Rm. 108 Sp.Ed. 6th

AF862053

Teacher's Edition

CONCEPTS AND CHALLENGES

EARTH SCIENCE

Fourth Edition

Leonard Bernstein ◆ Martin Schachter ◆ Alan Winkler ◆ Stanley Wolfe

Stanley Wolfe
Project Coordinator

PEARSON

Shoreview, Minnesota ◆ Boston, Massachusettes ◆ Chandler, Arizona ◆ Glenview, Illinois ◆ Upper Saddle River, New Jersey

ISBN-13: 978-0-7854-6764-9
ISBN-10: 0-7854-6764-5

1 2 3 4 5 6 7 8 9 10 12 11 10 09 08

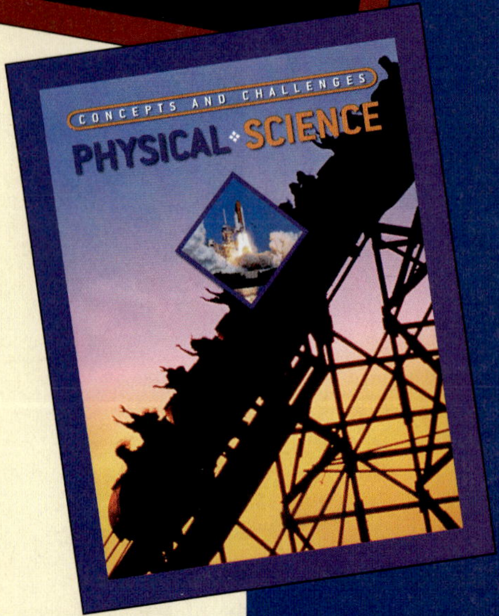

Concepts and Challenges, Fourth Edition

Science Textbooks Your Students Will Be Able to Read and Understand

Concepts and Challenges—the complete program proven to make scientific discovery accessible to all learners—offers you and your students:

- ✔ Hands-on inquiry
- ✔ Skill-building
- ✔ Teacher support for every lesson

Concepts and Challenges delivers the unique, two-page lesson format that teachers have always relied on to provide the right balance of challenge and accessibility. Now, science teachers have more of what they need:

- Updated Life, Earth, and Physical Science textbooks
- Full-length labs and hands-on activities in every chapter
- Special features that build skills and integrate content
- Point-of-use teacher lesson plans
- Teaching options for every learning style
- A separate Laboratory Manual that supports every chapter in the Student Edition
- Customizable Teacher's Resources online
- A companion website that connects students with the age-appropriate Fact Monster science database

Earth Science Table of Contents

The *Concepts and Challenges* Program

This complete program gives you and your students solutions for success.

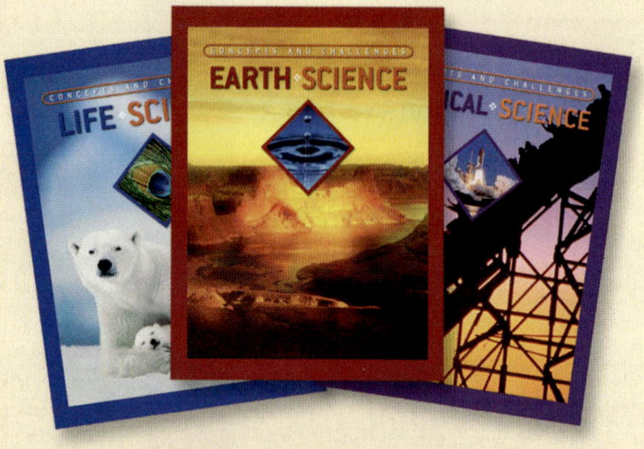

Student Edition

The Student Edition provides accessible text that your students will be able to read independently. Every chapter includes investigations, lab activities, integrated science connections, and technology applications. All lessons are presented in the acclaimed two-page format of "concept" and "challenge" that ensures learning and discovery for every student. *See pages vi–xiii.*

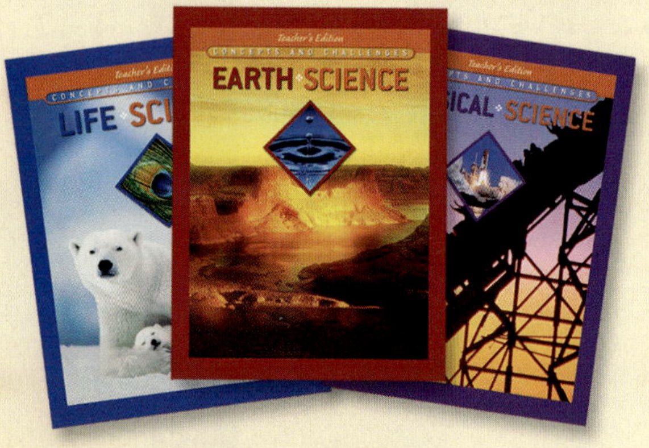

Teacher's Edition

The wraparound Teacher's Edition puts everything you need right at the point of use, including four-step lesson plans, an abundance of teaching options and enhancement activities, and chapter planning pages with information to help you prepare for each chapter. *See pages xiv–xvii.*

Laboratory Manual

Designed for appeal and ease of use, the student Laboratory Manual furnishes a full, formal, hands-on laboratory program correlated to the text. The Teacher's Laboratory Guide and Answers provides alternate materials, tips for success, safety warnings, and expected outcomes.

ConceptsandChallenges.com

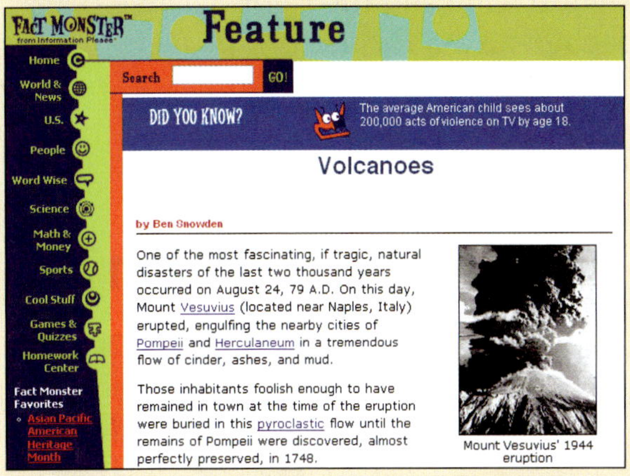

The student portal provides students access to the robust Fact Monster database—a reference website with resources tailored to the needs and interests of science students. The level-appropriate content includes an encyclopedia, almanacs, a dictionary, a thesaurus, scientific articles, and more. The site also features a comprehensive Homework Center that supports writing and research skills development, and provides students with test-taking tips. *See page xix.*

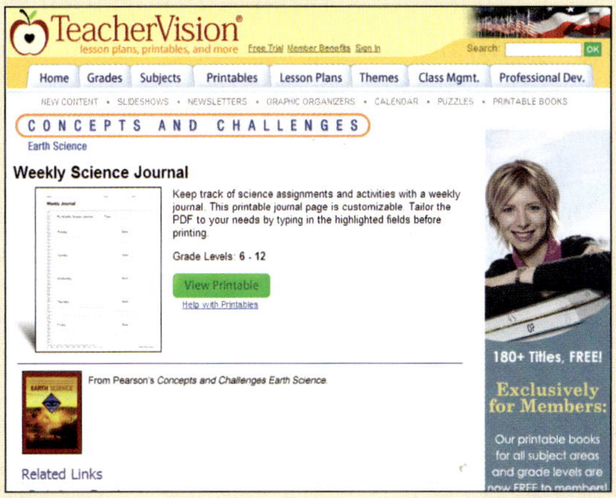

The Teacher's Resources online provide the opportunity to extend and support your classroom teaching with lesson review and enrichment worksheets, English and Spanish lesson notes, scoring rubrics, graphic organizers, comprehensive assessments, and more. Also available in print. *See page xviii.*

Concepts and Challenges in Science — Program Components

Concepts and Challenges in Life Science

Student Edition	0-785-46768-8
Teacher's Edition	0-785-46766-1
Laboratory Manual	0-130-23850-3
Teacher's Laboratory Guide and Answers	0-130-23861-9
Teacher's Resources in Print	0-130-24283-7
Color Transparencies	0-130-24286-1

Concepts and Challenges in Earth Science

Student Edition	0-785-46763-7
Teacher's Edition	0-785-46764-5
Laboratory Manual	0-130-23847-3
Teacher's Laboratory Guide and Answers	0-130-23848-1
Teacher's Resources in Print	0-130-24282-9
Color Transparencies	0-130-24285-3

Concepts and Challenges in Physical Science

Student Edition	0-785-46767-X
Teacher's Edition	0-785-46765-3
Laboratory Manual	0-130-23855-4
Teacher's Laboratory Guide and Answers	0-130-23856-2
Teacher's Resources in Print	0-130-24284-5
Color Transparencies	0-130-24287-X

SUPPORTING COMPONENTS:

Science Teacher Tech Pack:
0-785-46867-6

Discovery Channel School DVD Library:
0-785-46868-4

For more information, call 800-321-3106 or Fax 800-393-3156

Two-Page Lessons: A Proven Success

Consistent two-page lesson format provides the right pacing and the right balance of content for the right level of accessibility. This teacher-tested format is a proven success with students.

Investigate Activities help you jump start selected lessons, heighten student interest, and develop inquiry skills.

Objectives aid students in goal setting and furnish an important reference point throughout the lesson.

Key Terms identify and define vocabulary words students need to know before beginning the lesson.

Short Reading Passages give students frequent places to pause and check for understanding.

Comprehension Questions provide a basis for outlining lessons and make class discussions easy.

1-3 What is the structure of Earth?

INVESTIGATE

Modeling Earth's Layers

HANDS-ON ACTIVITY

1. Cut a hard-boiled egg in half with the shell still on it.
   ⚠ CAUTION: Be very careful when working with sharp objects.
2. On three small pieces of paper or self-stick labels, write "Yolk," "White," and "Shell." Tape the papers to three thin straws or coffee stirrers or fold the self-stick labels around them. Use the straws or stirrers to label each part of the egg.

THINK ABOUT IT: Like the egg, Earth has a core, material around the core, and a thin but hard outer skin. Which parts of the model are like Earth's outer skin, middle, and core?

Objective

Describe Earth's major parts: the crust, the mantle, and the core.

Key Terms

crust: solid, thin outer layer of Earth

mantle (MAN-tuhl)**:** thick layer of rock below the crust

core: innermost region of Earth

A Slightly Flat Sphere A perfect sphere has no real top. Every point on the surface of a sphere is the same distance from the center. Earth is a sphere, but not a perfect one. It is slightly squashed, or flat, at its poles. Also, Earth bulges out a little around its middle. Diameter is the distance across a sphere through its center. Earth is about 12,756 km in diameter at the equator. The diameter of Earth from pole to pole is about 12,664 km.

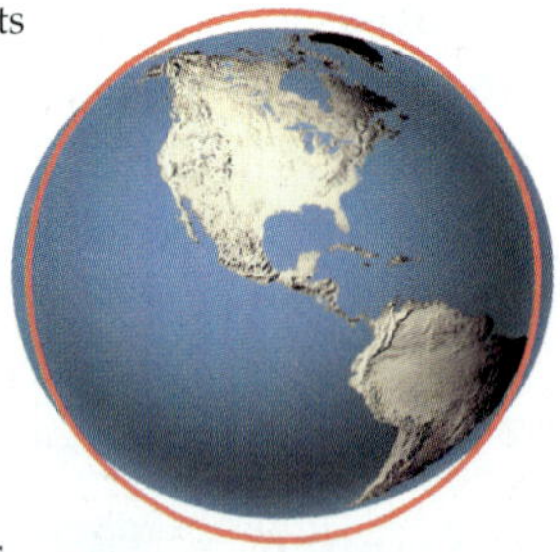

▲ **Figure 1-8** The outline of Earth does not fit perfectly into the outline of a sphere.

1 EXPLAIN: How is Earth not a perfect sphere?

Earth's Crust What is below Earth's surface? Like the hard-boiled egg above, Earth has three basic layers: the crust, the mantle, and the core. The **crust** is the thin, outer layer of Earth. It forms the upper lithosphere. Beneath the oceans, the crust is from 5 to 10 km thick. Beneath the continents, the crust is from 32 to 70 km thick. The thickest crust is beneath mountains.

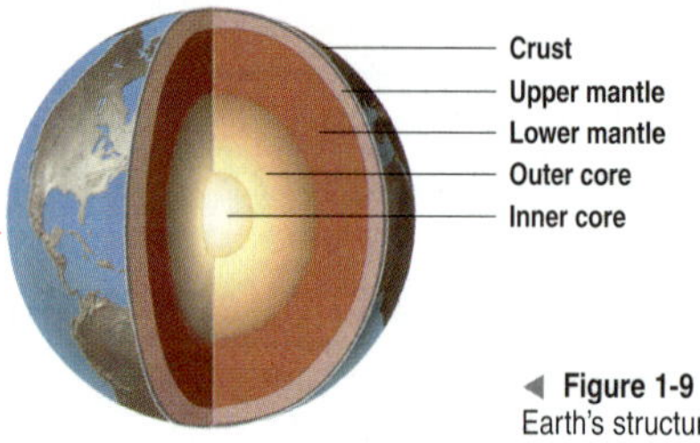

◀ **Figure 1-9** Earth's structure

2 CONTRAST: How is ocean crust different from continental crust?

Earth's Mantle The layer of Earth beneath the crust is called the **mantle.** Earth's mantle ranges from about 1,800 to 2,900 km thick. More than two-thirds of the mass of Earth is in the mantle.

The lithosphere includes all of the crust and the upper part of the mantle. Below this, the rock flows like a very thick liquid. Most of the mantle is made up of rock that flows very slowly.

3 OBSERVE: Where is the mantle located?

Highly detailed and functional **illustrations** reinforce student comprehension.

Checking Concepts questions supply a quick and easy way to review and assess learning at the end of the lesson.

Earth's Core The innermost region of Earth is the **core.** Scientists think that the core is made mostly of iron. The core has two parts. The outer core is liquid. It is about 2,250 km thick. The inner core is solid, with a radius of about 1,300 km.

THE LAYERS OF EARTH	
Layer	**Thickness**
Crust	5–70 km
Mantle	2,900 km
Outer Core	2,250 km
Inner Core	1,300 km

▲ **Figure 1-10**

4 CALCULATE: Which is thicker, the mantle or the two core layers combined?

CHECKING CONCEPTS

1. The crust is thickest beneath __________.
2. The core is made mostly of __________.
3. You can stand on Earth's layer called the __________.
4. The diameter of Earth at the equator is about __________ km.

THINKING CRITICALLY

5. SEQUENCE: Arrange Earth's layers in order, from thickest to thinnest.
6. ANALYZE: A drilling rig can go down only 6 km. Where would you set it up to drill into the mantle? Why?
7. INFER: Which layer of Earth includes the continents?

BUILDING MATH SKILLS

Calculating The radius, or *r*, of Earth is the point measured from its center to its surface. Use the formula below to find the approximate circumference, or C, of Earth. Pi, or π, is equal to 3.14. $C = 2\pi r$.

How Do They Know That?

EARTH IS A SPHERE

For thousands of years, people believed Earth was flat. This seemed logical. If you stand in a field or sail on the ocean, you can see for many kilometers. Everything appears flat. About 2,500 years ago, around 500 B.C., the Greek philosopher Pythagoras suggested that Earth was a sphere. His idea was probably based on observing the night sky. The heights of stars in the sky varied, depending on where you stood on Earth's surface. The Moon appeared to be round. Possibly, Earth had the same shape.

Many years later, another Greek philosopher, Aristotle, supported Pythagoras' theory. While observing lunar eclipses, Aristotle noted that a curved shadow fell on the Moon. He believed this to be Earth's shadow. Aristotle concluded that Earth was round.

Today, pictures from space show us that Earth is indeed a sphere.

▲ **Figure 1-11** Pythagoras (c. 580 B.C. to c. 500), Greek philosopher and mathematician

Thinking Critically How did Pythagoras' observation of the heights of stars support his theory?

CHAPTER 1: The Structure of Earth 21

Critical Thinking questions challenge students to use higher order thinking skills.

Challenge Features such as Building Math Skills give students the chance to hone essential skills, such as using the calculator for computation. Other Challenges are Health and Safety Tips, Interpreting Visuals, Web InfoSearch, Building Science Skills, and Designing an Experiment.

Lesson Features such as How Do They Know That? provide students with interesting articles and activities that make the lesson content more meaningful. Here, for example, historical and current scientific support is given for basic science concepts. Other features are Real-Life Science, Science and Technology, People in Science, and Integrating the Sciences.

Sample pages from ***Concepts and Challenges in Earth Science***, pages 20–21

Standards-Based Activities

Promote Scientific Inquiry

To help you engage students in scientific inquiry, experiments and hands-on activities are built into each chapter. These exercises have been carefully designed to meet the National Science Education Standards for scientific inquiry and discovery. A lesson may have any of the following features:

- **Investigate** heightens student interest and develops inquiry skills.
- **Hands-On Activity** gives students hands-on practice in making observations, interpreting information, thinking critically and logically, and communicating their findings. Students also use basic lab equipment, such as a graduated cylinder or a microscope, to gather and engage with data.
- **Designing an Experiment** asks students to plan a method for solving a problem. In the process, they gain valuable experience in clarifying questions, collecting evidence, interpreting data, and generating and critiquing explanations. Use the rubric from the Teacher's Resources to assess it.
- **Web InfoSearch** activities help students sharpen their research and technology skills via the Internet.

8-1 What are erosion and deposition?

INVESTIGATE

Preventing Erosion
HANDS-ON ACTIVITY

1. Fill a pie plate with a pile of sand or soil.
2. See if you can find a way to keep the soil from washing away when water is poured over it. You can try using small wooden sticks, strips of paper, pebbles—anything you can think of.

THINK ABOUT IT: Do any of the methods prevent the soil from washing away? If so, how?

STEP 2

Clipped from page 178

DESIGNING AN EXPERIMENT

Design an experiment to solve the following problem. Include a hypothesis, variables, a procedure with materials, and a type of data to study. Also tell how you would record your data.

PROBLEM: Using a board, sand, and a liquid, how can you demonstrate that glaciers change the shape of the land?

Clipped from page 191

Hands-On Activity

MODELING GROUNDWATER USE

You will need gravel or aquarium stones, one aquarium tank, one small plastic tub, sandy soil, a watering can, and clean spray pumps from bottles.

1. Put a 7.5-cm layer of gravel or stones in the aquarium. Cover with a layer of sandy soil.
2. Using the watering can, let it rain on your aquarium until you have at least 5 cm of groundwater.
3. Insert the spray pump into the gravel. Start pumping it into the plastic container. Record your observations. Note that if the soil is too sandy, it will clog your pump.
4. Add a few drops of food coloring to the watering can and water again. Repeat Step 3 and record your observations.

▲ **STEP 3** Pump the water into the plastic container.

Practicing Your Skills

5. **OBSERVE:** What happens to the water level as you pump?
6. **INFER:** How would adding food coloring to one water source affect the other?

Clipped from page 213

Web InfoSearch

Oases An oasis is a very rich piece of land in the middle of a desert. Oases vary in size from less than 1 square kilometer to several square kilometers. The land for hundreds of kilometers around is dry and barren. However, an oasis has plenty of plant life and a natural spring. Many people live in oases.

SEARCH: Use the Internet to find out more about oases. Why do they have water? What is it like to live there? You can start your search at www.conceptsandchallenges.com. Some key search words are **desert, oasis,** and **geology.**

◀ **Figure 9-10** Oasis in an Egyptian t

Clipped from page 213

Make Connections to All the Sciences and Other Subjects

To help you make connections to topics in life and physical science, as well as technology, mathematics, reading, and health, look for the following features in every chapter:

- **Science and Technology** integrates technological applications into the lessons.
- **Building Science Skills** challenges students to practice applying science process skills in the context of a lesson.
- **Health and Safety Tips** promotes student awareness of the connections between science and personal health.
- **Building Reading Skills** promotes content literacy through exercises that build vocabulary, explore word relationships, and apply definitions.
- **Integrating the Sciences** uses a high-interest topic as a springboard for connecting different branches of science.

Science and Technology

LASER DATING OF SEDIMENTS

A laser is a very strong, focused beam of light. The light has only one wavelength, or color. You may have several appliances in your house that use laser beams. CD and DVD players use laser beams to read the stored sounds and images on the discs. In Earth science, lasers are used to estimate the age of sediments.

How can lasers be used to estimate the age of sediments? Scientists bombard sediment layers with lasers. The lasers free electrons from the sediments. Electrons are negatively charged particles trapped in sediments. When the electrons are freed, they give off light. Scientists measure how much light is given off. Older sediments give off more electrons than younger sediments do. Laser dating is used to determine the ages of sediments that are up to 700,000 years old. It can also be used to determine the age of river sediments and glacial deposits.

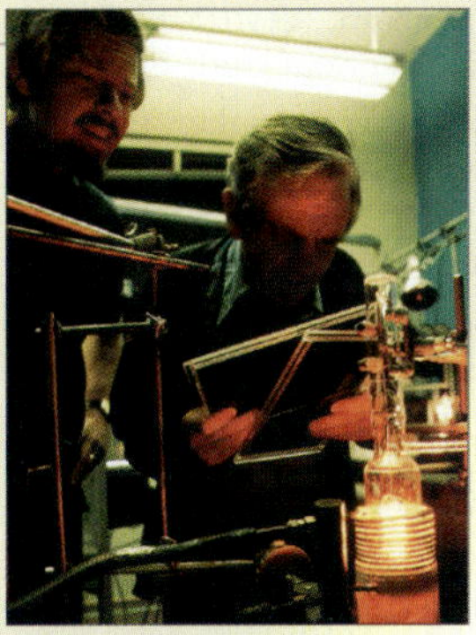

▲ **Figure 4-16** Scientists use lasers to date

Thinking Critically Why might scientists want to find out the age of glaciers?

Clipped from page 99

HEALTH AND SAFETY TIP

Radioactive elements emit radiation. Large amounts of radiation are harmful to living things. Radiation harms cells. Find out what the universal symbol for radiation danger looks like. Draw it on a sheet of paper. What two colors are used in the symbol?

Clipped from page 99

BUILDING READING SKILLS

Vocabulary There are many landforms other than the ones talked about so far. Find out where the names of the following landforms come from. Then, tell if and how their names relate to how they are formed: *butte, peneplain, monadnock, mesa.*

Clipped from page 117

BUILDING SCIENCE SKILLS

Classifying Below is a list of some important artificial satellites. Find out the function of each. Then, classify each according to its function. On a piece of paper, write the words *Communications Satellite, Weather Satellite, Navigation Satellite,* and *Scientific Satellite.* Place each satellite under its correct heading.

Early Bird	*Explorer I*	*Intelsat*	*Tiros*
Echo	*Landsat*	*Nimbus*	*Transit*

◀ **Figure 16-31** *Tiros*

Clipped from page 401

Integrating Life Science

TOPICS: mammals, habitats

BAT CAVE IN CARLSBAD CAVERNS

▲ **Figure 8-15** Bats leave their caves at dusk.

In Carlsbad, New Mexico, the skies sometimes fill with dark clouds just before sunset. These clouds are not made of smoke or dust but of huge numbers of bats. The bats live inside Carlsbad Caverns, clinging to cave walls. Carlsbad is one of the world's largest natural caverns. It provides a perfect habitat for bats.

The most common bat species found in the cave is the Mexican free-tailed bat. Mostly females, the Mexican free-tailed bats that live in the caverns stay there about half the year. Young bats, called pups, are born there. The darkness protects them from predators. At sunset, the bats leave the caverns to hunt insects for food. Clouds of departing bats may have led to the discovery of the caves in the 1800s.

The caverns became a national monument in 1923. Tourists are not allowed into Bat Cave. Some bat species hibernate there in cold weather.

Thinking Critically Why do you think some bats hibernate?

Clipped from page 187

The Big Idea Integrates Science!

Connect the science disciplines with this engaging two-page feature in each chapter. Set in a real-world context, The Big Idea helps your students synthesize what they're learning into one big idea.

Topic of integration is clearly identified at the top of the page.

High-interest graphic instantly captures students' attention with colorful illustrations and photos.

Short, easy-to-read article summarizes the science connection.

Integrating Technology

THE Big IDEA

What products are made from oil?

It might not seem like crayons, guitar strings, footballs, and bubble gum have anything in common, but they do. Each of these products is made from crude oil.

Ancient peoples recognized the value of this natural resource. The Chinese used oil to waterproof their homes. Egyptians wrapped their dead in rags saturated with oil to preserve them. Native Americans added a drop or two of oil to medicines, and ancient Romans applied oil to cuts and wounds. Thanks to technology, today there are more than 500,000 different uses for this natural resource.

The first step in creating these products is separating crude oil into its different parts. This is called fractional distillation and takes place in an oil refinery. Pipes carry raw crude oil from oil wells to the refinery. The crude oil is heated to about 430°C. This causes the liquid to turn into vapor. The vapor is then pumped into a tall tower divided into different levels. Each of these levels is kept at a certain temperature.

As the evaporated crude oil moves through the tower, its substances condense at different temperatures. The heaviest parts condense in the warmer, lower levels of the tower. The lightest parts condense in the cooler, higher levels of the tower. The separated substances are drawn from each level and sent to other locations, where they are used to make many products.

Look at the illustrations on these two pages. Then, follow the directions in the Science Log to learn more about "the big idea." ✦

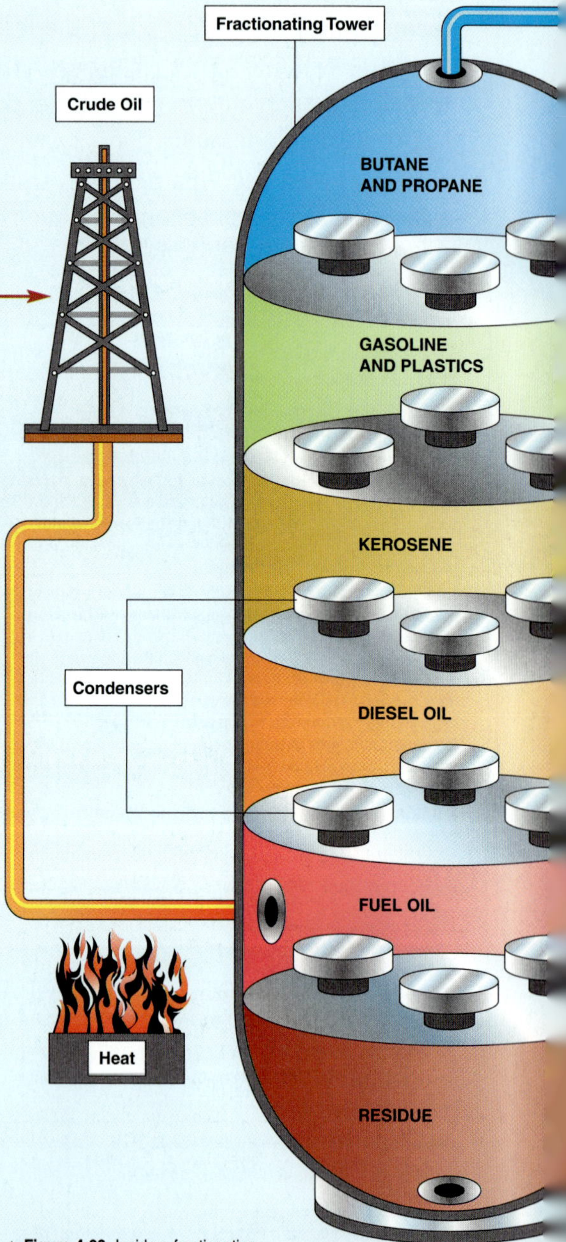

▲ **Figure 4-22** Inside a fractionating tower, different substances contained in crude oil condense at different temperatures.

104

In the Teacher's Edition

The Big Idea helps you guide students through researching on the Internet to complete their Science Log Writing Activity.

THE BIG IDEA

What products are made from oil? Have students start their research for their Science Log Writing Activity at www.conceptsandchallenges.com. Be sure students use search engines and websites that are reliable and acceptable. You may also wish to provide copies of the *Big Idea Planner* and *Big Idea Science Log Entry* study tools found in the Teacher's Resources.

Call-outs provide students with intriguing, bite-size pieces of information.

Butane and Propane The lightest parts of evaporated crude oil condense in the coolest level of the tower. Butane used in lighters and propane used in barbecue grills is collected here.

Gasoline and Plastics The gasoline that cars run on comes from substances that condense in this level. Chemicals and plastics used to make clothing, sporting equipment, and telephones start at this level.

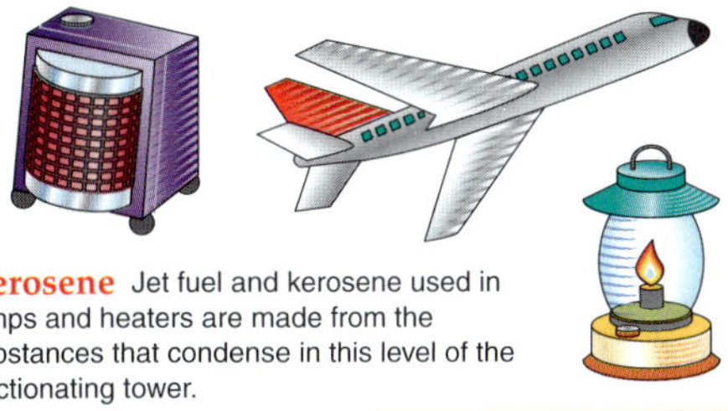

Kerosene Jet fuel and kerosene used in lamps and heaters are made from the substances that condense in this level of the fractionating tower.

Diesel Oil Trucks and trains run on diesel fuel made from substances that condense at this level.

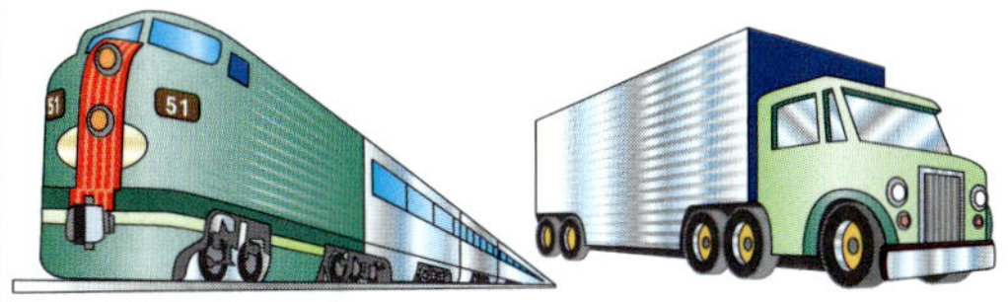

Fuel Oil Heating oil is used to keep many homes warm. Certain fuels power ships and factories. All are made from substances that condense in this level.

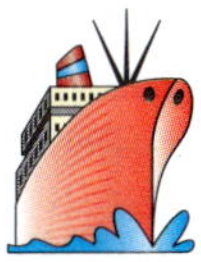

Residue Lubricants, paraffin, and asphalt are made from the residues that condense in the lowest level of the tower.

WRITING ACTIVITY

Science Log

List all of the crude oil products that you have used this week. Name at least five. Then, write several paragraphs entitled "A Day Without Oil Products." Describe in these paragraphs what your day would be like without these items. Start your search for information using www.conceptsandchallenges.com.

Science Log gives students a valuable writing experience as well as the opportunity to relate the lesson content to their own lives and to prior knowledge. Students can complete the exercise in their own Science Log which they build from lesson to lesson.

CHAPTER 4: The Rock Record **105**

Sample pages from ***Concepts and Challenges in Earth Science***, pages 104–105

Hands-On Approach in Every Chapter

A stand-alone Lab Activity is built into each chapter of the Student Edition. Directly correlated to chapter content, each easy-to-follow lab helps students apply a systematic approach to the scientific method. See the Teacher's Edition for point-of-use support.

Background reviews relevant chapter concepts to ensure understanding and to set the stage for the activity.

Materials provides an accessible list of what is needed to complete the lab.

Purpose clearly identifies the objective and provides an important reference point for students.

Procedure walks students step by step through the experiment with clear, easy-to-follow instructions.

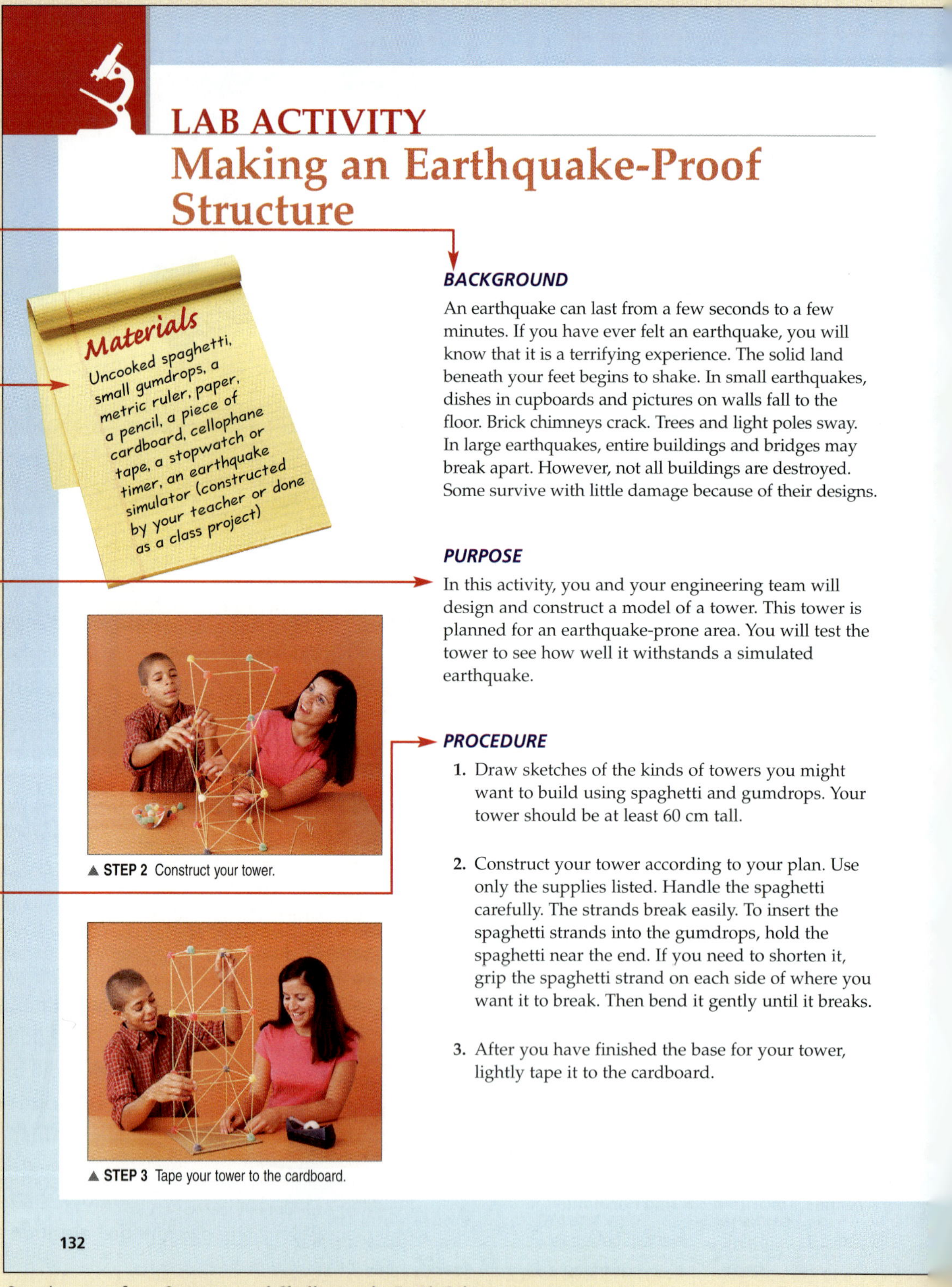

Sample pages from ***Concepts and Challenges in Earth Science***, pages 132–133

In the Teacher's Edition

Prepare the Lab provides you with everything you need to know to plan your lab.
Set-Up Time indicates the approximate time you should allot for setting up the lab.
Lab Time indicates the approximate time it will take students to complete the lab.

1 *Prepare the Lab*

Making an Earthquake-Proof Structure

SET-UP TIME: LAB TIME:

4. When the tower is completed, draw a picture of it. Label the height and width of the tower on your drawing. Then take it for testing to the earthquake simulator.

5. Time how long your building remains standing at different "strengths" of shaking. Copy the chart in Figure 5-36. Record your observations and times on it.

▲ **STEP 4** Test your tower with the simulator.

Visual aids, including photos, graphs, and tables, help clarify different procedures in the lab activity.

Data Collection Charts provide students with a format for recording their observations and results.

	P-Wave Test	S-Wave Test	L-Wave Test
Did your tower survive the test?			
Where did the tower shake the most?			

▲ **Figure 5-36** Use a copy of this chart to record your observations.

Conclusion questions encourage critical thinking and the application of the scientific method. The Teacher's Resources provide rubrics and study tools as support.

CONCLUSIONS

1. OBSERVE: What happened to your tower when it was tested? Where did it vibrate? Did it break or survive the test?

2. OBSERVE: Did any towers from other groups perform better than yours? Compare their height and other features.

3. INFER: How could you build a stronger tower?

CHAPTER 5: T

In the Laboratory Manual

Laboratory Challenges provide more investigations for every chapter.

See page v for details.

Planning Made Easy

Concepts and Challenges wraparound lesson plans enable you to customize lessons for diverse learning styles. A consistent four-step lesson plan puts the activities and answers right where you need them!

Step 1 Introduce furnishes tools for setting up the lesson and linking your students' prior knowledge to the lesson content.

Step 2 Teach delivers support for presenting the lesson, including discussion guidelines, ways to accommodate different learning styles, answers to questions on the student pages, and ideas for demonstrations that reinforce the concepts.

Demonstration helps you show students the concepts in a visual and tactile way.

Fast Facts provides in-depth background to support and extend the lesson.

Reading Strategy builds in techniques to help students become more proficient readers.

6-2 *(pp. 140–141)*

1 *Introduce*

STUDY HINT Write the title of each section in the lesson on the board. Have students copy these titles in their notebooks, using an outline format. As students read each section, have them write the topic sentence of each paragraph beneath each section title.

Linking Prior Knowledge Ask students: *What causes volcanoes and earthquakes in the Mid-Atlantic Ridge?* (the formation of new crust) *What do you think happens to old crust when new crust forms?* Explain that in this lesson, students will learn what happens to old crust when new crust forms. After reading the lesson, have students review their answers. Were they correct?

2 *Teach*

Demonstration Use a hard-boiled egg to represent seafloor spreading. Explain that Earth is like the egg. Just as the eggshell buckles in some places when it is cracked, exposing the egg white, Earth's mantle also buckles. When this happens, new ocean floor and mountains form, and earthquakes occur. **learning style: visual**

Discussion Using Figure 6-7, discuss the mid-ocean ridges and the process of seafloor spreading. Make sure that students understand that some of the tallest mountains and longest mountain ranges are under the ocean.
learning style: visual

Answers to questions

1. NAME: oceanic and continental
2. IDENTIFY: under the ocean
3. DEFINE: deep crack running down the center of the ridges
4. DESCRIBE: It cools and forms new crust.

6-2 Why is the seafloor spreading?

Objective
Relate seafloor spreading to the forming of new oceanic crust.

Key Terms
mid-ocean ridge: ocean-floor feature resembling a mountain ridge on land
rift valley: flat area between two ridges that is formed by spreading plates
seafloor spreading: process that forms new seafloor

Two Kinds of Crust There are two kinds of crust on Earth. One kind is continental crust. This crust makes up Earth's continents. The other kind is called oceanic (oh-shee-AN-ihk) crust. Oceanic crust, which makes up the ocean floor, is denser than continental crust.

1 NAME: What are the two kinds of crust?

Mid-Ocean Ridges The most obvious features of the crust that makes up the ocean floor are the **mid-ocean ridges.** These ridges are part of a series of underwater mountain ranges that run along the floors of all the oceans.

Some of the longest mountain ranges and tallest peaks on Earth are part of oceanic crust. These mountains form a loose chain around the world. Some peaks in the chain rise more than 3,000 m above the ocean floor. In a few places, the peaks rise above the surface of the ocean to form islands. Iceland is a mountain peak of the Mid-Atlantic Ridge. The Mid-Atlantic Ridge is in the middle of the Atlantic Ocean.

2 IDENTIFY: Where are some of the longest mountain ranges on Earth found?

Rift Valleys In the late 1940s, scientists began to map the mid-ocean ridges. They discovered deep cracks, or rifts, in the center of some parts of some ridges. Magma pouring out of the rift had hardened and formed broad underwater valleys with steep sides. These valleys are called **rift valleys.** There are many earthquakes and volcanoes in rift valleys.

On land, spreading plates form steep, wide valleys in the continental crust, as shown in Figure 6-7. The East African Rift System extends for 6,400 km, from Jordan in southwestern Asia through eastern Africa to Mozambique. The valley averages 48 km to 64 km in width. Other rift valley floors are below sea level and form deep lakes.

3 DEFINE: What is a rift valley?

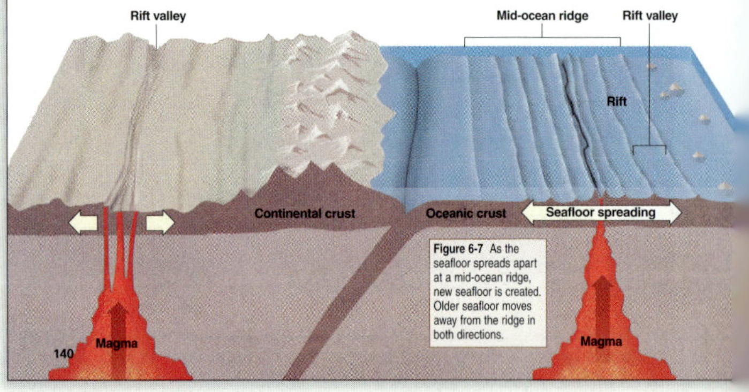

Figure 6-7 As the seafloor spreads apart at a mid-ocean ridge, new seafloor is created. Older seafloor moves away from the ridge in both directions.

140

Teaching Options

FAST FACTS *Continental Rift Valleys*

The two rift valleys that are on continents, rather than in oceans, are the Red Sea and the East African Rift System. The San Andreas Fault in California is also a spreading center but is composed of short segments rather than one long fault.

The Red Sea has already filled with seawater but does not yet completely separate the two landmasses of the Arabian Peninsula and northeast Africa.

The East African Rift System is in the early stages of spreading. If movement continues in the same way, east Africa will separate from the rest of the continent. The rift is marked by a chain of lakes, most of them deep and fjordlike, and seas such as the Dead Sea, whose surface is 400 m (1,312 ft) below sea level.

Planning Guide

The planning guide in each chapter contains:

- ✔ A suggested number of days to teach the chapter
- ✔ Every lesson, Lab Activity, and Big Idea
- ✔ The National Science Education Standards
- ✔ Skills and features in each lesson including
  - Building Science Skills
  - Hands-On Activity
- ✔ Projects and activities, including
  - Chapter Project
  - Cooperative Learning
  - More on the Web
- ✔ Achieving Science Literacy
  - Reading Strategy
  - Writing Hint
  - Tips for Using Technology
- ✔ Meeting Individual Needs
  - ESL/ELL Strategy
  - Enrichment Activity

Formation of New Seafloor Deep-sea drills have been used to bring up samples of oceanic crust. Most of these samples turn out to be younger than samples of continental crust. In addition, the oceanic crust near the mid-ocean ridges is younger than oceanic crust farther from the center of the ridge. The youngest oceanic crust is found in the center of the ridge.

Below the mid-ocean ridges, magma rises through the crust. As the magma cools, it forms new crust on both sides of a rift. The new crust pushes the seafloor apart at the ridges. Scientists called the process **seafloor spreading.** Seafloor spreading is partly what drives the movements of the continents, or continental drift.

DESCRIBE: What happens as magma rises through the crust of the mid-ocean ridges?

CHECKING CONCEPTS

1. Which kind of crust is made up of heavy, dense material?
2. What is the Mid-Atlantic Ridge?
3. How does magma come out of a chamber to form a ridge and valley?
4. Where did scientists find the youngest crust on the ocean floor?

THINKING CRITICALLY

5. MODEL: Draw a diagram that illustrates seafloor spreading. Label your diagram.
6. CONCLUDE: How do you think seafloor spreading helps explain the theory of continental drift?

BUILDING SCIENCE SKILLS

Modeling To make a model of seafloor spreading, do the following:

STEP 1 Tape two sheets of red construction paper together along their short ends. Then push two desks together.

STEP 2 Fold the sheets of construction paper together along the tape.

STEP 3 Push the open ends of the construction paper up through the crack between the desks until some paper falls on each side. Keep pushing the paper up.

7. CONCLUDE: What does the crack between the desks represent? What does the construction paper represent? How could you make your model more accurate?

Real-Life Science

THE SUMATRA-ANDAMAN EARTHQUAKE

On December 26, 2004, an undersea earthquake rocked the Indian Ocean off the coast of Sumatra. The magnitude of the quake registered at 9.1 on the Richter scale. It was the 3rd largest earthquake since 1900. It literally shook the planet.

During the earthquake, an underwater landslide displaced a large volume of water. This triggered a series of giant waves, called tsunamis, that devastated the coasts of the Indian Ocean. These waves were over 100 meters high. Nearly 230,000 people were killed or missing. This incident is now known as the Asian Tsunami.

▲ **Figure 6-8** A giant tsunami hits the shore.

Thinking Critically What effects might an undersea earthquake have on the ocean floor?

CHAPTER 6: Plate Tectonics **141**

COOPERATIVE LEARNING

Modeling Seafloor Spreading
Have students work in pairs to complete the Building Skills activity. After students determine how to make their models more accurate, have them present their revised model to the class.
learning style: tactile

• TEACHING RESOURCES •
Online or Print

Teacher's Resources

Lesson Review: Chapter 6, p. 3

Enrichment: Chapter 6, p. 5

Seafloor Spreading: Visuals, p. 5

MORE ON THE WEB

Tsunami Aftermath Have students research online, starting at www.conceptsandchallenges.com, to learn more about the devastation caused by the Asian Tsunami. Then have students write a short report about their findings.

QUICK CONVERSIONS

	Metric /SI	English
Chain peaks	3,000 m	9,843 ft
Great Rift Valley		
length	6,400 km	3,977 mi
width	48–64 km	30–40 mi

3 Assess

CHECKING CONCEPTS ANSWERS

1. oceanic
2. mountain chain that runs down the middle of the Atlantic Ocean
3. through a crack in the crust
4. along the mid-ocean ridges

THINKING CRITICALLY ANSWERS

5. MODEL: Students' models should resemble Figure 6-7.
6. CONCLUDE: Possible answer: Seafloor spreading causes the continents to move.

4 Close

BUILDING SCIENCE SKILLS ANSWER

Modeling

7. CONCLUDE: The crack between the two desks represents the rift valley in the mid-ocean ridge; the construction paper represents oceanic crust; by placing construction paper on either side so it gets pushed out toward the edge of the desk. learning style: tactile

Real-Life Science

THE SUMATRA-ANDAMAN EARTHQUAKE

Background Called a mega-thrust event, (subduction of one tectonic plate beneath another), this earthquake occurred as thrust-faulting where the India plate subducts under the Burma plate. Records were set for the longest fault rupture in history – about 1300 km (780 miles) – and the longest duration of shaking at about 10 minutes.

The initial earthquake caused the fault to slip nearly 15 meters near Banda Aceh, Sumatra. This event resulted in a huge displacement of water forming a deadly tsunami in the Indian Ocean. This tsunami caused more casualties than any on record, affecting South Asia and East Africa. Efforts to warn the Indian Ocean coasts about a possible tsunami were difficult because there were no buoys in place in 2004 to register tsunami threats.

Discussion Ask students: How might the number of deaths have been different if the Indian Ocean had a Tsunami Warning System in place? (Warnings would have reached coastal people well in advance of the waves, allowing them time to move to higher ground.)

Thinking Critically Answer
Possible answers include causing underwater landslides and changes in rifts and valleys.

CHAPTER 6: Plate Tectonics **T141**

Step 3 Assess saves you time by supplying point-of-use answers to every question in the Checking Concepts and Thinking Critically sections.

Step 4 Close presents activities that challenge students to apply what they have learned in the lesson.

Teaching Resources refers you to worksheets and support materials that reinforce and enrich the lesson. You'll also find references to relevant lab manual pages.

More on the Web is a handy reminder that students can find additional information on the day's topic on the Internet.

Quick Conversions takes the guesswork out of converting from metric to English measurement with easy-to-read charts and tables.

Sample pages from ***Concepts and Challenges in Earth Science, TE***, pages T140–T141

Successful Teaching Strategies

Concepts and Challenges already maximizes comprehension through its extensive visual support, highly focused content, and strong reinforcement of key vocabulary terms. In addition, a wide variety of hands-on activities motivate tactile learners.

Focusing on Reading

The Teacher's Edition provides you with many innovative tools that can be used to successfully teach struggling students.

- **Linking Prior Knowledge** builds on what students know by reminding them of material already covered or their own real-life experiences.

Linking Prior Knowledge Ask students what they think makes day differ from night. Also, ask them why the amount of daylight and darkness changes throughout the year.

Clipped from page T326

- **Reading Strategy and Writing Hint** help students develop their language arts skills, which can improve their reading and writing in the content area.

READING STRATEGY Have students create concept webs in their notenooks. (They can use the Fishbone Map Concept Map on the Teacher's R The topic should be "Earth's mo As studentsr ead thelesson, they arms to their webs on which the Key Terms or other details about Earth's motions.

Clipped from page T326

WRITING HINT Tell students that they can improve their spelling skills by keeping a dictionary on hand when writing. Encourage them to refer to the dictionary when doing any type of work that requires writing, even just answering questions.

Clipped from page T402

- **Demonstration** makes complex concepts come to life for students who have difficulty visualizing these concepts.

Demonstration Use two paperback books to demonstrate what can happen when two tectonic plates meet. Show how one book moves under the other and how the upper book folds upward. Point out that the upward fold represents a mountain.
learning style: visual

Clipped from page T150

- **Reteach** provides innovative ways of presenting lesson concepts so that students do not have to rely on print. Many of these activities also include suggestions for using a variety of graphic organizers, which can be found in the Teacher's Resources.

RETEACH

Reviewing Newton's Third Law of Motion Have students identify the action force in this activity. (the fizzing antacid tablet) Ask: *What is the reaction force?* (the flight of the rocket) *What is another name for reaction force?* (thrust) Have students write three cause-and-effect statements about the lab activity. Students can use the Cause-and-Effect graphic organizer in the Teacher's Resources.
learning styles: auditory, linguistic

Clipped from page T396

- **Cooperative Learning** offers activities that encourage students of different abilities and levels of knowledge to share their knowledge and to work together to improve understanding of core content.

COOPERATIVE LEARNING

Telling Time Using Figure 14-5, locate and identify the time zones in the continental United States. Display a large U.S. map and have students take turns reading aloud the names of any two places in the United States in different time zones. Tell half the students to write a sample time for the first place on the board. Have the rest of the class determine the time for the second place.
learning styles: visual, auditory

Clipped from page T328

Focusing on ESL/ELL Students

Concepts and Challenges recognizes the need to customize teaching for English-language learners. ESL/ELL Strategies, which appear in each chapter of the Teacher's Edition, provide insights into English vocabulary and increase comprehension of scientific concepts. Specific techniques that are cited in these strategies ensure that learning is taking place. The techniques may also remove potential language barriers for English-language learners at beginning levels of proficiency. The ESL/ELL Strategies help students by:

- **Recommending peer tutoring**, which encourages communication with peers and gives students an opportunity to ask questions and receive answers without the pressures of having to perform in front of the whole class.
- **Suggesting simple projects or activities** that reinforce comprehension of scientific concepts.
- **Using cooperative learning** to encourage interactions within a group. This helps students gain greater comprehension of the content while improving communication skills.
- **Providing vocabulary and linguistic clues** to aid understanding and recall of scientific terminology.
- **Having students use study tools and graphic organizers** to visually reinforce the comprehension of concepts. For example, a Venn diagram can be used to visually compare and contrast concepts.
- **Using graphic organizers and other study tools** to provide a focus for activities and assessment. Graphic organizers allow students to present complex processes and show what they know using the language skills they do possess. The Teaching Resources box, in the Teacher's Edition, provides references to a complete collection of study tools and graphic organizers. Look for activities and suggestions using any of the pages listed below.

> ***ESL/ELL STRATEGY***
>
> ***Peer Tutoring*** Peer-tutored hands-on activities help English-language learners in two ways. Peers encourage students to communicate, while hands-on activities help make scientific concepts easier to understand without the barriers of difficult scientific languages. Use the Investigate with your ELL students.

Clipped from page T150

Venn Diagram

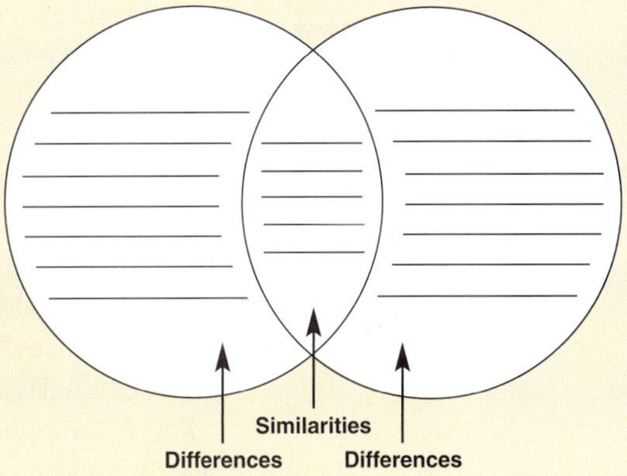

Clipped from Study Tools in Teacher's Resources

• *TEACHING RESOURCES* •
Online or Print

Teacher's Resources

Big Idea Planner	Lab Activity Report
Big Idea Science Log Entry	Making Connections–KWL
Cause-and-Effect Diagram	Project Sketch
Concept Map	Spider Map
Database Planner	Spreadsheet
Designing An Experiment	Thumbnail Sketch
Fishbone Map	Venn Diagram
Flowchart	Web Page Planner
Folder	Weekly Journal
KWL chart	Writing Activity Outline

Customize With Technology

Teacher's Resources Online

Customizable and printable pages make it easy for you to adapt materials to meet the needs of students with varied ability levels.

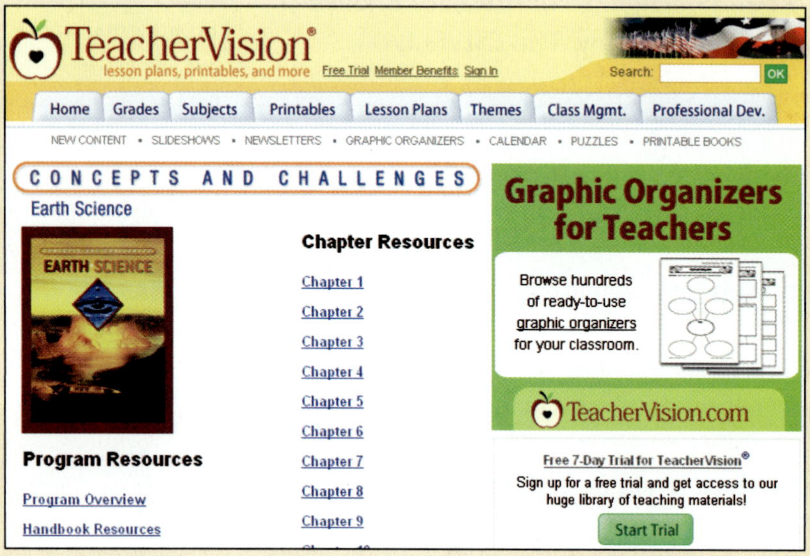

On the **conceptsandchallenges.com website,** you can select a section and then easily navigate to the page you want. Click the "View Printable" button to see the page. Many of the pages can be customized to meet your needs; simply type the appropriate information in the text fields before printing.

A variety of **chapter support** pages are provided, including: Lesson Reviews, Enrichment Activities, Key Term Reviews, English and Spanish Lesson Notes, and Chapter Tests. In addition, Diagnostic, Midterm, and Final Exams provide cumulative assessment, while a wealth of Teaching Tools, Study Tools, and Visuals help you meet individual needs. The online answer key password is ccearth09. These resources are also available in print.

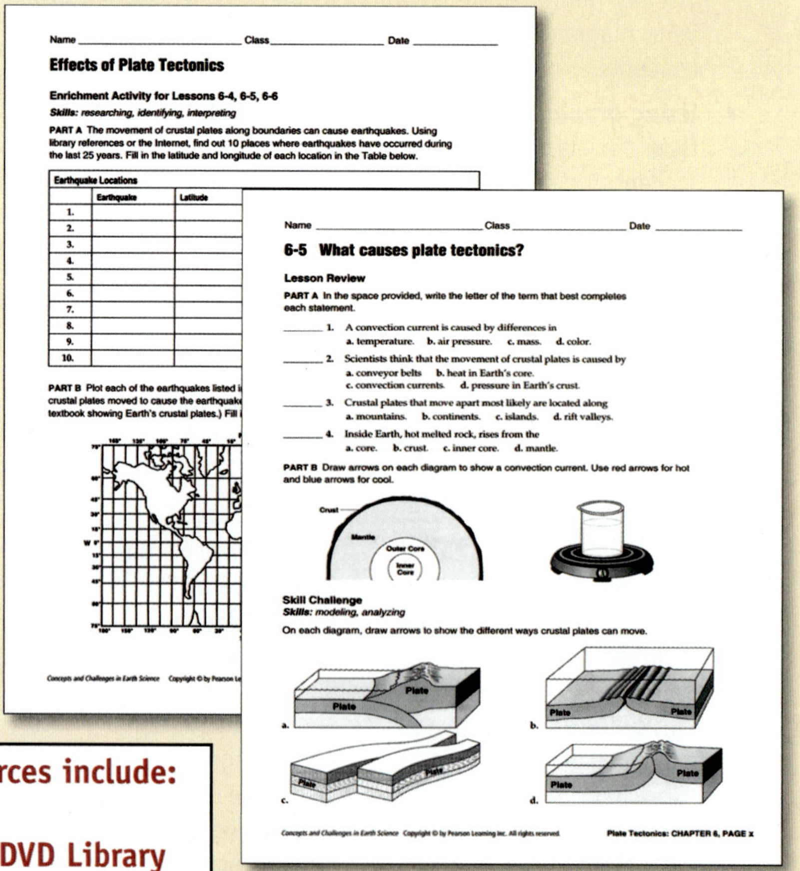

Name ______ Class ______ Date ______

Effects of Plate Tectonics

Enrichment Activity for Lessons 6-4, 6-5, 6-6

Skills: *researching, identifying, interpreting*

PART A The movement of crustal plates along boundaries can cause earthquakes. Using library references or the Internet, find out 10 places where earthquakes have occurred during the last 25 years. Fill in the latitude and longitude of each location in the Table below.

Earthquake Locations		
	Earthquake	Latitude
1.		
2.		
3.		
4.		
5.		
6.		
7.		
8.		
9.		
10.		

PART B Plot each of the earthquakes listed i… crustal plates moved to cause the earthquak… textbook showing Earth's crustal plates.) Fill i…

Name ______ Class ______ Date ______

6-5 What causes plate tectonics?

Lesson Review

PART A In the space provided, write the letter of the term that best completes each statement.

_____ 1. A convection current is caused by differences in
a. temperature. b. air pressure. c. mass. d. color.

_____ 2. Scientists think that the movement of crustal plates is caused by
a. conveyor belts b. heat in Earth's core.
c. convection currents. d. pressure in Earth's crust.

_____ 3. Crustal plates that move apart most likely are located along
a. mountains. b. continents. c. islands. d. rift valleys.

_____ 4. Inside Earth, hot melted rock, rises from the
a. core. b. crust. c. inner core. d. mantle.

PART B Draw arrows on each diagram to show a convection current. Use red arrows for hot and blue arrows for cool.

Skill Challenge
Skills: *modeling, analyzing*

On each diagram, draw arrows to show the different ways crustal plates can move.

Sample pages from ***Concepts and Challenges in Earth Science, Teacher's Resources***

Supporting technology resources include:
- **Science Teacher Tech Pack**
- **Discovery Channel School DVD Library**

See page v.

Student Research Online

Conceptsandchallenges.com connects students with Pearson's Fact Monster science database—a search engine that provides articles, facts, definitions, and more. Students can use it as a starting point to search for information on the Web and to research their Science Log writing activities. Students can also utilize the Reference Desk to use the Fact Monster atlas, almanac, dictionary, or encyclopedia exclusively.

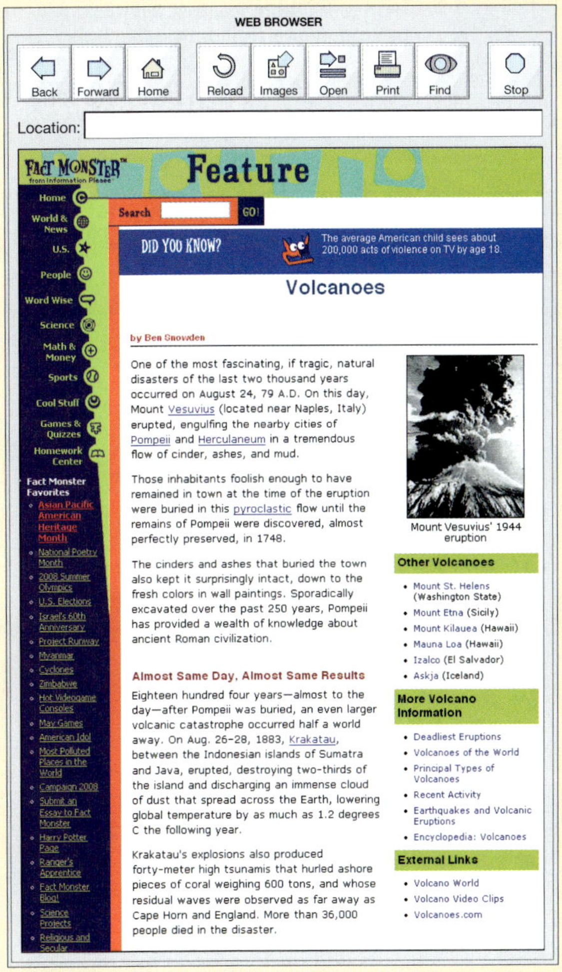

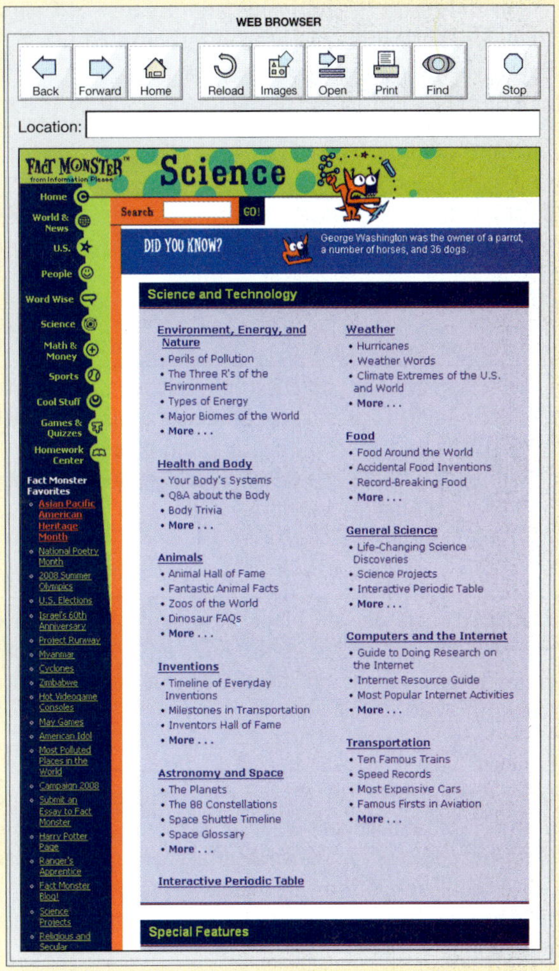

By entering keyword information or related topics, students can explore new concepts beyond the textbook. The interdisciplinary database provides opportunities to connect concepts between the sciences, as well as to history, geography, and current events.

Go Online
active art
For: Seismic Waves activity
Visit: PHSchool.com
Web Code: cfp-1022

Students can interact with the art of seismic waves online.

Active Art allows students to interact online with visuals related to selected lesson topics.

Earth Science
Content Standards D Grades 5–8

1. STRUCTURE OF THE EARTH SYSTEM

Standard	Chapter Correlation
a. The solid earth is layered with a lithosphere; hot, convecting mantle; and dense, metallic core. Code: D1a	**Chapter 1** 1-2, 1-3
b. Lithospheric plates on the scales of continents and oceans constantly move at rates of centimeters per year in response to movements in the mantle. Major geological events, such as earthquakes, volcanic eruptions, and mountain building, result from these plate motions. Code: D1b	**Chapter 5** 5-1, 5-2, 5-3, 5-4, 5-5, 5-6, 5-7, 5-8, 5-9, 5-10, BI, Lab **Chapter 6** 6-1, 6-2, 6-3, 6-4, 6-5, 6-6
c. Land forms are the result of a combination of constructive and destructive forces. Constructive forces include crustal deformation, volcanic eruption, and deposition of sediment, while destructive forces include weathering and erosion. Code: D1c	**Chapter 3** Lab **Chapter 5** 5-1, 5-2, 5-3, 5-4, 5-5, 5-6, 5-7, 5-9, Lab, BI **Chapter 8** 8-1, 8-2, 8-3, 8-4, 8-5, 8-7, 8-8, 8-9, 8-10, 8-11, Lab **Chapter 10** 10-8, 10-9, Lab
d. Some changes in the solid earth can be described as the "rock cycle." Old rocks at the earth's surface weather, forming sediments that are buried, then compacted, heated, and often recrystallized into new rock. Eventually, those new rocks may be brought to the surface by the forces that drive plate motions, and the rock cycle continues. Code: D1d	**Chapter 2** 2-6 **Chapter 3** 3-1, 3-2, 3-4, 3-6, 3-8, Lab, BI **Chapter 7** 7-1, 7-2, 7-3, Lab
e. Soil consists of weathered rocks and decomposed organic material from dead plants, animals, and bacteria. Soils are often found in layers, with each having a different chemical composition and texture. Code: D1e	**Chapter 4** 4-6 **Chapter 7** 7-4, 7-5, 7-6, BI **Chapter 8** Lab **Chapter 9** 9-2 **Chapter 15** 15-5
f. Water, which covers the majority of Earth's surface, circulates through the crust, oceans, and atmosphere in what is known as the "water cycle." Water evaporates from Earth's surface, rises and cools as it moves to higher elevations, condenses as rain or snow, and falls to the surface where it collects in lakes, oceans, soil, and in rocks underground. Code: D1f	**Chapter 1** 1-2 **Chapter 8** 8-10 **Chapter 9** 9-1, 9-2, 9-3, 9-4, 9-5, BI, Lab **Chapter 10** 10-1, 10-3 **Chapter 11** 11-5 **Chapter 12** 12-1, 12-5, 12-6, 12-9 **Chapter 15** 15-2, 15-8, 15-9, Lab
g. Water is a solvent. As it passes through the water cycle, it dissolves minerals and gases and carries them to the oceans. Code: D1g	**Chapter 7** 7-2, Lab, 7-3 **Chapter 8** 8-3 **Chapter 10** 10-3
h. The atmosphere is a mixture of nitrogen, oxygen, and trace gases that include water vapor. The atmosphere has different properties at different elevations. Code: D1h	**Chapter 1** 1-2 **Chapter 11** 11-1, 11-2, 11-3, 11-4, 11-5, 11-6, 11-7 **Chapter 12** 12-1, 12-2 **Chapter 15** 15-7, BI
i. Clouds, formed by the condensation of water vapor, affect weather and climate. Code: D1i	**Chapter 9** 9-1 **Chapter 12** 12-1, 12-2, 12-3, 12-4, 12-5, 12-6, 12-9, Lab, BI

1. STRUCTURE OF THE EARTH SYSTEM *(continued)*

Standard	Chapter Correlation
j. Global patterns of atmospheric movement influence local weather. Oceans have a major effect on climate, because water in the oceans holds a large amount of heat. Code: D1j	**Chapter 10** 10-4, 10-5 **Chapter 11** 11-7, 11-8, BI **Chapter 12** 12-7, 12-8, 12-10, 12-11 **Chapter 13** 13-1, 13-2, 13-3, 13-4, 13-5, Lab
k. Living organisms have played many roles in the Earth system, including affecting the composition of the atmosphere, producing some types of rocks, and contributing to the weathering of rocks. Code: D1k	**Chapter 3** 3-2, 3-3, 3-4, 3-5, 3-6, 3-7 **Chapter 4** 4-1, 4-2, 4-3, 4-4, 4-5, 4-6, Lab, BI **Chapter 7** 7-2, 7-4 **Chapter 10** 10-7, 10-9, BI **Chapter 12** 12-1 **Chapter 13** BI **Chapter 15** 15-1, 15-2, 15-3, 15-6

2. EARTH'S HISTORY

Standard	Chapter Correlation
a. The Earth processes we see today, including erosion, movement of lithospheric plates, and changes in atmospheric composition, are similar to those that occurred in the past. Earth history is also influenced by occasional catastrophes, such as the impact of an asteroid or comet. Code: D2a	**Chapter 4** 4-5 **Chapter 5** 5-1, 5-2, 5-3, 5-4, 5-5, BI **Chapter 6** 6-1, 6-2, 6-4, 6-5, 6-6, Lab, BI **Chapter 7** 7-1, 7-2, Lab **Chapter 8** 8-1, 8-2, 8-3, 8-7, 8-8, 8-9, 8-10, BI, Lab **Chapter 9** 9-4, 9-5, Lab
b. Fossils provide important evidence of how life and environmental conditions have changed. Code: D2b	**Chapter 4** 4-1, 4-2, 4-3, 4-4, 4-5, 4-6, Lab **Chapter 6** 6-1, 6-3

3. EARTH IN THE SOLAR SYSTEM

Standard	Chapter Correlation
a. The Earth is the third planet from the Sun in a system that includes the Moon, the Sun, seven other planets and their moons, and smaller objects, such as asteroids and comets. The Sun, an average star, is the central and largest body in the solar system. Code: D3a	**Chapter 14** 14-1 **Chapter 16** 16-1, 16-2, 16-6 **Chapter 17** 17-1, 17-3, 17-4, 17-5, 17-6, 17-7, 17-8, 17-9, 17-10, Lab **Chapter 18** 18-4, 18-6
b. Most objects in the solar system are in regular and predictable motion. Those motions explain such phenomena as the day, the year, phases of the moon, and eclipses. Code: D3b	**Chapter 14** 14-1, 14-3, 14-4, 14-5, 14-6, 14-7, 14-8, Lab **Chapter 18** 18-11
c. Gravity is the force that keeps planets in orbit around the Sun and governs the rest of the motion in the solar system. Gravity alone holds us to Earth's surface and explains the phenomena of the tides. Code: D3c	**Chapter 10** 10-6 **Chapter 16** 16-7, Lab **Chapter 17** 17-1, 17-2, 17-3, BI
d. The Sun is the major source of energy for phenomena on Earth's surface, such as growth of plants, winds, ocean currents, and the water cycle. Seasons result from variations in the amount of the Sun's energy hitting the surface, due to the tilt of Earth's rotation on its axis and the length of the day. Code: D3d	**Chapter 4** 4-6 **Chapter 11** 11-3, 11-4, 11-8 **Chapter 14** 14-3, 14-4, BI **Chapter 18** 18-3, 18-6, 18-7, 18-8

For correlations to selected states, visit **www.conceptsandchallenges.com** and enter the Teacher site.

Earth Science
Materials and Equipment

The following charts provide comprehensive lists of the materials needed to complete all of the Investigates, Hands-On Activities, and Lab Activities in the Student Edition of *Concepts and Challenges Earth Science*. This list assumes that the students are grouped according to the suggestions given in the Teacher's Edition for each activity. For a list of material suppliers, **see page xxviii.**

	Amount*	Investigate	Hands-On Activity	Lab Activity
Consumables				
Adding machine paper	1 roll	4-5		
Aluminum foil	1 roll			Ch. 13
Antacid tablets, fizzing	30 tablets			Ch. 16
Aquarium hose, narrow, 60-cm lengths	30**			Ch. 9
Assorted items (sticks, paper strips, or pebbles)		8-1		
Baking soda	1 large box	2-7		
Balloons, average size	40		16-7	Ch. 11
Balloons, average size, round	30	18-1		
Ballpoint pen covers	30		11-9	
Balls, plastic-foam, 4-in. diameter	15**	14-1		
Bottles, 2-L plastic	10†			Ch. 12
Bottles, glass, narrow-necked	10†			Ch. 11
Calcite or limestone samples	30	2-5		
Candy, hard	60 pieces	7-1		
Cardboard,				
8.5 in. × 11 in.	100 sheets		11-9, 17-2	Ch. 5, 11
15 cm × 15 cm	10 sheets†			
20 cm × 28 cm	75 sheets**	16-4	18-6	
Cardboard boxes, large	4			Ch. 8
Cement, waterproof, 4-oz size	3 tubes**			Ch. 9
Cereal flakes, crushed	2 18-oz. boxes**		5-6	
Chalk, white	60 pieces		7-3	
Clay powder	1 bag	3-4		
Coffee stirrers	15**	1-3		
Computer paper, continuous-feed	90 sheets		16-6	
Construction paper, assorted colors	200 sheets	6-6	6-1, 15-4, 18-2	
Container, very large plastic, with lid	8††	4-1		
Cornstarch	5 boxes**	6-4		

* Unless otherwise noted, the amount is for 30 students working individually.
** Amount for 30 students working in pairs
† Amount for 30 students working in groups of three
†† Amount for 30 students working in groups of four (8 groups)
††† Amount for 4 groups

Materials and Equipment *(continued)*

	Amount*	Investigate	Hands-On Activity	Lab Activity
Consumables *(continued)*				
Cotton balls	1 bag	17-10	12-3	
Craft sticks	360**		8-4	Ch. 7
Cups, plastic	130	10-2, 10-4	HB-3, 7-3	Ch. 2
Cups, clear plastic	30	7-6, 9-1, 11-5		
Cups, plastic-foam	135		3-5, 11-7	Ch. 6
Detergent powder	30 teaspoons**	15-6		
Diffraction gratings	10†		18-2	
Egg cartons, plastic-foam	8			Ch. 2
Eggs, hard-boiled	15**	1-3		
Eggs, uncooked	15**	10-3		
Film canisters with inner-locking caps	30			Ch. 16
Filters, blue and red	30 small pieces of each	17-6		
Fishing line	90 meters		12-7	
Food coloring, assorted colors	10 bottles**	10-4, 15-8		Ch. 2, 6, 7
Glass pieces, small	15**		2-4	
Gloves, plastic	30 per activity	2-7		Ch. 1
Glue	6 bottles	17-10	6-1, 11-9	Ch. 10, 11
Granite samples	30	2-5		
Grapefruit	15**	1-5		
Grapes	1 bag**	10-2		
Graph paper 8.5 in. × 11 in.	40 sheets †		14-3	Ch. 11
Gravel (or aquarium stones)	6 bags		9-3	
Gravel (or pebbles)	1 bag	3-4	7-6	
Gumdrops, small	5 large bags**			Ch. 5
Houseplants, small, potted	30		12-1	
Hydrochloric acid, dilute	1 bottle	2-5		
Ice cubes	1 bag	2-7	HA-4, 9-1, 12-4	
Isopropyl alcohol	1 bottle		12-3	
Knives, butter	15**		2-4	
Knives, plastic	55	1-3	4-3, 5-1	Ch. 3
Labels, self-stick	45**	1-3		
Leaves and grass clippings	2 bags		8-4	
Light bulbs	10†		18-2	
Lunch bags, paper	30	12-9		
Magnesium sulfate	25 tablespoons			Ch. 2
Matches, wooden striking	1 box†			Ch. 12
Meat trays, plastic-foam	30**		HB-3	
Modeling clay, assorted colors	80 packages**	7-6, 5-1, 4-3		Ch. 1, 4, 3
Nails, steel	15**		2-4	
Paint sticks, wooden	20†			Ch. 10
Paper clips, 1 in. size	1 box (30)	18-11		
Paper towels	1 roll per activity	1-5, 1-7, 2-7, 15-6	7-3, 8-7	

Materials and Equipment *(continued)*

	Amount*	Investigate	Hands-On Activity	Lab Activity
Consumables *(continued)*				
Pebbles	1 cup† 8 cups		3-5 10-7	
Peppercorns, whole (or peas)	30	17-10		
Petroleum jelly, med. size	1 jar**		15-7	
Plaster of Paris	4 cups† 1 box** 1 cup** 2 boxes		3-5 5-6 17-3	 Ch. 4
Plastic bags, clear	30		12-1	
Plastic wrap, clear	1 roll		9-1	Ch. 8
Plates, paper, regular dinner 5-in. size 10-in. size	60** 30 30	 17-10 17-10	5-6	Ch. 13
Poster board, 2 in. × 6 in.	30 pieces	17-6		
Rock samples (collected by students)	10 per student		3-8	
Rubber bands, heavy	10†			Ch. 11
Rubber bands, small, thin	30	18-11		
Salt	1 box	10-3		Ch. 2
Sand	20 bags	3-4, 5-5, 11-3	7-6, 3-5, 8-7, 10-7	Ch. 8, 9, 10, 13, 17
Soil dry and dark from the ground peat-based with added fertilizers	10 bags 1 bag 1 bag† 1 bag 1 bag	8-1 11-3 7-5 7-5 7-5	8-4, HB-3, 7-6, 10-7	
Shoeboxes	30		17-3	
Soap bars	30	8-3	8-7	
Sod	4 squares**		HB-3	
Soda water	2 2-L bottles**	10-2		
Soil, fine	1 bag	3-4		
Soil, sandy	6 bags		9-3	
Spaghetti, uncooked	10 boxes**			Ch. 5
Stickers	300		3-8	
Sticks, wooden	15**	14-1		
Stirrers	45	3-4, 10-4		
Storage boxes, large clear plastic	15**			Ch. 9
Straws, regular size, any color	50		8-4, 16-6, 16-7	Ch. 11
Straws, regular size (clear plastic)	40	11-5		Ch. 3
Streamers	3 rolls	17-10		
String	3 rolls	18-11	17-2	
Sugar	25 tablespoons			Ch. 2
Sugar cubes	11 boxes**			Ch. 4, 7
Tape, clear cellophane	6 rolls for each activity	17-10	11-4, 16-6, 16-7	Ch. 5, 11, 16
Tape, duct, 2 in. × 60 yd	1 roll			Ch. 8

Materials and Equipment *(continued)*

	Amount*	Investigate	Hands-On Activity	Lab Activity
Consumables ***(continued)***				
Tape, masking, 2 in. × 60 yd	1 roll		12-6,18-2	
Thread	3 spools		11-4	Ch. 1
Tissue paper	15 sheets**		11-4	
Toothpicks	1 box**		15-7	Ch. 4
Tracing paper	30 sheets		6-1, 14-7	
Tubes, cardboard	10†		18-2	
Twist ties	30		12-1, 16-7	
Vegetable oil	15 teaspoons**	15-6		
Vinegar	1 gallon		7-3	
Aquarium tank	1		10-5	
Baking pans	30		8-4	
Balance scales	30	3-1		
Balls, plastic-foam, 4-in. diameter	15**			Ch. 14
Beakers, 100 mL 250 mL	 30** 30**	 10-3	 HA-4	 Ch. 7
Bicycle hand pumps	10†			Ch. 12
Binoculars	30 pairs		18-6	
Blocks, wooden	30		8-2	
Boards, wooden	30		8-2	
Books	30**		HB-3	
Bottles, 2-L plastic	15**			Ch. 9
Bottles, small, plastic, with screw-on caps	30		11-7	
Bowls, small with rounded bottoms	10†		4-3	
Bowls, med. mixing	15**	15-6		
Boxes, 3 different sizes	15 sets**		HA-4	
Calculators	30	16-1		Ch. 15, 18
Cans, empty metal (no labels)	30		12-4	
Clamps	30		18-6	
Coal, anthracite or lignite	15 pieces**	15-3		
Colored pencils	30 sets	3-1		Ch. 3, 11
Compasses, directional	30		11-9	
Containers, clear, 1-gallon	10†	15-8		
Containers, plastic with lids	30**	6-4,10-4	8-7, 10-7	
Containers, plastic, clear	30		9-1	Ch. 6
Corks	5		10-5	
Crayons, assorted colors	10 small boxes†		18-2	
Drill	1			Ch. 9
Dropper bottles	15**			Ch. 7, 17

Materials and Equipment *(continued)*

	Amount*	Investigate	Hands-On Activity	Lab Activity
Non-Consumables *(continued)*				
Earthquake simulator	1			Ch. 5
Electric hair dryer	1			Ch. 8
Eyedroppers	30	5-8, 10-4		
Field guide to rocks	1		3-8	
Flashlights, small large	30 10	14-1, 16-4	14-3, 18-3 18-3	
Funnels	30		11-7	Ch. 9
Glasses, large	15**			Ch. 15
Globes	15**	1-5		
Graduated cylinders	30		12-1	Ch. 15
Grass sod	10 small squares†			Ch. 13
Hammers	30		7-3	
Hand lenses	30 15** 10†	3-1 15-3 7-5	3-8 3-3, 15-7	Ch. 2
Heat sources	15**		HA-4	
Hot plates	30	2-7		
Index cards, 4 in. × 6 in.	30		12-3	
Index cards, 5 in. × 8 in.	15**	16-4		
Jars, clear plastic with lids 12-oz. size small wide-mouthed with lids	30 15** 30 60	3-4 7-1	 12-6 	 Ch. 4
Knives	15**	1-5		
Lab aprons	30 per activity		HB-3, 2-5, 5-6, 7-3	Ch. 1, 2, 3, 4, 6, 7
Lamp, clip or table, with a 150-watt bulb	1			Ch. 14
Lamps	30	11-3	11-4	
Lenses, convex (or magnifying glasses)	60		16-3	
Marbles	30		8-2	
Measuring cups, 2-cup size with metric units labeled	15** 10†	 15-8	5-6 	
Microscope slides	60 15**	2-7 	 15-7	
Microscopes	30	2-7		
Mixing tubs	15**			Ch. 4
Modeling clay	8 packages** 1 package** 2 packages**	6-6 14-1 16-4		
Objects, small, assorted	30 sets**		HA-1	
Pails, plastic or metal, 2-gal size	15**			Ch. 9
Pebbles, assorted sets of 15	15 sets**		HA-2	
Pennies	30	3-1	2-4	
Pens, ballpoint	30	17-10		Ch. 2
Pens, felt-tip	15**	1-5		

Materials and Equipment *(continued)*

	Amount*	Investigate	Hands-On Activity	Lab Activity
Non-Consumables *(continued)*				
Pens, marking	30	18-11, 14-1	7-6	
Petri dishes, plastic	20†			Ch. 17
Picture, Hubble Deep Field				Ch. 18
Pictures, cratered bodies in our solar system				Ch. 17
Pictures, three-dimensional of Mars		17-6		
Pie plates, aluminum,8-in. diameter	30	5-5, 5-8, 8-1, 11-3		
Plastic tubs, large small	30 30		9-3 9-3	
Protractors	15**	14-1		
Pushpins	60		17-2	
Ring stands	30		18-6	
Rock samples, assorted basalt diorite gabbro two different sizes	 30 sets 15** 15** 15** 30 of each size	 3-1	 3-3 3-3 3-3 17-3	
Rocks, dark-colored (or asphalt pieces)	10 sets†			Ch. 13
Rocks, small	30			Ch. 8
Rubber bands, large	30**		HA-1	
Rulers, metric	30 per activity	4-5, 16-4, 18-11	HA-4, 1-9, 5-1, 5-6, 11-4, 14-7, 16-6, 17-2, 17-3	Ch. 1 Ch. 3, 5, 8
Rules, with metric and inch markings	30		12-6	
Safety goggles	30 per activity	2-7, 5-5, 7-1	HA-4, HB-3, 2-5, 7-3, 12-3	Ch. 2, 6, 7, 8, 12, 15 Ch. 16, 17
Sandpaper	30 pieces		8-2	
Scissors	30 pairs per activity	16-4, 17-6	HB-3, 6-1, 11-4, 11-9, 18-2	Ch. 2, Ch. 3, 8, 11, 16
Seashells or snail shells	15** small			Ch. 4
Shoeboxes, with lids	30**		HA-1	
Spoons, large	15**			Ch. 4
Spray bottles	30		9-3	
Stopwatches	30		12-3	Ch. 5, 6, 10
Storage boxes, plastic, large size	10†			Ch. 10
Tablespoons	30	3-4	3-5	Ch. 2, 16
Teaspoons	30	15-6	5-6	
Thermometers, mercury with Fahrenheit and Celsius units	30	11-3	HA-4, 12-4, 15-4	Ch. 12, 13, 15
Tweezers or small spoons	30	2-7		
Washers, metal	30		10-5	
Watches with second hands	15**			Ch. 15
Watering cans	30		HB-3, 8-4, 9-3	
Wax pencils	15**		HA-4	

Enhance Your Students' Performance

These additional titles can offer you more choices to help meet the diverse needs of your students and your teaching situation.

Event-Based Science

Engage your students in real-world, interdisciplinary tasks that motivate them to learn more about Earth science. Students are "hooked" after they read and watch authentic media coverage of a real-life event such as Hurricane Andrew or the Valdez oil spill. Extend your *Concepts and Challenges* lesson with the following Event-Based Science modules:

Asteroid!	***Oil Spill!***	***Tornado!***
Earthquake!	***Flood!***	***Toxic Leak!***
Gold Rush!	***Volcano!***	***Hurricane!***

For more information on these titles, visit **www.pearsonschool.com**.

Lab Suppliers and Addresses

American Science & Surplus
3605 Howard Street
Skokie, IL 60076
Phone: (800)934-0870
Fax: (800)934-0722
Email: info@sciplus.com
Web: http://www.sciplus.com

Arbor Scientific
P.O. Box 2750
Ann Arbor, MI 48106
Phone: (800)367-6695
Fax: (734)913-6201
Email: mail@arborsci.com
Web: http://www.arborsci.com

Carolina Biological Supply Co.
2700 York Road
Burlington, NC 27215
Phone: (800)334-5551
Fax: (800)222-7112
Email: carolina@carolina.com
Web: http://www2.carolina.com

Connecticut Valley
Biological Supply Co., Inc.
82 Valley Road, P.O. Box 326
Phone: (800)628-7748
Fax: (800)355-6813
Email: conval@ctvalleybio.com
Web: http://www.ctvalleybio.com

Edmund Scientific
60 Pearce Avenue
Tonawanda, NY 14150
Phone: (800)728-6999
Fax: (800)828-3299
Email: scientifics@edsci.com
Web: http://scientificsonline.com

Fisher Science Education
4500 Turnberry
Hanover Park, IL 60133
Phone: (800)955-1177
Fax: (800)955-0740
Email: info@fisheredu.com
Web: http://www.fishersci.com

Frey Scientific
P.O. Box 8101
100 Paragon Parkway
Mansfield, OH 44903
Phone: (800)225-3739
Fax: (877)256-3739
Email: customercare@freyscientific.com
Web: http://www.freyscientific.com

NASCO Science
901 Janesville Avenue, P.O. Box 901
Fort Atkinson, WI 53538
Phone: (800)558-9595
Fax: (920)563-8296
Email: custsevr@eNASCO.com
Web: http://wwwnascofa.com/science

Nature Watch
5312 Derry Avenue Unit R
Agoura Hills, CA 91301
Phone: (800)228-5816
Fax: (800)228-5814
Email: info@nature-watch.com
Web: http://www.nature-watch.com

Sargent Welch Scientific Company
P.O. Box 5229
Buffalo Grove, IL 60089
Phone: (800)727-4368
Fax: (800)676-2540
Email: sarwel@sargentwelch.com
Web: http://www.sargentwelch.com

Science Kit @ Boreal Laboratories
777 East Park Drive
P.O. Box 5003
Tonawanda, NY 14150
Phone: (800)828-7777
Fax: (800)828-3299
Email: sk@sciencekit.com
Web: http://www.sciencekit.com

WARD'S Natural Science Establishment, Inc.
5100 West Henrietta Road
Rochester, NY 14692
Phone: (800)962-2660
Fax: (800)635-8439
Email: customer-service@warsci.com
Web: http://www.warsci.com

CONCEPTS AND CHALLENGES

EARTH ❖ SCIENCE

Fourth Edition

Leonard Bernstein ◆ Martin Schachter ◆ Alan Winkler ◆ Stanley Wolfe

Stanley Wolfe
Project Coordinator

PEARSON

About the Cover: Earth science is the study of Earth and its history. The images of Glen Canyon and Lake Powell in Utah and dripping water represent some of the things you will be learning in this book. Erosion caused by water helped to create Glen Canyon. The rock layers of the Canyon provide Earth scientists with an opportunity to study the geologic history of this region of North America. What do you think are some of the other things you will study in Earth science?

Staff Credits: Amanda Aranowski, Doug Bauernschmidt, Melania Benzinger, Karen Blonigen, Sarah Brandel, Katie Colón, Barbara Drewlo, Daren Hastings, Helen Higgins, Mariann Johanneck, Becky Johnson, Mary Kaye Kuzma, Vivian Lemanowski, Charles Luey, Mary Lukkonen, Paul Ramos, Dan Ray, Marie Schaefle, Christopher Tures, Mike Vineski, Sue Will

ISBN-13: 978-0-7854-6763-2
ISBN-10: 0-7854-6763-7

1 2 3 4 5 6 7 8 9 10 12 11 10 09 08

PEARSON

1-800-321-3106
www.pearsonschool.com

Acknowledgments

Science Consultants

Gregory L. Vogt, Ed.D.
Associate Professor
Colorado State University
Fort Collins, CO

Stephen T. Lofthouse
Pace University
New York, NY

Laboratory Consultants

Sean M. Devine
Science Teacher
Ridge High School
Basking Ridge, NJ

Vincent R. Dionisio
Science Teacher
Clifton High School
Clifton, NJ

Reading Consultant

Sharon Cook
Consultant
Leadership in Literacy

Internet Consultant

Janet M. Gaudino
Seventh Grade Science Teacher
Montgomery Middle School
Skillman, NJ

ESL/ELL Consultant

Elizabeth Jimenez
Consultant
Pomona, CA

Content Reviewers

Sharon Danielsen (pp. 320-321, 334-335)
Site Manager
Darrin Fresh Water Institute
Rensselaer Polytechnic Institute
Troy, NY

Art DeGaetano (Chs. 12, 13)
Associate Professor
Cornell University
Ithaca, NY

Ivan Dmochowski (pp. 364-365)
Helen Hay Whitney Postdoctoral Scholar
California Institute of Technology
Pasadena, CA

Marian B. Jacobs Ph.D. (Chs. 1, 2, 3, 4)
Columbia University
New York, NY

Dr. Charles Liu (Chs. 5, 7, 8, 16, 17, 18)
Astrophysicist
Department of Astrophysics and Hayden Planetarium
American Museum of Natural History
New York, NY

Terry Moran (pp. 30-31, 152-153, 202-203)
Moran Research Service
Harvard, MA

Dr. Nathan M. Reiss (Ch. 11)
Professor Emeritus
Department of Environmental Sciences
Rutgers University
New Brunswick, NJ

Dr. Gerald Schubert (Ch. 11)
Department of Earth and Space Sciences
University of California—Los Angeles
Los Angeles, CA

Dr. Dirk Schulze-Makuch (Chs. 9, 10)
Department of Geological Sciences
University of Texas at El Paso
El Paso, TX

Seth Stein (Ch. 6)
Professor
Geological Sciences
Northwestern University
Evanston, IL

Hugh P. Taylor, Jr. (Chs. 14, 15)
Robert P. Sharp Professor of Geology
Division of Geological and Planetary Sciences
MS 100-23
California Institute of Technology
Pasadena, CA

Dr. Raymond C. Turner (Chs. 14, 15)
Alumni Distinguished Professor Emeritus of Physics
Department of Physics and Astronomy
Clemson University
Clemson, SC

Todd Woerner (pp. 84–85, 172–173, 268–269, 412–413)
Department of Chemistry
Duke University
Durham, NC

Teacher Reviewers

Peggy L. Cook
Lakeworth Middle School
Lakeworth, FL

Claudia Toback
Consultant/Mentor
Staten Island, NY

Contents

Scientific Skills and Investigations Handbooks

UNIT 1 INTRODUCTION TO EARTH

Chapter 3 Rocks and Their Origins . . . 65

Chapter 4 The Rock Record . . . 89

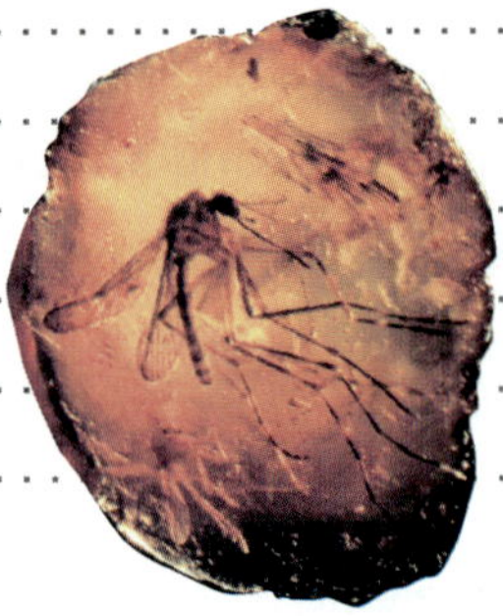

UNIT 2 THE LITHOSPHERE

Chapter 5 The Changing Landscape . . . 109

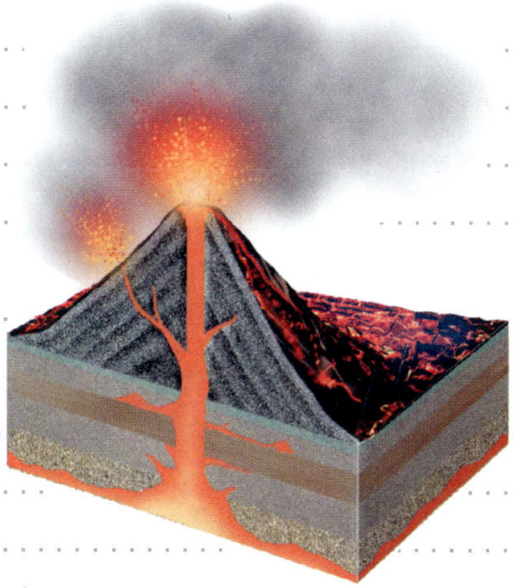

Chapter 6 Plate Tectonics . . . 137

Chapter 7 Weathering and Soil ... 157

Chapter 8 Erosion ... 177

UNIT 3 THE HYDROSPHERE

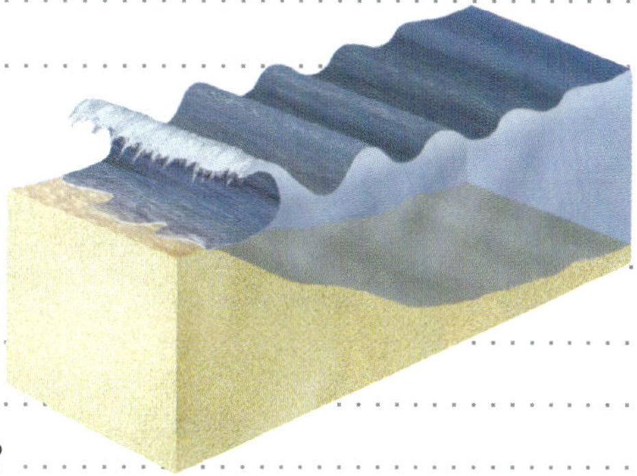

UNIT 4 *The Atmosphere*

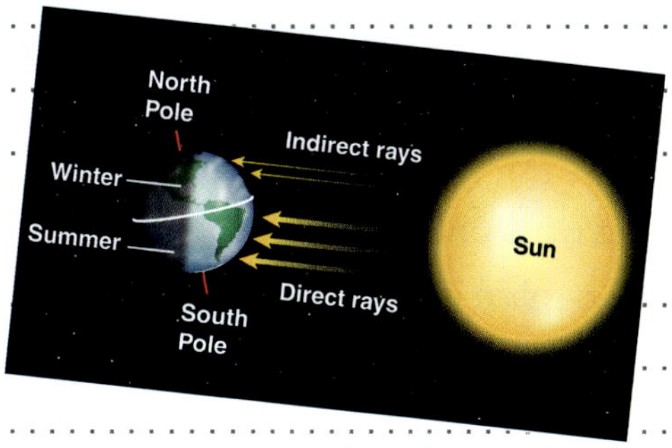

North Pole
Indirect rays
Winter
Summer
Sun
Direct rays
South Pole

UNIT 6 EARTH AND SPACE

Chapter 17 The Solar System . . . 407

Chapter 18 The Sun and Other Stars . . . 435

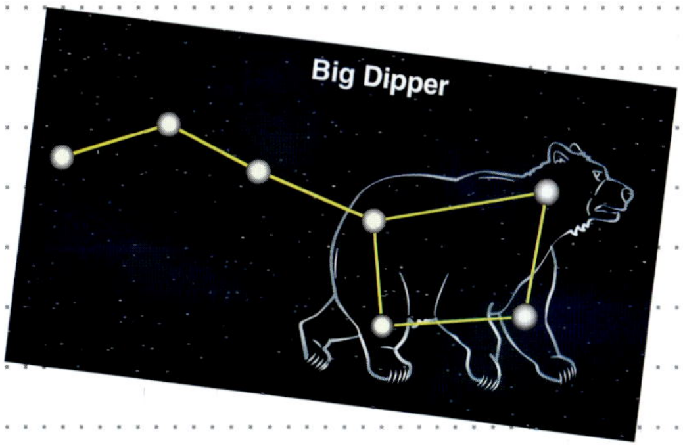

Appendices

Earth Science Features

Hands-On Activity

How Do They Know That?

People in Science

Integrating the Sciences

Real-Life Science

Science and Technology

INVESTIGATE

Web InfoSearch

What are scientific skills?

People are naturally curious. They want to understand the world around them. They want to understand what causes earthquakes and where is the best place to search for useful minerals. The field of science would probably not exist if it were not for human curiosity about the natural world.

People also want to be able to make good guesses about the future. They want to be able to track severe storms such as hurricanes and to find ways to protect their homes against flooding.

Scientists use many skills to explore the world and gather information about it. These skills are called science process skills. Another name for them is science inquiry skills.

Science process skills allow you to think like a scientist. They help you identify problems and answer questions. Sometimes they help you solve problems. More often, they provide some possible answers and lead to more questions. In this book, you will use a variety of science process skills to understand the facts and theories in Earth science.

Science process skills are not only used in science. You compare prices when you shop and you observe what happens to foods when you cook them. You predict what the weather will be by looking at the sky. In fact, science process skills are really everyday life skills that have been adapted for problem solving in science.

1 NAME: What is the name for the skills scientists use to solve problems?

▲ **Figure 1** Scientists use science process skills to understand how gravity affects the way crystals grow and materials mix, how caves form and change, how the land is built up and then torn down, and what Earth's place is in the universe.

Contents

Teaching Options

ASSESSMENT PLANNER

For assessment in the Student Edition, see the following pages:

Performance-Based Assessment
Hands-On Activity: pp. 2, 3, 6, 7

• TEACHING RESOURCES • Online or Print

Teacher's Resources

Weekly Journal: Study Tools, p. 19

Writing Activity Rubric: Teaching Tools, p. 9

Developing Science Skills: Handbook Resources, p. 1

Handbook Key Term Review: Handbook Resources, pp. 9–10

Handbook A: Developing Skills

What are scientific skills?

Handbook Overview

In this chapter, students will identify the science process skills and learn how to use these skills to help them think like scientists. They will read about the tools scientists use to collect data and how scientists classify data. Students will learn how length, area, weight, volume, and temperature are measured and will learn how scientists analyze data and communicate their results to other scientists. Finally, students will come to understand the importance of using data to make scientific predictions.

About Figure 1 Scientific skills help scientists understand the various processes that shape Earth's surface. Caves, such as the one shown in the top photograph, are formed when rain falls and collects over time on top of limestone. Compounds in the water dissolve the limestone and carry parts of it away. Over a period of years, underground chambers are formed. Crystals, such as those in the center picture, which are sometimes formed when liquids freeze, can be reheated and then quickly cooled. Canyons, such as the one shown in the bottom picture, are created by the erosive action of a river. Canyons are more likely to form in dry regions, where river erosion can cut through dry rock.

Answer to question

1 NAME: science process skills

Handbook Project

SCIENCE PROCESS SKILLS JOURNAL

MATERIALS: Weekly Journal reproducible found in the Teacher's Resources, pen or pencil

As students read Handbook A, ask them to keep a journal of the science process skills that they use in their daily lives. Students should make journal entries every day for a week and include at least five of the seven skills discussed in this handbook. Encourage students to write about their experiences at home in addition to their experiences in school. Invite students to share their projects with the class.
learning style: linguistic

1 Observing and Comparing

STUDY HINT Before beginning this lesson, write the title of each section on the board. Have students copy these titles in their notebooks, using an outline format. As student read each section, they should write notes in their outlines.

Demonstration Ask students what it means to observe, mentioning that observing is not limited to what they can see. Then, point to an object, such as a rock, and allow 2 minutes for students to observe it. You might pass the object around so that students can touch it.

Discussion As students read the description of each science skill, ask them to identify examples of how they use the skill in their daily lives. Emphasize that these skills are used in all academic disciplines, not only in science.

Answers to questions

2 **LIST:** sight, hearing, touch, smell, taste

3 **COMPARE/CONTRAST:** Possible response: Both valleys are surrounded by high hills. The valley made by a river has rolling hills that are covered with vegetation. The valley made by a glacier has steep, rocky hills with sharp edges.

4 **INFER:** Accept any reasonable response such as measuring the amount of vibration caused by a subway train.

Hands-On Activity
MAKING OBSERVATIONS

TIME: 15 minutes

PURPOSE: identifying, analyzing

PREPARATION: Gather materials before class. Include objects such as feathers, marbles, dice, jacks, and rocks. If you do not have enough materials for each student to have a box, make several boxes and have students work in small groups.

Practicing Your Skills Answers

6. **IDENTIFY:** observing
7. **IDENTIFY:** the sense of hearing
8. **ANALYZE:** Possible answer: Both. I used direct observation when I listened to the sounds that the object made in the box. I used indirect observation when I used my past experiences to try to picture what object was in the box.

1 Observing and Comparing

Making Observations An important part of solving any problem is observing, or using your senses to find out what is going on around you. The five senses are sight, hearing, touch, smell, and taste. When you look at a pebble and feel its smoothness, you are observing. When you observe, you pay close attention to everything that happens around you.

Scientists observe the world in ways that other scientists can repeat. This is a goal of scientific observation. It is expected that when a scientist has made an observation, other people will be able to make the same observation.

2 **LIST:** What are the five senses?

Comparing and Contrasting Part of observing is comparing and contrasting. When you compare data, you observe the characteristics of several things or events to see how they are alike. When you contrast data, you look for ways that similar things are different from one another.

▲ **Figure 2** River and glacial cut valleys look similar. However, you can see many differences from the valley floor.

3 **COMPARE/CONTRAST:** How are valleys carved by running water and valleys carved by glaciers similar? How are they different?

Using Tools to Observe Sometimes an object is too small to see with your eyes alone. You need a special tool to help you make observations. One tool that scientists use to observe is the seismograph. A seismograph detects earthquakes by measuring the vibrations of Earth's crust.

▲ **Figure 3** Seismologist checking a seismograph

4 **INFER:** Besides detecting earthquakes, what other use does a seismograph have?

Hands-On Activity
MAKING OBSERVATIONS

You and a partner will need 2 shoeboxes with lids, 2 rubber bands, and several small objects.

1. Place several small objects into the shoebox. Do not let your partner see what you put into the shoebox.
2. Cover the shoebox with the lid. Put a rubber band around the shoebox to keep the lid on.
3. Exchange shoeboxes with your partner.
4. Gently shake, turn, and rattle the shoebox.
5. Try to describe what is in the shoebox without opening it. Write your descriptions on a sheet of paper.

Practicing Your Skills

6. **IDENTIFY:** What science process skill did you use?
7. **IDENTIFY:** Which of your senses was most important to you?
8. **ANALYZE:** Direct observation is seeing something with your eyes or hearing it with your ears. Indirect observation involves using a model or past experience to make a guess about something. Which kind of observation did you use?

Teaching Options

• TEACHING RESOURCES •
Online or Print

Teacher's Resources

Venn Diagram: Study Tools, p. 10

Writing Activity Outline: Study Tools, pp. 20–21

COOPERATIVE LEARNING

Comparing and Contrasting Practice Students can work with partners to practice the skill of comparing and contrasting. Give pairs of students two objects or pictures of objects to compare and contrast. Give each student a copy of the Venn Diagram from the Teacher's Resources. Have students write similarities of earthquakes and volcanoes where the circles overlap and write the differences between the two in individual circles.
learning style: visual

2 Classifying Data

Key Term

data (DAYT-uh): information you collect when you observe something

Collecting and Classifying Data The information you collect when you observe something is called **data.** The data from an experiment or from observations you have made are first recorded, or written down. Then, they are classified.

When you classify data, you group things together based on how they are alike. This information often comes from making comparisons as you observe. You may classify by size, shape, color, use, or any other important feature. Classifying data helps you recognize and understand the relationships between things. Classification makes studying large groups of things easier. For example, Earth scientists use classification to organize the different types of rocks and minerals.

5 EXPLAIN: How can you classify data?

Hands-On Activity

ORGANIZING ROCKS

You will need 15 pebbles of different colors, textures, and shapes.

1. Lay the pebbles out on a table. Classify the pebbles into two categories based on texture: *Smooth* or *Rough*.
2. Look at the pebbles you classified as smooth. Divide these pebbles into new categories based on similar colors.
3. Repeat Step 2 for the pebbles you classified as rough.

Practicing Your Skills

4. ANALYZE: How did you classify the pebbles? What other ways could you classify the pebbles?
5. EXPLAIN: Why is a classification system useful?

3 Modeling and Simulating

Key Terms

model: tool scientists use to represent an object or process

simulation: computer model that usually shows a process

Modeling Sometimes things are too small to see with your eyes alone. Other times, an object is too large to see. You may need a model to help you examine the object. A **model** is a good way to show what a very small or a very large object looks like. A model can have more details than what may be seen with just your eyes. It can be used to represent a process or an object that is hard to explain with words. A model can be a three-dimensional picture, a drawing, a computer image, or a diagram.

6 DEFINE: What is a model?

Simulating A **simulation** is a kind of model that shows a process. It is often done using a computer. You can use a simulation to predict the outcome of an experiment. Scientists use simulations to study everything from the insides of a volcano to the development of a tornado.

▲ **Figure 4** This student is discovering how volcanoes are created through successive layers of erupted lava.

7 DEFINE: What is a simulation?

MORE ON THE WEB

Coming to Your Senses At www.conceptsandchallenges.com, students can learn more about the five senses. Students can then form into groups to prepare oral presentations on the senses. You may also choose from a number of activities to use with your class.

• TEACHING RESOURCES •
Online or Print

Teacher's Resources

Group Activity Rubric: Teaching Tools, p. 6

Parts of a Volcano: Visuals, p. 4

ESL/ELL STRATEGY

Setting Goals Suggest that before reading each section, students read the section title and the question at the end and then set a goal. For the first section, for example, students could set the following goal: *I am going to read to find out about the five senses and how they are used in making observations.*

2 Classifying Data

STUDY HINT Ask students to write the terms *data* and *classify* on separate lines in their notebooks. As students read, they should write the definitions of these terms.

Discussion Explain to students that classifying is a way to organize information. Have students mention ways that they can classify rocks. (Possible answers include size, color, texture, or place or origin.) **learning style: auditory**

Answer to question

5 EXPLAIN: by size, shape, color, use, or any other important feature

Hands-On Activity

ORGANIZING ROCKS

TIME: 15 minutes

PURPOSE: analyzing, explaining

PREPARATION: Gather materials before class. Choose pebbles that can easily be classified as smooth or rough, and be sure to include a variety of colors. Divide the pebbles into sets so that students can work in pairs or in small groups.

Practicing Your Skills Answers

4. ANALYZE: by texture and color; pebbles can also be classified by size, shape, or weight
5. EXPLAIN: It makes it easier to study large groups of things.

3 Modeling and Simulating

Demonstration Show students various models that you have in class, such as the model of a volcano in the Teacher's Resources. Ask students why models might be better than pictures when studying certain topics. (A model is three-dimensional, so it gives a better idea of the shape and dimensions of an object.)

Discussion Ask students to suggest some scenarios in which simulation would be useful for studying Earth science. Students might say that geologists could learn new mining techniques that way.

Answers to questions

6 DEFINE: a tool that scientists use to represent an object or a process

7 DEFINE: a computer model that usually shows a process

4 Measuring

STUDY HINT Read the Key Terms aloud with students and ask them to give examples of items that might be measured in terms of meters (a running track), grams (a box of cereal), and so on. Have students write the terms in their notebooks and notes or questions that pertain to each term as they read. Allow time for students to share their notes and ask questions.

Demonstration Show students a metric ruler. Point out centimeter and millimeter divisions. Demonstrate the proper way to measure, using a metric ruler.
learning style: visual

Discussion Remind students that the inch, foot, and yard are measurements in the English system. Point out that no simple relationship exists among these units. Perform a sample calculation on the board such as changing 60 inches into feet. (5 feet; 12 inches = 1 foot) Then, introduce the metric system. Write 1,500 mm on the board. Show students how this number can be changed to centimeters (150) or meters (1.5) by moving the decimal point to the left. **learning styles: auditory, visual**

Answers to questions

8 IDENTIFY: Accept any of the two following systems: the metric, English, and the SI systems.

9 CALCULATE: 300 cm

READING STRATEGY Call attention to Figure 5 and remind students that prefixes are word parts that, when added to words, change the meanings of the words. Encourage students to think of other words that they know that include these prefixes, such as *centipede* and *decimal*. Ask how knowing the meanings of prefixes might help students become better readers.

4 Measuring

Key Terms

unit: amount used to measure something

meter: basic unit of length or distance

mass: amount of matter in something

gram: basic unit of mass

volume: amount of space an object takes up

liter: basic unit of liquid volume

meniscus: curve at the surface of a liquid in a thin tube

temperature: measure of the amount of heat energy something contains

Two Systems of Measurement When you measure, you compare an unknown value with a known value using standard units. A **unit** is an amount used to measure something. The metric system is an international system of measurement. Examples of metric units are the gram, the kilometer, and the liter. In the United States, the English system and the metric system are both used. Examples of units in the English system are the pound, the foot, and the gallon.

There is also a more modern form of the metric system called SI. The letters *SI* stand for the French words *Système International*. Many of the units in the SI are the same as those in the metric system.

The metric and SI systems are both based on units of 10. This makes them easy to use. Each unit in these systems is ten times greater than the unit before it. To show a change in the size of a unit, you add a prefix to the unit. The prefix tells you whether the unit is larger or smaller. For example, a centimeter is ten times bigger than a millimeter.

PREFIXES AND THEIR MEANINGS	
kilo-	one thousand (1,000)
hecto-	one hundred (100)
deca-	ten (10)
deci-	one-tenth (1/10)
centi-	one-hundredth (1/100)
milli-	one-thousandth (1/1,000)

◀ Figure 5

8 IDENTIFY: What are two measurement systems?

Units of Length Length is the distance from one point to another. In the metric system, the basic unit of length or distance is the **meter.** A meter is about the length from a doorknob to the floor. Longer distances, such as the distances between cities, are measured in kilometers. A kilometer is 1,000 meters. Centimeters and millimeters measure shorter distances. A centimeter is 1/100 of a meter. A millimeter is 1/1,000 of a meter. Figure 6 compares common units of length. It also shows the abbreviation for each unit.

SI/METRIC UNITS OF LENGTH	
1,000 millimeters (mm)	1 meter (m)
100 centimeters (cm)	1 meter
10 decimeters (dm)	1 meter
10 millimeters	1 centimeter
1,000 meters	1 kilometer (km)

▲ Figure 6

Length can be measured with a meter stick. A meter stick is 1 m long and is divided into 100 equal lengths by numbered lines. The distance between each of these lines is equal to 1 cm. Each centimeter is divided into ten equal parts. Each one of these parts is equal to 1 mm.

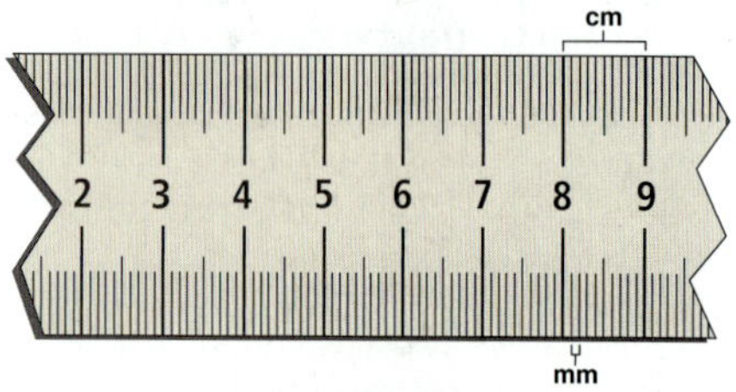

▲ Figure 7 A meter stick is divided into centimeters and millimeters.

9 CALCULATE: How many centimeters are there in 3 meters?

4

Teaching Options

EXTENDING THE LESSON

Measurements From the Past Discuss with students some of the measurements that people used in the past. (Alternatively, students could research past units of measurement on their own.) At one time, for example, the distance from a king's nose to his hand equaled 1 yard, and the length of a king's foot equaled 1 foot. Elicit from students why this system of measurement created problems. (Possible answer: The measuring system changed each time a new king was crowned.) Use this discussion to point out the importance of having a standard system of measurement.
learning style: auditory

RETEACH

Metric Unit Practice If students need more practice with metric units, use Figure 6 as the basis for review. Copy the chart by writing each of the units of measurement on separate index cards. Students can work in pairs or small groups to match equivalent measurements. They can check their answers against Figure 6.
learning styles: visual, tactile

Measuring Area Do you know how people find the area of the floor of a room? They measure the length and the width of the room. Then, they multiply the two numbers. You can find the area of any rectangle by multiplying its length by its width. Area is expressed in square units, such as square meters (m^2) or square centimeters (cm^2).

$$\text{Area} = \text{length} \times \text{width}$$

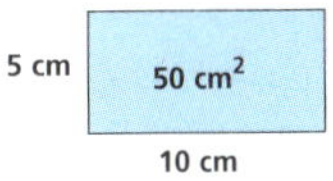

◀ **Figure 8** The area of a rectangle equals length times width.

10 CALCULATE: What is the area of a rectangle 2 cm × 3 cm?

Mass and Weight The amount of matter in something is its **mass.** The basic metric unit of mass is called a **gram (g).** A paper clip has about 1 g of mass. Mass is measured with an instrument called a balance. A balance works like a seesaw. It compares an unknown mass with a known mass.

One kind of balance that is commonly used to measure mass is a triple-beam balance. A triple-beam balance has a pan. The object being measured is placed on the pan. The balance also has three beams. Weights, called riders, are moved along each beam until the object on the pan is balanced. Each rider gives a reading in grams. The mass of the object is equal to the total readings of all three riders.

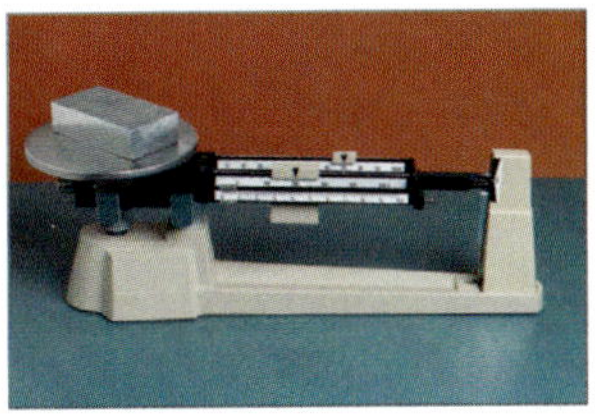

◀ **Figure 9** A triple-beam balance

Mass and weight are related; however, they are not the same. The weight of an object is a measure of Earth's pull of gravity between Earth and that object. Gravity is the force that pulls objects toward the center of Earth. The strength of the pull of gravity between two objects depends on the distance between the objects and how much mass they each contain. So, the weight changes as its distance from the center of Earth changes.

11 IDENTIFY: What instrument is used to measure mass?

Volume The amount of space an object takes up is its **volume.** You can measure the volume of liquids and solids. Liquid volume is usually measured in **liters.** Soft drinks in the United States often come in two-liter bottles.

A graduated cylinder is used to measure liquid volume. Graduated cylinders are calibrated, or marked off, at regular intervals. Look at Figure 10. It shows a graduated cylinder. On this graduated cylinder, each small line is equal to 0.05 mL. The longer lines mark off every 0.25 mL up to 5.00 mL. However, every graduated cylinder is not calibrated in this manner. They come in different sizes up to 2,000 mL, with different calibrations.

Always read the measurement at eye level. If you are using a glass graduated cylinder, you will need to read the mark on the graduated cylinder closest to the bottom of the meniscus. A **meniscus** is the curve at the surface of a liquid in a thin tube. A plastic graduated cylinder does not show a meniscus.

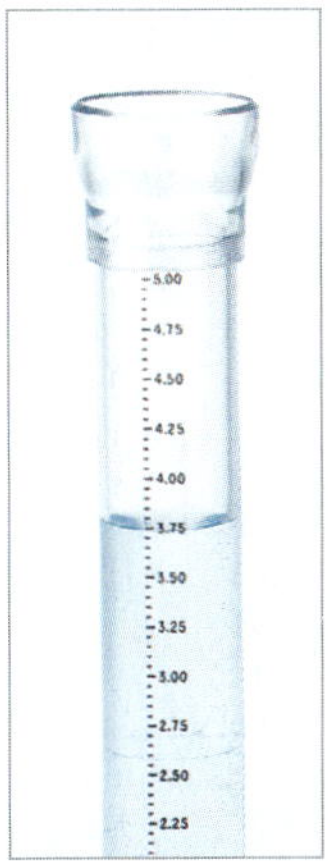

▲ **Figure 10** This glass graduated cylinder shows a meniscus.

COOPERATIVE LEARNING

Volume Displacement Experiment
Ask students how they could estimate the volume of an object if they had the following materials: a graduated cylinder, a cup, some liquid, and an object. Students can work in pairs or small groups to design an experiment. If students need help, lead them to realize that they should measure the volume of the liquid and the object. Then, they should remove the object, remeasure the volume, and subtract to find the difference. If time allows, groups of students can use displacement to approximate the volume of several small objects.
learning styles: tactile, visual

• *TEACHING RESOURCES* • *Online or Print*

Teacher's Resources

Measuring Length and Area: Handbook Resources, p. 2

Measuring Mass and Weight: Handbook Resources, p. 3

Measuring Volume: Handbook Resources, p. 4

Laboratory Manual

Lab. Skills Worksheets 1, 2, and 3

Demonstration Distribute 1-gram masses to the class. Have students lift these masses. Using the 1-gram masses for comparison, challenge students to estimate the mass (in grams) of books, pencils, and other common objects. Demonstrate by using a balance to check estimations.
learning styles: visual, tactile

Demonstration Display a 1-liter flask and explain that its volume is equal to 1,000 cubic cm. Remind students that 1/1000 of a liter is equal to 1 mL and, therefore, 1 mL is equal to 1 cubic cm.

You may also want to demonstrate reading the meniscus to students. Fill a glass graduated cylinder with a liquid. Lower your head until your eye is level with the meniscus. The line closest to the bottom of the meniscus should be recorded as the volume of the liquid. Allow students to observe the meniscus. **learning style: visual**

Discussion Introduce the square centimeter and square meter as units of area. Using a meterstick, measure the length and width of the classroom. Then, have students calculate the area of the classroom.
learning styles: auditory, visual

Reinforcement Make sure students understand that mass and weight are not the same measurement. Ask students to describe the difference between mass and weight. (*Mass* is the amount of matter in something, while an object's *weight* is the measure of Earth's pull of gravity between Earth and that object.) Then, ask students to explain in their own words why area is measured in square units. (Area involves two dimensions—length and width—so it must be measured in square units.)

Answers to questions

10 CALCULATE: 6 cm^2

11 IDENTIFY: a balance

Discussion Use Figure 12 to point out the differences between area and volume. The total area of the cubes at the front of the box is 6 cm^2, while the volume of the entire box is 12 cm^3. Volume is recorded in cubic units because it involves three dimensions.
learning style: auditory

Answer to question

12 CALCULATE: 12 mL

Hands-On Activity

CALCULATING AREA AND VOLUME

TIME: 15 minutes

PURPOSE: analyzing, calculating

PREPARATION: Gather a variety of boxes and label them with letters so that students can record their measurements. Remind students that volume is measured in cubic units, while area is measured in square units. Draw a table on the board with the columns *Box*, *Length (cm)*, *Width (cm)*, *Height (cm)*, *Volume (cm^3)*, and *Surface Area (cm^2)*. Ask students to copy the table into their notebooks and use it to record their measurements. Remind students that 1 cubic cm is equal to 1 mL.

COOPERATIVE LEARNING: Students can work in pairs or small groups for this activity.

Practicing Your Skills Answers

4. **ANALYZE:** Answer will depend on the boxes that are provided for the group.
5. **CALCULATE:** Answer will depend on the boxes that are provided for the group. Students should calculate the volume of each box and convert their answers to milliliters.
6. **ANALYZE:** Answer will depend on the boxes that are provided for the group.

Demonstration Demonstrate for students how a thermometer works. Display two beakers, one with ice and one with hot water. Place the thermometer in the ice and leave it until the temperature reading on the thermometer drops. Then, place the thermometer in the hot water and have students observe the liquid in the thermometer. Ask students to explain why the liquid rises in the tube.
learning style: visual

The volume of solid objects is often measured in cubic centimeters. One cubic centimeter equals 1 milliliter (mL).

Look at Figure 11. Each side of the cube is 1 cm long. The volume of the cube is 1 cubic centimeter (cm^3). Now, look at the drawing of the box in Figure 12. Its length is 3 cm. Its width is 2 cm. Its height is 2 cm. The volume of the box can be found by multiplying length by width by height. In this case, volume equals 3 × 2 × 2. Therefore, the volume of the box is 12 cm^3.

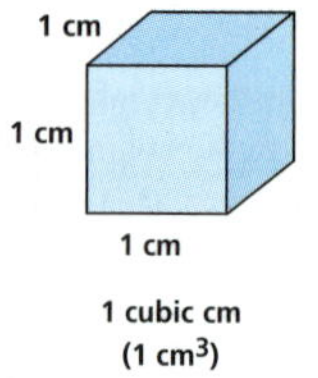

◀ **Figure 11** A cube

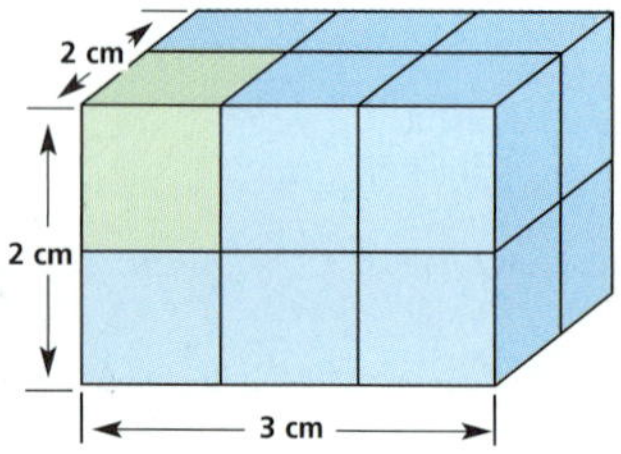

▲ **Figure 12** The volume of a box equals length times width times height.

$$V = L \times W \times H$$

If you have a box that is 10 cm on each side, its volume would be 1,000 cm^3. A liter is the same as 1,000 cm^3. One liter of liquid will fill the box exactly.

12 CALCULATE: How many milliliters of water would fill a 12 cm^3 box?

Hands-On Activity

CALCULATING AREA AND VOLUME

You will need 3 boxes of different sizes, paper, and a metric ruler.

1. Measure the length, width, and height of each box in centimeters. Record each measurement in your notes.
2. Calculate the volume of each box. Record each volume in your notes.
3. Find the surface area of each box. Record each area in your notes.

Practicing Your Skills

4. **ANALYZE:** Which of the three boxes has the largest volume?
5. **CALCULATE:** How many milliliters of liquid would fill each box?
6. **ANALYZE:** What is the surface area of the largest box?

Temperature **Temperature** is a measure of the amount of heat energy something contains. An instrument that measures temperature is called a thermometer.

Most thermometers are glass tubes. At the bottom of the tube is a wider part, called the bulb. The bulb is filled with liquid. Liquids that are often used include mercury, colored alcohol, or colored water. When heat is added, the liquid expands, or gets larger. It rises in the glass tube. When heat is taken away, the liquid contracts, or gets smaller. The liquid falls in the tube. On the side of the tube is a series of marks. You read the temperature by looking at the mark on the tube where the liquid stops.

Temperature can be measured on three different scales. These scales are the Fahrenheit (F) scale, the Celsius (C) scale, and the Kelvin (K) scale. The Fahrenheit scale is part of the English system of measurement. The Celsius scale is usually used in science. Almost all scientists, even in the United States, use the Celsius scale. Each unit on the Celsius scale is a degree Celsius (°C). The degree Celsius is the metric unit of temperature. Water freezes at 0°C. It boils at 100°C.

Teaching Options

ONGOING ASSESSMENT

Assess students' understanding of units of measurement by naming one and having students classify it as a measure of mass, length, or volume. You might instead name two of them, such as *decimeter* and *deciliter*, and ask students how the terms are similar and how they are different. (Both describe one-tenth, but one is a unit of length and one is a unit of volume.)

RETEACH

Estimating Measurements Ask students to use a ruler with centimeter markings to measure the width of their thumbs. Then, have them measure objects by using their thumb widths. They can check their estimates by measuring with rulers. Ask students to name things that they could measure if they knew the length of their feet (for example, the width and length of the classroom and the width of the hall outside the classroom).
learning styles: visual, tactile

Scientists working with very low temperatures use the Kelvin scale. The Kelvin scale is part of the SI measurement system. It begins at absolute zero, or 0K. This number indicates, in theory at least, a total lack of heat.

COMPARING TEMPERATURE SCALES			
	Kelvin	Fahrenheit	Celsius
Boiling point of water	373K	212°F	100°C
Human body temperature	310K	98.6°F	37°C
Freezing point of water	273K	32°F	0°C
Absolute zero	0K	−459.67°F	−273.15°C

▲ Figure 13

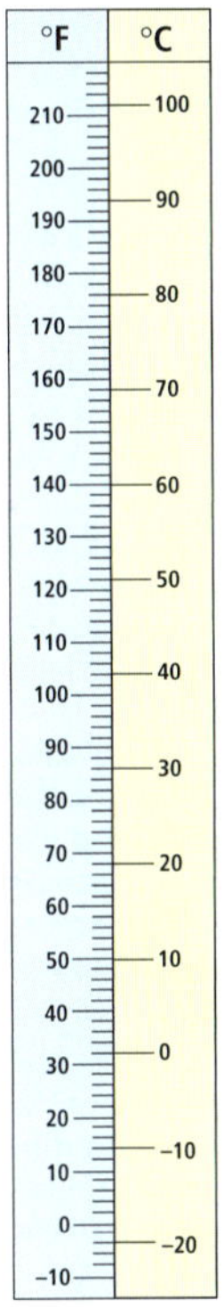

◀ **Figure 14** The Fahrenheit and Celsius scales

13 NAME: What are the three scales used to measure temperature?

Hands-On Activity

READING A THERMOMETER

You will need safety goggles, 2 beakers, a heat source, ice, water, a wax pencil, a ruler, and a standard Celsius thermometer.

1. Boil some water in a beaker.
   ⚠ CAUTION: Be very careful when working with heat. Place your thermometer in the beaker. Do not let the thermometer touch the sides or the bottom of the beaker. Wait until the mercury rises as far as it will go. Record the temperature.
2. Fill a beaker with ice water. Place the unmarked thermometer into this beaker. Wait until the mercury goes as low as it will go. Record the temperature.

▲ **STEP 1** Record the temperature of the boiling water.

Practicing Your Skills

3. IDENTIFY: What is the temperature at which the mercury rose as high as it would go?
4. IDENTIFY: What is the temperature at which the mercury went as low as it would go?

ESL/ELL STRATEGY

Working Together Have English-language learners work with students proficient in English to complete the two Hands-On Activities in this spread. Have one student do the setup and the other record the results.

MORE ON THE WEB

The History of the Metric System Have students use the Internet to learn more about the history of the metric system, including its modern form as the Système Internationale (SI).

• TEACHING RESOURCES • Online or Print

Teacher's Resources

Rulers—Inches and Centimeters: Visuals, p. 15

Temperature Scale: Visuals, p. 16

Measuring Temperature: Handbook Resources, p. 5

Discussion Using Figure 13, read out the temperatures of items listed on one of the scales. Have students identify the temperature readings on the other two scales. Then, have students refer to Figure 14. Ask: *Which temperature is higher—90°F or 40°C?* (40°C) *Which temperature would you most likely find on a winter day in the northern United States—20°F or 20°C?* (20°F) Have students make up questions and answers of their own.

Tell students that the most accurate thermometers are filled with mercury, which is a poisonous substance. Because mercury can be absorbed through the skin, a person should wear protective gloves to clean up any mercury that spills out of a broken thermometer. Mercury can be cleaned with paper towels. Caution students to handle thermometers carefully and notify you of any broken thermometers.

Answer to question

13 NAME: Kelvin, Fahrenheit, and Celsius

Hands-On Activity

READING A THERMOMETER

TIME: 15 minutes

PURPOSE: identifying

PREPARATION: Be sure to gather the materials before class. Demonstrate the experiment for students before they begin, emphasizing safety as you boil the water.

COOPERATIVE LEARNING: You can have students work in pairs. Ask them to set up the experiment.

DISCUSSION: Make sure that you check their setups before they begin boiling water. Remind them that they should read the thermometer at eye level to get an accurate reading.

Practicing Your Skills Answers

3. IDENTIFY: 100°C
4. IDENTIFY: 0°C

5 Analyzing Data and Communicating Results

Discussion Draw students' attention to the table and the graph. Point out that each graphic organizer shows the same data. Ask students: *Which is easier to use and understand? Why?* **learning style: auditory**

Answers to questions

14 **EXPLAIN:** They make data easier to compare and interpret.

15 **LIST:** books, magazines, newspaper articles, Web sites, and graphs

6 Making Predictions

Demonstration Give a concrete example to help students understand what it means to infer. Place a rock in a paper bag and, as you experiment with the bag, say: *Someone placed an object in this paper bag. I picked up the bag, and it was heavy. I shook the bag, and the item did not move very much. I touched the bag, and I felt something that was round and smooth. I inferred that the object inside was a rock.* Ask students how they can tell that you are inferring and not predicting. Encourage them to describe situations where they have inferred and predicted.

Discussion Discuss how the processes of inferring and predicting are different. Point out that an inference does not necessarily involve a future event.

Answer to question

16 **CONTRAST:** When you infer, you form a conclusion. When you predict, you state ahead of time what you think is going to happen. Predictions are sometimes based on inferences.

Handbook Project Wrap-Up

SCIENCE PROCESS SKILLS JOURNAL

Students can share their journals with the class. You can use the Individual Activity Rubric found in the Teacher's Resources to assess students' projects. Fill in the rubric with the additional information below. For this project, students should have:

- made a journal entry every day.
- included five skills discussed in the handbook.

5 Analyzing Data and Communicating Results

Key Term

communication: sharing information

Analyzing Data When you organize information, you put it in a logical order. In scientific experiments, it is important to organize your data. Data collected during an experiment are not very useful unless they are organized and easy to read. It is also important to organize your data if you plan to share the results of your experiment.

Scientists often organize information visually by using data tables, charts, graphs, and diagrams. By using tables, charts, graphs, and diagrams, scientists can display a lot of information in a small space. They also make it easier to compare and interpret data.

Tables are made up of rows and columns. Columns run up and down. Rows run from left to right. Tables usually show numerical data. Information in the table can be arranged in time order. It can also be set up to show patterns or trends. A table showing wind speed can reveal the effects the speed of wind will have on land. Figure 15 shows a table of gases in the atmosphere.

GASES IN THE ATMOSPHERE

Gas	Percentage
Oxygen	21
Carbon dioxide	0.04
Nitrogen	78
Water vapor, helium, and other gases	0.02
Argon	0.94

▲ Figure 15

Graphs, such as bar graphs, line graphs, and circle graphs, often use special coloring, shading, or patterns to represent information. Keys indicate what the special markings represent. Line graphs have horizontal (x) and vertical (y) axes to indicate such things as time and quantities.

14 **EXPLAIN:** How do tables and graphs help you analyze data?

Sharing Results When you talk to a friend, you are communicating, or sharing information. If you write a letter or a report, you are also communicating but in a different way. Scientists communicate all the time. They communicate to share results, information, and opinions. They write books and magazine or newspaper articles. They may also create Web sites about their work. This is called written **communication.**

Graphs are a visual way to communicate. The circle graph in Figure 16 is showing the same information from Figure 15. The circle graph presents the information in a different way.

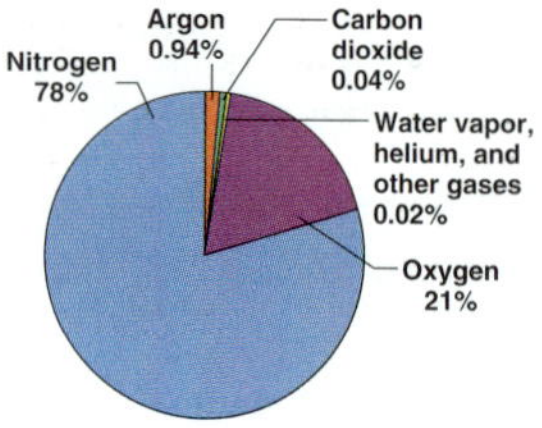

▲ Figure 16 Circle graphs are a good way to show parts of a whole.

15 **LIST:** What are some ways to communicate the results of an experiment?

6 Making Predictions

Key Terms

infer: to form a conclusion

predict: to state ahead of time what you think is going to happen

Thinking of Possibilities When you **infer** something, you form a conclusion. This is called making an inference. Your conclusion will usually be based on observations or past experience. You may use logic to form your statement. Your statement might be supported by evidence and perhaps can be tested by an experiment. An inference is not a fact. It is only one possible explanation.

When you **predict,** you state ahead of time what you think will happen. Predictions about future events are based on inferences, evidence, or past experience. The two science process skills of inferring and predicting are very closely related.

16 **CONTRAST:** What is the difference between inferring and predicting?

Teaching Options

EXTENDING THE LESSON

Displaying Data Use classroom data as the basis for a table and for a graph. For example, you might ask students to tell what their favorite school lunches are. Record students' responses on the board. Then, lead students as they construct a table and a circle graph based on the data.
learning style: visual

• TEACHING RESOURCES • Online or Print

Teacher's Resources

Graph: Visuals, p. 3

Graph Paper: Visuals, p. 4

Analyzing Data and Communicating Results: Handbook Resources, p. 6

Laboratory Manual

Lab. Skills Worksheet 4: Organizing and Analyzing Data, pp. 9–10

Lab. Skills Worksheet 5: Graphing, pp. 11–15

How do you conduct a scientific investigation?

By now, you should have a good understanding of the science process skills. These skills are used to solve many science problems. There is also a basic procedure, or plan, that scientists usually follow when conducting investigations. Some people call this procedure the scientific method.

The scientific method is a series of steps that can serve as a guide to solving problems or answering questions. It uses many of the science process skills you know, such as observing and predicting.

Not all experiments use all of the steps in the scientific method. Some experiments follow all of them, but in a different order. In fact, there is no one right scientific method. Each problem is different. Some problems may require steps that another problem would not. However, most investigations will follow the same basic procedure.

1 DESCRIBE: What is the scientific method?

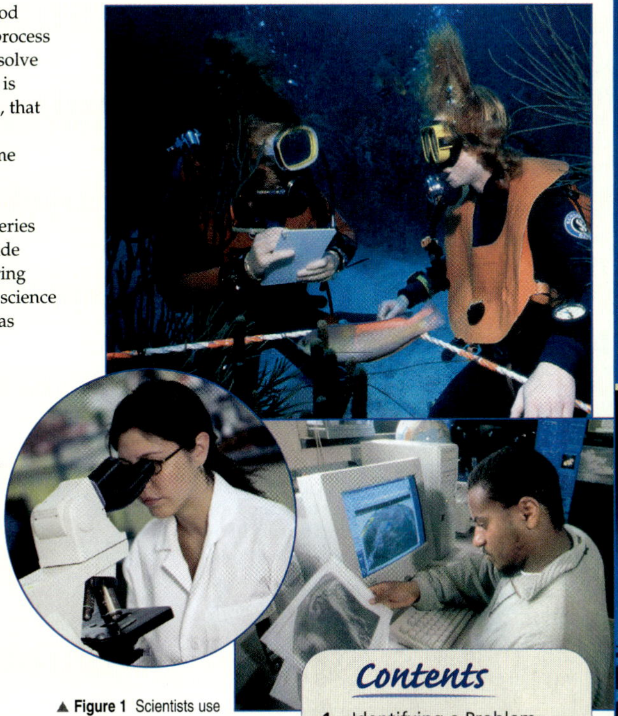

▲ **Figure 1** Scientists use the scientific method to guide experiments.

Contents

ASSESSMENT PLANNER

For assessment in the Student Edition, see the following pages:

Alternative Assessment
Building Skills: pp. 10, 12, 13

Performance-Based Assessment
Hands-On Activity: p. 12

• TEACHING RESOURCES • Online or Print

Teacher's Resources
Conducting Scientific Investigations: Handbook Resources, p. 7

Handbook B: Conducting Investigations

How do you conduct a scientific investigation?

Handbook Overview

In this part of the handbook, students will identify the steps of the scientific method. They will learn how to identify a problem and form a hypothesis as well as design and carry out an experiment. They will find out different ways that they can record and analyze data. Finally, they will become familiar with the final steps in the scientific method, which are stating a conclusion and writing a report. Students will also learn in this part of the handbook some basic safety procedures to be followed in the lab.

About Figure 1 Many people, including the ones shown here, use the scientific method in everyday life to help solve problems. The people shown in the photographs are most likely carrying out experiments, which is the third step in the scientific method.

Answer to question

1 DESCRIBE: a series of steps that can serve as a guide to solving problems or asking questions

Handbook Project

CONDUCTING EXPERIMENTS REFERENCE

MATERIALS: posterboard or paper (depending upon project chosen), markers or crayons, pen and pencil, computer

As students read Handbook B, they should create a resource to use throughout the year when they conduct experiments. Students should clearly record the steps of the scientific process and explain how to complete each. They might choose to illustrate the steps in a real experiment, such as the one described in this handbook. They should also include a copy of their completed lab report with the important parts highlighted and labeled. Students' resources could be in the form of pamphlets, illustrated posters, or a class Web site. Encourage students to use this resource throughout the year as they conduct and report on scientific experiments. **learning style: linguistic**

1 Identifying a Problem and Doing Research

Demonstration Show or describe various research resources and discuss with students what kind of information they would find in each one. A Web site, for example, might give up-to-the-minute information. An encyclopedia is a good source for general information about a topic. **learning styles: visual, linguistic**

Discussion Discuss with students why they would want to research a science problem before conducting experiments of their own. (Other people may have already investigated the problem. Learning all you can about a topic helps you to focus your research.) Give examples of scientific problems that students can restate as questions to be researched.

Answer to question

2 **IDENTIFY:** doing research to gather evidence

BUILDING SCIENCE SKILLS

Researching Background Information Provide resources for students' research. Remind them that using indexes can help them utilize their research time more effectively.

2 Forming a Hypothesis

Discussion Discuss with students the difference between a guess, a hypothesis, and a theory. Point out that a hypothesis is based on observations or prior knowledge or experience, while a guess may not have any basis or foundation. A theory is a set of hypotheses that have been tested many times. **learning style: auditory**

Answer to question

3 **DESCRIBE:** A scientist thinks of possible explanations for a set of observations or suggests possible answers to a scientific question.

BUILDING SCIENCE SKILLS

Developing a Hypothesis Guide students as they narrow the hypothesis. If students need help getting started, suggest possible hypotheses and ask students if these hypotheses are narrow enough to test.

1 Identifying a Problem and Doing Research

Starting an Investigation Scientists often state a problem as a question. This is the first step in a scientific investigation. Most experiments begin by asking a scientific question. That is, they ask a question that can be answered by gathering evidence. This question is the reason for the scientific investigation. It also helps determine how the investigation will proceed.

Have you ever done background research for a science project? When you do this kind of research, you are looking for data that others have already obtained on the same subject. You can gather research by reading books, magazines, and newspapers, and by using the Internet to find out what other scientists have done. Doing research is the first step of gathering evidence for a scientific investigation.

2 **IDENTIFY:** What is the first step of a scientific investigation?

BUILDING SCIENCE SKILLS

Researching Background Information Suppose you notice that a river running through your town looks brown on some days and clear on others. You also notice that when the river turns brown, it has usually rained the day before. You wonder if rain and the brown color of the river water are related.

To determine if the river water color is related to rainfall, look for information on rivers in encyclopedias, in geology books, and on the Internet. Put your findings in a report.

▲ **Figure 2** Water in river after a heavy rain.

2 Forming a Hypothesis

Key Terms

hypothesis: suggested answer to a question or problem

theory: set of hypotheses that have been supported by testing over and over again

Focusing the Investigation Scientists usually state clearly what they expect to find out in an investigation. This is called stating a hypothesis. A **hypothesis** is a suggested answer to a question or a solution to a problem. Stating a hypothesis helps to keep you focused on the problem and helps you decide what to test.

To form their hypotheses, scientists must think of possible explanations for a set of observations or they must suggest possible answers to a scientific question. One of those explanations becomes the hypothesis. In science, a hypothesis must include something that can be tested.

A hypothesis is more than just a guess. It must consider observations, past experiences, and previous knowledge. It is an inference turned into a statement that can be tested. A set of hypotheses that have been supported by testing over and over again by many scientists is called a **theory.** An example is the theory that explains how living things have evolved, or changed, over time.

A hypothesis can take the form of an "if…then" statement. A well-worded hypothesis is a guide for how to set up and perform an experiment.

3 **DESCRIBE:** How does a scientist form a hypothesis?

BUILDING SCIENCE SKILLS

Developing a Hypothesis If you are testing how river water and rainfall are related, you might write down this hypothesis:

Runoff is one cause of the river water turning brown.

Your hypothesis is incomplete. It is not enough to link water color and rainfall. You need to explain what materials make the river water brown and how rainfall causes those materials to get into the water. Revise the hypothesis above to make it more specific.

Teaching Options

RETEACH

Making a Flowchart Help students create a flowchart that lists the steps in the scientific method. This flowchart should have space for six steps and some details about each one. After students read this section, they should write the step in their flowcharts. They can continue filling in the chart as they read the rest of Handbook B. Give each student a copy of the Flowchart, which can be found in the Teacher's Resources.

• TEACHING RESOURCES •
Online or Print

Teacher's Resources
Flowchart: Study Tools, p. 6

3 Designing and Carrying Out an Experiment

Key Terms

variable: anything that can affect the outcome of an experiment

constant: something that does not change

controlled experiment: experiment in which all the conditions except one are kept constant

Testing the Hypothesis Scientists need to plan how to test their hypotheses. This means they must design an experiment. The plan must be a step-by-step procedure. It should include a record of any observations made or measurements taken.

All experiments must take variables into account. A **variable** is anything that can affect the outcome of an experiment. Room temperature, amount of sunlight, and water vapor in the air are just some of the many variables that could affect the outcome of an experiment.

4 DEFINE: What is a variable?

Controlling the Experiment One of the variables in an experiment should be what you are testing. This is what you will change during the experiment. All other variables need to remain the same. In this experiment, you will vary the type of earth.

A **constant** is something that does not change. If there are no constants in your experiment, you will not be sure why you got the results you did. An experiment in which all the conditions except one are kept constant is called a **controlled experiment.**

Some experiments have two setups. In one setup, called the control, nothing is changed. In the other setup, the variable being tested is changed. Later, the control group can be compared with the other group to provide useful data.

5 EXPLAIN: Explain how a controlled experiment is set up.

Designing the Procedure Suppose you now want to design an experiment to determine what makes river water brown. You have your hypothesis. You decide your procedure is to construct a slightly tilted model of the river, the town, and the land upstream from the town. You will send water down the river and measure the color and clarity of the water. Next, you will create artificial rain and again check the color and clarity of the water.

Does it matter how much rain you add to your model? Does it matter how heavy the rainfall is? Does 3 inches (7.5 cm) of rainfall in 5 minutes have the same effect on your model as 3 inches in 1 hour?

In designing your experiment, you need to identify the variables. The amount of water and the rate at which you apply it to your model are variables that could affect the outcome of your experiment. Another important variable for your experiment is the steepness of the river. To be sure of your results, you will have to conduct your experiment several times. Each time you will alter just one variable while keeping the other variables just the same.

Finally, you should decide on the data you will collect. How will you measure the color and clarity of the water of the river? You might make a color chart that you lower into the river water to see how the sediment in the water changes its color.

The hands-on activity on page 12 is an example of an experiment you might have designed.

◀ **Figure 3** In your experiment, you will elevate the trays with books to test soil runoff.

6 EXPLAIN: How do constants and variables affect an experiment?

Handbook B: Conducting Investigations

COOPERATIVE LEARNING

Designing an Experiment Ask students to work in pairs or small groups to design simple experiments that will test a hypothesis. Students should identify the question for research, state the hypothesis they would test, and tell how they would test it. Be sure that students identify both variables and constants in the experiment.

• TEACHING RESOURCES • Online or Print

Teacher's Resources

Designing an Experiment: Study Tools, p. 14

Designing an Experiment Rubric: Teaching Tools, p. 7

3 Designing and Carrying Out an Experiment

STUDY HINT Ask students to write a topic sentence for each of the sections as they read them. They can also write some supporting details about each.

Demonstration Tell students that you are going to test the hypothesis that heavier objects fall through liquid faster than lighter objects do. Set up a clear container, such as a fish tank, and fill it with a liquid. Drop several objects of different weights from the same place above the liquid. Ask: *What were the constants in your experiment?* (the type of liquid and the height from which objects were dropped) *What were the variables?* (the weights of the various objects) **learning style: visual**

Discussion Explain that in a controlled experiment, only one variable is tested at a time. Ask students to explain why it is important to have a control in an experiment. (By changing only one variable, scientists can be sure that any differences between the control and the experimental setup were caused by the variable being tested.) **learning style: auditory**

Answers to questions

4 DEFINE: anything that can affect the outcome of an experiment

5 EXPLAIN: An experiment is set up in which all the conditions except one are kept constant.

6 EXPLAIN: The data that you wish to collect determines the variables in the experiment. All the variables except one have to be kept constant to be sure that the desired question is being tested.

Hands-On Activity

CARRYING OUT AN EXPERIMENT

TIME: 30 minutes

PURPOSE: observing, comparing, explaining, inferring

PREPARATION: Students can work on this activity in pairs or small groups. Have them identify the controls (the make-up and preparation of the tray, and amount of water added to the samples) and variables (the soil and the sod).

Practicing Your Skills Answers

9. **OBSERVE:** Answers will vary. It should take less water added to the soil to get a full glass of water than it takes to get a full glass of water from the sod.
10. **OBSERVE:** Answers will vary. See answer to question 9 above.
11. **COMPARE:** the glass of water collected from the tray with the soil
12. **EXPLAIN:** The sod absorbs the water, and the water that runs off does not take much of the sod with it.
13. **INFER:** plant sod close to the edge of a river

4 Recording and Analyzing Data

STUDY HINT Give each student a copy of the Concept Map from the Teacher's Resources. Have students write the words *Dealing With Data* in the center circle. As students read, they can fill in the rest of the concept map.

Demonstration Take students on a "data walk" through another text. Show them different ways that data can be displayed. **learning styles: visual, auditory**

Discussion Tell students that collecting data is not the end of an experiment. Once data is collected, it must be organized and analyzed. Ask students to name some ways in which data might be organized. (charts, graphs, tables, and reports)

Answer to question

7 LIST: tables, charts, graphs, reports

BUILDING SCIENCE SKILLS

Analyzing Data Suggest that students organize the data shown in the notebook in Figure 4.

Hands-On Activity

CARRYING OUT AN EXPERIMENT

You will need 2 styrofoam meat trays from the grocery store, garden soil, grass sod, 2 plastic drinking glasses, 2 books, sprinkling can, scissors, and water. You should wear an apron and safety goggles.

1. Cut a small drain notch from the center of one end of each tray.
   ⚠ CAUTION: Be careful when using scissors.
2. Fill one tray with about a 1-inch layer (2.5 cm) of garden soil. Leave 2 inches (5 cm) of the notched end of the tray empty of soil.
3. Fill the second tray with a layer of sod except for 2 inches at the end with the notch.
4. Place the notched end of each tray at the edge of a table so that the trays extend over the edge a short distance.
5. Elevate the other end of the trays with books.
6. Label the cups *Soil* and *Sod*.
7. Sprinkle the soil tray with water. Keep sprinkling until the water runs off the surface of the soil and pours out the notch drain. Collect a glass of runoff water.
8. Repeat Step 7 with the sod tray.

Practicing Your Skills

9. **OBSERVE:** How much water did you have to sprinkle on the soil tray in order to collect a full glass?
10. **OBSERVE:** How much water did you have to sprinkle on the sod tray in order to collect a full glass?
11. **COMPARE:** Which glass had the dirtiest water?
12. **EXPLAIN:** What caused the difference in water clarity in the two glasses?
13. **INFER:** What would be the best way to reduce soil runoff in rivers?

4 Recording and Analyzing Data

Dealing With Data During an experiment, you must keep careful notes about what you observe. For example, you might need to note how long the rain fell on the trays before water began running off. How fast did the water run off each tray? This is important information that might affect your conclusion.

At the end of an experiment, you will need to study the data to find any patterns. Much of the data you will deal with is written text such as a report or a summary of an experiment. However, scientific information is often a set of numbers or facts presented in other, more visual ways. These visual presentations make the information easier to understand. Tables, charts, and graphs, for instance, help you understand a collection of facts on a topic.

After your data have been organized, you need to ask what the data show. Do they support your hypothesis? Do they show something wrong in your experiment? Do you need to gather more data by performing another experiment?

7 LIST: What are some ways to display data?

BUILDING SCIENCE SKILLS

Analyzing Data You made the following notes during your experiment. How would you display this information?

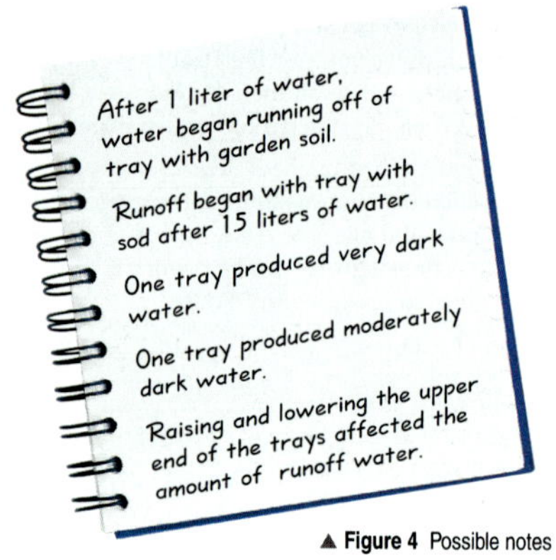

▲ **Figure 4** Possible notes

12

Teaching Options

RETEACH

For the Building Science Skills activity, students might draw pictures of the trays and what they looked like when water was added. Students could also make a sequence chart or graphic organizer. This exercise should show students that data can be organized in different ways.

• *TEACHING RESOURCES* •
Online or Print

Teacher's Resources
Concept Map: Study Tools, p. 4

Laboratory Manual
Lab. Skills Worksheet 4: Organizing and Analyzing Data, pp. 13–14

ESL/ELL STRATEGY

Sequence Words Remind English-language learners that words such as *first*, *next*, *then*, and *later* signal sequence. Sometimes these words are included to show a sequence of events. Other times, they are implied but not stated openly in the text. Ask students to summarize the steps in the scientific method that they have learned so far. Have them use sequence words to indicate the order of the steps.

5 Stating a Conclusion

Drawing Conclusions A conclusion is a statement that sums up what you have learned from an experiment. When you draw a conclusion, you need to decide whether the data you collected supported your hypothesis. You may need to repeat an experiment several times before you can draw any conclusions from it. Conclusions often lead you to ask new questions and plan new experiments to answer them.

8 EXPLAIN: Why might it be necessary to repeat an experiment?

BUILDING SCIENCE SKILLS

Stating a Conclusion Review your hypothesis statement regarding the effect of surface material on rainwater runoff. Then, review the data you obtained during the experiment.

- Was your hypothesis correct? Use your observations to support your answer.
- Which surface reduced soil runoff better?

▲ **Figure 5** Throughout this program, you may use forms like these to organize your lab reports.

6 Writing a Report

Communicating Results Scientists keep carefully written records of their observations and findings. These records are used to create a lab report. Lab reports are a form of written communication. They explain what happened in the experiment. A good lab report should be written so that anyone reading it can duplicate the experiment. It should contain the following information:

- A title
- A purpose
- Background information
- Your hypothesis
- Materials used
- Your step-by-step procedure
- Your observations
- Your recorded data
- Your analysis of the data
- Your conclusions

Your conclusions should relate back to the questions you asked in the "purpose" section of your report. Also, the report should not have any experimental errors that might have caused unexpected results. For example, did you follow the steps in the correct order? Did an unexpected variable interfere with your results? Was your equipment clean and in good working order? This explanation of possible errors should also be part of your conclusions.

9 EXPLAIN: Why is it important to explain possible errors in your lab report?

BUILDING SCIENCE SKILLS

Writing a Lab Report Write a lab report to communicate to other scientists your discoveries about soil runoff. Your lab report should include a title, your hypothesis statement, a list of materials you used, the procedure, your observations, and your conclusions. Try to include one table of data in your report.

MORE ON THE WEB

Learning to Write Online At www.conceptsandchallenges.com, students can visit the Homework Center to learn how to write better. Students can use the online guides to learn to write paragraphs, essays, and reports.

• TEACHING RESOURCES • Online or Print

Teacher's Resources
Lab Activity Report: Study Tools, pp. 16–17

Laboratory Manual
Lab. Skills Worksheet 6: Writing a Laboratory Report, pp. 19–20

ESL/ELL STRATEGY

Making a Chart If English-language learners have trouble making a table for the Building Science Skills activity, help them by creating one on the board that they can copy into their reports. A chart might center on the amount of water collected from the run-off of each tray.

5 Stating a Conclusion

Demonstration Return to your earlier demonstration, in which you dropped items in liquid to prove that the heavier items fell more quickly. Have students draw conclusions about the experiment. Ask: *Was the hypothesis correct? If you were conducting the experiment, would you repeat the steps? Why or why not?* **learning styles: visual, tactile**

Discussion Point out that the conclusion of an experiment may not be the end. An experimenter may find that there is more work to be done if a hypothesis was not proved.

Answer to question

8 EXPLAIN: The initial results may not lead to a conclusion. An experiment may need to be repeated several times before the conclusions are clear.

BUILDING SCIENCE SKILLS

Stating a Conclusion Students should conclude that less soil runoff occurs when water is added to sod than when water is added to dirt.

6 Writing a Report

Demonstration Show students examples of lab reports. Remind students that a report could end up in paragraph form, list form, or some other form. Ask students what would happen if various pieces of information were missing. (Other scientists could not perform the experiment themselves. If the data were missing, it might be hard to tell if the conclusion was correct.) **learning styles: visual, auditory**

Discussion Review the parts of a report as given in the bulleted list. Discuss with students why each component is necessary for a detailed and organized report.

Answer to question

9 EXPLAIN: Errors could produce unexpected results, which could result in the hypothesis not being proven.

BUILDING SCIENCE SKILLS

Writing a Lab Report Remind students to use the parts of a report listed on page 13. Students could turn the parts on the list such as *title*, *purpose*, *background*, *materials*, and *procedure* into topic heads in their report.

LAB SAFETY

STUDY HINT As students read each lab safety rule, make sure that they also note the icon to the left of the rule. Getting familiar with the icons will help students recall the proper safety procedure as they do the lab activities.

Demonstration Describe for students various situations in the lab that could be dangerous. For example, you should never pick up a jar of chemicals without wearing a lab apron or safety goggles. When you cut something, you should always point the knife away from you instead of toward you. **learning style: auditory**

Discussion Read through the safety information with students. Point out the need for lab safety rules, reiterating that students could be injured by not following safety procedures in the lab. As you go through each rule, be sure that students understand the rule and the reasons behind it. **learning style: auditory**

Reinforcement To help students remember the safety rules, have them work with partners or in small groups to make posters that feature safety reminders. Assign one or two rules to each group. Have groups write the rules in their own words and include illustrations. Display students' work around the room as a reminder. **learning style: visual**

Handbook Project Wrap-Up

CONDUCTING EXPERIMENTS REFERENCE

Students can explain their references to the rest of the class. You can use the Individual Activity Rubric found in the Teacher's Resources to assess students' projects. Fill in the rubric with the additional information below. For this project, students should have:

- created a resource reference.
- included a lab report.

LAB SAFETY

Working in a science laboratory can be both exciting and meaningful. However, you must always be aware of safety precautions when carrying out experiments. There are a few basic rules that should be followed in any science laboratory:

- Read all instructions carefully before the start of an experiment. Follow all instructions exactly and in the correct order.
- Check your equipment to make sure it is clean and working properly.
- Never taste, smell, or touch any substance in the lab that you are not told to do so. Never eat or drink anything in the lab. Do not chew gum.
- Never work alone. Tell a teacher at once if an accident occurs.

Experiments that use chemicals or heat can be dangerous. The following list of rules and symbols will help you avoid accidents. There are also rules about what to do if an accident does occur. Here are some rules to remember when working in a lab:

1. Do not use glass that is chipped or metal objects with broken edges. Do not try to clean up broken glassware yourself. Notify your teacher if a piece of glassware is broken.
2. Do not use electrical cords with loose plugs or frayed ends. Do not let electrical cords cross in front of working areas. Do not use electrical equipment near water.
3. Be very careful when using sharp objects such as scissors, knives, or tweezers. Always cut in a direction away from your body.
4. Be careful when you are using a heat source. Use proper equipment, such as tongs or a ringstand, when handling hot objects.
5. Confine loose clothing and hair when working with an open flame. Be sure you know the location of the nearest fire extinguisher. Never reach across an open flame.
6. Be careful when working with poisonous or toxic substances. Never mix chemicals without directions from your teacher. Remove any long jewelry that might hang down and end up in chemicals. Avoid touching your eyes or mouth when working with chemicals.
7. Use extreme care when working with acids and bases. Never mix acids and bases without direction from your teacher. Never smell anything directly. Use caution when handling chemicals that produce fumes.
8. Wear safety goggles, especially when working with an open flame, chemicals, and any liquids.
9. Wear lab aprons when working with substances of any sort, especially chemicals.
10. Use caution when handling or collecting plants. Some plants can be harmful if they are touched or eaten.
11. Use caution when handling live animals. Some animals can injure you or spread disease. Handle all live animals as humanely as possible.
12. Dispose of all equipment and materials properly. Keep your work area clean at all times.
13. Always wash your hands thoroughly with soap and water after handling chemicals or live organisms.
14. Follow the ⚠ CAUTION and safety symbols you see used throughout this book when doing labs or other activities.

14

Teaching Options

ONGOING ASSESSMENT

State the rules for students in random order, leaving some as they are but substituting words or rewording others to make them incorrect. Ask students to make a thumbs up sign if the rule is correct or a thumbs down sign if it is not. (Examples of incorrect statements: *You need only safety goggles when working with chemicals* or *It's okay to reach across an open flame as long as your hair is tied back.*)

• TEACHING RESOURCES • Online or Print

Teacher's Resources

Individual Activity Rubric: Teaching Tools, p. 5

Practicing Lab Safety: Handbook Resources, p. 8

Handbook Key Term Review: Handbook Resources, pp. 9–10

Chapter 1 The Structure of Earth

PLANNING GUIDE

◆ **TEACHING THE CHAPTER** This chapter should take approximately 9–12 days to complete instruction and assessment.

	Skills and Features	Projects and Activities	Achieve Science Literacy	Meet Individual Needs
Chapter 1 Opener p. T15	RESEARCH	• Chapter Project		
Lesson 1-1 **What is Earth science?** pp. T16–T17	IDENTIFY, CLASSIFY, ANALYZE • Web InfoSearch	• Extending the Lesson	• Study Hint	• ESL/ELL Strategy Teacher's Resources* Enrichment Activity
Lesson 1-2 **What are the main parts of Earth?** pp. T18–T19 Standards: D1a, D1f, D1h	NAME, INTERPRET, ANALYZE • Building Reading Skills • Integrating Life Science	• Cooperative Learning • More on the Web	• Study Hint	• Reteach • ESL/ELL Strategy Teacher's Resources* Enrichment Activity
Lesson 1-3 **What is the structure of Earth?** pp. T20–T21 Standard: D1a	CONTRAST, CALCULATE, SEQUENCE • Investigate • Building Math Skills • How Do They Know That?	• Cooperative Learning • More on the Web	• Study Hint	Teacher's Resources* Enrichment Activity Earth's Structure
Lesson 1-4 **What are maps?** pp. T22–T23	HYPOTHESIZE, MODEL, COMPARE • Building Science Skills • Science and Technology	• Cooperative Learning • Extending the Lesson • More on the Web	• Study Hint	• ESL/ELL Strategy Teacher's Resources* Enrichment Activity Concept Map Town Map
Lesson 1-5 **What is a map projection?** pp. T24–T25	DEFINE, INFER • Investigate • Web InfoSearch • How Do They Know That?	• Extending the Lesson	• Study Hint • Ongoing Assessment	• Reteach Teacher's Resources* Enrichment Activity Writing Activity Outline
Lesson 1-6 **What are latitude and longitude?** pp. T26–T27	COMPARE, MEASURE, LOCATE • Building Science Skills • Hands-On Activity	• Extending the Lesson • More on the Web • Integrating Mathematics	• Reading Strategy • Ongoing Assessment	Teacher's Resources* Enrichment Activity
Lesson 1-7 **How do you read a map?** pp. T28–T29	OBSERVE, ANALYZE, MODEL • Real-Life Science	• Cooperative Learning • More on the Web • Integrating Mathematics	• Study Hint	Teacher's Resources* Enrichment Activity Town Map

*www.conceptsandchallenges.com, password: ccearth09

Planning Guide continues on next page.

Standards: For details on the correlation to National Science Standards see pages *xx–xxi*.

	Skills and Features	Projects and Activities	Achieve Science Literacy	Meet Individual Needs
Big Idea **How did maps help the United States grow?** pp. T30–T31	RESEARCH, COMMUNICATE, ANALYZE • Science Log: Writing Activity	• Big Idea • Close Activity	• Tips for Using Technology	Teacher's Resources* Big Idea Planner
Lesson 1-8 **What is a topographic map?** pp. T32–T33	CONTRAST, EXPLAIN, RELATE • Web InfoSearch • People in Science	• Extending the Lesson • Active Art	• Study Hint	• Reteach • ESL/ELL Strategy Teacher's Resources* Enrichment Activity
Lesson 1-9 **How do you read a topographic map?** pp. T34–T35	OBSERVE, ANALYZE, READ A TOPOGRAPHIC MAP • Hands-On Activity	• Extending the Lesson • More on the Web • Lab Challenge	• Study Hint	• Reteach Teacher's Resources* Enrichment Activity
Lab Activity **Making a Topographic Map** pp. T36–T37	ANALYZE, OBSERVE, INFER • Set-Up Time: 10 min • Lab Time: 35 min	• Extending the Lesson • Lab Skills Worksheet • Extend the Lab Activity	• Tips for Using Technology	• ESL/ELL Strategy Teacher's Resources* Lab Activity Report
Chapter 1 Challenges pp. T38–T40	• Chapter Summary • Key Term Challenges • Content Challenges • Concept Challenges	• Chapter Project Wrap-Up	• Preparing Students for Standardized Tests	• ESL/ELL Strategy Teacher's Resources* Chapter Self-Check Weekly Journal

*www.conceptsandchallenges.com, password: ccearth09

Teacher Notes

Chapter 1 The Structure of Earth

▲ **Figure 1-1** Earth as seen from space

This is an image of Earth created by an artificial satellite in space. Artificial satellites are sent into orbit for many purposes. For example, some satellites are used to study our planet and other objects in space. They help us to map Earth's surface and determine the size and shape of Earth. They also are used to study the land, oceans, and air. Computers can be used to reveal even more detail.

▶The satellite image in Figure 1-1 shows some physical features of Earth. Which ones can you identify?

Contents

Teaching Options

ASSESSMENT PLANNER

For assessment in the Student Edition, see the following pages:

Content Assessment
Checking Concepts: pp. 17, 19, 21, 23, 25, 27, 29, 33, 35
Thinking Critically: pp. 17, 19, 21, 23, 25, 27, 29, 33, 35
Interpreting Visuals: pp. 29, 35
Chapter Challenges: pp. 38–40

Alternative Assessment
Building Skills: pp. 19, 21, 23, 27
Web InfoSearch: pp. 17, 25, 33
Science Log: p. 31

Performance-Based Assessment
Hands-On Activity: pp. 27, 35
Lab Activity: pp. 36–37

• TEACHING RESOURCES • Online or Print

Teacher's Resources
Lesson Review: Ch. 1, pp. 2, 3, 4, 6, 7, 8, 9, 12, 13
Enrichment: Ch. 1, pp. 5, 10, 14
Key Term Review: Ch. 1, p. 15
Chapter 1 Test: pp. 16–17

Laboratory Manual: pp. 19–20, 21–24

Chapter 1 The Structure of Earth

Chapter Overview

In this chapter, students will learn that Earth science is the study of Earth and its history and that this field includes four main branches. They will learn that Earth's surface and the mixture of gases directly above it are often divided into three main parts: the lithosphere, hydrosphere, and atmosphere. They will find out about Earth's interior. Lastly, they will read maps and find places by latitude and longitude.

About Figure 1-1 This photograph was taken from a satellite orbiting above Earth's surface. Although humans have known for centuries that Earth is a sphere, photographs taken from above Earth provided the first opportunity to directly view Earth's curved horizon.

Answer Accept any of the following: landmasses, oceans, clouds, islands, shorelines, or mountains.

Linking Prior Knowledge

Ask students to list ways in which they have used maps in the past, perhaps as part of a social studies class or to find their way around unfamiliar parts of a city or town. Ask students what kinds of features they have seen marked on maps.

Chapter Project

MAP DISPLAY

MATERIALS: maps, bulletin board, posterboard, markers

Have students bring to class as many different kinds of maps as they can find, such as political and physical world maps, state or city maps, road maps, land-use maps, natural resources maps, topographic maps, and navigational maps. Display the maps on the bulletin board. Provide students with posterboard and markers to make labels for the maps in the display. Use maps from the display when teaching students how to read maps. Have them point to special features, such as color, patterns, keys or legends, and scales.
learning styles: tactile, visual

1-1 *(pp. 16–17)*

1 *Introduce*

STUDY HINT Before beginning this lesson, have students scan the book's complete Table of Contents and write some general notes about what Earth science is.

Linking Prior Knowledge Ask students to describe the shape of Earth. Also ask: *What kinds of matter make up Earth?*

2 *Teach*

Demonstration Invite an Earth scientist to come to your classroom to demonstrate some of the tools he or she uses on the job. **learning style: visual**

Discussion As you describe each branch of Earth science, refer to the descriptions in Figure 1-2. Point out that astronomy is considered a part of Earth science because Earth is an astronomical body, as are the other planets in our solar system. **learning style: visual**

Answers to questions

1. **IDENTIFY:** geology, oceanography, meteorology, and astronomy
2. **DESCRIBE:** one part of a subject
3. **EXPLAIN:** Possible answers: It provides early warning of dangerous weather conditions. It also provides information that people can use in deciding how to dress or in making plans for travel or outdoor activities.

Reinforcement Divide the class into pairs of students who have similar career interests. Have partners work together to search for information about the Earth science career they chose.

1-1 What is Earth science?

Objective

Identify and describe the four main branches of Earth science.

Key Terms

Earth science: study of Earth and its history

specialist (SPEHSH-uhl-ihst)**:** person who studies or works on only one part of a subject

The Study of Earth Science The three major fields of science are Earth science, life science, and physical science. **Earth science** is the study of Earth and its history. It is also the study of changes on Earth and Earth's place in the universe.

As seen from space, Earth is much more than rock and soil. In fact, the most obvious features are the clouds and oceans. In addition, our environment is not just land, water, and air. Rather, it is air interacting with land, land interacting with water, water interacting with air, and so on.

Earth science is like a jigsaw puzzle made up of four pieces. Each piece of the puzzle is a main branch of Earth science. The four main branches of Earth science are geology, oceanography, meteorology, and astronomy. Together, they make up the field of Earth science. Look at Figure 1-2 below to learn something about each of the four main branches.

1. **IDENTIFY:** What are the four main branches of Earth science?

Specialists in Earth Science A **specialist** is a person who studies or works on only one part of a subject. This one part of the subject is called a **specialty**.

There are many Earth science specialties. Most Earth scientists are specialists. For example, some oceanographers study only waves and tides. Others study the makeup of ocean water. Still others study the ocean bottom or deep-sea vents.

2. **DESCRIBE:** What is meant by a specialty?

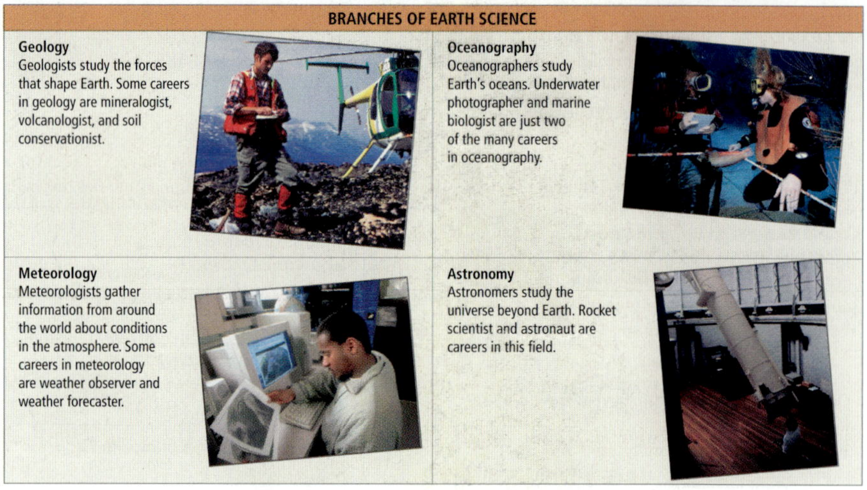

▲ Figure 1-2

Teaching Options

! FAST FACTS *Mining Geologist*

A mining geologist maps geological structures in the mines and drills for core samples for chemical analysis. The mining geologist's main responsibility is to ensure that any ore being mined from Earth is of high quality. He or she is also responsible for developing strategies to avoid wasting the material being mined.

Mining geologists usually spend some of their time in the field. They may work in remote areas that they get to by helicopter, in a four-wheel-drive vehicle, or on foot. Most geologists have a college degree in geology or Earth science.

ESL/ELL STRATEGY

Reinforcing Vocabulary Have students look again at Figure 1-2. Point out that the four branches of Earth science end in *y*. When *y* is on the end of a word with two or more syllables, the *y* usually sounds like a long *e*. Have students look through the lesson to find other words that end in *y* and that have the same sound. (study, history, specialty, only, many, early, property, electricity, every)

The Importance of Earth Science Earth science is an important part of our everyday lives. The observations and discoveries made by Earth scientists affect us in many ways.

For example, weather information is collected by Earth scientists called meteorologists. Meteorologists are able to give early warnings about severe storms such as hurricanes and tornadoes. Storm warnings can help save lives and limit property damage. Geologists help locate oil and coal supplies. These fuels are used to heat homes, run cars, and generate electricity. Oceanographers study the ocean and map currents. These help in containing oil spills and the cleaning of polluted waters.

Exploring space and experiments done in space have led to new medicines and technologies. They have also led to improvements in radios, televisions, and telephones. The development of the space shuttle led to improvements in fire-resistant materials.

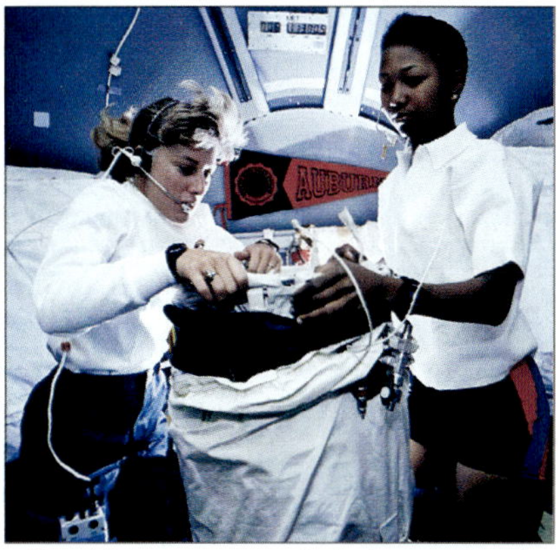

▲ Figure 1-3 Astronauts carry out Earth science experiments in space that help us on Earth.

3 EXPLAIN: Why is the study of weather important to our lives every day?

CHECKING CONCEPTS

1. The study of Earth and how it changes is called ___________.
2. A mineralogist studies the branch of Earth science called ___________.
3. The study of Earth's oceans is called ___________.
4. The study of space is part of the branch of Earth science called ___________.
5. Another name for a weather forecaster is a ___________.

THINKING CRITICALLY

6. CLASSIFY: In which field of Earth science would you study each of these subjects?
   a. coral reefs
   b. the Grand Canyon
   c. hurricanes
   d. the planet Jupiter
   e. soil quality
7. ANALYZE: What is one way, other than weather, in which the study of Earth science affected you today?

Web InfoSearch

Earth Science Careers Choose an Earth science career from the ones that are mentioned in Figure 1-2 or another one that you have heard about.

SEARCH: Use the Internet to find out more information about the career you chose. How much education is required? What tools are generally used on the job? What is the average salary for a person who has some experience? What fields of study in this career would interest you most? What on-the-job activities would you enjoy most? Start your search at www.conceptsandchallenges.com. Some key search words are **Earth science careers, meteorology, geology, oceanography,** and **astronomy.**

◀ Figure 1-4 As a geologist, you might study rock formations like this one.

3 Assess

CHECKING CONCEPTS ANSWERS

1. Earth science
2. geology
3. oceanography
4. astronomy
5. meteorologist

THINKING CRITICALLY ANSWERS

6. CLASSIFY:
   a. oceanography
   b. geology
   c. meteorology
   d. astronomy
   e. geology
7. ANALYZE: Possible answer: The bus I took to school uses a fuel that geologists helped locate.

4 Close

Web InfoSearch

Science Careers Students should select an Earth Science career from the ones listed in Figure 1-2 on page 16. Have students use the Internet to find out about how much education is required, average salary, travel requirements, and tools used on the job. In addition, have students indicate the fields of study that interest them within their chosen career and describe the on-the-job activities that they would enjoy most. Some key search words are **Earth science careers, meteorology, geology, oceanography,** and **astronomy.** Students can begin their search at www.conceptsandchallenges.com.

EXTENDING THE LESSON

Earth Science Specialties Geology specialists include gemologists, paleontologists, and petrologists. Gemologists study and appraise precious and semiprecious stones, such as diamonds, rubies, and jade. Paleontologists study fossils of living organisms that have been left behind in rocks and sediment, such as petrified wood or dinosaur bones. Petrologists are experts on the history, structure, chemical composition, and classification of rocks. Have students bring to class examples or illustrations of the materials studied by each of these specialists.

learning styles: visual, tactile

• TEACHING RESOURCES • Online or Print

Teacher's Resources

Lesson Review: Chapter 1, p. 2

Enrichment: Chapter 1, p. 5

1-2 *(pp. 18–19)*

1 *Introduce*

STUDY HINT Have students scan the lesson to note the parts of Earth. Ask students to describe how the three parts of Earth are alike and how they are different.

Linking Prior Knowledge Ask students to name the continent on which the United States is located. Have students name the oceans that border the East Coast and West Coast of North America. Point out that the continents are landmasses that make up part of Earth's lithosphere. The oceans make up part of Earth's hydrosphere.

2 *Teach*

Demonstration Display a globe. Point out that a globe is a model of Earth. Have students use the globe to point out features that make up the lithosphere (continents, islands, ocean floor, bottoms of lakes and streams) and the hydrosphere (oceans, seas, lakes, rivers, glaciers, polar ice caps). Ask students to explain why certain colors are used on different parts of the globe. (possible answers include to distinguish land from water and to denote political boundaries) **learning style: visual**

Discussion Ask students which objects in Figure 1-5 are parts of Earth's lithosphere. (the seven continents) Ask which objects on the map are parts of Earth's hydrosphere. (Atlantic, Pacific, and Indian Oceans) Explain that the lithosphere includes islands and the ocean floor. Tell students that the hydrosphere includes lakes, streams, glaciers, and polar ice caps. Ask what the atmosphere is. (the gases that surround Earth) Have students look back at the photograph of Earth in Figure 1-1 on page 15. Explain that clouds are made of water and are part of the atmosphere. **learning styles: auditory, visual**

Answers to questions

1. **NAME:** the lithosphere
2. **NAME:** the Pacific Ocean, Atlantic Ocean, and Indian Ocean
3. **IDENTIFY:** nitrogen and oxygen

1-2 What are the main parts of Earth?

Objective
Name and describe the three main parts of Earth: the lithosphere, the hydrosphere, and the atmosphere.

Key Terms
sphere (SFEER): round, three-dimensional object
lithosphere (LIHTH-oh-sfeer): solid part of Earth
hydrosphere (HY-droh-sfeer): part of Earth that is water
atmosphere (AT-muh-sfeer): envelope of gases surrounding Earth

The Lithosphere A **sphere** is a round, three-dimensional object. Earth is spherical in shape. The prefix *litho-* means "stone." The solid part of Earth is called the **lithosphere.** The ground you walk on is part of the lithosphere. Mountains are raised parts of the lithosphere. Valleys, found between mountains, are low areas of the lithosphere.

The lithosphere includes the continents as well as the land under the oceans. Continents are very large landmasses. The seven continents found on Earth today are shown in Figure 1-5.

1. **NAME:** What is the solid part of Earth called?

The Hydrosphere The part of Earth that is water is called the **hydrosphere.** The hydrosphere includes all of the liquid water and ice on Earth. Saltwater makes up about 97 percent of Earth's water. Most of the saltwater is found in the oceans. There are three major oceans on Earth: Atlantic, Pacific, and Indian. These are all interconnected to form one large world ocean. The remaining 3 percent is freshwater. Freshwater is found mostly in rivers, lakes, glaciers, and the polar ice caps.

▲ **Figure 1-6** Much of the freshwater on Earth is found frozen in glaciers. This glacier is in Alaska.

2. **NAME:** List the bodies of saltwater that are shown in Figure 1-5.

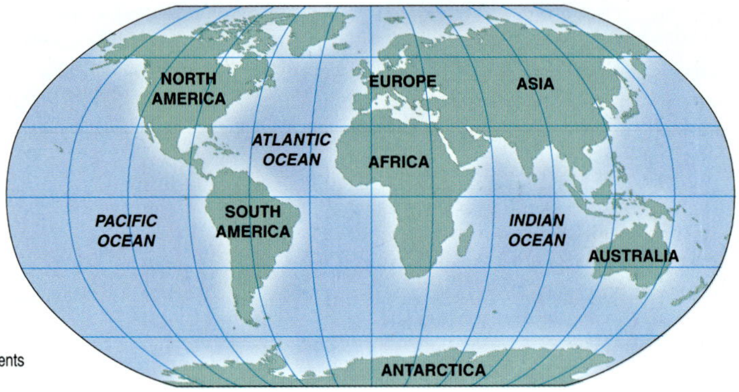

Figure 1-5 ▶ Earth's continents and oceans

Teaching Options

COOPERATIVE LEARNING
Explaining Terms Have students work in pairs to define in their own words the terms *lithosphere, hydrosphere*, and *atmosphere.* Invite students to use drawings or pictures cut out of magazines to illustrate their definitions.
learning styles: auditory, visual

RETEACH
Using a Map Display a large world map in the classroom. Have volunteers locate on the map a continent, an ocean, or other feature that you name. Ask students to state whether the feature is part of the lithosphere or part of the hydrosphere.
learning styles: auditory, visual

ESL/ELL STRATEGY
Key Term Review The Key Terms all end in *sphere*. Have students look at the definitions and see what they all refer to that could provide clues to the meaning of each term.

The Atmosphere The envelope of gases surrounding Earth is called the **atmosphere.** The air you breathe is part of the atmosphere. Nitrogen and oxygen make up most of the atmosphere. Organisms need these and other gases to live. The atmosphere also helps protect living things by blocking harmful rays given off by the Sun. These rays, called ultraviolet rays, cause sunburn, skin cancer, and eye damage.

3 IDENTIFY: What are the two main gases in Earth's atmosphere?

CHECKING CONCEPTS

1. The ground you walk on is part of the __________.
2. The air you breathe is part of the __________.
3. The __________ includes rivers, lakes, and oceans.
4. Earth is a round, three-dimensional object known as a __________.
5. Saltwater makes up about __________ percent of the hydrosphere.
6. There are __________ continents on Earth.
7. There are __________ major oceans on Earth.

THINKING CRITICALLY

8. INTERPRET: Use Figure 1-5 to answer the following questions:
   a. Which is the smallest continent?
   b. Which is the largest continent?
9. ANALYZE: Draw a pie graph that shows the percentages of saltwater and freshwater on Earth.

BUILDING READING SKILLS

Using Prefixes Prefixes are word parts that appear at the beginning of words. Prefixes have consistent meanings. Knowing the definition of a prefix can help you figure out the meaning of a particular word. Two prefixes that are used often in this lesson are *litho-* and *hydro-*. Find out the meanings of these prefixes. (Hint: *Litho* is defined in this lesson.) Then, figure out and write down the definitions of the following words that use these prefixes: *hydroelectric, hydrologist, lithology,* and *lithographer.* Circle the part of the definition that relates to each prefix. Put each word in a sentence.

Integrating Life Science

TOPIC: biosphere

LIFE ON EARTH

The part of Earth that supports all living things is called the biosphere. Parts of the lithosphere, the hydrosphere, and the atmosphere make up the biosphere.

The biosphere is a very narrow zone. You may think living things are almost everywhere. Some kinds of clams live in deep-sea vents on the ocean floor. Some spiders live high in the atmosphere. Bacteria are found deep in the crust. The biosphere extends 8 to 10 km into the atmosphere. Yet compared to the entire volume of Earth, the part where life exists is very small.

Thinking Critically Why do you think life is found only in the biosphere?

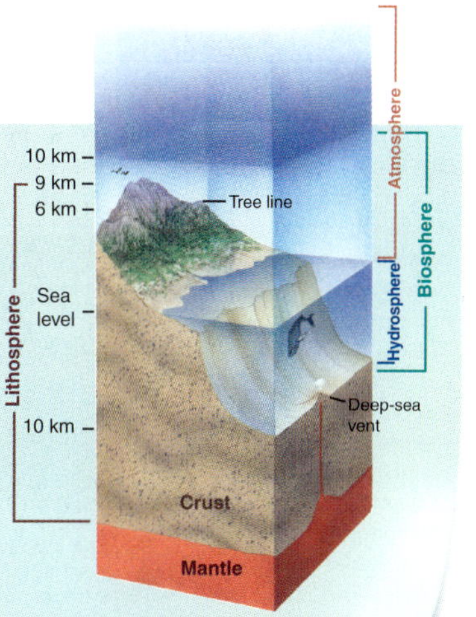

▲ **Figure 1-7** The biosphere is the part of Earth where life exists.

MORE ON THE WEB

Investigating the Atmosphere Have students use the Internet to investigate what the different layers of the atmosphere are called, starting their search at www.conceptsandchallenges.com.

QUICK CONVERSIONS

	Metric/SI	English
Earth's biosphere	6 km	3.7 mi
	9 km	5.6 mi
	10 km	6.2 mi

• *TEACHING RESOURCES* • *Online or Print*

Teacher's Resources

Lesson Review: Chapter 1, p. 3

Enrichment: Chapter 1, p. 5

3 *Assess*

CHECKING CONCEPTS ANSWERS

1. lithosphere
2. atmosphere
3. hydrosphere
4. sphere
5. 97
6. 7
7. 3

THINKING CRITICALLY ANSWERS

8. INTERPRET: **a.** Australia **b.** Asia
9. ANALYZE: The graph should show 97 percent salt water and 3 percent fresh water

4 *Close*

BUILDING READING SKILLS

Using Prefixes The prefix *litho-* means "stone" or "rock." The prefix *hydro-* means "water." *Hydroelectric* means "producing electricity by water power." A *hydrologist* is a scientist who studies Earth's waters. *Lithology* is the study of the physical characteristics of rocks. A *lithographer* is a person who makes prints from a flat stone or metal plate.

Integrating Life Science

TOPIC: biosphere

LIFE ON EARTH

Background The biosphere is made up of ecosystems. An ecosystem is made up of all the living and nonliving features that are present in a particular environment. For example, the living features of a forest ecosystem include trees and other plants, animals, and bacteria and fungi in the soil. The nonliving features of a forest ecosystem include soil, air, groundwater, rain, snow, and sunlight. The nonliving features of any ecosystem include land (lithosphere), water (hydrosphere), and air (atmosphere).

Discussion Point out to students that, even though the biosphere extends from high in the atmosphere to the deepest ocean floor, it is a very thin zone compared to the thickness of the entire Earth.

Thinking Critically Answer
Possible answer: Other regions do not contain some of the things essential to life, such as water, air, or nutrients.

1-3 *(pp. 20–21)*

1 Introduce

INVESTIGATE

TIME: 10–15 minutes

PURPOSE: Students will create a model of Earth's layers.

MATERIALS: hard-boiled egg, plastic knife, 3 thin straws or coffee stirrers, 3 self-stick labels or small pieces of paper, pen or pencil

PROCEDURE: Have students work in pairs to complete this activity. Caution students to use care when working with sharp objects such as knives. **learning style: tactile**

THINK ABOUT IT: The eggshell models Earth's crust. The egg white is like the middle layer, and the yolk is like the core.

STUDY HINT Explain to students that the use of models helps in the investigation of things in the real world. Figure 1-9 is an example of a model of Earth's structure.

Linking Prior Knowledge Before beginning this lesson, have students review the parts of Earth (Lesson 1-2). (lithosphere, hydrosphere, and atmosphere) Point out that those are Earth's outer parts. Explain that this lesson is about the inner structure of Earth.

2 Teach

Discussion As you describe each layer of Earth's interior, explain to students that the boundaries between these layers are not perfect circles, as shown in Figure 1-9. Point out that the thickness of each layer varies, as indicated by the numerical values given in the text. **learning style: auditory**

Answers to questions

1. **EXPLAIN:** It bulges in the middle and is slightly squashed at the poles.
2. **CONTRAST:** Ocean crust is thinner. It is only 5–10 km (3–6 mi) thick, compared to the continental crust, which is 32–70 km (19–43 mi) thick.
3. **OBSERVE:** beneath Earth's crust
4. **CALCULATE:** the combined core layers

1-3 What is the structure of Earth?

INVESTIGATE

Modeling Earth's Layers

HANDS-ON ACTIVITY

1. Cut a hard-boiled egg in half with the shell still on it.
   ⚠ CAUTION: Be very careful when working with sharp objects.
2. On three small pieces of paper or self-stick labels, write "Yolk," "White," and "Shell." Tape the papers to three thin straws or coffee stirrers or fold the self-stick labels around them. Use the straws or stirrers to label each part of the egg.

THINK ABOUT IT: Like the egg, Earth has a core, material around the core, and a thin but hard outer skin. Which parts of the model are like Earth's outer skin, middle, and core?

STEP 2

Objective

Describe Earth's major parts: the crust, the mantle, and the core.

Key Terms

crust: solid, thin outer layer of Earth

mantle (MAN-tuhl)**:** thick layer of earth below the crust

core: innermost region of Earth

A Slightly Flat Sphere A perfect sphere has no real top. Every point on the surface of a sphere is the same distance from the center. Earth is a sphere, but not a perfect one. It is slightly squashed, or flat, at its poles. Also, Earth bulges out a little around its middle. The diameter is the distance across a sphere through its center. Earth is about 12,756 km in diameter at the equator. The diameter of Earth from pole to pole is about 12,713 km.

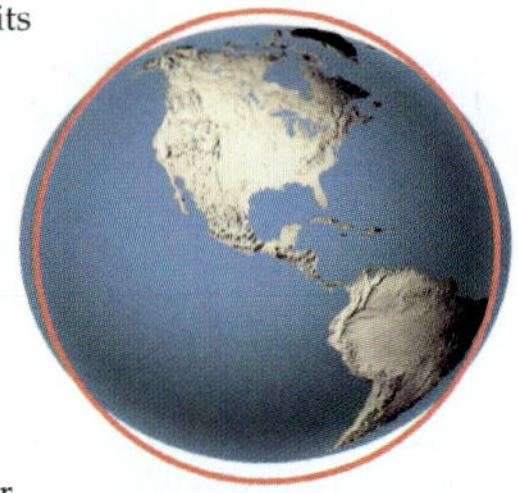

▲ **Figure 1-8** The outline of Earth does not fit perfectly into the outline of a sphere.

1. **EXPLAIN:** How is Earth not a perfect sphere?

Earth's Crust What is below Earth's surface? Like the hard-boiled egg above, Earth has three basic layers: the crust, the mantle, and the core. The **crust** is the thin, outer layer of Earth. It forms the upper lithosphere. Beneath the oceans, the crust is 5 to 10 km thick. Beneath the continents, the crust is 25 to 50 km thick. The thickest crust is beneath mountains.

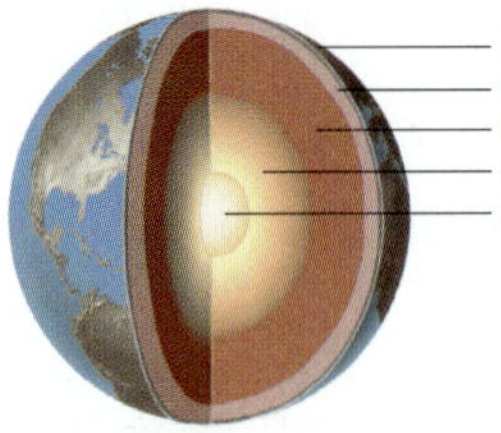

◄ **Figure 1-9** Earth's structure

2. **CONTRAST:** How is ocean crust different from continental crust?

Earth's Mantle The layer of Earth beneath the crust is called the **mantle.** Earth's mantle ranges from about 1,800 to 2,900 km down. More than two-thirds of the mass of Earth is in the mantle.

The lithosphere includes all of the crust and the upper part of the mantle. Below this, the rock flows like a very thick liquid. Most of the mantle is made up of rock that flows very slowly.

3. **OBSERVE:** Where is the mantle located?

Teaching Options

COOPERATIVE LEARNING

Another Way to Model Earth's Layers Have students work in pairs to create and label a model of Earth's layers, using different colors of modeling clay. Tell students that they are to create a model appropriate for an educational exhibit at a school or museum. Invite them to look up additional facts about each of Earth's layers to include in their exhibit. Provide students with modeling clay of several colors, thin straws or toothpicks, tape, markers, and index cards. **learning style: tactile**

MORE ON THE WEB

Investigating Earth's Crust Have students investigate the two types of crust that surround the outer layer of Earth. They can begin their search by visiting www.conceptsandchallenges.com. Students can use the Internet to learn about differences in composition, thickness, and age of the layers.

Earth's Core The innermost region of Earth is the **core.** Scientists think that the core is made mostly of iron. The core has two parts. The outer core is liquid. It is about 2,250 km thick. The inner core is solid, with a radius of about 1,300 km.

THE LAYERS OF EARTH

Layer	Thickness
Crust	5–70 km
Mantle	2,900 km
Outer Core	2,250 km
Inner Core	1,300 km

▲ Figure 1-10

4 **CALCULATE:** Which is thicker, the mantle or the two core layers combined?

CHECKING CONCEPTS

1. The crust is thickest beneath ___________.
2. The core is made mostly of ___________.
3. You can stand on Earth's layer called the ___________.
4. The diameter of Earth at the equator is about ___________ km.

THINKING CRITICALLY

5. **SEQUENCE:** Arrange Earth's layers in order, from thickest to thinnest.
6. **ANALYZE:** A drilling rig can go down only 6 km. Where would you set it up to drill into the mantle? Why?
7. **INFER:** Which layer of Earth includes the continents?

BUILDING MATH SKILLS

Calculating The radius, or *r*, of Earth is the point measured from its center to its surface. Use the formula below to find the approximate circumference, or *C*, of Earth. Pi, or π, is equal to approximately 3.14. $C = 2\pi r$.

How Do They Know That?

EARTH IS A SPHERE

For thousands of years, people believed Earth was flat. This seemed logical. If you stand in a field or sail on the ocean, you can see for many kilometers. Everything appears flat. About 2,500 years ago, around 500 B.C., the Greek philosopher Pythagoras suggested that Earth was a sphere. His idea was probably based on observing the night sky. The heights of stars in the sky varied, depending on where you stood on Earth's surface. The Moon appeared to be round. Possibly, Earth had the same shape.

Many years later, another Greek philosopher, Aristotle, supported Pythagoras's theory. While observing lunar eclipses, Aristotle noted that a curved shadow fell on the Moon. He believed this to be Earth's shadow. Aristotle concluded that Earth was round.

▲ **Figure 1-11** Pythagoras (about 580 B.C. to about 500 B.C.), Greek philosopher and mathematician

Today, pictures from space show us that Earth is indeed a sphere.

Thinking Critically How did Pythagoras's observation of the heights of stars support his theory?

3 *Assess*

CHECKING CONCEPTS ANSWERS

1. the continents
2. iron
3. crust
4. 12,756

THINKING CRITICALLY ANSWERS

5. **SEQUENCE:** mantle, outer core, inner core, and crust
6. **ANALYZE:** Possible answer: in the ocean where the crust is only 5–10 km deep
7. **INFER:** the crust

BUILDING MATH SKILLS ANSWER

Calculating If you choose the radius at the equator, $C = 40{,}054$ km. If you choose the radius at the poles, $C = 39{,}764$ km.

4 *Close*

How Do They Know That?

EARTH IS A SPHERE

Astronomy Connection If students are not familiar with how a lunar eclipse occurs, use a flashlight, a tennis ball, and a globe to demonstrate how Earth's shadow falls on the Moon. In a darkened room, have a student aim the flashlight (which represents the Sun) toward the globe (Earth), which is held by a second student. Have a third student hold the tennis ball (Moon) in the shadow of the globe, positioning it so that the curved edge of the globe's shadow falls on the ball.

Discussion Tell students that Pythagoras and his followers are believed to have been the first thinkers to propose that Earth is round. Aristotle lived about 200 years after Plato, from about 384 to 322 B.C. He proposed that Earth was at the center of a sphere-shaped universe. It was Nicolaus Copernicus, a Polish astronomer who lived from A.D. 1473 to 1543, who first theorized that Earth revolves around the Sun once each year.

Thinking Critically Answer
Pythagoras theorized that if Earth were curved, its curvature would change a person's position relative to the star, and the height of the star would appear to change.

QUICK CONVERSIONS

Earth's Dimensions	Metric/SI	English
Diameter		
At equator	12,756 km	7,916 mi
From pole to pole	12,664 km	7,869 mi
Crust thickness		
Beneath oceans	5–10 km	3–6 mi
Beneath continents	32–70 km	19–43 mi
Mantle thickness	1,800–2,900 km	1,118–1,802 mi
Core thickness		
Outer core	2,250 km	1,398 mi
Inner core	1,300 km	808 mi

• TEACHING RESOURCES •
Online or Print

Teacher's Resources
Lesson Review: Chapter 1, p. 4
Enrichment: Chapter 1, p. 5
Earth's Structure: Visuals, p. 1

1-4 *(pp. 22–23)*

1 Introduce

STUDY HINT Have students list the types of information that can be obtained from maps. As they go through the lesson, have them note beside each item on their lists what kinds of maps would be best for communicating that type of information.

Linking Prior Knowledge Have students review the ways in which they used globes in the past. Ask a volunteer to describe how the class used a globe while studying Lesson 1-2. (to identify features that make up the lithosphere and hydrosphere)

2 Teach

Demonstration Bring a model airplane, car, train, bridge, or ship to class. Remind students that a model represents a real object. Ask students to explain how the model is similar to and different from the actual object. **learning style: visual**

Discussion Display a globe. Explain that a globe shows the shapes and relative sizes of the features of Earth. Guide students to understand that distances on the globe's surface are related to distances on Earth's surface. Explain to students that maps are similar to globes, except that maps show more detail and are presented on a flat surface rather than on a curved one.
learning style: auditory

Reinforcement Have students use the Concept Map from the Teacher's Resources. Ask students how concept maps can be used to map items other than Earth's surface features, such as ideas or processes.
learning style: auditory

Answers to questions

1. **DEFINE:** a three-dimensional model of Earth's surface
2. **IDENTIFY:** many things, such as details of Earth's surface or locations of or distances between places on Earth
3. **COMPARE:** a map of the United States, because it shows a larger area than a map of Florida, and maps of smaller areas contain fewer distortions

1-4 What are maps?

Objectives

Explain why a globe is the most useful type of Earth map. Describe what happens when the round Earth is shown on a flat map.

Key Terms

globe: three-dimensional model of Earth's surface

map: flat model of Earth's surface

distortion (dih-STOR-shuhn)**:** error in shape, size, or distance

Globes A **globe** is a three-dimensional model of Earth's surface. A model is something that represents a real object such as a ship or an airplane. A globe correctly shows the shapes and sizes of features on Earth. For this reason, a globe is the best model of Earth. Most globes, however, are too small to show much detail. Larger globes are too big to handle. Imagine trying to carry a globe to school every day!

◀ **Figure 1-12** Globes are good models of Earth's surface. However, they can be hard to carry around.

1 **DEFINE:** What is a globe?

Maps A **map** is a flat model, or drawing, of Earth's surface. It is two-dimensional. There are many kinds of maps.

Some maps show the whole Earth. Other maps show only part of Earth. Maps that show a small part of Earth can show more details than maps that show more of Earth.

Maps can show many different things. They can show the locations of and distances between places on Earth. Some maps show city streets. Other maps show buildings in a town. Some maps even show the weather or types of soil in an area.

▲ **Figure 1-13** Navigation systems in cars display maps that help you find your way in a strange city.

2 **IDENTIFY:** What features are shown on maps?

Distortion When a round surface such as Earth is shown on a flat map, errors occur. If the shapes on the map are correct, the distances may be wrong. If the distances are correct, the shapes may be wrong. These errors in shape, size, or distance are called **distortions.** Nearly all maps have some distortions. Maps of smaller areas have fewer distortions.

▲ **Figure 1-14** Maps can vary in how much detail they show.

3 **COMPARE:** Which would have more distortions, a map of the United States or a map of Florida? Explain.

22

Teaching Options

ESL/ELL STRATEGY

Peer Interaction Have English-proficient students pair up with English-language learners to examine different types of maps and to compare the information that the maps provide.

MORE ON THE WEB

Using Maps on the Internet
Have students use the Internet to learn about some of the mapping tools available online.

EXTENDING THE LESSON

Comparing Map and Image
Provide students with a *Landsat* image of their local area and a map that covers roughly the same area. Have them compare the level of detail and type of information provided by the two items. Ask students to report their findings to the rest of the class. *Landsat* images can be obtained from Customer Services, U.S. Geological Survey, EROS Data Center, 47914 252nd Street, Sioux Falls, SD 57198-0001. Telephone: 1-800-252-4547.
learning styles: visual, auditory

CHECKING CONCEPTS

1. A map is a __________ model of Earth.
2. The best model of Earth is a __________.
3. Errors on a map are called __________.

THINKING CRITICALLY

4. **EXPLAIN:** Why is a globe the best model of Earth?
5. **HYPOTHESIZE:** When would a flat map be more useful than a globe?
6. **APPLY:** Globes and maps can be used to find the distances between different places. Maria wanted to know how far it was from her house to her friend's house in the next town. She tried to use a globe to find out. Why might Maria have a problem using a globe to find the distance between the two towns?

BUILDING SCIENCE SKILLS

Modeling and Comparing Using an $8\frac{1}{2} \times 11$-inch sheet of paper, make a map of your classroom.

1. Measure 2 cm from each edge of the paper to draw the walls of the room.
2. Fold the paper into quarters and then open it. Use these folds as guides.
3. Draw as many objects as you can in each part. Draw where you and your teacher sit.
4. Have a classmate locate places in the room on your map. Does your map have distortion? Indicate where in your map any distortion might be and why.

Science and Technology
EARTH-OBSERVATION SATELLITES

Landsat is the oldest U.S. land-surface observation satellite system. The satellites orbit from pole to pole. They use remote sensing to collect information about Earth's surface. Remote sensing is the ability to gather information about an object or event without actually being in physical contact with it. A *Landsat* satellite can provide images with great detail. Images from *Landsat* have been used to map wildfire hazards in Yosemite National Park, track lava flows in Hawaii, and observe population growth in large U.S. cities.

▲ **Figure 1-15** This satellite image shows Salt Lake City in Utah.

The National Aeronautics and Space Administration, or NASA, launched the first *Landsat* satellite more than 30 years ago. Before it was retired, it had recorded over 300,000 images. Since then, at least six more *Landsats* have been launched. Most of the launches were successful.

Other observation satellite systems are *SarSat* and *NavStar*. *NavStar* is made up of 24 satellites that orbit Earth over different areas. The system is used today to keep track of airplane traffic. It is also used to help during search and rescue missions. *SarSat* is an international search and rescue satellite system.

Thinking Critically Why do you think it is important to track wildfire hazards in a national park?

COOPERATIVE LEARNING

Correcting Distortion Have students complete the Building Science Skills activity. After students determine where distortions occur in their maps, have them work with partners to make their maps more accurate. **learning style: visual**

• *TEACHING RESOURCES* • *Online or Print*

Teacher's Resources

Lesson Review: Chapter 1, p. 6

Enrichment: Chapter 1, p. 10

Concept Map: Study Tools, p. 4

Town Map: Visuals, p. 2

3 *Assess*

CHECKING CONCEPTS ANSWERS

1. two-dimensional, or flat
2. globe
3. distortions

THINKING CRITICALLY ANSWERS

4. **EXPLAIN:** A globe more accurately shows the shapes and relative sizes of Earth's surface features.
5. **HYPOTHESIZE:** Possible answers: Maps can show more small details than globes can, and maps are easier to carry around.
6. **APPLY:** The distance between the two houses is probably too small to show on a globe. A map that included both towns would provide a better way to measure the distance.

4 *Close*

BUILDING SCIENCE SKILLS

Modeling and Comparing Discuss the variety of ways in which students could represent classroom features on their maps. Have students identify distortions in their maps by comparing their completed maps to the actual classroom. **learning style: visual**

Science and Technology
EARTH-OBSERVATION SATELLITES

Discussion Before students read the feature, ask them if they know what a satellite is. Discuss students' responses. Ask students if they can name some satellites. Explain that a satellite is an object in space that revolves around another object. Point out that there are natural satellites, such as the Moon, and artificial satellites, such as *Landsat,* which were made by humans. Ask students: *How do you think people use the information gathered by the* Landsat *and* NavStar *satellites?*

Thinking Critically Answer
Possible answer: Park managers would want to know if wildfire hazards are near campgrounds, hiking trails, buildings, or other visitor areas. They could use the information to help prevent forest fires or be better prepared to rescue park visitors during a fire.

1-5 *(pp. 24–25)*

1 Introduce

INVESTIGATE

TIME: 10–15 minutes

PURPOSE: Students will demonstrate the distortions that take place when a curved surface is flattened.

MATERIALS: globe, grapefruit, felt-tip pen, knife, paper towels

PROCEDURE: Have students work in pairs to complete this activity. Instruct them to discuss what happens after they complete Step 3.
learning styles: visual, tactile

THINK ABOUT IT: The shapes of the continents are distorted. By flattening two parts to match up better, you distort the other parts even more.

STUDY HINT Before beginning this lesson, have students copy the lesson titles in their notebooks, using an outline format and leaving a few blank lines between the titles. As students read each section, have them write the topic sentence for each paragraph beneath each section title. Have students use the Writing Activity Outline found in the Teacher's Resources.

Linking Prior Knowledge Remind students that maps, like globes, are models of Earth, or parts of Earth. Ask students to recall any distortions in their maps of the classroom.

2 Teach

Discussion Write *Map Projection* on the board. From this head, draw arrows to each of the following heads: *Mercator Projection*, *Polar Projection*, and *Conic Projection*. Have students copy the heads in their notebooks. Describe each type of projection and how it is made. Tell students to copy this information in their notebooks under the proper head. Refer to Figures 1-16, 1-17, and 1-18 on pages 24 and 25 as you describe how each one is made.
learning styles: auditory, visual

Answers to questions

1 **DEFINE:** a flat map that represents part or all of Earth's curved surface

2 **INFER:** a polar projection

1-5 What is a map projection?

INVESTIGATE

Making a Map Projection
HANDS-ON ACTIVITY

1. Look at a globe. With a felt-tip pen or permanent marker, roughly outline the continents on the surface of a grapefruit.
2. Carefully cut the grapefruit into quarters. Gently peel the skin off the grapefruit. Try not to tear the four pieces of peel into smaller pieces.
3. Lay the four pieces of peel flat on a table. Match up your pieces to form a flat map.

THINK ABOUT IT: What happens to the continents when you try to flatten the pieces? How can you adjust them to make them match better?

Objectives

Explain what a map projection is. Name three kinds of map projections.

Key Term

map projection (proh-JEHK-shuhn): drawing of Earth's surface, or part of it, on a flat surface

From Round to Flat A **map projection** is a flat map that represents all or part of Earth's curved surface. Imagine a clear plastic globe with a light inside. If paper is wrapped around the lighted globe, the outlines of the continents will be projected onto the paper. A mapmaker can trace these outlines to make a flat map. The result is called a map projection.

1 DEFINE: What is a map projection?

Types of Map Projections There are several types of map projections. Although no type is completely accurate, each kind is useful.

Gerardus Mercator (1512–1594) was a Flemish cartographer, or mapmaker. He drew a huge world map in 1569 that used the system of projection now named for him. Mercator was one of the most respected geographers of his time.

Mercator projections are made by wrapping a sheet of paper into a tube around a globe. An example is shown in Figure 1-16. In a Mercator projection, the distances between land areas and the sizes of land areas near the poles are distorted.

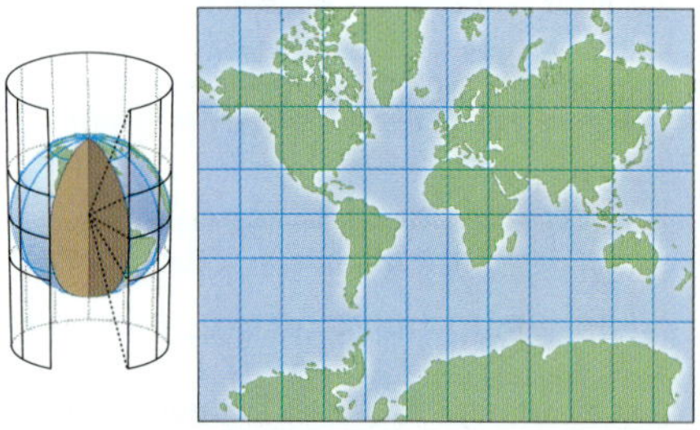

▲ **Figure 1-16** A Mercator projection

A polar projection, shown in Figure 1-17, is made by holding a flat sheet of paper on one pole of a globe. It shows little distortion near that pole. Farther from that pole, however, both direction and distance are distorted.

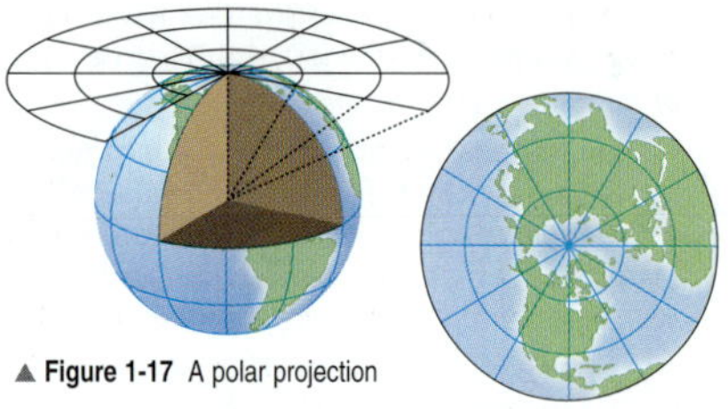

▲ **Figure 1-17** A polar projection

A conic (KAHN-ik) projection is shown in Figure 1-18. It is made by shaping a sheet of paper into a cone and placing it over a globe. When several conic projections are put together, the relative shapes and sizes of land areas on the map are almost the same as those on a globe.

Teaching Options

EXTENDING THE LESSON

Examining Early Maps Have students examine facsimiles of early maps, such as those Columbus used. Ask students to find out how early mapmakers gathered the information they used to make their maps. Have students obtain a copy of an early map or make a drawing of one and explain its characteristics to the class. **learning style: visual**

ONGOING ASSESSMENT

Supply copies of old maps and have students compare them to more modern maps of the same region.

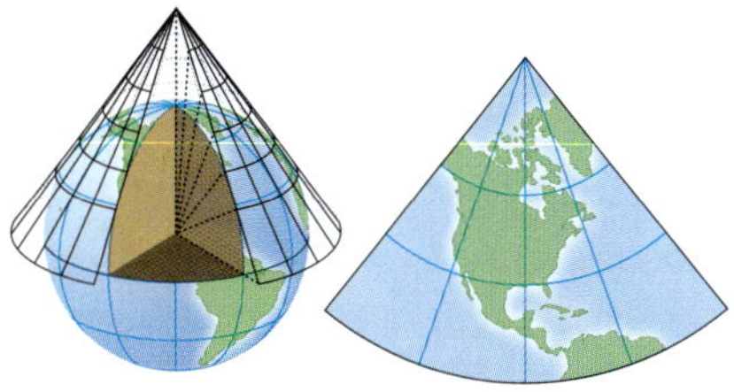

▲ **Figure 1-18** A conic projection

2 **INFER:** Which kind of map projection would you use to explore Antarctica?

CHECKING CONCEPTS

1. What is a flat map of a curved surface called?
2. How is a Mercator projection made?
3. How is a conic projection made?
4. How do mapmakers show Earth's surface on a flat map?
5. How is a polar projection made?

THINKING CRITICALLY

6. **INFER:** Why do you think a conic projection and a polar projection are called by these names?

Web InfoSearch

Mercator Projections Mercator projections are very helpful to navigators at sea. They are widely used to create navigation charts. On Mercator projections, lines of latitude and longitude are straight, parallel lines. This allows a navigator to plot a straight course.

SEARCH: Use the Internet to find out more. How does a sailor use one to plot a course? What problems would occur if the sailor were heading for the poles? Start your search at www.conceptsandchallenges.com. Some key search words are **sailing projections** and **Mercator projections.**

How Do They Know That?

LOCATING FEATURES ON EARTH'S SURFACE

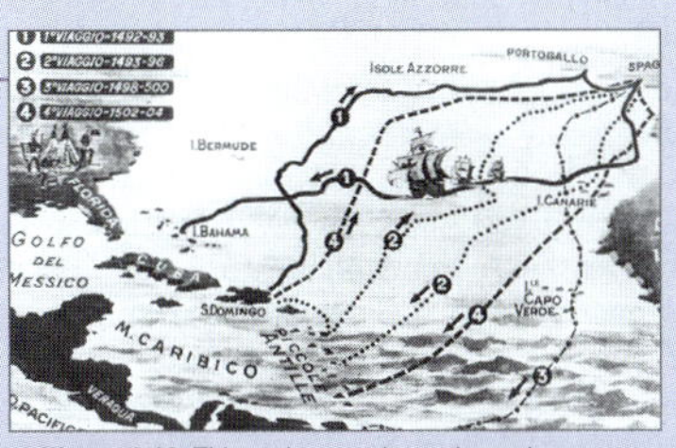

▲ **Figure 1-19** This early map shows the various routes Columbus took to reach and return from the New World.

When do you think the first maps were made? The earliest known maps are more than 4,000 years old. These early maps were carved on clay tablets. Maps during this time showed rivers, mountains, and human settlements. They may have been used to record property lines and help build roads, canals, and towns.

The first people to show Earth as a sphere on maps were the Greeks. About A.D. 100, Ptolemy (TAHL-uh-mee), a Greek-born Egyptian astronomer, mathematician, and geographer, wrote a book based on Greek ideas called *Geography*. In this book, he described how to make maps and globes. Columbus may have used Ptolemy's maps when he set out on his voyage to the New World.

During the fifteenth and sixteenth centuries, the explorations of Columbus, Da Gama, and Magellan led to more accurate maps of Earth. The most famous mapmaker during this time was Mercator. Today, aerial photographs, satellites, and computers are all used to make detailed, accurate maps of Earth's surface.

Thinking Critically How has technology helped us make more accurate maps of Earth's surface?

RETEACH

Showing Projections Display a Mercator projection and a globe to show students the distortion on this type of projection. Show the relative sizes of different landmasses on the globe and on the Mercator projection. Next, show students an example of a polar projection and a conic projection. Point out the kinds of distortions associated with each type of projection.

learning style: visual

• *TEACHING RESOURCES* • *Online or Print*

Teacher's Resources

Lesson Review: Chapter 1, p. 7

Enrichment: Chapter 1, p. 10

Writing Activity Outline: Study Tools, pp. 20–21

3 *Assess*

CHECKING CONCEPTS ANSWERS

1. a projection
2. A sheet of paper is shaped into a tube around a globe.
3. A sheet of paper is shaped into a cone and placed over a globe.
4. A sheet of paper is wrapped around a lighted globe. The outlines of the continents are projected onto the paper and are traced.
5. A sheet of paper is held over one pole of a globe.

THINKING CRITICALLY ANSWER

6. **INFER:** A conic projection is named for the cone-shaped paper on which the map is drawn. A polar projection has that name because the map is traced onto paper held over one of Earth's poles.

4 *Close*

Web InfoSearch

Mercator Projections Have students use the Internet to research how a sailor uses Mercator projections to plot a course. Some key search words are **sailing projections** and **Mercator projections.** Students can begin their search for information at www.conceptsandchallenges.com.

How Do They Know That?

LOCATING FEATURES ON EARTH'S SURFACE

Technology Connection Have students think of two ways of making a map of their community. The first method would involve walking around and recording the locations of streets and buildings. The second method would be to take photographs from a helicopter. Guide students to understand that technologies such as airplanes, satellites, and computers make it easier and faster to gather geographical information.

Thinking Critically Answer
Possible answer: Technology provides better views of Earth's features. Computers can help enhance details and help mapmakers use the information gathered by satellite images and photographs.

1-6 *(pp. 26–27)*

1 Introduce

Linking Prior Knowledge Display a globe or a map of the world. Ask students to explain why there are lines running around the globe, or from one side of the map to the other, and from pole to pole. Discuss students' responses.

2 Teach

Demonstration Display a large map of the world. Point out the equator and the lines of latitude and longitude. Show students how latitude and longitude can be used to find locations on the map. Explain that the Northern Hemisphere is north of the equator and the Southern Hemisphere is south of the equator. Explain that lines of latitude can be used to estimate distance, regardless of the scale of the map or globe. One degree of latitude equals about 111 km, or about 69 mi. One minute of latitude is about 1.8 km, or about 1.12 mi.
learning style: visual

Discussion Make sure that students understand the importance of including direction with measurements for latitude and longitude. To emphasize this, ask students what place is at 90° latitude. Students should realize that both the North Pole and South Pole are at 90° latitude. Ask students what place is at 90° north latitude. (North Pole) **learning style: auditory**

Reinforcement Give students practice using latitude and longitude to find locations on the map. Give the latitude and longitude of a location and have a volunteer find it on the map. After the student finds the location, have him or her state whether it is in the Northern or Southern Hemisphere.
learning styles: visual, auditory

READING STRATEGY Tell students that a mnemonic is a technique that can help people remember words or definitions. Ask students to think of a mnemonic to help them remember the difference between latitude and longitude.

1-6 What are latitude and longitude?

Objective

Explain how hemispheres, latitude, longitude, meridians, and parallels are all related.

Key Terms

parallel (PAR-uh-lehl)**:** horizontal line on a map or globe that circles Earth from east to west at intervals starting at the equator

meridian (muh-RIHD-ee-uhn)**:** line on a map or a globe running from the North Pole to the South Pole along Earth's surface

latitude (LAT-uh-tood)**:** distance in degrees north or south of the equator

longitude (LAHN-juh-tood)**:** measurement in degrees east or west of the prime meridian

Parallels and Meridians If you look at a globe or a map of the world, you will notice a series of lines drawn east and west and another series drawn north and south. These reference lines help people describe locations anywhere on Earth's surface. One handy reference is the equator. The equator is an imaginary line that runs around the middle of Earth's surface. It divides Earth into two half-spheres, or hemispheres — the Northern Hemisphere and the Southern Hemisphere. As shown in Figure 1-20, angles drawn from Earth's center to its surface are used to construct other east-west lines on Earth's surface. **Parallels** are lines on maps and globes that circle Earth east to west parallel to the equator.

Another set of lines on maps and globes, called meridians, run north and south across Earth's surface. A **meridian** is a half-circle that extends from the North Pole to the South Pole. Figure 1-21 shows how meridians are constructed.

1 COMPARE/CONTRAST: How are parallels and meridians similar? How are they different?

Latitude Parallels and meridians are used to describe locations on Earth's surface. Parallels describe latitude. **Latitude** is the distance in degrees north and south of the equator. The latitude of the equator is 0°.

The North Pole is at 90° north latitude. The South Pole is at 90° south latitude. Figure 1-20 shows how the locations of latitude lines are determined.

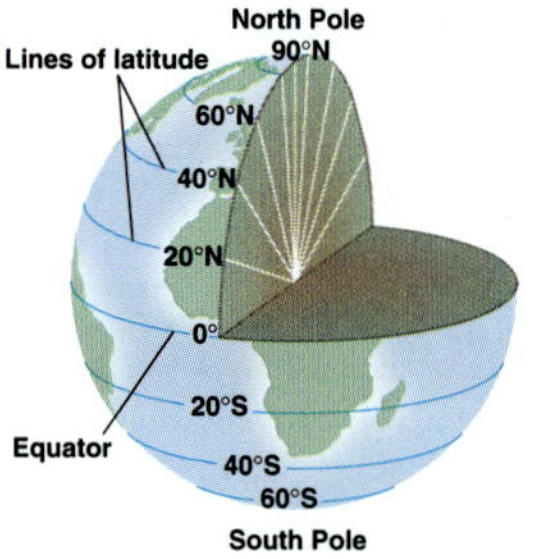

◄ **Figure 1-20** Lines of latitude run horizontally around the globe.

2 MEASURE: What is the latitude of the equator?

Longitude Meridians are used to describe longitude. **Longitude** is the distance in degrees east and west of the 0° meridian. Since there is no equator dividing Earth into north-south halves, scientists had to invent one. This 0° meridian, which runs through Greenwich, England, is called the prime meridian. On the other side of Earth from the prime meridian is the 180° meridian. Called the International Date Line, this is the place where the calendar day changes. Figure 1-21 shows how longitude lines are drawn.

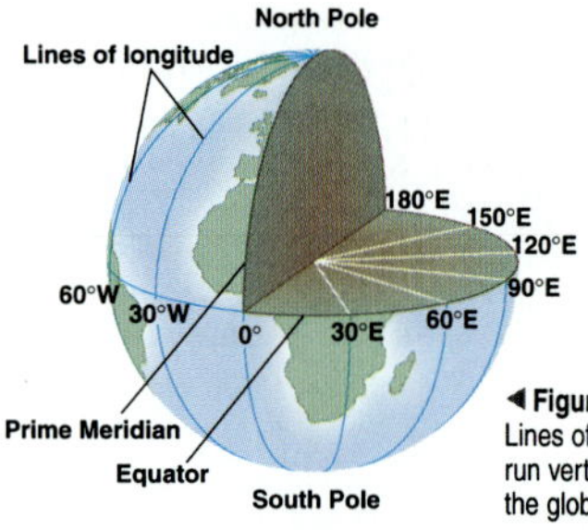

◄ **Figure 1-21** Lines of longitude run vertically around the globe.

3 NAME: What is the name of the meridian that passes through Greenwich, England?

26

Teaching Options

INTEGRATING MATHEMATICS

Circular Reasoning Longitude and latitude coordinates are measured in degrees. There are 360° in a circle. Show students how angles are used to create lines of longitude. Draw a circle on the board with a point at the center. Use a straightedge to draw a line through the center from one side to the other. Draw another line from the top to the bottom edge of the circle, also through the center. Point out that the lines crossing the center point make four 90° angles, and that $90° \times 4 = 360°$. Have students think of placing the center of this circle at the North Pole and the edges of the circle at the equator. Have them refer to Figure 1-21 to see how this would work. Lines of longitude are formed by the lines that radiate from the center of the circle.

ONGOING ASSESSMENT

Ask students to draw a globe, then add and label lines of latitude and longitude.

Locating Places Every place on Earth has its own latitude and longitude. For example, Washington, D.C., is very close to the point where the parallel for 39° north meets the meridian for 77° west. So, Washington's location can be described as 39° north latitude and 77° west longitude. This number can be abbreviated as 39°N, 77°W.

4 **COMPARE:** What describes the location of a place?

CHECKING CONCEPTS

1. The parallel around the middle of Earth is the __________.
2. Each half of Earth's surface is called a __________.
3. Points on a sphere or angles in a circle are measured in units called __________.
4. Lines running horizontally on a map or globe are called __________.
5. Lines running from pole to pole on a map or globe are called __________.

THINKING CRITICALLY

6. **EXPLAIN:** How can you locate a particular city on a map of Earth's surface?
7. **NAME:** In what place on Earth's surface is there a latitude of 0° and a longitude of 0°?
8. **INFER:** What do you think the hemispheres formed by the prime meridian and the International Date Line are called?
9. **INFER:** What is the North Pole's latitude and longitude?

BUILDING SCIENCE SKILLS

Locating Use a globe or a map to identify the cities with the following latitudes and longitudes.

IDENTIFYING CITIES		
City	Latitude	Longitude
A	38° N	77° W
B	30° N	90° W
C	35° N	140°E

▲ Figure 1-22

Hands-On Activity

FINDING PLACES ON A MAP

1. Find the following cities on the map:
   New Orleans, LA
   Pittsburgh, PA
   San Antonio, TX
   Jacksonville, FL
   Springfield, IL
   Denver, CO
2. Write the latitude and longitude of each city. Estimate as closely as possible.

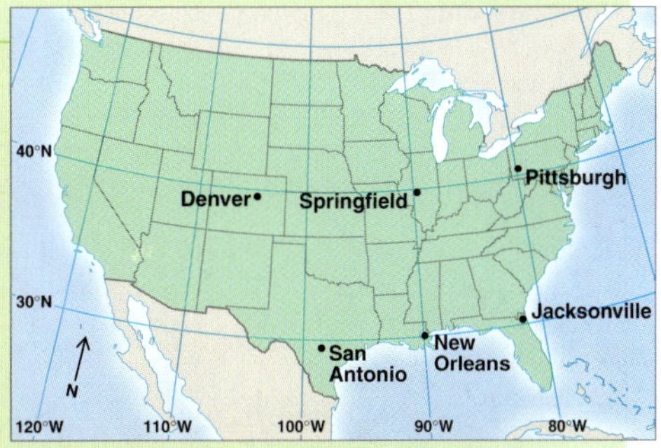

▲ **Figure 1-23** Latitudes and longitudes in the United States.

Practicing Your Skills

3. **EXPLAIN:** Why do you need to know both latitude and longitude to locate places on a map?
4. **IDENTIFY:** What is the approximate latitude and longitude of the city where you live?

EXTENDING THE LESSON

Planning a Journey Have students plan a trip to Europe, Asia, South America, Africa, or around the world. Ask them to include at least five cities or other specific sites in their itinerary. Have them find the latitude and longitude of each stop on their journey. **learning style: visual**

MORE ON THE WEB

Identifying Confluence Have students use the Internet to identify ten different places in the United States where latitude and longitude intersect. Students can begin at www.conceptsandchallenges.com.

• TEACHING RESOURCES • Online or Print

Teacher's Resources

Lesson Review: Chapter 1, p. 8

Enrichment: Chapter 1, p. 10

Answers to questions

1 **COMPARE/CONTRAST:** Both are lines found on the globe. Parallels run parallel to the equator. Meridians run from pole to pole.

2 **MEASURE:** 0°

3 **NAME:** prime meridian

4 **COMPARE:** its latitude and longitude

3 *Assess*

CHECKING CONCEPT ANSWERS

1. equator
2. hemisphere
3. degrees
4. parallels, or latitude
5. meridians, or longitude

THINKING CRITICALLY ANSWERS

6. **EXPLAIN:** Find its latitude and longitude.
7. **NAME:** where the prime meridian and the equator cross each other
8. **INFER:** Eastern and Western hemispheres
9. **INFER:** 90°N, 0°

4 *Close*

BUILDING SCIENCE SKILLS ANSWERS

Locating City A: Washington, D.C.; City B: New Orleans, LA; City C: Tokyo, Japan

Hands-On Activity

FINDING PLACES ON A MAP

TIME: 15–20 minutes

PURPOSE: explaining, identifying

2. Answers should be close to the following: New Orleans: 30°N, 90°W; Pittsburgh: 41°N, 80°W; San Antonio: 29°N, 98°W; Jacksonville: 30°N, 81°W; Springfield: 39°N, 89°W; Denver: 39°N, 105°W

Practicing Your Skills Answers

3. **EXPLAIN:** Where lines of latitude and longitude cross is a unique point that exactly locates a place. Just latitude or longitude gives you one line, and you cannot tell at which point on the line a place is.
4. **IDENTIFY:** You can ask students to choose the nearest large city.

1-7 *(pp. 28–29)*

1 Introduce

STUDY HINT Have students find the important words in the topic sentence of each paragraph. Ask them to make a list of these words and write a summary of the lesson based on the list.

Linking Prior Knowledge Review the different kinds of maps that students learned about (Lesson 1-4). Ask what kinds of information maps provide. Ask students if they have had any experience navigating with a map, perhaps while someone drove a car. Invite students to discuss any difficulties or successes they have had using maps.

2 Teach

Demonstration Display a large wall map and have students identify the compass directions. Show students how to use the map's compass rose or north arrow to determine directions. Point out two locations on the map. Have students determine the direction from one location to the other. Repeat this activity, using several locations. **learning style: visual**

Discussion Ask students why truck drivers or people visiting an unfamiliar area should be able to read a road map. Discuss students' responses. Point out that in order to read a map, a person needs to understand the map's scale and symbols. **learning style: auditory**

Reinforcement Emphasize the importance of orienting oneself by first locating north on a map. Show students a variety of maps and have volunteers practice finding north on each one. **learning styles: visual, auditory**

Answers to questions

1. **OBSERVE:** compass rose or north arrow
2. **CALCULATE:** 2 cm ($1/5 = x/10$; $10/5 = 2$)
3. **ANALYZE:** Symbols are a shorthand way of showing the locations of most objects on a map.
4. **IDENTIFY:** black

1-7 How do you read a map?

Objectives

Use the scale on a map. Interpret map symbols.

Key Terms

scale: feature that relates distances on a map to actual distances on Earth's surface

symbol: drawing on a map that represents a real object

legend (LEHJ-uhnd)**:** list of map symbols and their meanings

Showing Direction Most maps show directions. If you look at the map on this page, you will see an arrow marked N. This arrow shows the direction north. On most maps, north (N) is at the top of the map. From this you can infer that south (S) is at the bottom, east (E) is to your right, and west (W) is to your left.

Another way of showing direction on a map is using a compass rose. Most places on a map are not exactly north, south, east, or west. In Figure 1-24, you can see that Snake River flows both south and north and east and west. The direction between south and west is called southwest (SW). The direction between north and east is called northeast (NE).

A compass rose is a circle divided into eighths. It shows the four directions and directions in between the four.

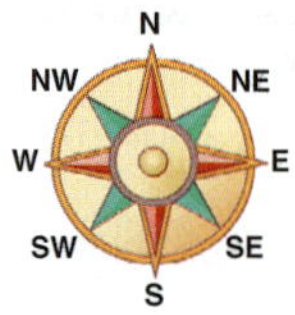

▲ **Figure 1-25** Compass rose

1. **OBSERVE:** On a map, what shows the direction?

Scale A map's **scale** shows how the distance on a map compares with real distance on Earth's surface. Look at the scale in Figure 1-24. It shows that 1 cm on the map equals 5 km on Earth's surface. Measure the distance on the map from Point A to Point B. The distance is 4.4 cm. To find the actual distance between Point A and Point B, multiply 4.4 × 5, because each centimeter on the map represents 5 km on Earth. On Earth's surface, the actual distance from Point A to Point B is 22 km.

2. **CALCULATE:** The main street in a town is 10 km long. How many centimeters would the street be on a map that uses the scale 1 cm = 5 km?

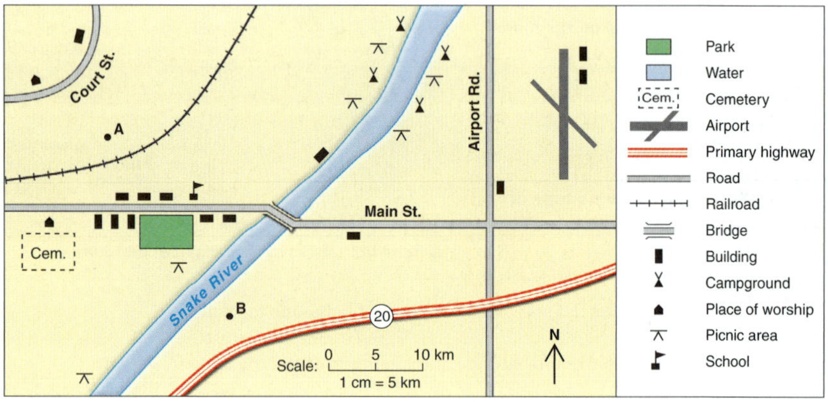

▲ **Figure 1-24** Town map

Teaching Options

! FAST FACTS *Compasses*

A magnetic compass contains a magnetized needle housed in a clear case. A liquid compass is a magnetic compass filled with a liquid that reduces vibrations that can make readings difficult.

A gyrocompass contains a gyroscope instead of a magnetic needle. The spinning wheel of the gyroscope is attached to an axle that keeps the wheel horizontal. The gyroscope indicates north because it aligns itself with the axis of Earth's rotation, which runs north to south and from pole to pole.

MORE ON THE WEB

Getting From Here to There
Have students use the Internet to plan a car trip with their family. Then have students share the trip and route with the class.

QUICK CONVERSIONS

Point A to Point B	Metric/SI	English
Measured distance	3.0 cm	1.2 in.
Actual distance	15.0 km	9.3 mi

Map Symbols Simple drawings on a map called **symbols** are used to show the location of real objects. Many different symbols are used on maps. These symbols are listed in a table called a **legend.** Different maps have different legends. The legend of each map explains the symbols used on that map. See the legend in Figure 1-24 for some common map symbols.

3 ANALYZE: Why are symbols used on a map?

Color Color often has meaning on maps. Black is usually used for anything made by people. Symbols for buildings, railroads, and bridges are colored black. Blue is used to show bodies of water. Forests and parks are green.

4 IDENTIFY: What color would be used to show your school on a map?

CHECKING CONCEPTS

1. The direction northwest would be shown on a map by the letters __________.
2. The __________ on a map shows distance.
3. Simple drawings on a map used to show real objects are called __________.
4. The symbols used on a map are listed in a __________.
5. A river would most likely be colored __________ on a map.

THINKING CRITICALLY

6. CLASSIFY: What types of information do the scale and the legend on a map provide?
7. MODEL: Draw a simple map of the area around your school. Include a scale and a legend of all the symbols you think belong there.

INTERPRETING VISUALS

Use Figure 1-24 to answer the following questions.

8. OBSERVE: How many buildings are shown on the map?
9. OBSERVE: What kinds of buildings and roads are shown?
10. ANALYZE: What direction is to your left on the map?
11. ANALYZE: In what two directions does Main Street run?

Real-Life Science
ORIENTEERING

Orienteering is a fun sport. The goal of the game is to find specific areas marked on a given course. As you move through the course, you race against other people looking for the same marked points as you. To win, you must be able to read a map quickly and use a compass. Classes in map and compass reading are offered at some orienteering meets.

Before an orienteering meet begins, the course is set up. Markers are left at different checkpoints along the course. Each orienteering participant is given a map that identifies each checkpoint. The first person to cross the finish line with each checkpoint marker wins.

You can participate in orienteering meets no matter where you live. Courses have been set up in cities as well as in wooded areas. You do not need any special equipment except a map and a compass.

Thinking Critically What checkpoints might there be along a city orienteering course?

▲ **Figure 1-26** Orienteering courses can be found in many places.

3 Assess

CHECKING CONCEPTS ANSWERS

1. NW
2. scale
3. symbols
4. legend
5. blue

THINKING CRITICALLY ANSWERS

6. CLASSIFY: A map's scale shows the distance between objects. Map legends explain symbols on a map.
7. MODEL: Students' maps should show an understanding of how maps represent actual objects, directions, and distances.

INTERPRETING VISUALS ANSWERS

8. OBSERVE: 17
9. OBSERVE: Unidentified buildings, a school, two places of worship, a primary highway, local roads. Some students may also include the railroad.
10. ANALYZE: west
11. ANALYZE: east and west

4 Close

Real-Life Science
ORIENTEERING

Background Bring a compass to class and show students how it works. Have students take turns placing the compass on a map lying on a flat table. Show how to orient the map and compass so that the compass needle points to north on the map. Have students turn the compass housing without picking it up from the table. They will observe that the needle of the compass remains pointing to the north as the compass housing moves. Explain to students that if they want to travel north, they should match up the compass needle with the large direction-of-travel arrow on the compass. As long as the needle is lined up with the arrow, they will be going north. Travel in other directions requires lining up the compass needle in other ways.

Discussion Ask students how a compass would be useful in orienteering. (You need to know that you are headed in the right direction.)

Thinking Critically Answer
Possible answers: street intersections, public buildings, and parking lots

COOPERATIVE LEARNING

Measuring Distance Have students work in pairs to measure distances between objects shown in Figure 1-24. Students will need a straightedge marked in centimeters and should convert centimeters to kilometers. Start them off with the Snake River Bridge to Airport Road (2.5 cm [.98 in.] or 12.5 km [7.8 mi]).
learning style: logical/mathematical

• TEACHING RESOURCES • Online or Print

Teacher's Resources
Lesson Review: Chapter 1, p. 9
Enrichment: Chapter 1, p. 10
Town Map: Visuals, p. 2

INTEGRATING MATHEMATICS

Using Algebra To help students use scales on maps, you may want to show them how to set up simple algebraic equations. Using the example in the text, write $1\ cm/5\ km = x\ cm/10\ km$ on the board. Have students solve for x. ($x = 2$) You may also want to ask students to convert distance in kilometers to distance in miles. (Multiply kilometers by 0.6 to convert to miles.)

The Big Idea

(pp. 30–31)

1 Introduce

Objective Students will learn how Lewis and Clark collected valuable information about the western part of the United States during their journey. Students will choose one place that the explorers visited and write about it in their Science Logs.

Linking Prior Knowledge Ask students to discuss what they know about the early exploration of the United States, including the journey of Lewis and Clark. Have students review what they learned about reading maps (Lesson 1-7), map directions, and symbols. Ask students to identify directions and symbols used on the map on pages 30 and 31.

2 Teach

Discussion After students have read the article and callouts, encourage them to use reference materials to look for more information on the places shown on the map before deciding what to write about in their Science Logs. Ask students what methods of travel into unexplored areas were available at the time of the Lewis and Clark expedition. (small boat, horseback, foot) Ask students why the expedition began by following the course of the Missouri River. (A large river made a good landmark for the explorers to use in mapping the new territory. Also, it was probably easier to travel by boat than over land.) Use the background information in Fast Facts to provide students with more details concerning the life and times of Meriwether Lewis. **learning style: auditory**

THE BIG IDEA

How did maps help the United States grow? Have students start their research for their Science Log Writing Activity at www.conceptsandchallenges.com. Be sure students use search engines and websites that are reliable and acceptable. You may also wish to provide copies of the *Big Idea Planner* and *Big Idea Science Log Entry* study tools found in the Teacher's Resources.

Integrating History

THE Big IDEA

How did maps help the United States grow?

Look at a modern map of the United States. You will see a vast stretch of land extending from the Atlantic Ocean to the Pacific Ocean and from Canada to the Gulf of Mexico. A map of what the United States looked like in 1800 would look very different. The country then was only about one-third of its current size. It consisted of lands that extended westward from the Atlantic Ocean to the Mississippi River.

President Thomas Jefferson dreamed of an America that would stretch from sea to sea. In 1803, he bought from France all of the lands between the Mississippi River and the Rocky Mountains. This so-called Louisiana Purchase doubled the size of the country. The United States paid only $15 million for this land.

No maps of the area existed. Jefferson decided to send explorers to map the region and report on local wildlife and climate. He asked his personal secretary, Meriwether Lewis, to lead the expedition. Lewis asked his friend, William Clark, to join him.

In 1804, the two men left St. Louis with 28 others. The group, known as the Corps of Discovery, arrived back home 2½ years later. The maps, charts, and journals they brought back opened the door to westward expansion.

Look at the illustrations that appear on these two pages. Then, follow the directions in the Science Log to learn more about "the big idea." ✦

Spring 1805

Lewis climbs to the top of a hill and views the Rocky Mountains for the first time.

Winter 1806

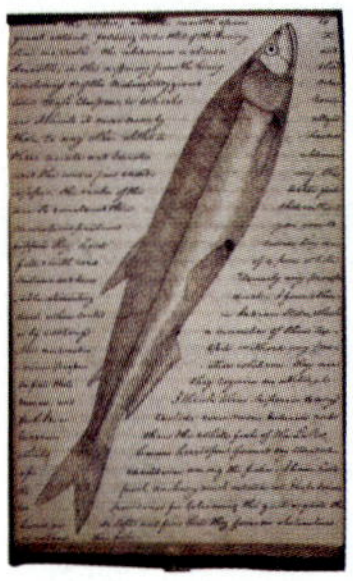

The expedition spends the harsh winter months at Fort Clatsop in present-day Oregon. Lewis and Clark make sketches of wildlife and of the Native Americans living in this previously unexplored area. This is a sample from their journal.

WRITING ACTIVITY

Science Log

You are traveling with the Corps of Discovery. On your journey, you observe many strange new animals and eat new foods. You view the Rocky Mountains for the first time. Write down your observations. Draw a simple map with a legend of the place you are describing. Start your search at www.conceptsandchallenges.com.

Teaching Options

! FAST FACTS Meriwether Lewis

Meriwether Lewis was born in Virginia in 1774. In school, he showed a talent for botany and zoology and was a keen observer of nature. In 1801, President Jefferson invited Lewis to serve as his personal secretary. Jefferson soon set Lewis to the task of preparing for an exploratory expedition to the Pacific Ocean. Over the next three years, Lewis estimated the expedition's travel expenses, purchased supplies, and consulted with scientists about the proposed route. During the expedition, Lewis served as naturalist, making careful notes of plants and animals that the expedition encountered.

In 1808, two years after the completion of the Lewis and Clark expedition, Jefferson appointed Lewis governor of the Louisiana Territory. However, Lewis was not well suited to the job and was beset by political troubles, financial problems, and episodes of depression. He died mysteriously in September 1809 on a trip to Washington, D.C., to answer complaints about his performance as governor. Whether his death was a murder or a suicide has not been determined.

Winter 1805

The Corps of Discovery sets up winter quarters near present-day Bismarck, North Dakota. During this time, Toussaint Charbonneau joins the group as an interpreter. His wife, a Shoshone named Sacagawea (sac-a-juh-WEE-ah), also joins the expedition as a guide. She is honored on the coin shown.

Fall 1804

The explorers leave the trail and meet the Arikara. Clark records information about this Plains group in his journal. He notes that the Arikara are farmers who live year round in earthen lodges.

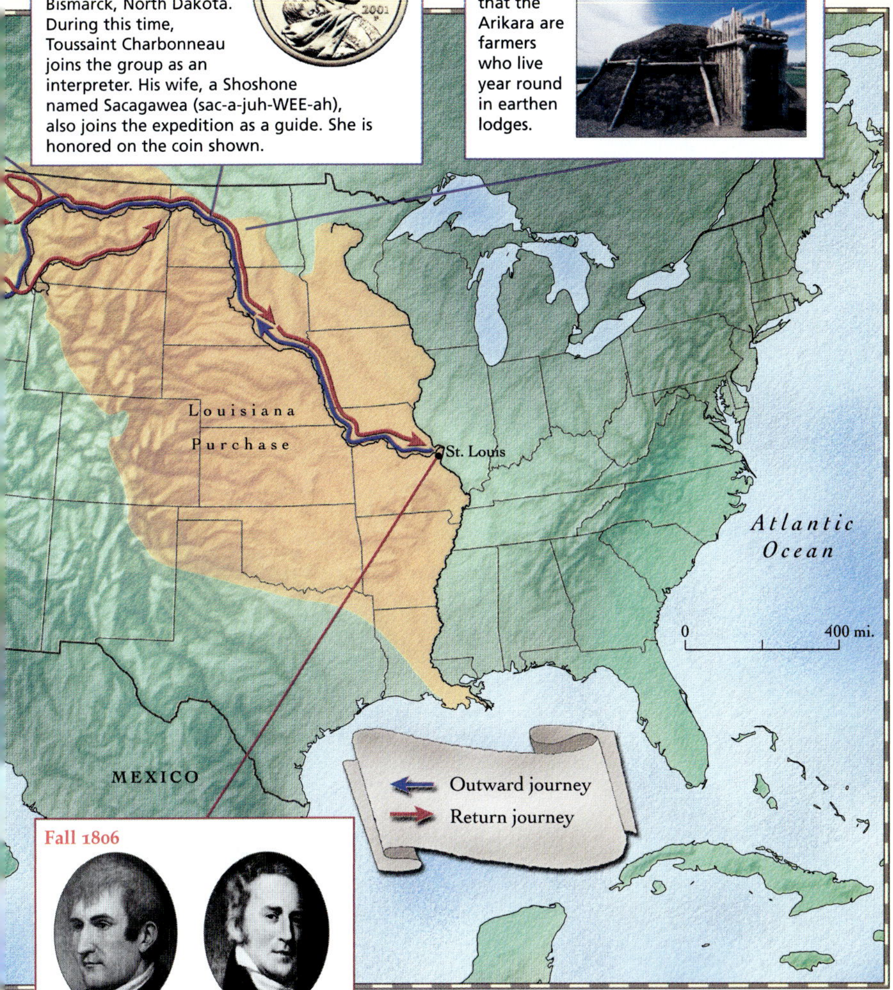

Fall 1806

M. Lewis W. Clark

The explorers arrive back in St. Louis on September 23. The whole town turns out to give them a heroes' welcome.

▲ **Figure 1-27** This map of the early United States shows the travel route of Lewis and Clark. The maps Lewis and Clark created led to the expansion of the western United States.

3 Assess

Use the Writing Activity Rubric from the Teacher's Resources to assess students' Science Logs. Fill in the rubric with the additional information below. For this assignment, students should have:

- drawn a map of the place they wrote about, including map symbols and a legend.
- communicated, in paragraph form, accurate information about the place.

4 Close

Picturing a Place Invite students to describe how the place they wrote about in their Science Logs looks today compared with when it was first observed by the Corps of Discovery. Have students write their descriptions in a paragraph and illustrate them. Ask students to find out how Lewis's and Clark's journals and maps helped encourage settlement of the area and how long it took before settlers went there to live. **learning style: visual**

Writing an Explorer's Journal Have pairs of students use the information from their Science Logs to write several entries in a daily journal kept by the Corps of Discovery. Their entries should include details about geological features encountered on each day's journey. Put together all of the entries into a diary and have students draw pictures to illustrate it.

Tips for Using Technology

Using Graphics Software to Draw a Map Students can use graphics creation software to draw a simple map. Most graphics creation software uses similar tools to let you create graphics and illustrations. Some of these tools include shape creation tools, text tools, and pencil tools.

The pencil, or drawing tool, allows you to draw in freehand. Depending on your graphics software, your mouse pointer will probably change to a pencil. Then, you can click and drag the pointer in your illustration to draw.

• TEACHING RESOURCES • Online or Print

Teacher's Resources

Big Idea: Chapter 1, p. 11

Big Idea Planner: Study Tools, pp. 11–12

Big Idea Science Log Entry: Study Tools, p. 13

Writing Activity Rubric: Teaching Tools, p. 9

1-8 *(pp. 32–33)*

1 *Introduce*

STUDY HINT Have students use the Key Terms in this lesson to write a paragraph describing what is shown on a topographic map.

Linking Prior Knowledge Have students discuss any experiences they may have had using very detailed maps to guide them while hiking in parks, forests, or wilderness areas. Point out that not all maps are of settled areas with streets and roads.

2 *Teach*

Discussion Define topography as the physical features, or shape, of Earth's surface. Ask students what is meant by physical features. (mountains, valleys, lakes, streams, and so on)
learning style: auditory

Demonstration Ask students how they would go about creating a map of a hiking trail that passed over hills, through valleys, and across streams. Discuss students' responses. Show students a topographic map, or have them study the topographic map in Figure 1-28. **learning style: visual**

Reinforcement Draw a horizontal line across the board. Label the line *Sea Level.* Point out that the average level of the water in the oceans is called sea level. Define *elevation.* Draw a line above and below the line on the board to represent points above and below sea level. **learning style: visual**

Answers to questions

1. **DEFINE:** the height of a point on Earth above or below sea level
2. **IDENTIFY:** mountains, plains, and plateaus
3. **DESCRIBE:** places on a map with the same elevation
4. **EXPLAIN:** a map with contour lines that shows the surface features of the land

Go Online *active art*
For: Topographic Map activity
Visit: PHSchool.com
Web Code: cfp-2014
Students can interact with the art of topographic maps online.

1-8 What is a topographic map?

Objective
Describe what is shown on a topographic map.

Key Terms
elevation (ehl-uh-VAY-shuhn): height of a point on Earth above or below sea level
topography (tuh-PAHG-ruh-fee): general form and shape of the land on Earth's surface
contour (KAHN-toor) **line:** line drawn on a map that connects all points having the same elevation

Elevation The level of the water in the oceans is about the same height everywhere on Earth. The average height for water in the oceans is called sea level. The height of a place on land is measured from sea level. For example, the height of Mount McKinley, Alaska, is more than 6 km above sea level. The height of land above or below sea level is called **elevation.**

1 **DEFINE:** What is elevation?

Topography The general form and shape of the land on Earth's surface is called **topography.** Examples of topography on Earth's surface are mountains, valleys, plains, and plateaus. Mountains have high elevations and steep slopes. Valleys are low areas bordered by high ground. Plains are broad, flat areas that are usually only a little above sea level. Plateaus are broad, flat areas of land that have an elevation of at least 300 m above sea level.

2 **IDENTIFY:** What are three types of topography on Earth's surface?

Contour Lines The elevations of surface features can be shown on a map using contour lines. A **contour line** is a line that passes through all points on a map that have the same elevation. When you look at the contour lines on a map, you can identify the shape, or contour, of the land. Every point on a contour line has the same elevation.

3 **DESCRIBE:** What do contour lines on a map show?

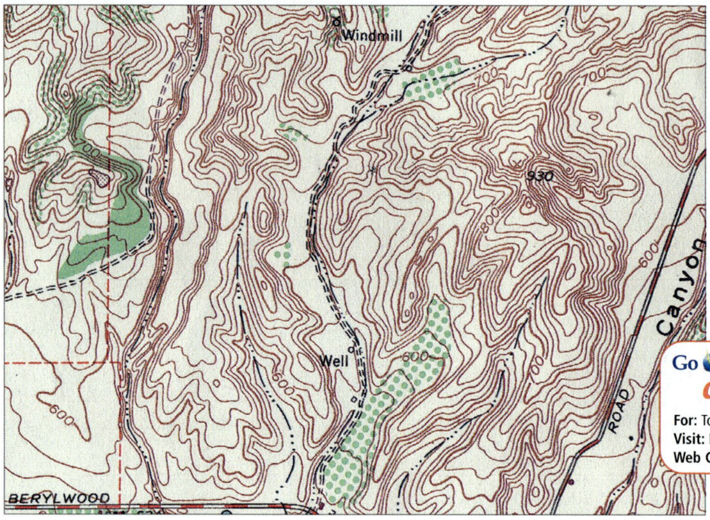

◀ **Figure 1-28** A topographic map, like the one to the left, shows surface features such as differences in elevation.

Go Online *active art*
For: Topographic Map activity
Visit: PHSchool.com
Web Code: cfp-2014

32

Teaching Options

! FAST FACTS *Topographic Map Symbols*

The margins of most topographic maps provide the latitude and longitude of the region the map covers. Contour lines are usually printed in shades of brown. However, not all contour lines are labeled with their elevation. Only index contour lines have elevation labels. Index contour lines are printed in a darker shade of brown. In Figure 1-28, index contour lines occur at elevations of 1,000, 1,250, 1,500, 1,750, and 2,000 m (1,094, 1,367, 1,640, 1,914, and 2,187 yd).

Green signifies wooded areas and blue represents water. Large streams and lakes are outlined and filled in with blue. Streams that are less than about 7.5 m (8.2 yd) wide are a single blue line. Streams that are dry for part of the year appear as a blue dashed line. Springs and water wells are represented by a blue circle.

Red represents hard-surface roads, such as highways. Red-tinted areas on a map indicate that the area contains many roads and buildings not shown in detail.

Black is used for objects made by humans as well as for place names and other labels. Footpaths or trails are dashed black lines. Black squares represent buildings. The symbols for campground, gravel pit, and picnic area also appear in Figure 1-28.

Topographic Maps A map with contour lines shows the surface features, or topography, of the land. That is why these maps are called topographic maps. The contour lines found on a topographic map connect places on the map having the same elevation. Figure 1-28 is a topographic map of the Grand Canyon area in Arizona.

4 EXPLAIN: What is a topographic map?

CHECKING CONCEPTS

1. What is sea level?
2. What is topography?
3. What is a contour line?
4. What is the height of a point above or below sea level called?

THINKING CRITICALLY

5. CONTRAST: What is the difference between sea level and elevation?
6. EXPLAIN: A contour line is marked 50 m. What does this mean?
7. INFER: Why are plateaus sometimes called "plains in the air"?
8. RELATE: Why is a map with contour lines called a topographic map?

Web InfoSearch

Contour Plowing Farmers planting on hilly ground are careful to plow their land in a special way. They plow along strips of land that are at the same elevation. In this way, all the furrows follow the slope of the hill. Water collects in the furrows and soaks into the soil. The furrows prevent the water from running down the hill and carrying away valuable topsoil. This kind of plowing is called contour plowing.

SEARCH: Use the Internet to find out more about contour plowing. In what countries is it used? Why? Start your search at www.conceptsandchallenges.com. Some key search words are **agriculture, farming, topsoil,** and **contour tilling.**

People in Science

SURVEYOR

A surveyor measures land to find the exact location of certain physical features on Earth's surface. Engineers employ surveyors to measure the heights of objects, the slope of the land, and the distances between physical features. With this information, engineers design roads, bridges, and homes.

Ken Stigner began his surveying career in the military. He prepared maps for the construction of roads and bridges. He started as a member of a crew taking measurements ("chain man"). Then, he became an instrument operator. Later, he was made party chief. This meant he had his own crew.

▲ **Figure 1-29** Surveyors use many different tools, such as this transit, when working on building projects.

Surveyors come from all education levels. Some have little education and just want to work outdoors. Although on-the-job training is possible, it helps to have a background in mathematics. It requires many years of experience to obtain a surveying license. Some states now require a four-year degree in surveying to qualify.

Thinking Critically Why do you think the military may need to build roads and bridges?

3 Assess

CHECKING CONCEPTS ANSWERS

1. the average level of water in the oceans
2. the general form and shape of land on Earth's surface
3. a line that connects all points on a map that have the same elevation
4. elevation

THINKING CRITICALLY ANSWERS

5. CONTRAST: Sea level is the average level of water in the oceans. Elevation is the height of land above or below sea level.
6. EXPLAIN: The points it connects are 50 m (165 ft) above sea level.
7. INFER: Plateaus are flat like plains, but they are at least 300 m (984 ft) above sea level.
8. RELATE: It shows the surface features, or topography, of the land.

4 Close

Web InfoSearch

Contour Plowing Have students use the Internet to determine the areas of the world that use contour tilling, or plowing. Then, have students write a short report saying why it is used there. Some key search words are **agriculture, farming, topsoil,** and **contour tilling.** Students can begin their search at www.conceptsandchallenges.com.

People in Science

SURVEYOR

Geometry Connection Surveyors need a thorough understanding of geometry and trigonometry. For example, triangulation is a surveying technique in which a survey area to be measured is divided into a series of triangles.

Discussion The device shown in Figure 1-29 is a surveyor's level. It is used to measure elevation and includes a telescope with crosshairs and a spirit level. The spirit level provides a horizontal baseline for measurements. Surveyors use tape measures and laser devices to measure distances.

Thinking Critically Answer Military personnel and heavy equipment need to be moved from place to place.

EXTENDING THE LESSON

Maps in the Field Give students an opportunity to use a contour map to follow their progress while hiking on a well-marked trail in a nearby park or forest area. **learning style: kinesthetic**

RETEACH

Locating Contour Lines Have students find the contour lines on the map. Point out the different contour lines and ask students to name the elevations. Emphasize that contour lines connect places on the map with the same elevations. Also point out the scale, compass rose, and symbols on the map.

• TEACHING RESOURCES • Online or Print

Teacher's Resources

Lesson Review: Chapter 1, p. 12

Enrichment: Chapter 1, p. 14

ESL/ELL STRATEGY

Word Derivations Point out that *graph* comes from a Greek word that means "to write." Ask students which of the Key Terms this information will help them understand. Ask what other words they can list that end in *graph*. (photograph, autograph)

1-9 *(pp. 34–35)*

1 *Introduce*

STUDY HINT Ask students to relate the Key Term *contour line* from the previous lesson to the Key Term *contour interval* from this lesson.

Linking Prior Knowledge Review with students the definition of the terms *elevation* and *contour lines* (Lesson 1-8). Remind students that a topographic map indicates the steepness or flatness of the terrain shown on the map.

2 *Teach*

Demonstration Have volunteers use a tape measure to establish a line of sight that crosses several objects of varying heights in the classroom, such as your desk, a student desk, a bookshelf, and a windowsill. Have other students use a tape measure or meterstick to measure the height (or highest elevation) of each object crossed by the line. On the board, create a line graph that shows the variation in elevation along the line. Use the floor as a baseline, giving it an elevation of zero. Point out to students that this is the kind of information used to create a topographic map.

Discussion Refer students to the contour lines of the island shown in Figure 1-30. Point out that the top portion of the figure is a side view of the island. The lower portion is a top view, looking down on the island. The top view is the perspective shown on topographic maps. Next, refer students back to the topographic map in Figure 1-28. Explain that when contour lines are close together, the land has a steep slope. Ask students: *What kind of slope do you think the land has if the contour lines are far apart?* (gentle) **learning style: visual**

Reinforcement Give students practice determining contour intervals. Write on the board pairs of numbers that represent labels on contour lines. Ask students to calculate the contour intervals. Make sure that students understand why large contour intervals are used for mountainous areas and small intervals are used for flat areas. Some students may have difficulty understanding the relationship.
learning style: logical/mathematical

1-9 How do you read a topographic map?

Objective
Explain how to read a topographic map.

Key Term
contour interval (KAHN-toor IHN-tuhr-vuhl)**:** difference in elevation between one contour line and the next

Contour Intervals A topographic map shows the elevation of features on Earth's surface. Look at the topographic map shown in Figure 1-30. Every point on a contour line is at the same elevation.

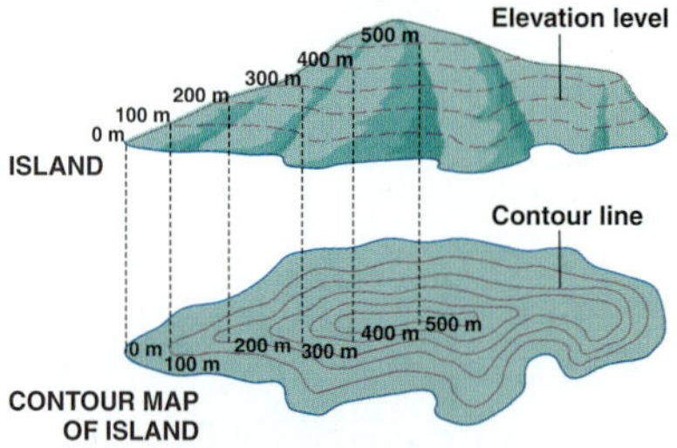

▲ **Figure 1-30** This drawing shows how an island's different elevations would look on a topographic map.

The places where contour lines are close together, the land has a steep slope. The places where the lines are far apart, the land has a gentle slope.

The difference in elevation between one contour line and the next is called a **contour interval.** If two contour lines near each other are marked 500 and 600 m, the contour interval between them is 100 m. Mapmakers use different contour intervals for different maps. A large contour interval is used for mountainous areas. A small contour interval is used for flat areas.

1 OBSERVE: What is the contour interval for the map in Figure 1-30?

Relief Map A map that uses color, shading, or contour lines to indicate the different heights of features is called a relief map. On a relief map, like the one in Figure 1-31, one color might be used to show high mountains. Another color might be used to show areas near sea level. Still other colors might be used to show elevations in between or for areas below sea level.

2 EXPLAIN: How is color sometimes used to show elevation on a relief map?

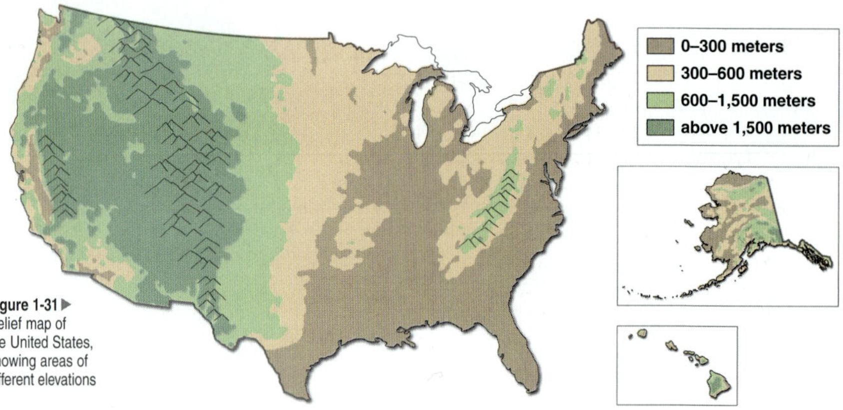

Figure 1-31 ▶ Relief map of the United States, showing areas of different elevations

Teaching Options

! FAST FACTS *River Valleys on a Topographic Map*

Contour lines are usually V-shaped where they cross river valleys. A river flows from higher elevations to lower elevations. Therefore, contour lines approaching rivers must go upstream in order to stay at the same elevation when crossing the rivers. Then, the lines must head downstream to stay at the same elevation. The resulting pattern is a series of V-shaped contour lines pointing upstream.

MORE ON THE WEB

Using Mapping Technology
Have students use the Internet to explore the development of the Global Positioning System (GPS) and how GPS satellites are used in today's society.

RETEACH

Looking at Local Topography
Topographic maps for all areas of the United States are available from the U.S. Geological Survey. Obtain maps of your local area for students to study. Have students describe the kinds of terrain shown on each.
learning styles: visual, auditory

CHECKING CONCEPTS

1. Where contour lines are close together, the land has a __________ slope.
2. The difference in elevation between two neighboring contour lines is called the __________.
3. A map of a flat area would usually have a __________ contour interval.
4. Instead of contour lines, some maps use __________ to show elevation.

THINKING CRITICALLY

Use Figures 1-30 and 1-31 to answer the following questions.

5. OBSERVE: What is the highest elevation shown in Figure 1-30?
6. ANALYZE: Is the land shown in Figure 1-30 flat or hilly? How do you know?
7. OBSERVE: What color is used in Figure 1-31 to show elevations from 0 to 300 m?
8. ANALYZE: What is the elevation of the land shown in dark green in Figure 1-31?

INTERPRETING VISUALS

Reading a Topographic Map The symbol X on a topographic map is called a benchmark. A benchmark is a mark on a map that shows a landmark with a known height. The landmark is used as a reference point in relation to the height of other things on the map. A depression is shown on a topographic map by using short, straight lines pointing toward the center of the depression. Look at Figure 1-32. Which point on the map is the benchmark? What is the elevation at the benchmark? Which point shows a depression?

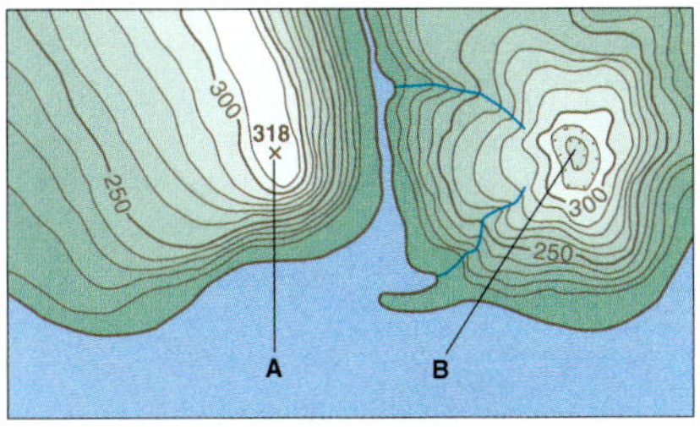

▲ **Figure 1-32** Elevation is shown in meters.

Hands-On Activity

INTERPRETING A TOPOGRAPHIC MAP

You will need a pencil, a sheet of paper, and a metric ruler. Examine the map in Figure 1-33. Then, answer the following questions.

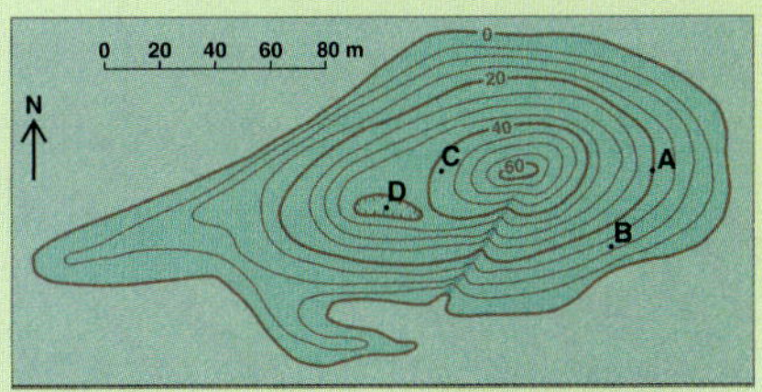

▲ **Figure 1-33** The contour lines on this map show several land features.

Practicing Your Skills

1. IDENTIFY: What contour interval is used on the map in Figure 1-33?
2. CALCULATE: What are the elevations of Points *A*, *B*, and *C*?
3. CALCULATE: How many meters is the highest point on the map?
4. MEASURE: How long in meters is the river?
5. INFER: The hill shown on the map has three sides that are steep and one side that is gentle. How can you tell which are the steep sides and which is the gentle side? What is the compass direction of the gentle side of the hill?
6. ANALYZE: Which letter is in a depression? How do you know?

EXTENDING THE LESSON

Scaling Contour Intervals Tell students that contour intervals are based on the scale of the map. A 1:1,000,000 scale has to have very large contour intervals or the contour lines would all blend together. A 1:15,000 scale can have smaller intervals because the lines will be far enough apart to read.
learning style: logical/mathematical

QUICK CONVERSIONS

	Metric/SI	English
Elevations	300 m	984 ft
	600 m	1,969 ft
	1,500 m	4,922 ft

• *TEACHING RESOURCES* • *Online or Print*

Teacher's Resources
Lesson Review: Chapter 1, p. 13
Enrichment: Chapter 1, p. 14

Laboratory Manual
Lab. Challenge: How can a three-dimensional landform be modeled from a topographic map? pp. 21–24

Answers to questions

1. OBSERVE: 100 m
2. EXPLAIN: Different colors can be used to indicate different elevations.

3 *Assess*

CHECKING CONCEPTS ANSWERS

1. steep
2. contour interval
3. small
4. color

THINKING CRITICALLY ANSWERS

5. OBSERVE: 500 m
6. ANALYZE: hilly (The elevation increases from 0 m to 500 m.)
7. OBSERVE: brown
8. ANALYZE: above 1,500 m

INTERPRETING VISUALS ANSWERS

Reading a Topographic Map Point A is a benchmark. The elevation at the benchmark is 318 m (1,043 ft). Point B shows a depression.

4 *Close*

Hands-On Activity

INTERPRETING A TOPOGRAPHIC MAP

TIME: 10–15 minutes
PURPOSE: observing, measuring
COOPERATIVE LEARNING: Have students work in pairs to complete this activity.

1. IDENTIFY: 5 m
2. CALCULATE: A: 20 m; B: 10 m; C: about 42 m
3. CALCULATE: 60 m
4. MEASURE: 50 m
5. INFER: The contour interval lines are farther apart on the gentler side of the hill. The compass direction is west.
6. ANALYZE: D. The short, straight lines on the contour line indicate a depression.

Lab Activity

(pp. 36–37)

1 Prepare the Lab

Making a Topographic Map

SET-UP TIME: **LAB TIME:**

BACKGROUND: Earth looks like a smooth sphere when seen from a distance. Close up, however, its solid surface is covered with mountains, valleys, cliffs, plains, crevasses, and plateaus. To understand the processes that created these landforms, geologists must first understand the shape of the surface. The surface provides clues as to what lies beneath.

PURPOSE: Students will construct a contour map from a clay landform.

ALTERNATIVE MATERIALS: Floral clay or modeling dough can be used to create the landforms. You may also substitute homemade modeling dough; recipes can be found on the Internet. Monofilament fishing line or dental floss can be used in place of thread.

SCIENCE SKILLS: Students will **model** a landform and **observe** and **collect data** to create a two-dimensional map.

ADVANCE PLANNING: Make cleanup easier by covering work surfaces with newspaper and having paper towels and running water on hand.

2 Guide the Lab

PROCEDURE: If you prefer, you can divide the class into groups of two or three students. Show students how to hold the thread to slice their landform models. The thread should work like a wire cheese slicer. Make sure that students create guide holes through the clay before slicing the model. Provide students with a copy of the Lab Activity Report found in the Teacher's Resources for students to record their observations and conclusions.

TROUBLESHOOTING: Be sure that students shape their mountains so that there are some steep and some gentle sides. Students' maps and models may not match precisely. Caution students to handle clay layers gently so they do not change shape. Remind students to keep the thread level when making the slices.

LAB ACTIVITY
Making a Topographic Map

Materials
- Lab apron
- Gloves (optional)
- Modeling clay
- Thread (about 30 cm long)
- Plain paper
- Pencil
- Metric ruler

BACKGROUND

Earth's surface is covered with mountains, valleys, plateaus, and broad plains. Topographic maps show these features. They contain contour lines. These lines show the look and shape of the land.

PURPOSE

In this activity, you will shape a clay model of a mountain. You will then construct a topographic map of your model.

▲ **STEP 2** Shape your clay to form a mountain.

▲ **STEP 3** Push a pencil down through the top of your mountain.

PROCEDURE

1. Put on your lab apron.
2. Place a sheet of plain paper on your desk. Put a lump of clay in the middle of the paper. Shape the clay to form a mountain and a valley. Make the mountain 8 cm high.
3. Push a pencil through the top of your mountain all the way down to the paper to make a mark. Pull the pencil out. Make a second hole about 2 cm from the first hole. Make a mark on the paper. Pull out the pencil.
4. Stand a ruler beside your mountain. Mark your clay at a height of 7 cm.
5. Hold both ends of the thread tightly so it is parallel to the table. Slice the top of the mountain off at the 7 cm mark you just made. Remove it and set it aside.
6. Now, measure and mark a 6 cm height on the mountain. Using the thread again, slice the mountain off at the 6 cm mark. Set this piece aside.

36

Teaching Options

EXTENDING THE LESSON

Modeling From a Map Challenge students to create a three-dimensional model based on a small section of a local topographic map. (You may want to use a copy machine to enlarge part of the map, making it easier to work with.) Provide students with a map, modeling clay, and sculpting tools (such as plastic knives) to create their models. Students will need to decide on a scale for their models that relates to the scale on the map. Invite students to make labels for specific landmarks, using paper, marking pens, and straight pins. **learning style: tactile**

ESL/ELL STRATEGY

Pairing Students English-language learners may have difficulty following the procedure. Pair an English-language learner with an English-proficient student during this lab. The English-proficient student can help the English-language learner in carrying out the procedure, as well as help in interpreting the results.

7. Repeat Steps 4 and 5 for heights of 5, 4, 3, 2, and 1 cm.

8. Use your pencil to trace a line around the bottom slice of the mountain. This is a contour line.

9. Remove the bottom slice from the paper. Place the slice that was on top of the bottom slice on the paper. Line up both holes in the slice with the pencil marks on the paper. Trace around this slice and remove it.

▲ **STEP 7** Slice off marked sections of the mountain.

10. Trace around each of the remaining four slices using the same paper. Be sure to line up both holes before you trace.

11. Rebuild your clay mountain by stacking all the pieces back together in order from largest to smallest. Compare your mountain to the topographic map you just drew.

▲ **STEP 10** Trace around your slices.

▲ **STEP 11** Rebuild your clay mountain.

CONCLUSIONS

1. **ANALYZE:** What do the contour lines on the topographic map show?
2. **OBSERVE:** What do widely spread contour lines tell you about the shape of the land?
3. **OBSERVE:** What do closely spaced contour lines tell you about the shape of the land?
4. **INFER:** How do you think geologists use topographic maps?

SAFETY TIP: Remind students to pick up clay bits from the floor immediately. Have students wash their hands after working with the clay.

EXPECTED OUTCOME: Students will observe the relationship between the spacing of the contour lines and the steepness of the slope in their landform models.

LAB HINT: The slice lines seen from above the reassembled mountains should match the contour lines on students' topographic maps.

3 Conclude the Lab

1. **ANALYZE:** the form and shape of the land
2. **OBSERVE:** The land is gently sloping.
3. **OBSERVE:** The land is steeply sloping.
4. **INFER:** to model the Earth's surface and get some clues about how it formed

Use the Laboratory Activity Rubric from the Teacher's Resources to assess students' lab activities. Fill in the rubric with the additional information below. For this activity, students should have:

- shaped a clay model of a mountain.
- constructed a topographic map.

4 Extend the Lab

Discuss how to use contour lines on a topographic map to determine the direction a river flows. Contour lines form a V as they cross rivers. The point of the V aims upstream. Show this by making a clay model with a gentle slope. Gouge out a steep-walled stream valley in the slope. Slice the model with thread to see the shape of the contour lines.
learning styles: visual, tactile

Tips for Using Technology

Using Maps on CD-ROMS
A variety of maps and atlases are now available on CD-ROMs. Students can use these resources to search for maps of particular locations. In many cases, they can zoom in for detailed views of an area and zoom back out for views of the surrounding area. Students can print out maps of selected areas to use in reports and presentations.

• TEACHING RESOURCES • Online or Print

Teacher's Resources
Lab Activity Report: Study Tools, pp. 16–17

Laboratory Activity Rubric: Teaching Tools, p. 8

Laboratory Manual
Lab. Skills Worksheet 6: Writing a Laboratory Report, pp. 19–20

Challenges (pp. 38–40)

Chapter Summary

Review Before students begin the Challenges, review the summary with them. Ask volunteers to mention something new they learned while studying each lesson.

Key Term Challenges

MATCHING

1. mantle (1-3)
2. atmosphere (1-2)
3. scale (1-7)
4. hydrosphere (1-2)
5. core (1-3)
6. map projection (1-5)
7. distortion (1-4)

IDENTIFYING WORD RELATIONSHIPS

8. A contour line links together points on a map that have the same elevation. (1-8)
9. A map projection is made by wrapping a sheet of paper around a lighted globe and then tracing the outlines. (1-5)
10. The crust is the solid, thin outer layer of Earth that forms the upper part of the lithosphere. (1-3)
11. Contour intervals indicate the topography, or form and shape, of the land shown on a map. (1-8)
12. Lines of latitude run parallel to the equator. A line of latitude is a type of line called a parallel. (1-6)
13. Lines of longitude are also called meridians. They run from pole to pole. (1-6)
14. A legend explains the symbols used on a map. (1-7)
15. A globe is a sphere-shaped model of Earth. (1-4)

Chapter 1 Challenges

Chapter Summary

Lesson 1-1

- **Earth science** is the study of Earth and its history. The four branches of Earth science are geology, oceanography, meteorology, and astronomy.

Lesson 1-2

- A **sphere** is a round, three-dimensional object. The **lithosphere** is the solid part of Earth. The **hydrosphere** is made of liquid water or ice. The **atmosphere** is an envelope of gases surrounding Earth.

Lesson 1-3

- Earth has three layers. The **crust** is the outer layer. The **mantle** is below the crust. Its upper part is solid rock. The rock below this flows like a thick liquid. The **core** in the center has an outer liquid layer and an inner solid layer.

Lessons 1-4 and 1-5

- A **globe** is a spherical model of Earth. A **map** is a flat model of Earth. All maps have **distortions.**
- A **map projection** shows Earth's surface, or part of it, on a sheet of paper. Three examples of map projections are Mercator, polar, and conic.

Lessons 1-6 and 1-7

- The equator divides Earth's surface into northern and southern **hemispheres.** The distance a place is north or south of the equator is its **latitude.**
- **Meridians** are lines that run from pole to pole. The distance in degrees a place is east or west of the prime meridian is its **longitude.**
- A map's **scale** compares map distances with real distances. Map **symbols** represent real objects. The symbols are explained in a **legend.** Colors can be used to show different features.

Lesson 1-8

- Height above or below sea level is called **elevation.** The study of the form and shape of Earth's surface is called **topography.** Relief maps show elevation often using colors or shading.

Lesson 1-9

- Topographic maps use **contour lines** to show elevation. A **contour interval** is the difference in elevation between contour lines.

Key Term Challenges

atmosphere (p. 18)	lithosphere (p. 18)
contour interval (p. 34)	longitude (p. 26)
contour line (p. 32)	mantle (p. 20)
core (p. 20)	map (p. 22)
crust (p. 20)	map projection (p. 24)
distortion (p. 22)	meridian (p. 26)
Earth science (p. 16)	parallel (p. 26)
elevation (p. 32)	scale (p. 28)
globe (p. 22)	specialist (p. 16)
hydrosphere (p. 18)	sphere (p. 18)
latitude (p. 26)	symbol (p. 28)
legend (p. 28)	topography (p. 32)

MATCHING **Write the Key Term from above that best matches each description.**

1. layer of Earth below the crust
2. layer of gases surrounding Earth's surface
3. shows distance on a map
4. all the liquid water and ice on Earth
5. innermost region of Earth
6. flat drawing of part of Earth's surface
7. error in shape, distance, or size on a map

IDENTIFYING WORD RELATIONSHIPS **Explain how the words in each pair are related. Write your answers in complete sentences.**

8. contour line, elevation
9. map projection, globe
10. crust, lithosphere
11. contour interval, topography
12. latitude, parallel
13. longitude, meridian
14. symbol, legend
15. sphere, globe

38

Teaching Options

PREPARING STUDENTS FOR STANDARDIZED TESTS

Reading Strategy: Tell students that when matching up definitions with Key Terms, they should be careful to match up parts of speech. In other words, match up nouns with nouns and verbs with verbs.

Writing Strategy: Be sure that students write complete sentences for the Identifying Word Relationships questions. Encourage them to try to explain both terms using one sentence. This will help them understand the relationship between the terms.

Interpreting Visuals Strategy: Remind students that maps contain many shorthand ways of presenting information, including symbols, legends, color, and place name abbreviations.

ESL/ELL STRATEGY

Working With Maps To become proficient at using maps, have English-language learners work with maps of places they are more familiar with. Perhaps they can prepare rough topographic maps of the area in which they were born or recently came from.

Content Challenges TEST PREP

MULTIPLE CHOICE Write the letter of the term or phrase that best completes each statement.

1. The study of weather is called
   a. geology.
   b. oceanography.
   c. meteorology.
   d. specialist.
2. Earth's crust is part of the
   a. atmosphere.
   b. mantle.
   c. hydrosphere.
   d. lithosphere.
3. Rivers, lakes, and streams are part of Earth's
   a. core.
   b. mantle.
   c. hydrosphere.
   d. atmosphere.
4. The two main gases in the atmosphere are
   a. oxygen and nitrogen.
   b. oxygen and hydrogen.
   c. hydrogen and nitrogen.
   d. oxygen and helium.
5. The thickest layer of Earth is the
   a. crust.
   b. mantle.
   c. outer core.
   d. inner core.
6. Globes are
   a. two dimensional.
   b. one dimensional.
   c. three dimensional.
   d. flat.
7. A map projection made by holding a sheet of paper to one pole of a globe is called a
   a. Mercator projection.
   b. polar projection.
   c. conic projection.
   d. relief projection.
8. Lines on maps or globes that run from the North Pole to the South Pole are called
   a. meridians.
   b. latitudes.
   c. longitudes.
   d. contour lines.
9. The North Pole is at
   a. 0° latitude.
   b. 10° longitude.
   c. 90° north latitude.
   d. 90° north longitude.
10. The distance in degrees east or west of the prime meridian is a place's
   a. longitude.
   b. latitude.
   c. equator.
   d. parallel.

FILL IN Write the term or phrase that best completes each statement.

11. Elevation is the distance of a point above or below ________.
12. The symbols used on a map are listed in a ________.
13. Latitude and longitude are measured in ________.
14. The longest parallel on Earth's surface is the ________.
15. A narrow field of study within a larger field is called a ________.
16. Maps of ________ areas have the fewest distortions.
17. The layer of Earth between the core and the crust is the ________.
18. The hydrosphere is the part of Earth that is ________.

ALTERNATIVE ASSESSMENT

Layers of Earth Have students create a poster display or model that depicts the layers of Earth. Have students present their display or model to the class. Students should describe how they made their model and explain the information it provides.

CUMULATIVE ASSESSMENT

You may wish to administer the pretest examination found in the Teacher's Resources at this point. Distribute copies of the Scantron Sheet, also found in the Teacher's Resources, for students to record their answers.

PORTFOLIO ASSESSMENT

Making Student Portfolios Portfolio Assessment is designed to evaluate a student's performance over an extended period of time. It also provides an opportunity for students who are generally not good test-takers to demonstrate their learning in another way. Students can demonstrate their comprehension of the concepts in this chapter by making a portfolio. The Chapter Self-Check, the Big Idea Planner, and the Weekly Journal are some of the reproducibles in the Teacher's Resources that they can include in their portfolios. You can use the Portfolio Assessment Rubric also found in the Teacher's Resources to assess students' portfolios.

Content Challenges

MULTIPLE CHOICE

1. c (1-1)
2. d (1-3)
3. c (1-2)
4. a (1-2)
5. b (1-3)
6. c (1-4)
7. b (1-5)
8. a (1-6)
9. c (1-6)
10. a (1-6)

FILL IN

11. sea level (1-8)
12. legend (1-7)
13. degrees (1-6)
14. equator (1-6)
15. specialty (1-1)
16. small (1-4)
17. mantle (1-3)
18. water (1-2)

Concept Challenges

WRITTEN RESPONSE

1. **EXPLAIN:** The longitude of Greenwich, England, is 0°. Greenwich is on the prime meridian. (1-6)
2. **INFER:** A place's exact location is identified by where the lines of latitude and longitude cross. If you did not know both latitude and longitude, you would know only one line on which the place is located. (1-6)
3. **ANALYZE:** Two contour lines cannot overlap because each contour line links only places with the same elevation. Different contour lines show different elevations. Contour lines can be very close together, however, if the slope is steep. (1-8)
4. **EXPLAIN:** The polar projection is made by placing a flat piece of paper on one of the poles on a globe and tracing a map onto the paper. The map is most accurate at the point of contact. It becomes distorted as you move farther from the point of contact. (1-5)
5. **INFER:** Because Earth is slightly flattened at the North and South poles, it is slightly larger when measured around the equator. (1-3, 1-6)

INTERPRETING VISUALS

6. **OBSERVE:** blue (1-7)
7. **CALCULATE:** 760 km (1-7)
8. **INTERPRET:** about 1,140 km; about 380 km (1-7)
9. **INTERPRET:** northeast (1-7)
10. **ANALYZE:** 75°W, 40°N (1-6)

Chapter Project Wrap-Up

MAP DISPLAY

Groups of students can present their map displays to the rest of the class. You can use the Individual Activity Rubric found in the Teacher's Resources to assess students' projects. Fill in the rubric with the additional information below. For this project, students should have:

- displayed several varieties of maps.
- discussed special features of maps.

Concept Challenges TEST PREP

WRITTEN RESPONSE **Answer each of the following questions in complete sentences.**

1. **EXPLAIN:** What is the longitude of Greenwich, England? Why?
2. **INFER:** Why must both latitude and longitude be known to find the location of a place on a map?
3. **ANALYZE:** Why do two contour lines never cross each other?
4. **EXPLAIN:** Why does a polar projection have less distortion nearer the poles than it has farther from the poles?
5. **INFER:** Is the diameter of Earth larger when measured from pole to pole or when measured at the equator? Explain.

INTERPRETING VISUALS **Use Figure 1-34 to answer the following questions.**

6. **OBSERVE:** What color on the map is used to show bodies of water?
7. **CALCULATE:** What real distance on the map is shown by 2 cm?
8. **INTERPRET:** How many kilometers is it from Miami, Florida, to Austin, Texas? From Charlotte, North Carolina, to Louisville, Kentucky?
9. **INTERPRET:** In what direction is Chicago, Illinois, from San Diego, California?
10. **ANALYZE:** What is the latitude and longitude of Philadelphia, Pennsylvania?

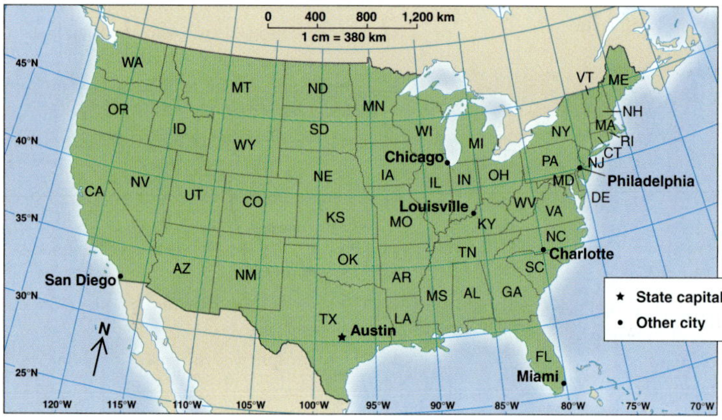

◄ Figure 1-34

Teaching Options

• TEACHING RESOURCES •
Online or Print

Teacher's Resources

Key Term Review: Chapter 1, p. 15

Chapter 1 Test: pp. 16–17

Individual Activity Rubric: Teaching Tools, p. 5

Scantron Sheet: Teaching Tools, p. 12

Chapter Self-Check: Teaching Tools, p. 11

Weekly Journal: Study Tools, p. 19

Portfolio Assessment Rubric: Teaching Tools, p. 10

PLANNING GUIDE

◆ **TEACHING THE CHAPTER** This chapter should take approximately 8–11 days to complete instruction and assessment.

	Skills and Features	Projects and Activities	Achieve Science Literacy	Meet Individual Needs
Chapter 2 Opener p. T41	IDENTIFY	• Chapter Project		
Lesson 2-1 **What are elements and compounds?** pp. T42–T43 Standards: B1b, B1c	ANALYZE, IDENTIFY, EXPLAIN • Web InfoSearch • How Do They Know That?	• Cooperative Learning • Integrating Chemistry	• Study Hint	Teacher's Resources* Enrichment Activity
Lesson 2-2 **What are chemical formulas?** pp. T44–T45 Standards: B1b, B1c	IDENTIFY, ANALYZE, INTERPRET • Hands-On Activity	• Cooperative Learning • Extending the Lesson	• Study Hint • Reading Strategy	• Reteach • ESL/ELL Strategy Teacher's Resources* Enrichment Activity
Lesson 2-3 **What are minerals?** pp. T46–T47 Standards: B1b, B1c	INTERPRET, DESCRIBE, ORGANIZE A COLLECTION • Building Science Skills	• Cooperative Learning • Extending the Lesson • More on the Web	• Study Hint	• Reteach Teacher's Resources* Enrichment Activity
Lesson 2-4 **How do we identify minerals?** pp. T48–T49 Standards: B1a, B1c	IDENTIFY, ANALYZE, INFER • Designing an Experiment • Hands-On Activity	• More on the Web • Lab Challenge	• Study Hint • Ongoing Assessment	• Reteach Teacher's Resources* Enrichment Activity
Lesson 2-5 **What are some other ways to identify minerals?** pp. T50–T51 Standard: B1a	EXPLAIN, CALCULATE, ANALYZE • Investigate • Designing an Experiment • Integrating Physical Science	• Extending the Lesson • Lab Challenge	• Study Hint	• ESL/ELL Strategy Teacher's Resources* Designing an Experiment
Lesson 2-6 **What are crystals, cleavage, and fracture?** pp. T52–T53 Standard: D1d	DEFINE, CONTRAST, CLASSIFY • Building Science Skills • People in Science	• More on the Web • Active Art	• Study Hint • Ongoing Assessment	

*www.conceptsandchallenges.com, password: ccearth09

Planning Guide continues on next page.

Standards: For details on the correlation to National Science Standards see pages *xx–xxi*.

Chapter 2 Minerals and Their Properties
(Planning Guide continued)

	Skills and Features	Projects and Activities	Achieve Science Literacy	Meet Individual Needs
Lesson 2-7 **How are minerals formed?** pp. T54–T55 Standards: B1a, B1c	CONTRAST, INFER, CLASSIFY • Investigate • Building Science Skills • Integrating Life Science	• More on the Web • Integrating Life Science	• Study Hint • Ongoing Assessment	• Reteach Teacher's Resources* Enrichment Activity
Lab Activity **Growing a Crystal Garden** pp. T56–T57 Standard: B1a	OBSERVE, ANALYZE, INFER • Set-Up Time: 15 min • Lab Time: 25–30 min	• Integrating Chemistry • Extend the Lab Activity • Extending the Lesson	• Tips for Using Technology	Teacher's Resources* Lab Activity Report Flowchart
Lesson 2-8 **How are minerals used?** pp. T58–T59	IDENTIFY, DEFINE, LIST • Science and Technology	• Extending the Lesson • More on the Web	• Study Hint	Teacher's Resources* Enrichment Activity Concept Map
Big Idea **How can the element carbon form different substances?** pp. T60–T61 Standard: B1c	RESEARCH, COMMUNICATE, ANALYZE • Science Log: Writing Activity	• Big Idea • Close Activity	• Tips for Using Technology	Teacher's Resources* Big Idea Planner
Chapter 2 Challenges pp. T62–T64	• Chapter Summary • Key Term Challenges • Content Challenges • Concept Challenges	• Chapter Project Wrap-Up	• Study Hint • Preparing Students for Standardized Tests	• ESL/ELL Strategy Teacher's Resources* Chapter Self-Check Weekly Journal

*www.conceptsandchallenges.com, password: ccearth09

Teacher Notes

Chapter 2 Minerals and Their Properties

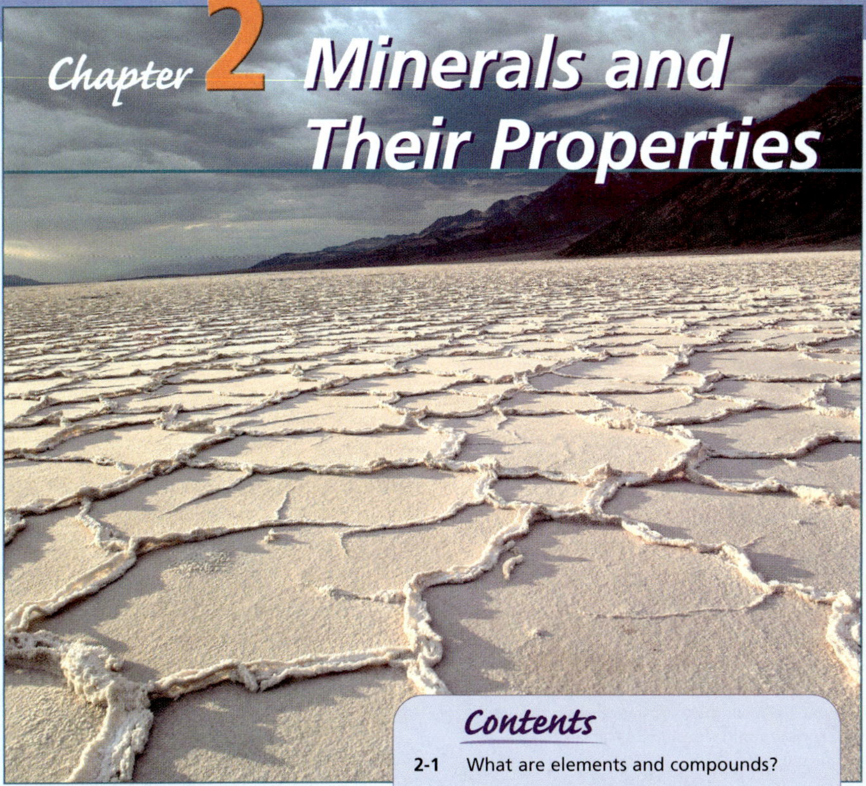

▲ **Figure 2-1** Death Valley, in southeastern California, has large salt flats.

Death Valley is a hot, dry place. Salt flats make up parts of the valley floor. Salt flats often form in pits. These pits sometimes fill with runoff from the mountain. Then, the water slowly seeps into the ground or evaporates forming salt flats. Death Valley attracts tourists and scientists because of its unusual properties.

► Water carries dissolved minerals from the mountains to Death Valley below. What common substances do you think are left behind as the water evaporates?

Contents

Teaching Options

ASSESSMENT PLANNER

For assessment in the Student Edition, see the following pages:

Content Assessment
Checking Concepts: pp. 43, 45, 47, 49, 51, 53, 55, 59
Thinking Critically: pp. 43, 45, 47, 49, 51, 53, 55, 59
Interpreting Visuals: pp. 45, 59
Chapter Challenges: pp. 62–64

Alternative Assessment
Building Skills: pp. 47, 53, 55
Web InfoSearch: p. 43
Science Log: p. 61

Performance-Based Assessment
Hands-On Activity: pp. 45, 49
Designing an Experiment: pp. 49, 51
Lab Activity: pp. 56–57

• TEACHING RESOURCES • Online or Print

Teacher's Resources
Lesson Review: Ch. 2, pp. 2, 3, 5, 6, 8, 9, 10, 11
Enrichment: Ch. 2, pp. 4, 7, 12
Key Term Review: Ch. 2, p. 14
Chapter 2 Test: pp. 15–17
Laboratory Manual: pp. 25–28, 29–32

Chapter 2 Minerals and Their Properties

Chapter Overview

In this chapter, students will learn about atoms, elements, and compounds and how they combine to form minerals. Students will also discover how to identify minerals by their physical properties. Finally, students will find out how the element carbon occurs as coal, graphite, and diamond.

About Figure 2-1 Death Valley is the hottest, driest spot in North America. Depressions filled with salt water sometimes form lakes called playa lakes. If this water evaporates, a flat bed made of salts and clay called a playa is left. The photo shows a playa heavily encrusted with salts.

Answer Most students will know or be able to deduce from the photo that common table salt (NaCl) is left behind when these waters evaporate. Other common substances left behind are mud and clay.

Linking Prior Knowledge

For this chapter, students should recall:

- the three main parts of Earth, especially the lithosphere (Lesson 1-2).
- Earth's structure, especially the crust (Lesson 1-3).

Chapter Project

MINERAL JEOPARDY

MATERIALS: paper, pens or pencils

Divide students into four groups. Have each group prepare a set of 25 answers to questions about minerals and their properties. The group should break their questions into five different categories. For example, one category might be Mineral Hardness, with an answer being "the hardest mineral known." The categories can later be put on the board, and someone from the group can read the answers aloud to the rest of the class. Students respond to each answer with an appropriate question (here, *What is diamond?*). Each question can be worth points, and the team that gets the most points wins the game.

2-1 *(pp. 42–43)*

1 *Introduce*

STUDY HINT Have students carefully read the Key Terms and their definitions. Then, help students order the terms from simplest to most complex substance. (atom, element, molecule, compound) Challenge students to come up with a mnemonic device to help them remember this hierarchy. For example, aemc—Andy eats many carrots.

Linking Prior Knowledge Refer students to Figure 2-2 and ask the following questions: *Which of the elements listed can be found in Earth's atmosphere?* (oxygen and helium) *Which elements are most likely to make up Earth's lithosphere?* (carbon, aluminum, silver, and copper)

2 *Teach*

Demonstration Have samples of some common elements, such as iron, copper, sulfur, zinc, and lead. Ask students to observe and describe each element. Reiterate that most elements are solids at room temperature. Have students contrast this information with the mercury shown in Figure 2-3, which is a liquid at room temperature. **learning style: visual**

Discussion Refer students to Figure 2-4 as you ask the following questions: *What are the parts of an atom?* (neutrons, protons, and electrons) *What two atoms are shown here?* (oxygen and hydrogen) *How many atoms of hydrogen and how many atoms of oxygen make up a water molecule?* (2 hydrogen, 1 oxygen)

Answers to questions

1. **ANALYZE:** Possible answers include that atoms are the smallest parts of an element or that atoms make up elements.
2. **ANALYZE:** Most compounds are made up of molecules.
3. **IDENTIFY:** sodium and chlorine

2-1 What are elements and compounds?

Objectives

Define element and compound. Identify and describe some common compounds.

Key Terms

atom: smallest part of an element that can be identified as that element

element (EHL-uh-muhnt)**:** simple substance that cannot be broken down into simpler substances by ordinary chemical means

compound (KAHM-pownd)**:** substance made up of two or more elements that are chemically combined

molecule (MAHL-ih-kyool)**:** smallest part of a substance that has all the properties of that substance

Atoms and Elements All matter is made up of particles too small to be seen. These particles are called atoms. **Atoms** are the smallest parts of an element that can be identified as that element. An **element** is a simple substance that cannot be broken down into simpler substances by ordinary chemical means.

There are more than 100 different elements. Elements can be solids, liquids, or gases. Most elements are solids at room temperature. Only two elements, bromine and mercury, are liquid at room temperature. Oxygen, hydrogen, and nitrogen are gases. Figure 2-2 shows the states of some common elements at room temperature.

STATES OF COMMON ELEMENTS			
Element	**Natural State**	**Element**	**Natural State**
Carbon	Solid	Oxygen	Gas
Aluminum	Solid	Helium	Gas
Silver	Solid	Bromine	Liquid
Copper	Solid	Mercury	Liquid

▲ Figure 2-2

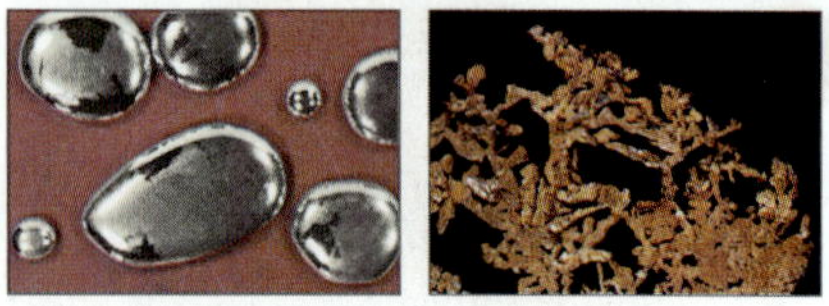

▲ **Figure 2-3** Mercury (left) and copper in their natural states

1. **ANALYZE:** How are atoms and elements related?

Compounds Many common substances are made up of combinations of elements. Water is made up of hydrogen and oxygen. Sugar is made up of hydrogen, oxygen, and carbon. Water and sugar are examples of compounds. A **compound** is a substance made up of two or more elements that are chemically combined.

Most compounds are made up of a string of atoms called **molecules.** A molecule is the smallest part of a substance that has all the chemical properties of that substance. The atoms in a molecule are chemically combined.

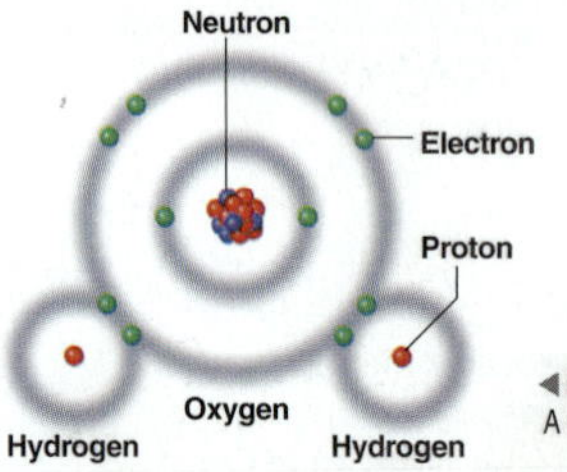

◀ **Figure 2-4** A water molecule

2. **ANALYZE:** How are molecules and compounds related?

More Than Its Parts When two or more elements are combined chemically, a compound is formed. The elements in the compound are in definite proportion to each other. Most compounds have physical and chemical properties that are

Teaching Options

INTEGRATING CHEMISTRY

Compounds and Mixtures Make sure that students understand that a compound is a chemical combination of two or more elements. Tell them that a mixture is a physical combination of two or more substances. Unlike a compound, the substances in a mixture do not combine chemically. Demonstrate the difference between a compound and a mixture with a small bowl of tap water and a bowl of paper clips, sand, and marbles. Explain that water is a compound. It is impossible to separate the hydrogen from the oxygen by simple physical means. Students might suggest boiling as a means of separation. Explain that boiling will cause the liquid water to evaporate, or change into a gas we call steam, but the steam is still made up of hydrogen and oxygen. Have students hypothesize ways to remove one component of the mixture. (Students might suggest picking out the paper clips or the marbles by hand or using a magnet to separate the paper clips from the rest of the mixture. The sand could be removed from the mixture by sifting.) **learning styles: visual, logical/mathematical**

different from the physical and chemical properties of the elements that formed them.

Sodium is a soft metal that can be cut with a knife. Chlorine is a green gas. Both sodium and chlorine are poisonous elements. Combined, they form a solid compound that can be eaten. You know this compound, sodium chloride, as table salt.

▲ **Figure 2-5** Sodium (left) is part of the compound known as table salt (right).

3 IDENTIFY: What elements make up table salt?

CHECKING CONCEPTS

1. About how many elements are there?
2. What form are most elements found in at room temperature?
3. What is a compound?
4. What is a molecule?

THINKING CRITICALLY

5. EXPLAIN: The compound water is made from the gases oxygen and hydrogen. How is water different from the elements that make it up?
6. ANALYZE: How are atoms and molecules related?
7. IDENTIFY: What elements do the compounds sugar and water have in common?

Web InfoSearch

Alloys That Remember Alloys are metallic compounds or solutions made up of two or more elements. Some metal alloys can return to their original shape after being twisted, bent, or knotted. These are called shape memory alloys.

SEARCH: Use the Internet to find out more about alloys. What elements are they made from? What are they used for? Start your search at www.conceptsandchallenges.com. Some key search words are **shape memory alloys** and **intelligent materials.**

How Do They Know That?

SOME ELEMENTS ARE RADIOACTIVE

A radioactive element gives off energy when particles inside its atoms break apart. Most of what is known today about radioactive elements can be traced to the work of Marie Curie and her husband, Pierre Curie.

The Curies worked with an ore called pitchblende. They observed that the pitchblende gave off unusual rays. The Curies soon discovered that these rays were being produced by two elements contained in the pitchblende. The Curies and another scientist, Henri Becquerel, shared the Nobel Prize in 1903 in physics for discovering these new elements. The elements were named polonium and radium. Polonium was named for Poland, the homeland of Marie Curie.

In 1906, Pierre Curie died in a tragic accident. However, Marie Curie continued her work with radium and polonium. In 1911, she received a second Nobel Prize, in chemistry.

▲ **Figure 2-6** Marie Curie (1867–1934) was born in Poland. However, she did her experiments in a laboratory in France.

Thinking Critically What do you think the energy inside an atom is used for?

3 Assess

CHECKING CONCEPTS ANSWERS

1. more than 100
2. solid
3. a substance made up of two or more elements that are chemically combined
4. the smallest part of a substance that has all the chemical properties of that substance

THINKING CRITICALLY ANSWERS

5. EXPLAIN: Oxygen and hydrogen are gases. They combine to form a liquid, water.
6. ANALYZE: Molecules are made up of atoms.
7. IDENTIFY: hydrogen and oxygen

4 Close

Web InfoSearch

Alloys That Remember Have students investigate what elements shape memory alloys are made from and what they are used for. Some key search words are **shape memory alloys** and **intelligent materials.** Students can begin their search at www.conceptsandchallenges.com.

How Do They Know That?

SOME ELEMENTS ARE RADIOACTIVE

Technology Connection Inform students of the practical uses of radioactive elements—as fuels (nuclear energy), to irradiate foods and drugs, to diagnose and treat some kinds of cancer, to study certain chemical and biological reactions, and as critical parts of most smoke detectors.

Discussion Explain that the protons and neutrons of an atom are held together by a force called the strong force. When a radioactive element breaks down, or decays, vast amounts of energy, or alpha, beta, and gamma particles, are released from the nucleus. Such particles, as well as the energy emitted from radioactive elements, can damage the DNA within living cells. When DNA gets damaged, the cells may die or mutate.

Thinking Critically Answer
The energy inside an atom (nuclear energy) serves to hold the atom together.

COOPERATIVE LEARNING

Modeling Molecules Provide pairs of students with different-sized plastic-foam balls, white glue, and toothpicks. Have students re-create the water molecule shown in Figure 2-4. Check models for accuracy. The ball that represents oxygen should be larger than the balls that represent hydrogen. Also, the two balls that represent hydrogen should be the same size. Finally, the model atom should be joined at approximately the same angles as shown. Some groups may draw the subatomic particles—neutrons, protons, and electrons—on each ball. **learning styles: visual, tactile**

• TEACHING RESOURCES •
Online or Print

Teacher's Resources
Lesson Review: Chapter 2, p. 2
Enrichment: Chapter 2, p. 4

2-2 (pp. 44–45)

1 Introduce

STUDY HINT Have students read the lesson objectives and scan the page for examples of chemical symbols and chemical formulas. Stress that symbols are often only one or two letters, whereas formulas are generally longer and often include numbers.

Linking Prior Knowledge Have students read the definitions of the Key Terms. All of the definitions contain Key Terms from Lesson 2-1. Ask students to recall the main concepts of Lesson 2-1.

2 Teach

Demonstration Write the following sentences on the board, one directly beneath the other: *The dog bit the cat. The dog bit the car.* Make sure that the letters align exactly. Point out how changing just one letter changes the whole sentence. Have students find pairs of similar chemical formulas from the text. These include water and hydrogen peroxide (H_2O and H_2O_2), and carbon monoxide and carbon dioxide (CO and CO_2). Have students tell the number of atoms in a molecule of each of these compounds.

Discussion Have students suggest reasons why chemical symbols are used. Lead them to conclude that these abbreviations save time and space, and, because they are standard symbols, people all over the world can understand and use them.

Reinforcement Make sure that students understand that when there is no subscript, the number 1 is understood to be the subscript.

Answers to questions

1 **IDENTIFY:** The names of these elements come from languages other than English.

2 **ANALYZE:** Chemical symbols are used to write chemical formulas.

3 **INTERPRET:** the number of atoms of each element in one molecule of a compound

2-2 What are chemical formulas?

Objectives

Interpret chemical symbols. Recognize common chemical formulas of some familiar compounds.

Key Terms

chemical (KEHM-ih-kuhl) **symbol:** shorthand way of writing the name of an element

chemical formula: formula that shows the elements that make up a compound

subscript (SUHB-skrihpt)**:** number in a chemical formula that shows how many of each atom are in one molecule of a compound

Chemical Symbols A **chemical symbol** is a shorthand way of writing the name of an element. Each element has its own chemical symbol. The chemical symbol is used to represent one atom of that element.

In 1813, a Swedish chemist named Jons Jakob Berzelius (buhr-ZEE-lee-uhs) suggested using the first letter of the Latin name of an element as its chemical symbol for classification. When more than one element has the same first letter, a second letter is added. That second letter is a lowercase letter.

CHEMICAL SYMBOLS			
Element	Symbol	Element	Symbol
Oxygen	O	Hydrogen	H
Nitrogen	N	Carbon	C
Aluminum	Al	Helium	He
Iron	Fe	Lead	Pb
Sodium	Na	Mercury	Hg

▲ Figure 2-7

Some elements do not use the first letters of their English names as their symbols. The symbols for these elements may come from the names of the elements in a different language.

Look at Figure 2-7 for some examples. The chemical symbol for sodium is Na. This comes from the Latin name for sodium, which is *natrium*.

1 IDENTIFY: Why do some elements have unusual letters for their symbols?

Chemical Formulas We just learned that a chemical symbol is a shorthand way of writing the name of an element. Chemical symbols are most often used to write chemical formulas. A **chemical formula** shows the elements that make up a compound. Sodium chloride is the chemical name for table salt. The chemical formula for sodium chloride is NaCl. By looking at the formula, you can see that table salt is made up of sodium and chlorine. The chemical symbol for sodium is Na. For chlorine, it is Cl.

2 ANALYZE: How are chemical symbols and chemical formulas related?

Subscripts Hydrogen peroxide is a compound made from hydrogen (H) and oxygen (O). The chemical formula for hydrogen peroxide is H_2O_2.

The number 2 after the H and the O is called a **subscript.** Subscripts show how many of each atom are in a molecule of a compound. Subscripts are written slightly below the line.

The chemical formulas for some compounds are shown in Figure 2-8. The chemical formula for one compound of iron sulfide is FeS_2. This chemical formula shows that a molecule of this compound contains two atoms of sulfur (S) and one atom of iron (Fe). When there is no subscript, it means that there is only one atom of that element in the molecule.

CHEMICAL FORMULAS FOR COMMON COMPOUNDS	
Compound	Formula
Water	H_2O
Carbon monoxide	CO
Carbon dioxide	CO_2
Iron sulfide	FeS_2
Calcium carbonate	$CaCO_3$
Aluminum oxide	Al_2O_3
Ammonium hydroxide	NH_4OH
Silica	SiO_2
Sodium phosphate	$NaPO_4$
Hydrogen peroxide	H_2O_2

▲ Figure 2-8

44

Teaching Options

COOPERATIVE LEARNING

It's Elementary! Have pairs of students make ten flash cards, one for each element listed in Figure 2-7. On one side of the card, students should write the name of the element. On the other side, they should write the chemical symbol. Have students use the cards to learn these common elements and their symbols. **learning styles: visual, interpersonal**

EXTENDING THE LESSON

Using the Periodic Table Have students refer to Appendix E in their text for a simple, up-to-date periodic table so that they can see the symbols for all of the elements.

ESL/ELL STRATEGY

Finding Gold Ask students who were born in other countries to give the word for *gold* in their native languages, or have students find the word for *gold* in languages other than English. Write these words on the board to show the usefulness of standard chemical symbols.

▲ **Figure 2-9** The chemical compound iron sulfide (FeS_2), also known as pyrite, has two atoms of sulfur in each molecule.

3 INTERPRET: What do subscripts show?

CHECKING CONCEPTS

1. Chemical symbols are shorthand for __________.
2. Subscripts are written slightly __________ the line.
3. A chemical symbol stands for one __________ of an element.
4. The first letter of a chemical symbol is always a __________ letter.

THINKING CRITICALLY

5. ANALYZE: Why do you think scientists need shorthand ways of writing elements and compounds?

INTERPRETING VISUALS

Use Figure 2-8 to answer the following questions.

6. CALCULATE: How many atoms make up one molecule of carbon dioxide?
7. ANALYZE: How does a molecule of carbon monoxide differ from a molecule of carbon dioxide?
8. IDENTIFY: How many different elements are in calcium carbonate?

Hands-On Activity

INTERPRETING CHEMICAL FORMULAS

1. Copy the table in Figure 2-10 onto a clean sheet of paper.
2. On your paper, fill in the information that is missing from the table. Use the tables from this lesson and Lesson 2-1 to help you identify the elements in each compound.

Practicing Your Skills

3. EXPLAIN: What information does a chemical formula contain?
4. COMPARE/CONTRAST: How does the chemical formula for aluminum oxide differ from the chemical formula for aluminum sulfide?
5. INFER: Why do scientists use chemical formulas?

CHEMICAL FORMULAS AND THEIR MEANINGS

Chemical Formula	Name of Compound	Elements in Compound	Relative Numbers of Atoms in Compound
H_2O	Water	Hydrogen and oxygen	2 atoms of hydrogen for every 1 atom of oxygen
CO_2	Carbon dioxide	Carbon and oxygen	1 atom of carbon for every 2 atoms of oxygen
CO	Carbon monoxide		
NaCl	Sodium chloride		
Al_2O_3	Aluminum oxide		
Al_2S_3	Aluminum sulfide		
H_2SO_4	Sulfuric acid		

▲ **Figure 2-10**

READING STRATEGY Chemical symbols are standard abbreviations. Ask students to think of other abbreviations they encounter everyday. Have students compile individual lists of abbreviations they come across in their readings and the meanings of those abbreviations.

3 *Assess*

CHECKING CONCEPTS ANSWERS

1. the names of elements
2. below
3. atom
4. capital

THINKING CRITICALLY ANSWER

5. ANALYZE: Abbreviations take up less space and are quicker to write than the actual chemical names. Because most abbreviations used in science are standard, they can be understood by scientists all over the world.

INTERPRETING VISUALS ANSWERS

6. CALCULATE: three
7. ANALYZE: Carbon dioxide has two atoms of oxygen. Carbon monoxide has one.
8. IDENTIFY: three—calcium (Ca), carbon (C), and oxygen (O)

4 *Close*

Hands-On Activity

INTERPRETING CHEMICAL FORMULAS

TIME: 15 minutes

PURPOSE: interpreting, explaining

COOPERATIVE LEARNING: Have pairs of students work together to complete the table in Figure 2-10.

Practicing Your Skills Answers

3. EXPLAIN: A chemical formula shows the elements in a compound as well as the number of atoms of each element in one molecule of that compound.
4. COMPARE/CONTRAST: Aluminum oxide has three atoms of oxygen. Aluminum sulfide has three atoms of sulfur.
5. INFER: Chemical formulas are quick, space-saving notations that communicate the same information to anyone in the world. Chemical formulas are also helpful for scientists writing out lengthy equations.

• TEACHING RESOURCES •
Online or Print

Teacher's Resources

Lesson Review: Chapter 2, p. 3

Enrichment: Chapter 2, p. 4

RETEACH

Symbol Scavenger Hunt Have students look for product labels that display chemical symbols and formulas. Good choices are ingredient labels on toiletries and medications, and content labels on cleaning products. Have students make tables that list each symbol or formula, the product on which it was found, and the names of the elements in each formula.

2-3 (pp. 46–47)

1 Introduce

STUDY HINT Tell students that definitions can often be simplified by picking out key words or phrases specific to the item being defined. Have students determine whether or not a given substance is a mineral by finding out if it is natural, inorganic, solid, and has a definite chemical makeup.

Linking Prior Knowledge Review Earth's structure and features of Earth's crust (Lesson 1-3).

2 Teach

Demonstration Obtain some samples of the minerals quartz, feldspar, mica, and a sample of a coarse-grained granite, which is a rock. Have students observe the minerals and identify each mineral within the rock. Provide hand lenses or magnifying glasses if possible. Use this demonstration to reinforce the concept that rocks and minerals are different. Rocks are mixtures of one or more minerals.
learning styles: visual, tactile

Discussion Refer students to Figure 2-12. Have volunteers identify each mineral as either an element or a compound. For each compound, have students determine the kind and number of atoms that make up the compound.
learning styles: visual, logical/mathematical

Reinforcement Have students study Figure 2-11. Tell them that the percentages that make up a pie graph must always add up to 100 percent. Draw a five-sectioned pie graph on the board. Place percentage values in four of the sections, and have students use math to determine the percentage value of the fifth section.
learning style: logical/mathematical

Answers to questions

1 DESCRIBE: Earth's crust is made of elements such as oxygen, silicon, potassium, sodium, calcium, iron, and aluminum, and many compounds.

2 INTERPRET: sodium (Na) and chlorine (Cl)

2-3 What are minerals?

Objectives
Define mineral. Identify and describe some minerals found in Earth's crust.

Key Term
mineral (MIHN-uhr-uhl): naturally occurring, inorganic solid formed from elements or compounds and having a definite chemical makeup and regular atomic structure

Earth's Crust The crust is the outer layer of Earth. It is made up of large chunks of solid rock and smaller pieces of rock, sand particles, and soil. Sand and soil are rock that has been broken into small pieces. All materials in Earth's crust are made up of elements and compounds.

About 75 percent of Earth's crust is made up of the elements oxygen and silicon. When oxygen and silicon combine, silica is formed. Sand is composed mostly of silica in the form of quartz.

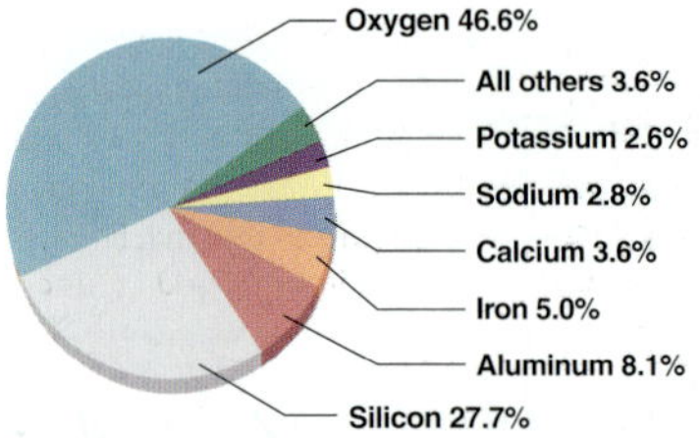

▲ Figure 2-11 Elements found in Earth's crust

Look at Figure 2-11 above. It shows the elements that make up most of Earth's crust. It also shows what percentage of Earth's crust is made up of each element. These elements are found combined into many different compounds.

1 DESCRIBE: What does Earth's crust consist of?

Natural Solids A **mineral** is a naturally occurring solid. It is formed from elements or compounds in Earth's crust. All minerals are inorganic. This means that they are not formed from living things or from the remains of living things.

A mineral may be either an element or a compound. Each mineral has a definite chemical makeup. Gold and silver are examples of minerals that are made up of only one element. Mostly, however, minerals are combinations of elements. For example, the mineral quartz is a compound. Quartz is made up of silicon and oxygen. The chemical formulas for some commonly found minerals are shown in Figure 2-12.

SOME MINERALS AND THEIR CHEMICAL FORMULAS			
Mineral	Formula	Mineral	Formula
Gold	Au	Calcite	$CaCO_3$
Silver	Ag	Halite	NaCl
Copper	Cu	Galena	PbS
Quartz	SiO_2	Pyrite	FeS_2

▲ Figure 2-12

2 INTERPRET: What are the elements that make up halite?

Rock-Forming Minerals The rocks that make up Earth's crust are made up of minerals. Scientists have identified more than 2,000 minerals. However, fewer than 20 of these minerals are commonly found in Earth's crust. These common minerals are called rock-forming minerals. Most rock-forming minerals are compounds.

▲ Figure 2-13 Sand is made up mostly of silicon and oxygen. Silicon makes up a large part of Earth's crust.

Teaching Options

COOPERATIVE LEARNING

Metallic Minerals Divide the class into seven groups. Assign each group one continent. Have the groups use Figure 2-14 on page 47 to help them find information about the major metallic minerals mined on their continents. Students should find that many useful metals do not occur as elements but are extracted from minerals. Have groups cite at least one use for each mineral found.
learning styles: visual, linguistic

EXTENDING THE LESSON

Researching Rocks and Minerals of the Ages Human history is often divided into ages based on materials used by society for technological or industrial advances. Such divisions include the Stone Age, the Bronze Age, and the Iron Age. Have small groups of students research one of these periods in history. Students can illustrate their findings with posters. Have one student from each group give an oral report on the findings. Display the posters.
learning styles: visual, kinesthetic

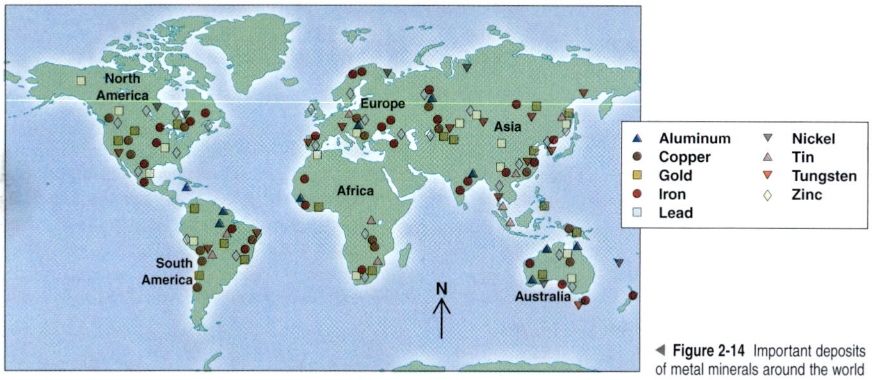

◀ **Figure 2-14** Important deposits of metal minerals around the world

In a rock, the minerals are usually found as mixtures. In a mixture, two or more substances come together. However, the substances do not chemically combine. They retain their original properties. In the case of rocks, each mineral retains its color, texture, and chemical makeup.

3 EXPLAIN: Why are rocks called mixtures?

Metal Minerals Minerals that can be recovered for use are called mineral resources. Aluminum, iron, copper, and silver are some metal minerals. Metals are useful because they can be stretched into wire, flattened into sheets, and hammered or molded without breaking. Tools, machinery, and even the steel girders used in buildings were all made of materials found inside Earth's crust. Metals can be found in many places around the world. See Figure 2-14 for some examples.

4 NAME: What are some metal minerals?

✓ CHECKING CONCEPTS

1. About three-fourths of Earth's crust is made up of oxygen and __________ .
2. Quartz is made up mostly of __________ .
3. Calcite has __________ atom of calcium for every three atoms of oxygen.
4. Gold is both an element and a __________ .

THINKING CRITICALLY

Use Figure 2-11 to answer questions 5 through 8.

5. OBSERVE: Which element makes up almost half of Earth's crust?
6. IDENTIFY: What are two elements of Earth's crust that are found in only small amounts?
7. CALCULATE: What percentage of Earth's crust is made up of iron?
8. SEQUENCE: What five elements make up most of Earth's crust, from most common to least?
9. INTERPRET: What two ways does the map legend in Figure 2-14 use to show metal mineral resources?
10. INTERPRET: Based on Figure 2-14, what is the most abundant metal mineral found around the world?

BUILDING SCIENCE SKILLS

Organizing a Collection There are more than 2,000 different minerals. However, you are likely to find less than 100 of these. You can buy specimens from museums and rock and mineral shows. Or, you can look for your own minerals on mountains, in rock quarries, or at the beach. All you need is goggles, a hammer, chisel or a geologist's pick, and collecting sack. Record the date, location, and name of any mineral you find. Use a field guide to help you identify it. Put your samples in labeled boxes. ⚠ CAUTION: Do not trespass on private land. Also, be aware that some national parks and beaches do not permit the taking of minerals.

3 EXPLAIN: The substances (minerals) that make up rocks do not combine chemically. They retain their original properties. Components of any mixture can be separated by physical means.

4 NAME: aluminum, iron, copper, and silver

3 Assess

✓ CHECKING CONCEPTS ANSWERS

1. silicon
2. oxygen
3. one
4. mineral

THINKING CRITICALLY ANSWERS

5. OBSERVE: oxygen
6. IDENTIFY: Accept any two of the following elements: potassium, sodium, calcium, or iron.
7. CALCULATE: 5.0%
8. SEQUENCE: oxygen (46.6%), silicon (27.7%), aluminum (8.1%), iron (5.0%), calcium (3.6%)
9. INTERPRET: symbols and colors
10. INTERPRET: iron

4 Close

BUILDING SCIENCE SKILLS

Organizing a Collection Students should wear long pants and long-sleeved shirts or jackets to protect skin from cuts or scrapes. Tell students not to wear open-toed shoes. In addition to field guides, high school Earth science texts and college introductory geology books are also good reference sources to help students identify mineral samples. **learning styles: visual, linguistic**

RETEACH

Minerals in Our LIves Tell students that minerals appear in a multitude of everyday items. Divide the class into small groups and have each group list minerals found in their homes, as well as items in which the minerals are found. Encourage students to list more than one item next to each mineral. For example, aluminum may be found in cans, foil, or cooking utensils. Iron may be found in machine or appliance parts, or in electrical equipment. Students in each group can compare their lists.
learning styles: tactile, visual

• TEACHING RESOURCES • Online or Print

Teacher's Resources
Lesson Review: Chapter 2, p. 5
Enrichment: Chapter 2, p. 7

MORE ON THE WEB

Career Connection Have students use the Internet to learn about the field of geochemistry. They can begin their search at www.conceptsandchallenges.com. Then, have students write a job description for a geochemist.

2-4 *(pp. 48–49)*

1 Introduce

STUDY HINT Have students write in their notebooks the physical properties that will be discussed in this lesson: color, streak, luster, and hardness. As they read the lesson, have them list specific examples next to each property.

Linking Prior Knowledge Discuss everyday uses of the word *property*. For example, property can be a piece of land or an object owned by an individual. Have volunteers list a few of their personal properties. Explain that in Earth science, a property is a characteristic of a rock or mineral that can be observed or measured.

2 Teach

Demonstration Obtain samples of some of the minerals discussed—gold (in jewelry), pyrite, purple quartz (amethyst), yellow quartz (citrine), rose quartz, colorless quartz, smoky quartz, and milky quartz—to demonstrate how color is not always a reliable property for identifying minerals. Mineral samples can be obtained from most scientific supply companies.
learning style: visual

Discussion Have a student use adjectives to describe an object in the classroom without naming it. Have a second student write the adjectives on the board. Challenge the class to identify the object based on the observed properties.
learning styles: auditory, interpersonal

Reinforcement Have a volunteer explain the relationship between a mineral's color and its streak. (The color of a mineral can vary. Its streak on an unglazed ceramic tile, however, is always the same color.)

Answers to questions

1. LIST: color, streak, luster, and hardness
2. DESCRIBE: white
3. IDENTIFY: Accept any two of the following: metallic, nonmetallic, waxy, glassy, or dull.
4. ANALYZE: diamond

2-4 How do we identify minerals?

Objective

Describe the properties that can be used to identify minerals.

Key Terms

physical property: observable characteristic that describes an object

streak: color of the powder left by a mineral

luster: how a mineral's surface reflects light

hardness: property of a mineral that relates to how much the mineral resists being scratched

Physical Properties The characteristics of an object that can be observed or measured are its **physical properties.** The element iron is gray. Objects made out of iron are attracted to a magnet. These are two physical properties of iron. Physical properties can be used to help identify minerals. Some physical properties of minerals are color, streak, luster, and hardness.

1 LIST: What are some physical properties of minerals?

Color and Streak Color is an important property of minerals. However, most minerals cannot be identified by color alone. Many minerals are the same color. Gold and pyrite are both brassy yellow. Other minerals are found in more than one color. Quartz, for example, can be purple, yellow, pink, or colorless. It may also be brown, black, or white. Two minerals that are found only in a single color are malachite and azurite. Malachite is always green. Azurite is always blue.

▲ **Figure 2-15** Malachite is a green mineral.

Streak is the color of the powder left by a mineral. You can find the streak of a mineral by rubbing the mineral across a square of unglazed ceramic tile. A mineral may be found in many different colors. Its streak, however, is always the same color.

Chalk is made up of the mineral calcite. When you write with a piece of naturally occurring chalk, the calcite leaves a white powder that you can see.

▲ **Figure 2-16** Streaks of powder are made by scratching two forms of the mineral hematite on an unglazed ceramic tile.

2 DESCRIBE: What is the streak of calcite?

Luster The way a mineral's surface reflects light is called **luster.** A mineral can have either a metallic or a nonmetallic luster. Minerals with a metallic luster shine like new coins. Other minerals may look waxy, glassy, or dull. Quartz has a glassy luster. Calcite has a glassy or a dull luster.

▲ **Figure 2-17** Silver (left) is a metal valued for its high luster, among other properties. Fluorite has a dull, waxy, or glassy luster.

3 IDENTIFY: What are two kinds of luster?

Teaching Options

RETEACH

Streak Tell students that streak is the color of the powder of a mineral, which is generally very consistent from specimen to specimen for a given mineral. Streak is important because color alone can be a misleading property for identifying minerals. One mineral may exist in two forms with different colors, while two different minerals may appear identical in color.) Provide pairs of students with unglazed bathroom tiles from a home improvement store and have them test hematite, pyrite, calcite, and gypsum for streak.
learning styles: auditory, tactile

ONGOING ASSESSMENT

Provide small groups of students with samples of talc, gypsum, calcite, fluorite, and quartz and the materials from the Hands-On Activity on page 49. Have students describe the color and luster of each sample. Then, have them determine the hardness of each mineral by testing it against the other minerals. Have students summarize their findings and list the mineral samples in order from softest to hardest.
learning styles: tactile, visual

Hardness The property of a mineral that relates to how much it resists being scratched is called **hardness.** In 1812, Friedrich Mohs, an Austrian mineralogist, worked out a scale of hardness for minerals.

Mohs' scale ranks ten minerals in hardness from 1 to 10. As the numbers increase, the hardness of the minerals also increases. A mineral with a high number can scratch any mineral that has a lower number. However, a mineral with a lower number cannot scratch a mineral that has a higher number.

MOHS' SCALE OF HARDNESS			
Mineral	**Hardness**	**Mineral**	**Hardness**
Talc	1	Orthoclase	6
Gypsum	2	Quartz	7
Calcite	3	Topaz	8
Fluorite	4	Corundum	9
Apatite	5	Diamond	10

▲ Figure 2-18

ANALYZE: What mineral is the hardest on Mohs' scale?

CHECKING CONCEPTS

1. What property allows one mineral to scratch another?
2. What is luster?
3. How is the streak of a mineral found?

THINKING CRITICALLY

4. IDENTIFY: What are some physical properties of the mineral diamond?
5. INFER: Why is streak a better way to identify a mineral than color?

DESIGNING AN EXPERIMENT

Design an experiment to solve the following problem. Include a hypothesis, variables, a procedure with materials, and the type of data to study. Be sure to say how you will record your data.

PROBLEM: Joan had two blue minerals. One left a blue streak. The other left a white streak. What other tests could Joan do to identify the minerals?

Hands-On Activity
PERFORMING HARDNESS TESTS

You will need a penny, a butter knife, a piece of glass, and an iron nail.

1. A field hardness scale, like the one in Figure 2-19, can help you determine a mineral's hardness. If a mineral scratches glass, it is harder than glass. If glass scratches the mineral, the mineral is softer than glass.
2. First, try to scratch the glass with your fingernail. ⚠ CAUTION: Do *not* scratch your fingernail with the glass.
3. To find the hardest material, try scratching each material mentioned with the others. Look carefully where you scratched. Are there any marks?

FIELD HARDNESS SCALE	
Hardness	**Test**
1	Scratched easily with fingernail
2	Scratched by fingernail (2.5)
3	Scratched by a penny
4	Scratched easily by a butter knife; does not scratch glass
5	Hard to scratch with a butter knife; barely scratches glass (5.5)
6	Scratched by an iron nail (6.5); easily scratches glass
7	Scratches an iron nail and glass

▲ Figure 2-19

Practicing Your Skills

4. SEQUENCE: List the objects in order from hardest to softest.
5. IDENTIFY: Which material will scratch all of the others?
6. CONCLUDE: How can you find the hardness of a mineral using the materials mentioned?

MORE ON THE WEB

Learning More About Minerals Have students use the Internet to create a set of ten mineral flash cards of their choice. Students should research their minerals, using different Web sites. Then, students should draw a sketch of the mineral on one side of the card and provide information about the mineral on the other side. Students can begin their research at www.conceptsandchallenges.com.

• TEACHING RESOURCES • Online or Print

Teacher's Resources

Lesson Review: Chapter 2, p. 6

Enrichment: Chapter 2, p. 7

Designing an Experiment Rubric: Teaching Tools, p. 7

Laboratory Manual

Lab. Challenge: How are magnetism, density, and the acid test used to identify minerals? pp. 25–28

3 Assess

CHECKING CONCEPTS ANSWERS

1. hardness
2. the way a mineral's surface reflects light
3. by scratching the mineral against an unglazed ceramic tile

THINKING CRITICALLY ANSWERS

4. IDENTIFY: Diamonds are the hardest substances on Earth, ranking 10 on Mohs' scale of hardness. Diamonds also are usually transparent and colorless.
5. INFER: A mineral's color can vary, but its streak on an unglazed ceramic tile is always the same.

4 Close

DESIGNING AN EXPERIMENT

Use the Designing an Experiment Rubric from the Teacher's Resources to assess students' experiments. Fill in the rubric with the additional information below. For this assignment, students should have:

- suggested that Joan compare the luster of the two minerals and determine the hardness of each by using different materials and Mohs' scale.
- proposed that the results of each test be recorded in a data table.

Hands-On Activity
PERFORMING HARDNESS TESTS

TIME: 15–20 minutes

PURPOSE: observing, comparing

SAFETY TIPS: Make sure that students read the directions carefully, especially the caution. Have them report any cuts to you immediately.

COOPERATIVE LEARNING: Have students work in pairs.

Practicing Your Skills Answers

4. SEQUENCE: iron nail, butter knife, glass, penny
5. IDENTIFY: the iron nail
6. CONCLUDE: Each mineral can be tested to determine its field hardness. If a mineral scratches the material, the mineral is harder than the material. If the mineral does not scratch the material, the mineral is softer than the material.

2-5 *(pp. 50–51)*

1 Introduce

INVESTIGATE

TIME: 10 minutes

PURPOSE: Students will observe, compare, and test mineral samples for calcium carbonate.

MATERIALS: safety goggles, lab aprons, calcite or limestone, granite, dilute hydrochloric acid (a pH of between 1 and 2)

SAFETY TIPS: Even weak acids can cause burns. Instruct students to avoid inhaling the acid or splashing it around their eyes. Flush any spills on the skin with plenty of cool tap water.

PROCEDURE:

1. Using a small hammer, carefully break each specimen so that students can apply acid to a fresh surface.
2. Emphasize to students that only a few drops of acid are needed per sample for this investigation.

THINK ABOUT IT: Coral skeletons are made of calcium carbonate (calcite) and will effervesce, or bubble, when exposed to hydrochloric acid.

STUDY HINT Before you begin this lesson, have students write the title of each section in their notebooks. As students review each section, help them find the topic sentence of each paragraph.

Linking Prior Knowledge Ask students to describe the properties that can be used to identify minerals (Lesson 2-4). (color, streak, luster, hardness) Tell students that there are other ways to identify minerals that they will learn in this lesson.

2 Teach

Discussion Ask students to list uses for magnetic materials. (to generate electricity, to hold other magnetic material such as some metals in place, in special trains, in TVs, in floppy computer disks, in medicine [MRIs], in stereo speakers, and in electric motors)

Reinforcement Review the definitions of the terms *mass* and *volume*. *Mass* is "the amount of matter in something." *Volume* is "the space occupied by something." Explain the formula, if necessary, by repeating that to calculate density, an object's mass is divided by its volume.
learning style: logical/mathematical

2-5 What are some other ways to identify minerals?

INVESTIGATE

Testing for Calcium Carbonate

HANDS-ON ACTIVITY

1. Obtain a sample of calcite or limestone and a sample of granite.
2. Put on goggles and a lab apron. Use an eyedropper to place a few drops of weak hydrochloric acid on each sample. Observe and record what happens to each.
   ⚠ **CAUTION:** Acids can cause serious burns.
3. Rinse both samples under warm running water and wash your hands.

THINK ABOUT IT: Coral has a similar chemical composition to limestone. How do you think it would react to the acid?

Objective

Explain how density, magnetism, and an acid test can be used to identify minerals.

Key Terms

density (DEHN-suh-tee)**:** amount of matter in a given volume

magnetism (MAG-nuh-tihz-uhm)**:** force of attraction or repulsion associated with magnets

acid test: test that helps identify minerals containing calcium carbonate

Density All matter has mass and volume. **Density** is the amount of matter in a given volume. You can calculate the density of an object by using the following formula:

Density = Mass ÷ Volume

As an example, suppose that you have a sample of the mineral talc. Your sample has a mass of 26 g and a volume of 10 cm^3. If you divide 26 g by 10 cm^3, you will get a density of 2.6 g/cm^3.

▲ **Figure 2-20** Talc is a soft, light mineral.

Density is a property of minerals that never changes. For this reason, density can be used to help identify minerals. Density is measured in grams per cubic centimeter, or g/cm^3. Most minerals have a density that measures between 2 and 3 g/cm^3.

1 EXPLAIN: How can density be used to identify minerals?

Magnetism **Magnetism** is a force of attraction or repulsion shown by matter that is magnetic. Magnetism is displayed when objects made of iron, nickel, or cobalt are attracted to a magnet. Steel, which is an alloy of iron, also shows magnetism. Magnetite is a mineral that contains iron. If you held a small chunk of magnetite near a magnet, the magnetite would be attracted to the magnet. Magnetite can also sometimes act as a magnet itself. Magnetite shows the property of magnetism.

▲ **Figure 2-21** Magnetite is also called lodestone. This mineral attracts the steel in paper clips.

2 EXPLAIN: How could you use a magnet to determine if an unknown mineral is magnetite?

Teaching Options

EXTENDING THE LESSON

Determining Density Distribute small samples of quartz, string, and a graduated cylinder. Help students determine the mass of their samples, in grams, using a triple-beam balance. Tell students to record this number. Then, have them half fill the graduated cylinder with water. Have students record this number in milliliters. Tell students to tie the string securely around the quartz and slowly lower the quartz into the water. Have students check the new volume from the graduated cylinder and record the number. Instruct students to subtract the volume of the water alone from the volume of water with quartz in it to determine the density. Have students use the following formula: density = mass/new volume − original volume. The average density of quartz is 2.65 g/cm^3, and 1 cm^3 = 1 mL.

ESL/ELL STRATEGY

Magnets Write the following words vertically down a sheet of paper: *magnet, magnetic, magnetism,* and *magnetite.* Point out that the base word is *magnet.* Have students circle *magnet* in each word. Then, ask students to write a definition for each term. Demonstrate magnetism by allowing students to hold a small bar magnet near a piece of magnetite.

The Acid Test Some minerals contain calcium carbonate ($CaCO_3$). Knowing that a mineral contains this compound can help you identify the mineral. To test for calcium carbonate in a mineral, you place a drop of a weak acid on a small piece of the mineral, as you did in the activity at the start of this lesson. If the surface of the mineral fizzes, the mineral contains calcium carbonate. This is an **acid test.** Calcite and dolomite are two minerals that contain calcium carbonate.

3 EXPLAIN: What is the acid test?

CHECKING CONCEPTS

1. What is density?
2. Which common mineral is attracted to a magnet?
3. What kinds of minerals can be identified by the acid test?

THINKING CRITICALLY

4. IDENTIFY: What two properties must be known to calculate the density of a mineral?
5. CALCULATE: What is the density of an object with a mass of 49 g and a volume of 7 cm^3?
6. ANALYZE: How can the property of magnetism be used to help you identify two gray minerals?

DESIGNING AN EXPERIMENT

Design an experiment to solve the following problem. Include a hypothesis, variables, a procedure, and the type of data to study. List the materials you need, and tell how you would record your data.

PROBLEM: How can the minerals quartz and calcite be identified?

Integrating Physical Science

TOPICS: metallic elements, density

FINDING THE DENSITY OF GOLD

Pure gold scratches easily. It is often mixed with silver or copper to make it harder. Jewelers measure the purity of gold in karats, or parts per 24. Twenty-four-carat gold is pure gold. Fourteen-carat gold is 14/24 pure, or 58 percent gold.

Have you ever wondered whether a piece of yellow, shiny jewelry was really gold? Color and luster are important properties of minerals. However, other properties must be identified for gold to be considered genuine.

Legend has it that a king asked a goldsmith to make him a 1 kg crown of gold. The goldsmith replaced some of the gold with silver. The goldsmith then made the king a crown with a mass of 1 kg.

▲ **Figure 2-22** Archimedes (above) may have tested a crown similar to the one shown here.

The king thought the goldsmith had stolen some of the gold but could not prove it. He asked Archimedes, a scientist and mathematician, for help. Archimedes calculated the density of the crown. He compared it with the density of gold. They were different. The crown was not pure gold. The goldsmith had stolen some of the king's gold.

Thinking Critically What are some properties of gold other than color and luster?

QUICK CONVERSIONS		
Properties of Talc	**Metric/SI**	**English**
Mass of talc sample	26 g	.91 oz
Volume of talc sample	10 cm^3	0.6 $in.^3$
Average density of talc samples	2.6 g/cm^3	1.51 $oz/in.^3$

• TEACHING RESOURCES •
Online or Print

Teacher's Resources

Lesson Review: Chapter 2, p. 8

Designing an Experiment: Study Tools, p. 14

Designing an Experiment Rubric: Teaching Tools, p. 7

Laboratory Manual

Lab. Challenge: How is specific gravity measured? pp. 29–32

Answers to questions

1 EXPLAIN: Every mineral has a density that does not change with the size or shape of the mineral.

2 EXPLAIN: Place the magnet close to the mineral to see if the mineral is attracted to the magnet. Of course, some other minerals are magnetic, but they are not as common as magnetite.

3 EXPLAIN: The acid test is a reaction between a weak acid and a mineral that contains calcium carbonate.

3 Assess

CHECKING CONCEPTS ANSWERS

1. amount of matter in a given volume
2. magnetite
3. minerals that contain calcium carbonate, such as limestone and calcite

THINKING CRITICALLY ANSWERS

4. IDENTIFY: mass and volume
5. CALCULATE: $49\ g \div 7 cm^3 = 7\ g/cm^3$
6. ANALYZE: If one of the samples is magnetite, it will attract a magnet.

4 Close

DESIGNING AN EXPERIMENT

Use the Designing an Experiment Rubric from the Teacher's Resources to assess students' experiments. Fill in the rubric with the additional information below.

For this assignment, students should have:

- suggested dropping a weak acid, such as hydrochloric acid, on each mineral.
- proposed that the results of each test be recorded in a data table.

Integrating Physical Science

TOPICS: metallic elements, density

FINDING THE DENSITY OF GOLD

Background Common alloys are brass, bronze, steel, pewter, tin, and the "gold" used in jewelry. Have objects made from alloys available for students to compare and contrast.

Discussion Ask students why a crown of pure gold might be highly impractical to wear. (too heavy, too soft)

Thinking Critically Answer
density (19.3 g/cm^3), cubic structure, streak (yellow), and hardness (2.5 to 3)

2-6 *(pp. 52–53)*

1 *Introduce*

STUDY HINT Ask students to write one similarity and one difference between the terms *cleavage* and *fracture*. Tell students that comparing and contrasting terms can help them remember definitions.

Linking Prior Knowledge Ask anyone in the class who has had a fractured bone to describe the injury. The description should explain that a fracture is an uneven break in the bone. Inform students that some minerals also fracture.

2 *Teach*

Demonstration Give each student about 1/8 teaspoon of kosher salt and a hand lens. Challenge students to identify the crystal system to which salt crystals belong. (isometric) **learning styles: visual, logical/mathematical**

Discussion Refer students to Figure 2-23. Explain that the intersecting lines represent the axes of a crystal. Point out that there are three axes in the isometric system that form right angles. Provide students with thin wooden skewers or plastic sticks and allow them to model the axes in each crystallographic system.

Reinforcement Use a butter knife to split samples of mica, which is a mineral that cleaves. Then, wrap a piece of quartz in a cloth towel and strike it with a hammer. Unwrap the quartz and have students observe how the quartz fractures. Ask students to contrast mica and quartz. **learning styles: tactile, visual**

Answers to questions

1 **DEFINE:** a solid material in which atoms are arranged in a regular pattern

2 **DEFINE:** the tendency of a mineral to split along a plane

3 **CONTRAST:** Cleavage is the tendency of a mineral to split into pieces with smooth, flat surfaces. Fracture is the tendency of a mineral to break into pieces with uneven surfaces.

2-6 What are crystals, cleavage, and fracture?

Objectives

Define crystal and show how it is related to minerals. Distinguish between fracture and cleavage in minerals.

Key Terms

crystal (KRIHS-tuhl): solid substance with its atoms arranged in a regular, three-dimensional pattern

cleavage (KLEEV-ihj): tendency of some minerals to split along smooth, flat surfaces called planes

fracture (FRAK-chuhr): tendency of some minerals to break into pieces with uneven surfaces

Crystal Shapes All minerals are made up of crystals. A **crystal** is a solid material in which the atoms are arranged in a regular pattern. This pattern is repeated over and over in a regular internal structure. This structure determines the external shape of the crystal. The crystals that make up a given mineral always have the same shape, but they may differ in size.

Crystal shape is a property of minerals. There are six basic crystal systems that describe crystal structure. Scientists use X-rays to study the structures of crystals. They can use the structure of a crystal to help identify the mineral.

1 **DEFINE:** What is a crystal?

Cleavage Cleavage is a property of some minerals. **Cleavage** is the tendency of a mineral to split along smooth, flat surfaces called planes. Mica is a mineral that splits into flat sheets. Mica looks like layers of thin plastic.

▲ **Figure 2-24** Mica (top) and feldspar (bottom)

Different minerals have different kinds of cleavage. However, the way the mineral splits, or shows cleavage, is always the same for that mineral. Mica cleaves into thin, flat sheets. Galena splits into small cubes. Feldspar shows what is called steplike cleavage.

2 **DEFINE:** What is cleavage in minerals?

Fracture A mineral shows **fracture** when it breaks into pieces with uneven surfaces. The fracture may be even, uneven, hackly, or concoidal. Hackly minerals have sharp, jagged surfaces. Copper shows a hackly structure. Concoidal minerals have ringed ridges or indentations like those of seashells. Obsidian shows concoidal

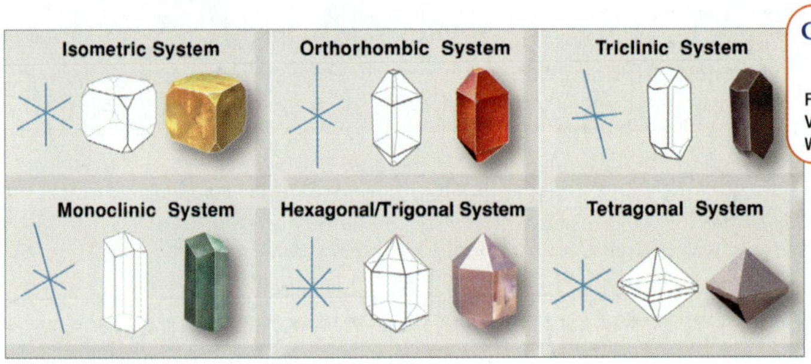

◀ **Figure 2-23** The six basic crystal systems

Go Online *active art*
For: Crystal Systems activity
Visit: PHSchool.com
Web Code: cfp-1041

Teaching Options

! FAST FACTS *Properties of Crystals*

It is not always easy to tell the difference between crystal and glass just by looking. Although crystal may sometimes visually resemble glass, the two items have very little in common structurally. Crystals are solids. Glass is neither a solid nor a liquid but exists in a vitreous, or glassy, state. Crystals have an orderly atomic structure. Glass has no regular atomic structure. Crystal has a definite external geometrical shape bounded by plane faces. Glass does not tend to assume a definite external shape. Crystal has physical properties that vary with direction. Glass has properties that are the same in all directions.

ONGOING ASSESSMENT

Quiz students either orally or in a written test by having them define or give examples of the following: color, streak, luster, hardness, density, magnetism, acid test, crystal structure, cleavage, and fracture.

fracture. Some minerals break into splinters or fibers. One example of this is asbestos.

HOW MINERALS BREAK APART			
Cleavage	Examples	Fracture	Examples
Pieces with smooth, flat surfaces	Mica, galena, feldspar	Pieces with rough, jagged, or uneven surfaces	Copper, obsidian, asbestos

▲ Figure 2-25

3 CONTRAST: How are fracture and cleavage different?

CHECKING CONCEPTS

1. How do scientists study crystal structure?
2. What kind of cleavage does mica show?
3. What is fracture?
4. What do the surfaces of a mineral with fracture look like?

THINKING CRITICALLY

5. MODEL: Halite forms cubic crystals. Draw the crystal form of a piece of halite.
6. IDENTIFY: Steplike cleavage is a property of what mineral?
7. INFER: What property of minerals splits the mineral into pieces with uneven surfaces?
8. DESCRIBE: How does galena look when split?

BUILDING SCIENCE SKILLS

Classifying Minerals have either cleavage or fracture. Find out whether each mineral listed shows cleavage or fracture. Then, make a chart listing each mineral in the appropriate box.

pyroxene	quartz
anatase	corundum
periclase	calcite
amphibole	opal
brookite	hematite
thorianite	talc

People in Science

GEMSTONE CUTTER

A gemstone cutter works with beautiful and valuable minerals called gems. The gem cutter may take the gemstone, such as a diamond, as it is found in nature and split it into two or three parts. Then, a grinding wheel and other tools are used to polish the dull stone. This creates flat surfaces called facets (FAS-ihts) that make the stone sparkle.

Most gem cutters cut gems for jewelry. Some create sculptures. Lawrence Stoller is a gem artist. He was taught gem-cutting by Glenn Lehrer. Stoller and Lehrer together developed many of the tools and technologies used to cut large stones.

▲ **Figure 2-26** Gemstone cutters use grinding wheels to polish gemstones.

Stoller has cut many of the world's largest gemstones. He is best known for his giant sculpted crystals. Some of these are in museums. Stoller also cuts tools for physicists and surgeons.

Gemstone cutting requires skill, focus, and patience. A mistake can destroy a costly stone. Most gem cutters learn their trade on the job.

Thinking Critically If you were to do gem cutting as a hobby, why would you use stones such as agate and quartz?

MORE ON THE WEB

Learning More About Fractures Have students use the Internet to explore the difference between conchoidal, subconchoidal, uneven, jagged, splintery, and earthy fractures. Students should describe each type of fracture and name a mineral that fractures that way. Students can begin searching at www.conceptsandchallenges.com.

• *TEACHING RESOURCES* • *Online or Print*

Teacher's Resources
Lesson Review: Chapter 2, p. 9

Go Online *active art*

For: Crystal Systems activity
Visit: PHSchool.com
Web Code: cfp-1041

Students can interact with the art of crystal systems online.

3 *Assess*

CHECKING CONCEPTS ANSWERS

1. using X-rays
2. thin, flat sheets
3. the tendency of some minerals to break into pieces with uneven surfaces
4. They are rough and jagged.

THINKING CRITICALLY ANSWERS

5. MODEL: Drawings should resemble the cube shown in Figure 2-23.
6. IDENTIFY: feldspar
7. INFER: fracture
8. DESCRIBE: like small cubes

4 *Close*

BUILDING SCIENCE SKILLS

Classifying Cleavage—pyroxene, anatase, periclase, amphibole, brookite, calcite, and talc. Fracture—thorianite, quartz, corundum, opal, and hematite

People in Science

GEMSTONE CUTTER

Background When people shop for diamonds, they usually look for the 4 C's: color, cut, clarity, and carat. Although most diamonds are colorless, some are more yellow or gray than others. Regarding the cut of a diamond, more facets add to the quality of the mineral. Impurities in diamonds can give them a dirty look. Finally, everything else being equal, large diamonds (high carats) are more valuable than smaller ones.

Real-Life Connection Arrange for a gemstone cutter to come to your class or take the class to a gemstone cutter's place of business. Ask the person to show students some of the tools used to cut and polish gems. The gemstone cutter might also show students samples of rough minerals. If possible, purchase a grinding wheel and agates with which the students can experiment.

Thinking Critically Answer
Agate and quartz are inexpensive and relatively easy to cut and polish.

2-7 *(pp. 54–55)*

1 *Introduce*

INVESTIGATE

TIME: 15–20 minutes

PURPOSE: Students will observe, compare, and contrast the formation of crystals.

MATERIALS: safety goggles, gloves, baking soda, small spoon, tweezers, microscope slides, hot plate, paper towels, ice cubes, microscope or hand lens

PROCEDURE: Have the hot plate warmed up, the ice cubes evenly arranged on a plate or paper towel, and two clean microscope slides ready. Use tweezers to move the slides on and off the hot plate.

SAFETY TIPS: Remind students to keep the hot plate set on low and to avoid touching it.

THINK ABOUT IT: The crystals that cooled slowly are generally larger and more developed.

STUDY HINT Have students use the lesson objective and the Key Terms to form an outline of the lesson. Have students fill in the outline as they review the lesson.

Linking Prior Knowledge Explain precipitation as it applies to the weather. Have students define the term as it applies to minerals.

2 *Teach*

Discussion Write each of the processes listed in the Key Terms on the board. Use colored chalk to divide the words into parts as follows: crystal/lization, e/vapor/ation, and precipita/tion. Use the dividing lines to help students differentiate among the three processes.

Answers to questions

1. **RELATE:** Rocks are made of minerals.
2. **CONTRAST:** Crystals that cool slowly are generally larger and more developed than crystals that cool quickly.
3. **IDENTIFY:** When water rich in dissolved substances evaporates, the mineral crystals are left behind and combine to form rock.
4. **DESCRIBE:** Chimneys are vents in the ocean floor that release a hot solution of minerals dissolved in water.

2-7 How are minerals formed?

INVESTIGATE

Forming Crystals Fast and Slow

HANDS-ON ACTIVITY

STEP 1

1. Put on safety goggles and gloves. Using tweezers or a spoon, put a small amount of baking soda on each of two microscope slides. Carefully place the slides on a hot plate set on low. ⚠ CAUTION: Be careful when using a hot plate.
2. When the baking soda melts, carefully move one slide aside with tweezers and put it on a paper towel to cool. Place the second slide on some ice cubes.
3. When the baking soda on both slides is hard, examine the slide using a hand lens or a microscope.

THINK ABOUT IT: How are the crystals on the two slides different?

Objective

Describe the main ways minerals are formed.

Key Terms

crystallization: formation of crystals caused by processes such as cooling and evaporation

evaporation (ee-vap-uh-RAY-shuhn)**:** process by which a liquid changes into a gas

solution: mixture in which the particles of one substance are evenly mixed with the particles of another substance

precipitation (pree-sihp-uh-TAY-shuhn)**:** process that occurs when elements and compounds leave a solution and crystallize out as solids

Minerals in the Crust Earth's crust is made up of rocks. Rocks, in turn, are made up of minerals. Most minerals are compounds in crystal form. In some rocks, the crystals are very large and easy to identify. In other rocks, the crystals are very tiny.

1 RELATE: How are minerals and rocks related?

Magma Cooling Molten rock, or magma, is a very hot mixture of many different minerals. Magma is found in deep underground chambers. As long as the magma stays hot, it remains liquid. Eventually, though, the magma cools. If it cools slowly underground, large mineral crystals start forming. This is called **crystallization.** Eventually, the mineral crystals grow together to form rock.

Sometimes, cracks form in the rock above magma chambers. Pressure causes the magma to move up to the surface. When it comes out, the magma, now called lava, cools more quickly than it would have underground. Soon, the lava hardens into rock that contains small crystals.

2 CONTRAST: How do crystals that cool very slowly differ from crystals that cool rapidly?

Evaporation The second way minerals form is by **evaporation.** In this process, water evaporates, or changes to a gas. The water in lakes and oceans contains dissolved minerals. As the water evaporates, only the mineral crystals remain. The crystals grow together to form rock.

3 IDENTIFY: How do crystals form from evaporated water?

Precipitation Sometimes the elements in a mineral dissolve in water heated by magma. These dissolved minerals form solutions. A **solution** is a mixture in which the particles of one substance are evenly mixed with the particles of another substance. When the solution begins to cool, the elements and compounds leave the solution and crystallize out as solids. This process is **precipitation.**

Crystals can also form when water has more dissolved minerals than it can usually hold. For example, warm water can dissolve more halite (salt) than cold water can. The Great Salt Lake of Utah has very salty water. A very hot and dry climate causes evaporation and salt to crystallize out.

54

Teaching Options

INTEGRATING LIFE SCIENCE

Sea-Vent Compounds Sulfur compounds are among the many substances that escape from deep-sea vents. Through chemosynthesis, a process that replaces photosynthesis, certain marine bacteria use these compounds to produce food and oxygen. These bacteria, in turn, are eaten by other ocean organisms such as tube worms.

RETEACH

Crystal Formation Make a solution of salt and water. Put about 1/2 cup of kosher salt into a cup of hot water and stir the solution until all of the salt dissolves. Pour the salt water into a shallow dish and put it on a sunny windowsill where it will not be disturbed. Have volunteers take turns observing the solution with a magnifying glass. When the water has evaporated, salt crystals will cover the plate. Lead students to conclude that the crystals formed as the result of evaporation of the water.

learning style: visual

Many minerals form where there are cracks in the ocean floor. First, ocean water seeps down through the cracks. The water is heated by magma in the crust. The heated water dissolves the minerals in the crust. This hot solution is released through vents called chimneys, as shown in Figure 2-27. When the hot solution hits cold seawater, minerals crystallize out and then fall to the ocean floor.

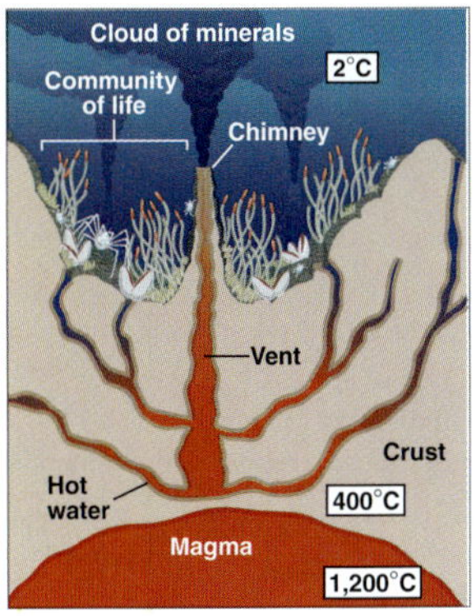

◄ **Figure 2-27** A deep-sea vent is a crack in the ocean floor.

4 **DESCRIBE:** What are chimneys?

CHECKING CONCEPTS

1. What are minerals?
2. What causes big crystals to form?
3. What causes tiny crystals to form?
4. What happens as water containing dissolved minerals evaporates from lakes and oceans?

THINKING CRITICALLY

5. **INFER:** What black, glassy mineral might you get if molten lava cooled so rapidly that crystals did not have time to form?
6. **INFER:** Sometimes rock has both large and small mineral crystals. How can you explain this?

BUILDING SCIENCE SKILLS

Classifying Find out what is meant by the following terms: *silicates, carbonates, sulfates, halides,* and *oxides.* Write a definition for each term. Then, find examples of two minerals that can be put into each group.

Integrating Life Science

TOPICS: medicine, health

PROTEIN CRYSTALS

The human body contains more than 100,000 different proteins. Proteins perform many important functions. For example, they allow muscles to contract. They help bones grow and form the attachments for muscles. They also help fight diseases.

Proteins are made up of very complex molecules. These molecules are strung out like long, twisted ribbons. Scientists can crystallize the proteins and analyze the molecules. However, Earth's gravity distorts the crystals as they form.

▲ **Figure 2-28** Protein crystals (right) grown in space by Russian cosmonaut Vladimir G. Titan (above)

In flight, the space shuttle acts as a gravity-free environment. Scientists have performed protein crystal experiments on the shuttle. This has allowed them to grow large protein crystals that are not distorted by gravity. These crystals may be effective at fighting diseases such as cancer.

Thinking Critically The space shuttle can only remain in space for a few weeks. The International Space Station (ISS) will have a permanent orbiting laboratory. Would the ISS be better for growing crystals? Why or why not?

MORE ON THE WEB

Observing Mineral Formation Students can use the Internet to research Carlsbad Caverns National Park. Students can later create a travel brochure of their own design based on the information they have gathered on the Internet.

• *TEACHING RESOURCES* • *Online or Print*

Teacher's Resources

Lesson Review: Chapter 2, p. 10

Enrichment: Chapter 2, p. 12

ONGOING ASSESSMENT

Have students draw a picture of a crystal, demonstrating one of the six basic crystal shapes. Instruct students to trade drawings with a partner. Ask students to identify the crystal system depicted in the drawing without referring to the chart. **learning style: visual**

3 *Assess*

CHECKING CONCEPTS ANSWERS

1. Minerals are compounds in crystal form.
2. slow rate of the mineral cooling
3. fast rate of the mineral cooling
4. Mineral crystals are left behind.

THINKING CRITICALLY ANSWERS

5. **INFER:** obsidian
6. **INFER:** Rocks are made up of different minerals with various cooling rates.

4 *Close*

BUILDING SCIENCE SKILLS

Classifying Silicates are minerals that contain silicon and oxygen and often other elements or compounds. Quartz, amphibole, talc, feldspar, mica, and pyroxene are a few of the many silicate minerals. Carbonates contain calcium carbonate and often other elements or compounds. Calcite and dolomite are carbonates. Sulfates contain sulfur and oxygen, plus other elements or compounds. Barite and gypsum are sulfates. Halides contain the element halogen and one or more other elements or compounds. Halite and fluorite are halides. Oxides are minerals in which oxygen combines with one or more metals. Oxides include corundum and hematite.

Integrating Life Science

TOPICS: medicine, health

PROTEIN CRYSTALS

Life-Science Connection Proteins are complex molecules that contain the elements carbon, hydrogen, oxygen, and nitrogen. A molecule of protein is made up of subunits called amino acids. The body needs 20 amino acids to construct proteins. About 10 of these acids can be made in the body's cells. The others must be obtained from foods.

Discussion Have students list sources of protein in their diets. Answers might include milk, eggs, meats and fish, nuts, and cheese.

Thinking Critically Answer Students who answer yes might argue that growing crystals aboard the ISS is better because it is a permanent lab.

Lab Activity

(pp. 56–57)

1 *Prepare the Lab*

Growing a Crystal Garden

SET-UP TIME: **LAB TIME:**

ONE WEEK LATER

BACKGROUND: All rocks are made up of minerals. With a few exceptions, the minerals are in crystal form. Some rocks, such as limestone, are made up entirely of one mineral. Most limestone forms along ocean bottoms, where calcite is dissolved in the water. Tiny calcite crystals grow and interlock to become limestone. Other rocks, such as granite, are mixtures of several minerals. Granite forms as magma slowly cools beneath Earth's surface. Quartz, feldspar, and other minerals make up granite.

PURPOSE: Students will grow small crystals and study them under magnification to observe their shapes.

ALTERNATIVE MATERIALS: Consult library books to find different crystal-growing recipes. Small plastic bathroom cups can be substituted for egg carton cups.

SCIENCE SKILLS: Students will **observe** the growth of crystals and **communicate** their findings by making drawings of the crystal shapes.

ADVANCE PLANNING: Obtain the materials needed for this activity. You must use the plastic-foam egg cartons, not cardboard. You may wish to prepare the crystal solutions yourself to avoid having students work with hot water. The solutions can be stored in thermos bottles until needed.

2 *Guide the Lab*

PROCEDURE: Have each student make a crystal garden. The crystal recipes make enough for about six to eight gardens. You may wish to designate certain students to do the mixing for others. Discuss the procedures. If students prepare their own crystal growing solutions, discuss safe handling of the hot water. Provide students with a copy of the Lab Activity Report from the Teacher's Resources for students to record their observations and conclusions.

LAB ACTIVITY
Growing a Crystal Garden

Materials

Goggles, lab apron, sugar, salt, magnesium sulfate (Epsom salts), hot food coloring, hot water, plastic cups, plastic foam egg carton, scissors, tablespoon, ballpoint pen, magnifying glass

▲ **STEP 1** Cut the egg carton to form three cups.

▲ **STEP 4** Stir until all chemicals are dissolved.

BACKGROUND

Most rocks contain minerals in crystal form. A crystal has a regular geometric shape, such as a cube. Some minerals form under water. Minerals dissolved in the water can crystallize out as the water evaporates or cools. Eventually, the crystals grow together and become minerals or rocks.

PURPOSE

In this activity, you will dissolve minerals in water to form solutions. As the solutions evaporate, the minerals will recrystallize.

PROCEDURE

1. Cut the egg carton so that you have three egg cups joined together. Leave a little of the carton lid above the hinge to write your name.
   ⚠ CAUTION: Be careful when using scissors.
2. Put on your safety goggles and lab apron.
3. Look at the crystal recipes listed in Figure 2-29. Place the right amount of salt for the first recipe in one of the plastic cups.
4. Add the right amount of hot water to the cup. Stir until all the chemicals dissolve.
   ⚠ CAUTION: Be careful when using hot water.
5. Add food coloring to the solution and stir.
6. Pour one tablespoon of your crystal solution into one egg cup. Share the rest of your solution with others doing the experiment.

56

Teaching Options

INTEGRATING CHEMISTRY

Solutions A solution is a type of matter in which one substance is dissolved in another. The substance being dissolved is the solute. The substance that dissolves the solute is the solvent. In this lab, the solids (and the food coloring) are the solutes. Water is the solvent. A saturated solution is a solution that contains all of the solute that it can hold at that temperature. A supersaturated solution contains more solute than a saturated solution at that temperature. Many crystals form from supersaturated solutions. When these solutions cool, some of the dissolved solids come out of, or precipitate from, the solution, and crystals begin to form.

EXTENDING THE LESSON

Evaporation The process that produces crystals from solutions, called evaporation, is an integral part of the water cycle. Water is constantly moving, either evaporating from the oceans into the atmosphere where it forms clouds or falling to Earth as precipitation. Have students write a paragraph on the similarities between the processes at work in the experiment and those at work in the water cycle.

7. Mix and pour the other crystal solutions into the other cups. Do not touch the egg cups for one week unless instructed to do so by your teacher.

8. One week later, carefully pour any remaining crystal solutions in your egg cups into the sink. Let the crystals dry.

9. Examine the crystals with a magnifying glass. Record your observations in a notebook.

▲ **STEP 7** Pour one tablespoon of each crystal solution into an egg cup.

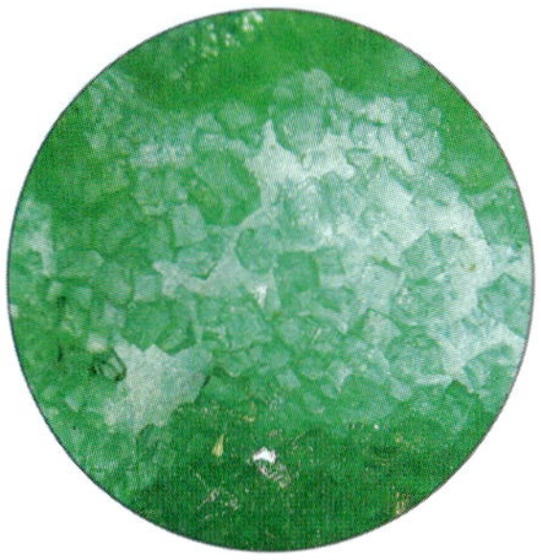

◀ **STEP 9** Examine the crystals with a magnifying glass.

Crystal Recipes

Halite (salt)	*Sugar*	*Epsom salts*
5 tbs salt	*5 tbs sugar*	*5 tbs Epsom salts*
6 tbs hot water	*5 tbs hot water*	*5 tbs hot water*
Green food coloring	*Red food coloring*	*Blue food coloring*

▲ **Figure 2-29** Use these recipes to make your crystal solutions.

CONCLUSIONS

1. OBSERVE: What do you see?
2. OBSERVE: What are the shapes of the crystals?
3. ANALYZE: Why are the crystals attached to each other?
4. INFER: How do your results resemble the way some kinds of rocks form at the bottom of the ocean?

TROUBLESHOOTING: Dissolve the chemicals as much as possible in the water. Rafts of crystals may form over the surface of the egg cups during crystallization. The rafts slow the evaporation process. Gently break the rafts so that the crystals settle to the bottom of the cups.

SAFETY TIP: Check the safety cautions on the packages of any chemicals you use. Do not boil the water. Use a hot plate and raise the water temperature to about 60°C (140°F). If you plan to have students mix their own chemicals, transfer the hot water to a thermos bottle and bring it to students' workstations.

EXPECTED OUTCOME: Students will grow crystals from water and chemical solutions and observe the crystal forms produced.

LAB HINT: If a large container of crystal solution is allowed to evaporate over many weeks, a thick mass of interlocking crystals will form at the bottom. This models the rock-forming process.

3 Conclude the Lab

1. OBSERVE: hundreds of small crystals
2. OBSERVE: Halite forms small cubes. Sugar forms long, tapered crystals with at least six sides. Epsom salts also form multifaceted crystals.
3. ANALYZE: Small crystals grew larger and joined with other crystals.
4. INFER: Dissolved substances in seawater crystallize out when the water contains too many dissolved substances.

Use the Laboratory Activity Rubric from the Teacher's Resources to assess students' lab activities. Fill in the rubric with the additional information below. For this activity, students should have:

- dissolved minerals in water to form solutions.
- examined and analyzed their resulting crystals.

4 Extend the Lab

Display individual mineral crystals, such as quartz and calcite, and other rocks with large, visible crystals. Have students examine the rocks under magnification. Also, display rocks such as lava and obsidian. Students will not be able to detect any crystals in these rocks because the lava that formed them cooled so rapidly that crystals did not have time to form.

Tips for Using Technology

Presenting Lab Conclusions

Have interested students use presentation software to organize and present the results of this lab to the class. Finished products should include a data table and sketches of the different type of crystals. Students can use the Flowchart found in the Teacher's Resources to outline their presentations.
learning styles: visual, logical/mathematical

• TEACHING RESOURCES • Online or Print

Teacher's Resources

Lab Activity Report: Study Tools, pp. 16–17

Laboratory Activity Rubric: Teaching Tools, p. 8

Flowchart: Study Tools, p. 6

2-8 *(pp. 58–59)*

1 Introduce

STUDY HINT Have students copy the boldface heads—Mineral Uses, Ores, Gemstones and Gems—from the lesson onto a sheet of paper. After they have read the information presented, challenge students to make a concept map that links these terms.

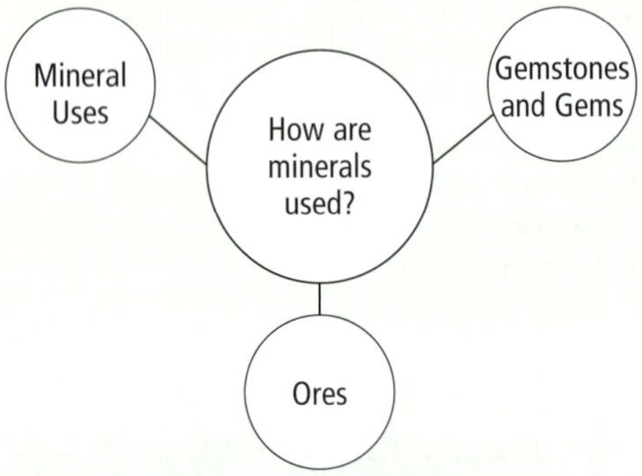

Linking Prior Knowledge Ask how many students know their birthstones. If anyone is wearing jewelry with a birthstone, have him or her show it to the class.

2 Teach

Demonstration Before students start the lesson, bring in some of the items listed in the right column of Figure 2-30. Challenge students to guess the minerals used in each item. If they are unable to do so, have them check the table. If possible, display hand specimens of each of the minerals listed next to the items in which they are found.
learning styles: tactile, visual

Discussion Explain that many of Earth's resources, such as ores and gemstones, form deep beneath the surface. Sometimes the rocks that hold these resources are exposed at the surface by forces within Earth. In many cases, however, these resources need to be extracted from underground mines. Mines at or near the surface are called strip mines. Ask students to speculate about the advantages and disadvantages of these two types of mines. (Underground mines are hard to get to and can collapse. Strip mines are more accessible but can harm the environment.)
learning style: linguistic

2-8 How are minerals used?

Objective
State some common uses of minerals.

Key Terms
ore: mineral that is mined because it contains useful metals or nonmetals
gem: stone that has been cut and polished

Mineral Uses Many products are made from minerals. Diamond crystals are used to make jewelry. Because they are so hard, diamonds are also used to make cutting and drilling tools.

Gypsum is a mineral that is commonly used to make plaster. Many walls and ceilings are made of plaster. Some minerals and their uses are shown in Figure 2-30.

MINERALS AND THEIR USES	
Mineral	**Uses**
Quartz	Glass, sandpaper, telephone, radio
Feldspar	Porcelain, china, dishes
Mica	Insulators, toasters, irons, motors
Talc	Talcum powder, crayons, soap
Calcite	Building materials, medicine
Graphite	Pencil lead

▲ **Figure 2-30**

Tiny amounts of minerals are needed as part of your diet to keep your body healthy and working properly. Calcium and phosphorus are two examples of minerals needed by your body. These minerals are needed for strong bones and teeth. The body uses the mineral iron to make new red blood cells.

The minerals your body needs are found in food substances. Calcium, for example, is found in milk and other dairy products. Iron is found in red meat and some vegetables. The minerals that are used in dietary supplements are usually found in compounds.

1 **IDENTIFY:** What are two minerals used by the human body?

Ores Minerals in the form of compounds contain useful metals and nonmetals. An **ore** is a rocky material that is mined because it contains useful metals or nonmetals that can be extracted from it.

Some examples of ores are hematite, bauxite, and halite. Hematite contains iron, which is used in the production of steel. Bauxite is mined to extract aluminum. Halite contains salt.

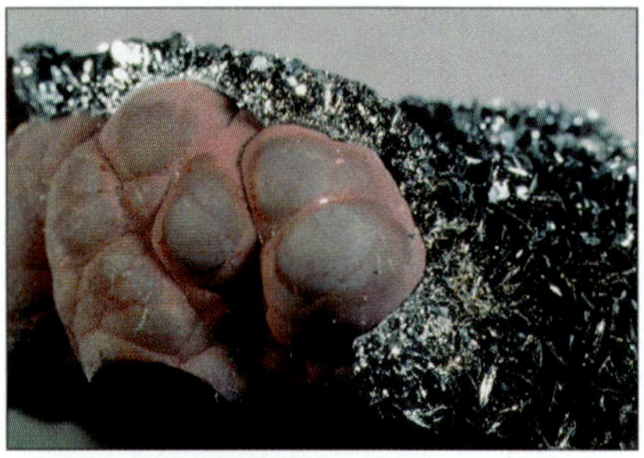

▲ **Figure 2-31** Hematite iron ore

2 **DEFINE:** What is an ore?

Gemstones and Gems Some minerals are valued because they are beautiful and long lasting. These minerals are called gemstones. Rare and beautiful gemstones are called precious stones. Gemstones that are more common are called semiprecious stones. Because they are beautiful and long lasting, many precious and semiprecious stones are used to make jewelry.

▲ **Figure 2-32** Ruby (left) from corundum and emerald (right) from beryl are both precious stones used in making jewelry.

58

Teaching Options

! FAST FACTS Birthstones

Wearing birthstones dates back to the fifteenth or sixteenth century. Some gemstones came to be worn because of specific powers associated with them. A different gemstone was worn each month, and people thought that each gemstone had heightened powers during certain months. For example, many people believed that the opal, October's birthstone, could prevent fainting. Amethyst, the birthstone for February, was thought to prevent intoxication. Topaz, November's birthstone, reportedly gave the wearer strength. Garnet, January's birthstone, protected the wearer from nightmares.

Month	Birthstone	Month	Birthstone
January	garnet	July	ruby
February	amethyst	August	peridot
March	aquamarine	September	sapphire
April	diamond	October	opal
May	emerald	November	topaz
June	pearl	December	turquoise

Gemstones that have been cut and polished are called **gems.** Diamonds, rubies, sapphires, and emeralds are precious stones that are often made into gems. Opal, amethyst, turquoise, topaz, garnet, aquamarine, and jade are some semiprecious stones that are made into gems.

▲ **Figure 2-33** Opal (above) and turquoise (left) are both semiprecious stones used in making jewelry.

3 IDENTIFY: What are three precious stones?

CHECKING CONCEPTS

1. Calcium and __________ are both needed by the body for strong bones and teeth.
2. Bauxite is an __________ of aluminum.
3. Plaster is made from __________.
4. Very rare and beautiful minerals are called __________ stones.
5. Halite is an ore of __________.

THINKING CRITICALLY

6. LIST: What are four minerals that are commonly taken from ores?

INTERPRETING VISUALS

Use Figure 2-30 to answer the following questions.

7. LIST: Which minerals are used to make items found in the bathroom?
8. IDENTIFY: What is the most common use of graphite?
9. IDENTIFY: Which mineral do objects that require heat insulation usually contain?

Science and Technology

GEMS FROM THE LABORATORY

Precious stones are rare. They are prized for their color, luster, and hardness. Any combination of these properties makes a precious stone desirable and costly.

Many gems are now made in laboratories. For example, crystals of quartz are heated until they change color. They are then sold as topaz. Such stones are called treated gems. A three-piece sandwich of heat-treated aquamarine inside of green glass can be substituted for an emerald. This is an example of an assembled gem.

▲ **Figure 2-34** Synthetic gems, which are made in laboratories, can be very difficult to tell from natural gems. Above is a synthetic diamond.

Corundum is the second hardest of all minerals. It can be specially treated to form synthetic rubies and sapphires. These look just like the naturally occurring gemstones.

Thinking Critically Manufactured gems are hard to tell from the real thing. Should they be as costly as the natural gems they resemble?

Answers to questions

1 **IDENTIFY:** Accept any two of the following: calcium, phosphorus, or iron.

2 **DEFINE:** a rocky material that is mined because it contains useful metals or nonmetals

3 **IDENTIFY:** Accept any three of the following: diamonds, rubies, sapphires, or emeralds.

3 *Assess*

CHECKING CONCEPTS ANSWERS

1. phosphorus
2. ore
3. gypsum
4. precious
5. salt

THINKING CRITICALLY ANSWER

6. **LIST:** hematite, bauxite, halite, and gemstones

4 *Close*

INTERPRETING VISUALS ANSWERS

7. **LIST:** quartz, feldspar, talc, and calcite
8. **IDENTIFY:** as pencil lead
9. **IDENTIFY:** mica

Science and Technology

GEMS FROM THE LABORATORY

Discussion Pose the following scenario: Suppose you were born in May and wanted to buy a natural emerald. How could you make sure that the gem was formed in nature and not in a lab? If necessary, have students recall from Lesson 2-6 that scientists can use X rays to determine crystal structure. (The gem could be X-rayed. If it was an assembled gem, as described in the feature, then its X-ray pattern would be different from the pattern of a natural emerald.)
learning style: logical/mathematical

Thinking Critically Answer
Accept either response as long as students can justify it. Most students will probably state that a buyer should be told that the gem is either naturally occurring or synthetic.

EXTENDING THE LESSON

Classifying Gems Have pairs of students research the chemical formulas for the gems listed in this lesson. Challenge students to classify each mineral into one of the following groups: native elements, silicates, carbonates, sulfates, oxides, or halides. (Native elements are any of a class of naturally occurring minerals with a molecular structure of only one element, such as copper or sulfur.)
learning styles: linguistic, logical/mathematical

• TEACHING RESOURCES • *Online or Print*

Teacher's Resources

Lesson Review: Chapter 2, p. 11

Enrichment: Chapter 2, p. 12

Concept Map: Study Tools, p. 4

MORE ON THE WEB

Famous Diamonds The Hope Diamond is one of the most famous diamonds in the world. Have students go online to read about the history of the Hope Diamond.

The Big Idea
(pp. 60–61)

1 *Introduce*

Objective Students will understand that atomic structure determines how carbon can form different substances. Students will write in their Science Logs about imaginary journeys through three substances made of carbon: coal, graphite, and diamond.

Linking Prior Knowledge Ask students to recall the relationship between elements and minerals (Lessons 2-1 and 2-3). (An element is a simple substance that cannot be broken down into simpler substances by ordinary means. Minerals are naturally occurring inorganic solids with definite chemical compositions. Minerals are made up of one or more elements. Carbon is an element. Diamond is a mineral. Coal is an organic substance, so it cannot be classified as a mineral.) Have students recall that atoms are the smallest part of an element and that minerals are formed from elements (Lessons 2-1 and 2-3).

2 *Teach*

Discussion After students have read the information on these two pages, have them choose facts to include in their imaginary journeys. These facts should include the structure of each substance and how each substance forms.

Tell students that carbon is a nonmetallic element with four electrons in its outermost shell. This arrangement allows carbon to easily combine with many other elements and itself. Carbon is found in more compounds than any other single element. It is always present in organic compounds. More than 90 percent of all compounds on Earth are organic compounds. Organic compounds are vital to all living things and are found in carbohydrates such as sugars, starch, and cellulose. These are used by plants and animals to store energy.

Integrating Chemistry

THE Big IDEA

How can the element carbon form different substances?

How are a lump of coal, the point of a pencil, and a sparkling diamond alike? All three substances are made up mostly of carbon atoms. However, the carbon atoms are arranged in different ways.

Coal is a carbon-rich organic material. It took millions of years to form. Overgrown swamps once covered huge areas of land, and the climate was mild and rainy. As the plants died, they became buried. Pressure and heat caused the deep deposits of buried plants to change into coal.

Graphite is a mineral made up entirely of carbon. The atoms in graphite are arranged in sheets. They are held together by weak chemical bonds. When these bonds break, as they do under light pressure, some of the carbon atoms flake off. That is why graphite is often used in pencils. Pressing on the pencil leaves behind a trail of carbon. Graphite is also used to lubricate machines.

Diamond is also a mineral made up entirely of carbon atoms. However, the diamonds crystallize only under extreme pressure. In nature, they are created by geological processes deep within Earth. Other processes, such as the wearing away of surface rock, bring the diamonds to the surface.

The atoms that make up diamonds are tightly bonded together. Its internal crystal structure makes diamond the hardest natural substance known.

Look at the illustrations on these two pages. Then, follow the directions in the Science Log to learn more about "the big idea." ✦

Coal Swamp
Coal is not a mineral but an organic material. It is formed from the remains of living things, mostly plants. Most coal was formed in huge, overgrown swamps during the Carboniferous Period, 300 to 350 million years ago.

WRITING ACTIVITY

Science Log
Imagine that you are a scientist exploring how carbon atoms combine to form different substances. Your research team gets on board a special vehicle that can travel through solids such as coal, graphite, and diamond. Write a few paragraphs describing your observations. Start your search at www.conceptsandchallenges.com.

Teaching Options

! FAST FACTS More on Carbon Substances

Coal
The first stage in coal formation is peat—a substance composed of water, decaying twigs, leaves, and branches. Peat is about 50 percent carbon. Lignite, the second stage, is a soft brown coal about 70 percent carbon. Bituminous coal is the third stage in coal formation. This substance is a dense, black, brittle coal containing about 85 percent carbon. If intense heat and pressure are applied to bituminous coal, it becomes anthracite—a hard coal about 90 percent carbon.

Graphite
Graphite is a hexagonal mineral with a metallic luster. It cleaves rather than fractures, has a hardness of between 1 and 2 on Mohs' scale, and gives a black streak on an unglazed ceramic tile. Graphite is used in the manufacture of steel, brass, bronze, and other alloys. It is also used as a lubricant and forms the "lead" of pencils.

Diamond
Diamonds, which belong to the isometric crystal system, are the hardest substances known. These minerals have a brilliant luster and perfect cleavage. Most diamonds form in kimberlite pipes (a collection of biotite-peridotite) deep beneath Earth's surface. Today, most diamonds are mined in Brazil and South Africa.

Types of Coal

Coal forms from the remains of living things. It is not a mineral and does not have a regular atomic structure. Coal is classified according to the amount of carbon it contains. Anthracite (hard) coal is made up mostly of carbon. It is found in rocks that are more than 65 million years old. There is also bituminous coal and lignite, which are softer forms of coal and contain more impurities such as sulfer.

Graphite

Graphite is made up of layers of carbon. The layers are loosely bonded to each other. This internal crystal structure makes graphite a very soft substance.

Diamond

Diamond is made up of tightly packed carbon atoms. The Hope diamond is a world-famous deep blue gem that weighs more than 45 carats. You can see this magnificent combination of carbon atoms by visiting the Smithsonian Institution in Washington, D.C.

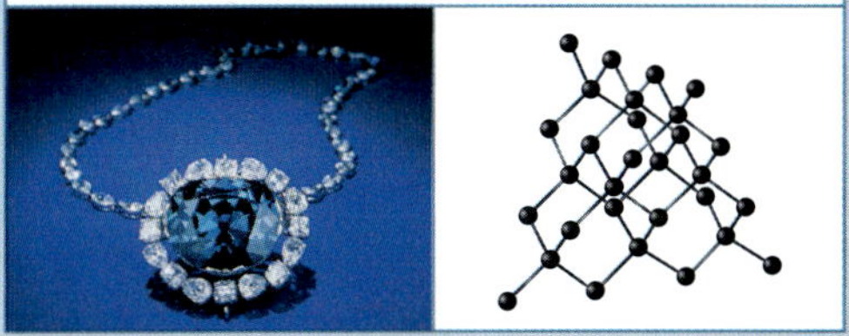

▲ **Figure 2-35** All of these forms of carbon can be mined from the earth. Diamonds come from a mine like this one in South Africa.

Use the background information in Fast Facts to help guide students in writing their Science Logs.

THE BIG IDEA

How can the same element form different substances? Have students start their research for their Science Log Writing Activity at www.conceptsandchallenges.com. Be sure students use search engines and websites that are reliable and acceptable. You may also wish to provide copies of the *Big Idea Planner* and *Big Idea Science Log Entry* study tools found in the Teacher's Resources.

3 Assess

Use the Writing Activity Rubric from the Teacher's Resources to assess students' Science Logs. Fill in the rubric with the additional information below. For this assignment, students should have:

- accurately explained the differences among the internal structures of coal, graphite, and diamond.
- communicated, in paragraph form, information about the three different carbon substances.

4 Close

A Day in the Life of . . . Have students work in pairs to write about the various stages and processes that occur for a diamond ring to be made. Students should include diamond formation, mining, processing, and the final construction of the ring. Students' writing should include accurate, scientific information about diamonds. Suggest that students illustrate their stories and then display them or read them aloud to classmates.
learning styles: linguistic, interpersonal

Tips for Using Technology

Using a Search Engine for Finding Information on Carbon Atoms Remind students that an important tool for finding information on the Internet is the search engine. After a user types in a key word or phrase, the search engine searches the Internet for information that matches the request. Students can begin their searches at www.conceptsandchallenges.com. Have students scan the text and the callouts from the Big Idea for key words to use in their search.

• TEACHING RESOURCES • Online or Print

Teacher's Resources

Big Idea: Chapter 2, p. 13

Big Idea Planner: Study Tools, pp. 11–12

Big Idea Science Log Entry: Study Tools, p. 13

Writing Activity Rubric: Teaching Tools, p. 9

Challenges (pp. 62–64)

Chapter Summary

Review Before students begin the Challenges, review the definitions of the Key Terms with them. In this chapter, it may be possible to assign one term to each student, who will then recite the definition to the rest of the class.

STUDY HINT Encourage students to answer the challenge questions without referring back to the text. Studying the Chapter Summary points may prompt students to recall related information and help them formulate answers on their own.

Key Term Challenges

MATCHING

1. physical property (2-4)
2. mineral (2-3)
3. luster (2-4)
4. hardness (2-4)
5. crystal (2-6)
6. density (2-5)
7. magnetism (2-5)
8. streak (2-4)

IDENTIFYING WORD RELATIONSHIPS

9. Ores are rocky materials that contain valuable minerals. Gems are made from gemstones, which are found in ores. (2-8)
10. Color is how a mineral appears in normal light. Streak is the color of the powder left by a mineral when rubbed against an unglazed ceramic tile. (2-4)
11. Cleavage and fracture describe different ways a mineral can break. (2-6)
12. A compound is a substance made of elements. A molecule is the smallest part of a substance that has all the chemical properties of that substance. Most compounds are made of molecules. (2-1)
13. An element is a substance made of microscopic particles called atoms. (2-1)
14. Chemical symbols are abbreviations for elements. (2-2)
15. Evaporation and crystallization are both processes that result in the formation of mineral crystals. (2-7)

Chapter 2 Challenges

Chapter Summary

Lessons 2-1 and 2-2

- Matter is made up of **atoms.** An **element** has only one kind of atom. Elements can be solids, liquids, or gases.
- A **compound** is two or more elements chemically combined. A **molecule** is the smallest part of a compound showing the compound's properties.
- **Chemical symbols** stand for element names. **Chemical formulas** stand for compound names.

Lessons 2-3, 2-4, and 2-5

- **Minerals** are solids formed from elements or compounds. Most rock-forming minerals are made up of compounds. Sand contains the mineral quartz.
- **Physical properties** help identify minerals. Color alone cannot be used to identify a mineral. The color of the powder left by a mineral is called a **streak.**
- Minerals have a metallic or nonmetallic **luster.** The property of a mineral to resist being scratched is called **hardness.**
- **Density** is the amount of matter in a given volume. Every mineral has its own density. Some minerals show **magnetism.** An **acid test** can identify minerals that contain the compound calcium carbonate.

Lessons 2-6 and 2-7

- Minerals are made up of **crystals.** Crystals form in one of six basic crystal systems.
- **Cleavage** is the splitting of minerals into pieces with smooth, flat surfaces. There are different kinds of cleavage. **Fracture** is the splitting of minerals into jagged or uneven pieces.
- **Crystallization** and **evaporation** are some of the processes that form minerals.

Lesson 2-8

- Minerals have many uses. The human body uses minerals. A mineral that contains a useful metal or nonmetal is called an ore.
- Gemstones are rare, beautiful, and long-lasting minerals. Cut and polished **gems** are used in jewelry and art.

Key Term Challenges

acid test (p. 50)	gem (p. 58)
atom (p. 42)	hardness (p. 48)
chemical formula (p. 44)	luster (p. 48)
chemical symbol (p. 44)	magnetism (p. 50)
cleavage (p. 52)	mineral (p. 46)
compound (p. 42)	molecule (p. 42)
crystal (p. 52)	ore (p. 58)
crystallization (p. 54)	physical property (p. 48)
density (p. 50)	precipitation (p. 54)
element (p. 42)	solution (p. 54)
evaporation (p. 54)	streak (p. 48)
fracture (p. 52)	subscript (p. 44)

MATCHING **Write the Key Term from above that best matches each description.**

1. characteristic used to describe objects
2. solid material formed from elements and compounds in Earth's crust
3. the way a mineral reflects light
4. property of a mineral to resist being scratched
5. solid substance with a definite pattern in its atoms
6. amount of mass in a given volume
7. natural force that few minerals show
8. color of a mineral's powder

IDENTIFYING WORD RELATIONSHIPS **Explain how the words in each pair are related. Write your answers in complete sentences.**

9. gem, ore
10. streak, color
11. cleavage, fracture
12. compound, molecule
13. atom, element
14. chemical symbol, element
15. evaporation, crystallization

62

Teaching Options

PREPARING STUDENTS FOR STANDARDIZED TESTS

Reading Strategy: Tell students that it is important to read every possible answer choice in a test. Some choices may be partly correct, but only one response is completely correct.

Writing Strategy: Stress the importance of making a brief outline of the topics to be covered when composing a paragraph or an essay. The outline will aid in generating an organized, concise answer.

Interpreting Visuals Strategy: Remind students to always study a chart carefully before answering the questions that pertain to it. Then, challenge students to answer the questions without referring back to the chart.

ESL/ELL STRATEGY

Word Bank Have students create a word bank of Key Terms from this list. Suggest that students classify related words into groups. This can be done by scanning the lessons to see which terms were presented in each lesson.

Content Challenges TEST PREP

MULTIPLE CHOICE Write the letter of the term or phrase that best completes each statement.

1. The smallest part of a compound that has all the properties of that substance is
   a. an atom.
   b. a molecule.
   c. an element.
   d. a formula.

2. CO_2 is
   a. an atom.
   b. a chemical symbol.
   c. a chemical formula.
   d. an element.

3. All materials in Earth's crust are made up of
   a. elements and compounds.
   b. elements and minerals.
   c. silicon and oxygen.
   d. crystals.

4. The softest mineral on Mohs' hardness scale is
   a. quartz.
   b. diamond.
   c. feldspar.
   d. talc.

5. The luster of pyrite can be described as
   a. metallic.
   b. glassy.
   c. dull.
   d. nonmetallic.

6. An acid test is used to test minerals for
   a. quartz.
   b. calcium carbonate.
   c. crystals.
   d. calcium sulfate.

7. The formula for density is
   a. Density = mass ÷ volume.
   b. Density = volume ÷ mass.
   c. Density = volume × mass.
   d. Density = mass × weight.

8. Scientists study the crystal structure of minerals using
   a. a hand lens.
   b. X-rays.
   c. the acid test.
   d. luster.

9. A mineral that breaks into thin, flat sheets is
   a. feldspar.
   b. mica.
   c. asbestos.
   d. quartz.

10. Bodies of water often contain minerals that are
   a. dissolved.
   b. precipitated.
   c. supercooled.
   d. solutions.

TRUE/FALSE Write *true* if the statement is true. If the statement is false, change the underlined term to make the statement true.

11. Most elements are liquids at room temperature.
12. A chemical formula stands for one atom of an element.
13. Mohs' scale is used to compare the luster of minerals.
14. There are six basic crystal systems.
15. A mineral with cleavage breaks into pieces with uneven edges.
16. A substance that is not formed from living things or the remains of living things is inorganic.
17. On Mohs' hardness scale, a mineral with a high number cannot scratch a mineral with a lower number.
18. One molecule of CO_2 has one atom of carbon and two atoms of oxygen.

Content Challenges

MULTIPLE CHOICE

1. b (2-1)
2. c (2-2)
3. a (2-3)
4. d (2-4)
5. a (2-4)
6. b (2-5)
7. a (2-5)
8. b (2-6)
9. b (2-6)
10. a (2-7)

TRUE/FALSE

11. false; solids (2-1)
12. false; symbol (2-2)
13. false; hardness (2-4)
14. true (2-6)
15. false; fracture (2-6)
16. true (2-3)
17. false; can (2-4)
18. true (2-2)

ALTERNATIVE ASSESSMENT

Relating Information Have students take the boldface terms from a single lesson and write a short paragraph that relates the terms. **learning style: linguistic**

PORTFOLIO ASSESSMENT

Making Student Portfolios Students can demonstrate their comprehension of the concepts in this chapter by making a portfolio. Have students make suggestions as to what they would like to include. The Chapter Self-Check, the Big Idea Planner, and the Weekly Journal are some of the reproducibles from the Teacher's Resources that they can include in their portfolios. You can use the Portfolio Assessment Rubric also found in the Teacher's Resources to assess students' portfolios.

Concept Challenges

WRITTEN RESPONSE

1. **ANALYZE:** Different minerals can have some of the same properties, such as color or hardness. Thus, it is important to test as many properties as possible to correctly identify the mineral. (2-4, 2-5, 2-6)
2. **APPLY:** Accept all justifiable answers. If cost is a consideration, 12K or 14K gold is cheaper than 18K gold. If a rich color is desired, 18K gold is the best choice. (2-5)
3. **INFER:** The acid test only works with minerals that contain calcium carbonate. (2-5)
4. **IDENTIFY:** Quartz, topaz, corundum, and diamond will scratch glass because they all have a hardness greater than 6. (2-4)
5. **EXPLAIN:** Magnets are attracted to magnetite. Placing a magnet next to the unknown mineral will determine whether or not the mineral is attracted to the magnet and is magnetite. (2-5)

INTERPRETING VISUALS

6. gold, silver, and diamond (2-3)
7. iron (Fe) (2-2)
8. pyrite, gold, silver, and quartz (2-6)
9. pyrite, magnetite, hematite, gold, silver, quartz, and corundum (2-4)
10. corundum and diamond (2-4)
11. testing both for magnetic attraction or testing the streak of both minerals (2-4, 2-5)
12. silver, gold, magnetite, hematite, pyrite, quartz, corundum, and diamond (2-4)
13. Both have a colorless streak. Testing for hardness would distinguish the two minerals. (2-4)
14. Corundum ranks 9 on Mohs' scale and can scratch any mineral with a smaller value, including hematite. (2-4)
15. magnetite, hematite, quartz, and corundum (2-2)
16. Magnetite has one more atom of iron per molecule than hematite does and one more atom of oxygen. (2-2)

Chapter Project Wrap-Up

MINERAL JEOPARDY

You can use the Group Activity Rubric found in the Teacher's Resources to assess students' projects. Fill in the rubric with the additional information below. For this project, students should have:

- determined appropriate categories for their answers.
- prepared questions for each category.

Concept Challenges TEST PREP

WRITTEN RESPONSE Answer each of the following questions in complete sentences.

1. **ANALYZE:** Why do you think it is important to test for more than one physical property of a mineral to identify it?
2. **APPLY:** If you were going to buy gold jewelry, what number carat of gold would you choose? Why?
3. **INFER:** Why would the acid test not be useful to identify a diamond?
4. **IDENTIFY:** Glass has a hardness of about 6. What minerals on Mohs' scale will scratch glass?
5. **EXPLAIN:** How would you test an unknown mineral to see if it was magnetite?

INTERPRETING VISUALS Use Figure 2-36 to answer the following questions.

6. Which minerals listed in the table are elements?
7. What element does pyrite have in common with magnetite?
8. Which minerals fracture?
9. Which minerals have a metallic luster?
10. What are the two hardest minerals listed?
11. What property would be most useful in distinguishing between black hematite and magnetite?
12. What are the minerals, in order, from softest to hardest?
13. Why might streak not be a useful property for telling apart quartz and diamond?
14. Which substance can scratch the other—corundum or hematite? Why?
15. Which minerals contain oxygen?
16. How does the chemical formula of magnetite differ from the chemical formula of hematite?

MINERAL PROPERTIES

Mineral	Chemical Formula	Color	Streak	Luster	Hardness	Fracture or Cleavage
Pyrite	FeS_2	Yellow	Greenish, brownish black	Metallic	6–6.5	Fracture
Magnetite	Fe_3O_4	Black	Black	Metallic	5–6	Fracture
Hematite	Fe_2O_3	Reddish brown to black	Light to dark red	Metallic	5.5–6.5	Fracture
Gold	Au	Gold	Yellow	Metallic	2.5–3	Fracture
Silver	Ag	Silver white	Silver to light gray	Metallic	2.5	Fracture
Diamond	C	Colorless, pale yellow, black	Colorless	Nonmetallic	10	Cleavage
Quartz	SiO_2	Colorless, purple, yellow, pink	Colorless	Nonmetallic	7	Fracture
Corundum	Al_2O_3	Brown, red, blue	White	Nonmetallic	9	Fracture

▲ Figure 2-36

64

Teaching Options

• *TEACHING RESOURCES* •
Online or Print

Teacher's Resources

Key Term Review: Ch. 2, p. 14

Chapter 2 Test: pp. 15–17

Group Activity Rubric: Teaching Tools, p. 6

Chapter Self-Check: Teaching Tools, p. 11

Big Idea Planner: Study Tools, pp. 11–12

Weekly Journal: Study Tools, p. 19

Portfolio Assessment Rubric: Teaching Tools, p. 10

PLANNING GUIDE

◆ **TEACHING THE CHAPTER** This chapter should take approximately 8–11 days to complete instruction and assessment.

	Skills and Features	Projects and Activities	Achieve Science Literacy	Meet Individual Needs
Chapter 3 Opener p. T65	OBSERVE, EXPLAIN	• Chapter Project		
Lesson 3-1 **What are rocks?** pp. T66–T67 Standard: D1d	IDENTIFY, INFER, CONTRAST • Investigate • Web InfoSearch • People in Science	• Cooperative Learning	• Reading Strategy • Ongoing Assessment	• Reteach Teacher's Resources* Enrichment Activity
Lesson 3-2 **How are igneous rocks formed?** pp. T68–T69 Standards: D1d, D1k	INFER, PREDICT, GRAPH • Web InfoSearch • How Do They Know That?	• Extending the Lesson		• Reteach Teacher's Resources* Enrichment Activity Concept Map
Lesson 3-3 **How are igneous rocks classified?** pp. T70–T71 Standard: D1k	OBSERVE, INFER, CLASSIFY • Building Science Skills • Hands-On Activity	• Extending the Lesson • More on the Web	• Study Hint • Writing Hint	Teacher's Resources* Enrichment Activity Concept Map
Lesson 3-4 **How are sedimentary rocks formed?** pp. T72–T73 Standards: D1d, D2e, D1k	DESCRIBE, IDENTIFY, EXPLAIN • Investigate • Designing an Experiment • Integrating Life Science	• Cooperative Learning • Extending the Lesson • More on the Web	• Study Hint	• Reteach Teacher's Resources* Designing an Experiment
Lesson 3-5 **How are sedimentary rocks classified?** pp. T74–T75 Standard: D1k	IDENTIFY, DESCRIBE, CLASSIFY • Web InfoSearch • Hands-On Activity	• Integrating Chemistry	• Tips for Using Technology • Study Hint • Reading Strategy	Teacher's Resources* Enrichment Activity Concept Map Flowchart
Lesson 3-6 **How are metamorphic rocks formed?** pp. T76–T77 Standards: D1d, D1k	EXPLAIN, DESCRIBE, PREDICT • Web InfoSearch • Integrating Physical Science	• Extending the Lesson	• Study Hint • Reading Strategy • Ongoing Assessment	• ESL/ELL Strategy

*www.conceptsandchallenges.com, password: ccearth09

Planning Guide continues on next page.

Standards: For details on the correlation to National Science Standards see pages *xx–xxi*.

	Skills and Features	Projects and Activities	Achieve Science Literacy	Meet Individual Needs
Lesson 3-7 How are metamorphic rocks classified? pp. T78–T79 Standard: D1k	CLASSIFY, COMPARE, RESEARCH • Building Science Skills • Real-Life Science	• Cooperative Learning • Extending the Lesson • Lab Challenge	• Study Hint • Reading Strategy	Teacher's Resources* Enrichment Activity
Lesson 3-8 What is the rock cycle? pp. T80–T81 Standard: D1d	INFER, APPLY, EXPLAIN • Hands-On Activity	• Cooperative Learning • Extending the Lesson • Active Art	• Study Hint	• Reteach Teacher's Resources* Rock Cycle
Lab Activity Modeling What Lies Beneath Earth's Surface pp. T82–T83 Standards: D1c, D1d	OBSERVE, INFER • Set-Up Time: 20 min • Lab Time: 45 min	• Lab Skills Worksheet • Extend the Lab Activity	• Tips for Using Technology	• ESL/ELL Strategy Teacher's Resources* Lab Activity Report
Big Idea What drives the rock cycle? pp. T84–T85 Standard: D1d	RESEARCH, COMMUNICATE, ANALYZE • Science Log: Writing Activity	• Big Idea • Cooperative Learning • Close Activity	• Writing Hint	Teacher's Resources* Big Idea Planner
Chapter 3 Challenges pp. T86–T88	• Chapter Summary • Key Term Challenges • Content Challenges • Concept Challenges	• Chapter Project Wrap-Up	• Study Hint • Preparing Students for Standardized Tests	• ESL/ELL Strategy Teacher's Resources* Chapter Self-Check Weekly Journal

*www.conceptsandchallenges.com, password: ccearth09

Teacher Notes

Chapter 3 Rocks and Their Origins

▲ **Figure 3-1** The Grand Canyon in northwestern Arizona

The Grand Canyon was cut into rock by the Colorado River. It is more than 1.6 km deep in places and from 6 to 29 km wide. The sides of the canyon are mostly layers of rock. The top has many tall peaks and other interesting landforms. As you can see, much of the canyon is red. However, each rock layer has slightly different coloring—beige and gray; green and pink; brown, silver-gray, and violet.

▶The rocks of the Grand Canyon show much about the history of the area. Why do you think the rock layers are different colors?

Contents

Teaching Options

ASSESSMENT PLANNER

For assessment in the Student Edition, see the following pages:

Content Assessment
Checking Concepts: pp. 67, 69, 71, 73, 75, 77, 79, 81
Thinking Critically: pp. 67, 69, 71, 73, 75, 77, 79, 81
Interpreting Visuals: p. 81
Chapter Challenges: pp. 86–88

Alternative Assessment
Building Skills: pp. 71, 79
Web InfoSearch: pp. 67, 69, 75, 77
Science Log: p. 85

Performance-Based Assessment
Hands-On Activity: pp. 71, 75, 81
Designing an Experiment: p. 73
Lab Activity: pp. 82–83

• TEACHING RESOURCES • Online or Print

Teacher's Resources
Lesson Review: Ch. 3, pp. 2, 3, 5, 6, 7, 8, 9, 11
Enrichment: Ch. 3, pp. 4, 10
Key Term Review: Ch. 3, p. 13
Chapter 3 Test: pp. 14–15
Laboratory Manual: pp. 15–18, 33–36

Chapter 3 Rocks and Their Origins

Chapter Overview

In this chapter, students will learn about the three main types of rocks: igneous, sedimentary, and metamorphic. In addition, they will learn how to classify rocks. Finally, students will explore the rock cycle.

About Figure 3-1 Most of the rock in the Grand Canyon is sedimentary rock. The current top, or youngest, layer of the Grand Canyon is the Kaibab limestone, which is about 250 million years old. The Kaibab limestone is composed mostly of sandy limestone with some sandstone and shale. This layer probably was formed near the coast of a shallow sea. The intrusion of sandstone and shale indicates that at times the layer was above the surface of the water but near its edge.

Answer The different colors are caused by the different minerals and sediments that make up the rocks. The rock layers were laid down at different times, and at those times, the mineral composition must have varied.

Linking Prior Knowledge

For this chapter, students should recall:

- the layers of Earth and the composition of the mantle (Lesson 1-3).
- the definition of minerals (Lesson 2-3).
- what crystal are (Lesson 2-6).
- how minerals are formed (Lesson 2-7).

Chapter Project

ROCK TESTS

MATERIALS: Materials will vary with students' tests.

Ask students to make up two tests to check the characteristics of three different rocks of the same type, for example, three different sedimentary rocks (limestone, sandstone, conglomerate). Students should explain what each test is and how it tests for the characteristic. Some ideas might include hitting each rock with a hammer, shaking it in water, and putting acid on it. Students can also use the Lab Activity Report from the Teacher's Resources.
learning style: tactile

3-1 *(pp. 66–67)*

1 Introduce

INVESTIGATE

TIME: 10–15 minutes

PURPOSE: Students will observe some characteristics of rocks.

MATERIALS: sample rocks, penny, hand lens, drawing paper, triple-beam balance scale, ruler, colored pencils

PROCEDURE: Students should use colored pencils to make realistic drawings of the rocks. Have students check rock weights with a balance scale.

THINK ABOUT IT: Most rocks contain mixtures of minerals. The color of the rocks, sizes of the mineral grains, and textures of each rock should be more apparent through the hand lens.

Linking Prior Knowledge Ask students to recall what the layers of Earth are (Lesson 1-3). (crust, mantle, core) Ask: *Which is the solid layer that contains rocks?* (crust)

2 Teach

Discussion Display examples of an igneous rock, a sedimentary rock, and a metamorphic rock. Tell students that petrologists classify rocks. Write the three main types of rocks on the board: igneous, sedimentary, metamorphic. Pass around the sample of each type of rock and ask students to describe it. Ask the last person holding the rock to stand next to its proper classification on the board. Tell students that stones are fragments of rocks.
learning styles: visual, tactile

Reading Strategy To help students remember that petrologists are scientists who study rocks, have them find the meaning of the prefix *petro-* in a dictionary. (rock or stone)

Answers to questions

1. **IDENTIFY:** Rocks are made up of one or more minerals.
2. **DEFINE:** the grouping of objects that have common features
3. **IDENTIFY:** igneous, sedimentary, and metamorphic

3-1 What are rocks?

INVESTIGATE

Observing Rocks
HANDS-ON ACTIVITY

1. Collect some samples of rocks. Draw pictures of them. Describe them in as much detail as you can.
2. Try to scratch all the rocks with a penny. What happens?
3. Use a ruler to measure the rocks and a triple-beam balance to find their mass.
4. Look for different materials within each rock.
5. Look at the rocks through a hand lens. Write down what you see.

THINK ABOUT IT: Do any of the rocks contain mixtures of materials? How do the rocks look different through the hand lens?

Objective
Identify and describe the three classes of rocks.

Key Terms
rock: mixture of minerals, generally cemented together

igneous (IHG-nee-uhs) **rock:** rock formed by the crystallization of hot melted rocks or minerals

sedimentary (sehd-uh-MEHN-tuh-ree) **rock:** rock formed from pieces of other rocks that are cemented together

metamorphic (meht-uh-MAWR-fihk) **rock:** rock formed when existing rocks are changed by heat and pressure

Rocks in Earth's Crust Earth's crust is made up of many kinds of rocks. **Rocks** are a mixture of minerals usually cemented together.

Small rocks found lying on the ground are fragments of larger rocks. Other rocks are so big that they are part of the landscape, as in Monument Valley in Arizona.

Petrologists (pet-RAH-low-jists) are Earth science specialists who study rocks and minerals and their origins. They also study rock composition and how rocks change. They can work in deserts, ocean or tropical environments, or arctic areas. They may do studies in areas affected by erosion, earthquakes, or volcanoes.

▲ **Figure 3-2** Monument Valley in northeastern Arizona

All rocks are made up of one or more minerals. There are more than 2,000 different minerals in Earth's crust. However, only a few of these minerals are found in most rocks.

1 **IDENTIFY:** What are rocks made up of?

Classification of Rocks The grouping of objects that have certain features in common is called classification. Scientists often classify things to make them easier to study. Biologists classify things as living or nonliving. Chemists classify elements as metals or nonmetals. Petrologists classify rocks and minerals found in Earth's crust. They also classify the types of structures rocks form.

2 **DEFINE:** What is classification?

66

Teaching Options

ONGOING ASSESSMENT

Have students prepare a list of questions that they would send to a geologist. Then, arrange for students to write letters containing their questions and send them to a geologist at a government agency, such as the U.S. Geological Survey (USGS). Alternatively, students could use library or Internet sources to find the answers.
learning styles: auditory, linguistic

COOPERATIVE LEARNING

Classifying Items Divide students into pairs. Have one student in each pair place items into a large box. Then, remove all the items and ask the other student in the pair to arrange the items in a different order. Arrangements may include those done by color, size, or use, or alphabetically by name. List each type of arrangement on the board. Point out that the arrangements are different methods of classification. Tell students that rocks can also have different classifications.
learning style: tactile

Classes of Rocks Petrologists classify rocks by the way they form. Some rocks form from melted minerals that cool and harden. These rocks are classified as **igneous rocks.** Rocks that form when small pieces of minerals and rocks, or the remains of living things, become cemented or compacted together are classified as **sedimentary rocks.** Other rocks form from existing rocks, which are slowly changed by heat and pressure. These rocks are classified as **metamorphic rocks.**

IDENTIFY: What are the three classes of rocks?

CHECKING CONCEPTS

1. Rocks are made up of __________.
2. There are more than __________ different minerals in Earth's crust.
3. There are __________ classes of rocks.
4. Rocks are classified as igneous, __________, or metamorphic.
5. When melted minerals cool, __________ rocks form.
6. Scientists who study rocks and minerals are called __________.

THINKING CRITICALLY

7. INFER: How would you group rocks to make identifying them easier?
8. CONTRAST: How do sedimentary rocks differ from metamorphic rocks?

Web InfoSearch

Stone Monuments A monument is a structure built to honor a person or an event. Many monuments are made of stone, or rock. Different kinds of stone are used for different kinds of monuments. Often, the color, texture, and grain size influence the choice of stone.

SEARCH: Use the Internet to find out about famous stone monuments in the United States. What kinds of stone are they most often made of? Why did the builders choose those kinds of stone? Start your search at www.conceptsandchallenges.com. Some key search words are **national monument, landmarks,** and **statues.**

People in Science

GEOLOGIST

Geologists do many different kinds of work. Some help build roads, dams, and pipelines. Some search for fossil-fuel deposits. Others explore areas for valuable minerals. A petrologist is one type of geologist.

Dr. Michael W. Howell is a marine geologist. He studies deep-sea sediments to learn more about changes in the ocean over time. He analyzes how these changes affected past climates. Dr. Howell also teaches geology at the University of South Carolina. In 2001, he traveled to the Marion Plateau, off the coast of Australia. He was part of an ocean-drilling expedition run by the Ocean Drilling Program.

▲ **Figure 3-3** A marine geologist studies deep-sea sediments.

A geologist can work in industry, for the government, or as a college teacher. To become a geologist, you must complete at least four years of college. Some geologists also pursue postgraduate degrees or degrees from technical institutes.

Thinking Critically Why would a geologist want to drill into the ocean floor?

3 Assess

CHECKING CONCEPTS ANSWERS

1. minerals
2. 2,000
3. three
4. sedimentary
5. igneous
6. petrologists or geologists

THINKING CRITICALLY ANSWERS

7. INFER: Possible answer: You could classify rocks by color, texture, chemical composition, or size of mineral grains.
8. CONTRAST: Sedimentary rocks form when pieces of minerals and rocks or the remains of living things become cemented or compacted together. Metamorphic rocks are formed from existing rocks that were slowly changed by heat and pressure.

4 Close

Web InfoSearch

Stone Monuments Have students use the Internet to find out about famous stone monuments in the United States. What kinds of stone are they most often made of? Why did the builders choose those kinds? Have students guess how the type of stone chosen is related to its use. Some key search words are **national monument, landmarks,** and **statues.** Have students begin their search at www.conceptsandchallenges.com.

RETEACH

Understanding Classifying

Remind students that classification is the grouping of things that are alike. Describe how some things are classified. You may wish to use the departments in large stores or the aisles in supermarkets as examples. Next, have students name some things that they classify, such as CDs or foods. Tell students that they will learn more about different types of rocks and how they are classified in the following lessons.

learning styles: tactile, visual

• TEACHING RESOURCES •
Online or Print

Teacher's Resources

Lesson Review: Chapter 3, p. 2

Enrichment: Chapter 3, p. 4

People in Science

GEOLOGIST

Discussion Ask: *What do geologists do?* List students' responses on the board.

Real-Life Connection Ask students if there is anything in their home that is made of rock. Show students photographs of kitchen counters made of granite. Tell them that granite is a very strong igneous rock and therefore a good building material. Help students to understand that rocks are valuable natural resources.

Thinking Critically Answer

Possible answers: to study deep-sea sediments or to learn how the sediments have changed over time

3-2 *(pp. 68–69)*

1 *Introduce*

Linking Prior Knowledge Ask students how minerals are formed (Lesson 2-7). Remind them that rocks are made of combinations of minerals.

2 *Teach*

Demonstration Carefully cut in half a 10-cm plastic-foam ball. Use different-colored markers to outline the layers of Earth on the flat surface of one-half of the ball. Have students identify each layer. **learning style: visual**

Discussion Earth is composed mostly of igneous rock with a very thin layer of weathered material. The most abundant minerals in igneous rock belong to a group called silicates, which are compounds that contain the elements silicon and oxygen. Granite, which is silica-rich, makes up most of the continental crust. It is a strong, hard rock that can be highly polished. For this reason, it is often used for buildings and monuments. Today, people use granite for kitchen countertops.

Refer students to Figure 3-4 and have them identify the temperature of Earth at each layer. Point out that the temperature between the crust and the mantle is high enough to melt many minerals to form igneous rocks such as granite.

Answers to questions

1 **INFER:** because they are formed from hot, molten rock

2 **DEFINE:** molten rock inside Earth

3 **DESCRIBE:** Magma rises through cracks in rocks.

Reinforcement Most likely, students have studied or seen pictures of volcanoes. Help students make the connection between volcanoes and lava.

3-2 How are igneous rocks formed?

Objective

Identify two ways that igneous rocks are formed.

Key Terms

molten (MOHL-tuhn) **rock:** melted minerals

magma (MAG-muh)**:** molten rock inside Earth

pluton (PLOO-tahn)**:** large body of igneous rock that can form into different shapes when magma cools inside Earth's crust

lava (LAH-vuh)**:** magma that reaches Earth's surface

Heat Inside Earth If you dig about 300 m beneath Earth's surface, the temperature increases about 10°C. As you dig deeper, the temperature gets hotter. The deepest oil wells are drilled about 6 km into Earth's crust. Here, the temperature is more than 60°C. At about 60 km below Earth's surface, in the mantle, it is more than 1,000°C. At about 3,000 km, into the outer core, it is more than 2,200°C. The center of Earth may be as hot as 5,000°C.

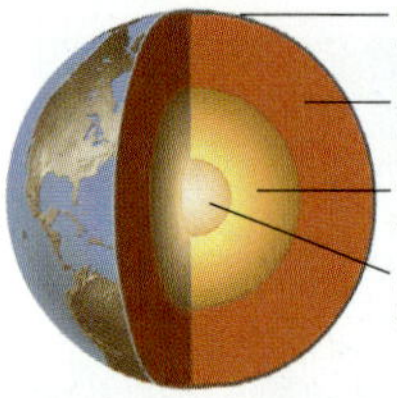

▲ **Figure 3-4** Temperatures inside Earth

The temperature of the upper mantle is hot enough to melt minerals and form **molten rock.** When the molten rock cools, it becomes solid, forming igneous rocks.

▲ **Figure 3-6** Basalt is from molten rock that cooled quickly.

1 **INFER:** Igneous means "formed by fire." Why are rocks formed from magma called igneous?

Magma and Igneous Rocks Molten rock inside Earth is called **magma.** There are large pools of magma inside Earth's upper mantle and lower crust. Sometimes, magma rises through cracks in rocks into the upper part of Earth's crust. Here, the temperature is cooler than the temperature inside the mantle. As magma rises through cracks in Earth's crust, it cools and crystallizes. As it hardens, igneous rock is formed. Igneous rock can take many thousands of years to form from magma.

▼ **Figure 3-5** Large bodies of igneous rock can form above or below ground.

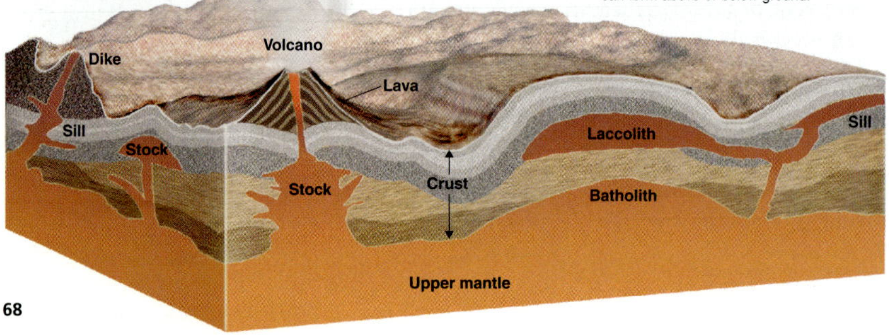

68

Teaching Options

! FAST FACTS *Rocks in Oceanic Crust*

Basalt and gabbro, each about 50 percent silica, are the primary rocks of oceanic crust. Gabbro forms from magma that is high in iron and magnesium. Gabbro is not common in most places, but large formations can be found in Finland. Basalt forms when the same magma that forms gabbro reaches Earth's surface. Much of the bedrock of Washington and Oregon is basalt that formed when ancient lava flowed from fissures in Earth's crust. The black sand beaches of Hawaii are made of eroded basalt.

QUICK CONVERSIONS		
	Metric/SI	**English**
Crust	300 m/10°C	328 yd/50°F
Mantle	60 km/1,000°C	37 mi/1,832°F
Outer core	3,000 km/4,100°C	1,864 mi/7,412°F

Large bodies of igneous rock with different shapes are formed when magma cools inside Earth's crust. These bodies are called batholiths, stocks, laccoliths, dikes, and sills. A general name for all of these bodies of igneous rock is **pluton.** An example of each kind of pluton is shown in Figure 3-5.

2 **DEFINE:** What is magma?

Lava and Igneous Rocks Sometimes magma rises through cracks in rocks and reaches Earth's surface. Magma that reaches Earth's surface is called **lava.** Lava cools upon contact with air or water. Cooling makes "red hot" lava harden into igneous rock. Igneous rocks form quickly from lava.

3 **DESCRIBE:** How does lava rise to Earth's surface?

CHECKING CONCEPTS

1. Igneous rocks formed from __________ form quickly.
2. Magma that reaches Earth's surface is called __________.
3. Melted minerals form __________ rock.
4. Large pools of __________ are found deep inside Earth.

THINKING CRITICALLY

5. **PREDICT:** If you were to dig a hole 2 m deep into Earth's surface, would you be likely to find molten rock? Why or why not?
6. **GRAPH:** Make a bar graph comparing the temperatures of the crust, the mantle, the outer core, and the inner core.

Web InfoSearch

Plutons Plutons form inside Earth. However, you can see plutons on Earth's surface in some areas of the United States.

SEARCH: Use the Internet to find out more about plutons. Where can you see them on Earth's surface? How are they exposed? Start your search at www.conceptsandchallenges.com. Some key search words are **plutons, igneous rock,** and **volcano.**

How Do They Know That?

THE ORIGIN OF ROCKS

James Hutton (1726–1797) was born in Edinburgh, Scotland. He was trained to be a doctor. However, after Hutton inherited a farm from his father, he became interested in the natural forces at work on Earth's surface. At the time, scientists thought that all rocks were sedimentary rocks. Hutton had a different idea. He said that many rocks, such as granite, were once molten, or melted, rock. Hutton also showed that mountains could be formed by the upward push of igneous rocks. Hutton published his ideas in a book called *Theory of the Earth* in 1795.

Based on his studies of Earth, Hutton proposed a theory about the history of Earth. Hutton's theory said that Earth was shaped in the past by the same forces, such as erosion and sedimentation, that were still at work. Hutton's theory was the beginning of the modern science of geology. Hutton is sometimes called the father of geology.

▲ **Figure 3-7** James Hutton, the father of geology

Thinking Critically Why do you think scientists once thought that all rocks were sedimentary?

3 Assess

CHECKING CONCEPTS ANSWERS

1. lava
2. lava
3. molten
4. magma

THINKING CRITICALLY ANSWERS

5. **PREDICT:** No, because the temperature at 2 m deep is not hot enough to melt rocks.
6. **GRAPH:**

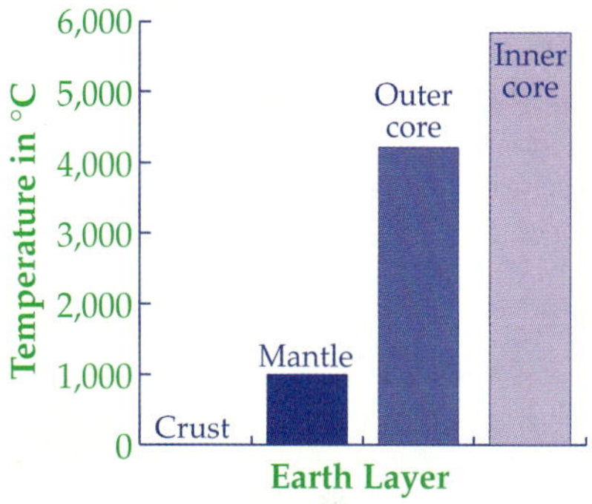

4 Close

Web InfoSearch

Plutons Have students use the Internet to locate plutons that are exposed on Earth's surface and determine how they got there. Some key search words are **plutons, igneous rock,** and **volcano.** Have students begin their search at www.conceptsandchallenges.com.

How Do They Know That?

THE ORIGIN OF ROCKS

Technology Connection A well can be drilled into a bed of cracked rocks. Water can then be pumped into the well and heated. The resulting steam can be used to generate electricity. Heated rocks are found everywhere beneath the surface. However, the wells must sometimes be drilled very deep and thus can be expensive.

Discussion Explain that James Hutton's theory is called *uniformitarianism.* Ask students to brainstorm other forces that shape Earth. (volcanoes and earthquakes)

Thinking Critically Answer They could directly observe the formation of sedimentary rocks.

RETEACH

Clarifying Concepts Some students may be confused by the relationships among molten rock, magma, lava, and igneous rock. Have students use the Concept Map from the Teacher's Resources to show the relationships.
learning style: tactile

EXTENDING THE LESSON

Evolution and Geology Explain that Charles Darwin's theory of evolution was greatly influenced by the ideas of geologists such as James Hutton. Darwin realized that to survive, organisms would have to adapt over time to the vast changes on Earth's surface.

• TEACHING RESOURCES • Online or Print

Teacher's Resources

Lesson Review: Chapter 3, p. 3

Enrichment: Chapter 3, p. 4

Concept Map: Study Tools, p. 4

3-3 (pp. 70–71)

1 Introduce

STUDY HINT Before beginning this lesson, have students scan the lesson for the science process skill words. Have them identify each skill used in the lesson and then review the definition for each skill in the Skills Handbook at the front of the Student Edition.

Linking Prior Knowledge Remind students that a mineral is a natural solid formed from elements and compounds in Earth's crust (Lesson 2-3).

2 Teach

Demonstration Hand out samples of granite, quartz, feldspar, and mica. Have students examine the samples, using a hand lens. Tell students to identify the minerals that make up the granite. (quartz, feldspar, and mica) Then, give students samples of obsidian and rhyolite to examine with the granite. Tell students to note the glassy texture of obsidian, the coarse texture of granite, and the fine texture of rhyolite. **learning style: visual**

Discussion Refer students to Figure 3-8. Tell them that the table lists the six minerals most commonly found in igneous rocks. Read aloud the names of the six minerals so that students can hear the pronunciations. Explain that igneous rocks are distinguished by their minerals.
learning styles: visual, auditory

Answers to questions

1. **OBSERVE:** by their size, shape, and color
2. **INFER:** The faster the cooling rate, the smaller is the crystal.
3. **CLASSIFY:** Granite has a coarse texture, obsidian has a glassy texture, and basalt and rhyolite have a fine texture.

Reinforcement Provide students with copies of the Concept Map from the Teacher's Resources. Have them write the main idea *Igneous rocks have mineral crystals of varying sizes* in the center box. Help them fill in the surrounding boxes with information about how crystals of different sizes form.
learning style: visual

3-3 How are igneous rocks classified?

Objective
Identify and describe igneous rocks by their minerals and textures.

Key Term
texture (TEHKS-chuhr)**:** size of the crystals in an igneous rock

A Combination of Minerals Igneous rocks are made up of many different kinds of minerals. However, there are only six minerals that are commonly found in igneous rocks. These six common minerals are listed in Figure 3-8.

MINERALS IN IGNEOUS ROCKS		
Quartz	Mica	Amphibole
Feldspar	Olivine	Pyroxene

◀ **Figure 3-8**

Often, igneous rocks can be identified by their minerals. Granite is an igneous rock mostly made up of the minerals known as quartz, feldspar, and mica. Look at the photograph of granite in Figure 3-9. See if you can distinguish the different minerals that make up granite.

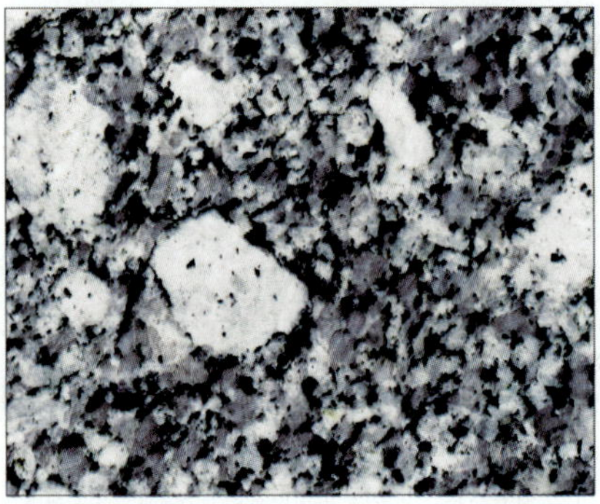

▲ **Figure 3-9** Granite is an igneous rock.

1 **OBSERVE:** How do the quartz and feldspar crystals in granite differ?

Crystal Size Igneous rocks have mineral crystals of varying sizes. Crystal size depends on the amount of time it takes the molten rock to cool. The largest crystals form when the rock cools very slowly. Igneous rocks formed from magma have large mineral crystals.

Small crystals take less time to form. Most igneous rocks formed from lava have very small crystals.

Sometimes lava cools so quickly that there is no time for crystals to form. For this reason, a few igneous rocks do not have any crystals. Obsidian is an example of an igneous rock that does not have any crystals.

▲ **Figure 3-10** Obsidian looks like glass because it rarely contains crystals.

2 **INFER:** How are cooling rate and crystal size related?

Texture The size of the crystals in an igneous rock is used to describe its **texture.** Texture can be used to identify different igneous rocks that are made up of the same minerals.

For example, you can tell if a rock is granite or rhyolite by looking at its texture. Granite has large crystals that you can see and feel. Rhyolite has very small crystals that are hard to see. Basalt is another commonly found rock with small crystals.

Teaching Options

! FAST FACTS Rocks and Crystals

Rock crystal, a variety of quartz, was once thought to be ice frozen so hard that it could not thaw. The word *crystal* is derived from the Greek word *kyros,* meaning "icy cold."

Some igneous rocks are made up of both large and small crystals of the same or different minerals. This type of texture is called *porphyritic.* A porphyritic texture indicates that the rock cooled under erratic conditions. First, the molten rock began cooling slowly, deep beneath Earth, forming large crystals. Then, conditions changed. The magma may have been forced upwards or an eruption may have occurred. This caused the magma to cool rapidly, forming finer-grained crystals.

Igneous rocks are classified according to their textures. Igneous rocks with large crystals have a coarse texture. Igneous rocks with small crystals have a fine texture. Igneous rocks that do not have crystals have a glassy texture.

3 CLASSIFY: Classify granite, obsidian, basalt, and rhyolite according to their textures.

CHECKING CONCEPTS

1. The six most common minerals in igneous rock are __________, feldspar, mica, olivine, amphibole, and pyroxene.
2. Granite is made up of the minerals quartz, __________, and mica.
3. An igneous rock that has very small mineral crystals most likely was formed from __________.
4. The texture of an igneous rock with no crystals is described as __________.
5. The texture of rhyolite would be described as __________.

THINKING CRITICALLY

6. ANALYZE: A rock collector found five igneous rocks. The rocks had mineral crystals with the following sizes: Rock A = 2 mm; Rock B = 7 mm; Rock C = 4.5 mm; Rock D = 10 mm; Rock E = 0.5 mm.
   a. Which rock probably formed from lava?
   b. Which rocks could have formed from magma?
   c. Which rock probably took the longest time to cool?

BUILDING SCIENCE SKILLS

Classifying Scientists often classify igneous rocks according to where they were formed. Igneous rocks that were formed at Earth's surface are called extrusive igneous rocks. Igneous rocks that were formed inside of Earth are called intrusive igneous rocks. Research the roots and prefixes of the terms *intrusive* and *extrusive.* Then, using that information and the information in this lesson, classify granite, obsidian, basalt, and rhyolite as intrusive or extrusive igneous rocks.

Hands-On Activity

CLASSIFYING IGNEOUS ROCKS BY CRYSTAL SIZE

You will need a hand lens; samples of the igneous rocks diorite, gabbro, and basalt; a sheet of paper; and a pencil.

1. Copy the table in Figure 3-11 onto a sheet of paper.
2. Carefully examine each of your rock samples with the hand lens.
3. Record your observations in your table.

Classifying Igneous Rocks

Rock	Crystal Size	Texture
Diorite		
Gabbro		
Basalt		

▲ Figure 3-11

Practicing Your Skills

4. INFER: Which of these igneous rocks formed deep inside Earth? Explain your answer.
5. INFER: Which of these rocks formed on Earth's surface? Explain your answer.
6. CLASSIFY: Using information from this lesson, add granite, obsidian, and rhyolite to your table.

▲ **STEP 2** Carefully examine the rock sample with a hand lens.

WRITING HINT Classifying almost always involves some comparing and contrasting, which are important science skills. Tell students that when they are asked to classify objects or other things, as they are in Building Science Skills, they should first prepare a list of similarities and differences.

3 Assess

CHECKING CONCEPTS ANSWERS

1. quartz
2. feldspar
3. lava
4. glassy
5. fine

THINKING CRITICALLY ANSWERS

6. ANALYZE:
   a. Rock E
   b. Rocks A, B, C, D
   c. Rock D

4 Close

BUILDING SCIENCE SKILLS

Classifying *Intrusive* comes from the Latin word *intrudere,* meaning "to thrust in." *Extrusive* comes from the Latin word *extrudere,* meaning "to thrust out." Granite is an intrusive igneous rock. Obsidian, basalt, and rhyolite are extrusive igneous rocks. **learning style: linguistic**

Hands-On Activity

CLASSIFYING IGNEOUS ROCKS BY CRYSTAL SIZE

TIME: 10–15 minutes

PURPOSE: observing, inferring, classifying

COOPERATIVE LEARNING: Have students working in pairs take turns, first as observers and then as recorders of the observations.

Practicing Your Skills Answers

4. INFER: Gabbro and diorite formed deep inside Earth. They have large crystals and coarse textures.
5. INFER: Basalt formed on Earth's surface. It has small crystals and a fine texture.
6. CLASSIFY: Check students' tables. Granite has large crystals and a coarse texture. Obsidian has no crystals and a glassy texture. Rhyolite has small crystals and a fine texture.

MORE ON THE WEB

Identifying Igneous Rocks
Have students review the formation and identification of igneous rocks by visiting www.conceptsandchallenges.com.

EXTENDING THE LESSON

Seeing Crystals Show students a polarized-light photomicrograph of a thin section of rock. Have them research how and why thin sections are made. (Geologists slice off thin sections to better study crystal shapes and sizes.) **learning style: visual**

• TEACHING RESOURCES • Online or Print

Teacher's Resources

Lesson Review: Chapter 3, p. 5

Enrichment: Chapter 3, p. 10

Concept Map: Study Tools, p. 4

3-4 (pp. 72–73)

1 Introduce

INVESTIGATE

TIME: 10–15 minutes

PURPOSE: Students will model the formation of sedimentary rock layers.

MATERIALS: 12-oz clear plastic jar with lid, sand, gravel or pebbles, clay powder, fine soil, water, measuring spoon, stirrer

PROCEDURE: Students should let the jar sit for at least 5 minutes. Have students recheck the mixture to see if layers formed.

THINK ABOUT IT: The water will get clearer over time as gravity causes sediments to settle to the bottom. The denser, heavier materials, such as gravel and sand, will settle to the bottom.

STUDY HINT Remind students to take notes as they read a lesson. Tell them to put the information in their own words so that they will understand it later.

Linking Prior Knowledge Ask students to describe some of the organisms that live in the ocean. Ask what they think happens when these organisms die. (They often sink to the bottom of the ocean.)

2 Teach

Discussion Display a piece of concrete. Ask students if they have ever seen a cement truck pouring concrete. Describe what concrete is made of (sand or gravel in cement) and explain how it is formed. Compare the formation of concrete to the formation of sedimentary rock.

Answers to questions

1 **RESTATE:** rock particles carried and deposited by water, wind, or ice

2 **DESCRIBE:** when water and air are squeezed out from between the sediment layers or when dissolved minerals in the water cement the sediments together

3 **IDENTIFY:** When organisms die, their remains settle to the ocean floor. Over time, calcium carbonate in the shells and skeletons dissolves and comes out of solution to cement the materials together, forming solid rock.

3-4 How are sedimentary rocks formed?

INVESTIGATE
Forming Sedimentary Rock Layers
HANDS-ON ACTIVITY

1. Fill a 12-oz clear plastic jar with water.
2. Measure several spoonfuls each of sand, gravel, clay powder, and fine soil. Put them in the jar.
3. Stir them all together. Observe the color of the mixture.
4. Put a lid on the jar. Let the water sit for a while. Check it again to see the layers that have formed.

THINK ABOUT IT: Do you think the water will remain muddy? Why or why not? How can you tell which materials will end up at the bottom?

◀ **Figure 3-12** Observe the layers that form in your jar.

Objective

Describe two ways that sedimentary rocks are formed.

Key Term

sediment (SEHD-uh-muhnt)**:** rock particles carried and deposited by water, wind, or ice

Sediments What happens when you mix soil and sand in a jar of water? The particles of soil turn to mud and, together with the sand, settle to the bottom of the jar in layers. The sand settles first because it is heavier than the mud. The mud settles on top of the sand. Mud and sand are sediments. **Sediments** are rock particles that are carried and deposited by wind, water, or ice.

1 **RESTATE:** What are sediments?

Natural Cement Many sedimentary rocks form in much the same way as concrete. Concrete is made up of sand, gravel, and cement. Have you ever seen trucks pouring concrete for a building? Inside the trucks, sand, gravel, cement, and water are mixed. A chemical reaction then occurs that causes the sand, gravel, and cement to bind together into concrete. As the water evaporates, the concrete hardens.

How do sedimentary rocks form? Most sedimentary rocks are formed in water. These rocks form from sediments that settle to the bottom of lakes, rivers, or oceans. Over millions of years, the sediments pile up in layers. Some sedimentary rock forms from sediments deposited by wind or glaciers.

The layers of sediment may be hundreds of meters thick. As more sediment is added to the layers, the lower layers of sediment become tightly packed under the pressure and weight of the new layers. The older sediments become solid rock when water and air are squeezed out from between the sediment layers. The sediments may also become solid rock when dissolved minerals in the water cement the sediments together.

2 **DESCRIBE:** When does sediment become solid rock?

Sedimentary Rock From Living Things Some types of sedimentary rocks form from the remains of living things. For example, the shells or skeletons of sea animals contain the compound calcium carbonate. When the organisms die, their remains are left on the bottom of the ocean. Over millions of years, the shells and skeletons build up on the ocean floor. Some of the calcium carbonate dissolves and comes out of solution to cement the minerals together. This forms solid rock.

Teaching Options

RETEACH

Forming Sediment Rub two pieces of sandstone over a sheet of white paper. Ask students to describe what happens. (Grains of sand will collect on the paper.) Tell students that the sand was once sediment that settled to the bottom of a body of water. Over the years, the weight of layers of sediment formed the sandstone.
learning style: visual

EXTENDING THE LESSON

Modeling Sedimentary Rock Formation Have students work in small groups to model the formation of sandstone or conglomerate by pouring white glue over a small pile of sand or pebbles and applying pressure. Allow the models to stand for about 10 minutes. Tell students that the models are similar to sedimentary rocks that formed as dissolved minerals cemented sediments together. Ask students what the glue represents in the models. (dissolved minerals) Explain that the durability of sedimentary rock depends on the cementing minerals.
learning styles: visual, tactile

Coquina limestone is a sedimentary rock formed from shells. Chalk limestone also forms from shells. However, you cannot see the shells in it because they were formed from microscopic organisms.

▲ **Figure 3-13** Coquina (left) and chalk (right)

3 IDENTIFY: How can shells of marine animals form sedimentary rock?

CHECKING CONCEPTS

1. Why does sand settle to the bottom of a jar faster than mud does?
2. What is sediment?
3. What sedimentary rock is made up of pieces of shells?
4. What are two types of sedimentary rock?

THINKING CRITICALLY

5. IDENTIFY: What materials combine together to bind sedimentary rock?
6. EXPLAIN: Why does sedimentary rock form in layers?
7. EXPLAIN: Why must water evaporate or be pressed out for sedimentary rock to form?
8. COMPARE/CONTRAST: How are the rocks chalk and coquina alike and how are they different?
9. EXPLAIN: How does weight affect the way in which sediments form layers?

DESIGNING AN EXPERIMENT

Design an experiment to solve the following problem. Include a hypothesis, variables, a procedure, and a type of data to study. Also, tell how to record the data.

PROBLEM: How can you separate salt from sand?

Integrating Life Science

TOPICS: fossils, microorganisms

THE WHITE CLIFFS OF DOVER

The city of Dover is located along the southeastern coast of England. Dover is known for its beautiful white cliffs. These cliffs are composed of deposits of natural chalk and are more than 100 million years old.

How were the White Cliffs of Dover formed? The natural chalk comes from the fossil shells of microscopic animals called foraminifera (for-AM-ih-NIH-fer-ah). Close examination of the chalk deposits reveals the shells of the foraminifera that formed them.

The shells of foraminifera are made up of calcium carbonate ($CaCO_3$). When the foraminifera die, their shells sink to the ocean floor. These shells form fine sediments on the ocean floor. Gradually, the sediments build up until they are hundreds of meters thick. They are then compacted to form solid rock.

▲ **Figure 3-14** The White Cliffs of Dover

▲ **Figure 3-15** Foraminifera

Thinking Critically What compresses the shells of the foraminifera?

3 Assess

CHECKING CONCEPTS ANSWERS

1. Sand particles are heavier.
2. rock particles carried and deposited by wind, water, or ice
3. coquina
4. coquina and chalk limestone

THINKING CRITICALLY ANSWERS

5. IDENTIFY: dissolved minerals
6. EXPLAIN: because the sediments are deposited at different times or because one sediment is denser than another
7. EXPLAIN: The sediments could not bind together to form solid rock in the presence of water.
8. COMPARE/CONTRAST: They are both limestones formed from shells. The shells are visible in coquina limestone, but not in chalk limestone.
9. EXPLAIN: The densest, heaviest sediments form the bottom layers because they settle first.

4 Close

DESIGNING AN EXPERIMENT

Use the Designing an Experiment Rubric from the Teacher's Resources to assess students' experiments. Fill in the rubric with the information below. For this assignment, students should have:

- mentioned that salt will dissolve in water but sand will not.
- mentioned putting the mixture in water, filtering out the sand, and then evaporating the water to recover the salt.

Integrating Life Science

TOPICS: fossils, microorganisms

THE WHITE CLIFFS OF DOVER

Background Forams, or foraminifera, are one-celled protists. They belong to the group of protists called protozoans and are classified in the same phylum as the amoeba.

Discussion Ask students: *What class of rock are these cliffs?* (sedimentary)

Thinking Critically Answer
As more and more shells fall to the bottom of the ocean, the weight of overlying layers compresses the layers underneath.

RETEACH

Demonstrating Compression Have students squeeze a mud ball between two pieces of cardboard. Students will notice that the mud becomes rocklike when the water is squeezed out. Have students try different amounts of pressure and note the results.
learning styles: tactile, visual

COOPERATIVE LEARNING

Examining Sedimentary Rocks Divide the class into small groups. Distribute samples of coquina and chalk limestone to each group. Have students compare and contrast the two types of sedimentary rocks.

• TEACHING RESOURCES • Online or Print

Teacher's Resources
Lesson Review: Chapter 3, p. 6
Designing an Experiment: Study Tools, p. 14

MORE ON THE WEB

Rocks on Mars Have students use the Internet to investigate the discovery of sedimentary rock on Mars. Then, have students write news articles detailing their findings.

3-5 *(pp. 74–75)*

1 Introduce

STUDY HINT Ask students to keep a running list of different ways that sedimentary rock can be classified.

Linking Prior Knowledge Ask students to tell what they know about beach sand. For example, do they know that sand is formed from the action of waves? Waves continually break shells and rocks into smaller and smaller fragments (weathering). Waves are agents of erosion. The resulting sediments eventually form sedimentary rocks.

2 Teach

Demonstration Evaporate the water from a saltwater solution, either by gently heating the solution or by placing it in sunlight. Show students the salt that forms. Have students classify the salt as clastic or nonclastic. (nonclastic)
learning styles: visual, auditory

Discussion Place a copy of the Concept Map from the Teacher's Resources on an overhead projector. Write the head *Clastics* in the center circle. Then, write the subheads *Conglomerate, Sandstone,* and *Shale* in each of three boxes surrounding the center circle. Compare the three types of clastics and make notes under each subhead.
learning styles: visual, linguistic

Reinforcement Ask students to describe how limestone forms. Make sure that they understand that limestone can form from the remains of living things and chemical precipitation. **learning style: linguistic**

Answers to questions

1. **DEFINE:** Clastics are sedimentary rocks made up of fragments of broken rock that have been eroded, transported by wind, water, or ice, and deposited elsewhere.
2. **IDENTIFY:** conglomerates, sandstones, and shales
3. **NAME:** halite
4. **DESCRIBE:** from the remains of plants that are compacted over a long time

3-5 How are sedimentary rocks classified?

Objective
Identify and describe the two main groups of sedimentary rocks.

Key Terms
clastic (KLAS-tihk) **rock:** sedimentary rock made up of fragments of rock
nonclastic rock: sedimentary rock made up of dissolved minerals or the remains of living things

Groups of Sedimentary Rocks There are two groups of sedimentary rocks. One group is made up of fragments of rock that have been eroded, transported by water, wind, or ice, and deposited elsewhere. These sedimentary rocks are called **clastic rocks.** Another group of sedimentary rock is made up of dissolved minerals or the remains of plants and animals. These sedimentary rocks are **nonclastic rocks.**

1 **DEFINE:** What are clastic rocks?

Particle Size Clastic rocks can be classified according to the sizes and shapes of the sediments that formed them. Conglomerates (kuhn-GLAHM-uhr-itz) are clastic rocks that are made up of rounded pebbles and gravel.

Sandstones are made of small grains of sand. Shales are clastic rocks made of silt and clay. Silt and clay are the smallest kinds of sediment.

2 **IDENTIFY:** Name three groups of clastic rocks.

▲ **Figure 3-16** Sandstone (top) and shale (bottom)

Dissolved Minerals Most nonclastics form from dissolved minerals. Rocks formed in this way are chemical rocks. When water evaporates from salty lakes and shallow seas, the salts are left behind. These salts form a mineral called halite. Rock salt is a sedimentary rock made up of halite. Some kinds of limestone also form from dissolved minerals. One dissolved mineral that forms limestone is calcite.

Some caves have icicle-shaped deposits of limestone hanging from their ceilings. The cave shown in Figure 3-17 is an example. These deposits are called stalactites (stuh-LAK-tyts). The same cave may also have cone-shaped deposits of limestone rising from its floor. Deposits that rise from the floor are called stalagmites (stuh-LAG-myts).

▼ **Figure 3-17** Icicle-shaped deposits of limestone can be seen in this cave in Lincoln County, New Mexico.

Teaching Options

INTEGRATING CHEMISTRY

The Chemical Makeup of Coal Plants have a high content of cellulose, a complex molecule with a formula of $C_6H_{10}O_5$. During coal formation, chemical changes lower the amount of oxygen and hydrogen of the cellulose molecule. As coal matures through the four stages, the ratio of carbon to oxygen and hydrogen increases. Lignite is about 70 percent carbon, bituminous coal is about 85 percent carbon, and anthracite contains the most carbon, about 90 to 95 percent. When coal burns, impurities remain behind as ash or are vaporized as oxides. The higher the carbon content of the coal, the cleaner the coal burns.

The ratio of elements such as nitrogen and sulfur in coal molecules varies. Bituminous coal has up to 10 percent sulfur, more than other forms of coal. When coal burns, the sulfur is released as sulfur dioxide gas. If the gas reacts with water, sulfuric acid is produced, causing acid rain. Acid rain damages structures made of limestone or marble, common building materials. Have students research methods that are being used to reduce the air pollution caused by burning fossil fuels such as coal and summarize their findings in a report. **learning style: linguistic**

Stalactites and stalagmites are formed from dripping water containing dissolved calcium carbonate. Sometimes, a stalactite and stalagmite may join to form a structure called a column.

3 NAME: What mineral makes up rock salt?

Plants and Animals Some nonclastics form from the remains of plants and animals. Rocks formed from the remains of living things are organic rocks. Coals are nonclastic rocks formed from the remains of plants. Coal forms when the remains of plants are compacted for a long time. Coquina and chalk are both nonclastic limestones made up of seashells.

4 DESCRIBE: How is coal formed?

CHECKING CONCEPTS

1. Nonclastics are formed from the remains of __________ and __________.
2. Silt and __________ are two kinds of sediment in shale.
3. Caves often have structures called __________ hanging from their ceilings.
4. Coal is a __________ sedimentary rock.

THINKING CRITICALLY

5. CLASSIFY: Which group of nonclastic rock is most likely to form from plants buried for a long time?
6. CLASSIFY: Which group of clastic rock might form from the largest kinds of sediment?

Web InfoSearch

Spelunking Exploring caves is called spelunking. People who explore caves are called spelunkers. There are many caves in the United States to explore. If you go spelunking, follow all safety rules. For example, never go spelunking without an experienced spelunker.

SEARCH: Use the Internet to find out more about spelunking. List at least four safety rules you would need to know before entering a cave. Also, list the equipment you would need. Start your search at www.conceptsandchallenges.com.
Some key search words are **spelunking, cave exploring,** and **stalactite.**

Hands-On Activity

FORMING CLASTIC ROCK

You will need safety goggles, three plastic foam cups, a tablespoon (tbsp), plaster of Paris, sand, pebbles, and water.

1. In one cup, stir together 2 tbsp each of sand and plaster of Paris. Add 1 tbsp of water. Stir.
2. In the second cup, stir together 1 tbsp each of pebbles, sand, and water. Add 2 tbsp of plaster of Paris and stir. Spoon this mixture on top of the mixture in the first cup.
3. In the third cup, prepare the same mixture as in Step 1. Spoon it on top of the other two layers in the first cup. Set the cup aside to harden.
4. The next day, peel away the cup from the clastic rock you created. Scrape away the surface from the side of the rock.

▲ **STEP 3** Spoon the mixture on top of the other layers.

Practicing Your Skills

5. DESCRIBE: What does the side of your rock look like?
6. OBSERVE: What kinds of particles does each layer have?

Tips for Using Technology

Using Presentation Software to Explain Spelunking Have students use presentation software to develop multimedia presentations about spelunking. Students may want to download photographs and illustrations from the Internet and incorporate these into their slide show. Students can then print out their slide show and use it as a handout to follow during their presentations.
Have students use the Flowchart graphic organizer from the Teacher's Resources to plan their presentations.

• TEACHING RESOURCES • Online or Print

Teacher's Resources

Lesson Review: Chapter 3, p. 7

Enrichment: Chapter 3, p. 10

Concept Map: Study Tools, p. 4

Flowchart: Study Tools, p. 6

READING STRATEGY Tell students that the word part *-sed* comes from the Latin word *sedere,* which means "to settle." Ask students how knowing this word origin can help them understand and remember the meaning of *sedimentary.*

3 Assess

CHECKING CONCEPTS ANSWERS

1. plants, animals
2. clay
3. stalactites
4. nonclastic

THINKING CRITICALLY ANSWERS

5. CLASSIFY: coal
6. CLASSIFY: conglomerates

4 Close

Web InfoSearch

Spelunking Have students use the Internet to find out more about spelunking, or cave exploring. Have students list at least four safety rules they need to know before entering a cave. Have students also list spelunking equipment. Some key search words are **spelunking, cave exploring,** and **stalactite.** Have students begin their search at www.conceptsandchallenges.com.

Hands-On Activity

FORMING CLASTIC ROCK

TIME: 15 minutes

PURPOSE: modeling, describing

SAFETY TIP: Tell students to wipe up spills immediately to avoid accidents. Have students wear safety goggles when mixing the plaster.

COOPERATIVE LEARNING: Working in groups, one student can measure the materials while the other mixes. All students should record their observations in their notebooks.

Practicing Your Skills Answers

5. DESCRIBE: Possible answer: It has different layers.
6. OBSERVE: Students should observe different-sized particles in each layer.

3-6 (pp. 76–77)

1 Introduce

STUDY HINT Before beginning this lesson, write the title of each section in the lesson on the board. Have students copy these titles in their notebooks, using an outline format. As students read each section, they should write a summary of each paragraph beneath the section.

Linking Prior Knowledge Make sure that students recall from Lesson 1-3 that the mantle has two parts: a solid upper part and a lower, more fluid, part. Ask: *Are temperatures hotter in the crust or mantle?* (mantle)

2 Teach

Demonstration Take a piece of paper and loosely form it into a wad or ball. Then put the paper on the table and press it flat with your hand. Have students discuss how and why pressure changed the shape of the paper. **learning style: visual**

Discussion Explain to students that in addition to heat, tremendous pressure inside Earth causes rocks to change. Ask students what is meant by the term *pressure.* Discuss students' responses. Point out that pressure is expressed in units of force per unit area. Stronger forces result in greater pressure. **learning style: auditory**

Answers to questions

1. **EXPLAIN:** because heat and pressure inside Earth changes the minerals in metamorphic rocks
2. **EXPLAIN:** extreme pressure
3. **IDENTIFY:** all kinds of rock, even existing metamorphic rocks

READING STRATEGY Prepare students for the concept of the rock cycle by having them reread the subsection entitled *Magma.* Students should begin to grasp that rocks continually change from one type to another.

3-6 How are metamorphic rocks formed?

Objective
Describe two ways in which metamorphic rocks form.

Changed Rocks You can make bread from water, flour, salt, and yeast. First, you mix the ingredients. Then, you bake the mixture in an oven. After baking the mixture, you will not recognize any of the ingredients. The heat of the oven changes the water, flour, salt, and yeast.

The inside of Earth is like an oven. Heat and pressure inside Earth "bakes" rocks and changes the minerals in them. These chemically changed rocks are called metamorphic rocks.

1. **EXPLAIN:** Why are metamorphic rocks called changed rocks?

Heat and Pressure Minerals in all types of rocks go through chemical changes when the rocks are heated. Minerals go through chemical changes at temperatures of between 200°C and 800°C. Minerals do not change chemically when the temperature is below 100°C. Above 800°C, the minerals melt into magma, or molten rock.

Rocks buried deep inside Earth's crust are also affected by forces. These forces cause pressure, which changes the form of minerals. Extreme pressure can flatten minerals into layers, as shown in Figure 3-18.

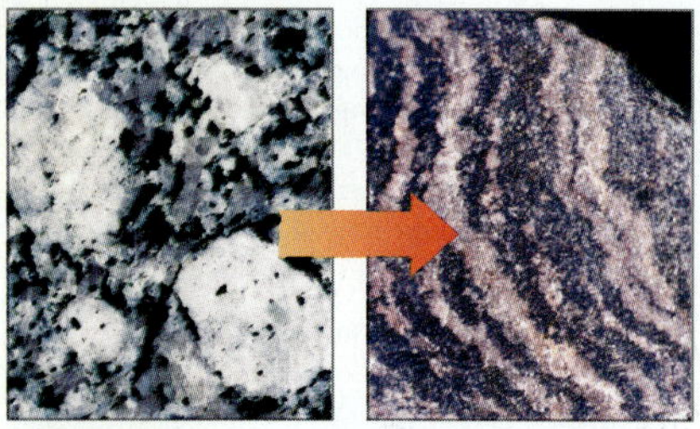

▲ **Figure 3-19** Pressure may cause minerals in rock to form into bands in metamorphic rock. Here granite (left), an igneous rock, becomes gneiss (right), a metamorphic rock.

2. **EXPLAIN:** What can flatten minerals into layers?

Magma Metamorphic rocks are sometimes formed when existing rocks come into contact with or mix with magma. Magma can move into cracks in deeply buried sedimentary rocks. It may also flow between the layers of sedimentary rocks. The heat and chemical solutions in the magma cause the minerals inside the sedimentary rocks to change. The heat from magma can also change minerals in igneous rocks and turn one kind of metamorphic rock into a different kind of metamorphic rock.

3. **IDENTIFY:** What kinds of rocks are changed by coming into contact with magma?

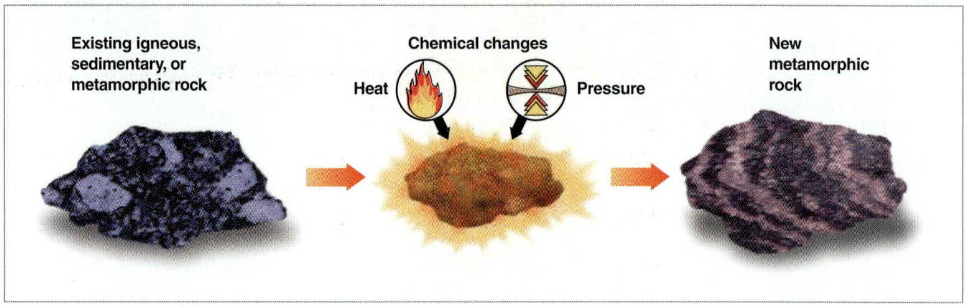

▲ **Figure 3-18** Heat and pressure cause chemical changes in rock.

76

Teaching Options

QUICK CONVERSIONS		
Changes in Rock	**Metric/SI**	**English**
Chemical changes do not occur at	100°C	212°F
Chemical changes occur at	100–800°C	212–1,472°F
Minerals melt at	800°C+	1,472°F+

EXTENDING THE LESSON

Building Blocks Have students research the type of rocks used locally in buildings such as schools, banks, churches, hospitals, and homes. Tell students to write their findings in a report.

ESL/ELL STRATEGY

Studying Word Origins Explain to students that the word *metamorphic* is related to the word *metamorphosis.* The prefix *meta-* means "change" or "transformation," and *morphosis* refers to the way in which an organism or part of it changes form or develops. Students may already be familiar with the term *metamorphosis* from life science, where it refers to animals such as butterflies. Here the term *metamorphic* is used to refer to a nonliving entity (rock).

✓ CHECKING CONCEPTS

1. Igneous and sedimentary rocks can be changed into metamorphic rocks by __________ and __________.
2. Minerals in deeply buried rocks are flattened by great __________.
3. The minerals in rocks can be changed by the heat and chemical solutions in __________.
4. Pressure is caused by __________ that push against an object.
5. When rocks are heated to 200°C, the minerals in the rocks go through a __________ change.

THINKING CRITICALLY

6. DESCRIBE: What happens to the minerals in rocks that are heated to a temperature of less than 100°C?
7. DESCRIBE: What happens to the minerals in rocks that are heated to a temperature of 950°C?
8. PREDICT: Would the minerals in a rock buried 2 m beneath Earth's surface be significantly affected by pressure? Explain.

Web InfoSearch

Stonemasonry In stonemasonry, stonemasons build structures using many types of stone from all around the world. Some of the most popular building stones are granite, marble, and slate. Stonemasons measure, cut, and carve stone to a specific shape or form. The stones are used in new and restored buildings. For restored buildings, the masons must often keep the new stonework as close as possible to the original stonework.

SEARCH: Use the Internet to find out more about stonemasonry. Are there different types of stonemasons? What training do you need to perform the job? What skills and interests should you have? After finishing your research, write a help wanted ad for a stonemason to restore old stone fireplaces. Start your search at www.conceptsandchallenges.com. Some key search words are **stonemasonry, fireplace restoration,** and **masonry.**

Integrating Physical Science

TOPICS: force, pressure

CALCULATING PRESSURE

The definition of pressure is force per unit of area. Pressure changes rocks and affects the atmosphere and oceans. The amount of pressure applied to an object can be calculated using the following formula:

Pressure = force ÷ area

In this formula, force is measured in units called newtons (N). A 1 kg weight exerts a downward force of 9.8 N. Area is measured in square units.

You can calculate pressure using the formula above. Suppose a person exerts a force of 120 N while pressing down on something with the heel of his or her shoe. The heel of the shoe has an area of 60 cm^2. What would the pressure be? Of course, in the real world, the entire weight of a person may not be concentrated on his or her heel.

▲ **Figure 3-20** Footprints in sand show the force of pressure.

Thinking Critically When would you not put all your weight into stepping down?

ONGOING ASSESSMENT

To show students that metamorphic rocks differ from the rocks that formed them, have students compare a piece of dried clay and a similar piece of kiln-fired clay. (If a kiln is not available, heat the clay in a flowerpot or crucible.) Ask: *Which piece of clay represents the sedimentary rock called shale?* (the dried clay) *Which piece represents the metamorphic rock made from shale?* (the kiln-fired clay)

• TEACHING RESOURCES • Online or Print

Teacher's Resources

Lesson Review: Chapter 3, p. 8

3 Assess

✓ CHECKING CONCEPTS ANSWERS

1. heat, pressure
2. pressure
3. magma
4. forces
5. chemical

THINKING CRITICALLY ANSWERS

6. DESCRIBE: They do not change chemically when the temperature is below 100°C.
7. DESCRIBE: The minerals melt.
8. PREDICT: No; The pressure 2 m beneath Earth's surface is not great enough.

4 Close

Web InfoSearch

Stonemasonry Have students use the Internet to find out more about stonemasonry. Some key search words are **stonemasonry, fireplace restoration,** and **masonry.** Then, have them write a want ad for a stonemason. Students can begin their search for information www.conceptsandchallenges.com.

Integrating Physical Science

TOPICS: force, pressure

CALCULATING PRESSURE

Real-Life Connection According to Pascal's principle, pressure applied to a fluid is transferred unchanged throughout the fluid. Hydraulic machines used to lift cars take advantage of this principle. They usually contain two cylinders of different sizes filled with a liquid such as oil. When pressure is applied to the smaller cylinder, the inner piston descends and pressure is transmitted to the larger cylinder. Because the larger cylinder has a greater area, the total force on the large piston increases. In this way, a relatively small amount of pressure can be used to lift or power heavier objects.

Discussion The pressure applied by the heel of the shoe would be 2 N/cm^2. Ask volunteers to show their calculations on the board. ($120\ N \div 60\ cm^2 = 2 N/cm^2$)

Thinking Critically Answer

when you are standing on your toes

3-7 (pp. 78–79)

1 Introduce

STUDY HINT Remind students that tables and charts often compare information. Ask: *What information is being compared in this lesson?* (the textures of metamorphic rocks)

Linking Prior Knowledge Review with students what they learned about crystals (Lesson 2-6). Ask: *What are crystals?* (natural solid substances that have a definite shape) *What determines the shape of a crystal?* (the arrangement of its atoms)

2 Teach

Demonstration Place a large stack of assorted-color construction paper on your desk. Then, lift several sheets of paper from the stack. Ask: *What type of metamorphic rock could this represent?* (foliated) Have students explain their reasoning. (Foliated metamorphic rocks are banded and tend to separate in layers along the bands.) **learning style: visual**

Discussion Describe how metamorphic rocks are classified. Compare foliated and nonfoliated metamorphic rocks. Ask students which type of metamorphic rocks have minerals arranged in parallel layers or bands. (foliated) **learning style: linguistic**

Answers to questions

1. CLASSIFY: according to their textures
2. COMPARE: Both are foliated metamorphic rocks formed from shale or granite. Schist is also formed from basalt.
3. IDENTIFY: quartzite

READING STRATEGY Remind students that the text uses certain strategies to help make information more accessible, such as heads and subheads. Have students scan the lesson and list the heads and subheads.

3-7 How are metamorphic rocks classified?

Objective

Explain the difference between foliated and nonfoliated metamorphic rocks.

Key Terms

foliated (FOH-lee-ay-tuhd)**:** texture of a metamorphic rock that has mineral crystals arranged in bands

nonfoliated: texture of a metamorphic rock that does not have mineral crystals arranged in bands

Classifying Metamorphic Rocks Petrologists classify metamorphic rocks based on the texture of the rock. Metamorphic rocks have two kinds of textures, foliated and nonfoliated. The texture of a metamorphic rock is determined by the arrangement of its mineral crystals.

CLASSIFICATION OF METAMORPHIC ROCKS

Original Rock	Metamorphic Rock	Texture
Granite, shale	Gneiss	Foliated
Granite, shale, basalt	Schist	Foliated
Shale	Slate	Foliated
Sandstone	Quartzite	Nonfoliated
Limestone	Marble	Nonfoliated

▲ **Figure 3-21**

1 CLASSIFY: How are metamorphic rocks grouped?

Banded Metamorphic Rocks The mineral crystals in some metamorphic rocks are arranged in bands. The texture of a metamorphic rock with minerals arranged in bands is called **foliated.** Foliated rocks tend to break along their mineral crystal bands.

Foliated metamorphic rocks are formed when existing rocks are placed under great heat and pressure. Under this great heat and pressure, the minerals in the rocks melt and recrystallize into minerals that form bands.

Gneiss (NICE) is a foliated metamorphic rock. Gneiss is formed when granite is put under great heat and pressure.

Schist is another rock that has bands of recrystallized minerals. Schist is formed from shale, basalt, or granite. Slate is also formed from shale.

▲ **Figure 3-22** Shale (left) becomes slate (right).

2 COMPARE: How are gneiss and schist similar?

Metamorphic Rocks Without Bands The texture of a metamorphic rock that does not have its minerals arranged in bands is **nonfoliated.** Nonfoliated metamorphic rocks do not break in layers.

Marble and quartzite are nonfoliated metamorphic rocks. Marble is formed from limestone. Marble contains large mineral crystals of calcite.

Quartzite is formed from sandstone put under great heat and pressure. Quartzite has large crystals of quartz.

▲ **Figure 3-23** Limestone (left) becomes marble (right).

3 IDENTIFY: Which metamorphic rock has large crystals of quartz formed from sand?

78

Teaching Options

COOPERATIVE LEARNING

Classifying Rocks Have small groups examine samples of gneiss, schist, slate, quartzite, and marble to observe differences among metamorphic rocks. Tell students to classify the rocks as foliated or nonfoliated. Ask them if there are other ways to describe the rocks' similarities and differences to help establish other subclassification systems. **learning styles: visual, logical/mathematical**

EXTENDING THE LESSON

Making Tables Have students make a list of all the sedimentary and igneous rocks mentioned in this chapter. Tell them to use library references to find out what type of metamorphic rocks can be formed from each of the sedimentary and igneous rocks on their lists. Have students write their findings in a table similar to the one in Figure 3-21. **learning style: linguistic**

CHECKING CONCEPTS

1. What are the two textures of metamorphic rock?
2. What causes banding in metamorphic rock?
3. How do foliated rocks break?
4. A metamorphic rock without banding has what kind of texture?

THINKING CRITICALLY

Use Figure 3-21 to answer the following questions.

5. INTERPRET: Which metamorphic rocks are formed when granite is subjected to high heat and pressure?
6. INTERPRET: Which metamorphic rock is formed from limestone?
7. INTERPRET: Which metamorphic rocks can be formed from shale?
8. INTERPRET: From which sedimentary rock is quartzite formed?
9. CLASSIFY: Is slate a foliated or a nonfoliated metamorphic rock? Why?

BUILDING SCIENCE SKILLS

Researching Metamorphic rocks, such as marble and slate, are economically important. Use the Internet or reference books to find out where in the United States deposits of marble and slate can be found. Determine how these varieties of rock are different from those found elsewhere, such as in Italy. Also, find out how the impurities in the rocks give them their special colors. Finally, list some ways in which marble and slate are important to business and industry.

▲ **Figure 3-24** India's Taj Mahal is made of white marble from India.

Real-Life Science

MARBLE IN ART AND ARCHITECTURE

Marble is often used in architecture and sculpture because it is beautiful to look at and lasts a long time. The Lincoln Memorial was built to honor our sixteenth president, Abraham Lincoln. Construction began in 1914. It was not completed until 1922.

The Piccirilli brothers, both well-known marble sculptors, were chosen to carve the statue. It is 5.8 m tall and contains 28 blocks of white Georgia marble. The statue itself took four years to complete. The outside of the memorial contains marble from Colorado and Tennessee. The ceiling looks clear. The marble used in it was soaked in a waxy, whitish mixture known as paraffin to make it shine.

Another famous memorial made of marble is the Taj Mahal in India. It was built by an emperor in the 1600s to honor his wife.

Are there buildings or statues in your town that contain marble? Take a walk around your community and see if you can find some.

Thinking Critically What kind of rock is marble?

▲ **Figure 3-25** The Lincoln Memorial in Washington, D.C.

• TEACHING RESOURCES • Online or Print

Teacher's Resources

Lesson Review: Chapter 3, p. 9

Enrichment: Chapter 3, p. 10

Laboratory Manual

Lab. Challenge: How are metamorphic rocks classified? pp. 33–36

3 Assess

CHECKING CONCEPTS ANSWERS

1. foliated and nonfoliated
2. great heat and pressure
3. along their mineral crystal bands
4. nonfoliated

THINKING CRITICALLY ANSWERS

5. INTERPRET: gneiss, schist
6. INTERPRET: marble
7. INTERPRET: gneiss, schist, and slate
8. INTERPRET: sandstone
9. CLASSIFY: Slate is foliated because its mineral crystals are arranged in bands.

4 Close

BUILDING SCIENCE SKILLS

Researching Marble consists mainly of recrystallized calcite or dolomite. Its colorful streaks and swirls are produced by inclusions of minerals such as mica, talc, pyrite, and quartz. Marble is used in sculpture and for ornamental stone on buildings and monuments.

Marble quarries are found throughout the world. In the United States, several grades of marble come mainly from the Appalachian Mountain areas, especially Vermont, Georgia, and Tennessee.

Slate results from the metamorphism of rocks such as shale or from volcanic ash. Slate's ability to be split into large, thin, flat sheets makes it a good material for floor tiles, roofing, the foundation of billiard tables, chalkboards, and electrical insulation. In the United States, the most important sources of slate are Virginia, Pennsylvania, and Vermont.

Real-Life Science

MARBLE IN ART AND ARCHITECTURE

Background Marble comes in a variety of colors, textures, and patterns. Because all types of marble can be cut and polished, marble is favored by sculptors and architects. The most highly valued marble is found in the Carrara quarry in Tuscany, Italy. It was the rock that Michelangelo used in the 1500s to sculpt the *Pieta* and other statues.

Thinking Critically Answer
a nonfoliated metamorphic rock

3-8 (pp. 80–81)

1 Introduce

STUDY HINT Point out to students that in a cycle, the direction of the arrows is very important. Arrows point from one step to the next. Between steps a change occurs.

Linking Prior Knowledge Have students briefly describe their daily activities over the course of a typical week. Students should see a cyclical nature to their activities, for example, they rise, eat breakfast, and go to school. Tell students that cycles, such as seasons, are also common in nature.

2 Teach

Demonstration Place an ice cube in a beaker and melt it over a hot plate. Ask students what part of the rock cycle this models. (the melting of solid rock into magma) Ask what type of rock would form if the "magma" became solid again. (igneous) **learning style: visual**

Discussion Refer students to the diagram of the rock cycle in Figure 3-26. Explain the many ways that rocks are changed. Tell students to note the cross-links in the cycle. Make sure that students understand that many rocks never make it through the complete rock cycle. Point out that rocks, like air and water, are recycled; the same materials are used repeatedly in different ways. **learning styles: visual, auditory**

Answers to questions

1. **DEFINE:** a series of events that happen over and over again during a certain length of time
2. **INFER:** through weathering and erosion
3. **OBSERVE:** by cooling to form solid rock

3-8 What is the rock cycle?

Objective

Explain the rock cycle.

Key Terms

cycle: series of events that happen over and over again

rock cycle: series of natural processes by which rocks are slowly changed from one kind of rock to another kind of rock

Nature's Cycles A series of events that happen over and over again for a certain length of time is called a **cycle.** In nature, there are many different cycles. If you look around you, you will probably find many different examples of nature's cycles.

The rise and fall of the ocean tides is a natural cycle that occurs twice each day. The movement of Earth around the Sun is a natural cycle that takes a year to complete. The Moon going through its phases is another cycle.

1 **DEFINE:** What is a cycle?

The Rock Cycle Rocks on Earth change over time. Some of the changes in rocks take place inside Earth. There, tremendous heat and pressure can slowly change rocks from one kind of rock into another. Other changes in rocks take place at the surface. On Earth's surface, rocks may be changed by rain, ice, wind, or the action of chemicals in air or water. The series of natural processes by which rocks are slowly changed from one kind of rock to another kind of rock is called the **rock cycle.** Look at the diagram of the rock cycle in Figure 3-26. It shows the ways that rocks are changed from one rock type to another.

2 **INFER:** How do you think ice, rain, and wind can change rocks?

A Closer Look at the Rock Cycle Look again at the diagram of the rock cycle. Notice that all three classes of rock eventually lead to sedimentary rock. The three classes of rocks can also be changed into metamorphic rocks or back into magma. However, only igneous rocks form directly from magma.

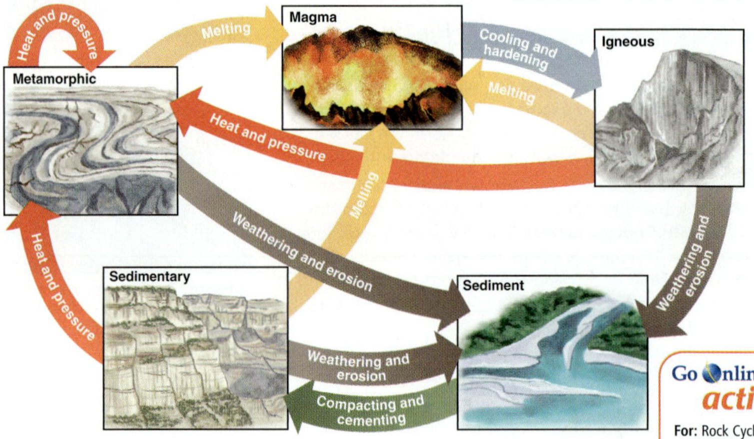

▲ **Figure 3-26** The rock cycle

Go Online **active art**
For: Rock Cycle activity
Visit: PHSchool.com
Web Code: cfp-1056

Teaching Options

RETEACH

Describing the Rock Cycle Hold up pictures of different kinds of igneous, sedimentary, and metamorphic rocks. For each picture, have students describe what may happen to the rock as it goes through the rock cycle. Tell them to use Figure 3-26 as a guide. Then, have students draw their own version of the rock cycle, using their own drawings or magazine clippings to illustrate it. **learning styles: visual, tactile**

COOPERATIVE LEARNING

Reviewing the Rock Cycle Make copies of the diagram of the rock cycle, but do not label the arrows. Have pairs of students take turns filling in the arrows with the proper processes, such as weathering or melting. **learning styles: visual, interpersonal**

EXTENDING THE LESSON

Modeling the Rock Cycle Have students design an experiment that models at least one part of the rock cycle. Check students' designs for accuracy and safety. Students should share their models with the class.

At Earth's surface, rocks are broken apart by the forces of nature—moving water, ice, plant action, and wind. The particles of rock may then settle in lakes or oceans to form sediment that eventually makes up sedimentary rock.

Deeply buried igneous and sedimentary rocks can change into metamorphic rock or melt back into magma. Intense heat and pressure can change any rock into metamorphic rock. High temperatures can change any rock back into magma.

3 **OBSERVE:** How does magma become igneous rock?

CHECKING CONCEPTS

1. What changes rock inside Earth?
2. What can all rocks be broken down into?
3. What kind of rock does magma cool and harden into?
4. What two conditions slowly change one type of rock into another?

THINKING CRITICALLY

5. **COMPARE:** How is the rock cycle like other natural cycles on Earth?
6. **INFER:** Why can only igneous rocks be formed directly from magma?
7. **APPLY:** Look around your neighborhood. What examples can you find of the rock cycle in action? What do you think caused the rocks to change?

INTERPRETING VISUALS

Use Figure 3-26 to answer the following questions.

8. **EXPLAIN:** How can igneous or metamorphic rock eventually become sedimentary rock?
9. **IDENTIFY:** What processes change sedimentary rock into metamorphic rock?
10. **NAME:** What processes or forces change metamorphic rock into igneous rock?

Hands-On Activity
CLASSIFYING ROCKS BY TYPE

You will need 10 different rocks, a hand lens, stickers, paper, pencil, and (optional) a field guide to rocks or other reference books.

1. Collect 10 different rocks from around your town. Label each rock with a numbered sticker.
2. Look at each rock with a hand lens. Record your observations on a sheet of paper.
3. Using your notes, classify each rock as sedimentary, igneous, or metamorphic.
4. Using reference materials, identify as many of your rock samples as you can. Write the name of each rock in your notes.

▲ **STEP 2** Use a hand lens to study each rock.

Practicing Your Skills

5. **CLASSIFY:** How many rocks did you classify as igneous? How many rocks did you classify as sedimentary? How many rocks did you classify as metamorphic? Explain your choices.
6. **EXPLAIN:** How many rocks were you not able to identify? Why weren't you able to identify these?

Go Online
active art

For: Rock Cycle activity
Visit: PHSchool.com
Web Code: cfp-1056

Students can interact with the art of the rock cycle online.

• *TEACHING RESOURCES* •
Online or Print

Teacher's Resources

Lesson Review: Chapter 3, p. 11

Rock Cycle: Visuals, p. 3

3 Assess

CHECKING CONCEPTS ANSWERS

1. heat and pressure
2. sediments of many sizes and shapes
3. igneous
4. time and the environment

THINKING CRITICALLY ANSWERS

5. **COMPARE:** Like natural cycles, the rock cycle is a series of events that happen over and over again.
6. **INFER:** because igneous rocks are the solid form of magma
7. **APPLY:** Possible answers: a crumbling stone wall caused by weathering, or sediment in streams or rivers caused by erosion

INTERPRETING VISUALS ANSWERS

8. **EXPLAIN:** through weathering and erosion
9. **IDENTIFY:** heat and pressure
10. **NAME:** melting, then cooling and hardening

4 Close

Hands-On Activity
CLASSIFYING ROCKS BY TYPE

TIME: 15–30 minutes

PURPOSE: classifying, explaining

COOPERATIVE LEARNING: Have students compare their rocks with the rocks of others. Ask students to determine the kinds of rocks they have in common and the kinds that are different.

Practicing Your Skills Answers

5. **CLASSIFY:** Answers should include a numerical response for igneous, sedimentary, and metamorphic and an explanation for each classification. An example of a possible explanation may be identifying sandstone by its small grains of sand, which classifies it as sedimentary.
6. **EXPLAIN:** Answers will vary. Further tests may be needed to identify some rock samples. Reasons for not identifying may include: rock comes in too wide a variety of colors, grain size is hard to see, and no banding.

Lab Activity

(pp. 82–83)

1 Prepare the Lab

Modeling What Lies Beneath Earth's Surface

SET-UP TIME: **LAB TIME:**

BACKGROUND: Geologists examine the positions of rock layers to determine the age and geologic history of an area. This is known as relative dating. Relative dating does not give the exact age of rocks. However, it does allow geologists to determine which layers predate the others.

PURPOSE: Students will construct model profiles and interpret core samples.

SCIENCE SKILLS: Students will use the available materials to **construct models** and profile **graphs** of rock layers. They will take core samples from the models of other students and **interpret** the samples. Students will record their observations and **compare and contrast** their profile graphs to actual models.

ADVANCE PLANNING: Gather the necessary materials beforehand. Cut wax paper into 12-inch sheets. Have a trash can available so students can safely dispose of their core samples and models.

2 Guide the Lab

PROCEDURE: Place all the available materials on a table. You may wish to have students work in small groups to complete the activity. Discuss the procedure before beginning. Check students' profile lines on the models and the 20-cm line on the paper before allowing them to remove the core samples. Provide students with a copy of the Lab Activity Report found in the Teacher's Resources for students to record their observations and conclusions.

TROUBLESHOOTING: Have students make a key code that describes which color of clay represents each type of rock. You may want to have everyone use the same color code so that students can easily interpret other groups' models. For best results, make sure that students push the straws straight into the model hills, rather than at an angle.

LAB ACTIVITY
Modeling What Lies Beneath Earth's Surface

Materials
Lab apron, three different colors of modeling clay, plastic picnic knife, plain paper, wax paper, colored pencils or markers, metric ruler, clear plastic straws, scissors.

BACKGROUND

Beneath Earth's surface lies a complex structure of rock layers. These layers provide clues as to how Earth's crust formed. Geologists drill deep into these layers of rock. The drills are like apple corers. They bring up rock cores for study.

PURPOSE

In this activity, you will work with a group to create a model of a hill made up of several rock layers. You will then trade your model for the model created by another group. Using clear plastic straws, you will determine what lies beneath the surface and make a profile graph.

▲ **STEP 3** Add layers to your hill using different colors of clay.

▲ **STEP 5** Make a line straight across your model.

PROCEDURE

1. Put on your lab apron.
2. Each color of clay should represent a different type of rock. For example, yellow could be sedimentary rock; blue, metamorphic rock; and red, igneous rock. Place a lump of modeling clay on a sheet of wax paper. Shape the clay to form a gentle hill.
3. Add additional layers to the hill using colored clay. Create a gentle hill 4 cm high and 20 cm wide. The top layer should cover all the layers below.
4. Exchange your model with one made by another group.
5. Use the ruler and a pencil to make a light line straight across the model at its widest point (20 cm). This is your profile line.

Teaching Options

! FAST FACTS *Rock Strata*

Earth's rocks range in age from more than 4 billion years to less than 1 million years. An exposed cliff often vividly illustrates strata, which are rock layers made up of different types of rocks. Generally, the oldest strata are found at the bottom of a cliff and the youngest at the top.

QUICK CONVERSIONS

Model Hill	Metric/SI	English
Height	4.0 cm	1.5 in.
Width	20.0 cm	7.9 in.

ESL/ELL STRATEGY

Noting the Icons English-language learners should pay close attention to the icons at the beginning of the Procedure. These icons indicate the precautions that should be taken during the lab. Have students review page 14 to identify the meanings of each icon.

6. Draw a straight line 20 cm long on a sheet of paper.

7. Place the end of a straw along the profile line on one end of the hill. Twirl the straw and slowly push it through the clay to the tabletop. Remove the straw. It will have a core sample inside. Cut the straw a short distance above the clay core sample.
⚠ CAUTION: Be careful when using scissors.

8. Place the lower edge of the core sample on the 20 cm line that you drew. The core should be in the same place on the line as it was on the model.

9. Using the remainder of the straw, take other core samples a few centimeters apart along the profile line until you have sampled the length of the hill.

10. Use colored pencils or markers to fill in the spaces on the paper between your core samples. This will complete the profile diagram.

11. Slice the clay model in half along the profile line. Compare the clay model to the profile diagram.

▲ **STEP 7** Slowly push the straw through the clay to the tabletop.

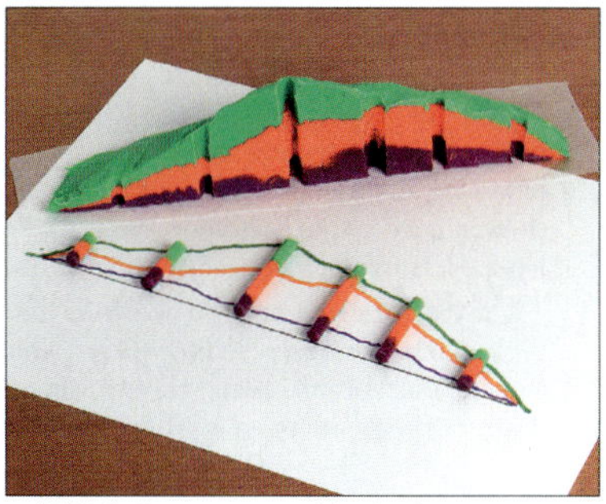

▲ **STEP 11** Compare your model to the diagram you made.

CONCLUSIONS

1. OBSERVE: How close was the profile graph you drew to the model?

2. OBSERVE: How many core samples did you take? Could you have used fewer cores?

3. INFER: How do profile graphs help geologists determine the history of a region on Earth?

SAFETY TIP: Caution students to take care when using scissors and plastic knives. Remind students not to eat or drink anything in the lab.

EXPECTED OUTCOME: Students' profile diagrams should correspond to the actual layers of their models.

LAB HINT: Remind students that precise measurements are necessary to obtain accurate results.

3 Conclude the Lab

1. OBSERVE: If students were careful, their profile graphs should be reasonably accurate.
2. OBSERVE: The more core samples they took, the greater the accuracy of the profile graphs would be.
3. INFER: Geologists can study the different rock layers under the surface and infer how they formed and what changes took place in a particular area over time.

Use the Laboratory Activity Rubric from the Teacher's Resources to assess students' lab activities. Fill in the rubric with the additional information below. For this activity, students should have:

- created a model of a hill made up of several rock layers.
- made a profile graph.

4 Extend the Lab

On the profiles, have students label the rock layers of their models as igneous, sedimentary, or metamorphic. Next to each layer, have students give an example of an igneous, a sedimentary, or a metamorphic rock. Students should include a brief paragraph on their profiles, stating how each of the three types of rocks forms.

Tips for Using Technology

Using Graphics Software to Draw a Profile Diagram After students have completed the activity, have them use the graphics program on a computer to draw their profile diagrams. Students should use different colors for each layer and include a color key. Print out the diagrams and display them in class. Most graphics creation software has similar tools to let the user create graphics and illustrations. Some of these tools include shape creation tools, text tools, and pencil tools.

• TEACHING RESOURCES • Online or Print

Teacher's Resources
Lab Activity Report: Study Tools, pp. 16–17
Laboratory Activity Rubric: Teaching Tools, p. 8

Laboratory Manual
Lab. Skills Worksheet 5: Graphing, pp. 15–18

The Big Idea

(pp. 84–85)

1 *Introduce*

Objective Students will come to understand the physics of the rock cycle in forces such as gravity and processes such as melting. Students will hypothesize how Earth's surface would be different if the rock cycle did not exist.

Linking Prior Knowledge Ask students to recall that gravity is the attractive force that exists between all objects. Gravity is related to mass. The greater the mass of an object, the greater is its gravitational pull. Because Earth has a large mass, all objects on Earth are pulled toward the center of the planet. This explains why sediments sink to the ocean floor and rock layers press down upon the layers below.

Have students recall the layers of Earth and how minerals are formed (Lessons 1-3 and 2-7).

2 *Teach*

Discussion Make a three-column chart on the board. Head the columns *Gravity*, *Pressure*, and *Heat*. Have volunteers read aloud the boxed information on pages 84 and 85. Have the class identify which processes involve gravity, pressure, or heat, and write the information in the correct column. Save the chart on the board for students to use when writing in their Science Logs. **learning styles: visual, logical/mathematical**

Use the background information in Fast Facts to help guide students in writing their Science Log entries.

THE BIG IDEA

What drives the rock cycle? Have students start their research for their Science Log Writing Activity at www.conceptsandchallenges.com. Be sure students use search engines and websites that are reliable and acceptable. You may also wish to provide copies of the *Big Idea Planner* and *Big Idea Science Log Entry* study tools found in the Teacher's Resources.

Integrating Physical Science

THE Big IDEA

What drives the rock cycle?

The rocks in Earth's crust are constantly being changed. These changes are caused by forces pushing and pulling rocks in all directions.

Have you ever been in a tug-of-war contest? If so, then you know that you pull on a rope together with other members of your team. Your team exerts a force on the rope. If that force is greater than the force exerted by the other team, the rope is pulled toward you, and your team wins the contest!

Forces are exerted on Earth's rocks, too. The effects of these push-pull forces create pressure. This pressure causes bits and pieces of rocks and minerals to join together into sedimentary rock. Metamorphic rocks form from pressure deep beneath Earth's surface.

Gravity is also a force at work in the rock cycle. Gravity causes sediments in water to sink to the bottom. Also, gravity causes the weight of rock layers on top to press down on rock layers below. This, in turn, causes compaction.

Heat is a form of energy. Heat can change rock. When the minerals in rocks are under great heat and pressure, they may melt and combine chemically to form metamorphic rock.

Removing heat from a substance also causes changes in rocks. When a substance cools, it loses heat energy. Its molecules slow down and move closer together. The substance hardens to form a solid. Igneous rocks form this way. As the molten rock moves towards Earth's surface, it cools. Molecules in the molten rock slow down, and minerals crystallize to form igneous rock.

Look at the illustrations on these two pages. Then, follow the instructions in the Science Log to learn more about "the big idea." ✦

Igneous Rocks
Lava loses heat energy to the air, causing its molecules to slow down and move closer together. Solid igneous rock, such as basalt, forms.

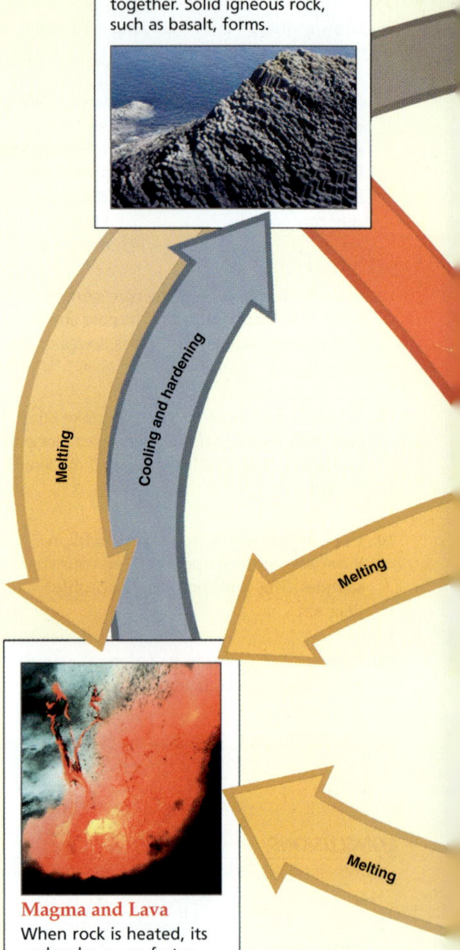

Magma and Lava
When rock is heated, its molecules move faster and farther apart until it changes from a solid to a liquid. Melted rock underground is called magma. Above ground it is called lava.

▲ **Figure 3-27** Pressure, gravity, heating, and cooling cause rocks to continually change form throughout the rock cycle.

84

Teaching Options

! FAST FACTS *The Physics and Chemistry of Rocks*

The motion of the molecules that make up a substance, such as magma or rock, determines the state of the substance. The molecules in a solid rock are generally stationary and do not move around much. The molecules in a liquid, such as magma, can slide past one another. They are not as strongly attracted to each other as those in a solid. In a gas, the molecules move freely and independently. Little attractive force exists between the molecules in a gas.

Heat and pressure can affect the chemical composition of rocks as well as their physical state (solid or liquid). For example, igneous rocks form from magma that contains many different elements. When magma cools, certain elements combine to form chemical compounds that crystallize as minerals. The first minerals to crystallize as magma cools are those that have the highest melting points. As these early-formed crystals grow, they remove specific elements from the magma, thus changing the magma's overall composition. Further cooling of the magma causes other minerals with progressively lower melting points to crystallize. If certain minerals settle out of the magma when they crystallize, they will be scarce in the igneous rock that forms when the remaining magma cools.

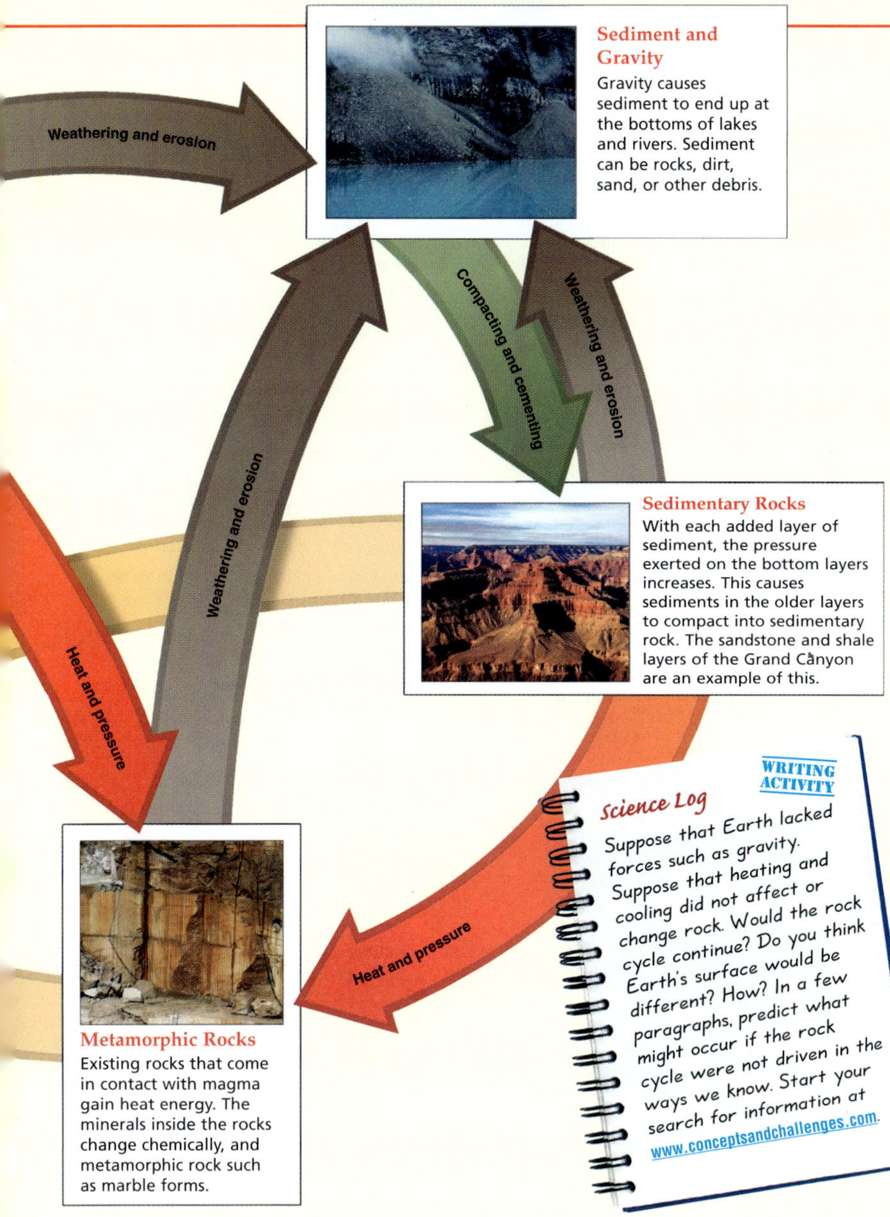

WRITING ACTIVITY

Science Log

Suppose that Earth lacked forces such as gravity. Suppose that heating and cooling did not affect or change rock. Would the rock cycle continue? Do you think Earth's surface would be different? How? In a few paragraphs, predict what might occur if the rock cycle were not driven in the ways we know. Start your search for information at www.conceptsandchallenges.com.

WRITING HINT Tell students that, when making predictions, as students will do for this Science Log, they can say almost anything, since no one can prove them wrong. However, to have a prediction be taken seriously, students must follow logic and support their hypothesis with scientific fact. Point out that the most successful science fiction stories start with a firm foothold in scientific fact.

3 Assess

Use the Writing Activity Rubric from the Teacher's Resources to assess students' Science Logs. Fill in the rubric with the additional information below. For this assignment, students should have:

- identified what drives the rock cycle.
- predicted how the cycle would be different if it were not driven by current means.

4 Close

Comparing Logs Have pairs of students share their Science Logs. Ask the partners to list how their ideas about the rock cycle are the same and how their ideas are different. Discuss students' lists as a class. Give students an opportunity to revise their Science Logs and incorporate relevant points raised by others.

COOPERATIVE LEARNING

Collecting Rocks Have students bring in rocks to start a class rock collection. Tell students that rock and mineral collecting are among the oldest and still most popular hobbies. Ask students to infer why. (Possible answer: Rocks have always been around, so it is natural that people would pick them up and be interested in them.)

• TEACHING RESOURCES • Online or Print

Teacher's Resources

Big Idea: Chapter 3, p. 12

Writing Activity Rubric: Teaching Tools, p. 9

Big Idea Planner: Study Tools, pp. 11–12

Big Idea Science Log Entry: Study Tools, p. 13

Challenges *(pp. 86–88)*

Chapter Summary

Review Hint Before students begin the Challenges, review the objectives of each lesson. If time permits, have students work in pairs to write at least one relevant fact related to each objective.

STUDY HINT Have pairs of students test each other on the classifications of different rocks and the different ways that rocks are classified. Then, have students test each other on the different processes at work in the rock cycle.

Key Term Challenges

MATCHING

1. igneous rock (3-2)
2. lava (3-2)
3. texture (3-3)
4. nonclastic rock (3-5)
5. foliated (3-7)
6. nonfoliated (3-7)
7. metamorphic rock (3-6)
8. cycle (3-8)

FILL IN

9. sedimentary rock (3-4)
10. clastic rocks (3-5)
11. magma (3-2)
12. rock cycle (3-8)
13. sediments (3-4)
14. pluton (3-2)

Chapter 3 Challenges

Chapter Summary

Lesson 3-1

- **Rocks** are made up of one or more minerals. The three types of rock are **igneous, sedimentary,** and **metamorphic.**

Lessons 3-2 and 3-3

- **Igneous rocks** form from **magma.** Magma is **molten rock** inside Earth.
- **Lava** is magma that has reached Earth's surface.
- The minerals in an igneous rock can be used to identify the rock.
- Crystal size is influenced by the rate at which molten rock cools. Crystal size in igneous rock determines its **texture.** Igneous rocks can be classified according to their textures.

Lessons 3-4 and 3-5

- **Sediment** is small pieces of rocks and minerals and the remains of living things that settle to the bottom of water. **Sedimentary rock** is formed when sediments are cemented together by dissolved minerals.
- **Clastics** and **nonclastics** are two groups of sedimentary rock. Clastics are classified according to the size and shape of their sediments. Nonclastics form from minerals dissolved in water or from the remains of plants and animals.

Lessons 3-6 and 3-7

- **Metamorphic rock** is formed when great heat and pressure inside Earth chemically change the minerals in rocks. The heat and chemical solutions in magma can also change the minerals in rocks.
- Metamorphic rocks have two kinds of textures. The mineral crystals in **foliated** metamorphic rocks are arranged in bands. The texture of a metamorphic rock that does not have bands is called **nonfoliated.**

Lesson 3-8

- The **rock cycle** is a series of natural processes by which rocks slowly change from one kind of rock to another kind of rock.
- All rocks can change into another type of rock or back to magma. Heat and pressure can change any rock into metamorphic rock.

Key Term Challenges

clastic rock (p. 74)	nonclastic rock (p. 74)
cycle (p. 80)	nonfoliated (p. 78)
foliated (p. 78)	pluton (p. 68)
igneous rock (p. 66)	rock (p. 66)
lava (p. 68)	rock cycle (p. 80)
magma (p. 68)	sediment (p. 72)
metamorphic rock (p. 66)	sedimentary rock (p. 66)
molten rock (p. 68)	texture (p. 70)

MATCHING **Write the Key Term from above that best matches each description.**

1. rock formed from molten material
2. magma that reaches Earth's surface
3. size of crystals in a rock
4. sedimentary rocks from dissolved minerals
5. texture of metamorphic rock with mineral crystals arranged in bands
6. texture of metamorphic rock that does not have mineral crystals arranged in bands
7. rock formed when another rock is changed by heat and pressure
8. process that occurs over and over

FILL IN **Write the Key Term from above that best completes each statement.**

9. A rock that forms when rocks and minerals are cemented together is classified as a __________.
10. Conglomerates belong to a group of sedimentary rocks called __________.
11. Molten rock inside Earth is called __________.
12. The process by which rocks slowly change from one kind of rock to another is the __________.
13. Rock materials that settle in water are called __________.
14. A large body of igneous rock formed by magma cooling is called a __________.

86

Teaching Options

PREPARING STUDENTS FOR STANDARDIZED TESTS

Reading Strategy: Tell students to answer those questions that they know first, then come back and reread the remaining questions.

Writing Strategy: Remind students to make sure that they answer both parts of a two-part question.

Interpreting Visuals Strategy: Tell students to carefully note arrows or connecting lines on visuals because these elements indicate relationships between the items shown.

ESL/ELL STRATEGY

Summarizing Learning English-language learners will benefit from stating processes in their own words. Have pairs of students work together to summarize the rock cycle and each process in the rock cycle.

Content Challenges TEST PREP

MULTIPLE CHOICE **Write the letter of the term or phrase that best completes each statement.**

1. All rocks are made up of one or more
   - **a.** sediments.
   - **b.** minerals.
   - **c.** magmas.
   - **d.** metals.
2. A scientist who studies and classifies rocks is a
   - **a.** chemist.
   - **b.** geographer.
   - **c.** biologist.
   - **d.** petrologist.
3. The kind of rock formed when melted minerals cool and harden is
   - **a.** molten rock.
   - **b.** sedimentary rock.
   - **c.** igneous rock.
   - **d.** metamorphic rock.
4. A rock formed when an existing rock is changed by heat and pressure is
   - **a.** molten rock.
   - **b.** sedimentary rock.
   - **c.** igneous rock.
   - **d.** metamorphic rock.
5. Rocks formed from once-living things are usually classified as
   - **a.** molten rocks.
   - **b.** igneous rocks.
   - **c.** sedimentary rocks.
   - **d.** metamorphic rocks.
6. Magma that reaches Earth's surface cools and hardens into
   - **a.** igneous rock.
   - **b.** sedimentary rock.
   - **c.** molten rock.
   - **d.** metamorphic rock.
7. The only kind of rock that can form directly from magma is
   - **a.** sedimentary rock.
   - **b.** molten rock.
   - **c.** igneous rock.
   - **d.** metamorphic rock.
8. Crystal size in an igneous rock is determined by the rate at which the magma in the rock
   - **a.** cools.
   - **b.** melts.
   - **c.** reaches Earth's surface.
   - **d.** freezes.
9. Silt and sand are examples of
   - **a.** molten rock.
   - **b.** magma.
   - **c.** lava.
   - **d.** sediments.

TRUE/FALSE **Write *true* if the statement is true. If the statement is false, change the underlined term to make the statement true.**

10. Most nonclastic sedimentary rocks are made up of mud.
11. Magma that reaches Earth's surface is called lava.
12. Metamorphic rocks are formed when existing rocks are changed by heat and pressure.
13. The texture of a sedimentary rock that does not have minerals arranged in bands is described as nonfoliated.
14. Limestone is formed from dissolved halite.
15. Igneous rocks can be identified by their minerals.
16. Igneous rocks formed from lava usually have large crystals.
17. Marble is an igneous rock formed when limestone is heated.
18. Slate is a foliated metamorphic rock.

Content Challenges

MULTIPLE CHOICE

1. b (3-1)
2. d (3-1)
3. c (3-2)
4. d (3-6)
5. c (3-5)
6. a (3-2)
7. c (3-8)
8. a (3-3)
9. d (3-4)

TRUE/FALSE

10. false; dissolved minerals or the remains of plants and animals (3-5)
11. true (3-2)
12. true (3-6)
13. false; metamorphic (3-7)
14. false; minerals (3-5)
15. true (3-3)
16. false; small (3-3)
17. false; metamorphic (3-7)
18. true (3-7)

ALTERNATIVE ASSESSMENT

Drawing Diagrams Students may sketch a diagram of the rock cycle in their Science Logs. Make sure that students label all parts of the diagram without referring to the textbook or their notes.

PORTFOLIO ASSESSMENT

Making Student Portfolios Students can demonstrate their comprehension of the concepts in this chapter by making a portfolio. The Chapter Self-Check, the Big Idea Planner, and the Weekly Journal are some of the reproducibles from the Teacher's Resources that they can include in their portfolios. Students can also suggest items from the chapter, such as their Big Idea Science Log Entry or Lab Activity Report, to include. You can use the Portfolio Assessment Rubric also found in the Teacher's Resources to assess students' portfolios.

Concept Challenges

WRITTEN RESPONSE

1. **EXPLAIN:** Rocks can change types, depending on the forces that act on them, but they can never be destroyed. (3-8)
2. **CLASSIFY:** Pumice is an igneous rock. (3-3)
3. **EXPLAIN:** Molten rock inside Earth is called magma. Magma that reaches Earth's surface is called lava. (3-2)
4. **CONTRAST:** Igneous rocks formed from lava usually have smaller crystals or no crystals at all because they cooled much more quickly than igneous rocks that formed from magma. (3-3)
5. **CLASSIFY:** Slate is foliated metamorphic rock because it has bands and breaks into sheets along the bands. (3-7)

INTERPRETING VISUALS (3-8)

6. igneous, sedimentary, and metamorphic
7. It cools and hardens.
8. weathering and erosion
9. They become magma.
10. through heat and pressure
11. igneous and metamorphic
12. igneous
13. igneous, sedimentary, and other metamorphic
14. It never ends. It is a continuous cycle.

Chapter Project Wrap-Up

ROCK TESTS

You can use the Individual Activity Rubric found in the Teacher's Resources to assess students' projects. Fill in the rubric with the additional information below. For this project, students should have:

- explained the tests they plan to use.
- prepared a summary of their results.

Concept Challenges TEST PREP

WRITTEN RESPONSE **Answer each of the following questions in complete sentences.**

1. **EXPLAIN:** The law of conservation of matter states that matter cannot be created or destroyed. However, matter can be changed from one form to another. Explain how the rock cycle supports the law of conservation of matter.
2. **CLASSIFY:** Pumice is a kind of rock often formed from volcanic lava. In what class of rocks should pumice be classified?
3. **EXPLAIN:** What is the difference between magma and lava?
4. **CONTRAST:** How do igneous rocks formed from lava differ from igneous rocks formed from magma?
5. **CLASSIFY:** Slate is a metamorphic rock that breaks into sheets along its crystal bands. Is slate foliated or nonfoliated metamorphic rock? Explain your answer.

INTERPRETING VISUALS **Use Figure 3-28 below to answer the following questions.**

6. What are the three types of rock?
7. How does magma form igneous rock?
8. What processes break down igneous rocks into sediments?
9. What happens when the minerals in igneous, sedimentary, and metamorphic rocks melt?
10. How do igneous rocks change into metamorphic rocks?
11. What are two kinds of rock that can form sedimentary rock?
12. What is the only kind of rock that can form directly from magma?
13. What kinds of rocks can form metamorphic rocks?
14. When does the rock cycle end? Explain.

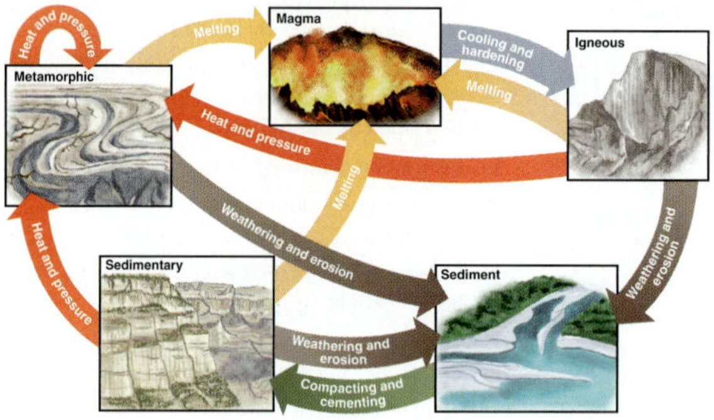

◀ **Figure 3-28** The rock cycle

Teaching Options

• TEACHING RESOURCES •
Online or Print

Teacher's Resources

Key Term Review: Chapter 3, p. 13

Chapter 3 Test: pp. 14–15

Chapter Self-Check: Teaching Tools, p. 11

Big Idea Planner: Study Tools, pp. 11–12

Weekly Journal: Study Tools, p. 19

Portfolio Assessment Rubric: Study Tools, p. 10

PLANNING GUIDE

◆ **TEACHING THE CHAPTER** This chapter should take approximately 6–9 days to complete instruction and assessment.

	Skills and Features	Projects and Activities	Achieve Science Literacy	Meet Individual Needs
Chapter 4 Opener p. T89	RESEARCH, ILLUSTRATE	• Chapter Project		
Lesson 4-1 **How are organisms preserved?** pp. T90–T91 Standards: D1k, D2b	DEFINE, HYPOTHESIZE, PREDICT • Investigate • Web InfoSearch	• Cooperative Learning • Extending the Lesson	• Study Hint • Ongoing Assessment	• ESL/ELL Strategy Teacher's Resources* Enrichment Activity
Lab Activity **Modeling a Fossil Formation** pp. T92–T93 Standards: D1k, D2b	OBSERVE, ANALYZE, APPLY • Set-Up Time: 15 min • Lab Time: 90 min over 2 days	• Extending the Lesson • Extend the Lab Activity	• Tips for Using Technology • Ongoing Assessment	• ESL/ELL Strategy Teacher's Resources* Lab Activity Report Database Planner Weekly Journal
Lesson 4-2 **Why do scientists study fossils?** pp. T94–T95 Standards: D1k, D2b	EXPLAIN, ANALYZE, OBSERVE • How Do They Know That?	• Cooperative Learning • Extending the Lesson • More on the Web • Integrating Life Science	• Study Hint • Reading Strategy	Teacher's Resources* Enrichment Activity Fishbone Map
Lesson 4-3 **How do fossils help us date rocks?** pp. T96–T97 Standards: D1k, D2b	INFER, DESCRIBE, MODEL • Building Science Skills • Hands-On Activity	• More on the Web • Cooperative Learning	• Study Hint	• Reteach Teacher's Resources* Enrichment Activity Concept Map
Lesson 4-4 **How is a rock's absolute age determined?** pp. T98–T99 Standards: D1k, D2b	DESCRIBE, INFER, EXPLAIN • Health and Safety Tip • Science and Technology	• More on the Web • Integrating Physical Science • Lab Challenge	• Study Hint	
Lesson 4-5 **What is the geologic time scale?** pp. T100–T101 Standards: D1k, D2a, D2b	DESCRIBE, INFER, MODEL • Investigate • Building Science Skills • Integrating Life Science	• Extending the Lesson	• Study Hint	Teacher's Resources* Enrichment Activity

*www.conceptsandchallenges.com, password: ccearth09 *Planning Guide continues on next page.*

Standards: For details on the correlation to National Science Standards see pages *xx–xxi*.

	Skills and Features	Projects and Activities	Achieve Science Literacy	Meet Individual Needs
Lesson 4-6 **What are fossil fuels?** pp. T102–T103 Standards: D1e, D1k, D2b, D3d	IDENTIFY, NAME, INFER • People in Science	• Extending the Lesson • More on the Web • Integrating Physical Science	• Study Hint	• Reteach
Big Idea **What products are made from oil?** pp. T104–T105 Standard: D1k	RESEARCH, COMMUNICATE, ANALYZE • Science Log: Writing Activity	• Big Idea • Close Activity	• Tips for Using Technology	• Reteach Teacher's Resources* Big Idea Planner
Chapter 4 Challenges pp. T106–T108	• Chapter Summary • Key Term Challenges • Content Challenges • Concept Challenges	• Chapter Project Wrap-Up	• Study Hint • Preparing Students for Standardized Tests	• ESL/ELL Strategy Teacher's Resources* Chapter Self-Check Weekly Journal

*www.conceptsandchallenges.com, password: ccearth09

Teacher Notes

Chapter 4 The Rock Record

▲ **Figure 4-1** A plant fossil imprint in sedimentary rock

Fossils are important tools for reconstructing the distant, or geologic, past. They give us clues as to what life was like on Earth millions of years ago, when the environment was different. Knowing the different species that existed at different times can help us guess what the climate and land were like at those times. Fossils can also help scientists match rocks of similar ages from different locations.

▶How might the land during earlier ages have been different from today?

Contents

Teaching Options

ASSESSMENT PLANNER

For assessment in the Student Edition, see the following pages:

Content Assessment

Checking Concepts: pp. 91, 95, 97, 99, 101, 103

Thinking Critically: pp. 91, 95, 97, 99, 101, 103

Interpreting Visuals: pp. 95, 103

Chapter Challenges: pp. 106–108

Alternative Assessment

Building Skills: pp. 97, 101

Web InfoSearch: p. 91

Science Log: p. 105

Performance-Based Assessment

Hands-On Activity: p. 97

Lab Activity: pp. 92–93

• TEACHING RESOURCES • Online or Print

Teacher's Resources

Lesson Review: Ch. 4, pp. 2, 3, 5, 6, 7, 9

Enrichment: Ch. 4, pp. 4, 8

Key Term Review: Ch. 4, p. 11

Chapter 4 Test: pp. 12–13

Laboratory Manual: pp. 13–14, 37–38

Chapter 4 The Rock Record

Chapter Overview

In this chapter, students will learn about fossils and how fossil remains are preserved. Students will see how to determine the relative and absolute ages of a rock. They will read and interpret the geologic time scale and learn about fossil fuels.

About Figure 4-1 Fossils are the remains or imprints of living things that lived long ago. Fossils help scientists find links between the plants and animals of today and their ancient ancestors. By studying fossils and the land in which they are found, scientists can trace how the land and the living things on it have changed over time.

Answer Possible answer: Some areas that are now dry land were once under water, and some modern oceans and seas may have once been dry land. Some of today's mountain ranges were once the floors of shallow seas.

Linking Prior Knowledge

For this chapter, students should recall:

- what extrusions and intrusions are (Lesson 3-2).
- how sedimentary rocks form (Lesson 3-4).

Chapter Project

FOSSIL BULLETIN BOARD

MATERIALS: reference works, drawing and writing materials, scissors, paper, stapler

As students read the chapter, they should create a bulletin board display showing a visual summary of Earth's geologic history. Have students research each period and together decide which events seem to be most significant during each period. Encourage students to come up with descriptive titles for each period. For example, the Devonian could be called "When Fish First Swam," and the Cretaceous could be called "Goodbye to the Dinosaurs." Students should draw or find images, make labels, and write captions or short summaries to include on the display. **learning styles: visual, linguistic**

4-1 (pp. 90–91)

1 Introduce

INVESTIGATE

TIME: 15 minutes

PURPOSE: Students will use a time capsule to model the burial and preservation of fossil remains.

MATERIALS: very large plastic container with lid, various objects that fit in container, paper, pencil

PROCEDURE: Have students read the activity and think about the message they want to send to people who will open the time capsule in the future. Suggest that they bring in inexpensive personal items that have some meaning to them (such as buttons, photographs, or souvenirs) for inclusion in the time capsule. **learning style: tactile**

THINK ABOUT IT: Students should explain the reasoning for including certain objects and their choice of burial location.

STUDY HINT Before beginning this lesson, have a volunteer read aloud the definition of *fossil* from the Key Terms list. Then, write the heads *Rock, Ice and Tar,* and *Amber* on the board. Have students use the text to identify fossils that belong under each head.

Linking Prior Knowledge Ask students why people freeze food. Make sure that students understand that freezing helps preserve food by preventing decay. Explain that fossils are preserved in a variety of ways, one of which is freezing.

2 Teach

Demonstration Slowly pour thick maple syrup into a flat pan. Explain that syrup is made from the sap of a maple tree. Allow a volunteer to touch the syrup and describe it to the class. Discuss the properties that syrup shares with sap, which make both ideal for trapping and preserving small organisms. **learning style: tactile**

Discussion Show students photographs or actual samples of fossils. Point out that fossils are the remains or traces of organisms that lived long ago. Identify each type of fossil. Make sure that students understand the difference between a mold and a cast. **learning style: visual**

4-1 How are organisms preserved?

INVESTIGATE

Making a Time Capsule

HANDS-ON ACTIVITY

1. A time capsule is one way to preserve objects and information for people of the future. To make your own time capsule, get a very large plastic container with a lid.
2. Put an object in the container. Have five of your classmates each place an object in it.
3. Write something down about each of the objects in the container. Put the note inside.
4. Close the lid.
5. Find an appropriate place to bury the container in the ground.

THINK ABOUT IT: Why did you include the objects you did in your time capsule? Why did you bury the time capsule where you did?

Objectives

Define fossil. Describe how organisms are preserved in rock and elsewhere.

Key Terms

fossil (FAHS-uhl)**:** remains or traces of an organism that lived long ago

mold: imprint or hollow in rock that is shaped like and made by an organism

cast: mold filled with hardened sediments

amber: hardened tree sap

What Are Fossils? The remains or traces of organisms that lived long ago are called **fossils.** A fossil can be a bone, a whole skeleton, an imprint, a shell, or even the body of an ancient species.

▲ **Figure 4-2** *Tyrannosaurus rex* was a very large, carnivorous dinosaur that lived more than 65 million years ago.

Many species that lived long ago are extinct. That means they are no longer found alive on Earth. Dinosaurs are extinct. However, their remains have been found in many places.

1 DEFINE: What are fossils?

Fossils in Rock Most fossils are found in sedimentary rock that was once under water. It takes millions of years for fossils to form. A fossil begins to form when an organism dies and is buried quickly by sediment. The soft parts of the organism decay. Usually, only the hard parts, such as shells and bones, are left. Through chemical processes, the sediments slowly harden into rock.

Sometimes, as the organism decays, it leaves a hollow opening in a rock. This hollow opening is called a **mold.** Later, the mold may fill up with sand or mud that later hardens. This material forms a **cast.**

Some fossils are only imprints, or marks, of the organisms in the rock. For example, footprints are made when an animal steps into soft mud. The mud hardens into rock, preserving the footprint. Scientists have also found many fish and leaf imprints.

▲ **Figure 4-3** Some fish imprints show the bones of the fish.

2 NAME: What kind of rock contains the most fossils?

Teaching Options

COOPERATIVE LEARNING

Visiting a Museum If possible, arrange for the class to visit a natural history museum or have students use the Internet to take a virtual museum tour. Have pairs of students observe the different types of fossils in the museum and list ten different fossils they observed. For each fossil, have students record the era and the environment in which the organism lived.
learning style: visual

EXTENDING THE LESSON

Researching Dinosaur Extinction Have students research theories of why dinosaurs became extinct. Students can use library resources or the Internet and present their findings to the class in oral reports. **learning style: auditory**

ESL/ELL STRATEGY

Analyzing Vocabulary Explain to English-language learners the different meanings of the nouns *mold* and *cast.* Have students research the derivations of these words to better understand and remember them.

Fossils in Ice and Tar Freezing helps preserve things by preventing decay. The bodies of about 50 elephantlike extinct animals called mammoths have been found frozen in Siberian and Alaskan soil and ice. Woolly rhinoceroses have also been found.

▲ **Figure 4-4** Museum model of a woolly mammoth

Animal remains have also been found in ancient pits of thick, sticky tar. These tar pits often become covered by water. Animals that went to them to drink were trapped in the tar. Other animals that tried to eat the trapped animals also became trapped. The animals sank into the tar and were preserved. In southern California, the bones of extinct animals such as saber-toothed cats and dire wolves have been found in the LaBrea tar pits. Parts of extinct camels, mastodons, vultures, and bison have also been found there. Tarlike oil deposits in Poland contained entire specimens of woolly rhinoceroses.

▲ **Figure 4-5** The skeleton of a saber-toothed cat

3 IDENTIFY: How are animals trapped in tar pits?

Fossils in Amber

A clear, sticky sap flows from some kinds of trees. This hardened tree sap is called **amber.** Millions of years ago, many insects and leaf fragments became trapped in the sticky sap. More sap covered these organisms. The sap hardened, and the organisms were perfectly preserved inside.

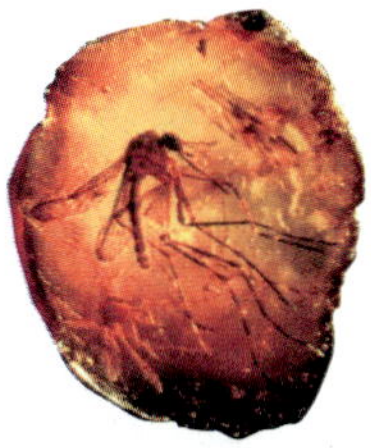

▲ **Figure 4-6** Insects are often found perfectly preserved in amber.

4 DEFINE: What is amber made from?

CHECKING CONCEPTS

1. What four substances help preserve the remains of organisms?
2. What animal parts are most often preserved?
3. Which extinct animals were preserved in ice?
4. What else besides plants are usually preserved in amber?

THINKING CRITICALLY

5. INFER: How do we know that dinosaurs existed?
6. HYPOTHESIZE: Why are fossils rarely found in igneous rocks?
7. PREDICT: Which of the following would most likely be found as a fossil: a clam shell, a flower petal, or an ancient worm? Why?

Web InfoSearch

Petrified Fossils Sometimes fossils are preserved by a process called petrification. To *petrify* means to "turn to stone."

SEARCH: Use the Internet to find out more about petrification. How does the process work, and exactly what does it do to the organism? Start your search at www.conceptsandchallenges.com. Some key search words are **fossils, petrified forest,** and **petrification.**

Answers to questions

1 DEFINE: the remains or traces of organisms that lived long ago

2 NAME: sedimentary rock

3 IDENTIFY: The animals are trapped in the tar when they try to drink the water that covers tar pits or when they try to pull out other trapped animals to eat.

4 DEFINE: hardened tree sap

3 Assess

CHECKING CONCEPTS ANSWERS

1. sediments, frozen soil and ice, tar, and sap
2. hard parts, such as shells and bones
3. mammoths and wooly rhinoceroses
4. insects

THINKING CRITICALLY ANSWERS

5. INFER: We infer their existence from their fossil remains.
6. HYPOTHESIZE: The conditions under which igneous rocks form do not favor preservation of the remains of organisms.
7. PREDICT: a clam shell, because it is hard

4 Close

Web InfoSearch

Petrified Fossils Have students use the Internet to find out more about petrification. How does the process work, and exactly what does it do to the organism? Some key search words are **fossils, petrified forest,** and **petrification.** Have students begin their search for information at www.conceptsandchallenges.com.

ONGOING ASSESSMENT

Have students explain how the fossil in Figure 4-6 formed.

• TEACHING RESOURCES •
Online or Print

Teacher's Resources

Lesson Review: Chapter 4, p. 2

Enrichment: Chapter 4, p. 4

Lab Activity

(pp. 92–93)

1 Prepare the Lab

Modeling a Fossil Formation

SET-UP TIME: 🕒 **LAB TIME: 90 MINUTES OVER 2 DAYS**

BACKGROUND: Sedimentary rocks provide a record of what Earth was like millions of years ago. Some of these rocks contain fossils. A fossil is the remains or traces of an organism that lived long ago. Fossils tell geologists what ancient organisms may have looked like and provide clues about the kind of environment in which the organisms lived.

PURPOSE: Students will model basic fossil-forming processes.

ALTERNATIVE MATERIALS: You could substitute premixed patching plaster for plaster of Paris and modeling dough for modeling clay. You could substitute homemade modeling dough; recipes for it can be found on the Internet.

SCIENCE SKILLS: Students will **observe** how a shell leaves an impression in clay and **model** fossil formation.

ADVANCE PLANNING: Have paper towels available for cleaning up.

2 Guide the Lab

PROCEDURE: You may wish to divide the class into groups of two or three students. Discuss the procedures. Tell students that they will conduct an experiment and later make their own fossils. Explain the difference between a mold and a cast. Provide students with a copy of the Lab Activity Report found in the Teacher's Resources for students to record their observations and conclusions.

LAB ACTIVITY

Modeling a Fossil Formation

Materials
Safety goggles, lab apron, modeling clay, sugar cubes, small jar or disposable plastic food tub, wax paper (15 cm square), small seashells or snail shells, plaster of Paris, water, mixing tub, large spoon, toothpicks

▲ **STEP 2** Press a sugar cube into the clay.

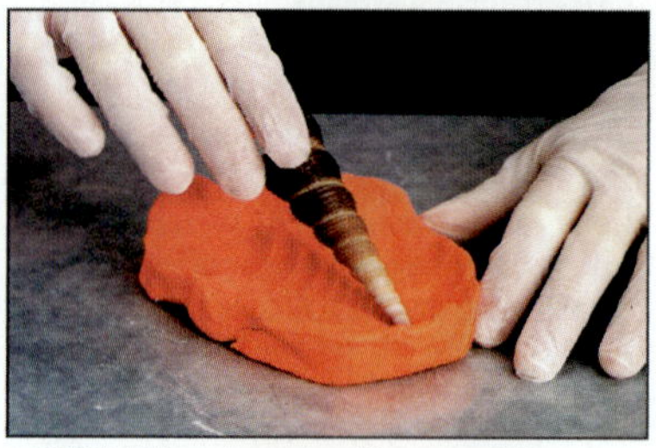

▲ **STEP 5** Make an impression with your seashell.

BACKGROUND

Many sedimentary rocks contain the remains of ancient animal and plant life. These remains are called fossils. A seashell, bone, or leaf is buried in sediment. Its remains may leave an impression in the sediment. Fossil impressions are called molds. Later, more sediment fills in the mold and hardens. These fossils are called casts.

PURPOSE

In this activity, you will experiment with creating fossil molds and casts.

PROCEDURE

1. Put on your lab apron and goggles.
2. Place the square of wax paper on your table and spread a small lump of clay on it. Press a sugar cube into the clay.
3. Remove the clay with the sugar cube from the wax paper and place them in a container with cold water. Set it aside and leave it overnight.
4. Place a larger lump of clay on the wax paper. Flatten it out. Pinch up the edges to form a shallow dish.
5. Press a small seashell or snail shell into the clay to make an impression. Gently remove the shell. You can make more impressions with different shells if there is extra clay.

92

Teaching Options

EXTENDING THE LESSON

Modeling More Fossils Once students have modeled fossil formation in rocks, have small groups outline a procedure for modeling another type of fossil formation. For example, they might describe how to preserve a leaf in ice. Once you have approved their procedure, allow students to carry out their plan. Have each group describe how their model is like a fossil.
learning style: tactile

ESL/ELL STRATEGY

Reviewing the Procedure Review the procedure with students and allow them to ask questions. Show English-language learners an example of the expected outcome of the activity to clarify what they should be aiming for.

ONGOING ASSESSMENT

Have students write a journal entry explaining why geologists spend so much time studying fossils.

6. Mix the plaster of Paris with just enough water to make it easy to pour.

7. Gently spoon plaster into your dish to fill it. You must leave the plaster to set overnight. Place a small name tag next to your clay dish to identify it.

8. After a full day has passed, examine your sugar cube experiment and answer the questions below. Then, gently remove the clay from the outside of your plaster fossil. If any bits stick to the model, you can clean them off with toothpicks. Compare your fossil to the shell that made it. Copy the chart in Figure 4-7 and record your observations.

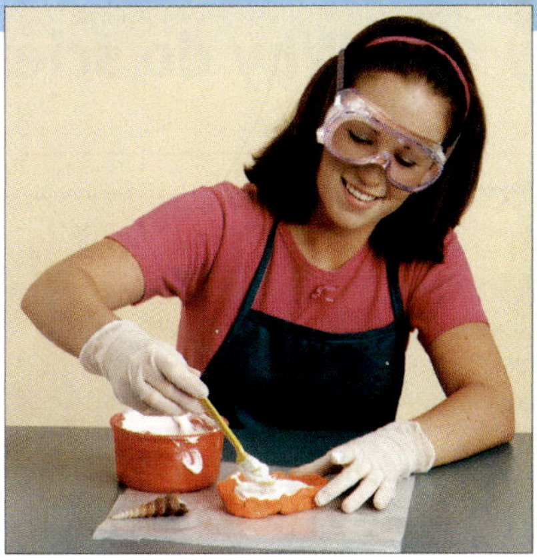

▲ **STEP 7** Spoon enough plaster of Paris into the clay dish to fill it.

▲ **STEP 8** Compare your plaster fossil shell with the shell from which it was made.

Fossil Observations

Sugar cube	Shell

▲ **Figure 4-7** Use a copy of this chart to record your observations.

CONCLUSIONS

1. OBSERVE: What happened to the sugar cube?
2. ANALYZE: Why did the change take place?
3. APPLY: How are fossils of life forms made in rocks?
4. IDENTIFY: What two kinds of fossils did you make?

TROUBLESHOOTING: Students may have difficulty making molds of their shells. Shells should be pressed straight down into the clay without overlap.

Make sure that the clay rim is pinched upward to keep the plaster from spilling. You may want to mix the plaster of Paris yourself. It must be used within about 10 to 15 minutes after you mix it.

SAFETY TIP: Clean up plaster spills immediately. Flush plaster-stained carpeting and clothing with cold water until the residue is gone.

EXPECTED OUTCOME: Students will learn how organisms leave impressions in sediment that eventually form fossils. They will learn the difference between fossil casts and fossil molds.

LAB HINT: As the fossils begin to set, students can write their names in the plaster with the tip of a pencil.

3 Conclude the Lab

1. **OBSERVE:** It disappeared and left its shape in the clay.
2. **ANALYZE:** The sugar dissolved in the water.
3. **APPLY:** Dead animals and plants are buried by sediment. When the sediment hardens into rock, the shape of the plant or animal remains.
4. **IDENTIFY:** mold and cast

Use the Laboratory Activity Rubric from the Teacher's Resources to assess students' lab activities. Fill in the rubric with the additional information below. For this activity, students should have:

- created fossil molds and casts.
- recorded their observations on a chart.

4 Extend the Lab

Have students create other fossil models by pressing leaves, dead beetles, chicken bones, and other materials into clay. Invite students to bring in and display fossils that they have.

Tips for Using Technology

Using Database Software for Observations Have students use database software to record their observations on modeling a fossil formation. Databases allow students to enter and edit information, store and organize a lot of information easily, and search for specific information. Students can also use database software to share their data with the class. Remind students to enter information into the file carefully. Have students use the Database Planner graphic organizer from the Teacher's Resources to plan their databases.

• TEACHING RESOURCES • Online or Print

Teacher's Resources

Lab Activity Report: Study Tools, pp. 16–17

Laboratory Activity Rubric: Teaching Tools, p. 8

Database Planner: Study Tools, p. 22

Weekly Journal: Study Tools, p. 19

4-2 *(pp. 94–95)*

1 Introduce

STUDY HINT Before beginning this lesson, read the Key Terms aloud so students can hear their pronunciations.

Linking Prior Knowledge Ask students to name and describe some of the features of organisms that they think can be preserved as fossils, for example, body structure, size, injuries it suffered, stomach contents, and so on. Then, ask what scientists might infer about the lives of the organisms by studying fossils.

2 Teach

Discussion Write the title of the lesson on the board. Ask students why they think scientists study fossils. Discuss all student responses. Describe why fossils are clues to Earth's history. **learning style: auditory**

Reinforcement Refer students to Figure 4-9, showing the evolution of the horse. Have students use the Fishbone Map from the Teacher's Resources to show how the horse has changed through time. Encourage them to note differences in the size of each horse and in their skulls, teeth, and legs. **learning style: visual**

Answers to questions

1. **LIST:** Accept any four of the following: bones, shells, teeth, tracks, burrows, coprolites, or gastroliths.
2. **EXPLAIN:** The horse has changed in size, structure of the legs, shape of the teeth, and number of toes.
3. **INFER:** The locations of certain fossils show that Earth's climate has changed.

READING STRATEGY As students read through this lesson, ask them to keep in mind the question that begins it. It may be very obvious why life scientists study fossils. They help us understand the history of life on Earth. However, students should ask themselves why Earth scientists would [e] interested in fossils. If they can answer [t]hey will understand the lesson.

4-2 Why do scientists study fossils?

Objective

Explain how fossils are clues to Earth's history.

Key Terms

coprolite (KAHP-roh-lyt)**:** fossilized dung or the stomach contents of ancient animals

gastrolith (GAS-troh-lihth)**:** stone used to grind food

The History of Life on Earth Fossils show that many kinds of organisms lived at different times in Earth's history. Many species are now extinct. Dinosaurs lived between 65 million and 220 million years ago. There were hundreds of species of dinosaurs. Not one dinosaur lives on Earth today. Saber-toothed cats, giant sloths, giant deer, and tiny shelled animals called trilobites are also examples of extinct animals.

▲ **Figure 4-8** Giant giraffes with antlers once roamed Earth.

Animal bones, shells, and teeth give us clues to past life. So do animal tracks. Another type of fossil that helps us to understand life in the past are animal burrows. These holes or tubes were made by animals in sediment, wood, or rock. They were later filled with sediment and became preserved. Some of the oldest known fossils of this type are worm burrows.

Coprolites are fossilized dung or the stomach contents of ancient animals. Coprolites can tell us what the organisms ate. Some ancient reptiles, including dinosaurs, may have swallowed stones that were used during digestion to grind food. These fossilized stones, which are highly polished, are called **gastroliths.**

1. **LIST:** What are four kinds of fossils?

Changes in Living Things Fossils show that plants and animals have changed over time. For example, many fossils of horses have been found. Look at Figure 4-9. These fossils show changes in the animal's size and the number of toes. There was once a four-toed horse about the size of a dog. Fossils of horses also show that changes occurred in their teeth and legs. Today, horses have hooves, not toes, and are fairly large. Other fossils show that modern-day animals such as elephants, giraffes, and camels are related to species that are now extinct.

2. **EXPLAIN:** How has the horse changed over millions of years?

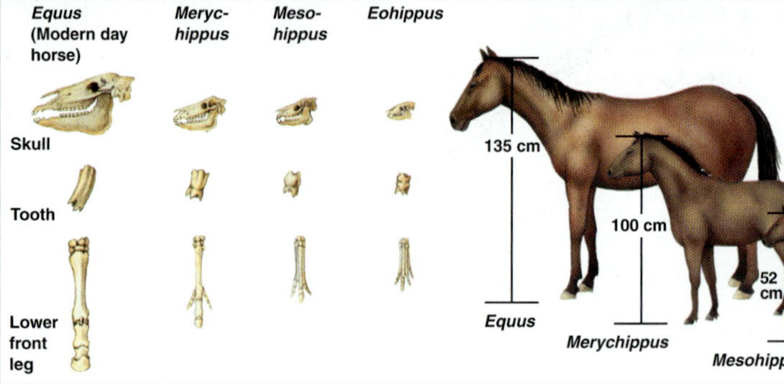

Figure 4-9
Fossil record of horses

Teaching Options

INTEGRATING LIFE SCIENCE

Locating Endangered Species Explain that many organisms living on Earth today are near extinction. These organisms are called endangered species. Endangered species can be plants, mammals, reptiles, amphibians, birds, or insects. Have students write to your state's wildlife agency for information on endangered species in the United States. Have them trace an outline map of the United States and write the name of each organism in the area where it lives. Suggest that students use a different color for each type of organism. Have them make a map key that shows what color indicates each type of organism.

EXTENDING THE LESSON

Discovering Early Fossils The Handyman discussed in the feature on page 95 is classified as *Homo habilis*. Have students research the discoveries of other early human fossils. Tell students to find out about the genus Australopithecus and the species *Homo erectus*. Have students present their findings orally. **learning style: linguistic**

Past Climates and Topography Fossils show that Earth's climate and surface have changed over time. Fossils of alligator-like animals have been found in Canada. Alligators usually live in warm climates. The fossils show that Canada once had a warmer climate than it has today.

Fossils of ferns have been found in Antarctica. These indicate that Antarctica was once very warm. Today, Antarctica is covered with ice and snow. Fossils of coral have been found in Arctic ice.

Fossils of ocean animals have been found in the Andes Mountains in South America. The Andes are 4,000 m above sea level. Scientists believe that the Andes were once under the ocean.

3 **INFER:** How do we know that climates change through the ages?

CHECKING CONCEPTS

Write an E for each extinct organism. Write an N for each living organism.

1. Mammoth
2. Saber-toothed cat
3. Horse
4. Alligator
5. Trilobite
6. Elephant

THINKING CRITICALLY

7. **ANALYZE:** What do fossils tell us about Earth's surface and climate?
8. **ANALYZE:** What do fossils tell us about living things?

INTERPRETING VISUALS

Use Figure 4-9 to answer the following questions.

9. **OBSERVE:** How many toes did the earliest horses have?
10. **OBSERVE:** Which horse has the largest skull?
11. **ANALYZE:** How has the size of the horse changed over time?

How Do They Know That?

ANCIENT HUMANLIKE SPECIES

▲ **Figure 4-10** Mary and Louis Leakey

Imagine finding a humanlike skull that is millions of years old. In the late 1950s, the Leakeys did just that. The Leakeys were a family of anthropologists. Anthropologists are scientists who study the history and development of humans and humanlike species.

Mary Leakey and her husband Louis studied fossils for more than 50 years. Much of their work involved looking for fossils of early humans or humanlike species. The Leakeys looked mostly in Tanzania, in the Olduvai Gorge.

In 1959, Mary Leakey found part of a humanlike skull that was more than 1.75 million years old. Over the years, Mary, Louis, and their son, Richard, found other humanlike fossils. Because he found stone tools at one site, Richard Leakey called the humanlike species he found nearby "Handyman."

Louis died in 1972. Mary died in 1996. Richard continues to study fossils.

Thinking Critically What did the Leakeys infer when they found ancient stone tools near the humanlike fossils?

MORE ON THE WEB

Careers in Science Have students learn more about paleontologists by visiting www.conceptsandchallenges.com. Then, have students write a job description for a paleontologist, using the information they gather on the Internet.

• TEACHING RESOURCES • Online or Print

Teacher's Resources

Lesson Review: Chapter 4, p. 3

Enrichment: Chapter 4, p. 4

Fishbone Map: Study Tools, p. 5

COOPERATIVE LEARNING

Playing a Memory Game Have students create pairs of flash cards with related terms from this lesson. For example, one card could have "three-toed horse," and its companion could have "Mesohippus." With all cards facedown, have students match each card with its companion by memory.

3 *Assess*

CHECKING CONCEPTS ANSWERS

1. E
2. E
3. N
4. N
5. E
6. N

THINKING CRITICALLY ANSWERS

7. **ANALYZE:** Fossils show that Earth's surface and climate have changed over time. Warm environments, for example, have turned cold. Some mountain crust was once ocean floor.
8. **ANALYZE:** Fossils show that living things have changed over time. For example, some types of organisms have grown larger. Others have gotten smaller. Still others have become extinct.

INTERPRETING VISUALS ANSWERS

9. **OBSERVE:** 4
10. **OBSERVE:** *Equus*
11. **ANALYZE:** The size has increased.

4 *Close*

How Do They Know That?

ANCIENT HUMANLIKE SPECIES

Discussion Ask students to locate Tanzania on a classroom map of the world. Have them also locate Olduvai Gorge if possible.

Explain that Olduvai Gorge is a rift in Earth's surface, similar to but smaller than the Grand Canyon. Students will learn how such rifts form in Chapter 7. Olduvai Gorge was once a large lake. Millions of years ago, ash from the eruptions of surrounding volcanoes, along with sediments carried by streams, were deposited in the lake bed. These sediments contained remains of ancient organisms, including early humans. Then, an earthquake drained the lake. Later, the Olduvai River eroded through the sediment layers, forming Olduvai Gorge and exposing the fossil-containing rock layers. Louis Leakey first explored Olduvai Gorge in 1931.

Thinking Critically Answer

They inferred that the humanlik ... ies that lived in that area could use ...

4-3 *(pp. 96–97)*

1 Introduce

STUDY HINT Before beginning this lesson, have students read the heads in the lesson. Help them speculate on what might be covered under each head.

Linking Prior Knowledge Have students review how sediments are deposited and how sedimentary rocks form (Lesson 3-4). Emphasize that sediments are deposited one layer on top of the other.

2 Teach

Discussion Describe the law of superposition and the relative ages of rocks. Make sure that students understand that unless it is disturbed, a rock layer is older than the one above it and younger than the one below it. Refer students to Figure 4-11 and ask them to describe the relative age of rock layer B. (It is older than layer A and younger than layer C.) **learning style: visual**

Demonstration Demonstrate how rock layers are deposited by filling a clear plastic shoebox first with a layer of sand, next with a layer of gravel, and then with soil. Ask students: *Which layer is on the bottom?* (sand) *Which is on the top?* (soil) Relate this to the law of superposition and how it can be used to determine the relative age of the three layers in the shoebox.
learning style: visual

Reinforcement Describe how index fossils help scientists determine the relative ages of rock layers. Have students use the Concept Map from the Teacher's Resources to list the criteria needed for a fossil to be classified as an index fossil. For students who have difficulty grasping the concept, clarify why an organism must have lived during only a short part of Earth's history to be an index fossil. **learning style: logical/mathematical**

Answers to questions

1 **OBSERVE:** A

2 **INFER:** It is older than layer B and younger than layer D.

3 **EXPLAIN:** to determine the relative ages of rock layers

4-3 How do fossils help us date rocks?

Objective
Describe how the relative ages of fossils and rock layers can be determined.

Key Terms
law of superposition: states that each undisturbed sedimentary rock layer is older than the layer above it
relative age: age of an object compared to the age of another object
index fossil: fossil used to help determine the relative age of rock layers

Reading Rock Layers Rock sediments are carried from one place and deposited in another by water and wind. These sediments pile up layer upon layer. The bottom layer is deposited first. Each layer is deposited on top of other layers. The sediments are pressed together and harden into sedimentary rock layers that form beds of rock.

The **law of superposition** states that each undisturbed rock layer is older than the layer above it. This law is used by scientists to read rock layers.

Where would you expect to find the oldest layer? Usually, the bottom layer is the oldest layer. The youngest layer is the top layer.

▲ **Figure 4-11** Rock formed in layers will usually follow the law of superposition.

1 **OBSERVE:** Which rock layer is the youngest in the rock bed illustrated in Figure 4-11?

Relative Age Using the law of superposition, scientists can tell the **relative age** of a rock layer. Relative age is the age of an object compared to the age of another object. Relative age does not tell the exact age of a rock layer.

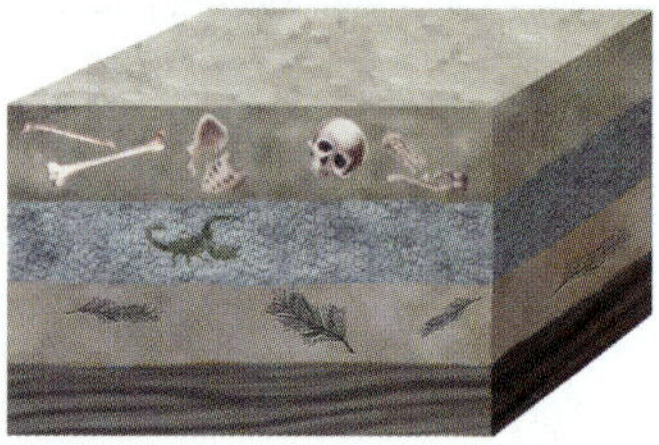

▲ **Figure 4-12** The relative age of a rock layer tells scientists only that one rock layer is older or younger than another rock layer.

2 **INFER:** What is the relative age of rock layer C in Figure 4-12?

Index Fossils Fossils that can be used to help determine the relative age of rock layers are called **index fossils.** To be an index fossil, a species must have existed for only a short time in Earth's history. Many fossils of this short-lived species need to be recovered from rock layers. These fossils must have a wide geographic range. The fossil organisms must also be unique.

Many kinds of trilobites and graptolites are used as index fossils. Trilobites appeared about 590 million years ago and lived until 250 million years ago. They were small, shelled animals that lived in the ocean. There were many different species of trilobites.

◀ **Figure 4-13** Trilobite and graptolite

Teaching Options

COOPERATIVE LEARNING

Determining Relative Age To help students understand what the relative age of a rock means, find out the relative ages of students. Have students write the month and year they were born. Then, have them stand in a line. Have each student show his or her date of birth to the student on either side. Then, have students say whether they are older or younger than the students on their right and left. Point out that students are not saying their exact age aloud. They are describing their age relative to the students standing next to them.
learning style: logical/mathematical

MORE ON THE WEB

Examining Index Fossils Have students use the Internet to learn more about index fossils. Have students share how index fossils can help us determine the relative age of sedimentary rock layers?

Trilobites had three body parts. They also had a skeleton made of chitin (KY-tihn). Chitin is what your fingernails are made of. If a sedimentary rock contains trilobite fossils, it must have formed between 590 and 250 million years ago, when trilobites were still living. Graptolites appeared about 500 million years ago and lived until 335 million years ago.

Index fossils can also be used to date rock layers from two different parts of the world. Suppose rock layers found in different places contain fossils of the same species of trilobite. Scientists can infer that the layers are about the same age.

EXPLAIN: What are index fossils used for?

CHECKING CONCEPTS

1. The youngest rock layer is usually the __________ layer.
2. Trilobites are examples of __________ fossils.
3. If one rock layer is older than another, you know the __________ ages of the rock layers.

THINKING CRITICALLY

4. DESCRIBE: What four qualities make up an index fossil?
5. MODEL: Draw a rock bed with four layers. In one layer, there are 25 trilobite fossils. In another, there are 37 graptolite fossils. The oldest layer is about 600 million years old. Label the oldest and youngest layers and the trilobite and graptolite layers.

BUILDING SCIENCE SKILLS

Modeling Two igneous rock formations can be used to find relative age. These formations are called extrusions and intrusions. They are always younger than the sedimentary rock beds in which they are found. You learned earlier what an extrusion and an intrusion are. Draw a model of a rock bed that has an intrusion and an extrusion. Color the oldest layer in your drawing red. Color the youngest layer yellow.

Hands-On Activity

MODELING HOW SEDIMENTS ARE DEPOSITED

You will need several different colors of clay, a bowl, and a cheese slicer or plastic knife.

1. Make a stack of different colored layers of clay. Each layer should be about the diameter and thickness of a pancake.
2. Turn the stack into a dome by pressing it over a small rounded object, such as a bowl.
3. With the cheese slicer or plastic knife, carefully slice down from the top and across through the dome so you can see a cross section of the stack. Examine the layers exposed.
4. Push your rock layers close together and then press them down.

▲ **STEP 3** Examine the layers you have exposed.

Practicing Your Skills

5. IDENTIFY: Which layer was deposited first?
6. IDENTIFY: Which layer is the oldest?
7. APPLY: How does this model show how sediments are usually deposited?
8. APPLY: Is the oldest layer on the top or the bottom now?

3 *Assess*

CHECKING CONCEPTS ANSWERS

1. top
2. index
3. relative

THINKING CRITICALLY ANSWERS

4. DESCRIBE: The species existed for only a short time in Earth's history; many fossils of the species are found in rock layers; the species was widespread on Earth; the species must be unique.
5. MODEL: Students' drawings should show the rock layers with trilobite fossils and graptolite fossils above the layer that is 600 million years old.

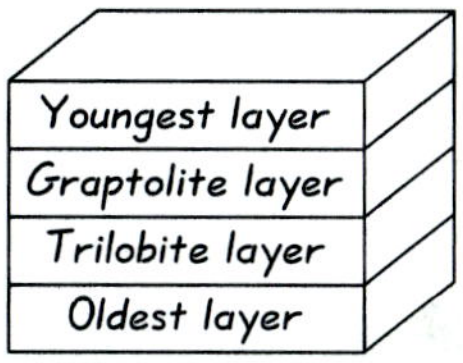

BUILDING SCIENCE SKILLS

Modeling Students' drawings should reflect the fact that an extrusion forms on Earth's surface and an intrusion forms below Earth's surface.
learning style: visual

4 *Close*

Hands-On Activity

MODELING HOW SEDIMENTS ARE DEPOSITED

TIME: 15 minutes

PURPOSE: modeling, comparing, identifying, applying

SAFETY TIP: Caution students to be careful when using the cheese slicer or knife.

COOPERATIVE LEARNING: Have students work in small groups. Have a student from each group describe to the class the relative age of each rock layer in their model.

Practicing Your Skills Answers

5. IDENTIFY: the bottom layer
6. IDENTIFY: the bottom layer
7. APPLY: Sediments pile up layer upon layer, just as the clay layers were stacked one on top of the other.
8. APPLY: the bottom

RETEACH

Newest or Oldest? Show students a clear bottle or container filled with five or six layers of different colored sand. Ask students which layer of sand is the newest or youngest. Guide students in recognizing that the layers of sand are similar to layers of rock. The oldest sand or rocks are on the bottom, while the newest sand or rocks are on the top.

• TEACHING RESOURCES •
Online or Print

Teacher's Resources

Lesson Review: Chapter 4, p. 5

Enrichment: Chapter 4, p. 8

Concept Map: Study Tools, p. 4

4-4 *(pp. 98–99)*

1 Introduce

STUDY HINT Before beginning this lesson, have a volunteer read question 1 from page 98 of the text aloud. Then guide students to use the other three questions to predict how absolute age might be determined.

Linking Prior Knowledge Ask students what proof they have of their ages and how they could determine exactly how old they are. If necessary, ask: *What information is on your birth certificate?* Have students explain why their age in years is not the same as their relative age.

2 Teach

Demonstration Place 16 pennies on a desk to represent the original amount of carbon-14 in a sample. Replace 8 pennies with 8 nickels and explain that after one half-life, half the carbon-14 decays to a new substance, nitrogen-14 (the nickels). **learning style: visual**

Discussion Explain how the decay rates of radioactive elements are used to determine the absolute ages of fossils. Describe the carbon-14 radioactive dating method. Note that this method is used only for fossils less than 50,000 years old. Emphasize that uranium is not used for dating organic materials or fossils. It is only used for dating inorganic materials such as rocks.

Reinforcement Have students stand in a group. Tell them that together they represent an animal bone and that each of them individually is a carbon-14 atom. Tell students that every minute will represent 5,800 years, the half-life of carbon-14. To demonstrate the gradual decay of carbon-14, have students sit down when you tap their shoulder. Tap students at regular intervals so that after 1 minute, half of them have sat down; after 2 minutes, half of the remainder have sat down, and so on. **learning styles: tactile, visual**

4-4 How is a rock's absolute age determined?

Objectives

Define absolute age in rocks and fossils. Describe ways used to measure it.

Key Terms

absolute age: true age of a rock or fossil

half-life: length of time it takes for one-half the amount of a radioactive element to change into a stable element

Absolute Age To measure your age, you count the number of years from the time you were born. This number is your age. The true age of a rock layer is called its **absolute age.** Absolute age tells scientists the approximate number of years ago a rock layer formed.

1 DESCRIBE: What is meant by absolute age?

Natural Clocks The process of a radioactive element changing into another element is called radioactive decay. Radioactive elements were first discovered in 1896. They are elements that emit, or give off, particles and energy. As a radioactive element gives off particles and energy, it decays. This causes new elements to form that may or may not be radioactive.

The rate at which radioactive decay happens can be measured. Each radioactive element decays at a regular, steady rate. Radioactive elements are like natural clocks.

2 EXPLAIN: What is radioactive decay?

Half-life The **half-life** of a radioactive element is the time it takes for one-half of the mass of a sample of a radioactive element to decay. Each radioactive element has a different half-life.

Uranium is a radioactive element. One form of it, called U-238, slowly decays into lead. If you begin with 6 kg of U-238, 3 kg will decay into lead after 4.5 billion years. After another 4.5 billion years, only 1.5 kg of U-238 would be left. By comparing the amount of the radioactive element in a rock to its decay element, scientists can find the absolute age of a rock.

3 INFER: If a rock has equal amounts of lead and U-238 in it, how old would the rock be?

Carbon-14 Carbon-14 is used to date the remains of living things. As long as living things are alive, they take in carbon-14. Carbon-14 is a radioactive form of carbon. It decays into nitrogen gas. The half-life of carbon-14 is about 5,800 years.

This rock sample contains 6 kg of uranium. | About 4.5 billion years later it has 3 kg of uranium left. | About 9 billion years later it has 1.5 kg of uranium left.

◄ **Figure 4-14** Radioactive decay of uranium

98

Teaching Options

! FAST FACTS *Dating Lucy With a Laser*

Lucy, the famous fossil of a humanlike creature called *Australopithecus afarensis*, was discovered in Africa in 1974. Scientists have determined Lucy's age at 3.2 million years by using a laser-dating technique called *potassium-argon dating* to date the ancient volcanic ash layers in which her fossils were found.

In potassium-argon dating, single crystals of volcanic material above and below Lucy's fossils were melted by using a laser. Gases released as the crystals melted contained an isotope known as argon-40, the decay element of radioactive potassium-40. The gases were fed into a machine called a mass spectrometer, which counted the number of argon-40 atoms. Using the traditional technique of using the half-life of potassium-40 (1.25 billion years) and comparing the number of potassium-40 atoms to the number of argon-40 atoms, scientists were able to determine the age of the material in which Lucy was buried. From that information, they could infer that Lucy was the same age. Scientists who used this technique to date Lucy believe that the figure of 3.2 million years is accurate to within about 32,000 years.

Carbon-14 is used to find out the absolute ages of wood, bones, and so on. It is also used to date "young" samples of fossils. If a sample is more than 50,000 years old, this process cannot be used, since almost all of the carbon-14 has already decayed into nitrogen-14.

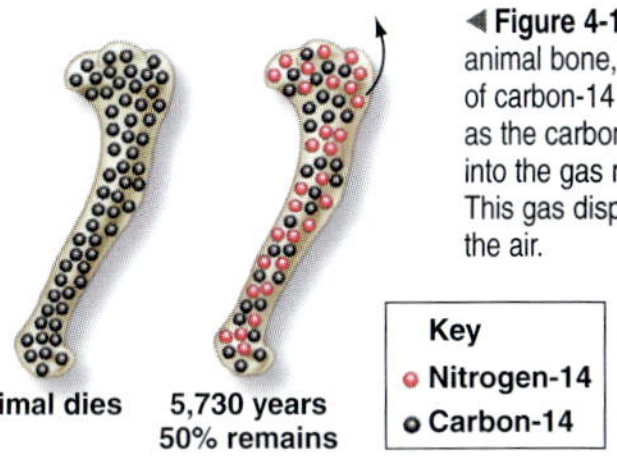

◀ **Figure 4-15** In the animal bone, the amount of carbon-14 decreases as the carbon-14 decays into the gas nitrogen-14. This gas disperses into the air.

4 DESCRIBE: What is carbon-14 used for?

CHECKING CONCEPTS

1. The decay element of U-238 is __________.
2. Radioactive uranium gives off particles and __________.
3. If you read that a fossil is 350-400 billion years old, you are reading its__________ age.
4. The radioactive element __________ is used to find out the absolute age of once-living things.

THINKING CRITICALLY

5. INFER: A radioactive element has a half-life of 70 million years. How much of a 10 g sample will be unchanged after 140 million years?
6. INFER: Would you use carbon-14 or uranium to find the age of a tooth you think is 20,000 years old? Why?

HEALTH AND SAFETY TIP

Radioactive elements emit radiation. Large amounts of radiation are harmful to living things. Radiation harms cells. Find out what the universal symbol for radiation danger looks like. Draw it on a sheet of paper. What two colors are used in the symbol?

Science and Technology

LASER DATING OF SEDIMENTS

A laser is a very strong, focused beam of light. The light has only one wavelength, or color. You may have several appliances in your house that use laser beams. CD and DVD players use laser beams to read the stored sounds and images on the discs. In Earth science, lasers are used to estimate the age of sediments.

How can lasers be used to estimate the age of sediments? Scientists bombard sediment layers with lasers. The lasers free electrons from the sediments. Electrons are negatively charged particles that can be trapped in sediments. When the electrons are freed, they give off light. Scientists measure how much light is given off. Older sediments give off more electrons than younger sediments do. Laser dating is used to determine the ages of sediments that are up to 700,000 years old. It can also be used to determine the age of river sediments and glacial deposits.

▲ **Figure 4-16** Scientists use lasers to date sediments.

Thinking Critically Why might scientists want to find out the age of glaciers?

INTEGRATING PHYSICAL SCIENCE

Radioactive Decay Two common types of radioactive decay are alpha and beta decay. In alpha decay, the nucleus of a radioactive element gives off an alpha particle, made up of two protons and two neutrons. In beta decay, an atom gives off either an electron or a positron.

QUICK CONVERSIONS		
	Metric/SI	English
Uranium mass	6.0 kg 3.0 kg 1.5 kg	13.2 lb 6.6 lb 3.3 lb

• TEACHING RESOURCES • Online or Print

Teacher's Resources
Lesson Review: Chapter 4, p. 6

Laboratory Manual
Lab. Challenge: How can half-life be modeled? pp. 37–38

MORE ON THE WEB

Radioactive Decay Have students investigate how radioactive decay works and study isotopes and their role in that decay. Students can begin their investigation at www.conceptsandchallenges.com.

Answers to questions

1 DESCRIBE: the true age of a rock or fossil

2 EXPLAIN: A radioactive element changes into another more stable element by giving off particles and energy.

3 INFER: 4.5 billion years

4 DESCRIBE: to date the remains of living things

3 Assess

CHECKING CONCEPTS ANSWERS

1. lead
2. energy
3. absolute
4. carbon-14

THINKING CRITICALLY ANSWERS

5. INFER: 2.5 g (half of half)
6. INFER: carbon-14, because it is used to date young fossils, or those less than 50,000 years old

4 Close

HEALTH AND SAFETY TIP ANSWER

The colors are black and yellow.

Science and Technology

LASER DATING OF SEDIMENTS

Discussion Ask students how sediments bombarded with lasers are similar to and different from radioactive elements. (Both give off particles and energy, and both can be used to determine absolute age. However, radioactive elements decay naturally, whereas sediments must be bombarded to emit electrons.) Ask: *Could laser dating be used to find the ages of sediments containing graptolites? Why or why not?* (No. Graptolites lived more than 700,000 years ago, which is the upper limit of laser dating.)

Thinking Critically Answer
Possible answer: to find out about past climates or world climate changes

4-5 (pp. 100–101)

1 Introduce

INVESTIGATE

TIME: 15 minutes

PURPOSE: Students will create a time line that resembles the geologic time scale.

MATERIALS: roll of adding machine paper, pencil, ruler

PROCEDURE: Have students work in pairs to brainstorm important events in their lives, such as achievements, awards, trips, illnesses, and so on. Each partner should then create a timeline. Remind students to begin their time lines with their birth date and include each birthday.
learning style: visual

THINK ABOUT IT: Possible answers: Students could divide their lives into developmental periods such as infancy, toddler years, childhood years, and teen years. Students who have lived in different places could divide their lives into periods spent in various locations.

STUDY HINT Before beginning this lesson, have students write the section-ending questions from pages 100 and 101 of the text and then scan the text for the answers.

2 Teach

Discussion Draw students' attention to Figure 4-17. Explain that the geologic time scale outlines some of the major events in Earth's history. Point out the major divisions on the scale. Read aloud the names of the eras, periods, and epochs so that students can hear their pronunciations. Tell students to note the major organisms that dominated each of the four eras. **learning styles: visual, auditory**

Reinforcement Write the names of the four geologic eras on the board. Have students look at Figure 4-17. Help students come up with two or three sentences that describe each era. **learning style: verbal**

Answers to questions

1. DESCRIBE: 4.6 billion years
2. DESCRIBE: an outline of the major divisions in Earth's history
3. NAME: eras, periods, and epochs

4-5 What is the geologic time scale?

INVESTIGATE

Making a Time Line of Your Life

HANDS-ON ACTIVITY

1. List important events that have occurred in your life.
2. Draw a long line across a sheet of paper to represent your life. Break it up into years. You can use a scale of 2 cm for every year.
3. Write each event in the correct year along the time line.
4. Now, divide the time line into parts that describe major periods in your life, such as preschool years, elementary school years, and middle school years.

THINK ABOUT IT: If you had to break up your time line into different groupings or divisions, what might they be?

Objective

Read and interpret the geologic time scale.

Key Term

geologic (jee-uh-LAHJ-ihk) **time scale:** outline of the major divisions in Earth's history

Age of Earth The oldest rocks found on Earth are about 4 billion years old. Moon rocks are older.

Scientists think that Earth and the Moon were formed at about the same time. However, because of the rock cycle, Earth's earliest rocks are gone. Based on the age of Moon and Earth rocks, scientists think that Earth may be about 4.6 billion years old.

1. DESCRIBE: About how old is Earth?

Outlining Earth's History By studying rocks and fossils, geologists have developed a geologic time scale. The **geologic time scale** outlines the major divisions in Earth's history. It also outlines the kinds of organisms that lived on Earth in the past. The geologic time scale begins when Earth was formed. It continues until the present day.

2. DESCRIBE: What is the geologic time scale?

GEOLOGIC TIME SCALE

Era	Period	Epoch	Approximate Start Date (millions of years ago)	Organisms First Appeared
Cenozoic	Quaternary	Recent	0.025	Modern humans
		Pleistocene	1.75	Mammoths
	Tertiary	Pliocene	14	Large carnivores
		Miocene	26	Many land mammals
		Oligocene	40	Primitive apes
		Eocene	55	Early horses
		Paleocene	65	Primates
Mesozoic	Cretaceous		130	Flowering plants
	Jurassic		180	Dinosaurs
	Triassic		225	Conifers
Paleozoic	Permian		275	Seed plants
	Carboniferous		345	Reptiles
	Devonian		405	Insects, amphibians
	Silurian		435	Fishes
	Ordovician		480	Algae, fungi
	Cambrian		600	Invertebrates
Precambrian			4,600	Bacteria, blue-green bacteria

◀ Figure 4-17

100

Teaching Options

! FAST FACTS *Mass Extinctions*

Throughout Earth's history, large groups of organisms have become extinct at the same time in events known as mass extinctions. Earth scientists have identified five mass extinctions. No one knows exactly what caused these mass extinctions, but the major reason seems to be glacial buildup, which caused lowering of sea temperatures, dropping of sea level, and elimination of continental shelf environments. Volcanic activity and meteor, asteroid, or comet impact—all of which could have blanketed Earth in a thick cloud cover—are also sometimes blamed.

In the first known mass extinction, toward the end of the Precambrian era, 70 percent of marine species disappeared. These were the only forms of life that existed at the time. The Cambrian period saw at least four mass extinctions. Other mass extinctions occurred during the Ordovician and at the end of the Devonian. As a result of these extinctions, many species of marine invertebrates disappeared, such as trilobites, graptolites, brachiopods, and bryozoans. During the Permian mass extinction, as many as 95 percent of all marine species died out, including early jawed fish. The most famous mass extinction occurred, however, about 65 million years ago, at the end of the Cretaceous period. This event wiped out about 85 percent of all species on the planet, including the dinosaurs.

Divisions of Geologic Time A year is divided into units called months, weeks, and days. Geologic time is also divided into units. The major unit is an era. Each era in geologic history lasted for millions of years. There are four eras. The eras are divided into periods. The more recent periods are divided into epochs.

The divisions of geologic time are based on important changes that occurred on Earth at these times. Some examples are the extinction of an important group of organisms and afterward the appearance of new categories of animals.

3 NAME: What are the divisions of geologic time?

CHECKING CONCEPTS

1. How many eras make up the geologic time scale?
2. How old are the oldest Earth rocks ever found?
3. What does the geologic time scale model?
4. What is the largest unit of geologic time?
5. What is the most recent period of geologic time?

THINKING CRITICALLY

6. INFER: Why are only the most recent periods divided into epochs?
7. HYPOTHESIZE: If the oldest Earth rocks are only 4 billion years old, why do scientists think that Earth is 4.6 billion years old?

BUILDING SCIENCE SKILLS

Modeling Use adding machine tape to make your own time line. Have each centimeter equal 10 million years. Calculate the length of each era in centimeters. Draw lines on the tape to mark each era. Mark each of the events listed below. Add five other events to your time line using Figure 4-17.

- First fossils appeared 3.4 billion years ago.
- First fish appeared 450 million years ago.
- First reptiles appeared 290 million years ago.
- Age of dinosaurs began 200 million years ago.
- First mammals appeared 200 million years ago.
- First birds appeared 150 million years ago.
- First humanlike animals appeared 2 million years ago.
- Last major ice age ended 10,000 years ago.

Integrating Life Science

TOPICS: extinction, biodiversity

THE DIVERSITY OF LIFE

Earth is teeming with life. However, scientists estimate that for every species existing today, at least 100 others are now extinct.

Earth has changed greatly over its history. Mountains have been built up. Volcanoes have formed. There have been ice ages. The Sun's output has varied over time. Each time a great change has occurred, species that could not adapt died out. Others replaced them. Today, Earth may have up to 100 million species.

▲ **Figure 4-18** An asteroid striking Earth is believed to have led to the extinction of the dinosaurs.

Scientists call this vast variety of organisms biodiversity. Together, the various species keep natural cycles, such as the water and nitrogen cycles, running smoothly. Each species helps balance the ecosystem of which it is a part. For example, some species break down wastes. Others add oxygen to the air. The loss of even a few species can result in the destruction of an ecosystem.

Thinking Critically How do pollution and the spread of civilization threaten biodiversity?

EXTENDING THE LESSON

Recognizing a Controversy Explain that some scientists recognize only three eras on the geologic time scale, classifying the Precambrian as an eon rather than an era. The eon is the largest unit of geologic time. Ask students why scientists' conclusions may not always agree.

• TEACHING RESOURCES • Online or Print

Teacher's Resources

Lesson Review: Chapter 4, p. 7

Enrichment: Chapter 4, p. 8

3 Assess

CHECKING CONCEPTS ANSWERS

1. four
2. 4 billion years old
3. the major divisions in Earth's history
4. era (Eon is also an acceptable answer.)
5. Quaternary

THINKING CRITICALLY ANSWERS

6. INFER: Possible answer: More information is available for the most recent period so that finer distinctions can be made.
7. HYPOTHESIZE: If we assume that the Moon and Earth formed at the same time and the oldest Moon rocks are 4.6 billion years old, then Earth must also be 4.6 billion years old.

4 Close

BUILDING SCIENCE SKILLS

Modeling Have students work in pairs or small groups. Check students' time lines. Be sure that they correctly position the listed events as well as four or five other events on their time lines. Possible events from Figure 4-17 include appearance of flowering plants about 120 million years ago, first insects about 410 million years ago, and large carnivores about 10 million years ago. **learning style: visual**

Integrating Life Science

TOPICS: extinction, biodiversity

THE DIVERSITY OF LIFE

Discussion When an organism adapts to a new condition, its structure or behavior changes in a way that contributes to the survival of the species. If an organism cannot adapt or does not already possess a characteristic that helps it survive in new conditions, it dies, and, over time, the species becomes extinct.

Ask students why people should be concerned about human-caused extinctions when extinctions are so common anyway.

Thinking Critically Answer

Pollution poisons the land, air, and water, often interfering with the ability of a species to reproduce. Civilization cuts down on available living space for species.

4-6 *(pp. 102–103)*

1 *Introduce*

STUDY HINT Have students rewrite the lesson objective in the form of a question. As they read the lesson, have them write the answer to the question.

Linking Prior Knowledge Ask students to describe what they think a fuel is and what it is used for. Lead students to think of fuel in terms of energy. Have them give examples of fuels that people have in their homes or use in their cars. Accept all logical responses, including foods that fuel the body and fuels for cooking and heating.

2 *Teach*

Demonstration Introduce the lesson by turning the classroom lights on and off several times. Ask students if they know what energy source is used to produce electricity. **learning style: visual**

Discussion Point out that most of our electricity is produced by the burning of fossil fuels. Define fossil fuels as natural fuels that come from the remains of living things. Tell students that coal, natural gas, and petroleum are fossil fuels.
learning style: auditory

Reinforcement Remind students that what all fossils fuels have in common is that they form from remains of living things. Both living things and fossil fuels store energy in the form of carbon compounds. Bacteria, heat, and pressure chemically change the complex carbon compounds in the remains of living things into simple carbon compounds called hydrocarbons.
learning style: auditory

Answers to questions

1. **DEFINE:** fuels made from the remains of living things
2. **IDENTIFY:** light and heat
3. **NAME:** peat, soft coal, and hard coal
4. **NAME:** microorganisms that were buried in sediments

4-6 What are fossil fuels?

Objective
Describe how coal, oil, and natural gas were formed.

Key Term
fossil fuel (FYOO-uhl)**:** natural fuel that was formed from the remains of living things

Fossils as Fuel A fuel is a substance that gives off energy when it is burned. A fuel may be a solid, a liquid, or a gas. **Fossil fuels** are fuels that were formed from fossils.

Coal and natural gas are fossil fuels. The liquid fuel known as petroleum is also a fossil fuel. Most of the energy that heats our homes and runs our appliances comes from fossil fuels.

1 **DEFINE:** What are the fossil fuels?

▲ **Figure 4-19** During the coal-forming age, much of Earth was covered by huge swamps.

Hydrocarbons Fossil fuels are made up mostly of hydrocarbons. Hydrocarbons are compounds made up of the elements hydrogen and carbon. Hydrocarbons store energy. This energy comes from the sunlight absorbed by plants and microorganisms that lived millions of years ago. When hydrocarbons are burned, they give off this energy as light and heat.

2 **IDENTIFY:** What two forms of energy are given off when hydrocarbons are burned?

The Formation of Coal Coal is formed in swamps. Swamps are areas of shallow water with a lot of plant life. When the plants in the swamps die, they are covered by water and by sediments such as mud. Bacteria, pressure, and heat slowly cause chemical changes to take place in the plants. After many millions of years, the decaying plant material changes to peat. Peat is the first stage in coal formation. After another long period of time, peat changes into soft coal. Pressure changes soft coal into hard coal. Soft and hard coal are mostly carbon. They give off a lot of heat when they burn.

3 **NAME:** What are the three basic stages of coal formation?

The Formation of Oil and Natural Gas Geologists think that petroleum, or crude oil, and natural gas were formed from decaying microorganisms. When these microorganisms died, they were covered with sediments that became sedimentary rock.

Like coal, petroleum and natural gas were formed by the actions of bacteria, heat, and pressure. Petroleum moved with water through the cracks and pores in rock. When it reached a rock layer it could not pass through, the mixture of petroleum and water began to collect. Petroleum is usually found in pools that collect in layers of shale or sandstone. The petroleum floats on top of the water.

102

Teaching Options

INTEGRATING PHYSICAL SCIENCE

Alkanes The hydrocarbons that make up petroleum are called *alkanes*. Alkanes in gasoline vary in terms of the number of carbon atoms per molecule as well as in their ability to burn smoothly and resist knocking. Knocking refers to a noise that a car's engine may make as gasoline is burned. Gasoline is given an octane rating reflecting its resistance to knocking. The higher the octane rating, the more resistant the gasoline is to knocking. Ask students to look for octane ratings on the pumps the next time they go to a gas station.

RETEACH

Explaining Fossil Fuels On the board, draw a graphic organizer similar to the Concept Map in the Teacher's Resources. Use *Fossil Fuels* as the main concept and *Coal*, *Natural Gas*, and *Petroleum* as the subconcepts. In the boxes, describe how each type of fossil fuel forms.

EXTENDING THE LESSON

Researching Fossil Fuels Have students research the locations of the world's major reserves of fossil fuels. Instruct them to present their findings in the form of a word map.
learning style: auditory

If natural gas is also present, it is on top of the oil. To get to the oil and natural gas, it is necessary to drill into Earth's crust.

4 NAME: What types of ancient organisms formed oil and natural gas?

CHECKING CONCEPTS

1. What compounds are found in fossil fuels?
2. What are three fossil fuels?
3. What three things act on decaying organisms to help form fossil fuels?
4. In what kind of rock layers would you most likely find petroleum and natural gas?

THINKING CRITICALLY

5. INFER: Why are coal, petroleum, and natural gas called fossil fuels?
6. IDENTIFY: What is the source of the energy stored in hydrocarbons?
7. INFER: Why is natural gas found above petroleum in an oil deposit?

INTERPRETING VISUALS

A pictograph is a graph that uses a picturelike symbol to show an idea. Use the pictograph below to do the following exercises.

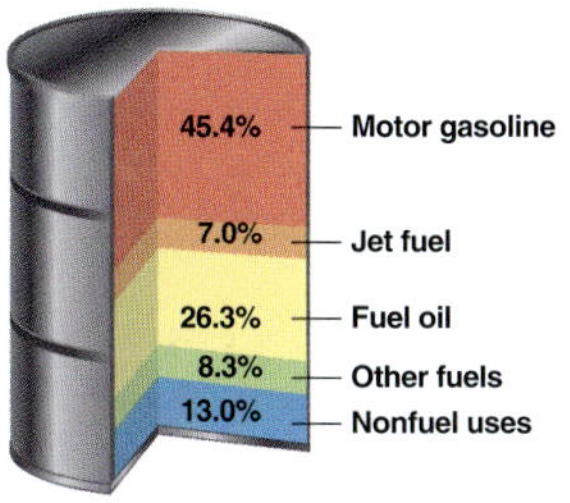

▲ **Figure 4-20** Uses of Petroleum

8. Place the uses of petroleum in order from greatest use to least use.
9. Give the percentage of petroleum that is used to make all fuels.
10. What total percentage of petroleum fuel goes toward transportation needs?

People in Science

OIL-RIG OPERATOR

An oil-rig operator often works at sea. The drilling structure is a tall steel frame called a jacket. It is built onshore and carried out to sea on a barge. Then, it is anchored to the seafloor. The machines used to extract the oil are on a platform on top of the jacket. The construction crew sets the derrick, which holds the drilling equipment, over the spot where the well is to be drilled.

The workers live on the platform. They may be away from home for up to three weeks at a time. They run the drill, pumps, the power plant, and the control room. The drilling crew changes bits that are dull or cannot penetrate the rock. Most drilling crews have one driller, one or more derrickmen, and several other workers. Many oil-rig workers start out as roustabouts. Roustabouts do the heavy, unskilled labor. Many companies train roustabouts to do more skilled work.

Oil-rig workers must be physically fit. They must also usually take what is called the Basic Offshore Survival and Firefighting training course.

▲ **Figure 4-21** An oil rig in operation off the coast of California

Thinking Critically What kind of labor might roustabouts do?

MORE ON THE WEB

The Trans-Alaskan Pipeline
Have students use the Internet to research the construction of the Trans-Alaskan oil pipeline. Students should create an information brochure about the pipeline, including pictures. You may want to assist students in printing pictures from the Internet. Artistic students can draw their own pictures.

• TEACHING RESOURCES • Online or Print

Teacher's Resources
Lesson Review: Chapter 4, p. 9

3 Assess

CHECKING CONCEPTS ANSWERS

1. hydrocarbons
2. coal, natural gas, and petroleum
3. bacteria, pressure, and heat
4. shale or sandstone (sedimentary rock)

THINKING CRITICALLY ANSWERS

5. INFER: They are fuels formed from fossils.
6. IDENTIFY: sunlight
7. INFER: Gas is less dense than liquid oil.

INTERPRETING VISUALS ANSWERS

8. motor gasoline, fuel oil, nonfuel uses, other fuels, jet fuel
9. 87%
10. 52.4% (motor gasoline plus jet fuel)

4 Close

People in Science

OIL-RIG OPERATOR

Real-Life Connection If possible, arrange for an oil-rig operator or someone else who works in the oil industry to visit the class. Ask the operator to describe his or her responsibilities and what a typical work day involves. Have students prepare questions ahead of time to ask the visitor. If bringing in an oil-rig worker is not practical, show students a sample of motor oil. Tell students that the motor oil has been refined, or purified, to make it usable and that petroleum is pumped from wells in a cruder form. Explain that some crude oil resembles motor oil but that other crude oil is black and thicker, sometimes like tar. Encourage students to speculate on the challenges of getting crude oil from beneath the ocean floor to the surface.

Discussion Ask students what dangers they think an oil-rig operator may face on the job. Suggest that students look for clues in the feature to dangers when oil-rig operators do heavy labor or firefighting.

Thinking Critically Answer
Accept any of the following: setting up, lifting, and maintenance of drilling and other equipment on an oil rig.

The Big Idea

(pp. 104–105)

1 Introduce

Objective Students will be able to identify products made from oil and describe the process of fractional distillation. In their Science Logs, students will list all the crude-oil products they used in the past week and describe what their day would be like without these products.

Linking Prior Knowledge Have students recall what kind of substance crude oil is and how it forms (Lesson 4-6).

2 Teach

Discussion As a class, brainstorm a list of common petroleum products. Be sure that students list all the products shown in Figure 4-22 on pages 104 and 105 as well as any others they can think of. Examples of petroleum products not mentioned in the text include paint, dyes, explosives, ink, insecticides, medicines, fertilizers, and cosmetics. **learning style: auditory**

Use the background information in Fast Facts to help guide students in understanding the Big Idea.

THE BIG IDEA

What products are made from oil? Have students start their research for their Science Log Writing Activity at www.conceptsandchallenges.com. Be sure students use search engines and websites that are reliable and acceptable. You may also wish to provide copies of the *Big Idea Planner* and *Big Idea Science Log Entry* study tools found in the Teacher's Resources.

Integrating Technology

THE Big IDEA

What products are made from oil?

It might not seem like crayons, guitar strings, footballs, and bubble gum have anything in common, but they do. Each of these products is made from crude oil.

Ancient peoples recognized the value of this natural resource. The Chinese used oil to waterproof their homes. Egyptians wrapped their dead in rags saturated with oil to preserve them. Native Americans added a drop or two of oil to medicines, and ancient Romans applied oil to cuts and wounds. Thanks to technology, today there are more than 500,000 different uses for this natural resource.

The first step in creating these products is separating crude oil into its different parts. This is called fractional distillation and takes place in an oil refinery. Pipes carry raw crude oil from oil wells to the refinery. The crude oil is heated to about 430°C. This causes the liquid to turn into vapor. The vapor is then pumped into a tall tower divided into different levels. Each of these levels is kept at a certain temperature.

As the evaporated crude oil moves through the tower, its substances condense at different temperatures. The heaviest parts condense in the warmer, lower levels of the tower. The lightest parts condense in the cooler, higher levels of the tower. The separated substances are drawn from each level and sent to other locations, where they are used to make many products.

Look at the illustrations on these two pages. Then, follow the directions in the Science Log to learn more about "the big idea." ✦

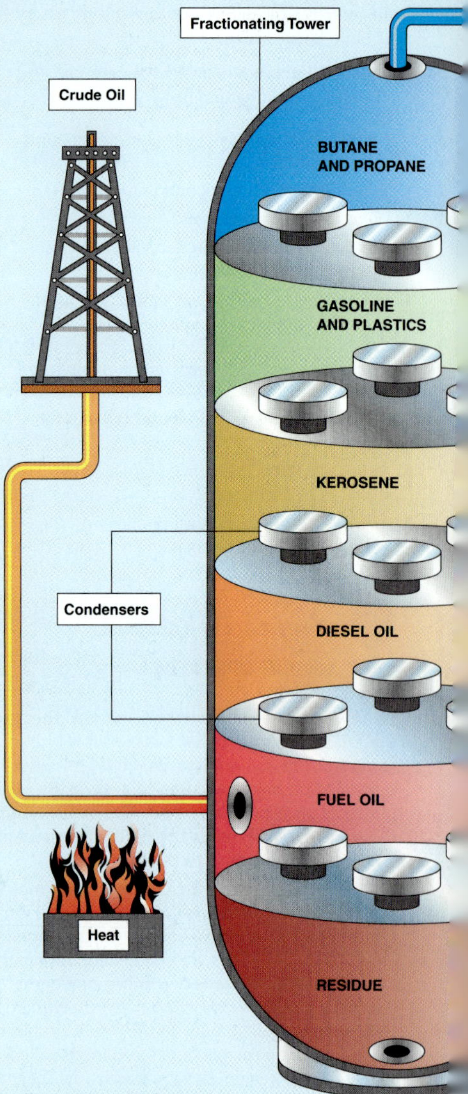

▲ **Figure 4-22** Inside a fractionating tower, different substances contained in crude oil condense at different temperatures.

104

Teaching Options

! FAST FACTS *The Chemistry of Fractional Distillation*

Fractional distillation is the process of separating a mixture of liquids that have different boiling points. Petroleum is such a mixture. When heated, the different types of hydrocarbons separate out. Hydrocarbon molecules with fewer carbon atoms boil at lower temperatures than those molecules with more carbon atoms. Upon boiling, the liquid evaporates, or changes from a liquid to a gas or vapor. Once cooled, the gas condenses, changing to a liquid. Each substance condenses at a different temperature, so the substances can be drawn off separately as they condense.

RETEACH

Demonstrating Distillation To demonstrate the processes that occur during fractional distillation, boil water in a clear cooking pot covered with a matching lid. Explain that crude oil, like water, turns to vapor when heated. The vapor rises and condenses as it cools on the inside of the pot lid, just as evaporated crude oil rises in the tower and condenses as it cools in the upper levels. **learning style: visual**

Butane and Propane The lightest parts of evaporated crude oil condense in the coolest level of the tower. Butane used in lighters and propane used in barbecue grills are collected here.

Gasoline and Plastics The gasoline that cars run on comes from substances that condense in this level. Chemicals and plastics used to make clothing, sporting equipment, and telephones start at this level.

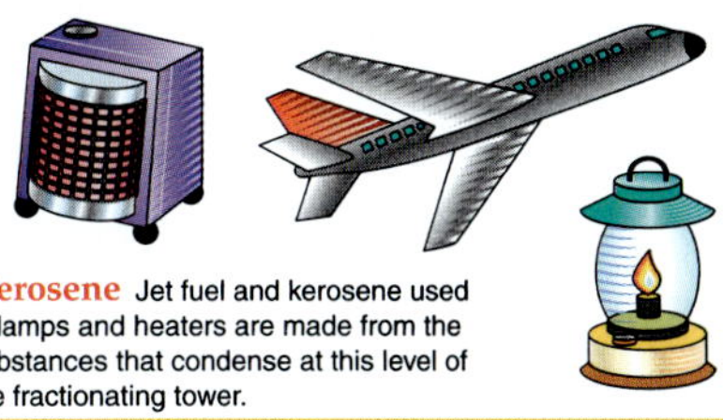

Kerosene Jet fuel and kerosene used in lamps and heaters are made from the substances that condense at this level of the fractionating tower.

Diesel Oil Trucks and trains run on diesel fuel made from substances that condense at this level.

Fuel Oil Heating oil is used to keep many homes warm. Certain fuels power ships and factories. All are made from substances that condense at this level.

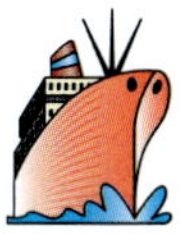

Residue Lubricants, paraffin, and asphalt are made from the residues that condense in the lowest level of the tower.

WRITING ACTIVITY

Science Log

List all of the crude oil products that you have used this week. Name at least five. Then, write several paragraphs entitled "A Day Without Oil Products." Describe in these paragraphs what your day would be like without these items. Start your search for information using www.conceptsandchallenges.com.

3 Assess

Use the Writing Activity Rubric from the Teacher's Resources to assess students' Science Logs. Fill in the rubric with the additional information below. For this assignment, students should have:

- listed all the crude oil products that they used in the past week.
- described in several paragraphs what their day would be like without crude oil products.

4 Close

Diagramming Oil Products Have students work in groups to create a mural about one of the crude oil products in Figure 4-22 on pages 104 and 105. The mural should show the journey the product takes and the changes it undergoes, starting with its formation from ancient organisms and ending with its final use in a home, school, vehicle, and so on. Display the murals in the classroom.

Tips for Using Technology

Using the Internet to Download Graphics Students can download graphics from the Internet to add to their Science Logs. Students may use saved images in reports and presentations.

On a Microsoft® Windows®-based computer, right-click on the image you want to save. Choose *Save As...* or *Save Image As...* (depending on your operating system). A *Save* dialog box will appear. Choose the location into which you would like to save the file. The file name should automatically be filled in.

• TEACHING RESOURCES • Online or Print

Teacher's Resources

Big Idea: Chapter 4, p. 10

Big Idea Planner: Study Tools, pp. 11–12

Big Idea Science Log Entry: Study Tools, p. 13

Writing Activity Rubric: Teaching Tools, p. 9

Challenges *(pp. 106–108)*

Chapter Summary

Review Have a different volunteer read each point of the Chapter Summary aloud to the class.

STUDY HINT Have students review the Chapter Summary and summarize each lesson in one sentence.

Key Term Challenges

MATCHING

1. geologic time scale (4-5)
2. mold (4-1)
3. relative age (4-3)
4. absolute age (4-4)
5. half-life (4-4)

FILL IN

6. fossil fuels (4-6)
7. cast (4-1)
8. amber (4-1)
9. fossils (4-1)
10. index fossil (4-3)
11. gastrolith (4-2)
12. law of superposition (4-3)

Chapter 4 Challenges

Chapter Summary

Lesson 4-1

- **Fossils** are the remains or traces of organisms that lived many years ago.
- Fossils form in sedimentary rock. Members of some extinct species have been preserved in frozen soil or ice.
- The bones of many animals trapped in tar pits were preserved.
- Some insects and plant parts have been perfectly preserved in **amber.**

Lesson 4-2

- Fossils show that many kinds of organisms lived at different times in Earth's history. They show that some living things have changed over millions of years. They also show that Earth's climate and surface have changed over million of years.

Lesson 4-3

- Scientists use the **law of superposition** to tell the **relative age** of a rock layer.
- **Index fossils** can be used to help find the relative age of rock layers in the same area or from two different parts of the world.

Lesson 4-4

- The specific age of a rock or a fossil is called its **absolute age.**
- Radioactive elements are like natural clocks. The **half-life** of radioactive elements can be used to find the absolute age of a rock or fossil.
- Carbon-14 is used to find the absolute age of the remains of living things.

Lesson 4-5

- Some scientists estimate that Earth is more than 4.6 billion years old.
- The **geologic time scale** is a record of the major divisions in Earth's history.
- The major divisions of geologic time are eras, periods, and epochs.

Lesson 4-6

- Petroleum, coal, and natural gas are **fossil fuels.**
- When hydrocarbons are burned, they give off energy as light and heat.

Key Term Challenges

absolute age (p. 98)
amber (p. 90)
cast (p. 90)
coprolite (p. 94)
fossil (p. 90)
fossil fuel (p. 102)
gastrolith (p. 94)
geologic time scale (p. 100)
half-life (p. 98)
index fossil (p. 96)
law of superposition (p. 96)
mold (p. 90)
relative age (p. 96)

MATCHING **Write the Key Term from above that best matches each description.**

1. outline of the major divisions in Earth's history
2. imprint in a rock that has the shape of an extinct organism
3. age of an object compared to the age of another object
4. specific age of a rock or fossil
5. length of time it takes for one-half the amount of a radioactive element to change into another element

FILL IN **Write the Key Term from above that best completes each statement.**

6. Oil, coal, and natural gas are three kinds of __________.
7. A mold that has been filled with sediments forms a __________.
8. The bodies of entire insects have been found preserved in hardened tree sap called __________.
9. The traces or remains of living things that lived long ago are called __________.
10. A trilobite can be used as an __________.
11. A __________ was a stone some ancient animals swallowed to help grind up their food.
12. Scientists use the __________ to tell the relative age of rocks and fossils.

106

Teaching Options

PREPARING STUDENTS FOR STANDARDIZED TESTS

Reading Strategy: Remind students to reread the test directions several times before answering any questions.

Writing Strategy: Explain that, when writing complete sentences, students should check to make sure that each sentence has a subject, a predicate, and correct punctuation.

Interpreting Visuals Strategy: Use the geologic time scale on page 108 to review with students the way in which charts and tables are organized. Point out the boldface title and column heads.

ESL/ELL STRATEGY

Answering Questions As students answer the Challenge questions, move around the classroom, remaining accessible to those who may have questions and need help.

Content Challenges TEST PREP

MULTIPLE CHOICE Write the letter of the term or phrase that best completes each statement.

1. A species that once lived on Earth but is no longer found alive is called
   a. a fossil.
   b. an imprint.
   c. a cast.
   d. extinct.
2. Most fossils are found in
   a. metamorphic rock.
   b. sedimentary rock.
   c. mud.
   d. igneous rock.
3. Footprints are a kind of fossil called
   a. a mold.
   b. a cast.
   c. an imprint.
   d. amber.
4. The bodies of woolly mammoths and woolly rhinoceroses have been found preserved in
   a. amber.
   b. sedimentary rocks.
   c. ice.
   d. petrified forests.
5. Three ways in which the entire bodies of organisms are preserved are
   a. ice, tar, and amber.
   b. ice, molds, and casts.
   c. tar, molds, and casts.
   d. molds, petrified wood, and amber.
6. Graptolites and trilobites are two kinds of
   a. radioactive elements.
   b. molds.
   c. casts.
   d. index fossils.
7. Peat is the first stage in the formation of
   a. natural gas.
   b. tar.
   c. coal.
   d. petroleum.
8. The law of superposition states that each undisturbed rock layer is older than the one
   a. beside it.
   b. above it.
   c. below it.
   d. to its left.
9. Fossils show how Earth's climate and topography have
   a. changed over time.
   b. stayed about the same over time.
   c. improved over time.
   d. worsened over time.

TRUE/FALSE Write *true* if the statement is true. If the statement is false, change the underlined term to make the statement true.

10. Shells and nails are often the only parts of organisms that are preserved.
11. Elephantlike animals called saber-toothed cats have been discovered preserved in ice.
12. The LaBrea tar pits are located in southern California.
13. The bodies of entire insects have been found preserved in amber.
14. When fossil fuels are burned, they give off energy in the forms of heat and electricity.
15. Fossil fuels are formed when bacteria, heat, and pressure act on decaying plants and animals.
16. Fossils of alligatorlike animals found in Canada indicate that the climate of Canada was always cold.

Content Challenges

MULTIPLE CHOICE

1. d (4-1)
2. b (4-1)
3. c (4-1)
4. c (4-1)
5. a (4-1)
6. d (4-3)
7. c (4-6)
8. b (4-3)
9. a (4-2)

TRUE/FALSE

10. false; bones (4-1)
11. false; mammoths (4-1)
12. true (4-1)
13. true (4-1)
14. false; light (4-6)
15. true (4-6)
16. false; once warmer (4-2)

ALTERNATIVE ASSESSMENT

Illustrating Fossil Formation Have students draw diagrams that show three ways in which fossils are preserved.

PORTFOLIO ASSESSMENT

Making Student Portfolios Students can demonstrate their comprehension of the concepts in this chapter by making a portfolio. The Chapter Self-Check, the Big Idea Planner, and the Weekly Journal are some of the reproducibles from the Teacher's Resources that they can include in their portfolios. It is important that students include work that they are proud of or that they have put great effort into completing. You can use the Portfolio Assessment Rubric also found in the Teacher's Resources to assess students' portfolios.

Concept Challenges

WRITTEN RESPONSE

1. **INFER:** It was pushed up from the ocean floor. (4-2)
2. **DESCRIBE:** By comparing the amount of radioactive uranium in a rock to its decay element, lead, the absolute age of the rock can be found, using uranium's half-life. (4-4)
3. **DESCRIBE:** The geologic time scale is an outline of the major divisions in Earth's history. Its divisions include eras, periods, and epochs. (4-5)
4. **EXPLAIN:** After 50,000 years, most of the carbon-14 in the sample has already decayed into nitrogen-14. (4-4)
5. **ANALYZE:** I would use uranium, because it has a half-life of 4.5 billion years. (4-4)

INTERPRETING VISUALS (4-5)

6. Cenozoic era
7. 180 million years ago
8. Triassic period
9. Carboniferous period
10. Precambrian era

Chapter Project Wrap-Up

FOSSIL BULLETIN BOARD

Students can present their bulletin board display to another class and explain each fossil and how it is preserved. You can use the Individual Activity Rubric found in the Teacher's Resources to assess students' projects. Fill in the rubric with the additional information below. For this project, students should have:

- displayed images of fossils and fossil organisms and labeled them appropriately.
- explained accurately the process by which fossils form.

Concept Challenges TEST PREP

WRITTEN RESPONSE **Answer each of the following questions in complete sentences.**

1. **INFER:** Suppose you find fossils of clams in rock high on a mountain. What can you infer about the mountain?
2. **DESCRIBE:** How is U-238 used for the specific dating of rocks?
3. **DESCRIBE:** What is the geologic time scale? Give examples in your explanation.
4. **EXPLAIN:** Why is carbon-14 not useful for dating rocks that are more than 50,000 years old?
5. **ANALYZE:** What method would you use to date a sedimentary rock that you think is about 1 million years old? Why?

INTERPRETING VISUALS **Use Figure 4-23 to answer the following questions.**

6. What is the name of the most recent era in geologic time?
7. How long ago did dinosaurs roam Earth's surface?
8. During which period did conifers such as pine trees first appear?
9. During which period did snakes first appear?
10. Which era lasted the longest?

GEOLOGIC TIME SCALE

Era	Period	Epoch	Approximate Start Date (millions of years ago)	Organisms First Appeared
Cenozoic	Quaternary	Recent	0.025	Modern humans
		Pleistocene	1.75	Mammoths
	Tertiary	Pliocene	14	Large carnivores
		Miocene	26	Many land mammals
		Oligocene	40	Primitive apes
		Eocene	55	Early horses
		Paleocene	65	Primates
Mesozoic	Cretaceous		130	Flowering plants
	Jurassic		180	Dinosaurs
	Triassic		225	Conifers
Paleozoic	Permian		275	Seed plants
	Carboniferous		345	Reptiles
	Devonian		405	Insects, amphibians
	Silurian		435	Fishes
	Ordovician		480	Algae, fungi
	Cambrian		600	Invertebrates
Precambrian			4,600	Bacteria, blue-green bacteria

▲ Figure 4-23

Teaching Options

• TEACHING RESOURCES • Online or Print

Teacher's Resources

Key Term Review: Ch. 4, p. 11

Chapter 4 Test: pp. 12–13

Individual Activity Rubric: Teaching Tools, p. 5

Chapter Self-Check: Teaching Tools, p. 11

Big Idea Planner: Study Tools, pp. 11–12

Weekly Journal: Study Tools, p. 19

Portfolio Assessment Rubric: Teaching Tools, p. 10

PLANNING GUIDE

◆ **TEACHING THE CHAPTER** This chapter should take approximately 10–13 days to complete instruction and assessment.

	Skills and Features	Projects and Activities	Achieve Science Literacy	Meet Individual Needs
Chapter 5 Opener p. T109	MODEL	• Chapter Project		
Lesson 5-1 **What processes change Earth's crust?** pp. T110–T111 Standards: D1b, D1c, D2a	DEFINE, DESCRIBE, INFER • Hands-On Activity	• Extending the Lesson • More on the Web	• Study Hint	
Lesson 5-2 **What are mountains?** pp. T112–T113 Standards: D1b, D1c, D2a	SEQUENCE, CLASSIFY, RESEARCH AND ORGANIZE • Building Science Skills • People in Science	• Extending the Lesson • More on the Web • Integrating Language Arts	• Study Hint	• ESL/ELL Strategy
Lesson 5-3 **How are mountains classified?** pp. T114–T115 Standards: D1b, D1c, D2a	DESCRIBE, CLASSIFY, MODEL • Building Science Skills • Real-Life Science	• Extending the Lesson • More on the Web	• Study Hint • Reading Strategy	• Reteach
Lesson 5-4 **What are plains and plateaus?** pp. T116–T117 Standards: D1b, D1c, D2a	OBSERVE, SEQUENCE, CLASSIFY • Building Reading Skills • Integrating Life Science	• More on the Web • Integrating Physical Science	• Reading Strategy • Ongoing Assessment	
Lesson 5-5 **What are volcanoes?** pp. T118–T119 Standards: D1b, D1c, D2a	DEFINE, CONTRAST, DESCRIBE • Investigate • Web InfoSearch • Science and Technology	• Active Art	• Reading Strategy	• Reteach • ESL/ELL Strategy Teacher's Resources* Enrichment Activity Parts of a Volcano
Lesson 5-6 **How are volcanoes classified?** pp. T120–T121 Standards: D1b, D1c	DESCRIBE, INFER, PREDICT • Web InfoSearch • Hands-On Activity	• Extending the Lesson	• Study Hint • Ongoing Assessment	Teacher's Resources* Enrichment Activity

*www.conceptsandchallenges.com, password: ccearth09

Planning Guide continues on next page.

Standards: For details on the correlation to National Science Standards see pages *xx–xxi*.

	Skills and Features	Projects and Activities	Achieve Science Literacy	Meet Individual Needs
Big Idea **How do volcanic eruptions affect ecosystems?** pp. T122–T123 Standards: D1b, D1c, D2a	RESEARCH, COMMUNICATE, ANALYZE • Science Log: Writing Activity	• Big Idea • Close Activity	• Tips for Using Technology	• ESL/ELL Strategy Teacher's Resources* Big Idea Planner
Lesson 5-7 **What are earthquakes?** pp. T124–T125 Standards: D1b, D1c	DEFINE, IDENTIFY, ANALYZE • People in Science	• Cooperative Learning • Extending the Lesson • More on the Web • Active Art	• Study Hint • Reading Strategy	Teacher's Resources* Enrichment Activity
Lesson 5-8 **What are the different kinds of seismic waves?** pp. T126–T127 Standard: D1b	COMPARE, OBSERVE, ANALYZE • Investigate	• Extending the Lesson • More on the Web • Lab Challenge	• Writing Hint	• Reteach Teacher's Resources* Enrichment Activity
Lesson 5-9 **How do earthquakes cause damage?** pp. T128–T129 Standards: D1b, D1c	INFER, SEQUENCE, RESEARCH • Building Science Skills • Real-Life Science	• More on the Web	• Study Hint • Writing Hint	Teacher's Resources* Enrichment Activity
Lesson 5-10 **What is the Ring of Fire?** pp. T130–T131 Standard: D1b	IDENTIFY, DESCRIBE, INTERPRET • Web InfoSearch • How Do They Know That?	• Extending the Lesson	• Study Hint • Ongoing Assessment	• ESL/ELL Strategy
Lab Activity **Making an Earthquake-Proof Structure** pp. T132–T133 Standards: D1b, D1c	OBSERVE, INFER • Set-Up Time: 25–30 min • Lab Time: 40–45 min	• Extend the Lab Activity • Lab Skills Worksheet	• Tips for Using Technology	• ESL/ELL Strategy Teacher's Resources* Lab Activity Report
Chapter 5 Challenges pp. T134–T136	• Chapter Summary • Key Term Challenges • Content Challenges • Concept Challenges	• Chapter Project Wrap-Up	• Study Hint • Preparing Students for Standardized Tests	• ESL/ELL Strategy Teacher's Resources* Chapter Self-Check

*www.conceptsandchallenges.com, password: ccearth09

Chapter 5 The Changing Landscape

▲ **Figure 5-1** Mount Fuji in Japan

According to legend, the volcano on Mount Fuji was formed in 286 B.C. by an earthquake. It appears to be a single volcano. However, it is actually made up of three separate volcanoes: Komitake, Ko Fuji, and Shin Fuji. The most recent is Shin Fuji ("New Fuji"). Shin Fuji first became active about 10,000 years ago. It still smolders and erupts from time to time.

▶Mount Fuji is called a volcanic cone. Why do you think it is called this?

Contents

Chapter 5 The Changing Landscape

Chapter Overview

In this chapter, students will learn about three major landforms and some of the processes that slowly create and change them. Students will also find out about volcanoes and earthquakes. Finally, students will experiment with building earthquake-resistant structures.

About Figure 5-1 Mount Fuji is classified as a composite cone. Composite cones are kinds of volcanic cones. They form when the flow of thick (viscous) lava alternates with the violent ejection of pieces of volcanic rocks and ash. Composite cones have steep sides.

Answer because it resembles an overturned funnel or cone

Linking Prior Knowledge

For this chapter, students should recall:

- the three layers of Earth (Lesson 1-3).
- maps and how elevation maps are used (Lessons 1-7 to 1-9).
- the three classes of rock, how they form, and how they are classified (Lessons 3-1 to 3-4).

Chapter Project

LANDFORM MODEL

MATERIALS: different colors (red, gray, brown, black) of modeling clay, red or orange aquarium sand, small aluminum pie plates, plastic knives

Divide the class into six groups. Have each group make two models of one of the volcanoes described on page 120 or a model of a mountain being built. One model should show the intact landform, and the other should be a three-dimensional cross section of it. Encourage students to be creative, but insist on scientific accuracy. Have each group write a short paragraph that identifies the type of volcano or the mountain and explains how it formed. Display the models in the classroom or for the school, if possible.

learning styles: tactile, visual

Teaching Options

ASSESSMENT PLANNER

For assessment in the Student Edition, see the following pages:

Content Assessment
Checking Concepts: pp. 111, 113, 115, 117, 119, 121, 125, 127, 129, 131
Thinking Critically: pp. 111, 113, 115, 117, 119, 121, 125, 127, 129, 131
Interpreting Visuals: pp. 111, 125, 127
Chapter Challenges: pp. 134–136

Alternative Assessment
Building Skills: pp. 113, 115, 117, 129
Web InfoSearch: pp. 119, 121, 131
Science Log: p. 123

Performance-Based Assessment
Hands-On Activity: pp. 111, 121
Lab Activity: pp. 132–133

• TEACHING RESOURCES • Online or Print

Teacher's Resources
Lesson Review: Ch. 5, pp. 2, 3, 4, 5, 6, 7, 10, 11, 12, 14
Enrichment: Ch. 5, pp. 9, 13
Key Term Review: Ch. 5, p. 15
Chapter 5 Test: pp. 16–18
Laboratory Manual: pp. 13–14, 39–42

5-1 *(pp. 110–111)*

1 Introduce

STUDY HINT Have students carefully read the Key Terms and their definitions. Ask them to group the terms based on their definitions. Students should be able to state that anticlines and synclines are folds and that fractures and faults are breaks.

Linking Prior Knowledge Ask students to name the three classes of rocks and briefly describe how they form (Lesson 3-1). Have students recall how sedimentary rock forms layers (Lesson 3-4).

2 Teach

Demonstration Use silicon putty to demonstrate how different forces result in either folds or faults. Make a rectangle from the putty and gently push on the rectangle from the sides. Tell students that you just formed a fold. Form another rectangle and very quickly pull on the putty to make it snap. Stress the difference in the amount of force and the length of time it was applied in each case. Explain that to produce folds, rocks are exposed to slow, steady forces. To produce faults, rocks are subjected to relatively abrupt forces.
learning style: visual

Discussion Refer students to Figure 5-3. Point out that the principal direction of movement in normal, thrust, and reverse faults is vertical, while the principal direction of movement along a strike-slip fault is horizontal. If possible, use wooden blocks to model the movements shown in the figure. **learning style: visual**

Answers to questions

1. **DEFINE:** a bend, wrinkle, or curve in a rock or rock layers
2. **DESCRIBE:** a fracture along which horizontal movement takes place
3. **INFER:** Each time movement occurs along a fault, it puts pressure on the surrounding area. Thus, an earthquake can occur along any one of the faults in an area.

5-1 What processes change Earth's crust?

Objective

Compare and contrast the processes of folding and faulting in Earth's crust.

Key Terms

anticline (AN-tih-klyn): upward fold
syncline (SIHN-klyn): downward fold
fracture (FRAK-chuhr): break in a rock
fault: break in Earth's crust along which movement occurs

Folding Earth's surface is always changing. Some changes occur very suddenly. Others take hundreds or thousands of years, or even longer. Over millions of years, pressure in Earth's crust can cause rock layers to bend, curve, and wrinkle. This is called folding. Folding occurs when rock layers are squeezed from the sides. The rocks may crack under the pressure. However, the layers stay together.

Folds in rock look like waves. An upward fold is called an **anticline.** A downward fold is called a **syncline.** Some folds are small enough to be seen in a single rock. Others are very large. You often see anticlines and synclines along roads cut through rock.

▲ **Figure 5-2** Rock layers can fold upward or downward.

1 **DEFINE:** What is a fold?

Faulting Pressure deep inside Earth can also break rocks. A break in a rock is called a **fracture.** When movement of rock takes place along a fracture, it is called a **fault.** Four common kinds of faults are shown in Figure 5-3.

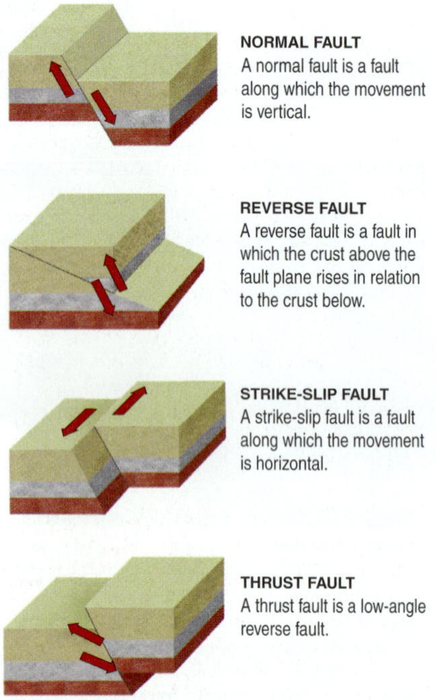

▲ **Figure 5-3** Common types of faults

Faulting causes rocks to move up and down or side to side a tiny distance. Each time movement occurs along a fault, the pressure between rock layers along the fault eases. Then, the pressure builds up again. Movement occurs again when this pressure is released.

2 **DESCRIBE:** What is a strike-slip fault?

Teaching Options

! FAST FACTS Fold Shapes

Folds can take various forms. Four are shown below.

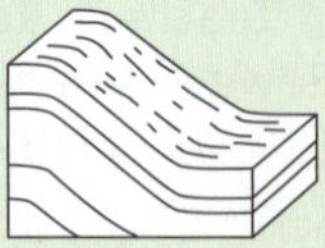

1. A *monocline* is a steep, steplike fold, with upper and lower bends in a set of rock layers.

2. A *pericline* is an anticline in the form of an elongated dome.

3. A lopsided anticline with one side folded over the other is an *overfold.*

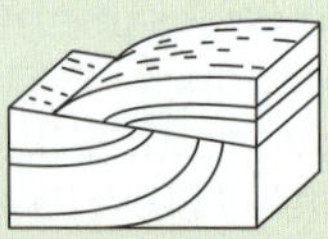

4. A *nappe* is an extreme overfold that has sheared through so that the upper side is forced forward.

San Andreas Fault There is a huge strike-slip fault that runs through a large part of California. This fault is called the San Andreas Fault. The San Andreas Fault extends from the Gulf of California through San Francisco Bay. The fault reaches more than 30 km down into Earth's crust.

▲ **Figure 5-4** The San Andreas Fault is a strike-slip fault.

Movement along this fault caused the great San Francisco Earthquake of 1906. In 1989, another earthquake hit Santa Cruz and the Bay area. In 1994, an earthquake in the Northridge area put further stress on the San Andreas Fault and other faults nearby. Areas with networks of faults, such as southern California, have many earthquakes.

3 INFER: Why would a network of faults have many earthquakes?

✓ CHECKING CONCEPTS

1. What is a fracture?
2. Which way does a syncline fold?
3. What do rock layers that change shape but do not break form?
4. What do rock layers that break but do not move form?
5. What do rock layers that break and move form?

THINKING CRITICALLY

6. INFER: What slowly changes Earth's crust?
7. INFER: What suddenly changes Earth's crust?

INTERPRETING VISUALS

Use Figure 5-3 to answer the following questions.

8. IDENTIFY: Along which fault is the movement horizontal?
9. COMPARE/CONTRAST: How is the movement along a reverse fault similar to and different from that along a normal fault?
10. INFER: Why do you think the thrust fault was given its name?

Hands-On Activity

MODELING THE FOLDING OF ROCK LAYERS

You will need 4 different colors of modeling clay, a plastic knife, a metric ruler, and a sheet of wax paper 30 cm long.

1. Flatten each color of clay into a strip about 9 cm wide and 24 cm long. Each strip should be about 1 cm thick.
2. Using the plastic knife, trim the clay strips so that they form a rectangle 8 cm × 23 cm.
3. Stack the four strips of clay on the sheet of wax paper.
4. Push both ends of the stacked clay layers toward the center.
5. Sketch the folded layers. Label the layers by color.
6. Label the anticlines and synclines on your drawing.

▲ **STEP 4** Push both ends of the clay together.

Practicing Your Skills

7. MODEL: What do the clay layers represent?
8. ANALYZE: What caused the clay layers to fold?
9. OBSERVE: Did any cracks appear in the clay layers? If so, what are they called?

3 *Assess*

✓ CHECKING CONCEPTS ANSWERS

1. a break in a rock
2. downward
3. folds
4. fractures
5. faults

THINKING CRITICALLY ANSWERS

6. INFER: processes such as folding
7. INFER: processes such as faulting and earthquakes

INTERPRETING VISUALS ANSWERS

8. IDENTIFY: strike-slip
9. COMPARE/CONTRAST: Movement along reverse and normal faults is principally vertical. In a normal fault, the rocks above the fault plane move down relative to the rocks below the fault plane. In a reverse fault, rocks above the fault plane move up relative to the rocks below the fault plane.
10. INFER: In a thrust fault, rocks are pushed, or thrust, up over the fault plane.

4 *Close*

Hands-On Activity

MODELING THE FOLDING OF ROCK LAYERS

TIME: 15 minutes

PURPOSE: modeling, observing, analyzing

SAFETY TIP: Remind students to use the knives only as instructed and to report any cuts to you immediately.

ALTERNATIVE MATERIALS: If you wish to reuse the clay, have students put wax paper between adjacent layers. You might want to provide rolling pins to flatten the clay.

COOPERATIVE LEARNING: Have pairs of students apply force to a single rectangle and compare the folds that form.

Practicing Your Skills Answers

7. MODEL: different rock layers
8. ANALYZE: the force applied to the sides
9. OBSERVE: If the clay is somewhat dry, cracks called fractures might have formed. Fractures are not faults, as there is no movement along these breaks.

EXTENDING THE LESSON

Using Miners' Terms Have students research what the terms *hanging wall* and *footwall* mean. (Miners once used these terms to describe the rocks on a fault plane. The miners hung their lanterns on the hanging wall and stood on the footwall.) Have students trace the normal, thrust, and reverse faults from Figure 5-3 onto a sheet of paper and identify the hanging wall and footwall in each.

QUICK CONVERSION		
	Metric/SI	English
Depth of San Andreas Fault	32.0 km	19.9 mi

• *TEACHING RESOURCES* • *Online or Print*

Teacher's Resources

Lesson Review: Chapter 5, p. 2

MORE ON THE WEB

Researching Other Faults Have students research the San Andreas Fault, the Denali Fault, and the New Madrid Fault online. Then, have them write a short report about a famous earthquake associated with one of these faults. Have students visit www.conceptsandchallenges.com to begin their search.

5-2 *(pp. 112–113)*

1 *Introduce*

STUDY HINT Remind students that textbooks use certain strategies to help make information more accessible. Ask students to scan the lesson and list the strategies used.

Linking Prior Knowledge Have students recall what they learned about maps and how elevation is shown on a map (Lessons 1-7 to 1-9).

2 *Teach*

Demonstration Use a clear plastic storage box with a lid, a ruler, a black marker, colored water, and clay to demonstrate how the elevation of a mountain is plotted on a topographic map. Using the clay, make a mountain that is a little smaller than the depth of the box and put it into the box. Use the ruler and the marker to mark off the side of the box every 2 cm (0.75 in.). Have a volunteer pour colored water into the box until it reaches the 2-cm mark. Put the lid on the box and trace the water level. Mark this contour *2 cm*. Repeat this for the 4-cm mark and the remaining marks. Allow students to observe the map on the lid.
learning style: visual

Discussion Draw diagrams that show a mountain in its various stages of development. Explain that deposition, which occurs as a result of the processes of weathering (the physical and chemical breaking down of rocks into smaller pieces) and erosion (the carrying away of these bits of rocks) are responsible for changing the shape and elevation of a mountain or mountain system.
learning styles: visual, auditory

Answers to questions

1. **DEFINE:** a landform that reaches more than a few hundred meters above the surrounding land
2. **OBSERVE:** The Eurasian-Melanesian is in Europe and Asia, and the Circum-Pacific rings the Pacific Ocean along the coasts of Asia and North and South America.
3. **CLASSIFY:** old

5-2 What are mountains?

Objective
Identify the world's major mountain systems.

Key Terms
landform: physical feature on Earth's solid surface
elevation (el-uh-VAY-shuhn)**:** height of a point on Earth above or below sea level
summit: highest point on a mountaintop

Defining Mountains A mountain is a landform that reaches a high elevation. A **landform** is any physical feature on Earth's solid surface. **Elevation** is the height of a point on Earth above or below sea level. For a hill or a mound of land to be classified as a mountain, its **summit,** or highest point, must be more than a few hundred meters above the land around it.

1 **DEFINE:** What is a mountain?

Mountain Systems Most mountains do not stand alone. They are part of a group of mountains. A group of mountains with the same general shape and structure is called a mountain range. For example, Mount St. Helens in Washington State is part of the Cascade mountain range. Groups of mountain ranges form into what are called mountain systems.

In the eastern part of the United States is the Appalachian mountain system. The Blue Ridge and Great Smoky mountain ranges are part of the Appalachian system.

▲ **Figure 5-5** The Blue Ridge Mountains are in the Appalachian mountain system.

Some scientists break up mountain systems into two major groups, called belts. These belts, shown in Figure 5-6, are called the Eurasian-Melanesian belt and the Circum-Pacific belt.

2 **OBSERVE:** Where are the major mountain belts?

Mountain Development Over millions of years, mountains change and take on different forms. Mountains change as the crust is either built up or worn away by various processes. This process is called mountain development.

Some mountains have steep slopes. The peaks are sharp and jagged. The valleys are narrow. These mountains are often called "young" because they are formed from new crust.

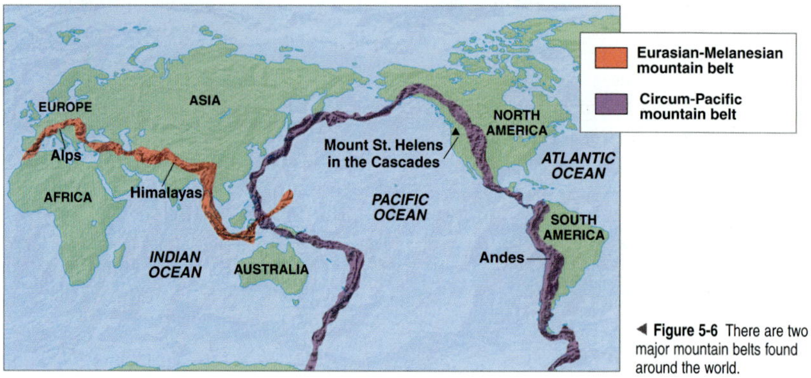

◄ **Figure 5-6** There are two major mountain belts found around the world.

112

Teaching Options

EXTENDING THE LESSON

Locating Mountain Peaks Using references, have students find at least three peaks and their elevations in each range shown in Figure 5-6. Students should then locate these ranges on a classroom map, using pushpins and labels. **learning style: visual**

ESL/ELL STRATEGY

Personalizing the Lesson Ask students what they think of when they hear the word *mountain.* Students from relatively low-lying areas may have concepts that differ from students that come from mountainous regions.

Mountain	Country	Range It Is Part of
Everest	Nepal-Tibet border	Himalayas
Logan	Canada	St. Elias Mountains
McKinley	United States	Alaska Range
Nebo	Jordan	Abarim Mountains
Whitney	United States	Sierra Nevada

Mountain	Height	
	Metric/SI	English
Everest	8,850 m	29,035 ft
Logan	5,951 m	19,524 ft
McKinley	6,194 m	20,320 ft
Nebo	806 m	2,644 ft
Whitney	4,418 m	14,494 ft

As time passes, a mountain's peak is worn down by weather conditions. These make the peak more rounded. The slopes become less steep. Mountains like these are sometimes called "mature" mountains.

Some mountains continue to be worn away for a very long time. No buildup occurs. These "old" mountains are almost flat and have no jagged peaks. They have many rolling hills, and the valleys between these mountains are wide.

3 CLASSIFY: The Appalachians have very rounded peaks. Do you think they are young or old?

CHECKING CONCEPTS

1. The highest point of a mountain is its __________.
2. There are narrow valleys between __________ mountains.
3. Mountain systems make up mountain __________.
4. If a mountain contains many rolling hills, the mountain is __________.
5. A mountain's stage depends on the __________ of the mountain.

THINKING CRITICALLY

6. SEQUENCE: Place the following terms in order, from smallest to largest: mountain belt, single mountain, mountain range, mountain summit, mountain system.
7. CLASSIFY: A mountain has high jagged peaks and steep slopes. Its valleys are narrow. Is the mountain a young mountain or an old mountain? Explain your answer.
8. INTERPRET A VISUAL: What are two mountain ranges found in the Eurasian-Melanesian Belt shown in Figure 5-6?

BUILDING SCIENCE SKILLS

Researching and Organizing Mount McKinley, Mount Whitney, Mount Everest, Mont Blanc, Mount Nebo, and Mount Logan are some famous mountains. Find out the height in meters or kilometers of each mountain. Identify the country in which the mountain is located. Also, find out the name of the mountain range in which each mountain is located. Organize the information into a table.

People in Science

MOUNTAIN EXPLORERS

On May 29, 1953, two mountain climbers reached the top of Mount Everest. They had climbed to 8,848 meters above sea level. Edmund Hillary from New Zealand and Tenzing Norgay, a Sherpa tribesman from Nepal, were the first mountain climbers to reach the summit of Mount Everest. Mount Everest is the highest mountain in the world. Many people had tried before them, but no one had succeeded in climbing to the top. Heavy snow, strong winds, and thin air make climbing the cliffs of Mount Everest very difficult. Sir Edmund's book, *High Adventure*, describes his experiences during the 1953 climb.

▲ **Figure 5-7** Edmund Hillary (left) and Tenzing Norgay pose for pictures right after climbing to the top of Mount Everest.

In 1957, Hillary set forth on an expedition to the South Pole. On January 4, 1958, he reached the South Pole by tractor. In 1977, he led the first jet boat expedition up the Ganges River to find its source. When the boat could go no farther, Hillary climbed the Himalayas to reach the source of the river.

Thinking Critically Why do you think Hillary wanted to climb Mount Everest?

3 Assess

CHECKING CONCEPTS ANSWERS

1. summit
2. young
3. belts
4. old
5. age

THINKING CRITICALLY ANSWERS

6. SEQUENCE: mountain summit, single mountain, mountain range, mountain system, mountain belt
7. CLASSIFY: The mountain range is young. Its peak would be rounded from weathering if it were an old mountain.
8. INTERPRET A VISUAL: the Alps and the Himalayas

4 Close

BUILDING SCIENCE SKILLS

Researching and Organizing Encourage students to use an atlas to find accurate elevations of the peaks. The table in Teaching Options on page T112 shows possible answers. Note that there is also a Mount Nebo in Utah.

People in Science

MOUNTAIN EXPLORERS

Real-Life Connection Tell students what they would need to go hiking in a hilly nature area. Each student would need a filled water bottle, a healthy snack, a first-aid kit, and appropriate shoes for the hike.

Discussion Have volunteers describe their experiences hiking in mountainous regions. Prompt students to describe how the air temperature, the plant and animal life, and their own circulatory and respiratory systems changed with an increase in elevation. (Air temperatures decrease with elevation. Different types of trees and other plants and animals often thrive at different altitudes. The body's respiratory and circulatory systems work harder as a person ascends a mountain because oxygen levels decrease as elevation increases.)

Thinking Critically Answer
Students should be able to deduce from the title of this feature as well as the text that Hillary was an explorer, interested in great challenges, seeing new things, and making discoveries about the world.

MORE ON THE WEB

People in Science Have students learn more about Tenzing Norgay by visiting www.conceptsandchallenges.com. They can take turns acting as reporter and as Norgay. Have the reporter interview Norgay, asking him about the climb and how his life changed because of it.

• TEACHING RESOURCES • Online or Print

Teacher's Resources
Lesson Review: Chapter 5, p. 3

INTEGRATING LANGUAGE ARTS

Reading About High Adventures Have students read Hillary's book *High Adventure* and write a book report. Suggest that interested students find out more about Hillary's other adventures and books. **learning style: linguistic**

5-3 *(pp. 114–115)*

1 Introduce

STUDY HINT Have students sketch or trace the three figures that correspond to the photographs on these two pages. Below each sketch, have students label the type of mountain shown. As you teach this lesson, have students write facts about each mountain type below the appropriate sketch.

Linking Prior Knowledge Have students recall how igneous rocks form (Lesson 3-2). Ask: *What is magma?* (molten rock inside Earth) *What is magma that reaches Earth's surface called?* (lava)

2 Teach

Demonstration Use four or five shoeboxes with their lids taped on and as many volunteers to demonstrate how the relative uplift or dropping of rocks creates fault-block mountains. Use Figure 5-10 as an example of how to align the boxes. Have another student point out the fault planes in the model. Have students tilt the boxes and repeat their mountain-forming demonstration. Challenge the rest of the class to identify each fault as a normal fault or a reverse fault.
learning styles: visual, tactile

Discussion Discuss the fault-block demonstration. Ask students about the direction of the forces that form fault-block mountains. (They are primarily horizontal and push on the bodies of rock from the sides.) Ask about the direction of forces that form dome mountains. (They are vertical and push up against the overlying rocks.) Have students discuss whether folded mountains are formed more like fault-block or dome mountains. (dome mountains)
learning style: auditory

Answers to questions

1. **EXPLAIN:** by how they are formed
2. **DESCRIBE:** Continents collide, squeezing rocks from opposite sides.
3. **LIST:** the Grand Tetons and the Sierra Nevada

5-3 How are mountains classified?

Objectives

Identify three types of mountains. Describe how they are formed.

Key Terms

folded mountain: mountain formed by the folding of rock layers

dome mountain: mountain formed when upfolds in rocks create a rounded structure that looks like a bowl turned upside down

fault-block mountain: mountain formed when normal faults uplift a block of rock

▲ **Figure 5-8** Mount Everest, in the Himalayas, is a folded mountain.

Classifying Mountains Mountains are classified based on how they are formed. There are three main types of mountains. These are volcanic, folded, and fault-block mountains.

Volcanic mountains are formed from lava or debris, such as ash or rocks, thrown out of a volcano. Mount Fuji, shown in the photo at the beginning of this chapter, is a volcanic mountain. Volcanic mountains are built up from eruptions that occur over thousands or even millions of years. They will be discussed in more detail in Lesson 5-6.

1 **EXPLAIN:** How are mountains classified?

Folded Mountains When rock layers are subject to certain forces, they can become bent in a process called folding. Mountains formed by the folding of rock layers are called **folded mountains.**

Earth's continents move slowly on top of the upper mantle. Most folded mountains form when the continents collide.

The movements of the continents squeeze rock layers together. Over millions of years, pressure builds up. The rock layers of the crust buckle and fold.

Large upfolds, or anticlines, form folded mountains. The Himalayas in south central Asia and the Urals in Russia are folded mountains. The Appalachian Mountains in the United States are also folded mountains.

When some mountains fold upward they produce a shape that looks like a bowl that has been turned upside down. These mountains are called **dome mountains.**

The rock layers in a dome mountain dip down and out in all directions from a center point. As the top layers of unfolded sedimentary rock are worn away, older igneous or metamorphic rock beneath is exposed. The Black Hills of South Dakota are an example. The upper rock layers of these dome mountains have been worn away unevenly. This has resulted in many separate peaks being formed.

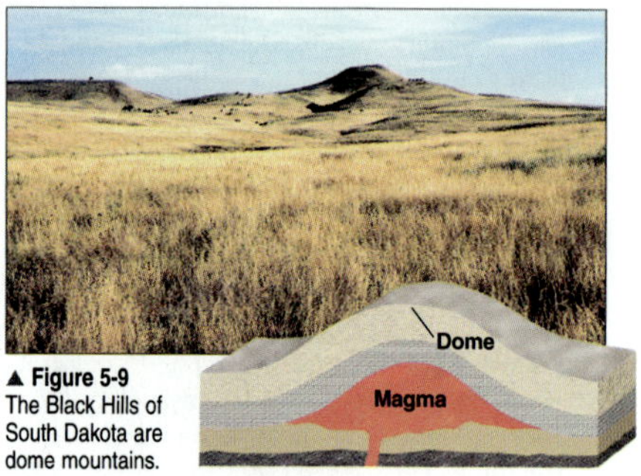

▲ **Figure 5-9** The Black Hills of South Dakota are dome mountains.

2 **DESCRIBE:** How are folded mountains formed?

Teaching Options

RETEACH

Classifying Mountains Have students classify the mountain ranges (not the individual peaks given in the text) that they identified in the Building Science Skills activity on page 113 as folded, fault-block, or dome mountains. learning style: logical

EXTENDING THE LESSON

Identifying More Landforms Have students find out about horsts and grabens. A *horst* is an uplifted block of rock bound by faults. A *graben* is a narrow valley that forms when fault blocks of rock drop below the surrounding landscape. Students should find examples of each. (The Great Rift Valley in East Africa is a series of horsts and grabens. The Imperial Valley in California is a graben.) Have students report their findings to the class and use a world map to point out where the examples are located. learning style: visual

Fault-Block Mountains Fractures can form in Earth's crust. These fractures may break the crust into large blocks. Sometimes faulting lifts these large blocks. One side of the fault slips past the crust on the other side. If the blocks are pushed up enough, a mountain is formed. Mountains formed in this way are called **fault-block mountains.**

The Grand Tetons in Wyoming are fault-block mountains. The Sierra Nevada in California are also fault-block mountains.

▲ **Figure 5-10** The Sierra Nevada are fault-block mountains.

3 LIST: Name two fault-block mountain ranges in the United States.

CHECKING CONCEPTS

1. Which type of mountain is usually formed by colliding continents?
2. Which type of mountain is formed by vertical pressure?
3. Which type of mountain is formed by upward thrusts of Earth's crust?

THINKING CRITICALLY

4. CLASSIFY: Classify each mountain range listed as dome, folded, or fault-block. Explain your choices.
   a. Appalachian Mountains
   b. Grand Tetons
   c. Himalayas

BUILDING SCIENCE SKILLS

Modeling On a sheet of paper, make a series of drawings to show the formation of a dome mountain. Use arrows to show the movement of magma and pressure. Color your drawings. Use a key to show what the colors mean.

Real-Life Science

MOUNTAIN CLIMBING

Mountain climbing, or mountaineering, is a sport enjoyed around the world. It can range from simple trail hiking to more difficult rock and ice climbing. Professional climbers prefer folded mountains. These are some of the world's tallest mountains. They feature very rugged and challenging terrains. The Himalayas are folded mountains. They contain all but one of the 14 tallest mountains on Earth.

Fault-block mountains often have high cliffs. This provides climbers with special challenges. Dome and volcanic mountains tend to be easier to climb. However, they attract climbers of all skill levels because of their unique terrains and beautiful scenery.

As with all sports, one must follow some basic safety rules. Climbing gear such as helmets and ropes must be suitable for the climber. A person should never climb alone. A qualified rock climbing instructor should train and supervise all climbing activities.

Thinking Critically Why would dome mountains be easier to climb than other mountains?

▲ **Figure 5-11** Special gear is needed for climbing snow-covered mountains.

MORE ON THE WEB

Researching Mountaineering Students can use the Internet to learn about the sport of mountaineering. Have small groups of students research the safety, equipment, and techniques involved. Each group should create an oral presentation on the subject.

• TEACHING RESOURCES • Online or Print

Teacher's Resources

Lesson Review: Chapter 5, p. 4

READING STRATEGY Before students read the lesson, have them look at the head *Folded Mountains* and predict what shape these mountains have.

3 Assess

CHECKING CONCEPTS ANSWERS

1. folded mountain
2. dome mountain
3. fault-block mountain

THINKING CRITICALLY ANSWERS

4. CLASSIFY: **a.** The Appalachians are folded mountains that formed when the North American continent collided with another continent. **b.** The Grand Tetons are fault-block mountains that formed when large blocks of faulted rocks were forced upward. **c.** The Himalayas are folded mountains that were formed as India collided with the Asian continent.

4 Close

BUILDING SCIENCE SKILLS

Modeling Students' drawings should be sequenced as follows: flat layers of undeformed rock, magma pool forming under the layers of rock, and gradual uplift and doming of rocks over the magma pool.

Real-Life Science

MOUNTAIN CLIMBING

Real-Life Connection Invite an expert rock climber to the class to discuss some of his or her experiences and also to demonstrate climbing safety equipment. If this is not possible, ask a salesclerk at a local sporting goods store to show students basic rock-climbing equipment.

Discussion Have students brainstorm to compile a list of reasons why one should never climb alone. Reasons might include that if a person fell, a partner could go for help or if a climber got lost, the partner could help find the way. Stress the old adage that there's safety in numbers.

Thinking Critically Answer Possible answer: Dome mountains are smaller and have less rugged terrain than folded and fault-block mountains.

5-4 *(pp. 116–117)*

1 Introduce

Linking Prior Knowledge Have students recall the layers of Earth (Lesson 1-3). Ask: *What is the solid outer layer of Earth called?* (the crust)

2 Teach

Demonstration Display a relief map of the United States (which can be found in the Appendix of this book and most atlases) so that students can compare the major landforms of the country. Draw an east-west cross section through roughly the middle of the United States to demonstrate how, in general, relief changes. (The Appalachians are old mountains in the eastern United States that rise hundreds of meters above sea level. Much of the central part of the country is the Great Plains, which is an essentially flat feature. Elevation increases sharply in Colorado at the foothills of the Rockies. Much of California is dotted with mountains and plateaus that border the sea.)
learning style: visual

Discussion Discuss several definitions of the word *plain*, such as "without flavor," "with little or no color," and "simple." Have students use these meanings to formulate a working definition of a geological plain. (an area with little change in elevation) Ask students to suggest adjectives that contrast the characteristics of plains and plateaus. (Plains might be described as flat, featureless, and low-lying. Plateaus might be described as elevated landforms that sometimes have towering cliffs and deep canyons.)
learning styles: logical/mathematical, auditory

Answers to questions

1. **NAME:** coastal plains and interior plains
2. **DESCRIBE:** an elevated landform that often has canyons
3. **OBSERVE:** Possible answers: mountains, plains, or plateaus

READING STRATEGY When students have finished reading, ask them to compare and contrast plains and plateaus.

5-4 What are plains and plateaus?

Objectives

Compare and contrast plains and plateaus.
Describe how they form.

Key Terms

plain: large, flat area just above sea level
plateau: large, flat area at a high elevation

Plains Large, flat areas that are just above sea level are called **plains.** There are coastal plains and interior plains. Coastal plains are located in coastal areas. Interior plains are located inland. All plains slope gently over great distances.

▲ **Figure 5-12** The Great Plains of the United States are interior plains.

Plains are formed in several ways. One way is for land of uneven elevation to be worn down by weather conditions. The Great Plains of the United States were formed this way.

Another way for plains to form is for material to be deposited in a large body of water. Then, something causes the water level to drop or the land to rise. A flat, dry area of land remains.

1 **NAME:** What are the two kinds of plains?

Plateaus **Plateaus** have much higher elevations than plains. Yet, plateaus are also large, flat areas, like plains. Most plateaus are located inland. Plateaus that are near oceans have cliffs that face the ocean.

Many plateaus have canyons. A canyon is a steep-sided valley formed by a river. The Colorado Plateau has many canyons carved out by the Colorado River. This same river also formed the Grand Canyon.

▲ **Figure 5-13** The Colorado Plateau was carved out by the Colorado River.

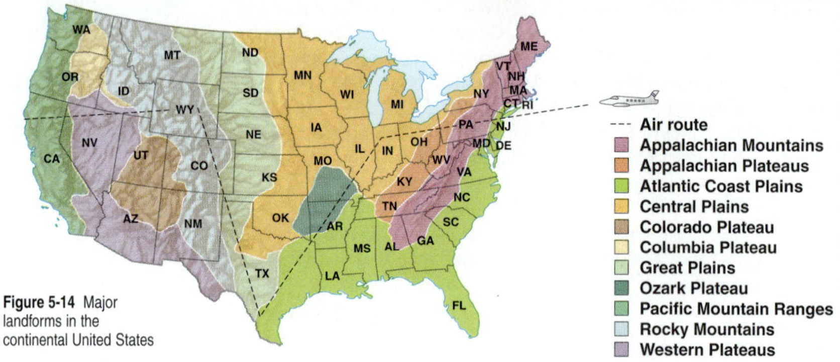

Figure 5-14 Major landforms in the continental United States

116

Teaching Options

! FAST FACTS *The Grand Canyon*

The Grand Canyon, shown on page 65, is a huge gorge, 1.6 km (1 mi) deep in places. It cuts through mostly sedimentary rocks that overlie layers of metamorphic and igneous rocks. The oldest rocks in the canyon are about 2 billion years old. The youngest, on the canyon's rim, are about 225 million years old. The rocks of the Grand Canyon are part of the Colorado Plateau.

Weathering causes the variation in the shape of the canyon's slopes. Limestone and some sandstone weather to form steep walls. Shale weathers more easily and so forms more gradual slopes. Cliffs form when soft, less-resistant rock erodes and undercuts overlying resistant rocks. The subtle hues of the valley walls are largely because of different amounts of iron in the rocks..

INTEGRATING PHYSICAL SCIENCE

Making Bread Write a simple bread recipe on the board and have students identify chemical and physical changes that occur when making bread. (Physical changes include mixing the ingredients, increasing volume, and decreasing density. Chemical changes include converting sugar to gas.) Invite interested students to follow the recipe at home to make bread.

The same forces that build mountains form plateaus. Large areas of the crust are pushed upward. Some plateaus are formed by lava pouring out of a volcano or a chain of volcanoes. The lava cools and hardens. This forms a large, raised, flat area. The Columbia Plateau in Washington State was formed in this way.

2 **DESCRIBE:** What is a plateau?

Landform Regions Mountains, plains, and plateaus are the three main kinds of landforms. Landforms make up the regions of the world. Figure 5-14 shows the different landform regions of the United States.

3 **OBSERVE:** In what landform region do you live?

CHECKING CONCEPTS

1. What are the three main kinds of landforms?
2. Which landform has the lowest elevation?
3. What lava plateau is found in the United States?
4. What river formed the Grand Canyon?

THINKING CRITICALLY

Use Figure 5-14 to answer the following questions.

5. **COMPARE/CONTRAST:** How are plains and plateaus alike and how are they different?
6. **SEQUENCE:** You are taking an airplane flight across the United States, going from east to west. Your plane will follow the air route shown in the illustration. List the landform regions of the United States in the order that you will pass over them as you make the trip.
7. **CLASSIFY:** Choose five states. In which landform region or regions is each state located?

BUILDING READING SKILLS

Vocabulary There are many landforms other than the ones talked about so far. Find out where the names of the following landforms come from. Then, tell if and how their names relate to how they are formed: *butte, peneplain, monadnock, mesa.*

Integrating Life Science

TOPICS: agriculture, biomes

AMERICA'S BREADBASKET

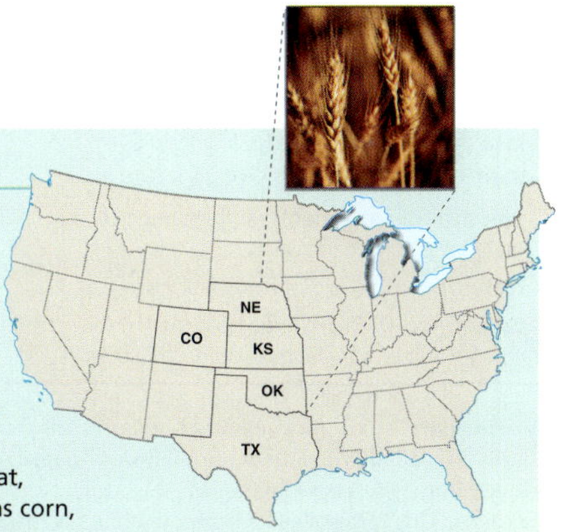

▲ **Figure 5-15** The five breadbasket states, outlined, grow crops such as wheat (inset).

Most of the Great Plains, shown in light green on Figure 5-14, are flat. There are few hills or mountains. The states in the Great Plains have rich soil that is good for growing crops, especially Colorado, Kansas, Nebraska, Oklahoma, and Texas. Many grains used to make bread, such as oat, wheat, and rye, are grown there. Other food crops such as corn, potatoes, soybeans, and alfalfa are also grown there. Beef cattle and sheep graze on the grasslands. Hogs are raised on farms. In the northern part of the breadbasket, dairy farms are more common.

The Great Plains are grasslands, also known as prairies. Grasslands are usually found in the centers of continents, where rainfall is relatively low. These regions generally have hot summers and very cold winters. Tallgrass prairies have grasses more than a meter high. Many of the grains grown in the Great Plains are tallgrasses.

Thinking Critically Why are plains good grazing areas?

MORE ON THE WEB

Researching Plateaus Have students use the Internet to learn about three plateaus in the United States. Students can use the online dictionary at www.conceptsandchallenges.com to find the definitions of *monsoon*, *flora*, and *petroleum*.

• TEACHING RESOURCES • Online or Print

Teacher's Resources
Lesson Review: Chapter 5, p. 5

ONGOING ASSESSMENT

Have students use a physical map of the world to locate several examples of plains, plateaus, and mountains that are not discussed in the text. Students should list the landforms and the countries in which they are found. **learning style: visual**

3 *Assess*

CHECKING CONCEPTS ANSWERS

1. mountains, plains, and plateaus
2. a plain
3. the Columbia Plateau
4. the Colorado River

THINKING CRITICALLY ANSWERS

5. **COMPARE/CONTRAST:** Both are large and flat. Plains are essentially featureless and low-lying. Plateaus are elevated with cliffs and deep canyons.
6. **SEQUENCE:** Atlantic Coastal Plains, Appalachian Mountains, Appalachian Plateau, Central Plains, Ozark Plateau, Atlantic Coast Plains, Great Plains, Rocky Mountains, Western Plateaus, Pacific Mountain Ranges
7. **CLASSIFY:** Encourage students to choose a variety of locations.

4 *Close*

BUILDING READING SKILLS

Vocabulary A butte is an elevated, often isolated, flat-topped hill or mountain with steep slopes. *Butte* is French for "knoll" or "rounded hill." A peneplain is a low, nearly featureless, gently rolling expanse of land produced by erosion. *Pene-* is Latin for "almost." A monadnock is a knob of rock that protrudes above a peneplain. The term comes from the place where such a structure was first described—Mount Monadnock in New Hampshire. A mesa is an elevated, flat-topped landmass smaller than a plateau but larger than a butte. *Mesa* is Spanish for *table*. Buttes, peneplains, and mesas result from erosion. A monadnock is a rock that resists erosion.

Integrating Life Science

TOPICS: agriculture, biomes

AMERICA'S BREADBASKET

Background The conditions that make the states in the breadbasket ideal for farming include thick, fertile soils; adequate yearly precipitation; and mild temperatures during the growing season. The dominant plants are grasses and perennial wildflowers.

Discussion Work with students to list the breadbasket foods that they eat regularly.

Thinking Critically Answer
The abundant grass on plains can support large herds of grazing animals.

5-5 *(pp. 118–119)*

1 Introduce

INVESTIGATE

TIME: 10 minutes

PURPOSE: Students will use sand to model volcano formation.

MATERIALS: coarse sand, aluminum pie plates, safety goggles

PROCEDURE: You can have students work in pairs to complete this activity. Explain that after Steps 2 and 3, each pair should discuss their findings.

SAFETY TIPS: Remind students to wear safety goggles during this activity.

THINK ABOUT IT: The sand represents volcanic materials. The shape of the model is typical of a volcanic cone.

Linking Prior Knowledge Ask students to describe Earth's mantle (Lesson 1-3). (The mantle has two parts—an upper solid part and a lower part that flows like a liquid.)

2 Teach

Discussion Ask students to describe a volcano. Most will probably describe a cone-shaped mountain shooting red-hot lava and volcanic rocks high into the air. Although this description is true of some volcanoes, explain that a volcano is any opening in the crust from which lava flows. **learning style: auditory**

READING STRATEGY Have students find the following words in the lesson: *volcano, volcanism,* and *volcanic*. Have students use a dictionary to identify what part of speech each term is and use each term in a sentence. Encourage students to construct their own sentences rather than relying on those in the text.

Answers to questions

1. **DEFINE:** magma that flows onto Earth's surface
2. **CONTRAST:** A crater is the pit at the top of a volcanic cone. A caldera is a large hole that forms when the walls of a crater fall back into the volcanic vent.
3. **DESCRIBE:** shaking, cracks in crust, hot gases, and lava

5-5 What are volcanoes?

INVESTIGATE

Modeling a Volcano

HANDS-ON ACTIVITY

1. Take a handful of sand and make a fist over an aluminum pie plate. Dribble a little of it out at a time onto the plate. Do not touch the sand or move the plate.
2. Take another handful of sand and make a fist. Again, dribble the sand out onto the pie plate over the first pile. Keep taking handfuls of sand until you have a good-sized pile in the plate.
3. Note the shape the sand takes as it builds up on the plate.

THINK ABOUT IT: What does the sand represent? How is the shape like that of a typical volcano?

STEP 2

Objective

Describe volcanism.

Key Terms

volcanism (VAHL-kuh-nihz-uhm): movement of magma inside Earth

lava: magma that reaches Earth's surface

vent: volcano opening from which lava flows

volcano (vahl-KAY-noh): vent and the volcanic material around it

crater: pit at the top of a volcanic cone

caldera (kal-DER-uh): large hole that forms when the roof of a magma dome collapses

Volcanism Any movement of magma inside Earth is called **volcanism.** Sometimes magma flows between rock layers of the crust and hardens. This forms a sill. Magma that cuts across rock layers and hardens forms a dike.

Sometimes magma breaks through the crust and flows onto Earth's surface. Then, it is called **lava.** The opening that lava flows through is called a **vent.** Dust, ash, and rock particles are often thrown out of the vent.

The word **volcano** really refers to the volcanic mountain, or cone, that forms from the materials that collect around the vent. It also refers to the vent itself.

◀ **Figure 5-16** Parts of a volcano

Go Online **active art**
For: Composite Volcano Eruption activity
Visit: PHSchool.com
Web Code: cfp-1033

1 **DEFINE:** What is lava?

Craters and Calderas At the top of a volcanic cone there may be a deep pit. This pit, called a **crater,** forms as material is blown out of the volcano's vent.

The top of a volcano may also explode, emptying the vent. This can turn the mountain into a huge, hollow shell. The walls of the crater may fall back into the vent. This partly filled gaping hole is called a **caldera.**

Some calderas fill with water from rain or snow. They form large lakes such as Crater Lake in Oregon. Crater Lake is 9.6 km long and 8 km wide. It is about 600 m deep.

Teaching Options

! FAST FACTS *Volcanic Material*

Rock fragments ejected from a volcano are called *tephra.* Tephra are classified by size. Dust refers to fragments that are smaller than 0.06 mm (0.02 in.) in diameter. Ash ranges in size from 0.06 mm to less than 2 mm (0.08 in.). Cinders are tephra that are larger than 2 mm but less than 64 mm (2.5 in.) in diameter. Large tephra, called bombs and blocks, can be as big as a car or a small house.

ESL/ELL STRATEGY

Labeling a Volcano On an overhead transparency, draw a volcanic cone with lava flowing down its flanks. Draw pointers to the magma chamber, vent, and crater. Define each of the parts. **learning style: visual**

QUICK CONVERSIONS

	Metric/SI	English
Height of Parícutin	400 m	1,312 ft
Width of Parícutin	5 km	3.1 mi
Crater Lake length	9.6 km	5.97 mi
Crater Lake width	8.0 km	4.97 mi
Crater Lake depth	600 m	1,969 ft
Distance between Indonesia and Australia	3,540 km	2,220 mi

▲ **Figure 5-17** Crater Lake in Oregon is a caldera that filled with water.

2 CONTRAST: What is the difference between a crater and a caldera?

The Story of Parícutin In 1943, a farmer in Mexico saw a mountain form! First, the ground started shaking. A few weeks later, cracks, called fissures, appeared and widened. Hot gases came out of the fissures. Soon, lava began to seep out. After several months, a volcanic cone about 400 m high and 5 km wide had formed. The volcano was called Parícutin (pah-ree-koo-TEEN).

3 DESCRIBE: What signs were there that a volcano was forming at Parícutin?

CHECKING CONCEPTS

Explain the differences in each word pair.

1. magma, lava
2. caldera, crater
3. vent, volcano
4. mountain, volcanic cone

THINKING CRITICALLY

5. INFER: How are volcanoes like windows to the inside of Earth?

Web InfoSearch

Krakatoa Krakatoa (krah-kah-TOH-uh) is a volcano found near Indonesia. In August 1883, it erupted in a huge volcanic explosion that was heard in Australia, thousands of kilometers away.

SEARCH: Use the Internet to find out more about Krakatoa. How did it change the surrounding islands? Start your search at www.conceptsandchallenges.com. Some key search words are **Krakatoa, Krakatoa island,** and **Krakatoa Volcano.**

Science and Technology

PREDICTING VOLCANIC ERUPTIONS

On May 18, 1980, Mount St. Helens in Washington State blew its top! The force of the blast destroyed much of the area around the mountain. Scientists knew years before that the volcano in the Cascade Mountains might erupt. However, they could not predict when.

New technologies have now made such predictions possible. Seismometers alert scientists when a volcano rumbles. Tiltmeters and geodimeters measure the tiniest swelling of a volcano, caused by movement of magma. Correlation spectrometers (COSPECS) measure gases such as sulfur and carbon dioxide. These gases are released in increasing amounts before an eruption.

▲ **Figure 5-18** Mount St. Helens erupted in 1980.

Earth-observation satellites also help in predicting volcanic eruptions. Instruments on the satellites monitor Earth's surface temperature. This data can alert scientists to any volcanic changes that may signal trouble.

Thinking Critically Why is it important to predict volcanic eruptions?

3 *Assess*

CHECKING CONCEPTS ANSWERS

1. Magma is molten material beneath Earth's surface. Lava is molten material that flows onto Earth's surface.
2. A crater is the pit at the top of a volcanic cone. A caldera is a large hole that forms when the walls of a crater collapse.
3. A volcano is an accumulation of volcanic material that builds up around an opening called a vent.
4. Many volcanoes are cone-shaped mountains. Not all mountains, however, are volcanoes.

THINKING CRITICALLY ANSWER

5. INFER: They allow us to see the materials below the surface of Earth.

4 *Close*

Web InfoSearch

Krakatoa Have students use the Internet to find out how Krakatoa changed the surrounding islands. Some key search words are **Krakatoa, Krakatoa island,** and **Krakatoa Volcano.** Students can begin their search at www.conceptsandchallenges.com.

Science and Technology

PREDICTING VOLCANIC ERUPTIONS

Background Because the magma cools below the surface, moving magma does not always result in an eruption. Monitoring every site is not possible because there are thousands of volcanoes, some erupting only every few hundred or thousand years. Monitoring is expensive. However, scientists should be able to predict many eruptions.

Discussion Have students hypothesize what kinds of movements are measured by each instrument mentioned. (A seismometer measures seismic [earthquake] waves. A tiltmeter measures any change from the horizontal in a surface. A geodetic distance meter measures changes in the ground that might be caused by magma movement.)

Thinking Critically Answer
to minimize injuries and loss of life and also to reduce property damage

RETEACH

Modeling Lava Use a needle and a self-sealing plastic bag nearly filled with water to model the flow of lava from a volcano. Make sure the bag is sealed. Poke a hole near the top of the bag. Ask students to hypothesize what will happen if you gently squeeze the bag. Ask what will happen if you squeeze the bag with much greater force. (The water will flow with greater force from the hole and probably also from the opening. The bag might even pop.) Ask students to compare this to the upward movement of magma in a volcanic vent.
learning style: visual

• *TEACHING RESOURCES* •
Online or Print

Teacher's Resources

Lesson Review: Chapter 5, p. 6

Enrichment: Chapter 5, p. 9

Parts of a Volcano: Visuals, p. 4

Go Online
active art

For: Composite Volcano Eruption activity
Visit: PHSchool.com
Web Code: cfp-1033

Students can interact with the art of volcanic eruption online.

5-6 *(pp. 120–121)*

1 *Introduce*

STUDY HINT Have students summarize the main idea of each section in a sentence or two.

Linking Prior Knowledge Review the formation of igneous rock from lava (Lesson 3-2). Ask: *What happens to lava when it flows out onto the surface?* (It cools and hardens.)

2 *Teach*

Demonstration Use a warm bottle of club soda to which you have added about a dozen raisins to demonstrate an explosive eruption. With the cap in place, shake the bottle vigorously for a minute. With everyone wearing safety goggles and well out of your way, open the bottle over a sink and have students observe the gases, liquids, and solids that "erupt" from the model volcano. Stress that all three states of matter—solids, liquids, and gases—are released during any type of volcanic eruption. **learning style: visual**

Discussion Point out that because the volcanoes shown in Figures 5-19 to 5-21 are not drawn to scale, the cinder and composite cone volcanoes appear larger than the shield volcano. Inform students that shield volcanoes are the widest of the three types of volcanoes. **learning style: visual**

Reinforcement Show videos of erupting volcanoes. Challenge students to identify the type of eruption as well as the type of volcano responsible for the eruption. **learning style: visual**

Answers to questions

1 **NAME:** quiet and explosive

2 **DESCRIBE:** Lava from quiet eruptions accumulates and hardens.

3 **DESCRIBE:** Dust, ash, and rock particles thrown violently from a volcano accumulate to form a cone.

4 **INFER:** Lava layers form in quiet eruptions. Layers of dust, ash, and rock form in explosive eruptions.

5 **INFER:** cinder cones

5-6 How are volcanoes classified?

Objective

Identify and describe the three kinds of volcanic cones.

Key Terms

shield cone: volcanic cone made up of layers of hardened lava

cinder cone: volcanic cone made up of rock particles, dust, and ash

composite (kuhm-PAHZ-iht) **cone:** volcanic cone made up of alternating layers of lava and rock particles

Classifying by Eruption Volcanic eruptions may be classified as quiet or explosive. During a quiet eruption, lava flows freely through a vent or a fissure. Explosive eruptions shoot rocks, lava, gases, ash, and dust high into the air. Different kinds of volcanic eruptions form different types of volcanic cones. Volcanic mountains are really just large volcanic cones.

1 **NAME:** What are the two basic kinds of volcanic eruptions?

Shield Cones A **shield cone** is a type of volcano, or volcanic cone. It is made up of layers of hardened lava. A shield cone forms from quiet eruptions. Lava flows over a large area and hardens.

Layers of lava build up to form the cone. The cone has a wide base. The sides of the cone have gentle slopes.

Mauna Loa in the Hawaiian Islands is one of the largest shield cone volcanoes in the world. It is more than 4 km above sea level.

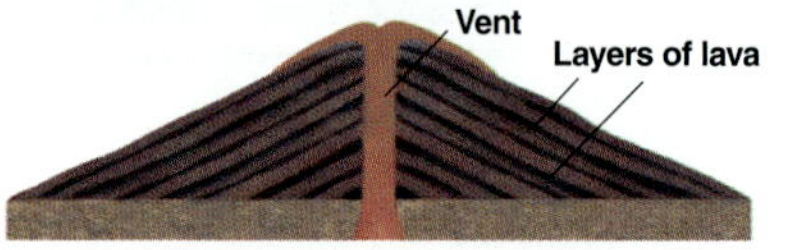

▲ **Figure 5-19** Shield cone volcano

2 **DESCRIBE:** How does a shield cone form?

Cinder Cones Explosive eruptions form **cinder cones.** Dust, ash, and rock particles are thrown out of the vent and settle to form the cone. Cinder cone volcanoes have steep sides and narrow bases. The rock particles are loose and roll down the slope. Parícutin is a cinder cone volcano.

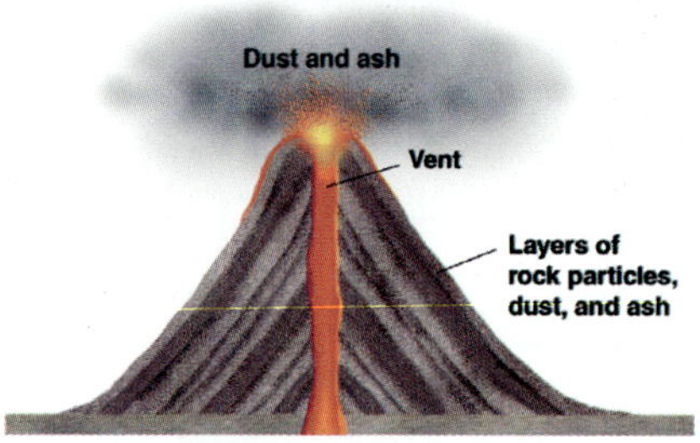

▲ **Figure 5-20** Cinder cone volcano

3 **DESCRIBE:** How does a cinder cone form?

Composite Cones A **composite cone** is made up of layers of lava and rock particles. It is formed from a series of quiet and explosive eruptions. A quiet eruption results in the lava forming a wide base. An explosive eruption adds a layer of dust, ash, and rock particles. Then, another quiet eruption adds more lava. Eventually, a very high, wide volcanic cone with steep sides is formed. Mount St. Helens and Mount Hood, in Oregon, are composite cones.

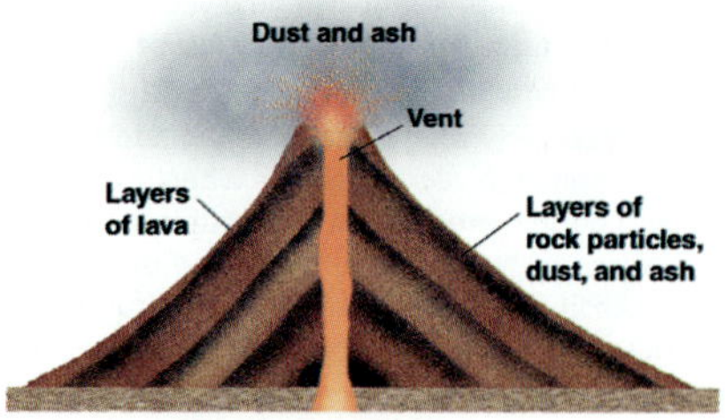

▲ **Figure 5-21** Composite cone volcano

4 **INFER:** How could you tell what kind of eruption formed a layer of a composite cone?

Teaching Options

EXTENDING THE LESSON

Comparing Volcano Size and Shape Have students repeat the Investigate activity from page 118, using gravel and very fine sand. Have students compare the angle that each slope makes with the table. (The pile of very fine sand should have the gentlest slopes, followed by the pile of coarse sand. The gravel should form a pile with relatively steep slopes.) Have students compare their results to the volcanoes shown in the diagrams on page 120.
learning styles: visual, tactile

ONGOING ASSESSMENT

Have students write a sentence or two that explains the relationship between mountains and volcanic cones. (A mountain is a landform that rises high above the surrounding area. Volcanic cones form mountains. However, not all mountains are volcanic cones.)
learning style: linguistic

Volcanoes in Space Io is a bright red and yellow moon of Jupiter. Scientists think Io's color is caused by volcanic action. Io is the first moon or body other than Earth on which scientists have seen active volcanoes. The volcanoes on Io are very powerful. They shoot out many metric tons of material high into Io's atmosphere each month.

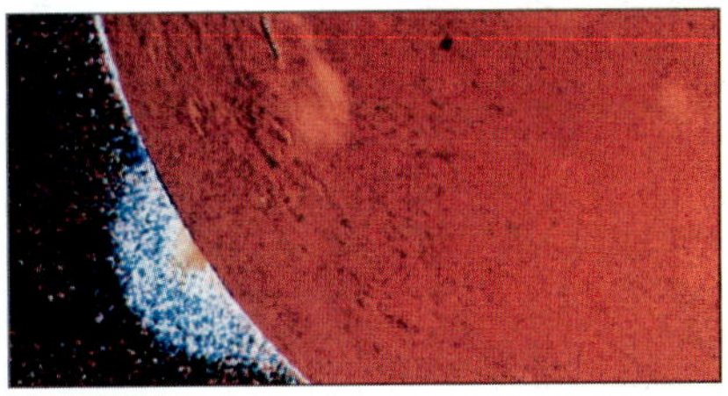

▲ **Figure 5-22** Red-hot gases shoot out from this volcanic eruption on Io.

5 INFER: What types of cones do you think Io's volcanoes are forming?

CHECKING CONCEPTS

1. Cinder cones are formed by __________ eruptions.
2. There are __________ kinds of volcanic cones.
3. Volcanic cones with gentle slopes and wide bases are __________ cones.
4. Shield cones are formed by __________ eruptions.

THINKING CRITICALLY

5. SEQUENCE: Place these volcanic cones in order from steepest to flattest: shield, cinder, and composite.
6. PREDICT: What type of eruption do you think each of these volcanoes might have in the future: Mauna Loa, Mount Hood, and Parícutin?

Web InfoSearch

Volcanoes of the World Both active and inactive volcanoes can be found in many places around the world.

SEARCH: Use the Internet to make a list of five volcanoes and find out about each of them. Are they active or inactive? Have they erupted in recent years? Start your search at www.conceptsandchallenges.com. Some key search words are **volcano, Ring of Fire,** and **volcano observatory.**

Hands-On Activity
MODELING VOLCANIC CONES

You will need a lab apron, goggles, crushed cereal flakes, 2 paper plates, a metric ruler, plaster of Paris, a measuring cup, and a teaspoon.

1. In the measuring cup, mix 3 teaspoons of plaster of Paris with about 75 mL of water. Be sure the mixture is not too runny.
2. Hold the measuring cup about 15 cm over a paper plate. Pour the plaster of Paris slowly onto the plate. Let it harden.
3. Clean the cup. Fill it halfway with crushed cereal flakes. Hold it about 15 cm above the second paper plate. Slowly pour the cereal onto the plate.

▲ **STEP 3** Slowly pour the cereal onto the plate.

Practicing Your Skills

4. INFER: **a.** What does the plaster of Paris represent? **b.** What does the cereal represent?
5. MEASURE: What is the width of each base?
6. ANALYZE: Which kind of volcanic cone does each model represent? Explain.

QUICK CONVERSIONS		
	Metric/SI	English
Plaster of Paris	75.0 mL	2.5 oz
Height over plate	15 cm	6 in.

• TEACHING RESOURCES •
Online or Print

Teacher's Resources
Lesson Review: Chapter 5, p. 7
Enrichment: Chapter 5, p. 9

3 Assess

CHECKING CONCEPTS ANSWERS

1. explosive
2. three
3. shield
4. quiet

THINKING CRITICALLY ANSWERS

5. SEQUENCE: cinder cone, composite cone, shield cone
6. PREDICT: Mauna Loa is a shield volcano that erupts quietly. Mount Hood is a composite cone. An eruption from this volcano will either be a quiet lava flow or an explosive eruption of ash, dust, and rocks. Parícutin is a cinder cone that erupts violently.

4 Close

Web InfoSearch

Volcanoes of the World Have students use the Internet to make a list of five volcanoes and find out about them. Ask students: *Are they active or inactive? Have they erupted in recent years?* Some key search words are **volcano, Ring of Fire,** and **Volcano Observatory.** Students can begin their search at www.conceptsandchallenges.com.

Hands-On Activity
MODELING VOLCANIC CONES

TIME: 15–20 minutes

PURPOSE: modeling, inferring, measuring, analyzing

COOPERATIVE LEARNING: Have students work in pairs. One student can do Steps 1 and 2, and the other Step 3.

Practicing Your Skills Answers

4. INFER: **a.** lava **b.** volcanic dust, ash, and rock particles
5. MEASURE: The cone formed by the plaster of Paris should be wider than the cone of cereal crumbs.
6. ANALYZE: The cone made of plaster of Paris models a shield volcano, which is made of hardened lava. The cereal cone models a cinder cone, which is made of volcanic rock particles.

The Big Idea

(pp. 122–123)

1 Introduce

Objective Students will find out how a volcanic eruption can affect the ecosystem in the area of a volcano. They will compare current conditions with the same ecosystem after a major eruption.

Linking Prior Knowledge Have students recall that Mount St. Helens is a composite cone, made of alternating layers of lava and volcanic debris (Lesson 5-6). Have a volunteer remind the class that the lava is ejected during quiet eruptions and the ash, dust, and rock particles are ejected during explosive eruptions.

2 Teach

Discussion After students have read page 122, have them expand on how an ecosystem develops after a major volcanic eruption. Prompt students, if necessary, to state that the first organisms to return are generally lichen and mosses, followed by other simple plants. Eventually, complex plants populate the fertile soil. Animals that feed on these plants then return to the ecosystem. Finally, animals that feed on other animals return to the area. Have students find out how the plants and animals in the area of Mount St. Helens were affected by the 1980 eruption. Ask students what scientists who were studying the area found. Have students present their findings in an oral report to the class.
learning style: auditory

Use the background information in Fast Facts to help guide students in preparing their Science Logs.

THE BIG IDEA

How do volcanic eruptions affect ecosystems? Have students start their research for their Science Log Writing Activity at www.conceptsandchallenges.com. Be sure students use search engines and websites that are reliable and acceptable. You may also wish to provide copies of the *Big Idea Planner* and *Big Idea Science Log Entry* study tools found in the Teacher's Resources.

Integrating Life Science

THE Big IDEA

How do volcanic eruptions affect ecosystems?

Ecosystems are made up of living and nonliving things in balance with each other. A volcanic eruption may disrupt that balance. Gases with temperatures of more than 1,000°C shoot into the air. Ash released by the eruption forms thick clouds that can block out sunlight. Huge chunks of rock are hurled onto the surrounding land. Many plants and animals are killed. Even the soil and water can be damaged. Hundreds of years may pass before the ecosystem returns to its former state.

On May 18, 1980, Mount St. Helens in Washington State erupted. Nearly 7,000 land animals were killed in the blast. Millions of fish died when the ash and dust settled in streams and lakes. The number of trees blown down by the eruption would have provided enough wood to build 300,000 homes.

By 1982, life had begun returning to the area. Ecological succession had started. Succession is the often predictable series of changes an environment undergoes over time. Pioneer plants such as mosses and lichens began growing on bare rocks. Wind and birds deposited seeds that soon sprouted. A few years later, larger plants began to grow. Animals began returning to the area. Ecologists studied the changes. From them, we have learned more about how ecosystems and volcanoes interact with each other.

Look at the illustrations that appear on these two pages. Then, follow the directions in the Science Log to learn more about "the big idea." ✦

Ecosystem Thrives
In early 1980, Mount St. Helens is a thriving ecosystem. The volcano has been dormant since 1857. However, later that year, it shows signs of erupting. Magma begins pushing up inside the mountain. This causes the north face of the mountain to bulge out.

Volcano Erupts
On May 18, 1980, an earthquake causes a landslide on Mount St. Helens. This triggers the eruption. Ash and steam rush out of the volcano at speeds of up to 1,000 km/h. The blast kills 57 people.

Ecosystem Destroyed
For at least a month after the explosion, the ecosystem is stripped of most of its life forms. The blast damages the land and kills wildlife for about 200 sq km around. A larger area is covered by dust and ash. The height of the mountain is decreased by about 400 m.

Teaching Options

! FAST FACTS Who Survived?

Among the survivors of the 1980 Mount St. Helens eruption were fireweed, spiders, millipedes, centipedes, ants, and beetles. Some animals such as gophers survived, but most of the plant life did not. Because of the lack of food, the larger animals migrated from this area shortly after the eruption.

Three areas of destruction were noted. In the blast zone, trees were completely destroyed. In the blow-down zone, trees were leveled. In the outermost zone, or scorch zone, trees still stood but had been burned by the high temperatures caused by volcanic debris flows. Just beyond the scorch zone, many plant and animal species survived but with fewer individuals.

QUICK CONVERSIONS

	Metric/SI	English
Gas temperatures during an eruption	1,000°C	1,832°F
Speed of ash and gas flows	1,000 km/h	621 mph

ESL/ELL STRATEGY

Working With Peers Pair an English-language learner with an English-proficient student. Have them explain the article and the captions in their own words to each other.

▲ **Figure 5-23** Mount St. Helens erupting

WRITING ACTIVITY

Science Log

Find out about the current conditions at Mount St. Helens, Mount Pinatubo, or another volcano of your choice. Write a brief news article about it. Compare conditions today to what conditions were like just after an eruption. You can start your search for information about volcanoes at www.conceptsandchallenges.com.

Life Returns

By fall of 1982, life starts returning to the area in the form of pioneering plants, such as moss. Pioneer plants are also called bryophytes. Their root-like rhizoids help break down rocks to form soil. As the plants die and decay, they add nutrients to the soil. This helps other plants to grow.

Ecosystem Thrives Again

By the summer of 1987, much of the area near Mount St. Helens is coming back to life. A new ecosystem is developing. It is similar to but also different from the one that was there before. Here, we see Indian paintbrush and purple wildflowers blooming on recovering slopes near the mountain.

Providing a Model Have newspapers available so that students can see the general format of a newspaper article before they begin their own articles for their Science Logs. Some students might wish to include drawings to accompany their articles. **learning style: visual**

3 Assess

Use the Writing Activity Rubric from the Teacher's Resources to assess students' Science Logs. Fill in the rubric with the additional information below. For this assignment, students should have:

- accurately explained the conditions in the same ecosystem before and after a volcanic eruption.
- written a newspaper article, using two or more information sources.

4 Close

A Day in the Life of . . . Have students write about the adaptations of one plant or animal living in an ecosystem near an active volcano that allow that organism to survive in its environment. Suggest that students illustrate their reports.

Tips for Using Technology

Using the Internet to Find Information Remind students that an important tool for finding information on the Internet is a search engine. A search engine allows the user to type in a keyword or phrase. The engine then searches the Internet for those sites that contain the keyword or phrase.

When using search words more than one word long, suggest to students that they surround them in quotation marks, such as "Ring of Fire." This will instruct the search engine to search for Web sites that have that exact phrase.

• TEACHING RESOURCES • Online or Print

Teacher's Resources

Big Idea: Chapter 5, p. 8

Big Idea Planner: Study Tools, pp. 11–12

Big Idea Science Log Entry: Study Tools, p. 13

Writing Activity Rubric: Teaching Tools, p. 9

5-7 *(pp. 124–125)*

1 Introduce

STUDY HINT This lesson contains many Key Terms. Have students make sketches to help them visualize the relationships among the terms.

Linking Prior Knowledge Have students recall visiting the beach or standing in a wave pool. Ask a volunteer to describe what happens as a wave passes. (The person is lifted up and dropped back down.) Inform students that the energy released by an earthquake also travels as waves.

2 Teach

Demonstration Explain that some earthquake waves travel the same way as sound waves. Demonstrate this by striking a tuning fork and placing it on the surface of a pan of water. Have students observe and describe the waves (the concentric circles) that form. Make an analogy between this demonstration and the seismic waves that travel from an earthquake's focus, as shown in Figure 5-24.
learning style: visual

Reinforcement To help students distinguish between the focus and the epicenter of an earthquake, have them look up the definition of the prefix *epi-* in a dictionary. (on, upon, over, on the outside)
learning style: linguistic

Answers to questions

1. **DEFINE:** sudden, strong shaking movements of Earth's crust
2. **DEFINE:** the point beneath Earth's surface where an earthquake originates
3. **NAME:** release of energy generated by faults locking
4. **DESCRIBE:** an instrument that detects and measures earthquakes

READING STRATEGY Have students locate all of the words in this lesson that contain the word part *seismo-*. (*seismograph, seismogram, seismology,* and *seismologists*) Have students look up what this word part means.

5-7 What are earthquakes?

Objectives

Explain what causes earthquakes. Describe what happens during an earthquake.

Key Terms

earthquake: sudden, strong movement of Earth's crust

focus (FOH-kuhs): point beneath Earth's surface where an earthquake starts

epicenter (EHP-ih-sehnt-uhr): place on Earth's surface directly above the focus

seismic (SYZ-mihk) **wave:** earthquake wave

seismograph (SYZ-muh-graf): instrument that detects and measures earthquakes

Tremors and Shakes Earth's crust is always moving. However, most of the movements are so small that you do not even feel them. Movements of the crust that you may or may not feel are called tremors. There may be more than six million tremors each year. Sudden, strong, shaking movements of Earth's crust are called **earthquakes.** Earthquakes can cause a lot of damage.

1. **DEFINE:** What is an earthquake?

Focus and Epicenter Earthquakes begin deep inside Earth's crust. The point beneath Earth's surface where an earthquake starts is called the **focus.** The place on Earth's surface that is directly above the focus is called the **epicenter.** The surface of Earth shakes the hardest at the epicenter.

2. **DEFINE:** What is the focus of an earthquake?

Cause of Earthquakes Earthquakes are associated with faulting. Usually, the rocks on both sides of a fault are squeezed together very tightly. However, they have no place to go. Geologists say that the fault is "locked." The pressure on the rocks increases. Eventually, the rocks break at their weakest point. Rocks first slip and move near the focus. As the rocks slip, they release energy in the form of waves, or vibrations. These vibrations are called **seismic waves,** or earthquake waves.

Seismic waves travel out from the focus in all directions. Imagine throwing a pebble into a pond. At the point where the pebble hits the water, waves move outward in all directions. Earthquake waves move out from the focus in the same way.

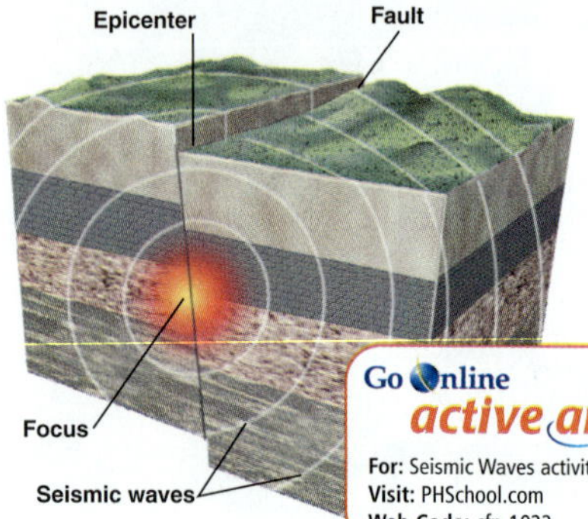

Go Online active art
For: Seismic Waves activity
Visit: PHSchool.com
Web Code: cfp-1022

▲ **Figure 5-24** Seismic waves travel from the focus in all directions.

3. **NAME:** What is the main cause of earthquakes?

Measuring Earthquakes A **seismograph** is an instrument that detects and measures earthquakes. A seismograph can even measure very small tremors that most people cannot feel. It records the movements in Earth's crust on a sheet of paper. This record is called a seismogram. The seismogram has wavy lines. The higher the wavy lines are on the seismogram, the stronger the earthquake.

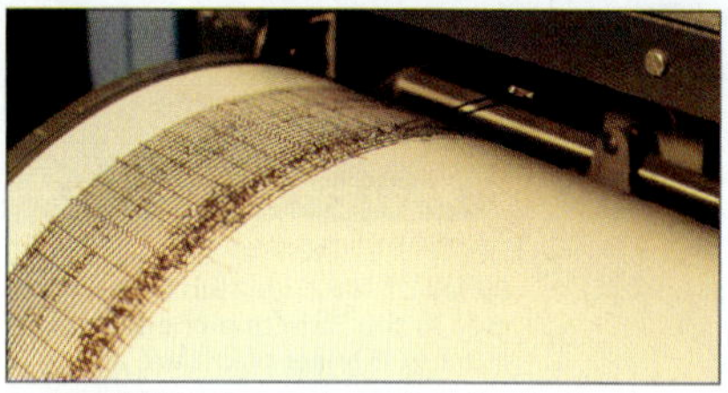

▲ **Figure 5-25** A seismograph records earthquake strength.

4. **DESCRIBE:** What is a seismograph?

Teaching Options

EXTENDING THE LESSON

Researching Local Faults Have students use library resources and the Internet to find out the location of any major faults in their local area or state. Have students search newspapers to find the date and location of the most recent earthquake along those faults. Have them write their own news article about the event. **learning style: linguistic**

COOPERATIVE LEARNING

Preparing for Earthquakes Have students role-play what might happen to them during a moderate earthquake. Inform students that thousands of earthquakes occur each day, but only a few are strong enough to be felt by people.
learning styles: kinesthetic, auditory

Go Online active art
For: Seismic Waves activity
Visit: PHSchool.com
Web Code: cfp-2014

Students can interact with the art of seismic waves online.

CHECKING CONCEPTS

1. What are small, often unfelt shifts of Earth's crust called?
2. What are strong, shaking movements of Earth's crust called?
3. Where does the ground shake the hardest during an earthquake?
4. How do the vibrations known as seismic waves travel?
5. What internal process causes most earthquakes?

THINKING CRITICALLY

Explain the difference between the words in each of the following pairs.

6. focus, epicenter
7. earthquake, tremor
8. seismograph, seismogram

INTERPRETING VISUALS

Use Figure 5-26 below to answer the following questions.

9. **IDENTIFY:** What are diagrams A and B called?
10. **ANALYZE:** Which shows the stronger earthquake?

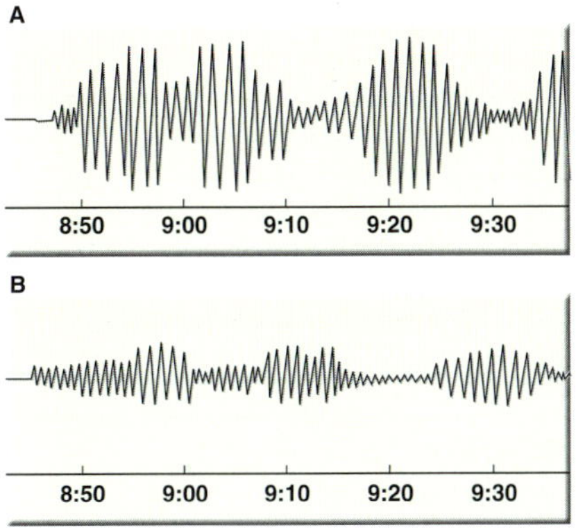

▲ **Figure 5-26**

People in Science

SEISMOLOGIST

The study of earthquakes is a specialty in the field of Earth science. This specialty is called seismology. Earthquake scientists are known as seismologists.

Seismologists study areas of Earth's surface where earthquakes tend to occur. After an earthquake, seismologists will study the ground along the faults. When an earthquake occurred near San Francisco, California, in 1989, seismologists studied the San Andreas Fault. They wanted to learn more about how Earth's crust moves along the fault.

▲ **Figure 5-27** A seismologist checks a seismogram.

Seismologists collect a lot of data about earthquakes. They take measurements and examine the damage caused by earthquakes. They analyze seismograms, which are records of the movement of Earth during an earthquake. Using the data, seismologists can try to predict where earthquakes will most likely happen. They may even someday be able to find ways to prevent earthquakes.

Thinking Critically What can seismologists learn from studying the damage caused by an earthquake?

3 *Assess*

CHECKING CONCEPTS ANSWERS

1. tremors
2. earthquakes
3. near the epicenter
4. outward from the focus in all directions
5. faulting

THINKING CRITICALLY ANSWERS

6. The focus is the point beneath Earth's surface where an earthquake originates. The epicenter is the point on the surface directly above the focus.
7. An earthquake is a strong, shaking movement of the crust. A tremor is a vibration of the crust that may or may not be felt by people.
8. A seismograph is an instrument that measures and records the energy released by an earthquake. A seismogram is the record produced by a seismograph.

INTERPRETING VISUALS ANSWERS

9. **IDENTIFY:** seismograms
10. **ANALYZE:** A, or the top seismogram

4 *Close*

People in Science

SEISMOLOGIST

Real-Life Connection Arrange for a seismologist to visit your class. Have each student prepare one question to ask the scientist. If a seismologist is unable to visit the class, obtain some seismograms from a seismology center or lab and have students interpret them.

Discussion After students have read the information on seismologists, ask: *How do you think the work of a seismologist may help to save lives?* (By predicting earthquakes before they happen, a seismologist may help to prevent or reduce loss of lives.)

Thinking Critically Answer
By studying the damage caused by an earthquake, seismologists can determine the strength (Richter-scale value) of an earthquake and how different structures withstand earthquakes.

MORE ON THE WEB

Measuring Earthquakes
Introduce students to the Richter scale and the Mercalli scale. Then, have students find the five largest or deadliest earthquakes since 1900 and determine the Richter and Mercalli measurements for each of the earthquakes. Students can begin their search at www.conceptsandchallenges.com.

• TEACHING RESOURCES • Online or Print

Teacher's Resources
Lesson Review: Chapter 5, p. 10
Enrichment: Chapter 5, p. 13

5-8 *(pp. 126–127)*

1 Introduce

INVESTIGATE

TIME: 5–10 minutes

PURPOSE: Students will use a model to observe wave motion.

MATERIALS: small pie plates, water, eyedroppers

PROCEDURE: Stress that before students drop the water from the eyedropper, the water in the pan must be completely still. If the water in the pan is sloshing even a little, students will have a difficult time making their observations.
learning style: visual

THINK ABOUT IT: The waves spread out in concentric circles from the point where the water was dropped into the pan.

Linking Prior Knowledge Write the following words on the board: *primary*, *secondary*, and *surface*. Ask volunteers to tell the meaning of each word. Tell students that each word describes a type of wave produced by an earthquake.

2 Teach

Demonstration Use a spring toy and a student to show how seismic waves travel through Earth. Gently stretch the spring between yourself and the student. Have the student gather about ten coils and quickly release them. Explain that this is how P-waves move through rocks. Instruct the student to hold the spring still while you set it in motion vertically with a quick flick of your wrist. Explain that this is how S-waves travel through rock.
learning styles: tactile, visual

Discussion Make sure that students understand the characteristics of the three main types of seismic waves. Ask: *What are the fastest waves and how do they move?* (P-waves are the fastest waves. They push and pull rock particles as they move through them.) *How do S-waves move through rocks?* (S-waves move from side to side as they travel through rocks.) *What are L-waves and how do they affect Earth's surface?* (L-waves are the slowest-moving waves. They move Earth's surface up and down, much like an object moves as a water wave passes.) **learning style: auditory**

5-8 What are the different kinds of seismic waves?

INVESTIGATE

Modeling How Seismic Waves Move
HANDS-ON ACTIVITY

1. Fill a shallow pan or pie plate with water. You may also use a white plastic tub or a plastic foam bowl that has a rim.
2. Use an eyedropper to add a few drops of water to the center of the pan.
3. Watch carefully for the waves on the bottom of the plate. You will more easily see the shadows of the waves on the bottom than the waves themselves.

THINK ABOUT IT: If you drop the water into the middle of the pan, the waves will spread out the way certain earthquake waves travel. How did they spread out?

Objective

Describe three kinds of seismic waves.

Key Terms

P-wave: fastest earthquake wave

S-wave: second earthquake wave to be recorded at a seismograph station

L-wave: surface wave

Primary Waves The fastest-moving seismic waves are primary waves, or **P-waves.** P-waves are push-pull waves. They cause the particles in materials to move back and forth in place. The wave itself moves out from the focus. The particles move together and apart along the direction of the wave. P-waves move through solids, liquids, and gases.

1 **DESCRIBE:** What are P-waves like?

Secondary Waves The second waves that are recorded by a seismograph are called secondary waves, or **S-waves.** S-waves move more slowly than P-waves. S-waves travel through solids only. S-waves cause the particles in materials to move from side to side. The particles move at right angles to the direction in which the waves are traveling.

2 **COMPARE:** Which move faster, P-waves or S-waves?

Surface Waves Surface waves are also called long waves, or **L-waves.** L-waves are the slowest-moving waves. They are the last waves to be recorded. L-waves can travel through solids, causing Earth's surface to rise and fall like ocean waves. They cause great damage by bending and twisting Earth's surface.

3 **NAME:** What is another name for L-waves?

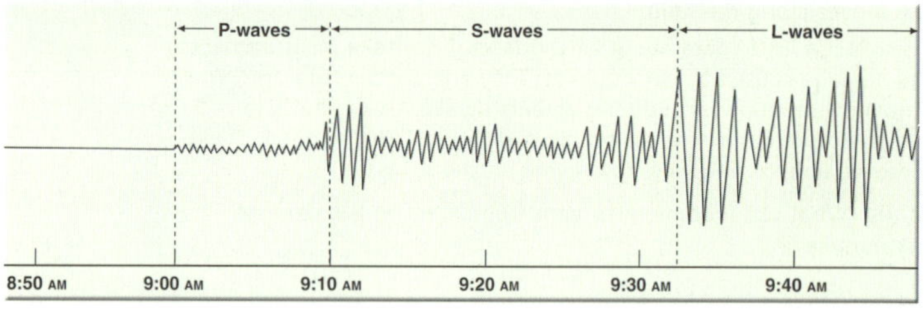

◄ **Figure 5-28** The different types of earthquake waves are recorded on a seismogram.

Teaching Options

RETEACH

Explaining Seismograms Refer students to Figure 5-28. Ask: *What is shown in the figure?* (a seismogram) *What made this record?* (a seismograph) *Which are the P-waves?* (the smallest lines on the seismogram) *Which waves are the last to be recorded?* (L-waves) *When were the first P-waves of this earthquake recorded?* (at about 9:00) *How much time elapsed between the first P-wave and the first L-wave?* (about 32 minutes) *What can be determined from a seismogram?* (the travel times of seismic waves, strength of the earthquake, and distance to the epicenter)
learning styles: visual, auditory

EXTENDING THE LESSON

Graphing Data Have students work in small groups to graph the data in Figure 5-29. Before they start their graphs, make sure that they realize that the distances should be plotted on the *x*-axis and the travel times on the *y*-axis. Have students describe the trend of the data. (Travel time increases with an increase from an earthquake's epicenter.)
learning style: logical/mathematical

Studying Earthquake Waves A seismogram tells scientists a great deal about earthquakes. For example, it can indicate how far away the epicenter of an earthquake is and the force of the earthquake.

Look at the seismogram in Figure 5-28. When did the first P-wave arrive? The time shows 9:00 AM When did the first S-wave arrive? It arrived at 9:10 AM There was a 10-minute difference between the arrival times of the P-waves and the S-waves. Using this time difference, called an S-P interval, scientists can tell how far the epicenter is from the seismograph. Figure 5-29 shows how these intervals change depending on distance.

S-P INTERVALS AND DISTANCE	
S-P Interval	Distance From Epicenter
1 min	700 km
2 min	1,200 km
3 min	1,800 km
4 min	2,500 km
5 min	3,400 km

▲ Figure 5-29

4 DESCRIBE: What two things can a seismogram indicate?

Finding the Epicenter To find the epicenter, seismograms from three stations are needed. Look at Figure 5-30. A red circle has been drawn on the map around each station.

Each station is at the center of its circle. The diameter of the circle is based on the travel time measured. There is only one point where all three circles cross. The epicenter is near this point.

5 OBSERVE: Where is the epicenter of the earthquake shown on the map?

CHECKING CONCEPTS

1. There are ________ kinds of seismic waves.
2. The slowest-moving waves are ________.

THINKING CRITICALLY

3. DESCRIBE: Why are P-waves called push-pull waves?
4. ANALYZE: Why do you think surface waves are also called long waves?

INTERPRETING VISUALS

Use Figure 5-29 to answer the following question.

5. ANALYZE: If the difference in travel time of P-waves and S-waves is 4 minutes, how far from the epicenter is the seismograph station that recorded it?

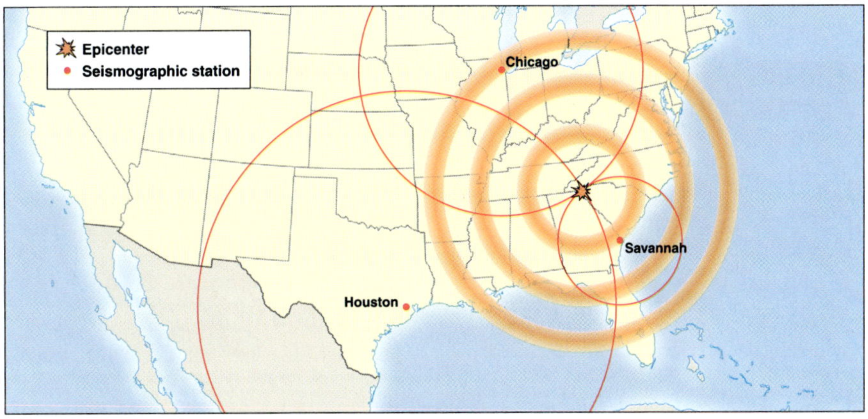

▲ Figure 5-30 Seismograms help scientists locate the epicenter of an earthquake.

WRITING HINT Ask students to write descriptive captions for each of the three kinds of waves shown in Figure 5-28.

Reinforcement Have students refer to Figure 5-30 to explain why data from at least three seismic stations are needed to accurately plot the epicenter of an earthquake. (Having data from only one station produces one circle. The epicenter of the earthquake can be anywhere along the circumference of the circle. Data from two stations results in two possible locations for the earthquake. Data from three stations is needed to pinpoint the epicenter of the earthquake because the three circles will intersect at only one point.)
learning style: logical/mathematical

Answers to questions

1 DESCRIBE: P-waves travel quickly and push and pull on particles.

2 COMPARE: P-waves move faster.

3 NAME: surface waves or long waves

4 DESCRIBE: distance from an earthquake's epicenter and the strength of an earthquake

5 OBSERVE: near the borders of North Carolina, South Carolina, and Georgia (Have students refer to the relief map of the United States in the Appendix of their books to help them answer this question.)

3 Assess

CHECKING CONCEPTS ANSWERS

1. three
2. surface waves or L-waves

THINKING CRITICALLY ANSWERS

3. DESCRIBE: P-waves cause particles in materials to be pushed together and pulled apart.
4. ANALYZE: They take the longest time to reach the seismic recording station.

INTERPRETING VISUALS ANSWER

5. ANALYZE: 2,500 km

MORE ON THE WEB

Simulating Seismic Waves
Have students use an online seismic wave simulator to see how simulated earthquake waves affect buildings at ground level as well as below ground level. Have students visit www.conceptsandchallenges.com for more details about seismic waves.

QUICK CONVERSIONS

	Metric/SI	English
Distance from epicenter	700 km	435 mi
	1,200 km	746 mi
	1,800 km	1,118 mi
	2,500 km	1,553 mi
	3,400 km	2,113 mi

• TEACHING RESOURCES • Online or Print

Teacher's Resources
Lesson Review: Chapter 5, p. 11
Enrichment: Chapter 5, p. 13

Laboratory Manual
Lab. Challenge: How does a seismograph detect and measure earthquake waves? pp. 39–42

5-9 (pp. 128–129)

1 Introduce

STUDY HINT Ask students to turn the chart in Figure 5-31 into a paragraph of text. Let them read their paragraphs aloud.

Linking Prior Knowledge Most students have probably heard the Richter scale mentioned on the news or seen the term in a newspaper article. Have students guess why the *R* in Richter is capitalized. (The scale is named for an American seismologist, Charles Richter.)

2 Teach

Demonstration Build a small city out of wooden blocks on a cart with wheels. Have students observe what happens to the city during a minor earthquake, when you gently shake the cart. Have students compare this to what might happen to the city during a major earthquake, as you push and pull with moderate force on the cart. **learning style: visual**

Discussion Show as many photos as possible of varying degrees of damage caused by earthquakes. Have students try to sequence the photos from the least damaging earthquake to the most. Have students brainstorm what the damage caused by an earthquake depends on. (the population density near the epicenter, the time of day at which the earthquake strikes, the types of buildings in the stricken area, the type of ground on which these structures are built, the preparedness of the people in the affected area, whether or not the earthquake caused water and gas mains to rupture, and so on)

Reinforcement Read various statements from the Earthquake Checklist on page 129. Have students try to determine—without looking at the table if possible—if the statement is from the "Before," "During," or "After" part. **learning style: auditory**

WRITING HINT Tell students that any kind of writing can benefit from careful rereading and revising where necessary. Students should be encouraged to write at least two drafts before finalizing a written report.

5-9 How do earthquakes cause damage?

Objective

Recognize the power of earthquakes and the damage they can cause.

Key Terms

Richter (RIHK-tuhr) **scale:** scale that measures the energy released by an earthquake

tsunami (tsoo-NAH-mee)**:** large ocean wave caused by an earthquake

The Richter Scale In 1935, a geologist named Charles Richter developed a scale to measure the energy released by earthquakes. The **Richter scale** shows an earthquake's relative strength, or magnitude. On the Richter scale, an earthquake is given a number, usually between 1 and 9. The stronger the earthquake, the higher the number. For each increase in number, the earthquake is said to release ten times more energy.

An earthquake measuring 7 or more on the Richter scale can cause a great deal of damage. Earthquakes that measure 2.5 or less on the Richter scale are usually not felt by people. The largest earthquakes recorded so far have measured around 9.5 on the scale.

RICHTER SCALE	
Magnitude	**Effects Near Epicenter**
Less than 2.0	Generally not felt but recorded
2.0–2.9	Potentially perceptible
3.0–3.9	Felt by some
4.0–4.9	Felt by most
5.0–5.9	Damaging shocks
6.0–6.9	Destructive in settled regions
7.0–7.9	Major earthquakes; inflict serious damage
More than 8.0	Great earthquakes; destroy communities near epicenter

▲ **Figure 5-31** The strength of an earthquake is called its magnitude. Different magnitudes cause different levels of damage.

1 EXPLAIN: What is the Richter scale?

Earthquake Damage Many new buildings are built to be "earthquake-proof." This means they are not likely to collapse during an earthquake. Older buildings, however, may be destroyed completely. Tall buildings might sway or even topple over. Earthquakes can also damage electrical lines, telephone lines, and water pipes. Explosions and fires are caused by broken electric and gas lines.

▲ **Figure 5-32** Earthquakes can destroy homes and buckle roads.

2 EXPLAIN: What is meant by "earthquake-proof"?

Tsunamis A large ocean wave that is caused by an earthquake is called a **tsunami.** A tsunami forms when an earthquake's epicenter is on the ocean floor. Out in the open ocean, tsunamis do not get very high. Near the shore, as water depth decreases, wave heights increase. A tsunami near shore may be more than 30 m high. When a tsunami hits land, it can kill many people and cause a lot of damage.

▲ **Figure 5-33** Tsunamis can cause great damage to coastal cities.

3 DESCRIBE: What is a tsunami?

Teaching Options

! FAST FACTS *Tsunamis*

Tsunami is a Japanese word meaning "harbor wave." A tsunami can result from an earthquake, a landslide, or a volcanic eruption on the ocean floor. The speed of a tsunami can be as much as 800 km (497 mi) per hour. A tsunami originating in Japan can reach Hawaii in less than 8 hours. Earthquakes that produce tsunamis are usually greater than 6.5 on the Richter scale. Tsunamis generally occur in the Pacific Ocean because of the high level of seismic activity near the plate boundaries there.

The International Tsunami Information Center (ITIC) in Hawaii was established in 1965 by the Intergovernmental Oceanographic Commission (IOC) of the United Nations Educational, Scientific, and Cultural Organization (UNESCO). This organization provides educational and public service information about earthquakes and tsunamis, earthquake warnings, and other related information. Teaching materials are available for all grade levels.

Predicting Earthquakes Scientists use past earthquakes to help predict future earthquakes. Small movements in Earth's crust are a signal that an earthquake may soon occur. Scientists look at the ground in the area to see if it has moved up or down. Laser field stations record the smallest movements along faults. Lasers are very strong, focused light beams. The laser beam is aimed at a reflector. By measuring the time it takes the beam to hit the reflector and come back, scientists can find out if there has been any movement along a fault. Using past earthquake information, scientists have developed a seismic risk map. The map shows where earthquakes may occur and the kind of damage they may cause.

4 INFER: Why is it helpful to be able to predict an earthquake?

CHECKING CONCEPTS

1. A strong earthquake on the Richter scale measures __________ or more.
2. Earthquakes less than __________ on the Richter scale are usually not felt.
3. Buildings built to stand during an earthquake are called __________.
4. A map that shows where an earthquake may happen is called a __________ map.

THINKING CRITICALLY

5. SEQUENCE: Place the following earthquakes in order from weakest to strongest on the Richter scale.
   - **a.** San Francisco, California, 1906, 8.3
   - **b.** Santa Cruz, California, 1989, 7.0
   - **c.** Mexico City, Mexico, 1985, 8.1
   - **d.** New York, New York, 1984, 5.0
   - **e.** Tokyo, Japan, 1923, 8.3
   - **f.** Sumatra, Indonesia, 1994, 7.2

BUILDING SCIENCE SKILLS

Researching The Mercalli scale is also used to measure earthquakes. Find out about the Mercalli scale. Make a poster that shows the Mercalli scale. How does this scale measure earthquakes? How is it different from the Richter scale?

Real-Life Science

EARTHQUAKE SAFETY

Earthquake Checklist

BEFORE: Be prepared
- ☐ **1.** Always have a supply of ready-to-eat canned food and bottled water on hand.
- ☐ **2.** Have a portable radio and extra batteries also on hand.
- ☐ **3.** Learn how to turn off the electricity, gas, and water in your house.

DURING: Stay calm
- ☐ **1.** If you are indoors, stay indoors. Protect yourself from falling materials by standing in a doorway or taking cover under a desk or large table.
- ☐ **2.** Stay away from glass, especially windows.
- ☐ **3.** If you are outdoors, move away from buildings and overhead electrical and telephone wires.
- ☐ **4.** If you are in a car, stop as long as you are away from buildings, bridges, tunnels, and so on. Stay in the car until the shaking stops. Get out of a tunnel or off a bridge if you can.

AFTER: Be careful
- ☐ **1.** Check the gas, water, and electricity. Look for fires or fire hazards. If you smell gas, open windows, and turn off the gas. Leave the building. Contact the gas company or police. Do not go back into the building. If water pipes are broken, turn off the main water valve. If there are electrical shorts, turn off the electricity at the main fuse box or circuit breaker.
- ☐ **2.** Do not use the telephone except for emergencies.
- ☐ **3.** Turn on a radio to get emergency information. Use the television if you can.
- ☐ **4.** Do not enter badly damaged buildings.
- ☐ **5.** Do not go sightseeing.

Thinking Critically Why is it important to be prepared before an earthquake happens?

MORE ON THE WEB

Researching Tsunamis Have students use the Internet to learn how to calculate the threat of tsunamis, plus ways that those in the path of a tsunami can protect themselves.

Simulating Tsunamis Have students use the Internet to view computer simulations of tsunamis. Have students visit www.conceptsandchallenges.com for more details about tsunamis.

• TEACHING RESOURCES • Online or Print

Teacher's Resources

Lesson Review: Chapter 5, p. 12

Enrichment: Chapter 5, p. 13

Answers to questions

1 EXPLAIN: a scale that rates the relative strength, or magnitude, of an earthquake

2 EXPLAIN: Earthquake-proof structures are less likely to fall over or suffer extensive damage during an earthquake.

3 DESCRIBE: a wave that forms when an earthquake occurs on the ocean floor and grows much larger as it approaches the shore

4 INFER: Successfully predicting earthquakes can prevent or reduce damage and loss of life.

3 Assess

CHECKING CONCEPTS ANSWERS

1. 7.0
2. 3.0
3. earthquake-proof
4. seismic risk

THINKING CRITICALLY ANSWER

5. SEQUENCE: d, b, f, c, e or a, a or e (e = a)

4 Close

BUILDING SCIENCE SKILLS ANSWER

Researching The Mercalli scale is a measure of earthquake intensity, or the damage done to structures by an earthquake. This scale uses Roman numerals and a description of the damage to rate an earthquake. This scale differs from the Richter scale, which measures the energy released by an earthquake.

Real-Life Science

EARTHQUAKE SAFETY

Discussion Ask students: *Have you ever felt an earthquake? If so, what was your experience? Did you feel safe?* Lead a discussion about the importance of earthquake safety. Review the Earthquake Checklist as a class. Invite students to write additional safety tips on the board.

Thinking Critically Answer
Earthquakes occur without warning. Preparing beforehand is the best way to make sure that you will be safe.

5-10 *(pp. 130–131)*

1 *Introduce*

STUDY HINT Have students frequently refer to Figure 5-34 as they read this lesson. Ask them to explain the map and locate places discussed in the text.

Linking Prior Knowledge Have students compare Figure 5-6, which shows Earth's major mountain belts, with Figure 5-34. Point out the similarities between the locations of major mountain belts and regions of seismic and volcanic activity.

2 *Teach*

Demonstration Obtain some volcanic ash and volcanic cinders from a scientific supply company. Allow students to rub the ash and the cinders between their fingers and to observe each type of volcanic debris with a magnifying glass. Tell students that a mixture of ash, cinders, and hot, poisonous gases, which is called a pyroclastic flow, covered and destroyed Pompeii when Mount Vesuvius erupted over 2,000 years ago.
learning styles: kinesthetic, visual

Discussion Ask students where in the world they could travel in order to study earthquakes. Have them find the location in Figure 5-34. Then, ask them where they would go to study active volcanoes. When someone comes up with the same place for both questions, ask whether earthquakes cause volcanoes or vice versa. Make sure students understand that although major seismic zones often coincide with zones of active volcanism, earthquakes and volcanoes usually do not cause each other.
learning style: auditory

Answers to questions

1. STATE: three
2. IDENTIFY: the Ring of Fire
3. DESCRIBE: an underwater chain of volcanic mountains
4. DESCRIBE: collisions between large parts of Earth's crust

5-10 What is the Ring of Fire?

Objective

Identify the three major volcanic and earthquake zones on Earth.

Key Term

Ring of Fire: major earthquake and volcanic zone that almost forms a circle around the Pacific Ocean

Zones of Activity Most big earthquakes and volcanic eruptions on Earth occur in three areas, or zones. These zones are called the Ring of Fire, the Mid-Atlantic Ridge, and the Eurasian Belt. A lot of movement and activity happen in these places. There are many active volcanoes. Volcanoes that have erupted at least once within recorded history are called active volcanoes. There are more than 500 active volcanoes on land. There are many more under the oceans.

1. STATE: How many major volcanic and earthquake zones are there?

Ring of Fire One major volcanic and earthquake zone is the **Ring of Fire.** The Ring of Fire almost forms a circle or ring around the Pacific Ocean. This is how it got its name. Most of the active volcanoes on landmasses are located in the Ring of Fire. Many earthquakes also occur in this area. The western coasts of North and South America are part of the Ring of Fire.

2. IDENTIFY: What is the name of the earthquake and volcanic zone around the edge of the Pacific Ocean?

Mid-Atlantic Ridge A second major volcanic and earthquake zone is located in the Atlantic Ocean. This zone is called the Mid-Atlantic Ridge. The Mid-Atlantic Ridge is a long underwater chain of volcanic mountains. In this zone, earthquakes and volcanoes are caused by the formation of new parts of Earth's crust. Iceland is part of this zone. Iceland is a volcanic island.

3. DESCRIBE: What is the most important feature of the Mid-Atlantic Ridge?

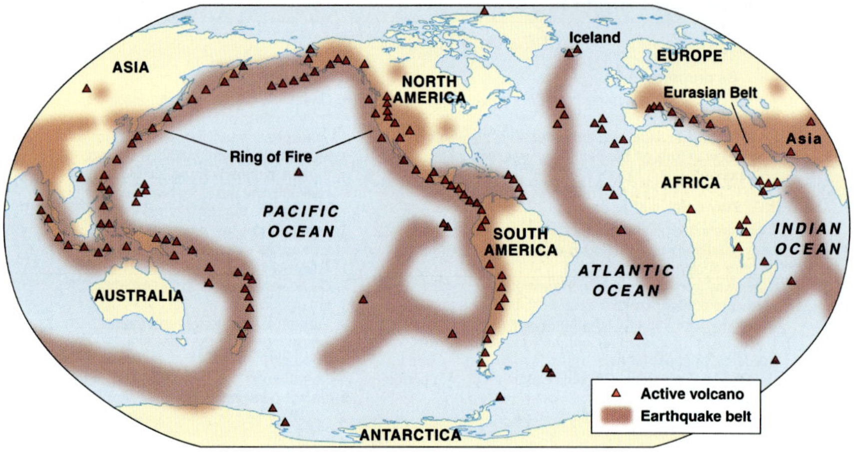

▲ **Figure 5-34** There are three major areas worldwide where earthquakes and volcanic eruptions are common.

Teaching Options

ONGOING ASSESSMENT

Provide students with black and white copies of Figure 5-34. Have them use three different colors of markers to identify the three different volcanic and earthquake zones. Tell students to write one important fact for each zone on the copy. learning style: tactile

EXTENDING THE LESSON

Living in Pompeii Have students find out what has been learned about the everyday life of ancient Romans from studying Pompeii. Students will discover that buildings, jewelry, coins, utensils, and many other objects were preserved by the volcanic ash. Have them locate images or draw the buildings, decorations, and other artifacts of Pompeian life. Make a bulletin board display of the images. learning style: visual

Eurasian Belt Many countries in Europe have had big earthquakes. Many also have active volcanoes. These countries are in the third zone, called the Eurasian Belt. The mountains along this belt were formed when parts of Earth's crust collided.

4 **DESCRIBE:** What formed the Eurasian-Melanesian Belt?

CHECKING CONCEPTS

1. The Ring of Fire nearly surrounds the ___________ Ocean.
2. The second major earthquake and volcanic zone is located in the middle of the ___________ Ocean.
3. The third major volcanic and earthquake zone is the ___________.

THINKING CRITICALLY

4. **INTERPRET:** Use Figure 5-34 and a map of the United States to name two states that are part of the Ring of Fire.

Web InfoSearch

New Madrid Between 1811 and 1812, a series of major earthquakes hit New Madrid, Missouri. They caused flooding. They even changed the course of the Mississippi River. Scientists did not know why the earthquakes happened in New Madrid. Then, in the 1970s, they discovered three faults in the area. The faults were buried deep beneath Earth's surface. Scientists think that movement along these faults caused the earthquakes.

SEARCH: Use the Internet to find out more about this event. What did the earthquakes measure on the Richter scale? How were the New Madrid earthquakes different from earthquakes in the western United States? Start your search at www.conceptsandchallenges.com. Some key search words are **earthquake, New Madrid 1811,** and **New Madrid seismic zone.**

How Do They Know That?

THE STORY OF POMPEII

Mount Vesuvius (vuh-SOO-vee-uhs) is one of the few active explosive volcanoes in Europe. In A.D. 79, Vesuvius erupted in a giant explosion. It rained hot, wet ashes and cinders on the city of Pompeii (pahm-PAY). The entire city was completely covered. Some of the people of Pompeii escaped. However, many others died from the hot ash and poisonous fumes in the air.

In the 1500s, people were digging a tunnel when they came across some of the remains of Pompeii. The ash and cinders had preserved the city. Today, more than half of Pompeii has been uncovered and restored.

▲ **Figure 5-35** The restored city of Pompeii

Mount Vesuvius has not erupted since 1944. However, the longer the volcano is quiet, the more explosive the next eruption may be. Hundreds of thousands of people live on or near the fertile slopes of Vesuvius. Scientists constantly monitor the volcano. They measure escaping gases and slight ground movements. Scientists hope to predict an eruption two or three weeks ahead of time. This would give people time to leave the area.

Thinking Critically Why would people want to live on the slope of the volcano?

ESL/ELL STRATEGY

Working With Peers Pair an English-language learner with an English-proficient student. Have each student relate two facts about Pompeii to each other, and then have them discuss with each other the point of the article.

• *TEACHING RESOURCES* • *Online or Print*

Teacher's Resources

Lesson Review: Chapter 5, p. 14

3 *Assess*

CHECKING CONCEPTS ANSWERS

1. Pacific
2. Atlantic
3. Eurasian Belt

THINKING CRITICALLY ANSWER

4. **INTERPRET:** Possible answers include California, Oregon, and Washington.

4 *Close*

Web InfoSearch

New Madrid The faults responsible for the New Madrid earthquakes are subsurface faults that remained inactive until the 1811–1812 earthquakes struck. Students can begin their search at www.conceptsandchallenges.com using the key search words **earthquake, New Madrid 1811,** and **New Madrid Seismic Zone.**

How Do They Know That?

THE STORY OF POMPEII

Background Vesuvius is an active volcano on the mainland of Europe. It lies in southern Italy near the shore of the Bay of Naples and the city of Naples. The eruption in A.D. 79 was not the only one. In 1631, a volcanic eruption destroyed many nearby towns and killed more than 3,000 people. In 1906, an eruption that lasted 10 days caused the death of more than 2,000 people. Smaller eruptions have occurred in 1913, 1926, 1929, and 1944. Today, vineyards and orchards cover the fertile slopes of the volcano.

Discussion Ask students to describe why visiting Pompeii could be like taking a trip back in time. (The volcanic ash and cinders preserved the city as it existed almost 2,000 years ago.)

Have students recall the 1980 eruption of Mount St. Helens. Tell them that this eruption produced about a third of the amount of debris produced by the A.D. 79 eruption of Vesuvius.

Thinking Critically Answer
Possible answer: Volcanic ash and other debris contain minerals that make the soil around a volcano very fertile.

Lab Activity

(pp. 132–133)

1 Prepare the Lab

Making an Earthquake-Proof Structure

SET-UP TIME: **LAB TIME:**

BACKGROUND: Internal waves continually shake the rocks that make up Earth's crust. Most of these vibrations are produced when slabs of Earth's crust and upper mantle move toward, away from, or horizontally past each other. Many of these vibrations are nearly undetectable. Some, however, are major earthquakes that destroy buildings and result in many deaths.

PURPOSE: Students will design and construct a tower and subject it to forces produced by an earthquake simulator.

ALTERNATIVE MATERIALS: Commercial classroom earthquake simulators are available from science and technology education catalogs. An electrical simulator can be made from a percussion massager. The massager should have a speed control to simulate different earthquake magnitudes.

SCIENCE SKILLS: Students will **model** a tall building and **contrast** how it withstands earthquakes.

ADVANCE PLANNING: Separate the tower supplies into plastic bags for each group. Give each group 16 gumdrops and 36 strands of spaghetti.

Make an earthquake simulator from four 1-in. cubes of foam rubber that are fixed to a flat surface with drops of hot glue. Arrange the cubes in a square pattern slightly smaller than the cardboard squares used for tower building. P-waves can be simulated by placing a 1-in. × 1-in. × 12-in. piece of foam along the back of a counter.

2 Guide the Lab

PROCEDURE: Divide the class into groups with two or three students per group. Provide a set of tower construction materials for each group. Students will place their towers on the foam cubes of the earthquake simulator and tap on one corner of the base. The amount and frequency of the tapping should be changed during the test. This simulates L-waves. To simulate P-waves, students press the cardboard base against the foam

LAB ACTIVITY
Making an Earthquake-Proof Structure

Materials

Uncooked spaghetti, small gumdrops, a metric ruler, paper, a pencil, a piece of cardboard, cellophane tape, a stopwatch or timer, an earthquake simulator (constructed by your teacher or done as a class project)

BACKGROUND

An earthquake can last from a few seconds to a few minutes. If you have ever felt an earthquake, you will know that it is a terrifying experience. The solid land beneath your feet begins to shake. In small earthquakes, dishes in cupboards and pictures on walls fall to the floor. Brick chimneys crack. Trees and light poles sway. In large earthquakes, entire buildings and bridges may break apart. However, not all buildings are destroyed. Some survive with little damage because of their designs.

PURPOSE

In this activity, you and your engineering team will design and construct a model of a tower. This tower is planned for an earthquake-prone area. You will test the tower to see how well it withstands a simulated earthquake.

PROCEDURE

1. Draw sketches of the kinds of towers you might want to build using spaghetti and gumdrops. Your tower should be at least 60 cm tall.
2. Construct your tower according to your plan. Use only the supplies listed. Handle the spaghetti carefully. The strands break easily. To insert the spaghetti strands into the gumdrops, hold the spaghetti near the end. If you need to shorten it, grip the spaghetti strand on each side of where you want it to break. Then bend it gently until it breaks.
3. After you have finished the base for your tower, lightly tape it to the cardboard.

▲ **STEP 2** Construct your tower.

▲ **STEP 3** Tape your tower to the cardboard.

132

Teaching Options

! FAST FACTS *Flexible Moorings*

Flexible moorings are standard building parts in most earthquake-prone areas today. The moorings, which are made of steel plates filled with rubber and steel disks, are attached to the foundation of the building and to the ground. During an earthquake, the moorings move back and forth with the seismic waves. The rubber disks absorb most of the seismic vibrations, so the building gently sways rather than crumbling during an earthquake.

ESL/ELL STRATEGY

Summarizing Procedures English-language learners may benefit from stating procedures in their own words. Have English-language learners paraphrase the procedures outlined in this lab activity to English-proficient students.

4. When the tower is completed, draw a picture of it. Label the height and width of the tower on your drawing. Then take it for testing to the earthquake simulator.

5. Time how long your building remains standing at different "strengths" of shaking. Copy the chart in Figure 5-36. Record your observations and times on it.

▲ **STEP 4** Test your tower with the simulator.

	P-Wave Test	S-Wave Test	L-Wave Test
Did your tower survive the test?			
Where did the tower shake the most?			

▲ **Figure 5-36** Use a copy of this chart to record your observations.

CONCLUSIONS

1. **OBSERVE:** What happened to your tower when it was tested? Where did it vibrate? Did it break or survive the test?
2. **OBSERVE:** Did any towers from other groups perform better than yours? Compare their height and other features.
3. **INFER:** How could you build a stronger tower?

strip on the counter and tap the edge of the cardboard, causing it to press into the foam and then rebound. S-waves can be simulated by turning the cardboard 90° and repeating the procedure. More information about earthquake waves can be found on pages 126 and 127 (Lesson 5-8). Provide students with a copy of the Lab Activity Report found in the Teacher's Resources for students to record their observations and conclusions.

TROUBLESHOOTING: Caution students to handle the spaghetti strands gently when inserting them into the gumdrops. Students should hold the strands near the end being inserted and push firmly without forcing.

SAFETY TIP: Tell students not to eat the gumdrops. They will be handled extensively during construction. Consider having students wear eye goggles when testing their structures. Spaghetti fragments may break off the collapsing structures.

EXPECTED OUTCOME: Students will observe how their structures withstand vibration tests and identify weaknesses in their designs.

LAB HINT: Make sure that students work together to assemble their structures. Spaghetti strands will easily break if not supported during assembly.

3 Conclude the Lab

1. **OBSERVE:** Answers will depend on how the towers were built.
2. **OBSERVE:** Answers will depend on how the towers were built.
3. **INFER:** Some possible ways to improve a tower are to use more flexible materials that bend rather than break or to use better materials that give the tower more support.

Use the Laboratory Activity Rubric from the Teacher's Resources to assess students' lab activities. Fill in the rubric with the additional information below. For this activity, students should have:

- designed and constructed a model of a tower.
- tested the tower to see how it would withstand a simulated earthquake.

4 Extend the Lab

Collect news stories and pictures of damage caused by recent earthquakes. Have groups of students examine pictures of buildings that have collapsed and try to analyze where structural weaknesses existed.

Tips for Using Technology

Using Graphics Software to Create Illustrations Most graphics creation software use similar tools to let you create illustrations. One of these tools is the pencil tool. Students can use this tool to sketch their towers before beginning the lab.

The pencil, or drawing tool, allows the user to draw in freehand in an illustration. This can be useful for sketching something. Depending on the graphics software, the mouse pointer will probably change to a pencil. Then, click and drag the pointer in the illustration to draw as if holding a pencil.

• *TEACHING RESOURCES* •
Online or Print

Teacher's Resources

Lab Activity Report: Study Tools, pp. 16–17

Laboratory Activity Rubric: Teaching Tools, p. 8

Laboratory Manual

Lab. Skills Worksheet 4: Organizing and Analyzing Data, pp. 13–14

Challenges *(pp. 134–136)*

Chapter Summary

Review Assign students one or two Key Terms each to explain to the rest of the class.

STUDY HINT Have students refer to the Chapter Summary statements to help them answer the Challenge questions.

Key Term Challenges

MATCHING

1. seismic wave (5-7)
2. S-wave (5-8)
3. elevation (5-2)
4. tsunami (5-9)
5. Ring of Fire (5-10)
6. vent (5-5)
7. seismograph (5-7)
8. folded mountain (5-3)

APPLYING DEFINITIONS

9. An anticline is an upward fold. A syncline is a downward fold. (5-1)
10. A focus is the point below Earth's surface where an earthquake originates. The epicenter is the point on the surface directly above the focus. (5-7)
11. A crater is a pit on top of a volcanic cone. A caldera forms when a crater falls back into a vent or when a volcano explodes. (5-5)
12. A fracture is any break in a rock. A fault is a large fracture along which movement takes place. (5-1)
13. A composite cone is a volcano that is made of alternating layers of lava and volcanic rock particles. The lava erupts during quiet explosions, and the particles are ejected during explosive eruptions. A shield cone is made of hardened lava extruded during quiet eruptions. (5-6)
14. A folded mountain is formed by the folding of rock layers. A fault-block mountain forms from upthrusts of the crust. (5-3)

Chapter 5 Challenges

Chapter Summary

Lesson 5-1

- Folds form when rock bends. A **fracture** is a break in a rock. Faulting causes rocks to move.

Lessons 5-2 and 5-3

- Mountains can form by rock layers folding, large blocks of crust being lifted during faulting, and by molten rock pushing up crust. Most mountains are part of ranges, systems, or belts.

Lesson 5-4

- The three main landforms are **plains, plateaus,** and mountains. Plateaus are large, flat plains with high elevations. The same forces that build mountains also form plateaus.

Lesson 5-5

- **Volcanism** is magma moving inside Earth. A **volcano** is a **vent** with **lava,** dust, ash, and rocks.
- A **crater** is a pit on top of a volcanic cone. A **caldera** forms when crater walls fall back into a vent or when a volcano explodes.

Lesson 5-6

- Volcanoes form different volcanic cones. These include **shield, cinder,** and **composite cones.**

Lesson 5-7

- An **earthquake** is a sudden, strong movement of Earth's crust. It starts at a **focus. Seismic waves** travel from there in all directions. Most earthquakes are associated with faulting. A **seismograph** detects and measures earthquakes.

Lessons 5-8 and 5-9

- **P-waves** are the fastest waves. **S-waves** are second. **L-waves** are the slowest waves.
- **Seismographs** help locate **epicenters** and record an earthquake's force. The **Richter scale** indicates the energy released by an earthquake.
- A **tsunami** is a giant wave caused by an earthquake on or under the ocean floor.

Lesson 5-10

- The three major volcanic and earthquake zones are the **Ring of Fire,** the Mid-Atlantic Ridge, and the Eurasian Belt.

Key Term Challenges

anticline (p. 110)
caldera (p.118)
cinder cone (p. 120)
composite cone (p. 120)
crater (p. 118)
dome mountain (p. 114)
earthquake (p. 124)
elevation (p. 112)
epicenter (p. 124)
fault (p. 110)
fault-block mountain (p. 114)
focus (p. 124)
folded mountain (p. 114)
fracture (p. 110)
L-wave (p. 126)
landform (p. 112)
lava (p. 118)
P-wave (p. 126)
plain (p. 116)
plateau (p. 116)
Richter scale (p. 128)
Ring of Fire (p. 130)
S-wave (p. 126)
seismic wave (p. 124)
seismograph (p. 124)
shield cone (p. 120)
summit (p. 112)
syncline (p. 110)
tsunami (p. 128)
vent (p. 118)
volcanism (p. 118)
volcano (p. 118)

MATCHING **Write the Key Term from above that best matches each description.**

1. earthquake wave
2. second earthquake wave to be recorded at a seismograph station
3. height above or below sea level
4. ocean wave caused by an earthquake
5. major earthquake and volcanic zone around the Pacific Ocean
6. opening from which lava flows
7. instrument that records earthquake activity
8. type of mountain formed when rock layers fold upward due to vertical pressure

APPLYING DEFINITIONS **Explain the difference between the words in each pair. Write your answers in complete sentences.**

9. anticline, syncline
10. epicenter, focus
11. crater, caldera
12. fracture, fault
13. composite cone, shield cone
14. folded mountain, fault-block mountain

Teaching Options

PREPARING STUDENTS FOR STANDARDIZED TESTS

Reading Strategy: Tell students to carefully read each question and to watch for words such as *not* and *only* when choosing the correct answer.

Writing Strategy: Many tests ask for responses that are complete sentences. Tell students to review their written responses to see that each sentence begins with a capital letter and ends with a period. Also, students should check that each sentence has a subject and a predicate.

Interpreting Visuals Strategy: Remind students to examine all of the visuals that pertain to a particular question or set of questions.

ESL/ELL STRATEGY

Word Bank Have students add Key Terms from this list to their word banks. Suggest that students classify the words into groups based on similarities. This can easily be done by quickly scanning the lessons to see which terms were presented within a particular lesson. Some examples of classifications are movements and landforms.

Content Challenges TEST PREP

MULTIPLE CHOICE Write the letter of the term or phrase that best completes each statement.

1. Folding and faulting are caused by
   a. heat.
   b. earthquakes.
   c. volcanoes.
   d. pressure.
2. The peaks of a young mountain are
   a. sharp and jagged.
   b. worn.
   c. rounded.
   d. flat.
3. A group of mountains with the same general shape and structure make up a mountain
   a. range.
   b. system.
   c. cascade.
   d. belt.
4. Magma that reaches Earth's surface is called
   a. crater.
   b. lava.
   c. magma.
   d. volcanic ash.
5. Magma that flows between rock layers and hardens forms a
   a. dike.
   b. composite cone.
   c. sill.
   d. cinder cone.
6. A volcanic cone made up of rock particles, dust, and ash is a
   a. fault-block cone.
   b. composite cone.
   c. shield cone.
   d. cinder cone.
7. An instrument that detects and measures earthquakes is a
   a. seismogram.
   b. focus.
   c. seismic map.
   d. seismograph.
8. Earthquake waves that travel along Earth's surface are
   a. P-waves.
   b. S-waves.
   c. L-waves.
   d. N-waves.
9. One example of a landform on Earth's surface is
   a. the Pacific Ocean.
   b. the San Andreas Fault.
   c. the Colorado Plateau.
   d. the continent of North America.

TRUE/FALSE Write *true* if the statement is true. If the statement is false, change the underlined term to make the statement true.

10. The second seismic waves to be recorded by a seismograph are P-waves.
11. Parícutin in Mexico is an example of a cinder cone.
12. A volcanic cone with layers of lava and rock particles is a shield cone.
13. A funnel-shaped pit at the top of a volcanic cone is called a crater.
14. A large, flat landform with a high elevation is a plateau.
15. Mountains, plateaus, and sills are the three main landforms of Earth's crust.
16. The Great Plains of the United States are coastal plains.

ALTERNATIVE ASSESSMENT

Relating Information Have students write a paragraph or two based on the chapter title, "The Changing Landscape." They should explain some forces and processes that change Earth's landscape.

PORTFOLIO ASSESSMENT

Making Student Portfolios Portfolio Assessment is designed to evaluate a student's performance over an extended period of time. It also provides an opportunity for students who are generally not good test-takers to demonstrate their learning in another way. Encourage students to select their own work for inclusion in their portfolios. The mountain formation drawings from Lessons 5-3 and Mercalli scale poster from Lesson 5-9 may be included in their portfolios. You can use the Portfolio Assessment Rubric found in the Teacher's Resources to assess students' portfolios.

Content Challenges

MULTIPLE CHOICE

1. d (5-1)
2. a (5-2)
3. a (5-2)
4. b (5-5)
5. c (5-5)
6. d (5-6)
7. d (5-7)
8. c (5-8)
9. c (5-2)

TRUE/FALSE

10. false; S-waves (5-8)
11. true (5-6)
12. false; composite cone (5-6)
13. true (5-5)
14. true (5-4)
15. false; plains (5-4)
16. false; interior (5-4)

Concept Challenges

WRITTEN RESPONSE

1. **EXPLAIN:** The two cones can be distinguished by the steepness of their slopes and by composition. A shield cone is gently sloping and made of lava. A cinder cone is steeply sloping and made of rock particles called cinders. (5-6)
2. **CONTRAST:** A mountain has sloping sides, whereas a plateau is a flat landform that has been uplifted. (5-4)
3. **CONTRAST:** A sill forms when magma intrudes into parallel rock layers. A dike forms when magma intrudes across existing rock layers. (5-5)
4. **ANALYZE:** Usually one that occurs on land would do more damage. However, it would depend on the population density of the affected area. An earthquake on land could do more damage if it struck a densely populated area. Earthquakes that originate in the ocean can generate tsunamis, which can pose great threats to nearby coastal communities. (5-9)
5. **PREDICT:** Tsunamis are likely to flood coastal areas and nearby inland locations. (5-9)

INTERPRETING VISUALS

6. A is a cloud of dust and ashes from an eruption. B is a crater. C is the volcanic cone. D is a volcanic vent. (5-5)
7. shield (5-6)
8. cinder (5-6)
9. composite (5-6)
10. A shield cone is made of lava that flows out of a vent during quiet eruptions. A cinder cone is made of rock particles called cinders that are ejected during explosive eruptions. (5-6)

Chapter Project Wrap-Up

LANDFORM MODEL

Students' projects can be exhibited around the classroom. Students can present their models and summarize their paragraphs. You can use the Individual Activity Rubric found in the Teacher's Resources to assess students' projects. Fill in the rubric with the additional information below. For this project, student should have:

- created scientifically accurate models of an intact landform and a cross section of it.
- prepared a logical and well-organized paragraph.

Concept Challenges TEST PREP

WRITTEN RESPONSE **Answer each of the following questions in complete sentences.**

1. **EXPLAIN:** How could you use observations to distinguish between a cinder cone and a shield cone?
2. **CONTRAST:** How does a mountain differ from a plateau?
3. **CONTRAST:** How does a dike differ from a sill?
4. **ANALYZE:** Which is likely to do more damage, an earthquake that occurs on land or an earthquake that occurs beneath the ocean? Explain.
5. **PREDICT:** What kind of damage is likely to be caused by a tsunami?

INTERPRETING VISUALS **Use Figures 5-37 to 5-40 to complete the following.**

6. Identify each of the lettered parts in Figure 5-37.
7. What kind of volcanic cone is shown in Figure 5-38?
8. What kind of volcanic cone is shown in Figure 5-39?
9. What kind of volcanic cone is shown in Figure 5-40?
10. How does the makeup of the volcanic cone in Figure 5-38 differ from that of the volcanic cone in Figure 5-39?

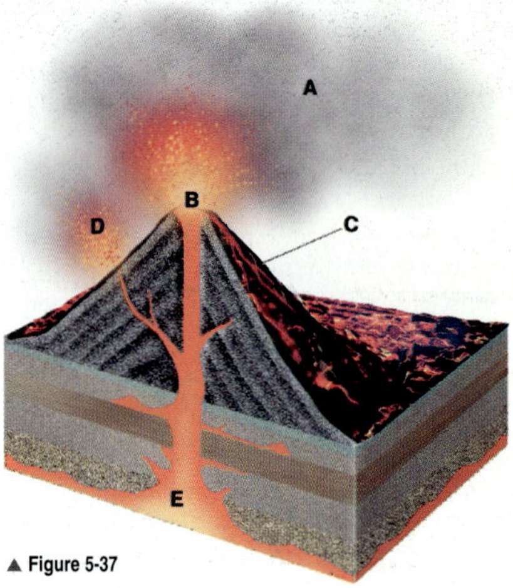

▲ Figure 5-37

▲ Figure 5-38

▲ Figure 5-39

▲ Figure 5-40

Teaching Options

• TEACHING RESOURCES •
Online or Print

Teacher's Resources

Key Term Review: Chapter 5, p. 15

Chapter 5 Test: pp. 16–18

Portfolio Assessment Rubric: Teaching Tools, p. 10

Individual Activity Rubric: Teaching Tools, p. 5

Chapter Self-Check: Teaching Tools, p. 11

PLANNING GUIDE

◆ **TEACHING THE CHAPTER** This chapter should take approximately 6–9 days to complete instruction and assessment.

	Skills and Features	Projects and Activities	Achieve Science Literacy	Meet Individual Needs
Chapter 6 Opener p. T137	APPLY	• Chapter Project		
Lesson 6-1 **What is continental drift?** pp. T138–T139 Standards: D1b, D2a, D2b	EXPLAIN, RELATE, INFER • Hands-On Activity	• Cooperative Learning • Extending the Lesson • More on the Web • Active Art	• Study Hint • Reading Strategy	
Lesson 6-2 **Why is the seafloor spreading?** pp. T140–T141 Standards: D1b, D2a	IDENTIFY, MODEL, CONCLUDE • Building Science Skills • Science and Technology • Real-LIfe Science	• Cooperative Learning	• Study Hint	Teacher's Resources* Enrichment Activity Seafloor Spreading
Lesson 6-3 **What evidence supports seafloor spreading?** pp. T142–T143 Standards: D1b, D2b	DEFINE, PROVE, CALCULATE • Investigate • Web InfoSearch	• Cooperative Learning • Integrating Physical Science	• Study Hint • Ongoing Assessment	Teacher's Resources* Enrichment Activity
Lesson 6-4 **What is plate tectonics?** pp. T144–T145 Standards: D1b, D2a	IDENTIFY, NAME, MODEL • Investigate • How Do They Know That?	• Cooperative Learning • Extending the Lesson • More on the Web • Lab Challenge	• Study Hint	• ESL/ELL Strategy Teacher's Resources* Enrichment Activity
Lesson 6-5 **What causes plate tectonics?** pp. T146–T147 Standards: D1b, D2a	DEFINE, STATE, INFER • Designing an Experiment • People in Science	• More on the Web	• Writing Hint • Ongoing Assessment	Teacher's Resources* Enrichment Activity Designing an Experiment
Lab Activity **Modeling Convection Currents** pp. T148–T149 Standards: D1a, D2a	OBSERVE, MODEL, ANALYZE • Set-Up Time: 10–15 min • Lab Time: 25–30 min	• Integrating the Sciences • Extend the Lab Activity	• Tips for Using Technology	• ESL/ELL Strategy Teacher's Resources* Lab Activity Report

*www.conceptsandchallenges.com, password: ccearth09 *Planning Guide continues on next page.*

Standards: For details on the correlation to National Science Standards see pages *xx–xxi*.

	Skills and Features	Projects and Activities	Achieve Science Literacy	Meet Individual Needs
Lesson 6-6 **What are some effects of plate tectonics?** pp. T150–T151 Standards: D1b, D2a	DESCRIBE, IDENTIFY, INFER • Investigate • Integrating Physical Science	• More on the Web • Integrating Physical Science	• Study Hint	• ESL/ELL Strategy Teacher's Resources* Enrichment Activity
Big Idea **How does plate tectonics affect countries around the world?** pp. T152–T53 Standards: D1a, D2a	RESEARCH, COMMUNICATE, ANALYZE • Science Log: Writing Activity	• Big Idea • Close Activity	• Tips for Using Technology	Teacher's Resources* Big Idea Planner
Chapter 6 Challenges pp. T154–T156	• Chapter Summary • Key Term Challenges • Content Challenges • Concept Challenges	• Chapter Project Wrap-Up	• Study Hint • Preparing Students for Standardized Tests	• ESL/ELL Strategy Teacher's Resources* Chapter Self-Check Weekly Journal

*www.conceptsandchallenges.com, password: ccearth09

Teacher Notes

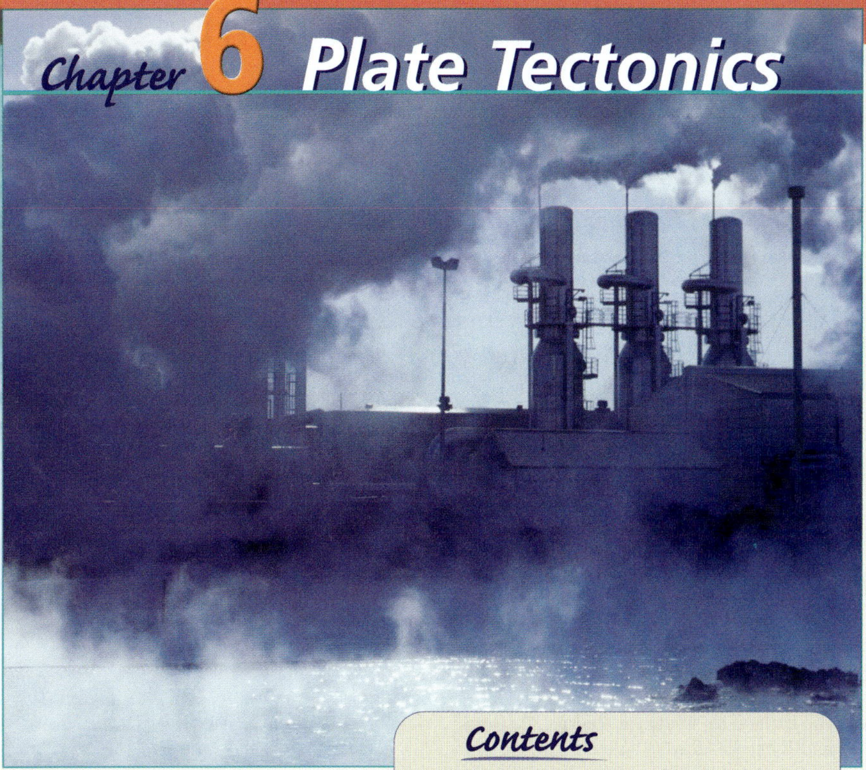

▲ **Figure 6-1** This power plant uses steam to generate electricity.

Earth's core heats the mantle. This contributes to the process known as plate tectonics. Rocks deep in the crust above the mantle also become hot in some spots. The power plant you see in the background has drilled into this hot rock. When cold water from the lagoon is pumped down into the hot rock, steam forms. The steam is then brought to the surface to power electrical generators. The warm waste water of this plant is released into the Blue Lagoon.

▶ During the summer, Iceland is very cold. Why are people still able to swim in the lagoon?

Contents

Teaching Options

ASSESSMENT PLANNER

For assessment in the Student Edition, see the following pages:

Content Assessment

Checking Concepts: pp. 139, 141, 143, 145, 147, 151
Thinking Critically: pp. 139, 141, 143, 145, 147, 151
Interpreting Visuals: p. 145
Chapter Challenges: pp. 154–156

Alternative Assessment

Building Skills: p. 141
Web InfoSearch: p. 143
Science Log: p. 153

Performance-Based Assessment

Hands-On Activity: p. 139
Designing an Experiment: p. 147
Lab Activity: pp. 148–149

• TEACHING RESOURCES • Online or Print

Teacher's Resources

Lesson Review: Ch. 6, pp. 2, 3, 4, 6, 7, 8

Enrichment: Ch. 6, pp. 5, 9

Key Term Review: Ch. 6, p. 11

Chapter 6 Test: pp. 12–14

Laboratory Manual: pp. 43–44

Chapter 6 Plate Tectonics

Chapter Overview

In this chapter, students will learn about the theory of continental drift and the evidence that supports this theory. Students will understand that seafloor spreading forms new oceanic crust and examine evidence for seafloor spreading. They will learn what causes plate tectonics. Finally, students will learn to recognize some of the effects of plate tectonics.

About Figure 6-1 The water in Iceland's Blue Lagoon is heated by geothermal energy. Earth's heat warms the water. A power plant pumps the very hot water to the surface and uses it to produce electricity. By the time the water reaches the Blue Lagoon, it is hot and free of pollutants.

Answer Heat from inside Earth warms the water.

Linking Prior Knowledge

For this chapter, students should recall:

- the names of the seven continents (Lesson 1-2).
- the three layers of Earth (Lesson 1-2).
- the causes of volcanoes and earthquakes on the Mid-Atlantic Ridge (Lesson 5-10).

Chapter Project

PLATE TECTONICS COLLAGE

MATERIALS: magazines, posterboard, glue

As students read the chapter, they should create a collage depicting some of the visible effects of plate tectonics. Have students look through magazines to find pictures of volcanoes, mountains, areas affected by earthquakes, and islands that formed over hot spots. Students can create their collages by gluing the pictures on posterboard. **learning styles: visual, tactile**

6-1 *(pp. 138–139)*

1 Introduce

STUDY HINT Before beginning this lesson, read the lesson objective with students. Have them review the Key Terms and their definitions.

Linking Prior Knowledge Ask students to name the seven continents. List the names of the continents on the board. Have students point out each continent on a globe or world map.

2 Teach

Demonstration Bring in seven pieces of a jigsaw puzzle that fit together. Have students join the pieces. Then, have them break the seven pieces apart. Relate this activity to the breaking apart of Pangaea to form the seven continents.
learning style: tactile

Discussion Display a large world map. Direct students' attention to the similar configurations of the east coast of South America and the west coast of Africa. Explain that the shapes of the coastlines support the theory of continental drift. Use Figure 6-5 to discuss other evidence supporting continental drift. Point out that mountain ranges and rocks on different continents seem to match.
learning style: visual

Answers to questions

1. **IDENTIFY:** Alfred Wegener
2. **DESCRIBE:** They seem to fit together like pieces of a jigsaw puzzle.
3. **EXPLAIN:** He used them to show that Africa and South America were once part of the same continent.
4. **EXPLAIN:** Some mountains on different continents seem to match.

READING STRATEGY Tell students that a passage can be summarized in one statement, called the main idea. Suggest that students take notes as they read this lesson. They should write the main ideas of this lesson in their notes.

6-1 What is continental drift?

Objective

Explain the theory of continental drift and the evidence that supports it.

Key Terms

continental (kahnt-ihn-EHNT-uhl) **drift:** theory that the continents were at one or more times a single landmass that broke apart and eventually moved into the positions they are in today

Pangaea (pan-JEE-uh)**:** single, giant landmass, or continent, that later broke apart

Continental Drift Most scientists think that millions of years ago, Earth was very different from the way it is today. There was only one supercontinent that had water all around it. Over time, the continent began to break apart. Pieces of it slowly drifted apart and then came together again. Eventually, those pieces became today's seven continents.

The theory that the seven modern continents split apart one or more times from a single landmass was suggested by German scientist Alfred Wegener in 1915. Wegener called his theory **continental drift.** He named the giant landmass **Pangaea,** meaning "all lands." Others added to Wegener's theory. One scientist suggested that Pangaea had separated into two landmasses: Laurasia (law-RAY-zhuh) in the north and Gondwanaland (gahnd-WAH-nuh-land) in the south. The Tethys (TEE-thihs) Sea was in the middle. Laurasia later broke apart to form North America, Europe, and Asia. Gondwanaland became South America, Africa, Australia, India, and Antarctica.

1 IDENTIFY: Who first stated the modern theory of continental drift?

A Giant Jigsaw Puzzle To support his theory, Wegener showed how the continents could fit together. Look at the shapes of the coastlines of South America and Africa in Figure 6-5. They seem to fit together like jigsaw puzzle pieces. Other places can be found that might once have fit together as well.

2 DESCRIBE: In what ways do the continents seem to fit together?

Fossil Evidence Coastline shapes were only one of many clues Wegener used to support his theory of continental drift. Fossils of once-living organisms also provided support for Wegener's theory.

Wegener studied the fossils of *Mesosaurus* (meh-soh-SAWR-uhs). *Mesosaurus* fossils were found in both Africa and South America. *Mesosaurus* was an animal that lived in fresh water. How could it have swum across the salty Atlantic Ocean? Wegener believed that it had not. Instead, he believed that it had lived at a time when there was only one large landmass. Later, that landmass broke apart, and some of the fossils remained on each part.

3 EXPLAIN: How did Wegener use *Mesosaurus* fossils to help explain his continental drift theory?

More Evidence Today, most scientists support the theory of continental drift. Here is some other evidence scientists have uncovered.

- Some mountain ranges on different continents seem to match. A mountain range along the eastern United States and Canada is similar to ranges in Greenland and northwestern Europe. Figure 6-5 shows where the mountains match.

Go Online **active art**
For: Continental Drift activity
Visit: PHSchool.com
Web Code: cfp-1015

▲ **Figure 6-2** Alfred Wegener

▲ **Figure 6-3** Pangaea, 225 million years ago

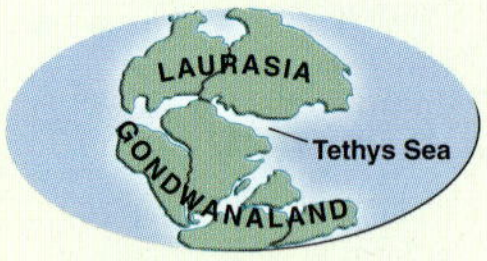

▲ **Figure 6-4** Gondwanaland and Laurasia, 180–200 million years ago

Teaching Options

COOPERATIVE LEARNING

Presenting the Evidence Have a volunteer explain the theory of continental drift to the class. Have a second student describe the evidence supporting continental drift.
learning style: auditory

EXTENDING THE LESSON

Discussing More Evidence The remains of tropical life-forms have been found near the North and South poles. Ask students to explain how such life-forms could have survived the cold climates of the polar regions. Lead students to conclude that at one time, the polar regions were closer to the equator and that over long periods of time, they drifted to their present polar positions.
learning style: auditory

- The ages and kinds of rocks along the edge of one continent match those of rocks along the edge of another continent. Even the sizes of the diamonds found in Brazilian and West African mines are the same.

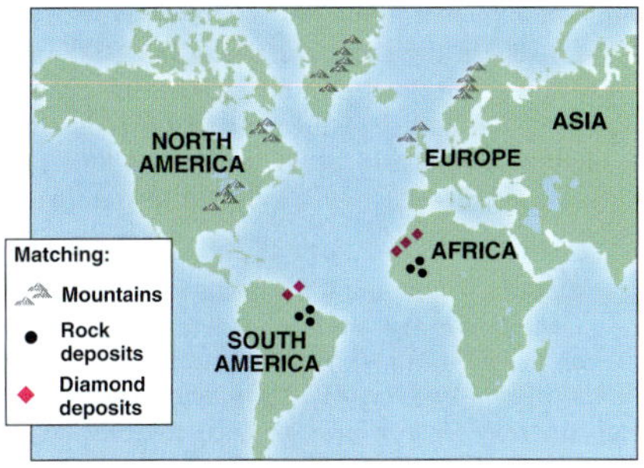

▲ **Figure 6-5** Evidence from matching mountains, rocks, and diamonds supports Wegener's theory.

4 EXPLAIN: How do similarities among certain mountain ranges support the theory of continental drift?

CHECKING CONCEPTS

1. What is the theory of continental drift?
2. Which continents seem to fit together?
3. What are three things that scientists use to support the theory of continental drift?
4. How do the shapes of coastlines today support the theory of continental drift?

THINKING CRITICALLY

5. **RELATE:** How does the idea of Pangaea support the theory of continental drift?
6. **INFER:** Fossils of the plant *Glossopteris* (glaw-SAWP-tuh-rihs) have been found on four continents. The continents are Africa, Asia, Australia, and South America. *Glossopteris* dates back more than 245 million years. It was a fern plant that may have produced spores, which are seedlike structures similar to those of a flowering plant. What might *Glossopteris* fossils tell scientists about these continents?

Hands-On Activity
MODELING PANGAEA

You will need a sheet of tracing paper, scissors, glue, and two sheets of construction paper.

1. Trace and label the continents shown on the map.
2. Glue the tracing paper onto a sheet of construction paper.
3. Carefully cut out the continents and Greenland. Do not cut Europe and Asia apart. Keep them as one continent. ⚠ **CAUTION:** Be careful when using scissors.
4. Arrange the pieces to form Pangaea. Glue the model of Pangaea onto another sheet of construction paper.

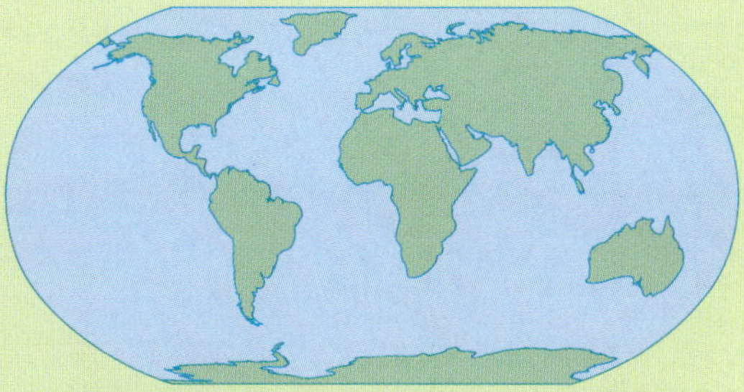

▲ **Figure 6-6** Earth's current landmasses are divided into seven continents.

Practicing Your Skills

5. **EXPLAIN:** Which continents seem to fit together?
6. **IDENTIFY:** Label your model. Then write a caption for it.
7. **COMPARE:** How is the reconstructed Pangaea like a jigsaw puzzle?

MORE ON THE WEB

Modeling Pangaea Have students use the Internet to learn more about Alfred Wegener's theory on the supercontinent of Pangaea. Students can begin their search at www.conceptsandchallenges.com.

Go Online active art
For: Continental Drift activity
Visit: PHSchool.com
Web Code: cfp-2014

Students can interact with the art of continental drift online.

• TEACHING RESOURCES •
Online or Print

Teacher's Resources
Lesson Review: Chapter 6, p. 2

3 Assess

CHECKING CONCEPTS ANSWERS

1. The continents were once part of a giant landmass that split apart.
2. Possible answer: Africa and South America
3. Accept any three of the following: the shapes of coastlines; fossil evidence; mountain ranges on different continents that seem to match; the matching ages and kinds of rocks along the edges of continents.
4. The shapes of some coastlines seem to fit together.

THINKING CRITICALLY ANSWERS

5. **RELATE:** Possible answer: Pangaea was one large continent that apparently split into smaller landmasses. Continental drift explains how that could happen.
6. **INFER:** The continents were connected 245 million years ago.

4 Close

Hands-On Activity
MODELING PANGAEA

TIME: 15–30 minutes

PURPOSE: modeling, comparing, explaining, identifying

SAFETY TIP: Caution students to be careful when using scissors.

ALTERNATIVE MATERIALS: See More on the Web for a larger map to trace.

COOPERATIVE LEARNING: Have volunteers show and explain their models to the class.

Practicing Your Skills Answers

5. **EXPLAIN:** Africa and South America; North America and Europe; Africa and Asia; Africa, Antarctica, and Australia
6. **IDENTIFY:** Models should include labels on the continents and a caption such as "The seven continents on Earth once were part of a single landmass called Pangaea."
7. **COMPARE:** The continents making up Pangaea fit together like the pieces of a jigsaw puzzle.

6-2 *(pp. 140–141)*

1 Introduce

STUDY HINT Write the title of each section in the lesson on the board. Have students copy these titles in their notebooks, using an outline format. As students read each section, have them write the topic sentence of each paragraph beneath each section title.

Linking Prior Knowledge Ask students: *What causes volcanoes and earthquakes in the Mid-Atlantic Ridge?* (the formation of new crust) *What do you think happens to old crust when new crust forms?* Explain that in this lesson, students will learn what happens to old crust when new crust forms. After reading the lesson, have students review their answers. Were they correct?

2 Teach

Demonstration Use a hard-boiled egg to represent seafloor spreading. Explain that Earth is like the egg. Just as the eggshell buckles in some places when it is cracked, exposing the egg white, Earth's mantle also buckles. When this happens, new ocean floor and mountains form, and earthquakes occur. **learning style: visual**

Discussion Using Figure 6-7, discuss the mid-ocean ridges and the process of seafloor spreading. Make sure that students understand that some of the tallest mountains and longest mountain ranges are under the ocean.
learning style: visual

Answers to questions

1. **NAME:** oceanic and continental
2. **IDENTIFY:** under the ocean
3. **DEFINE:** deep crack running down the center of the ridges
4. **DESCRIBE:** It cools and forms new crust.

6-2 Why is the seafloor spreading?

Objective
Relate seafloor spreading to the forming of new oceanic crust.

Key Terms
mid-ocean ridge: ocean-floor feature resembling a mountain ridge on land
rift valley: flat area between two ridges that is formed by spreading plates
seafloor spreading: process that forms new seafloor

Two Kinds of Crust There are two kinds of crust on Earth. One kind is continental crust. This crust makes up Earth's continents. The other kind is called oceanic (oh-shee-AN-ihk) crust. Oceanic crust, which makes up the ocean floor, is denser than continental crust.

1 **NAME:** What are the two kinds of crust?

Mid-Ocean Ridges The most obvious features of the crust that makes up the ocean floor are the **mid-ocean ridges.** These ridges are part of a series of underwater mountain ranges that run along the floors of all the oceans.

Some of the longest mountain ranges and tallest peaks on Earth are part of oceanic crust. These mountains form a loose chain around the world. Some peaks in the chain rise more than 3,000 m above the ocean floor. In a few places, the peaks rise above the surface of the ocean to form islands. Iceland is a mountain peak of the Mid-Atlantic Ridge. The Mid-Atlantic Ridge is in the middle of the Atlantic Ocean.

2 **IDENTIFY:** Where are some of the longest mountain ranges on Earth found?

Rift Valleys In the late 1940s, scientists began to map the mid-ocean ridges. They discovered deep cracks, or rifts, in the center of some parts of some ridges. Magma pouring out of the rift had hardened and formed broad underwater valleys with steep sides. These valleys are called **rift valleys.** There are many earthquakes and volcanoes in rift valleys.

On land, spreading plates form steep, wide valleys in the continental crust, as shown in Figure 6-7. The East African Rift System extends for 6,400 km, from Jordan in southwestern Asia through eastern Africa to Mozambique. The valley averages 48 km to 64 km in width. Other rift valley floors are below sea level and form deep lakes.

3 **DEFINE:** What is a rift valley?

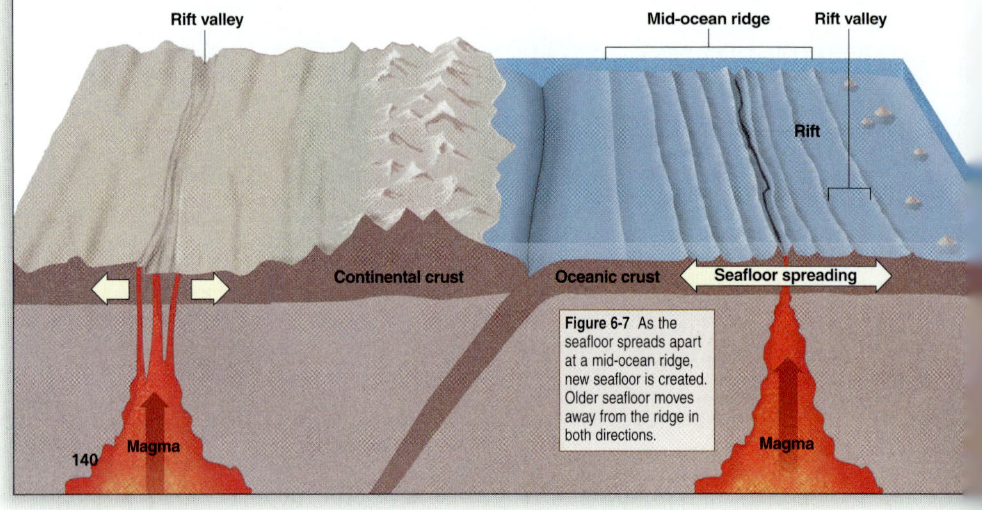

Figure 6-7 As the seafloor spreads apart at a mid-ocean ridge, new seafloor is created. Older seafloor moves away from the ridge in both directions.

Teaching Options

! FAST FACTS *Continental Rift Valleys*

The two rift valleys that are on continents, rather than in oceans, are the Red Sea and the East African Rift System. The San Andreas Fault in California is also a spreading center but is composed of short segments rather than one long fault.

The Red Sea has already filled with seawater but does not yet completely separate the two landmasses of the Arabian Peninsula and northeast Africa.

The East African Rift System is in the early stages of spreading. If movement continues in the same way, east Africa will separate from the rest of the continent. The rift is marked by a chain of lakes, most of them deep and fjordlike, and seas such as the Dead Sea, whose surface is 400 m (1,312 ft) below sea level.

Formation of New Seafloor Deep-sea drills have been used to bring up samples of oceanic crust. Most of these samples turn out to be younger than samples of continental crust. In addition, the oceanic crust near the mid-ocean ridges is younger than oceanic crust farther from the center of the ridge. The youngest oceanic crust is found in the center of the ridge.

Below the mid-ocean ridges, magma rises through the crust. As the magma cools, it forms new crust on both sides of a rift. The new crust pushes the seafloor apart at the ridges. Scientists called the process **seafloor spreading.** Seafloor spreading is partly what drives the movements of the continents, or continental drift.

4 **DESCRIBE:** What happens as magma rises through the crust of the mid-ocean ridges?

CHECKING CONCEPTS

1. Which kind of crust is made up of heavy, dense material?
2. What is the Mid-Atlantic Ridge?
3. How does magma come out of a chamber to form a ridge and valley?
4. Where did scientists find the youngest crust on the ocean floor?

THINKING CRITICALLY

5. **MODEL:** Draw a diagram that illustrates seafloor spreading. Label your diagram.
6. **CONCLUDE:** How do you think seafloor spreading helps explain the theory of continental drift?

BUILDING SCIENCE SKILLS

Modeling To make a model of seafloor spreading, do the following:

STEP 1 Tape two sheets of red construction paper together along their short ends. Then push two desks together.

STEP 2 Fold the sheets of construction paper together along the tape.

STEP 3 Push the open ends of the construction paper up through the crack between the desks until some paper falls on each side. Keep pushing the paper up.

7. **CONCLUDE:** What does the crack between the desks represent? What does the construction paper represent? How could you make your model more accurate?

Real-Life Science

THE SUMATRA-ANDAMAN EARTHQUAKE

On December 26, 2004, an undersea earthquake rocked the Indian Ocean off the coast of Sumatra. The magnitude of the quake registered at 9.1 on the Richter scale. It was the 3rd largest earthquake since 1900. It literally shook the planet.

During the earthquake, an underwater landslide displaced a large volume of water. This triggered a series of giant waves, called tsunamis, that devastated the coasts of the Indian Ocean. These waves were over 100 meters high. Nearly 230,000 people were killed or missing. This incident is now known as the Asian Tsunami.

▲ **Figure 6-8** A giant tsunami hits the shore.

Thinking Critically What effects might an undersea earthquake have on the ocean floor?

COOPERATIVE LEARNING

Modeling Seafloor Spreading

Have students work in pairs to complete the Building Skills activity. After students determine how to make their models more accurate, have them present their revised model to the class.
learning style: tactile

MORE ON THE WEB

Tsunami Aftermath Have students research online, starting at www.conceptsandchallenges.com, to learn more about the devastation caused by the Asian Tsunami. Then have students write a short report about their findings.

• TEACHING RESOURCES • *Online or Print*

Teacher's Resources

Lesson Review: Chapter 6, p. 3

Enrichment: Chapter 6, p. 5

Seafloor Spreading: Visuals, p. 5

QUICK CONVERSIONS

	Metric /SI	English
Chain peaks	3,000 m	9,843 ft
Great Rift Valley		
length	6,400 km	3,977 mi
width	48–64 km	30–40 mi

3 *Assess*

CHECKING CONCEPTS ANSWERS

1. oceanic
2. mountain chain that runs down the middle of the Atlantic Ocean
3. through a crack in the crust
4. along the mid-ocean ridges

THINKING CRITICALLY ANSWERS

5. **MODEL:** Students' models should resemble Figure 6-7.
6. **CONCLUDE:** Possible answer: Seafloor spreading causes the continents to move.

4 *Close*

BUILDING SCIENCE SKILLS ANSWER

Modeling

7. **CONCLUDE:** The crack between the two desks represents the rift valley in the mid-ocean ridge; the construction paper represents oceanic crust; by placing construction paper on either side so it gets pushed out toward the edge of the desk. **learning style: tactile**

Real-Life Science

THE SUMATRA-ANDAMAN EARTHQUAKE

Background Called a mega-thrust event, (subduction of one tectonic plate beneath another), this earthquake occurred as thrust-faulting where the India plate subducts under the Burma plate. Records were set for the longest fault rupture in history – about 1300 km (780 miles) – and the longest duration of shaking at about 10 minutes.

The initial earthquake caused the fault to slip nearly 15 meters near Banda Aceh, Sumatra. This event resulted in a huge displacement of water forming a deadly tsunami in the Indian Ocean. This tsunami caused more casualties than any on record, affecting South Asia and East Africa. Efforts to warn the Indian Ocean coasts about a possible tsunami were difficult because there were no buoys in place in 2004 to register tsunami threats.

Discussion Ask students: How might the number of deaths have been different if the Indian Ocean had a Tsunami Warning System in place? (Warnings would have reached coastal people well in advance of the waves, allowing them time to move to higher ground.)

Thinking Critically Answer
Possible answers include causing underwater landslides and changes in rifts and valleys.

6-3 *(pp. 142–143)*

1 Introduce

INVESTIGATE

TIME: 10–15 minutes

PURPOSE: Students will demonstrate seafloor spreading.

MATERIALS: sheet of paper with "Mid-Ocean Ridge" written on it, additional sheets of paper

PROCEDURE: Have three students come to the front of the room and stand side by side. Have the student in the middle hold the sheet of paper with "Mid-Ocean Ridge" written on it. After 1 minute, students on the outside of the row write "1 minute" on a sheet of paper. Have two more students join the line, one on either side of the middle student, thus pushing the original two students apart. Have the five students stand for another minute and then record how long they have been standing. Continue this activity until all students are standing.
learning styles: visual, tactile

THINK ABOUT IT: The students who have been standing for the shortest time are nearest the student who represents the mid-ocean ridge, just as the oceanic crust near the mid-ocean ridges is younger than the crust farther away. The growing line of students models the spreading of the seafloor.

STUDY HINT Have students write the lesson title and objective in their notebooks. As students read the lesson, tell them to write the sentence or sentences that provide the information that meets the objective.

2 Teach

Demonstration Bring a compass to class. Show students that the needle of a compass always points north.
learning style: visual

Discussion To introduce this lesson, write the following numbers on the board: *4.6 billion years* and *160 million years.* Explain that Earth is about 4.6 billion years old, but that the oldest rocks on the ocean floor are only about 160 million years old. Have students try to explain the discrepancy in time. Discuss all responses. (As a hint, tell students that Earth is not getting bigger.) Then, describe subduction zones. Refer students to Figure 6-9 to illustrate your description.
learning style: auditory

6-3 What evidence supports seafloor spreading?

INVESTIGATE

Modeling Seafloor Spreading

HANDS-ON ACTIVITY

1. Three students stand in a row. The student in the middle holds a sheet of paper with "Mid-Ocean Ridge" written on it.
2. After one minute, the other two students write "1 minute" on their sheets of paper.
3. Two students join the row on either side of the person in the middle. They are the "lava" that forms the "new crust."
4. After another minute, everyone records how long they have been standing.
5. Repeat this process until all students are standing in the row.

THINK ABOUT IT: Compare the numbers on the sheets of paper. How does your model show seafloor spreading? What part of the new seafloor will be the youngest part?

Objective

Describe some evidence of seafloor spreading.

Key Terms

trench: deep canyon

subduction (suhb-DUHK-shuhn) **zone:** place where old oceanic crust is forced back down into an ocean trench

Trenches On the ocean floor, there are deep canyons called **trenches.** Trenches are the deepest parts of the oceans. They may be more than 10,000 m deep. Most trenches are found in the Pacific Ocean along the coasts of continents.

Some trenches are found near chains of islands. The volcanic region known as the Ring of Fire in the Pacific Ocean contains many examples of these.

1 DEFINE: What is a trench?

Disappearing Crust The oldest rocks found on the ocean floor are only about 160 million years old. Yet, Earth is about 4.6 billion years old. How is this possible? As new crust is made in one place, old crust is destroyed someplace else. If crust were not being destroyed, our planet would keep growing. Old crust on the ocean floor ends up deep down in ocean trenches. There, it melts and becomes magma again.

2 STATE: What happens to older crust?

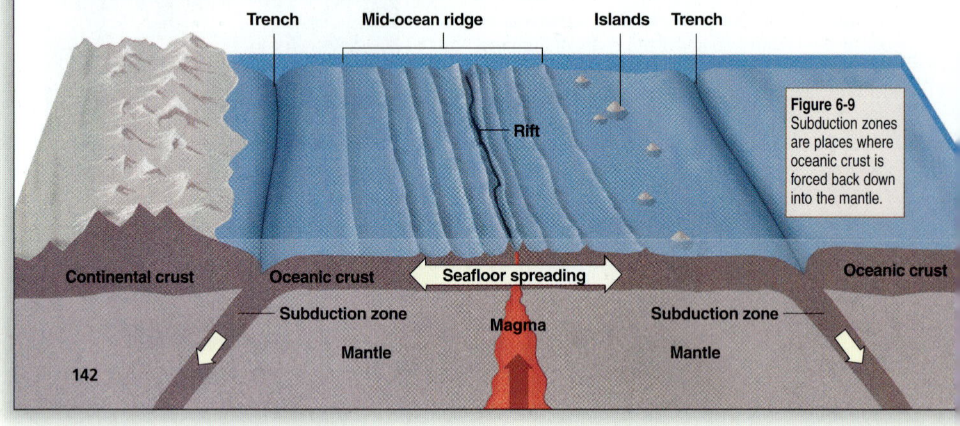

Figure 6-9 Subduction zones are places where oceanic crust is forced back down into the mantle.

Teaching Options

INTEGRATING PHYSICAL SCIENCE

Looking at Magnets Discuss the magnetic properties of ocean-floor rocks. Point out that Earth is like a giant bar magnet with a magnetic field. The part of a magnet where the magnetic forces are the strongest is called a magnetic pole. Every magnet has two poles. In a bar magnet, the poles are at the ends. Tell students that a compass can be any magnet that can turn freely. It will swing around so that its poles point north and south. If students are not familiar with how a compass works, have them create one with a needle, a magnet, and a bowl of water and check its direction against a pocket compass. **learning style: tactile**

ONGOING ASSESSMENT

Have students explain Figure 6-9 to other students.

Subduction Zones The areas in which oceanic crust is forced back down into ocean trenches are called **subduction zones,** shown in Figure 6-9. A lot of volcanic activity and many earthquakes occur near subduction zones. The Ring of Fire has many subduction zones. In these, the older oceanic crust is forced back down into the mantle. It is in these areas that trenches are formed.

3 DEFINE: What are subduction zones?

Ages of Fossils In 1968, the crew of the research ship *Glomar Challenger* made an important discovery. They drilled holes at various sites along the Mid-Atlantic Ridge and brought up samples of crust. Scientists studied the fossils in the samples. They discovered that the rock samples closest to the rift valley held the youngest fossils. The fossils farthest from the rift valley were older. The fossil evidence supports the idea that new oceanic crust is forming along the ridge. It also supports the theory of continental drift.

4 IDENTIFY: Where were the oldest fossils found?

Magnetic Fields There is another important piece of evidence that supports the theory of seafloor spreading. It was found in rocks near the Mid-Atlantic Ridge.

Earth has a magnetic field. If you have ever used a compass, you know that a compass needle always points toward Earth's magnetic north pole. Some minerals are magnetic. When magnetic minerals form, their particles, like the compass needle, line up with Earth's magnetic field.

Earth's magnetic poles have switched places many times. If the poles were to switch again, your compass would point south instead of north. When rocks with magnetic minerals are formed, their minerals point to where Earth's magnetic north pole was when the minerals formed. Scientists found bands of rock in the Mid-Atlantic Ridge with minerals pointing toward present-day north. These alternated with bands of rock pointing toward present-day south. Both sides of the ridge showed the same pattern.

You cannot see the magnetic bands on the ocean floor. They can only be detected with scientific instruments. However, this pattern is evidence that new ocean floor is being created and that the seafloor is spreading in places.

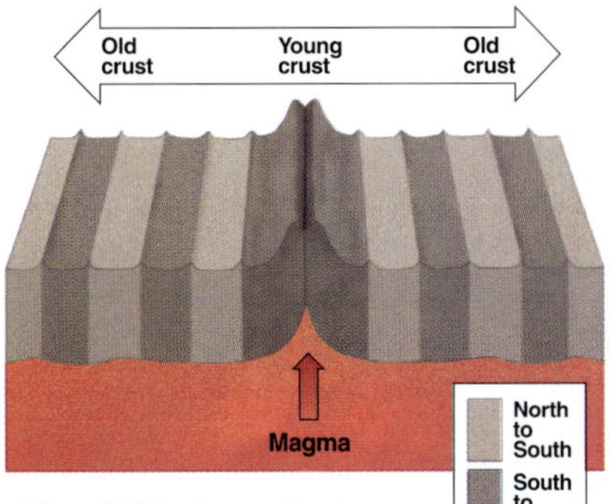

▲ **Figure 6-10** A mid-ocean ridge shows alternating bands of magnetic particles.

5 PROVE: What evidence shows that Earth's magnetic field has changed?

CHECKING CONCEPTS

1. The deepest parts of the ocean are __________.
2. The oldest rocks found on the ocean floor are only __________ million years old.
3. A compass needle always points __________.

THINKING CRITICALLY

4. INFER: Why is there a lot of volcanic and earthquake activity at subduction zones?
5. CALCULATE: Seafloor spreading adds about 2.5 cm of new material to Iceland each year. How much wider could Iceland be in 150 years?

Web InfoSearch

Challenger Deep The deepest place on Earth is Challenger Deep. Challenger Deep is a trench 11,033 m deep. It is part of the Mariana Trench in the Pacific Ocean. If Mount Everest were put into it, its top would be covered by 1,600 m of water.

SEARCH: Use the Internet to find out more about Challenger Deep. What else do we know about it? Start your search at www.conceptsandchallenges.com. Some key search words are **Mariana Trench, Mount Everest,** and **Ring of Fire.**

Reinforcement Some students may not easily grasp the relationship between seafloor spreading and the alignment of magnetic particles in rocks. Use the Integrating Physical Science activity on page T142 to explain this concept.
learning style: auditory

Answers to questions

1 DEFINE: a deep canyon

2 STATE: It ends up in ocean trenches.

3 DEFINE: places where crust is forced into ocean trenches

4 IDENTIFY: in the rocks farthest from the rift valley

5 PROVE: bands of rock with minerals pointing north alternating with bands of rock with minerals pointing south

3 Assess

CHECKING CONCEPTS ANSWERS

1. trenches
2. about 160
3. north

THINKING CRITICALLY ANSWERS

4. INFER: because crust is moving at subduction zones
5. CALCULATE: 375 cm, or 3.75 m

4 Close

Web InfoSearch

Challenger Deep Have students read the passage on page 143. Have them write two questions that they cannot answer from the information in the passage. Then, have students work in pairs to search the Internet to find the answers. Students can begin their search for information at www.conceptsandchallenges.com. Key search words are **Mariana Trench, Mount Everest,** and **Ring of Fire.**

COOPERATIVE LEARNING

Modeling the Ocean Floor Using Figure 6-9 as a guide, have students work in small groups to make models of the ocean floor near a subduction zone. Tell students to use different colors of clay to make their models.
learning style: tactile

• TEACHING RESOURCES • Online or Print

Teacher's Resources

Lesson Review: Chapter 6, p. 4

Enrichment: Chapter 6, p. 5

QUICK CONVERSIONS		
	Metric/SI	**English**
Ocean depths	10,000 m 11,033 m 1,600 m	32,808 ft 36,198 ft 5,249 ft

6-4 *(pp. 144–145)*

1 Introduce

INVESTIGATE

TIME: 5–10 minutes

PURPOSE: Students will model Earth's crust to demonstrate that plates are movable.

MATERIALS: 250 mL of cornstarch, 100 mL of room-temperature water, plastic container

PROCEDURE: Have students work in pairs to complete this activity. Explain that after Steps 3 and 4, each pair should discuss their findings. **learning style: tactile**

THINK ABOUT IT: It makes the crust movable.

STUDY HINT Have students scan the Skills Handbook lesson on science skills at the front of the Student Edition. Have them identify each skill used in the lesson. Students should then review the definition given for each skill on page 4 of the text.

Linking Prior Knowledge Before beginning this lesson, have students review the structure of Earth (Lesson 1-2). Ask students to name the three layers of Earth. (crust, mantle, and core)

2 Teach

Demonstration To model floating plates, put a piece of soap in a pan of water. Put a few cotton balls or other light objects on the soap. Compare the water to the solid rock of the mantle that flows like a thick liquid. Compare the soap to the tectonic plates and the cotton balls to the continents and oceans.
learning style: visual

Discussion Refer students to Figure 6-11, which shows tectonic plates. Randomly read out the names of the tectonic plates and have students locate them on the map.
learning style: visual

Answers to questions

1. **IDENTIFY:** 21
2. **NAME:** mantle
3. **NAME:** seafloor spreading and continental drift

6-4 What is plate tectonics?

INVESTIGATE

Modeling Earth's Mantle

HANDS-ON ACTIVITY

1. Begin by measuring 250 mL of cornstarch. Pour it into a container.
2. Measure 100 mL of room-temperature water. Pour it over the cornstarch, and stir until blended smoothly.
3. Tap a fingertip on the mixture you made. What happens?
4. Slowly press the mixture with your fingertip. What happens?

THINK ABOUT IT: Earth's mantle is made from solid rock. Like your mixture, the mantle rock can also flow like a very thick liquid. How do you think this property affects the crust on top of the mantle?

Objectives

Name some crustal plates. Describe the theory of plate tectonics.

Key Terms

tectonic plate: large, solid piece of Earth's surface

theory of plate tectonics (tehk-TAHN-ihks): theory that Earth's crust is broken into plates that float on the upper part of the mantle

Tectonic Plates Earth's lithosphere is made up of crust and the solid, uppermost part of the mantle. The mantle is the layer of rock below the crust. The lithosphere is broken up into large pieces called **tectonic plates.**

There are 7 large tectonic plates and about 14 smaller ones. The largest plate is the Pacific plate. Figure 6-11 below shows some of the tectonic plates.

1. **IDENTIFY:** About how many tectonic plates have been identified?

Floating Plates Part of the mantle contains rock that flows like a thick liquid. This part, called the athenosphere, is just below the more solid rock of the uppermost mantle. Tectonic plates float on the athenosphere like rafts floating on a lake. The continents and oceans are carried along on the plates like passengers on a raft.

2. **NAME:** On which part of Earth do the tectonic plates float?

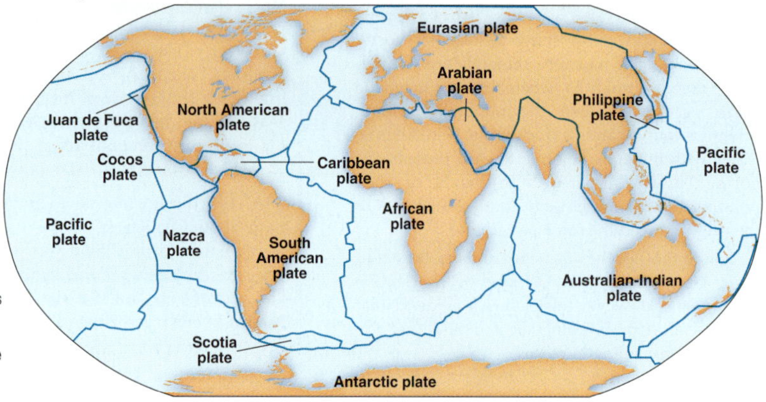

Figure 6-11 ▶ This map shows the seven large tectonic plates and some of the small ones.

Teaching Options

COOPERATIVE LEARNING

Modeling Earth's Plates Have students work in pairs to create models of Earth's plates. Students can use the models to show how Pangaea Ultima will form.
learning style: tactile

ESL/ELL STRATEGY

Restating English-language learners will benefit from restating theories and processes in their own words. Have pairs of students work together to summarize the theory of plate tectonics as presented in the lesson. **learning style: auditory**

EXTENDING THE LESSON

Crustal Components All of the continents have four main crustal components: *shields* (relatively level land with very old crust), *stable platforms* (horizontal land covered with a thin layer of sedimentary rock), *sedimentary basins* (broad, deep depressions filled with sedimentary rock that formed in shallow seas), and *folded mountain belts* (made of younger sedimentary rock). Have interested students investigate one or more of these crustal components and write an explanation of it.

Theory of Plate Tectonics Using information that supports seafloor spreading and continental drift, scientists have developed the **theory of plate tectonics.** The theory of plate tectonics combines the theories of continental drift and seafloor spreading. The theory of plate tectonics explains how and why the continents move. It states that Earth's lithosphere is broken into plates that float on the upper mantle. The continents move because they are carried along on the moving plates.

B NAME: What two theories does plate tectonics combine?

CHECKING CONCEPTS

1. Which layer of Earth is below the crust?
2. How many large tectonic plates are there?
3. What is the uppermost mantle rock like?
4. What is the theory of plate tectonics?

THINKING CRITICALLY

5. MODEL: Draw a picture of a raft on a lake with people on the raft. Using the lake, the people, and the raft, label your picture with the following terms: tectonic plate, athenosphere, continents.

INTERPRETING VISUALS

Use Figure 6-11 to answer the following.

6. On which plate is Australia?
7. Name one plate that the United States is on.
8. What are three large plates and two of the smaller plates?
9. Do you think the present number, sizes, and shapes of tectonic plates will remain the same as they are today? Explain your answer.

How Do They Know That?

PREDICTING PANGAEA ULTIMA

Many geologists study how Earth's continents were formed. They can use that information to answer questions such as whether we can predict volcanic eruptions and earthquakes. Geologists use many techniques for studying Earth's movements. One way is to bounce laser beams from stations on Earth off satellites. The time it takes for the beams to return tell them how far the crust has moved since previous measurements.

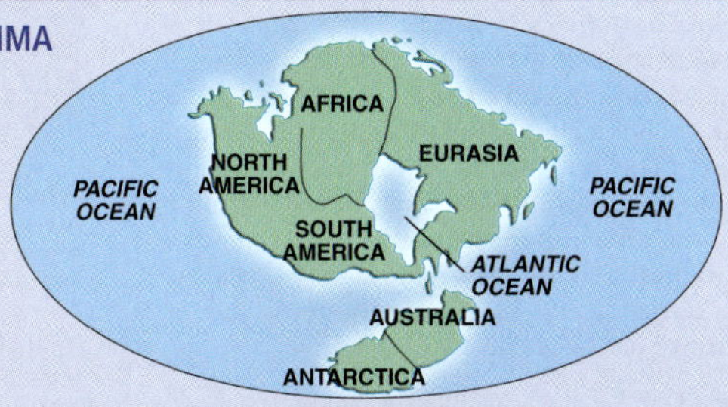

▲ **Figure 6-12** Pangaea Ultima

Dr. Christopher Scotese is a geologist at the University of Texas in Arlington, Texas. Dr. Scotese creates maps of Earth's continents and oceans from computer and satellite data. He uses the data to help him predict what the continents might look like in the future. According to Dr. Scotese, most of today's continents could rejoin as one giant continent in about 250 million years. The Pacific Ocean will become much larger. The Atlantic Ocean will become a small landlocked sea. Dr. Scotese has named his future world Pangaea Ultima.

Thinking Critically How might Africa's new position affect the wildlife there?

3 *Assess*

CHECKING CONCEPTS ANSWERS

1. mantle
2. seven
3. It is made up of solid rock that flows like a thick liquid.
4. Earth's lithosphere is broken into plates that float and move.

THINKING CRITICALLY ANSWER

5. MODEL: Students should label the people "continents," the raft "tectonic plate," and the lake "athenosphere."

INTERPRETING VISUALS ANSWERS

6. Australian-Indian plate
7. Possible answers include North American, Pacific, and Juan de Fuca.
8. Accept any three of the following: Eurasian, Pacific, Australian-Indian, African, South American, Antarctic, and North American. Accept any two of the smaller plates.
9. Possible answer: probably not, as plates will run into each other and form new, larger landmasses

4 *Close*

How Do They Know That?

PREDICTING PANGAEA ULTIMA

Technology Connection In addition to making models, computers are used by scientists to store data; calculate and measure; create charts, tables, and graphs; and run simulations. Scotese used a computer model to depict how Pangaea Ultima will look.

Scientists also used a computer model to depict how the world looked around 750 million years ago, before the supercontinent Pangaea formed. Scientists call this earlier supercontinent Rodinia. In the Rodinia model, North America was connected to Antarctica. When Rodinia broke up, the continents moved farther apart. About 260 million years ago, they moved back together to form Pangaea.

Discussion Have students compare the diagram of Pangaea Ultima with a map of the world today. Discuss changes in the placement of the continents and the oceans on Scotese's model.

Thinking Critically Answer
Possible answer: Some species might become endangered because they would have to face more competition and new enemies.

MORE ON THE WEB

Defining Terms Have students use the Internet to find out what the term *isostasy* means and how it relates to plate tectonics. Students should write their findings in a report. Students can begin their search at www.conceptsandchallenges.com.

• *TEACHING RESOURCES* • *Online or Print*

Teacher's Resources
Lesson Review: Chapter 6, p. 6
Enrichment: Chapter 6, p. 9

Laboratory Manual
Lab. Challenge: How does the thickness of tectonic plates vary? pp. 43–44

6-5 (pp. 146–147)

1 Introduce

Linking Prior Knowledge Before discussing convection currents as they relate to crustal movement, discuss how some room heaters use convection. Students may be familiar with these heating systems. Steam or hot water moves through a heater. It heats some of the air in the room. The warm air then rises. Cooler air sinks and moves toward the heater. This occurs in an endless cycle of rising and sinking air.

2 Teach

Demonstration Use two plastic-foam sheets (such as blueboard insulation) of different thicknesses to simulate plates. Use a thick sheet for a continental plate and a thinner sheet for an oceanic plate. Shape the leading edges into a point. Show students that the oceanic plate slides under the continental plate.
learning style: visual

Discussion Referring to Figure 6-14, point out that molten material is pouring onto the surface at one point. Ask where rock is moving down into Earth.
learning style: visual

Reinforcement Describe the different ways that tectonic plates move. Refer students to the diagrams in this lesson that show the three types of tectonic plate movement.
learning style: visual

Answers to questions

1. **DEFINE:** the movement of a gas or liquid caused by changes in temperature
2. **STATE:** giant convection currents in Earth's mantle
3. **LIST:** together, apart, and past each other

WRITING HINT Suggest that students write answers in complete sentences. This helps improve the quality of their thinking as well as of their writing.

6-5 What causes plate tectonics?

Objective

Describe the causes of plate tectonics.

Key Terms

convection current: movement of a gas or a liquid caused by changes in temperature

plate boundary: place where two plates meet

Convection Currents A **convection current** is the movement of a gas or a liquid caused by differences in temperature and density. For example, warm air rises and cool air sinks. When a pan of water boils on a stove, the water near the bottom of the pan heats first. The less dense hot water rises. The cooler and denser water near the top of the pan sinks.

1 DEFINE: What is a convection current?

Inside the Mantle Scientists think that giant convection currents in Earth's upper mantle cause the tectonic plates to move. The mantle rock close to Earth's core is hot and has some properties of a thick liquid. The mantle rock farther from the core is cooler. The hot mantle rock rises. The cooler mantle rock sinks. As the cooler rock moves closer to the core, it heats up and rises. This process repeats in an endless cycle, as shown in Figure 6-13. The plates are carried along like packages on a moving conveyor belt or like people on a raft.

2 STATE: What causes the movement of plates?

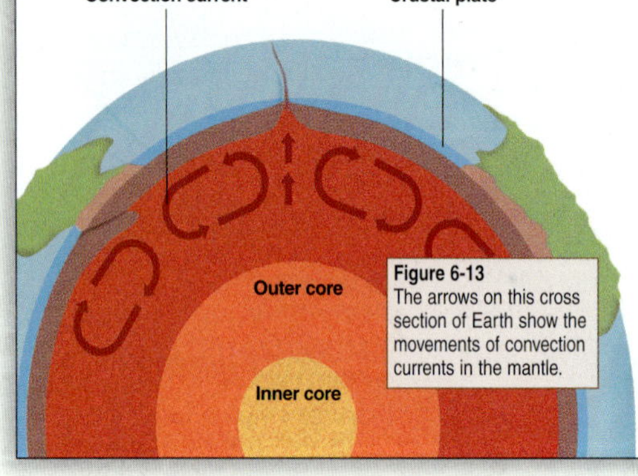

Figure 6-13 The arrows on this cross section of Earth show the movements of convection currents in the mantle.

Plate Boundaries and Movements The place where two plates meet is called a **plate boundary.** Scientists have identified three different kinds of plate boundaries: transform, convergent, and divergent.

At transform boundaries, two plates slide past each other. The movement is not smooth. This sliding movement causes earthquakes, as shown in Figure 6-20 on page 150.

At a convergent boundary, the plates move toward each other. When oceanic crust meets continental crust, the denser oceanic crust slides under the continental crust. This downward movement of the oceanic crust produces a subduction zone, as shown in Figure 6-14.

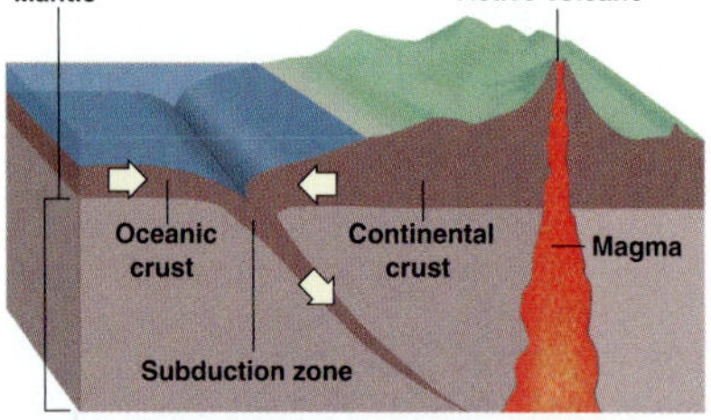

▲ **Figure 6-14** When oceanic crust meets continental crust at a convergent boundary, a subduction zone forms. This is often an area of volcanic activity.

If two continental plates meet, the crust crumples upward. This kind of movement is shown in Figure 6-15.

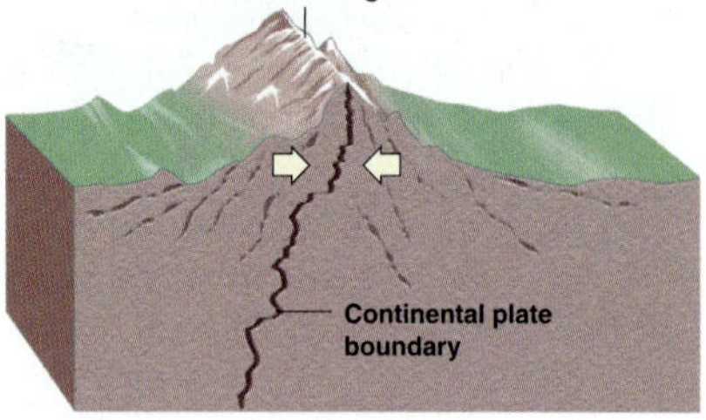

▲ **Figure 6-15** When two continental crusts meet at a convergent boundary, they crumple upward. Mountains may be formed.

Teaching Options

! FAST FACTS *Support for Plate Tectonics*

Arthur Holmes was a geologist and a contemporary of Alfred Wegener's. In 1930, he offered support to Wegener's theory of continental drift. Holmes said that currents of heat inside Earth could provide the power to break apart and move the continents. Holmes's theory helped revive interest in Wegener's theory, which had never been taken very seriously. However, Holmes did not have evidence to support his theory.

As a result, few scientists paid attention to Holmes's theory of convection currents either. Then, the discovery of mid-ocean ridges brought new attention to his theory. In 1960, a geologist named Harry Hess proposed that the ocean floors move, starting at the mid-ocean ridges. Hess said that magma rises from the mantle and pushes older rock to both sides of the ridge. Evidence supporting Hess's theory of seafloor spreading was also used to support Wegener's theory of continental drift and Holmes's theory of convection currents.

Arthur Holmes is also known for helping to create the geologic time scale that is used today.

At a divergent boundary, the plates move apart. These plates sometimes form rift valleys, as shown in Figure 6-16.

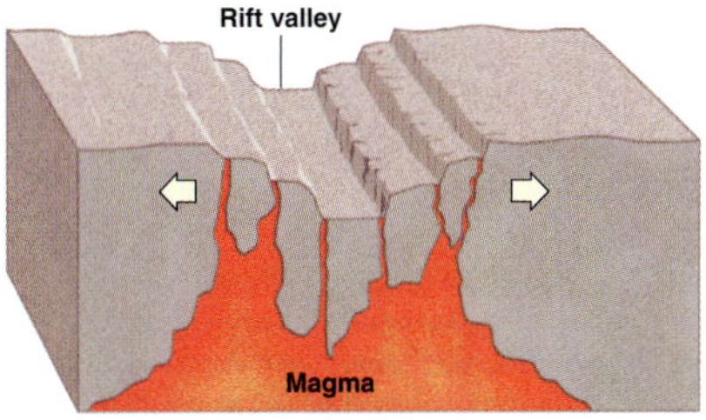

▲ **Figure 6-16** A rift valley may form at a divergent boundary.

3 LIST: What are three ways plates move?

✓ CHECKING CONCEPTS

1. What causes the movement of plates?
2. How is crustal rock heated inside the mantle?
3. At which type of boundary do plates move apart?
4. What happens when two plates move toward each other?

THINKING CRITICALLY

5. **INFER:** What land feature may be formed when two continental plates meet?

DESIGNING AN EXPERIMENT

Design an experiment to solve the following problem. Include a hypothesis, variables, a procedure with materials, and a type of data to study. Also say how you would record your data.

PROBLEM: Scientists believe that rift valleys are created by the movement of continental plates away from each other. As the plates pull apart, a valley forms between them. Design a small-scale experiment to show how a rift valley forms.

People in Science

VOLCANOLOGIST

Volcanologists are scientists who study volcanoes. They collect data and samples, and observe volcanoes. A volcanologist must know geology, physics, chemistry, and math.

Volcanologists have very dangerous jobs because they walk on active volcanoes! Often called lava hunters, they travel the world to observe volcanoes in action.

Two famous French lava hunters were Maurice and Katia Krafft. Their goal was to educate people on the dangers of volcanic eruptions.

In many parts of the world, warnings about possible eruptions are often ignored by local people. For example, mudflows from the 1985 Nevado del Ruiz eruption in Colombia killed 25,000 people. The Kraffts produced videos about how to survive an eruption.

▲ **Figure 6-17** Lava hunters must wear protective gear when near an active volcano.

Together, the Kraffts observed and filmed more active volcanoes than any other volcanologists. Sadly, they were killed by an explosion of hot ash and gases from an eruption of Mount Unzen in Japan on June 3, 1991.

Thinking Critically Why do you think people often ignore warnings of possible volcanic eruptions?

MORE ON THE WEB

Career Connection Students can learn more about the job description and job requirements of volcanologists by searching the Internet. Have students write their findings in a report.

ONGOING ASSESSMENT

Have students make posters depicting what occurs at each of the different plate boundaries.

• TEACHING RESOURCES • Online or Print

Teacher's Resources

Lesson Review: Chapter 6, p. 7

Enrichment: Chapter 6, p. 9

Designing an Experiment: Study Tools, p. 14

Designing an Experiment Rubric: Teaching Tools, p. 7

3 Assess

✓ CHECKING CONCEPTS ANSWERS

1. giant convection currents in Earth's mantle
2. It is pushed into the mantle at subduction zones.
3. divergent
4. One may slide under the other, they may slide past each other, or they may crumple upward.

THINKING CRITICALLY ANSWER

5. **INFER:** a mountain

4 Close

DESIGNING AN EXPERIMENT

Use the Designing an Experiment Rubric from the Teacher's Resources to assess students' experiments. Fill in the rubric with the additional information below. For this assignment, students should have:

- designed a small-scale experiment that shows how a rift valley forms.
- included, in the procedure, a way to record observations of the experiment.

People in Science

VOLCANOLOGIST

Real-Life Connection If you live near a volcano, it is important to plan an evacuation route and have a backup route in case the volcano erupts. Because eruptions can lead to other disasters, such as earthquakes, flash floods, landslides, and severe thunderstorms, it is important to have a disaster kit handy.

A good disaster kit should include a flashlight, a battery-operated radio, a first-aid kit, canned food, bottled water, and extra batteries. Every family member should have a pair of goggles and a disposable breathing mask to wear while outdoors to protect him or her from toxic fumes and ash.

Discussion Ask students: *Would you like to become a volcanologist?* Have them explain their answers to the class. (See More on the Web for additional information.)

Thinking Critically Answer
Possible answer: They do not think the volcano will erupt because it has been dormant for so many years.

Lab Activity

(pp. 148–149)

1 Prepare the Lab

Modeling Convection Currents

SET-UP TIME: **LAB TIME:**

BACKGROUND: Earth is a giant heat engine that transfers energy from its core to the surface through slow-moving convection currents.

PURPOSE: Students will use hot water to model how differences in temperatures cause convection currents.

ALTERNATIVE MATERIALS: Small coffee cups can be used in place of plastic cups. If you do not have running water in your classroom, bring hot (not boiling) water in a thermos and cold water in plastic milk bottles. Use liquid food coloring. If you only have the gel form, you will have to water it down.

SCIENCE SKILLS: Students will **collect data** and **communicate** their findings in written and sketch form. They will use this experiment to **model** what takes place inside Earth and **infer** how this relates to crustal movements.

ADVANCE PLANNING: Have an empty bucket on hand for collecting water after the experiment.

2 Guide the Lab

PROCEDURE: Divide the class into groups of two or three students. Discuss the procedure. Emphasize that students must avoid disturbing the water in the plastic box. Discuss data collection and reporting. Have students make their data charts before starting the experiment. Provide students with a copy of the Lab Activity Report found in the Teacher's Resources to record their observations and conclusions.

TROUBLESHOOTING: If there are no convection currents, test your water temperature to make sure that there is a 25°C (77°F) difference between the hot and cold water.

If the food coloring spreads out too much, have students wipe off the tip of the dropper bottle between uses. Make sure that the water in the box is undisturbed.

LAB ACTIVITY
Modeling Convection Currents

Materials

Safety goggles, lab aprons, 5 plastic or plastic foam cups, 1 clear plastic food storage box, food coloring (red or blue), hot water, cold water, stopwatch

BACKGROUND

Earth's continents are moving on tectonic plates that float on the mantle. The part of the mantle they float on flows like a thick liquid. It is very hot and is under great pressure. The mantle's heat comes from deep inside Earth's core. Convection currents carry the heat to the mantle.

PURPOSE

In this activity, you will observe convection currents forming in water as a model of how convection currents are formed in Earth's mantle.

PROCEDURE

1. Put on your safety goggles and lab apron.
2. Place four plastic cups upside down on a table top to make pillars that will support the food storage box at its corners.
3. Fill the food storage box about half full with cold water from a faucet, and place it on the pillars.
4. Wait five minutes for the water to stop moving completely. Try not to bump the table.
5. While waiting for the water to settle, copy the chart in Figure 6-19 onto your paper. Make six rows on your chart. Title the chart *Observing Convection Currents.*
6. Fill the last cup with very hot water from a faucet and gently slide this cup under the middle of the box. Do not bump the box or the pillars. ⚠ CAUTION: Be extremely careful when handling the hot water.
7. Gently place four to five drops of the food coloring in the water near the center and bottom of the box.

▲ **STEP 6** Carefully place the cup of hot water under the plastic box.

▲ **STEP 7** Add the food coloring to the bottom of the box.

148

Teaching Options

INTEGRATING THE SCIENCES

Chemistry In cooking, convection currents help to mix ingredients in simmering soups and sauces.

Meteorology Convection currents strongly influence wind patterns. In summer, hot air above the land is driven upward by convection currents. Cool breezes from the oceans fill the void. This process is reversed in winter.

Technology Heated air or liquid expands and becomes less dense. Consequently, it becomes buoyant and rises. Hot-air balloons can rise because their air bags capture hot air rising.

Ask students to think about situations at home that use convection currents. Have them describe the source of the temperature difference and the fluid that flows.

ESL/ELL STRATEGY

Filling in Data Charts When filling in a data chart, it may be helpful for English-language learners to add illustrations as well as describe their observations.

8. Begin timing the experiment. Once each minute for five minutes, observe and record what you see on your chart. Write your observations in the third column. In the last column, draw pictures of what you see. Draw in arrows to show any motions you observe.

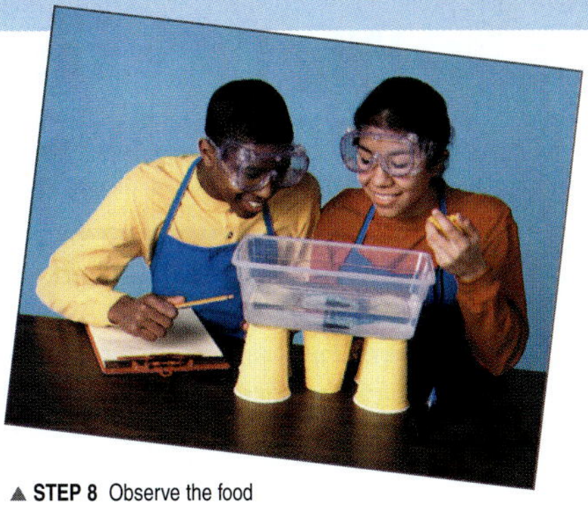

▲ **STEP 8** Observe the food coloring as the water in the box is heated. Record what you see.

◀ **Figure 6-18** The food coloring shows a convection current forming.

Observing Convection Currents

Observation	Time	Written Description	Diagram
1	0 min.		
2	1 min.		

▲ **Figure 6-19** Make a chart like this one with six rows. Leave enough room to write and draw in each row.

CONCLUSIONS

1. OBSERVE: What happened in the experiment?
2. MODEL: What does the cold water in the experiment represent?
3. MODEL: What does the hot water in the experiment represent?
4. INFER: How might this modeling experiment help to explain why the continents are moving?
5. ANALYZE: How are convection currents within Earth similar to the ones you created in your model?

SAFETY TIP: Caution students not to get food coloring in their eyes or on their clothing. Have soap and water or towelettes available for cleaning hands. You do not have to use very hot water as long as the difference between the hot and cold water is about 25°C (77°F).

EXPECTED OUTCOME: Students will immediately observe upward moving currents in the plastic container. They will observe that the currents rise above the cup with the hot water and spread out in all directions when they reach the surface. Students will also observe that cool water falls around the edges.

LAB HINT: Remind students to consider all variables when analyzing the results of the experiment.

3 Conclude the Lab

1. **OBSERVE:** The food coloring rose from the bottom of the container and spread across the surface of the water. The water cooled and dropped back down. Gradually, a doughnut shape formed.
2. **MODEL:** the cooler upper layers of Earth's mantle
3. **MODEL:** Earth's hot core
4. **INFER:** Convection currents bring heat up from Earth's interior. This softens the mantle rock and reduces friction with the tectonic plates above.
5. **ANALYZE:** They move because of differences in temperature.

Use the Laboratory Activity Rubric from the Teacher's Resources to assess students' lab activities. Fill in the rubric with the additional information below. For this activity, students should have:

- observed convection currents in a model.
- carefully recorded their observations.

4 Extend the Lab

Discuss the role that gravity plays in convection currents. (Convection currents form on Earth because fluids expand or become less dense when heated. This causes them to become buoyant and rise in the cooler surrounding fluid.) The force of gravity pulls the denser liquids toward Earth's center.

Ask students to predict what would happen if they could conduct their experiment on the International Space Station. (Because the effects of gravity are greatly reduced in space, convection currents would also be reduced.)

Tips for Using Technology

Using Graphics Tools to Draw a Diagram Most graphics creation software uses similar tools to create graphics and illustrations. Students can make a diagram of a convection current by clicking on some of these tools, such as shape-creation tools, text tools, and pencil tools. Shape tools allow students to create different types of shapes in their illustrations. The pencil, or drawing, tool allows students to draw in freehand in their illustrations. The text tool allows students to enter text into their illustrations, to label parts of the illustration, or to give it a title.

• TEACHING RESOURCES • Online or Print

Teacher's Resources

Lab Activity Report: Study Tools, pp. 16–17

Laboratory Activity Rubric: Teaching Tools, p. 8

6-6 *(pp. 150–151)*

1 Introduce

INVESTIGATE

TIME: 5–10 minutes

PURPOSE: Students will model mountain formation.

MATERIALS: clay, construction paper

PROCEDURE: Have pairs of students work together. Make sure that they try the activity several times.

THINK ABOUT IT: The collision of the plates pushes the crust upward. This upward movement creates mountains.

STUDY HINT Discuss the lesson objective with the class and have students review the Key Terms and their definitions.

2 Teach

Demonstration Use two paperback books to demonstrate what can happen when two tectonic plates meet. Show how one book moves under the other and how the upper book folds upward. Point out that the upward fold represents a mountain.
learning style: visual

Discussion Bring in pictures showing a mountain, a volcano, or an area affected by an earthquake, as well as pictures of the Hawaiian Islands. Display the pictures in the front of the room. Point out that each picture shows an effect of plate tectonics. Describe how the movement of tectonic plates results in earthquakes, mountain building, volcanoes, and the formation of some islands. You may want to refer students to Figures 6-14 through 6-16 and Figure 6-20, showing the different ways that tectonic plates move.
learning style: visual

Answers to questions

1. **DESCRIBE:** earthquakes
2. **IDENTIFY:** the Himalaya mountains
3. **DESCRIBE:** by melting rock or through cracks in the rock
4. **DESCRIBE:** as the Pacific plate moved over a hot spot

6-6 What are some effects of plate tectonics?

INVESTIGATE

Modeling Mountain Formation
HANDS-ON ACTIVITY

1. Flatten two different pieces of clay into two squares or rectangles.
2. Place each piece of clay on a different sheet of construction paper. Line the short edge of the clay along the short edge of the paper.
3. Push the pieces of clay gently together. What happens?
4. Repeat Steps 1 to 3 several times.

THINK ABOUT IT: Based on your clay model, how do you think some mountains are formed?

Objective

Explain how plate tectonics causes changes on Earth's surface.

Key Terms

magma chamber: underground pocket of molten rock

hot spot: place where magma reaches the surface of a tectonic plate

Earthquakes In some areas, two tectonic plates slide past each other at a transform boundary. The San Andreas Fault, in California, is one example. Many earthquakes are caused by the movement of plates at a fault.

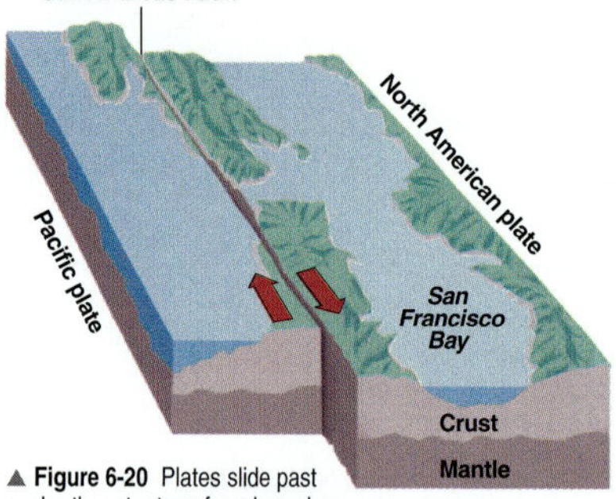

▲ **Figure 6-20** Plates slide past each other at a transform boundary.

The San Francisco earthquakes of 1906 and 1989 were caused by the movement of tectonic plates at the San Andreas Fault. This movement is shown in Figure 6-20.

1 **DESCRIBE:** What can happen when two plates slide past each other at a fault?

Mountain Building When two plates collide, the force of the collision may push oceanic crust under continental crust. Such a collision causes the continental crust to crumple. It may even be pushed upward to form new mountains. Mountains along the western coasts of North and South America were formed in this way. They are young mountains that are still being pushed upward.

Two plates carrying continents may collide without one plate being pushed down under the other. The Himalayas were formed in this way. The plate carrying India collided with the Eurasian plate. The edges of the two plates buckled upward, forming the Himalayas.

2 **IDENTIFY:** What mountain range was formed by the collision of two continental plates?

Volcanoes At subduction zones, oceanic plates dip back down into the hot mantle. The heat from the mantle melts the rocks in the subducting plate, forming magma. The magma collects in underground pockets called **magma chambers.** Hotter than the surrounding rock, the magma melts the solid rock around it and rises to the surface.

150

Teaching Options

INTEGRATING PHYSICAL SCIENCE

Friction Rocks move along a fault, causing friction between the opposite sides of the fault. The way the rocks move depends on how much friction there is. In many places along the San Andreas Fault, high friction causes the plates to lock. Stress builds up, an earthquake releases it, and the plates slide past each other.

Attach a brick or rock to an elastic shock cord and have students drag it along a board. They should discover that the elastic stretches for a while, then the weight moves with a jerk. Discuss how this compares to a slip fault and what the sudden movements would cause in real life. (earthquakes)
learning style: tactile

Magma also moves through cracks in rock. When it reaches the surface, a volcano forms. Mount St. Helens, in Washington State, formed in this way.

3 DESCRIBE: How does magma reach the surface?

Islands Some islands are formed by plate tectonics. The Hawaiian Islands are a chain of volcanic islands in the Pacific Ocean. The islands formed one after the other as the Pacific plate drifted over a hot spot. A **hot spot** is a place where magma works its way to the surface within a plate. As the plate moves over the hot spot, new layers of volcanic mountains are built up.

The islands that are farthest northwest are the oldest. Kauai (koo-EYE) is the oldest of the large islands. Five million years ago, it was the only Hawaiian island. The islands to the southeast are the youngest. The big island of Hawaii is now over the hot spot. It is still being formed.

4 DESCRIBE: How were the Hawaiian Islands formed?

CHECKING CONCEPTS

1. What is a hot spot?
2. What is a magma chamber?
3. How were the Himalayas formed?
4. In what direction is the oceanic crust beneath the Hawaiian Islands moving?

THINKING CRITICALLY

5. INFER: What are some of the possible surface features along the San Andreas Fault?
6. INFER: If railroad tracks were built over the boundary of two plates, such as in Figure 6-20, what would happen to the railroad tracks as the plates moved?
7. INFER: How can the chain of islands that make up Hawaii be used to show the direction the plate beneath them is moving?

Integrating Physical Science

TOPICS: alternate energy sources, heat

GEOTHERMAL ENERGY

The opening page of this chapter shows Iceland's Blue Lagoon. The Blue Lagoon is heated by geothermal energy.

Geothermal energy is energy stored by heated rock deep inside Earth's crust. The heat from Earth's liquid core causes the movement of rock below Earth's surface. Iceland is a hot spot in the Mid-Atlantic Ridge. Deep pockets of water below the Blue Lagoon are heated by hot rocks. Steam forms and is pumped to the surface. The power plant uses the steam to turn a turbine, which generates electricity. Heated water is also pumped directly into homes to provide heat. Most commercial energy used in Iceland today also comes from geothermal energy.

▲ **Figure 6-21** The Blue Lagoon in Iceland is heated by geothermal energy.

The movement of rock below Earth's surface is part of plate tectonics. Iceland is affected by plate tectonics in many different ways. It has about 200 volcanoes. Only 70 have erupted in the last 10,000 years. Islands off the coast of Iceland were created by volcanoes. Iceland has many hot springs and geysers. Also, earthquakes are common in Iceland.

Thinking Critically How are islands created by volcanoes?

MORE ON THE WEB

Researching Geothermal Energy Have students research on the Internet about the benefits and types of geothermal energy. Students should focus on the reliability of geothermal energy and some of the environmental and financial costs associated with it.

• TEACHING RESOURCES • Online or Print

Teacher's Resources

Lesson Review: Chapter 6, p. 8

Enrichment: Chapter 6, p. 9

ESL/ELL STRATEGY

Peer Tutoring Peer-tutored hands-on activities help English-language learners in two ways. Peers encourage students to interact, while hands-on activities help make scientific concepts easier to understand without the barriers of difficult scientific language. The Investigate features can be especially useful with your English-language learners.

3 Assess

CHECKING CONCEPTS ANSWERS

1. a place where magma reaches the surface within a tectonic plate
2. an underground pocket of molten rock
3. When the plate carrying India collided with the Eurasian plate, the edges of the two plates buckled upward.
4. northwest

THINKING CRITICALLY ANSWERS

5. INFER: steep slopes (caused by erosion), ridges, and troughs
6. INFER: The railroad tracks would be offset from their original location.
7. INFER: The youngest island is southeast of the oldest island, so the plate is moving northwest over the stationary hot spot.

4 Close

Integrating Physical Science

TOPICS: alternate energy sources, heat

GEOTHERMAL ENERGY

Physical-Science Connection California, Nevada, Utah, and Hawaii all have power plants that use geothermal energy to generate electricity. Worldwide, however, geothermal energy accounts for very little energy production because not many places on Earth are suitable for producing it.

Geothermal power has some limitations and dangers. The amount of steam from a reservoir is limited, and so the steam must be managed and conserved. In other words, it may not be a truly renewable resource. Geothermal plants may cause environmental problems as well. The wells can leak pollutants such as sulfides into the groundwater. Gases from underground can also pollute the air. Finally, geothermal reservoirs are often in sensitive wilderness areas that should not be disturbed.

Discussion Ask students: *Is geothermal energy a better energy choice than fossil fuels? Why or why not?*

Thinking Critically Answer
When an underwater volcano erupts, lava flows out and hardens into rock. Each new eruption adds more layers of rock. Eventually, the land rises above the surface to form an island.

The Big Idea

(pp. 152–153)

1 Introduce

Objective Students will identify areas worldwide that are affected by plate tectonics. Students will choose one place to research and write about in their Science Logs.

Linking Prior Knowledge Have students recall what happens at the different types of plate boundaries (Lessons 6-5 and 6-6).

2 Teach

Discussion After students have read the article and callouts, encourage them to use reference materials to look for pictures of some of the places on the map before deciding on a place to write about in their Science Logs.

Use the background information in Fast Facts to help guide students in choosing a topic for their Science Logs.

THE BIG IDEA

How do plate tectonics affect places around the world? Have students start their research for their Science Log Writing Activity at www.conceptsandchallenges.com. Be sure students use search engines and websites that are reliable and acceptable. You may also wish to provide copies of the *Big Idea Planner* and *Big Idea Science Log Entry* study tools found in the Teacher's Resources.

Integrating Geography

THE Big IDEA

How does plate tectonics affect countries around the world?

Like Earth science, geography is the study of Earth's surface. The word *geography* comes from two Greek words. *Geo* means "Earth," and *graphia* means "writing." Geography, then, means writing about, or describing, Earth. However, a geographer looks at Earth's surface differently from the way an Earth scientist does. Most geographers are not usually concerned about how the features were formed.

Geography has two main branches. One branch, called physical geography, describes the physical features of Earth's surface, such as mountains, valleys, plains, oceans, and rivers. It is this branch that is most closely related to Earth science. The other branch, called human geography, describes the people who live in or near those places and the cultures of the people who live there.

Iceland is not the only country affected by plate tectonics. Countries all over the world have many interesting land features that were formed by plate tectonics. Geographers are as interested in those features as Earth scientists are. The callouts that appear on these two pages point out some places on Earth where plate tectonics have left a mark on the geography of a region.

Look at the illustration that appears on these two pages. Then, follow the directions in the Science Log to find out more about "the big idea." ✦

Western North America
All along the San Andreas Fault, which runs through California, earthquakes are common.

Galápagos Rift
Along the Galápagos Rift, deep-sea vents and strange sea life have been discovered.

Baños, Ecuador
In December 1999, the Tungurahua volcano, also known as the Black Giant, erupted. It killed people and ruined farmland.

152

Teaching Options

FAST FACTS More on Plate Tectonics Around the World

Western North America
In January 1994, a very damaging earthquake struck the San Fernando Valley in northern Los Angeles.

Galápagos Rift
Many unusual species of animals can be found only on or near the Galápagos Islands.

Baños, Ecuador
The Black Giant was one of the most popular tourist attractions in Ecuador before it erupted. There is still a stream of burning lava pouring from the volcano.

Iceland
Seafloor spreading increases the size of Iceland by about 2.5 cm per year.

The Alps
The converging African plate and Eurasian plate created the Alps and the Mediterranean Sea. The mountains helped Switzerland stay neutral during World War II.

East African Rift System
Animals such as the simian fox and the giant mole rat of the Ethiopian Highlands are not found anywhere outside of the East African Rift System. The vast plains, formed by volcanic ash, have the greatest concentration of large animals on Earth.

Iceland
Iceland is a land of fire and ice. It is home to many glaciers as well as the Blue Lagoon, hot springs, earthquakes, geysers, and active volcanoes. All of these features are caused by plate tectonics.

The Alps
This chain of mountains, formed millions of years ago, crosses Europe. It is one of the most popular skiing and hiking locations on Earth.

East African Rift System
This long system of cliffs and valleys in eastern Africa is due to the rifting apart of the African plate, which will someday become two plates. The Rift System can be clearly seen from as far away as the Moon!

◄ **Figure 6-22** This satellite photo of Earth shows some of the major plates.

Plate Boundaries
- Transform
- Convergent
- Divergent

WRITING ACTIVITY

Science Log
What place on this map would you like to visit? Do some research, and write a report describing why you would like to go there. Then, explain what effect plate tectonics has on the area and what you would expect to see because of it. Start your search at www.conceptsandchallenges.com.

3 Assess

Use the Writing Activity Rubric from the Teacher's Resources to assess students' Science Logs. Fill in the rubric with the additional information below. For this assignment, students should have:

- explained how plate tectonics affected the region they wrote about, including its effect on landscape and the implications for people living in the region.
- communicated, in paragraph form, accurate information about the region.

4 Close

Writing Travel Brochures Have pairs of students create a travel brochure for one of the regions featured on the map. Their brochures should include why plate tectonics makes the region an interesting place to visit. Students can present their completed brochures to the class.

Tips for Using Technology

Using the Internet to Find Information Remind students that one important tool for finding information on the Internet is a search engine. After a user types in a keyword or phrase, such as "East African Rift System," the engine searches the Internet for those sites that contain the keyword or phrase.

Students can start their search at www.conceptsandchallenges.com. Tell them to use some of the information in the Big Idea callouts to help them get started.

• TEACHING RESOURCES • Online or Print

Teacher's Resources

Big Idea: Chapter 6, p. 10

Big Idea Planner: Study Tools, pp. 11–12

Big Idea Science Log Entry: Study Tools, p. 13

Challenges *(pp. 154–156)*

Chapter Summary

Review Before students begin the Challenges, review the summary with them. Give each pair of students one summary point to describe in their own words.

STUDY HINT Have students use index cards to create flash cards for each Key Term. Have students write the Key Term on one side of the index card and the definition on the other side. Have pairs of students review the Key Terms together.

Key Term Challenges

MATCHING

1. mid-ocean ridge (6-2)
2. seafloor spreading (6-2)
3. trench (6-3)
4. tectonic plate (6-4)
5. Pangaea (6-1)
6. magma chamber (6-6)
7. convection current (6-5)
8. rift valley (6-2)

FILL IN

9. hot spot (6-6)
10. continental drift (6-1)
11. subduction zone (6-3)
12. theory of plate tectonics (6-4)

Chapter 6 Challenges

Chapter Summary

Lesson 6-1

- Most scientists think that millions of years ago there was one giant continent called **Pangaea.**
- The theory of **continental drift** states that Pangaea split apart to eventually form today's continents.

Lesson 6-2

- Magma rising through the crust forms new crust on each side of the **mid-ocean ridges.** This process, which also creates **rift valleys,** is called **seafloor spreading.**

Lesson 6-3

- The direction in which the magnetic particles in rocks point is evidence that the ocean floor is spreading out from both sides of the Mid-Atlantic Ridge.
- Old oceanic crust is forced into ocean **trenches,** where it is changed to magma. **Subduction zones** are places where old oceanic crust is forced back down into an oceanic trench.

Lesson 6-4

- The crust of Earth and the upper, solid part of the mantle are broken into **tectonic plates** that float on the athenospere, the part of the mantle that can flow. This theory is called **plate tectonics.**

Lesson 6-5

- Scientists think that **convection currents** in Earth's mantle cause the movement of tectonic plates.
- The tectonic plates move toward or away from each other at **plate boundaries.** They may also slide past each other.

Lesson 6-6

- Earthquakes often occur where two plates slide past each other. When two plates collide, sometimes the crust crumples up to form mountains.
- Volcanoes often form near subduction zones. This is because of the melting that occurs as the plate subducts into the hotter mantle.
- Some islands are formed when plates move over **hot spots.**

Key Term Challenges

continental drift (p.138)
convection current (p.146)
hot spot (p.150)
magma chamber (p.150)
mid-ocean ridge (p.140)
Pangaea (p.138)
plate boundary (p.146)
rift valley (p.140)
seafloor spreading (p.140)
subduction zone (p.142)
tectonic plate (p.144)
theory of plate tectonics (p.144)
trench (p.142)

MATCHING **Write the Key Term from above that best matches each description.**

1. underwater mountain chain
2. process that forms new seafloor
3. long valley
4. large piece of the solid part of Earth's surface
5. single giant landmass that scientists think existed millions of years ago
6. underground pocket of molten rock
7. movement of a gas or a liquid caused by changes in temperature
8. flat area between two ridges that is formed by spreading plates

FILL IN **Write the Key Term from above that best completes each statement.**

9. The place where magma reaches the surface within a tectonic plate is called a __________.
10. Wegener's idea that the continents were once part of a giant landmass that broke apart and moved into the positions they are in today is called __________.
11. The place where old crust is forced into a trench is a __________.
12. The idea that Earth's crust is broken into pieces that float on the mantle is the __________.

154

Teaching Options

PREPARING STUDENTS FOR STANDARDIZED TESTS

Reading Strategy: Tell students to try to eliminate obviously wrong choices quickly from multiple-choice questions. This will improve their odds of guessing correctly.

Writing Strategy: Tell students to pay attention to spelling, grammar, and organization of thoughts. This will help others understand what they are trying to say.

Interpreting Visuals Strategy: Remind students to read captions carefully to fully understand what they are looking at.

ESL/ELL STRATEGY

Compound Words Suggest that students make note of any compound words in the Key Terms list. Ask them to write definitions for the individual words. Then, have students explain how the meaning of each word relates to the meaning of the compound word.

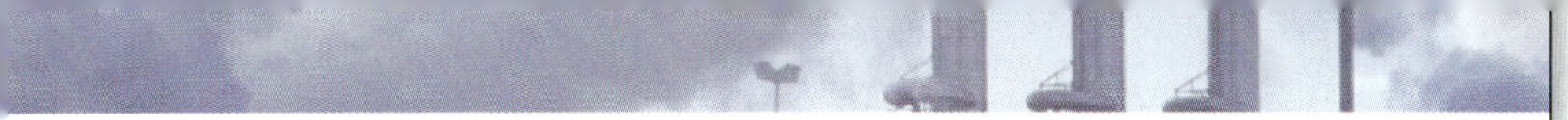

Content Challenges TEST PREP

MULTIPLE CHOICE **Write the letter of the term or phrase that best completes each statement.**

1. Wegener called the giant landmass that later formed the seven continents
   a. *Mesosaurus.*
   b. Pangaea.
   c. Gondwanaland.
   d. Tethys.
2. The oldest rocks found on the ocean floor are about
   a. 160 million years old.
   b. 160 billion years old.
   c. 4.6 million years old.
   d. 4.6 billion years old.
3. Scientists estimate that the age of Earth is
   a. 160 million years.
   b. 160 billion years.
   c. 4.6 million years.
   d. 4.6 billion years.
4. The deepest parts of the ocean floor are
   a. rift valleys.
   b. trenches.
   c. mid-ocean ridges.
   d. subduction zones.
5. When continental crust is pushed upward,
   a. a trench forms.
   b. a mountain forms.
   c. a rift valley forms.
   d. an island forms.
6. Tectonic plates sliding past each other are most likely to cause
   a. an earthquake.
   b. a mountain to form.
   c. a volcano to form.
   d. an island to form.
7. An animal fossil that is used to support the theory of continental drift is
   a. Tethys.
   b. Gondwana.
   c. *Mesosaurus.*
   d. Panthalassa Laurasia.
8. In a model of Pangaea, the coastline of South America seems to fit together with the coastline of
   a. Australia.
   b. northern Asia.
   c. North America.
   d. Africa.
9. Plates are made up of continental crust and
   a. islands.
   b. oceanic crust.
   c. magma.
   d. convection currents.

TRUE/FALSE **Write *true* if the statement is true. If the statement is false, change the underlined term to make the statement true.**

10. Peaks of mid-ocean ridges that rise above the surface of the ocean form continents.
11. Earthquakes and volcanic activity often take place along mid-ocean ridges.
12. The oldest crust on the ocean floor is found in the center of mid-ocean ridges.
13. Continental crust is younger than oceanic crust.
14. The theory that Earth's crust is broken into plates that float on the mantle is called continental drift.
15. The movement of tectonic plates is caused by convection currents.

Content Challenges

MULTIPLE CHOICE

1. b (6-1)
2. a (6-3)
3. d (6-3)
4. b (6-3)
5. b (6-6)
6. a (6-6)
7. c (6-1)
8. d (6-1)
9. b (6-4)

TRUE/FALSE

10. false; islands (6-2)
11. false; rift valleys (6-2)
12. false; youngest (6-2)
13. false; older (6-2)
14. false; theory of plate tectonics (6-4)
15. true (6-5)

ALTERNATIVE ASSESSMENT

Relating Information Have students write a Science Log entry relating the theory of plate tectonics to the formation of mountains, volcanoes, or trenches.

PORTFOLIO ASSESSMENT

Making Student Portfolios Students can demonstrate their comprehension of the concepts in this chapter by making a portfolio. The Lesson Reviews and Weekly Journal are some of the reproducibles from the Teacher's Resources that they can include in their portfolios. They can also include their computer-generated diagram of convection currents from the Big Idea. You may wish to make certain assignments planned portfolio items. Tell students ahead of time, so they can give it their best efforts. You can use the Portfolio Assessment Rubric also found in the Teacher's Resources to assess students' portfolios.

Concept Challenges

WRITTEN RESPONSE

1. **APPLY:** Tectonic plates are sliding past each other, causing the continents to move. (6-6)
2. **EXPLAIN:** At mid-ocean ridges, new oceanic crust is being formed. At subduction zones, oceanic crust is being destroyed. (6-2, 6-3)
3. **INFER:** Possible answer: The heat from inside Earth could melt the rocks, thereby turning them into magma. (6-4)
4. **EXPLAIN:** It explains how and why the continents move. It states that Earth's crust is broken into tectonic plates that float on the lower mantle. The continents move because they are carried along on the moving plates. (6-4)
5. **ANALYZE:** Old crust is destroyed at about the same rate as new crust is being made. (6-3)
6. **CALCULATE:** Earth is 4.6 billion years old. To find the number, subtract 150 million from 4.6 billion: (6-3)

$$\begin{array}{r} 4{,}600{,}000{,}000 \\ -\quad 150{,}000{,}000 \\ \hline 4{,}450{,}000{,}000 \end{array} \text{ or 4.45 billion}$$

INTERPRETING VISUALS (6-1)

7. Europe and North America
8. matching diamonds and matching rocks
9. The diagram looks like all the landmasses could have been attached.
10. The modern coastlines of South America and Africa fit together like jigsaw-puzzle pieces.

Chapter Project Wrap-Up

PLATE TECTONICS COLLAGE

Students' projects can be exhibited around the room. Have students present their collages and explain what each picture represents. You can use the Individual Activity Rubric from the Teacher's Resources to assess students' projects. Fill in the rubric with the additional information below. For this project, students should have:

- included pictures of volcanoes, mountains, areas affected by earthquakes, and islands that formed over hot spots.
- organized a logical and neat overall presentation.

Concept Challenges TEST PREP

WRITTEN RESPONSE **Answer each of the following questions in complete sentences.**

1. **APPLY:** How is the San Andreas Fault explained by the theory of plate tectonics?
2. **EXPLAIN:** How does the activity at subduction zones balance the activity at mid-ocean ridges?
3. **INFER:** How could mineral deposits formed along mid-ocean ridges become magma?
4. **EXPLAIN:** How does the theory of plate tectonics combine the idea of continental drift and seafloor spreading?
5. **ANALYZE:** Why doesn't Earth get bigger as a result of seafloor spreading?
6. **CALCULATE:** The oldest rocks found on the ocean floor are only about 160 million years old. For how many years of Earth's history are we missing ocean floor samples?

INTERPRETING VISUALS **Use Figure 6-23 to answer the following questions.**

7. Which two continents have similar mountain ranges?
8. What evidence on the continents of South America and Africa supports continental drift?
9. How does the evidence shown in the map support the idea of continental drift?
10. How do the shapes of the coastlines of South America and Africa support the idea of continental drift?

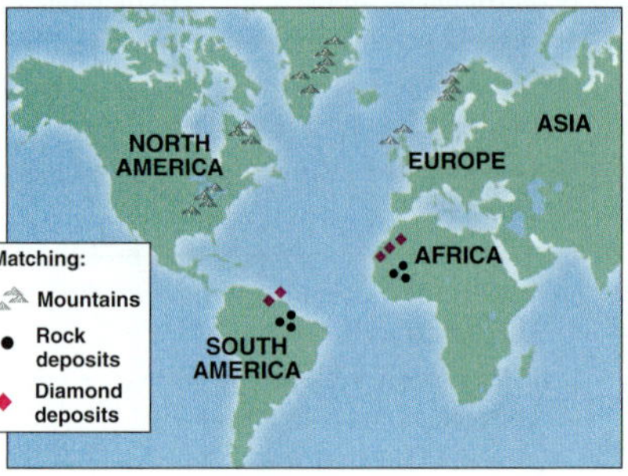

▲ Figure 6-23

Teaching Options

• TEACHING RESOURCES •
Online or Print

Teacher's Resources

Key Term Review: Chapter 6, p. 11

Chapter 6 Test: pp. 12–14

Individual Activity Rubric: Teaching Tools, p. 5

Weekly Journal: Study Tools, p. 19

Portfolio Assessment Rubric: Teaching Tools, p. 10

Chapter Self-Check: Teaching Tools, p. 11

PLANNING GUIDE

◆ **TEACHING THE CHAPTER** This chapter should take approximately 6–9 days to complete instruction and assessment.

	Skills and Features	Projects and Activities	Achieve Science Literacy	Meet Individual Needs
Chapter 7 Opener p. T157	LIST	• Chapter Project		
Lesson 7-1 **What is mechanical weathering?** pp. T158–T159 Standards: D1d, D2a	CLASSIFY, NAME, HYPOTHESIZE • Investigate • Building Science Skills • Real-Life Science	• Extending the Lesson • More on the Web • Integrating Life Science	• Study Hint	• ESL/ELL Strategy
Lesson 7-2 **What is chemical weathering?** pp. T160–T161 Standards: D1d, D1g, D1k, D2a	EXPERIMENT, EXPLAIN, INFER • Building Science Skills • Science and Technology	• More on the Web	• Study Hint • Reading Strategy • Ongoing Assessment	• ESL/ELL Strategy Teacher's Resources* Enrichment Activity Concept Map
Lab Activity **Modeling the Weathering of Sedimentary Rocks** pp. T162–T163 Standards: D1c, D1d, D1g, D2a	OBSERVE, INFER, HYPOTHESIZE • Set-Up Time: 15 min • Lab Time: 45 min	• Integrating Physical Science • Extend the Lab Activity	• Tips for Using Technology	• ESL/ELL Strategy Teacher's Resources* Lab Activity Report
Lesson 7-3 **What factors affect the rate of weathering?** pp. T164–T165 Standards: D1c, D2a, D1d, D1g, D1k	IDENTIFY, PREDICT, LIST • Web InfoSearch • Hands-On Activity		• Study Hint	• Reteach Teacher's Resources* Enrichment Activity
Lesson 7-4 **How does soil form?** pp. T166–T167 Standard: D1e	IDENTIFY, DESCRIBE, NAME • Web InfoSearch • Integrating Life Science	• Extending the Lesson • Integrating Environmental Science • Lab Challenge	• Study Hint	• ESL/ELL Strategy • Reteach Teacher's Resources* Enrichment Activity Flowchart

*www.conceptsandchallenges.com, password: ccearth09

Planning Guide continues on next page.

Standards: For details on the correlation to National Science Standards see pages *xx–xxi*.

	Skills and Features	Projects and Activities	Achieve Science Literacy	Meet Individual Needs
Lesson 7-5 **How do soils differ?** pp. T168–T169 Standard: D1e	DESCRIBE, CONTRAST, IDENTIFY • Investigate • Web InfoSearch	• Cooperative Learning • Extending the Lesson	• Study Hint • Writing Hint • Ongoing Assessment	Teacher's Resources* Enrichment Activity
Lesson 7-6 **What is a soil profile?** pp. T170–T171 Standard: D1e	IDENTIFY, DESCRIBE, RELATE • Building Reading Skills • Hands-On Activity	• Extending the Lesson • More on the Web • Active Art	• Reading Strategy	• Reteach Teacher's Resources* Soil Profile
Big Idea **How does the chemistry of soil affect plant growth?** pp. T172–T173 Standard: D1e	RESEARCH, COMMUNICATE, ANALYZE • Science Log: Writing Activity	• Big Idea • Extending the Lesson • Close Activity		• Reteach Teacher's Resources* Big Idea Planner
Chapter 7 Challenges pp. T174–T176	• Chapter Summary • Key Term Challenges • Content Challenges • Concept Challenges	• Chapter Project Wrap-Up	• Study Hint • Preparing Students for Standardized Tests	• ESL/ELL Strategy Teacher's Resources* Chapter Self-Check Weekly Journal

*www.conceptsandchallenges.com, password: ccearth09

Teacher Notes

Chapter 7 Weathering and Soil

▲ **Figure 7-1** The Great Sphinx at Giza in Egypt

Ancient peoples carved stone statues to honor their gods and heroes. They believed the statues would last forever. Today, we know that rocks, statues, and even whole mountains do break down over time. They are worn away in a process known as weathering.

▶The sphinx is a mythical creature. It has a lion's body and a human head. What signs of weathering show on this limestone carving of the Great Sphinx?

Contents

Teaching Options

ASSESSMENT PLANNER

For assessment in the Student Edition, see the following pages:

Content Assessment
Checking Concepts: pp. 159, 161, 165, 167, 169, 171
Thinking Critically: pp. 159, 161, 165, 167, 169, 171
Chapter Challenges: pp. 174–176

Alternative Assessment
Building Skills: pp. 159, 161, 171
Web InfoSearch pp. 165, 167, 169
Science Log p. 173

Performance-Based Assessment
Hands-On Activity: pp. 165, 171
Lab Activity: pp. 162–163

• TEACHING RESOURCES • Online or Print

Teacher's Resources
Lesson Review: Ch. 7, pp. 2, 3, 4, 6, 7, 9
Enrichment: Ch. 7, pp. 5, 8
Key Term Review: Ch. 7, p. 11
Chapter 7 Test: pp. 12–13
Laboratory Manual: pp. 45–48

Chapter 7 Weathering and Soil

Chapter Overview

In this chapter, students will learn how Earth's surface is worn away by mechanical and chemical weathering. They will describe the factors that affect the rate of weathering. They will explain how soil forms and describe different soils. Finally, they will identify and describe the parts of a soil profile.

About Figure 7-1 The Sphinx was carved in place thousands of years ago from soft, natural limestone. For much of its existence, the Sphinx was covered in sand, which trapped water from rain, runoff, and floods. This moisture promoted chemical weathering of the statue. When the statue was exposed to the air, the wind sandblasted it, causing mechanical weathering.

Answer Possible answer: the Sphinx's missing nose and worn facial features and the absence of original details on its body

Linking Prior Knowledge

For this chapter, students should recall:

- that minerals are made of different elements and compounds (Lesson 2-3).
- how minerals react with acid (Lesson 2-5).
- that different minerals are made up of different chemicals, so they react differently (Lesson 3-2).
- how rocks are broken down (Lesson 3-8).

Chapter Project

WEATHERING LOG

MATERIALS: paper, pencil, posterboard, markers

Have students keep a log as they go around their community looking for signs of weathering. Ask them to keep a list of what they find: weathered buildings or statues, rusted cars, broken sidewalks, and so on. Have them indicate if the weathering was chemical or mechanical and suggest what they think might have caused it. Tell them to create a chart based on their notes, which they can put on a poster and illustrate if they wish. **learning styles: visual, tactile**

7-1 *(pp. 158–159)*

1 Introduce

INVESTIGATE

TIME: 10–15 minutes

PURPOSE: Students will model mechanical and chemical weathering.

MATERIALS: safety goggles, water, 2 jars with lids, 2 pieces of hard candy

PROCEDURE: Caution students to wear safety goggles. Be sure that students tightly secure the jar lids. Instruct them to vigorously shake one of the jars while they leave the other undisturbed. Have students leave the undisturbed jar overnight and then observe whether their prediction was correct.
learning styles: visual, tactile

THINK ABOUT IT: The candy in the shaken jar dissolved. The candy in the undisturbed jar is still whole. Eventually, this candy will also dissolve. This experiment models both types of weathering that will be discussed in this chapter.

STUDY HINT Tell students to set up two columns in their notebooks, one headed *Mechanical Weathering* and the other *Chemical Weathering*. As they read this lesson and the next, they should list examples of each kind of weathering in the appropriate column.

Linking Prior Knowledge Review the rock cycle (Lesson 3-8). Ask: *What are some ways rocks are broken down?* (wind, rain, and ice) Tell students that they will learn more about these processes in this lesson.

2 Teach

Discussion Ask students if they have ever noticed cracked or uplifted sidewalks in areas near trees. Have students discuss what they think caused the cracking and uplifting. Point out that they were caused by the action of tree roots.
learning style: auditory

Answers to questions

1. **NAME:** chemical and mechanical
2. **CLASSIFY:** mechanical weathering
3. **IDENTIFY:** Water expands when it freezes.
4. **CLASSIFY:** mechanical weathering

7-1 What is mechanical weathering?

INVESTIGATE
Modeling Weathering
HANDS-ON ACTIVITY

1. Pour 1 cup (250 mL) of tap water into each of two jars.
2. Add 1 piece of hard candy to each jar and close the lids.
3. Put one jar where it will not be disturbed.
4. Shake the other jar until the candy dissolves.

THINK ABOUT IT: What happened to the candy in the jar you shook? How is it different from the other candy? What do you think will eventually happen to the candy in the other jar?

Objective

Explain the ways that Earth's surface is worn away by mechanical weathering.

Key Terms

weathering: breaking down of rocks and other materials on Earth's surface

chemical (KEHM-ih-kuhl) **weathering:** weathering that changes the chemical makeup of rocks

mechanical (muh-KAN-ih-kuhl) **weathering:** weathering in which the chemical makeup of rocks does not change

ice wedging: mechanical weathering caused by the freezing and melting of water

Weathering The breaking down of rocks and other materials on Earth's surface is called **weathering.** Newly made bricks are bright red. They have sharp corners and edges. Old bricks are darker red. They have rounded corners and edges. What changes the bricks' color and shape? The process of weathering.

▼ **Figure 7-2** Over long periods of time, weathering changes the shape of mountains.

There are two kinds of weathering. **Chemical weathering** occurs when the chemical makeup of the rocks changes. The sizes and shapes of the rocks may also change.

Mechanical weathering involves only physical changes, such as size and shape. The chemical makeup of the rocks does not change. Mechanical weathering occurs as a result of temperature changes, ice wedging, and root action.

1 **NAME:** What are the two different types of weathering?

Temperature Changes Rocks may be broken apart by changes in temperature. Heat makes things expand, or become larger. Cooling makes things contract, or become smaller. During the day, heat can cause the outside of a rock to expand. The inside of the rock stays cool. At night, the outside of the rock cools and contracts. The repeated heating and cooling of the rock's surface each day may cause pieces of it to flake or peel off.

2 **CLASSIFY:** What kind of weathering do changes in temperature cause?

Ice Wedging The repeated freezing and melting of water causes **ice wedging.** Water enters cracks in rocks. When the temperature drops below freezing (0°C), water freezes and expands. The expanding water is like a wedge, making the crack wider. After the repeated freezing and melting of water, the rock breaks apart. Potholes in streets are caused by ice wedging.

3 **IDENTIFY:** What property of water causes it to act like a wedge when it freezes in cracks?

Teaching Options

INTEGRATING LIFE SCIENCE

Digestive Breakdown Explain that during the digestive process, both the mechanical and chemical breakdown of food takes place. Mechanical digestion physically breaks down food into small pieces. Chemical digestion breaks large food molecules into smaller ones. Have students create a chart that lists the various types of mechanical weathering and gives similar digestive processes.
learning styles: auditory, visual

ESL/ELL STRATEGY

Clarifying Weathering If possible, bring a new brick and an old brick to class. Display the bricks and ask students to describe each. Identify weathering as the reason bricks—and rocks—change shape and color. Point out the signs of weathering on the old brick.
learning style: visual

QUICK CONVERSIONS

Statue Statistics	Metric/SI	English
Height	4 m	13 ft
Mass	12 t	26,455 lb

Root Action Plant roots in search of minerals and water can grow into cracks in rocks. Look at the sidewalk near a tree. Is the sidewalk cracked or raised? Tree roots may have caused the cracks by lifting up the sidewalk. The pressure of the growing root can make the cracks in rocks larger. As the roots grow, they can break the rocks apart.

▲ **Figure 7-3** Growing plant roots can lift up sidewalks.

4 **CLASSIFY:** Does root action cause mechanical or chemical weathering?

CHECKING CONCEPTS

1. Is breaking a rock into pieces chemical or mechanical weathering? Explain your choice.
2. What kind of weathering causes potholes?
3. What causes ice wedging?

THINKING CRITICALLY

4. **HYPOTHESIZE:** Why should trees not be planted near underground water pipes?
5. **ANALYZE:** The breaking apart of rocks by growing roots is sometimes called root-pry. Is this is a good name? Explain.
6. **PREDICT:** Would you find potholes caused by weathering in tropical regions? Why or why not?

BUILDING SCIENCE SKILLS

Classifying Use the definitions of chemical change and mechanical change to classify each of the following.

a. A rock breaks into many pieces.
b. A match burns.
c. A sidewalk is cracked by a tree root.

Real-Life Science
THE STATUES OF EASTER ISLAND

▲ **Figure 7-4** The statues of Easter Island show the effects of weathering.

Easter Island is in the South Pacific Ocean. Its name comes from the fact that a Dutch explorer landed there on Easter Day in 1722. Tourists visit Easter Island to see its most famous attraction—the statues. The statues are carved in the form of humans. A typical statue might stand 4 m tall and weigh more than 12 metric tons. The statues were carved from hardened volcanic ash. This material was probably taken from the volcanic cone Rano Raraku, on the southern coast of the island.

There are about 800 to 1,000 of these statues on Easter Island. Several other Pacific island cultures have statues like these. It is believed that all of these statues were carved for religious purposes.

Perhaps one-third of the statues on Easter Island are still standing. Weathering has caused the stone to slowly crumble away. Efforts are being made to preserve the statues.

Thinking Critically How do you think these statues could be preserved?

MORE ON THE WEB

Modeling Weathering Have students create and design their own "famous" landmark that has been affected by weathering. Have each student draw pictures of the landmark, give the history, and explain how weathering has shaped it. To begin their research, have students visit www.conceptsandchallenges.com.

• TEACHING RESOURCES • Online or Print

Teacher's Resources
Lesson Review: Chapter 7, p. 2

EXTENDING THE LESSON

Weathering Survey Explain that potholes form as a result of mechanical weathering of asphalt, and that sidewalks often break or lift up because of tree roots. Have students survey the streets in their neighborhoods and record the number of potholes, sidewalk cracks, and uplifted sections of pavement they observe. **learning style: visual**

3 Assess

CHECKING CONCEPTS ANSWERS

1. mechanical, because the chemical makeup of the rock does not change
2. mechanical weathering
3. the repeated freezing and melting of water

THINKING CRITICALLY ANSWERS

4. **HYPOTHESIZE:** Tree roots can crack the ground and pipes.
5. **ANALYZE:** yes, because roots grow into cracks in rocks and pry them apart
6. **PREDICT:** no, because potholes are caused by repeated freezing and melting of water

4 Close

BUILDING SCIENCE SKILLS ANSWERS

Classifying **a.** mechanical change **b.** chemical change **c.** mechanical change

Real-Life Science
THE STATUES OF EASTER ISLAND

Archaeology Connection The statues of Easter Island are about 600 years old. They are thought to represent the spirits of ancestors or high-ranking men on Easter Island. Of the existing statues, 288 were hauled from a quarry and erected at various sites, 92 are lying in transit, and 397 are still in the quarry where they were carved. It is not known why so many were never moved from the quarry. The smallest statue is 1.15 m tall (3.76 ft). The largest is 21.92 m tall (71.93 ft) and weighs 160 to 182 metric tons (176 to 201 tons).

Discussion Have students use information in the feature to calculate how many of the original statues on Easter Island remain standing today. (around 300) Discuss the types of weathering that likely eroded the statues of Easter Island. **learning style: logical/mathematical**

Thinking Critically Answer
Possible answer: by covering the statues to protect them from further weathering or coating them with a protective seal of some type

7-2 *(pp. 160–161)*

1 Introduce

STUDY HINT Ask students to think of situations relevant to what they are reading in the lesson and write them in their notebooks.

Linking Prior Knowledge Have students recall what they learned about minerals (Lesson 2-5). Ask: *How do acids react with calcite, limestone, and marble?* (The substances fizz.)

2 Teach

Demonstration Show students a glass of club soda. Ask: *What is in the bubbles in club soda?* (carbon dioxide) Use litmus paper to prove that club soda (carbonic acid, H_2CO_3) is acidic. Demonstrate the effect of carbonic acid on calcite by placing a small piece of calcite in club soda. Have students note the formation of small bubbles on the calcite as the acid reacts with it.
learning style: visual

Discussion Point out that, not only does water chemically weather rocks during hydrolysis, but water is needed for carbonation. Discuss the role that water plays in each of the processes mentioned.
learning style: auditory

Answers to questions

1. **NAME:** oxygen, water, and acids
2. **DEFINE:** the chemical change that occurs when oxygen combines with another substance
3. **EXPLAIN:** Using water, it chemically changes substances.
4. **IDENTIFY:** carbonic acid

READING STRATEGY On the board, write the Key Term *hydrolysis*. Point out that this term consists of the word parts *hydro-* and *-lysis*. Have students use a dictionary to find the definition of each word part. (*hydro-*, "having to do with water"; *-lysis*, "breaking down or decomposition")

7-2 What is chemical weathering?

Objective
Explain how Earth's surface is worn away by chemical weathering.

Key Terms
oxidation (ahk-sih-DAY-shuhn): chemical change that occurs when oxygen reacts with another substance
hydrolysis (hy-DRAHL-uh-sihs): chemical reaction that occurs when minerals with little water content react with water
carbonation (kahr-buh-NAY-shuhn): chemical reaction that occurs when carbonic acid reacts with certain minerals

Kinds of Chemical Weathering In chemical weathering, substances in water cause substances in rocks to dissolve. This action weakens the structure of the rocks.

Chemical changes occur when minerals in rocks are broken down into other substances. They also occur when minerals are added to or removed from rocks. Chemical weathering is usually caused by reactions with oxygen, water, or acids.

▲ **Figure 7-5** The Statue of Liberty contains a large amount of copper. Its thin copper skin reacts with oxygen in the air to form a green coating.

1. **NAME:** What substances cause most chemical weathering?

Oxidation One kind of chemical weathering is **oxidation.** Oxidation occurs when oxygen combines with another substance. New substances called oxides are formed. Rust is an iron oxide (Fe_2O_3). Many rocks, such as pyrite and magnetite, have minerals that contain iron. When these rocks are exposed to the air, the iron undergoes oxidation. The rocks weaken and may crumble.

▲ **Figure 7-6** Like copper statues, rocks can be oxidized. Oxidized rocks change color. Here, the iron in the rock has reacted with oxygen in the air. The red streaks are oxidized iron, or rust.

2. **DEFINE:** What is oxidation?

Water Most chemical weathering is caused by water. Water reacting chemically with a mineral that has little water content is called **hydrolysis.** Many minerals in rocks undergo hydrolysis. For example, feldspar may combine chemically with water and change to clay. When heated or dried even a little, minerals formed by hydrolysis may fall apart.

3. **EXPLAIN:** What does hydrolysis do?

Acids Carbon dioxide can dissolve in rain to form weak carbonic acid. When this acid comes into contact with certain minerals, a chemical change called **carbonation** occurs. The mineral calcite is changed by this process. Limestone and marble are made of calcite. Buildings made of these materials are often weathered by carbonation.

160

Teaching Options

! FAST FACTS *Dry Weathering*

Water can cause weathering, but so can a lack of water. As water evaporates from the cracks inside of rocks during dry conditions, dissolved salts crystallize out of solution. As the crystals increase in size, they can split rocks.

ONGOING ASSESSMENT

Have students draw a concept map that includes the following terms: *mechanical, carbonation, weathering, ice wedging, root action, chemical, hydrolysis, temperature changes,* and *oxidation.* An example of a concept map can be found in the Teacher's Resources.

ESL/ELL STRATEGY

Word Connections Point out to English-language learners the connection between the words *oxygen* and *oxidation* and between the words *carbon* and *carbonation*. Also discuss the use of the *-tion* suffix in relation to these weathering processes. For more support, look these up in a bilingual dictionary.

▲ **Figure 7-7** Many buildings and building decorations are made of limestone. Limestone can be damaged by carbonation.

Some green plants produce weak acids. Mosses, which grow on rocks, are an example. The acids can wear away a rock's surface. As acid seeps into cracks in a rock, it can break the rock apart. Decaying organisms also produce acids, which contribute to chemical weathering. Chemical weathering changes the chemical composition of materials. This can weaken the material's internal structure.

4 IDENTIFY: What acid forms from carbon dioxide and water?

CHECKING CONCEPTS

1. How are minerals in a rock oxidized?
2. What does carbonation do to minerals?
3. How can plants cause chemical weathering?
4. What process will most likely chemically weather an iron-rich rock?

THINKING CRITICALLY

5. INFER: Why do you think structures containing iron are painted?
6. CLASSIFY: Identify the kind of chemical weathering described in each case.
   a. A marble statue's nose can no longer be seen.
   b. A bicycle fender rusts.

BUILDING SCIENCE SKILLS

Experimenting Place one piece of steel wool outside your home. Make sure it is exposed to the weather. Place another piece in your room. Wait two weeks. Check the pieces often. What changes did you observe? Are there any differences in the two pieces of steel wool? If so, why?

Science and Technology

PREVENTING CORROSION

Corrosion is the chemical weathering of materials such as metals. One common kind of corrosion is rust. Rust damages iron and steel. It weakens these metals and changes their color. Rusting affects highways and bridges, underground pipes and tanks, and building pipes.

One of the best ways to slow rusting is to paint metals. Paint stops water from corroding the metals. Building steel is often coated with the metal zinc to protect it. Zinc resists corrosion. Steel coated with zinc is called galvanized steel.

▲ **Figure 7-8** Rusting made this bridge unsafe.

Rust is also a big problem in cars and trucks. Rust destroys more cars each year than accidents do. New cars come with some corrosion protection. A wax-based spray can be applied to a car to give it added protection. A car can also be sprayed each year with an oil-based product that contains antirust agents. These methods help keep out the moisture that can cause corrosion.

Thinking Critically What might happen if water pipes in a building rusted?

MORE ON THE WEB

Identifying Weathering Processes Have students use the Internet to learn more about the different processes of mechanical and chemical weathering. To start their research you can have students visit www.conceptsandchallenges.com.

• TEACHING RESOURCES • Online or Print

Teacher's Resources

Lesson Review: Chapter 7, p. 3

Enrichment: Chapter 7, p. 5

Concept Map: Study Tools, p. 4

3 Assess

CHECKING CONCEPTS ANSWERS

1. When minerals containing iron are exposed to air, they combine with oxygen.
2. It chemically changes them.
3. Some plants produce weak acids that can seep into cracks and break rocks apart.
4. oxidation

THINKING CRITICALLY ANSWERS

5. INFER: Painting the iron helps prevent rusting.
6. CLASSIFY: **a.** carbonation **b.** oxidation

4 Close

BUILDING SCIENCE SKILLS ANSWER

Experimenting Possible answer: The steel wool rusted. The steel wool placed outside rusted more because it was exposed to more moisture.

Science and Technology

PREVENTING CORROSION

Technology Connection Have interested students research rustproofing of new cars. Have students find out what substances are used and why. (waxes and oils that coat the surface, preventing oxygen and water from reaching the metal) Students should learn how much rustproofing adds to the cost of a new car. Have them report their findings to the class.
learning style: auditory

Discussion Bring in pictures of rusted objects, such as bridges, cars, and steel girders. Discuss the problem of rusting. Describe some of the measures that are taken to prevent rusting.
learning style: visual

Thinking Critically Answer
The pressure in the pipes could cause the pipes to burst, resulting in flooding.

Lab Activity

(pp. 162–163)

1 Prepare the Lab

Modeling the Weathering of Sedimentary Rocks

SET-UP TIME: **LAB TIME:**

BACKGROUND: The land surface of Earth is constantly changing. Rain, wind, ice, chemicals, plants, and gravity act upon rocks and soil to wear them away. The beginning of this change is a process called weathering, the breakdown of rocks and other materials by chemical and mechanical action. The product of this process is sediment that later may become new rock.

PURPOSE: Students will observe a model of chemical weathering and describe its effects.

ALTERNATIVE MATERIALS: Drinking glasses can be used in place of beakers. Sedimentary rock samples made by students in the activity on page 75 can be used in place of the sugar cubes. However, the chemical weathering of these samples will be much slower. Instead of the dropper bottles, set up a very slow siphon with a 2-L bottle and a hose clamp. The bottle should be elevated above the sample and the clamp adjusted so that a slow drip of water hits the surface of the sample. Put a catch bucket below for runoff water. Noticeable changes will take place overnight.

SCIENCE SKILLS: Students will **model** a sample of sedimentary rock and **observe** and **measure** the changes that take place when the sample is subjected to chemical weathering.

ADVANCE PLANNING: Display rocks that show signs of weathering. Bring in pictures of weathering, such as broken curbs, worn stone steps, sidewalk cracks with weeds, and worn tombstones.

2 Guide the Lab

PROCEDURE: Begin the lab by displaying weathered rocks. Ask students to speculate on what might have caused the weathering. Discuss the weathering in the pictures you brought in.

LAB ACTIVITY

Modeling the Weathering of Sedimentary Rocks

Materials

Safety goggles
Lab apron
30 Sugar cubes
4 Craft sticks
Beaker
Dropper bottle
Water
Red or blue food coloring

▲ **STEP 3** Stack the sugar cubes in two rows.

BACKGROUND

If you ever walk through an old cemetery, you will see that many markers are hard to read. Walk along an old sidewalk, and you will see many cracks. Stone steps leading to an old building will be thinner in the middle than at the edges. Old stone statues look worn and cracked. Weathering has caused these changes. Weathering also changes rocks that make up the surface of Earth.

PURPOSE

In this activity, you will create a model of a small section of Earth's crust out of sugar cubes. Sugar cubes are made of small crystals joined together.

PROCEDURE

1. Copy the chart in Figure 7-9. In it, describe the appearance of a sugar cube.
2. Place the four craft sticks side by side across the rim of the beaker. ⚠ CAUTION: Be careful when working with glass.
3. Carefully place two side-by-side rows of five sugar cubes each on the sticks. The rows should touch each other.
4. Gently stack two more rows of cubes on top of the first rows.
5. Put on your safety goggles and a lab apron. Fill the dropper bottle with water mixed with one drop of food coloring.

162

Teaching Options

INTEGRATING PHYSICAL SCIENCE

Solutions Explain to students that when a solid (such as the sugar cubes) dissolves in a liquid (such as water), a substance called a solution forms (such as the sugar water). The solid that dissolves is called a solute, and the liquid is called a solvent. When a solute dissolves, it does not actually disappear. Its molecules spread evenly throughout the solvent. Water is sometimes referred to as the "universal solvent" because it can dissolve many different solutes. Have students give examples of other solutions, naming the solutes and solvents involved in their formation.
learning style: auditory

ESL/ELL STRATEGY

Noting the Icons English-language learners should pay close attention to the icons at the beginning of the Procedure part of the lab activity. The icons indicate the precautions that should be taken during the lab.

▲ **STEP 6** Squeeze drops of water on the sugar cubes.

▲ **STEP 8** Observe what happens each time you add more water to your sugar cube structure.

6. Moving up and down the rows, squeeze single drops of water on the center of each cube. Repeat this step four times until each cube has received five drops.

7. Observe what happens to the sugar cubes. Write your observations in the chart.

8. Repeat Step 6 and 7 at least five times. Describe the results each time.

Number of water drops/cube	Total number of water drops/cube	Observations	Sugar cube appearance

▲ **Figure 7-9**

CONCLUSIONS

1. OBSERVE: What happened to the cubes?
2. OBSERVE: Which of the sugar cubes changed first?
3. INFER: What caused the sugar cubes to change?
4. INFER: Why did the walls between the rows of sugar cubes start dissolving?
5. HYPOTHESIZE: Are you modeling mechanical or chemical weathering?

Show students how to stack the sugar cubes on the craft sticks. Make sure that students understand that water is a chemical (mild acid) that can dissolve many materials, given enough time. Also make sure that students understand that the sugar cubes are not chemically changed by water, though they are dissolved. Provide students with a copy of the Lab Activity Report found in the Teacher's Resources to record their observations and conclusions.

TROUBLESHOOTING: Students may focus on dropping the water and miss some of the effects they are creating. Remind students to pause for observations after adding five drops of water to each of the top sugar cubes. Check to see that the stacks of cubes are close together.

SAFETY TIP: Make sure that students do not eat the sugar cubes. The cubes will be handled extensively during the activity and will be dirty. Furthermore, the sugar water collecting in the beaker will be unsanitary.

EXPECTED OUTCOME: Students will observe changes in the sugar cubes as water dissolves the crystals.

LAB HINT: Students can observe the faces on the two rows of sugar cubes by gently sliding away the craft sticks supporting one of the rows. After, students can slide the sticks back into position.

3 Conclude the Lab

1. OBSERVE: The cubes began dissolving.
2. OBSERVE: those in the top layer
3. INFER: Chemical weathering; The water began dissolving the sugar.
4. INFER: Water from both sides joined in the middle and flowed to the bottom.
5. HYPOTHESIZE: chemical weathering

Use the Laboratory Activity Rubric from the Teacher's Resources to assess students' lab activities. Fill in the rubric with the additional information below. Students should have:

- filled in the chart with data collected during the activity.
- correctly answered the questions.

4 Extend the Lab

Challenge students to collect evidence of chemical weathering around the school. They can sketch what they discover, take pictures using cameras, or take electronic pictures to transfer to classroom computers for multimedia presentations.

Tips for Using Technology

Using Word Processing to Record Data Students can use the table function of a word processing program to record their data. A simple word processing program allows them to do basic tasks, such as key entering and editing text. A more complex word processing program allows students to make tables, draw pictures, do mathematical calculations, and make graphs. Have students use word processing software to create a table to record their lab activity data. Students can print out their document to share with the class.

• TEACHING RESOURCES • Online or Print

Teacher's Resources

Lab Activity Report: Study Tools, pp. 16–17

Laboratory Activity Rubric: Teaching Tools, p. 8

7-3 *(pp. 164–165)*

1 *Introduce*

STUDY HINT Before beginning this lesson, have students preview it by reading the title and boldface subheads. Have them predict the answer to the title question.

Linking Prior Knowledge Remind students that they learned that different minerals are different chemicals, so they react differently with acid in the atmosphere or in water (Lesson 2-3). Because rocks are made of minerals with different properties and reactions, rocks also weather in different ways.

2 *Teach*

Demonstration Use a sugar cube to illustrate the effect of surface area on the rate of weathering. Break the cube in half. Crush one half, and drop both halves into separate beakers of water. Stir the mixture, and have students note the speed at which each sample dissolves. (The sample that was crushed dissolves faster.) Hold up another sugar cube and have students count its surfaces (six). Break the cube in half and have students note the two new surfaces. Guide students to see that, as you break the cube into smaller and smaller pieces, you expose more surfaces and speed up the rate of dissolving. Compare this process to weathering.
learning style: visual

Discussion Point out the three factors that affect the rate of weathering: climate, surface area, and rock composition. Make sure that students understand why weathering occurs quickest in hot, wet climates. Review the role that water plays in chemical weathering.
learning style: auditory

Answers to questions

1. **IDENTIFY:** the amount of water in the air and the temperature
2. **DESCRIBE:** The more surface area a rock has, the more quickly it will weather.
3. **ANALYZE:** minerals
4. **LIST:** carbonic acid

7-3 What factors affect the rate of weathering?

Objective
Identify three factors that affect the rate of weathering.

Key Term
acid rain: rain containing acids produced by water chemically combining with certain gases

Climate The amount of water in the air and the temperature of an area are both part of an area's climate. Moisture speeds up chemical weathering. The more water there is in the air, the faster something weathers. Weathering occurs fastest in hot, wet climates. It occurs very slowly in hot, dry climates. Without liquid water, most chemical weathering cannot occur. Without temperature changes, ice wedging cannot occur. In very cold, dry areas, there is little weathering.

1 **IDENTIFY:** What two climate factors affect the rate of weathering?

Surface Area Most weathering occurs on exposed surfaces of rocks and minerals. Picture a large block of granite. It has six exposed surfaces. If the block is cut in half, two more surfaces are exposed. If each half is cut in half, four more surfaces are exposed. The more surface area a rock has, the more quickly it will weather. So, the smaller pieces of granite will weather faster than the large block of granite.

▲ **Figure 7-10** A block has more surface area when it is cut into smaller pieces.

2 **DESCRIBE:** How does surface area affect the weathering of rocks?

Rock Composition Some minerals that hold rocks together weather more quickly than others. Quartz, for example, resists weathering. Feldspar weathers quite easily. So, rocks rich in quartz will weather slowly, and rocks rich in feldspar will weather much faster. Some rocks, such as limestone, have large amounts of calcite in them. These rocks weather very quickly, often by carbonation. Rocks with iron generally weather faster than rocks without iron.

3 **ANALYZE:** What substances in rocks cause them to weather at different rates?

Chemical Reaction Carbon dioxide in the air becomes dissolved in rainwater. This forms carbonic acid, which is a weak acid. As the rainwater moves through soil, the carbonic acid dissolves calcite, a mineral found in marble and limestone. Limestone caves show the effects of this kind of weathering.

◄ **Figure 7-11** Formations in limestone caves show the effects of chemical weathering. The calcite solution drips into the cave, forming stalagtites, such as those shown here, and stalagmites. This formation is in Luray Caverns in Virginia's Shenandoah Valley.

Weathering is usually a slow process. However, pollution tends to speed it up. Factories and cars release carbon dioxide and other gases such as sulfur and nitrogen into the air. These become dissolved in the rainwater. Rain containing acids is called **acid rain.** Acid rain, which contains droplets of nitric and sulfuric acid, causes rocks and minerals to weather faster. The effects of acid rain can be seen on buildings and statues.

4 **LIST:** What acids are often present in acid rain?

164

Teaching Options

! FAST FACTS *Differential Weathering*

Rocks do not all weather at the same rate. Hard rocks, such as granite, weather more slowly than softer rocks, such as limestone. The rate at which minerals weather is directly related to the rate at which they crystallize from cooling magma. Minerals that form quickly at high temperatures and pressures are less resistant to chemical weathering than are minerals that form slowly at lower temperatures.

Even rocks that have the same chemical composition can weather at different rates. Rocks that form by slow cooling will form larger crystals that interlock like puzzle pieces. They tend to be more resistant to weathering. Faster cooling rock of the same material will not form the same large crystals. If cooling happens very rapidly, no crystals may form at all.

CHECKING CONCEPTS

Choose the word *increase* or *decrease* for each question.

1. The rate of weathering will __________ with moisture.
2. A rock that contains more quartz than feldspar will __________ the rate of weathering.
3. As a rock is broken into smaller pieces, the rate of weathering will __________.
4. In very polluted air, the weathering of a statue will probably __________.

THINKING CRITICALLY

5. LIST: What factors affect the rate of weathering?
6. PREDICT: Which would most likely have more weathered building stones: a city, a farm, or a desert? Explain.

Web InfoSearch

Cleopatra's Needle In 1881, a large stone monument called Cleopatra's Needle was moved from Egypt to New York City. It had stood for more than 3,000 years in Egypt. Writings carved on its surface had changed very little. Since its move to New York City, most of the writing has worn away.

▲ **Figure 7-12** Cleopatra's Needle

SEARCH: Use the Internet to find out more about Cleopatra's Needle. What happened in New York City? Why? Start your search at www.conceptsandchallenges.com. Some key search words are **Cleopatra's Needle** and **Cleopatra's New York.**

Hands-On Activity

OBSERVING THE EFFECTS OF WEATHERING ON SURFACE AREA

You will need 2 plastic cups, a hammer, 2 pieces of chalk, a paper towel, vinegar, safety goggles, and a lab apron.

1. Break a piece of chalk in half. Place one half of the chalk into a plastic cup.
2. Wrap the other piece of chalk inside of a paper towel. Break it into small pieces with a hammer. Place these pieces into the second cup. ⚠ CAUTION: Wear your safety goggles and apron.
3. Pour enough vinegar into each cup to cover the chalk. Observe what happens.

▲ **STEP 3** Cover the chalk with vinegar.

Practicing Your Skills

4. COMPARE: Of the two pieces of chalk, which has the greater surface area?
5. IDENTIFY: What kind of weathering takes place when the pieces of chalk are placed in vinegar?
6. INFER: How is the rate of weathering affected by breaking up the chalk?

RETEACH

Explaining Weathering Display several samples showing weathered and unweathered rock surfaces. Challenge students to explain what caused the samples to weather. Point out that as soon as a surface becomes exposed to air, it begins to weather.

learning style: visual

• *TEACHING RESOURCES* •
Online or Print

Teacher's Resources

Lesson Review: Chapter 7, p. 4

Enrichment: Chapter 7, p. 5

3 *Assess*

CHECKING CONCEPTS ANSWERS

1. increase
2. decrease
3. increase
4. increase

THINKING CRITICALLY ANSWERS

5. LIST: climate, surface area, and rock composition
6. PREDICT: a city, because it has the most moisture and pollution

4 *Close*

Web InfoSearch

Cleopatra's Needle Have students use the Internet to find out more about Cleopatra's Needle. Ask students: *What happened to the monument in New York? Why?* Some key search words are **Cleopatra's Needle** and **Cleopatra New York.** Students can begin their search at www.conceptsandchallenges.com.

Hands-On Activity

OBSERVING THE EFFECTS OF WEATHERING ON SURFACE AREA

TIME: 15 minutes

PURPOSE: observing, comparing, identifying, inferring

SAFETY TIP: Caution students to wear safety goggles.

ALTERNATIVE MATERIALS: Jars or plastic containers can be substituted for plastic cups.

Practicing Your Skills Answers

4. COMPARE: the hammered piece
5. IDENTIFY: chemical weathering
6. INFER: The more pieces the chalk is broken into, the faster will be the rate of weathering.

7-4 (pp. 166–167)

1 Introduce

STUDY HINT Have students read the captions and art labels in the lesson before reading the text to preview what is coming and to focus on what is important.

Linking Prior Knowledge Ask students who have worked in a garden to describe their experiences with soil. Have them use their observations to infer the contents of soil.

2 Teach

Discussion Bring in pictures of organisms that live in or on the soil, such as mosses, lichen, bacteria, fungi, earthworms, ants, moles, and prairie dogs. Display the pictures in the front of the classroom. Ask students if they know what all these organisms have in common. Discuss students' responses. Point out that each kind of organism helps form soil in some way. Discuss the ways the organisms contribute to soil formation.
learning styles: visual, auditory

Answers to questions

1. **IDENTIFY:** bedrock
2. **DESCRIBE:** the organic material in soil
3. **LIST:** silt, sand, clay, weathered pieces of rocks and minerals, humus, air, and water
4. **NAME:** potassium, phosphorus, and nitrogen

Reinforcement Review weathering. Ask students if they consider weathering to be a constructive or a destructive process. Guide them to see that soil production is a constructive result of weathering.
learning style: auditory

7-4 How does soil form?

Objective
Explain how soil is a product of weathering and organic processes.

Key Terms
bedrock: solid rock that lies beneath the soil
humus (HYOO-muhs)**:** decaying remains of plants and animals
soil: mixture that includes silt, sand, and clay

Formation of Soil If you dig a hole through the soil, sooner or later you will hit a layer of solid rock that is called **bedrock.** Bedrock is the parent material of soil.

In some areas, bedrock is at Earth's surface. Soil formation begins when bedrock is broken down by weathering. Weathering breaks the parent material into smaller and smaller pieces. Over time, the weathered rock is broken down into soil particles.

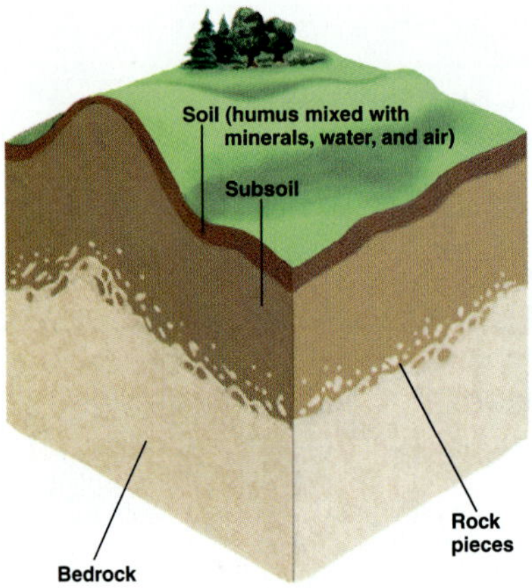

▲ **Figure 7-13** Bedrock is the parent material of soil.

1 **IDENTIFY:** From what parent material does soil form?

Living Things and Soil Many different kinds of organisms live in or on the soil. Living things also help to form soil. Some, such as mosses and lichens, form acids that help break down rocks. Bacteria in the soil cause dead plants and animals to decay. Fungi, such as mushrooms, also help break down dead animals and plants. Acids formed by decay speed up soil formation. The decayed remains of plants and animals form humus. **Humus** is the organic material in soil.

Animals that live in the soil help to form it. Earthworms, ants, prairie dogs, and moles burrow through the ground. As they do, they help break apart large pieces of soil. The burrowing also lets more water and air into the soil. The water helps speed up the breakdown of rock.

▲ **Figure 7-14** Prairie dogs live in burrows in the ground. In digging out their tunnels, they help form soil.

2 **DESCRIBE:** What is humus?

Soil Silt, sand, and clay help make up the mixture known as **soil.** Weathered pieces of rocks, minerals, and humus are also in soil. Humus has a lot of minerals in it, especially quartz. These mixtures, which lie above bedrock, are important to plants. Soil that is rich in humus is very dark in color. Air and water are also in soil. They fill the spaces between soil particles.

3 **LIST:** What things make up soil?

Teaching Options

INTEGRATING ENVIRONMENTAL SCIENCE

Quantifying Decomposition The speed of vegetation decomposition depends on type of litter, the resident microorganisms and fauna, and the weather. If you have a secure space available outside, suggest to students one way to measure speed of decomposition. Weigh several small nylon mesh bags, put a measured mass of dry litter in each one, and tie or glue them closed. Bury them outdoors in marked sites, or anchor them to a surface that is exposed to the weather. After a month, retrieve one bag, let it dry out, and weigh it to see how much the litter has decomposed. Each successive month, retrieve another bag and weigh it. Have students graph the results.
learning styles: tactile, visual

RETEACH

Sequencing the Process of Soil Formation Have students use the Flowchart from the Teacher's Resources to sequence the steps in soil formation. **learning style: logical/mathematical**

ESL/ELL STRATEGY

Reversing Compound Words Tell English-language learners that reversing the parts of the compound word *bedrock* may help them understand it better: *rockbed.*

The Importance of Soil Soil is important to plants. It supplies plants with water and support. Most plants need the nutrients, or chemical elements, in soil to grow properly. Important nutrients such as potassium, phosphorus, and nitrogen are part of soil.

Soil is also important to people. Houses, cities, and roads are built on soil. Food crops are grown in soil. Many animals eat plants and could not survive without them. Other animals hunt and kill plant-eating animals for food. Without soil, few organisms could live on Earth.

4 **NAME:** What are three important nutrients needed by plants?

CHECKING CONCEPTS

1. The parent material of soil is __________.
2. The organic material in soil is __________.
3. Air and __________ fill the spaces between soil particles.
4. The chemical elements in soil that are needed by plants are called __________.
5. The process that forms soil is __________.

THINKING CRITICALLY

6. **INFER:** Why is bedrock called the parent material of soil?
7. **ANALYZE:** How do ants help to speed up the formation of soil?
8. **DESCRIBE:** Explain how soil meets the needs of plants for support, water, and nutrients.

Web InfoSearch

Lichens Not many organisms can live on bare rock. Rocks hold little water and are battered by heat and wind. Lichens, however, can live and grow on bare rocks. Lichens are fungi that form a permanent partnership with certain algae, bacteria, or both. Each organism is helped by the partnership. This is called mutualism.

SEARCH: Use the Internet to find out why lichens are important. How do they help to form soil? Start your search at www.conceptsandchallenges.com. Some key search words are **lichens, lichen soil,** and **lichen uses.**

Integrating Life Science

TOPICS: food chain, ecology

LIFE IN THE SOIL

The soil is teeming with living things. Some organisms mix the soil and make room in it for air and water. Others make humus.

Fallen, decaying leaves form a loose layer in the soil called litter. Humus forms during decomposition. Decomposers break down the litter and the remains of dead organisms and return their nutrients to the soil. Mushrooms, protists, insects, bacteria, and worms are the main soil decomposers. Earthworms carry humus down to the subsoil. Mice, moles, prairie dogs, and gophers break up hard soil. Animal waste also adds nitrogen to the soil.

Thinking Critically What is mixed into the soil when prairie dogs break it up?

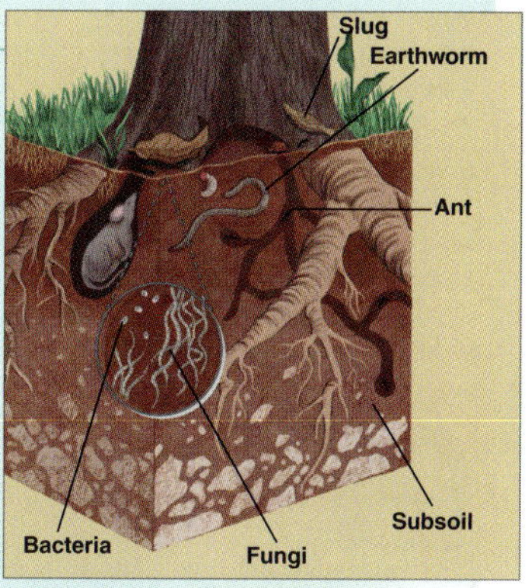

▲ **Figure 7-15** The soil teems with life.

3 Assess

CHECKING CONCEPTS ANSWERS

1. bedrock
2. humus
3. water
4. nutrients
5. weathering

THINKING CRITICALLY ANSWERS

6. **INFER:** Soil is formed from bedrock.
7. **ANALYZE:** They help break apart large pieces of soil. They also let more water into the soil, which breaks down rock.
8. **DESCRIBE:** Plant roots grow in the soil. Plants take in water and nutrients from the soil.

4 Close

Web InfoSearch

Lichens Have students use the Internet to find out more about lichens. Ask: *Why are they important? How do they help to form soil?* Some key search words are **lichens, lichen soil,** and **lichen uses.** Students can begin their search at www.conceptsandchallenges.com.

EXTENDING THE LESSON

Researching Soil Organisms Earthworms play an important role in soil formation. Have interested students use library resources to find out about the characteristics of earthworms and what they can accomplish. Tell students to make a poster of a surprising fact about earthworms, such as how much soil they turn over in a year (about 40 tons per acre), how many can be found in a fertile field (a million per acre), or how many hearts a worm has (five pairs of enlarged blood vessels that serve as hearts). Display the posters. **learning style: visual**

• *TEACHING RESOURCES* •
Online or Print

Teacher's Resources

Lesson Review: Chapter 7, p. 6

Enrichment: Chapter 7, p. 8

Flowchart: Study Tools, p. 6

Laboratory Manual

Lab. Challenge: How do pore space and porosity affect the movement of water through soil? pp. 45–48

Integrating Life Science

TOPICS: food chain, ecology

LIFE IN THE SOIL

Ecology Connection Soils have their own intricate food webs. Dead plant and animal tissue is the principal source of food for most soil organisms. The primary consumers include microorganisms, mites, wood lice, and earthworms. In turn, the primary consumers become food for small predators and parasites, such as bacteria, fungi, mites, termites, protozoa, springtails, spiders, and snails. At the top of the soil food web are tertiary consumers, such as ants, spiders, centipedes, mites, and scorpions.

Discussion Discuss the different types of organisms that affect soil. Have students list all the organisms in the feature and describe how they interact with soil. **learning style: auditory**

Thinking Critically Answer

air

7-5 (pp. 168–169)

1 Introduce

INVESTIGATE

TIME: 10 minutes

PURPOSE: Students will observe and describe characteristics of soils.

MATERIALS: soil from the ground, peat-based soil, fertilized soil, paper plates, magnifying glass

PROCEDURE: Before class, purchase a peat-based soil and one with fertilizer added. Have students work in small groups and label each paper plate with the soil type. Then, distribute soil samples to each group. After students have examined the soils, allow them to read the soil packaging for content information. Be sure that students wash their hands after handling the soil.
learning style: visual

THINK ABOUT IT: Possible answers: They differ in color and size of particles. Peat and fertilizers were added to the purchased soils to improve their nutrient content.

STUDY HINT Have students summarize in a paragraph or two what they see on the map in Figure 7-17. Ask them to draw some conclusions or inferences from the information on the map.

Linking Prior Knowledge Ask students what they know about the chemical formulas of minerals (Lesson 2-3). Emphasize that minerals have different chemical compositions and, therefore, different properties.

2 Teach

Demonstration Display large samples of limestone and granite along with samples of soil formed from each type of rock. Discuss how the kinds of minerals in soil depend on the parent material. Point out that a limestone-rich soil develops from limestone. learning style: visual

Answers to questions

1. DESCRIBE: the size of the soil particles
2. CONTRAST: Transported soils have been carried away from the place they formed. Residual soils form in place.

7-5 How do soils differ?

INVESTIGATE

Examining Soil Composition
HANDS-ON ACTIVITY

1. Obtain soil from the ground, and buy a peat-based soil and one with fertilizers added. Keep the packaging that shows the composition of the soil from the store.
2. Put the three different soils on paper plates.
3. Using your eyes alone, examine the soils.
4. Using a magnifying glass, examine the soils again.

THINK ABOUT IT: How are the various soils different? What was added to the soil you bought from the store? Why were these added?

Objective

Describe differences in soils in terms of texture, mineral content, and where the soil was formed.

Key Terms

texture (TEHKS-chuhr): size of soil particles

transported (trans-POR-tihd) **soil:** soil moved away from the bedrock from which it was formed

residual (rih-ZIJ-oo-uhl) **soil:** soil remaining on top of the bedrock from which it formed

leaching (LEECH-ing)**:** removing or washing away of the minerals in soil

Texture Soils have different textures. **Texture** describes the size of the soil particles. Sandy soils have a coarse texture. Sand particles are between 0.06 mm and 2 mm in diameter. Some soils contain silt. Silt particles are only between about 0.008 mm and 0.016 mm in diameter. Silty soils are finer than sandy soils.

▲ **Figure 7-16** Soil comes in three basic textures: fine, medium, and coarse.

The smallest particles in soil are clay. Clay particles are 0.004 mm or smaller. Soils with a lot of clay have a fine texture. When these soils are wet, they feel sticky or muddy.

1 DESCRIBE: What is meant by soil texture?

Bedrock and Soil Sometimes soil is carried away from the place in which it is formed. Soil can be moved by running water, glaciers, wind, or waves. The soil is then deposited in a new place. This soil is called **transported soil.**

Transported soil may differ from the bedrock, or parent material, that is beneath it. The process that moves the soil from one place to another may also act to change it chemically.

Many soils stay on top of the bedrock from which they were formed. These are called **residual soils.** Residual soils have a chemical makeup similar to that of their parent material.

2 CONTRAST: How do transported and residual soils differ?

Minerals in Soil Weathering rates depend on the mineral makeup of the soil. The kinds and amounts of minerals in soil vary. It all depends on the parent material. For example, the minerals in the rock granite are different from those in limestone. Weathered granite produces one kind of soil. Weathered limestone produces a different kind of soil.

Teaching Options

EXTENDING THE LESSON

Classifying Soil Have students collect soil samples from around their neighborhood. Caution them to wear gloves and get permission to dig. Remind them to wash their hands thoroughly with soap and water afterwards. Have them examine the soil and make a chart listing the following information: overall appearance, kind and size of soil particles, shape of rock particles, and presence of organic material. Help students use the information they gather to determine the type of soil in the area.
learning styles: visual, tactile

QUICK CONVERSIONS

Particle size	Metric/SI	English
Sand	0.06–2.0 mm	0.002–0.08 in.
Silt	0.008–0.016 mm	3.15×10^{-4} in.–6.30×10^{-4} in.
Clay	0.004 mm	1.57×10^{-4} in.

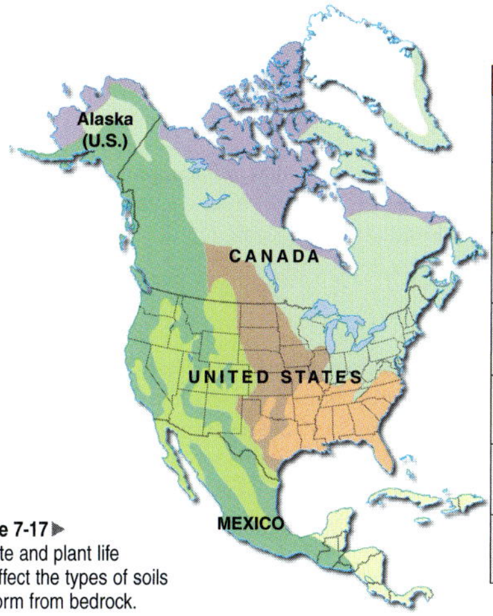

Figure 7-17 ▶ Climate and plant life can affect the types of soils that form from bedrock.

SOILS IN NORTH AMERICA
Tundra soils Cold climate year-round; thin soil with little humus
Northern forest soils Cool, wet climate; ranges from thick to thin soils
Prairie soils Cool, dry, grasslands climate; topsoil thick and rich
Mountain soils Cold temperatures cause topsoil to thin; slow chemical weathering can cause erosion
Desert soils Dry climate with little chemical weathering; often thin, sandy soil; low in humus
Southern forest soils Warm, wet climate; soils may be low in humus
Tropical soils Wet, tropical climate; soils may be low in humus

Chemical weathering also affects the minerals in soil. Minerals are removed from soils by plants and by leaching. **Leaching** is the removing or washing away of minerals by water. The minerals in the soil are dissolved and washed deeper into the soil or are washed away completely.

On farmland, these minerals need to be replaced. Crops need minerals to grow properly. Farmers add fertilizers to the soil to replace minerals used up by plants or washed away by leaching.

3 **DESCRIBE:** How are minerals removed from soil?

Climate and Soil An area's climate helps to determine the type of soil formed from bedrock. Scientists classify different types of soil into groups based partly on a region's climate. Look at Figure 7-17 above to see where each of the major soil types is found in North America.

4 **IDENTIFY:** Which soil type is found on your part of the continent?

CHECKING CONCEPTS

1. The size of particles in soil describes the __________ of the soil.
2. If a soil is very different from the bedrock beneath it, the soil is probably a __________ soil.
3. A soil with a fine texture probably has a lot of __________ in it.
4. The movement of minerals by water deeper into the soil is called __________.

THINKING CRITICALLY

5. **INFER:** Which types of soil described in Figure 7-17 would you expect to form most slowly?
6. **HYPOTHESIZE:** How might chemical weathering affect the minerals in soil that is forming?

Web InfoSearch

Soil Conservation One job of the federal government's Natural Resources Conservation Service (NRCS) is to help protect natural resources. NRCS soil conservationists work with farmers and ranchers to prevent soil erosion, protect wildlife, and keep water supplies clean.

SEARCH: Use the Internet to find out how you can work for the NRCS. Start your search at www.conceptsandchallenges.com. Some key search words are **National Resources Conservation Service, NRCS,** and **NRCS jobs.**

3 **DESCRIBE:** by leaching

4 **IDENTIFY:** If students' local soil differs from the types of soil on the map, point that out.

Reinforcement To help students understand and remember the definition of *transported soil,* relate it to the more common usage of the word *transportation.* Ask students to name different means of transportation. Point out that soil can also be transported. The means of soil transportation are running water, glaciers, wind, and waves.
learning style: auditory

WRITING HINT Point out the use of the word *affects* in the first sentence on page 169. Remind students not to confuse *affect* with *effect* when writing. Have students look up both words in a dictionary, and then use each word in a sentence.

3 Assess

CHECKING CONCEPTS ANSWERS

1. texture
2. transported
3. clay
4. leaching

THINKING CRITICALLY ANSWERS

5. **INFER:** Possible answer: tundra soils or desert soils because they are cold or dry and thus weathering occurs more slowly
6. **HYPOTHESIZE:** It might chemically change some of the minerals and could also cause a soil to become mineral-poor.

4 Close

Web InfoSearch

Soil Conservation Have students use the Internet to find out how they can become soil conservationists for the NRCS. Some key search words are **Natural Resources Conservation Service, NRCS,** and **NRCS jobs.**

COOPERATIVE LEARNING

Forming Questions Tell students to prepare as if a soil scientist were coming to class. Students should form small groups and prepare questions beforehand that they would like to ask the scientist.
learning style: auditory

• TEACHING RESOURCES • Online or Print

Teacher's Resources
Lesson Review: Chapter 7, p. 7
Enrichment: Chapter 7, p. 8

ONGOING ASSESSMENT

Have students refer to Figure 7-17 to rank four regions in terms of how fast soil is likely to form in them, from fastest to slowest: northern forests, southern forests, desert, and mountains. (southern forests, northern forests, mountains, desert)

7-6 *(pp. 170–171)*

1 Introduce

Linking Prior Knowledge Remind students that they learned how soil forms (Lesson 7-4). Ask them to describe the texture differences (Lesson 7-5).

2 Teach

Discussion Using the soil profile in Figure 7-19, guide students through a step-by-step development of a soil profile. At one time, this spot had bare bedrock. It weathered into smaller pieces, which eventually became the C horizon. Plants and organisms that can live in stony soil got a toehold and started to add organic material to the top of the soil. The material decomposed, adding nutrients to the finer material and forming topsoil. Weathering continued at all levels, producing more fine particles and leaching chemicals from the topsoil, until a B horizon could develop. This process, from bedrock to mature soil, can take thousands of years.
learning styles: visual, auditory

Reinforcement Show students photographs of a mature soil and an immature soil. Compare the two soils. Make sure that students understand that an immature soil has only two horizons. Ask students to identify what horizon is missing in an immature soil.
(B horizon) **learning style: visual**

Answers to questions

1. **LIST:** stream banks and road cuts
2. **DESCRIBE:** a soil that has only two horizons
3. **IDENTIFY:** It is a mixture of humus and tiny rock particles.
4. **COMPARE:** The B horizon contains a lot of clay.
5. **DESCRIBE:** at the bottom

READING STRATEGY Ask: *Why do you think the word* horizon *is used to describe the layers in a soil profile?* Have students look up the word in a dictionary to see how it applies here.

7-6 What is a soil profile?

Objective

Identify and describe the parts of a soil profile.

Key Terms

horizon (huh-RY-zuhn)**:** soil layer

soil profile (PROH-fyl)**:** all the layers that make up the soil in an area

Soil Layers Soil forms in layers called **horizons.** A soil horizon is a layer of soil that differs in color and texture from the layers on top of it or below it.

▲ **Figure 7-18** The layers of soil are easy to see here.

Most soils have at least three horizons. Each one is named with a capital letter. The A horizon is the top layer of soil. However, this layer is actually covered with a thin coating of organic material called the O horizon. The O horizon is made up mostly of plant litter, such as loose and decaying leaves. It is also teeming with microscopic life and the remains of plants and animals.

The A and O horizon together make up the top layer of soil. The B horizon is beneath the A horizon. The C horizon is the bottom layer. All the soil layers together are called a **soil profile.**

You can see soil profiles in stream banks and in road cuts. A road cut is land that has been dug out to build a road.

1. **LIST:** Name two places where you might see a soil profile.

Mature and Immature Soils A soil that has at least three horizons is called a mature soil. A mature soil can take thousands of years to form or even longer. Some soils have only two horizons. A soil that has two horizons is called immature soil. After a long time, weathering may turn an immature soil into a mature soil.

2. **DESCRIBE:** What is immature soil?

The A Horizon The A horizon is found below the O horizon, the thin layer of organic material. The A horizon is usually the topsoil, a mixture of humus and rock particles. Humus gives topsoil its dark color.

Topsoil is rich in nutrients. Most organisms that live in the soil are found in the A horizon. Plants grow best in topsoil.

3. **IDENTIFY:** What is topsoil made up of?

The B Horizon The B horizon, called subsoil, is found beneath the A horizon. Subsoil is mostly clay, small pieces of weathered rock, and minerals. Leaching moves minerals from the A into the B horizon. Because the subsoil has a lot of clay in it, it is much firmer than the topsoil. For this reason, only the roots of very large plants grow into the B horizon.

4. **COMPARE:** Why is the B horizon firmer than the A horizon?

Go Online **active art**
For: Soil Layers activity
Visit: PHSchool.com
Web Code: cfp-2022

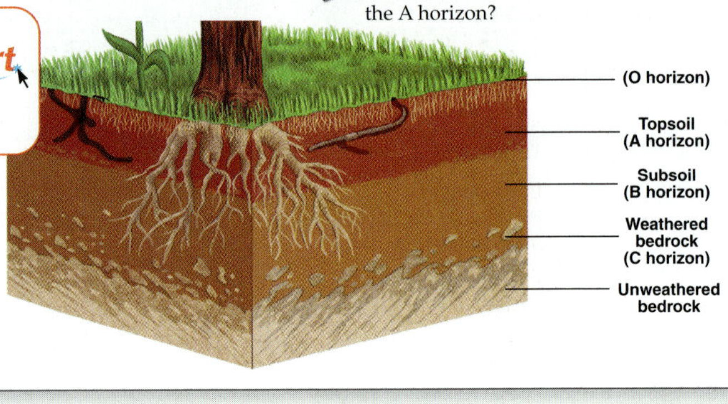

Figure 7-19 ▶ A common soil profile

170

Teaching Options

! FAST FACTS *Soil Types*

Classifying Soil Some soil scientists, such as those in the U.S. Department of Agriculture, categorize soils according to their properties. Other soil scientists use a system based on climate zones.

Soil Fertility In hot deserts, water brings dissolved salts and other materials to the surface of the soil. When the water evaporates, the dissolved materials stay behind, forming a whitish crust on the soil. Alternatively, the heavy rainfall of some areas causes minerals and other substances in the soil to flow downward through the soil, redepositing them at lower soil levels. This downward movement of materials affects soil fertility at different levels.

RETEACH

Comparing Horizons Have students construct charts comparing soil horizons. Remind students to label the column heads on their charts. Tell students to use their charts when studying this lesson. **learning style: visual**

The C Horizon The C horizon is the bottom layer of a soil profile. The C horizon is made up of large pieces of rock. Weathering of these large pieces of rock is the first step in forming soil. At the bottom of the C horizon is solid bedrock. The large pieces of rock in the C horizon come from the weathering of bedrock.

5 DESCRIBE: Where is the C horizon located in a soil profile?

CHECKING CONCEPTS

1. Soil forms in layers called ___________.
2. The B horizon is the ___________.
3. The number of horizons in immature soil is ___________.
4. The A horizon is made up of a dark-colored soil called ___________.
5. If a soil has at least three horizons, it is a ___________ soil.
6. In the C horizon, the large pieces of rock are weathered ___________.

THINKING CRITICALLY

7. RELATE: What is the relationship between soil horizons and a soil profile?
8. CONTRAST: What is the difference between mature and immature soils?

BUILDING READING SKILLS

Prefixes Find out the meaning of the prefix *sub-*. Then, use it to help you define the words in the list below. Write a definition for each word. Circle the word or words in the definition that relate to the meaning of the prefix *sub-*.

- subheading
- submarine
- submerge
- subsoil
- substance
- subway

Hands-On Activity

MODELING A SOIL PROFILE

You will need a clear plastic cup, modeling clay, sand, gravel, soil, and a marking pen.

1. In a clear plastic cup, arrange layers of modeling clay, sand, gravel, and soil to make a model of a soil profile. Let the modeling clay represent the bedrock.
2. Use a marking pen to identify each layer of your soil profile. Label the A, B, and C horizons, and the bedrock.
3. Draw a diagram of your soil profile. Compare your drawing with the soil profile in Figure 7-20.

▲ **Figure 7-20** Be sure to label each layer of your soil profile.

Practicing Your Skills

4. IDENTIFY: Which material did you use to represent the A horizon? The B horizon? The C horizon?
5. CLASSIFY: Does your soil profile show a mature soil or an immature soil? Explain.
6. IDENTIFY: **a.** Which horizon of a soil profile is the topsoil? **b.** Which horizon is the subsoil?

3 Assess

CHECKING CONCEPTS ANSWERS

1. horizons
2. subsoil
3. two
4. topsoil
5. mature
6. bedrock

THINKING CRITICALLY ANSWERS

7. RELATE: All the soil layers, or horizons, together are called a soil profile.
8. CONTRAST: Mature soils have at least three horizons. Immature soils have only two horizons.

4 Close

BUILDING READING SKILLS ANSWERS

Prefixes The prefix *sub-* means "under" or "beneath." A subhead is a head that comes beneath the main head. *Submarine* means "under the ocean." An object is submerged when it is underwater. *Subsoil* is under the topsoil. The word *substance* is from a French word that means "to stand under something." A *subway* is an underground passage.

Hands-On Activity

MODELING A SOIL PROFILE

TIME: 15 minutes

PURPOSE: modeling, classifying, identifying

SAFETY TIP: Have students wash their hands after working with the soil and sediment.

ALTERNATIVE MATERIALS: Clear jars can be used instead of plastic cups. Students might want to wear gloves while working with soil.

Practicing Your Skills Answers

4. IDENTIFY: Students should use soil to represent the A horizon, sand to represent the B horizon, and gravel to represent the C horizon.
5. CLASSIFY: a mature soil, because it has three horizons
6. IDENTIFY: **a.** the A horizon **b.** the B horizon

MORE ON THE WEB

Creativity With Soil Have students use the Internet to learn more about soil. Then, have small groups write a song, story, or poem about soil. Groups can present their final product to the class, or they can submit it to a website that features other student entries. To start their research, students can visit www.conceptsandchallenges.com.

EXTENDING THE LESSON

Observing Soil Profiles Have students observe a road cut in their neighborhood. Tell them to draw a soil profile of it.
learning style: visual

• TEACHING RESOURCES • Online or Print

Teacher's Resources

Lesson Review: Chapter 7, p. 9

Soil Profile: Visuals, p. 6

Go Online active art

For: Soil Layers activity
Visit: PHSchool.com
Web Code: cfp-2022

Students can interact with the art of soil layers online.

The Big Idea

(pp. 172–173)

1 *Introduce*

Objective Students will describe how the chemistry of soil affects plant growth. They will research the soil pH requirements of plants and record the information in their Science Logs.

Linking Prior Knowledge Review the importance of soil to living things (Lesson 7-4). Ask students if they have any experience with gardening. Discuss what materials are added to soil to make it better suited for growing plants. Be sure that students understand what nutrients are.

2 *Teach*

Demonstration Demonstrate how soils are tested for pH. Soil test kits can be obtained from scientific supply houses or local nurseries. Indicators such as chlorphenol red, litmus, phenol red, and bromthymol blue have different pH ranges and can be used in testing soils. Have students collect small samples of soil from different areas. If possible, use soil found under evergreens and soil from open, grassy fields. Caution students about using gloves, getting permission to dig, and washing their hands thoroughly with soap and water afterwards.
learning style: visual

Discussion Explain that the pH scale is used to determine whether a substance is an acid or a base. A pH value below 7 describes an acidic solution. A pH value above 7 describes a basic solution. Make sure that students understand that the higher the pH, the more basic is the substance. The lower the pH, the more acidic is the substance.
learning style: logical/mathematical

Integrating Chemistry

THE Big IDEA

How does the chemistry of soil affect plant growth?

To carry out life processes, plants must obtain certain nutrients. Plants get these essential nutrients from the soil. But not all soil is alike. The chemical makeup of soil differs from place to place. This difference affects how easily nutrients can be absorbed by a plant.

For example, some soils are more acidic than others. Chemists measure acid content using the **pH scale**. The pH scale consists of numbers between 0.0 and 14.0. A pH of 7 means that the substance is neutral. A pH above 7 means that the substance is basic, or alkaline. A pH below 7 means that the substance is acidic. Neutral solutions are neither basic nor acidic.

Most plants grow well in slightly acidic soil, with a pH of between 6 and 7. Soils with pH levels between 4 and 5 often have high amounts of aluminum, iron, and manganese, which harm plants. Soils with pH levels between 7 and 8 often contain a lot of calcium, potassium, magnesium, and sodium. Few plants will grow in soils with a pH below 4 or above 8.

The acidity of soil can be checked with a pH meter. It can also be tested using an **indicator kit**. Indicators change color in acidic and basic solutions. After testing, the soil pH can be adjusted by adding certain materials. To raise the pH, the soil can be "sweetened" with lime. Lime is finely ground limestone. To lower the pH, peat moss or sulfur can be added.

Look at the illustrations on these two pages. Then, follow the directions in the Science Log to learn more about "the big idea." ✦

Growing Plants in Acidic Soil
Rhododendrons grow well in acidic soil with a pH between 4.0 and 6.0.

Growing Plants in Alkaline Soil
Cabbage grows well in slightly alkaline soil with a pH of about 7.5.

172

Teaching Options

! FAST FACTS Soil Colors

Soil comes in many different colors, including brown, red (such as the Red River in Oklahoma and Texas), gray, yellow (such as the Yellow River in China), and white. The color is determined by the soil's chemistry. For example, gray soil occurs when a soil has a high water table or water settles above an impermeable layer. This turns the iron content in it into colorless, water-soluble ferrous iron. Different-colored soils were used as coloring agents in many early human cultures.

EXTENDING THE LESSON

Defining Terms Have students find the meaning of the terms *acid* and *base* in a chemistry textbook. Have them show, using numbers or a graph, how much more acidic a substance that has a pH of 3 is than substances with a pH of 4, 5, and 8. **learning style: visual**

RETEACH

Measuring pH Provide students with safety goggles, gloves, and litmus paper or other indicators. Have them determine whether common substances such as milk, ammonia, lemon juice, and vinegar are acidic or basic.
learning style: visual

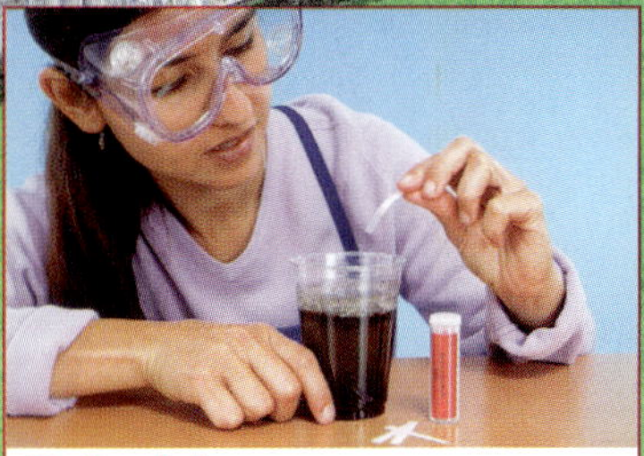

Testing Soil

Red litmus paper turns blue in basic soils. Blue litmus paper turns red in acidic soils.

Changing the pH of Soil

Adding lime will raise the pH. This will make the soil more basic, or alkaline. Peat moss will lower it, or make it more acidic.

WRITING ACTIVITY

Science Log

Make a two-column chart. Label one column "Plant" and the other "pH Level." Record the following items in the first column: broccoli, blueberry, carrot, cauliflower, corn, lima bean, pea, pumpkin, and spinach. Find out the pH of the soil these plants grow best in and complete the chart. Start your search for information using www.conceptsandchallenges.com.

Composting Soil

As organic matter decays, such as in compost, acids are produced. Therefore, most garden soils become acidic over time.

Figure 7-21 The acidity of soil determines the kinds of plants that can grow well in it. Different plants growing next to one another in a garden should have the same pH needs.

THE BIG IDEA

How does the chemistry of soil affect plant growth? Have students start their research for their Science Log Writing Activity at www.conceptsandchallenges.com. Be sure students use search engines and websites that are reliable and acceptable. You may also wish to provide copies of the *Big Idea Planner* and *Big Idea Science Log Entry* study tools found in the Teacher's Resources.

3 Assess

Use the Writing Activity Rubric from the Teacher's Resources to assess students' Science Logs. Fill in the rubric with the additional information below. For this assignment, students should have:

- determined the pH of the soil in which the listed plants grow best.
- created a two-column chart.

4 Close

Writing Gardening Brochures Have students work in small groups to create a brochure explaining to home owners how to choose the best plants to grow in their soil. The brochures should explain how soil acidity affects plant growth and should include colorful illustrations. Students can pass out copies of their brochures to friends and family members.

• TEACHING RESOURCES • Online or Print

Teacher's Resources

Big Idea: Chapter 7, p. 10

Big Idea Planner: Study Tools, pp. 11–12

Big Idea Science Log Entry: Study Tools, p. 13

Writing Activity Rubric: Teaching Tools, p. 9

Challenges (pp. 174–176)

Chapter Summary

Review Before students begin the Challenges, have pairs of students review the Key Terms and lesson summaries together.

STUDY HINT Have students classify the Key Terms into groups, based on whether they are related to weathering or to soil. Then, have students write a paragraph on each concept (weathering and soil), using all of the relevant Key Terms.

Key Term Challenges

MATCHING

1. bedrock (7-4)
2. texture (7-5)
3. weathering (7-1)
4. acid rain (7-3)
5. leaching (7-5)
6. horizon (7-6)

IDENTIFYING WORD RELATIONSHIPS

7. These soils both form from bedrock. Residual soil stays where it forms, and transported soil is moved from where it forms. (7-5)
8. Mechanical weathering changes the shape of rocks, and chemical weathering changes the chemical makeup of rocks. (7-1)
9. Both processes involve water that causes weathering. (7-1, 7-2)
10. Both processes cause chemical weathering. (7-2)
11. Horizons make up a soil profile. (7-6)
12. Humus is one ingredient in soil. (7-4)
13. Acid rain speeds weathering. (7-3)
14. Leaching is a type of chemical weathering. (7-5)

Chapter 7 Challenges

Chapter Summary

Lesson 7-1

- **Weathering** is the breaking down of materials on Earth's surface. There is **mechanical weathering** and **chemical weathering.**
- Changes in temperature can break rocks apart. **Ice wedging** is a kind of mechanical weathering caused by the repeated freezing and melting of water. Pressure from the roots of growing trees can also cause mechanical weathering.

Lesson 7-2

- Chemical weathering takes place when rocks break down as a result of chemical changes. **Oxidation** is the chemical reaction of oxygen with other substances.
- Most chemical weathering is from **hydrolysis,** which is water reacting with water-poor substances.
- **Carbonation** breaks down minerals in rocks. Plants also produce acids that cause chemical weathering.

Lesson 7-3

- Weathering is affected by moisture and temperature. The more surface area of rocks exposed, the faster the rocks will weather.
- The rate of weathering is affected by the minerals that make up rocks. **Acid rain** speeds the weathering process.

Lessons 7-4 and 7-5

- Soil forms from **bedrock. Humus** is the organic material in soil. Burrowing organisms help break up large pieces of soil.
- **Soil** is a mixture that supplies plants with water, support, and nutrients. Soils have different **textures.**
- **Transported soils** are soils that were moved from where they formed. **Residual soils** stay on top of the bedrock from which they formed.
- Different soils contain different kinds and amounts of minerals. Minerals are removed from soils by plants and by **leaching.**

Lesson 7-6

- Soil forms in layers called **horizons.** These horizons make up a **soil profile.** Mature soil has at least three horizons: A, B, and C.

Key Term Challenges

acid rain (p. 164)
bedrock (p. 166)
carbonation (p. 160)
chemical weathering (p. 158)
horizon (p. 170)
humus (p. 166)
hydrolysis (p. 160)
ice wedging (p. 158)
leaching (p. 168)
mechanical weathering (p. 158)
oxidation (p. 160)
residual soil (p. 168)
soil (p. 166)
soil profile (p. 170)
texture (p. 168)
transported soil (p. 168)
weathering (p. 158)

MATCHING **Write the Key Term from above that best matches each description.**

1. layer of rock beneath soil
2. size of soil particles
3. breaking down of rocks and materials of Earth's surface
4. rain containing harmful acids
5. washing away of minerals in soil
6. soil layer

IDENTIFYING WORD RELATIONSHIPS **Explain how the words in each pair are related. Write your answers in complete sentences.**

7. residual soil, transported soil
8. chemical weathering, mechanical weathering
9. ice wedging, hydrolysis
10. oxidation, carbonation
11. horizon, soil profile
12. humus, soil
13. acid rain, weathering
14. leaching, chemical weathering

174

Teaching Options

PREPARING STUDENTS FOR STANDARDIZED TESTS

Reading Strategy: Remind students that when they answer multiple-choice questions, they should read all the possible answers to a question, even when the first answer seems correct. Explain that an "all of the above" option may be the best choice.

Writing Strategy: Tell students that when they describe a cause or an effect, they can go beyond the keywords *cause* and *effect* and use *because, reason, thus, as a result,* and *consequently.*

Interpreting Visuals Strategy: When labeling diagrams, students should think about what each part of the diagram is showing to help them recall the label to be used.

ESL/ELL STRATEGY

Fill-In Questions Have English-language learners turn the fill-in sentences into simple questions, for example, for question 10 on page 175, "Potholes in the street are caused by what natural process?" This may help them come up with the missing words.

Content Challenges TEST PREP

MULTIPLE CHOICE **Write the letter of the term or phrase that best completes each statement.**

1. Hydrolysis, carbonation, and oxidation are kinds of
   - **a.** leaching.
   - **b.** mechanical weathering.
   - **c.** chemical weathering.
   - **d.** acid rain.
2. Subsoil is located in
   - **a.** the O horizon.
   - **b.** the A horizon.
   - **c.** the B horizon.
   - **d.** the C horizon.
3. A soil that has only two horizons is classified as
   - **a.** a sweet soil.
   - **b.** a sour soil.
   - **c.** a mature soil.
   - **d.** an immature soil.
4. Topsoil is located in
   - **a.** no horizon.
   - **b.** the A horizon.
   - **c.** the B horizon.
   - **d.** the C horizon.
5. Most chemical weathering is caused by the process called
   - **a.** acid rain.
   - **b.** carbonation.
   - **c.** oxidation.
   - **d.** hydrolysis.
6. Chemical weathering occurs the fastest in
   - **a.** hot, dry climates.
   - **b.** cold, dry climates.
   - **c.** cold, wet climates.
   - **d.** hot, wet climates.
7. Soil carried away from where it was formed is called
   - **a.** residual soil.
   - **b.** mature soil.
   - **c.** transported soil.
   - **d.** immature soil.
8. The smallest particles in soil are
   - **a.** silt.
   - **b.** sand.
   - **c.** clay.
   - **d.** weathered rock.

FILL IN **Write the term or phrase that best completes each statement.**

9. Mechanical weathering does not change the __________ of rocks.
10. Potholes in streets are caused by __________.
11. Root action and ice wedging are kinds of __________ weathering.
12. Oxidation forms new substances called __________.
13. Carbon dioxide can dissolve in rain to form __________.
14. A rock containing quartz weathers __________ than a rock containing calcite.
15. Bedrock is the __________ material of soil.
16. The organic material found in soil is called __________.
17. Most plants need the __________ found in soil to grow properly.
18. Sandy soils have a __________ texture.
19. Farmers add __________ to the soil to replace minerals washed away by leaching or used by plants.
20. An area's __________ helps to determine what kind of soil is formed from bedrock.

Content Challenges

MULTIPLE CHOICE

1. c (7-2)
2. c (7-6)
3. d (7-6)
4. b (7-6)
5. d (7-2)
6. d (7-3)
7. c (7-5)
8. c (7-5)

FILL IN

9. chemical makeup (7-1)
10. ice wedging (7-1)
11. mechanical (7-1)
12. oxides (7-2)
13. carbonic acid (7-2)
14. slower (7-3)
15. parent (7-4)
16. humus (7-4)
17. nutrients (7-4)
18. coarse (7-5)
19. fertilizers (7-5)
20. climate (7-5)

ALTERNATIVE ASSESSMENT

Describing Soil Have students write a paragraph explaining what soil is and how it forms.

PORTFOLIO ASSESSMENT

Making Student Portfolios Students can demonstrate their comprehension of the concepts in this chapter by making a portfolio. The Chapter Self-Check and the Weekly Journal are some of the reproducibles in the Teacher's Resources that they can include in their portfolios. They can also include their Chapter Project Weathering Log or Big Idea gardening brochure. Encourage students to select their own work for inclusion in their portfolios. You can use the Portfolio Assessment Rubric also found in the Teacher's Resources to assess students' portfolios.

Concept Challenges

WRITTEN RESPONSE

1. **EXPLAIN:** Heat and water speed up the process of weathering. Therefore, weathering would be slow in a cold, dry climate. (7-3)
2. **PREDICT:** Carbonation would affect a rock because of the reaction of carbonic acid with calcite. (7-2)
3. **EXPLAIN:** Living things burrow into soil, breaking apart the large pieces. The burrows let more water into the soil, which helps speed up the breakdown of rock. (7-4)
4. **INFER:** It removes from the soil some nutrients that plants need. (7-5)

INTERPRETING VISUALS (7-6)

5. a soil profile
6. a horizon
7. O horizon
8. A horizon or topsoil
9. humus and rock particles
10. subsoil
11. bedrock
12. Mature; The soil has three horizons.

Chapter Project Wrap-Up

WEATHERING LOG

Students can publish their journal information on a Web page, either as a class or individually. You can use the Individual Activity Rubric found in the Teacher's Resources to assess students' projects. Fill in the rubric with the additional information below. For this project, students should have:

- listed at least ten examples of weathering.
- presented a logical explanation for the cause of each example.

Concept Challenges TEST PREP

WRITTEN RESPONSE **Answer each of the following questions in complete sentences.**

1. **EXPLAIN:** Why is weathering slow in a cold, dry climate?
2. **PREDICT:** What kind of chemical weathering is most likely to affect a rock containing calcite? Why?
3. **EXPLAIN:** How do living things help to form soil?
4. **INFER:** How does leaching affect plant life?

INTERPRETING VISUALS **Use Figure 7-22 below to answer the following questions.**

5. What is shown in the diagram?
6. What is each layer of soil called?
7. What is the name of the layer indicated by the letter *K*?
8. What is the name of the layer indicated by the letter *L*?
9. What kind of soil makes up the layer labeled *L*?
10. What is the soil that makes up layer *M* called?
11. What material makes up the layer labeled *N*?
12. Is this mature or immature soil? Explain your answer.

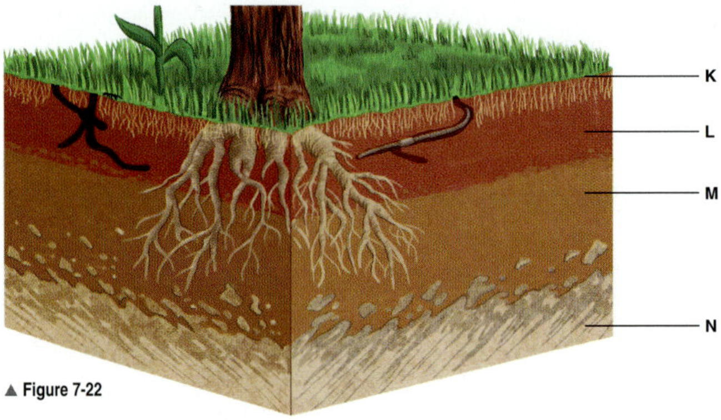

▲ Figure 7-22

Teaching Options

• TEACHING RESOURCES •
Online or Print

Teacher's Resources

Key Term Review: Chapter 7, p. 11

Chapter 7 Test: pp. 12–13

Individual Activity Rubric: Teaching Tools, p. 5

Chapter Self-Check: Teaching Tools, p. 11

Big Idea Planner: Study Tools, pp. 11–12

Weekly Journal: Study Tools, p. 19

• TEACHING RESOURCES •
Online or Print

Teacher's Resources

Portfolio Assessment Rubric: Teaching Tools, p. 10

Chapter 8 Erosion

PLANNING GUIDE

◆ **TEACHING THE CHAPTER** This chapter should take approximately 11–14 days to complete instruction and assessment.

	Skills and Features	Projects and Activities	Achieve Science Literacy	Meet Individual Needs
Chapter 8 Opener p. T177	MODEL	• Chapter Project		
Lesson 8-1 **What are erosion and deposition?** pp. T178–T179 Standards: D1c, D2a	RESEARCH, WRITE, DEFINE • Investigate • Building Science Skills • Real-Life Science	• Extending the Lesson • More on the Web • Integrating Life Science		• Reteach Teacher's Resources* Concept Map Flowchart
Lesson 8-2 **How does gravity cause erosion?** pp. T180–T181 Standards: D1c, D2a	EXPLAIN, CONTRAST, IDENTIFY • Web InfoSearch • Hands-On Activity	• Extending the Lesson • Active Art	• Reading Strategy • Ongoing Assessment	• Reteach Teacher's Resources* Flowchart
Lesson 8-3 **How does running water cause erosion?** pp. T182–T183 Standards: D1c, D1f, D1g, D2a	MODEL, RESEARCH, PREDICT • Investigate • Building Science Skills • Integrating Environmental Science	• Extending the Lesson • More on the Web • Lab Challenge		• ESL/ELL Strategy • Reteach Teacher's Resources* Enrichment Activity
Lesson 8-4 **What landforms does running water create?** pp. T184–T185 Standard: D1c	IDENTIFY, LOCATE, INFER • Hands-On Activity	• Cooperative Learning • More on the Web • Integrating Environmental Science	• Study Hint • Writing Hint	Teacher's Resources* Enrichment Activity Running Water and Landforms
Lesson 8-5 **What causes caves, caverns, and sinkholes?** pp. T186–T187 Standards: D1c, D1g	DEFINE, IDENTIFY, EXPLAIN • Building Geography Skills • Integrating Life Science	• Cooperative Learning • More on the Web	• Study Hint	Teacher's Resources* Enrichment Activity
Lesson 8-6 **What is a glacier?** pp. T188–T189 Standard: D1f	DEFINE, NAME, IDENTIFY • Web InfoSearch • How Do They Know That?	• Extending the Lesson • Integrating Physical Science	• Study Hint • Ongoing Assessment	

*www.conceptsandchallenges.com, password: ccearth09 *Planning Guide continues on next page.*

Standards: For details on the correlation to National Science Standards see pages *xx–xxi*.

	Skills and Features	Projects and Activities	Achieve Science Literacy	Meet Individual Needs
Lesson 8-7 **How do glaciers cause erosion?** pp. T190–T191 Standards: D1c, D2a	EXPLAIN, HYPOTHESIZE, COMPARE • Designing an Experiment • Hands-On Activity	• More on the Web	• Study Hint	• ESL/ELL Strategy • Reteach Teacher's Resources* Designing an Experiment
Lesson 8-8 **What landforms do glaciers create?** pp. T192–T193 Standards: D1c, D2a	DEFINE, NAME, INFER • Web InfoSearch • People in Science	• Extending the Lesson	• Study Hint • Reading Strategy • Writing Hint • Ongoing Assessment	
Lesson 8-9 **How do ocean waves cause erosion?** pp. T194–T195 Standards: D1c, D1f, D2a	DESCRIBE, IDENTIFY, HYPOTHESIZE • Building Math Skills • Health and Safety Tip • Real-Life Science	• Cooperative Learning • More on the Web	• Study Hint • Writing Hint	Teacher's Resources* Enrichment Activity
Lesson 8-10 **What landforms do ocean waves create?** pp. T196–T197 Standards: D1c, D1f, D2a	MODEL, DESCRIBE, EXPLAIN • Integrating Physical Science	• Cooperative Learning • Extending the Lesson • More on the Web	• Study Hint • Reading Strategy	• ESL/ELL Strategy • Reteach Teacher's Resources* Enrichment Activity Flowchart
Lesson 8-11 **How does wind cause erosion?** pp. T198–T199 Standard: D1c	CLASSIFY, INFER, DEFINE • Science and Technology	• More on the Web	• Study Hint • Writing Hint	Teacher's Resources* Enrichment Activity
Lab Activity **Modeling the Effects of Wind on Sand** pp. T200–T201 Standards: D1c, D1e, D2a	OBSERVE, ANALYZE • Set-Up Time: 45 min • Lab Time: 45 min	• Cooperative Learning • Extending the Lesson • Lab Skills Worksheet • Extend the Lab Activity	• Tips for Using Technology	• ESL/ELL Strategy Teacher's Resources* Lab Activity Report Weekly Journal
Big Idea **How did erosion change American history?** pp. T202–T203 Standard: D2a	RESEARCH, COMMUNICATE, ANALYZE • Science Log: Writing Activity	• Big Idea • Extending the Lesson • Close Activity	• Tips for Using Technology	Teacher's Resources* Big Idea Planner Thumbnail Sketch
Chapter 8 Challenges pp. T204–T206	• Chapter Summary • Key Term Challenges • Content Challenges • Concept Challenges	• Chapter Project Wrap-Up	• Study Hint • Preparing Students for Standardized Tests	• ESL/ELL Strategy Teacher's Resources* Chapter Self-Check Weekly Journal

*www.conceptsandchallenges.com, password: ccearth09

Chapter 8 Erosion

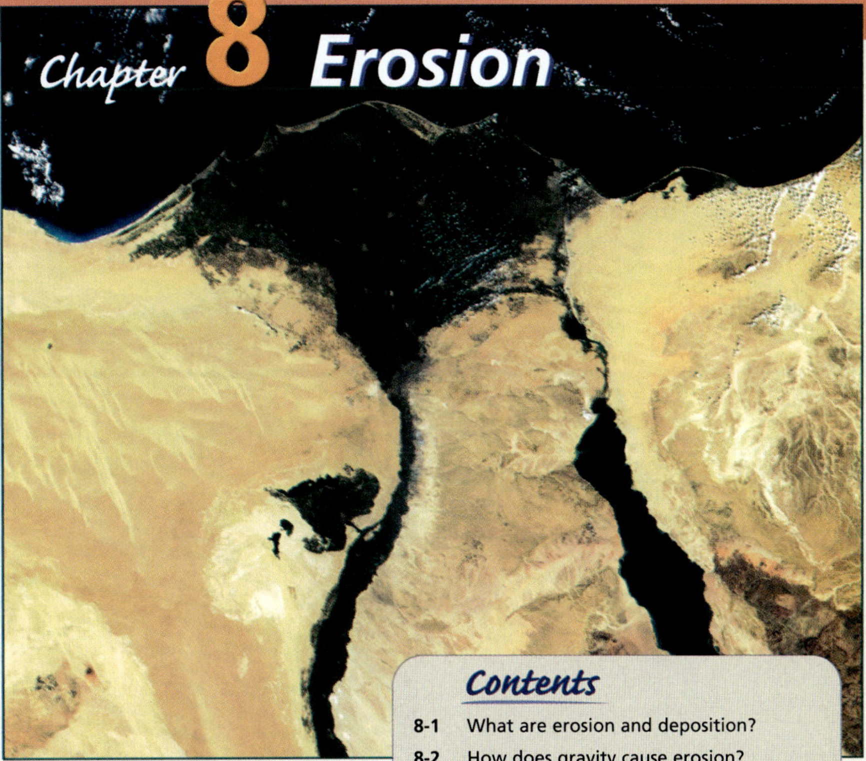

Figure 8-1 The delta of the Nile River is near Cairo, Egypt. At the delta, the river divides into several tributaries.

The flooding of rivers can cause erosion and property damage. With many rivers, such as the Nile, flooding also enriches the soil. The Nile is the longest river in the world. About 5,000 years ago, the first great African civilization developed in the Nile Valley, where Egypt is today. As an agricultural state, Egypt relied on the yearly flooding of the Nile. The rich soil around the Nile made Egypt a very successful civilization. Egypt ruled much of northeastern Africa for thousands of years.

▶What benefits did the yearly flooding of the Nile bring to Egypt?

Contents

Teaching Options

ASSESSMENT PLANNER

For assessment in the Student Edition, see the following pages:

Content Assessment

Checking Concepts: pp. 179, 181, 183, 185, 187, 189, 191, 193, 195, 197, 199
Thinking Critically: pp. 179, 181, 183, 185, 187, 189, 191, 193, 195, 197, 199
Interpreting Visuals: pp. 185, 197, 199
Chapter Challenges: pp. 204–206

Alternative Assessment

Building Skills: pp. 179, 183, 187, 195
Web InfoSearch: pp. 181, 189, 193
Science Log: p. 203

Performance-Based Assessment

Hands-On Activity: pp. 181, 185, 191
Designing an Experiment: p. 191
Lab Activity pp. 200–201

• TEACHING RESOURCES • Online or Print

Teacher's Resources

Lesson Review: Ch. 8, pp. 2, 3, 4, 5, 6, 8, 9, 10, 11, 12, 13
Enrichment: Ch. 8, pp. 7, 14
Key Term Review: Ch. 8, p. 16
Chapter 8 Test: pp. 17–19
Laboratory Manual: pp. 13–14, 49–52

Chapter 8 Erosion

Chapter Overview

In this chapter, students will learn about erosion and deposition. Students will list five agents of erosion and explain how these agents cause erosion and deposition to create landforms. They will name and describe features of karst topography. In addition, they will identify two kinds of glaciers.

About Figure 8-1 The Nile River delta is one of the world's largest. At its widest point, it stretches 250 km (160 mi) from east to west. Cairo, the capital and largest city in Egypt, is located near the southern end of the delta. Alexandria, Egypt's principal seaport, is located near the western end of the delta. Other major deltas include the Mississippi River delta in the United States and the combined delta of the Ganges River and Brahmaputra River in India.

Although the caption states that at the delta, the Nile divides into tributaries, these areas are technically known as *dis*tributaries, since true tributaries feed water into a river, while here the Nile is losing water.

Answer It deposited a nutrient-rich layer of black soil, enabling farmers to raise abundant crops.

Linking Prior Knowledge

For this chapter, students should recall:

- erosion in the rock cycle (Lesson 3-8).
- what causes mechanical weathering (Lesson 7-1).
- what causes chemical weathering (Lesson 7-2).

Chapter Project

MODEL OF A LANDFORM CREATED BY EROSION

MATERIALS: modeling clay, sand

Divide the class into groups and have each prepare a model of a different type of landform created by erosion, such as a sand dune, a glacier-carved valley, or a cavern. Have each group research actual locations of the landform they are modeling.
learning styles: visual, tactile

8-1 (pp. 178–179)

1 Introduce

INVESTIGATE

TIME: 10 minutes

PURPOSE: Students will try to devise a method for preventing erosion.

MATERIALS: pie plate; sand or soil; various items such as sticks, paper strips, and pebbles

ALTERNATIVE MATERIALS: sponges, cotton balls, strips of window screen or fabric

PROCEDURE: Instruct students to pile up the sand or soil on the pie plate, experiment with one erosion-prevention method, drain the water from the plate into a pail, and repeat the procedure several times, using other erosion-prevention methods.

THINK ABOUT IT: None of the methods should completely prevent erosion. However, some methods may reduce the rate of erosion by slowing the flow of water or by helping to anchor the sand or soil.

Linking Prior Knowledge Review mechanical and chemical weathering with students (Lessons 7-1 and 7-2).

2 Teach

Discussion Remind students that weathering is the breakdown of materials on Earth's surface. Ask students to recall the two types of weathering. (mechanical and chemical) Make sure that students understand that weathering and erosion are two different processes.
learning style: auditory

Reinforcement Write the following terms on the board: *Running Water, Glaciers, Wind, Waves,* and *Gravity.* Have students write an explanation that tells what these five things have in common. (They are the five agents of erosion.) **learning style: linguistic**

Answers to questions

1. **DEFINE:** the process by which weathered material is moved from place to place
2. **LIST:** running water, glaciers, wind, waves, and gravity
3. **DEFINE:** the process by which weathered materials are carried by agents of erosion and dropped in new places

8-1 What are erosion and deposition?

INVESTIGATE

Preventing Erosion
HANDS-ON ACTIVITY

1. Fill a pie plate with a pile of sand or soil.
2. See if you can find a way to keep the soil from washing away when water is poured over it. You can try using small wooden sticks, strips of paper, pebbles—anything you can think of.

THINK ABOUT IT: Do any of the methods prevent the soil from washing away? If so, how?

STEP 2

Objectives

Define erosion and deposition. List five agents of erosion.

Key Terms

erosion (ee-ROH-zhuhn)**:** process by which weathered material is removed and carried from a place

deposition (dehp-uh-ZIHSH-uhn)**:** process by which material carried by erosion is dropped in new places

Weathering and Erosion The rocks that make up Earth's crust are broken down by weathering. Weathering, however, is not the only force that acts upon Earth's surface. After rocks are broken down, they may be moved from one place to another. The process of moving weathered material from place to place is called **erosion.**

1 DEFINE: What is erosion?

Agents of Erosion Through erosion, parts of Earth's surface are worn away. One example is the Grand Canyon in Arizona. The Grand Canyon is a gash in Earth's surface almost 2 km deep. At the bottom of the Grand Canyon is the Colorado River. As the river flows, it carries away small pieces of weathered rock. Over millions of years, enough rock bits were carried away to carve out the Grand Canyon.

▲ **Figure 8-2** Rivers carry debris such as weathered rocks to new places.

Running water, glaciers, wind, waves, and gravity are five common agents of erosion. Running water and glaciers carry with them large amounts of rock and soil. Waves move sand onto and off of a beach. Wind picks up sand and moves dirt and dust. Gravity causes soil and rocks to move down a slope.

2 LIST: What are five common agents of erosion?

Teaching Options

INTEGRATING LIFE SCIENCE

Running Water Running water at the bottom of the Grand Canyon picks up and carries away bits of rocks to carve the canyon. Explain that organisms that live in the running water of such a river environment must be specially adapted to the rate of the water's flow in order not to be carried away. Most fish that live in fast-flowing water have streamlined bodies and powerful swimming muscles. Illustrate this point by showing photographs of trout or salmon swimming upstream. **learning styles: auditory, visual**

MORE ON THE WEB

Writing About Erosion Have students learn more about Bryce Canyon in Utah by visiting the Internet. Then, have them write about their favorite features of the canyon while also noting any presence of erosion.

RETEACH

Weathering, Erosion, and Deposition Have students make a flowchart, concept map, or other graphic organizer to illustrate how weathering, erosion, and deposition are related. Refer to the Teacher's Resources for examples of graphic organizers.

Deposition Materials removed by erosion are laid down in new places. The process by which weathered materials are carried by agents of erosion and dropped in new places is called **deposition.** Deposition builds landforms on Earth.

▲ **Figure 8-3** This sandbar was built by deposition.

DEFINE: What is deposition?

CHECKING CONCEPTS

1. The Grand Canyon was formed by the process of __________.
2. There are __________ common agents of erosion.
3. Running water, wind, __________, gravity, and glaciers are agents of erosion.
4. The dropping of rocks carried by a glacier is an example of __________.

THINKING CRITICALLY

5. **CONTRAST:** What is the difference between erosion and deposition?
6. **INFER:** A roadside sign on a mountain road says, "Beware of Falling Rocks." Which agent of erosion is probably at work in the area?
7. **ANALYZE:** Are shells on the beach an example of erosion or deposition? How do you think the shells were carried there?

BUILDING SCIENCE SKILLS

Researching and Writing The federal government spends millions of dollars each year repairing damage caused by erosion. Research the types of damage that are caused by erosion in your area. Explain what is being done to stop the damage and whether you think the effort is worth the money spent.

Real-Life Science

RECLAIMING THE LAND

Coal was formed millions of years ago from buried plants and other organisms. Today, more than half of our country's electric power comes from burning coal.

In the method of mining coal called strip mining, plants and topsoil are removed to reach the coal inside the rock. The exposed coal is called a coal seam. This process leaves a gash in Earth's surface. The newly exposed broken rock is now subject to weathering and begins to erode.

▲ **Figure 8-4** A coal seam is exposed in this strip mine.

Until the 1970s, there were no federal laws about abandoned mines. Many were simply left behind by the companies that had dug them. However, the Surface Mining and Reclamation Act of 1977 changed that. Today, mining companies must put down new layers of topsoil. They must replant former crops or grass. Dry ground must be irrigated. Finally, the site is studied for many years. This is to help make sure that the land weakened by mining returns to its original state.

Thinking Critically Why is it necessary to put down new topsoil?

EXTENDING THE LESSON

Researching Landforms Ask students who have visited an area where the landscape has been shaped by erosion to describe that landscape to the class. Examples of such areas in the United States include the Grand Canyon, Carlsbad Caverns, Badlands National Park, and Niagara Falls. Challenge interested students to research how one of these landforms was created and to prepare an oral report on their findings.
learning styles: visual, auditory

• TEACHING RESOURCES • Online or Print

Teacher's Resources

Lesson Review: Chapter 8, p. 2

Concept Map: Study Tools, p. 4

Flowchart: Study Tools, p. 6

3 Assess

CHECKING CONCEPTS ANSWERS

1. erosion
2. five
3. waves
4. deposition

THINKING CRITICALLY ANSWERS

5. **CONTRAST:** Erosion is the process by which weathered material is moved from one place to another. In deposition, materials carried by agents of erosion are dropped in new places.
6. **INFER:** gravity
7. **ANALYZE:** deposition; by waves

4 Close

BUILDING SCIENCE SKILLS

Researching and Writing Erosion of soil reduces the fertility of croplands. Planting trees, which act as windbreaks, and practicing farming methods that slow water flow and anchor the soil help reduce soil erosion. Shoreline erosion is controlled by building wall-like gabion groins and jetties that reach from the shore into shallow water. By blocking waves and currents, they prevent sand from drifting down the shore.

Real-Life Science

RECLAIMING THE LAND

Real-Life Connection Strip mining is known as open-cut, open-cast, or surface mining when it refers to the mining of valuable nonmetals other than coal and metal ores such as iron and copper. In this process, earthmoving equipment is used to strip away surface materials to expose the ore.

Discussion Show students photographs of strip mines from environmental textbooks such as *Living in the Environment* by G. Tyler Miller, Jr. Have students describe what they see in the photographs and predict the effects that the various agents of erosion would have on the landscapes.

Thinking Critically Answer
In strip mining, topsoil is removed to reach the coal below the surface. Restoring topsoil allows for replanting of vegetation in hopes of returning the land to its original condition.

8-2 *(pp. 180–181)*

1 Introduce

Linking Prior Knowledge Ask students to tell what they know about gravity and give examples or a demonstration of gravity's effect on objects on Earth.

2 Teach

Demonstration Demonstrate mass movement by placing several pieces of gravel on one end of a book lying flat on a desk. Slowly lift that end of the book until the gravel slides down onto the desk. Ask: *What force causes this movement?* (gravity)
learning style: visual

Discussion Ask students if they have ever seen a road sign warning "Beware of Falling Rocks" or noticed rocks at the base of slopes, hills, or mountains. Point out that such signs occur in areas in which there is a danger of rocks falling on the road, and that this is an example of erosion caused by gravity. Explain that the movement of weathered material caused by gravity is called mass movement.
learning style: auditory

Answers to questions

1. **EXPLAIN:** Gravity pulls things downward, causing them to move from areas of higher elevation to areas of lower elevation.
2. **DEFINE:** downhill movement of weathered materials caused by gravity
3. **CONTRAST:** Landslides are sudden movements of rocks down a hill, whereas mudflows are rapid movements of mud down a hill.
4. **IDENTIFY:** earthflow and creep

READING STRATEGY Remind students that comparing means telling how two or more things are alike, and contrasting means telling how they are different. Encourage students to compare and contrast the slow and sudden forms of mass movement.

8-2 How does gravity cause erosion?

Objective
Describe how erosion can be caused by gravity.

Key Terms
mass erosion: downhill movement of weathered materials caused by gravity
talus: pile of rocks and rock particles that collects at the base of a slope

The Effects of Gravity On Earth, the force of gravity pulls all things toward Earth's center. This downward pull of gravity can cause materials to move from areas of high elevation to areas of low elevation.

Most erosion begins with the force of gravity pulling materials toward the center of Earth. For example, the force of gravity can cause rocks and glaciers to move down mountain slopes. Gravity also causes rivers to flow toward the oceans.

1 EXPLAIN: How does gravity cause erosion?

Mass Movement The downhill movement of weathered materials caused by gravity is called **mass erosion.** Mass erosion can occur quickly or slowly. Materials moved by gravity come to rest in piles at the base of a slope. A pile of these materials is called **talus.** The talus may be carried away by running water or other agents of erosion.

▲ **Figure 8-5** Talus that was moved down the mountain by gravity

2 DEFINE: What is mass erosion?

Landslides and Mudflows The sudden mass erosion of rocks down a hill is called a landslide. Rocks on a hill can be loosened by earthquakes, volcanic eruptions, or even heavy rains. The force of gravity may suddenly pull the loosened rocks down the slope of the hill. The moving rocks in a landslide can seriously damage anything that gets in their path.

A mudflow is the rapid mass erosion of mud down a hillside. Mudflows usually happen in dry, mountainous regions after a heavy rainfall.

Like landslides, mudflows can damage property. In the United States, mudflows commonly occur in the hillside communities of southern California.

▲ **Figure 8-6** Mudflow caused this property damage.

3 CONTRAST: How do mudflows differ from landslides?

Earthflow and Creep Earthflow and creep are both slower forms of mass erosion. Earthflow is the slow movement of soil, rocks, and plant life down a slope. Loose earth and extra moisture help cause it. Earthflow can cause part of a hillside to look like it is sliding away.

Creep is even slower. It is caused by gravity pulling soil and rocks down a gentle slope. Creep can be very hard to see. However, bent trees or bent telephone poles are important clues that it is happening.

180

Teaching Options

! FAST FACTS More on Landslides

Geologists recognize several types of landslides. A *rockslide* is a landslide in which masses of bedrock slip down a slope. If a landslide includes huge blocks of rock that tilt back as they slip, it is called *slumping.* In a *rockfall,* pieces of rock tumble from a cliff through the air.

Large landslides can cause sudden, dramatic changes in a landscape. For example, in 1959, a powerful earthquake in western Montana triggered a landslide of material into the Madison River Canyon. This material blocked the river, causing a lake to form behind the blockage. The lake, named Earthquake Lake, reached a depth of more than 58 m (190 ft) within a few weeks.

RETEACH

Mass Movements Write the following head on the board: *Mass Movement.* From this head, draw arrows to each of the following heads: *Rapid Movement* and *Slow Movement.* From *Rapid Movement,* draw arrows to the following heads: *Landslides* and *Mudflows.* From *Slow Movement,* draw arrows to the following heads: *Earthflow* and *Creep.* Have students copy the chart in their notebooks. **learning style: visual**

▲ **Figure 8-7** Creep is hard to see happening. However, its effects, such as bent trees, are obvious.

4 IDENTIFY: What are two examples of slow mass erosion?

CHECKING CONCEPTS

1. The force of gravity can cause rocks and __________ to move down mountain slopes.
2. Gravity causes rivers to flow toward the __________.
3. The sudden movement of rocks down a hillside is called a __________.
4. Weathered material that collects at the base of a slope is called __________.

THINKING CRITICALLY

5. CLASSIFY: Identify each description as evidence of a landslide, mudflow, earthflow, or creep. **a.** A hillside slowly slides down toward the bottom. **b.** A car driving on a mountain road must swerve to avoid falling rocks. **c.** Mud from a heavy rain blocks a road at the base of a mountain. **d.** The fenceposts on a slope are tilting slightly.

Web InfoSearch

Nevado del Ruiz November 13, 1985: The Nevado del Ruiz volcano in Colombia, South America, erupts. The explosion sends lahars rushing down nearby valleys. Lahars are mixtures of water, rock, sand, and mud. The lahars rip huge boulders, trees, and houses from the ground and carry them away. Over 23,000 people are killed.

SEARCH: Use the Internet to find out more about landslides. How do they cause erosion? How do lahars form? Start your search at www.conceptsandchallenges.com. Some key search words are **landslide, lahars,** and **Nevado del Ruiz.**

Hands-On Activity

MODELING GRAVITY ON A SLOPE

Friction is a force that opposes motion. Friction affects how easily material that has been eroded travels down a slope. For this activity, you will need a board, a block of wood, a marble, and sandpaper.

1. Place a board flat on your desk. Place a marble on the board and slowly tip the board up. Observe what happens.
2. Place a block of wood on the board. Slowly lift up one end of the board.
3. Now, cover the wood block with sandpaper. Use tape to hold the sandpaper in place. Repeat Step 2.

▲ **STEP 3** Repeat Step 2 using the wood block covered with sandpaper.

Go Online *active art*
For: Mass Movement activity
Visit: PHSchool.com
Web Code: cfp-2031

Practicing Your Skills

4. EXPLAIN: What do you think caused the differences in your observations?
5. RELATE: Which type of gravity erosion does each of the setups model?

EXTENDING THE LESSON

Gathering Mass Movement Data Have students look for examples of mass movement in their area. Tell students to record the type of mass movement they observe and describe how it has affected buildings, roads, and other structures in the area. Students should organize their data in a chart that they display in class.
learning style: visual

ONGOING ASSESSMENT

Give descriptions of each type of mass movement introduced in the lesson and have students take turns identifying those mass movements.

• TEACHING RESOURCES • Online or Print

Teacher's Resources
Lesson Review: Chapter 8, p. 3
Flowchart: Study Tools, p. 6

Go Online *active art*
For: Mass Movement activity
Visit: PHSchool.com
Web Code: cfp-2031
Students can interact with the art of mass movement online.

3 *Assess*

CHECKING CONCEPTS ANSWERS

1. glaciers or soil
2. oceans
3. landslide
4. talus

THINKING CRITICALLY ANSWERS

CLASSIFY:
5. **a.** earthflow
   b. landslide
   c. mudflow
   d. creep

4 *Close*

Web InfoSearch

Nevado del Ruiz Have students use the Internet to find out more about the Nevado del Ruiz volcano. Ask: *How do landslides cause erosion? How do lahars form?* Some key search words are **landslide, Nevado del Ruiz,** and **lahar.** Students can begin their search for information at www.conceptsandchallenges.com.

Hands-On Activity

MODELING GRAVITY ON A SLOPE

TIME: 15 minutes

PURPOSE: modeling, explaining, relating

DISCUSSION: Suggest that students vary the angle of tilt of the board to see how easily the different materials travel down a slope. Materials that start sliding at a less-steep angle move down a slope more easily.

SAFETY TIP: Remind students to immediately pick up any marbles dropped on the floor.

ALTERNATIVE MATERIALS: A small ball or other round object can be used in place of a marble.

Practicing Your Skills Answers

4. EXPLAIN: Possible answer: When the sandpaper is in place, there is more friction. Thus, the materials do not travel down the slope as easily as when the sandpaper is absent.
5. RELATE: Possible answer: They model earthflow and a landslide.

8-3 *(pp. 182–183)*

1 Introduce

INVESTIGATE

TIME: 15 minutes

PURPOSE: Students will model and observe erosion by running water.

MATERIALS: bars of soap, sink with running water

PROCEDURE: Divide the class into groups based on the number of faucets in your classroom. If there is only one sink in your classroom, consider performing this activity as a demonstration.

THINK ABOUT IT: Possible answers: The soap would dissolve. Eventually, it would be gone. The process could be speeded up by increasing the water pressure or slowed by decreasing it.

Linking Prior Knowledge Ask students to describe the way that water in a river looks after a heavy rain. Relate the muddy quality of the water to running water's capacity to pick up and carry soil. **learning style: auditory**

2 Teach

Discussion Have students locate the Mississippi River on a map. Have them identify several tributaries and where all the water in the river system eventually flows. (The most important tributaries are Illinois, Ohio, Chippewa, Black, Wisconsin, Saint Croix, Iowa, Des Moines, Rock, Missouri, Arkansas, Red and White rivers. The Mississippi River flows into the Gulf of Mexico.) **learning style: visual**

Answers to questions

1. **LIST:** rivers, streams, and runoff
2. **PREDICT:** the hillside, because areas with steep slopes have the most runoff
3. **IDENTIFY:** the Mississippi, and the Missouri rivers.
4. **EXPLAIN:** As gullies are eroded, they become wider and form channels.

8-3 How does running water cause erosion?

INVESTIGATE

Modeling Erosion by Water

HANDS-ON ACTIVITY

1. Turn on a faucet and let water slowly drip from it.
2. Place a bar of soap in a dry place. Place another bar of the exact same soap under the dripping faucet. Let the faucet drip for 10 minutes.
3. Turn off the faucet and compare the two bars of soap.

THINK ABOUT IT: What did you think would happen to the soap when you placed it under the faucet? Why? What do you think it would look like if you left it there longer? How can you speed up or slow down the process?

STEP 3

Objective

Describe how running water causes erosion.

Key Terms

runoff: water from rain or snow that flows into streams and rivers from surface areas

tributary (TRIHB-yoo-tehr-ee)**:** smaller stream that flows into the main stream of a river system

Running Water Running water changes more of Earth's surface than any other agent of erosion. Rivers, streams, and runoff are forms of running water. **Runoff** is water from rain or snow that flows over Earth's surface. Runoff empties into streams and standing bodies of water such as lakes and ponds.

1 LIST: What are three forms of running water?

Runoff and Erosion As runoff flows over Earth's surface, the running water carries away soil particles. The faster the water moves, the more soil the water may carry away. For example, gullies in a hillside are formed by erosion from runoff.

The amount of rainfall, plant growth, and shape of the land affect the amount of runoff in an area. During heavy rains, there is a lot of runoff. However, plants help to stop erosion from runoff.

The roots of plants take in some water and hold soil particles in place. For this reason, areas with a lot of plant growth have less runoff than areas with little plant growth. Areas with steep slopes have the most runoff. The greater the amount of runoff, the more erosion there will be.

2 PREDICT: Would there be more erosion from runoff on a hillside or on flat land? Explain.

River Systems A river is made up of a main stream and all the smaller streams that flow into the main stream. The smaller streams that flow into the main stream of a river system are called **tributaries,** or branches. The Mississippi River is the main stream of a large river system.

▲ **Figure 8-8** The Mississippi River has many tributaries.

Teaching Options

EXTENDING THE LESSON

Relating River Terms Explain to students that an analogy shows a similarity in two or more different things. Provide the following example: *Organic remains are to sedimentary rocks as magma is to igneous rocks.* Ask students to compare a river system to a tree by writing analogies that include the following terms: *main stream, tree trunk, branches, tributaries.* (Possible answer: Tributaries are to a river system as branches are to a tree.)
learning style: logical/mathematical

ESL/ELL STRATEGY

Modeling Water Erosion Have students make a stream table with a 1-m rain gutter. Attach a piece of wire mesh to one end of the gutter. Fill the gutter with gravel, sand, and clay. Place the gutter on a slight incline and pour water on the upper end. (A tin can with small holes will allow the water to fall slowly.) Have students observe how the running water produces grooves called gullies. Point out that gullies can lead to river formation. **learning style: visual**

The Mississippi River has many tributaries. One of them, the Missouri River, is almost as long as the Mississippi River itself. Runoff from the surrounding land feeds the many tributaries of a river system.

3 **IDENTIFY:** What are two rivers in the Mississippi River system?

Formation of a River A river system usually forms in mountains or in hills. The place where a river starts is called its source. When water cannot soak into the ground, runoff flows down the slope of the hill or mountain. The running water erodes the soil and cuts gullies into the slope. A small channel forms. Channels are the paths the streams follow. The small stream may join other small streams to form a larger stream. Larger streams may flow together to form the tributaries of the main stream, or river, of a river system.

4 **EXPLAIN:** How do channels form?

CHECKING CONCEPTS

1. How does runoff cause erosion?
2. What causes gullies in a hillside?
3. What is a tributary?
4. What is a river system?
5. Where do river systems usually form?

THINKING CRITICALLY

6. **PREDICT:** Can planting trees on a hillside help to prevent running water from eroding the land? Explain.
7. **SEQUENCE:** Place the following terms in the correct order to show how rivers can form: channel, small stream, river, large stream, and tributary.
8. **INFER:** How does snow help to form rivers?

BUILDING SCIENCE SKILLS

Modeling and Researching Draw and label a diagram comparing a river and its tributaries to a tree and its branches. Indicate what part of a tree is like the main stream of a river. Then, do research to find out the names of at least one tributary of each of the following rivers: the Nile, the Colorado, the Thames, and the Amazon. Tell where each river is located.

Integrating Environmental Science

TOPICS: pollution, agriculture

PESTICIDE RUNOFF

Pesticides are poisons used to kill insects and other pests. Farmers use them to protect their crops. Sometimes, though, the pesticides are washed away by water, such as during heavy rainstorms. The pesticides can then travel long distances, polluting streams and water in the ground. This can be dangerous for fish and other organisms. In some areas, pesticides are starting to reach poisonous levels.

One solution to this problem is farming without chemical fertilizers and pesticides. More and more farmers are combining programs of soil conservation with methods of natural pest control. One pest-control technique is to release natural predators into the farm fields.

Thinking Critically Who might be harmed by poisoned groundwater?

3 *Assess*

CHECKING CONCEPTS ANSWERS

1. by carrying away soil particles
2. runoff flowing down the slope
3. a smaller stream that flows into the main stream of a river system
4. a main stream and all the streams that flow into it
5. in mountains or in hills

THINKING CRITICALLY ANSWERS

6. **PREDICT:** Yes. The roots of plants absorb water and hold onto soil particles.
7. **SEQUENCE:** channel, small stream, large stream, tributary, river
8. **INFER:** Snow melts to form runoff, which erodes soil to form streams that eventually grow to form rivers.

4 *Close*

BUILDING SCIENCE SKILLS

Modeling and Researching Possible answers for tributaries: *Nile* (Africa): Blue Nile, White Nile, Atbara; *Colorado* (western United States): San Juan, Little Colorado, Gila, Virgin, Gunnison, Dolores, Green; *Thames* (England): Churn, Coln, Windrush, Evenlode, Cherwell, Ock, Thame, Kennet, Loddon, Cole, Wey, Mole, Blackwater, Wandle; *Amazon* (South America): Ucayali, Maranon, Rio Negro, Xingu.

Integrating Environmental Science

TOPICS: pollution, agriculture

PESTICIDE RUNOFF

Life Science Connection Some pesticides are quickly broken down into harmless chemicals after they are applied. Other pesticides, such as the insecticide DDT, remain toxic for a long time and accumulate in the tissues of organisms. When these organisms are consumed, the pesticides are passed up the food chain, becoming more concentrated at each higher level.

Discussion Discuss what students think of when they hear the word *pollution.*

Thinking Critically Answer
Possible answer: animals and people who drink the poisoned water

MORE ON THE WEB

Finding Water on Mars Have students use the Internet to learn about the possibility of water on Mars. Students can also view photographs of channels and gullies that have been found on the surface of the planet.

• TEACHING RESOURCES • Online or Print

Teacher's Resources
Lesson Review: Chapter 8, p. 4
Enrichment: Chapter 8, p. 7

Laboratory Manual
Lab. Challenge: What factors affect erosion caused by running water? pp. 49–52

RETEACH

Identifying Tributaries Provide students with a photocopy of a photograph, diagram, or map that shows a river system. Have them use a colored marker to outline the main stream and a different colored marker to outline its tributaries. **learning style: visual**

8-4 *(pp. 184–185)*

1 Introduce

STUDY HINT Have students write each head in the lesson and one sentence summarizing the main idea of the section below the head.

Linking Prior Knowledge Ask students who have witnessed flooding in person, on television, or in the movies to describe what they saw. Have students predict what they think happens when a river overflows its banks. **learning style: auditory**

2 Teach

Demonstration Use a stream table to demonstrate the forming of a delta. Additional information about stream tables can be found in the ESL/ELL Strategy in Lesson 8-3 on page T182. Have students note the amount of sediment that forms at the bottom of the stream table.
learning style: visual

Discussion Write the following terms on the board: *valley, delta,* and *floodplain.* Identify each term as a landform created by running water. Describe each landform and how it is created. Refer students to Figures 8-10 and 8-11 to observe photographs of a delta and a valley.
learning styles: visual, auditory

Answers to questions

1 **IDENTIFY:** a landform created by running water

2 **LOCATE:** at the mouth of a river

3 **DEFINE:** flat areas on both sides of a river where fertile soil is deposited during floods

WRITING HINT Tell students that if they are answering a question that does not require a full sentence, they should not start their answer with a capital letter or end with a period. However, to improve their writing skills, tell students to try, as often as possible, to use full sentences that express complete thoughts and use correct punctuation.

8-4 What landforms does running water create?

Objective

Describe three landforms created by running water.

Key Terms

abrasion (uh-BRAY-zhuhn)**:** wearing away of rock particles by wind and water

delta: triangular deposit of sediment located at the mouth of a river

sediment (SEHD-uh-muhnt)**:** rock particles carried and deposited by water, wind, or ice

floodplain: flat area on the side of a river where sediments are deposited during floods

Valleys The bed, or bottom, of a river is often solid rock. Small pieces of rock bounce along the bed of a fast-moving river. **Abrasion** is the wearing away of rock particles by scraping. Abrasion occurs as pieces of rock scrape and then settle into the riverbed. Over time, the bed is cut deeper into the rock, and a valley is formed. A valley is a landform that can be created by running water. Fast-moving rivers can cut very deep V-shaped valleys.

▲ **Figure 8-10** Fast-moving rivers can cut deep valleys.

1 **IDENTIFY:** What is a valley?

Delta A triangular-shaped deposit of muddy land called a **delta** often forms at the mouth, or end, of a river. The word *delta* is a Greek letter. The Greek letter delta looks like a triangle.

How does a delta form? When a fast-moving river empties into a larger body of water, it slows down. Pieces of rock and soil carried by the now slow-moving water are deposited. The particles of soil and rock that settle to the bottom, or bed, of a river are called **sediments.** Sediments in the water are deposited at the mouth of the river. This forms the triangle of land known as the delta.

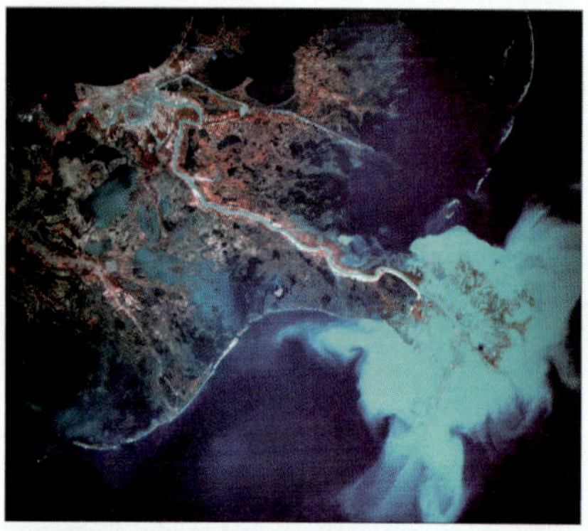

▲ **Figure 8-11** The sediments in the Mississippi River delta appear light blue in this false-color image.

2 **LOCATE:** Where does a delta form?

Floodplains During periods of heavy rain, some rivers overflow their banks, or sides. When a river floods its banks, fertile soil carried along by the river is deposited on the sides of the river. The fertile soil is deposited in low-lying flat areas called **floodplains.**

The soil deposited in floodplains is sometimes carried to the river by tributaries that feed the river. Other times, the soil is picked up by the river itself as it flows along.

Teaching Options

INTEGRATING ENVIRONMENTAL SCIENCE

Dams and Soil Fertility The yearly flooding of the Nile built up the plains of Egypt. These floods brought fertile soil from land in the south to Egypt, making the Nile Valley one of the most fertile regions in the world. In 1968, the Aswan High Dam was built in Aswan, Egypt. This dam stopped the annual flooding of the Nile, and thus the natural addition of nutrients to the soil. Egyptian farmers must now add fertilizers to the soil to maintain its fertility for growing crops. Have students research the reasons for building the Aswan High Dam and its advantages and disadvantages.
learning styles: auditory, interpersonal

COOPERATIVE LEARNING

Building Vocabulary Have students work in pairs or small groups to find in a dictionary the following terms for types of valleys: *canyon, gorge, glen, ravine,* and *chasm.* Have students write the definitions and use an encyclopedia or other reference source to name an example of each type of valley. **learning styles: interpersonal, linguistic**

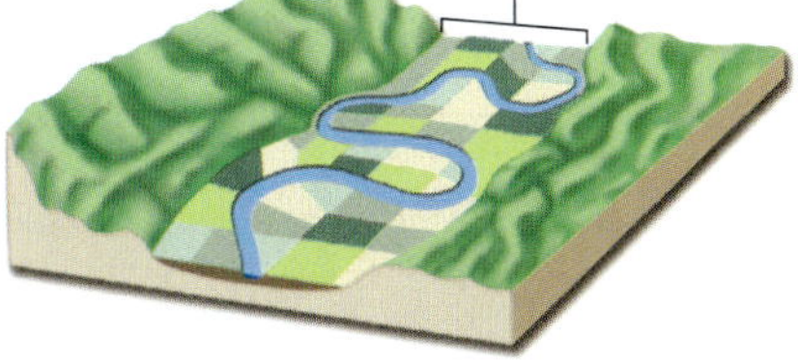

▲ **Figure 8-12** Water spills onto the floodplain.

3 DEFINE: What are floodplains?

CHECKING CONCEPTS

1. What is the bed of a river made of?
2. What happens when pieces of rock bounce along the bed of a river?
3. What is sediment?
4. How is a delta formed?
5. Where does the fertile soil in a floodplain come from?

THINKING CRITICALLY

6. INFER: Why is the triangle of sediments deposited at the mouth of a river called a delta?
7. INFER: Why does a river usually flow more slowly at its mouth than at its source?
8. INFER: Why are floodplains valued for planting crops?

INTERPRETING VISUALS

Use Figure 8-12 to answer the following questions.

9. COMPARE: How does the level of the floodplain compare to the rest of the area?
10. HYPOTHESIZE: What would Figure 8-12 look like after a heavy rain?
11. INFER: What would happen to the farms and houses in the floodplain after a heavy rain?
12. HYPOTHESIZE: How could flooding possibly be prevented here?

Hands-On Activity

MODELING LANDFORMS CREATED BY WATER EROSION

You will need sand or soil, a baking pan, several ice cream sticks, leaves and grass clippings, a straw, and a watering can.

1. Make a mountain of sand or soil at one end of a shallow baking pan.
2. Tilt the pan on one side.
3. Mark off centimeters on the edges of several ice cream sticks. Place the sticks around the mountain, burying them at least 3 cm.
4. Cover the right side of the mountain with leaves and grass clippings.
5. Using a straw, try blowing away the sides of the mountain.
6. Now, use a watering can to make it rain on the mountain.

▲ **STEP 6** Use a watering can to make it rain on the mountain.

Practicing Your Skills

7. OBSERVE: Judging from the marks on your sticks, which side is eroding faster?
8. ANALYZE: How do the leaves and grass clippings help prevent erosion? If these plants had roots, would it help to prevent erosion even more? Why?

MORE ON THE WEB

Writing About Deposition
Have students use the Internet to learn about the Mekong Delta in Vietnam. Then, have students write a report on the delta, the people that live near it, and what their lives are like.

• TEACHING RESOURCES • Online or Print

Teacher's Resources

esson Review: Chapter 8, p. 5

Enrichment: Chapter 8, p. 7

Running Water and Landforms: Visuals, p. 7

3 *Assess*

CHECKING CONCEPTS ANSWERS

1. solid rock or particles of soil and rock
2. Abrasion occurs.
3. particles of soil and rock that settle to the bed of a river
4. When a river empties into a larger body of water, it slows down and deposits its sediments to form a delta.
5. It is deposited by the river.

THINKING CRITICALLY ANSWERS

6. INFER: The Greek letter delta looks like a triangle.
7. INFER: because the slope of the river is steeper at the source
8. INFER: The soil of the floodplains is very fertile.

INTERPRETING VISUALS ANSWERS

9. COMPARE: It is lower and flatter.
10. HYPOTHESIZE: The water would be overflowing the riverbanks.
11. INFER: They could be flooded.
12. HYPOTHESIZE: by damming the river upstream to control its flow or piling sandbags along the banks

4 *Close*

Hands-On Activity

MODELING LANDFORMS CREATED BY WATER EROSION

TIME: 15 minutes

PURPOSE: modeling, observing, analyzing

SAFETY TIP: Have paper towels handy for students to immediately wipe up spills.

ALTERNATIVE MATERIALS: Any kind of markable stick or straw can be used.

DISCUSSION: Ask students: *What properties of plants do you think would be the most useful in preventing erosion?* (roots that spread near the soil to hold soil in place and large leaves or branches that form canopies to filter and control rainfall)

Practicing Your Skills Answers

7. OBSERVE: the left side
8. ANALYZE: The leaves and grass clippings help to hold the soil in place. If the plants had roots, they could further slow erosion by helping to absorb the water and by holding soil particles.

8-5 *(pp. 186–187)*

1 Introduce

STUDY HINT Have students summarize the main idea of this lesson. Then, have them give three details to support the main idea.

Linking Prior Knowledge Before beginning this lesson, have students review how sedimentary rocks are formed (Lesson 3-4) and chemical weathering (Lesson 7-2). Ask students: *What is chemical weathering?* (weathering that changes the chemical makeup of rocks) Also, have students review limestone (Lesson 3-4). Ask students: *What is limestone?* (a sedimentary rock made up of the remains of dead sea creatures)

2 Teach

Demonstration To demonstrate chemical weathering similar to that associated with karst topography, drop diluted hydrochloric acid on a sample of limestone. Allow students to observe the bubbling that occurs when the acid reacts with the limestone. **learning style: visual**

Discussion Display pictures showing karst topography. Ask students how they think each land feature in the pictures was formed. Then, discuss each feature of karst topography. Make sure that students understand that karst topography results from the chemical weathering of limestone. **learning style: visual**

Answers to questions

1. **IDENTIFY:** irregular topography with land that is bare and rocky, has few or no plants, has no lakes or streams, and has underground rivers, caves, and large holes
2. **EXPLAIN:** Limestone is chemically weathered by carbonic acid.
3. **DEFINE:** a series of connected caves
4. **PREDICT:** A natural bridge might be formed.
5. **DEFINE:** holes that are formed when a cavern roof collapses

8-5 What causes caves, caverns, and sinkholes?

Objective

Name and describe some features of karst topography.

Key Terms

karst topography: land that has sinkholes, caverns, and underground rivers

cave system: series of connected underground caves

sinkhole: large hole in the ground formed when the roof of a cavern collapses

Karst Topography In some places in the world, the topography, or general shape and look of the land, is irregular. The land is bare and rocky. There is little or no plant growth. There are no lakes or streams on the surface. Instead, there are underground pools of water, caves, and large holes in the ground. This kind of topography is called **karst topography.**

Karst topography gets its name from a region in Yugoslavia where these kinds of features are common. Karst topography can be seen in some areas of the United States, including parts of Florida and Kentucky.

1 IDENTIFY: What features are typical of karst topography?

Chemical Weathering of Limestone Karst topography results from the action of water in the ground on limestone. In many areas of the world, layers of limestone rock are found under a thin layer of other rocks.

As water in the ground seeps through the layers of rock, the water mixes with carbon dioxide in the rocks. This forms carbonic acid. When carbonic acid reaches the limestone layers, carbonation takes place. The acid reacts with the limestone. The limestone is chemically weathered, or broken down, by the acid.

2 EXPLAIN: How does carbonic acid affect limestone?

Caverns As carbonic acid seeps into small cracks and holes in underground limestone deposits, it reacts with the limestone. Over the years, the holes increase in size and form caves. A series of connected caves forms a **cave system.** Underground water often pools in these caverns.

▲ **Figure 8-13** Many caverns have underground pools of water.

3 DEFINE: What is a cave system?

Natural Bridges When water empties out of a cavern, part of the roof of the cavern may collapse. The remaining section of the roof may form a natural bridge. Natural bridges can also be formed by erosion.

◄ **Figure 8-14** Natural Bridge, near Lexington, Virginia, is over 100 million years old. Made of limestone, this natural wonder is about 66 m tall and between 15 m and 45 m wide.

Teaching Options

FAST FACTS *Stalactites and Stalagmites*

When water containing dissolved calcite drips from the ceiling of a cavern, some of the calcite solidifies on the ceiling. As the calcite builds up, it forms a stalactite. To help students remember the term, tell them that a stalactite "sticks tight" to the ceiling. When drops of water fall on the floor of the cavern, the deposited calcite forms a stalagmite. A stalactite will commonly extend downward and a stalagmite will extend upward until the two meet, forming a single column, or pillar, of calcite. Have students use Figure 8-13 to help them draw and label stalactites and stalagmites in a cavern. **learning style: visual**

COOPERATIVE LEARNING

Modeling Karst Topography Have students work in small groups to make relief maps of an area with karst topography. Have students use clay to model caverns, sinkholes, and natural bridges. Display the models in the classroom. **learning style: tactile**

A natural bridge can also form when a surface stream runs through a crack that is beneath the surface. It then flows back onto the surface. The weathering and erosion is a continuing process that causes rock to wear away. The stream wears away the rock, and the crack gets bigger. Scientists think that Natural Bridge in Virginia formed as a result of a stream wearing away rock.

4 **PREDICT:** What can happen when part of a cavern roof collapses?

Sinkholes Sometimes the whole cavern roof may collapse. When that happens **sinkholes,** or sinks, are formed. A sinkhole may serve as an entrance to a cavern. Tourists enter sinks to visit places such as Carlsbad Caverns in New Mexico and Howe Caverns in New York. If a sinkhole fills with water, a sinkhole lake results.

5 **DEFINE:** What are sinkholes?

CHECKING CONCEPTS

1. The reaction of carbonic acid on limestone is a form of __________.
2. Karst topography gets its name from a region in __________.
3. A __________ is a series of underground connected caves.
4. A large opening from the surface into a cavern is a __________.

THINKING CRITICALLY

5. **EXPLAIN:** Why does karst topography develop where the underground rock is limestone?
6. **HYPOTHESIZE:** Why would karst topography not develop in a desert?
7. **INFER:** Why would you find few surface streams in a karst topography region?

BUILDING GEOGRAPHY SKILLS

Mapping Trace an outline of a map of the United States on a sheet of paper. Look up the locations of the following caverns: Carlsbad Caverns, New Mexico; Howe Caverns, New York; Mammoth Cave, Kentucky; Luray Caverns, Virginia; and Wind Cave, South Dakota. Plot the location of each cavern on your outline map. Have you ever visited any of them? If so, describe what the caverns looked like.

Integrating Life Science

TOPICS: mammals, habitats

BAT CAVE IN CARLSBAD CAVERNS

In Carlsbad, New Mexico, the skies sometimes fill with dark clouds just before sunset. These clouds are not made of smoke or dust but of huge numbers of bats. The bats live inside Carlsbad Caverns, clinging to cave walls. Carlsbad is one of the world's largest natural caverns. It provides a perfect habitat for bats.

▲ **Figure 8-15** Bats leave their caves at dusk.

The most common bat species found in the cave is the Mexican free-tailed bat. Mostly females, the Mexican free-tailed bats that live in the caverns stay there about half the year. Young bats, called pups, are born there. The darkness protects them from predators. At sunset, the bats leave the caverns to hunt insects for food. Clouds of departing bats may have led to the discovery of the caves in the 1800s.

The caverns became a national monument in 1923. Tourists are not allowed into Bat Cave. Some bat species hibernate there in cold weather.

Thinking Critically Why do you think some bats hibernate?

3 Assess

CHECKING CONCEPTS ANSWERS

1. chemical weathering
2. Yugoslavia
3. cavern
4. sinkhole

THINKING CRITICALLY ANSWERS

5. **EXPLAIN:** The limestone reacts with the carbonic acid in the groundwater and is dissolved to form structures common in karst topography.
6. **HYPOTHESIZE:** There is not enough water.
7. **INFER:** Most of the runoff flows into caves or sinkholes.

4 Close

BUILDING GEOGRAPHY SKILLS

Mapping Check students' outline maps. Have students who have visited caverns describe what the caverns looked like. If students have photographs of their trip, have them show the pictures to the class, and tell any unusual things that they noticed such as the constant, year-round temperatures in caves or unusual formations.

Integrating Life Science

TOPICS: mammals, habitats

BAT CAVE IN CARLSBAD CAVERNS

Real-Life Connection Carlsbad Caverns are the largest known caverns and have not yet been fully explored. The caverns include one of the largest known underground chambers in the world. This national monument's spectacular features and unique fauna make it one of the most popular tourist attractions in New Mexico.

Discussion Show students photographs of Mexican free-tailed bats or other bats. Have students describe the animals' characteristics. Ask students to infer why bats have large ears. Then, explain that bats produce high-pitched sounds and navigate by listening to echoes as those sounds are reflected off objects. Explain how this can be especially effective in a cave setting.

Thinking Critically Answer
Possible answers: because food is scarce in cold weather or to conserve energy

MORE ON THE WEB

Researching Cave Life Have students write a short report that explains the unique appearance of the different species that live in Mammoth Cave. Students can learn about Mammoth Cave and its varied wildlife by visiting the Internet.

• TEACHING RESOURCES • Online or Print

Teacher's Resources
Lesson Review: Chapter 8, p. 6
Enrichment: Chapter 8, p. 7

QUICK CONVERSIONS

Natural Bridge Dimensions	Metric/SI	English
Height	66 m	220 ft
Width	27 m	89 ft

8-6 (pp. 188–189)

1 Introduce

STUDY HINT Before beginning this lesson, have students predict an answer to the lesson title. Then, have them read the lesson to see if their prediction was correct and revise their answer as needed.

Linking Prior Knowledge Remind students of how rivers form (gullies with water deposits) and that a glacier is a frozen river. Have students recall experiences of packing snow to make snowballs or snow sculptures. Ask students: *As more snow and pressure is added, how does the snow change?* (Snow compacts to form ice.) After students read page 188, recall students' experiences and compare the making of a snowball or a snow sculpture to glacier formation.

2 Teach

Demonstration Float an ice cube in a clear glass of water. Have students note that most of the ice cube is below the surface of the water. Only a small part of the ice cube is visible above the surface. Compare the ice cube to an iceberg. **learning style: visual**

Discussion Make sure that students understand that there are two kinds of glaciers: valley glaciers and continental glaciers. Show students pictures of valley glaciers and continental glaciers. **learning style: visual**

Answers to questions

1. **DEFINE:** a moving river of ice and snow
2. **NAME:** valley glaciers and continental glaciers
3. **EXPLAIN:** Only a small part of an iceberg is visible above the water, so ships might not see it in time and crash into it.
4. **IDENTIFY:** a period of very cold temperatures in which glaciers spread over Earth's surface

8-6 What is a glacier?

Objectives

Explain how a glacier is formed. Name two kinds of glaciers.

Key Terms

glacier (GLAY-shuhr): moving river of ice and snow

ice cap: large sheet of ice found near Earth's poles

iceberg: large piece of a glacier that enters the ocean

Ice Rivers A **glacier** is a moving river of ice and snow. Glaciers form where the temperature is close to or below freezing most of the year. In these places, the snow does not melt after heavy snowfalls. More snow falls on top of snow that is already on the ground. The snow gets deeper and deeper. Ice forms at the bottom of this deep layer of snow.

Gravity and the weight of the snow on top of the ice cause the ice to move forward. When more snow melts than falls, the glacier begins to recede.

▲ **Figure 8-16** Glaciers look like what they are—rivers of ice.

1 **DEFINE:** What is a glacier?

Kinds of Glaciers There are two kinds of glaciers. Glaciers that form in mountains and move slowly downhill are called valley glaciers or alpine glaciers.

Other glaciers form near Earth's poles. These glaciers form from very large sheets of ice called **ice caps.** Another name for these glaciers is continental glaciers.

2 **NAME:** What are two kinds of glaciers?

Icebergs When a glacier reaches the ocean, a large piece of it may break off and float away. This is called an **iceberg.** Large amounts of sediment may be frozen into an iceberg. As the iceberg melts, these sediments sink to the ocean bottom.

Most of an iceberg is below the surface of the water. Only a small part of the iceberg is visible above the water's surface.

Many ships have crashed into icebergs. To protect ships, the U.S. Coast Guard keeps track of where icebergs are and the direction in which they are heading. Sometimes icebreakers are used to break up icebergs and clear the way for ships.

▲ **Figure 8-17** Iceberg in the ocean near Antarctica

3 **EXPLAIN:** How do icebergs threaten ships?

Ice Ages An ice age is a period of very cold temperatures. Glaciers grow and spread during ice ages. There have been many ice ages in Earth's history. The last one ended about 8,000 to 15,000 years ago.

During the last ice age, the sheet of ice around the North Pole grew larger. The expanding ice cap eventually covered much of North America.

4 **IDENTIFY:** What is an ice age?

Teaching Options

INTEGRATING PHYSICAL SCIENCE

Flowing Ice Have students line a long, shallow box with plastic wrap. Then, have them fill the container with water and place it in a freezer. Have students remove the ice block from the container and return it to the freezer, supporting each end on objects, leaving the middle unsupported. Students should then place a weight on the middle of the ice block. Ask students to check the ice daily and note the physical change. The middle of the ice block should bend downward over time. This is caused by the weight on top of the block. Explain that the process is similar to the forward movement of a glacier, caused by the weight of snow and ice at the top. **learning style: visual**

ONGOING ASSESSMENT

Have students make a two-column chart with the headings *Glaciers* and *Icebergs* to compare and contrast glaciers and icebergs. Then, have students use their charts to tell how glaciers and icebergs are alike and how they are different. **learning style: linguistic**

CHECKING CONCEPTS

1. Another name for a moving river of ice and snow is a __________.
2. A glacier that forms in mountains is a __________ glacier.
3. A glacier that forms near Earth's poles is a __________ glacier.
4. An __________ is a period during which glaciers spread over Earth's surface.
5. The last ice age ended about __________ years ago.

THINKING CRITICALLY

6. **INFER:** How are glaciers "rivers of ice"?
7. **PREDICT:** Even a small increase in average temperatures around the world could cause the ice caps to melt. This melting would raise the sea level. How would this affect coastal cities?
8. **INFER:** What happens to sea level during an ice age? Explain your answer.

Web InfoSearch

Titanic The British steamer *Titanic* was the largest ship in the world when it was built. It was thought to be unsinkable. In April 1912, the *Titanic* set sail on its first voyage. As it sailed toward New York, it struck an iceberg and sank. Of the more than 2,200 people on board, only 705 were rescued.

SEARCH: Use the Internet to find out more about the *Titanic.* Why did it sink? Why didn't the crew see the iceberg in time? Start your search for information at www.conceptsandchallenges.com. Some key search words are ***Titanic,* RMS *Titanic,*** and **titanic—movie.**

▲ **Figure 8-18** *Titanic*

How Do They Know That?

GLACIERS GROW AND SHRINK

Jean Louis Agassiz (AG-uh-see) (1807–1873) was a Swiss naturalist, geologist, and teacher. Agassiz earned degrees in philosophy and medicine. His early research was on fossil fishes. In 1836, Agassiz began to study glaciers. He built a shack on top of a glacier to study the glacier's movement. Agassiz was the first person to discover that the center of a glacier moves faster than the sides.

Agassiz observed that as glaciers moved, rocks carried by the glacier carved scratches into other rocks the glacier passed over. Using these observations, Agassiz showed that sheets of ice had once covered large areas of land. From his research, Agassiz also concluded that Earth had gone through several ice ages.

▲ **Figure 8-19** Jean Louis Agassiz giving a lecture

In 1846, Agassiz moved to the United States. He taught at Harvard University and founded Harvard's Museum of Natural History. At his death, a boulder carried by a glacier was placed near his grave as a memorial.

Thinking Critically What might have led Agassiz to conclude that there had been several ice ages?

• TEACHING RESOURCES •
Online or Print

Teacher's Resources
Lesson Review: Chapter 8, p. 8

EXTENDING THE LESSON

Ice Ages Have interested students research the four major ice ages and write their findings in a report. Tell students to include a description of when the ice ages occurred and the conditions that led to each. Major ice ages occurred from about 800–600 million years ago, 460–430 million years ago, 350–250 million years ago, and 4 million–approximately 10,000 years ago. Factors involved in causing ice ages include small changes in the path of Earth's revolution around the Sun that reduce the amount of solar energy reaching Earth's surface; continental drift, which changes Earth's distribution of land; changes in ocean currents; and reduction of heat-trapping carbon dioxide gas in the atmosphere.

3 Assess

CHECKING CONCEPTS ANSWERS

1. glacier
2. valley
3. continental
4. ice age
5. 8,000 to 15,000

THINKING CRITICALLY ANSWERS

6. **INFER:** They flow very slowly, like rivers, from a higher point to a lower point.
7. **PREDICT:** A rise in sea level could cause flooding in coastal cities.
8. **INFER:** Sea level is lowered because more of Earth's water is frozen as glacial ice.

4 Close

Web InfoSearch

Titanic Have students use the Internet to learn about the *Titanic. Why did the* Titanic *sink? Why didn't the crew see the iceberg in time?* Some key search words are ***Titanic,* RMS *Titanic,*** and **titanic −movie.** Explain to students that the minus sign (−) in front of *movie* helps avoid search results about the movie *Titanic.* Students can begin their search at www.conceptsandchallenges.com.

How Do They Know That?

GLACIERS GROW AND SHRINK

Physical-Science Connection All ice within a glacier does not flow at the same rate. Because of frictional drag with the bedrock floor, lower portions of a glacier flow more slowly than upper portions. Similarly, frictional drag with the valley walls results in glacial flow being fastest in the center of the glacier and slowest along the sides.

Discussion Ask students to suggest a method for measuring glacial movement. Then, explain that the first measurements of glacial movement involved placing stakes in a straight line across the top of an alpine glacier and then periodically recording their positions.

Thinking Critically Answer
Possible answer: Agassiz observed evidence, such as scratches from sediment, of ice sheets from different time periods.

8-7 (pp. 190–191)

1 Introduce

STUDY HINT Have students write the lesson objectives in the form of questions. As students read the lesson, have them look for and record the answers to the questions.

Linking Prior Knowledge Have students review the rock cycle and recall that erosion is one of the forces that changes rocks (Lesson 3-8). Ask: *What changes rocks on Earth's surface?* (ice, rain, wind, and chemicals in air and water)

2 Teach

Demonstration To demonstrate erratics, place a weight on top of an ice cube. Elicit that when the ice cube melts, the water will evaporate, and the weight, which was supported by the ice, will be left sitting on the table. Relate this process to the deposition of erratics. **learning style: visual**

Discussion Review the definition of *glacier*. (a moving river of ice and snow) Ask students to infer how something as slow-moving as a glacier, which flows very slowly compared to river water, might cause erosion. Accept all reasonable responses. **learning style: auditory**

Reinforcement If possible, distribute samples of rocks with grooves, or striations, caused by glacial erosion, or show students photos of such rocks. Have students examine the rocks or photos. Point out that the grooves were formed when small pieces of rock were carved away by gravel and rocks frozen into a glacier. Then, have students retell in their own words how the grooves formed.
learning styles: visual, auditory

Answers to questions

1. **EXPLAIN:** Pieces of rock carried by the glacier scrape and carve away bedrock.
2. **EXPLAIN:** a glacier scraping away the floor and sides of a valley
3. **DESCRIBE:** The sediments are deposited in a new place.
4. **EXPLAIN:** by a glacier

8-7 How do glaciers cause erosion?

Objectives

Explain how a glacier causes erosion. Describe two results of glacial deposition.

Key Terms

hanging valley: small glacial valley above a main valley

till: rock material deposited by a glacier

erratic (ih-RAT-ihk)**:** boulder left behind by a retreating glacier

Erosion by Abrasion Gravel and pieces of rock are frozen into the ice at the bottom and along the sides of a glacier. As the glacier moves over bedrock, small pieces of the bedrock are scraped and carved away by the rocks set into the glacier. This rock abrades the valley as the glacier moves forward.

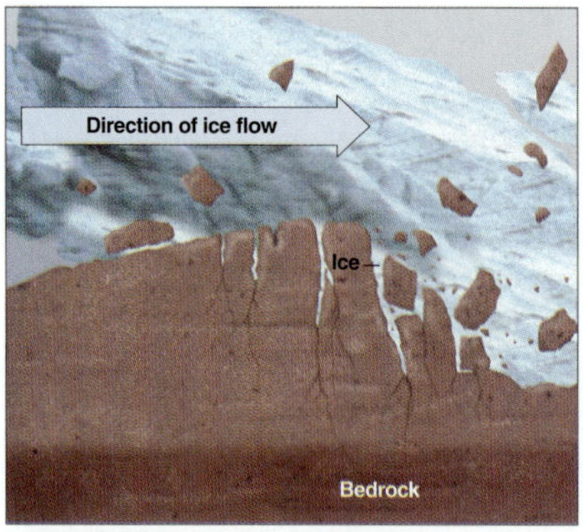

▲ **Figure 8-20** Glaciers abrade, or scrape away, pieces of bedrock as they flow over it.

1 **EXPLAIN:** How does a glacier cause abrasion?

U-shaped Valleys As a glacier scrapes away the floor and sides of a valley, the valley becomes U-shaped. Sometimes small side glaciers flow into the main glacier. When the side glaciers melt, they leave small glacier valleys. These valleys are called **hanging valleys.**

▲ **Figure 8-21** This hanging valley is located in New Zealand.

2 **EXPLAIN:** What causes a U-shaped valley to form?

Glacial Deposition Frozen into a glacier are large rocks and sediments. As the glacier moves, these materials are carried to new places. If a glacier moves into a warmer area, the ice around the edges begins to melt. The glacier appears to be shrinking. Rocks and sediments that were frozen in the ice are left behind. This loose material deposited by the glacier is called **till.**

3 **DESCRIBE:** What happens to sediments as a glacier melts?

Erratics A boulder left behind by a retreating glacier is called an **erratic.** Central Park in New York City has many erratics. These boulders are not like the bedrock beneath the surface. The bedrock they came from is located many kilometers away.

▲ **Figure 8-22** Erratics in Central Park, New York City

Teaching Options

! FAST FACTS More Landforms Created by Glaciers

Alpine glaciers create a variety of landforms besides moraines, drumlins, and lakes. At the top of an alpine glacier, the ice erodes a bowl-shaped depression called a *cirque*. Two glaciers eroding opposite sides of a mountain eventually produce a sharp-crested ridge known as an *arête*. A saddle-shaped depression in such a ridge is called a *col*. A sharp peak called a *horn* results when glaciers erode several sides of a mountain; the Matterhorn in Switzerland is an example of a horn. When a glacier at sea level recedes, the sea invades the glacier valley, forming a *fjord*.

RETEACH

Modeling Glacial Deposition Have small groups of students use a pen to poke a few holes in the sides of a plastic container. Have them fill about half the container with ice or snow, then cover the ice or snow with sand, dirt, and gravel. Allow the ice or snow to melt. (Students can place the container in a larger container to trap the water.) Have students compare the deposit left when the ice or snow melts to glacial till.
learning styles: visual, tactile

The boulders are too large and too heavy to have been moved by running water. How did they get to Central Park? They were carried there by a glacier. The boulders were frozen into the ice and left behind when the glacier melted.

4 EXPLAIN: How are erratics moved?

CHECKING CONCEPTS

1. What is frozen into the ice at the bottom and sides of a glacier?
2. What happens to the shape of a valley as a result of glacial erosion?
3. What happens when small side glaciers melt and disappear?
4. What happens when a glacier moves into a warmer area?
5. What is the loose material deposited by a glacier called?

THINKING CRITICALLY

6. HYPOTHESIZE: How can you tell if a valley was formed by a glacier?
7. EXPLAIN: Why is the main valley of a glacier lower than a hanging valley left behind by a side glacier?
8. COMPARE: How is erosion caused by a glacier similar to erosion caused by a river?

DESIGNING AN EXPERIMENT

Design an experiment to solve the following problem. Include a hypothesis, variables, a procedure with materials, and a type of data to study. Also tell how you would record your data.

PROBLEM: Using a board, sand, and a liquid, how can you demonstrate that glaciers change the shape of the land?

Hands-On Activity

OBSERVING ABRASION CAUSED BY GLACIERS

You will need safety goggles, sand, water, a bar of soap, a plastic container, and paper towels.

1. Put some sand into a plastic container.
2. Fill the container with water.
3. Place the container in a freezer until the water turns to ice.
4. Remove the ice from the container.
5. Hold the piece of ice with paper towels. Move the ice, sand side facing it, over the bar of soap.
6. Press some sand into the soap. Repeat Step 5.

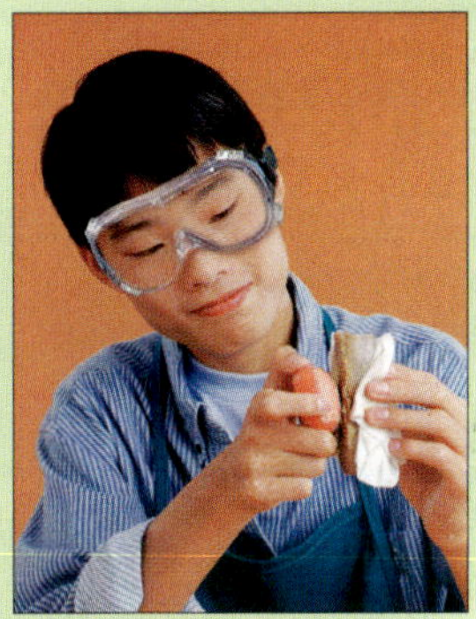

▲ **STEP 5** Rub the soap against the sand-coated ice.

Practicing Your Skills

7. OBSERVE: What happened to the soap when you moved the ice and sand over it?
8. OBSERVE: Did the sand in the soap mix with the sand in the ice?
9. OBSERVE: What marks, if any, were left in the soap by the sand?
10. MODEL: How does this activity show how a glacier can erode the land?

MORE ON THE WEB

Researching Valley Glaciers
Have students use the Internet to learn about valley glaciers. Then, have students write a short report explaining the differences between valley glaciers and hanging valleys.

• TEACHING RESOURCES • Online or Print

Teacher's Resources

Lesson Review: Chapter 8, p. 9

Designing an Experiment: Study Tools, p. 14

Designing an Experiment Rubric: Teaching Tools, p. 7

ESL/ELL STRATEGY

Multiple Meanings Remind students that many words have more than one meaning and that certain words, such as *till* and *erratic*, have a meaning that is unique to science.

3 Assess

CHECKING CONCEPTS ANSWERS

1. sediments and pieces of rock
2. The valley becomes U-shaped.
3. Hanging valleys form.
4. Its edges begin to melt, and rocks and sediments frozen in the ice are deposited.
5. till

THINKING CRITICALLY ANSWERS

6. HYPOTHESIZE: It will be U-shaped.
7. EXPLAIN: because the main valley continues to be eroded by the glacier
8. COMPARE: Like a river, glaciers carve away bedrock, forming valleys.

4 Close

DESIGNING AN EXPERIMENT

Use the Designing an Experiment Rubric from the Teacher's Resources to assess students' experiments. Fill in the rubric with the additional information below. For this assignment, students should have:

- included in the procedure a board, sand, and a liquid.
- demonstrated that glaciers change the shape of the land.

Hands-On Activity

OBSERVING ABRASION CAUSED BY GLACIERS

TIME: 10 minutes each on two different days

PURPOSE: observing, modeling

DISCUSSION: Ask students to note the direction of the scratches or grooves on the soap. (They should be parallel in the direction of the movement of the ice.) Ask students to predict what will happen when the ice melts. (The sand will be deposited, and the water from the melted ice will either run off or evaporate.)

SAFETY TIP: Tell students to immediately pick up dropped soap or wipe up spills.

Practicing Your Skills Answers

7. OBSERVE: Pieces of the soap were scraped away.
8. OBSERVE: yes
9. OBSERVE: grooves
10. MODEL: Just as the ice and sand scrape the soap, glaciers with sediment frozen in them scrape the land.

8-8 *(pp. 192–193)*

1 Introduce

STUDY HINT Have students read the lesson title. Then, have them sketch Figure 8-23 in their notebooks. As they read the lesson, they should draw and label on their sketches the landforms that are shown in the figure.

Linking Prior knowledge Review with students how glaciers cause erosion and deposition. Tell students that in this lesson, they will learn how these processes produce certain landforms.

2 Teach

Discussion Refer students to Figure 8-23. Point out the moraine, U-shaped valley, drumlin, glacial lake, and kettle lake. Discuss how glaciers form each of these landforms. **learning style: visual**

Answers to questions

1 **NAME:** terminal moraine and lateral moraine

2 **DEFINE:** an oval-shaped mound of till

3 **DEFINE:** a lake formed when a mass of glacial ice melts

READING STRATEGY Write the words *terminal* and *lateral* on the board and circle the *-al* in each word. Explain that this adjective-forming suffix means "of," or "like." Have students use this definition and information from the text to determine the meanings of *terminal* and *lateral*. (of the end; of the side) Have students identify other words in the lesson that include the same suffix. (glacial, physical)

WRITING HINT Remind students that when they explain how something happened, they should use time order words such as *before, first, next, after,* and *then.*

8-8 What landforms do glaciers create?

Objective

Describe two landforms created by glaciers.

Key Terms

moraine (muh-RAYN)**:** ridge of till deposited by a melting glacier

drumlin: oval-shaped mound of till

kettle lake: lake formed by a retreating glacier

Moraines Glaciers are enormous and well-insulated blocks of ice. It takes a very long time for them to melt. When a glacier melts, it deposits till. The till builds up a long, low ridge. This ridge of till is called a **moraine.** When till is deposited at the front edge of a glacier, a terminal moraine is formed. When till is deposited along the sides of a glacier, a lateral moraine is formed.

1 **NAME:** What are two kinds of moraines?

Drumlins Sometimes when a glacier moves, it leaves behind oval-shaped mounds of till. These mounds are called **drumlins.** The tip of a drumlin points in the direction the glacier was moving.

Drumlins often form in groups. Drumlins can be seen in New York and in the farmlands of Vermont, among other places.

2 **DEFINE:** What is a drumlin?

Glacial Lakes During the last ice age, glaciers formed in many river valleys. These glaciers weathered and eroded the valleys and made them deeper. As the glaciers moved, till carried by them was left behind. Then the glaciers melted. The water of a melting glacier is called meltwater. Meltwater filled the valleys with water and formed lakes. Glacial lakes are usually narrow, long, and deep. The Great Lakes and New York's Finger Lakes are glacial lakes.

▲ **Figure 8-24** Grenell Glacier Lake in Montana

Sometimes, a moving glacier leaves behind huge masses of ice. These ice masses can become covered with sediments. When the ice melts, it leaves a large hole in the ground. The sediments in the meltwater fall to the bottom. The hole fills with meltwater, and a lake is formed. A lake formed in this way is called a **kettle lake.** Lake Ronkonkoma in New York is a kettle lake.

3 **DEFINE:** What is a kettle lake?

▼ **Figure 8-23** Many different landforms are carved by glaciers.

Teaching Options

! FAST FACTS *Glaciers and Midwest Landforms*

Landforms over much of the midwestern United States were formed by glaciers that advanced and retreated across this region during the last ice age. These glaciers covered large parts of what are now Illinois, Indiana, Iowa, Minnesota, Wisconsin, and other states. As the glaciers retreated, they left behind thick deposits of clay, sand, and gravel, which formed flat plains. The moraines produced by these glaciers became low hills and ridges.

Glaciers early in the last ice age spread very far south. As a result, glacial deposits in southern areas of the Midwest were exposed to erosive forces longer than deposits in northern areas. That explains why eroded landforms, such as steep-sided valleys and cliffs, are more common in the southern areas.

EXTENDING THE LESSON

Glacial Lakes Display a map of the Great Lakes. Tell students that at one time, the lakes were river valleys and that the rivers became blocked by a glacier. Encourage students to find out more about this and prepare a report with diagrams to share with the class. **learning styles: auditory, linguistic**

CHECKING CONCEPTS

1. A ridge formed by a glacier is a __________.
2. The material deposited at the front edge of a glacier forms a __________ moraine.
3. Glacial lakes are usually long and __________.
4. A lake formed when a block of ice from a glacier melts is called a __________ lake.
5. Melting glaciers deposit __________.

THINKING CRITICALLY

6. INFER: What can a scientist learn about a glacier by studying moraines?
7. ANALYZE: Why was the name *kettle* given to this kind of glacial lake?
8. ANALYZE: The tip of a drumlin points south. In what direction was the glacier that formed the drumlin moving?
9. INFER: Why do you think glacial lakes are usually narrow, long, and deep?

Web InfoSearch

The Ross Ice Shelf More than 65 percent of the fresh water on Earth is locked up in ice. One of the world's largest bodies of floating ice is the Ross Ice Shelf in Antarctica. In 2001, a huge iceberg, about 160 km long and 35 km wide, was found in the sea of Antarctica. It had broken off from the Ross Ice Shelf. Scientists placed tracking devices and other scientific instruments on the iceberg. They are concerned that it could block shipping lanes in the area.

SEARCH: Use the Internet to find out the size of the piece that broke off the Ross Ice Shelf. Where did the Ross Ice Shelf get its name? Start your search at www.conceptsandchallenges.com. Some key search words are **Ross Ice Shelf, ice shelf, Ross Sea, icebergs,** and **Antarctic icebergs.**

People in Science

PALEOCLIMATOLOGIST

Paleoclimatologists (pale-ee-oh-KLEYE-muh-TAHL-uh-jihsts) are scientists who study past climates. They do not use thermometers or barometers. They look for clues to past climates in ocean sediment, coral, glaciers, ice caps, and tree rings.

Lonnie Thompson, a paleoclimatologist who teaches at Ohio State University, led an expedition to Nevado Sajama in the late 1990s. This is Bolivia's highest mountain. Drilling equipment, scientists, technicians, and a hot-air balloon crew were all taken to the top. During the expedition, frozen ice cores were removed from the summit glacier. They were taken away, still frozen, by the hot-air balloon crew. The cores would be studied to learn more about the climate 14,000 years ago.

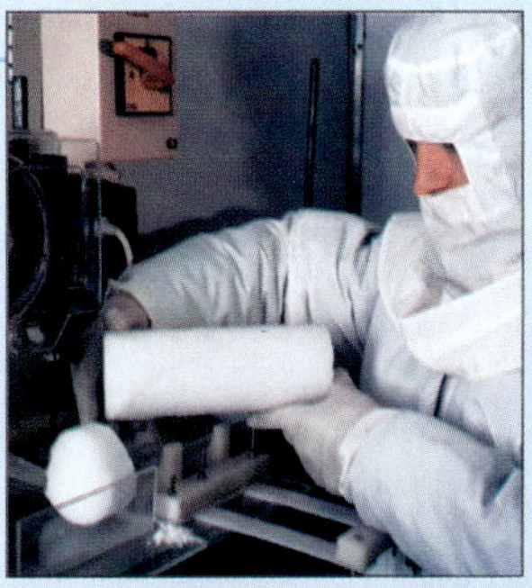

▲ **Figure 8-25** A paleoclimatologist saws off a section of an ice core for further study.

To be a paleoclimatologist, you need a background in mathematics and science. For fieldwork, you must be in good physical shape. Also, paleoclimatologists are often gone from home for a long time.

Thinking Critically Why are the usual weather tools useless to a paleoclimatologist?

3 Assess

CHECKING CONCEPTS ANSWERS

1. moraine
2. terminal
3. narrow or deep
4. kettle
5. till

THINKING CRITICALLY ANSWERS

6. INFER: how far the glacier reached before retreating or where the glacier formed
7. ANALYZE: A kettle is a rounded, hollow container filled with a liquid such as water. Kettle lakes have a similar shape.
8. ANALYZE: south
9. INFER: Glacial lakes formed in river valleys that were made deeper by glacial erosion.

4 Close

Web InfoSearch

The Ross Ice Shelf Have students use the Internet to learn about the Ross Ice Shelf. Ask students: *Where did it get its name? What are some other ice shelves on Earth?* Some key search words are **Ross Ice Shelf, ice shelf, Ross Sea, icebergs,** and **Antarctica icebergs.** Students can begin their search at www.conceptsandchallenges.com.

QUICK CONVERSIONS		
Ross Ice Shelf Iceberg Dimensions	Metric/SI	English
Length	160 km	99 mi
Width	35 km	22 mi

• TEACHING RESOURCES • Online or Print

Teacher's Resources
Lesson Review: Chapter 8, p. 10

ONGOING ASSESSMENT

Hold up pictures of landforms created by glaciers. Have students identify each landform and describe how it formed.

People in Science

PALEOCLIMATOLOGIST

Real-Life Connection Although *paleo-* means "ancient," data gathered by paleoclimatologists are not just applied to the past. Past climatic patterns can be used to predict future climatic trends.

Discussion On a world map, point out the location of Bolivia and its proximity to the equator. Ask students: *How can glacial ice exist in this hot, equatorial country?* (On mountaintops, temperatures remain cold enough to support glacial ice.)

Thinking Critically Answer
Paleoclimatologists study past climates. Most weather tools measure current conditions, such as air pressure and temperature.

8-9 *(pp. 194–195)*

1 Introduce

STUDY HINT Have students write a paragraph explaining the differences among sea cliffs, sea arches, sea stacks, and wave-cut terraces. The paragraph should also indicate what they have in common. Students should make simple sketches to accompany the explanations.

Linking Prior Knowledge Have students recall mechanical weathering (Lesson 7-1). Then, elicit features of wave action from students' experiences at the beach. Ask students to describe the force of waves. Point out that when a wave breaks on a person, it can knock the person down. Explain that the force of waves striking the shoreline also affects rocks.

2 Teach

Discussion If possible, arrange for your class to go on a field trip to an ocean area to observe shoreline features caused by wave erosion, or show students a video on wave erosion. Afterward, have students describe the features that they observed. **learning style: visual**

Answers to questions

1. **DESCRIBE:** by carrying away weathered sand and rock particles
2. **DEFINE:** a steep rock face caused by erosion along a rocky shoreline
3. **IDENTIFY:** a sea stack

WRITING HINT Remind students that, when writing a sentence, they should make sure that the verb agrees with its subject. Write the following sentence on the board: *The force of ocean waves breaks rocks into small pieces*. Point out that the subject (*force*) is singular, so the verb (*breaks*) is also singular. The prepositional phrase *of ocean waves* that comes between the subject and verb does not affect subject-verb agreement.

8-9 How do ocean waves cause erosion?

Objectives

Describe how ocean waves cause erosion. List shoreline features caused by wave erosion.

Key Terms

wave-cut terrace: flat section of rock formed by the erosion of a sea cliff

sea arch: gap formed when waves cut completely through a section of rock

sea stack: column of rock remaining after the collapse of a sea arch

Wave Erosion A **wave** is an up-and-down movement of water. Ocean waves are formed when wind blows over the water. Tides, storms, and earthquakes also cause waves.

The force of ocean waves striking a shoreline can break rocks into small pieces. The rock pieces grind against one another. This grinding motion causes abrasion. Abrasion wears down the rock particles until they become sand. The sand is then carried away by the waves.

▲ **Figure 8-26** Waves crash against a shoreline in Oregon.

Waves also cause chemical weathering of the rocks along a shoreline. As waves meet the shoreline, salt water is forced into cracks in rocks. The chemical action of salt water causes the rocks to break down and makes the cracks larger. Broken pieces of rock are then carried away by the waves.

1. **DESCRIBE:** How do waves cause shoreline erosion?

Sea Cliffs, Caves, and Terraces Waves pound into the rocks on a rocky shoreline. The rocks are broken down into small pieces. The broken rock is carried away by the waves. A sea cliff is formed. A sea cliff is a steep rock face caused by wave erosion. Soft rock is eroded more quickly than hard rock. When waves hollow out the soft rock in a sea cliff, a sea cave is formed.

Over time, the bottom of a sea cliff may be slowly worn away. A flat section of rock may remain below the surface of the water. This flat platform is called a **wave-cut terrace.**

2. **DEFINE:** What is a sea cliff?

Sea Arches and Sea Stacks When waves cut completely through a section of rock, a **sea arch** is formed. The sea arch looks like a natural bridge. In time, the top of the sea arch may fall into the water. The remaining columns of rock are called **sea stacks.** Sea stacks were once the sides of an arch.

▲ **Figure 8-27** A sea arch in Quebec, Canada

3. **IDENTIFY:** What is left when the top of a sea arch falls into the water?

194

Teaching Options

! FAST FACTS *Controlling Coastal Erosion*

Seawalls and groins are artificial structures designed to control erosion along coastlines. A *seawall* is an embankment that protects the shore by absorbing or reflecting the energy of crashing waves. A *groin* is a barrier of concrete or large boulders that extends into the ocean along a beach. Sand accumulates on one side of the groin, preventing erosion of the beach on that side. However, beach erosion on the other side of the groin is accelerated.

COOPERATIVE LEARNING

Modeling Wave Erosion Have students work in small groups to make clay models of wave erosion. Have each group make a different model. Tell students to show what happens when waves erode soft rock on the shore. **learning style: tactile**

CHECKING CONCEPTS

1. What are two causes of ocean waves?
2. What are two ways waves break down rocks?
3. What is the term for a steep rock face caused by erosion along a rocky shoreline?
4. What is formed when part of a sea cliff is hollowed out?
5. What is the shoreline feature that looks like a natural bridge called?

THINKING CRITICALLY

6. **ANALYZE:** What processes or events other than wave erosion might change the shape of a shoreline?
7. **HYPOTHESIZE:** The shoreline of Cape Cod, Massachusetts, is made up of loose glacial deposits. Would you expect this shoreline to erode quickly or slowly? Explain.

BUILDING MATH SKILLS

8. **CALCULATE:** Barrier island beaches are easily eroded by storm waves. An entire shoreline can be reshaped in a day. If a shoreline erodes at 1.5 m per year, how much erosion would occur over 25 years?

HEALTH AND SAFETY TIP

Hurricanes are the most powerful storms on Earth. They erode coastlines, damage homes, and cause injury. If a hurricane watch is issued, be sure you have canned goods, gasoline for the car, and first-aid kits on hand. Put fresh batteries in flashlights and portable radios. Tape your windows or close your shutters to avoid flying glass. If a hurricane warning is issued, leave the area if you live near the shore. If you stay, don't go out in the storm. Fill your bathtub with water. Tie down anything that can be blown away. After the storm, watch out for downed power lines and broken gas mains.

Real-Life Science

COASTAL EROSION AND STORMS

Hurricanes batter the U.S. Atlantic coastline each year. During the 1990s, there were more than 150 hurricanes reported. Twenty of those hurricanes were major storms. In 2005, the southern coast of the United States was hit by Hurricane Katrina and Hurricane Rita. These storms caused millions of dollars of damage and forced millions of people from their homes. Hurricane Katrina caused over 200 billion dollars of damage and killed 1,500 people. It stands as the most destructive storm in U.S. history.

▲ **Figure 8-28** Hurricane damage in New Orleans, Louisiana.

Large storms cause rapid changes along a coastline. However, slow changes are always occurring in coastal areas. Beaches are moved, sand grain by sand grain, by wind and waves. A beach washed away in one place may build up in another place. Sand dunes are slowly moved by the wind. Houses built on sand dunes or hills near a shore may be washed away.

People have tried to protect shorelines by building breakwaters and jetties. However, these only slow down coastal erosion. They cannot prevent it.

Thinking Critically Artificial structures can reduce erosion. However, shoreline changes are natural, and some people believe that shorefront development should be limited. What do you think?

MORE ON THE WEB

Effects of Coastal Erosion
Have students search on the Internet about coastal erosion in Louisiana. Then have them write about how the reduced coastline might affect Louisiana and other states in the future.

• TEACHING RESOURCES • Online or Print

Teacher's Resources

Lesson Review: Chapter 8, p. 11

Enrichment: Chapter 8, p. 14

3 Assess

CHECKING CONCEPTS ANSWERS

1. Accept any two of the following: wind, tides, storms, and earthquakes.
2. abrasion and chemical weathering
3. a sea cliff
4. a sea cave
5. a sea arch

THINKING CRITICALLY ANSWERS

6. **ANALYZE:** Possible answers include chemical weathering, mechanical weathering, sea-level changes, and human activities such as construction.
7. **HYPOTHESIZE:** quickly, because the shoreline is made up of loose deposits that will be carried away by many different forces of erosion

BUILDING MATH SKILLS ANSWER

8. **CALCULATE:** 1.5 m per year × 25 years = 37.5 m

4 Close

HEALTH AND SAFETY TIP

The National Hurricane Center in Florida tracks every Atlantic hurricane from its beginning. Scientists analyze data collected by ocean-based sensors, aircraft, satellites, and other sources to determine each hurricane's strength and predict its path. This allows authorities to evacuate communities, reducing the loss of life.

Real-Life Science

COASTAL EROSION AND STORMS

Real-Life Connection With the growth of leisure time in industrial nations during the nineteenth century, beaches became increasingly popular recreational sites. Every year millions of people vacation at the beach. Many hotels are built along the shore for the convenience of their guests. These beachfront properties also face the risk of being damaged by erosion.

Discussion Ask students: *What are some risks of owning beachfront property?* (It can be damaged and eroded by waves.) *What can be used to reduce shoreline erosion?* (jetties and breakwaters)

Thinking Critically Answer
Ask students to justify their opinions. Hold a class debate about whether or not shorefront development should be limited.

8-10 *(pp. 196–197)*

1 Introduce

STUDY HINT Have students prepare a three-column chart with the following heads: *Name of Landform, What It Looks Like*, and *How It Forms*. Tell students to fill in the first column with the Key Terms. As students read the lesson, have them fill in the remaining columns for each landform.

Linking Prior Knowledge Have students review what they learned about particle sizes (Lesson 3-5). Then, have students recall their experiences at the beach and describe the variety of materials they observed that make up the beach, such as rounded pebbles, white sand, or black sand. Explain that the color and texture of beaches vary, depending on the type of weathered material available for waves to transport and deposit along the shore. **learning style: auditory**

2 Teach

Discussion Obtain travel brochures and books that include photographs of a variety of beaches. Show the photographs to students and have them describe and compare the beaches. Emphasize the differences in beaches.
learning styles: auditory, visual

Answers to questions

1. **EXPLAIN:** Sand and rock particles are deposited on a shoreline by waves.
2. **DEFINE:** the movement of water parallel to a shoreline
3. **DESCRIBE:** by sand being deposited along the curve of a beach
4. **DESCRIBE:** a sand bar that reaches above the surface of the water

READING STRATEGY Tell students that scanning is a method of searching text for a particular piece of information. Explain that when you scan, you do not read every word but sweep your eyes quickly across the page, looking for keywords that signal the information you seek. For example, to find where examples of spits occur, students might scan page 196 for the keywords *spits* and *examples.*

8-10 What landforms do ocean waves create?

Objective

Describe three shoreline features created by wave deposition.

Key Terms

longshore current: movement of water parallel to a shoreline

spit: long, narrow deposit of sand connected at one end to the shore

sand bar: long, offshore underwater deposit of sand

Beaches A beach is an offshore stretch of land covered by sediment, sand, or pebbles. Waves carry rock particles and other material away from a shoreline. A beach is formed when sand and rock particles are deposited on a shoreline by waves.

Materials that form beaches may vary in size and color. Pebble beaches are found along some shorelines. Along the east and west coasts of the United States, weathered quartz forms white sand beaches. Weathered volcanic rock forms black sand. Hawaii has some black sand beaches. Some Florida beaches are made up of broken shells.

▲ **Figure 8-29** This black sand beach in Hawaii is made of weathered volcanic rock.

1 **EXPLAIN:** How is a beach formed?

Longshore Currents Waves do not usually move in a straight line to the shore. They come in to shore at an angle. A **longshore current** is a current formed as waves approach a beach at an angle. The current itself runs parallel to the shoreline. Longshore currents can carry sand away from the beach.

2 **DEFINE:** What is a longshore current?

Spits A curved or hooked deposit of sand on a shoreline is called a **spit.** One end of a spit is always connected to the shore.

How is a spit formed? A longshore current in the ocean carries sand in a direction parallel to the beach. The sand keeps moving in a straight line until the shoreline of the beach changes direction. Then, the sand is deposited at the spot where the beach curves, forming a spit. Sandy Hook in New Jersey and Cape Cod in Massachusetts are both examples of spits.

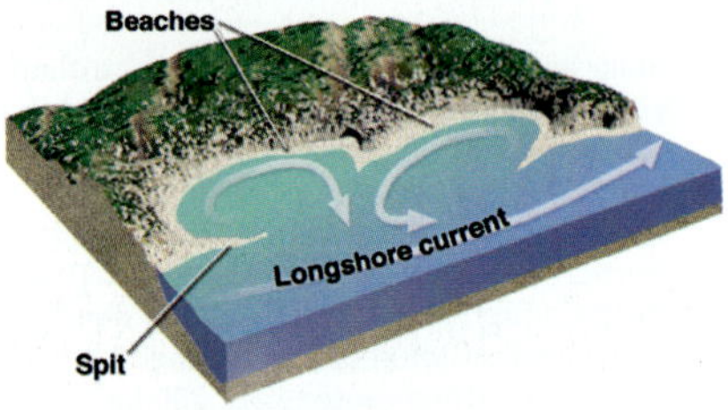

▲ **Figure 8-30** One end of a spit is always connected to the shore.

3 **DESCRIBE:** How is a spit formed?

Sand Bars Waves can carry a lot of sand away from an ocean shoreline. Most of the sand is dropped offshore. This deposit of sand builds up parallel to the shoreline.

A long, offshore underwater deposit of sand is called a **sand bar.** Sometimes, a sand bar completely crosses a bay, sealing it off from the open ocean. This type of sand bar is called a bay mouth bar.

Teaching Options

EXTENDING THE LESSON

Illustrating a Tombolo Have students do Internet or library research to find out what a tombolo is and how it forms. (A tombolo is a sand or gravel bar that connects an island with the mainland or another island. Tombolos form when sediments carried by a longshore current are deposited.) Then, have students illustrate the formation of a tombolo in a poster for display in the classroom. **learning style: visual**

COOPERATIVE LEARNING

Answering Questions Have students work in pairs to answer the Checking Concepts and Thinking Critically questions. **learning style: interpersonal**

ESL/ELL STRATEGY

Working With Peers Working with peers encourages English-language learners to communicate. Have students work in pairs, alternating asking each other questions about the lesson and summarizing important points aloud.

If a sand bar reaches above the water, a barrier beach or an island is formed. Miami Beach, Florida, and Padre Island in Texas are both built on barrier islands.

▲ **Figure 8-31** Padre Island is a barrier beach.

4 DESCRIBE: What is a barrier beach?

CHECKING CONCEPTS

1. A beach is formed when __________ and rock particles are deposited on a shoreline.
2. The material forming a beach may vary in size and __________.
3. Weathered __________ forms black sand.
4. A longshore current is formed when waves approach a beach at an __________.
5. A curved sand deposit that is connected to land is called a __________.

THINKING CRITICALLY

6. **EXPLAIN:** Why do some beaches appear different from other beaches?
7. **DESCRIBE:** How is a sand bar formed?

INTERPRETING VISUALS

Use Figure 8-30 to complete the following.

8. **MODEL:** Draw and label a diagram that shows the direction a longshore current is moving after it has formed near the shoreline.
9. **MODEL:** Draw and label a diagram that shows the relationship of a sand bar to a beach.

Integrating Physical Science

TOPICS: energy, waves

THE ENERGY OF WAVES

Energy is the ability to do work. Work is done when a force moves an object. The forces of wind and tide cause waves to form. These waves contain energy that can make objects move. Sand and pebbles are moved each time a wave breaks along the shoreline. The energy in waves does work.

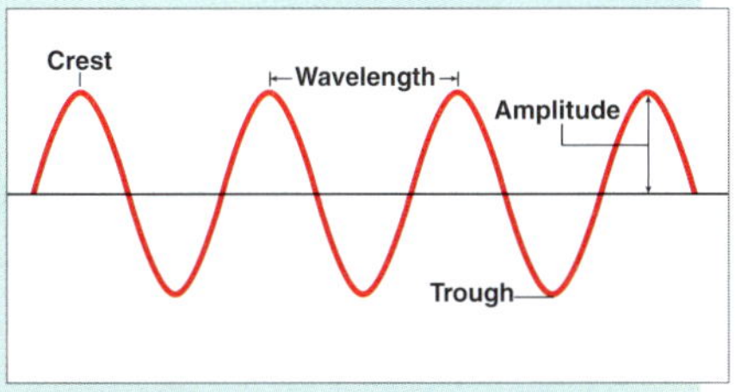

▲ **Figure 8-32** Energy in waves

The waves in water move up and down to form crests and troughs. Half of the difference in height between the crest and the trough is called the amplitude. The greater a wave's amplitude, the greater the energy carried by it. A wavelength is the distance between two consecutive crests.

You can feel the energy in waves if you stand in the water when the surf comes in. This energy may have traveled across the ocean. The water itself does not travel all that distance. Only the pattern of motion in the wave makes the entire trip. When the waves reach shore, they pass most of their energy onto the land at the water's edge.

Thinking Critically What happens to the rocks on the shoreline when they are hit by the energy in waves?

MORE ON THE WEB

Preventing Coastal Erosion
Have students use the Internet to learn how some communities are trying to prevent coastal erosion. Then, have students make a list of the effects that coastal erosion can have on a community and its environment.

• *TEACHING RESOURCES* •
Online or Print

Teacher's Resources
Lesson Review: Chapter 8, p. 12
Enrichment: Chapter 8, p. 14
Flowchart: Study Tools, p. 6

RETEACH

Sand Bars and Related Landforms Have students make a flowchart or other graphic organizer to show how sand bars, bay mouth bars, and barrier beaches or islands are related. Refer to the Teacher's Resources for examples of graphic organizers. **learning style: visual**

3 Assess

CHECKING CONCEPTS ANSWERS

1. sand
2. color
3. volcanic rock
4. angle
5. spit

THINKING CRITICALLY ANSWERS

6. **EXPLAIN:** Materials that form beaches may vary in size and color.
7. **DESCRIBE:** A sand bar is formed when waves carry sand away from a beach and deposit the sand offshore.

INTERPRETING VISUALS ANSWERS

8. **MODEL:** Students' diagrams should resemble Figure 8-30 and include arrows that indicate a current moving parallel to the shore.
9. **MODEL:** Students' diagrams should resemble Figure 8-30 but a sand bar offshore parallel to the shoreline should be added.

4 Close

Integrating Physical Science

TOPICS: energy, waves

THE ENERGY OF WAVES

Background Water waves, sound waves, earthquake waves, and waves that travel through a rope are all examples of mechanical waves. In order to travel, mechanical waves require a medium, such as water, air, or rock. Particles of the medium move in response to a passing wave. However, they do not move along with the wave.

Discussion Ask students: *What is the high point of the wave called?* (crest) *What is the low point called?* (trough)

Thinking Critically Answer
They become eroded.

8-11 *(pp. 198–199)*

1 Introduce

STUDY HINT Have students divide the concepts in this lesson into processes (deflation and abrasion) and things created by these processes (sand dunes and loess).

Linking Prior Knowledge Ask students if they have ever felt sand hitting their face or dust in their eyes on a windy day. Ask them why this happens. (On a windy day, the fast-moving air picks up sand and dust and carries it along.) To get students to think of wind as a force, remind them of windmills.

2 Teach

Demonstration To introduce this lesson, pile a cup of dry sand at the end of a cardboard box. Gently blow through a straw onto the sand. Have students describe what they observe, and then compare their observations to wind moving sand and dust particles from Earth's surface. CAUTION: Safety goggles should be worn for this demonstration. Be sure that the sand is blown in a direction away from students. **learning style: visual**

Discussion Introduce the term *deflation* and ask students why they think deflation occurs most often in deserts, in plowed fields, and on beaches. (There is little or no vegetation holding the soil and sand down in these locations.) **learning style: auditory**

Answers to questions

1. **EXPLAIN:** Wind carries away loose particles from Earth's surface.
2. **EXPLAIN:** Sand particles carried by the wind hit rock and soil surfaces, wearing them away.
3. **INFER:** from the north
4. **DEFINE:** thick deposits of wind-blown dust

WRITING HINT Adjectives usually come before the word they describe. Point out the use of the phrase *loose materials* on page 198. Explain that *loose* is an adjective describing the noun *materials*. Then, have students write two sentences using adjectives to describe a dune or other landform.

8-11 How does wind cause erosion?

Objective

Describe two kinds of wind erosion and deposition.

Key Terms

deflation (dee-FLAY-shuhn)**:** removal of loose material from Earth's surface by wind

loess (LOH-ehs)**:** thick deposits of wind-blown dust

Deflation Wind has energy. Wind can move a sailboat across a lake and turn the blades of windmills. Wind can also move loose materials, such as sand and dust particles. The removal of loose materials from Earth's surface by the wind is called **deflation.** Deflation most often occurs in deserts, in plowed fields, and on beaches.

1 **EXPLAIN:** How can wind be an agent of erosion?

Abrasion Sand particles blown about by the wind can wear away rocks. Have you ever rubbed sandpaper against a piece of wood? The sand on the sandpaper wears away the surface of the wood.

The same thing happens when sand and rock particles are blown over exposed rock or soil. Sand particles carried by the wind are bounced along close to the ground. As the sand particles hit rocks and exposed soil, some of the surface of the rock and soil is worn away or abraded.

2 **EXPLAIN:** How is blowing sand like sandpaper?

Sand Dunes Particles in fast-moving wind have kinetic energy. When a rock or other barrier blocks the wind, the wind slows down. As the wind slows down, it drops the sand it carries the same way moving rivers drop their sediment. The sand builds up and forms a mound called a sand dune. Sand dunes are found in deserts and on beaches.

A sand dune has two sides. The side facing the wind is the windward side. The windward side has a gentle slope. Sand is blown up the windward side and over the top, or crest, to the other side. The side away from the wind is called the slipface. The slipface has a steep slope.

3 **INFER:** The north side of a sand dune has a gentle slope. From which direction is the wind blowing?

▲ **Figure 8-33** Sand dunes form from sand deposited by slow-moving wind.

Teaching Options

! FAST FACTS Wind and Weathered Materials

Wind transports weathered material in one of two main ways, depending on the size of the material. Sand grains are carried as *bed load*, which rolls and bounces along Earth's surface. Smaller dust particles (silt and clay) are swept high above the surface and carried by the wind as *suspended load*.

Most bedload and suspended load is deposited near its source area. However, high winds are able to transport large amounts of dust over great distances. During the 1930s, sediment picked up by the wind in Kansas was carried all the way to New England and into the North Atlantic. Dust traced to the Sahara in Africa has been carried by the wind across the Atlantic to the distant West Indies.

Ask students to infer how scientists might trace the source of loess deposits. (Possible answer: comparing the composition of loess deposits with possible source deposits) **learning styles: linguistic, logical/mathematical**

Loess Wind carries silt and clay higher and farther than it carries sand. Thick deposits of wind-blown dust, called **loess,** may build up many kilometers away. Loess deposits are found in parts of the Mississippi River valley, Washington State, and Oregon.

▲ **Figure 8-34** A loess deposit in Missouri

4 DEFINE: What is loess?

CHECKING CONCEPTS

1. The removal of loose materials from Earth's surface by the wind is called __________.
2. Wind erosion is common in deserts and __________ fields.
3. The __________ of a sand dune has a steep slope.
4. Deposits of wind-blown __________ are called loess.

THINKING CRITICALLY

5. INFER: Why are rocks placed around the bottom of telephone poles in the desert?
6. CLASSIFY: Is abrasion an example of chemical or mechanical weathering? Explain.

INTERPRETING VISUALS

7. Label the parts of the sand dune in Figure 8-35 below.

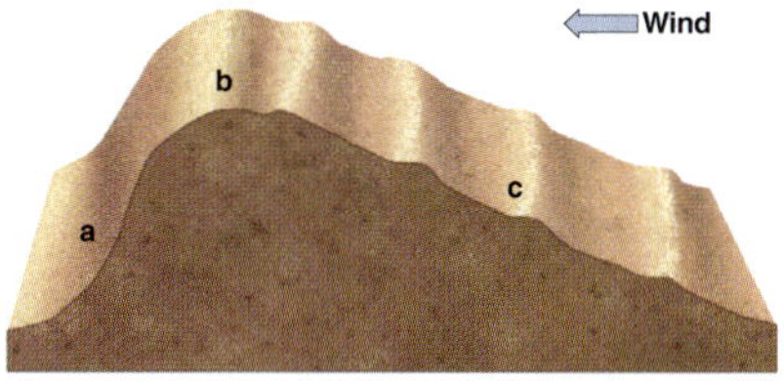

▲ **Figure 8-35**

Science and Technology
WIND EROSION LAB

Wind erosion is worst in very dry regions such as northern Africa. In North America, it is a serious problem for the Great Plains states.

During the 1930s, a very long, dry spell led to dust storms and widespread soil destruction in the Great Plains. The damage caused by the wind led the U.S. Department of Agriculture to set up a research program on wind erosion with Kansas State University (KSU). Today, the KSU laboratory is officially known as the Wind Erosion Research Unit (WERU). Researchers there use wind tunnels to perform tests. Scientists conduct field studies on the mechanics of wind erosion. They also predict wind erosion and the environmental impacts of erosion.

▲ **Figure 8-36** An approaching dust storm

Thinking Critically What are some possible environmental impacts of wind erosion?

MORE ON THE WEB

Preventing Wind Erosion
Have students find out more about wind erosion online.

Viewing Wind Erosion
Students can use the Internet to see how scientists use wind tunnels to research wind erosion. Students can view online images of wind tunnel research in action. Then, have them write a short report on wind erosion prevention and the equipment used to do research on it.

• TEACHING RESOURCES • Online or Print

Teacher's Resources
Lesson Review: Chapter 8, p. 13
Enrichment: Chapter 8, p. 14

3 Assess

CHECKING CONCEPTS ANSWERS

1. deflation
2. plowed
3. slipface
4. dust

THINKING CRITICALLY ANSWERS

5. **INFER:** to prevent the telephone poles from being hit and worn away by the sand
6. **CLASSIFY:** Abrasion is an example of mechanical weathering because it changes only the physical properties of rocks and not the chemical makeup.

INTERPRETING VISUALS ANSWERS

7. **a.** slipface
   b. crest
   c. windward side

4 Close

Science and Technology
WIND EROSION LAB

Environmental-Science Connection
Desertification is the spreading of desert conditions. People and their activities can cause desertification. Expanding development and overgrazing can strip land of vegetation that protects soil from wind erosion. As a result, wind carries away the exposed soil, leaving behind desert land unable to support plant growth or an expanding human population.

Discussion As a class, discuss whether or not wind erosion research still has merit during relatively wet spells without dust storms. Make sure that students support their opinions with facts.

Thinking Critically Answer
Possible answers may include loss of fertile soil, inability to grow crops (drought), inability for plants and animals to survive, and destruction of the ecosystem.

Lab Activity

(pp. 200–201)

1 Prepare the Lab

Modeling the Effects of Wind on Sand

SET-UP TIME: **LAB TIME:**

BACKGROUND: Though not as powerful as flowing water and glacial ice, wind is nevertheless a significant agent of erosion. Strong winds can pick up dust and sand and transport it great distances. Furthermore, the suspended dust and sand in wind will wear away rock walls in desert canyons and scour out dry valleys.

PURPOSE: Students will observe and describe some of the effects of wind erosion by using a cardboard wind tunnel.

ALTERNATIVE MATERIALS: A cardboard wardrobe box can be substituted for smaller boxes, such as copy paper boxes.

SCIENCE SKILLS: Students will **model** a sandy desert and **observe** and **identify** the changes that take place when the desert is subjected to wind erosion.

ADVANCE PLANNING: A single wind tunnel is sufficient for the entire class to use. Small groups can take turns using it. If you prefer, construct the tunnel before class and have it ready for use. Move classroom tables around so that the exhaust end of the tunnel points at an open space about 3 m (3 yd) long. Spread a plastic sheet or tarp over the floor to catch sand and dust blown through the tunnel.

2 Guide the Lab

PROCEDURE: Construct the wind tunnel by removing the bottoms of the boxes and taping them together with duct tape to form a long tunnel. Stand the tunnel on its short side and use tape as a brace if necessary to keep the boxes from sagging. Cut 25-cm squares to make viewing windows at opposite sides of the box near the downwind end of the tunnel. Cover the windows with clear plastic wrap. Place a blow dryer at the other end of the tunnel and direct its airflow down the tunnel. The dryer should be powerful and have adjustable speeds.

LAB ACTIVITY

Modeling the Effects of Wind on Sand

Materials

Safety goggles, 4 large cardboard boxes, duct tape, scissors, clear plastic wrap, electric hair dryer, sand, small rocks, metric ruler

BACKGROUND

On a windy day, you will occasionally get a small particle of dust or sand in your eye. The wind picked up the particle and transported it perhaps great distances before your eye got in the way. Wind is a powerful agent of erosion. A strong wind can pick up sand and dust particles and toss them great distances. It can also wear away rocks. When flying sand and dust strike a rock cliff, tiny chips are broken off. When the wind stops, the particles fall and pile up in layers as sediment.

PURPOSE

In this activity, you will study the force of wind as an agent of erosion.

▲ **STEP 1** Tape the boxes together with duct tape.

▲ **STEP 3** Cover the windows with clear plastic wrap and tape them down.

PROCEDURE

1. First, you will construct a wind tunnel. Remove the bottoms of four large boxes. Tape the boxes together with duct tape to form a long tunnel.
2. Stand the tunnel on its short side. Use tape to keep the boxes from sagging if necessary.
3. Cut two 25 cm square viewing windows on opposite sides of the last box. Cover the windows with clear plastic wrap and hold them in place with tape. ⚠ CAUTION: Be careful when working with scissors.
4. Copy the chart in Figure 8-37. Put on your goggles.
5. Place a small sand pile inside the wind tunnel. Measure the height of the sand pile and record this number in your chart.
6. Hold a hair dryer at the end of the tunnel nearest the windows. Direct its stream of air into the tunnel. The dryer should be powerful and have adjustable speeds.

200

Teaching Options

EXTENDING THE LESSON

Writing About Wind Erosion Have interested students write a journal entry describing a field trip to the desert. Their journal entry should include a description of wind erosion and deposition and landforms students would be likely to encounter. Students can take turns reading their journal entries to the class.
learning style: auditory

COOPERATIVE LEARNING

Exchanging Roles Have students in each group take turns performing each part of the lab activity so that every student has an opportunity to pile or spread the sand, operate the hair dryer, and observe the results.
learning styles: visual, tactile

ESL/ELL STRATEGY

Summarizing Procedures English-language learners may benefit from stating the procedures in their own words. Have these students explain the procedures in their own words to English-proficient students.

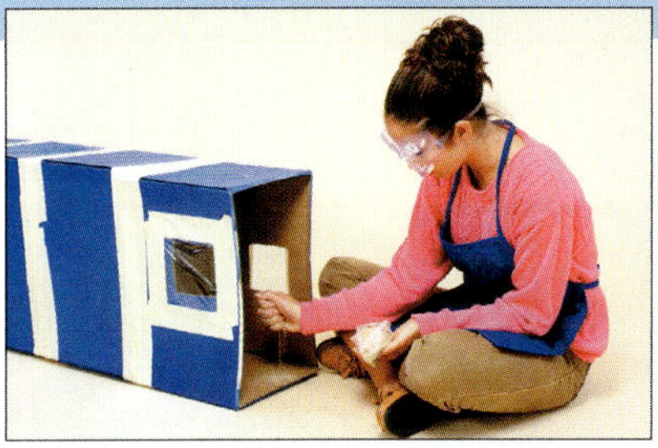

▲ **STEP 5** Create a small pile of sand inside the tunnel.

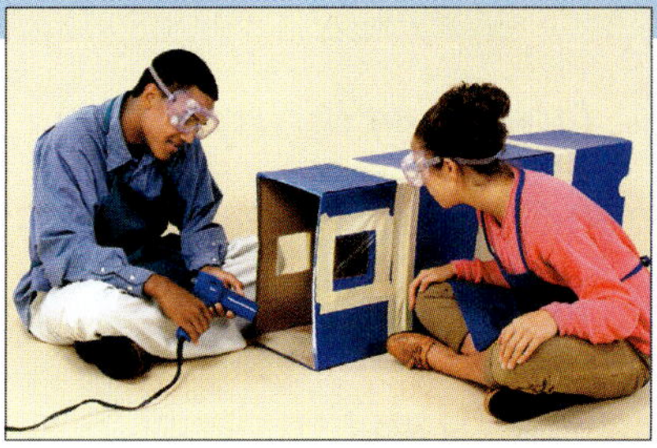

▲ **STEP 6** Direct the stream of air into the tunnel.

7. Turn on the hair dryer for 1 minute and observe what happens. Try different dryer speeds for the best effect. Measure how high the pile is now. Record this number.

8. Return the sand to a pile. Push the top of the pile down to form a crater. Turn on the hair dryer for 1 minute and observes what happens. Record your observations.

9. Spread a thin layer of sand across the tunnel. Place several small rocks in different places on the sand. Turn on the hair dryer for 1 minute and observe what happens. Record your observations.

Wind and Sand

Test Performed	*Data or Observations*	*Explanation*
1.		
2.		
3.		

▲ **Figure 8-37** Use a copy of this chart to record your data.

CONCLUSIONS

1. OBSERVE: What happened to the sand when the wind began blowing?
2. OBSERVE: What happened to the crater in your second experiment?
3. OBSERVE: What happened to the sand in front of and behind the small rocks in your third experiment?
4. ANALYZE: What does this tell you about how wind erodes land?

Begin the lab by discussing how a wind tunnel works. Divide the class into groups of three or four students. Discuss the procedures. Show students where to place the sand and other objects in the tunnel. Also, discuss safety procedures. Provide students with a copy of the Lab Activity Report found in the Teacher's Resources for students to record their observations and conclusions.

TROUBLESHOOTING: It may be necessary to aim the dryer slightly downward to get a strong wind blowing across the sand. Experiment with different speed settings for the best effects.

SAFETY TIP: Make sure that students wear eye protection. Do not permit students to stand in the exhaust of the tunnel. Using prewashed sandbox sand will reduce dust.

EXPECTED OUTCOME: Students will observe sand movements and wind shadows.

LAB HINT: Depending upon classroom lighting conditions, it may be difficult to see some of the erosion effects in the sand. A flashlight may help students get a better view. Have a small broom and dustpan on hand to collect the sand for reuse.

3 Conclude the Lab

1. OBSERVE: The top layer of the sand began moving with the wind.
2. OBSERVE: The crater began filling in.
3. OBSERVE: Sand was removed from the front of the rocks (scouring) and dropped behind (wind shadow).
4. ANALYZE: Wind picks up the sand and carries it. The sand chips away at any rocks it strikes while airborne.

Use the Laboratory Activity Rubric from the Teacher's Resources to assess students' lab activities. Fill in the rubric with the additional information below. For this activity, students should have:

- studied wind as an agent for erosion.
- recorded their observations in a chart.

4 Extend the Lab

Show students pictures of the surface of Mars. Mars has a very thin atmosphere, and sometimes strong winds blow across its surface. Dark wind shadows behind boulders and craters are visible in orbital spacecraft pictures. Surface pictures show accumulations of sand in low spots.

Tips for Using Technology

Using Word Processing Software to Write About Erosion
Have students use word processing software to write about erosion for the Extending the Lesson on page T200. A simple word processing program allows them to do basic tasks such as entering and editing text. A more complex word processing program allows students to make tables, draw pictures, do math and make graphs. Students can print their word processing document to share with the class.

• TEACHING RESOURCES • Online or Print

Teacher's Resources

Lab Activity Report: Study Tools, pp. 16–17

Laboratory Activity Rubric: Teaching Tools, p. 8

Weekly Journal: Study Tools, p. 19

Laboratory Manual

Lab. Skills Worksheet 4: Organizing and Analyzing Data, pp. 13–14

The Big Idea

(pp. 202–203)

1 Introduce

Objective Students will be able to describe how the Dust Bowl erosion changed American history. Students will research and write a description of Black Sunday.

Linking Prior Knowledge Have students review Lesson 8-11 and the Lab Activity conclusions on page 201 to recall how wind causes erosion.

2 Teach

Discussion Ask students: *What factors caused the Dust Bowl?* (poor farming methods and drought) *What is the relationship between soil moisture and wind erosion?* (Soil without moisture is more prone to erosion.) Refer students to the information in Fast Facts to help them identify some results of severe erosion on the landscape. **learning style: auditory**

THE BIG IDEA

How did erosion change American history? Have students start their research for their Science Log Writing Activity at www.conceptsandchallenges.com. Be sure students use search engines and websites that are reliable and acceptable. You may also wish to provide copies of the *Big Idea Planner* and *Big Idea Science Log Entry* study tools found in the Teacher's Resources.

Integrating History

THE Big IDEA

How did erosion change American history?

In 1917, the United States entered World War I. Americans throughout the country were called to help their country. While some served as soldiers, others back home produced items for the troops. Farmers in the Midwest set to work growing increased amounts of wheat that could be made into food for the soldiers. They cleared millions of acres of grassland. When the war ended, the farmers continued growing wheat. The soil's thick mass of tough roots kept the soil in place and held the moisture.

However, the practice of planting the same crop over and over again was hard on the soil. The nutrients in the once fertile soil disappeared. The soil began to lose its ability to hold water, and the topsoil quickly dried out.

In 1931, a severe drought hit the region. The lack of rain caused the already dry soil to turn to dust. Strong winds blew the dusty topsoil away in huge black clouds. The area became known as the Dust Bowl.

As the drought continued, more dust storms occurred. Crops were destroyed. Livestock caught in the storms became sick from inhaling the soil-filled air. Homes and farm machinery were damaged.

Many farmers left the area. They traveled westward to California in the hope of starting new, successful farms. They believed the state's mild climate and long growing season would guarantee success. By 1940, the number of people who had moved out of the Great Plains states grew to a record number of 2.5 million. No other event in American history has led to a greater migration of people from one part of the country to another.

Look at the illustrations that appear on this page and the next. Then, follow the directions in the Science Log to learn about "the big idea."✦

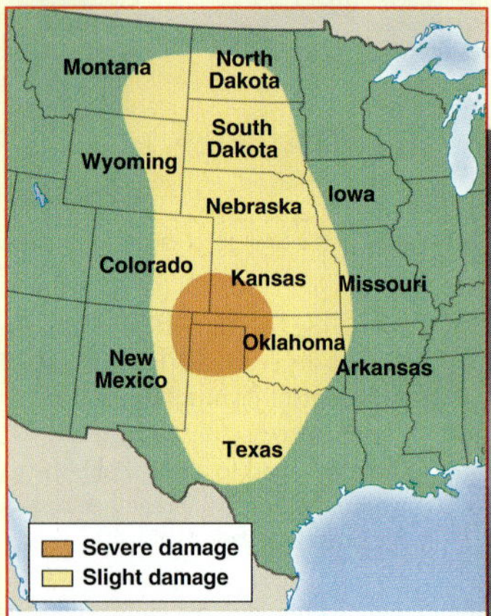

The Damage
About 50 million acres of land in the Midwest states were damaged by the dust storms of the 1930s.

The People
This famous photograph was taken by Dorothea Lange. Lange traveled around the Midwest during the 1930s. She is known for her striking photographs of poverty and hopelessness. Both were results of the drought and the Great Depression.

202

Teaching Options

! FAST FACTS Badlands

In dry areas of sedimentary rock with little or no vegetation, the land has been carved into ridges, pinnacles, and gullies. These areas are called *badlands.* When rain falls in the badlands, the rainwater runs off very quickly because of the scarcity of vegetation. The water erodes the soft, exposed rock, forming steep canyons.

In the United States, the best-known examples of badlands are found in the western parts of North Dakota, South Dakota, and Nebraska. Some of these sites are protected in Theodore Roosevelt National Park in North Dakota and Badlands National Park in South Dakota.

EXTENDING THE LESSON

Writing About the Dust Bowl Have students write a book report on the novel *The Grapes of Wrath* by John Steinbeck. Tell students that the story is a fictional account of the impact of the drought and other events of the 1930s on farmers. Have students present their book reports to the class. **learning styles: linguistic, auditory**

The Migration

Dust-covered farms were abandoned by one out of every four farmers in the Midwest. More than 200,000 farmers moved to California.

▲ **Figure 8-38** Daytime skies would darken, as huge black clouds swept over farmland.

WRITING ACTIVITY

Science Log

On April 14, 1935, one of the most severe storms of the Dust Bowl struck the Midwest. Afterwards, people referred to the day as "Black Sunday." Find firsthand accounts of that day. Then, write a brief description of it as if you had been there yourself. Also, find out how farmers in the old Dust Bowl states are doing now. Begin your search for information by going to www.conceptsandchallenges.com.

The Soil Conservation Service

Government agencies, such as the Soil Conservation Service, helped farmers practice soil conservation methods.

3 Assess

Use the Writing Activity Rubric from the Teacher's Resources to assess students' Science Logs. Fill in the rubric with the additional information below. For this assignment, students should have:

- researched firsthand accounts and written a description of Black Sunday.
- communicated in paragraph form a description of how erosion changed history.

4 Close

Writing an Article Have students write a newspaper account of how erosion caused the Dust Bowl of the 1930s. Articles should include all of the features of a real newspaper article, such as a headline, byline, date and place, as well as a description of soil erosion and how it led to the Dust Bowl.

You can also have small groups of students prepare a class presentation on measures that farmers have taken since the 1930s to conserve soil and prevent severe dust storms. These measures include planting grasses in large areas, rotating crops, contour plowing, terracing, and planting trees as windbreaks. Students' presentations could be posters or oral reports with visual aids.

learning styles: linguistic, visual

Tips for Using Technology

Using Desktop Publishing to Produce a Newspaper Have students use desktop publishing software to create a newspaper that features their firsthand Dust Bowl accounts. Desktop publishing software allows students to combine text and art on a page. Most word processing software programs include tools for desktop publishing. Students can use these to give text special formats, make borders around the text, and plan a page layout. Have students use the Thumbnail Sketch graphic organizer from the Teacher's Resources to plan their desktop publishing layouts.

• TEACHING RESOURCES • Online or Print

Teacher's Resources

Big Idea: Chapter 8, p. 15

Big Idea Planner: Study Tools, pp. 11–12

Big Idea Science Log Entry: Study Tools, p. 13

Thumbnail Sketch: Study Tools, p. 24

Challenges *(pp. 204–206)*

Chapter Summary

Review Hint Have students group terms to make them more meaningful. For example, they can list all processes together and all landforms together.

STUDY HINT Assign a different Key Term to each student. Have students write a definition of their term and use it in a sentence.

Key Term Challenges

MATCHING

1. runoff (8-3)
2. karst topography (8-5)
3. tributary (8-3)
4. wave-cut terrace (8-10)
5. glacier (8-6)
6. erratic (8-7)
7. hanging valley (8-7)
8. sand bar (8-10)
9. mass erosion (8-2)

APPLYING DEFINITIONS

10. Talus is deposited by mass erosion. Till is deposited by a glacier. (8-2, 8-7)
11. A sea arch is a gap formed when waves cut through a section of rock. A sea stack is a column remaining when a sea arch collapses. (8-9)
12. Deflation involves the removal of loose material by wind. Deposition involves the dropping of rock material in a new place. (8-1, 8-11)
13. Erosion is the process by which weathered material is removed and deposited elsewhere. Weathering is the process by which rock is broken down. (8-1)
14. Loess is a kind of sediment deposited by wind. Sediment is deposited soil and rock particles. (8-4, 8-11)
15. A floodplain is a flat area on the sides of a river where flood sediments are deposited. A delta is a triangular deposit of sediment at the mouth of a river. (8-4)

Chapter 8 Challenges

Chapter Summary

Lessons 8-1 and 8-2

- **Erosion** removes material from a place. **Deposition** drops material in new places.
- **Mass erosion** is the downhill movement of weathered materials due to gravity. Landslides, mudflows, and creep are mass movements.

Lesson 8-3

- **Runoff** causes erosion. The amount of rainfall, plant growth, and topography all affect erosion.
- A river system has a main stream and **tributaries.**

Lessons 8-4 and 8-5

- Rock and sand carried by wind or water weathers rocks and soil. This is **abrasion.**
- A **delta** forms when sediments are deposited by slow-moving water at the mouth of a river. **Floodplains** deposit fertile soil on both sides.
- Caves, underground rivers, and **sinkholes** are features typical of **karst topography.** They are formed by the chemical weathering of limestone.
- **Cave systems** are made up of several connected caves.

Lessons 8-6 and 8-7

- A **glacier** is a moving river of ice and snow. An **iceberg** breaks off of continental glaciers.
- A **hanging valley** forms when a small side glacier melts. **Till** is loose material left by a glacier. **Erratics** are boulders left by glaciers.

Lessons 8-8 and 8-9

- Till left by a glacier forms terminal or lateral **moraines. Drumlins** are oval mounds of till.
- Glacial lakes form when valley glaciers melt. **Kettle lakes** form when ice is buried, then melts.
- Waves can break up rocks on shore. An eroded rocky shoreline may have sea cliffs, sea caves, **sea arches, sea stacks,** or **wave-cut terraces.**

Lessons 8-10 and 8-11

- A **longshore current** is formed when waves approach a beach at an angle. Longshore currents create **spits** and **sand bars.**
- **Deflation** removes loose material from Earth's surface. Sand dunes form when wind is blocked. Wind-blown dust forms **loess** deposits.

Key Term Challenges

abrasion (p. 184)
cave system (p. 186)
deflation (p. 198)
delta (p. 184)
deposition (p. 178)
drumlin (p. 192)
erosion (p. 178)
erratic (p. 190)
floodplain (p. 184)
glacier (p. 188)
hanging valley (p. 190)
iceberg (p. 188)
ice cap (p. 188)
karst topography (p. 186)
kettle lake (p. 192)
loess (p. 198)
longshore current (p. 196)
mass erosion (p. 180)
moraine (p. 192)
runoff (p. 182)
sand bar (p. 196)
sea arch (p. 194)
sea stack (p. 194)
sediment (p. 184)
sinkhole (p. 186)
spit (p. 196)
talus (p. 180)
till (p. 190)
tributary (p. 182)
wave-cut terrace (p. 194)

MATCHING **Write the Key Term from above that best matches each description.**

1. water from rain or snow that flows on Earth's surface
2. land that has sinkholes, caverns, and underground pools of water
3. smaller stream that flows into the main stream of a river system
4. flat section of rock formed by the erosion of a sea cliff
5. moving river of ice and snow
6. boulder left behind by a retreating glacier
7. small glacial valley above a main valley
8. underwater deposit of sand
9. downhill movement of weathered materials caused by gravity

APPLYING DEFINITIONS **Explain the difference between the words in each pair. Write your answers in complete sentences.**

10. talus, till
11. sea arch, sea stack
12. deflation, deposition
13. erosion, weathering
14. loess, sediment
15. floodplain, delta

204

Teaching Options

PREPARING STUDENTS FOR STANDARDIZED TESTS

Reading Strategy: When reading a question, tell students to focus on keywords, such as numbers, that tell them how many parts must be included in their answer.

Writing Strategy: Remind students to check their writing for incorrect use of words that sound similar or commonly confused homophones that sound the same but have different meanings, such as *than* and *then*, *affect* and *effect*, *coarse* and *course*, and *weather* and *whether*.

Interpreting Visuals Strategy: Remind students to pay attention to direction when interpreting visuals. Have students carefully read any labels associated with direction arrows.

ESL/ELL STRATEGY

Compound Words Tell students that some of the Key Terms are compound words, such as *runoff* and *sinkhole*. Explain that knowing the definitions of the individual words can help them understand the meaning of the compound word.

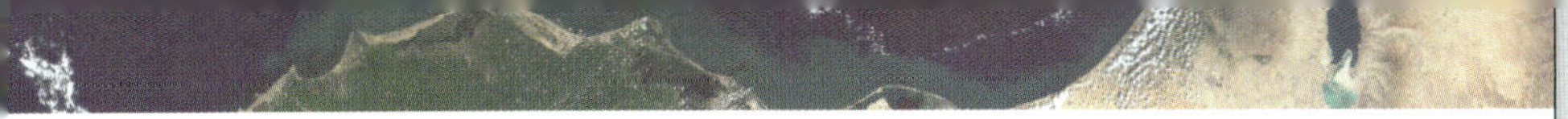

Content Challenges TEST PREP

MULTIPLE CHOICE Write the letter of the term or phrase that best completes each statement.

1. Mudflows, creep, landslides, and earthflow are caused by
   a. wind.
   b. waves.
   c. running water.
   d. gravity.
2. Weathered material is removed from a place by
   a. abrasion.
   b. deposition.
   c. erosion.
   d. moraines.
3. The sudden movement of rocks down a hillside is
   a. an earthflow.
   b. a mudflow.
   c. creep.
   d. a landslide.
4. Tilted telephone or fence poles on a hillside is evidence of
   a. creep.
   b. earthflow.
   c. runoff.
   d. glaciers.
5. Gullies in a hillside are caused by
   a. creep.
   b. talus.
   c. runoff.
   d. glaciers.
6. Sediments that are carried by a glacier are called
   a. till.
   b. talus.
   c. moraines.
   d. drumlins.
7. The Great Lakes and New York's Finger Lakes are
   a. kettle lakes.
   b. drumlins.
   c. oxbow lakes.
   d. glacial lakes.
8. Sea cliffs, sea stacks, and sea arches are caused by
   a. glaciers.
   b. wind.
   c. waves.
   d. gravity.
9. A piece of glacier that breaks off and goes into the ocean is
   a. a spit.
   b. an iceberg.
   c. a drumlin.
   d. a valley glacier.
10. Glaciers that form in mountains are
    a. valley glaciers.
    b. ice caps.
    c. continental glaciers.
    d. moraines.

FILL IN Write the term or phrase that best completes each statement.

11. The rocks that make up Earth's crust are removed and then carried away to new places by __________.
12. Deposition builds __________ on Earth's surface.
13. A piece of ice that has broken off from a glacier is called an __________.
14. Landslides and __________ are forms of rapid mass erosion.
15. A delta usually forms at the __________, or end, of a river.

ALTERNATIVE ASSESSMENT

Describing Landforms and Deposits On small strips of paper, write the names of the landforms and deposits introduced in the chapter and place the strips in a box. Have students take turns drawing one strip and describing the indicated landform or deposit and how it forms.

PORTFOLIO ASSESSMENT

Making Student Portfolios Portfolio Assessment is designed to evaluate a student's performance over an extended period of time. Students can demonstrate their comprehension of the concepts in this chapter by making a portfolio and including in it something they are especially proud of. The Weekly Journal, the Big Idea Planner, and the Chapter Self-Check are some of the reproducibles in the Teacher's Resources that they can include in their portfolios. You can use the Portfolio Assessment Rubric also found in the Teacher's Resources to assess students' portfolios.

Content Challenges

MULTIPLE CHOICE

1. d (8-2)
2. c (8-1)
3. d (8-2)
4. a (8-2)
5. c (8-3)
6. a (8-7)
7. d (8-8)
8. c (8-10)
9. b (8-6)
10. a (8-6)

FILL IN

11. erosion (8-1)
12. landforms (8-1)
13. iceberg (8-6)
14. mudflows (8-2)
15. mouth (8-4)

Concept Challenges

WRITTEN RESPONSE

1. **ANALYZE:** The glacier was moving north. You know this because the tips of drumlins always point in the direction the glacier was moving. (8-8)
2. **PREDICT:** Sea level would increase, or rise, as the glaciers melted. (8-6)
3. **INFER:** Deflation occurs mostly in deserts, in plowed fields, and on beaches because there is not much vegetation to secure the soil and sand. (8-11)
4. **INFER:** Possible answer: Dust particles are smaller or lighter than sand particles. (8-11)
5. **INFER:** Slopes are steep in these areas. Therefore, heavy, wet materials easily slide down a slope. (8-2)
6. **PREDICT:** Runoff would increase because areas with little plant growth have more runoff than areas with a lot of plant growth. (8-3)
7. **ANALYZE:** A glacier probably deposited it. (8-7)
8. **INFER:** Sand moves up the windward side and accumulates on the slipface, making the angle of the slope steeper as more sand is deposited. (8-11)
9. **EXPLAIN:** Fast-moving water is more abrasive and can carry away more sediment than slow-moving water. (8-3)
10. **DESCRIBE:** A glacier carves away the land in places by eroding rock materials and builds up the land in other places by depositing rock materials. (8-7, 8-8)

INTERPRETING VISUALS

11. a river and its floodplain (8-4)
12. water (8-4)
13. a sand dune (8-11)
14. wind (8-11)
15. It was caused by deposition. A sand dune forms when wind carrying sand slows down and deposits the sand. (8-11)

Chapter Project Wrap-Up

MODEL OF A LANDFORM CREATED BY EROSION

Have students present their models to the class and explain what they represent. You can use the Individual Activity Rubric found in the Teacher's Resources to assess students' models. Fill in the rubric with the additional information below. For this project, students should have:

- prepared a model that includes features of that landform and an explanation of what creates it.
- given a logical class presentation.

Concept Challenges TEST PREP

WRITTEN RESPONSE **Answer each of the following questions in complete sentences.**

1. **ANALYZE:** If the tip of a drumlin points north, in which direction was the glacier that formed the drumlin moving? Explain how you know.
2. **PREDICT:** What would happen to the sea level if Earth's temperature rose and caused glaciers to melt?
3. **INFER:** Why does deflation occur mostly in deserts, in plowed fields, and on beaches?
4. **INFER:** Why can wind carry dust farther than it can carry sand?
5. **INFER:** Why do most mudflows occur in dry mountainous regions after a heavy rainfall?
6. **PREDICT:** How would runoff in a forest area change if all the trees were cut down?
7. **ANALYZE:** You find a large boulder that is unlike the bedrock in that area. It is too heavy to have been carried by running water. How did the boulder get there?
8. **INFER:** Why does the windward side of a sand dune have a gentle slope while the slipface has a steep slope?
9. **EXPLAIN:** Why do fast-moving rivers cut deep valleys?
10. **DESCRIBE:** In what ways does a glacier change a landscape?

INTERPRETING VISUALS **Use Figures 8-39 and 8-40 below to answer the following questions.**

11. What is shown in Figure 8-39?
12. Which agent of erosion causes what is shown in Figure 8-39?
13. What is the name of the landform shown in Figure 8-40?
14. What natural force created the landform shown in Figure 8-40?
15. Was the landform shown in Figure 8-40 caused by erosion, deposition, or both? Explain your answer.

▲ Figure 8-39

▲ Figure 8-40

206

Teaching Options

• TEACHING RESOURCES •
Online or Print

Teacher's Resources

Key Term Review: Chapter 8, p. 16

Chapter 8 Test: pp. 17–19

Individual Activity Rubric: Teaching Tools, p. 5

Chapter Self-Check: Teaching Tools, p. 11

Big Idea Planner: Study Tools, pp. 11–12

Weekly Journal: Study Tools, p. 19

• TEACHING RESOURCES •
Online or Print

Teacher's Resources

Portfolio Assessment Rubric: Teaching Tools, p. 10

PLANNING GUIDE

◆ **TEACHING THE CHAPTER** This chapter should take approximately 5–8 days to complete instruction and assessment.

	Skills and Features	Projects and Activities	Achieve Science Literacy	Meet Individual Needs
Chapter 9 Opener p. T207	WRITE	• Chapter Project		
Lesson 9-1 **What is the water cycle?** pp. T208–T209 Standards: D1f, D1i	CLASSIFY, STATE, INTERPRET • Building Science Skills • Hands-On Activity	• Cooperative Learning • Extending the Lesson • Active Art	• Study Hint • Reading Strategy • Ongoing Assessment	Teacher's Resources* Enrichment Activity
Lesson 9-2 **What is groundwater?** pp. T210–T211 Standards: D1e, D1f	DESCRIBE, DEFINE, INFER • People in Science	• More on the Web	• Reading Strategy	• Reteach Teacher's Resources* Enrichment Activity Groundwater
Lesson 9-3 **What are wells, springs, and geysers?** pp. T212–T213 Standards: D1f, D2a	DEFINE, EXPLAIN, INFER • Web InfoSearch • Hands-On Activity		• Study Hint • Ongoing Assessment	Teacher's Resources* Enrichment Activity
Lesson 9-4 **What are rivers and streams?** pp. T214–T215 Standards: D1f, D2a	DEFINE, INFER, IDENTIFY • Web InfoSearch	• Cooperative Learning • Extending the Lesson	• Study Hint	• Reteach Teacher's Resources* Enrichment Activity
Lab Activity **Modeling the Life Cycle of a Stream** pp. T216–T217 Standards: D1f, D2a	OBSERVE, ANALYZE, INFER • Set-Up Time: 1 hour • Lab Time: 45 min	• Extending the Lesson • Lab Skills Worksheet • Extend the Lab Activity	• Tips for Using Technology	Teacher's Resources* Lab Activity Report Spreadsheet

*www.conceptsandchallenges.com, password: ccearth09

Planning Guide continues on next page.

Standards: For details on the correlation to National Science Standards see pages *xx–xxi*.

	Skills and Features	Projects and Activities	Achieve Science Literacy	Meet Individual Needs
Lesson 9-5 **What are lakes and ponds?** pp. T218–T219 Standards: D1f, D2a	DEFINE, DESCRIBE, INFER • Designing an Experiment • Real-Life Science	• More on the Web • Lab Challenge	• Study Hint • Writing Hint	• ESL/ELL Strategy Teacher's Resources* Enrichment Activity Designing an Experiment
Big Idea **Why are wetlands important to the environment?** pp. T220–T221 Standards: D1f, D1k	RESEARCH, COMMUNICATE, ANALYZE • Science Log: Writing Activity	• Big Idea • Close Activity	• Tips for Using Technology	Teacher's Resources* Big Idea Planner
Chapter 9 Challenges pp. T222–T224	• Chapter Summary • Key Term Challenges • Content Challenges • Concept Challenges	• Chapter Project Wrap-Up	• Study Hint • Preparing Students for Standardized Tests	• ESL/ELL Strategy Teacher's Resources* Chapter Self-Check Weekly Journal

*www.conceptsandchallenges.com, password: ccearth09

Teacher Notes

Chapter 9 Inland Waters

▲ **Figure 9-1** Riding the rapids is one way to explore rivers. It is also a fun pastime for many people.

Running water is important to people. It helps shape the land on which we live. We depend on rivers for transportation, farming, energy, and food. We also use rivers for recreational activities, such as boating and rafting. Many people enjoy riding the rapids. A rapid forms where a streambed has a steep slope.

▶How is running water important in transportation?

Contents

Teaching Options

ASSESSMENT PLANNER

For assessment in the Student Edition, see the following pages:

Content Assessment
Checking Concepts: pp. 209, 211, 213, 215, 219
Thinking Critically: pp. 209, 211, 213, 215, 219
Interpreting Visuals: p. 211
Chapter Challenges: pp. 222–224

Alternative Assessment
Building Skills: p. 209
Web InfoSearch: pp. 213, 215
Science Log: p. 221

Performance-Based Assessment
Hands-On Activity: pp. 209, 213
Designing an Experiment: p. 219
Lab Activity: pp. 216–217

• TEACHING RESOURCES • Online or Print

Teacher's Resources
Lesson Review: Ch. 9, pp. 2, 3, 4, 6, 7
Enrichment: Ch. 9, pp. 5, 8
Key Term Review: Ch. 9, p. 10
Chapter 9 Test: pp. 11–12
Laboratory Manual: pp. 13–14, 53–56

Chapter 9 Inland Waters

Chapter Overview

In this chapter, students will trace the steps in the water cycle. Students will explain how groundwater collects in soil. They will describe how groundwater reaches Earth's surface. They will name and describe the three stages in the life cycle of a river or stream. Finally, they will describe how lakes and ponds form and change.

About Figure 9-1 The photograph shows adventurers riding river rapids. Rapids, along with waterfalls, occur where a stream profile drops suddenly. This is usually because of variations in the erodibility of the bedrock into which a stream channel is cutting.

Answer Possible answer: The running water of rivers provides pathways for boats and barges to transport goods and people from one location to another.

Linking Prior Knowledge

For this chapter, students should recall:

- the names of the main parts of Earth (Lesson 1-2).
- the definition of runoff (Lesson 8-3).
- how glacial lakes form (Lesson 8-8).
- how waves cause erosion (Lesson 8-9).

Chapter Project

WATER CYCLE GAZETTE

MATERIALS: paper, scissors, glue, tape, magazines, newspapers, markers

As students read the chapter, they should work in small groups to create a newspaper called the *Water Cycle Gazette.* They should write news articles about events relating to water in the atmosphere, on Earth's surface, and in the ground. Suggest that students' newspapers include articles and advertisements, comics, and any other features that might be found in a newspaper. Encourage students to be creative but insist that their work be scientifically accurate.
learning styles: visual, linguistic

9-1 *(pp. 208–209)*

1 Introduce

STUDY HINT Have students note that the first three Key Terms are processes that occur during the last Key Term, *water cycle.* Have them make up their own drawings of the water cycle based on the definitions given.

Linking Prior Knowledge Review with students the main parts of Earth (Lesson 1-2). Discuss the role that water plays in each part.

2 Teach

Demonstration Demonstrate to students the relationship between state and temperature. Heat a container of water to the boiling point. Use a thermometer to record the temperature at which the water boils (100°C/212°F). Write this temperature on the board. Ask students to identify what they observe in the air above the boiling water. (steam) Explain that steam is water in the gaseous state. Then, place ice cubes in a container. Point out to students that ice forms from liquid water at 0°C/32°F. Ask students to predict what will happen to the ice if it is left out at room temperature. (The ice will melt.) Elicit from students what causes this change in state. (an increase in temperature, or absorption of heat)
learning style: visual

Discussion Review the meaning of the term *precipitation*. Have students list the four main kinds of precipitation. (rain, snow, sleet, hail) Ask students how the forms of precipitation differ. (Rain falls to Earth as a liquid. Snow, sleet, and hail fall to Earth as solids.) Discuss the impact of the atmosphere's temperature on the kind of precipitation that forms.
learning style: auditory

Answers to questions

1. **DESCRIBE:** a collection of water droplets or ice crystals
2. **LIST:** Accept any two of the following: rain, snow, hail, or sleet.
3. **STATE:** the repeated movement of water between Earth's surface and the atmosphere

9-1 What is the water cycle?

Objective

Trace the steps in the water cycle.

Key Terms

evaporation (ee-vap-uh-RAY-shuhn)**:** changing of a liquid to a gas

condensation (kahn-duhn-SAY-shuhn)**:** changing of a gas to a liquid

precipitation (pree-sihp-uh-TAY-shuhn)**:** water that falls to Earth's surface from the atmosphere

water cycle: repeated pattern of water movement between Earth and the atmosphere

Evaporation and Condensation Liquids change to gases by a process called **evaporation.** Most of Earth's surface is covered with water. When liquid water absorbs enough heat energy from the Sun, it changes into the gas water vapor. Air always contains some water vapor through evaporation. We become more aware of water vapor when the air is humid.

The changing of a gas to a liquid is called **condensation.** When air containing water vapor is cooled, the water vapor loses heat. If enough heat is lost, the water vapor changes to a liquid.

Water vapor condenses into tiny water droplets. These water droplets join to form clouds. If the temperature reaches freezing, ice forms. A cloud is a collection of water droplets or ice crystals.

1. **DESCRIBE:** What is a cloud?

Precipitation Water that falls to Earth from the atmosphere is called **precipitation.** Rain and snow are two forms of precipitation. As the water droplets in a cloud grow bigger, they become too heavy to stay in the air. Gravity pulls them toward Earth. The water falls as rain. If the rain passes through very cold air, the water may turn to a solid and fall to Earth as snow, hail, or sleet.

2. **LIST:** What are two forms of precipitation?

The Water Cycle Water is always changing state. As water evaporates from Earth's surface, it changes from a liquid to water vapor. In the atmosphere, the water vapor condenses to a liquid. This forms clouds. Finally, the water falls to Earth as precipitation. The repeated movement of water between Earth and the atmosphere is called the **water cycle.** The water cycle is also known as the hydrologic cycle.

About 97 percent of Earth's water is salt water. The rest is fresh water.

Go Online **active art**
For: Water Cycle activity
Visit: PHSchool.com
Web Code: cfp-4024

▲ **Figure 9-2** In the water cycle, water continually moves between the atmosphere and Earth.

Teaching Options

COOPERATIVE LEARNING

Interviewing Professionals Have pairs of students interview by phone, letter, or e-mail an employee of the local water company or watershed management agency to find out the origin of their community's fresh water. Each pair of students can present its findings to the class in an oral report. **learning styles: auditory, interpersonal**

EXTENDING THE LESSON

Researching Sublimation Have students find out what sublimation and transpiration are and the roles these processes play in the water cycle. (Sublimation is the process in which a substance changes directly from a solid to a gas. Transpiration is the release of water from plant leaves. Both sublimation of ice and transpiration participate in the water cycle by adding water vapor to the atmosphere.) Have students summarize their findings in two paragraphs.
learning style: linguistic

About 76 percent of fresh water is frozen into ice caps near the North and South poles. A tiny amount is in the air in the form of water vapor. Only about 23 percent of Earth's fresh water is readily available for use by living things. That water is found in lakes, rivers, and streams, and below Earth's surface.

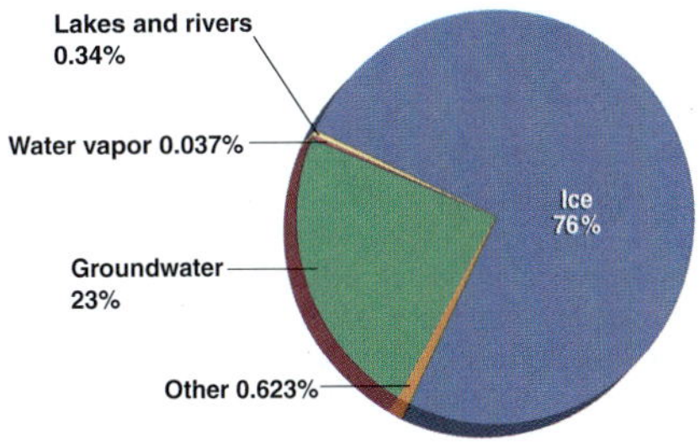

▲ **Figure 9-3** Fresh water distribution on Earth

STATE: What is the water cycle?

CHECKING CONCEPTS

1. What happens when water absorbs enough heat?
2. During condensation, what happens to gas?
3. What makes up a cloud?
4. What processes make up the water cycle?

THINKING CRITICALLY

5. **HYPOTHESIZE:** How might the water cycle be affected when dust in the air blocks sunlight?
6. **INTERPRET:** Based on Figure 9-3, what would you say is the largest source of fresh water available for use by living things?

BUILDING SCIENCE SKILLS

Classifying Write *Solid*, *Liquid*, and *Gas* across the top of a sheet of paper. Put each of the following terms under its correct state in your chart: rain, snow, water vapor, ocean, river, ice cap, and glacier.

Hands-On Activity

OBSERVING THE STATES OF WATER

You will need safety goggles, a glass, plastic wrap, a rubber band, hot water, and ice cubes.

1. Fill a glass halfway with hot tap water. Put a piece of plastic wrap tightly over the top of the glass. Use a rubber band to secure it. ⚠ **CAUTION:** Be careful not to burn yourself with the hot water.
2. Place 5 or 6 ice cubes on the plastic wrap as shown in Figure 9-4.
3. Allow the glass to stand for about 5 minutes. Observe what happens.
4. Pick up the plastic wrap. Feel the side that was closest to the hot water.

▲ **Figure 9-4** Set up your experiment as shown.

Practicing Your Skills

5. **IDENTIFY:** In what state was the hot water? The ice cubes? What state could not be seen?
6. **ANALYZE:** What happened to the ice? What caused this?
7. **ANALYZE:** What is on the underside of the plastic wrap? How did it get there?
8. **INFER:** Which state of matter are glaciers?

READING STRATEGY Many words, such as *precipitation*, have multiple meanings. Tell students that the meteorological meaning of precipitation is "substances such as rain and snow that fall from the atmosphere." The chemical meaning is "the process of separating out of a solution."

3 Assess

CHECKING CONCEPTS ANSWERS

1. It evaporates.
2. It changes to a liquid.
3. water droplets or ice crystals
4. evaporation, condensation, and precipitation

THINKING CRITICALLY ANSWERS

5. **HYPOTHESIZE:** Possible answer: If heat from the Sun could not reach Earth, water would not evaporate.
6. **INTERPRET:** shallow groundwater

4 Close

BUILDING SCIENCE SKILLS ANSWER

Classifying *Solid:* snow, ice cap, glacier; *Liquid:* rain, ocean, river; *Gas:* water vapor

Hands-On Activity

OBSERVING THE STATES OF WATER

TIME: 15 minutes

PURPOSE: observing, analyzing, inferring

SAFETY TIP: Caution students to be careful not to spill hot water on skin or clothing.

ALTERNATIVE MATERIALS: A plastic container can be used in place of a glass.

DISCUSSION: Help students recall the states of matter for water.

Practicing Your Skills Answers

5. **IDENTIFY:** The hot water was a liquid. The ice cube was a solid. The gas could not be seen.
6. **ANALYZE:** It melted. Heat from the hot water melted the ice cube.
7. **ANALYZE:** Water droplets were on the underside of the plastic wrap. The water droplets formed as steam (water vapor) condensed on the underside of the plastic wrap.
8. **INFER:** solid

ONGOING ASSESSMENT

Have students write a steps-in-a-process description of the water cycle. Suggest that they number each step in the correct order. **learning styles: linguistic, logical/mathematical**

• TEACHING RESOURCES • Online or Print

Teacher's Resources

Lesson Review: Chapter 9, p. 2

Enrichment: Chapter 9, p. 5

Go Online active art

For: Water Cycle activity
Visit: PHSchool.com
Web Code: cfp-4024

Students can interact with the art of the water cycle online.

9-2 *(pp. 210–211)*

1 Introduce

Linking Prior Knowledge Review the definition of runoff with students (Lesson 8-3). Explain that water that does not flow along Earth's surface as runoff but sinks into the ground is called groundwater.

2 Teach

Demonstration Demonstrate the water table by pouring gravel into an empty aquarium or clear plastic storage box. Give the gravel an irregular surface. Pour some water into the aquarium and have students observe how it saturates the lower portion of the gravel. Add more water and have students observe how the level of the water rises. Have students locate the water table in the container. **learning style: visual**

Discussion Bring a sponge to class. Have students describe the characteristics of the sponge. Elicit from students that a sponge has many tiny holes, or pores, that allow it to absorb water. Relate this characteristic of the sponge to the ability of soil to hold water. **learning style: visual**

Answers to questions

1. **DEFINE:** water that collects in the pores between rock and soil particles and sinks into the ground
2. **DESCRIBE:** when they are loosely packed and have many pores
3. **DESCRIBE:** through connected pores between soil particles
4. **DEFINE:** the upper level of saturated soil and rock

READING STRATEGY Have students break the compound word *groundwater* into two words. (ground, water) Ask students to define the two words. Discuss how combining these words makes a new meaning for the compound word. Then, have students identify other compound words in this lesson. (*freshwater, cannot, bedrock, topsoil, underground, background, outdoors*) **learning styles: linguistic, auditory**

9-2 What is groundwater?

Objective
Explain how groundwater collects in soil.

Key Terms
pore: tiny hole or space
groundwater: water that collects in pores in soil and sinks into the ground
water table: upper layer of saturated rock and soil

Pores and Groundwater Some of the rain or snow that falls to Earth soaks into the soil. The water collects in the spaces, or **pores,** between bits of rock and soil. Water that collects between the bits of rock and soil and sinks into the ground is called **groundwater.** About 23 percent of Earth's freshwater supply is stored as groundwater.

1 DEFINE: What is groundwater?

Properties Affecting Groundwater Different kinds of rock and soil can hold different amounts of groundwater. Loosely packed rock or soil has many pores. It can hold a lot of groundwater.

Tightly packed rock or soil does not have many pores. It holds very little groundwater. Rock and soil containing particles that are all the same size can hold a lot of water. Suppose the soil particles are of all different sizes. Then, the smaller particles can fill up the pores. This kind of rock or soil has little room for water.

2 DESCRIBE: When can rocks and soil hold a lot of groundwater?

Movement of Groundwater Groundwater moves easily through rocks and soil with large, connected pores. However, if the pores are not connected, the water cannot sink deeper into the ground.

3 DESCRIBE: How does groundwater travel?

The Water Table Groundwater eventually reaches bedrock, where it almost stops moving. The pores in the rock above begin to fill up. When the pores are completely filled, the rock or soil is saturated (SACH-uh-rayt-ihd). The underground water level rises. The upper layer of saturated soil and rock is called the **water table.**

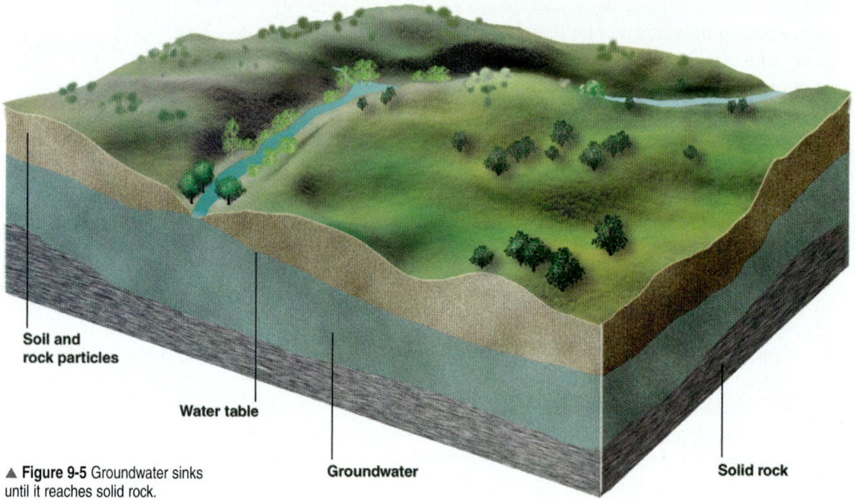

▲ **Figure 9-5** Groundwater sinks until it reaches solid rock.

Teaching Options

! FAST FACTS *The Ogallala Aquifer*

An aquifer is a layer of rock that is saturated with groundwater. The largest aquifer in the United States is the Ogallala Aquifer, which stretches beneath the High Plains from South Dakota to Texas. The Ogallala Aquifer covers an area of about 450,000 sq km (175,000 sq mi) and has an average thickness of 60 m (200 ft). Farmers in the High Plains states withdraw irrigation water from the aquifer through thousands of wells.

Water in the Ogallala Aquifer originally came from the Rockies and from precipitation that fell to the ground above the aquifer. Over thousands of years, the aquifer accumulated enough water to fill Lake Huron. To avoid depleting the aquifer, the amount of water that is withdrawn for irrigation must not exceed the amount that is added. However, because so much water is pumped out of the aquifer—chiefly for agricultural uses—the rate of water removal greatly exceeds the rate at which it is being replenished.

Because of the differences in soil density, not all of the water in the ground sinks down to the water table. Some of it stays near the surface, in the topsoil. Plant roots need this water because the roots of most plants do not reach the water table.

4 **DEFINE:** What is the water table?

CHECKING CONCEPTS

1. Groundwater collects in __________ between soil particles.
2. About __________ percent of fresh water is stored as groundwater.
3. Loosely packed soil holds a lot of __________.
4. The upper layer of saturated rock forms the __________.

THINKING CRITICALLY

5. **INFER:** Describe two ways in which your life would change if supplies of fresh water ran low.

INTERPRETING VISUALS

Use Figure 9-6 to answer the following question.

6. **ANALYZE:** Which soil sample can hold more groundwater, *A* or *B*? Why?

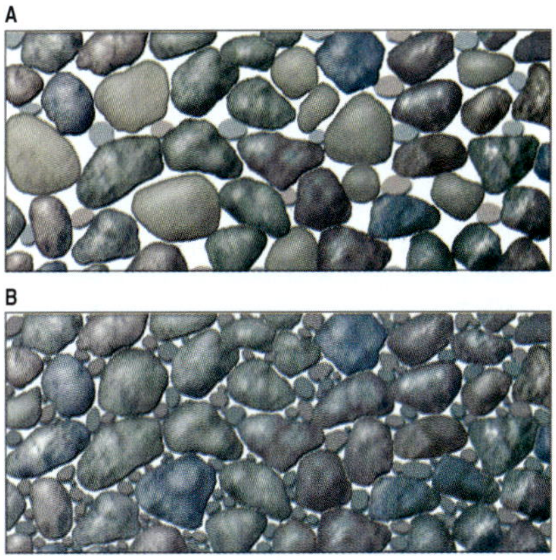

▲ Figure 9-6

People in Science

HYDROLOGIST

All living things need water to live. Water is also used for cooking, cleaning, bathing, and recreational activities. Industry and farming are dependent on water supplies.

Scientists who study Earth's water supplies are called hydrologists. Hydrologists work to keep us supplied with the fresh water we need. They may help select the best place to dig a well. They may draw up plans on how to route water from its source to a city. They help farmers irrigate their land by showing them how to use artificial canals or sprinkler systems. A hydrologist may also do research on how to keep water supplies clean. Many hydrologists work outdoors. A hydrologist must have a college degree and a background in geology, physics, chemistry, and mathematics.

▲ Figure 9-7 A hydrologist searches in rock openings for groundwater sources.

Thinking Critically A developer is planning to build new homes with private wells. How could a hydrologist help this builder?

3 *Assess*

CHECKING CONCEPTS ANSWERS

1. pores
2. 22
3. groundwater
4. water table

THINKING CRITICALLY ANSWER

5. **INFER:** Possible answer: All activities involving necessary water usage, including cooking, showering, and drinking, would need to be reduced, and unnecessary usage, such as watering lawns and flowers, would need to be stopped.

INTERPRETING VISUALS ANSWER

6. **ANALYZE:** Soil sample *A* can hold more groundwater because it contains more pores.

4 *Close*

People in Science

HYDROLOGIST

Real-Life Connection *Hydro* means "water." The science of hydrology deals with Earth's water—its properties, circulation, and distribution on, below, and above Earth's surface. Within recent years, the science of hydrology has expanded to include both the environmental and economic aspects of Earth's water.

Discussion Have students brainstorm a list of questions that they would ask a hydrologist. If possible, have a local hydrologist visit the class and allow students to take turns asking their questions. If this is not possible, have students send questions to a hydrologist via letter or e-mail.

Thinking Critically Answer
Possible answer: The hydrologist could help by locating the water table and the best place to dig the wells.

RETEACH

Modeling Pore Spaces Provide students with two plastic cups and marbles or beans of two different sizes. Have students use the materials to model loosely packed rock with many pore spaces and tightly packed rock with fewer pore spaces.
learning styles: tactile, visual

• TEACHING RESOURCES • Online or Print

Teacher's Resources
Lesson Review: Chapter 9, p. 3
Enrichment: Chapter 9, p. 5
Groundwater: Visuals, p. 8

MORE ON THE WEB

Learning How Wells Work
Have students use the Internet to learn more about how wells work. Students can begin their search at www.conceptsandchallenges.com.

9-3 *(pp. 212–213)*

1 Introduce

STUDY HINT Have students read the definitions of the Key Terms in this lesson. Then, have them make a labeled diagram that illustrates each Key Term based on their understanding of the definitions. After they have read the lesson, ask them to review their diagrams and to correct any mistakes that they may have made.

Linking Prior Knowledge To introduce this lesson, ask students where their supply of fresh water comes from. Discuss students' responses. Point out that in many communities, people get fresh water from the ground through wells.

2 Teach

Demonstration Partially fill a balloon with water and tie off the open end. Use a pin to poke a small hole in the balloon at the end opposite the tied end. Slowly squeeze the balloon to cause the water to squirt out. Tell students that squeezing the balloon puts the water under pressure and that this is similar to the water in an artesian well. Point out that a pump is not necessary to get water out of an artesian well because the water is under pressure.
learning styles: visual, auditory

Discussion Display a bottle of spring water. Ask students if they know where the water comes from. Some students may know that it comes from springs. Explain that spring water is a natural flow of groundwater at Earth's surface.
learning styles: visual, auditory

Answers to questions

1 **DEFINE:** a hole dug below the water table that fills with groundwater

2 **EXPLAIN:** pressure

3 **IDENTIFY:** on hillsides

4 **EXPLAIN:** the pressure of steam

9-3 What are wells, springs, and geysers?

Objective
Describe how groundwater reaches Earth's surface.

Key Terms
well: hole dug below the water table that fills with groundwater
spring: natural flow of groundwater to Earth's surface
geyser (GY-zuhr)**:** heated groundwater that erupts onto Earth's surface

Wells In many places, people get their fresh water from wells. A **well** is a hole dug into the soil to reach below the water table. Water enters an opening in a pipe placed in the well. The water can then be pumped to the surface.

The level of the water table changes from season to season. When there is little rain, the level of the water table drops. The pipe for a well must be set much deeper than the water table line. This will ensure that the well will not run dry during dry weather.

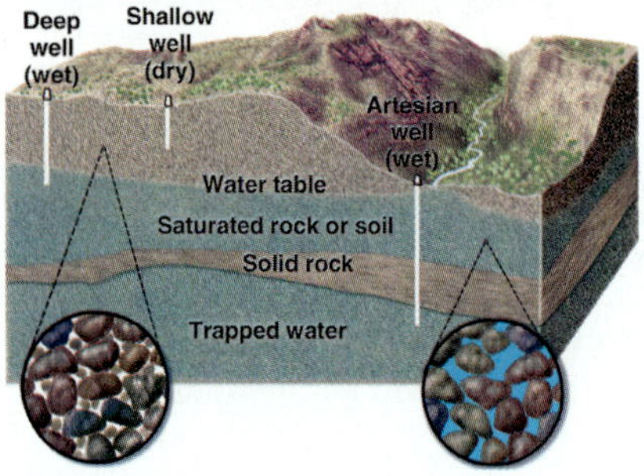

▲ **Figure 9-8** A well must reach below the water table.

1 **DEFINE:** What is a well?

Artesian Wells A pump is not necessary to get water out of artesian (ahr-TEE-zhen) wells. Water rises freely from them.

The water level in an artesian well can rise higher than ground level. The water source, called a trapped aquifer, is under pressure. The pressure makes the water rise. An artesian well may be located far from its original water source.

2 **EXPLAIN:** What makes water rise in an artesian well?

Springs A **spring** is a natural flow of groundwater that reaches Earth's surface. The side of a steep hill may dip below the water table. Water can then flow out of cracks in the rocks. This is why springs are usually found on hillsides. Spring water is often sold in stores as high-quality drinking water. Most spring water is cold. But when it is near an underground heat source, hot springs may form.

3 **IDENTIFY:** Where are most springs found?

Geysers Sometimes steam and boiling water shoot into the air, forming a geyser. A **geyser** is heated groundwater that erupts onto Earth's surface. Water in a deep hot spring may be heated above the boiling point of water. If the water is trapped and pressure is generated, the water becomes superheated. This superheated water turns to steam. The pressure of the steam forces the water above it out into the air. Geothermal areas containing geysers are found in Wyoming, New Zealand, and Iceland, among other places.

▲ **Figure 9-9** This geyser, called Old Faithful, is found in Yellowstone National Park in Wyoming.

4 **EXPLAIN:** What force causes geysers to erupt?

Teaching Options

! FAST FACTS *Yellowstone's Springs and Geysers*

Established in 1872, Yellowstone National Park was the first national park in the United States, extending over northwest Wyoming into Montana and Idaho. The area is a geological hot spot with nearly 10,000 hot springs and 200 geysers. The best-known geyser, although not the largest, is Old Faithful. Old Faithful erupts

- every 40 to 70 minutes.
- with approximately 41,640 L (11,000 gal) of water.
- to a height of about 46 m (150 ft).

CHECKING CONCEPTS

1. A well must be dug __________ than the water table.
2. Wells with no need for pumps are called __________ wells.
3. The temperature of spring water is usually __________.
4. Hot water erupting onto Earth's surface forms a __________.
5. Water heated above the boiling point is called __________ water.

THINKING CRITICALLY

6. **INFER:** What states of water are found in geysers?
7. **HYPOTHESIZE:** Why does the temperature of spring water stay the same all year?
8. **INFER:** How is the hot water and steam found in hot springs and geysers used to generate electricity from geothermal energy?

Web InfoSearch

Oases An oasis is a very rich piece of land in the middle of a desert. Oases vary in size from less than 1 square kilometer to several square kilometers. The land for hundreds of kilometers around is dry and barren. However, an oasis has plenty of plant life and a natural spring. Many people live in oases.

SEARCH: Use the Internet to find out more about oases. Why do they have water? What is it like to live there? You can start your search at www.conceptsandchallenges.com. Some key search words are **desert oasis** and **geology.**

◀ **Figure 9-10** Oasis in an Egyptian desert

Hands-On Activity

MODELING GROUNDWATER USE

You will need gravel or aquarium stones, one aquarium tank, one small plastic tub, sandy soil, a watering can, and clean spray pumps from bottles.

1. Put a 7.5 cm layer of gravel or stones in the aquarium. Cover with a layer of sandy soil.
2. Using the watering can, let it rain on your aquarium until you have at least 5 cm of groundwater.
3. Insert the spray pump into the gravel. Start pumping into the plastic container. Record your observations. Note that if the soil is too sandy, it will clog your pump.
4. Add a few drops of food coloring to the watering can and water again. Repeat Step 3 and record your observations.

▲ **STEP 3** Pump the water into the plastic container.

Practicing Your Skills

5. **OBSERVE:** What happens to the water level as you pump?
6. **INFER:** How would adding food coloring to one water source affect the other?

ONGOING ASSESSMENT

Have students draw a cross-sectional diagram showing an artesian well.
learning style: visual

• TEACHING RESOURCES • Online or Print

Teacher's Resources

Lesson Review: Chapter 9, p. 4

Enrichment: Chapter 9, p. 5

3 *Assess*

CHECKING CONCEPTS ANSWERS

1. deeper or lower
2. artesian
3. cold
4. geyser
5. superheated

THINKING CRITICALLY ANSWERS

6. **INFER:** gas and liquid
7. **HYPOTHESIZE:** The temperature of spring water is dependent on the temperature underground, not changes of season.
8. **INFER:** The hot water and steam is pumped up through wells and used to drive turbines to generate electricity.

4 *Close*

Web InfoSearch

Oases Have students use the Internet to find out more about oases. Have students answer these questions: *Why does an oasis have water? What is life like for people who live there?* Some key search words are **desert oasis,** and **geology.** Students can begin their search at www.conceptsandchallenges.com.

Hands-On Activity

MODELING GROUNDWATER USE

TIME: 15 minutes

PURPOSE: modeling, observing, inferring

DISCUSSION: Be sure that students place the bottom of the plastic tube of the spray pump in the gravel below the water level. Make sure that they understand that this tube represents a well dug below the water table.

SAFETY TIP: Remind students to wipe up spills immediately.

ALTERNATIVE MATERIALS: A large, clear plastic storage box can be used in place of an aquarium.

Practicing Your Skills Answers

5. **OBSERVE:** The water level becomes lower.
6. **INFER:** Food coloring would spread throughout the groundwater and into the pumped water.

9-4 *(pp. 214–215)*

1 *Introduce*

STUDY HINT Have students preview the lesson by examining the photographs and reading the captions. Ask students how they think this information relates to topics they have already studied or to this lesson about rivers and streams.

Linking Prior Knowledge To introduce this lesson, ask students to describe the different stages people go through during their lives. Discuss students' responses, which may include infancy, childhood, adolescence, adulthood, and old age. Tell students that rivers also have stages of development.

2 *Teach*

Discussion Point out the three stages of a river's development and describe the river at each stage. Associate youthful rivers with rapids. Then, explain that rapids are caused by the flow of a river over a rapidly sloping rocky bed. The rapids in the St. Lawrence River and at Sault St. Marie at the outlet of Lake Superior are among the most noted rapids in North America. **learning style: auditory**

Reinforcement Be sure that students understand that the terms *youth*, *maturity*, and *old age*, as they relate to rivers, do not have a chronological meaning. The important factor in determining the stage of a river is not time but the rate of flow of a river's water. A youthful river flows very quickly. As the rate of flow decreases, the river matures. The slowest, flattest rivers are considered old. **learning style: auditory**

Answers to questions

1. **DEFINE:** a large, natural channel containing flowing water
2. **IDENTIFY:** youth, maturity, old age
3. **INFER:** Niagara Falls

9-4 What are rivers and streams?

Objective

Describe the three stages in the life cycle of a river or stream.

Key Terms

rapids: part of a river where the current is swift

waterfall: steep fall of water, as of a stream, from a height

meander (mee-AN-duhr)**:** loop in a mature river

oxbow lake: curved lake formed when a bend in a river is cut off at both ends

What Is a River? A river is a large, natural channel containing flowing water. River water flows downhill because of the force of gravity. The main sources of the water in the river are runoff from rainwater, streams that flow into the river, groundwater from springs, and melting snow.

1 DEFINE: What is a river?

The Stages of a River Rivers go through three stages in their development. The three stages can be described as youthful, mature, or old. However, the stage of a river is not really determined by the age of a river in years. It depends on how fast the water in the river flows and how steep the slope is that it is on.

2 IDENTIFY: What are the three stages in the development of a river?

Youthful Rivers A youthful river has a steep slope and fast-moving water. The Yellowstone River and the Niagara River are examples of youthful rivers. The fast-moving water erodes the riverbed, or bottom, and forms a narrow, V-shaped valley. The river fills almost the whole valley from side to side.

Two features of a youthful river are **rapids** and **waterfalls.** As the moving water rushes over steep slopes and rocks, rapids are formed. Sometimes the slope drops straight down. Then, a waterfall is formed.

▲ **Figure 9-11** Niagara Falls is a waterfall on the Niagara River.

3 INFER: What evidence indicates that the Niagara River is a young river?

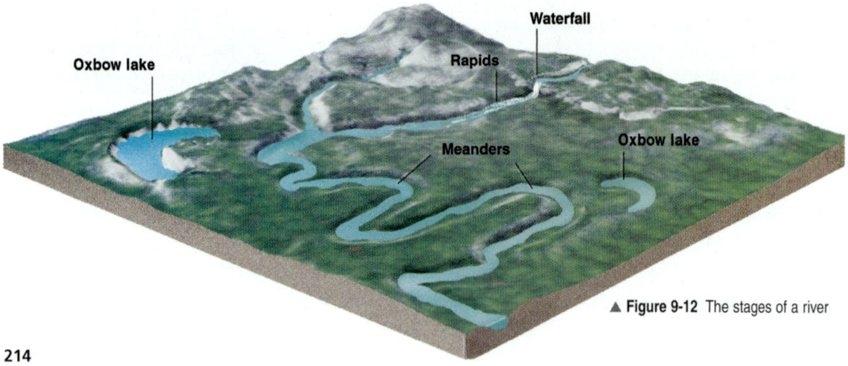

▲ **Figure 9-12** The stages of a river

Teaching Options

COOPERATIVE LEARNING

Describing an Oxbow Lake Have a volunteer describe to the class how an oxbow lake is formed, while classmates draw a diagram of a river with an oxbow lake. **learning styles: auditory, visual**

EXTENDING THE LESSON

Identifying the Stages of a River on a Topographic Map Provide students with topographic maps that show several rivers or streams at different stages of development. Explain that the lines on a topographic map are close together where the slope is steep and far apart where the slope is shallow. Ask students to identify portions of the rivers or streams that are youthful, mature, or old, based on the slope and the presence of rapids, waterfalls, or meanders. **learning style: visual**

Mature Rivers The waters of a mature river move more slowly than the waters of a youthful river. A mature river is on a shallower slope. It does not have rapids and waterfalls. Erosion widens the river and the valley floor. This creates gentle slopes. The bottom is flatter. The sides are smoother and rounder. The river winds back and forth in loops called **meanders.** The Missouri River and the Ohio River are mature rivers.

4 DEFINE: What are meanders?

Old Rivers Water moves very slowly in an old river. The Mississippi River is an old river. An old river has a nearly flat slope. Because of this, it floods easily. The slow river cannot get rid of the extra water, so it overflows.

▲ **Figure 9-13** A river floods its banks.

Flooding causes erosion and deposition along the meanders. Meanders do not extend to the valley walls. Over long periods of time, they change position. A meander may be cut off from the rest of the river. This forms a C-shaped lake called an **oxbow lake,** as shown in Figure 9-12.

5 INFER: Why does an old river flow very slowly?

Endangered Rivers Rivers can be endangered. How? Mostly it is the pollution that results from developing new energy sources that endangers rivers. This includes building dams for power, draining acid from coal mines, and drilling for oil and natural gas resources. Since 1986, a report has been issued each year listing America's most endangered rivers. In 2007, the most endangered river on the list was the Santa Fe. Figure 9-14 lists endangered rivers from earlier years.

ENDANGERED RIVERS

Year	River	Threat
2007	Santa Fe River	Dams
2005	Susquehanna River	Raw Sewage
2004	Colorado River	Dams
2003	Big Sunflower River	Flood control projects
2002	Missouri River	Dams
2000	Snake River	Dams

▲ **Figure 9-14**

6 EXPLAIN: What activities endanger a river?

CHECKING CONCEPTS

1. Use the terms *youthful, mature,* or *old* to identify the stage of each river described: **a.** has a steep slope; **b.** has a nearly flat slope; **c.** forms oxbow lakes; **d.** has many rapids; **e.** has meanders.

THINKING CRITICALLY

2. EXPLAIN: Why is a river with meanders sometimes said to "snake along"?
3. PREDICT: How can we tell if a lake near a river was once part of the river?

Web InfoSearch

The Nile River In the mid-1860s, the Royal Geographical Society sent the Scottish explorer Dr. David Livingstone to look for the source of the Nile River. Livingstone believed the source was Lake Tanganyika. He was gone a long time. No one heard from him. It was feared he was missing. A New York newspaper assigned Henry Morton Stanley, a journalist, to find him.

SEARCH: Use the Internet to find out more about this. What is the source of the Nile? Did Livingstone find it? What happened when Stanley finally met up with Livingstone? Start your search at www.conceptsandchallenges.com. Some key search words are **Nile River, Dr. David Livingstone,** and **H. M. Stanley.**

4 DEFINE: loops in a mature river

5 INFER: because the river's slope is nearly flat

6 EXPLAIN: polluting, developing new energy sources by building dams, draining acid from coal mines, and drilling for oil and natural gas

3 Assess

CHECKING CONCEPTS ANSWERS

1. **a.** youthful
   b. old
   c. old
   d. youthful
   e. mature, old

THINKING CRITICALLY ANSWERS

2. EXPLAIN: A river with meanders winds back and forth similar to the way a snake moves.
3. PREDICT: If it has a C shape, then it is an oxbow lake formed when a meander was cut off from the rest of a river.

4 Close

Web InfoSearch

The Nile River Have students use the Internet to learn more about the source of the Nile and the disappearance of Dr. David Livingstone. Ask students: *Did Dr. Livingstone eventually find the source? What happened when Stanley finally met up with him?* Some key search words are **Nile River, Dr. David Livingstone,** and **Henry Morton Stanley.** Students can begin their search at www.conceptsandchallenges.com.

RETEACH

Labeling Rivers Bring in pictures that illustrate features of youthful, mature, and old rivers. Write each stage of development (youthful, mature, old) on a separate index card. Display the pictures in front of the classroom and describe each stage of a river. Then, have students take turns placing the correct index card labels next to the river pictured. learning style: visual

• TEACHING RESOURCES • Online or Print

Teacher's Resources

Lesson Review: Chapter 9, p. 6

Enrichment: Chapter 9, p. 8

Lab Activity

(pp. 216–217)

1 Prepare the Lab

Modeling the Life Cycle of a Stream

SET-UP TIME: **LAB TIME:**

BACKGROUND: Streams are dynamic erosional features on Earth's surface. Wherever there is an excess of water that cannot be absorbed into the soil, the water begins running across the surface. Streams have life cycles that progress over thousands or millions of years. Many streams begin in high country and flow very rapidly. They carve steep, V-shaped valleys. Downstream, valleys broaden, and the stream widens to form a river. Eventually, the flow becomes very sluggish and winding as the river ultimately empties into a lake or an ocean. As new land is pushed up, the cycle begins again.

PURPOSE: Students will observe and describe the changes that take place in a stream as it matures.

ALTERNATIVE MATERIALS: A commercial classroom stream table, available from scientific supply houses, may be used in place of the storage box. Small stream tables can be made with plastic boxes, but for this lab, a longer flow distance is preferable.

SCIENCE SKILLS: Students will **model** a steep landscape of sand. After they **observe,** they will describe changes that take place over time.

ADVANCE PLANNING: A drain hole is needed at the lower end of the plastic box. Drill a small hole through the end wall near the bottom. The hole should be just large enough for a short piece of aquarium hose to be inserted. Seal the hose around the hole with silicone aquarium cement or another waterproof cement.

2 Guide the Lab

PROCEDURE: Set the stream tables on flat work surfaces. The lower end of each table should be near the end of the work surface so that the drain hose can hang downward toward the catch bucket. Elevate the upper end of the table about 5 cm, supporting the length of the box if needed.

LAB ACTIVITY

Modeling the Life Cycle of a Stream

Materials

Safety goggles, lab apron, large plastic under-the-bed storage box, 2-liter soft drink bottle, two pieces of narrow aquarium hose (about 60 cm each), drill, waterproof cement, bucket, water, sand, funnel

BACKGROUND

Water in mountain streams moves very fast. It carries with it boulders, pebbles, and sand. These cause erosion. Downstream, water moves slower and has only enough force to carry silt. As streams erode the land, they become more winding and move more slowly. This is the typical life cycle of a stream.

PURPOSE

In this activity, you will study the development of a stream as it flows over sand.

PROCEDURE

1. Have your teacher drill a small hole in one end of the plastic box, near the bottom. This hole should be just large enough to put one piece of your aquarium tubing into it. Seal the hose around the hole with waterproof cement. This is your stream table.
2. Fill the tray evenly with sand about 6 cm deep. Shape the sand into a mound near the upper end. The other end should be almost free of sand.
3. Set the plastic box on a table. The end with the hole should be near the end of the table so that the hose can hang down. Put the catch bucket directly below the hose end. Raise the upper end of the box about 5 cm, using two or three textbooks.
4. Fill the plastic bottle almost full with water. Raise it about 12 cm by resting it on top of another large box or several large books. Put one end of the second piece of hose into the bottle.
5. Push the hose almost entirely under the water in the bottle so that the tube fills. Now, pinch off the free end of the hose.

▲ **STEP 3** Raise the upper end of the box about 5 cm.

▲ **STEP 4** Insert the second hose into the bottle.

216

Teaching Options

! FAST FACTS

Rejuvenated Waters

Mature, meandering streams and rivers sometimes start cutting down into their beds as if they were youthful waterways. Called *rejuvenation*, this can occur when a sudden rise in rainfall boosts a river's or stream's flow, if land uplifts, or if a fall in sea level leaves a river's mouth above the sea. In any case, the river or stream will regrade its bed. This creates various distinct landforms, such as knickpoints (sharp steps in a waterway's profile, often with a rapid or waterfall), new terraces, meanders carved in bedrock, gorges and canyons, and natural bridges.

EXTENDING THE LESSON

Modeling Rivers Have students plan and summarize in a short report how they could use the materials in the activity to model each stage of a river. Review and approve students' plans before they make their models.
learning styles: tactile, logical/mathematical

6. Let go of the pinched end, and the water will start flowing. Make sure this end is lower than the water level in the bottle. Aim the water into the sand at the upper end of the stream table. Remember to keep the other end of the tube in the water.

7. Keep adding water to the sand until it is saturated. You will know it is saturated when water begins to flow over the surface. Watch what happens to the water and sand as you let the water continue to flow for 15 minutes. Use the funnel to add water to the bottle as needed.

8. Copy the chart in Figure 9-15. In your chart, describe the appearance of the stream after 10 minutes. Also sketch the shape of the valley.

▲ **STEP 6** Aim the water into the sand at the upper end of the stream table.

◀ **STEP 7** Watch what happens as water continues to flow over the sand.

Life Cycle of a Stream

	Appearance of stream	*Valley sketches*
Upper end		
Middle		
Lower end		

▲ **Figure 9-15** Use a copy of this chart to record your observations and draw your sketches.

CONCLUSIONS

1. **OBSERVE:** What happened to the sand when the water began flowing over it?
2. **OBSERVE:** What happened to the sand as it entered the "lake" at the lower end of the stream table?
3. **ANALYZE:** How does the stream change the land?
4. **INFER:** What will happen to the sand that ends up in a lake or an ocean?

Fill the box about 6 cm deep with sand. Mound the sand near the upper end and leave the lower end clear of sand. Place the water supply above the upper end of the table. Start the aquarium hose siphon by pushing the hose almost entirely under the water so that it fills. Pinch the end of the hose and pull it out until it is lower than the water level. Release the hose and the water will start flowing. The sand must become saturated with water before the stream flow will start. Provide students with a copy of the Lab Activity Report found in the Teacher's Resources for students to record their observations and conclusions.

TROUBLESHOOTING: Cover work areas with plastic sheets to protect the floor from spilled sand.

SAFETY TIP: Provide students with a place to rinse sand from their hands. Instruct them to clean up water spills immediately.

EXPECTED OUTCOME: Students will observe changes in the stream course as the water erodes the sand at the lower end of the stream table.

LAB HINT: The water flow can be adjusted with hose clamps.

3 Conclude the Lab

1. **OBSERVE:** Carried by the water, the sand began moving toward the lower end of the tray.
2. **OBSERVE:** The sand settled on the bottom.
3. **ANALYZE:** The stream cuts a valley into the sand and carries the sand downhill.
4. **INFER:** The sand will be deposited and will build up in layers.

Use the Laboratory Activity Rubric from the Teacher's Resources to assess students' lab activities. Fill in the rubric with the additional information below. For this activity, students should have:

- modeled a stream table.
- observed the changes that took place as the stream matured.

4 Extend the Lab

Show students pictures of different river systems, such as mountain streams, the Grand Canyon, and the lower end of the Mississippi River. Discuss what causes the differences in these rivers. (slope, underlying rock, local climate)

Tips for Using Technology

Using Spreadsheets to Record Data Have students use spreadsheet software to record their descriptions of the appearance of the stream. Spreadsheets are tables that display data such as numbers, text, dates, or all three. Students can use spreadsheets to format their data, analyze numerical data, and make graphs and charts to present data in a meaningful way. They can also use spreadsheets to present their data to the class. Have students use the Spreadsheet graphic organizer from the Teacher's Resources to plan their spreadsheets.

• TEACHING RESOURCES • Online or Print

Teacher's Resources

Lab Activity Report: Study Tools, pp. 16–17

Laboratory Activity Rubric: Teaching Tools, p. 8

Spreadsheet: Study Tools, p. 23

Laboratory Manual

Lab. Skills Worksheet: Organizing and Analyzing Data, pp. 13–14

9-5 *(pp. 218–219)*

1 Introduce

STUDY HINT Read the objective aloud and have students predict how lakes and ponds form and change. After students have read the lesson, have them revise their predictions as needed.

Linking Prior Knowledge Review with students how glacial lakes form (Lesson 8-8).

2 Teach

Demonstration Fill a shoebox with a level layer of sand several centimeters deep. Use a marble to represent a meteorite and drop the marble into the sand from a height of about 10 cm (4 in.). Remove the marble and ask: *How did the marble affect the surface?* (A depression formed.) *What might happen if the depression filled with water?* (A lake would form.) **learning style: visual**

Discussion Discuss with students any local lakes or ponds with which they are familiar. Have them describe features of the lake or pond that might indicate how old it is, such as water depth, in-filling with sediment, or marshy shorelines.
learning style: auditory

Answers to questions

1 **DEFINE:** a low spot in Earth's surface filled with still water

2 **DESCRIBE:** Erosion, glacial ice melt, land shifts during earthquakes, and meteorites hitting Earth form low spots that eventually fill with water.

3 **DEFINE:** artificial lakes formed behind dams

4 **DESCRIBE:** Lakes fill in with sediment and organic material and eventually become meadows or forests.

WRITING HINT Encourage students to use a dictionary to look up the spelling of words they are unsure of. Suggest that they keep a list of words that they often misspell and regularly review the list.
learning style: linguistic

9-5 What are lakes and ponds?

Objective
Describe how lakes and ponds form and how they change.

Key Terms
lake: low spot in Earth's surface filled with still water
pond: body of water similar to a lake but usually smaller and shallower
kettle lake: lake formed by a retreating glacier
reservoir: artificial lake

Still Waters Surface water that collects in hollows and low spots on Earth's continents forms **lakes** and **ponds.** Unlike the running waters of rivers and streams, lakes and ponds contain relatively still water. Ponds are similar to lakes but are usually smaller and shallower.

Over 98 percent of the surface water on the continents is found in lakes. Most lakes contain fresh water. However, some of the largest lakes in the world are salty.

Like oceans, lakes have shorelines that erode. Many lakes are small enough to see across. Some lakes are so big that you cannot see the other side. Most of the water collected in lakes and ponds originally came from rain, snow, or melting ice.

1 **DEFINE:** What is a lake?

Formation of Lakes Low spots on Earth's surface are created in many ways, including erosion. Also, when glaciers receded from North America, big chunks of ice were left behind. These became covered by soil and slowly melted. The soil on top fell in, and large kettle-shaped depressions were formed. Water filled the kettles, making **kettle lakes.**

Land that shifts during earthquakes sometimes forms low spots that become lakes. Large rocks from space, called meteorites, have also caused holes that have later become lakes.

2 **DESCRIBE:** How do lakes form?

Reservoirs Not all lakes are made by nature. People create lakes by building dams across rivers. A dam blocks the water until it reaches the top. The lake that is formed behind the dam is called a **reservoir.** As more water flows down the river, it is directed through a gate near the dam's top. The force of the running water can be used to power electric generators.

3 **DEFINE:** What are reservoirs?

Growing Old Lakes and ponds change over long periods of time. Look at Figure 9-16 as an example. At their peak, ponds are clear pools of water. In time, sand, silt, and dead leaves and branches may fill in the bottom. The water near the shore can become very shallow, and many plants may start growing there. Eventually, the shoreline may become a marsh.

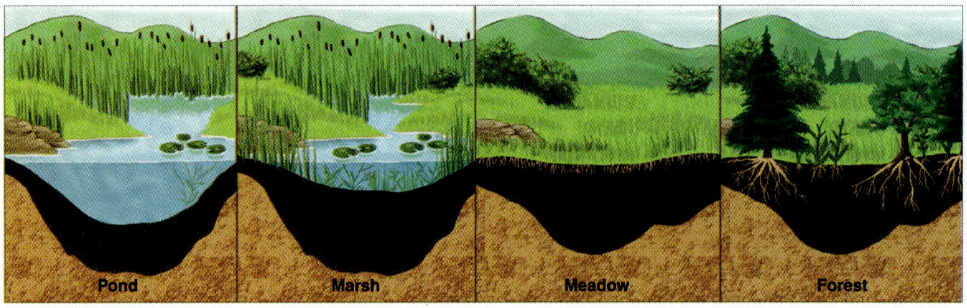

▲ **Figure 9-16** Life cycle of a pond

218

Teaching Options

! FAST FACTS *The Largest and Deepest Lakes*

The largest lake in the world is the Caspian Sea, a saltwater lake in southeastern Europe and southwestern Asia. It has an area of 371,000 sq km (143,000 sq mi), is slightly smaller than Montana, and has an average depth of about 170 m (560 ft). The deepest lake in the world is Lake Baikal in southeastern Russia. It is a freshwater lake with a maximum depth of about 1,600 m (5,200 ft).

The largest lake in the United States is Lake Superior, which is also the largest freshwater lake in the world. Lake Superior covers about 82,100 sq km (31,700 sq mi). Crater Lake in Oregon is the deepest lake in the United States, with a depth of 589 m (1,932 ft). It is also a freshwater lake.

QUICK CONVERSIONS

	Metric/SI	English
Volume of Great Lakes	2.3×10^{16} L	5.1×10^{14} gal
Average depth of Lake Superior	150 m	492 ft
Average depth of Lake Erie	19 m	62 ft

▲ **Figure 9-17** Lake George in New York State is fed by mountain streams and underground springs. It is a clear, deep lake with many fish and other wildlife.

When a pond is completely filled in, a meadow or a forest may grow on top. Only a slight dip in the land will show that a pond was once there. As old lakes and ponds fill in, new ones are being created elsewhere.

4 **DESCRIBE:** How do lakes change over time?

CHECKING CONCEPTS

1. Where does the water in lakes and ponds come from?
2. What can happen to a lake as it grows old?
3. How can people create artificial lakes?

THINKING CRITICALLY

4. **INFER:** What happens to wildlife when a lake grows old?

DESIGNING AN EXPERIMENT

Design an experiment to solve the following problem. Include a hypothesis, variables, a procedure with materials, and a type of data to study. Also include a way to record the data.

PROBLEM: A river is bringing sediment from upstream into a dam. The engineers need to prevent the reservoir from filling with sediment. How can they stop the sediment from filling the dam's reservoir?

Real-Life Science

FINDING FRESH WATER

You can see the Great Lakes in pictures of Earth taken from as far away as the Moon. The Great Lakes consist of five connected lakes along the border of the United States and Canada. The lakes are Superior, Michigan, Huron, Erie, and Ontario. Together, they contain a very large supply of fresh water. Scientists calculate that the five lakes combined hold 23,000,000,000,000,000 liters of fresh water. This is equal to about 20 percent of the world's total supply of fresh water. If this water were spread evenly over the United States, the country would be covered with almost 3 m of water.

▲ **Figure 9-18** Satellite image of the Great Lakes

The deepest Great Lake is Lake Superior. It averages 150 m deep. The shallowest is Lake Erie. It averages only 19 m deep. Water from the Great Lakes flows to the Atlantic Ocean. Niagara Falls connects Lake Ontario to Lake Erie.

Thinking Critically Cities around the country want to build pipelines to move Great Lake water to their growing communities. Do you think this is a good idea? Why or why not?

MORE ON THE WEB

Lake-Effect Snow Have students use the Internet to learn about what causes lake-effect storms. Then, have students create a news report on the crippling effect of a lake-effect snowstorm on Buffalo, New York.

ESL/ELL STRATEGY

Word Relationships Point out to English-language learners the relationship between the word *reserve* (to keep or set aside) and *reservoir*. Then have students create a word family poster with other words that share the root *reserve*.

• *TEACHING RESOURCES* •
Online or Print

Teacher's Resources

Lesson Review: Chapter 9, p. 7

Enrichment: Chapter 9, p. 8

Designing an Experiment: Study Tools, p. 14

Designing an Experiment Rubric: Teaching Tools, p. 7

Laboratory Manual

Lab. Challenge: What factors affect the speed of a river? pp. 53–56

3 *Assess*

CHECKING CONCEPTS ANSWERS

1. rain, snow, and melting ice
2. It can fill in with sediment and organic material and become a meadow or a forest.
3. by building dams across rivers

THINKING CRITICALLY ANSWER

4. **INFER:** Possible answer: Organisms that depend on the lake die or move on. New organisms move in when the lake becomes a meadow or forest.

4 *Close*

DESIGNING AN EXPERIMENT

Use the Designing an Experiment Rubric from the Teacher's Resources to assess students' experiments. Fill in the rubric with the additional information below. For this assignment, students should have:

- included a procedure that involves moving water and sediment.
- attempted to solve the problem of sediment deposition in a reservoir.

Real-Life Science

FINDING FRESH WATER

Real-Life Connection Fresh water is not the only resource supplied to the United States and Canada by the Great Lakes. Fish are also an important resource that must be shared and protected by both countries. The Great Lakes Fishery Commission, a joint venture of the United States and Canada, was established by the Convention on Great Lakes Fisheries in 1955. One of the commission's main responsibilities is to develop research programs on the lakes and to recommend measures that will promote maximum productivity of fish stocks.

Discussion Discuss the possible advantages and disadvantages of transporting water from the Great Lakes to distant cities around the country.

Thinking Critically Answer
Possible answers: It is a good idea because such a large percentage of Earth's freshwater supply is stored in the Great Lakes. It is not a good idea because it may lower the level of the lakes, with possible harmful effects to the lakes and the land surrounding them.

The Big Idea
(pp. 220–221)

1 *Introduce*

Objective Students will explain why wetlands are important. Students will research local wetlands and write an essay in their Science Logs about why they think a wetland should or should not be destroyed.

Linking Prior Knowledge Have students review the water cycle discussion (Lesson 9-1). Ask them to recall the discussions of the movement of groundwater (Lesson 9-2) and how lakes and ponds become marshes (Lesson 9-5).

2 *Teach*

Discussion Show students photographs of a variety of wetlands and characteristic vegetation, such as peat moss, mangrove trees, orchids, and cattails. After examining the photographs, have students make a three-column chart with the following heads: *What I Know About Wetlands, What I Want to Know About Wetlands,* and *What I Learned.* Encourage students to fill in the first two columns of the chart before reading the lesson and to fill in the third column after reading the lesson.
learning styles: visual, linguistic

Use the background information in Fast Facts to help guide students in writing their Science Logs.

THE BIG IDEA

Why are wetlands important to the environment? Have students start their research for their Science Log Writing Activity at www.conceptsandchallenges.com. Be sure students use search engines and websites that are reliable and acceptable. You may also wish to provide copies of the *Big Idea Planner* and *Big Idea Science Log Entry* study tools found in the Teacher's Resources.

Integrating Environmental Science

THE Big IDEA

Why are wetlands important to the environment?

How are swamps, bogs, and marshes alike? All are types of wetlands, or areas covered with shallow water.

For years, people thought that wetlands were simply a waste of good property. They drained the water and put down dry soil. Homes, roads, even shopping malls were then built on the newly landscaped areas. Almost half the original wetlands in the continental United States may have been destroyed in this way.

The destruction of wetlands has several harmful consequences. The chances for flooding increase. This is because wetlands act like giant sponges, soaking up runoff, groundwater, and precipitation. Another harmful result of wetland destruction is an increase in shoreline erosion. The roots of cordgrass and other saltwater plants commonly found in coastal wetlands anchor the soil. Without these natural anchors, bits of shoreline are broken off and carried away by waves.

Perhaps more importantly, filling in wetlands destroys the natural habitats of many plant and animal species. Inland wetlands are home to muskrats, beavers, and wood ducks. Coastal wetlands contain flounder, sea trout, and striped bass. They are also nurseries for many kinds of fish and crustaceans. Shrimp and crab populations are harmed every time a wetland is destroyed. Without wetlands, many of these animals could not survive.

Look at the illustrations on these two pages. Then, complete the Science Log to find out more about "the big idea." ✦

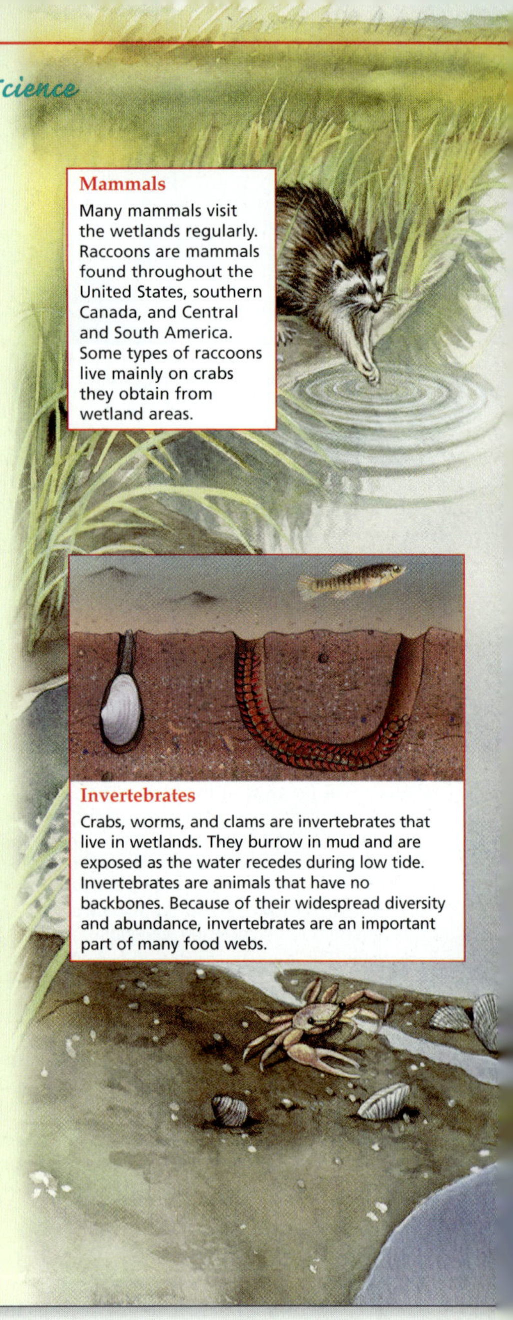

Mammals
Many mammals visit the wetlands regularly. Raccoons are mammals found throughout the United States, southern Canada, and Central and South America. Some types of raccoons live mainly on crabs they obtain from wetland areas.

Invertebrates
Crabs, worms, and clams are invertebrates that live in wetlands. They burrow in mud and are exposed as the water recedes during low tide. Invertebrates are animals that have no backbones. Because of their widespread diversity and abundance, invertebrates are an important part of many food webs.

220

Teaching Options

! FAST FACTS *More on Wetlands*

Wetlands are low-elevation ecosystems in which the water table is always at or near Earth's surface. Wetlands help to regulate the water cycle and filter the water supply. The poor drainage of wetlands creates unique habitats that serve as spawning and feeding grounds for many of the endangered species in the United States.

Wetlands include bogs, swamps, and marshes, which vary in terms of the types of soil and plant life present.

- A bog is a very old lake that has become acidic and overgrown with a thick mat of mostly peat moss.
- A swamp is a shallow body of water usually containing abundant plant growth dominated by high shrubs and trees, such as cypress.
- A marsh is a large, wet area of land where the dominant vegetation consists of grasses, rushes, and sedges (grasslike plants).

Fishes

Many kinds of fish live in or visit the wetlands. Killifish is a common name for over 100 different species of fish. During high tide, some kinds of killifish swim into the marsh in search of food. Killifish are important in controlling mosquito populations.

Birds

Waterfowl such as the great blue heron stop at coastal marshes along their migratory routes. They use the ecosystem as a place to rest, feed, and breed.

WRITING ACTIVITY

Science Log

Suppose your community was considering filling in a local wetland. How could this action affect the wildlife found in this wetland? How might this, in turn, affect the surrounding area? Do some research. Then, write an essay explaining why you think the wetland should or should not be destroyed. Start your search for information at www.conceptsandchallenges.com.

Figure 9-19 More than one-third of threatened and endangered species in the United States live only in wetlands.

3 Assess

Use the Writing Activity Rubric from the Teacher's Resources to assess students' Science Logs. Fill in the rubric with the additional information below. For this assignment, students should have:

- explained the importance of wetlands to the environment and determined how filling in a local wetland might affect its wildlife and the surrounding area.
- explained in an essay why they think the wetland should or should not be destroyed.

4 Close

Cooperative Learning Have pairs of students role-play an interview between a television reporter who wants to learn more about why wetlands are important and a wetlands scientist. Students should prepare a script for their interview and then perform it for the class.

Tips for Using Technology

Using Search Engines to Find Information Remind students that an important tool for finding information on the Internet is a search engine. A search engine allows the user to type in a keyword or phrase. The engine then searches thousands of sites on the Internet for those that contain the keyword or phrase.

When using search words that are more than one word long, suggest to students that they surround the words in quotation marks, such as "coastal wetlands." This will instruct the search engine to search for Web sites that have that exact phrase.

• TEACHING RESOURCES • Online or Print

Teacher's Resources

Big Idea: Chapter 9, p. 9

Big Idea Planner: Study Tools, pp. 11–12

Big Idea Science Log Entry: Study Tools, p. 13

Writing Activity Rubric: Teaching Tools, p. 9

Challenges (pp. 222–224)

Chapter Summary

Review Have students reread the summary and list any unfamiliar words. Have them look these words up in the chapter and write the definitions in their notebooks.

STUDY HINT Tell students to look for keywords in a definition that will help them remember the entire definition. After choosing their keywords, have them practice and test themselves until they can remember both the keywords and the complete definition.

Key Term Challenges

MATCHING

1. pore (9-2)
2. water table (9-2)
3. meander (9-4)
4. lake (9-5)
5. evaporation (9-1)
6. well (9-3)
7. geyser (9-3)
8. water cycle (9-1)

APPLYING DEFINITIONS

9. Evaporation is the changing of a liquid to a gas. Condensation is the changing of a gas to a liquid. (9-1)
10. Precipitation is water that falls to Earth from the atmosphere. The water cycle is the repeated pattern of water movement between Earth and the atmosphere. (9-1)
11. A spring is a natural flow of groundwater to Earth's surface. A geyser is heated groundwater that erupts onto the surface. (9-3)
12. Groundwater is water that collects in pores in soil. A well is a hole dug in the ground that fills with groundwater. (9-2, 9-3)
13. A lake is a larger and deeper body of water than a pond. (9-5)
14. A kettle lake forms in a depression left by melting glacial ice. An oxbow lake forms from a former river meander. (9-4, 9-5)
15. Waterfalls are steep falls of river water from a height. Rapids are part of a river where the current is swift. Both are features of youthful rivers. (9-4)

Chapter 9 Challenges

Chapter Summary

Lesson 9-1

- When liquid water takes in enough heat energy, it **evaporates,** or changes to a gas called water vapor. When water vapor loses enough heat, it **condenses,** or changes to a liquid.
- **Precipitation** is any form of water that falls from the atmosphere to Earth.
- The **water cycle** is the repeated movement of water between Earth and the atmosphere.

Lessons 9-2 and 9-3

- **Groundwater** collects in **pores** in soil. Soil with the same size particles can hold more groundwater than can soil with different-sized particles. Groundwater can move easily through soil with large, interconnected pores.
- The upper level of a layer of saturated rock is the **water table.** Some water remains near the surface of the soil.
- Groundwater can reach Earth's surface by means of a **well.** The level of the water table changes from season to season. Water rises freely out of an artesian well, so a pump is not needed.
- A **spring** is a natural flow of groundwater that reaches Earth's surface. Superheated groundwater breaks through to the surface as a **geyser.**

Lesson 9-4

- The three stages in the development of a river are described as youthful, mature, and old. A youthful river has a steep slope, fast-moving water, V-shaped valleys, and many rapids and waterfalls. A mature river has a shallower slope, is slow-moving, and forms **meanders.** An old river has a nearly flat slope and floods easily.

Lesson 9-5

- **Lakes** and **ponds** form in low areas of Earth's continents. The water in them comes mostly from rain, snow, or glacial ice and rivers and streams.
- A **reservoir** is an artificial lake.
- Lakes and ponds can age. This affects animals and plants that live in or near them.

Key Term Challenges

condensation (p. 208)	pore (p. 210)
evaporation (p. 208)	precipitation (p. 208)
geyser (p. 212)	rapids (p. 214)
groundwater (p. 210)	reservoir (p. 218)
kettle lake (p. 218)	spring (p. 212)
lake (p. 218)	water cycle (p. 208)
meander (p. 214)	waterfall (p. 214)
oxbow lake (p. 214)	water table (p. 210)
pond (p. 218)	well (p. 212)

MATCHING **Write the Key Term from above that best matches each description.**

1. tiny hole or space
2. upper layer of saturated rock and soil
3. bend in a mature river
4. low spot in Earth's surface filled with water from rain or snow
5. changing of a liquid to a gas
6. hole dug below the water table that fills with groundwater
7. heated groundwater that erupts onto Earth's surface
8. repeated pattern of water movement between Earth and the atmosphere

APPLYING DEFINITIONS **Explain the difference between the words in each pair. Write your answers in complete sentences.**

9. evaporation, condensation
10. precipitation, water cycle
11. spring, geyser
12. groundwater, well
13. lake, pond
14. kettle lake, oxbow lake
15. waterfalls, rapids

222

Teaching Options

PREPARING STUDENTS FOR STANDARDIZED TESTS

Reading Strategy: Inform students that multiple-choice answers that include words such as *never, none, only, all, always,* and *everyone* are usually not the best choice for the answer because most rules have exceptions.

Writing Strategy: When answering essay questions, remind students to be concise and to avoid using more words than necessary to answer the questions. However, students should check that they have used complete sentences throughout their essays.

Interpreting Visuals Strategy: Remind students to read all labels and captions carefully when interpreting visuals.

ESL/ELL STRATEGY

Definitions To help students learn new vocabulary terms, have them write each unfamiliar term, such as *oxbow lake* and *kettle lake*, on one side of an index card. Then, pair English-proficient students with English-language learners to write the definition for the unfamiliar term on the other side of the index card to use as flash cards to study for tests.

Content Challenges TEST PREP

MULTIPLE CHOICE Write the letter of the term or phrase that best completes each statement.

1. The process by which a gas changes to a liquid is called
   a. precipitation.
   b. evaporation.
   c. sublimation.
   d. condensation.
2. The main forms of precipitation are
   a. rain and sleet.
   b. sleet and hail.
   c. rain and hail.
   d. rain and snow.
3. About 23 percent of Earth's fresh water supply is stored
   a. as groundwater.
   b. in wells.
   c. in the oceans.
   d. in geysers.
4. In dry weather, the level of the water table in an area
   a. rises.
   b. drops.
   c. stays the same.
   d. increases, then decreases.
5. Old Faithful is
   a. a warm spring.
   b. an artesian well.
   c. a geyser.
   d. a hot spring.
6. The valley of a mature river is
   a. V-shaped with steep banks.
   b. V-shaped with gentle banks.
   c. narrow.
   d. steep.
7. The main threat to rivers is
   a. human recreation.
   b. overfishing.
   c. pollution.
   d. age.
8. A pump is usually used to get water from a
   a. geyser.
   b. spring.
   c. well.
   d. pond.
9. Artificial lakes are called
   a. reservoirs.
   b. oxbow lakes.
   c. kettle lakes.
   d. ponds.
10. Most of the fresh water that is found on Earth is found
    a. in groundwater.
    b. in lakes.
    c. in oceans.
    d. frozen in the polar ice caps.

FILL IN Write the term or phrase that best completes each sentence.

11. Water that evaporates from Earth and is changed to a gas is called __________.
12. A cloud is a collection of __________.
13. The pipe for a well should be __________ than the lowest level of the water table.
14. The temperature of spring water is usually __________.
15. A reservoir is a __________ created by a dam.
16. Like oceans, lakes have shorelines that __________.
17. The Yellowstone and Niagara Rivers are __________ rivers.
18. Ponds are similar to lakes but are usually __________.
19. Glaciers help to form __________ lakes.
20. In 2007, the most endangered river in the United States was the __________ River.

Content Challenges

MULTIPLE CHOICE

1. d (9-1)
2. d (9-1)
3. a (9-1)
4. b (9-2)
5. c (9-3)
6. b (9-4)
7. c (9-4)
8. c (9-3)
9. a (9-5)
10. d (9-1)

FILL IN

11. water vapor (9-1)
12. water droplets (9-1)
13. deeper (9-3)
14. cold (9-3)
15. lake (9-5)
16. erode (9-5)
17. youthful (9-4)
18. shallower or smaller (9-5)
19. kettle (9-5)
20. Santa Fe (9-4)

ALTERNATIVE ASSESSMENT

Diagramming the Water Cycle Have students draw a diagram that shows how water moves through the water cycle.

CUMULATIVE ASSESSMENT

Assessing Progress You may wish to administer the mid-term examination found in the Teacher's Resources at this point. Distribute copies of the Scantron Sheet, also found in the Teacher's Resources, for students to record their answers.

PORTFOLIO ASSESSMENT

Making Student Portfolios Students can demonstrate their comprehension of the concepts in this chapter by making a portfolio. The Chapter Self-Check, the Big Idea Planner, and the Weekly Journal are some of the reproducibles in the Teacher's Resources that they can include in their portfolios. Be sure that students know beforehand what they will be contributing to the portfolio so that they can put forth their best efforts. You can use the Portfolio Assessment Rubric also found in the Teacher's Resources to assess students' portfolios.

Concept Challenges

WRITTEN RESPONSE

1. **CALCULATE:** It should be at a depth of more than 15 m, since 15 m is the lower limit of the water table and the pipe must be much deeper than that. (9-2)
2. **CONTRAST:** Water from a hot spring flows freely from the spring, whereas water from a geyser erupts only under great pressure. (9-3)
3. **EXPLAIN:** With time, a lake that is clear and deep fills in, becoming shallow and eventually becoming a meadow or forest. (9-5)
4. **ANALYZE:** Youthful rivers have V-shaped valleys. Mature rivers have wider, flatter valley floors and smoother, rounder valley sides. (9-4)
5. **EXPLAIN:** If a meander becomes cut off from the river, it may form an oxbow lake. (9-4)
6. **ANALYZE:** Possible answer: Plants take in runoff or water in pores above the water table. (9-2)
7. **HYPOTHESIZE:** Possible answer: by using less electricity and fuels and passing laws that prevent pollution (9-4)
8. **HYPOTHESIZE:** Water would remain on or below Earth's surface and no longer enter the atmosphere. (9-1)

INTERPRETING VISUALS (9-1)

9. A
10. the changing of a liquid to a gas
11. D
12. from rain or snow that falls to Earth and soaks into the soil
13. C
14. Accept any two of the following: snow, sleet, or hail.
15. in the atmosphere or on Earth's surface
16. the changing of a gas to a liquid

Chapter Project Wrap-Up

WATER CYCLE GAZETTE

Students' projects can be photocopied and distributed in class as well as to other classes. You can use the Individual Activity Rubric found in the Teacher's Resources to assess students' projects. Fill in the rubric with the additional information below. For this project, students should have:

- organized information into a neat, scientifically accurate newspaper about the water cycle.
- included images illustrating concepts related to the water cycle.

Concept Challenges TEST PREP

WRITTEN RESPONSE **Answer each of the following questions in complete sentences.**

1. **CALCULATE:** If the level of a water table is 10 m below Earth's surface during the wet season and 15 m below the surface during the dry season, at what depth should a pipe for a well be placed? Why?
2. **CONTRAST:** How does a hot spring differ from a geyser?
3. **EXPLAIN:** How does a lake change over time?
4. **ANALYZE:** How can you tell the stage of a river by observing the shape of the river's valley?
5. **EXPLAIN:** How can a meander turn into an oxbow lake?
6. **ANALYZE:** How do plants get water if their roots do not go deep enough into the soil to reach the groundwater?
7. **HYPOTHESIZE:** How can people help rivers get off the endangered rivers list?
8. **HYPOTHESIZE:** What would happen to Earth's water supply if water did not evaporate?

INTERPRETING VISUALS **Use Figure 9-20 below to answer the following questions.**

9. What letter on the diagram shows evaporation taking place?
10. What is evaporation?
11. Which letter on the diagram shows groundwater?
12. Where does groundwater come from?
13. Which letter in the diagram shows precipitation?
14. What are three other forms of precipitation?
15. Where does condensation take place?
16. What is condensation?

▲ Figure 9-20

224

Teaching Options

• TEACHING RESOURCES •
Online or Print

Teacher's Resources

Weekly Journal: Study Tools, p. 19

Writing Activity Rubric: Teaching Tools, p. 9

Chapter Self-Check: Teaching Tools, p. 11

Developing Science Skills: Handbook Resources, p. 1

Handbook Key Term Review: Handbook Resources, pp. 9–10

Chapter 10 The Oceans

PLANNING GUIDE

◆ **TEACHING THE CHAPTER** This chapter should take approximately 9–12 days to complete instruction and assessment.

	Skills and Features	Projects and Activities	Achieve Science Literacy	Meet Individual Needs
Chapter 10 Opener p. T225	RESEARCH	• Chapter Project		
Lesson 10-1 **What is the world ocean?** pp. T226–T227 Standards: D1f, D1g	GRAPH, IDENTIFY, OBSERVE • Building Science Skills • People in Science	• More on the Web	• Study Hint • Reading Strategy • Writing Hint	• Reteach Teacher's Resources* Enrichment Activity Oceans
Lesson 10-2 **How are the oceans explored?** pp. T228–T229	CALCULATE, CONTRAST, EXPLAIN • Investigate • Web InfoSearch • How Do They Know That?	• Extending the Lesson • More on the Web	• Study Hint	• ESL/ELL Strategy Teacher's Resources* Enrichment Activity
Lesson 10-3 **What are some properties of the ocean?** pp. T230–T231 Standards: D1f, D1g	SEQUENCE, PREDICT, DESCRIBE • Investigate • Building Reading Skills	• Extending the Lesson • Lab Challenge	• Reading Strategy • Ongoing Assessment	
Lesson 10-4 **What are ocean currents?** pp. T232–T233 Standard: D1j	DEFINE, DESCRIBE, HYPOTHESIZE • Investigate • Health and Safety Tip • Real-Life Science	• More on the Web	• Writing Hint	• ESL/ELL Strategy • Reteach
Lesson 10-5 **What are ocean waves?** pp. T234–T235 Standard: D1j	MODEL, DEFINE, EXPLAIN • Web InfoSearch • Hands-On Activity	• Cooperative Learning • Extending the Lesson • Active Art	• Study Hint • Writing Hint	• Reteach
Lab Activity **Modeling the Shapes of Shorelines** pp. T236–T237 Standard: D1c	OBSERVE, ANALYZE, INFER • Set-Up Time: 25–30 min • Lab Time: 40–45 min	• Extend the Lab Activity	• Tips for Using Technology	• ESL/ELL Strategy Teacher's Resources* Lab Activity Report

*www.conceptsandchallenges.com, password: ccearth09

Planning Guide continues on next page.

Standards: For details on the correlation to National Science Standards see pages *xx–xxi*.

	Skills and Features	Projects and Activities	Achieve Science Literacy	Meet Individual Needs
Lesson 10-6 **What are the tides?** pp. T238–T239 Standard: D3c	HYPOTHESIZE, DEFINE, PREDICT • Real-Life Science	• Cooperative Learning • Extending the Lesson • More on the Web	• Study Hint	
Lesson 10-7 **What are ocean sediments?** pp. T240–T241 Standards: D1c, D2k	RESEARCH, IDENTIFY, DEFINE • Building Science Skills • Hands-On Activity	• Extending the Lesson • More on the Web	• Writing Hint	• Reteach
Lesson 10-8 **What are some ocean landforms?** pp. T242–T243 Standard: D1c	PREDICT, HYPOTHESIZE, IDENTIFY • Web InfoSearch • How Do They Know That?	• Extending the Lesson	• Study Hint • Reading Strategy • Ongoing Assessment	• Reteach Teacher's Resources* Enrichment Activity
Lesson 10-9 **What are coral reefs?** pp. T244–T245 Standards: D1c, D2k	IDENTIFY, INFER, HYPOTHESIZE • Designing an Experiment • Integrating Environmental Science	• More on the Web	• Study Hint	• Reteach Teacher's Resources* Enrichment Activity Designing an Experiment
Big Idea **Where is life found in the ocean?** pp. T246–T247 Standard: D2k	RESEARCH, COMMUNICATE, ANALYZE • Science Log: Writing Activity	• Big Idea • Close Activity • Active Art	• Tips for Using Technology	• ESL/ELL Strategy Teacher's Resources* Big Idea Planner
Chapter 10 Challenges pp. T248–T250	• Chapter Summary • Key Term Challenges • Content Challenges • Concept Challenges	• Chapter Project Wrap-Up	• Study Hint • Preparing Students for Standardized Tests	• ESL/ELL Strategy Teacher's Resources* Chapter Self-Check Weekly Journal

*www.conceptsandchallenges.com, password: ccearth09

Chapter 10 The Oceans

▲ **Figure 10-1** Coral reefs are home to a wide variety of life forms.

The oceans form the largest living environment on Earth. As on land, there are many smaller environments within the larger ocean environment. Coral reefs are an example. In addition to living things, the oceans contain valuable nonliving resources. They are great storehouses of dissolved minerals. For example, if all the water evaporated from the oceans, a layer of salt about 50 m thick would cover the ocean floor.

► Coral reefs are often described as underwater rain forests. Why?

Contents

Teaching Options

ASSESSMENT PLANNER

For assessment in the Student Edition, see the following pages:

Content Assessment

Checking Concepts: pp. 227, 229, 231, 233, 235, 239, 241, 243, 245
Thinking Critically: pp. 227, 229, 231, 233, 235, 239, 241, 243, 245
Interpreting Visuals: p. 239
Chapter Challenges: pp. 248–250

Alternative Assessment

Building Skills: pp. 227, 231, 241
Web InfoSearch: pp. 229, 235, 243
Science Log: p. 247

Performance-Based Assessment

Hands-On Activity: pp. 235, 241
Designing an Experiment: p. 245
Lab Activity: pp. 236–237

• TEACHING RESOURCES • Online or Print

Teacher's Resources

Lesson Review: Ch. 10, pp. 2, 3, 5, 6, 7, 8, 9, 10, 11

Enrichment: Ch. 10, pp. 4, 12

Key Term Review: Ch. 10, p. 14

Chapter 10 Test: pp. 15–17

Laboratory Manual: pp. 57–60

Chapter 10 The Oceans

Chapter Overview

In this chapter, students will learn about ocean currents, waves, and how the oceans are explored. They will see why tides occur. They will read about ocean sediments, landforms, and reefs. Students will also learn about the forces that shape shorelines and life found in the oceans.

About Figure 10-1 Only the surface of a reef contains live coral. Below are the exoskeletons of dead coral that have built up over hundreds or thousands of years. Coral reefs occur only in shallow, warm ocean waters.

Answer Possible answer: Like rain forests, coral reefs are located in Earth's warm, equatorial regions. They are also home to a wide variety of life.

Linking Prior Knowledge

For this chapter, students should recall:

- the definition of Earth's hydrosphere (Lesson 1-2).
- the concepts of erosion and deposition (Lesson 8-1).
- how ocean waves cause erosion and create landforms (Lesson 8-9).

Chapter Project

TREASURES OF THE SEA BOX

MATERIALS: plastic container, sand, shells, clay

Have students fill a large plastic container with sand. Leave half of it with only a shallow layer of sand. This end will represent the ocean floor. Have students decorate the outside of the box with pictures of sea life. Divide the class into five or six groups. Have each choose an appropriate subject, such as animals or plants of the sea, ocean resources, ocean landforms (including coral reefs, lagoons, volcanic islands), and the ocean floor. Each group should do research on its subject, producing a paper to present to the class. Each group should also make or find small plastic models appropriate to their topic and bury them in the sandbox. Coral reefs and the continental shelf can go in the shallow end of the box.
learning styles: visual, tactile

10-1 *(pp. 226–227)*

1 Introduce

STUDY HINT Ask students to explain why the term *world ocean* is used and how it relates to the main idea of the lesson.

Linking Prior Knowledge Ask students to discuss what they already know about Earth's oceans. List responses on the board. As you proceed with the lesson, compare students' responses with information in the text.

2 Teach

Demonstration Display a large map of the world. Point out some seas and gulfs on the map. Ask volunteers to identify which of the world's oceans each sea or gulf is connected to. Point out that some very large bodies of water are called seas, even if they are not connected to an ocean. (Examples include the Aral Sea, near southern Kazakhstan, and the Caspian Sea, between southeast Europe and western Asia.) **learning styles: visual, auditory**

Discussion Refer students to the map of the oceans in Figure 10-3. Discuss how geographers divide the world ocean into three major oceans. Have students locate the Atlantic, Pacific, and Indian oceans on the map. Describe the characteristics of each ocean. Ask students which ocean is the largest. (Pacific)
learning style: auditory

Answers to questions

1. **CALCULATE:** 25%
2. **IDENTIFY:** Atlantic, Pacific, Indian
3. **OBSERVE:** Possible answers include the Mediterranean Sea, North Sea, Caribbean Sea, and Gulf of Mexico.
4. **DEFINE:** the study of Earth's oceans

READING STRATEGY Remind students that in Lesson 1-8, they learned that *graph* comes from a Greek word that means "write." Ask students which of the Key Terms this information will help them understand and how. Have students list other words that contain *graph.*

10-1 What is the world ocean?

Objectives

Describe the world ocean. Explain what is meant by oceanography.

Key Terms

world ocean: body of salt water covering much of Earth's surface

oceanography (oh-shuh-NAHG-ruh-fee): study of Earth's oceans

The Water Planet About 70 percent of Earth's surface is covered with water. Most of this is salt water. This large body of salt water is known as the **world ocean.** Earth is the only planet in the solar system that has a covering of liquid water.

1 **CALCULATE:** About what percentage of Earth's surface is dry land?

Divisions of the World Ocean The world ocean is divided into three major bodies of salt water, also called oceans. These are the Atlantic Ocean, the Pacific Ocean, and the Indian Ocean. Geographers often mention two more oceans, the Arctic and the Antarctic. However, these are regarded by Earth scientists as parts of the Atlantic and Pacific oceans.

The word *sea* is sometimes used as another name for ocean. It also refers to a smaller body of water connected to or near an ocean.

Some inland bodies of water are also called seas. Most seas, though, are at least partly connected to an ocean. A gulf is a large area of ocean, larger than a bay, reaching into land.

▲ **Figure 10-2** Because of its great amounts of liquid water, Earth is often known as the blue planet.

2 **IDENTIFY:** What are the three major oceans?

Size and Depth The Pacific Ocean is the largest of the world's oceans. More than half of Earth's ocean water is in the Pacific Ocean. The Pacific is also Earth's deepest ocean. Its average depth is 4.3 km.

The Atlantic Ocean is the second largest ocean. Several seas and gulfs are part of the Atlantic Ocean. Its average depth is 3.3 km.

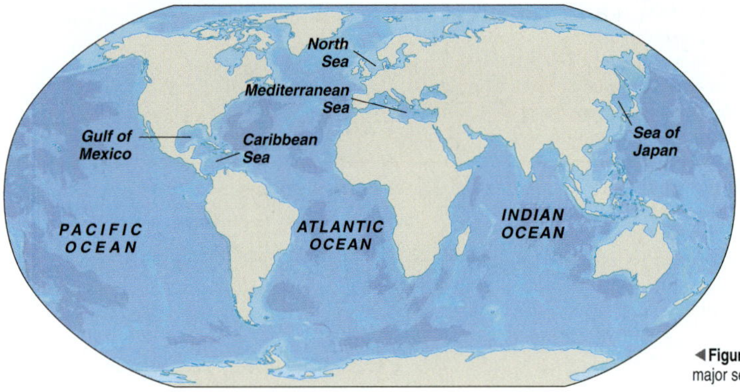

◀ **Figure 10-3** Oceans and major seas around the world

Teaching Options

RETEACH

Locating Oceans Provide students with blank world maps. Have them fill in the names of oceans, seas, gulfs, and other bodies of salt water, using an atlas or world map as an information source. Have students use a blue crayon or colored pencil to color the water and show how the bodies of water are connected.
learning style: visual

QUICK CONVERSIONS

Average Ocean Depth	Metric/SI	English
Pacific	4.3 km	2.6 mi
Atlantic	3.3 km	2.0 mi
Indian	3.8 km	2.3 mi

The Indian Ocean is the smallest ocean. It is deeper than the Atlantic Ocean, but not as deep as the Pacific. The average depth of the Indian Ocean is 3.8 km.

3 **OBSERVE:** Use Figure 10-3 to name two seas and a gulf that are part of the Atlantic Ocean.

Oceanography The study of Earth's oceans is called **oceanography.** Scientists who study the oceans and ocean life are called oceanographers. An oceanographer might specialize in the study of the oceans' depths, coral reefs, or the geography of the ocean floor.

4 **DEFINE:** What is oceanography?

CHECKING CONCEPTS

1. About how much of Earth's surface is covered with water?
2. How many oceans do geographers often name?
3. Which of Earth's oceans is the deepest?
4. About what fraction of Earth's ocean water is in the Pacific Ocean?
5. Why are Earth's oceans collectively called "the world ocean"?

THINKING CRITICALLY

6. **INFER:** Why is Earth often called the blue planet?
7. **HYPOTHESIZE:** Why would an oceanographer need to specialize?
8. **SEQUENCE:** List the three major oceans in order, from the smallest to the largest: Atlantic, Indian, Pacific.

BUILDING SCIENCE SKILLS

Graphing Create a bar graph that compares the relative depths of the three oceans.

People in Science

MARINE BIOLOGIST

Marine biologists are oceanographers who study life in the ocean. They work in laboratories, aboard research ships at sea, or in coastal areas. If they work in the ocean, they must be in good physical shape. For example, they may dive into the deep ocean to observe the interactions of living things with the environment.

Research is a large part of marine biology. Some marine biologists take samples of ocean water to test the amount of pollution in it. They may investigate how the pollution affects sea life.

▲ **Figure 10-4** This marine biologist is studying hammerhead sharks.

Research work is often done in special laboratories near the seacoasts. Among these laboratories are the Scripps Institution of Oceanography in California and the Marine Biological Laboratory of the Woods Hole Oceanographic Institute in Massachusetts. Marine biologists must know biology, physics, chemistry, and math.

Some of the most exciting discoveries in marine biology have been made on the ocean floor. Deep-sea, or hydrothermal, vents are cracks in the ocean floor that leak hot, acidic water. The discovery of them gave marine biologists whole new communities of organisms to study.

Thinking Critically What tool would a marine biologist use to study ocean bacteria?

MORE ON THE WEB

The Origins of Salt Water Have students search on the Internet to learn more about salt in the oceans. Then, have students create a children's book that illustrates how the oceans became salty.

• TEACHING RESOURCES • Online or Print

Teacher's Resources

Lesson Review: Chapter 10, p. 2

Enrichment: Chapter 10, p. 4

Oceans: Visuals, p. 9

WRITING HINT Emphasize that the word *sea* can mean "ocean," "body of salt water connected to an ocean," or "large inland water body." Remind students that any report they write on the subject should give the reader enough clues to understand which meaning is being used.

3 Assess

CHECKING CONCEPTS ANSWERS

1. 75% or three-quarters
2. five
3. the Pacific Ocean
4. more than half
5. They are all connected.

THINKING CRITICALLY ANSWERS

6. **INFER:** It is the only planet in our solar system covered with liquid water.
7. **HYPOTHESIZE:** Possible answer: There are so many aspects of the ocean to study that one person cannot master them all.
8. **SEQUENCE:** Indian, Atlantic, Pacific

BUILDING SCIENCE SKILLS ANSWER

Graphing The graphs should reflect deepest (Pacific) to shallowest (Atlantic).

4 Close

People in Science

MARINE BIOLOGIST

Real-Life Connection Water is heavier than air, so humans who dive to significant depths are under much more pressure than normal. If the diver ascends too quickly, the pressure on the body decreases too rapidly. Nitrogen gas present in the blood and other body fluids forms bubbles, much as a soft-drink forms bubbles when its can is opened and the pressure is released. These bubbles can stretch or break body tissues or interfere with blood circulation. This painful and possibly life-threatening condition is known as the bends.

Discussion Ask students: *Would you like to become a marine biologist?* Have them explain their answers. Ask students to describe the types of marine life that they would like to learn more about.

Thinking Critically Answer Possible answer: a microscope

10-2 *(pp. 228–229)*

1 Introduce

INVESTIGATE

TIME: 10–15 minutes

PURPOSE: Students will model buoyancy in soda water.

MATERIALS: drinking cup, soda water, grapes

PROCEDURE: Have students work in pairs. They will see the grapes rise to the surface and then wipe the bubbles off the surface to change the grapes' buoyancy. **learning styles: visual, tactile**

THINK ABOUT IT: Buoyancy makes the grapes float. When the bubbles are removed, the grapes are denser than the soda water and sink. Submersibles use changes in buoyancy to rise and fall in the water.

STUDY HINT Have students explain how sonar works, using Figure 10-5 for clues. Then, ask them to draw a relationship between sonar and submersibles.

Linking Prior Knowledge Point out that technology has helped humans to learn more about places that are familiar but hard to explore. Ask what some of these might be. (space, interior of Earth, ocean depths) Ask students to describe any movies or videos about the deep sea that they may have seen.

2 Teach

Discussion Ask students if they have ever heard their own echo. Explain that echoes are produced when sound waves bounce off solid objects. Explain how echo sounding is used to map the ocean floor. Refer students to Figure 10-5. Give different times in seconds to students and have them determine the depth.
learning styles: auditory, logical/mathematical

Answers to questions

1. **EXPLAIN:** by drilling into it
2. **DESCRIBE:** Possible answer: to map the ocean floor
3. **EXPLAIN:** They can reach greater depths and stay there for long periods of time.

10-2 How are the oceans explored?

INVESTIGATE

Modeling Buoyancy in the Ocean
HANDS-ON ACTIVITY

You will need a drinking cup, soda water, and grapes.

1. Fill the cup almost full with the soda water.
2. Immediately add five grapes to the cup, one at a time.
3. Wait and watch.

THINK ABOUT IT: What physical force makes the grapes rise to the surface? What happens when the bubbles are knocked off the surface of the grapes? Why? How might this effect help underwater submersibles explore the deep ocean?

Objective
Describe three ways scientists explore the oceans.

Key Terms
sonar: echo-sounding system that bounces sound waves off the ocean floor

submersible (suhb-MUHR-suh-buhl)**:** underwater research vessel

Deep-Sea Drilling Samples of rock from the ocean floor can be obtained by drilling. Studying these samples allows scientists to learn more about the ocean floor. The Deep Sea Drilling Project (DSDP) was an important ocean research program that ended in 1983. The research ship *Glomar Challenger* was specially built for the DSDP. It could drill 1 km into the crust beneath the ocean. The DSDP was replaced in 1985 by another program called the Ocean Drilling Program (ODP). A new ship, called the *JOIDES Resolution,* took over for the retired *Glomar Challenger*.

1. EXPLAIN: How do we get rock samples from the ocean floor?

Sonar Scientists can map the ocean floor by using sonar. The word *sonar* comes from the letters in sound navigation and ranging. **Sonar** is an echo-sounding system. Sound waves travel through water at about 1,500 m/s. A transmitter bounces a sound wave off the ocean floor. The returning sound wave, or echo, is picked up by a receiver. Scientists can measure the time it takes for the sound wave to return. This number can be used to calculate the depth of the ocean floor.

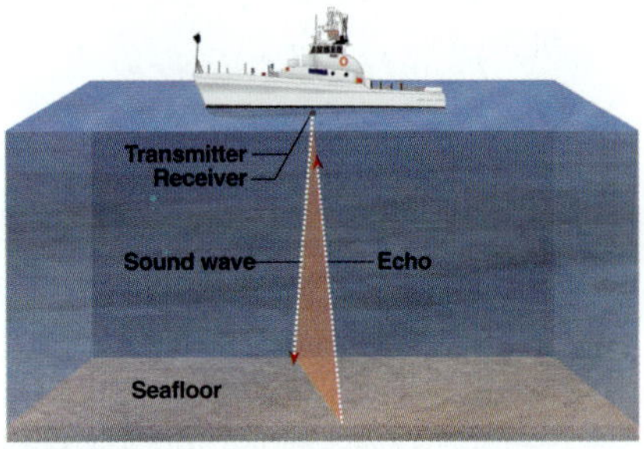

▲ **Figure 10-5** Sonar uses echoes of sound waves.

Suppose a sound wave makes a round trip in 10 seconds. The sound wave takes 5 seconds to reach the ocean floor. It takes another 5 seconds to bounce back to the ship. The depth of the ocean at that point is 1,500 m/s × 5 s = 7,500 m.

2. DESCRIBE: What is sonar used for?

Submersibles Scientists also study the oceans by traveling in underwater research vessels called **submersibles.** One of the first submersibles to be developed was a bathysphere (BATH-ih-sfeer). The bathysphere remained attached to the research ship. Therefore, its movements were limited.

Teaching Options

EXTENDING THE LESSON

Understanding Buoyancy Buoyancy is the upward pressure exerted on an object by the fluid in which it is immersed. The object is subject to pressure from all sides, but greatest on the bottom. The result is a force that pushes the object upward equal to the weight of the fluid being displaced. Because the Greek mathematician Archimedes was the first to discover this effect, it is also known as Archimedes' principle. Have students test the buoyancy of various small objects by placing them in water and describing what they see.

ESL/ELL STRATEGY

Waves in Motion Use a ripple tank to show that waves are reflected when they strike a surface. A glass tray placed on the stage of an overhead projector works well. Use a plastic ruler to set a wave in motion. Students will see the wave travel to the far end of the tray and bounce off. Draw an analogy between water waves and sound waves. Stress that the ripple tank is a model of sound waves and echoes. **learning style: visual**

Another kind of submersible is called a bathyscaph (BATH-ih-skayf). A bathyscaph is not attached to anything. It can hold one pilot and two scientists. Scientists in the bathyscaph *Alvin* discovered many unusual forms of life in deep-sea vents. Robot submersibles equipped with underwater cameras can reach great depths and stay there for long periods of time.

▲ **Figure 10-6** This deep-diving minisub is a bathyscaph.

EXPLAIN: How are robot submersibles better for deep-ocean research than other submersibles are?

CHECKING CONCEPTS

1. What was the Deep Sea Drilling Project?
2. What does the word *sonar* stand for?
3. What is a submersible?

THINKING CRITICALLY

4. **CONTRAST:** How are a bathysphere and a bathyscaph different?
5. **CALCULATE:** A sonar signal sent out returns in 55 seconds. How deep is the ocean floor in that spot?

Web InfoSearch

Robot Submersibles Many submersibles have robots aboard. These can both explore and take pictures of the ocean floor. In 1985, Dr. Robert D. Ballard used the robot submersible *Argo* to find the remains of the sunken *Titanic.* On board was a robotic camera called Jason, Jr.

SEARCH: Use the Internet to find out more about these amazing vehicles. What else have they found in exploring the ocean floor? Start your search at www.conceptsandchallenges.com. Some key search words are **submersibles, underwater exploration, *Titanic*,** and **Robert Ballard.**

How Do They Know That?

PEOPLE CAN BREATHE IN THE OCEAN

People were not designed to breathe underwater. Yet, different devices have been invented to help us do just that.

In 1939, the self-contained underwater breathing apparatus, or scuba, was designed for use by the U.S. military. In 1943, two Frenchmen, Jacques Cousteau and Emile Gagnan, invented the Aqua-lung. This device allowed a diver both to breathe and to move about freely underwater. A tank filled with compressed air was strapped to a diver's back. A hose carried air from the tank to the diver's mouth. A valve controlled air flow. Cousteau tested the Aqua-lung in more than 500 dives. Aqua-lungs are still used today. In the 1950s, Cousteau built an underwater diving station at the edge of the continental shelf. Divers lived and worked there. Cousteau himself traveled all over the world exploring the oceans. He died in 1997. His granddaughter Alexandra continues his work.

▲ **Figure 10-7** A scuba diver can breathe underwater by using tanks filled with air.

Thinking Critically How did the Aqua-lung help in the study of the oceans?

MORE ON THE WEB

Deep-sea Archaeology Have students use the Internet to read about Dr. Robert D. Ballard and his many ocean discoveries Then, have students write a brief biographical sketch of Dr. Ballard. Students can begin their search at www.conceptsandchallenges.com.

QUICK CONVERSIONS		
	Metric/SI	English
Drilling depth	4.0 km	2.5 mi
Sound-wave speed	1,500 m/s	4,920 ft/s
Ocean depth	7,500 m	24,600 ft

• *TEACHING RESOURCES* •
Online or Print

Teacher's Resources

Lesson Review: Chapter 10, p. 3

Enrichment: Chapter 10, p. 4

3 *Assess*

CHECKING CONCEPTS ANSWERS

1. a research program investigating the ocean floor
2. sound navigation and ranging
3. an underwater research vessel

THINKING CRITICALLY ANSWERS

4. **CONTRAST:** A bathysphere is a diving vessel attached to a ship. A bathyscaph is a small submarine that is not attached to anything.
5. **CALCULATE:** 1,500 m/s × 27.5 s = 41,250 m (25.6 mi)

Web InfoSearch

Robot Submersibles Have students read the passage and come up with two questions about robot submersibles that they cannot answer from the information in the passage. Have students then search the Internet to find the answers. They should begin their search for information at www.conceptsandchallenges.com using the key search words **submersibles, underwater exploration, *Titanic*,** and **Robert Ballard.**

4 *Close*

How Do They Know That?

PEOPLE CAN BREATHE IN THE OCEAN

Technology Connection One of the first successful diving suits was manufactured in 1837 by Augustus Siebe, who fitted a vented metal helmet to an airtight full-body suit worn by the diver. A hose ran from the helmet to a pump on the surface that supplied the diver with air.

Discussion Ask students how the Aqua-Lung was an improvement over underwater breathing systems that used a hose to carry air from the surface to the diver. (Using the Aqua-Lung, the diver could move more freely around obstacles and through small passageways.

Thinking Critically Answer
Possible answer: The Aqua-Lung made it possible to go deeper in the ocean than before. People could spend more time observing underwater life.

10-3 *(pp. 230–231)*

1 Introduce

INVESTIGATE

TIME: 10–15 minutes

PURPOSE: Students will determine the difference in density between fresh water and salt water.

MATERIALS: 250-mL beakers, salt, tap water, uncooked eggs

PROCEDURE: Have pairs of students work together. Caution students to handle eggs carefully.

THINK ABOUT IT: An egg in salt water will float. An egg in fresh water will not float. This result shows that salt water has a higher density and can therefore hold up the weight of the egg.

Linking Prior Knowledge Ask students what they know about salt. Point out that the table salt they use is the most abundant salt in the ocean. Salt has been used as a spice and food preservative for thousands of years, and one source of it is the ocean. Remind students of evaporation and the water cycle. Ask them why rain does not contain salt. (Evaporation involves only water molecules. When ocean water evaporates, the salt is left behind.)

2 Teach

Demonstration To model ocean water, mix 35 g (1.2 oz) of table salt in 1,000 g (35 oz) of water. Tell students that the solution is similar to ocean water. Ask how a better model of ocean water could be made. (by adding correct amounts of the other salts found in ocean water)
learning style: visual

Discussion Refer students to the circle graph in Figure 10-8 and discuss the salinity of ocean water. Ask students to identify the most common salt in ocean water. (sodium chloride) Point out that the salinity of ocean water varies slightly from place to place. Ask students what conditions they think could change the salinity of ocean water. (rivers flowing into the ocean, rainfall, melting icebergs, and evaporation) List students' responses on the board. Guide students to understand that pouring fresh water into ocean water lowers the salinity, while evaporation raises it.
learning style: auditory

10-3 What are some properties of the ocean?

INVESTIGATE

Testing the Density of Ocean Water

HANDS-ON ACTIVITY

1. Fill two 250 ml beakers or jars with 200 ml of tap water.
2. Add 3 tsp of salt to one beaker. Stir until the salt dissolves.
3. Place a whole, uncooked egg in each jar. Handle the eggs carefully to avoid breaking them. ⚠ **CAUTION:** Wash your hands when you are finished with this activity.

THINK ABOUT IT: What happened to the two eggs? What does this tell you about the difference between saltwater and freshwater?

Objective

Explain why ocean temperatures and salinity in the oceans vary.

Key Terms

salinity (suh-LIHN-uh-tee)**:** amount of dissolved salts in ocean water

thermocline (THUHR-muh-klyn)**:** layer of ocean water in which the temperature drops sharply with depth

The Salty Sea The water in Earth's oceans is saltwater. Saltwater contains more dissolved salts and other minerals than freshwater does.

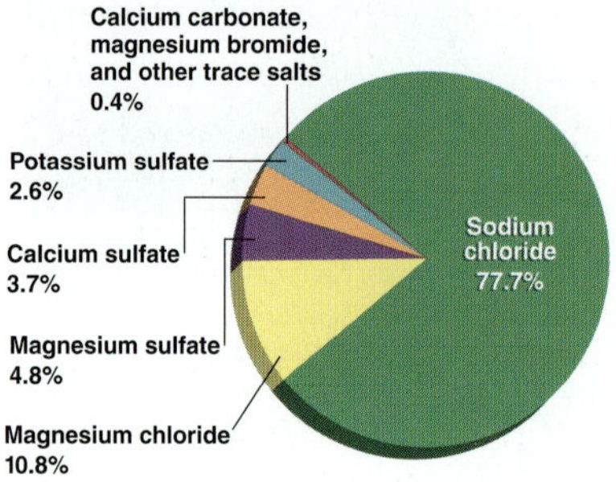

▲ **Figure 10-8** Minerals in saltwater

The amount of dissolved salts in ocean water is called **salinity.** Ocean water contains from 33 to 37 g of dissolved salt in every 1,000 g of water.

1 **DEFINE:** What is salinity?

Levels of Salinity The salinity of ocean water differs slightly from place to place. Freshwater from rivers, precipitation, and melting glaciers lowers ocean salinity.

During the day, water evaporates from the surface of the ocean. Evaporation leaves behind dissolved salts. This raises salinity. Salinity varies more at the surface than in deep ocean water.

2 **INFER:** Would rain increase or decrease the salinity in ocean water? Explain.

Temperature Layers Oceanographers recognize three "layers" of the ocean based on temperature. These layers are surface, thermocline, and deep.

Heat from the Sun warms ocean water. The water is warmest at the surface and coldest near the ocean floor. The surface layer is from 100 to 300 m deep. Constant winds and waves keep the water in the surface layer well mixed. As a result, the temperatures vary slightly in the ocean's surface layer.

Below the surface layer is the **thermocline.** In this layer, temperatures drop sharply with depth. The ocean below the thermocline, the deep layer, is even colder. Here the temperatures are usually below 5°C.

Teaching Options

EXTENDING THE LESSON

Desalinating Water Invite interested students to desalinate salt water by freezing it in an ice-cube tray, washing the frozen ice cubes with fresh water, and then allowing them to melt. Ask students to explain what they think happened to the salt. (It was left on the surface when the water froze into ice crystals.)
learning style: tactile

ONGOING ASSESSMENT

Refer students to Figure 10-9, showing different temperature layers in the ocean. Give a depth and have students describe what the water is like in that part of the ocean.

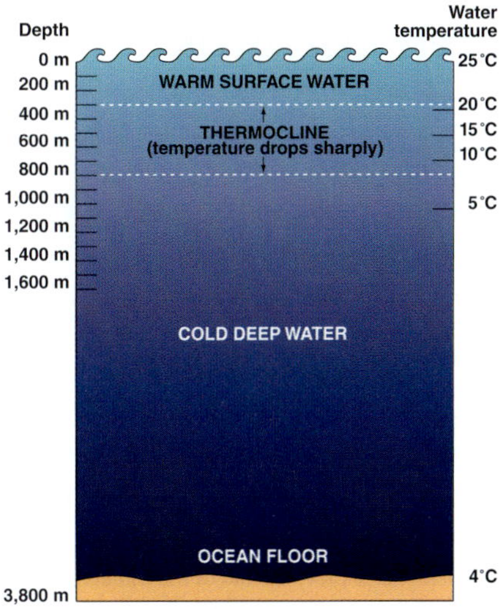

▲ **Figure 10-9** Ocean temperature decreases as depth increases.

3 IDENTIFY: What is the thermocline?

Desalination Only a small part of Earth's freshwater is **potable** (POHT-uh-buhl), or fit to drink. Fresh water is a valuable natural resource that is in short supply.

As the population of the world increases, more freshwater is needed. Most supplies of freshwater depend on precipitation. During dry periods, those supplies are reduced.

Scientists have discovered a way to use ocean water to meet the increasing need for water. Before people can use the supply of water available in the ocean, minerals and salts in the ocean water must be removed.

In many places, desalination (dee-sal-uh-NAY-shuhn) plants have been built to remove the salts from ocean water. These plants use several different methods to remove salts. The most common is to heat the water until it evaporates, leaving the salts behind. The water vapor is then condensed to recover fresh water. Another method is to freeze ocean water. When ocean water is frozen, the ice formed is free of salts. The ice is then cleaned and melted to provide freshwater. Currently, Saudi Arabia, Israel, Malta, and some U.S. states operate desalination plants.

▲ **Figure 10-10** A desalination plant changes saltwater into drinking water.

4 DESCRIBE: What happens when you desalinate water?

CHECKING CONCEPTS

1. The amount of dissolved salts in ocean water is called __________.
2. Adding freshwater to saltwater __________ the salinity of the water.
3. Ocean water is warmest at the __________.
4. There are __________ different temperature layers in the ocean.
5. Salinity varies more at the __________ than in deep ocean water.

THINKING CRITICALLY

6. **SEQUENCE:** Use Figure 10-8 to list the minerals found in oceans, from the greatest percentage to the least.
7. **PREDICT:** The Mediterranean Sea has a high rate of evaporation. Would it have a high or low salinity? Explain.

BUILDING READING SKILLS

Vocabulary Write the definitions of the following words that contain the prefix *thermo*, meaning "heat": thermocline, thermoelectric, thermograph, thermomagnetic, and thermometer. Circle the part of the definition that relates to the prefix.

QUICK CONVERSIONS		
	Metric/SI	**English**
Salt in ocean water	33–37 g 1,000 g	1.2–1.3 oz 2.2 lb
Water depths	100 m 800 m 1,600 m	330 ft 2,640 ft 5,280 ft
Ocean temperatures	25°C 10°C –4°C	77°F 50°F 25°F

• TEACHING RESOURCES • Online or Print

Teacher's Resources

Lesson Review: Chapter 10, p. 5

Laboratory Manual

Lab. Challenge: How do minerals form in ocean water? pp. 57–60

Answers to questions

1 DEFINE: the amount of dissolved salts in water

2 INFER: It would decrease it, because rain adds fresh water to the ocean water.

3 IDENTIFY: the layer of ocean water in which the temperature drops sharply with depth

4 DESCRIBE: Minerals and salts are removed.

READING STRATEGY Ask students what they think the prefix *de-* means. ("do the opposite of" or "remove") Have students identify a term used in this lesson that this information will help them understand. (desalination) Ask students to think of other words with this prefix. (Possible answers: *deflate, defog, deface, deactivate*) List these on the board.

3 Assess

CHECKING CONCEPTS ANSWERS

1. salinity
2. lowers
3. surface
4. three
5. surface

THINKING CRITICALLY ANSWERS

6. **SEQUENCE:** sodium chloride, magnesium chloride, magnesium sulfate, calcium sulfate, potassium sulfate, calcium carbonate, magnesium bromide, other trace salts
7. **PREDICT:** high, because evaporation raises salinity

4 Close

BUILDING READING SKILLS ANSWERS

Vocabulary *Thermo-* means "heat." *Thermocline* is a layer of ocean water in which temperature drops sharply. *Thermoelectric* refers to the direct relationship between heat and electricity. A *thermograph* is a thermometer that records variations in temperature. *Thermomagnetic* pertains to the relationship between heat and magnetism. A *thermometer* is an instrument for measuring temperatures, or heat energy.

10-4 *(pp. 232–233)*

1 Introduce

INVESTIGATE

TIME: 10 minutes

PURPOSE: Students compare density of water at different temperatures.

MATERIALS: plastic container, cup, warm water, ice water, eyedropper, food coloring, stirring stick

PROCEDURE: Have students work in pairs. Tell them to let the warm water stand for a few minutes so that it is quite still.

THINK ABOUT IT: The colored water sank, showing that cold water is denser than warm water.

Linking Prior Knowledge Ask students what they learned about convection currents (Lesson 6-5). Ask how these currents affect the movements of plates. Point out that, like density differences in the crust, ocean layers with different densities will make water flow.

2 Teach

Discussion Refer students to the map in Figure 10-11. Define ocean currents as streams of water flowing in the oceans. Identify winds as being the major cause of surface currents. Explain that currents move clockwise in the Northern Hemisphere and counterclockwise in the Southern Hemisphere. Ask students: *In which direction do currents move near the United States?* (clockwise) Describe warm and cold currents and have students locate warm and cold currents on the map. **learning styles: auditory, visual**

Answers to questions

1. **DEFINE:** a stream of water that flows in the ocean
2. **DESCRIBE:** differences in water temperature and salinity
3. **STATE:** wind

WRITING HINT Remind students that taking notes is the best way to remember unfamiliar material. Tell them to write only keywords or phrases, and new heads when the topic changes.

10-4 What are ocean currents?

INVESTIGATE

Comparing Densities of Cold and Warm Water

HANDS-ON ACTIVITY

1. Fill a plastic container half full with warm water. Wait for the water to stop moving.
2. Add several drops of food coloring to a cup of ice water and stir.
3. Using an eyedropper, gently dribble the colored water down the inside of the plastic container.

THINK ABOUT IT: What happened? Why? What does this tell you about the density of cold and warm water?

Objectives

Define current. Describe how surface currents and density currents are formed.

Key Terms

current: stream of water flowing in the oceans

density current: stream of water that moves up and down in ocean depths

Coriolis effect: bending of Earth's winds and ocean currents by Earth's rotation

On the Move The water in Earth's oceans is always moving. Have you ever heard of someone throwing a bottle containing a message into the ocean? Sometime later, the bottle is found far away.

How did it get there? It was carried by ocean currents. These **currents** are streams of water in the oceans. Some currents move along the ocean bottom. Some move up and down within the ocean depths. Currents can also flow along the surface.

1 **DEFINE:** What is a current?

Density Currents Differences in density can cause currents to move up and down in the ocean depths. This movement of water causes **density currents.**

Ocean currents can be warm or cold. Currents flowing from areas near the equator are warm. They bring warm water into cooler regions. These warm currents tend to warm the air over nearby land areas.

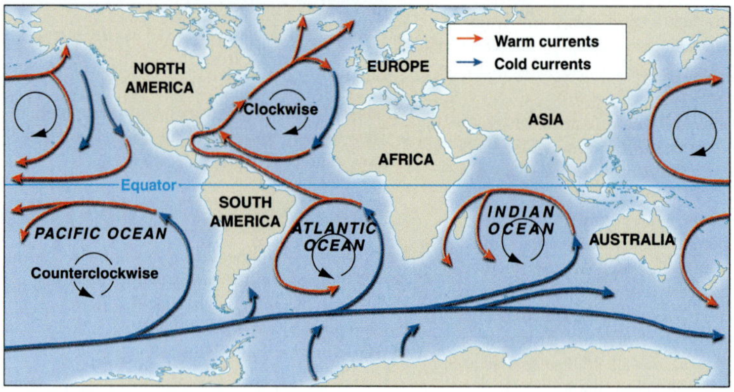

◀ **Figure 10-11** Different densities in ocean water cause density currents.

Teaching Options

! FAST FACTS ***Ocean Currents Around the World***

The cold *Peru Current* flows to the north along the western coast of South America. It is also known as the Humboldt Current. The warm *Kuroshio Current* flows northeast from the Philippine Sea to southeast Japan and on to the North Pacific. It is also called the Japan Current. The cold *Labrador Current* flows south from the Arctic Ocean along the coast of Labrador. There it turns east to join the *Gulf Stream.* This warm current starts in the Caribbean Sea, loops through the Gulf of Mexico, then flows northward along the Florida and Carolina coasts, and finally turns northeast and continues on to the north Atlantic.

RETEACH

Have a tank of fresh water at room temperature and a beaker of water at the same temperature. Prepare a saturated solution of salt water colored with food dye. Make the solution by adding salt to the water until no more salt will dissolve. Use a thermometer to show that both the salt water and fresh water are the same temperature. Slowly pour the colored salt solution into the tank of fresh water. Have students observe how the salt water sinks. Elicit from students that this shows that salt water is denser than fresh water. **learning style: visual**

Currents coming from areas near the poles are cold. They bring cold water into warmer regions and cool these areas. Cold water is denser than warm water. Cold water around the poles sinks to the ocean bottom. Water around the equator is warm. Warm water rises up toward the ocean surface.

Different amounts of salt in ocean water also cause density currents. Water with a lot of salt is denser than water with only a little salt. Dense, salty water sinks. Less salty water rises.

2 DESCRIBE: What conditions cause density currents in the ocean?

Surface Currents Winds cause most surface currents. Winds near the equator blow mainly from east to west. In the Northern Hemisphere, these winds blow from the northeast. In the Southern Hemisphere, these winds blow from the southeast. Earth's rotation causes the winds in the Northern Hemisphere to curve toward the west and the winds in the Southern Hemisphere to curve toward the east. This is called the **Coriolis effect.** Continents and large islands also influence ocean currents. As a result, surface currents move in huge circles. They move clockwise in the Northern Hemisphere and counterclockwise in the Southern Hemisphere.

3 STATE: What causes most surface currents?

CHECKING CONCEPTS

1. Ocean currents are streams of __________ in the oceans.
2. Most surface currents are caused by __________.
3. Surface currents from the __________ are cold currents.
4. Surface currents from the __________ are warm currents.

THINKING CRITICALLY

5. HYPOTHESIZE: Why do large land areas cause surface currents to change direction?
6. COMPARE: Which is probably denser, the water in the Arctic Ocean or the water in the Caribbean Sea? Explain your answer.

HEALTH AND SAFETY TIP

An undertow is a current that moves beneath and in a different direction from the surface current. Undertows can be very dangerous. This is one reason you should never swim alone. Interview a lifeguard to find out about other safety guidelines for swimming. Make a chart that outlines some of these guidelines.

Real-Life Science

TRAVELING ON CURRENTS

Knowing the direction and the strength of ocean currents is important to the shipping industry. Any good sailor knows that traveling with a current saves time.

In 1768, King George III asked Benjamin Franklin why it took mail ships longer to go from England to the colonies than to return to England. Franklin asked the captain of a whaling ship. He was told of a strong current of warm, salty water that flowed along the eastern coast of North America, then across the North Atlantic. Whaling ships rode the current going out but stayed outside of it when returning home. That ocean current is the Gulf Stream.

Thinking Critically How do you think ships today shorten their trips to Europe?

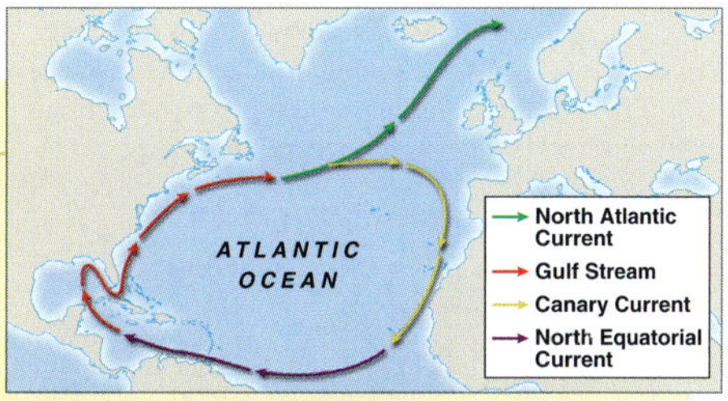

▲ **Figure 10-12** Currents can help ships cross the Atlantic Ocean faster.

3 *Assess*

CHECKING CONCEPTS ANSWERS

1. water
2. wind
3. poles
4. equator

THINKING CRITICALLY ANSWERS

5. HYPOTHESIZE: Ocean water is blocked by continents and large islands.
6. COMPARE: water in the Arctic Ocean, because it is colder

HEALTH AND SAFETY TIP

Water carried onto a beach by breaking waves returns to the sea by flowing under the waves. This returning water is called an undertow. Often, the returning water is channeled to a particular spot along the beach where the waves are not as strong. The resulting current is called a rip current or, erroneously, a rip tide. Swimmers caught in a rip current should swim parallel to the shore until they are out of the current. A person who tries to swim directly back to shore against the current will tire quickly and may drown.

4 *Close*

Real-Life Science

TRAVELING ON CURRENTS

Background Before motorized ships, sailors used ocean currents and prevailing winds to move from place to place. Just as there are prevailing surface currents in the ocean, there are prevailing wind currents in the atmosphere.

Discussion Ask students if they have ever floated down a stream or a river in a boat or on a flotation device such as an inner tube. Have students share their experiences with the class. Ask students to describe the difference between traveling with the current and traveling against the current. Point out that surface currents in the ocean are like the currents in a stream. Ask students why traveling in the same direction as the ocean current is easier. (It requires less energy.)

Thinking Critically Answer
Possible answer: by traveling along a current that flows toward Europe

MORE ON THE WEB

The Coriolis Effect At www.conceptsandchallenges.com, students can learn more about the movement of surface currents on Earth. Have students explain how the Coriolis effect influences currents.

• *TEACHING RESOURCES* •
Online or Print

Teacher's Resources
Lesson Review: Chapter 10, p. 6

ESL/ELL STRATEGY

Reading Coastal Currents Have students look at the map in Figure 10-11 and find the coast nearest to where they live. Ask in which direction the surface currents move on that coast.
learning style: visual

10-5 (pp. 234–235)

1 Introduce

STUDY HINT Have students turn the section heads into questions and try to answer the questions as a way to review the material.

Linking Prior Knowledge Remind students of what they learned about waves (Lessons 8-9 and 8-10). Discuss how waves affect the shore, and point out that their effects are not as dramatic farther out from shore.

2 Teach

Demonstration Float a cork or other small object in a large tub of water. Make waves by tapping a spoon on the surface of the water. Have students observe how the object moves. Make sure that students understand that wave motion is the transfer of energy between two points. Stress that matter (that is, the water) is not transferred. **learning styles: auditory, visual**

Discussion Draw a series of waves on the board, pointing out that waves are up-and-down motions. Figure 10-14 can serve as a guide. Use the drawing to illustrate wavelength, wave height, wave crest, and wave trough. **learning styles: auditory, visual**

Reinforcement Tell students that sea waves are water motions started mainly by the wind. Wave height and power depend on wind strength and the amount of unobstructed ocean over which the wind is blowing. In the open ocean, sea waves pass through the water without moving it forward. Closer to shore, wave crests are closer together. They pile up and become top-heavy, causing the water to fall forward. **learning style: auditory**

Answers to questions

1. **DEFINE:** a regular up-and-down movement of water
2. **EXPLAIN:** from the crest of a wave to the trough
3. **DESCRIBE:** in circles
4. **EXPLAIN:** The shallow waters near the shoreline drag on the wave.

10-5 What are ocean waves?

Objective

Identify the properties of an ocean wave.

Key Terms

wave: regular up-and-down movement of water

crest: highest point of a wave

trough (TRAWF): lowest point of a wave

How Waves Form When a wind blows across the water, waves are formed. A **wave** is a regular up-and-down movement of water. On a windy day at the beach, the ocean water gets rough. The waves are high when the wind is strong. On a calm day, the waves are not as high.

1. **DEFINE:** What is a wave?

Wave Shape A wave has a high point and a low point. The highest point, or top of a wave, is the **crest.** The lowest point of a wave is the **trough.** The height of a wave is the distance from crest to trough. Waves can reach heights of more than 15 m. However, eventually a wave reaches a point when it becomes too high and topples over. A white cap is then formed.

As you watch waves move across the water, you see one crest following another. The distance from one crest to another crest (or from one trough to another) is the wavelength of the wave.

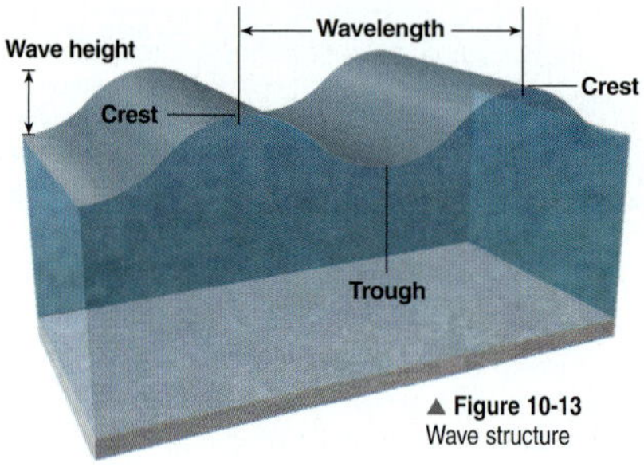

▲ Figure 10-13 Wave structure

2. **EXPLAIN:** How is wave height measured?

Water Movement in Waves In deep water, the water in a wave does not move forward as the wave moves. Only the energy in the wave moves forward.

You can see the movement of water by watching a floating object. As a wave moves by, the object moves slightly forward. As the wave passes, the object falls back about the same distance. The object appears to be moving up and down in the same place.

As a wave moves across the ocean, water particles in the wave move in circles. At the surface, the size of the circles is the same as the height of the waves.

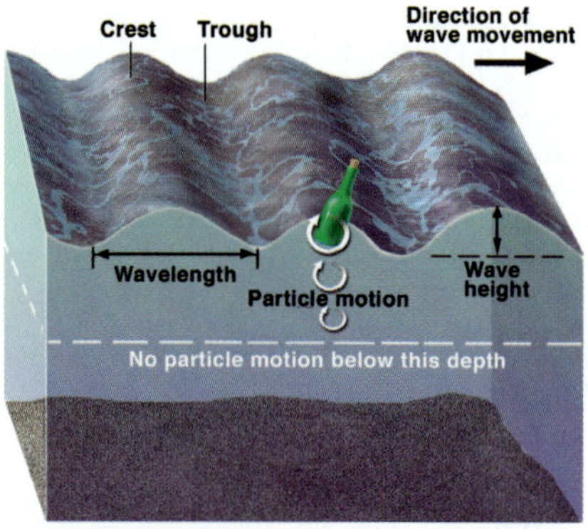

▲ Figure 10-14 Particle motions in waves

3. **DESCRIBE:** How do water particles move in a wave?

Breaking Waves As waves move through deep water, they are not affected much by the depth of the water. However, the shallow waters of the shoreline drag on the wave and cause it to slow down.

As a wave slows down, its height rises until a certain critical height is reached. At this critical point, the wave breaks. The energy contained in the wave up until then changes form. The wave no longer moves in an up-and-down motion. It advances up the shore as a sheet of water.

Teaching Options

EXTENDING THE LESSON

Exploring Other Waves Invite interested students to explore the similarities and differences among water waves, sound waves, and electromagnetic waves (such as light waves and radio waves). Ask students to sketch the information and report their findings to the rest of the class. **learning style: auditory**

RETEACH

Modeling Wave Action Challenge pairs or small groups of students to create their own method of modeling wave motion. Suggest that they use a rope or a line of students to create a dynamic model of a wave. Two students, each holding one end of a rope, can create a wave that moves down the length of the rope. A lineup of five or six students can use coordinated arm movements to create a wave that moves down the line.

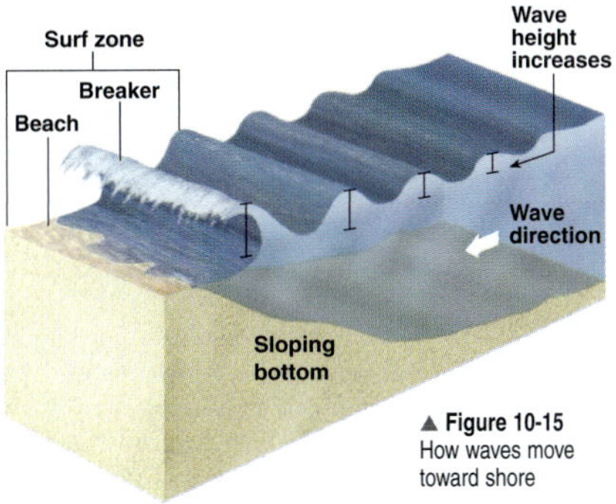

▲ **Figure 10-15** How waves move toward shore

4 **EXPLAIN:** Why does a wave slow up in shallow water?

CHECKING CONCEPTS

1. Most waves are caused by __________.
2. The top of a wave is the __________.
3. The distance between crests is the __________.
4. Only the __________ of a wave moves forward.

THINKING CRITICALLY

5. **MODEL:** Draw a wave and label it.
6. **INFER:** Early in the day, the water at the beach is fairly calm. Later in the day, the water begins to get rough. What do you think might be causing this?

Web InfoSearch

Tsunamis A tsunami (soo-NAH-mee) is an ocean wave. However, unlike other ocean waves, tsunamis are not caused by wind. Tsunamis carry a great deal of energy. They can be very destructive.

SEARCH: Use the Internet to find out more about tsunamis. What causes them? What happens as they approach the shore? To find out, start your search at www.conceptsandchallenges.com. Some key search words are **tsunamis, how tsunamis form,** and **underwater earthquakes.**

Hands-On Activity

MODELING WAVE MOTION

You will need an aquarium tank, metal washers, and four or five corks.

1. Fill the aquarium tank about 3/4 full of water.
2. Tie enough metal washers to a cork so that it floats about 3 cm from the tank bottom.
3. Repeat Step 2 with more corks so that they float 9 cm from the bottom, 15 cm from the bottom, and so on, until the last cork floats on the surface.
4. Make small, steady waves in the tank by moving your hand up and down in the water. Note what happens to each cork.
5. Repeat Step 4, but increase the height of the waves by moving your hand faster.

▲ **STEP 4** Make small, steady waves in the tank.

Go Online **active art**
For: Water Motion activity
Visit: PHSchool.com
Web Code: cfp-3031

Practicing Your Skills

6. **ANALYZE:** How does increasing the wave height affect the motion of each cork?
7. **OBSERVE:** What features of a wave did you observe in this activity?

COOPERATIVE LEARNING

Understanding Waves Divide the class into small groups. Have each group work together to draw a diagram showing several waves. Tell students to use the diagram showing the parts of a wave in Figure 10-14 as a guide. Tell them to label all the crests, troughs, wave heights, and wavelengths on their diagrams. Give students a scale to use, such as 5 cm (1 in.) = 1 m (40 in.). Have students use this scale to calculate the heights and lengths of the waves they drew. **learning style: visual**

• TEACHING RESOURCES • Online or Print

Teacher's Resources

Lesson Review: Chapter 10, p. 7

Go Online **active art**
For: Water Motion activity
Visit: PHSchool.com
Web Code: cfp-3031

Students can interact with the art of water motion online.

3 *Assess*

CHECKING CONCEPTS ANSWERS

1. wind
2. crest
3. wavelength
4. energy

THINKING CRITICALLY ANSWERS

5. **MODEL:** Students' diagrams should resemble Figure 10-13.
6. **INFER:** stronger winds

4 *Close*

Web InfoSearch

Tsunamis Have students use the Internet to find out more about tsunamis. Have students answer these questions: *What causes these ocean waves? What happens as they approach the shore?* Some key search words are **tsunamis, how tsunamis form,** and **underwater earthquakes.** Students can begin their search at www.conceptsandchallenges.com.

WRITING HINT Remind students that when they are ready to organize their research into a report, as they might do for the Web InfoSearch in this lesson, they should create a simple, rough outline of the material. For this, they will need to decide the order of topics or important events and how to begin and end their report. Stress that organizing this ahead of time will greatly simplify the task ahead.

Hands-On Activity

MODELING WAVE MOTION

TIME: 30–45 minutes

PURPOSE: analyzing, observing

Practicing Your Skills Answers

6. **ANALYZE:** The cork moves in larger circles.
7. **OBSERVE:** wavelength, wave height, and particle motion

Lab Activity

(pp. 236–237)

1 Prepare the Lab

Modeling the Shapes of Shorelines

SET-UP TIME: **LAB TIME:**

BACKGROUND: Oceans wrap around islands and continental landmasses. Far out to sea, the actions of wind and occasional seismic disturbances create waves that can travel thousands of miles. Over deep ocean floors, the waves appear small. Shallow waters, however, concentrate wave movement so that the waves crash on shorelines with substantial energy. This energy erodes and shapes shorelines.

PURPOSE: Students will observe and describe the changes that take place in a shoreline as waves crash upon it.

ALTERNATIVE MATERIALS: A plastic under-the-bed storage tray is used for the shoreline table. You could use a stream table, if one is available, by turning the tray so that the drain hose is at the upper end. Commercial classroom stream tables are available from scientific supply houses. Wooden paint stirrers make good wave generators. Plastic rulers can also be used for wave generators.

SCIENCE SKILLS: Students will **model** a shoreline and **observe** and **describe** the changes that waves create.

ADVANCE PLANNING: Obtain pictures of different kinds of shorelines from around the world. Look for postcards of wide beaches, rocky shores, sea cliffs, and so on. Display the pictures.

Cover the working areas with plastic sheets to protect the floor from spills. Set two or more shoreline tables on flat working surfaces. Elevate the upper ends of the tables about 5 cm (2 in.). Depending on the strength of the tray, support may be needed along its length. Fill the upper ends of these trays with sand about 6 cm (2.5 in.) deep. Arrange a place for students to rinse their hands during the lab. You may wish to create the wave generators in Step 2 beforehand.

LAB ACTIVITY

Modeling the Shapes of Shorelines

Materials
- Safety goggles
- Lab aprons
- Large plastic storage box
- Water
- Sand
- Wooden paint sticks
- Stop watch or clock
- Glue

▲ **STEP 1** Raise one end of the storage box about 5 cm.

▲ **STEP 2** Move your wave generator around in the water.

BACKGROUND

Large waves carry great amounts of energy. Waves toss around rocks, sand, and swimmers. Gentle waves also contain energy that moves sand and finer particles around. Waves erode and shape shorelines.

PURPOSE

In this activity, you will study the development of a shoreline as it is struck by waves.

PROCEDURE FOR WATER-ONLY SHORELINE TABLE

1. Use the plastic storage box for the shoreline table. Set the shoreline table on a flat working surface. Raise one end of the box about 5 cm. Fill the box part way with water.
2. Create a wave generator by gluing together two paint sticks in a T-shape. Move the sticks back and forth in the water for a minute to create gentle waves.
3. Observe what happens to the waves as they approach shallower water. How do they change shape? Copy the first column only of the chart in Figure 10-16 and draw a picture of what you observe. Be sure to label your drawing.
4. Create more powerful waves and again observe what happens to the shoreline.

PROCEDURE FOR WATER AND SAND SHORELINE TABLE

5. Use the same plastic storage box for the next shoreline table. Set the shoreline table on a flat working surface. Raise one end of the box about 5 cm.
6. Fill the raised end of the shoreline table with sand that is 6 cm deep. Then, fill the box part way with water.

236

Teaching Options

! FAST FACTS *More on Shorelines*

On a gently sloping shoreline, the sea erodes the lower part and dumps the sediment on the upper part. On a steeply sloping shoreline, the opposite occurs. Most beaches, however, are in a state of near-equilibrium, and the result is that change occurs very slowly.

If the sea level rises, the low-lying areas are impacted. Coastal plains can become drowned, valleys flooded, ridges converted to peninsulas, and uplands turned into islands. *Dalmatian coastlines* have mountain ridges that parallel the sea. *Fjords* are submerged inlets further gouged out by glaciers. They have sheer, high sides and a U-shaped profile.

ESL/ELL STRATEGY

Understanding Peninsulas To help English-language learners remember the word *peninsula,* explain to them that it comes from the Latin word *paeninsula,* which means "almost" (*paene*) and "island" (*insula*).

7. Again, move the wooden paint sticks back and forth in the water to act as wave generators. Create gentle waves in your shoreline table.

8. Observe what happens to the sand on the shoreline as waves strike it. Copy the whole chart below and draw pictures of what you observe. Be sure to label your drawings.

9. Scoop out sand to make a small bay. Create waves for one minute and observe what happens to the bay. Draw what you see.

10. Place a small pile of sand in the middle of the shoreline to make a small peninsula. A peninsula is a land area that projects out into water.

11. Create waves for one minute and observe what happens to the peninsula. Draw what you see.

▲ **STEP 9** Create waves and draw how the bay looks now.

Waves and Shorelines

Diagram of how waves change as they enter shallow water	Diagram of how waves changed the bay	Diagram of how waves changed the peninsula

▲ **Figure 10-16** Use a copy of this table on which to draw your diagrams.

CONCLUSIONS

1. OBSERVE: How did the waves change as they reached shallower water in the water-only shoreline table?

2. OBSERVE: What happened to the sand as waves broke on the shore of the water and sand shoreline table?

3. ANALYZE: How do waves change the shoreline?

4. INFER: What might happen to a shoreline that has both a bay and a peninsula?

2 Guide the Lab

PROCEDURE: Divide students into pairs or small groups. Point out the postcards of shorelines and ask students to speculate on why the shorelines are different from one another. Tell them that they will be experimenting to see how these differences arise. Designate half of the tables to be water only and half to be water and sand. Halfway through the activity, groups will switch shoreline tables. The water-only tables are for studying wave development. The sand-water tables are for studying shoreline erosion. Provide students with a copy of the Lab Activity Report found in the Teacher's Resources to record their observations and conclusions.

TROUBLESHOOTING: Water and sand will spill during the lab. Provide cleanup materials.

SAFETY TIP: The water will become dirty. Students will need a place to rinse their hands. Clean up water spills immediately.

EXPECTED OUTCOMES: Students will observe how waves concentrate their energy as they break on shorelines and how shorelines are eroded by wave action.

3 Conclude the Lab

1. OBSERVE: The waves got taller and splashed on the shore.
2. OBSERVE: The sand was pulled into the water as the waves drained back.
3. ANALYZE: Waves smooth out shorelines by moving the sand around.
4. INFER: Possible answer: The peninsula will be washed away, and the bay will be filled in with sand.

Use the Laboratory Activity Rubric from the Teacher's Resources to assess students' lab activities. Fill in the rubric with the additional information below. For this activity, students should have:

- created waves in a shoreline table.
- made observations about the development of the shoreline.

4 Extend the Lab

Play an environmental sounds CD of ocean waves crashing on a shore. Have students speculate on what they think is happening as the sounds change.

Tips for Using Technology

Using a Computer to Draw Diagrams Most graphics-creation software uses similar tools to create diagrams. Students may wish to use one of these graphics programs to create their lab diagrams. Shape tools allow them to create different types of shapes in their illustrations. The bucket tool allows them to fill a shape or an object that they have created with a color. The pencil, or drawing, tool allows them to draw in freehand in their illustrations. The text tool allows them to enter text into their illustrations.

• TEACHING RESOURCES • Online or Print

Teacher's Resources

Lab Activity Report: Study Tools, pp. 16–17

Laboratory Activity Rubric: Teaching Tools, p. 8

10-6 *(pp. 238–239)*

1 *Introduce*

STUDY HINT Have students read the Real-Life Science feature before beginning the lesson. Discuss the concepts presented. Ask students how they think this information relates to topics that they have already studied or to the lesson that they are about to study.

Linking Prior Knowledge Ask volunteers to tell what they know about ocean tides. Write students' responses on the board. As students go through the lesson, encourage them to make adjustments in their initial statements about tides.

2 *Teach*

Demonstration If possible, show students photographs of high and low tides for a region with a shallow shoreline and a steep shoreline. Use these photographs to show that the distance between the waterline at high and low tide can be great in regions with shallow shorelines.

Discussion Ask students to differentiate between currents and tides. Ask: *How are currents and tides caused?*(currents: wind and water density; tides: the Moon's gravitational pull on Earth) *What effect do they have?* (Currents bring warm water into cold water areas and vice versa. Tides change the level of the ocean along the shoreline.) **learning style: auditory**

Reinforcement Remind students that the gravitational attraction between two objects is directly related to the distance between them. This will help students understand why the side of Earth facing the Moon has a high tide. Point out that low tides are produced by the movement of water to high-tide locations. **learning style: auditory**

Answers to questions

1 **DEFINE:** regular changes in the levels of Earth's oceans

2 **IDENTIFY:** the gravitational pull of the Moon

3 **INFER:** twice

10-6 What are the tides?

Objective

Describe what causes and affects the tides.

Key Terms

tide: regular change in the level of Earth's oceans
flood tide: incoming, or rising, tide
ebb tide: outgoing, or falling, tide

Ocean Water Levels The water level of the ocean rises and falls throughout the day. Early in the day, ocean water rises and covers part of the beach. Later in the day, the ocean level falls. The beach is exposed. These regular changes in ocean water levels are called **tides.** A low water level is called low tide. A high water level is called high tide.

▲ **Figure 10-17** Low tide (above) and an incoming tide (below) in the same location

1 **DEFINE:** What are tides?

Causes of Tides You probably know that Earth's gravitational pull on the Moon keeps the Moon in orbit around us. But did you know that the Moon also pulls back on Earth and causes the tides? The Sun's gravitational pull also affects Earth's tides, but because it is so far away from us, not as strongly.

Earth's continents are slightly stretched by the pull of the Moon. Earth's oceans, which move more freely, are stretched even more. This stretching effect creates two bulges of water on Earth, one facing the Moon, the other directly opposite it. These bulges are the high tides. The bulging water also causes two areas with low tides between the high tides. As the Earth turns on its axis, the tide levels rise and fall.

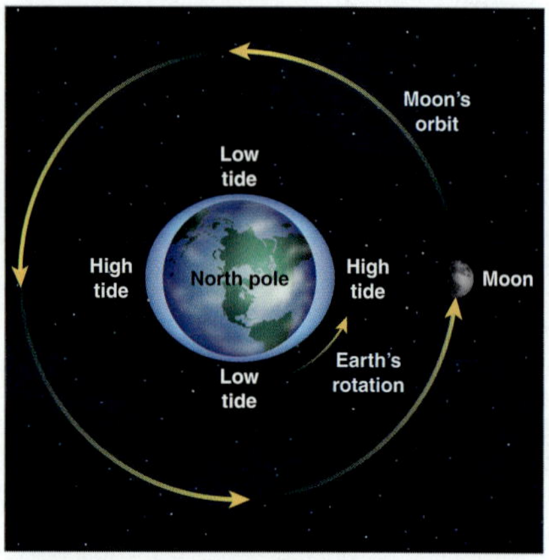

▲ **Figure 10-18** The Moon, along with the Sun, causes the tides. Whether it is high tide or low tide depends mostly on where you are on Earth relative to the Moon.

2 **IDENTIFY:** What mainly causes tides on Earth?

Changing Tides Some newspapers print tide tables. A tide table tells the times at which high tide and low tide will occur. If you look at a tide table, you will notice that it often shows two high tides and two low tides each day. The tides change about every 6 hours and 15 minutes.

Teaching Options

COOPERATIVE LEARNING

Living in the Intertidal Zone The intertidal zone along the ocean shore is underwater at high tide and exposed to air at low tide. Organisms living in the intertidal zone include algae, clams, oysters, and mussels. Have a group of students find out the variations in conditions that these organisms endure. Students should make graphs of the daily fluctuations in temperature and salinity, and describe these and other changes to the class. **learning styles: auditory, visual**

EXTENDING THE LESSON

Researching Tidal Patterns Have interested students find out where tides have a different pattern from two lows and two highs a day. They should find these places on a world map and point them out to the class as they describe the tidal pattern. Students may point to the Mediterranean Sea, Gulf of Mexico, or some Pacific Islands. **learning style: visual**

TIDE TABLE

Sunday	Monday	Tuesday	Wednesday
1 Low 4:45a	2 Low 5:38a	3 Low 6:27a	4 Low 7:11a
High 11:31a	High 12:32p	High 1:25p	High 2:10p
Low 4:30p	Low 5:24p	Low 6:15p	Low 7:02p
High 10:24p	High 11:10p	High 11:55p	
8 Low 2:47a	9 Low 3:31a	10 Low 4:18a	11 Low 5:10a
High 9:40a	High 10:13a	High 10:48a	High 11:23a
Low 4:26p	Low 4:58p	Low 5:30p	Low 6:05p
High 9:58p	High 10:46p	High 11:36p	

▲ **Figure 10-19** A newspaper gives the times for high and low tide.

Each quarter rotation of Earth causes a major change in tides. Water slowly floods the beach until high tide is reached. The incoming tide is called a **flood tide.** As Earth rotates another quarter turn, the water begins to leave the beach until low tide is reached. This outgoing tide is called an **ebb tide.**

3 INFER: How often each day do flood tides occur?

CHECKING CONCEPTS

1. What is the outgoing tide called?
2. What causes tides?
3. What is a flood tide?
4. How many times do the tides change each day?

THINKING CRITICALLY

5. INFER: Who would use a tide table? Why?
6. PREDICT: If low tide occurs at 6:30 AM, when will the next low tide probably occur?
7. HYPOTHESIZE: Suppose you docked a boat where the tides rise and fall about 1 m. How would this affect the use of the boat?

INTERPRETING VISUALS

Use Figure 10-19 to answer the following question.

8. INFER: Fishing is best on mornings with high tides. Which day should you go?

Real-Life Science

TIDES AND FISHING

Grunions are small silver fish that live in the Pacific Ocean off the coast of California. Grunions spawn and lay their eggs on sandy beaches from late February to early September, but only on nights of the highest tides. During this time, thousands of grunions cover the beaches. Many people gather on the beaches and catch the fish by hand. Some newspapers announce the nights the fish are expected to be on the beach.

▲ **Figure 10-20** Catching grunions on the beach is a common event in California.

Much of the fishing industry depends on an understanding of tides. Tides involve the movements of huge volumes of water. The water carries fresh oxygen. It also carries microorganisms, which serve as food for fish. Fish carried in with high tides can be easily trapped in nets later in the day.

The Bay of Fundy, between Nova Scotia and New Brunswick, Canada, is known worldwide for its tides. The difference between high and low tides can be over 10 m. This region is famous for its fishing industry.

Thinking Critically Why do you think the fish carried in with the high tides are easily trapped in the nets?

MORE ON THE WEB

Spring and Neap Tides Have students search the Internet to find out about spring tides and neap tides, which they will learn about in Chapter 17. Students should define these two terms and sketch pictures of the Sun, the Moon, and Earth for each type of tide.

• TEACHING RESOURCES • Online or Print

Teacher's Resources
Lesson Review: Chapter 10, p. 8

3 Assess

CHECKING CONCEPTS ANSWERS

1. an ebb tide
2. the gravitational pull of the Moon and, to a lesser extent, the Sun
3. an incoming high tide
4. generally three or four times

THINKING CRITICALLY ANSWERS

5. INFER: Possible answers include people who fish for species that appear at high or low tide, and sailors and ship captains, to help them come in to shore or go out to sea.
6. PREDICT: 7:00 P.M.
7. HYPOTHESIZE: Possible answer: At low tide, there might not be enough clearance between the ocean bottom and the bottom of the boat to allow the boat to float out to sea.

INTERPRETING VISUALS ANSWER

8. INFER: on Day 1 or 9

4 Close

Real-Life Science

TIDES AND FISHING

Life-Science Connection Other ocean-dwelling organisms also come onshore to mate, lay eggs, or bear young. Female sea turtles climb up certain beaches to bury their eggs in the sand. When the eggs hatch, the young turtles head for the water and are carried out to sea by high tides. Seals, sea lions, and elephant seals come to shore to mate and bear young. Sea birds such as gulls, puffins, and cormorants make nests on rocks and cliffs at the edges of the mainland or on islands.

Discussion Ask students to suggest why grunions are easy to catch by hand when they are spawning. (They cannot swim freely when they are in shallow water.) Tell students that tidal motion creates water currents that carry organisms from place to place. Fishers can time their fishing activities to take advantage of tidal currents moving in toward shore or out toward the open ocean.

Thinking Critically Answer

Possible answer: As the tide goes out, the water currents carry the fish into the nets.

10-7 *(pp. 240–241)*

1 *Introduce*

Linking Prior Knowledge In Lesson 8-1, students focused on erosion and deposition. Ask them to describe how they think these two processes affect the ocean. As you go though the lesson, prompt students to adjust their perceptions of the effects on the ocean.

2 *Teach*

Demonstration If possible, have students view slides of microscopic organisms that form the ooze on the ocean floor, such as foraminifera and diatoms. Have students view video clips taken from submersible vehicles that show the shower of particles that fall to the ocean floor. Tell students that this shower of particles is sometimes called marine snow.
learning style: visual

Discussion Describe the material that makes up the ocean floor. Remind students that three agents of erosion are running water, wind, and waves. Explain that the sediments found close to the shore are made up of rock particles eroded by water, wind, and waves. Then, define and discuss the formation of ooze.
learning style: auditory

Answers to questions

1. **LIST:** eroded rocks, the remains of living organisms, volcanic dust and ash, and meteoric dust
2. **EXPLAIN:** water, waves, and wind
3. **DEFINE:** lumps of minerals found on the ocean floor
4. **DEFINE:** ocean sediments formed from the remains of ocean organisms mixed with volcanic dust, clay, and water
5. **IDENTIFY:** sediment that collects near the edges of underwater slopes

WRITING HINT When students start their research for the Building Science Skills activity, remind them that they should look for more than one source of information. Point out that sources may be incorrect or incomplete.

10-7 What are ocean sediments?

Objective
Describe different kinds of ocean sediments.

Key Terms
nodule (NAHJ-ool)**:** mineral lump found on the ocean floor
ooze: ocean sediment that contains the remains of many ocean organisms

Sediment Formation Ocean sediments are formed by materials that collect on the ocean floor. Some sediments come from eroded land rocks. Others contain the remains of living things. Dust and ash from volcanoes sink to the ocean floor. Dust from space also falls and sinks to the ocean floor.

1 LIST: What makes up ocean sediments?

Eroded Rock Sediments Most of the sediments near the shore are eroded rock particles. Rivers carry rocks of all sizes to the ocean. Waves and wind also weather the rocks along the shoreline. These rock particles become ocean sediments. The sediments gradually spread out over the ocean floor. Large particles settle close to shore. Smaller particles settle farther from shore.

▲ **Figure 10-21** An arch can be formed by wind and waves eroding rock.

2 EXPLAIN: What weathers rock to form sediment?

Nodules Lumps of minerals called **nodules** are found on the ocean floor. Nodules form very slowly by a chemical process. Nodules are made up mostly of compounds of manganese, nickel, and iron. Small amounts of copper, lead, zinc, and silver are also found in ocean sediments. Most of the nodules lie thousands of meters deep in the ocean. Suction devices similar to huge vacuum cleaners have been used to collect them.

▲ **Figure 10-22** Nodules are lumps of minerals.

3 DEFINE: What are nodules?

Ooze Much of the ocean floor is made up of fine, soft sediments called **ooze.** At least 30 percent of ooze is the remains of ocean organisms. When ocean organisms die, their shells and skeletons sink to the ocean floor. These remains eventually fall apart or decompose. This material mixes with volcanic dust, clay, and water to form ooze.

▲ **Figure 10-23** Ooze may contain the decomposing skeletons of organisms such as these diatoms.

Teaching Options

RETEACH

Classifying Ocean Sediments
Have students prepare a three-column chart in their notebooks. Have them label the first column *Type of Sediment*, the second column *How It Forms*, and the third column *Other Features.* Ask students to fill in the chart for the three types of ocean sediments discussed in the lesson.
learning styles: visual, tactile

EXTENDING THE LESSON

Mining the Ocean Floor Have interested students research efforts being made to mine materials from the ocean floor. Tell students to find out about the problems associated with mining the ocean floor, such as the need to find ways to bring mined materials from great depths to the surface. Have students present their findings in a written report.

Much of the ooze on the ocean floor comes from tiny plants and animals such as the diatoms shown in Figure 10-23. Larger animals, such as clams and corals, also add to ooze.

4 DEFINE: What is ooze?

Underwater Canyons Beyond the shoreline of the ocean, the land gradually slopes downward. Ocean sediments collect near the edges of these slopes. From time to time, the sediments slide down the slope, pushing water ahead of them. Gradually, this underwater landslide erodes the slope in places and forms canyons.

5 IDENTIFY: What causes underwater landslides?

CHECKING CONCEPTS

1. Fallen materials on the ocean floor form __________.
2. The largest rock particles settle __________ to shore.
3. Nodules are made up mostly of manganese, nickel, and __________.
4. Ooze is formed from __________.
5. Rocks are carried to the ocean by __________.

THINKING CRITICALLY

6. INFER: Why do small rock particles settle farther from shore than larger rock particles?
7. INFER: How do sediments from a volcano on land get into the ocean?
8. EXPLAIN: A large rock is located in a river. Outline the steps that might lead to this rock eventually breaking apart and being deposited as sediment on the ocean floor.

BUILDING SCIENCE SKILLS

Researching Find out about other kinds of ocean sediment. Then, make a chart listing them all. Add those discussed here. Say where they come from and what structures they form.

Hands-On Activity

OBSERVING THE SETTLING OF OCEAN SEDIMENTS

You will need safety goggles, a plastic container with a lid, small pebbles, sand, soil, and water.

1. Place a small amount of pebbles, sand, and soil into a plastic container. Fill the container halfway with water. Then, put on the lid.
2. Shake the container gently for about 10 seconds.
3. Leave the particles undisturbed for several hours. Then, observe how the particles settled.

▲ **STEP 2** Shake the container for 10 seconds.

Practicing Your Skills

4. OBSERVE: Which particles settled on the bottom of the container? Which particles settled on top?
5. ANALYZE: Why did the particles settle out the way they did?
6. COMPARE: How is this like sediments settling out on the ocean floor?

MORE ON THE WEB

Writing About the Abyssal Plains Have students use the Internet to learn more about the effects of ocean sediment on the formation of the abyssal plains, which are ocean features that occur at great depths. Have students write a report based on their findings.

• *TEACHING RESOURCES* • *Online or Print*

Teacher's Resources
Lesson Review: Chapter 10, p. 9

3 *Assess*

CHECKING CONCEPTS ANSWERS

1. sediments
2. close
3. iron
4. the remains of living organisms mixed with volcanic dust, clay, and water
5. rivers

THINKING CRITICALLY ANSWERS

6. INFER: They can be carried farther by water. Heavy particles settle more quickly.
7. INFER: They are carried by rivers or wind.
8. EXPLAIN: Possible answer: Every winter, ice forms in cracks in the rock and splits it apart. Other rocks and particles tumble against the rock during floods, wearing it down and removing small particles. The small particles are carried downriver, getting smaller all the time, and eventually reach the ocean.

4 *Close*

BUILDING SCIENCE SKILLS ANSWERS

Researching From the land come clay and glacial till. From the upper ocean come skeletons of larger organisms, coral shells, snails, sponges, foraminifera, and other small animals with hard body parts. From the deep-ocean vents come sulfides that precipitate on the floor. Evaporites from the upper layers of water are salt, dolomite, and gypsum.

Hands-On Activity

OBSERVING THE SETTLING OF OCEAN SEDIMENTS

TIME: 15–20 minutes

PURPOSE: observing, analyzing, comparing

Practicing Your Skills Answers

4. OBSERVE: Small pebbles settled at the bottom, soil at the top.
5. ANALYZE: The heaviest materials settle quickly. The lightest settle slowly.
6. COMPARE: Large, heavy particles settle quickly, close to shore. Smaller particles settle more slowly, and so are carried farther from shore.

10-8 *(pp. 242–243)*

1 Introduce

STUDY HINT Three of the Key Terms refer to landforms. Have students create a chart or graphic showing how each of these landforms is created.

Linking Prior Knowledge Have students recall three of the major landforms found on Earth, which they studied in Chapter 5. (mountains, plains, plateaus) Ask students to review the definition of each landform and see how it relates to ocean landforms.

2 Teach

Demonstration Display a topographical world map that shows the ocean floor. Point out the continental shelf and the continental slope that surround each continent. Point out that the continental shelf is narrow in certain areas, such as along the west coast of the United States. It is wide in other regions, such as in the area between Siberia and Alaska. **learning style: visual**

Discussion Point out that the ocean floor has the same major landforms that are found on land. Refer students to the profile of the ocean floor in Figure 10-24. Have students locate the seamounts and the guyot. Remind students that the ocean floor also has mid-ocean ridges, which are like mountain ranges on land, and deep ocean trenches, which are like canyons. **learning styles: auditory, visual**

Answers to questions

1. COMPARE: the continental slope
2. IDENTIFY: plains and mountains
3. COMPARE: The trench is more than 2,000 m deeper than Mount Everest is high.
4. DEFINE: a seamount with a flattened top

READING STRATEGY Have students list alphabetically all of the labels that appear in Figure 10-24, with at least one blank line below each label for writing a sentence. As students read the lesson, tell them to make up sentences using these labels. Then, students should take their sentences and put them in order, to turn them into a summarizing paragraph.

10-8 What are some ocean landforms?

Objective

Describe the ocean floor.

Key Terms

continental shelf: part of a continent that slopes gently away from the shoreline

continental slope: part of a continent between the continental shelf and the ocean floor

trench: deep canyon on the ocean floor

seamount: volcanic mountain on the ocean floor

guyot (GEE-oh)**:** flat-topped, underwater seamount

Continental Margin The continental margin divides a continent from the ocean floor. The edges of the continent extend into the oceans. At first, the continent slopes gently down under the ocean. This area is called the **continental shelf.** In some places, the continental shelf is very narrow. In other places, it extends for more than 150 km from the edge of the continent. Beyond the continental shelf is the **continental slope.** The continental slope is steeper than the continental shelf. It ends at the ocean floor.

1 COMPARE: Which is steeper, the continental shelf or the continental slope?

The Ocean Floor Like Earth's surface, the ocean floor has different landforms. The flat parts are plains. Plains cover about half of the ocean floor. High mountain ranges run along the middle of the oceans. These ranges are the mid-ocean ridges. In some places, mid-ocean ridges rise above the ocean surface to form islands.

2 IDENTIFY: What two landforms are on the ocean floor?

Trenches The ocean floor has deep **trenches,** or underwater canyons. The deepest of these is the Mariana Trench in the Pacific Ocean. This trench is more than 11,000 m deep. Mount Everest, the tallest mountain on land, is about 8,900 m high. Mount Everest could fit into the Mariana Trench and still be more than 2,000 m below the ocean's surface.

3 COMPARE: How does the depth of the Mariana Trench compare with the height of Mount Everest?

Seamounts and Guyots There are many mountains scattered around the ocean floor. These are called **seamounts.** Seamounts were once active underwater volcanoes. The ones that reach above the ocean surface form volcanic islands. The Hawaiian Islands are the peaks of underwater volcanoes.

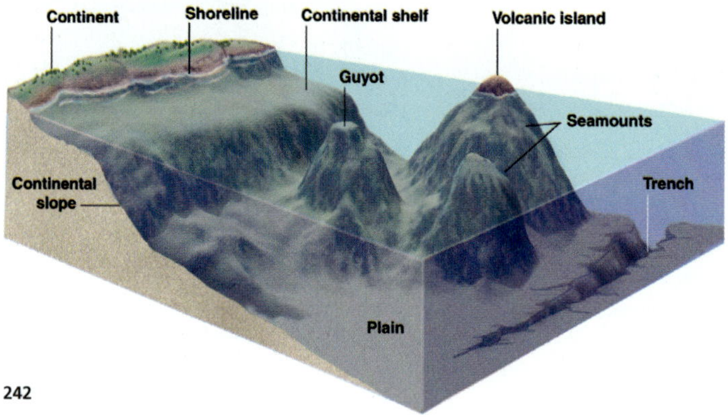

◀ **Figure 10-24** The ocean floor in many places is not smooth and flat. It contains many different landforms.

Teaching Options

RETEACH

Modeling the Ocean Floor Have students work in small groups to make models of the ocean floor, using different colors of clay. Tell students to use the diagram in Figure 10-24 as a guide. Display the models in the classroom. **learning style: tactile**

QUICK CONVERSIONS		
	Metric/SI	**English**
Width of continental shelf	150 km	90 mi
Marianas depth	11,000 m	36,000 ft
Everest height	8,900 m	29,200 ft
Depth	2,000 m	6,600 ft

ONGOING ASSESSMENT

Give students a map of the southeast Pacific Ocean. Ask them to sketch what the profile of the ocean floor might look like from the coast of South America to the Galápagos Islands. Tell them to label the main features, including the continental slope, continental shelf, plain, and seamounts.

Some underwater seamounts have flattened tops. These seamounts are called **guyots.**

4 DEFINE: What is a guyot?

CHECKING CONCEPTS

1. Two parts of the continental margin are the continental shelf and the continental __________.
2. Three landforms on the ocean floor are __________, mountains, and trenches.
3. Mountains on the ocean floor are called __________.
4. A __________ is an underwater seamount with a flat top.

THINKING CRITICALLY

5. PREDICT: What would happen to the continental shelf if the sea level dropped sharply?
6. HYPOTHESIZE: On the ocean floor, mountains are higher and plains are flatter than they are on Earth's surface. Why do you think this is so?

Web InfoSearch

Fishing on the Grand Banks The most productive fishing grounds in the world are found on continental shelves. The Grand Banks off the east coast of Canada is an example. Fishing boats from Canada, the United States, Japan, and Europe catch tons of fish off the Grand Banks every year.

SEARCH: Use the Internet to find out more about the Grand Banks. What kinds of fish are found there? Start your search at www.conceptsandchallenges.com. Some key search words are **fishing, Grand Banks,** and **continental shelves.**

◀ **Figure 10-25** Fishing on the Grand Banks

How Do They Know That?

MAPPING THE OCEAN FLOOR

Much of the ocean floor cannot be seen from the surface. Until the twentieth century, the best method to map it was to tie a heavy object to a piece of rope and lower it into the water. The amount of rope used showed how deep the bottom was.

Certain technologies have made the ocean floor more visible to us. These technologies allow us to map it in great detail. The main advance in mapping occurred when sonar was developed in the 1920s. Modern sonar systems produce very detailed images.

Another advance was the camera sled. This device takes photographs of the ocean floor while it is being towed across it. For the mapping of small areas, submersibles are used. Also, satellites provide very precise measurements of seafloor features.

Thinking Critically What seafloor features does sonar map?

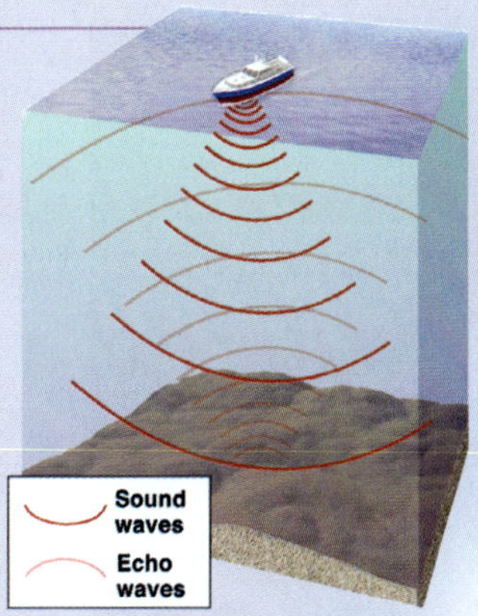

▲ **Figure 10-26** Sonar mapping of the ocean floor

EXTENDING THE LESSON

The Abyssal Plains Invite students to learn about abyssal plains, which are ocean features that occur at an average depth of 4,000 m (13,000 ft). They cover about 75 percent of the ocean floor. Have students find out the gradient of abyssal plains and what abyssal hills are. Students can also do the More on the Web activity described in the Teaching Options on page T241.
learning styles: visual, auditory

• TEACHING RESOURCES • Online or Print

Teacher's Resources

Lesson Review: Chapter 10, p. 10

Enrichment: Chapter 10, p. 12

3 Assess

CHECKING CONCEPTS ANSWERS

1. slope
2. plains
3. seamounts or mid-ocean ridges
4. guyot

THINKING CRITICALLY ANSWERS

5. PREDICT: It would become land, part of the continent.
6. HYPOTHESIZE: There is little weathering on the ocean floor. Plains are constantly being created by the rain of particles from above. Underwater mountains and plains are not eroded by wind, rain, snow, or glaciers.

4 Close

Web InfoSearch

The Grand Banks Have students use the Internet to learn about the kinds of fish found on the Grand Banks. Some key search words are **fishing, Grand Banks,** and **continental shelves.** Have students visit www.conceptsandchallenges.com.

How Do They Know That?

MAPPING THE OCEAN FLOOR

Background The surface of the ocean has slight bulges and depressions that reflect the shapes of landforms on the ocean floor below. The *TOPEX/Poseidon* satellite detects these small surface differences and uses them to create a detailed map of the tectonic structure of the ocean floor. Maps of the Mid-Atlantic Ridge and the fault network of the northwest Pacific Ocean have been made from these data.

Discussion Review the concept of sonar (Lesson 10-2). Make sure that students understand that sonar can be used to make a profile of the ocean floor. Point out that sonar can also be used to locate schools of fish or good fishing grounds.

Thinking Critically Answer
plains, seamounts, mid-ocean ridges, guyots, and trenches

10-9 *(pp. 244–245)*

1 *Introduce*

STUDY HINT Have students start a chart with four columns, headed *Fringing Reef, Barrier Reef, Lagoon,* and *Atoll.* As they read each section, have them write the characteristics of each formation in the correct column. They may also want to sketch what a typical formation looks like.

Linking Prior Knowledge Ask volunteers to speculate about what a coral reef is and how it forms.

2 *Teach*

Demonstration If possible, bring to class small pieces of dead coral for students to examine. Have students use a hand lens to look for tiny openings where coral once lived. If coral specimens are not available, display photographs of several types of stony coral. Include a few photographs taken with the aid of a microscope that show the tentacles of the coral animals extending out of their limestone exoskeletons. Point out the crevices and hollows in a coral reef that provide shelter or hiding places for fish, shrimp, and other organisms. **learning styles: visual, tactile**

Discussion Describe how barrier reefs, fringing reefs, and atolls are formed. Point out that fringing reefs form in shallow water near coastlines. Barrier reefs begin as fringing reefs next to coastal volcanoes. Over time, the volcano sinks into the ocean, but the coral reef continues to grow. Eventually, the volcano completely disappears, and only the reef is left. An atoll begins as a fringing reef that circles a volcanic island. After the volcano sinks into the ocean, only the ring of coral that forms the atoll is left. The water surrounded by an atoll is the lagoon. **learning style: auditory**

Answers to questions

1. **DESCRIBE:** living coral and the calcium carbonate skeletons of dead coral
2. **IDENTIFY:** around sunken volcanic islands
3. **DEFINE:** a ring-shaped coral reef around a lagoon

10-9 What are coral reefs?

Objective

Describe three kinds of coral reefs.

Key Terms

coral: small animals found in warm, shallow ocean waters

fringing reef: coral reef that is directly attached to a shore

barrier reef: coral reef that forms around a sunken volcanic island

lagoon: shallow body of water between a reef and the mainland

atoll (A-tawl)**:** ring-shaped coral reef around a lagoon

Building the Reefs Tiny animals called **coral** live in warm, shallow ocean waters. Corals absorb calcium from the water. They use it to make hard skeletons of limestone, or calcium carbonate, around their bodies. Corals often attach themselves to other coral, forming a colony. New corals grow on top of dead ones. In time, a coral reef is formed.

◀ **Figure 10-27** A coral reef in Egypt's Red Sea. The pink-orange fish are basslets. Both the basslets and the soft coral are typical of Red Sea coral reefs.

1 **DESCRIBE:** What is a coral reef made of?

Fringing Reefs and Barrier Reefs Two kinds of coral reefs are fringing reefs and barrier reefs. A **fringing reef** is a coral reef that is directly attached to a shore. It borders the coastline closely or is separated from it by only a narrow stretch of water. Fringing reefs are located along the east coast of Florida, among other places.

▲ **Figure 10-28** This fringing reef surrounds the South Pacific island of Bora Bora.

Barrier reefs form around sunken volcanic islands. The reef once extended out from the shore. As the island sank, the reef became separated from the mainland by a body of water. This body of water is a **lagoon.**

The Great Barrier Reef off the northeast coast of Australia is the world's largest barrier reef. It is more than 2,000 km long. In some places, it is almost 200 km wide. The Great Barrier Reef has been described by some as the largest structure ever built by living creatures. It actually consists of about 2,100 individual reefs and about 800 fringing reefs.

2 **IDENTIFY:** Where do barrier reefs form?

Atolls A ring-shaped coral reef around a lagoon is called an **atoll.** An atoll forms around a sunken volcanic island. Only the circular coral reef remains above the ocean's surface. In the center of the atoll is the lagoon. Ships are sometimes able to enter the lagoons through channels that connect the ocean and the lagoon. Most atolls are in the Pacific Ocean.

Teaching Options

! FAST FACTS *Where Coral Reefs Are Found*

Coral reefs occur in warm, clear water that is shallow enough to allow sunlight to penetrate. The water never falls below 16°C (61°F), which limits reefs to the tropical and subtropical oceans.

Many of the island chains in the south Pacific Ocean are composed of atolls, such as the Caroline Islands, the Marshall Islands, and the islands of Tuvalu. Atolls that occur in the Indian Ocean include the Maldives. Fringing reefs are along the coastlines of the Hawaiian Islands, the Pacific coast of Mexico, and the Caribbean coast. Barrier reefs include one off the island of Bermuda as well as the Great Barrier Reef off the northeast coast of Australia.

RETEACH

Classifying Reefs Make sure that students understand that there are two types of lagoons associated with coral reefs. A lagoon may lie between a barrier reef and the mainland or may be surrounded by a ring-shaped atoll. Point out that the formation of both atolls and barrier reefs begins along the edges of volcanoes. **learning style: auditory**

▲ **Figure 10-29** Aerial view of Kayangel Atoll in the western Pacific Ocean

3 DEFINE: What is an atoll?

CHECKING CONCEPTS

1. Where do the tiny animals known as corals live?
2. What do corals absorb from water to build reefs?
3. What is a reef colony?
4. What is found in the center of an atoll?
5. What are three kinds of coral reefs?

THINKING CRITICALLY

6. COMPARE/CONTRAST: How are fringing reefs and barrier reefs alike and different?
7. INFER: Why is the sand on a coral island white?
8. HYPOTHESIZE: Fossil coral reefs have been found beneath farmland far from the ocean. What might this discovery mean?

DESIGNING AN EXPERIMENT

Design an experiment to solve the following problem. Include a hypothesis, variables, a procedure with materials, and a type of data to study. Also include a way to record that data.

PROBLEM: Do the skeletons of coral contain calcium carbonate?

Integrating Environmental Science

TOPIC: ecosystems

ARTIFICIAL REEFS

Natural coral reefs support ecosystems. However, natural reefs take a long time to form. To help increase the fish population in the ocean, especially of those species used for food, people are building artificial reefs. These reefs also reduce beach erosion by blocking waves before they reach shore. The Japanese have been building artificial reefs for more than 300 years. The first artificial reef in the United States was built in 1830 off the coast of South Carolina.

Old oil rigs, old tires, sunken ships, concrete blocks, even old subway cars can serve as artificial reefs. Battle tanks and armored vehicles are part of a large artificial reef off the coast of Long Island in New York. Near Miami Beach, Florida, a 727 airliner was sunk to serve as a reef. Sea creatures such as coral, sea anemones (ah-NEM-oh-neez), mussels, barnacles, and sponges cling to or grow on the metal surfaces.

▲ **Figure 10-30** Artificial reefs can be built of almost anything, such as this sunken ship.

Thinking Critically How do reefs help increase the fish population?

MORE ON THE WEB

Threats to the Reefs On the Internet, students can learn about threats to coral reefs and conservation practices. Have students work in pairs to design coral reef conservation posters. The posters should address one of the issues facing the world's reefs and offer a possible solution to the problem.

• TEACHING RESOURCES • Online or Print

Teacher's Resources

Lesson Review: Chapter 10, p. 11

Enrichment: Chapter 10, p. 12

Designing an Experiment: Study Tools, p. 14

Designing an Experiment Rubric: Teaching Tools, p. 7

3 Assess

CHECKING CONCEPTS ANSWERS

1. warm, shallow ocean waters, inside hard skeletons made of limestone
2. calcium
3. an assembly of coral animals and their skeletons cemented together
4. a lagoon
5. fringing reefs, barrier reefs, and atolls

THINKING CRITICALLY ANSWERS

6. COMPARE/CONTRAST: Both lie along a mainland coast. Barrier reefs are separated from the mainland by a lagoon. Fringing reefs lie very close to it.
7. INFER: It consists of pulverized coral skeletons made of calcium carbonate. Calcium carbonate is white.
8. HYPOTHESIZE: The farmland was once beneath the ocean.

DESIGNING AN EXPERIMENT

Use the Designing an Experiment Rubric from the Teacher's Resources to assess students' experiments. Fill in the rubric with the added information below. For this assignment, students should have:

- mentioned that dead coral leave behind skeletons that accumulate to form a mass of coral.
- included, in the procedure, a test that uses weak acid to check for calcium carbonate.

4 Close

Integrating Environmental Science

TOPIC: ecosystems

ARTIFICIAL REEFS

Real-Life Connection Natural and artificial reefs supply living space and shelter for a wide variety of organisms. Many of these feed by filtering tiny food particles from the water. The species that live in a coral reef create a food web that supplies food for reef animals and larger fish that visit the reef.

Discussion Display photographs of artificial reefs. Provide some information about the food habits and behavior of the organisms pictured. Have students identify as many animals as they can.

Thinking Critically Answer
Possible answer: They provide plenty of food and living space for the fish.

The Big Idea

(pp. 246–247)

1 Introduce

Objective Students will distinguish between the pelagic and benthic zones of the ocean and recognize organisms adapted to life in each zone. They will see how different factors influence the ways that an ocean organism adapts to its environment.

Linking Prior Knowledge Ask groups of students to recall and list ocean organisms that they learned about (Lessons 10-7 and 10-9). Have them expand the lists from their own experience. Use the lists to create a master list for the class on the board. After the lesson, return to the master list and make additions or deletions as needed.

2 Teach

Discussion Describe plankton, nekton, and benthos for the class. After students have read the article and callouts, have them classify the organisms on the class master list as plankton, nekton, or benthos. If possible, allow students to view microscope slides of plankton.
learning styles: auditory, visual

THE BIG IDEA

Where is life found in the ocean? Have students start their research for their Science Log Writing Activity at www.conceptsandchallenges.com. Be sure students use search engines and websites that are reliable and acceptable. You may also wish to provide copies of the *Big Idea Planner* and *Big Idea Science Log Entry* study tools found in the Teacher's Resources.

Go Online active art
For: Ocean Food Web activity
Visit: PHSchool.com
Web Code: cfp-3042

Students can interact with the art of the ocean food web online.

Integrating Life Science

THE Big IDEA

Where is life found in the ocean?

Diatoms
Diatoms are one type of phytoplankton. When these single-celled organisms die, their glasslike shells sink to the ocean bottom, forming sediment.

Go Online active art
For: Ocean Food Web activity
Visit: PHSchool.com
Web Code: cfp-3042

Many different kinds of organisms live in the ocean. Living things in the ocean are found in two main zones. The waters lying above the ocean bottom are called the pelagic zone. The waters next to the ocean floor are called the benthic zone.

Plankton are organisms that float along the surface of the pelagic zone. The motion of wind, waves, and currents moves these microscopic organisms. Some rise and fall in response to the day/night cycle.

There are two kinds of plankton. **Phytoplankton** (fyt-oh-PLANK-tuhn) are floating, plant-like protists. These organisms use sunlight to make food from carbon dioxide and water. Phytoplankton are a source of food for floating animal-like protists called zooplankton.

Nekton are free-swimming ocean animals that move throughout the pelagic zone in search of food. Fishes, whales, dolphins, seals, and squids are nekton. Some nekton eat plankton. Others eat animals that have eaten plankton.

Organisms that live in the benthic zone are called benthos. Benthos are found in shallow waters along a coast and in the deepest parts of the ocean. Some benthos attach themselves to the ocean floor. They stay in that spot until they die. Mussels, barnacles, and some seaweeds are benthos that remain attached to the ocean floor. Some benthos bury themselves in sand or mud. Others, such as sea stars and crabs, move around on the ocean bottom.

Look at the illustrations on these two pages. Then, complete the Science Log to learn more about "the big idea."✦

Whales
Like all other mammals, whales have lungs and must come to the water surface to take in air. However, some whales, such as sperm whales, can dive to depths of 1,000 m in the pelagic zone.

Lantern Fish and Angler Fish
Organs that give off light are located along the bodies of the lantern fish (shown in photo below). This adaptation helps the animal survive the cold, dark benthic zone. Female angler fish, which live on or near the ocean floor, have structures that look like fishing rods or lures on top of their heads to attract prey. The illustration shows two types of angler fish.

246

Teaching Options

! FAST FACTS *More on Algae*

Algae are plantlike organisms that belong to the Kingdom Protista. They make their own food by photosynthesis. Ocean-dwelling algae include microscopic organisms that make up part of the plankton, as well as the larger seaweeds that grow in shallow water near coastlines.

Many human cultures, such as the Japanese, eat coastal seaweeds. Seaweed derivatives, such as alginates, are widely used in processed foods. Ask students to find out more about the substances derived from algae that are used in food and other products. Have interested students describe to the class the uses of some algae-derived substances. **learning style: auditory**

ESL/ELL STRATEGY

Working With Peers Working with students who are proficient in English can help English-language learners to communicate. Pair an English-language learner with an English-proficient student. Have both students explain the article and the captions in their own words to each other.

Dolphins

Just like other members of the whale family, dolphins are warm-blooded animals. Their body temperature stays about the same regardless of water temperature. Dolphins are strong swimmers and can dive to depths of 300 m.

Sea Stars

Sea stars, also known as starfish, can grow a new arm if they lose one. They have tiny tube feet on their arms that help them move about the ocean floor. These organisms live in the shallow part of the benthic zone.

Mackerel

The streamlined body of a mackerel is adapted to swimming throughout the pelagic zone. Mackerel generally live and travel in large groups called shoals.

Science Log — WRITING ACTIVITY

Available sunlight, water depth, temperature, salinity, and seafloor sediments all influence how an ocean organism adapts to its environment. Choose a marine organism. Research how its environment relates to how it lives. Write a report to be sent to an agency that monitors the environment, such as the Environmental Protection Agency (EPA). Start your search at www.conceptsandchallenges.com.

Figure 10-31 Sunlight and water temperature decrease as you move from the pelagic zone down to the benthic zone of the ocean. This affects the kinds of ocean life that live there.

3 Assess

Use the Writing Activity Rubric from the Teacher's Resources to assess students' Science Logs. Fill in the rubric with the additional information below. For this assignment, students should have:

- described their chosen organism and its relationship to its environment.
- communicated, in paragraph form, accurate information about the organism and its environment.

4 Close

Creating an Ocean-Life Exhibit Have students work in small groups to develop an exhibit that presents information about one or more types of ocean life. Students could create an informative poster or use labeled specimens of sea life, such as clamshells or dried sea stars, for their exhibit. Have each group present its exhibit to the class.

Tips for Using Technology

Downloading and Saving Graphics From the Internet Demonstrate for students how they can save graphic files while on the Internet. Students may use saved images in their letters.

On a Macintosh computer, click and hold the mouse button on the image. A menu will appear. Move the mouse pointer to *Download Image to Disk.* Let go of the mouse button. A Save dialog box will appear. Choose the location into which you would like to save the file. The file name should automatically be filled in. Click Save to save the image you have specified.

• TEACHING RESOURCES • Online or Print

Teacher's Resources

Big Idea: Chapter 10, p. 13

Big Idea Planner: Study Tools, pp. 11–12

Big Idea Science Log Entry: Study Tools, p. 13

Writing Activity Rubric: Teaching Tools, p. 9

Challenges (pp. 248–250)

Chapter Summary

Review Before students begin the Challenges, review the summary with them. Assign each student one summary point to describe in his or her own words.

STUDY HINT Have students choose ten of the Key Terms and use each term in a complete sentence.

Key Term Challenges

MATCHING

1. oceanography (10-1)
2. submersible (10-2)
3. salinity (10-3)
4. thermocline (10-3)
5. current (10-4)
6. wave (10-5)
7. density current (10-4)
8. nodule (10-7)
9. ooze (10-7)
10. trench (10-8)

APPLYING DEFINITIONS

11. The crest is the highest part of a wave. The trough is the lowest part of a wave. (10-5)
12. The continental shelf slopes gently from the coastline of a continent. The continental slope drops steeply from the continental shelf to the ocean floor. (10-8)
13. A guyot is a seamount with a flat top. (10-8)
14. An atoll is a ring-shaped island resulting from the growth of coral. (10-9)
15. A lagoon is an area bounded by a coral reef. A sea is a portion of the ocean that is partially surrounded by a landmass. (10-1, 10-9)
16. A barrier reef has a lagoon between it and the mainland. A fringing reef lies close to the shore. (10-9)

Chapter 10 Challenges

Chapter Summary

Lessons 10-1 and 10-2

- The world ocean is divided into three major oceans. The Pacific is the largest. The Atlantic is the second largest. The Indian is the smallest.
- Oceanographers use small **submersibles, sonar,** and ocean cores from drilling to study the oceans.

Lesson 10-3

- Saltwater contains many more dissolved salts and other minerals than freshwater does.
- The ocean has three different temperature layers. In the **thermocline,** the temperature drops sharply.

Lessons 10-4, 10-5, and 10-6

- **Currents** of water move through the oceans. Most surface currents are caused by wind. They move in different directions in each hemisphere. Ocean currents can be warm or cold. **Density currents** move up and down in the ocean depths.
- Waves are formed by wind. They have **crests** and **troughs.**
- Only the energy, not the water, in a wave moves forward. Waves form breakers near the shore.
- High and low tides change the ocean's surface level. They are caused mostly by the Moon's gravity. **Flood tides** and **ebb tides** are caused by the rotation of Earth.

Lesson 10-7

- Eroded rock particles form most of the sediments close to shore. Minerals also help form ocean sediments.
- Much of the ocean floor is made up of **ooze.**

Lessons 10-8 and 10-9

- The **continental shelf** and **continental slope** lead into the ocean floor. Deep canyons, or **trenches,** are found on the floor. The flat parts of the ocean floor are called plains.
- **Seamounts** are underwater volcanoes and mountains. A **guyot** is a flattened seamount.
- A coral reef is made up of the skeletons of living and dead **coral. Fringing reefs** are attached to a coastline. **Barrier reefs** form around partially sunken volcanic islands. **Atolls** are ring-shaped coral reefs around **lagoons.**

Key Term Challenges

atoll (p. 244)	nodule (p. 240)
barrier reef (p. 244)	oceanography (p. 226)
continental shelf (p. 242)	ooze (p. 240)
continental slope (p. 242)	salinity (p. 230)
coral (p. 244)	seamount (p. 242)
Coriolis effect (p. 232)	sonar (p. 228)
crest (p. 234)	submersible (p. 228)
current (p. 232)	thermocline (p. 230)
density current (p. 232)	tide (p. 238)
ebb tide (p. 238)	trench (p. 242)
flood tide (p. 238)	trough (p. 234)
fringing reef (p. 244)	wave (p. 234)
guyot (p. 242)	world ocean (p. 226)
lagoon (p. 244)	

MATCHING **Write the Key Term from above that best matches each description.**

1. study of Earth's oceans
2. underwater research vessel
3. amount of dissolved salts and minerals in ocean water
4. layer of ocean water in which temperature drops sharply
5. stream of water flowing in the ocean
6. regular up-and-down movement of water at the surface
7. stream of water that moves up and down in the ocean's depths
8. mineral lump found on the ocean floor
9. kind of ocean sediment
10. deep canyon on the ocean floor

APPLYING DEFINITIONS **Explain the difference between the words in each pair. Write your answers in complete sentences.**

11. crest, trough
12. continental shelf, continental slope
13. seamount, guyot
14. coral, atoll
15. lagoon, sea
16. barrier reef, fringing reef

248

Teaching Options

PREPARING STUDENTS FOR STANDARDIZED TESTS

Reading Strategy: Tell students to skip any questions that they have great difficulty answering. They should come back to those questions when they finish the others.

Writing Strategy: Tell students that comparing and contrasting terms involves more than simply giving their definitions. Students must write what connects or differentiates the terms.

Interpreting Visuals Strategy: If students have trouble understanding the information a diagram presents, point out that reading through the questions about the diagram may help them understand the diagram's meaning.

ESL/ELL STRATEGY

Remembering Ocean Landforms Have English-language learners work with students proficient in English to practice differentiating between the different landforms associated with the ocean. Students can use flash cards, for example, to quiz each other.

Content Challenges TEST PREP

MULTIPLE CHOICE Write the letter of the term or phrase that best completes each statement.

1. The three major oceans are the
   - a. Indian, Atlantic, Pacific.
   - b. Pacific, Antarctic, Indian.
   - c. Arctic, Pacific, Indian.
   - d. Atlantic, Pacific, Caribbean.
2. The study of the world's oceans is called
   - a. hydrology.
   - b. geography.
   - c. oceanography.
   - d. technology.
3. To find the depth of the ocean, scientists use
   - a. hydrology.
   - b. sonar.
   - c. bathyspheres.
   - d. bathyscaphs.
4. Ocean water is warmest
   - a. at the surface.
   - b. in the thermocline.
   - c. in the deep ocean.
   - d. at the continental margin.
5. Warm ocean currents come from areas near the
   - a. North Pole.
   - b. South Pole.
   - c. prime meridian.
   - d. equator.
6. The distance between crests is the
   - a. trough.
   - b. wavelength.
   - c. wave height.
   - d. density.
7. Nodules are mostly nickel, iron, and
   - a. manganese.
   - b. chlorine.
   - c. sodium.
   - d. volcanic dust.
8. The process by which a gas changes to a liquid is called
   - a. precipitation.
   - b. evaporation.
   - c. sublimation.
   - d. condensation.
9. Just beyond the shoreline of a continent you would find a
   - a. continental slope.
   - b. continental shelf.
   - c. deep sea trench.
   - d. guyot.
10. Salinity of ocean water can rise when
   - a. glaciers melt.
   - b. it rains.
   - c. water evaporates.
   - d. runoff from rivers increases.

FILL IN Write the term or phrase that best completes each sentence.

11. Most of the water on Earth's surface is __________ water.
12. Oceanographers are specialists who study the __________.
13. Underwater research vessels are called __________.
14. Heat from the __________ warms ocean water.
15. Tides are caused mainly by the __________.
16. The time of low water level is __________ tide.
17. A ring-shaped coral reef around a lagoon is called an __________.
18. A guyot is an underwater, flat-topped __________.

Content Challenges

MULTIPLE CHOICE

1. a (10-1)
2. c (10-1)
3. b (10-2)
4. a (10-3)
5. d (10-4)
6. b (10-5)
7. a (10-7)
8. d (10-3)
9. b (10-8)
10. c (10-3)

FILL IN

11. salt (10-1)
12. oceans (10-1)
13. submersibles (10-2)
14. Sun (10-3)
15. Moon (10-6)
16. ebb (10-6)
17. atoll (10-9)
18. seamount (10-8)

ALTERNATIVE ASSESSMENT

Relating Information Have students write a Science Log entry relating ocean currents, waves, and tides to the formation of shorelines. Have them describe how the presence of coral reefs affects the influence of currents, waves, and tides on the shoreline. You can use the Writing Activity Rubric found in the Teacher's Resources to assess students' entries.

PORTFOLIO ASSESSMENT

Making Student Portfolios Portfolio Assessment is designed to evaluate a student's performance over an extended period of time. Students can demonstrate their comprehension of the concepts in this chapter by making a portfolio. The Chapter Self-Check, the Big Idea Planner, and the Weekly Journal are some of the reproducibles in the Teacher's Resources that they can include in their portfolios. You may wish to make certain assignments planned Portfolio items. You can use the Portfolio Assessment Rubric also found in the Teacher's Resources to assess students' portfolios.

Concept Challenges

WRITTEN RESPONSE

1. **CALCULATE:** 1,500 m/s × 4 s = 6,000 m (10-2)
2. **EXPLAIN:** The bathysphere made it possible for humans to travel to deeper ocean regions. (10-2)
3. **PREDICT:** The salinity would decrease because melted polar ice would add fresh water to the oceans. (10-3)
4. **COMPARE:** A barrier reef has a large lagoon between it and the mainland. A fringing reef does not enclose a lagoon. (10-9)
5. **DESCRIBE:** Possible answers: marine biology, geology, geography (10-1)
6. **COMPARE/CONTRAST:** Both oceans and seas contain salt water, and all are interconnected. A sea is a smaller body of salt water that is partly surrounded by land but connected to one of the world's major oceans. (10-1)
7. **DESCRIBE:** Deep-sea drilling is a method of obtaining samples of crust from the ocean floor. (10-2)
8. **ANALYZE:** Weathering and erosion by wind, waves, tides, rain, and flowing water from streams and rivers contribute to changes in the shoreline. (10-7)

INTERPRETING VISUALS (10-8)

9. Possible answer: It looks like a shelf attached to the mainland continent.
10. guyot
11. The shoreline would move inland because portions of the coast would be covered with water.
12. shoreline, continental shelf, and continental slope

Chapter Project Wrap-Up

TREASURES OF THE SEA BOX

Have students explain the subject and content of the box. You can use the Individual Activity Rubric found in the Teacher's Resources to assess students' contributions. Fill in the rubric with the additional information below. For this project, students should have:

- selected items that represent the ocean-related subject they chose.
- written a paper and explained the subject they chose and the reason it belongs in a particular category.

Concept Challenges TEST PREP

WRITTEN RESPONSE **Answer each of the following questions in complete sentences.**

1. **CALCULATE:** If a sonar signal sent from a ship returns 8 seconds later, how deep is the ocean floor in that spot?
2. **EXPLAIN:** Why was a bathysphere considered an advancement in ocean exploration technology?
3. **PREDICT:** How would the salinity of the oceans be affected if increased temperatures caused the polar ice caps to melt into the oceans?
4. **COMPARE:** How can you tell the difference between a barrier reef and a fringing reef?
5. **DESCRIBE:** What are some specialties within the field of oceanography?
6. **COMPARE/CONTRAST:** How are oceans and seas alike and different?
7. **DESCRIBE:** What is the purpose of deep-sea drilling?
8. **ANALYZE:** What kinds of weathering and erosion change the shoreline?

INTERPRETING VISUALS **Use Figure 10-32 below to answer the following questions.**

9. How do you think the seafloor feature labeled *C* got its name?
10. What is the name of the landform labeled *D*?
11. How would the size of the shoreline change if sea levels rose due to global warming?
12. What parts of the continent, labeled *B*, *C*, and *G* on the diagram, extend into the ocean?

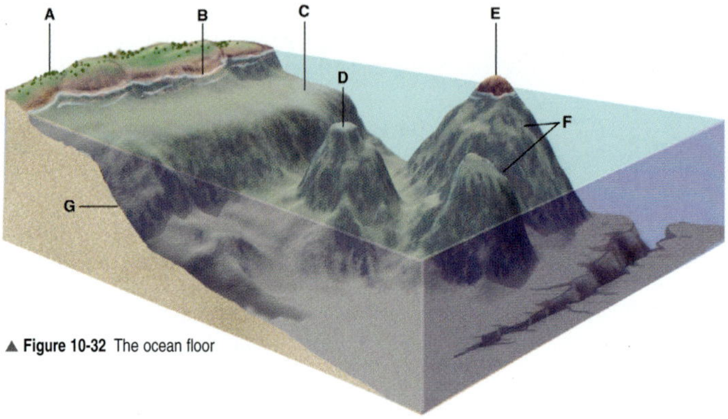

▲ **Figure 10-32** The ocean floor

Teaching Options

• TEACHING RESOURCES •
Online or Print

Teacher's Resources

Key Term Review: Chapter 10, p. 14

Chapter 10 Test: pp. 15–17

Individual Activity Rubric: Teaching Tools, p. 5

Chapter Self-Check: Teaching Tools, p. 11

Weekly Journal: Study Tools, p. 19

Portfolio Assessment Rubric: Teaching Tools, p. 10

• TEACHING RESOURCES •
Online or Print

Teacher's Resources

Writing Activity Rubric: Teaching Tools, p. 9

PLANNING GUIDE

◆ **TEACHING THE CHAPTER** This chapter should take approximately 9–11 days to complete instruction and assessment.

	Skills and Features	Projects and Activities	Achieve Science Literacy	Meet Individual Needs
Chapter 11 Opener p. T251	LABEL	• Chapter Project		
Lesson 11-1 **What is air?** pp. T252–T253 Standard: D1h	IDENTIFY, ANALYZE, STATE • Real-Life Science	• More on the Web • Integrating Life Science	• Study Hint • Reading Strategy • Ongoing Assessment	• Reteach Teacher's Resources* Enrichment Activity
Lesson 11-2 **What are the layers of the atmosphere?** pp. T254–T255 Standard: D1h	LIST, INFER, CALCULATE • Integrating Physical Science	• More on the Web	• Reading Strategy	Teacher's Resources* Enrichment Activity
Lesson 11-3 **How is Earth's surface heated?** pp. T256–T257 Standards: D1h, D3d, D4b	NAME, PREDICT, DESCRIBE • Investigate • Web InfoSearch • Science and Technology	• Extending the Lesson	• Study Hint	• ESL/ELL Strategy • Reteach
Lesson 11-4 **How does heat move through the atmosphere?** pp. T258–T259 Standards: D1h, D3d	EXPLAIN, DESCRIBE, INFER • Health and Safety Tip • Hands-On Activity	• Cooperative Learning • More on the Web • Lab Challenge	• Study Hint • Reading Strategy • Ongoing Assessment	
Lesson 11-5 **What is air pressure?** pp. T260–T261 Standards: D1f, D1h	EXPLAIN, DESCRIBE, RELATE • Investigate • Web InfoSearch	• Integrating Life Science	• Study Hint • Writing Hint • Ongoing Assessment	• Reteach
Lesson 11-6 **How is air pressure measured?** pp. T262–T263 Standard: D1h	CALCULATE, ANALYZE, DEFINE • Building Math Skills • How Do They Know That?	• Extending the Lesson • Active Art		• Reteach Teacher's Resources* Venn Diagram

*www.conceptsandchallenges.com, password: ccearth09

Planning Guide continues on next page.

Standards: For details on the correlation to National Science Standards see pages *xx–xxi*.

	Skills and Features	Projects and Activities	Achieve Science Literacy	Meet Individual Needs
Lab Activity **Constructing a Barometer** pp. T264–T265 Standard: D1h	OBSERVE, ANALYZE, INFER • Set-Up Time: 15–20 min • Lab Time: 15–20 min plus 5 min each day for 2 weeks	• Extending the Lesson • Extend the Lab Activity	• Tips for Using Technology	Teacher's Resources* Lab Activity Report
Lesson 11-7 **How do winds form?** pp. T266–T267 Standards: D1h, D1j	PREDICT, INFER, EXPLAIN • Web InfoSearch • Hands-On Activity		• Reading Strategy • Ongoing Assessment	Teacher's Resources* Enrichment Activity
Big Idea **How do pressure changes create global winds?** pp. T268–T269 Standards: D1h, D1j, D3d	RESEARCH, COMMUNICATE, ANALYZE • Science Log: Writing Activity	• Big Idea • Close Activity • Active Art	• Tips for Using Technology	Teacher's Resources* Big Idea Planner
Lesson 11-8 **What causes local winds?** pp. T270–T271 Standards: D1j, D2a, D3d	ANALYZE, DEFINE, MODEL • Building Reading Skills • Science and Technology	• Cooperative Learning	• Study Hint	• Reteach Teacher's Resources* Enrichment Activity
Lesson 11-9 **How is wind measured?** pp. T272–T273 Standard: D2a	CLASSIFY, INFER, DESCRIBE • Building Science Skills • Hands-On Activity	• Cooperative Learning • Extending the Lesson • More on the Web	• Study Hint	• ESL/ELL Strategy Teacher's Resources* Enrichment Activity
Chapter 11 Challenges pp. T274–T276	• Chapter Summary • Key Term Challenges • Content Challenges • Concept Challenges	• Chapter Project Wrap-Up	• Study Hint • Preparing Students for Standardized Tests	• ESL/ELL Strategy Teacher's Resources* Chapter Self-Check

*www.conceptsandchallenges.com, password: ccearth09

Chapter 11 Air

▲ **Figure 11-1** Colorful balloons dot the sky as they gracefully rise in the atmosphere.

Heating air causes it to rise. This is how the balloons in the picture can rise and float. The heating and rising of air is how air currents are formed. Air currents affect the water cycle and weather patterns. Air thins out the higher you go in the atmosphere. Therefore, hot-air balloons cannot go too high.

▶What do you think would happen if you tried to take a balloon where there is very little air?

Contents

Teaching Options

ASSESSMENT PLANNER

For assessment in the Student Edition, see the following pages:

Content Assessment
Checking Concepts: pp. 253, 255, 257, 259, 261, 263, 267, 271, 273
Thinking Critically: pp. 253, 255, 257, 259, 261, 263, 267, 271, 273
Interpreting Visuals: pp. 253, 255, 259
Chapter Challenges: pp. 274–276

Alternative Assessment
Building Skills: pp. 263, 271, 273
Web InfoSearch: pp. 257, 261, 267
Science Log: p. 269

Performance-Based Assessment
Hands-On Activity: pp. 259, 267, 273
Lab Activity: pp. 264–265

• TEACHING RESOURCES • Online or Print

Teacher's Resources

Lesson Review: Ch. 11, pp. 2, 3, 5, 6, 7, 8, 9, 11, 12

Enrichment: Ch. 11, pp. 4, 13

Key Term Review: Ch. 11, p. 14

Chapter 11 Test: pp. 15–17

Laboratory Manual: pp. 61–64

Chapter 11 Air

Chapter Overview

In this chapter, students will learn about the composition and structure of Earth's atmosphere. Students will also study heat transfer throughout the atmosphere and examine how atmospheric variables, such as air pressure and wind, are measured.

About Figure 11-1 People have been riding in hot-air balloons since 1783, when the first crewed flight occurred. Although scientists such as Evangelista Torricelli and Blaise Pascal had already discovered the relationship between air pressure and altitude, many early balloonists did not understand that air pressure decreases with increasing altitude. Several balloonists during the early 1800s floated too high and died of oxygen deprivation.

Answer Possible answer: The hot-air balloon would eventually cool, collapse, and fall back to Earth.

Linking Prior Knowledge

For this chapter, students should recall:

- elements and compounds (Lesson 2-1).
- how minerals are formed (Lesson 2-70).
- ocean currents (Lesson 10-4).

Chapter Project

LAYERS OF THE ATMOSPHERE POSTER

MATERIALS: posterboard, construction paper, glue, old magazines

Display a posterboard with the head *Layers of the Atmosphere.* Label the layers. As you study the layers of the atmosphere in Lesson 11-2, have students place something pictorial in each layer. They can make drawings or cut out pictures from magazines to glue onto the posterboard. As you go through the rest of the chapter, have students add to the display pictures that are relevant to each lesson. When the chapter is finished, divide the class into four groups. Each group should write and present a summary about one of the four major layers.

learning styles: visual, tactile

11-1 (pp. 252–253)

1 Introduce

STUDY HINT Have students preread the section titles of this lesson to help them see at a glance what is in air.

Linking Prior Knowledge Ask students to recall what they learned about elements and compounds (Lesson 2-1) and how they combine to form minerals (Lesson 2-7). Tell students that elements and compounds also make up the atmosphere. To emphasize the basic structure of matter, point out that some of these same elements are found in the human body.

2 Teach

Demonstration To demonstrate that air has mass and volume, fill a clear plastic container with water. Then, invert a clear glass and push it down into the water. Tell students to observe what happens. Ask them why the water cannot enter and fill up the glass. (Air in the glass prevents the water from entering.)
learning style: visual

Discussion Discuss the importance of nitrogen, oxygen, and carbon dioxide for living things. Ask students what happens if humans do not have air to breathe. (They suffocate and die.) Ask: *How do humans exist in environments that have no air, such as in outer space and underwater?* (They use life support systems.) Lead students to understand that life as we know it could not exist without these gases.

Answers to questions

1. **ANALYZE:** 0.002%
2. **DEFINE:** microscopic organisms that live in soil, water, and air
3. **INFER:** by breathing
4. **STATE:** from burning things and breathing out
5. **LIST:** tiny particles of dust, smoke, salt, and other chemicals

11-1 What is air?

Objectives
Describe air as matter. Identify and describe the main gases in air.

Key Terms
matter: anything that has mass and volume
atmosphere (AT-muh-sfeer): envelope of gases surrounding Earth
cellular respiration (rehs-puh-RAY-shuhn): process by which a cell releases energy from food molecules

A Mixture of Gases Air is a colorless, tasteless, odorless mixture of gases. Air is matter. **Matter** is anything that has mass and takes up space, or has volume.

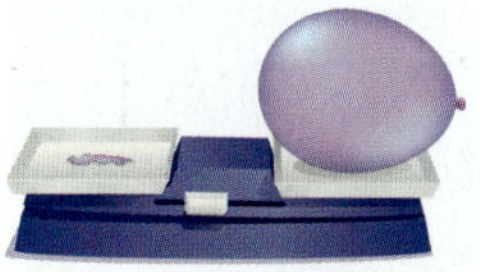

◀ **Figure 11-2** Air has mass and volume.

The **atmosphere,** or air, is the envelope of gases surrounding Earth. Air is made up mostly of nitrogen and oxygen. Air is also made up of other gases. Figure 11-3 shows the approximate percentages of the gases that make up air.

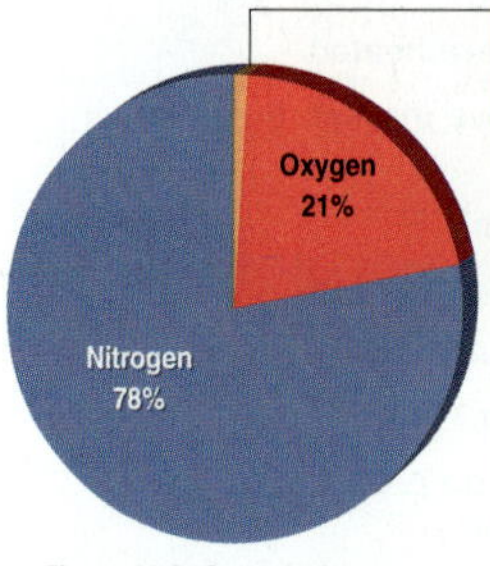

SOME OTHER GASES	
Gases	**Percentage by Volume**
Argon	0.934
Carbon dioxide	0.036
Neon	0.0018
Helium	0.00052
Hydrogen	0.00005
Krypton	0.00011
Methane	0.00002

▲ **Figure 11-3** Gases in the air

1 **ANALYZE:** What percentage of air is made up of helium, neon, and krypton?

Nitrogen About 78 percent of the atmosphere is nitrogen. All organisms need nitrogen. However, most living things cannot use nitrogen gas from the air.

Bacteria are microscopic organisms that live in soil, in water, and in the air. Some bacteria can change the nitrogen gas in the air into nitrogen compounds. Plants get the nitrogen they need by absorbing the nitrogen compounds made by these bacteria. Animals get the nitrogen they need by eating plants.

2 **DEFINE:** What are bacteria?

Oxygen About 21 percent of air is oxygen. Living things need oxygen to carry on cellular respiration. **Cellular respiration** is the process by which a cell releases energy from food molecules. The energy comes from the food made, eaten, or absorbed by an organism. Plants make their own food. Most animals must eat food. Some other organisms absorb food from their environment. Most living things get the oxygen they need from air.

3 **INFER:** How do people get the oxygen they need to carry on respiration?

Carbon Dioxide About 0.04 percent of air is made up of carbon dioxide. Carbon dioxide is released when things burn. Respiration produces carbon dioxide as a by-product. You get rid of carbon dioxide when you breathe out. Plants need carbon dioxide to make their own food.

4 **STATE:** How does carbon dioxide get into air?

Other Components in Air Most of the time, air looks and feels dry to us. However, it is never really completely dry. In addition to the gases mentioned above, air contains water vapor. Water vapor is water in the form of a gas. Steam is heated water vapor. Sometimes steam contains tiny droplets of water, which is what you see. The water vapor itself is invisible. The amount of water vapor in the air varies greatly from place to place and over time. In rain forests, up to 5 percent of the air may be water vapor.

Teaching Options

INTEGRATING LIFE SCIENCE

Converting to Food Atmospheric carbon dioxide plays an important role in photosynthesis. During photosynthesis, green plants use sunlight to convert carbon dioxide and water into nutrients and oxygen. Plants that photosynthesize are called producers because they make their own food. Ask students to list five plants that they eat for their nutritional value.

RETEACH

Demonstrating Air Mass
Re-create the demonstration shown in Figure 11-2. Use a triple-beam balance to find the mass of a deflated balloon. Then, blow up the balloon, tie it, and place it back on the balance. Have students find the mass of the blown-up balloon. Ask: *What is the mass of the volume of air in the balloon?* (the mass of the blown-up balloon minus the mass of the deflated balloon)
learning styles: visual, mathematical/logical

Water vapor plays an important role in weather. Clouds form as water vapor condenses out of air that has cooled. It forms tiny drops of liquid water or crystals of ice. If these drops or crystals grow large enough, they can fall as rain or snow.

In addition to the gases that make it up, air contains tiny particles of dust, smoke, salt, and other chemicals. You can see some of these particles. They are not part of the air itself. They float in the air, and sometimes they make the air unhealthy to breathe.

5 **LIST:** What is in air besides the gases?

CHECKING CONCEPTS

1. What two main gases make up air?
2. What is matter?
3. Why do living things need oxygen?
4. How does water vapor affect the weather?
5. What are two nongaseous particles in air?

THINKING CRITICALLY

6. **CONTRAST:** How is water vapor different from other gases that make up air?

INTERPRETING VISUALS

Use Figure 11-3 to answer the following questions.

7. **ESTIMATE:** What gas makes up most of air?
8. **CALCULATE:** What percentage of air is made up of methane, krypton, and hydrogen?
9. **IDENTIFY:** What gases make up the 1 percent of air not composed of nitrogen and oxygen?

Real-Life Science

BRONCHIAL ASTHMA

We all need air to live. For people who suffer from bronchial asthma, getting that air into their lungs can be a struggle. There are more than 14 million asthma sufferers in the United States alone.

An asthma attack usually starts with a swelling in the passages that carry air into our lungs when we breathe. One of the most common signs of an asthma attack is wheezing. Wheezing is the high-pitched noise caused by air moving through a narrowed air passage.

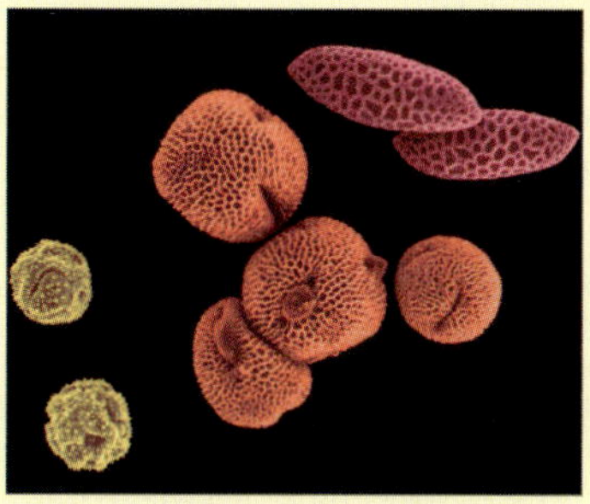

▲ **Figure 11-4** Three types of pollen grains

What causes an asthma attack? Usually, the air passages become irritated by tiny particles floating in the air. Often, these particles are things that a person with asthma is allergic to, such as pollen or animal dander. Household cleaners and cigarette smoke can also cause asthma attacks. The body's immune system releases a substance called histamine to fight the invaders. It is the histamine that causes the swelling.

Asthma cannot be cured. Its symptoms can be controlled, though, with treatment.

Thinking Critically How can the condition of asthma be triggered by particles suspended in air?

READING STRATEGY Have students review the discussion of the term *sphere* (Lesson 1-2). Have them make the connection to atmosphere.

3 *Assess*

CHECKING CONCEPTS ANSWERS

1. nitrogen and oxygen
2. anything that has mass and volume
3. for respiration
4. It forms clouds and rain or snow.
5. Accept any two of the following: dust, smoke, salt, or chemicals.

THINKING CRITICALLY ANSWER

6. **CONTRAST:** It varies greatly from place to place and over time.

INTERPRETING VISUALS ANSWERS

7. **ESTIMATE:** nitrogen
8. **CALCULATE:** 0.00031 percent
9. **IDENTIFY:** argon, carbon dioxide, helium, neon, krypton, methane, and hydrogen

4 *Close*

Real-Life Science

BRONCHIAL ASTHMA

Real-Life Connection Weather forecasters often provide information about air quality. When air quality is low, people with asthma and other respiratory problems are advised to stay indoors. The information about air quality is gathered by air pollution technicians who collect air samples from different places and run tests on the samples. The tests reveal the quantity and type of pollutants in the air. The results of the tests are sent to weather stations to be incorporated into weather forecasts.

Discussion Ask students: *How do air pollution technicians and weather forecasters work together?* (Weather forecasters report information gathered by air pollution technicians.)

Thinking Critically Answer
The body's immune system reacts to these particles by releasing histamine, which causes swelling in the air passages.

MORE ON THE WEB

Charting Atmospheric Gases
Have students use the Internet to learn about the most common gases found in the atmosphere. Then have students design a chart of the atmospheric gases, including their discovery, scientific and original names, and percentage found in the atmosphere.

• TEACHING RESOURCES •
Online or Print

Teacher's Resources
Lesson Review: Chapter 11, p. 2
Enrichment: Chapter 11, p. 4

ONGOING ASSESSMENT

Have students design a different experiment to show that air is matter. Students should list the materials they would use and write the steps of their procedure, including how they would record data. You may wish to have volunteers explain their experiments to the class.

11-2 *(pp. 254–255)*

1 *Introduce*

Linking Prior Knowledge Ask students if they have ever gone up a tall mountain or traveled in an airplane. Ask if they noticed any effect on their body as they traveled higher.

2 *Teach*

Discussion Tell students that airplane pilots change their flying altitude depending on the weather. Ask students to brainstorm why pilots increase altitude when faced with bad weather. Point out that pilots prefer to fly in the region of the atmosphere called the stratosphere because there is no weather in this layer.

Demonstration Use volunteers and a basketball to demonstrate how radio waves reflect off the ionosphere. Place yourself and four students in a circle. Tell students that the basketball represents reflected radio waves. You and the students represent the ionosphere, and the floor represents the surface of Earth. Bounce the basketball to one student so that the ball hits the ground once before reaching the student's hands. Have that student bounce the ball to another student, and so on around the circle.
learning styles: visual, kinesthetic

Reinforcement Have students write the following terms on four index cards: *troposphere, thermosphere, stratosphere, mesosphere.* Tell students that these terms name the main layers of the atmosphere. As you describe each layer, have students arrange their cards in the correct order from top to bottom: thermosphere, mesosphere, stratosphere, troposphere. Refer students to Figure 11-5 to check their cards. Have students identify one characteristic of each atmospheric layer on the index cards.
learning styles: auditory, visual

READING STRATEGY Read the list of Key Terms aloud so that students can hear their pronunciations. Have students identify the common root in each word. (sphere) Ask students how knowing this will help them better understand what they are reading.

11-2 What are the layers of the atmosphere?

Objective
Name and describe the layers of the atmosphere.

Key Terms
troposphere (TROH-puh-sfeer)**:** lowest layer of the atmosphere
stratosphere (STRAT-uh-sfeer)**:** second layer of the atmosphere
mesosphere (MEHZ-uh-sfeer)**:** third layer of the atmosphere
thermosphere (THER-muh-sfeer)**:** upper layer of the atmosphere

Parts of the Atmosphere The atmosphere begins at Earth's surface and goes more than 700 km up. Not all parts of the atmosphere are the same. The atmosphere is made up of four main layers. These layers are the troposphere, the stratosphere, the mesosphere, and the thermosphere.

1 LIST: What are the four main layers of the atmosphere?

The Troposphere The **troposphere** is the layer of the atmosphere closest to Earth. The air you breathe into your lungs is part of the troposphere. Most of the water vapor found in the atmosphere is in the troposphere. This water vapor forms clouds. Weather takes place in the troposphere.

The higher you go in the troposphere, the colder it gets. Near the top of the troposphere, the temperature stops getting colder. The boundary between the troposphere and the layer above is called the tropopause.

2 DEFINE: What is the tropopause?

The Stratosphere The **stratosphere** is the second layer of the atmosphere. The temperature of the air remains almost constant here. There is no weather in the stratosphere. Airplanes travel in this layer.

The upper stratosphere contains a layer of ozone. Ozone is a form of oxygen. Ozone prevents most of the ultraviolet light given off by the Sun from reaching Earth. Large amounts of ozone, however, are harmful to breathe and can irritate the lungs.

3 PREDICT: What might happen if the ozone layer of the stratosphere were destroyed?

The Mesosphere and Thermosphere Above the stratosphere is the third layer of the atmosphere, the **mesosphere,** which means "middle layer." In the mesosphere, temperatures begin to fall again. Above the mesosphere is the fourth region, the **thermosphere.** Here temperatures actually rise with height. Beyond the thermosphere is mostly empty space. This region is called the exosphere.

From the mesosphere up to the top of the thermosphere is a broad region of space that contains many charged particles called ions. Radio waves sent from Earth are reflected or bounced off of these ions. Because of ions, radio signals can be sent between distant parts of Earth. The area where the ions are most concentrated is called the ionosphere.

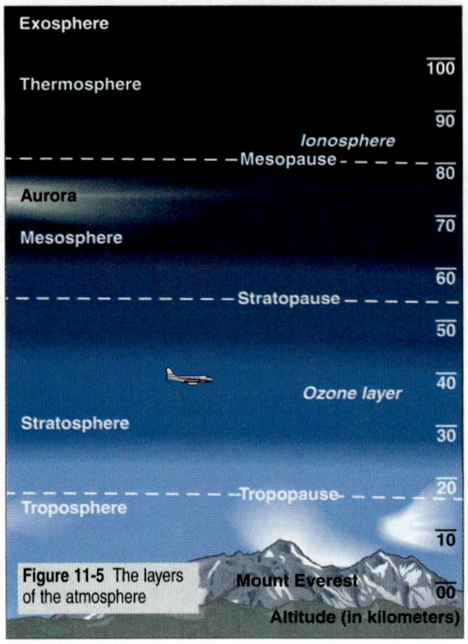

Figure 11-5 The layers of the atmosphere

Teaching Options

! FAST FACTS *CFCs and the Ozone Layer*

In the 1970s, scientists discovered a thinning in the ozone layer above Antarctica. Scientists hypothesize that this thinning is caused by pollutants in the atmosphere known as chlorofluorocarbons, or CFCs. In the presence of sunlight, CFCs break down into chlorine atoms, which in turn break down and destroy ozone molecules. CFCs were once widely used in aerosol spray cans and refrigerants. Many countries have since banned the use of CFCs.

QUICK CONVERSIONS

	Metric/SI	English
Troposphere	0–18 km	0–11 mi
Stratosphere	18–54 km	11–34 mi
Mesosphere	54–80 km	34–50 mi
Thermosphere	80 km+	50 mi+

Sometimes, in the night sky near the poles, you can see glowing colored light, usually green. These glowing lights, called auroras, are caused by charged particles from the Sun. The particles are interacting with matter in Earth's upper atmosphere. In the northern polar region, these events are known as the aurora borealis, or the northern lights. In the Southern Hemisphere, they are known as aurora australis.

▲ **Figure 11-6** The aurora borealis lights up the northern sky at night.

4 **DESCRIBE:** How is the ionosphere used in communications?

CHECKING CONCEPTS

1. The atmosphere is made up of __________ main layers.
2. The upper boundary of the troposphere is the __________.
3. Ozone is a form of __________.

THINKING CRITICALLY

4. **INFER:** Why do most airplanes travel in the stratosphere?
5. **COMPARE:** How do air temperatures differ in the troposphere and stratosphere?

INTERPRETING VISUALS

Use Figure 11-5 to answer the following questions.

6. **INFER:** In which layer of the atmosphere do we fly kites? Why?
7. **CALCULATE:** About how thick is the stratosphere?
8. **CALCULATE:** About how thick is the troposphere?

Integrating Physical Science

TOPICS: air pressure, force

GETTING A LIFT INTO THE AIR

Daniel Bernoulli (bur-NOO-lee) was a Swiss scientist known for his work with fluids. Bernoulli found that the faster a fluid moves, the less pressure it exerts. Pressure is the amount of force acting on a surface. Bernoulli further stated that the pressure of a moving stream of fluid or gas is lower than the pressure of any fluid or gas around it. This explains how air moving around a wing can produce a force called lift, allowing an airplane to fly.

Objects can be designed so air moves at different speeds around them. If the air moves faster above the object, pressure pushes the object upward.

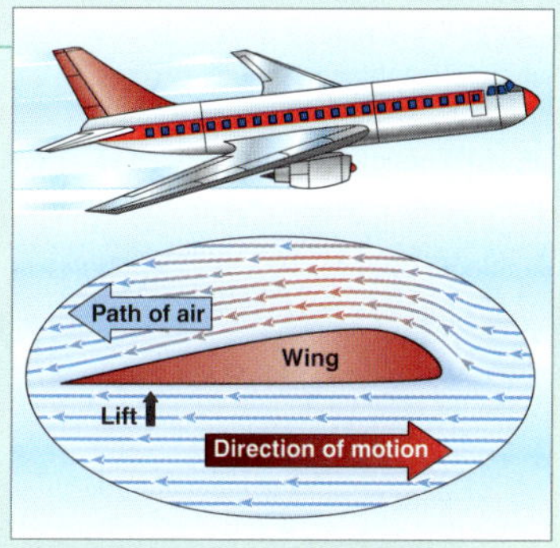

▲ **Figure 11-7** Airplane wings are curved on top, like bird wings.

Like a bird's wing, the top of an airplane wing is curved. Air that moves over the top of the wing must travel farther than air that moves along the bottom. The air moving over the top moves faster so its pressure is lower than the air pressure on the bottom. This difference in pressure creates the lift.

Thinking Critically Why are airplane wings curved on top?

MORE ON THE WEB

The Aurora Borealis Have students learn more about the aurora borealis by visiting www.conceptsandchallenges.com. Then, have students design an advertising brochure for the aurora borealis, describing what it is and where it can be observed.

• TEACHING RESOURCES • Online or Print

Teacher's Resources

Lesson Review: Chapter 11, p. 3

Enrichment: Chapter 11, p. 4

Answers to questions

1 **LIST:** troposphere, stratosphere, mesosphere, and thermosphere

2 **DEFINE:** boundary between the troposphere and the layer above

3 **PREDICT:** Large amounts of harmful ultraviolet radiation would be able to reach Earth.

4 **DESCRIBE:** Radio waves bounced off the ionosphere can be sent from one part of Earth to another.

3 *Assess*

CHECKING CONCEPTS ANSWERS

1. four
2. tropopause
3. oxygen

THINKING CRITICALLY ANSWERS

4. **INFER:** There is no weather in the stratosphere to interfere with travel.
5. **COMPARE:** The temperature of the troposphere decreases with altitude. The temperature of the stratosphere remains fairly constant.

INTERPRETING VISUALS ANSWERS

6. **INFER:** Troposphere; It is closest to Earth.
7. **CALCULATE:** 36 km
8. **CALCULATE:** 18 km

4 *Close*

Integrating Physical Science

TOPICS: air pressure, force

GETTING A LIFT INTO THE AIR

Real-Life Connection Show students a boomerang. Point out that the top is curved, like an airplane wing. A boomerang is an example of Bernoulli's principle. The boomerang gets its lift from the differences in pressure between its top and bottom parts. The overall shape of the boomerang affects its flight, causing it to spin back to the thrower.

Discussion Ask students to describe the relationship between speed and pressure in a fluid. (As the speed of a fluid increases, the pressure it exerts decreases.)

Thinking Critically Answer
The curve creates a difference in air pressure between the top and bottom of the wing, which creates lift.

11-3 (pp. 256–257)

1 Introduce

INVESTIGATE

TIME: 10–15 minutes

PURPOSE: Students will observe how different materials absorb light.

MATERIALS: aluminum plate, lamp, dark soil, sand, thermometer

PROCEDURE: Caution students not to touch the lamp; it will get hot. Also, make sure that students remove the thermometers from the pans before turning on the lamp. The thermometers should be reinserted into the soil and sand after the materials have been under the lamp for 10–15 minutes.

THINK ABOUT IT: Both materials increased in temperature, but the soil heated up more than the sand. The soil absorbs light more efficiently than does the sand. **learning style: visual**

STUDY HINT Have students study Figure 11-8 before they read the text to see if they can begin to answer the lesson question. Have them look for more information in the text.

2 Teach

Demonstration Demonstrate radiant energy with a radiometer, which can be purchased from a scientific supply house or a teacher's resources store. The vanes of a radiometer turn when hit by radiant energy. Place the radiometer in sunlight. Have students observe the turning vanes. Point out that radiant energy from the Sun causes the movement of the radiometer. **learning style: visual**

Discussion Using Figure 11-8 as a guide, discuss what happens to incoming solar radiation. Trace each path of solar radiation from the Sun to Earth's surface or atmosphere. Ask students why not all of the Sun's radiant energy reaches the surface of Earth. (The radiant energy is absorbed or reflected before reaching Earth's surface.) It may be helpful to make a transparency of the diagram for this discussion. **learning styles: visual, auditory**

Answers to questions

1 **NAME:** radiant energy

11-3 How is Earth's surface heated?

INVESTIGATE

Modeling the Absorption of Light

HANDS-ON ACTIVITY

1. Place an aluminum plate under a lamp whose light is directed downward.
2. Fill half the plate with dry, dark soil and half with dry sand.
3. Insert a thermometer into each side. Record the temperatures.
4. Turn on the lamp and let the plate stand for a while. Compare the new temperatures with the starting ones.

THINK ABOUT IT: How do the temperatures change? Why is one higher than the other?

Objective

Describe how energy from the Sun warms Earth's surface.

Key Terms

radiant (RAY-dee-uhnt) **energy:** energy given off by the Sun that can travel through empty space

radiation (ray-dee-AY-shuhn)**:** movement of the Sun's energy through empty space

Radiant Energy The Sun gives off **radiant energy.** If you go out into the sunlight, you can feel the radiant energy from the Sun warming your skin.

Light is a form of radiant energy. Radiant energy can travel across millions of kilometers of empty space. The movement of this energy through empty space is called **radiation.** Most of Earth's energy comes from the radiation given off by the Sun.

1 NAME: What kind of energy is sunlight?

Absorption of Energy Dark surfaces absorb light. When light is absorbed, or taken in, it is usually changed into heat.

Suppose you wrap two ice cubes with cloth. You wrap one ice cube in a dark-colored cloth and the other in light-colored cloth. You place both ice cubes in sunlight. Which one would melt first?

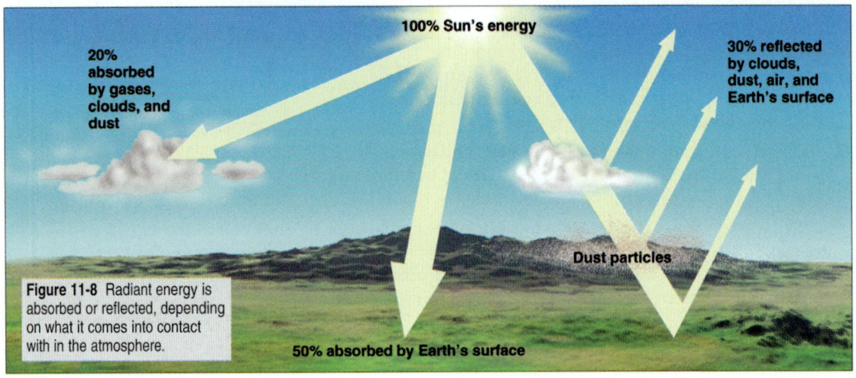

Figure 11-8 Radiant energy is absorbed or reflected, depending on what it comes into contact with in the atmosphere.

Teaching Options

EXTENDING THE LESSON

Researching Absorption Rates Tell students that water and land have different heat absorption rates. Have students research and write about these differences. Students should explain why the differences exist. (Water heats and cools more slowly than land. This is because water has a higher specific heat than land does. It takes more heat to raise the temperature of a given volume of water than it does to raise the temperature of the same volume of land.) **learning style: linguistic**

RETEACH

Feeling the Heat Have students place a dark-colored cloth and a white cloth in the Sun for several minutes. Have students feel the fabric to determine which is warmer and which is cooler. (The dark cloth will feel warmer.) Tell students that the darker color absorbs more light, which is changed into heat that they can feel. **learning style: tactile**

The ice cube wrapped in the dark cloth would melt faster. Surfaces that reflect light, such as white surfaces, remain cooler than surfaces that absorb light, such as dark surfaces.

2 DESCRIBE: What happens when light is absorbed by a surface?

Energy from the Sun Only a small part of the Sun's energy reaches Earth. Some of the Sun's energy is absorbed by the atmosphere. Clouds, dust particles, and water droplets in the atmosphere also absorb or reflect some of the Sun's energy. The energy that is reflected goes back into space. Some of the energy that passes through the atmosphere is absorbed by Earth's surface. This energy is changed into heat. As a result, Earth becomes warmer. The entire process is shown in Figure 11-10 on page 258.

3 STATE: What happens when the Sun's energy is absorbed by Earth's surface?

CHECKING CONCEPTS

1. Light is a type of __________.
2. The Sun's energy reaches Earth by __________.
3. When light is __________, it is changed into heat.
4. Clouds, __________, and water droplets can absorb or reflect the Sun's energy.

THINKING CRITICALLY

5. **PREDICT:** If there was no wind, would it be cooler on a cloudy or on a clear day? Why?

Web InfoSearch

Electromagnetic Radiation Electromagnetic radiation is energy in the form of waves of different frequencies. *Frequency* refers to how often the wave occurs in a given amount of time. Light is only one type of electromagnetic radiation.

SEARCH: Use the Internet to find out more about this topic. What are other types of electromagnetic radiation? In what technologies are they used? Start your search at www.conceptsandchallenges.com. Some key search words are **electromagnetic radiation, light waves,** and **X-rays.**

Science and Technology

SOLAR ENERGY

The Sun is a potentially huge source of energy. It could be used to meet all of the world's energy needs. Scientists are working to develop efficient ways to use solar energy.

At one power plant in California's Mojave Desert, rings of huge mirrors capture sunlight and reflect it to a central tower. A liquid runs through pipes in the tower. The liquid becomes hot. The hot liquid is then used to boil water to make steam. The steam turns a turbine and a generator. In this way, solar energy can be converted to produce electricity when people need it, even after sunset.

▲ **Figure 11-9** On this artist's drawing of a communications satellite, the large blue panels contain the solar cells.

Solar cells can also be used to turn solar energy directly into electricity. Solar cells are also called photovoltaic, or PV, cells. In the 1950s, PV cells were developed for use on U.S. spacecraft. Panels of solar cells power satellites in space. You might have a calculator or a watch that uses a solar cell. Some small, experimental cars are powered by solar cells.

Thinking Critically Why is it important to develop solar energy for heating homes?

2 DESCRIBE: It is changed into heat.

3 STATE: The energy is changed into heat, and Earth becomes warmer.

3 *Assess*

CHECKING CONCEPTS ANSWERS

1. radiant energy
2. radiation
3. absorbed
4. dust particles

THINKING CRITICALLY ANSWER

5. **PREDICT:** a cloudy day, because more solar radiation would be reflected back into space by the clouds

4 *Close*

Web InfoSearch

Electromagnetic Radiation Have students use the Internet to find out more about electromagnetic radiation. Ask students: *What are the other types of electromagnetic radiation? In what familiar technologies are they used?* Some key search words are **electromagnetic radiation, light waves,** and **X rays.** Students should begin their search at www.conceptsandchallenges.com.

Science and Technology

SOLAR ENERGY

Physical-Science Connection The different forms of energy include mechanical, electrical, chemical, nuclear, and radiant. Energy cannot be created, but it can change forms. Solar cells, for example, change radiant energy directly into electrical energy.

Discussion Have students describe how solar power plants convert energy from one form to another. For example, radiant energy from steam turns turbines and is converted into mechanical energy. The mechanical energy of the turbines is then converted into electrical energy by the generators. **learning style: auditory**

Thinking Critically Answer
Possible answer: to conserve fossil fuels and reduce air pollution

ESL/ELL STRATEGY

Suffixes Tell English-language learners that the suffix *-ion* means "act or process of." Radiation is the act or process of radiating, or sending out rays.

• *TEACHING RESOURCES* •
Online or Print

Teacher's Resources
Lesson Review: Chapter 11, p. 5

11-4 *(pp. 258–259)*

1 Introduce

STUDY HINT Have students compare and contrast the processes described in this lesson. They can put the information in a chart, a graphic organizer, or draw sketches to reinforce what they read.

Linking Prior Knowledge Ask students if they have ever walked barefoot across an asphalt parking lot or a sandy beach in the summer. Have them describe how the hot surface scorched their feet. Tell students that this is one example of how heat is transferred. After students have read the lesson, ask: *How is heat transferred from the hot asphalt or sand to your feet?* (by conduction)

2 Teach

Demonstration To reinforce the concept that heat energy is related to the motion of molecules, use marbles, a glass tray, and an overhead projector to model conduction. Cover about three-quarters of the bottom of the tray with a single layer of marbles. Agitate the marbles at one end of the tray. Have students note that the disturbance is "conducted" to the marbles at the other end. **learning style: visual**

Discussion Discuss how heat is transferred from a stove burner to a pot of boiling water. The burner radiates heat. This heat is transferred from the burner to the pot by conduction. Convection then causes the water in the pot to boil.
learning styles: auditory, visual

Answers to questions

1. **NAME:** by conduction
2. **DESCRIBE:** The energy is absorbed by the atmosphere and helps to heat it.
3. **DEFINE:** the movement of heat through a gas or liquid

READING STRATEGY In Lesson 6-5, students learned about the relationship between plate tectonics and convection currents. Have students review this lesson to increase their understanding of heat transfer through convection.

11-4 How does heat move through the atmosphere?

Objective
Explain how the atmosphere is heated.

Key Terms
conduction (kuhn-DUHK-shuhn): transfer of heat through matter by direct contact
convection (kuhn-VEK-shuhn): transfer of heat within a liquid or a gas

Conduction Heat moves through Earth's atmosphere in three main ways. These ways are conduction, radiation, and convection.

The troposphere, or lower layer of the atmosphere, is heated by a process that moves heat through matter. A metal pan placed over a flame will get hot. The metal molecules directly over the flame begin to move faster. They bump into the slower-moving molecules surrounding them and make them move faster. This is how heat moves through the metal pan.

Heat generally moves from an area of higher temperature to an area of lower temperature. This kind of movement of heat through matter is called **conduction.**

Sunlight absorbed by Earth's surface is changed into heat. This warms the surface. Air touches the warmed surface and is heated by conduction.

1. **NAME:** How does heat move through matter?

Radiation The atmosphere is also heated by the process of radiation. Radiant energy travels from the Sun through space in waves.

Most of the Sun's energy is short-wave radiation. This radiation passes easily through the atmosphere and strikes Earth's surface, where it is mostly absorbed and changed into heat energy. Heat energy is long-wave radiation. Earth's surface warms up and radiates most of the heat energy back into the atmosphere. There it is absorbed by gases. This warms the atmosphere.

2. **DESCRIBE:** What happens to the energy radiated by Earth?

Convection The transfer of heat within a liquid or a gas is called **convection.** When air is heated, it expands. As the warm air expands, it becomes lighter, because it becomes less dense. Warm air is lighter than cool air. Warm air rises. The cooler, denser air sinks.

3. **DEFINE:** What is convection?

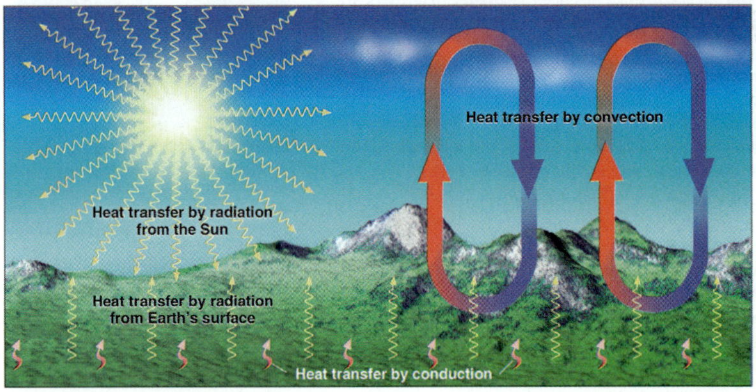

◀**Figure 11-10** There are three types of heat transfer: conduction, radiation, and convection.

Teaching Options

! FAST FACTS
Protecting Earth

If there were no atmosphere, Earth's temperature would soar by day and plummet at night, creating conditions too extreme to support life. This is because the atmosphere acts as a protective blanket, warming Earth's surface and helping to regulate the balance of radiation absorbed and reflected back into space.

COOPERATIVE LEARNING

Listing Facts Have each student make a set of five flash cards that list facts about the three methods of heat transfer. Then, have students work in pairs and take turns quizzing each other with the flash cards.
learning style: interpersonal

ONGOING ASSESSMENT

Have students make drawings showing the three different types of heat transfer. Students can use Figure 11-10 as a guide. Make sure that students include the appropriate labels and arrows on their drawings.

CHECKING CONCEPTS

1. When a solid is heated, its __________ move faster.
2. Solids are heated by __________.
3. Earth's radiant energy is __________-wave radiation.
4. Energy traveling in waves from the Sun is called __________ energy.
5. Heat moves through air by__________.

THINKING CRITICALLY

6. IDENTIFY: Which of the following is conduction, which is radiation, and which is convection?
   a. The water in a fish tank becomes warmer after the heater in it is turned on.
   b. The Sun warms your skin on a summer day.
   c. A glass bowl is warmed by the steaming rice it contains.
   d. A burning log in a fireplace causes the temperature in a room to go up.

INTERPRETING VISUALS

Use Figure 11-11 to answer the following questions.

7. EXPLAIN: Why do the balloons rise?
8. INFER: What does the flame do?

▲ **Figure 11-11** Hot-air balloons

HEALTH AND SAFETY TIP

Never grab the handle of a hot pot with your bare hands. Use a pot holder. Metal is a good conductor of heat. The pot holder is not. It keeps most of the heat from passing through to your hands.

Hands-On Activity

DETERMINING THE EFFECT OF TEMPERATURE ON AIR MOVEMENT

You will need scissors, a ruler, tissue paper, thread, cellophane tape, a desk lamp, and a partner.

1. Cut a 6-cm-long spiral from the tissue paper.
2. Cut a piece of thread 15 cm long.
3. Tape one end of the thread to the center of the paper spiral.
4. Turn on the desk lamp. Point the light up.
5. Have your partner hold the end of the thread. Position the paper spiral about 10 cm above the light. ⚠ CAUTION: Do not allow the paper to touch the light bulb.

▲ **STEP 5** Be sure the spiral does not touch the light bulb.

Practicing Your Skills

6. DESCRIBE: What happens to the spiral?
7. INFER: How do temperature differences affect what happens?
8. CONCLUDE: How is this an example of convection?

3 *Assess*

CHECKING CONCEPTS ANSWERS

1. molecules
2. conduction
3. long
4. radiant
5. convection

THINKING CRITICALLY ANSWER

6. IDENTIFY:
   a. convection
   b. radiation
   c. conduction
   d. radiation

INTERPRETING VISUALS ANSWERS

7. EXPLAIN: The air inside the balloon is heated. The hot air becomes less dense and rises, causing the balloon to rise.
8. INFER: It heats the air inside the balloon.

4 *Close*

HEALTH AND SAFETY TIP

Conduction can also pose a health threat when it carries too much heat out of the body. In a cold environment, a person loses body heat through conduction and may be at risk of hypothermia, a severe lowering of body temperature. The risk is greater in water than in air because water is 32 times better as a conductor of heat than air is. When outdoors in very cold weather, tell students to always wear warm clothing and be especially careful around open water.

Hands-On Activity

DETERMINING THE EFFECT OF TEMPERATURE ON AIR MOVEMENT

TIME: 10–15 minutes

PURPOSE: describing, inferring, concluding

DISCUSSION: Before beginning the activity, ask students for their predictions of how temperature will affect air movement. Discuss what they expect to see happening in this activity.

Practicing Your Skills Answers

6. DESCRIBE: It turns or twists.
7. INFER: The light bulb heats the air, which rises. The rising air causes the spiral to turn.
8. CONCLUDE: The heat from the light bulb moves the air, which is a gas.

MORE ON THE WEB

Writing About Radiometers
Have students use the Internet to learn how a radiometer works. Have students write a short report with their findings and include a labeled sketch of the apparatus. Have students visit www.conceptsandchallenges.com to begin their search.

• TEACHING RESOURCES • Online or Print

Teacher's Resources
Lesson Review: Chapter 11, p. 6

Laboratory Manual
Lab. Challenge: Which heats up faster—land or water? pp. 61–64

11-5 *(pp. 260–261)*

1 Introduce

INVESTIGATE

TIME: 10–15 minutes

PURPOSE: Students will observe the effects of air pressure.

MATERIALS: drinking straw, cup, water

PROCEDURE: Students should use clear plastic drinking straws.

THINK ABOUT IT: The water stayed in the straw. Air pressure on the bottom of the straw is greater than on top, so the water stays in the straw. Air pressure pushes downward and gravity pulls downward, forcing the water out. This shows that air exerts force or pressure in all directions on objects. When the pressure is released in one direction (in this case, the top of the straw because the finger covers the straw opening), the pressure in the other direction continues to exert force. In this case, the upward pressure exerts enough force on the bottom of the straw to hold the water in the straw.
learning styles: visual, tactile

STUDY HINT Have students write a summarizing sentence for each lesson section.

2 Teach

Demonstration Demonstrate that air has pressure. Place a piece of cardboard on top of a small cup filled with water. Hold the cardboard tight against the cup as you quickly turn the cup upside down. Ask students why the cardboard remains on the cup and the water remains in the cup. (The air pressure pushing up on the cardboard is greater than the air pressure pushing down on the water inside the cup.)
learning style: visual

Discussion To begin this lesson, ask: *What is air?* (a colorless, odorless, tasteless mixture of gases) Remind students that air is matter. Ask students to define matter. (anything that occupies space and has mass) Because of this mass, air has weight and exerts pressure.
learning styles: auditory, linguistic

Answers to questions

1 EXPLAIN: Air molecules are pulled toward Earth by gravity.

11-5 What is air pressure?

INVESTIGATE

Observing Air Pressure
HANDS-ON ACTIVITY

1. Place a drinking straw in a cup of water.
2. Put your finger over the top end of the straw.
3. Take the straw out of the water. Observe the water inside the straw.
4. Hold the straw over the cup and remove your finger. Observe what happens.

THINK ABOUT IT: What happened when you removed the straw from the water with your finger on top of it? What holds the water in the straw? Why does the water fall out when you take your finger away? How is this activity related to air pressure?

Objective
Explain air pressure and describe what affects it.

Key Terms
newton: metric unit of force
pressure: amount of force per unit of area

Weight and Pressure Weight is a force. If you hold a book in the palm of your hand, you feel the weight of the book pressing down. This force is measured in units called **newtons (N).** A 1-kg mass has a force of about 10 N.

The amount of force per unit of area is called **pressure.** When you hold the book in the palm of your hand, the book's weight is spread over your hand. Suppose the book's force is 10 N, and your hand has an area of 100 square cm, or 100 cm^2. The force on each square centimeter is then 10 N divided by 100 cm^2, or 0.1 N/cm^2. The pressure of the book on your hand is 0.1 N/cm^2.

$$\frac{\text{Force}}{\text{Area}} = \text{Pressure}$$

$$\frac{10\ \text{N}}{100\ \text{cm}^2} = 0.1\ \text{N/cm}^2$$

A force exerted over a small area causes more pressure than the same force applied over a large area. See Figure 11-12 for an example of this.

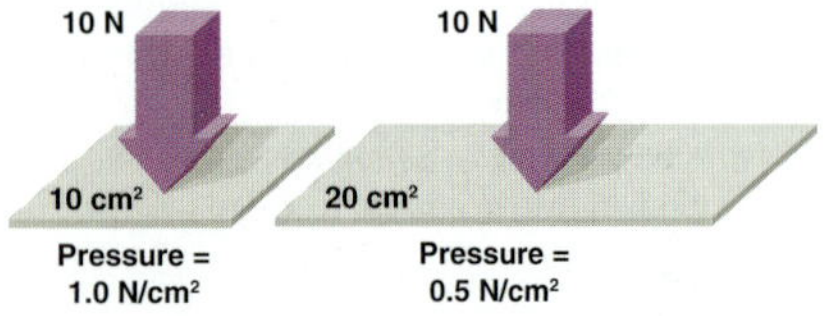

▲ **Figure 11-12** The same force exerted over a smaller area causes more pressure.

Air has weight. One liter of air weighs about 0.01 N at sea level. This is about the weight of a paper clip. The surface of Earth is at the bottom of the atmosphere. Air molecules are in constant motion and are pulled toward Earth's center by gravity. The force of all these moving molecules causes air pressure. Most of the air in the atmosphere is concentrated near Earth's surface. So air pressure is greatest near Earth's surface and decreases as altitude increases.

1 EXPLAIN: Why does air exert pressure on Earth's surface?

Elevation Air pressure changes with elevation, or height above sea level. The atmosphere is over 100 kilometers thick. The weight of all this air causes more pressure near the ground. This pushes the air molecules closer together. Near the top of the atmosphere, the air molecules remain farther apart. There is very little weight of air pressing down. Therefore, the air pressure is lower. The higher the elevation, the lower the air pressure.

Teaching Options

INTEGRATING LIFE SCIENCE

Breathing In and Out The mechanics of breathing are related to air pressure. When a person inhales, the volume of the chest cavity increases, and the air pressure within the chest cavity is lower than atmospheric air pressure. This causes air to rush into the lungs. When a person exhales, air pressure within the lungs is greater than atmospheric air pressure, so air is forced out of the lungs. Have students inhale and exhale into a paper bag to demonstrate this and note how the bag deflates when they inhale and inflates when they exhale. **learning style: visual**

ONGOING ASSESSMENT

Ask students to sketch on the board the relationship between air pressure and elevation. Also, have them draw how air pressure changes with temperature and humidity.
learning style: visual

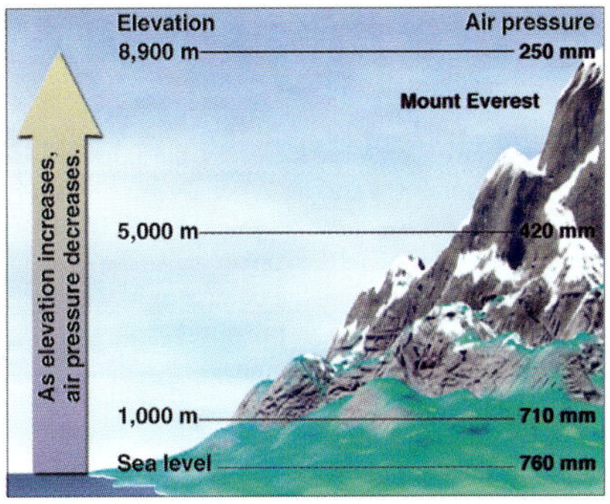

▲ **Figure 11-13** Elevation affects air pressure, which is measured in millimeters of mercury.

Air pressure decreases as distance above the surface increases. The air pressure on top of a mountain is less than the air pressure at sea level.

2 DESCRIBE: How does elevation affect air pressure?

Water Vapor The more water vapor in the air, the lower the air pressure. Water evaporates from lakes, rivers, and oceans. Living things give off water vapor. All of this water vapor goes into the air. The lighter molecules of water vapor replace some of the other gas molecules in air. Air with a lot of water vapor weighs less than dry air with less water vapor. Thus, moist air exerts less pressure. Air pressure goes down as the amount of water vapor in the air goes up.

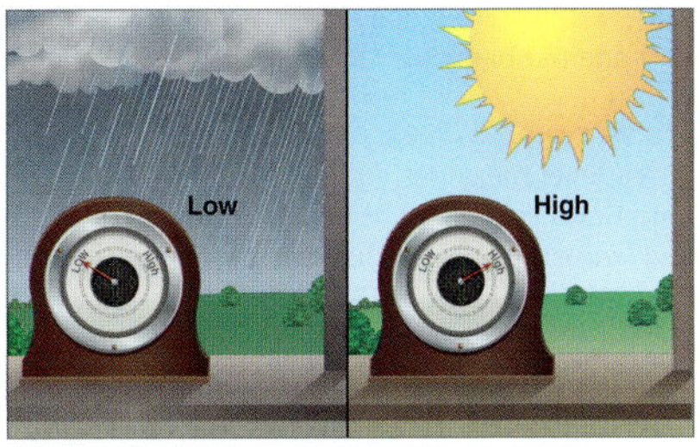

▲ **Figure 11-14** The weather conditions outside influence the air pressure.

3 DESCRIBE: How does water vapor affect air pressure?

Temperature Under ordinary conditions, the higher the temperature, the lower the air pressure. Heat makes air molecules move faster. As the molecules move faster, they spread apart. This makes the air less dense.

So warm air is less dense than cool air. In summer, when temperatures are higher, the air pressure is usually lower.

4 RELATE: How is temperature related to air pressure?

CHECKING CONCEPTS

1. Pressure is the amount of __________ on a unit of area.
2. Air pressure at sea level is __________ than air pressure on top of a mountain.
3. Air pressure __________ as elevation increases.
4. Warm air weighs __________ than cool air.

THINKING CRITICALLY

5. **CALCULATE:** A 5-N force pushes down on an area that is 10 cm^2. How much pressure does the force have?
6. **HYPOTHESIZE:** Why do ears "pop" in an airplane?

Web InfoSearch

The Magdeburg Hemispheres In 1654, Otto von Guericke, the mayor of a small German town called Magdeburg, did an experiment. He made a hollow metal sphere with two halves, or hemispheres, fitted tightly together. Air was pumped out of the sphere through a valve. This lowered the air pressure inside. The higher outside air pressure held the sphere together.

SEARCH: Use the Internet to find out more about this experiment. Why couldn't horses pull the two hemispheres apart? Start your search at www.conceptsandchallenges.com. Some key search words are **air pressure, von Guericke,** and **Magdeburg.**

2 DESCRIBE: Air pressure decreases with altitude.

3 DESCRIBE: Air pressure decreases as temperature increases.

4 RELATE: As water vapor in the air increases, air pressure decreases.

WRITING HINT Tell students that when writing formulas or equations, they should remember to use the first letter of the measurement called for in the equation. For example, use *F* with force and *P* with pressure.

3 Assess

CHECKING CONCEPTS ANSWERS

1. force
2. higher
3. decreases
4. less

THINKING CRITICALLY ANSWERS

5. **CALCULATE:** 0.5 N/cm^3
6. **HYPOTHESIZE:** Elevation affects air pressure. As elevation increases, air pressure decreases. When a person travels to a high elevation, as in an airplane, the inside of the person's ear has a higher pressure than the atmosphere. When the pressure equalizes inside and outside the ear, a popping sound is heard.

4 Close

Web InfoSearch

The Magdeburg Hemispheres In Guericke's experiment, a hollow metal sphere was formed with two halves, or hemispheres, fitted tightly together. Air was pumped out of the sphere through a valve. This lowered the air pressure inside. The higher outside air pressure held the sphere together. Ask students: *Why couldn't horses pull the two hemispheres apart?* Students can begin their search at www.conceptsandchallenges.com. Some key search words are **Von Guericke air pump, von Guericke horses,** and **Magdeburg hemisphere experiment.**

RETEACH

Modeling Air Molecules Use marbles, a glass tray, and an overhead projector to model how air pressure changes with altitude. Place the tray on the projector, then scatter a few marbles in the tray. Tell students that this illustrates air molecules high in the atmosphere. Add more marbles so that the tray is densely packed, then explain that this represents air molecules close to Earth's surface. Point out that these molecules exert a greater pressure than air molecules at higher altitudes.

learning style: visual

• TEACHING RESOURCES • Online or Print

Teacher's Resources

Lesson Review: Chapter 11, p. 7

11-6 (pp. 262–263)

1 Introduce

Linking Prior Knowledge Remind students that a thermometer is an instrument used to measure air temperature. Tell students that weather forecasters use a variety of instruments to measure atmospheric conditions. In this lesson, students will learn about barometers, which are used to measure air pressure.

2 Teach

Demonstration To show students that air pressure can hold up a column of liquid, do the following activity. Put a glass in a container of water. Turn the glass upside down. Slowly lift the glass out of the water. Have students describe what happens. (Water stays in the glass as long as the mouth of the glass is under the surface of the water.) Point out that air pressure pushes down on the water in the container. The pressure is transmitted upward into the glass and holds up the column of water. **learning style: visual**

Discussion Display an aneroid barometer. Use the diagram in Figure 11-16 to explain how the instrument works. Be sure that students understand that an aneroid barometer does not contain any liquid—mercury, in particular—but its scale still indicates air pressure in millimeters of mercury. **learning styles: visual, auditory**

Answers to questions

1. **DEFINE:** It measures air pressure.
2. **DESCRIBE:** The sides of the container bend in when air pressure increases and bend out when air pressure decreases. A pointer connected to the container records the changes in air pressure.
3. **ANALYZE:** 760 mm
4. **INFER:** so they know their altitude as they descend (and know when their parachutes should open)

11-6 How is air pressure measured?

Objective
Explain how a barometer measures air pressure.

Key Term
barometer (buh-RAHM-uht-uhr): instrument used to measure air pressure

Mercury Barometer Air pressure is measured with an instrument called a **barometer.** A mercury barometer is a glass tube filled with mercury. It is open at one end. The space at the closed end of the tube forms a vacuum. The open end of the tube sits in a container of mercury. Air pressure pushes down on the surface of the mercury in the container. The mercury is pushed up the vacuum. At sea level, air pressure can raise a column of mercury to a height of 760 mm. As the air pressure changes, the level of mercury in the tube rises or falls.

▲ **Figure 11-15** Near the ocean, a mercury barometer (on right) shows a reading of 760 mm.

1. **DEFINE:** What does a barometer do?

Aneroid Barometer Another kind of barometer is called an aneroid barometer. The word *aneroid* means "without liquid." An aneroid barometer is made of an airtight metal container. The sides of the container are very thin. They can bend in or out.

When the air pressure increases, the sides of the metal container bend in. When the air pressure decreases, the sides of the container bend out. A pointer is connected to the container. As the container changes shape, the pointer moves along a scale. The scale shows air pressure in millimeters of mercury. Some aneroid barometers, called barographs, keep a continuous record of air pressure.

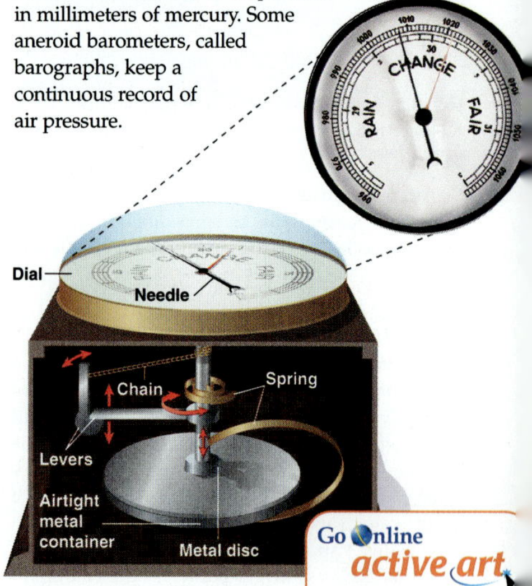

▲ **Figure 11-16** Aneroid barometer

Go Online active art
For: Measuring Air Pressure activity
Visit: PHSchool.com
Web Code: cfp-4012

2. **DESCRIBE:** How does an aneroid barometer work?

Measuring Air Pressure Standard air pressure at sea level measures 760 mm of mercury. This is sometimes called one atmosphere. Air pressure is also measured in millibars (mb). Standard air pressure is equal to 1,013.20 mb.

3. **ANALYZE:** How many millimeters of mercury equal 1,013.20 mb?

Measuring Altitude An altimeter is a device used to measure altitude. Pilots, scientists, surveyors, and mountain climbers all use altimeters.

At sea level, air pressure will raise a column of mercury 760 mm. As you go higher, air pressure decreases. The mercury column drops.

Teaching Options

EXTENDING THE LESSON

Finding Out More About Pascal Have interested students research and present brief oral reports about the other contributions to meteorology made by the French scientist Blaise Pascal. For example, Pascal was the first to hypothesize that pressure in a fluid is transmitted equally in all directions. (A fluid is any material that can flow and that takes the shape of its container. Liquids, and gases such as air, are fluids.) Pascal's discovery is known as Pascal's law.
learning style: linguistic

RETEACH

Aneroid Barometers vs. Mercury Barometers Remind students of the differences between an aneroid barometer and a mercury barometer. Then, have them write a brief paragraph in their notebooks comparing and contrasting the two barometers. Tell students to begin their paragraph with a topic sentence and to include sentences that tell how the barometers are alike and how they are different. Students can use a compare/contrast graphic organizer such as the Venn Diagram reproducible from the Teacher's Resources.

Mount Everest is about 8,900 m high. It is the highest point on Earth. The mercury column here is only about 250 mm high.

Skydivers and parachuters wear devices that contain altimeters. These let them see how high they are as they descend. This type of altimeter is actually an aneroid barometer.

▲ **Figure 11-17** Parachuters wear devices that contain altimeters.

4 **INFER:** Why do parachuters wear altimeters?

CHECKING CONCEPTS

1. Air pressure is measured with a __________.
2. Two kinds of barometers are mercury barometers and __________ barometers.
3. The mercury column in a mercury barometer rises or falls with changes in __________.
4. Air pressure is measured in __________.

THINKING CRITICALLY

5. **CALCULATE:** Mercury is 13.6 times more dense than water. If a container of water can hold 20 g of water, how much more massive would an equal volume of mercury be?
6. **CALCULATE:** The density of mercury is 13.6 g/cm^3. If air pressure at sea level supports a 760-mm column of mercury, how high a column of water can air pressure support at sea level?

BUILDING MATH SKILLS

Graphing At sea level, normal air pressure is 760 mm of mercury. It drops about 10 mm for every 123-m rise in elevation. First, calculate normal air pressure for the following places, which are above sea level: Denver, CO (1,600 m); Kansas City, MO (230 m); Mount Whitney, CA (4,400 m); Mount St. Helens, WA (2,950 m); Mount Washington, NH (1,916 m); the Empire State Building in New York City (380 m). Then, calculate air pressure in Death Valley, CA (85 m below sea level). Make a bar graph showing the air pressure for each location.

How Do They Know That?

MEASURING AIR PRESSURE

Evangelista Torricelli (eh-vahn-jeh-LEES-tah tawr-uh-CHEL-ee) (1608–1647) was an Italian scientist. During his lifetime, he made important improvements both in the microscope and in the telescope. His greatest accomplishment, however, was inventing the mercury barometer in 1643.

Torricelli filled a long glass tube with mercury. He placed his finger over one open end. (⚠ CAUTION: Mercury is poisonous to the touch. Do not attempt to repeat this experiment.) Then, he turned the tube upside down and placed it straight up in a container filled with mercury. When the mouth of the glass tube was under the surface of the mercury, Torricelli took his finger off the opening. The mercury in the tube dropped. It stopped at about 750 mm. At the top of the tube was a vacuum. Torricelli hypothesized that it was air pressure keeping the mercury in place.

Thinking Critically What was the air pressure in the city where Torricelli did his experiment?

▲ **Figure 11-18** Torricelli and his barometer

3 *Assess*

CHECKING CONCEPTS ANSWERS

1. barometer
2. aneroid
3. air pressure
4. millimeters or millibars

THINKING CRITICALLY ANSWERS

5. **CALCULATE:** 20 g × 13.6 = 272 g
6. **CALCULATE:** 760 mm × 13.6 = 10,336 mm

4 *Close*

BUILDING MATH SKILLS ANSWER

Graphing Students should first divide the elevation of each place by 123 m, then multiply by 10 mm. The resulting figure should be rounded, then either subtracted or added to 760 mm, depending on whether the location is above or below sea level.

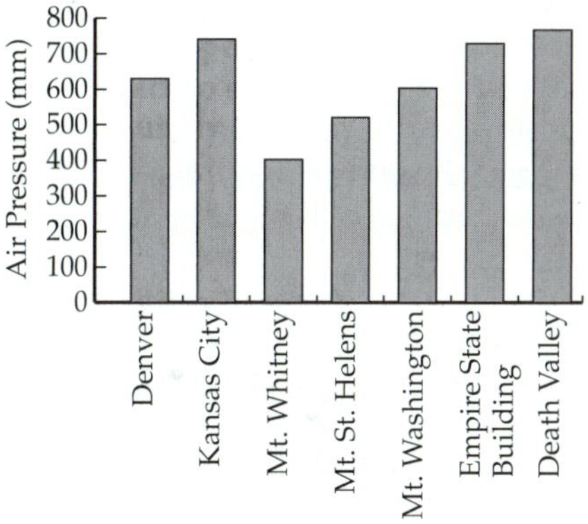

How Do They Know That?

MEASURING AIR PRESSURE

Physical-Science Connection French scientist Blaise Pascal built upon Torricelli's experiments with air pressure. Pascal fashioned an instrument much like Torricelli's and recorded air pressure first at a low-lying village and then at the top of a mountain. The mercury in the tube rose higher in the village than it did on the mountain. This experiment confirmed that air pressure varied with elevation.

Discussion Ask students: *Why did Pascal record the air pressure both in the village and on the mountaintop?* (to compare the readings to determine how air pressure varied with height)

Thinking Critically Answer
750 mm

QUICK CONVERSIONS

Air Pressure	Metric/SI	English
1,013.2 mb	760 mm of mercury	29.92 in. of mercury

Go Online active art

For: Measuring Air Pressure activity
Visit: PHSchool.com
Web Code: cfp-4012

Students can interact with the art of measuring air pressure online.

• TEACHING RESOURCES • Online or Print

Teacher's Resources

Lesson Review: Chapter 11, p. 8

Venn Diagram: Study Tools, p. 10

Lab Activity

(pp. 264–265)

1 Prepare the Lab

Constructing a Barometer

SET-UP TIME: LAB TIME:

PLUS **5 MINUTES EACH DAY FOR 2 WEEKS**

BACKGROUND: The weight of overlying layers of Earth's atmosphere constantly presses down on Earth's surface. Humans are accustomed to this pressure and are usually only affected by changes in air pressure. For example, at an altitude of 5,486 m, air exerts half the pressure it does at Earth's surface, and this change in altitude often causes people to experience light-headedness, shortness of breath, and overall weakness.

PURPOSE: Students will construct barometers and observe air pressure and related weather changes for a 2-week period. They will use their observations to forecast the weather.

ALTERNATIVE MATERIALS: Wooden bamboo skewers can be used in place of drinking straws.

SCIENCE SKILLS: Students will **construct** a barometer to **measure** and **record** changes in air pressure. They will **relate** these changes to weather conditions and use this information to **predict** the weather.

ADVANCE PLANNING: Students will need access to current weather reports for the duration of the lab. You can obtain these from a local newspaper or the Internet.

Place the barometers in a location where they will not be disturbed. If possible, obtain a commercial barometer so that students can use it to calibrate their own models. Also, display a thermometer in the classroom for comparison.

2 Guide the Lab

PROCEDURE: You can divide students into pairs or small groups. Tell students that they should calibrate their barometers by comparing their measurements to those recorded by a commercial barometer. Provide students with a copy of the Lab Activity Report found in the Teacher's Resources for students to record their observations and conclusions.

LAB ACTIVITY
Constructing a Barometer

Materials
Narrow-neck glass bottle, balloon, heavy rubber band, drinking straw, glue, 8-1/2-inch by 11-inch graph paper, tape, scissors, 8-1/2-inch by 11-inch cardboard square, colored pencils

BACKGROUND

Some days, the weather is clear and warm. Other times, it is stormy and cold. Weather forecasters try to predict changes in the weather. One of the tools they use is a barometer.

PURPOSE

In this activity, you will construct a barometer and use it to forecast the weather.

PROCEDURE

1. Tightly cover the mouth of the glass bottle with the rubber balloon. Cut the opening of the balloon as needed to fit over the mouth. Stretch the balloon so that it doesn't have any wrinkles. Use a rubber band to hold the balloon in place around the bottle's neck. ⚠ CAUTION: Be careful when using scissors.
2. Cut the tip of the drinking straw to make a pointer.
3. Put a small dab of glue on the other end of the straw and stick it to the balloon, as shown in photo. Hold the straw up until the glue sets.
4. Tape or glue the graph paper to the cardboard for support.
5. Using a pencil, darken one of the vertical lines in the center of the graph. This will become the barometer's scale.
6. When the glue holding the straw sets, begin calibrating the barometer. Put graph paper glued to the cardboard behind the barometer. Mark where the point of the straw is each day. You may want to mark the top of the barometer on graph paper also. You can use this as a reference point.

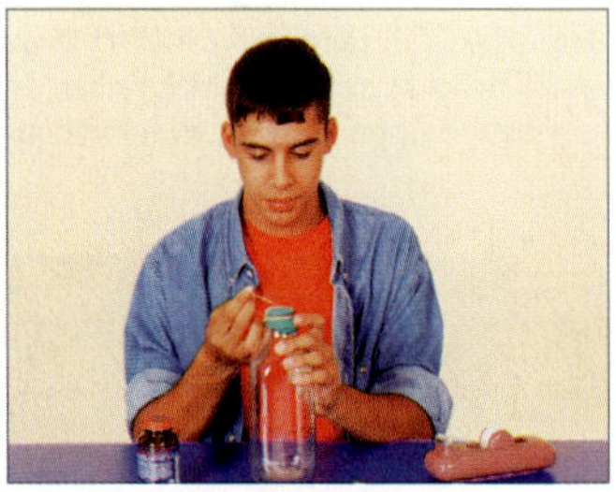

▲ **STEP 1** Hold the balloon in place with a rubber band.

▲ **STEP 3** Glue the straw to the balloon.

Teaching Options

! FAST FACTS *Pressure Tendencies*

When developing a forecast, a meteorologist closely examines any changes in air pressure that have occurred within the last 3 hours. These changes are referred to as *pressure tendency.* If air pressure changes less than 0.1 mb within a 3-hour period, it is considered steady. Changes within the 0.1 to 1.5 mb range are described as rising slowly or falling slowly. A change greater than 6.0 mb is described as rising or falling very rapidly.

EXTENDING THE LESSON

Adjusting Air-Pressure Readings Have interested students research how and why meteorologists adjust air-pressure readings at weather stations. (Meteorologists adjust the readings to account for fluctuations in temperature, latitude, and altitude among different weather stations. This allows readings from different stations to be compared directly.)
learning style: linguistic

7. Hold the graph paper behind the pointer end of the straw so that it is next to the line you drew. Mark the spot where the pointer is pointing. Ask your teacher to tell you what the actual pressure is. Write this next to the spot.

▲ **STEP 7** On your graph paper, mark where the straw is pointing.

8. Copy the chart in Figure 11-19. For the next two weeks, use your barometer to measure the air pressure each day. Mark where the barometer is pointing and the correct pressure. Use the chart to record what the weather outside is like each day.

9. Using your barometer, predict what the future weather will be like.

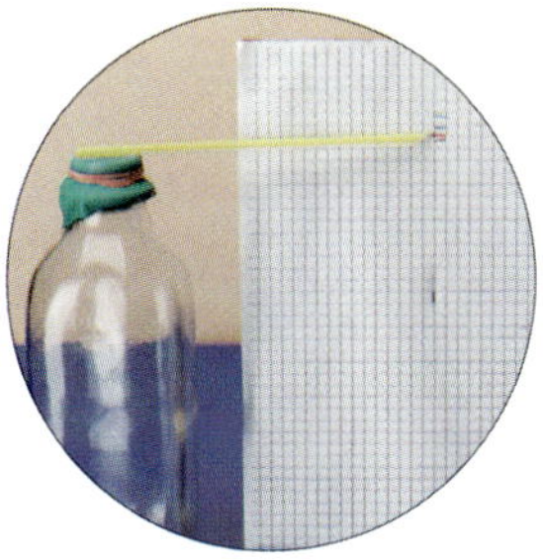

◄ **STEP 8** Using different colored pencils, mark where your barometer is pointing each day for two weeks.

Air Pressure and Weather

Date	Barometric Pressure	Weather for the Day

▲ **Figure 11-19** Use a copy of this chart to record your observations.

CONCLUSIONS

1. **OBSERVE:** What happened to the barometer pointer when the pressure became greater?
2. **OBSERVE:** What happened to the barometer pointer when the pressure decreased?
3. **ANALYZE:** What caused the pointer to move?
4. **INFER:** What do air pressure changes tell us about the coming weather?

TROUBLESHOOTING: Tell students to handle the bottles as little as possible to avoid warming the air inside. If your classroom has a thermostat, maintain as uniform a temperature as possible.

SAFETY TIP: Caution students to handle the glass bottles carefully and notify you immediately if a bottle breaks.

EXPECTED OUTCOME: Students will recognize the relationship between changes in air pressure and changes in the weather.

LAB HINT: If the barometers must be handled or moved, use mitts to grasp the bottles to prevent any transfer of heat.

3 Conclude the Lab

1. **OBSERVE:** It pointed up.
2. **OBSERVE:** It pointed down.
3. **ANALYZE:** The air inside the bottle was squeezed when air pressure increased, causing the rubber balloon to push into the bottle. This, in turn, caused the pointer to rise. The air inside the bottle expanded when air pressure decreased, and this pushed out the balloon, causing the pointer to fall.
4. **INFER:** Rising air pressure often indicates fair weather. Falling air pressure is most often associated with bad weather.

Use the Laboratory Activity Rubric from the Teacher's Resources to assess students' lab activities. Fill in the rubric with the additional information below. For this activity, students should have:

- constructed a barometer.
- used the barometer to predict the weather.

4 Extend the Lab

Have students compare their weather forecasts to actual weather conditions. Then, have students discuss the accuracy of their forecasts. If their forecasts varied from actual conditions, have them brainstorm reasons why. Lead students to understand that while air pressure is a good indicator of future weather, a forecast cannot be based entirely on one weather factor.

You can also have students graph the data they gathered during this investigation. (See Tips for Using Technology in the Teaching Options on this page.) Students can construct either a line graph or a bar graph. Have students look for patterns in the plotted data.
learning style: logical/mathematical

Tips for Using Technology

Using Word Processing Software to Make Graphs Have students use word processing software to graph the data they gathered during this investigation. A simple word processing program allows them to do basic tasks such as entering and editing text. However, a more complex word processing program allows students to do more. They can make tables, draw pictures, do math, and make graphs. Students can print out their graphs to share with the class.

• TEACHING RESOURCES • Online or Print

Teacher's Resources

Lab Activity Report: Study Tools, pp. 16–17

Laboratory Activity Rubric: Teaching Tools, p. 8

11-7 *(pp. 266–267)*

1 Introduce

Linking Prior Knowledge Ask students to recall what they learned about ocean currents (Lesson 10-4). Tell them that, like water, air moves when it is warmed or cooled. Remind students that convection is the movement of heat in a gas or liquid. Ask students what happens to air when it is warmed or cooled. (Warm air becomes less dense and rises. Cool air becomes denser and sinks.)

2 Teach

Demonstration If dry ice is available, use mitts and tongs to place a chunk of dry ice in a 200-mL beaker three-quarters full of water. Do not let the dry ice touch your skin. Place the beaker on a table. The dry ice will cool the surrounding air and cause the water vapor in the air to condense. Direct students' attention to the movement of the cloud of moisture over the rim of the beaker. Have students note how the mist moves down and outward over the surface of the table. Ask: *What causes the mist to move in this way?* (a convection current) Explain that similar horizontal movements of air over Earth are called winds.
learning styles: visual, auditory

Discussion Have students divide a notebook page into two columns. Have them write *Wind* at the top of the first column and *Air Currents* at the top of the second column. Ask students to use their charts to compare and contrast wind and air currents. Then, discuss students' charts. Make sure that students define *wind* as the horizontal movement of air along Earth's surface and understand that winds are caused by differences in air pressure.
learning style: auditory

Answers to questions

1 **PREDICT:** large

2 **DEFINE:** an up-and-down movement of air

READING STRATEGY After students have read the lesson, help them sharpen their critical thinking skills by asking: *How would you determine wind direction? What materials would you need?*

11-7 How do winds form?

Objective
Explain how winds form.

Key Terms
wind: horizontal movement of air
air current: up-and-down movement of air

Winds and Air Pressure Earth's atmosphere is nearly always in motion. Some air movements are weak and end quickly. Others are strong and last a long time.

Air is free to move in any direction. Its actual movements are determined by many factors. The horizontal movement of air along Earth's surface is called **wind.** Wind has often been described as "air in a hurry."

Air can also move vertically. Instead of being called wind, small vertical movements are usually called updrafts or downdrafts. Vertical movements are also important in the atmosphere. However, a much greater volume of air moves horizontally.

Winds form as cool, heavy air moves toward warm, light air. Cool air moves in under warm air. The cool air moves along the surface of Earth toward warmer air.

Winds are caused by differences in air pressure. Regions of cold, heavy air have high air pressure. These regions are called highs. Regions of warm, light air have low air pressure. These regions are called lows.

Air moves from regions of high pressure to regions of low pressure. Winds form when air moves. The speed of the wind depends on the differences in air pressure.

▲ **Figure 11-20** The winds that move these sailboards are produced by differences in air pressure.

1 **PREDICT:** Will the speed of a wind be greater if the difference in air pressure is large or small?

Air Currents Up-and-down movements of air are called **air currents.** Air currents are formed because the Sun does not heat all parts of Earth equally. Some areas of Earth are warmed more than other areas. As air over the warmer regions is heated, it expands and becomes less dense. As air over cooler regions is cooled, it becomes heavier, or denser. The cool air moves in under the warm air. It pushes the warm air upward. As the warm air mixes with the cool air, it becomes heavier and moves downward.

2 **DEFINE:** What is an air current?

▼ **Figure 11-21** How winds form

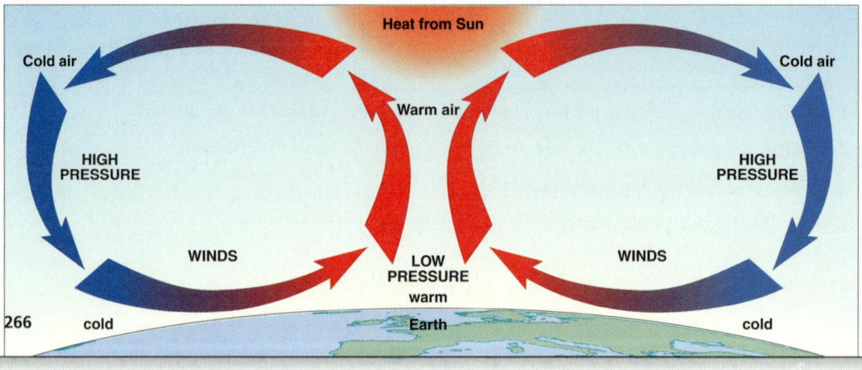

Teaching Options

! FAST FACTS *Wind Makers*

Winds result from three important forces acting together: pressure gradient forces, friction, and the Coriolis effect.

Pressure Gradient Forces The strength of winds is greater when high- and low-pressure areas are close together. The size of the pressure difference and the distance between the high- and low-pressure areas is called the *pressure gradient force.* The pressure gradient force between pressures of 1,023 mb and 1,000 mb over a distance of 150 km would be weak. However, the same pressure difference over a smaller distance, such as 30 km, would be stronger.

Friction As blowing winds rub against the surface of Earth, they slow down because of friction. Winds blowing over water encounter less friction than those blowing over land.

Coriolis Effect If Earth did not rotate on its axis, winds would blow straight from high-pressure areas to low-pressure areas. Earth's rotation causes winds to curve. In the Northern Hemisphere, the wind acts as if it is being pushed to the right. The reverse is true in the Southern Hemisphere. This curving motion is called the Coriolis effect. The faster the wind is moving, the stronger is the Coriolis effect.

CHECKING CONCEPTS

1. An air current is an __________ movement of air.
2. Cool air is more __________ than warm air.
3. As cool air moves in under warm air, it pushes the warm air __________.
4. Wind is the __________ movement of air along Earth's surface.
5. Wind speed depends on differences in __________.

THINKING CRITICALLY

6. INFER: The Sun does not heat the surface of Earth equally. What effect does this have on weather patterns?
7. CONTRAST: What are the differences between highs and lows?
8. EXPLAIN: Why does cold air push warm air upward?
9. INFER: **a.** Where on Earth would you expect to find the warmest and least dense air? Why? **b.** Where would you expect to find the coldest and densest air? Why?

Web InfoSearch

El Niño Changes in the temperatures of surface ocean water may be responsible for some global weather patterns. Such changes have been linked to El Niño events. El Niño is a warm ocean current that appears off the coast of Peru in December each year and may spread south. El Niño affects Earth's wind patterns and can cause extreme weather far from where it occurs.

SEARCH: Use the Internet to find out more about this. What happens during an El Niño event? Start your search at www.conceptsandchallenges.com. Some key search words are **El Niño, trade winds,** and **climate change.**

Hands-On Activity

OBSERVING AIR PRESSURE

You will need safety goggles, an oven mitt, a plastic foam cup, hot water, a funnel, and a small plastic bottle with a screw-on cap.

1. Your teacher will provide you with some hot water in a plastic foam cup. Carefully pour the water into the plastic bottle using a funnel.
   ⚠ CAUTION: Use an oven mitt to hold the bottle. Also, be careful when handling the cup of hot water.
2. Quickly pour the hot water out of the plastic bottle back into the plastic foam cup.
3. Immediately screw the cap on the plastic bottle.

▲ **STEP 2** Pour the hot water quickly into the cup.

Practicing Your Skills

4. OBSERVE: What happened to the plastic bottle after you screwed on the cap?
5. INFER: What can you say about the pressure of the warm air compared to the pressure of the cold air?
6. CONCLUDE: Based on your observations, should wind move from cold areas to warm or from warm areas to cold?
7. RELATE: How does this activity relate to the way winds form?

ONGOING ASSESSMENT

Provide students with copies of a map of the United States. Label the following cities and their corresponding air pressures on the map: Chicago, 770 mm; Boston, 750 mm; New York, 760 mm; Philadelphia, 765 mm. Have students determine the following: **a.** Will winds move from Chicago to New York or from New York to Chicago? (Chicago to New York); **b.** Will winds move from New York to Boston or from Boston to New York? (New York to Boston); **c.** Will winds move from Philadelphia to New York or from New York to Philadelphia? (Philadelphia to New York)

• TEACHING RESOURCES • Online or Print

Teacher's Resources

Lesson Review: Chapter 11, p. 9

Enrichment: Chapter 11, p. 13

3 *Assess*

CHECKING CONCEPTS ANSWERS

1. up-and-down
2. dense
3. upward
4. horizontal
5. air pressure

THINKING CRITICALLY ANSWERS

6. INFER: It causes air currents of different temperatures and pressures to form.
7. CONTRAST: Highs are cold, heavy air masses. Lows contain warm, light air.
8. EXPLAIN: Cold air is denser and moves in under lighter, warmer air.
9. INFER: **a.** the equator, because of the intensity of the sunlight there **b.** the poles, because there is less sunlight there

4 *Close*

Web InfoSearch

El Niño Have students use the Internet to find out what happens during an El Niño event. Ask: *How long does one last?* Some key search words are **El Niño, trade winds,** and **climate change.** Students can go to www.conceptsandchallenges.com.

Hands-On Activity

OBSERVING AIR PRESSURE

TIME: 10–15 minutes

PURPOSE: observing, inferring, concluding, relating

DISCUSSION: Ask students what happens to air when it is warmed or cooled. (Warmed air becomes lighter, rises, and exerts less air pressure. Cooled air becomes heavier, sinks, and exerts greater air pressure.) Ask students to predict what will happen to the plastic bottle when the air inside the bottle is warmer than the air outside the bottle. (The cooler, heavier air exerts greater pressure, causing the plastic bottle to collapse.)

Practicing Your Skills Answers

4. OBSERVE: Its sides bent inward.
5. INFER: The pressure of the warm air is lower.
6. CONCLUDE: from cold to warm areas
7. RELATE: It shows that winds form as cold air moves toward warm air.

The Big Idea

(pp. 268–269)

1 Introduce

Objective Students will learn how pressure causes global wind systems and where these wind systems are located. Students will also research and write about high-altitude jet streams.

Linking Prior Knowledge Ask students to recall what they learned about wind formation (Lesson 11-7). Have students relate differences in air pressure to the movement of global winds.

Have students review Lesson 11-8 and the discussion on local winds. Tell students that local winds can be thought of as small-scale versions of global wind patterns.

2 Teach

Discussion Point out that polar regions are areas of high pressure and warm equatorial regions are areas of low pressure. Remind students that winds move from areas of high pressure toward areas of low pressure. Describe how differences in air pressure at the poles and equator cause the vertically moving global winds. The doldrums and the horse latitudes are also vertical winds, not horizontal winds. Have students sketch a globe in their notebook and label the doldrums, horse latitudes, trade winds, prevailing westerlies, and polar easterlies on their drawings.
learning style: auditory

Demonstration Have students use a globe to demonstrate the movement of wind. Because of Earth's rotation, free-moving objects such as air and water are deflected to the right in the Northern Hemisphere and to the left in the Southern Hemisphere. Facing the globe, students can use their fingers to trace the movement of wind from the equator to the North Pole. Their fingers should trace a path that curves slightly to the right, or east. Then, have students stand above the globe and, looking directly down at the North Pole, trace the movement of wind from the pole to the equator. This time, their fingers should again curve slightly to the right, but in this case its direction is westerly.
learning styles: visual, tactile

Integrating Physical Science

THE Big IDEA

How do pressure changes create global winds?

We know that there is energy in winds. We use this energy to move a sailboat and turn a wind turbine. A cool breeze from the wind on a hot day is a pleasure. On a cold, windy day, we do not think it is a pleasure when we feel the wind's force on our bodies.

Because cool air is denser than warm air, it exerts greater pressure, or force, on a surface. That is why a pocket of cold air is called an area of high pressure. A pocket of less dense, warmer air is called an area of low pressure.

Winds blow from regions of high pressure to regions of low pressure. Warm air over the equator forms a region of low pressure. Bands of high pressure are found north and south of the equator. The cold air over the poles forms regions of high pressure. These differences in air pressure produce patterns of **global winds**. Global winds are large wind systems that circle Earth. At the equator, warm air rises and moves toward the poles. At the poles, cool air sinks and moves toward the equator.

Because Earth rotates, or spins like a top, winds curve as they move from high- to low-pressure regions. The winds moving toward the equator curve to the west in the Northern Hemisphere. The winds moving toward the poles curve to the east. In the Southern Hemisphere, the directions in which the wind curves are reversed. This is called the Coriolis effect.

Look at Figure 11-22 and the information on these two pages. Then, follow the directions in the Science Log to learn more about "the big idea."✦

Trade Winds
Some of the sinking cool air from the horse latitudes moves back toward the equator. It replaces the warm air rising in the doldrums. Earth's rotation causes this pocket of cool air to blow from east to west. These so-called trade winds are very reliable in terms of both speed and direction. Their name, trade winds, actually means "winds of commerce." European sailors and merchants in the sixteenth century realized that sailing with the trade winds was the quickest and safest route from Europe to America.

Doldrums
Warm air near the equator rises to form an area of low pressure. Because the region has little wind, sailors who traveled on tall ships named these winds the doldrums, which means "dull."

Prevailing Westerlies
Some of the sinking cool air from the horse latitudes moves toward Earth's poles. Because they are moving in a direction opposite to the trade winds, the prevailing westerlies are also called the antitrade winds.

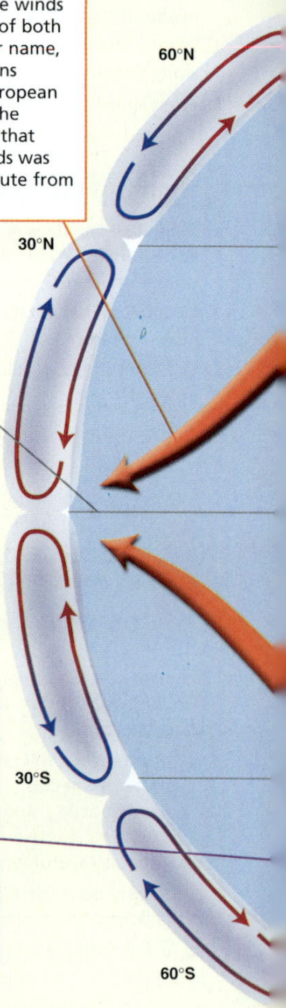

Figure 11-22 Global winds are large wind systems that travel around the world.

Teaching Options

! FAST FACTS Jet Streams

Jet streams are belts of high-speed air currents. Unlike surface winds, such as trade winds and prevailing westerlies, jet streams are high altitude winds located in the upper troposphere. Here are some additional jet stream facts:

- Their speeds sometimes reach 500 km/hr.
- There are two main belts of jets streams: the subtropical belt, located approximately between latitudes 20° and 30° North, and the polar jet stream, found approximately between latitudes 40° and 60° North. The position of both belts often varies.
- The polar jet stream in particular affects weather in the United States.

Go Online active art
For: Global Winds activity
Visit: PHSchool.com
Web Code: cfp-4023

Students can interact with the art of global winds online.

Polar Easterlies

Air surrounding Earth's poles is very cold and dense. These high pressure areas slowly move toward the equator. If they meet a mass of warm air moving upward from the equator, a weather condition known as a polar front forms.

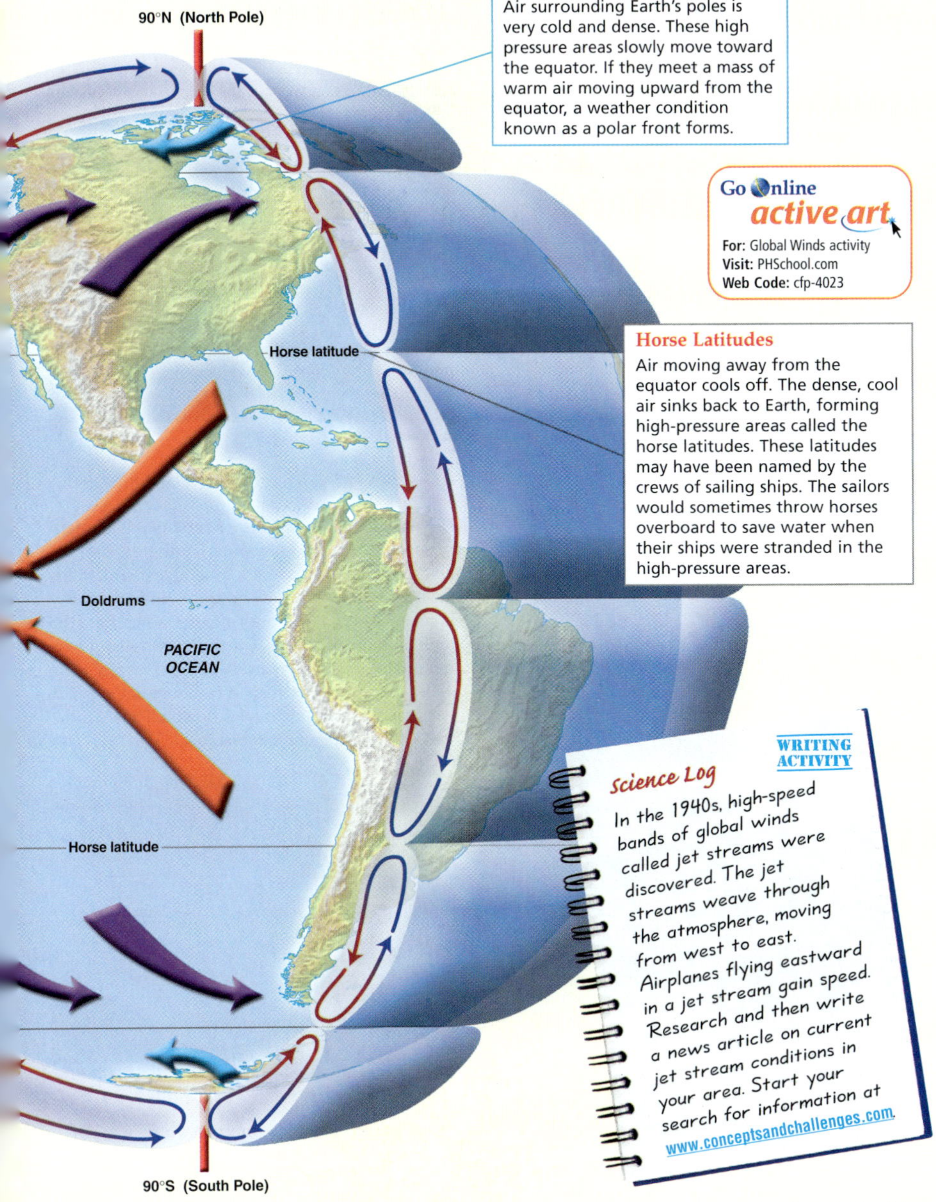

Go Online
active art
For: Global Winds activity
Visit: PHSchool.com
Web Code: cfp-4023

Horse Latitudes

Air moving away from the equator cools off. The dense, cool air sinks back to Earth, forming high-pressure areas called the horse latitudes. These latitudes may have been named by the crews of sailing ships. The sailors would sometimes throw horses overboard to save water when their ships were stranded in the high-pressure areas.

Science Log — WRITING ACTIVITY

In the 1940s, high-speed bands of global winds called jet streams were discovered. The jet streams weave through the atmosphere, moving from west to east. Airplanes flying eastward in a jet stream gain speed. Research and then write a news article on current jet stream conditions in your area. Start your search for information at www.conceptsandchallenges.com

Use the background information in Fast Facts to help guide students in conducting research for their Science Logs.

THE BIG IDEA

What winds travel around the world? Have students start their research for their Science Log Writing Activity at www.conceptsandchallenges.com. Be sure students use search engines and websites that are reliable and acceptable. You may also wish to provide copies of the *Big Idea Planner* and *Big Idea Science Log Entry* study tools found in the Teacher's Resources.

3 Assess

Use the Writing Activity Rubric from the Teacher's Resources to assess students' Science Logs. Fill in the rubric with the additional information below. For this assignment, students should have:

- recognized the relationship between air pressure and the formation of wind.
- identified how global wind systems form and where they are located.

4 Close

Illustrating Wind Patterns Divide students into small groups. Provide each group with posterboard and different-colored markers. Have students draw diagrams showing the movement of global winds. Tell students to use Figure 11-22 as a guide. Students should use a different-colored marker for each wind system and make a key that tells the color for each wind system. Display the posters around the classroom.

Tips for Using Technology

Searching the Internet for Information on Jet Streams
Remind students that one important tool for finding information on the Internet is a search engine. A search engine allows a user to type in a keyword or phrase. The search engine then searches thousands of sites on the Internet for those that contain the keyword or phrase. After a few seconds, the search engine lists many sites related to the topic.

• TEACHING RESOURCES • Online or Print

Teacher's Resources

Big Idea: Chapter 11, p. 10

Big Idea Planner: Study Tools, pp. 11–12

Big Idea Science Log Entry: Study Tools, p. 13

Writing Activity Rubric: Teaching Tools, p. 9

11-8 *(pp. 270–271)*

1 Introduce

STUDY HINT Have students study the diagrams in this lesson and describe what they show in relation to the lesson title and objective.

Linking Prior Knowledge Some students in your class have likely been to the ocean or to a large lake. Ask them if they noticed which way the wind was blowing during the day. Ask if the wind hit their faces or the backs of their heads when they faced the water. Tell students that usually the wind blows from the water to the land during the day. This is a local wind, which they will learn about in this lesson.

2 Teach

Demonstration Demonstrate the movement of warm air. Place a plastic pinwheel above the bare light bulb of a lamp. Turn the lamp on and have students observe what happens. Ask: *Why does the pinwheel turn?* (The air above the light was heated. The warm air then rose, causing the pinwheel to turn.) *Is this an example of conduction, convection, or radiation?* (convection) **learning style: visual**

Discussion Have students scan the wind energy feature in this lesson and make a list of other uses for wind. For example, wind is used to fly kites and sail boats. Compile responses in a list on the board.
learning styles: auditory, linguistic

Reinforcement To help students distinguish between land and sea breezes, tell them that winds are named for their source. Winds coming from the north are called northerly winds, and winds coming from the south are called southerly winds. A wind coming from the ocean is called a sea breeze, and a wind coming from the land is called a land breeze.
learning style: auditory

Answers to questions

1. **COMPARE:** land
2. **DESCRIBE:** from the valley to the mountaintop
3. **DEFINE:** winds that change direction with the seasons

11-8 What causes local winds?

Objective
Describe patterns of local winds.

Key Term
monsoon: wind that changes direction with the seasons

Sea and Land Breezes All winds are produced by temperature differences caused by unequal heating of Earth's surface. Local winds are simply small-scale winds produced by local changes in air pressure.

A breeze coming from the sea toward the land is a sea breeze. A breeze going from the land toward the sea is a land breeze. Land and sea breezes are local winds.

The Sun heats land faster than it does water. As a result, during the day, air over the land is warmer and lighter than air over the water. The cooler, heavier air over the ocean moves in toward the land. The warmer, lighter air over the land rises. The result is a sea breeze.

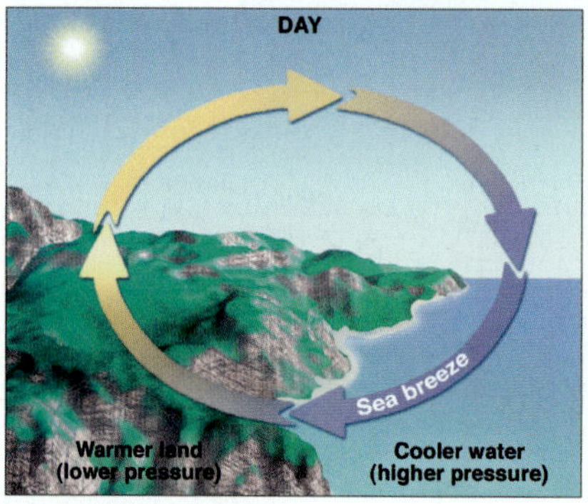

▲ **Figure 11-23** Sea breeze

At night, the land cools faster than the water. The air over the land becomes cooler than the air over the water. The heavier air over the land moves toward the water. The warmer, lighter air over the water rises. The result is a land breeze.

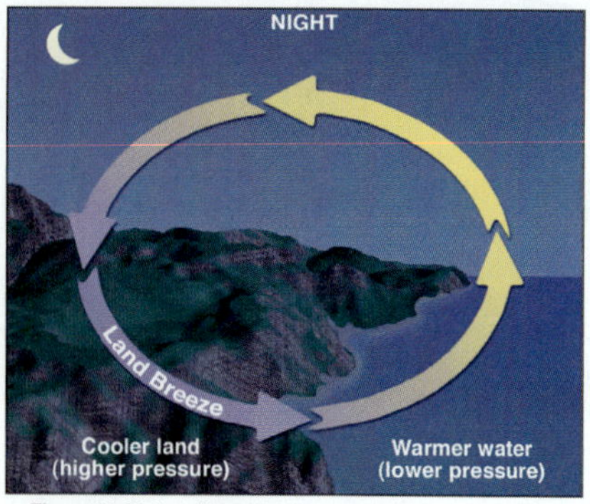

▲ **Figure 11-24** Land breeze

1 **COMPARE:** Which cools faster, land or water?

Mountain and Valley Breezes Mountain regions also have local winds. During the day, the air on a mountain slope is warmer than the air in the valleys. Warm air has low pressure. Air in the valley is cooler and has high pressure. Air moves from the high pressure of the valley to the low pressure of the mountain slope. This is a valley breeze. At night, the valleys are warmer than the mountaintops. The heavier mountain air moves downhill toward the valley. This is a mountain breeze.

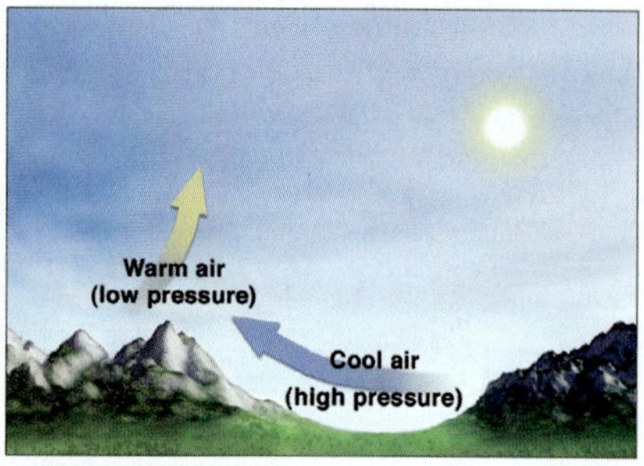

▲ **Figure 11-25** Valley breeze

2 **DESCRIBE:** In which direction do valley breezes move?

Teaching Options

COOPERATIVE LEARNING

Building a Kite Have students work in pairs or small groups to design and build kites. Tell students to use library references to gather ideas on kite construction. If possible, invite a kite-flying enthusiast to class and ask him or her to provide tips for students. Afterwards, you may want to arrange a field trip to a nearby open area to fly the kites.
learning styles: tactile, visual

RETEACH

Diagramming Local Breezes
Write *Local Winds* on the board. Write these subheads below: *Sea Breezes, Land Breezes, Mountain Breezes, Valley Breezes,* and *Monsoons.* Ask volunteers to draw arrows from the head to each subhead. Describe each local wind as its arrow is drawn. Have students copy the information in their notebooks.
learning styles: visual, auditory

Monsoons Parts of some continents have winds that change direction with the seasons. These winds are called **monsoons.**

In the summer, when the land is warmer than the water, winds move from the ocean toward the land. In the winter, when the land is colder than the water, the winds move from the land toward the ocean. Winds blow toward the ocean all winter. India is famous for its monsoons. The summer monsoon brings warm, moist air with heavy rains. The winter monsoon carries dry air. There is little rain in winter.

DEFINE: What are monsoons?

CHECKING CONCEPTS

1. A sea breeze blows toward the __________ from the ocean.
2. Land is heated __________ than water.
3. Movement of air from the land toward the ocean is called a __________ breeze.
4. A __________ breeze moves downhill toward a valley.
5. Winds that change direction with the seasons are __________.

THINKING CRITICALLY

6. **ANALYZE:** During part of the year, monsoons bring heavy rains and warm temperatures to many countries. Do these rainy seasons occur in the summer or winter? Explain.
7. **MODEL:** Draw and label a diagram of a mountain breeze. Use Figure 11-25 to help you.

BUILDING READING SKILLS

Vocabulary Mistral, foehn (FUHN), and chinook are the names of three local winds. Use a dictionary or other references to look up each of these winds. Write a brief description of each wind on a sheet of paper. Where does each wind occur?

Science and Technology

WIND ENERGY

For years, farmers have used windmills to pump water. Today, many communities are using the wind to produce electricity. Modern materials and engineering could combine to make wind energy an important source of power for the future. Wind energy is a promising source of power because the wind is free and no pollution is produced. However, the speed and direction of the winds are unpredictable.

In some places, large windmill farms provide power for electric generators. Windmill farms may include hundreds or even thousands of windmills. Over 6,000 wind generators located at Altamont Pass, east of San Francisco, California, are already in operation and producing electricity. One of the largest windmills was in Medicine Bow, Wyoming. This one windmill provided enough electricity for 1,200 homes. It was torn down in 2002.

Modern windmills do not look much like the windmills used to pump water on farms. Some of the newer windmills look more like airplane propellers or eggbeaters. Instead of wood, they are made of new, lightweight materials.

▲ **Figure 11-26** A large windmill farm near Palm Springs, California

Thinking Critically Why are fields with many windmills called wind farms?

• *TEACHING RESOURCES* •
Online or Print

Teacher's Resources
Lesson Review: Chapter 11, p. 11
Enrichment: Chapter 11, p. 13

3 *Assess*

CHECKING CONCEPTS ANSWERS

1. land
2. faster
3. land
4. mountain
5. monsoons

THINKING CRITICALLY ANSWERS

6. **ANALYZE:** They occur in the summer, when winds blowing from the ocean bring warm, moist air toward the land.
7. **MODEL:** Students' diagrams should show warm air in the valley and cool air on the mountaintop moving downhill.

4 *Close*

BUILDING READING SKILLS ANSWERS

Vocabulary *Mistral* winds are strong valley breezes formed by the Rhone Valley in the south of France. *Foehn* winds are mountain breezes first noticed in the Alps, and *chinook* winds are mountain breezes in the Rocky Mountains.

Science and Technology

WIND ENERGY

Real-Life Connection Wind is one of the oldest energy resources. In the Middle East, windmills were first used around A.D. 600. They spread to Europe by the 1100s. Today, large wind farms convert wind energy to mechanical energy and later to electrical energy. In the United States, California is the leading consumer of wind energy, with roughly 1 percent of its energy needs supplied by wind farms. However, experts predict that by 2020, wind energy will provide 10 percent of the world's energy needs.

Discussion Bring in a photograph of an old-fashioned windmill and have students compare it to the wind turbines in Figure 11-26. Discuss the uses of both old-fashioned windmills and modern wind turbines. Ask students how the windmills are different and how they are similar.

Thinking Critically Answer
Just as a farm may have many plants of the same type in one field, a wind farm has many wind turbines in one place.

11-9 *(pp. 272–273)*

1 Introduce

STUDY HINT Ask students to describe the difference between determining wind direction and measuring wind. Ask them to summarize in a sentence how weather balloons are useful for both activities.

Linking Prior Knowledge Some students may have a wind vane on their home or have seen wind vanes in illustrations. Ask them to describe the wind vanes. Explain how these instruments indicate wind direction.

2 Teach

Demonstration Make a wind vane, using the materials and directions given in the Hands-On Activity on page 273. Blow a fan at medium speed on the wind vane to demonstrate how the arrow points in the direction in which the wind is blowing.
learning style: visual

Discussion Display an anemometer. Ask students: *What does the anemometer measure?* (wind speed) Point out that the box under the cups generates a small amount of electricity. The anemometer is connected to a meter, which records the amount of electricity produced. The faster the cups turn, the greater the amount of electricity that is produced. By using an anemometer, an observer can determine wind speed.
learning styles: visual, auditory

Answers to questions

1. **NAME:** a northeast wind
2. **DESCRIBE:** Winds blow against cups in the anemometer, causing it to move. The faster and stronger the winds blow, the faster the anemometer moves. A meter measures the speed of the cups.
3. **IDENTIFY:** wind speed and direction

11-9 How is wind measured?

Objective

Explain how weather instruments are used to describe wind.

Key Terms

wind vane: instrument that indicates wind direction

anemometer (an-uh-MAHM-uht-uhr)**:** instrument that measures wind speed

Showing Wind Direction Both wind speed and wind direction are significant to weather watchers. A wind is named for the direction from which it comes. If a wind comes from the north, it is a north wind. If it comes from the east, it is an east wind.

The direction of a wind is determined with a wind vane. A **wind vane** shows the direction a wind is coming from. Many wind vanes are shaped like arrows. When the wind blows, the arrow turns and points into the wind. Often the direction of the wind is shown on a dial connected to the wind vane. The dial may indicate wind direction by points on a compass (E, W, N, S) or by degrees on a scale.

▲ **Figure 11-27** A wind vane indicates wind direction.

1 **NAME:** What is a wind that blows from the northeast named?

Measuring Wind Speed An **anemometer** is an instrument used to measure wind speed. It is often made of cups turned on their sides. These cups are attached to rods. Wind blowing against the cups causes the anemometer to turn. The faster and stronger the wind, the faster the anemometer turns. Anemometers usually have a meter attached to them that measures how fast the wind is blowing. It does this by measuring how fast the cups turn.

▲ **Figure 11-28** An anemometer measures wind speed.

2 **DESCRIBE:** How does an anemometer work?

Weather Balloons Scientists sometimes use weather balloons to determine wind speed and direction. A weather balloon is filled with the gas helium. Helium gas is lighter than air. As a result, balloons filled with helium gas rise into the air.

▲ **Figure 11-29** Weather balloons often carry instrument packages called radiosondes. These contain temperature, pressure, and humidity sensors.

272

Teaching Options

EXTENDING THE LESSON

High-Tech Instruments Have interested students research the latest methods of measuring wind speed and direction. For example, students can focus on lidar, or light radar systems, which use laser beams to measure wind speed. Sodar, or sound radar, relies on the Doppler effect to determine wind speed and direction. Encourage students to develop a visual presentation of their results.
learning styles: visual, linguistic

COOPERATIVE LEARNING

Measuring Wind Speed Set up an anemometer and a wind vane in an outside open area. Have small groups of students take turns recording wind speed and direction each day for one week. Assign each student in the group a task. One student can read the anemometer, another can read the wind vane, and a third can record the results. Keep an ongoing display of the measurements on a bulletin board. **learning styles: visual, tactile**

Winds high in the troposphere move a weather balloon along. Scientists can measure the speed of the wind by measuring the speed at which the balloon moves along. The direction it moves shows wind direction.

IDENTIFY: What do weather balloons measure?

CHECKING CONCEPTS

1. Winds are named based upon the __________ from which they come.
2. Wind direction is shown with a __________ .
3. Wind speed is measured with an __________.
4. Weather balloons are filled with __________.
5. Helium gas is __________ than air.

THINKING CRITICALLY

6. **INFER:** A weather vane points to the north. From which direction is the wind blowing?
7. **HYPOTHESIZE:** If carbon dioxide is heavier than air, can it be used in a weather balloon? Explain.
8. **INFER:** Why is it important for weather forecasters to know the speed and direction of the wind?
9. **INFER:** What happens to the cups on an anemometer when the wind gusts?

BUILDING SCIENCE SKILLS

Classifying Winds are classified by their speed or strength. They are also classified by the damage they do to homes and property. Place each wind below in order from calmest to strongest.

a. Hurricane: great damage is done to buildings
b. Strong breeze: hard to walk against the wind and open umbrellas
c. Calm: smoke goes straight up
d. Gale: branches are broken from trees; store windows break; TV antennas break
e. Moderate to fresh breeze: small trees sway; papers are blown around
f. Strong gale: trees are uprooted

Hands-On Activity
MAKING A WIND VANE

You will need cardboard, glue, a ballpoint pen cover, a compass, a pencil, and scissors.

1. Cut out two cardboard arrows about the same size. ⚠ CAUTION: Be careful when using scissors.
2. Clip a ballpoint pen cover in the middle between the two arrows. Then, glue the two arrows together.
3. Place the pen cover on the point of a pencil. Move the cover back and forth between the arrows until the arrow balances.
4. Go outside and hold up your vane where the wind will not be blocked by buildings or trees. Use a compass to see where the arrow points.

▲ **STEP 4** Hold your wind vane where the wind can blow it freely.

Practicing Your Skills

5. **INFER:** In what direction does your wind vane point?
6. **ANALYZE:** A strong wind would blow your cardboard vane away. What could you use to make a more sturdy wind vane?

MORE ON THE WEB

Sir Francis Beaufort Using the Internet, www.conceptsandchallenges.com, students can read about Sir Francis Beaufort and his contribution to the measurement of wind speed. Have students write a short biographical sketch about Beaufort's life and the development of the Beaufort Scale, which can be found in the Appendix of their texts.

• *TEACHING RESOURCES* •
Online or Print

Teacher's Resources

Lesson Review: Chapter 11, p. 12

Enrichment: Chapter 11, p. 13

ESL/ELL STRATEGY

Homophones Tell students that the word *vane* is a homophone for *vein* and *vain*. Explain the difference in the words' meanings.

3 Assess

CHECKING CONCEPTS ANSWERS

1. direction
2. wind vane
3. anemometer
4. helium gas
5. lighter

THINKING CRITICALLY ANSWERS

6. **INFER:** north
7. **HYPOTHESIZE:** no, because the balloon would not rise
8. **INFER:** Knowing wind speed and direction helps weather forecasters determine which distant weather systems will affect local areas and when.
9. **INFER:** The cups turn faster.

4 Close

BUILDING SCIENCE SKILLS ANSWER

Classifying c, e, b, d, f, a

Hands-On Activity
MAKING A WIND VANE

TIME: 15–30 minutes

PURPOSE: inferring, analyzing

COOPERATIVE LEARNING: You can have students work in pairs to construct the wind vane. Outside, one student can hold up the wind vane while the other uses the compass.

DISCUSSION: Ask students to stand. Then, have them turn and point out the four compass directions (north, east, south, west) and the intermediate compass directions (northeast, southeast, southwest, northwest). Discuss with students how a wind vane is similar to a compass. (A wind vane arrow points in a direction that provides the user with information, and the wind vane and compass share the same compass point names.) **learning style: kinesthetic**

Practicing Your Skills Answers

5. **INFER:** the direction from which the wind is blowing
6. **ANALYZE:** Possible answer: using more rigid materials, such as metal for the arrows and a metal rod in place of the pencil

Challenges (pp. 274–276)

Chapter Summary

Review Before students begin the Challenges, review the objectives of each lesson with them. Students can work in pairs. Each pair should come up with at least one relevant fact related to each objective.

STUDY HINT Have students use at least five of the Key Terms in a paragraph that summarizes the chapter.

Key Term Challenges

MATCHING

1. matter (11-1)
2. atmosphere (11-1)
3. cellular respiration (11-1)
4. thermosphere (11-2)
5. conduction (11-4)
6. monsoon (11-8)
7. troposphere (11-2)
8. wind (11-7)

IDENTIFYING WORD RELATIONSHIPS

9. Radiant energy is energy given off by the Sun. Radiation is the movement of the Sun's energy through space. (11-3)
10. Newton is a unit to measure force. Pressure is the amount of force per unit of area. (11-5)
11. Wind is the horizontal movement of air. An anemometer is used to measure wind speed. (11-7, 11-9)
12. Convection is the transfer of heat through a liquid or through a gas such as air. An air current is the up-and-down movement of air caused by convection. (11-4, 11-7)
13. Breezes and local winds are small-scale winds produced by local changes in air pressure. (11-8)
14. Wind is the horizontal movement of air. A wind vane is used to measure wind direction. (11-7, 11-9)
15. The stratosphere is the second layer of Earth's atmosphere. The troposphere is the layer nearest to Earth's surface. (11-2)

Chapter 11 Challenges

Chapter Summary

Lesson 11-1

- The **atmosphere** is an envelope of gases that surrounds Earth. About 78% of air is nitrogen, 21% is oxygen, and 0.04% is carbon dioxide.

Lesson 11-2

- The atmosphere is made up of four basic layers. The **troposphere** is the lowest layer. The **stratosphere** is the second layer. The ozone layer here protects us from harmful rays from the Sun. The third layer is the **mesosphere.** The fourth is the **thermosphere.** Within the mesosphere and thermosphere are concentrations of ions useful in radio communications.

Lessons 11-3 and 11-4

- The Sun gives off **radiant energy** that travels through space by **radiation.** When light energy is absorbed, it is changed into heat energy. Earth's surface changes light energy into heat energy.
- Heat moves through matter and the troposphere by **conduction.** It moves through the entire atmosphere by **convection.**

Lessons 11-5 and 11-6

- **Pressure** is the amount of force per unit of area. The weight of the air in the atmosphere exerts pressure on Earth's surface.
- Air pressure is affected by altitude, temperature, and water vapor.
- A **barometer** measures air pressure. Barometers can be mercury or aneroid.
- Air pressure is measured in millimeters of mercury or in millibars.

Lessons 11-7, 11-8, and 11-9

- **Air currents** are caused by the unequal heating of air. **Wind** is air moving horizontally along Earth's surface. Winds are caused by air pressure differences.
- Land breezes and sea breezes are local winds. Mountain regions have local winds called mountain breezes and valley breezes. A **monsoon** is a wind that changes direction with the seasons.
- A **wind vane** measures wind direction. An **anemometer** measures wind speed. Weather balloons measure wind speed and direction.

Key Term Challenges

air current (p. 266)
anemometer (p. 272)
atmosphere (p. 252)
barometer (p. 262)
cellular respiration (p. 252)
conduction (p. 258)
convection (p. 258)
matter (p. 252)
mesosphere (p. 254)
monsoon (p. 270)
newton (p. 260)
pressure (p. 260)
radiant energy (p. 256)
radiation (p. 256)
stratosphere (p. 254)
thermosphere (p. 254)
troposphere (p. 254)
wind (p. 266)
wind vane (p. 272)

MATCHING **Write the Key Term from above that best matches each description.**

1. anything with mass and volume
2. envelope of gases that surrounds Earth
3. process by which a cell releases energy from food molecules
4. upper layer of the atmosphere
5. movement of heat through matter
6. wind that changes direction with the seasons
7. lowest layer of the atmosphere
8. horizontal movement of air

IDENTIFYING WORD RELATIONSHIPS **Explain how the words in each pair are related. Write your answers in complete sentences.**

9. radiant energy, radiation
10. newton, pressure
11. wind, anemometer
12. convection, air current
13. breeze, local wind
14. wind, wind vane
15. stratosphere, troposphere

Teaching Options

PREPARING STUDENTS FOR STANDARDIZED TESTS

Reading Strategy: Tell students to quickly eliminate obviously wrong answers in multiple-choice questions. This will help them narrow their choices.

Writing Strategy: Tell students to have several sharpened pencils on hand when taking a test and to make sure that the pencils have erasers. Students should carefully reread their written responses and make any necessary corrections.

Interpreting Visuals Strategy: Tell students to read all of the questions about a visual before studying it. Knowing the questions can help students focus on the parts of the visual that contain the answers to the questions.

ESL/ELL STRATEGY

Rereading Questions Remind English-language learners to read multiple-choice questions carefully to make sure that they understand what is being asked. Students should reread the questions at least once to make certain that they have read the question correctly.

Content Challenges TEST PREP

MULTIPLE CHOICE **Write the letter of the term or phrase that best completes each statement.**

1. An altimeter measures
   - a. wind speed.
   - b. wind direction.
   - c. altitude.
   - d. rainfall.
2. Seventy-eight percent of the air is made up of
   - a. nitrogen.
   - b. oxygen.
   - c. helium.
   - d. carbon dioxide.
3. The layer of atmosphere closest to Earth is
   - a. the tropopause.
   - b. the stratosphere.
   - c. the ionosphere.
   - d. the troposphere.
4. The region of the atmosphere that reflects radio signals is the
   - a. tropopause.
   - b. stratosphere.
   - c. ionosphere.
   - d. troposphere.
5. Heat travels through empty space by
   - a. convection.
   - b. conduction.
   - c. evaporation.
   - d. radiation.
6. Earth's surface is heated by
   - a. ozone.
   - b. conduction.
   - c. ions.
   - d. radiation.
7. Air pressure is measured with
   - a. a barometer.
   - b. an altimeter.
   - c. a wind vane.
   - d. an anemometer.
8. Regions of cold, heavy air are called
   - a. highs.
   - b. lows.
   - c. convections.
   - d. monsoons.
9. A wind that blows from the southwest is a
   - a. southwest wind.
   - b. monsoon.
   - c. local wind.
   - d. northeast wind.
10. Sea breezes and land breezes are two kinds of
    - a. local winds.
    - b. monsoons.
    - c. jet streams.
    - d. global winds.

FILL IN **Write the term or phrase that best completes each sentence.**

11. Heat moves through __________ by conduction.
12. Heat moves through liquids and __________ by convection.
13. During respiration, living things give off __________ and water vapor as by-products.
14. Air is a __________ of gases.
15. Oxygen, argon, and __________ are the three main gases in air.
16. Beyond the atmosphere is the region known as the __________.
17. The ozone layer of the stratosphere protects Earth from __________ light from the Sun.
18. Ozone is a form of __________.
19. When air is heated, it __________.
20. Cool air is heavier and more __________ than warm air.

Content Challenges

MULTIPLE CHOICE

1. c (11-6)
2. a (11-1)
3. a (11-2)
4. c (11-2)
5. d (11-3)
6. b (11-4)
7. a (11-6)
8. a (11-7)
9. a (11-9)
10. a (11-8)

FILL IN

11. solids (11-4)
12. gases (11-4)
13. carbon dioxide (11-1)
14. mixture (11-1)
15. nitrogen (11-1)
16. exosphere (11-2)
17. ultraviolet (11-2)
18. oxygen (11-2)
19. expands (11-4)
20. dense (11-7)

ALTERNATIVE ASSESSMENT

Illustrating Local Winds Have students use a computer graphics program to create diagrams that illustrate local winds, such as sea breezes, land breezes, mountain breezes, and valley breezes. Print out students' diagrams and display the them around the classroom.

PORTFOLIO ASSESSMENT

Making Student Portfolios Portfolio Assessment is designed to evaluate a student's performance over an extended period of time. It also provides an opportunity for students who are generally not good test-takers to demonstrate their learning in another way. Encourage students to choose something to contribute. The Lesson Reviews, the Big Idea Planner, and the Chapter Self-Check are some of the reproducibles in the Teacher's Resources that they can include on their portfolios. You can use the Portfolio Assessment Rubric also found in the Teacher's Resources to assess students' portfolios.

Concept Challenges

WRITTEN RESPONSE

1. **ANALYZE:** Air pressure decreases because the air molecules are farther apart. (11-5)
2. **EXPLAIN:** It points in the direction from which the wind is blowing. (11-9)
3. **PREDICT:** Air pressure would be greater in the winter because as temperature decreases, air pressure increases. (11-5)
4. **EXPLAIN:** Mercury is denser than water. A barometer made with water would have to be much taller than a barometer made with mercury. (11-6)
5. **INFER:** A mountain breeze is named for the heavy, cool air that moves from a mountaintop into a valley at night. A valley breeze is named for the cool air that moves from a valley up a mountain during the day. (11-8)
6. **CONTRAST:** A wind vane indicates wind direction. An anemometer measures wind speed. (11-9)
7. **HYPOTHESIZE:** Dark-colored clothing would be warmer in winter because dark colors absorb more light and heat energy. (11-3)
8. **INFER:** Most wind vanes are shaped like arrows to better show the direction the wind is blowing by pointing into the wind and toward the compass direction for which the wind is named. (11-9)

INTERPRETING VISUALS (11-2)

9. the layers of the atmosphere
10. the thermosphere
11. troposphere, stratosphere, mesosphere, and thermosphere
12. the thermosphere
13. to 18 km
14. in the upper stratosphere
15. the troposphere
16. ions

Chapter Project Wrap-Up

LAYERS OF THE ATMOSPHERE POSTER

Display the completed poster in the classroom or common school area. Use the Individual Activity Rubric from the Teacher's Resources to assess students' projects. Fill in the rubric with the additional information below. For this project, students should have:

- contributed at least two pictures to the display.
- written and presented a summary that reflects information about one of the layers.

Concept Challenges TEST PREP

WRITTEN RESPONSE **Answer each of the following questions in complete sentences.**

1. **ANALYZE:** As elevation increases, what happens to air pressure? Explain why it does this.
2. **EXPLAIN:** How does a wind vane indicate wind direction?
3. **PREDICT:** Would air pressure on top of a mountain be greater in the summer or in the winter? Explain.
4. **EXPLAIN:** Why do you think mercury is used in barometers instead of water?
5. **INFER:** How do mountain and valley breezes get their names?
6. **CONTRAST:** How is a wind vane different from an anemometer?
7. **HYPOTHESIZE:** Why would dark-colored clothing be warmer in the winter than light-colored clothing?
8. **INFER:** Why are most wind vanes shaped like arrows?

INTERPRETING VISUALS **Use Figure 11-30 to answer the following questions.**

9. What is shown in the diagram?
10. In what layer of the atmosphere do auroras occur?
11. What are the names of the layers of the atmosphere?
12. What is the name of the upper part of the atmosphere?
13. How far does the troposphere extend?
14. Where is the ozone layer located?
15. What is the name of the layer of the atmosphere in which most weather takes place?
16. What kinds of particles make up most of the upper atmosphere?

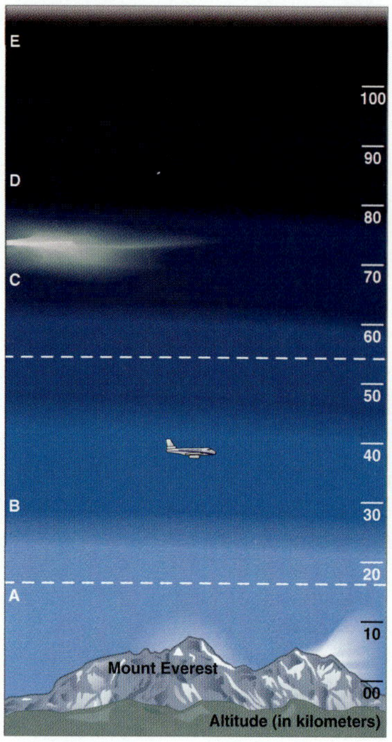

▲ Figure 11-30

Teaching Options

• TEACHING RESOURCES •
Online or Print

Teacher's Resources

Key Term Review: Chapter 11, p. 14

Chapter 11 Test: pp. 15–17

Individual Activity Rubric: Teaching Tools, p. 5

Chapter Self-Check: Teaching Tools, p. 11

Big Idea Planner: Study Tools, pp. 11–12

Portfolio Assessment Rubric: Teaching Tools, p. 10

Chapter 12 The Weather

PLANNING GUIDE

◆ **TEACHING THE CHAPTER** This chapter should take approximately 11–13 days to complete instruction and assessment.

	Skills and Features	Projects and Activities	Achieve Science Literacy	Meet Individual Needs
Chapter 12 Opener p. T277	PREDICT	• Chapter Project		
Lesson 12-1 **Where does water in the atmosphere come from?** pp. T278–T279 Standards: D1f, D1h, D1i, D1k	IDENTIFY, NAME, HYPOTHESIZE • Designing an Experiment • Hands-On Activity	• Extending the Lesson • More on the Web • Integrating Life Science • Lab Challenge	• Study Hint • Reading Strategy	Teacher's Resources* Designing an Experiment Enrichment Activity
Lesson 12-2 **What is humidity?** pp. T280–T281 Standards: D1h, D2i	EXPERIMENT, DEFINE, STATE • Building Science Skills • People in Science	• Cooperative Learning • Extending the Lesson • More on the Web	• Study Hint	Teacher's Resources* Enrichment Activity
Lesson 12-3 **How is relative humidity measured?** pp. T282–T283 Standard: D1i	CALCULATE, DEFINE, INFER • Hands-On Activity		• Study Hint • Reading Strategy • Ongoing Assessment	• ESL/ELL Strategy • Reteach Teacher's Resources* Enrichment Activity
Lesson 12-4 **What is the dew point?** pp. T284–T285 Standard: D1i	HYPOTHESIZE, APPLY, CONTRAST • Web InfoSearch • Hands-On Activity	• Integrating Physical Science	• Study Hint	• ESL/ELL Strategy Teacher's Resources* Enrichment Activity
Lesson 12-5 **How do clouds form?** pp. T286–T287 Standards: D1f, D1i	DESCRIBE, NAME, LIST • Web InfoSearch • Science and Technology	• Cooperative Learning • Extending the Lesson	• Study Hint • Writing Hint • Ongoing Assessment	Teacher's Resources* Enrichment Activity
Lab Activity **Making Cloud Models** pp. T288–T289 Standard: D1i	OBSERVE, INFER, ANALYZE • Set-Up Time: 15–20 min • Lab Time: 15–20 min	• Extend the Lab Activity	• Tips for Using Technology	• ESL/ELL Strategy • Reteach Teacher's Resources* Lab Activity Report Database Planner

*www.conceptsandchallenges.com, password: ccearth09

Planning Guide continues on next page.

Standards: For details on the correlation to National Science Standards see pages *xx–xxi*.

	Skills and Features	Projects and Activities	Achieve Science Literacy	Meet Individual Needs
Lesson 12-6 **What is precipitation?** pp. T290–T291 Standards: D1f, D1i	CONTRAST, CLASSIFY, DEFINE • Web InfoSearch • Hands-On Activity	• Extending the Lesson • Integrating Physical Science	• Study Hint • Ongoing Assessment	• ESL/ELL Strategy Teacher's Resources* Enrichment Activity
Lesson 12-7 **What are air masses?** pp. T292–T293 Standard: D1j	INFER, NAME, APPLY • How Do They Know That?	• Cooperative Learning • Extending the Lesson • More on the Web	• Study Hint • Reading Strategy	• Reteach Teacher's Resources* Enrichment Activity
Lesson 12-8 **What is a front?** pp. T294–T295 Standard: D1j	DEFINE, DESCRIBE, EXPLAIN • Web InfoSearch	• Cooperative Learning • Active Art	• Study Hint • Writing Hint • Ongoing Assessment	• ESL/ELL Strategy Teacher's Resources* Enrichment Activity Concept Map
Lesson 12-9 **What causes severe storms?** pp. T296–T297 Standards: D1f, D1i	DESCRIBE, INFER, CLASSIFY • Investigate • Web InfoSearch • Real-Life Science	• Cooperative Learning • Extending the Lesson • Integrating Physical Science	• Writing Hint	• ESL/ELL Strategy Teacher's Resources* Enrichment Activity
Big Idea **How do electrical charges cause lightning?** pp. T298–T299 Standard: D1i	RESEARCH, COMMUNICATE, ANALYZE • Science Log: Writing Activity	• Big Idea • Close Activity	• Tips for Using Technology	Teacher's Resources* Big Idea Planner
Lesson 12-10 **What is a station model?** pp. T300–T301 Standard: D1j	MODEL, INTERPRET, ANALYZE • Designing an Experiment • Hands-On Activity	• Extending the Lesson • More on the Web	• Reading Strategy • Ongoing Assessment	• ESL/ELL Strategy Teacher's Resources* Enrichment Activity Designing an Experiment
Lesson 12-11 **How do you read a weather map?** pp. T302–T303 Standard: D1j	RESEARCH, DEFINE, OBSERVE • Building Science Skills • Real-Life Science	• Cooperative Learning	• Ongoing Assessment	Teacher's Resources* Enrichment Activity
Chapter 12 Challenges pp. T304–T306	• Chapter Summary • Key Term Challenges • Content Challenges • Concept Challenges	• Chapter Project Wrap-Up	• Study Hint • Preparing Students for Standardized Tests	• ESL/ELL Strategy Teacher's Resources* Chapter Self-Check

*www.conceptsandchallenges.com, password: ccearth09

Chapter 12 Weather

▲ **Figure 12-1** Severe thunderstorms produce lightning and sometimes even tornadoes.

Every day, about 40,000 thunderstorms occur around the world. Most of these develop in the regions near the equator. The power of a thunderstorm can be enormous. Bright bolts of lightning flash across the sky. This is followed by the ear-splitting roar of thunder. Many thunderstorms bring heavy rains or strong winds. Sometimes there is even hail or tornadoes.

▶What evidence of a storm is there in this picture?

Contents

Teaching Options

ASSESSMENT PLANNER

For assessment in the Student Edition, see the following pages:

Content Assessment
Checking Concepts: pp. 279, 281, 283, 285, 287, 291, 293, 295, 297, 301, 303
Thinking Critically: pp. 279, 281, 283, 285, 287, 291, 293, 295, 297, 301, 303
Interpreting Visuals: pp. 283, 293, 301
Chapter Challenges: pp. 304–306

Alternative Assessment
Building Skills: pp. 281, 303
Web InfoSearch: pp. 285, 287, 291, 295, 297
Science Log: p. 299

Performance-Based Assessment
Hands-On Activity: pp. 279, 283, 285, 291, 301
Designing an Experiment: pp. 279, 301
Lab Activity: pp. 288–289

• TEACHING RESOURCES • Online or Print

Teacher's Resources

Lesson Review: Ch. 12, pp. 2, 3, 4, 5, 6, 8, 9, 10, 11, 14, 15

Enrichment: Ch. 12, pp. 7, 12, 16

Key Term Review: Ch. 12, p. 17

Chapter 12 Test: pp. 18–20

Laboratory Manual: pp. 65–68

Chapter 12 Weather

Chapter Overview

In this chapter, students will learn about the factors that affect weather, such as specific and relative humidity and dew point. Students will also learn about clouds and precipitation, as well as storm systems caused by the movements of air masses and fronts. Lastly, students will examine station models to interpret weather maps.

About Figure 12-1 You can estimate the distance to a thunderstorm by measuring the time between a lightning flash and the thunder it produces. It takes approximately 3 seconds for thunder to travel 1 km, or 5 seconds to travel 1 mi. Therefore, if you hear thunder 12 seconds after seeing lightning, you know the storm is about 4 km (2.5 mi) away. As a storm gets nearer, the time lag between lightning and thunder decreases. You know that a storm is directly overhead if you see lightning and hear thunder at the same time.

Answer Possible answer: lightning bolts and a dark cloud

Linking Prior Knowledge

For this chapter, students should recall:

- the processes of the water cycle (Lesson 9-1).
- how heat moves through the atmosphere (Lesson 11-4).
- the relationship between air pressure and wind (Lesson 11-7).

Chapter Project

PREPARING A WEATHER REPORT

MATERIALS: daily newspapers or Internet access

Have students collect U.S. weather maps over five consecutive days from a local newspaper or the Internet. Have them write a report on the weather around the nation over that period. They should also locate their own community on the maps and report the local weather conditions. Ask students to predict what the weather will be like on the sixth day. Have them check their predictions against a weather map published on the sixth day.
learning style: visual

12-1 (pp. 278–279)

1 *Introduce*

STUDY HINT Have students write a few sentences that tie together all the photographs shown in this lesson.

Linking Prior Knowledge Have students recall the processes involved in the water cycle, such as evaporation, precipitation, and condensation. Tell students that each of these processes plays a role in weather.

2 *Teach*

Demonstration Show students the relationship between heat and molecular motion. Put a glass of very hot water and a glass of very cold water side by side. Put a drop of food coloring into each glass. Tell students to observe the food coloring in each glass and note in which glass the food coloring spreads faster. (the glass containing hot water)
learning style: visual

Discussion Ask students if they know what happens to the water in hair or in clothes when hair dryers and clothes dryers are used. Discuss students' responses. Explain that the dryers heat the water molecules, which then change from a liquid to a gas. Define evaporation as the changing of a liquid to a gas.
learning styles: visual, auditory

Reinforcement Place a flat, transparent dish on an overhead projector. Put a few drops of water in the dish. Ask students to write a prediction of what will happen to the water. Turn the projector on for 5 minutes. Ask students to write their observation and an explanation of what happened to the water. (The water will have evaporated because of the heat from the projector.) **learning style: visual**

Answers to questions

1. **NAME:** water vapor
2. **IDENTIFY:** the oceans
3. **DESCRIBE:** heat
4. **NAME:** through transpiration

12-1 Where does water in the atmosphere come from?

Objective
Explain how water gets into the atmosphere.

Key Terms
evaporation (ee-vap-uh-RAY-shuhn): changing of a liquid to a gas
transpiration (tran-spuh-RAY-shuhn): process by which plants give off water vapor into the air

Water Vapor On a warm day, droplets of water may form on the outside of a window. The window surface is cooler than the outside air. The water comes from the air. Water in air is in the form of a gas called water vapor. Water vapor is an odorless, colorless gas. It mixes freely with other gases in the atmosphere. Most other gases change state only under extreme conditions. Water changes state at ordinary temperatures and pressures. It is because of these changes of state at ordinary temperatures that water leaves the oceans as a gas and returns again as a liquid. Water is also found in the atmosphere as clouds and fog. These are made up of tiny droplets of water or ice.

▲ **Figure 12-2** Up to 5 percent of the water in the rain forests may be in the form of water vapor.

The amount of water vapor in the air varies from place to place and over time. However, it has been estimated that there are a total of about 14 million tons of water vapor in the atmosphere.

1 **NAME:** What is water called when it is a gas?

Evaporation Water vapor enters the air through the process of evaporation. **Evaporation** is the changing of a liquid to a gas. Most of the water in the air evaporates from the oceans.

Every day, millions of tons of water evaporate from the surface of the oceans. Water also evaporates from lakes, rivers, puddles, and wet soil. Winds carry the water vapor in the air all over Earth's surface.

2 **IDENTIFY:** Where does most of the water in the air come from?

Heat and Evaporation Molecules in a liquid are always moving. Some are moving faster than others. Some of the fast-moving molecules near the surface escape from the liquid. They enter and become part of the air. This is evaporation.

When a liquid is heated, its molecules move faster. More of the molecules can escape from the surface. Evaporation occurs more rapidly in warm liquids. Water in a pan over a heater evaporates faster than does water in a pan on a table. Evaporation from the oceans occurs most rapidly around the equator, where the water is heated by the more direct rays of the Sun.

▲ **Figure 12-3** Near the equator, evaporation occurs rapidly.

3 **DESCRIBE:** What causes a liquid to evaporate faster?

Teaching Options

INTEGRATING LIFE SCIENCE

Looking for Stomata The tiny openings in the leaves of plants are called stomata. Stomata are surrounded by guard cells that control the stomata's opening and closing. Stomata help plants to conserve their water supply. To observe stomata, have students apply clear nail polish to the underside of a green leaf and allow the polish to dry. Students should place a piece of clear tape over the area, then remove the tape. Next, have them place the tape on a microscope slide and use a microscope to view the stomata. Have students make a drawing of their observations.
learning style: visual

EXTENDING THE LESSON

Structures of a Leaf Have students use biology textbooks as references to draw diagrams showing the structures of a leaf. Students should label their diagrams and describe the function of each leaf part. Have students circle the structures through which transpiration occurs.
learning styles: visual, linguistic

Living Things and Water Vapor If you blow on a cold mirror, moisture forms on it. When it is cold outside, you can see your breath. This is water vapor changing back into liquid water droplets. Breathing adds moisture to air. Plants also give off water vapor through a process called **transpiration.** Transpiration occurs in the leaves of plants through tiny openings called stomata.

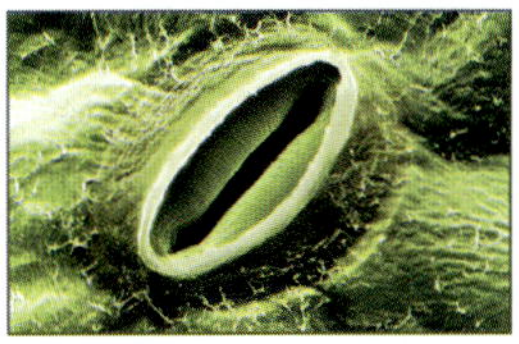

◀ **Figure 12-4** Stomata, or tiny openings, on a leaf release water vapor.

4 NAME: How do plants give off water vapor?

CHECKING CONCEPTS

1. About how much water vapor is in the air?
2. How do animals add water to the air?
3. What happens to the molecules in a liquid when they are heated?
4. What part of a plant gives off water vapor through openings called stomata?
5. What is the gaseous form of water called?

THINKING CRITICALLY

6. IDENTIFY: Where does most of the moisture in the air come from?
7. HYPOTHESIZE: If you wipe a damp cloth across a tabletop, the table is wet. What will probably happen to the water in a few minutes? Why?

DESIGNING AN EXPERIMENT

Design an experiment to solve the following problem. Include a hypothesis, materials needed, variables, a procedure, and a type of data to study. Also tell how you would record your data.

PROBLEM: How can you show that heated water evaporates faster than unheated water?

Hands-On Activity

OBSERVING TRANSPIRATION

You will need a small potted houseplant, a twist tie, a clear plastic bag, and a graduated cylinder.

1. Carefully cover the plant with the plastic bag. Use a twist tie to secure the bag near the bottom of the plant stem. ⚠ CAUTION: Plastic bags can be dangerous when placed over a person's head.
2. Set the plant in a place where it will get plenty of sunlight. Water the plant. Observe the inside of the plastic bag for five days. Record your observations each day.
3. On the fifth day, carefully remove the plastic bag. Measure the amount of water in the bag using the graduated cylinder. Record the amount of water that was collected in five days.

▲ **STEP 1** Cover the plant with a plastic bag.

Practicing Your Skills

4. OBSERVE: On which day did you first see water in the bag?
5. COMPARE: What happened to the amount of water in the plastic bag each day?
6. INFER: How much water did the plant give off in five days? Where did the water come from?

MORE ON THE WEB

Defining Hydrology Terms At www.conceptsandchallenges.com, students can explore the field of hydrology. Have them write a short report describing what hydrology is and what a hydrologist studies.

• TEACHING RESOURCES • Online or Print

Teacher's Resources

Lesson Review: Chapter 12, p. 2

Enrichment: Chapter 12, p. 7

Designing an Experiment: Study Tools, p. 14

Laboratory Manual

Lab. Challenge: How can evaporation be speeded up or slowed down? pp. 65–68

READING STRATEGY Point out to students that the Key Terms both end with *-ation.* Explain that some nouns (words that name things) are made by adding the suffix *-ation* to the end of a verb (an action word). Review the pronunciation of *evaporation* and *transpiration.* Then, have students list other words in this lesson with an *-ation* ending. (transportation, observation)

3 *Assess*

CHECKING CONCEPTS ANSWERS

1. about 14 million tons
2. by breathing
3. They move faster.
4. its leaves
5. water vapor

THINKING CRITICALLY ANSWERS

6. IDENTIFY: the surface of the oceans
7. HYPOTHESIZE: The water will evaporate because its molecules are always moving and the molecules will escape from the surface and enter the air.

4 *Close*

DESIGNING AN EXPERIMENT

Use the Designing an Experiment Rubric from the Teacher's Resources to assess students' experiments. Fill in the rubric with the additional information below.
For this assignment, students should have:

- designed a procedure that shows heated water evaporating faster than unheated water.
- explained how to record data from the experiment.

Hands-On Activity

OBSERVING TRANSPIRATION

TIME: 30 minutes

PURPOSE: observing, comparing, inferring

Practicing Your Skills Answers

4. OBSERVE: most likely on day one
5. COMPARE: It increased.
6. INFER: The amount of water that the plant gives off will vary. The water came from the process of transpiration.

12-2 *(pp. 280–281)*

1 *Introduce*

STUDY HINT Have students write a sentence using each of the Key Terms. Have students turn the sentences into questions that might be on a test of the lesson.

Linking Prior Knowledge Ask students if their skin has ever felt sticky on a hot summer day. Tell them that this sticky feeling is related to the amount of water vapor in the air, which is known as humidity. If you do not live in a humid area, ask students how the air feels before a storm. Have students infer whether the air outside is currently humid.

2 *Teach*

Demonstration Place a sponge in a pan of water and let it soak up as much water as possible. Point out that when a sponge cannot hold any more water, it is saturated. Compare the sponge to the air. Explain that when air is saturated, it holds as much water vapor as it can at that temperature.
learning styles: visual, auditory

Discussion Write the term *humidity* on the board and discuss what the word means to students. Ask if they are more comfortable during times of high humidity or low humidity. Tell students that humidity is the amount of water vapor in the air. Ask what they think high humidity means. (There is a lot of water vapor in the air.)
learning styles: auditory, linguistic

Reinforcement Remind students that air expands when heated and contracts when cooled. This will help them understand that as temperature goes down, the capacity of air for holding water goes down.

Answers to questions

1 **DEFINE:** amount of water vapor in the air

2 **STATE:** the temperature

3 **INFER:** 25°C

4 **NAME:** grams of water vapor/kg of air

12-2 What is humidity?

Objective

Explain how we measure the amount of water vapor in the air.

Key Terms

humidity amount of water vapor in the air

capacity (kuh-PAS-ih-tee)**:** amount of material something can hold

saturated (SACH-uh-rayt-ihd)**:** filled to capacity

specific humidity (hyoo-MIHD-uh-tee)**:** actual amount of water in the air

Humidity The amount of water vapor in a particular parcel of air is its **humidity.** High humidity means the air in that location contains a large amount of water vapor. Low humidity means that there is just a small amount of water vapor in the air in that location.

▲ **Figure 12-5** Water is always moving into or out of air.

1 DEFINE: What is humidity?

Capacity How much water can a 100-mL glass hold? If the glass is filled to the top, its capacity is 100 mL. **Capacity** is the amount of matter something can hold. The capacity of a 100-mL glass is always the same.

Air has a capacity for holding water vapor. Air's capacity for holding water vapor changes with the air temperature. As the temperature of the air goes down, air's capacity to hold water vapor goes down.

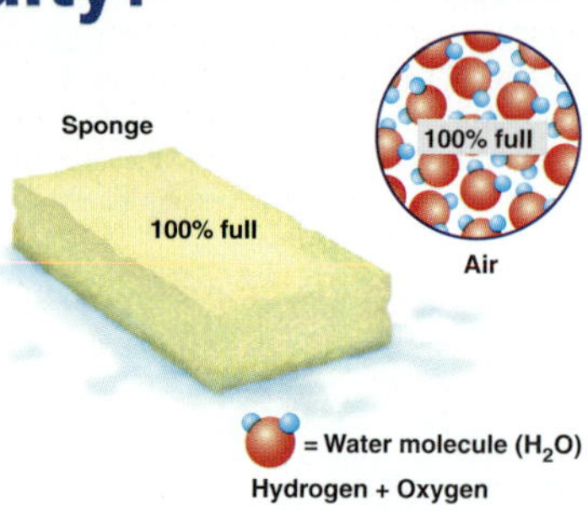

▲ **Figure 12-6** Air can be filled to capacity, just like this sponge.

2 STATE: What affects the capacity of air to hold water?

Saturated Air If you place a sponge in a pan of water, the sponge soaks up the water. Soon the sponge is filled with water. It cannot hold any more water. It is **saturated.** Air can be saturated, too. When air is saturated, it holds all the water vapor it can at a certain temperature. As the temperature goes up, the capacity of air for holding water goes up.

SATURATION OF AIR		
Temperature (°C)	(°F)	Grams of Water Vapor per Kilogram of Air
–40	–40	0.1
–30	–22	0.3
–20	–4	0.75
–10	14	2
0	32	3.5
5	41	5
10	50	7
15	59	10
20	68	14
25	77	20
30	86	26.5
35	95	35
40	104	47

▲ **Figure 12-7**

3 INFER: At which temperature will air hold more water vapor, 5°C or 25°C?

Teaching Options

EXTENDING THE LESSON

Measuring Humidity With a Hygrometer Bring a home-style hygrometer to class and tell students that it is one type of instrument used to measure humidity. Explain that the pointer on the front of the hygrometer is connected to a fiber inside the instrument that absorbs water from the air. Tell them that when the air is humid, the fiber absorbs more water and stretches, making the pointer move. Then, demonstrate this effect by holding a cup of boiling water beneath the hygrometer. **learning style: visual**

COOPERATIVE LEARNING

Studying Weather Instruments Have students work in pairs to gather from magazines or the Internet photographs of weather instruments. Students can create a collage of the instruments. Each instrument should be labeled with a brief caption describing what type of data it gathers.
learning style: visual

Specific Humidity The actual amount of water vapor that is in air is called **specific humidity.** Meteorologists express specific humidity as the number of grams of water vapor in 1 kg of air. Because specific humidity is measured by units of mass, it does not change with temperature or pressure. Only adding more water vapor to the air can change the specific humidity.

4 NAME: How is specific humidity expressed?

CHECKING CONCEPTS

1. The amount of __________ in the air is its humidity.
2. The amount of water vapor the air can hold is its __________.
3. The actual amount of water vapor in the air is called __________.
4. When air is __________, it holds all the water vapor it can at a given temperature.
5. The amount of water vapor that air can hold changes with __________.

THINKING CRITICALLY

6. INFER: The air temperature in Boise, Idaho, is 25°C. The air temperature in Austin, Texas, is 35°C. Which city's air could hold more water vapor? Why?
7. COMPARE: How is a sponge similar to air in terms of capacity?

BUILDING SCIENCE SKILLS

Experimenting Obtain three different brands of paper towels. Take one sheet from each. Put water in a measuring cup, graduated cylinder, or marked beaker. Slowly saturate each towel with water and check to see how much water each towel holds by how much is left in the cup. Which brand has the greatest capacity for holding water? Is it the most expensive brand? Compare how much water each towel can hold to the price for each towel.

People in Science

METEOROLOGIST

During an airplane flight, a pilot radios to the tower to find out weather conditions. The pilot needs to know wind speed and direction, visibility, and cloud conditions. The pilot depends on weather information gathered by meteorologists.

Meteorologists are scientists who study the atmosphere and try to forecast the weather. They gather information about the atmosphere from hundreds of different places and at many altitudes. Weather satellites and computers also help meteorologists gather and analyze data.

▲ **Figure 12-8** A meteorologist checks all kinds of weather data.

Meteorologists work for the National Weather Service, airports, news bureaus, and farming organizations. Some work as weather broadcasters for TV or radio stations.

To be a meteorologist, you need a college education and a background in mathematics and science. The military services also offer special training programs in meteorology.

Thinking Critically What kinds of data do weather satellites gather?

MORE ON THE WEB

Modeling Dehumidifiers
Dehumidifiers are devices that remove moisture from the air. Have students investigate online how dehumidifiers work and what they are used for. Then, have students sketch and label a model of a dehumidifier.

• TEACHING RESOURCES •
Online or Print

Teacher's Resources

Lesson Review: Chapter 12, p. 3

Enrichment: Chapter 12, p. 7

3 Assess

CHECKING CONCEPTS ANSWERS

1. water vapor
2. capacity
3. specific humidity
4. saturated
5. temperature

THINKING CRITICALLY ANSWERS

6. INFER: Austin, Texas, because warmer air can hold more water vapor
7. COMPARE: Both have a capacity to hold moisture, and when saturated cannot hold any more moisture.

4 Close

BUILDING SCIENCE SKILLS

Experimenting Answers will vary, but thicker towels will probably hold more moisture. Have students work in pairs to complete the experiment.

People in Science

METEOROLOGIST

Real-Life Connection Ask students to notice that the sound of a police vehicle or fire truck siren changes as it passes. It likely has a higher pitch as it approaches, then a lower pitch as it passes. This change in pitch is known as the Doppler effect. The Doppler effect describes the change in wave frequency that occurs when sound and other types of energy move toward or away from something. Tell students that Doppler radar is based on the Doppler effect and is used to plot the movements of storms. Doppler radar sends out waves, which are scattered by raindrops. The scattered waves are picked up by receivers, and the location of the storm can then be pinpointed.

Discussion Ask students why it is important to predict the weather. Help them see that the weather affects many aspects of their daily lives, such as participation in sporting events, clothing choices, and family vacations. In addition, severe weather that is predicted in advance can reduce loss of lives and damage to property.

Thinking Critically Answer
Possible answers include temperature, relative humidity, wind speed and direction, cloud cover, and air pressure.

12-3 *(pp. 282–283)*

1 *Introduce*

STUDY HINT Have students write the lesson objective in their notebooks. As students read the lesson, tell them to write three facts from the lesson that meet the objective.

Linking Prior Knowledge Ask students if their hair gets curlier or more frizzy on humid days. Tell them that hair reacts to changes in humidity. Then, ask them to recall that a hygrometer contains a fiber that changes length as humidity changes. (See Extending the Lesson on page 280.) Explain that in many hygrometers that fiber is a human hair.

2 *Teach*

Demonstration Fill a beaker to its capacity. Then, pour water out of the beaker a little at a time so that different amounts remain. Compare the amount of water in the beaker to the amount it can hold. Use fractions and percentages to describe how full the beaker is. Explain how this demonstration connects to relative humidity.
learning style: visual

Discussion If possible, bring a psychrometer to class. Show students the parts of the psychrometer and discuss how the instrument works. Relate the measurements it takes to the relative humidity chart in Figure 12-10. Use the psychrometer to find the relative humidity in the classroom.
learning styles: visual, auditory

Answers to questions

1. **CALCULATE:** 33%
2. **DEFINE:** an instrument used to measure relative humidity
3. **INFER:** It causes both to decrease.

READING STRATEGY Suggest to students that while reading a chart such as the one in Figure 12-10, they should use a ruler or sheet of paper to cover up the lines below the one they are looking at, so as not to become confused. Emphasize to them that this is especially important to do during a test.

12-3 How is relative humidity measured?

Objective

Explain relative humidity and how it is measured.

Key Terms

relative humidity: amount of water vapor in the air compared with the amount of water vapor the air can hold at capacity

psychrometer (sy-KRAHM-uht-uhr)**:** instrument used to find relative humidity

Relative Humidity The amount of water vapor in air compared with its capacity is **relative humidity.** The relative humidity of air that is saturated, or filled to its capacity, is 100 percent. Air is usually not filled to capacity. It may be filled to half its capacity. The relative humidity is then 50 percent.

The relative humidity of air changes as water vapor leaves the air and returns to the ocean or falls as precipitation. It also changes when water evaporates and goes into the air. If the amount of water vapor stays the same and the temperature drops, the relative humidity goes down. Relative humidity goes up if the temperature goes up.

1. **CALCULATE:** If air is filled to one-third its capacity, what is the relative humidity?

Measuring Relative Humidity A **psychrometer** is sometimes used to find relative humidity. Figure 12-9 shows a simple psychrometer made up of two thermometers. The bulb of one thermometer is covered with a damp piece of cloth. The other bulb is dry. Air is then passed over the psychrometer. This causes water on the cloth to evaporate. The dry thermometer measures air temperature. Water evaporating from the cloth cools the wet thermometer. The wet thermometer is cooled more when the relative humidity is low than when it is high. Relative humidity can be found by using the difference in temperature between dry- and wet-bulb thermometers and a chart such as the one shown in Figure 12-10.

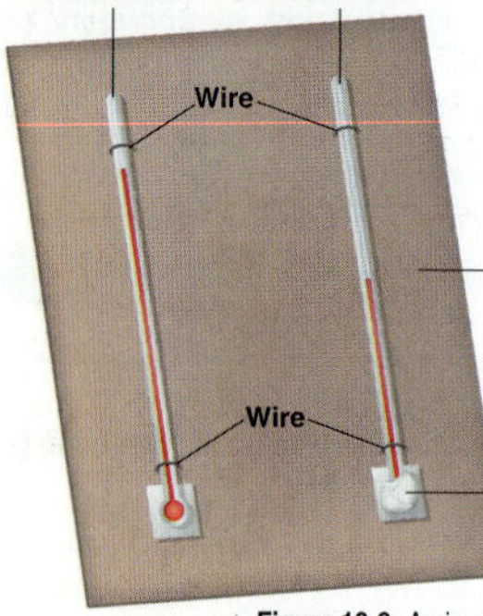

▲ **Figure 12-9** A simple psychrometer

Suppose the dry-bulb thermometer reads 25°C and the wet one reads 18°C. The difference is 7°C. Find 25°C along the top of Figure 12-10. Find 7°C down the left side. Where the two rows meet, you see the number 50. This means the relative humidity is 50 percent.

Difference Between Dry-Bulb and Wet-Bulb Thermometer (°C)	Temperature of Air from Dry-Bulb Thermometer (°C): −5	0	5	10	15	20	25	30	35
1	75	81	86	88	90	91	92	93	94
2	52	64	72	77	80	83	85	86	87
3	29	46	58	66	70	74	77	79	81
4	6	29	46	55	62	66	70	73	75
5		13	32	44	53	59	63	67	70
6			20	34	44	51	57	61	70
7			0	24	36	44	50	55	59
8				15	28	37	45	50	54

Relative Humidity (%)

▲ **Figure 12-10**

Another type of psychrometer is the **sling psychrometer**. This instrument can measure both relative humidity and dew point. The thermometers on this instrument are spun around until the temperature of the wet-bulb thermometer stops dropping.

2. **DEFINE:** What is a psychrometer?

282

Teaching Options

ESL/ELL STRATEGY

***Learning About* Meter** Show students a meterstick. Ask for the word *meter* in their native language. Tell students that *meter* is a Greek root that means "measure." In their science notebooks, have students make a list of words that end in *meter* that are related to measuring. Students should write the definition of each word. (for example, *centimeter, kilometer, millimeter, speedometer, barometer, thermometer,* and *psychrometer*)

ONGOING ASSESSMENT

Write examples of dry thermometer and wet thermometer readings on the board. Have students find the relative humidity for each example, using Figure 12–10.

QUICK CONVERSIONS	
Air Temperature	
Metric/SI	**English**
0°C	32°F
10°C	50°F
20°C	68°F
30°C	86°F

Humidity and Comfort As we perspire, moisture evaporates from our skin. The moisture evaporating from our bodies cools us. The faster the moisture evaporates, the cooler we feel. When the humidity is low, the rate of evaporation is faster. This makes us feel cooler. People living in a hot, dry climate may feel more comfortable than those in a cooler, moister climate. High relative humidity makes people feel uncomfortable. A temperature of 35°C with very low relative humidity may be quite comfortable. However, a temperature of 25°C with high relative humidity can be uncomfortable.

TEMPERATURE AND HUMIDITY IN THREE CITIES

City	Temperature	Humidity
A	35°C	17%
B	35°C	62%
C	25°C	44%

▲ Figure 12-11

3 INFER: What do you think an air conditioner does to the air temperature and humidity in a room?

CHECKING CONCEPTS

1. The relative humidity of air filled to capacity at a given temperature is __________ percent.
2. The relative humidity goes up if the air temperature __________.
3. A __________ has two thermometers and is used to find relative humidity.
4. The relative humidity of air changes as __________ leaves the air.

THINKING CRITICALLY

5. CALCULATE: Use Figure 12-10 to find the relative humidity when the psychrometer readings are as follows: **a.** wet bulb, 12°C; dry bulb, 20°C; **b.** wet bulb, 33°C; dry bulb, 35°C; and **c.** wet bulb, 22°C; dry bulb, 30°C.

INTERPRETING VISUALS

Use Figure 12-11 to answer the following question.

6. INFER: In which city would people be the most comfortable? The least comfortable?

Hands-On Activity

EVAPORATION AND COOLING

You will need safety goggles, isopropyl (rubbing) alcohol, water, an index card, a stopwatch, and two cotton balls.

1. Gather your materials and put on the goggles. Fan your arm with an index card. Describe how it feels.
2. Dip a small cotton ball into some water and rub it on your arm. Fan the spot for 25 seconds. Describe how it feels.
3. Dip another cotton ball into some alcohol and rub this on your other arm. Fan this spot for 25 seconds. Describe how it feels.

▲ STEP 1 Gather your materials.

Practicing Your Skills

4. OBSERVE: What happens to the water and to the alcohol when you fan your arm?
5. EXPLAIN: Why do evaporating liquids make your arm feel cooler?
6. INFER: Why do you feel cooler after you swim on a hot day?
7. COMPARE: Which evaporates faster, water or alcohol?
8. INFER: Why does sweating help to cool you down?
9. HYPOTHESIZE: Why do you feel warmer on a humid day?

RETEACH

Reviewing Math To help students develop a better understanding of relative humidity, you may wish to review fractions and percentages. Give students fractions to convert to percentages, such as 1/4 = 25% and 1/2 = 50%.

• *TEACHING RESOURCES* •
Online or Print

Teacher's Resources

Lesson Review: Chapter 12, p. 4

Enrichment: Chapter 12, p. 7

3 *Assess*

CHECKING CONCEPTS ANSWERS

1. 100
2. drops
3. psychrometer
4. water vapor

THINKING CRITICALLY ANSWERS

5. CALCULATE:
   a. 37%
   b. 87%
   c. 50%

INTERPRETING VISUALS ANSWER

6. INFER: People would be most comfortable in City C and least comfortable in City B.

4 *Close*

Hands-On Activity

EVAPORATION AND COOLING

TIME: 10–15 minutes

PURPOSE: inferring, observing, comparing, hypothesizing

DISCUSSION: Ask: *Would you feel cooler on a windy day or a calm day if the temperature and humidity were the same on both days? Why?* (You would feel cooler on the windy day because sweat evaporates more rapidly when air passes over your skin.) *Why is sweating ineffective at cooling your body if the relative humidity is 100 percent?* (At 100 percent relative humidity, the air is saturated with water vapor, so sweat will not evaporate from your skin.) Explain that people who work outside in hot, humid weather must be careful to avoid becoming overheated. The large amounts of sweat they produce will do little to keep them cool but will quickly cause them to become dehydrated.

Practicing Your Skills Answers

4. OBSERVE: They evaporate.
5. EXPLAIN: They remove heat from your skin.
6. INFER: Water evaporates from your body, cooling you off.
7. COMPARE: alcohol
8. INFER: The sweat evaporates from your skin.
9. HYPOTHESIZE: Sweat does not evaporate as quickly in humid air.

12-4 (pp. 284–285)

1 Introduce

STUDY HINT Have students summarize the main idea of each section in one sentence, then add two supporting details below it.

Linking Prior Knowledge Remind students that when air is saturated, it holds all the water vapor it can at a certain temperature. Tell them that when saturated air is cooled, the water vapor in the air changes from a gas to a liquid. This process is known as condensation.

2 Teach

Demonstration Place a soft-drink can in the refrigerator for several hours. Make sure that it is much colder than room temperature. Bring the can into the classroom. Have students observe the formation of water droplets on the sides of the can. Tell them that they are watching condensation occur as the air around the can grows colder and water vapor in the air condenses on it, or changes from a gas to a liquid. **learning style: visual**

Discussion Students are probably familiar with dew forming on grass. Have them describe how dew feels and looks. Then, discuss the dew point and why dew forms. Tell students that some insects, such as certain beetles, depend on dew for their water supply. They drink the dew that collects on the ground, grass, leaves, and other surfaces.
learning styles: auditory, linguistic

Reinforcement Some students may have difficulty understanding the formation of frost. You may have to review this concept more than once.

Answers to questions

1. **DESCRIBE:** A gas changes to a liquid.
2. **DEFINE:** drops of water that form on objects when condensation takes place
3. **HYPOTHESIZE:** because the freezing point of water is 0°C

12-4 What is the dew point?

Objective
Explain what happens when air temperature goes above or below the dew point.

Key Terms
condensation (kahn-duhn-SAY-shuhn)**:** changing of a gas to a liquid
dew point: temperature to which air must be cooled to reach saturation
frost: ice formed from condensation below the freezing point of water

Condensation Air always contains some water vapor. As the temperature of air drops, water vapor in the air changes from a gas to a liquid. The process of changing a gas to a liquid is called **condensation.**

Have you ever seen water form on the outside of a cold can or bottle? The water forms when water vapor in the air condenses into liquid water on the cold metal or glass.

1 **DESCRIBE:** What happens during the process known as condensation?

Dew Point The temperature of air at which condensation takes place is called the **dew point** of air. Condensation takes place when saturated air is cooled.

Warm air can hold more water vapor than cold air can. As the air cools, it can hold less and less water vapor. If the temperature of the air drops enough, the air becomes saturated. Its relative humidity reaches 100 percent.

If saturated air continues to be cooled, some of the water vapor in the air condenses. The water vapor changes to liquid water.

At night, the ground cools faster than the air. Air near the ground is then cooled by the ground. The temperature of the air may drop to or below its dew point. When this happens, condensation takes place. Drops of water called dew begin to form on grass and bushes. Dew may also form on the windows of cars.

▲ **Figure 12-12** Dew forms on leaves when condensation occurs.

2 **DEFINE:** What is dew?

Frost The freezing point of water is 0°C. At this temperature, water changes from a liquid to a solid. When the humidity is low, the dew point of the air may be lower than the freezing point of water. If the air temperature drops below the dew point, water vapor will come out of the air. However, the water vapor will change directly to ice instead of water. Ice that forms this way is called **frost.**

▲ **Figure 12-13** Frost forms if the air temperature drops below the freezing point and the dew point.

3 **HYPOTHESIZE:** Why doesn't frost form at temperatures above 0°C?

284

Teaching Options

! FAST FACTS *Fighting Frost*

In the United States, oranges and other citrus fruits are grown mainly in Florida and California. Often, citrus farmers have to deal with unseasonable frosts that can threaten their entire orchards. Heaters called smudge pots were once widely used in orange groves, but the heaters caused air pollution and blanketed the groves with soot. Most modern groves have huge fans to mix cold surface air with warmer air aloft. More efficient heaters are also used that mix the air through convection. In addition, in cold weather, crops are sometimes sprayed with water, which allows a coating of ice to form over the plants. The release of latent heat warms the plant, while the ice acts as a protective layer that insulates it.

ESL/ELL STRATEGY

Explaining Dew Point Another way to explain dew point is to say that dew forms when air has cooled to the point where water vapor begins to condense. Tell students that if the air is very cold, the water vapor turns into frost instead of dew. **learning style: auditory**

CHECKING CONCEPTS

1. What is the process of condensation?
2. When does the process of condensation take place?
3. What is the dew point?
4. What happens to the relative humidity when the air becomes saturated?
5. What is the freezing point of water in the Celsius temperature scale?

THINKING CRITICALLY

6. APPLY: Suppose condensation begins to form on the window of a car at 8°C. What is the dew point?
7. EXPLAIN: Why is the relative humidity of saturated air 100 percent?
8. CONTRAST: How do evaporation and condensation differ?
9. HYPOTHESIZE: How could you show that dew found on grass that is in direct sunlight evaporates quicker than dew found on grass that grows in the shade?

Web InfoSearch

Aircraft Deicing One of the most dangerous things that can happen to an airplane is icing. Icing is caused by the freezing of so-called supercooled water droplets. Icing usually occurs on the wings and upper body of an airplane. The extra weight of the ice can reduce the ability of the airplane to fly. Airports usually deice airplanes, or melt the ice, during dangerous weather conditions.

▲ **Figure 12-14** Plane being deiced

SEARCH: Use the Internet to find out how ice affects airplanes. Also, how do airports deice an airplane? Start your search at www.conceptsandchallenges.com. Some key search words are **airplane deicing, airplane safety,** and **airplane disasters.**

Hands-On Activity

MEASURING DEW POINT

You will need an empty can, water, a cup of ice cubes, and a thermometer.

1. Fill a can about two-thirds full of water.
2. Place a thermometer into the water. Do not let the bulb of the thermometer touch the bottom or sides of the can.
3. Add the ice cubes to the water. Observe the outside of the can. Record the temperature on the thermometer as soon as condensation appears on the outside of the can.

▲ **STEP 3** Observe the condensation on the outside of the can.

Practicing Your Skills

4. MEASURE: At what temperature did condensation appear on the can?
5. IDENTIFY: What are the water droplets that formed on the outside of the can called?
6. IDENTIFY: What is the name for the temperature at which condensation took place?
7. APPLY: What is the room's dew point?

INTEGRATING PHYSICAL SCIENCE

Movements of Water Molecules Water molecules contain opposing electrical charges, so although the molecules are in motion, they remain connected to one another because of the attraction between the charges. Heating causes the molecules in water to move more rapidly. As they speed up, some of the molecules break free and escape into the air above in the form of water vapor. When air cannot hold any more water vapor, it will condense.

• TEACHING RESOURCES • Online or Print

Teacher's Resources

Lesson Review: Chapter 12, p. 5

Enrichment: Chapter 12, p. 7

3 Assess

CHECKING CONCEPTS ANSWERS

1. the changing of a gas to a liquid
2. when saturated air is cooled
3. the temperature at which condensation takes place
4. It reaches 100%.
5. 0°C

THINKING CRITICALLY ANSWERS

6. APPLY: 8°C
7. EXPLAIN: The air is holding as much water vapor as it can.
8. CONTRAST: Evaporation is the changing of a liquid to a gas. Condensation is the changing of a gas to a liquid.
9. HYPOTHESIZE: Possible answer: Compare the amount of wetness of dewy grass in sunlight and dewy grass in shade by placing a clear plastic sheet over each area and observing the formation of moisture on the sheet.

4 Close

Web InfoSearch

Aircraft De-icing Have students use the Internet to learn how airports de-ice planes during dangerous weather conditions. Ask students: *How does ice affect airplanes? How do airports de-ice an airplane?* Some key search words are **airplane deicing, airplane safety,** and **ice airplane wing.** Students can begin their search at www.conceptsandchallenges.com.

Hands-On Activity

MEASURING DEW POINT

TIME: 10–15 minutes

PURPOSE: measuring, identifying

Practicing Your Skills Answers

4. MEASURE: Temperature readings will vary but will be much lower than room temperature.
5. IDENTIFY: dew
6. IDENTIFY: the dew point
7. APPLY: The temperature will vary, depending on the room.

12-5 *(pp. 286–287)*

1 Introduce

STUDY HINT Have students make a chart comparing the types of clouds discussed in this lesson. Charts should include shape, height, weather indications, and any other relevant information.

Linking Prior Knowledge To introduce this lesson, have a volunteer describe the clouds in the sky. Ask students if they know what clouds are made of or how they form. Discuss students' responses. Lead students to understand that clouds form when water droplets form around particles in the air.

2 Teach

Demonstration Bring dishwashing detergent or bubble bath to class. Pour some of the solution into a sink or bucket and add warm water. Have students observe the bubbles that form. Relate the different shapes of the bubbles to the many different shapes of clouds. Blow softly on the bubbles so that they break up. Relate this to upper-level winds that move clouds. **learning style: visual**

Discussion Ask students if they have ever looked for a familiar shape in a cloud. Tell them that scientists classify clouds by appearance. Feathery clouds are called cirrus, puffy clouds are called cumulus, and sheetlike clouds are called stratus. Clouds are also classified by where they form. Generally, cirrus clouds form at high altitudes, cumulus clouds form at middle altitudes, and stratus clouds form at low altitudes.

Answers to questions

1 **NAME:** condensation

2 **LIST:** cirrus, cumulus, and stratus

3 **DESCRIBE:** a cloud that forms near the ground

4 **COMPARE:** A cumulonimbus cloud is a type of cumulus cloud.

WRITING HINT Tell students that when they research a writing assignment, they should write notes on note cards to indicate where the information comes from, including a Web address.

12-5 How do clouds form?

Objectives

Describe how clouds form. Identify kinds of clouds.

Key Terms

cirrus (SIR-uhs) **cloud:** light, feathery cloud

cumulus (KYOO-myuh-luhs) **cloud:** big, puffy cloud

stratus (STRAY-tuhs) **cloud:** sheetlike cloud that forms layers across the sky

Cloud Formation From space, Earth sometimes seems to be covered with clouds. Clouds form from condensation in the atmosphere. Water droplets and ice form around dust and other particles in the air. Many billions of tiny water droplets and ice crystals form clouds. A cloud's shape is determined by how it formed.

1 **NAME:** What process forms clouds?

Kinds of Clouds There are three basic kinds of clouds. Light, feathery clouds are called **cirrus clouds.** They are made up of ice crystals. They sometimes form at heights above 10,000 m.

Big, puffy clouds are called **cumulus clouds.** They form from rising currents of warm air that build to great heights. The base is usually flat.

Sometimes the sky is covered with a layer of sheetlike clouds. These are **stratus clouds.** Stratus clouds form layer upon layer, usually at low altitudes.

2 **LIST:** What are the three basic kinds of clouds?

Fog A cloud that forms near the ground is fog. Fog forms from condensation. At night, the ground cools quickly. It cools the layer of air that lies above it. The air may be cooled to the dew point. If it is, the water vapor condenses and forms fog.

▲ **Figure 12-16** Fog rolls in off the ocean.

Thick blankets of fog can cover valleys or other low areas. Sometimes fog forms over rivers and lakes. This happens when cool air moves in over warm water.

3 **DESCRIBE:** What is fog?

◀ **Figure 12-15** Clouds come in three basic types: cirrus, cumulus, and stratus.

286

Teaching Options

COOPERATIVE LEARNING

Observing the Sky Have pairs of students make daily observations of clouds for a week. Tell students to try to identify the types of clouds they observe each day. Inform them that it is not always easy to identify clouds and that sometimes more than one type will be present. Students can make sketches of the clouds in their notebooks. **learning style: visual**

EXTENDING THE LESSON

Researching Clouds Many clouds have word parts added to their names. Have students research and make a list of cloud names that have word parts added and explain why these clouds have these names. (Possible cloud names: nimbostratus, altostratus, stratocumulus, cumulonimbus, cirrostratus, and cirrocumulus) **learning style: linguistic**

QUICK CONVERSION		
	Metric/SI	English
Cirrus clouds	10,000 m	10,936 yd

Cumulonimbus Clouds

▲ **Figure 12-17** Cumulonimbus clouds are associated with thunderstorms.

Cumulonimbus clouds are cumulus clouds that are often associated with thunderstorms. Updrafts can raise a cumulonimbus cloud to heights taller than Mount Everest. As the clouds rise, their tops become flattened against the tropopause, the top of the troposphere. This large, flat top resembles a blacksmith's anvil. An anvil is a good indication that a thunderstorm is coming.

4 **COMPARE:** How is a cumulonimbus cloud like a cumulus cloud?

CHECKING CONCEPTS

1. What are clouds made of?
2. What are light, feathery clouds called?
3. What are sheetlike, layered clouds called?
4. What are three places you might see fog?
5. What kinds of clouds are large and puffy?

THINKING CRITICALLY

6. **DESCRIBE:** Look outside at the sky. Are there clouds? What kind are they?
7. **MODEL:** Use cotton balls or wads of cotton to make models of the three kinds of clouds. Mount the clouds on a sheet of blue construction paper. Label the clouds.

Web InfoSearch

Thunderheads Another name for a cumulonimbus cloud is thunderhead. The sudden appearance of a thunderhead can mean that strong winds and heavy rains are on the way.

SEARCH: Use the Internet to find out more about thunderheads. How do they develop? How do meteorologists use them to learn more about the atmosphere? Start your search at www.conceptsandchallenges.com. Some key search words are **cumulus, cumulonimbus,** and **meteorology.**

Science and Technology

CLOUD SEEDING

Imagine being able to make it rain on very hot days. How about making it snow so that you could go sledding or skiing? There is a way to do this. Cloud seeding can sometimes make it rain or snow.

In cloud seeding, crystals of carbon dioxide or silver iodide are dropped on clouds from high-flying airplanes. Ice crystals form around these crystals. The ice crystals grow until they become heavy and fall from the clouds. Whether it rains or snows depends on the air temperature.

▲ **Figure 12-18** Cloud seeding can sometimes make it rain or snow.

Cloud seeding can help prevent severe storms by causing it to rain before a storm grows too large. It can also help during droughts. However, cloud seeding does not work all of the time. Better technologies need to be developed to improve its success rate.

Thinking Critically Some people think that cloud seeding is like stealing. Farmers in one area that have the clouds seeded may be taking water from farmers in another area. What do you think?

ONGOING ASSESSMENT

Have students use Figures 12-15 and 12-17 to explain how cumulus clouds and cumulonimbus clouds are similar and how they differ. (Both are big and puffy and usually have flat bases. Cumulonimbus clouds have flat tops and are associated with thunderstorms.) **learning style: verbal**

• TEACHING RESOURCES • Online or Print

Teacher's Resources

Lesson Review: Chapter 12, p. 6

Enrichment: Chapter 12, p. 7

3 Assess

CHECKING CONCEPTS ANSWERS

1. tiny water droplets and ice crystals
2. cirrus
3. stratus
4. valleys, rivers, and lakes
5. cumulus or cumulonimbus

THINKING CRITICALLY ANSWERS

6. **DESCRIBE:** Answers should identify clouds as cirrus, cumulus, or stratus.
7. **MODEL:** Models should accurately resemble the three basic types of clouds and be correctly labeled.

4 Close

Web InfoSearch

Thunderheads Have students use the Internet to learn more about thunderheads. Ask them: *How do they develop? What do they tell us about the atmosphere?* Some key search words are **cumulus, cumulonimbus,** and **meteorology.** Have students begin their search at www.conceptsandchallenges.com.

Science and Technology

CLOUD SEEDING

Real-Life Connection Cloud seeding can also be used to weaken severe storms. During the 1960s and 1970s, a U.S. project called Stormfury used seeding to weaken tropical hurricanes in the Caribbean Sea and Gulf of Mexico. Northern Mexico, however, receives much of its annual precipitation from these storms, which weaken naturally as they move inland.

Discussion Ask students why they think scientists developed cloud seeding. Lead students to understand that droughts can devastate crops. This can bankrupt farmers and drive up food prices.

Thinking Critically Answer
Possible answer: Water (rain) is a natural resource, and any special control exerted over this natural resource would give the controlling party an unfair advantage.

Lab Activity

(pp. 288–289)

1 Prepare the Lab

Making Cloud Models

SET-UP TIME: **LAB TIME:**

BACKGROUND: Earth's atmosphere is a complex mixture of many different materials, such as various gases, microscopic particles, and water. Although the quantity of water is relatively small, it is the most visible component of the atmosphere. Water condenses on tiny suspended particles, such as dust, pollution, and volcanic ash, to form droplets. Condensation is triggered by cooling. Cool air holds less water vapor than warm air does. As moist air rises, the air cools. Tiny droplets come together to form clouds. If more water vapor condenses, the drops become too heavy to remain suspended in the air, and they fall to Earth as precipitation.

PURPOSE: Students will investigate the relationship between water vapor, particles, and temperature change in Earth's atmosphere.

ALTERNATIVE MATERIALS: Birthday candles can be substituted for paper matches as the source of particles. Light the candles for students.

SCIENCE SKILLS: Students will **model** the formation of a cloud inside a chamber. They will **observe** changes in temperature and the quantity of particles. Finally, they will **relate** these changes to the formation of clouds.

ADVANCE PLANNING: Set up the cloud growth chambers in advance. Obtain enough plastic 2-L soft-drink bottles for each group of students. Clean the bottles and turn them upside down to dry. Also, obtain rubber tire valves from an auto supply store. Ask for TR418 snap-in tubeless tire valves (small .453″ rim hole). One valve is needed for each chamber. Drill a 3/8-inch hole in each cap and press a valve through the hole. The valve should seal itself to the cap. If the valve does not form an airtight seal, spread a little silicone aquarium sealant around the inside of the cap.

LAB ACTIVITY

Making Cloud Models

Materials
- Safety goggles
- 2-Liter soft drink bottle
- Bicycle hand pump
- Paper match
- Water
- Thermometer

BACKGROUND

Clouds are made from billions of tiny water droplets. The droplets form around nearly invisible particles of dust, volcanic ash, and pollution.

PURPOSE

In this activity, you will investigate what is needed to make a cloud in a bottle.

PROCEDURE

1. Put on your goggles. Copy the chart in Figure 12-19 onto a clean sheet of paper.
2. For the first experiment, carefully slide the thermometer into the bottle. Screw on the cap with the thermometer still inside.
3. Have your teacher make a small hole in the bottle top. The hole should be just big enough for the hose of the bicycle hand pump to fit. Attach the hose of the bicycle pump to the cap of the cloud bottle. Make sure you have no air leaking out once you have the hose inserted into the bottle. Pump 10 to 20 times. Record the temperature.
4. Remove the hose. Observe what happens when you unscrew the cap and let the air out. Record the temperature again.
5. For the second experiment, add about a teaspoon of water to the bottle. Cap the bottle and shake it for about a minute with your thumb over the hole. Record the temperature.

▲ **STEP 3** Pump air into the bottle.

▲ **STEP 5** Add a teaspoon of water to the bottle.

288

Teaching Options

! FAST FACTS Cloud Classification

English scientist Luke Howard (1772–1864) developed the first widely accepted cloud classification scheme in 1803. Howard based his system largely on cloud shapes. Because Latin was the language of scholars at that time, Howard used Latin terms to describe the shapes. *Cumulus* means "heap," *stratus* means "layer," and *cirrus* means "lock of hair." *Nimbus* referred to rain-bearing clouds. In Howard's system, combinations of these names could be used to identify more specific cloud types.

RETEACH

Identifying Clouds Distribute drawings of clouds and have students label the drawings. Ask students to describe each kind of cloud. Students can add this information to the drawings.
learning styles: visual, linguistic

ESL/ELL STRATEGY

Noting the Icons English-language learners should pay close attention to the icons at the beginning of the procedure. The icons indicate what precautions should be taken during the lab.

6. Attach the bicycle pump and pump the same number of times as in Step 3. Observe what happens when you unscrew the cap. Some water will still be in the bottle. Record the temperature.

7. For the third experiment, ask your teacher to light a paper match and blow it out. Your teacher should immediately drop the used match into the bottle to capture its smoke inside. Cap the bottle. Record the temperature.

8. Once again, attach the bicycle pump and pump the bottle as before. Observe what happens when you unscrew the cap. Record the temperature.

▲ **STEP 8** Attach the bicycle pump and pump the bottle again.

Making Clouds			
Experiment	Temperature Before	Temperature After	What Happened?
Air Only			
Water and Air			
Water, Smoke, and Air			

▲ **Figure 12-19** Use a copy of this chart for recording your observations.

CONCLUSIONS

1. OBSERVE: What happened to the temperature when you pumped air into the bottle? What happened when you released the pressure?
2. OBSERVE: In which experiment did you create the best cloud?
3. INFER: What are the parts of a cloud?
4. ANALYZE: How does air temperature affect cloud formation?

Tips for Using Technology

Using Databases for Cloud Observations Have students use database software to record their observations of cloud models. Databases allow students to enter and edit information, store and organize a lot of information easily, and search for specific information. Databases also allow students to arrange information by putting it in a specific order and to present information by using special layouts. Have students use the Database Planner graphic organizer from the Teacher's Resources to plan their databases.

• TEACHING RESOURCES • Online or Print

Teacher's Resources

Lab Activity Report: Study Tools, pp. 16–17

Laboratory Activity Rubric: Teaching Tools, p. 8

Database Planner: Study Tools, p. 22

2 Guide the Lab

PROCEDURE: Divide students into pairs or groups of three. Each pair or group should have its own cloud growth chamber. If there is only one bicycle pump, make sure that each pair or group has had a chance to complete Step 3 before using the pump again in Step 6. Provide students with a copy of the Lab Activity Report found in the Teacher's Resources for students to record their observations and conclusions.

TROUBLESHOOTING: The plastic-backed thermometers must be placed inside the bottles to measure temperature changes. Do the same for Steps 6 and 8.

SAFETY TIP: Review safety procedures for handling fire.

EXPECTED OUTCOME: Students will observe that clouds form best when the cloud growth chambers have both water vapor and particles from the smoke present.

LAB HINT: Make sure that students understand that the smoke from the match contains tiny soot particles.

3 Conclude the Lab

1. OBSERVE: The temperature rose. The temperature fell.
2. OBSERVE: the experiment with both water and smoke inside the bottle
3. INFER: particles and water
4. ANALYZE: Cool air holds less water than warm air. When the air cools, water collects around particles and forms tiny drops that become clouds.

Use the Laboratory Activity Rubric from the Teacher's Resources to assess students' lab activities. Fill in the rubric with the additional information below. For this activity, students should have:

- made a cloud in a bottle.
- analyzed how air temperature affects cloud formation.

4 Extend the Lab

Demonstrate condensation. Boil a kettle or beaker of water and hold a cool glass in the resulting steam. Droplets of water will form on the glass. Next, place ice water into a clean coffee can. Put the can in a shallow dish. Have students observe the can for several minutes and note the condensation on the outside.

12-6 *(pp. 290–291)*

1 *Introduce*

STUDY HINT Have students draw diagrams or sketches of how each type of precipitation forms based on the explanations in this lesson. Remind students to label the drawings.

Linking Prior Knowledge Remind students that clouds develop when microscopic water droplets and ice form around dust and other particles in the atmosphere. Tell them that the same process is involved in precipitation.

2 *Teach*

Demonstration Bring in potting soil, a flat pan, and a plastic bottle with an adjustable opening for controlling the spray of water. Fill the bottle with water at room temperature and pour the soil evenly into the pan. Have students observe as you spray the soil with the water, using a light drizzle on one end of the pan and a heavy spray on the other end. Relate the demonstration to different rates of rainfall. Let students touch the drenched soil. Ask them which type of rainfall would likely lead to flooding. (a heavy rainfall)
learning styles: visual, tactile

Discussion Identify rain, sleet, snow, and hail as the four kinds of precipitation. Describe the formation of each kind. Refer students to Figure 12-20 to illustrate your descriptions. Discuss the forms of precipitation that fall in your area. Allow students who have experienced hailstorms to describe the size of the hailstones.
learning styles: visual, auditory

Answers to questions

1. **LIST:** rain, sleet, snow, and hail
2. **CONTRAST:** Rain is made of droplets of water. Snow is made of crystals of ice.
3. **OBSERVE:** a ball of ice
4. **DEFINE:** an instrument used to measure rainfall

12-6 What is precipitation?

Objective
Identify and describe forms of precipitation.

Key Terms
precipitation (pree-sihp-uh-TAY-shuhn)**:** water that falls to Earth from the atmosphere
rain gauge (GAYJ)**:** device used to measure rainfall

Precipitation Conditions of the atmosphere vary greatly from place to place and season to season. This results in several different types of precipitation. **Precipitation** is water that falls to Earth from clouds. Precipitation may be a liquid or a solid. There are four major kinds of precipitation: rain, snow, sleet, and hail. All precipitation begins when water condenses in cool air.

1 **LIST:** What are the four kinds of precipitation?

Rain, Snow, and Sleet Droplets of water and crystals of ice that make up clouds are very small. They are kept up in the air by air currents. The droplets of water are always moving. They hit into each other. When they hit, they join together. When they become too heavy, they fall as rain. Snow falls when ice crystals grow too heavy.

A very heavy snowstorm with strong winds and low temperatures is called a blizzard. During a blizzard, it is difficult to see. The heavy falling snow can block highways, airports, and railroad tracks.

Sometimes rain falls through cold layers of air. Sleet is rain that freezes as it falls through a layer of cold air near the ground.

Occasionally, the freezing of the rain doesn't occur until the rain strikes a surface near the ground. In this event, the rain forms a thick sheet of ice. When these conditions occur, it is called an ice storm.

2 **CONTRAST:** How are rain and snow different?

Hail Hail is made up of lumps of ice. These lumps form as winds toss ice crystals up and down in a rain cloud. Each time the crystals move up, water freezes around them. These heavy lumps of ice eventually fall to the ground, often damaging property and crops. Hail usually occurs during strong thunderstorms.

3 **OBSERVE:** What does a hailstone look like?

Measuring Precipitation A **rain gauge** is a device used to measure rainfall. Rain gauges collect water in one spot. The amount of rain that falls is usually measured in millimeters.

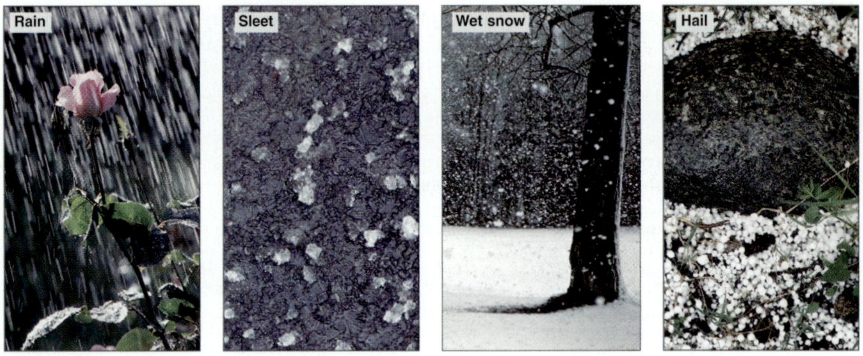

▲ **Figure 12-20** Precipitation can take many different forms.

Teaching Options

INTEGRATING PHYSICAL SCIENCE

Spreading Light A rainbow is made when raindrops act like natural prisms. Sunlight contains an almost equal mixture of all the visible frequencies of light. When a ray of sunlight enters and leaves a raindrop, it is refracted, or bent. This causes the white light to spread out into a spectrum of colors, forming a rainbow. Have students diagram this process, showing a light source (Sun), raindrop, refraction of light, and the resulting rainbow colors: red, orange, yellow, green, blue, indigo, and violet. **learning style: visual**

ONGOING ASSESSMENT

Have students write creative stories that describe how each of the different forms of precipitation fall to Earth and how each feels to the touch.

EXTENDING THE LESSON

Researching Storm Damage Have interested students research the damage that is caused by hail. Students will find that each year in the United States, several million dollars worth of crops are destroyed by hail.
learning style: linguistic

Snow can also be measured with a rain gauge. The snow is collected, melted, and then measured. Snow depth is measured with a meter stick.

4 **DEFINE:** What is a rain gauge?

CHECKING CONCEPTS

1. Precipitation may be a __________ or a solid.
2. There are __________ kinds of precipitation.
3. A liquid form of precipitation is __________.
4. Rain that freezes as it falls to Earth is called __________.
5. Snow depth is usually measured with a __________.

THINKING CRITICALLY

6. **CLASSIFY:** What state of matter—liquid or solid—is each kind of precipitation?
7. **CONTRAST:** What is the difference between sleet and freezing rain?
8. **INFER:** What effect do you think an ice storm might have on tree branches?

Web InfoSearch

Cloud Forests On the tops of mountains in Central America, hidden by mist, are the cloud forests. Trees in cloud forests are much shorter than lowland tropical trees. Their branches are usually covered with thick carpets of mosses, lichens, and ferns. Plants and animals in these high-elevation forests are like those in milder climates.

▲ **Figure 12-21** A cloud forest in Central America

SEARCH: Use the Internet to find out more about cloud forests. How are they alike and different from rain forests? What species are found in there? Start your search at www.conceptsandchallenges.com. Some key search words are **cloud forests, mountains,** and **Central America.**

Hands-On Activity

MAKING A RAIN GAUGE

You will need safety goggles, a wide-mouthed jar, tape, and a ruler with both metric and inch markings.

1. Stand a ruler inside a wide-mouthed jar. Tape the ruler to the side of the jar.
2. Place the jar in an open area outdoors.
3. After it rains, measure the water in the jar. If it snows, allow the snow to melt. Measure in both inches and millimeters, then write down the numbers.
4. Compare your results with those listed in your local newspaper.

▲ **STEP 3** Write down your measurements in both inches and millimeters.

Practicing Your Skills

5. **INFER:** Why should a rain gauge be placed in an open area?
6. a. **COMPARE:** How did your readings compare with those listed in the newspaper?
   b. **EXPLAIN:** Were your readings and the newspaper readings of the weather different? Explain.
7. **INFER:** Why should you allow snow to melt before measuring it?

3 Assess

CHECKING CONCEPTS ANSWERS

1. liquid
2. four (main)
3. rain
4. sleet
5. meterstick

THINKING CRITICALLY ANSWERS

6. **CLASSIFY:** Liquid—rain; solid—snow, sleet, and hail
7. **CONTRAST:** Sleet is rain that freezes as it falls. Freezing rain is rain that does not freeze until it hits the ground.
8. **INFER:** Possible answer: The ice might weigh the branch down and cause it to break.

4 Close

Web InfoSearch

Cloud Forests Have students use the Internet to learn more about cloud forests. Ask students: *How are cloud forests alike and different from rain forests? What species of plants live there?* Some key search words are **cloud forests, mountains,** and **Central America.** Students can begin their search at www.conceptsandchallenges.com.

Hands-On Activity

MAKING A RAIN GAUGE

TIME: 10–15 minutes

PURPOSE: measuring, inferring, comparing

Practicing Your Skills Answers

5. **INFER:** so it can collect precipitation
6. a. **COMPARE:** Answers will vary, but make sure that students are comparing data from the same area.
   b. **EXPLAIN:** Answers will vary, but check that the local readings are for your area.
7. **INFER:** so you can accurately measure the amount of water that has fallen

ESL/ELL STRATEGY

Identifying the Processes of the Water Cycle Remind students of the water cycle (Lesson 9-1). Draw a diagram of it on the board. Do not label the parts. Instead, use arrows to show the movement of water into the atmosphere (evaporation), the formation of clouds (condensation), and the movement of water to Earth's surface (precipitation). Based on what students have learned in this chapter so far, have them identify and describe each part of the water cycle. Then, have them draw and label the water cycle in their notebooks.

learning styles: visual, linguistic

• TEACHING RESOURCES • Online or Print

Teacher's Resources

Lesson Review: Chapter 12, p. 8

Enrichment: Chapter 12, p. 12

12-7 (pp. 292–293)

1 Introduce

STUDY HINT Have students turn each section review question into a statement, adding information from the section.

Linking Prior Knowledge Review with students the concept of latitude (Lesson 1-6).

2 Teach

Demonstration Place several ice cubes on a plate and put a beaker of hot water nearby. Have students place their hands 1 or 2 cm above the ice cubes and the beaker. Make sure that students do not spill or touch the water. Have students describe the temperature of the air above both the ice cubes and the hot water. They should notice that the air over the ice is cool, while the air over the water is warm. Relate the demonstration to how air masses take on the temperature of the area over which they travel. **learning style: tactile**

Discussion Define and describe an air mass. Describe the various types of air masses. Guide students to see that air masses above a region take on the temperature and moisture characteristics of that region. **learning style: auditory**

Reinforcement Refer students to Figure 12-22, which shows the movement of polar and tropical air masses across the United States. Read aloud the names of the air masses and have students trace the movement of the air masses.

Answers to questions

1. **INFER:** cold and moist
2. **NAME:** a continental polar air mass
3. **NAME:** over the oceans, the Caribbean, and the Gulf of Mexico

READING STRATEGY Students may have difficulty grasping the idea that air masses form over oceans. Have students look up the meaning of the word *maritime* (pertaining to the ocean or sea) and point to places on the map where these air masses occur. **learning style: linguistic**

12-7 What are air masses?

Objective

Describe different kinds of air masses.

Key Terms

air mass: large volume of air with about the same temperature and amount of moisture throughout

polar air mass: air mass that forms over cold regions

tropical (TRAHP-ih-kuhl) **air mass:** air mass that forms over warm regions

Air Masses One day, it may be rainy. The next, it may be cloudy or sunny. To understand why weather changes, you need to know about air masses. An **air mass** is a large volume of air with about the same temperature and amount of moisture throughout. Air masses form when air stays over an area for a while or moves slowly over an area more than 1,000 km across.

An air mass is affected by the region it covers. Air masses that form over land are dry. Air masses that form over water are moist. Air masses that form over warm regions are warm. Those that form over cold regions are cold.

1 **INFER:** What is an air mass formed over the ocean in a cold region like?

Polar Air Masses Cold air masses that form over cold regions are called **polar air masses.** There are two kinds of polar air masses. The United States is affected by continental polar (cP) air masses that form over Canada. These are called continental polar air masses because the air mass forms over land. Continental polar air masses that form over Canada in the winter are very cold and dry.

Air masses that form over oceans in cold regions are moist. These air masses are called maritime polar (mP) air masses. The United States is affected by maritime polar air masses that form over the northern Pacific and Atlantic oceans.

2 **NAME:** What is a cold, dry air mass called?

Tropical Air Masses Warm air masses that form near the tropics are called **tropical air masses.** There are two kinds of tropical air masses. If they form over water, they are moist and warm. These are called maritime tropical (mT) air masses. The United States is affected by maritime tropical air masses that form over the oceans, the Caribbean Sea, and the Gulf of Mexico. Continental tropical (cT) air masses are warm and dry. They form over tropical land areas such as northern Mexico.

3 **NAME:** Where do warm, moist air masses in the United States form?

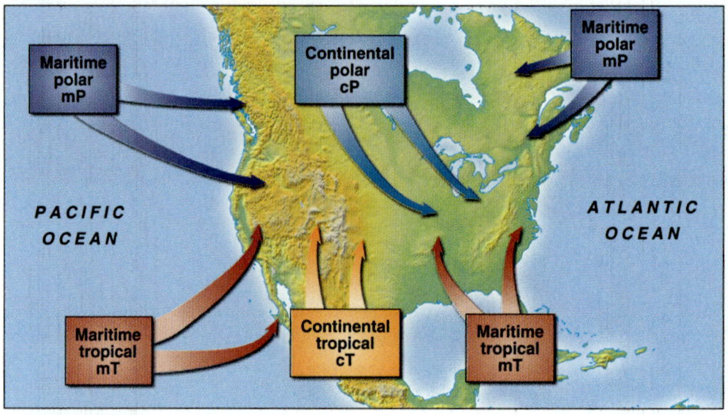

◄ **Figure 12-22** Air masses over the United States

Teaching Options

EXTENDING THE LESSON

Air-Mass Modification Air masses can be moved from place to place by upper-level winds. When an air mass moves over a new region, it takes on the temperature and moisture content of that new region. A maritime tropical air mass, for example, forms over warm oceans, and is thus warm and moist. The air mass may be pushed inland and become cooler and drier as it moves over land. This process is known as air-mass modification. Have interested students investigate air-mass modification and give another example of it.

COOPERATIVE LEARNING

Identifying Air Masses Provide pairs of students with maps of North America. Each map should be labeled with six dots corresponding to the locations of the six boxes in Figure 12-22. Have students take turns matching each dot with a type of air mass (maritime polar, continental tropical, and so on). Ask students to describe the air in that air mass (warm and dry, cold and moist, and so on) to their partners.
learning styles: visual, interpersonal

CHECKING CONCEPTS

1. What is an air mass?
2. What are the four main kinds of air masses?
3. What is a warm, dry air mass called?
4. Does a continental polar air mass form over land or water?
5. What is a cold, moist air mass called?

THINKING CRITICALLY

Use Figure 12-22 to answer the following questions. Abbreviations are used to show the four kinds of air masses.

6. **APPLY:** Which type of air mass would be represented by mT?
7. **APPLY:** Which type of air mass is represented by cP?
8. **INFER:** What do you think the letters for a maritime polar air mass would be?

INTERPRETING VISUALS

Use Figure 12-23 below to do the following exercise.

9. **INFER:** Seven air masses affect the weather of North America. Four are tropical and three are polar. The air masses are listed below. Match them to the letters on the map.

   Polar Pacific; polar Atlantic; tropical Pacific; tropical gulf; polar Canadian; tropical continental

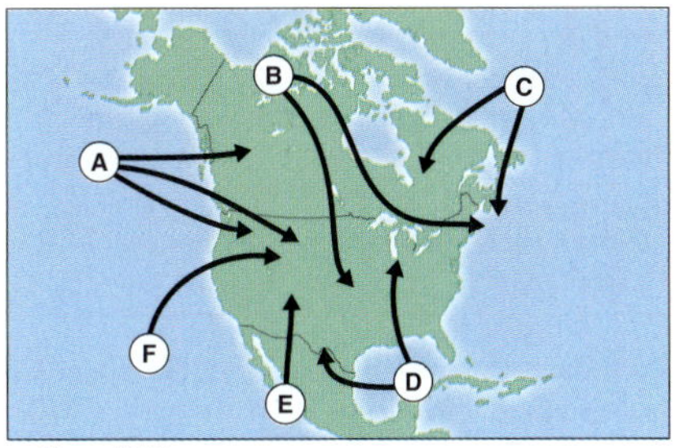

▲ Figure 12-23

How Do They Know That?

RIVERS OF AIR

Jet streams are rivers of fast-moving air. They can travel more than 300 km/h in a west-to-east direction. American pilots discovered the jet streams during World War II. Before then, planes were not equipped to reach the altitudes where jet streams occur. The World War II pilots were able to take advantage of the jet streams' high speeds to save fuel and time when traveling east. However, airplanes flying at high altitudes were slowed down when traveling west against the jet stream winds.

▲ Figure 12-24
Clouds in a jet stream

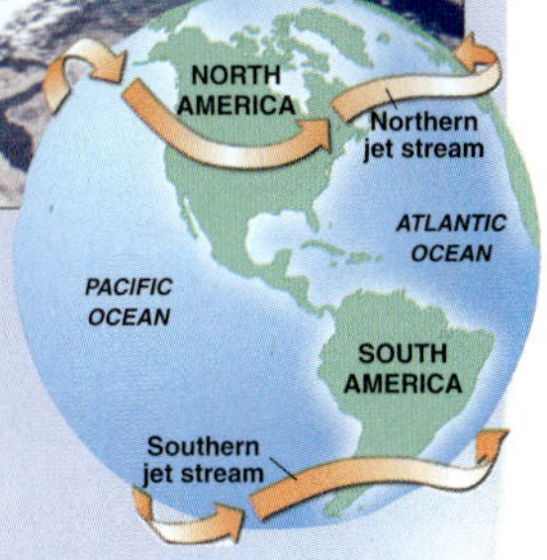

Jet streams form because of temperature and pressure differences. During winter, when there are greater variations in temperature, the jet streams are stronger.

Meteorologists often study jet streams with weather balloons. It appears that jet streams can strengthen and even change the movements of some weather systems. This affects the climate in some places. In India, for example, the jet stream is linked to the heavy monsoon rains.

Thinking Critically Why are jet streams weaker in the summer?

3 Assess

CHECKING CONCEPTS ANSWERS

1. a large volume of air that has the same temperature and amount of moisture throughout
2. a continental tropical air mass
3. land
4. continental polar, maritime polar, continental tropical, maritime tropical
5. maritime polar air mass

THINKING CRITICALLY ANSWERS

6. **APPLY:** a maritime tropical
7. **APPLY:** a continental polar
8. **INFER:** mP

INTERPRETING VISUALS ANSWER

9. **INFER:** polar Pacific: A; polar Atlantic: C; tropical Pacific: F; tropical gulf: D; polar Canadian: B; tropical continental: E

4 Close

How Do They Know That?

RIVERS OF AIR

Technology Connection Weather balloons are often used to carry a package of weather instruments aloft to measure upper-air conditions. This package is called a *radiosonde.* Sensors on the radiosonde can measure air temperature, pressure, and humidity. The readings are transmitted by radio signals to ground stations on the surface. By tracking the movement of the balloon, meteorologists can also measure wind speed and direction.

Discussion Like jet streams, monsoons are caused by temperature and pressure differences. In summer, the air is warmer over the land than over the ocean. This causes the denser, moist air from the ocean to blow inland, triggering heavy rains. In winter, the air over the ocean is warmer than the dry air over the land. This causes the denser, cool air to blow from land toward the ocean. Ask students if they have ever been to a beach. If so, they probably noticed a cool, refreshing breeze that flowed from the water to the land during the day. This breeze is like a small version of a summer monsoon.

Thinking Critically Answer
because the temperature range is not as great

MORE ON THE WEB

Describing Air Masses Have students search the Internet to learn more about the five different Northern Hemisphere air masses. Then, have students create a table that identifies the air mass and its symbol, location, and characteristics.

• *TEACHING RESOURCES* • *Online or Print*

Teacher's Resources

Lesson Review: Chapter 12, p. 9

Enrichment: Chapter 12, p. 12

RETEACH

Moving Air Masses Give students a map of North America and colored disks. Have them choose a color for each air mass, such as blue for maritime polar. Have students place the disks on the map to show the origin, then move the disks across the map to show how each air mass moves.
learning styles: tactile, visual

12-8 *(pp. 294–295)*

1 Introduce

STUDY HINT Have students create a concept map to better visualize the various fronts. A reproducible Concept Map can be found in the Teacher's Resources.

Linking Prior Knowledge Remind students that warm air is denser than cold air. Thus, warm air rises over cold air.

2 Teach

Demonstration A simple demonstration can be used to illustrate fronts. Fill a large, clear bottle halfway with water and fill the rest with motor oil. Ask students which is denser—the water or the oil. (the water) Place a cap on the bottle and turn it on its side. Direct students' attention to the movement of the liquids. Tell students that the boundary between the two fluids represents a front. Make the connection between the different densities of cold and warm air. Ask students: *Which liquid represents a cold air mass?* (the water) Point out the turbulence that occurs along the "front" where the two fluids meet. **learning style: visual**

Discussion Point out that the boundary between two different air masses is a front. Explain that fronts bring changes in weather. Discuss the characteristics of a cold air mass and those of a warm air mass. **learning style: auditory**

Reinforcement After the lesson, discuss the recent and current weather in your area. Ask students to determine which type of front—cold, warm, stationary, or occluded—has most recently passed through your area and how it changed the weather. **learning styles: auditory, logical/mathematical**

Answers to questions

1 EXPLAIN: Usually they bring changes in weather.

2 DESCRIBE: rain and cloudy skies

12-8 What is a front?

Objective

Describe the different kinds of fronts and the weather they cause.

Key Terms

front: boundary between air masses of different densities

cold front: forward edge of a cold air mass, formed when a cold air mass pushes under a warm air mass

warm front: forward edge of a warm air mass, formed when a warm air mass pushes over a cold air mass

Boundaries in Air A **front** is the boundary between two air masses of different densities. In the United States, air masses usually travel west to east. As they move, they meet up with other air masses.

Individual air masses do not usually mix. Instead, a front forms between them. A front can be hundreds of kilometers long. Fronts bring changes in the weather.

▲ **Figure 12-25** Fronts can be detected in satellite photos or sometimes by just looking up at the sky.

1 EXPLAIN: What do fronts bring?

Cold Fronts A **cold front** is the forward edge of a cold air mass. A cold front is formed when a cold air mass pushes its way underneath a warm air mass.

Cold air is denser than warm air. As a cold front moves through an area, the warm air over a region is pushed upward. Gusty winds are formed because of differences in air pressure. Cold fronts usually bring rain and cloudy skies. Once the cold front passes, the cold air mass moves in.

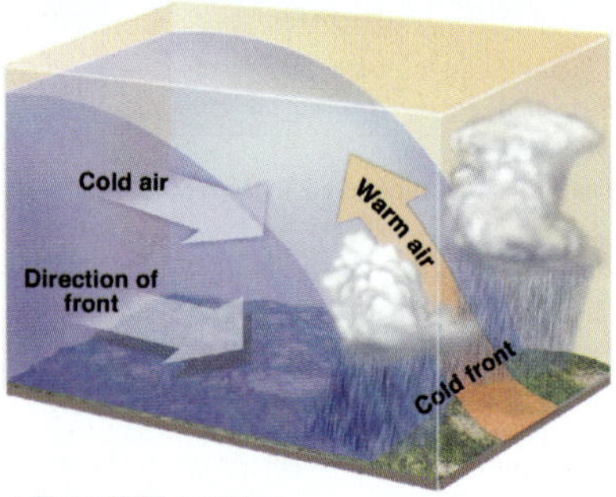

▲ **Figure 12-26** A cold front forms when cold air pushes warm air upward.

2 DESCRIBE: What kind of weather does a cold front bring?

Warm Fronts What happens when a warm front moves into an area? A **warm front** is the forward edge of a warm air mass. A warm front forms when a warm air mass pushes over a cooler air mass.

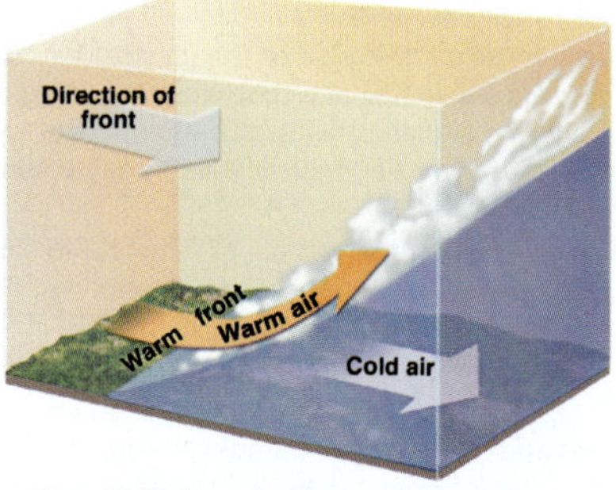

▲ **Figure 12-27** A warm front forms when a warm air mass pushes over cold air.

294

Teaching Options

! FAST FACTS More About Fronts

In weather forecasting, a front is often called a "trough" of low pressure. On surface pressure charts, fronts are shown in a two-dimensional form along a thermal gradient.

ESL/ELL STRATEGY

***Different Meanings of* Front**

Discuss the use of the term *front* in this lesson. Ask students how it is related to and different from the more common use of the word *front.*

ONGOING ASSESSMENT

Have students study the diagrams in Figures 12-26 and 12-27. In their notebooks, have students sequence in correct order the events that occur during the formation of a cold front and a warm front.

Warm air moves more slowly than cold air. Warm air rises above cold air. The rising warm air forms a gentle slope. The warm air slowly moves up and over the cold air. As the warm air rises, it cools. Cirrus clouds form high up, and precipitation may follow. Slow clearing and warmer temperatures show that a warm front has passed and the warm air mass has moved in.

3 DEFINE: What is a warm front?

Stationary Fronts Sometimes cold and warm air masses stay put for a while. They do not move. They remain stationary. This forms a stationary front. A stationary front brings very little change in the weather.

▲ **Figure 12-28** A stationary front brings little change in weather conditions.

4 DEFINE: What is a stationary front?

Occluded Fronts The most complex weather situation occurs with occluded fronts. In an occluded front, a warm air mass is between two cooler air masses.

Go Online
active art
For: Weather Fronts activity
Visit: PHSchool.com
Web Code: cfp-4031

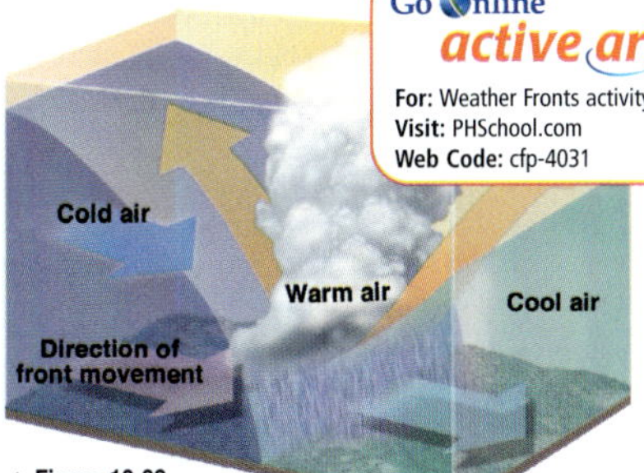

▲ **Figure 12-29** An occluded front traps a warm air mass between two cooler air masses.

In an occluded front, the cooler air masses move beneath the warm air mass and push it up. They then come together in the middle and may mix. This makes the temperature near the ground cooler. The warm air mass is cut off, or occluded, from the ground. As the warm air cools and its water vapor condenses, the weather often turns cloudy, rainy, or snowy.

5 DESCRIBE: What happens to the weather with an occluded front?

CHECKING CONCEPTS

1. What are four kinds of fronts?
2. How does a cold front form?
3. What kinds of clouds form when a warm front approaches?
4. Which kind of front causes very little change in weather?
5. What kind of weather is common with an occluded front?

THINKING CRITICALLY

6. **INFER:** A city gets a day of rain and on the next day skies clear. The temperature rises. Which kind of front has passed?
7. **INFER:** If a city has clear skies and warm temperatures for three days, which kind of front might be keeping the weather the same?

Web InfoSearch

Tracking the Weather Go to the Internet to find out the weather report for three days. Keep a record of the fronts that are coming to your area. Write down the weather conditions that are forecast. Using that information, forecast the weather for the next two days. Check your forecast with current weather reports.

SEARCH: Start your search at www.conceptsandchallenges.com. Some key search words are **weather** and **fronts.**

3 DEFINE: the forward edge of a warm air mass

4 DEFINE: an air mass formed when cool and warm air masses stay in one place for a while

5 DESCRIBE: It turns cloudy, rainy, or snowy.

WRITING HINT Remind students that record-keeping, such as they are required to do for the Web InfoSearch in this lesson, requires great accuracy and careful attention to detail. Suggest to them that they create a chart for themselves into which they enter their data, rather than just writing notes on loose pieces of paper. Then, tell students to double-check their information before they write it and use a pencil rather than a pen so that incorrect information can be changed easily.

3 Assess

CHECKING CONCEPTS ANSWERS

1. cold, warm, stationary, and occluded
2. A cold air mass pushes under a warm air mass.
3. cirrus
4. a stationary front
5. clouds, rain, or snow

THINKING CRITICALLY ANSWERS

6. **INFER:** a warm front
7. **INFER:** a stationary front

COOPERATIVE LEARNING

Recording the Weather Have a group of students set up a weather station equipped with a barometer, thermometer, psychrometer, and wind vane or anemometer. Students can alternate reading the instruments and recording the measurements. Have students compare their measurements with actual measurements reported on television or in the newspaper and report how fronts are part of the forecasting process.
learning styles: visual, tactile

• TEACHING RESOURCES •
Online or Print

Teacher's Resources

Lesson Review: Chapter 12, p. 10

Enrichment: Chapter 12, p. 12

Concept Map: Study Tools, p. 4

Go Online
active art
For: Weather Fronts activity
Visit: PHSchool.com
Web Code: cfp-4031

Students can interact with the art of weather fronts online.

4 Close

Web InfoSearch

Tracking the Weather Have students read the weather report online for three days and keep a record of fronts approaching the places where they live. Then, have students predict the weather for the next two days and check their predictions against current weather reports. Some key search words are **weather** and **fronts.** Students should begin by going to www.conceptsandchallenges.com.

12-9 *(pp. 296–297)*

1 Introduce

INVESTIGATE

TIME: 5–10 minutes

PURPOSE: Students will model a sound that is similar to thunder.

MATERIALS: paper lunch bags

PROCEDURE: Before students blow into the bags, caution them not to make holes or tears in the bags. Otherwise, the air will escape. After the bags are inflated, students should give the tops a twist to better hold the air inside.

THINK ABOUT IT: The bag burst. The noise is like thunder. The air contracted when the bag was hit, causing the vibrations that produced the sound. **learning styles: tactile, auditory**

Linking Prior Knowledge Review cumulonimbus clouds (Lesson 12-5). Have students recall their experiences with these clouds. Ask them what other weather features they remember witnessing when cumulonimbus clouds were present.

2 Teach

Demonstration Display a map of your school. Have students note safer areas of the building in the event of a hurricane or tornado. Discuss why it is dangerous to be near a window during a severe storm. (Strong winds can blow out the glass.) Use this opportunity to review storm warnings and safety procedures for your school. **learning style: visual**

Discussion Describe thunderstorms, hurricanes, and tornadoes, as well as their causes. Be sure that the differences between them are clear and that students realize how much more widespread is the damage that is usually caused by hurricanes. **learning style: auditory**

Answers to questions

1. **DESCRIBE:** a storm with thunder, lightning, and possibly heavy rains and strong winds
2. **DESCRIBE:** a small, violent, funnel-shaped cloud that spins
3. **DESCRIBE:** calm, with no wind or rain

12-9 What causes severe storms?

INVESTIGATE

Making a Boom
HANDS-ON ACTIVITY

1. Take a paper lunch bag and blow into it.
2. Close the open end of the bag with your hand, keeping the air inside.
3. With force, quickly hit the bag with your other hand.

THINK ABOUT IT: What happens when you hit the bag? What part of a storm does this noise remind you of? How do you think air was involved in making the sound?

STEP 3

Objective
Identify three kinds of severe storms.

Key Terms
thunderstorm: storm with thunder, lightning, and often heavy rain and strong winds

tornado (tawr-NAY-doh)**:** small, very violent, funnel-shaped cloud that spins

hurricane (HUR-ih-kayn)**:** tropical storm with very strong winds

Thunderstorms A **thunderstorm** is a storm with thunder, lightning, and often heavy rain and strong winds. Thunderstorms usually occur in summer, when a cold front forces warm, moist air to rise rapidly. Cumulus clouds build up to form cumulonimbus clouds. From these, heavy rain or hail falls.

▲ **Figure 12-30** A tornado acts like a giant vacuum when it touches the ground.

Lightning occurs when a cloud discharges electricity. The current causes the air to heat and expand explosively. It creates a wave of compressed air. The wave produces the boom you hear as thunder. You see lightning flash before you hear thunder because light travels faster than sound does.

1 **DESCRIBE:** What is a thunderstorm like?

Tornadoes A **tornado** is a funnel-shaped cloud that spins. Tornadoes cause very small, but very violent, storms. They frequently occur with severe thunderstorms. Tornadoes that form over water are called waterspouts.

Most tornadoes form during the spring or in early summer. Scientists are not sure how they form. They do know that a storm cloud may develop a small, spinning funnel that reaches down to the ground.

The funnel has very low air pressure. When it touches the ground, it acts like a giant vacuum cleaner, sucking everything up. As the funnel zigzags along, nearly all objects in its path are destroyed.

2 **DESCRIBE:** What is a tornado?

Hurricanes A **hurricane** is a tropical storm with strong winds. These winds spiral toward the center. The winds may be stronger than 300 km/h.

People who study hurricanes may fly over or even into them to gather information. Most are scientists who work for the National Weather Service (NWS) or the National Oceanic and Atmospheric Administration (NOAA).

296

Teaching Options

INTEGRATING PHYSICAL SCIENCE

Updrafts and Downdrafts Thunderstorms are associated with cumulonimbus clouds, which have strong updrafts and downdrafts. Friction between these opposing drafts strips electrons from some of the atoms in the cloud or near Earth's surface. Electrons have a negative charge, and atoms that lose electrons acquire a positive charge. Other atoms pick up the electrons and so acquire a negative charge. The two opposite charges attract, ultimately forming lightning. To demonstrate the electron attraction, have students rub an inflated balloon with a wool cloth and then attempt to stick the balloon to their clothing. **learning style: tactile**

COOPERATIVE LEARNING

Sharing Information Have the class as a group use a computer graphics program to develop a brochure that offers safety tips for each type of severe weather. This brochure can be distributed around the school.

ESL/ELL STRATEGY

Researching Eye Discuss the definition of the word *eye* for the calm center of a hurricane. Mention that the word was probably chosen because the eye resembles an eye shape.

From an airplane, the hurricane looks like bands of spinning clouds. The rain is heavy. The wind grows stronger and more destructive the closer you get to the eye of the storm. The eye is the calm, clear center of the storm. After the eye passes, the winds change direction.

▲ **Figure 12-31** Satellite photo of a hurricane

3 DESCRIBE: What is the eye of a hurricane like?

CHECKING CONCEPTS

1. When do thunderstorms usually happen?
2. What causes thunder?
3. What is the air pressure like in a tornado?
4. In what seasons do tornadoes usually form?

THINKING CRITICALLY

5. INFER: Not all the houses on a block may be damaged by a tornado. How can this be?
6. CLASSIFY: Which type of severe storm has each of these features? **a.** thunder and lightning **b.** funnel-shaped cloud **c.** giant cumulus clouds **d.** calm eye **e.** low air pressure

Web InfoSearch

Storms in History Many storms have changed history. Columbus and later the passengers on the *Mayflower* met with severe storms. During World War II, one storm sank three U.S. naval destroyers and many smaller boats.

SEARCH: Use the Internet to find out more about this. How did storms affect Columbus's journey, World War II, and the *Mayflower* voyage? Start your search at www.conceptsandchallenges.com. Some key search words are **hurricanes, World War II,** and **voyages of Columbus.**

Real-Life Science

TORNADO ALLEY

More tornadoes occur in the United States than anywhere else on Earth. During some years, more than 800 tornadoes are counted in the United States. Most of these tornadoes form in the Great Plains and the southwestern United States. Oklahoma and Kansas have more tornadoes than most other states. This area of the United States is called the Tornado Belt or Tornado Alley.

In Tornado Alley, tornadoes occur most often in April, May, and June. Usually, they strike during the middle or late afternoon.

▲ **Figure 12-32** Tornadoes cause a great deal of destruction where they touch down.

The National Weather Service warns communities when dangerous thunderstorms are approaching. During a tornado, the best place to go is a basement or the lowest floor of a building. Crouch under a table or another heavy piece of furniture. Stay away from windows and doors. After the tornado passes, watch out for dangerous debris and damaged structures.

Thinking Critically Why do most thunderstorms form in late afternoon?

3 Assess

CHECKING CONCEPTS ANSWERS

1. in summer, when a cold front and a warm front meet
2. a wave of compressed air
3. very low
4. late spring or early summer

THINKING CRITICALLY ANSWERS

5. INFER: Some houses may not be in the zigzagging path of the tornado.
6. CLASSIFY: **a.** thunderstorm **b.** tornado **c.** thunderstorm **d.** hurricane **e.** tornado

WRITING HINT Tell students that when describing a process such as storm development, they should always be sure to put things in the proper time order. If a detail is missing or is out of order, the explanation will be confusing. Tell students to use words such as *first, next,* and *then.*

4 Close

Web InfoSearch

Storms in History Many storms have changed the course of history. Ask students: *How did storms affect Columbus's journey, World War II, and the* Mayflower *voyage?* Some key search words are **Mayflower storms, World War II storms,** and **Columbus voyage storms.** Students can begin their search at www.conceptsandchallenges.com.

Real-Life Science

TORNADO ALLEY

Background Tell students that the National Weather Service advises that anyone caught outside during a tornado should lie flat in a nearby ditch or depression. Cars and mobile homes are dangerous and should be abandoned.

Discussion Ask students: *Why is it important to practice safety drills?* Lead students to understand that tornadoes can hit with little warning. Being prepared allows people to react quickly and make smart choices.

Thinking Critically Answer
This is the warmest time of day, and warmth contributes to storm formation.

EXTENDING THE LESSON

Microbursts Have interested students research microbursts and write their findings in a report. Their reports should state what microbursts are (brief, powerful gusts of wind caused by strong downdrafts), distinguish between the wet and dry types (wet microbursts produce rain that reaches the ground; dry microbursts do not), and explain why they are hazardous. (They can destabilize airplanes that are taking off or landing.)
learning style: linguistic

• TEACHING RESOURCES • Online or Print

Teacher's Resources
Lesson Review: Chapter 12, p. 11
Enrichment: Chapter 12, p. 12

The Big Idea

(pp. 298–299)

1 Introduce

Objective Students will be able to compare and contrast positive and negative electrical charges. They will be able to relate these charges to the formation of lightning. Students will then research and identify safety tips to follow during thunderstorms.

Linking Prior Knowledge Ask students to recall what they learned about thunderstorms (Lesson 12-9). Ask them to identify other characteristics of thunderstorms. (thunder, heavy rains, strong winds) Also ask students to identify the three basic parts of atoms (Lesson 2-1). (protons, neutrons, and electrons)

2 Teach

Discussion Ask students if lightning is always accompanied by thunder. Lead them to understand that thunder is an effect caused by lightning. The two always occur together. You may want to take this opportunity to discuss causes and effects.
learning style: auditory

Refer students to the feature in Lesson 12-9 that discusses safety tips in the event of a tornado. After they have completed the Science Log Writing Activity, they can compare and contrast safety tips for thunderstorms and tornadoes.

Reinforcement Obtain newspaper clippings describing the impact of lightning storms. Have volunteers read the articles to the class. After all the articles have been read aloud, place students in small discussion groups. Have them discuss the damage that was caused and how weather forecasting may have prevented or reduced loss of life or property damage.
learning styles: auditory, linguistic

Integrating Physical Science

THE Big IDEA

How do electrical charges cause lightning?

Every atom of matter is made up of smaller particles of matter. Some of these particles called protons have a positive electrical charge. Other particles called electrons have a negative electrical charge.

If the number of protons and electrons is equal, the atom is neutral. This means that the atom lacks any electrical charge.

Rubbing an atom can cause it to lose some of its electrons. An example is when you drag your feet across a rug on a dry day. As your feet rub against the rug, they pick up electrons. As a result, your body has more electrons than protons. It carries an electrical charge. If you touch a metal object, the extra electrons travel through your fingers to the metal. The shock you feel is actually the movement of electrons.

Clouds that float above Earth's surface are made of water and ice. Temperature differences caused by the freezing of water droplets result in electrons being transferred between atoms. The clouds build up an electrical charge that is released as lightning.

Lightning can travel within a single cloud, between two clouds, or from a cloud to the ground. It can even travel from the ground to a cloud. When the electricity travels through the air, it causes the air to warm and expand quickly. The rapid expansion and then contraction of air molecules is heard as thunder.

Look at the illustrations on these two pages. Then, follow the directions in the Science Log to find out more about "the big idea." ✦

Saint Elmo's fire
The sparks of light called Saint Elmo's fire occur during a storm when not enough charges have built up to form a lightning bolt. Instead, a mass of sparks appears above the ground. This phenomenon was first noted above ships' mast, and so was named for the patron saint of sailors.

Cloud to Ground
In cloud-to-ground lightning, electrical charges flow from a cloud to Earth's surface. Some violent storms produce as many as 100 electrical charges per second. The air around lightning can become as hot as 30,000°C.

298

Teaching Options

! FAST FACTS *Violent Winds*

Thunderstorms generate downdrafts. When a downdraft is very intense and focused on one relatively small area, it is called a *downburst.*

Downbursts can slam into the ground with the force of hurricane winds. They are the second greatest cause of airplane accidents. Since 1964, 29 major aviation accidents have been directly linked to the strong wind shears caused by downbursts. There are two types of downbursts: *microbursts* and *macrobursts.*

Microbursts affect small areas roughly 3 km (1.9 mi) wide and have wind speeds of more than 270 km/h (168 mph). Because microbursts are extremely difficult to detect by radar or other means, they are especially damaging.

Macrobursts affect areas roughly 5 km (3.1 mi) wide and have wind speeds of 216 km/h (134 mph). Although not as damaging as microbursts, macrobursts last three times longer than microbursts (about 30 minutes).

Figure 12-33 Lightning is a very large electrical current flowing through the air. Its energy produces a bright flash of light and often a loud clap of thunder.

THE BIG IDEA

How do electrical charges cause lightning? Have students start their research for their Science Log Writing Activity at www.conceptsandchallenges.com. Be sure students use search engines and websites that are reliable and acceptable. You may also wish to provide copies of the *Big Idea Planner* and *Big Idea Science Log Entry* study tools found in the Teacher's Resources.

3 Assess

Use the Writing Activity Rubric in the Teacher's Resources to assess students' Science Logs. Fill in the rubric with the additional information below. For this assignment, students should have:

- compared and contrasted positive and negative electrical charges and identified the relationship between electrical charges and the formation of lightning.
- communicated safety rules to follow in the event of a thunderstorm.

4 Close

Creating a Safety Drill Remind students that some storms strike quickly, and many people are caught outside. Have students work in small groups to develop a step-by-step safety drill to follow if a thunderstorm strikes during a field trip. Students should base their safety drills on the rules they developed for the Science Log. Their safety drill should include warning signs of approaching thunderstorms and action to take to ensure safety in a thunderstorm. Have students share their safety drills with the class. You may also wish to have some students create a Web page on the subject. See the Tips for Using Technology in the Teaching Options on this page.

Tips for Using Technology

Developing Web Pages on Lightning Safety Ask students to create a Web page that contains lightning safety information. Developing basic student Web pages can be an easy task, even if you are not at all familiar with HTML. Web sites such as www.Geocities.com, www.Angelfire.com, and www.Tripod.Lycos.com offer free Web site hosting. They also offer tools that make creating a Web page easier.

• TEACHING RESOURCES • Online or Print

Teacher's Resources

Big Idea: Chapter 12, p. 13

Big Idea Planner: Study Tools, pp. 11–12

Big Idea Science Log Entry: Study Tools, p. 13

Writing Activity Rubric: Teaching Tools, p. 9

12-10 *(pp. 300–301)*

1 *Introduce*

Linking Prior Knowledge When discussing the millibar unit used to show air pressure, remind students that millimeters of mercury are used with a barometer. Point out that the millibar is another unit of measurement used to describe air pressure.

2 *Teach*

Demonstration Cut black construction paper into circles that represent cloud cover at 50%, 70–80%, and 100%, using Figure 12-35 as a guide. Trace the circles using the lightbulb end of a flashlight. Then, have students observe while you shine the flashlight on a surface, such as a desk. Cover the lightbulb end of the flashlight with one of the circles and have students relate the demonstration to amount of cloud cover. Do the same for the remaining circles. Clear skies, or 0% cloud cover, can be demonstrated by shining the flashlight on the surface without covering its end with a circle. **learning style: visual**

Discussion Ask students why it is helpful to use symbols to display information. Lead them to understand that symbols are a kind of universal language. Meteorologists all over the world use the same types of symbols in station models. This overcomes language barriers.

Reinforcement Ask students to study Figure 12-34. Then, have them draw the symbols that most closely correspond to current weather conditions. Have them look at Figure 12-36 on page 302 and identify the cloud cover symbols on it. **learning style: visual**

Answers to questions

1. DEFINE: a record of weather conditions at a particular weather station
2. OBSERVE: 21–25 mph
3. ANALYZE: 1,018 mb

READING STRATEGY Make sure that students understand how to read Figure 12-35. Go over the symbols and discuss why they were chosen.

12-10 What is a station model?

Objective

Understand how to read a station model.

Key Terms

station model: record of weather information at a weather station

millibar (MIHL-ih-bahr)**:** unit of measurement for air pressure

Station Models A **station model** shows the weather conditions at a particular weather station. Station models such as the one in Figure 12-34 are used to depict weather conditions on a weather map. A station model uses symbols instead of words to describe weather. Each weather factor has a different symbol. Meteorologists, or weather scientists, use these symbols to make a station model.

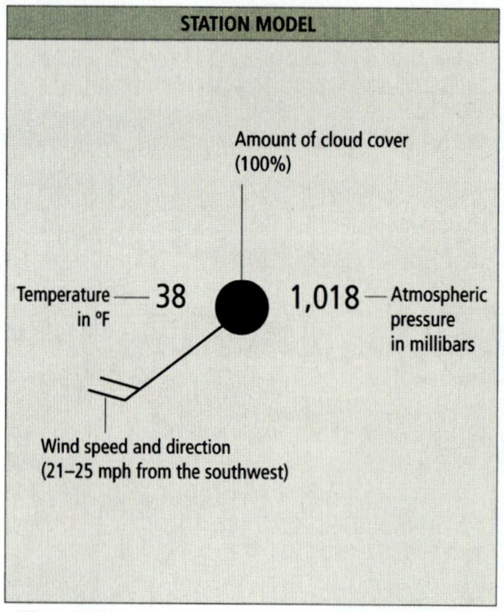

▲ Figure 12-34

1. DEFINE: What is a station model?

Reading Station Models Each station model has a circle, which represents cloud cover. The amount of shading in the circle tells how much cloudiness there is. It is given in percentages. Figure 12-35 shows these percentages and their symbols. Symbols are also used for rain, snow, and other conditions.

Wind speed and direction are indicated on a station model with an arrow shaft leading to the station circle and small "feathers" sticking out at the end of this shaft. Look at Figure 12-34 for an example. The shaft is slanted in the direction from which the wind is coming. In the illustration, this is northeast. The length and number of feathers represent wind speed. Wind direction symbols are also given.

STATION MODEL SYMBOLS

Wind Speed (mph)	Symbol	Cloud Cover (%)	Symbol
1–2		0	
3–8		10	
9–14		20–30	
15–20		40	
21–25		50	
26–31		60	
32–37		70–80	
38–43		90	
44–49		100	
50–54			
55–60			
61–66			
67–71			
72–77			

▲ Figure 12-35

2. OBSERVE: What is the wind speed on the station model shown in Figure 12-34?

Teaching Options

EXTENDING THE LESSON

Researching Flood Impact Have students contact the local county extension agency or research past news articles to determine how floods have impacted their area. If floods are not an occurrence in their region, have students use the Internet to research floods in one region of the U.S. for the previous year. **learning style: linguistic**

ONGOING ASSESSMENT

Using the symbols in Figure 12-35, construct two different station models on the board. Then, write a scrambled list of the weather conditions associated with each symbol in the station models. Have students take turns coming to the board and drawing a line to connect the correct symbol to the weather condition it represents. **learning style: visual**

ESL/ELL STRATEGY

Understanding Station Model Symbols Be sure that English-language learners understand how the symbols in Figure 12-35 relate to the station model shown in Figure 12-34. Say aloud different wind speeds and ask students how many feathers should be used to indicate each.

Air Pressure In the United States, air pressure is measured in **millibars** (mb). A bar is a unit of pressure. It is based on the pressure created by a column of mercury that rises 750 mm at a temperature of 0°C. A millibar (mb) is a thousandth of a bar. Air pressure at the ground, or standard atmospheric pressure, is 1,013 mb.

Normal air pressure at ground level can range from 980 to 1,040 mb. The air pressure during a storm is usually lower than normal. From the outer edge of a hurricane to its center, air pressure can drop from over 950 mb to less than 800 mb. The lowest pressures ever recorded in the Western Hemisphere are associated with these storms. The steep drop in pressure creates the rapid, inward-spiraling winds of a hurricane. High pressure, on the other hand, usually brings with it clear, sunny skies.

3 ANALYZE: What is the air pressure shown on the station model in Figure 12-34?

CHECKING CONCEPTS

1. What does a station model show?
2. How can you tell how cloudy it is by looking at a station model?
3. How is wind direction shown on a station model?
4. How is the pressure measured?

THINKING CRITICALLY

5. MODEL: Cut out a weather map from your local newspaper. Gather other weather information about your local area. Using the weather map and the information you are able to gather, draw a station model of your local area for that day.

INTERPRETING VISUALS

Use the station model shown in Figure 12-34 to answer the following questions.

6. INTERPRET: From which direction is the wind blowing?
7. INTERPRET: How would you describe the cloud cover?
8. INTERPRET: What is the wind speed?

DESIGNING AN EXPERIMENT

Design an experiment to solve the following problem. Include a hypothesis, variables, a procedure with materials, and a type of data to study. Also tell how you would record the data.

PROBLEM: Suzie wants to collect data on the percentage of cloud cover in her neighborhood over a week's time. She will use the following materials to do so: a crayon, a ruler, and a large mirror. How can Suzie collect data using these materials?

Hands-On Activity

READING A STATION MODEL

Examine the station models shown. Then answer the questions.

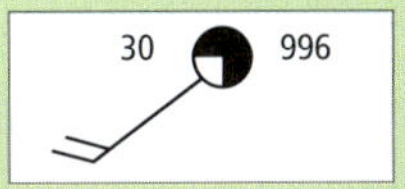

A

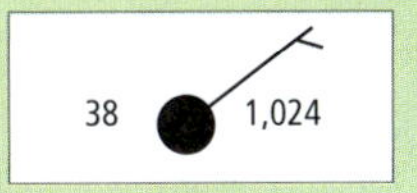

B

Practicing Your Skills

1. DESCRIBE: What is the sky condition for *A*?
2. MEASURE: What is the air pressure for *A* and *B*?
3. OBSERVE: What is the temperature for each?
4. **a.** OBSERVE: How fast is the wind blowing in each?
   b. IDENTIFY: From which direction is the wind blowing in each?
5. INFER: What kind of precipitation might be falling in *B*?
6. MODEL: Draw a station model that shows the conditions listed: **a.** air pressure, 1,018.8 mb; **b.** no cloud cover; **c.** temperature, 64°F; **d.** wind direction, from the southwest; **e.** wind speed, 35 mph.

MORE ON THE WEB

Interpreting Weather Maps Have the students go online to learn more about some of the most common weather map symbols. Then, have students design their own station model and attempt to interpret other students' maps.

• TEACHING RESOURCES • Online or Print

Teacher's Resources

Lesson Review: Chapter 12, p. 14

Enrichment: Chapter 12, p. 16

Designing an Experiment: Study Tools, p. 14

Designing an Experiment Rubric: Teaching Tools, p. 7

3 Assess

CHECKING CONCEPTS ANSWERS

1. the weather conditions for a particular weather station
2. by examining the amount of shading in the circle at the center of the station model
3. An arrow shaft is slanted in the direction from which the wind is coming.
4. in millibars

THINKING CRITICALLY ANSWER

5. MODEL: Check that students' models match the gathered information.

INTERPRETING VISUALS ANSWERS

6. INTERPRET: southwest
7. INTERPRET: 100%
8. INTERPRET: 21–25 mph

4 Close

DESIGNING AN EXPERIMENT

Use the Designing an Experiment Rubric in the Teacher's Resources to assess students' experiments. Fill in the rubric with the additional information below. For this assignment, students should have:

- described a method for collecting data, using the materials listed.
- developed a procedure for reading the data.

Hands-on Activity

READING A STATION MODEL

TIME: 10–15 minutes

PURPOSE: measuring, observing, inferring, modeling

COOPERATIVE LEARNING: Have pairs of students work together to answer the questions.

Practicing Your Skills Answers

1. DESCRIBE: 70–80% cloud cover
2. MEASURE: A: 996 mb; B: 1,024 mb
3. OBSERVE: A: 30°F; B: 38°F
4. **a.** OBSERVE: A: 21–25 mph; B: 3–8 mph
   b. IDENTIFY: A: southwest; B: northeast
5. INFER: rain
6. MODEL:

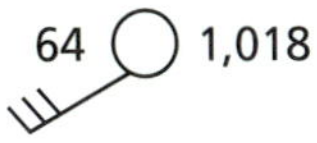

12-11 *(pp. 302–303)*

1 *Introduce*

Linking Prior Knowledge Remind students that the higher the temperature is, the lower the air pressure is (Lesson 11-5). Help students recall that air pressure is measured with a barometer (Lesson 11-6).

2 *Teach*

Demonstration Bring a large wall clock to class and remove the plastic or glass that covers its face. Use the hands of the clock to demonstrate clockwise and counterclockwise motion as you explain to students that in the Northern Hemisphere air moves around a high in a clockwise direction and around a low in a counterclockwise direction.
learning style: visual

Discussion Describe how to read a weather map. Using Figure 12-36, point out the weather conditions in different cities. Identify isobars, highs, and lows. Then, discuss the weather that occurs in high-pressure and low-pressure areas.
learning styles: visual, auditory

Reinforcement Prepare and distribute a data sheet listing weather factors such as temperature, pressure, wind direction and speed, dew point, cloud cover, and amount of precipitation for several cities around the United States. Also, distribute to students an outline map of the United States. Using an overhead projector and a transparency of the United States, construct a weather map with students, using Figure 12-36 as a guide. Have students mark their maps as you do the following on the transparency:

- Locate the cities and draw the symbol for cloud cover and precipitation.
- Add wind direction and wind speed.
- Add temperature.
- Locate various points of air pressure (including some points of equal pressure).

Construct a series of isobars on the transparency. Have students locate and mark one high (H) and one low (L).
learning style: visual

12-11 How do you read a weather map?

Objective

Read and interpret a weather map.

Key Terms

isobar (EYE-soh-bahr)**:** line on a weather map that connects points of equal air pressure

isotherm: line on a weather map that joins places that have the same temperatures

Weather Maps A weather map shows weather conditions for many places at one time. You can find the temperature, cloud cover, wind speed, and so on for many different locations. Data from more than 300 weather stations all over the country are assembled by the National Weather Service into weather maps each day. These maps are then used by newspapers and TV stations to prepare weather maps for their area.

1 DEFINE: What do weather maps show?

Reading a Weather Map Different weather maps can show different things. A key on a weather map helps you to read it. The pink lines that you see on the map in Figure 12-36 are called **isobars.** They connect points of equal air pressure. The pink numbers tell the pressure in millibars. Areas of high pressure are usually called highs. They are shown with an "H." Areas of low pressure are called lows. They are shown with an "L." On a different map, the pink lines might show isotherms. **Isotherms** are lines joining places that have nearly the same temperatures. Other symbols show fronts. Precipitation is shown as shading or with symbols like those in Figure 12-36.

2 OBSERVE: What was the temperature in Atlanta on the day this map was made?

Highs and Lows High-pressure regions have different weather from low-pressure regions. Highs usually bring clear skies. Lows usually bring cloudy skies and possibly precipitation.

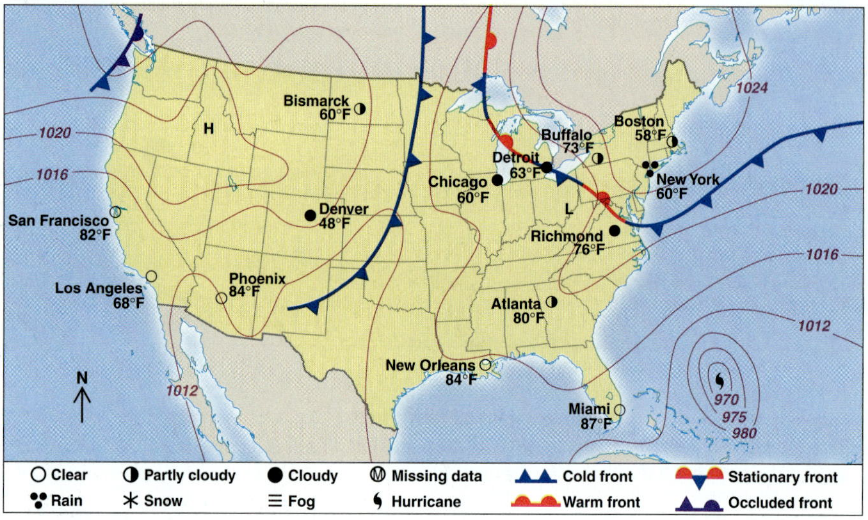

▲ **Figure 12-36** Weather map

302

Teaching Options

COOPERATIVE LEARNING

Weather Maps Most students have likely seen weather maps on television, in newspapers, or on the Internet. Have them work in small groups to list the kinds of information found on a weather map. After reading this lesson, have students revise their list.

ONGOING ASSESSMENT

Display a series of weather maps for a 2-week period. Have students describe how the highs and lows move across the country.
learning styles: visual, linguistic

Highs and lows usually move from west to east across the United States. As they pass through a region, they bring changes in the weather.

3 **DESCRIBE:** What type of weather does a low bring?

✓ CHECKING CONCEPTS

1. What are three weather conditions that you might find indicated on a weather map?
2. What do isobars show?
3. What are two ways to show precipitation on a weather map?
4. What do meteorologists usually call areas of low pressure?
5. What is the symbol weather maps use to indicate high pressure?
6. What do isotherms do?

THINKING CRITICALLY

Use Figure 12-36 to do the following exercise.

7. **IDENTIFY:** Find a city that has each of the following weather conditions.
   - **a.** clear skies
   - **b.** cloudy
   - **c.** partly cloudy
   - **d.** highest temperature
   - **e.** lowest temperature
   - **f.** low air pressure
   - **g.** inside a high

BUILDING SCIENCE SKILLS

Researching Many different symbols are used on weather maps. Do some research to find out some other weather symbols or different weather symbols for the same weather conditions. Draw and label the symbols on 3- × 5-inch index cards. What new symbols did you find? Make a list.

Real-Life Science

THE WEATHER PAGE IN A NEWSPAPER

Everyone is interested in the weather. Is it going to be sunny or stormy? Can we have our picnic? Will the ballgame be canceled? Weather has a big effect on our daily lives.

To find out the forecast, people often look in the newspaper. In it, you might find a four- to five-day local forecast, weather maps, a national forecast, and temperature graphs. Some newspapers contain a weather map for the local area. Others have a map of the country. Most newspapers have expected temperatures, in Fahrenheit. They also show fronts, precipitation data, and severe weather zones.

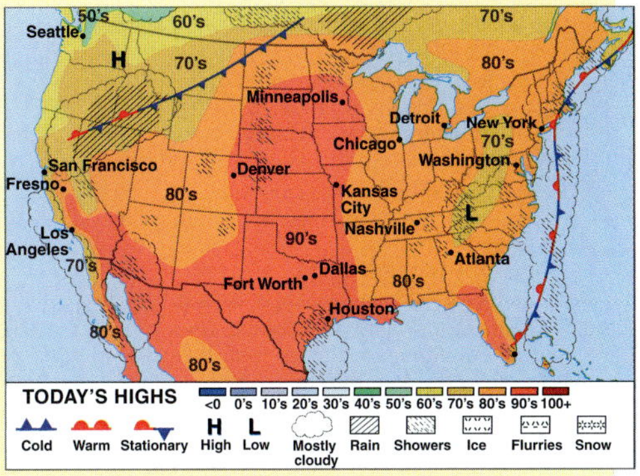

▲ **Figure 12-37** A typical newspaper weather map of the United States

Temperature zones may be shaded in different colors. High- and low-pressure areas are indicated. Symbols for the fronts, steady rain, showers, flurries, and snow are shown. Some papers give average temperatures for that month. Others have the temperatures for different cities around the world.

Thinking Critically Why do newspapers have so much weather data?

Answers to questions

1 **DEFINE:** weather conditions for many places at one time

2 **OBSERVE:** 80°F

3 **DESCRIBE:** cloudy skies and possibly precipitation

3 *Assess*

✓ CHECKING CONCEPTS ANSWERS

1. Possible answers: temperature, cloud cover, wind speed, and type of precipitation
2. places with the same air pressure
3. shading or symbols
4. lows
5. H
6. join places that have nearly the same temperature

THINKING CRITICALLY ANSWERS

7. **IDENTIFY:** Possible answers: **a.** New Orleans **b.** Chicago **c.** Atlanta **d.** Miami **e.** Denver **f.** Richmond **g.** Bismarck

4 *Close*

BUILDING SCIENCE SKILLS

Researching Collect students' index cards and display the different symbols on a bulletin board.

Real-Life Science

THE WEATHER PAGE IN A NEWSPAPER

Background Many weather forecasters on television are trained meteorologists. Before an actual news telecast begins, these professionals are busy behind the scenes. They spend time gathering information, analyzing satellite reports, organizing data in charts, and making weather predictions. In addition to a college degree, many meteorologists have backgrounds in communications and public speaking.

Discussion Have students discuss some ways that weather has affected their lives. Ask if they ever had to cancel plans because of the weather. If students are unable to come up with examples, point out that the clothes they wear each day are chosen with the weather in mind.

Thinking Critically Answer
to give people as much information as possible on a topic that is important to them

• TEACHING RESOURCES •
Online or Print

Teacher's Resources

Lesson Review: Chapter 12, p. 15

Enrichment: Chapter 12, p. 16

Challenges *(pp. 304–306)*

Chapter Summary

Review Before students begin the Challenges, review the objectives of each lesson with them. If time permits, have students work in pairs to come up with at least one relevant fact related to each objective.

STUDY HINT Have pairs of students take turns quizzing each other on key concepts in this chapter.

Key Term Challenges

MATCHING

1. transpiration (12-1)
2. saturated (12-2)
3. isobar (12-11)
4. tornado (12-9)
5. dew point (12-4)
6. capacity (12-2)
7. millibar (12-10)
8. station model (12-10)

APPLYING DEFINITIONS

9. Evaporation is changing of a liquid to a gas. Condensation is changing of a gas to a liquid. (12-1, 12-4)
10. Cirrus are light, feathery clouds. Stratus are clouds that form in layers. (12-5)
11. An air mass is a large volume of air that has the same temperature and amount of moisture throughout. A front is a boundary between different air masses. (12-7, 12-8)
12. A thunderstorm is a storm with thunder, lightning, and often heavy rain and strong winds. A hurricane is a tropical storm with very strong winds. (12-9)
13. Humidity refers to the presence of water vapor in air. Specific humidity is the actual amount of water vapor in air. (12-2)
14. A cold front is the forward edge of a cold air mass. A warm front is the forward edge of a warm air mass. (12-8)
15. Frost is ice formed from condensation below the freezing point of water. Precipitation is water that falls to Earth from the atmosphere. (12-4, 12-6)
16. An isobar is a line that connects points of equal pressure. An isotherm is a line that connects points of equal temperature. (12-11)

Chapter 12 Challenges

Chapter Summary

Lessons 12-1, 12-2, and 12-3

- The gaseous form of water is water vapor. Most of the water vapor in the air evaporates from the oceans. Living things also add water to air. Humidity is the amount of water in the air.
- Air is **saturated,** or filled to **capacity,** when it holds all the water vapor it can at a given temperature.
- **Specific humidity** is the actual amount of water vapor in air. **Relative humidity** is the amount of water vapor in the air compared with what air can hold at full capacity. Temperature changes cause the relative humidity to change. A **psychrometer** is used to find relative humidity.

Lessons 12-4, 12-5, and 12-6

- When saturated air is cooled, **condensation** takes place. Water vapor condenses at the **dew point.**
- Saturated air cooled close to the ground forms dew. **Frost** forms when the dew point is below water's freezing point.
- Clouds form from condensation in the atmosphere. The three basic kinds of clouds are **cirrus, cumulus,** and **stratus.** Fog is a cloud that forms near the ground.
- Types of **precipitation** are rain, snow, sleet, and hail. Precipitation is measured with a **rain gauge.**

Lessons 12-7 through 12-11

- An **air mass** is a large volume of air with about the same temperature and moisture level throughout. **Polar air masses** form over cold regions. **Tropical air masses** form near the equator. Either can form over land or water.
- A **front** is where two air masses meet. Moving fronts cause weather changes. A **cold front** is the forward edge of a cold air mass. A **warm front** is the forward edge of a warm air mass.
- **Thunderstorms** include lightning, thunder, strong winds, heavy rains, and sometimes tornadoes. **Tornadoes** have funnel-shaped clouds that touch down on Earth's surface. A **hurricane** is a tropical storm with strong, spiraling winds.
- **Station models** describe local weather conditions. **Millibars** are units of air pressure.
- Weather maps show the weather for an area.

Key Term Challenges

air mass (p. 292)	polar air mass (p. 292)
capacity (p. 280)	precipitation (p. 290)
cirrus cloud (p. 286)	psychrometer (p. 282)
cold front (p. 294)	rain gauge (p. 290)
condensation (p. 284)	relative humidity (p. 282)
cumulus cloud (p. 286)	saturated (p. 280)
dew point (p. 284)	specific humidity (p. 280)
evaporation (p. 278)	station model (p. 300)
front (p. 294)	stratus cloud (p. 286)
frost (p. 284)	thunderstorm (p. 296)
hurricane (p. 296)	tornado (p. 296)
isobar (p. 302)	transpiration (p. 278)
isotherm (p. 302)	tropical air mass (p. 292)
millibar (p. 300)	warm front (p. 294)

MATCHING **Write the Key Term from above that best matches each description.**

1. process in which plants give off water
2. filled to capacity
3. line on a weather map that connects points of equal air pressure
4. funnel-shaped cloud
5. temperature to which air must be cooled to reach saturation
6. amount of matter something can hold
7. unit of measurement for air pressure
8. record of weather data at a weather station

APPLYING DEFINITIONS **Explain the difference between the words in each pair. Write your answers in complete sentences.**

9. evaporation, condensation
10. cirrus cloud, stratus cloud
11. air mass, front
12. thunderstorm, hurricane
13. humidity, specific humidity
14. cold front, warm front
15. frost, precipitation
16. isobar, isotherm

304

Teaching Options

PREPARING STUDENTS FOR STANDARDIZED TESTS

Reading Strategy: Tell students to look for important words, such as *half* and *not,* in questions. These words can give students important clues about the specific amount or type of data being asked for.

Writing Strategy: Tell students to always use correct punctuation and grammar when writing.

Interpreting Visuals Strategy: Remind students to carefully read titles, captions, and labels in visuals and to put into their own words the information contained in them, to be sure that they understand what is being pictured.

ESL/ELL STRATEGY

Weather Maps Around the World Have students obtain weather maps from newspapers written in other languages and compare these to weather maps in U.S. newspapers. This should demonstrate to them the universal nature of weather forecasting.

Content Challenges TEST PREP

MULTIPLE CHOICE Write the letter of the term or phrase that best completes each statement.

1. Light, feathery clouds are called
   - **a.** cirrus clouds.
   - **b.** cumulus clouds.
   - **c.** stratus clouds.
   - **d.** nimbus clouds.
2. Each station model is marked by a
   - **a.** word.
   - **b.** circle.
   - **c.** square.
   - **d.** triangle.
3. In the United States, air pressure is measured in
   - **a.** millibars.
   - **b.** meters.
   - **c.** milliliters.
   - **d.** kilograms.
4. If air is filled to half its capacity, the relative humidity is
   - **a.** 100 percent.
   - **b.** 75 percent.
   - **c.** 50 percent.
   - **d.** 25 percent.
5. Fog forms over rivers and lakes when cool air moves in over
   - **a.** cold water.
   - **b.** warm water.
   - **c.** warm air.
   - **d.** cirrus clouds.
6. Air masses do not usually
   - **a.** form fronts.
   - **b.** move.
   - **c.** mix.
   - **d.** meet.
7. Tornadoes usually form during
   - **a.** early winter.
   - **b.** late fall.
   - **c.** early summer.
   - **d.** late winter.
8. Most of the water in the air evaporates from
   - **a.** the oceans.
   - **b.** the soil.
   - **c.** puddles.
   - **d.** lakes and rivers.

FILL IN Write the term or phrase that best completes each statement.

9. As temperature drops, the capacity of air for holding water goes __________.
10. Molecules in a liquid are always __________.
11. Air masses that form over oceans are __________.
12. Standard air pressure is about __________ millibars.
13. As the temperature of the air reaches the dew point, water vapor changes from a gas to a __________.
14. Lightning is caused when giant storm clouds discharge __________.
15. The way a cloud forms gives it its __________.
16. A front brings changes in the __________.
17. Relative humidity goes __________ if the temperature drops.
18. There is no wind or rain in the __________ of a hurricane.

ALTERNATIVE ASSESSMENT

Drawing Station Models Give students information about current weather conditions, such as temperature, wind speed and direction, cloud cover, and air pressure. Have students construct a station model based on this weather data. **learning styles: visual, logical/mathematical**

PORTFOLIO ASSESSMENT

Making Student Portfolios Students can demonstrate their comprehension of the concepts in this chapter by making a portfolio. The Lesson Reviews, the Big Idea Planner, and the Chapter Self-Check are some of the reproducibles in the Teacher's Resources that they can include in their portfolios. They can also include their weather prediction report for the Chapter Project or their individual contribution to group work in their portfolios. You can use the Portfolio Assessment Rubric also found in the Teacher's Resources to assess students' portfolios.

Content Challenges

MULTIPLE CHOICE

1. a (12-5)
2. b (12-10)
3. a (12-10)
4. c (12-3)
5. b (12-5)
6. c (12-7)
7. c (12-9)
8. a (12-1)

FILL IN

9. down (12-2)
10. moving (12-1)
11. moist (12-7)
12. 1,013 (12-10)
13. liquid (12-4)
14. electricity (12-9)
15. shape (12-5)
16. weather (12-8)
17. up (12-3)
18. eye (12-9)

Concept Challenges

WRITTEN RESPONSE

1. **RELATE:** Specific humidity is the actual amount of water vapor in the air. Relative humidity is the amount of water vapor in the air compared to the amount of water vapor the air can hold. (12-2, 12-3)
2. **CONTRAST:** Sleet is rain that freezes as it falls through cold layers of air, while hail is rain that freezes as it moves up and down in clouds. (12-6)
3. **EXPLAIN:** Cool air is denser than warm air and pushes the warm air upward. (12-8)
4. **EXPLAIN:** Evaporation of water from the cloth cools the wet thermometer more when the relative humidity is low. (12-3)
5. **APPLY:** Maritime polar air masses bring in cold, moist air. Continental polar air masses bring in cold, dry air. Maritime tropical air masses bring in warm, moist air. Continental tropical air masses bring in warm, dry air. (12-7)
6. **RELATE:** The changing of liquid water to a gas provides the water vapor in the atmosphere. When this water vapor changes back into a liquid or a solid, it falls to Earth as rain, snow, sleet, or hail. (12-1, 12-6)
7. **EXPLAIN:** It is measured in number of grams of water vapor in 1 kg of air. (12-2)
8. **EXPLAIN:** Rain forms when droplets of water in a cloud hit, join together, and fall. Snow forms when ice crystals do the same. Sleet forms when rain falls through a layer of cold air and freezes. Hail forms when water freezes around lumps of ice that are tossed up and down in a cloud by winds. (12-6)
9. **INFER:** Condensation occurred. (12-4)

INTERPRETING VISUALS (12-3)

10. 32%
11. 44%
12. 22°C

Concept Challenges TEST PREP

WRITTEN RESPONSE Answer each of the following questions in complete sentences.

1. **RELATE:** How are specific humidity and relative humidity related?
2. **CONTRAST:** What is the difference between sleet and hail?
3. **EXPLAIN:** Why does a cold air mass push under a warm air mass?
4. **EXPLAIN:** Why is the wet thermometer of a psychrometer cooled more when the relative humidity is low than when it is high?
5. **APPLY:** How do different air masses affect the weather in the United States?
6. **RELATE:** How does the process of changing a liquid to a gas relate to different forms of weather?
7. **EXPLAIN:** How is specific humidity measured?
8. **EXPLAIN:** How are each of the four kinds of precipitation formed?
9. **INFER:** The grass outside in the morning is wet, and it has not rained. What process probably took place to cause the dew?

INTERPRETING VISUALS Use Figure 12-38 below to answer the following questions.

10. What is the relative humidity when the dry-bulb thermometer reads 5°C and the wet-bulb thermometer reads 0°C?
11. What is the relative humidity when the dry-bulb thermometer reads 20°C and the wet-bulb thermometer reads 13°C?
12. If the relative humidity is 50 percent and the dry-bulb thermometer reads 30°C, what does the wet-bulb thermometer read?

Difference Between Dry-Bulb and Wet-Bulb Thermometer (°C)	Temperature of Air from Dry-Bulb Thermometer (°C)								
	–5	0	5	10	15	20	25	30	35
1	75	81	86	88	90	91	92	93	94
2	52	64	72	77	80	83	85	86	87
3	29	46	58	66	70	74	77	79	81
4	6	29	46	55	62	66	70	73	75
5		13	32	44	53	59	63	67	70
6			20	34	44	51	57	61	70
7			0	24	36	44	50	55	59
8				15	28	37	45	50	54

Relative Humidity (%)

▲ Figure 12-38

Teaching Options

Chapter Project Wrap-Up

PREPARING A WEATHER REPORT

You can use the Individual Activity Rubric found in the Teacher's Resources to assess students' reports. Fill in the rubric with the additional information below. For this project, students should have:

- written a report about the weather around the nation over a 5-day period and explained how they knew what the weather was like.
- predicted what the weather would be like on the sixth day and checked their predictions against a weather map published that day.

• TEACHING RESOURCES • Online or Print

Teacher's Resources

Key Term Review: Chapter 12, p. 17

Chapter 12 Test: pp. 18–20

Individual Activity Rubric: Teaching Tools, p. 5

Big Idea Planner: Study Tools, pp. 11–12

Chapter Self-Check: Teaching Tools, p. 11

Portfolio Assessment Rubric: Teaching Tools, p. 10

PLANNING GUIDE

◆ **TEACHING THE CHAPTER** This chapter should take approximately 5–8 days to complete instruction and assessment.

	Skills and Features	Projects and Activities	Achieve Science Literacy	Meet Individual Needs
Chapter 13 Opener p. T307	RESEARCH	• Chapter Project		
Lesson 13-1 **What is climate?** pp. T308–T309 Standard: D1j	COMPARE, INFER, DEFINE • Building Science Skills • How Do They Know That?	• Extending the Lesson • More on the Web	• Reading Strategy • Writing Hint	Teacher's Resources* Enrichment Activity
Lesson 13-2 **What factors determine climate?** pp. T310–T311 Standard: D1j	INFER, DESCRIBE, OBSERVE • Web InfoSearch • Real-Life Science		• Study Hint • Ongoing Assessment	• ESL/ELL Strategy • Reteach Teacher's Resources* Enrichment Activity Spider Map
Lesson 13-3 **What are climate zones?** pp. T312–T313 Standard: D1j	DEFINE, ANALYZE • Investigate	• More on the Web	• Study Hint • Ongoing Assessment	Teacher's Resources* Enrichment Activity Climate Zones
Lesson 13-4 **What are local climates?** pp. T314–T315 Standard: D1j	IDENTIFY, HYPOTHESIZE, EXPLAIN • Web InfoSearch • How Do They Know That?	• Extending the Lesson • Integrating Chemistry • Lab Challenge	• Study Hint	• ESL/ELL Strategy • Reteach Teacher's Resources* Enrichment Activity
Lab Activity **Investigating How Land Surface Affects Air Temperature** pp. T316–T317 Standard: D1j	OBSERVE, INFER, ANALYZE • Set-Up Time: 5–10 min • Lab Time: 40–45 min	• Cooperative Learning • Extending the Lesson • Lab Skills Worksheet • Extend the Lab Activity	• Tips for Using Technology	Teacher's Resources* Lab Activity Report Spreadsheet

*www.conceptsandchallenges.com, password: ccearth09

Planning Guide continues on next page.

Standards: For details on the correlation to National Science Standards see pages *xx–xxi*.

	Skills and Features	Projects and Activities	Achieve Science Literacy	Meet Individual Needs
Lesson 13-5 **How does climate affect living things?** pp. T318–T319 Standard: D1k	IDENTIFY, INFER, DEFINE • Integrating Environmental Science	• More on the Web	• Study Hint • Reading Strategy	• ESL/ELL Strategy
Big Idea **How do organisms adapt to climate?** pp. T320–T321 Standard: D1k	RESEARCH, COMMUNICATE, ANALYZE • Science Log: Writing Activity	• Big Idea • Close Activity	• Tips for Using Technology	• ESL/ELL Strategy Teacher's Resources* Big Idea Planner
Chapter 13 Challenges pp. T322–T324	• Chapter Summary • Key Term Challenges • Content Challenges • Concept Challenges	• Chapter Project Wrap-Up	• Study Hint • Preparing Students for Standardized Tests	• ESL/ELL Strategy Teacher's Resources* Chapter Self-Check Weekly Journal

*www.conceptsandchallenges.com, password: ccearth09

Teacher Notes

Chapter 13 Climate

▲ **Figure 13-1** Tree trunks reveal changes in climate over the years. This cross section is from a giant redwood.

Contents

Look at a cross section of a tree trunk. You should see a series of bands or rings. Starting in spring, the trunks of trees grow by producing bands of new material just beneath the bark. The bands produced in late summer look darker. This forms a pattern of rings. One combined ring represents one year of a tree's life. Rings provide a clue to climate. In good weather, there is more growth. That means the darker bands will be wider.

►About how old was the tree that produced the trunk in this picture?

Teaching Options

ASSESSMENT PLANNER

For assessment in the Student Edition, see the following pages:

Content Assessment
Checking Concepts: pp. 309, 311, 313, 315, 319
Thinking Critically: pp. 309, 311, 313, 315, 319
Interpreting Visuals: pp. 313, 319
Chapter Challenges: pp. 322–324

Alternative Assessment
Building Skills: p. 309
Web InfoSearch: pp. 311, 315
Science Log: p. 321

Performance-Based Assessment
Lab Activity: pp. 316–317

• TEACHING RESOURCES • Online or Print

Teacher's Resources

Lesson Review: Ch. 13, pp. 2, 3, 5, 6, 8

Enrichment: Ch. 13, pp. 4, 7

Key Term Review: Ch. 13, p. 10

Chapter 13 Test: pp. 11–13

Laboratory Manual: pp. 13–14, 69–72

Chapter 13 Climate

Chapter Overview

In this chapter, students will learn about the factors that influence climate. Students will explore the three main climate zones and how climate affects living things.

About Figure 13-1 Dendroclimatology is the science of using tree rings to study climate. The width of each ring indicates a tree's annual growth rate. This, in turn, is influenced by the amount of precipitation the tree received and the temperatures it experienced in a given year. The most useful records of past climates are found in trees that grow in relatively extreme conditions and are subject to a wide temperature range.

Answer about 80 years

Linking Prior Knowledge

For this chapter, students should recall:

- the concepts of latitude and longitude (Lesson 1-6).
- different types of ocean currents (Lesson 10-4).
- different types of precipitation (Lesson 12-6).

Chapter Project

TRAVEL BROCHURE

MATERIALS: paper, markers, research materials

Divide the class into three groups. Each group should produce a brochure for a famous tourist area from one of the three major climate zones. For example, one group can pick Alaska, another Florida or Hawaii, and a third New York. The brochure should list some of the major geologic sites in the area along with general climate information and weather. It should tell what kinds of clothes to pack and whether to expect precipitation in what months or parts of the day. It should also mention some local flora and fauna. The brochures can be created using a computer desktop publishing program or a pamphlet made from pictures cut from magazines.
learning styles: visual, tactile

13-1 (pp. 308–309)

1 Introduce

Linking Prior Knowledge Ask students to recall the two main types of precipitation (Lesson 12-6). (snow and rain) The type of precipitation and the amount that falls in an area help determine the climate of that area.

2 Teach

Demonstration Put some ice shavings or snow in a beaker. Have students mark the level of the ice shavings or snow. After the ice or snow has melted, have students note the level of the water. Tell them that meteorologists use the liquid volume of melted snow to determine how much precipitation an area receives.
learning style: visual

Discussion Ask students to describe the day's weather. They will probably mention wind conditions, humidity, and temperature, among other factors. Explain that weather is the day-to-day state of the atmosphere. Climate represents average weather conditions for an area over a long period. Stress that climate refers to more than just temperature and precipitation trends. Compare the climates of New York City and Miami to emphasize this fact.

Reinforcement Bring in newspapers that list daily high and low temperatures. Write the dates and the temperatures on the board. Have students calculate the average temperature range for each day by adding the high and low temperatures and dividing by 2. **learning style: logical/mathematical**

READING STRATEGY Remind students that tables, charts, and graphs compare information. Help students to understand what is being compared.

Answers to questions

1. **DEFINE:** the average weather conditions of an area from year to year
2. **DESCRIBE:** by adding 12 monthly averages and dividing by 12
3. **INFER:** New York City
4. **INFER:** Unlike Miami, New York City gets snow and hail, so it has the colder climate.

13-1 What is climate?

Objectives

Explain how weather and climate are related.
Identify the factors that determine climate.

Key Terms

weather: day-to-day conditions of the atmosphere

climate (KLY-muht): average weather conditions of an area over many years

Weather and Climate Air temperature, the appearance of the sky, winds, and the amount of moisture in the air all are part of weather. **Weather** is the day-to-day conditions of the atmosphere. The average weather conditions of an area from year to year is its **climate.** Climate describes the weather patterns of an area over time.

1. **DEFINE:** What is climate?

Average Temperatures Average monthly and yearly temperatures can be used to describe climate. The monthly average is found by adding together the daily average temperature for each day of the month and dividing by the number of days in the month. The yearly average temperature is a total of the twelve monthly averages divided by twelve.

2. **DESCRIBE:** How do you find daily average temperature?

Temperature Range How much the temperature changes during the year is also important in describing climate. This is temperature range. To find the temperature range, subtract the lowest monthly average temperature from the highest. In Eureka, California, for example, the average monthly temperature for July is 14°C. In January it is 9°C. In New York City, the average monthly temperature goes from 25°C in July to about −0°C in January. New York City has a greater temperature range than Eureka does. The temperature range for Eureka is 5°C. New York City has a temperature range of 31°C. Eureka and New York City have different climates partly because they have different temperature ranges.

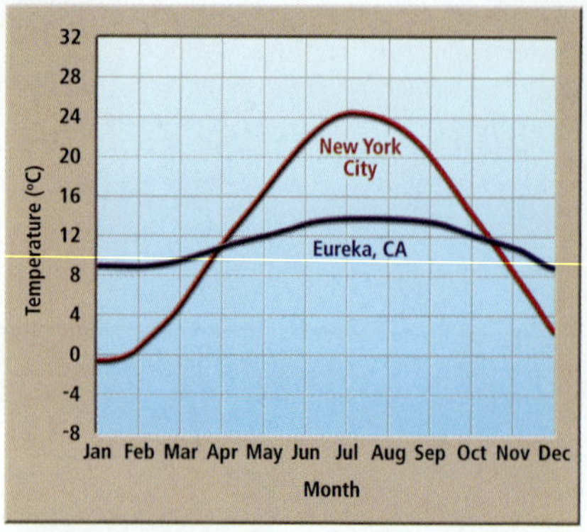

▲ **Figure 13-2** Average monthly temperatures for New York City and Eureka, California

3. **INFER:** Does New York City or Eureka have a colder climate?

Average Precipitation Average monthly precipitation is also used to describe climate. Precipitation is rain, snow, sleet, or hail. Average monthly precipitation is the average amount of water from rain, snow, sleet, or hail that falls in an area in a month. It is measured in centimeters or inches.

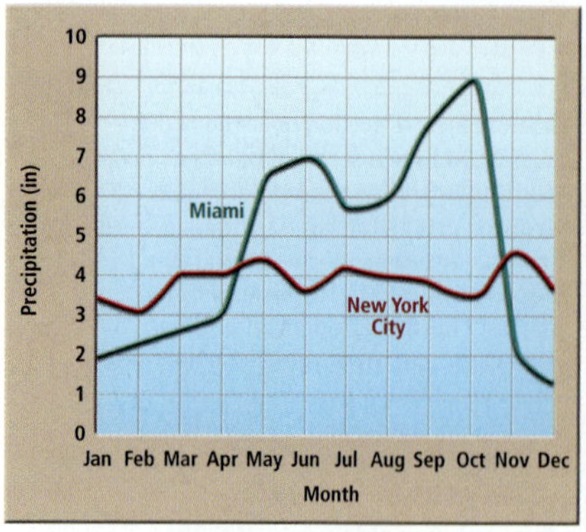

▲ **Figure 13-3** Average monthly precipitation for New York City and Miami, Florida, over a thirty-year period.

Teaching Options

! Fast Facts

Climate and Plants

A favorable climate allowed plants to spread onto land about 400 million years ago. During the last 80 million years, the relatively warm climate allowed flowering plants to diversify and become the dominant group. Today, there are about 250,000 species.

MORE ON THE WEB

Writing About Climate Changes Have students go online to explore the climate changes that have occurred in Antarctica over the past century. Have students write a newspaper article describing the effects on the lives of Antarctica's inhabitants and Earth's oceans.

EXTENDING THE LESSON

Researching Local Climates Have students research the climatic data for their area for the past year. Tell students to find out the average monthly temperatures, the average yearly temperature, the temperature range, the average monthly precipitation, and the average yearly precipitation. Have students organize their data in a table. **learning style: visual**

However, average precipitation is not enough to describe climate. Both Miami, Florida, and New York City get moderate amounts of precipitation yearly. However, Figure 13-3 shows that in New York City, precipitation falls steadily throughout the year. New York City gets rain, snow, sleet, and hail. In Miami, the most rain falls during a rainy season. The rainy season is May through October. Miami usually does not get snow or sleet. New York City and Miami have very different climates.

4 INFER: Which probably has a colder climate, Miami or New York City? Explain.

CHECKING CONCEPTS

1. The condition of the atmosphere today is the ___________.
2. The ___________ of an area describes the average weather conditions from year to year.
3. Temperature ___________ describes the yearly change in temperature.
4. Rain and snow are two different kinds of ___________.

THINKING CRITICALLY

5. CALCULATE: **a.** The high temperature in Chicago was 12°C. The low was 8°C. What was the temperature range that day? **b.** The high temperature in Houston was 25°C. The low was 21°C. What was the average temperature in Houston that day?
6. INFER: What two factors have the greatest effect on climate?

BUILDING SCIENCE SKILLS

Comparing and Contrasting In warm, dry climates, people usually wear loose-fitting, lightcolored, lightweight clothing. In cooler climates, they wear snug-fitting clothing and dark colors. Find out about the kinds of clothing that are appropriate to each kind of climate. What fabrics are best? How do the clothing styles of different cultures relate to their climates? Write a brief report on the relationship between clothing and climate.

How Do They Know That?

THE LOST COLONY OF GREENLAND

In A.D. 986, during a warm period, Erik the Red led an expedition to Greenland from Iceland. As a result, two small Norse settlements were established on the western coast. By the early twelfth century, there were about 10,000 people living there. They had cattle, sheep, and goats. There was plenty of wildlife. The colonists received supplies from Iceland and Scandinavia regularly.

Later that century, the weather in Greenland cooled sharply. This cooling of temperature was ahead of any climate change in Europe. The following century was even colder. These changes caused more frequent storms. More pack ice grew around the island. Visits from Icelanders decreased. By the late 1400s, both settlements had died out completely. One possible cause was malnutrition. This is the only recorded example of an established European society being completely wiped out.

▲ **Figure 13-4** Traces of the lost settlements of Greenland are still found there today.

Thinking Critically How did the change in climate cause the settlements to disappear?

QUICK CONVERSIONS		
	Celsius	**Fahrenheit**
Eureka temperatures		
July	14°C	57°F
January	9°C	48°F
Range	5°C	9°F
New York City temperatures		
July	25°C	77°F
January	−6°C	21°F
Range	31°C	54°F

• TEACHING RESOURCES •
Online or Print

Teacher's Resources

Lesson Review: Chapter 13, p. 2

Enrichment: Chapter 13, p. 4

WRITING HINT Remind students to organize the information they have gathered before writing a report. Preparing an outline is one way to organize. Each major section of the outline should have a topic sentence.

3 Assess

CHECKING CONCEPTS ANSWERS

1. weather
2. climate
3. range
4. precipitation

THINKING CRITICALLY ANSWERS

5. CALCULATE: **a.** 12°C − 8°C = 4°C
   b. (25°C + 21°C) ÷ 2 = 23°C
6. INFER: temperature and precipitation

4 Close

BUILDING SCIENCE SKILLS

Comparing and Contrasting Have students share their reports with the class.

How Do They Know That?

THE LOST COLONY OF GREENLAND

Background Between A.D. 1000 and A.D. 1250, Europe and other areas experienced a warming trend known as the Little Climatic Optimum. The warm temperatures allowed Norse settlers to explore and colonize previously harsh lands, such as Greenland. By the 1400s, however, temperatures began to fall, leaving Europe in the grip of the Little Ice Age, which lasted until the mid-1800s.

Discussion Ask students if they think the climate might change again in the future. Lead them to understand that Earth's climate has sometimes been very different from the climate of today and that it will likely be very different in the far future.

Thinking Critically Answer
Possible answer: The climate grew colder, and storms became more frequent. Supplies from Iceland and Scandinavia were cut off, leading to malnutrition among the settlers.

13-2 *(pp. 310–311)*

1 Introduce

STUDY HINT Have students create word webs in their notebooks. The topic should be *Climate Factors*. As students read the lesson, have them add arms to their webs on which they record keywords with their definitions and other relevant information about the word. A Spider Map from the Teacher's Resources can help students develop their webs.

Linking Prior Knowledge Remind students that lines of latitude run horizontally around Earth (Lesson 1-6). Ask students to identify the line of latitude that encircles the middle of Earth. (equator) Review the relationship between latitude and how much solar energy an area receives.

2 Teach

Demonstration Place a globe on a table. From about 25 cm (10 in.) away, shine a small flashlight directly over the equator. Have students note that the light is concentrated directly on the equator. Then, tilt the flashlight so that it shines on 60° north latitude. Students should observe that the light is now spread over a wider area and that its intensity has decreased. Have students relate the demonstration to variations in the intensity of sunlight received by different places on Earth.
learning style: visual

Discussion Point out that winds passing over ocean currents are either warmed or cooled, depending on the temperature of the current. The winds, in turn, heat or cool the land that they pass over.
learning style: auditory

Answers to questions

1. **RELATE:** Areas at low latitudes are warmer than areas at high latitudes.
2. **DEFINE:** height above sea level
3. **DESCRIBE:** Ocean currents raise or lower the temperatures of nearby land areas.
4. **OBSERVE:** the windward side

13-2 What factors determine climate?

Objective

Identify and describe the conditions that determine climate.

Key Terms

latitude (LAT-uh-tood)**:** distance in degrees north or south of the equator in degrees

altitude (AL-tuh-tood)**:** height above sea level

Latitude The climate of an area is affected by its latitude. **Latitude** is the distance in degrees north or south of the equator. Latitude determines how much heat energy an area gets from the Sun. At the equator, the Sun's rays fall almost directly on Earth. The closer an area is to the equator, the warmer is its climate. At higher latitudes, the Sun's rays strike Earth at more of an angle. The heat energy from the Sun is spread out, and the climate is colder. Global wind patterns at different latitudes also alter the number and kinds of storms in an area.

1 **RELATE:** How does latitude affect climate?

Altitude The height above sea level is called **altitude.** Air is warmer at sea level than it is at higher altitudes. The average air temperature drops about 1°C for every 100-m rise in altitude. Even near the equator, mountaintops are snow-covered all year.

▲ **Figure 13-5** This picture shows the kinds of vegetation that can be found at different altitudes, from mountaintop to sea level.

2 **DEFINE:** What is altitude?

Ocean Currents Ocean currents have an effect on the climate of areas along the seacoast. An ocean current is like a river of water within the ocean. Some ocean currents are warm. Other ocean currents are cold. Winds passing over ocean currents are either warmed or cooled by them. When these winds reach nearby land areas, they heat or cool the land.

3 **DESCRIBE:** How do ocean currents affect land temperatures?

Mountains When air passes over a mountain range, the air rises and cools. The side on which the air rises is called the windward side. Moisture condenses from the cooled air, causing it to either rain or snow.

As the air moves down the other side of the mountain, the leeward side, it is warmed and most of the moisture is removed. It rarely rains on the leeward side of a mountain. Many deserts, such as Death Valley, are found on the leeward sides of mountains. These areas are sometimes called rain shadow deserts.

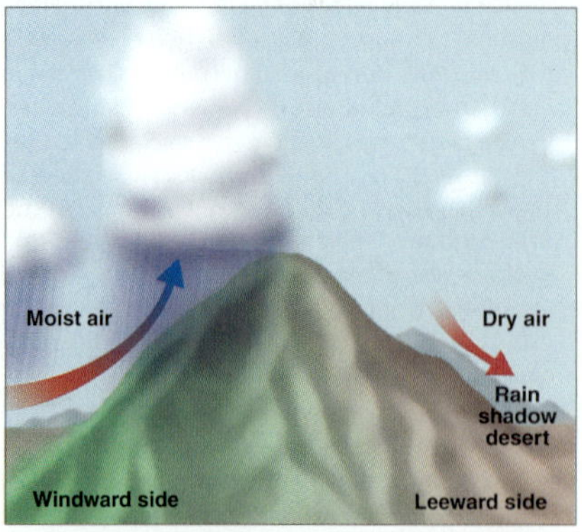

▲ **Figure 13-6** Different sides of a mountain often have different weather patterns.

4 **OBSERVE:** What side of a mountain faces the wind?

Teaching Options

RETEACH

The Effects of Mountains Use a copy machine to enlarge Figure 13-6, which shows how a mountain affects climate. Remind students that cold air can hold less water than warm air. Using the diagram, show students how mountains force air to rise up over them. Guide students to see how this rising air makes the windward side of a mountain moist and the leeward side dry.
learning style: visual

QUICK CONVERSIONS

	Metric/SI	English
Temperature drop	1°C	34°F
Altitude	100 m	328 ft
Rainfall		
Mt. Waialeale average	1,150 cm	453 in.
Cherrapunji average	1,125 cm	443 in.
Cherrapunji record	2,605 cm	1,026 in.
Seattle	93 cm	37 in.

ESL/ELL STRATEGY

Getting Familiar With the Rainiest Places Pronounce the place names in the feature and have students point out the places on a globe.

☑ CHECKING CONCEPTS

1. The closer an area is to the equator, the __________ its climate.
2. At __________ latitudes, the Sun's rays strike Earth at more of an angle.
3. Air at sea level is __________ than air at higher altitudes.
4. Land areas near cold-water currents usually have __________ temperatures.

THINKING CRITICALLY

5. CALCULATE: How much lower will the temperature be on top of a 1,500-m mountain than at sea level?
6. INFER: Miami, Florida, is at a lower latitude than San Francisco, California. Which city probably has a warmer climate? Explain.

Web InfoSearch

The Butterfly Effect Many factors influence the weather. A change in any one can affect the others in unpredictable ways. This is chaos theory. Weather patterns do repeat over the years. This is climate. However, past patterns are never repeated exactly. This is partly the result of the butterfly effect, which is the idea that a butterfly's wings stirring the air in one place can lead to large storms a month later far away.

SEARCH: Use the Internet to find out more about chaos theory. What besides the weather does it affect? Start your search at www.conceptsandchallenges.com. Some key search words are **chaos theory, climate change,** and **weather patterns.**

Real-Life Science
THE RAINIEST PLACES ON EARTH

You might think that the rain forests are the rainiest places on Earth. They do get more than 150 cm of rain per year. Some get up to 1,000 cm each year. Where could it rain more than that?

Actually, some of the rainiest places on Earth are found on windward mountain slopes. Mount Waialeale (wy-ahl-ay-AHL-ay) in Hawaii gets 1,150 cm per year, the highest average annual rainfall in the world. Cherrapunji (cher-uh-PUN-jee), India, is second, with about 1,125 cm a year. Cherrapunji holds the record for the most rainfall in one year, at 2,605 cm. Most of this rain fell in one month, July. Compare this with Seattle, Washington, which is known for its rainy weather. The average rainfall there is about 80 cm per year.

▲ **Figure 13-7** Mount Waialeale in Hawaii has the highest annual rainfall in the world.

Mountain areas get a lot of rain and snow. Runoff from mountains such as the Rockies can be a source of water for dry areas in the southwestern United States. Reservoirs in mountain areas can store spring runoff. This water can then be delivered to Los Angeles and other cities by a network of canals.

Thinking Critically Based on the climate, what do you think you would see on a hike up Mount Waialeale?

ONGOING ASSESSMENT

Have students collect national weather maps and tables from a newspaper for 1 week. Tell them to select five cities from the map and calculate the average daily temperature and daily temperature range for each city. They should also record the type of precipitation, if any. Students should compile their data in a table. After 1 week, tell students to write a summary that compares the weekly weather conditions of the cities.

• TEACHING RESOURCES • Online or Print

Teacher's Resources

Lesson Review: Chapter 13, p. 3

Enrichment: Chapter 13, p. 4

Spider Map: Study Tools, p. 9

3 Assess

☑ CHECKING CONCEPTS ANSWERS

1. warmer
2. higher
3. warmer
4. cooler

THINKING CRITICALLY ANSWERS

5. CALCULATE: (1,500 m ÷ 100 m/°C) = 15 Celsius degrees lower
6. INFER: Miami. Areas at low latitudes usually have warmer climates than areas at higher latitudes.

4 Close

Web InfoSearch

The Butterfly Effect Have students use the Internet to learn about the butterfly effect. Ask students: *What is chaos theory? What besides the weather does it affect?* Some key search words are **chaos theory, climate change,** and **weather patterns.** Students can begin their search at www.conceptsandchallenges.com.

Real-Life Science
THE RAINIEST PLACES ON EARTH

Real-Life Connection Cherrapunji, in northeastern India, receives much of its annual rainfall during the monsoon season.

Discussion Although Seattle is known for its rainy weather, it actually receives less total annual rainfall than either Miami or New York City. Ask students which city they think is the wettest: Seattle, Miami, or New York. Refer students back to Figure 13-3. Have them calculate average annual amounts of precipitation for New York City and Miami, then compare these amounts to Seattle's. (Average monthly amount for New York City: 9.9 cm [3.9 in.]; Average monthly amount for Miami: 11.4 cm [4.5 in.])

Thinking Critically Answer
Answers may vary. Some acceptable answers are: lush vegetation and plants; a thriving ecosystem.

13-3 (pp. 312–313)

1 Introduce

INVESTIGATE

TIME: 10–15 minutes

PURPOSE: Students will use a climatograph to compare and contrast climatic data.

MATERIALS: none

PROCEDURE: Have students make observations and come up with answers on their own. They should discuss their results with a partner and reach a consensus.

THINK ABOUT IT: Precipitation and temperature are related in Kansas City. As temperatures rise, so does precipitation.

STUDY HINT Display a globe while going over the definitions of the Key Terms. Point out which part of the globe each zone covers.

Linking Prior Knowledge Students learned that ocean currents move cool water into warm regions and warm water into cool regions (Lesson 10-4). Discuss what the climate zones would be like without this movement. (There would be greater climate differences between the zones.)

2 Teach

Demonstration Shine a bright light on a square piece of tin or a mirror. Have students observe how the light is reflected. Tell them that snow is also a good reflector of light. Polar climates are cold year-round partly because sunlight is reflected by ice and snow cover.
learning style: visual

Discussion Refer students to Figure 13-9, showing the three major climate zones on Earth. Describe the latitudes and temperatures of each zone. Have students make tables that list the characteristics of each climate zone. Ask when a table would be preferable to a narrative description of the three zones.
learning styles: visual, auditory

13-3 What are climate zones?

INVESTIGATE

Reading a Climograph

HANDS-ON ACTIVITY

1. Climographs allow scientists to compare different weather conditions to see if they are related. For example, scientists might compare hours of daylight and temperatures. Look at the climograph in Figure 13-8. In it, total rainfall and average monthly temperatures are compared.
2. Observe which month is warmest.
3. Observe which month has the most rainfall.

THINK ABOUT IT: Are precipitation and temperature related in Kansas City, Missouri? How can you tell?

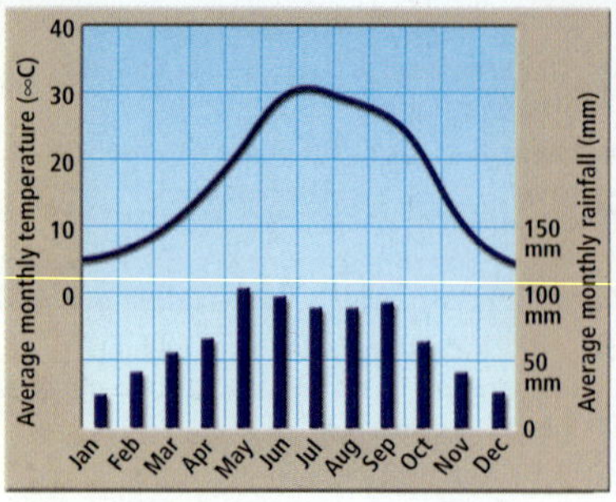

▲ **Figure 13-8** Climograph of Kansas City, Missouri

Objective

Identify and describe the three main climate zones.

Key Terms

tropical (TRAHP-ih-kuhl) **zone:** warm region near the equator

polar zone: cold region above 60°N and below 60°S latitude

middle-latitude zone: region between 30° and 60°N and S latitude

Latitudes and Climate Zones A climate zone is an area of Earth that has a certain temperature range and similar weather conditions. Figure 13-9 on the next page shows the three main climate zones. The warm zone near the equator is the **tropical zone.** It is between 30°N and 30°S latitude. The average monthly temperature is 18°C or higher. The coldest climate zones are the **polar zones.** These are above 60°N and below 60°S latitude. Temperatures do not go above 10°C. Between 30° and 60°N and S latitude are the **middle-latitude zones.** The temperature in the coldest months in these places averages no less than 10°C. The warmest month averages no colder than 18°C.

1 **DEFINE:** What is a climate zone?

Solar Energy and Climate Zones The biggest influence on climate and weather is energy from the Sun. The amount of solar energy received at a particular spot is determined by the tilt of Earth on its axis. This tilt influences the angle at which sunlight strikes Earth.

Climate conditions and weather are affected by how directly the Sun's rays strike an area. The Sun's rays are most direct and have the greatest effect at the equator. The places on Earth closest to the equator have the warmest climates.

Because Earth's poles get the least amount of sunlight, they are the coldest places on Earth. In between, it depends on the seasons. Other factors that can affect the climate of a region include topography, location of lakes and oceans, availability of moisture, global wind patterns, ocean currents, and location of air masses.

2 **INFER:** What is a state in the United States that has long, cold winters and short, warm summers?

Rainfall and Climate Zones In each climate zone, there are many smaller climate zones based on rainfall. In the tropics, climates may be very arid, or dry. They can also be very humid, or wet. In the polar zones, the climate is always cold and never humid. The middle-latitude zones have many different climates.

Teaching Options

FAST FACTS: *Koppen's Climate Classification System*

Climates are often classified using a system developed in 1918 by the Russian climatologist Wladimir Koppen (1846–1940). Originally, Koppen based his system on the different types of vegetation found in different areas. Later, he refined his system to include amount of precipitation and temperature.

Koppen divided climates into five main categories, all of which are represented by capital letters: tropical (A), dry (B), subtropical (C), temperate (D), and polar (E). These five climate classifications are subdivided into numerous categories according to variations in seasonal weather patterns, vegetation, amount of rainfall, and so on. The subdivisions are represented by additional capital and lowercase letters. For example, a tropical rain forest climate would be represented by the letters *Af*; a hot, dry, low-latitude climate would be represented by the letters *BWh*.

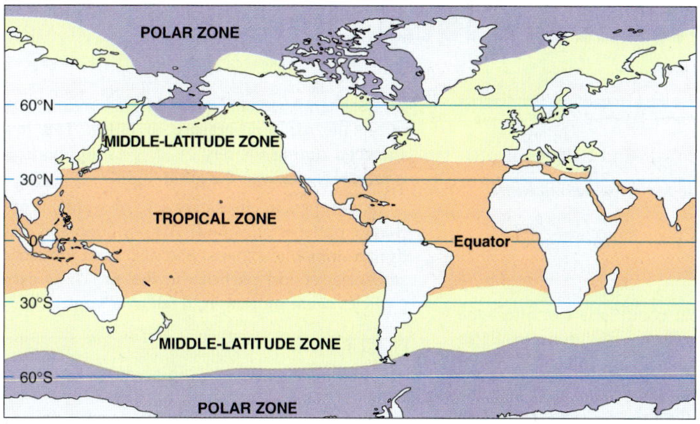

◀ **Figure 13-9** The three main climate zones are the tropical, middle-latitude, and polar zones.

The importance of climate cannot be overestimated. In addition to its effects on human life, climate determines the type of soil and vegetation found in a given area. This, in turn, determines how the land supports living things.

3 **ANALYZE:** What creates smaller climate zones within the larger ones?

CHECKING CONCEPTS

1. Earth is divided into __________ climate zones.
2. Places with the coldest temperatures are located in the __________ zone.
3. Places with the warmest temperatures are located in the __________ zone.
4. The middle-latitude zone has many different combinations of __________ and amounts of rainfall.
5. The middle-latitude zone is located between __________ N and S latitude.

THINKING CRITICALLY

6. **IDENTIFY:** Using Figure 13-9, indicate the lines of latitude that separate the tropical zone from the middle-latitude zones.

INTERPRETING VISUALS

Copy the table in Figure 13-10 onto a sheet of paper. Look at the three cities listed and their locations. Use the terms warm, moderate, *and* cold *to fill in the column titled "Type of Climate." Study the list of temperatures. Match each city with its temperature. Write the numbers in the table. Then, answer the questions.*

LATITUDE AND CLIMATE				
City	Location (Latitude)	Type of Climate	Average Temperature	
			January	July
Singapore	1∞N			
Point Barrow, AK	71∞N			
Boston, MA	42∞N			

▲ **Figure 13-10**

Coldest Month	Warmest Month
January average: 78°F	July average: 79°F
January average: 30°F	July average: 73°F
January average: −11°F	July average: 44°F

7. **EXPLAIN:** Why are regions closer to the equator (0° latitude) warmer than those farther from the equator?
8. **CALCULATE:** What is the yearly temperature range for each city listed in your table?

MORE ON THE WEB

Design a Virtual Garden Have students use the Internet to determine where certain climate zones are located. Students can design a virtual garden for their area. They should draw a model of their garden, labeling where their plant species would grow and explaining the characteristics of the plants that allow them to grow in this environment.

• TEACHING RESOURCES • Online or Print

Teacher's Resources

Lesson Review: Chapter 13, p. 5

Enrichment: Chapter 13, p. 7

Climate Zones: Visuals, p. 10

ONGOING ASSESSMENT

Have students find out the latitude and altitude of their community. Also, have them list factors that affect their climate, such as nearness to a body of water.

Reinforcement Have one student point to places on a world map or globe. Have another student determine the latitude of each place pointed to and identify the major climate zone in which it is located. Then, have students reverse roles. Students should figure out the temperature range of each place based on its climate zone. **learning styles: visual, auditory**

Answers to questions

1 **DEFINE:** an area of Earth with a given temperature range and similar weather conditions

2 **INFER:** Possible answer: Alaska

3 **ANALYZE:** rainfall

3 Assess

CHECKING CONCEPTS ANSWERS

1. three
2. polar
3. tropical
4. temperature
5. 30° and 60°

THINKING CRITICALLY ANSWER

6. **IDENTIFY:** 30°S and 30°N

INTERPRETING VISUALS ANSWERS

LATITUDE AND CLIMATE		
City	Location (Latitude)	Type of Climate
Singapore	1°N	warm
Pt. Barrow, AK	71°N	cold
Boston, MA	42°N	moderate
City	Average Temperature	
	January	July
Singapore	78°F	79°F
Pt. Barrow, AK	−11°F	44°F
Boston, MA	30°F	73°F

7. **EXPLAIN:** They receive more direct sunlight.
8. **CALCULATE:** Singapore: 1°F (−17°C); Pt. Barrow: 55°F (13°C); Boston: 43°F (6°C)

13-4 *(pp. 314–315)*

1 Introduce

STUDY HINT Before beginning this lesson, write the title of each section on the board. Have students copy these titles in their notebooks, using an outline format. As students read the lesson, they should write one or two facts for each section.

Linking Prior Knowledge Students learned that dark surfaces absorb heat more efficiently than light surfaces do (Lesson 11-3). Tell students that cities create microclimates partly because they contain many asphalt parking lots, sidewalks, and streets. These paved surfaces all absorb more heat than do grassy or forested surfaces. Students learned that humidity can change the comfort level of the air (Lesson 12-3). High humidity makes the air feel hotter, while low humidity makes the air feel cooler, even though the actual air temperature may be the same.

2 Teach

Demonstration To demonstrate how buildings can create microclimates, place one thermometer in an open area, such as a playground. Place a second thermometer against the school building so that it is in the shade for most of the day. Take students outside and have them read the temperatures of both thermometers. **learning style: visual**

Discussion Describe local climates and the conditions that form them. Identify altitude as the factor that most affects local climates. Then, describe microclimates. Ask students if they have ever noticed while listening to a weather report that the temperature in a city is always higher than in surrounding areas. Explain that cities are microclimates. Point out the things that affect the climate in a city, such as tall buildings, heavy traffic, and pollution. **learning style: auditory**

Answers to questions

1. **IDENTIFY:** altitude, large lakes, and forests
2. **EXPLAIN:** They act like mountains in that they can change rainfall patterns.
3. **PREDICT:** Most of the city would be underwater.

13-4 What are local climates?

Objective

Describe some factors that affect local climates.

Key Term

microclimate (MY-kroh-kly-muht): very small climate zone

Local Climate Local conditions can affect the climate of any area. These conditions result in small climate zones called local climates.

Altitude, or distance above sea level, has the greatest effect on local climates. Large lakes and forests also affect local climates. Like oceans, large lakes can warm or cool the temperature of the air. Forests are another factor. They slow down winds and add water vapor to the air. This increases humidity.

▲ **Figure 13-11** Large lakes can cool or warm the air temperature of nearby regions.

1. **IDENTIFY:** What are three factors that can affect local climates?

Microclimates Local climates can be broken down into even smaller climate zones. These very small climate zones are called **microclimates.** A microclimate can be as small as a schoolyard.

All cities have microclimates. The average temperature in a city is higher by several degrees than it is in surrounding areas. Heavy traffic in cities warms the air and raises average temperatures.

The energy used to heat and light buildings also raises the air temperature in cities. The heat absorbed during the day by streets and buildings is radiated back into the air at night.

Skyscrapers in cities may act as mountains and change patterns of rainfall. Air pollution and dust from automobile exhaust and also from industrial smokestacks add particles to the air. These extra particles cause more rain to fall over large cities.

▲ **Figure 13-12** Tall buildings, such as skyscrapers in New York City, create city microclimates.

2. **EXPLAIN:** How do skyscrapers affect the microclimate in cities?

Global Warming Climate has changed many times during Earth's history. These changes were mostly due to natural causes. Today, human activities may be changing Earth's climate. Increased air pollution may cause temperatures around the world to increase. This pattern of increased temperature is called global warming.

314

Teaching Options

INTEGRATING CHEMISTRY

Looking for Sources Tell students that the major greenhouse gases are carbon dioxide, nitrous oxides, methane, and chlorofluorocarbons. Have several students find out how these gases are produced naturally and in what quantities. Students should also find out how these gases are produced by human activities and in what quantities. Have them report their findings to the class. **learning style: auditory**

EXTENDING THE LESSON

Air Pollution Impurities in the air cause air pollution. The impurities can be particulates, or they can be tiny bits of solids, liquids, or gases. Both types of impurities can occur naturally or be caused by humans. Humans create pollutants mainly by burning fuels to run vehicles and factories and to create power, and by burning garbage. Have interested students research the types of pollutants commonly found in large cities. Students should present their data in a table, listing the pollutant, its source, and the damage it causes to living things.

Global warming could cause the polar ice caps to melt. Sea levels might rise as much as 60 m. New York City would be almost covered with water. Only the tops of very tall buildings would be above the water.

PREDICT: What would New York City look like if the ice caps melted?

✓ CHECKING CONCEPTS

1. What factor affects local climates the most?
2. Is the average temperature in a city lower or higher than that of surrounding areas? Why?
3. Can air pollution affect the amount of rainfall in an area? Explain.
4. What is global warming?

THINKING CRITICALLY

5. **HYPOTHESIZE:** Chicago is located in the north central United States. It is on one of the Great Lakes. How does this location affect the local climate?
6. **INFER:** Would you expect a city or its suburbs to get more snow? Explain.

Web InfoSearch

The Greenhouse Effect Earth's surface is surrounded by gases. Some of these gases trap heat from the Sun. Without these so-called greenhouse gases, Earth would be much colder. The warming that occurs is known as the greenhouse effect.

In the last few centuries, people have been burning wood, coal, oil, and gasoline for fuel. This has released huge amounts of carbon dioxide into the atmosphere. Carbon dioxide is a greenhouse gas. The burning of wood and fossil fuels is thought to be a major cause of global warming.

SEARCH: Use the Internet to find out more about the greenhouse effect. Will Earth continue to warm up? If so, what changes could occur? Start your search at www.conceptsandchallenges.com. Some key search words are **greenhouse effect, carbon dioxide, greenhouse gases,** and **global warming.**

How Do They Know That?

LONG-TERM CLIMATE TRENDS

Changes in the surface temperatures of ocean water can cause long-term changes in climate. These changes could explain sudden climate shifts, such as the Little Ice Age. The Little Ice Age was a cold period from about 1450 to 1850 that included many harsh winters in Europe.

Climate changes may occur because ocean water is always on the move. Scientists see the ocean as a kind of giant conveyor belt that moves and mixes water in a never-ending cycle around the globe.

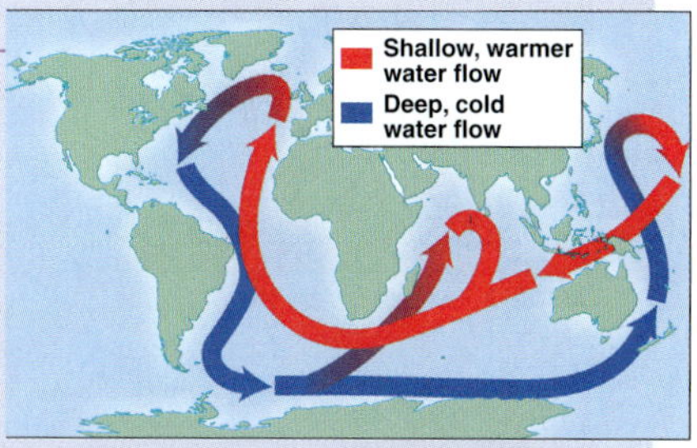

▲ **Figure 13-13** Ocean water makes a never-ending round-trip journey around the world.

Cold, salty water sinks into the deep ocean in the North Atlantic. The water flows south and then east around southern Africa. There, the water rises again to the surface. This water is then warmed in the Indian and Pacific oceans. Surface currents carry the warmed water back through the Pacific and South Atlantic oceans. The round trip takes between 500 and 2,000 years.

Thinking Critically How can ocean currents affect climate?

3 *Assess*

✓ CHECKING CONCEPTS ANSWERS

1. altitude
2. It is higher because heavy traffic warms the air, and tall buildings and paved surfaces absorb and radiate heat.
3. Yes. Water droplets can form around the particles and cause more rain to fall.
4. the pattern of increased temperatures worldwide

THINKING CRITICALLY ANSWERS

5. **HYPOTHESIZE:** It makes the local climate cooler and wetter.
6. **INFER:** Possible answer: Suburbs. City temperatures would be slightly higher, so they could get rain instead of snow.

4 *Close*

Web InfoSearch

The Greenhouse Effect Have students use the Internet to show how the greenhouse effect has developed. Some key search words are **greenhouse effect, carbon dioxide, greenhouse gases,** and **global warming.** Students should begin their search at www.conceptsandchallenges.com.

How Do They Know That?

LONG-TERM CLIMATE TRENDS

Real-Life Connection Scientists hypothesize that some long-term changes in climate may be caused by variations in sunspot cycles. The English scientist E. W. Maunder noted that during the Little Ice Age sunspot activity was extremely low. The term *Maunder minimum* is used to describe this dip in sunspot activity.

Discussion Ask students: *How are ocean currents similar to the global winds you learned about in Chapter 11?* (Cold, dense water sinks and moves toward the equator, where it warms and rises. Cold, dense air also sinks and moves toward the equator, where it warms and rises.)

Thinking Critically Answer
by changes in the surface temperatures of ocean currents

RETEACH

Exploring Microclimates Show students a map of their state that includes a key for the population size of cities and towns. Explain how to use the key. Point to various rural and urban areas on the map, and have students infer which places would have the highest temperatures.
learning style: visual

ESL/ELL STRATEGY

Micros Tell students that *micro-* means "very small." Give them the words *microbiology*, *microchip*, and *microcomputer*. Have them look up the meanings of these words.

• TEACHING RESOURCES •
Online or Print

Teacher's Resources
Lesson Review: Chapter 13, p. 6
Enrichment: Chapter 13, p. 7

Laboratory Manual
Lab. Challenge: How does the greenhouse effect contribute to global warming? pp. 69–72

QUICK CONVERSION

	Metric/SI	English
New York submerged	60 m	197 ft

Lab Activity

(pp. 316–317)

1 Prepare the Lab

Investigating How Land Surface Affects Air Temperature

SET-UP TIME: **LAB TIME:**

BACKGROUND: By 2025, an estimated 80 percent of the world's population will live in urban areas. This increase in urban populations will likely spark a corresponding increase in urban heat islands, a type of microclimate where the temperature of a city is as much as 5.5°C (10°F) higher than the temperature of surrounding areas. This higher temperature is caused by the large expanses of paved surfaces found in urban areas. In many cities, urban planners are designing ways to incorporate more green space into city landscapes. Large areas of vegetation significantly reduce the heat-island effect. In addition, some cities are experimenting with light-colored roofing materials. This reduces the heat-island effect because light colors reflect more heat than dark colors do.

PURPOSE: Students will investigate the relationship between different surface materials and rates of heat absorption.

ALTERNATIVE MATERIALS: Alcohol thermometers can be used in place of mercury thermometers.

SCIENCE SKILLS: Students will **measure** and **record** changes in temperature at three test sites. They will **compare** and **contrast** how different materials absorb heat and relate these differences to microclimates in cities.

ADVANCE PLANNING: Select an outside location for the test sites before beginning the lab. The test sites should be sunny and sheltered from strong winds.

2 Guide the Lab

PROCEDURE: You can divide the class into groups of three. Each group can set up and monitor its own test sites. Discuss the procedure for the lab and clarify any misunderstandings beforehand. Emphasize that students should follow a consistent method when measuring air temperature.

LAB ACTIVITY

Investigating How Land Surface Affects Air Temperature

Materials

- Safety goggles
- Lab apron, gloves.
- 3 Paper or foam plates
- 3 Thermometers
- Aluminum foil
- Dark rock or pieces of asphalt
- Grass sod
- Light-colored sand

BACKGROUND

Year round, cities are warmer than suburbs and rural areas. When land is cleared of trees to make way for roads, buildings, and parking lots, the local climate is changed.

PURPOSE

In this activity, you will investigate the relationship between air temperature and land surface.

PROCEDURE

1. Put on goggles. Prepare three plates for your experiment. Cover one plate with grass sod. Cover another plate with dark-colored rock or asphalt. Cover the last plate with light-colored sand.
2. Copy the chart in Figure 13-14.
3. Cover the end, or bulb, of each thermometer with aluminum foil. Take the plates outside and place them in the sunlight. Hold your thermometer 2.5 cm above the middle of each plate. In your chart record the temperatures.
4. Measure and record the temperatures again 5 minutes later.
5. Measure and record the temperatures after 10 minutes.

▲ **STEP 1** Pour sand into one plate.

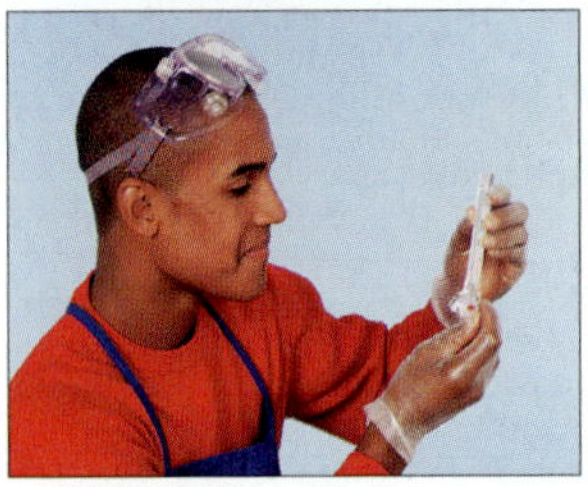

▲ **STEP 3** Use foil to cover your thermometer bulbs.

Teaching Options

! FAST FACTS *The Heat Index*

Because of the heat-island effect, summer temperatures in cities can reach dangerously high levels, particularly when the air is humid. Humidity reduces rates of evaporation, which in turn reduces the body's ability to cool itself. Severe heat waves can cause sunstroke, heat exhaustion, and death. To help people deal effectively with the dangers of excessive heat, the National Weather Service publishes the heat index. When the heat index is between 90°F and 105°F, extreme caution is advised for outdoor activities. A heat index above 105°F is considered extremely dangerous.

COOPERATIVE LEARNING

Reporting Data If, during the lab, students worked in groups of three, have each group give a brief oral report summarizing the data from a test site.
learning style: auditory

EXTENDING THE LESSON

Another Experiment Have students repeat the lab, using soil and water. Students will find that water heats up and cools down more slowly than soil.
learning style: tactile

6. Measure and record the temperatures after 15 minutes.

7. Take your plates out of the sunlight, so that they can start to cool. In your chart, below the first group of numbers, record the temperatures after 5, 10, and 15 minutes for each plate.

STEP 7 ▶ Take the temperatures above the three plates as they cool.

Air Temperature and Land Surfaces			
Time Outdoors	Temperature above dark surface	Temperature above grassy surface	Temperature above light surface
0 minutes			
5 minutes			
10 minutes			
15 minutes			
Cooling Times			
0 minutes			
5 minutes			
10 minutes			
15 minutes			

▲ **Figure 13-14** Use a copy of this chart to record the temperatures.

CONCLUSIONS

1. **OBSERVE:** Which plate had the highest temperature?
2. **OBSERVE:** Which plate had the lowest temperature? Which cooled fastest? Which stayed warmer the longest?
3. **INFER:** Why was there a difference?
4. **ANALYZE:** How could cities be changed to make them cooler?

Tips for Using Technology

Using Spreadsheets to Create a Table Have students use spreadsheet software to keep track of the data they are collecting for this activity. Spreadsheets allow students to enter and edit information, to arrange information by putting it in a specific order, and to present information by using special layouts such as graphs. Have students use the Spreadsheet graphic organizer in the Teacher's Resources to plan their spreadsheets.

• TEACHING RESOURCES • Online or Print

Teacher's Resources

Lab Activity Report: Study Tools, pp. 16–17

Laboratory Activity Rubric: Teaching Tools, p. 8

Spreadsheet: Study Tools, p. 23

Laboratory Manual

Lab. Skill Worksheet 4: Organizing and Analyzing Data, pp. 13–14

Provide students with a copy of the Lab Activity Report found in the Teacher's Resources to record their observations and conclusions.

TROUBLESHOOTING: Wind will cool the air over the plates. A piece of cardboard can be used to shelter the materials.

SAFETY TIP: If students are using mercury thermometers, caution them to handle the thermometers with care. If one should break, stress that they should not touch the mercury or broken glass. They should notify you immediately. Also, ask if any students are allergic to grass. Design an alternative procedure for these students.

EXPECTED OUTCOME: Students will conclude that the air above light-colored surfaces and grassy surfaces is cooler than the air above dark surfaces.

LAB HINT: If your school has grassy, concrete, and asphalt surfaces nearby, temperature data can be collected without constructing test sites.

3 Conclude the Lab

1. **OBSERVE:** In general, the air above the dark surfaces should be warmer than the air above the light surfaces.
2. **OBSERVE:** In general, the air above the light surfaces should be cooler than the air above the dark surfaces. The sand will probably cool the fastest and the asphalt the slowest.
3. **INFER:** Different surfaces absorb different amounts of sunlight.
4. **ANALYZE:** Possible answer: Cities could plant more trees and use light-colored paving materials.

Use the Laboratory Activity Rubric from the Teacher's Resources to assess students' lab activities. Fill in the rubric with the additional information below. For this activity, students should have:

- set up the experiment in order to observe air temperature.
- accurately recorded data and observations.

4 Extend the Lab

Discuss why people wear dark clothing in the winter and light-colored clothing in the summer. Lead students to understand how the heat absorption rates of different materials affect our lifestyles. Ask students to identify ways to save on heating and cooling costs, based on what they learned in this lab.

13-5 *(pp. 318–319)*

1 Introduce

STUDY HINT To reinforce the main idea of this lesson, have each student choose a biome and research one plant and one animal native to that biome. Students should be prepared to explain their choices.

Linking Prior Knowledge Ask students where a polar bear lives. Most will know that polar bears live only in cold climates. Tell them that many other organisms—both plants and animals—can live only in certain types of climates. In this lesson, they will learn how climate influences living things. Students learned about latitude and longitude (Lesson 1-6). Lines of latitude tell how far north or south of the equator a place is. The cold climates where polar bears live occur in the high northern latitudes.

2 Teach

Demonstration Show students a live cactus or a photo of a cactus. Point out the plant's spines and thick, leathery stem. Explain that these features are adaptations for life in the desert. Ask students how the adaptations help a cactus survive. (They help the plant retain water.) **learning style: visual**

Discussion Discuss biomes. Point out that precipitation and temperature are the two factors that determine a land biome. Emphasize that scientists classify biomes according to vegetation. The vegetation of a biome, in turn, influences the type of animals found in an area. **learning style: auditory**

Answers to questions

1. **IDENTIFY:** by the major types of plants growing in an area
2. **DEFINE:** the plants in an area
3. **NAME:** Some animals eat only certain plants and thus live only where those plants grow. Other animals can live only in climates that have a certain temperature range.

READING STRATEGY After reviewing Figure 13-15, have students summarize in their notebooks what the illustration is showing by listing three important points that it makes.

13-5 How does climate affect living things?

Objective

Identify ways in which plants and animals are affected by climate.

Key Terms

biome (BY-ohm): large region with a characteristic climate and plant and animal communities

vegetation (vehj-uh-TAY-shuhn): plants

Biomes A large region with a characteristic climate and certain kinds of living things is called a **biome.** Scientists identify a biome mainly by the major types of plants growing in the area. Because many animals eat only certain kinds of plants, each biome also has a particular group of animals and other organisms living there.

1 IDENTIFY: How do scientists identify a biome?

Climate and Vegetation Climate affects the vegetation in an area. **Vegetation** refers to the plants in an area. Rainfall and temperature are the climate factors that most affect vegetation.

Some kinds of plants need a lot of water. For example, many species of trees grow only in tropical rainy climates. Some plants grow where there is very little water. Cacti are an example. They grow in desert climates.

Many plants cannot survive very cold temperatures. Live oak trees, for example, may be killed by frost. Orange trees need both rainfall and warm temperatures to grow well.

2 DEFINE: What is vegetation?

Climate and Animals Animals that eat certain plants can live only in climates where those plants grow. For example, eucalyptus leaves are the main food of koalas. Koalas live only in areas where eucalyptus trees can grow.

▲ **Figure 13-16** Koalas eat mainly eucalyptus leaves.

Figure 13-15 ▶ Major land biomes found on Earth

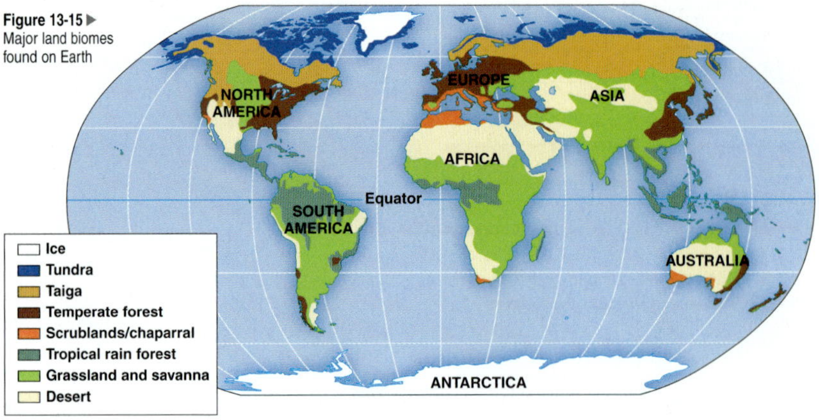

318

Teaching Options

! FAST FACTS *Irrigating Deserts*

Plants need sunlight, minerals, nutrients, and water to grow properly. Desert soil is rich in minerals and some nutrients. In addition, deserts receive ample sunlight. However, they are very dry. This problem is overcome in some places through irrigation, or the watering of land by artificial means.

There are four ways to irrigate land. Water can be run directly on a field or into narrow ditches called furrows. Huge sprinkler systems can also be used for irrigation. Plastic tubes can be placed next to plants; water then trickles out of the tubes. Underground pipes can carry water to plants.

Irrigation is widely used in the desert regions of the western and southwestern United States. The Imperial Valley in south-central California, for example, is irrigated. It is one of the richest farming areas in the United States.

Other animals are directly affected by temperature. The body temperature of cold-blooded animals is about the same as the air temperature. The animals would die if the air temperature were too low. Most snakes, for example, live in warm climates. They could not survive in a cold climate.

NAME: What are two ways animals are affected by climate?

CHECKING CONCEPTS

1. What is a biome?
2. How are scientists able to identify different biomes?
3. What are all the plants in a biome called?
4. What are two climate factors that can affect vegetation?
5. What kind of climate do most species of snakes live in?

THINKING CRITICALLY

6. INFER: Giant pandas are bearlike animals. They eat only bamboo shoots. Bamboo grows only in China and Tibet. Where do you think giant pandas live? Explain.
7. HYPOTHESIZE: There is a forest biome in the northern United States that is called the spruce-moose belt. How might this area have received this name?

INTERPRETING VISUALS

Use Figure 13-15 to answer the following questions.

8. IDENTIFY: Which type of biome covers the most land area?
9. ANALYZE: Which type of biome has hot and wet weather?
10. INFER: Which type of biome has hot and dry weather?

Integrating Environmental Science

TOPICS: natural resources, ecosystems

TROPICAL RAIN FORESTS

Tropical rain forests are very dense areas of vegetation. They have hot, humid climates. Many different kinds of animals and plants live in rain forests. Scientists estimate that in 10 square km, there may be more than 750 different species of trees and more than 1,000 different species of animals.

Many rain-forest plants are natural resources for medicines and other products. Some rain-forest animals are found nowhere else. Scientists are concerned because the rain forests are being cut down, burned, and used for farming and development. Once the nutrients in the soil are gone, farmers move to a new area. Many countries are trying to stop the destruction of the rain forests. They also hope to save some species that are in danger of extinction.

▲ **Figure 13-17** Tropical orchids produce the flavoring of vanilla from vanilla beans. The beans (right) are found in rain forests from Florida to South America.

Thinking Critically What are some ways governments can help stop the rain forests from being destroyed?

3 Assess

CHECKING CONCEPTS ANSWERS

1. a large region with a characteristic climate and plant and animal communities
2. mainly by the major types of plants in an area
3. the vegetation
4. rainfall and temperature
5. warm

THINKING CRITICALLY ANSWERS

6. INFER: China and Tibet, because those are the only places where giant pandas can find food
7. HYPOTHESIZE: There are many spruce trees and moose there.

INTERPRETING VISUALS ANSWERS

8. IDENTIFY: ice
9. ANALYZE: tropical rain forest
10. INFER: desert

4 Close

Integrating Environmental Science

TOPICS: natural resources, ecosystems

TROPICAL RAIN FORESTS

Life-Science Connection A tropical rain forest is made up of three layers. The canopy consists of the tops of trees, which form a dense layer. Many animals, such as lemurs and monkeys, spend most of their lives in the canopy. The second layer, the understory, is made up of smaller plants that have adapted to low light. The bottom layer, the forest floor, is covered by leaves and organic matter that rapidly decompose in the hot, humid environment.

Discussion Ask students: *Why must the plants in the understory be adapted to low light?* (because the forest canopy blocks out most of the sunlight)

Thinking Critically Answer
Possible answers include by passing and enforcing strict laws, teaching people about the importance of preserving the rain forest, and showing how people could make a better living from intact forests.

MORE ON THE WEB

Investigating Biomes At www.conceptsandchallenges.com, students can investigate the different types of biomes. Have students create a chart of seven biomes, describing where each biome is located, its climate, and examples of plants and animals that live in that biome.

• TEACHING RESOURCES • Online or Print

Teacher's Resources
Lesson Review: Chapter 13, p. 8

ESL/ELL STRATEGY

Making Connections Make the connection for English-language learners between the words *vegetables* and *vegetation*. *Vegetation* is the general term for all plants, while *vegetables* are edible plants with certain useful nutrients for humans and other animals.

The Big Idea
(pp. 320–321)

1 Introduce

Objective Students will learn how certain organisms have adapted to a particular climate.

Linking Prior Knowledge Ask students to recall what they know about a climate's effect on living things. Have them describe how animals are affected by climate. Have students review the different climate zones before reading the Big Idea.

2 Teach

Discussion Discuss the ways that people adapt to different environments. For example, people who live in tropical climates wear different types of clothing and live in different types of homes than do people who live in colder climates.
learning style: auditory

Demonstration Write *Adaptations* on the board. Write the following subheads under the main head: *Polar Bear*, *Camel*, and *Cactus*. Leave space between each subhead. Have students take turns coming up to the board and writing adaptations for each organism under the correct subhead.
learning style: linguistic

Reinforcement Tell students that adaptations can be structural or behavioral. A camel's long eyelashes and rounded hump are examples of structural adaptations. Migration and hibernation are examples of behavioral adaptations. Have students research an animal that shows structural adaptations. Then, have students draw the animal and its unique structural adaptations. **learning styles: tactile, visual**

Integrating Life Science

THE Big IDEA

How do organisms adapt to climate?

In nature, you would never see a palm tree and a cactus plant growing in the same location. Each type of plant is suited to a particular climate. A palm tree grows best in a warm, moist, tropical climate. A cactus grows best in a hot, dry, desert climate. A palm tree could not survive in the Arctic. A cactus could not survive in a tropical location.

Most living things have adaptations that help them survive in their climate zone. Adaptations are features that let an organism live and reproduce in its environment.

Look at the map and photographs that appear on these two pages. Then, follow the directions in the Science Log to learn more about "the big idea."✦

Bighorn Sheep
Bighorns are wild sheep found only in North America. Their feet are adapted to gripping slippery surfaces. This helps them leap across mountain ledges.

Sloth
Sloths spend most of their lives hanging from tree branches in the tropics. Green algae growing on its fur becomes camouflage for the sloth.

TROPICAL RAINY	TEMPERATE CONTINENTAL
Tropical wet	Humid continental
Tropical wet-and-dry	Subarctic
DRY	**POLAR**
Semiarid	Tundra
Arid	Ice cap
TEMPERATE MARINE	**HIGHLANDS**
Mediterranean	Highlands
Humid subtropical	
Marine west coast	

Penguin
Unlike other birds, penguins cannot fly. Their winglike flippers are well suited to their watery life. Thick layers of body fat help them survive in the cold polar regions.

▲ **Figure 13-18**
Each type of living thing has adaptations suited for life in its particular climate zone.

320

Teaching Options

! FAST FACTS Climates and Soils

Climate plays a large role in determining an area's soil type. Rainfall and temperature affect rates of weathering. In hot, wet climates, weathering occurs quickly. The resulting soils are usually thin. Most of the humus, or decayed organic material, is washed away by heavy rains. In warm, moist climates with changing seasons, soils are often thick. Grasslands and prairie soils in particular have rich layers of topsoil and are excellent for growing crops. In cold, dry climates, rates of weathering are slow. The ground is barren rock covered by an extremely thin layer of soil.

ESL/ELL STRATEGY

Working With Peers Working with students who are proficient in English can help English-language learners to communicate. Pair an English-language learner with an English-proficient student. Have both students explain the article and the captions in their own words to each other.

Polar Bear

Polar bears have light-colored fur that blends into the snow. Their heavy fur coats and layers of body fat keep the animals warm in frigid waters.

Arctic Lichen

Arctic lichens are fungi that partner with algae. Because lichens lack roots, they can survive the cycle of thawing and freezing that occurs in the permanently frozen ground that covers the Arctic.

Giant Rafflesia

Rain forests are home to the largest flowers in the world, the giant rafflesias. The plants that produce these huge flowers lack leaves and stems. They smell like rotting meat to attract the flies that pollinate them.

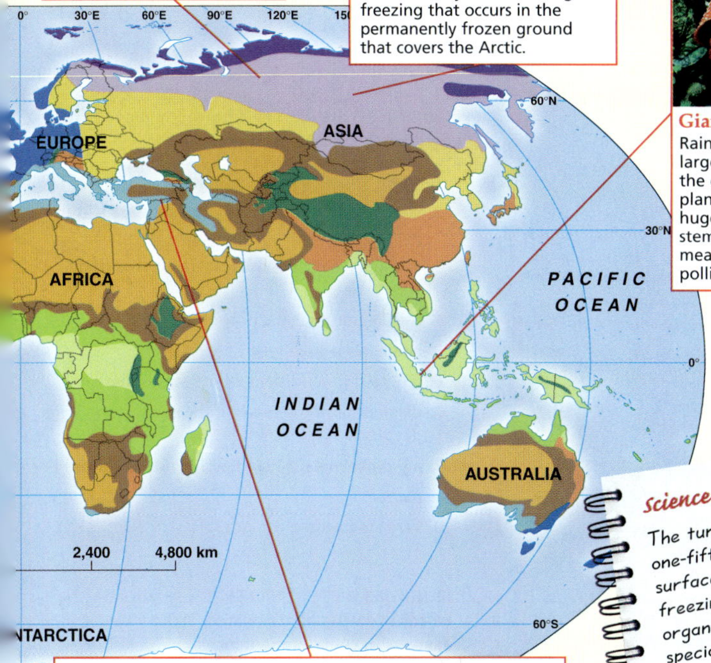

Camel

Camels have long eyelashes that protect their eyes from blowing sand. Thick eyebrows shade their eyes from the harsh Sun. The hump is a pad of fat used for energy when food is scarce. Because it does not sweat much, a camel can hold a lot of water in its body.

Science Log WRITING ACTIVITY

The tundra covers about one-fifth of Earth's land surface. To survive its freezing temperatures, organisms must have special adaptations. Find out what animals live in the tundra. Pick one and write an essay explaining how it can survive in this harsh environment. Start your search at www.conceptsandchallenges.com.

THE BIG IDEA

How do organisms adapt to climate? Have students start their research for their Science Log Writing Activity at www.conceptsandchallenges.com. Be sure students use search engines and websites that are reliable and acceptable. You may also wish to provide copies of the *Big Idea Planner* and *Big Idea Science Log Entry* study tools found in the Teacher's Resources.

3 Assess

Use the Writing Activity Rubric from the Teacher's Resources to assess students' Science Logs. Fill in the rubric with the additional information below. For this assignment, students should have:

- chosen an organism that has adapted to the tundra.
- communicated, in paragraph form, accurate information about the animal's adaptations to the harsh tundra environment.

4 Close

Exploring the Sea Have students work in pairs to research the adaptations of a marine organism of their choice. Tell students to search the Internet or look in magazines for pictures to incorporate into their reports. Have the class compile its research into a booklet about marine-life adaptations.

Tips for Using Technology

Using Word Processing Software to Write Essays Have students use word processing software to write their essay for the Science Log Writing Activity. A simple word processing program allows them to do basic tasks, such as entering and editing text. A more complex word processing program allows students to do more, such as draw or bring in pictures. Students can print out their word processing documents to share with the class.

• TEACHING RESOURCES • Online or Print

Teacher's Resources

Big Idea: Chapter 13, p. 9

Big Idea Planner: Study Tools, pp. 11–12

Big Idea Science Log Entry: Study Tools, p. 13

Writing Activity Rubric: Teaching Tools, p. 9

Challenges (pp. 322–324)

Chapter Summary

Review Before students begin the Challenges, review the objectives of each lesson with them. Have pairs of students write at least one relevant fact related to the objectives.

STUDY HINT Have students group the Key Terms that are closely related and write a summary paragraph about each group.

Key Term Challenges

MATCHING

1. tropical zone (13-3)
2. biome (13-5)
3. climate (13-1)
4. polar zone (13-3)
5. altitude (13-2)
6. weather (13-1)
7. latitude (13-2)
8. middle-latitude zone (13-3)
9. vegetation (13-5)
10. microclimate (13-4)

IDENTIFYING WORD RELATIONSHIPS

11. Altitude is the height above sea level. Latitude is the distance north or south of the equator in degrees. (13-2)
12. A biome is a large area of Earth that is characterized by the vegetation that grows there. (13-5)
13. Climate is the average weather conditions of an area from year to year. (13-1)
14. Microclimates are small climate zones within a local climate. (13-4)
15. The middle-latitude zone is the region north and south of the tropical zone. (13-3)

Chapter 13 Challenges

Chapter Summary

Lesson 13-1

- **Climate** is the average weather conditions in an area from year to year. Average monthly and yearly temperature, temperature range, and average precipitation are used to describe climate.

Lesson 13-2

- An area's climate is affected by its **latitude** and **altitude.**
- Ocean currents affect the temperatures of areas along coasts.
- The leeward sides of mountains have warmer, drier climates than the windward sides of mountains do.

Lesson 13-3

- A climate zone is an area with a given yearly temperature range and regular weather patterns. The three climate zones are the **tropical zone,** the **polar zone,** and the **middle-latitude zone.**
- The climates in each climate zone can be characterized by temperature and amount of rainfall.

Lesson 13-4

- Local climates are formed by local conditions. Very small climate zones are called **microclimates.** Cities are microclimates.
- Human activities and skyscrapers can affect the climate in an area. Global warming due to increased air pollution could cause temperatures to rise around the world.

Lesson 13-5

- A **biome** is a large region of Earth that has a characteristic climate and certain kinds of living things. It is characterized by the plant life found there.
- Climate affects the **vegetation** in an area.
- The animals and other organisms in an area are affected by climate.

Key Term Challenges

altitude (p. 310)
biome (p. 318)
climate (p. 308)
latitude (p. 310)
microclimate (p. 314)
middle-latitude zone (p. 312)
polar zone (p. 312)
tropical zone (p. 312)
vegetation (p. 318)
weather (p. 308)

MATCHING **Write the Key Term from above that best matches each description.**

1. warm region near the equator
2. large region with a characteristic climate and plant and animal communities
3. average weather conditions in an area from year to year
4. cold region near the poles
5. height above sea level
6. day-to-day conditions of the atmosphere
7. distance north or south of the equator in degrees
8. region between 30° and 60°N and S latitudes
9. plants
10. very small climate zone

IDENTIFYING WORD RELATIONSHIPS **Explain how the words in each pair are related. Write your answers in complete sentences.**

11. altitude, latitude
12. biome, vegetation
13. climate, weather
14. microclimate, local climate
15. tropical zone, middle-latitude zone

322

Teaching Options

PREPARING STUDENTS FOR STANDARDIZED TESTS

Reading Strategy: Remind students to look for context clues in the questions as they read. Information may be given in a question to help jog the student's memory.

Writing Strategy: Tell students to write in complete sentences, when appropriate, in their science notebooks. This practice will help prepare them for essays on standardized tests.

Interpreting Visuals Strategy: Tell students to pay attention to labels and legends. They often help explain the illustration.

ESL/ELL STRATEGY

Understanding Climate Zones To reinforce the concept of climate zones for English-language learners, have them create their own drawings. Have students color-code the climate zones and glue relevant magazine pictures in appropriate places, such as polar bears in the polar zone and orchids in the tropical zones.

Content Challenges TEST PREP

FILL IN **Write the term or phrase that best completes each statement.**

1. To find average temperature, you add two or more temperatures and divide the sum by the number of __________.
2. The amount of temperature change in an area is its temperature __________.
3. Rain, snow, sleet, and hail are forms of __________.
4. At higher latitudes, the Sun's rays strike Earth __________.
5. Areas at low latitudes have a __________ climate than areas at higher latitudes do.
6. Air is __________ at sea level than it is at higher altitudes.
7. Land areas near cold-water currents have __________ temperatures than land areas near warm-water currents do.
8. When air passes over a mountain range, it rises and __________.
9. It hardly ever rains on the __________ side of a mountain.
10. The leeward side of a mountain is the side of the mountain that faces __________ the wind.
11. The tropical zone is located between __________ north and south latitude.
12. Some scientists think that __________ will lead to global warming.
13. Energy from the __________ is the biggest factor influencing weather and climate.
14. Biomes are identified mainly by the type of __________ found in the area.
15. The type of vegetation found in an area depends on the area's __________.

TRUE/FALSE **Write *true* if the statement is true. If the statement is false, change the underlined term to make the statement true.**

16. The day-to-day condition of the atmosphere is called the climate.
17. The leeward side of a mountain gets more rain than the windward side.
18. Dry climates are also described as arid.
19. Antarctica is located in the tropical zone.
20. A city is an example of a microclimate.
21. Latitude, ocean currents, and mountains affect an area's climate.
22. Climate has never changed during Earth's history.
23. Each biome has certain kinds of plants and animals.
24. Most snakes live in cool climates.
25. Cacti grow best in tundra climates.
26. Average daily temperature and temperature range are both used to describe the climate of an area.
27. The temperature gets colder as you climb a mountain.
28. Global warming might cause sea levels to fall.
29. A rain shadow desert is sometimes found on the windward side of a mountain.
30. The tilt of Earth directly influences climate and weather.

Content Challenges

FILL IN

1. temperature readings (13-1)
2. range (13-1)
3. precipitation (13-1)
4. at more of an angle (13-2)
5. warmer (13-2)
6. warmer (13-2)
7. cooler (13-2)
8. cools (13-2)
9. leeward (13-2)
10. away from (13-2)
11. 0° and 30° (13-3)
12. air pollution (13-4)
13. Sun (13-3)
14. plants (13-5)
15. climate (13-5)

TRUE/FALSE

16. false; weather (13-1)
17. false; less (13-2)
18. true (13-3)
19. false; polar (13-3)
20. true (13-4)
21. true (13-2)
22. false; often (13-4)
23. true (13-5)
24. false; warm (13-5)
25. false; desert (13-5)
26. false; yearly (13-1)
27. true (13-2)
28. false; rise (13-4)
29. false; leeward (13-2)
30. true (13-3)

ALTERNATIVE ASSESSMENT

Making Tables Have students use the map on page 324 to make a table listing the different types of climates found on each continent. You may want to have students complete their tables before answering the Interpreting Visuals questions on that page, since the tables will help them answer some of those questions.

PORTFOLIO ASSESSMENT

Making Student Portfolios Students can demonstrate their comprehension of the concepts in this chapter by making a portfolio. The Chapter Self-Check, the Big Idea Planner, and the Weekly Journal are some of the reproducibles in the Teacher's Resources that they can include in their portfolios. You may wish to make certain assignments planned portfolio items, such as the Chapter Project or Big Idea Science Log. You can use the Portfolio Assessment Rubric also found in the Teacher's Resources to assess students' portfolios.

Concept Challenges

WRITTEN RESPONSE

1. **PREDICT:** It would have a cool climate. (13-2)
2. **INFER:** Climate conditions are usually warm and dry there. (13-2)
3. **INFER:** Earth's latitude zones would experience exactly the same intensity of sunlight year-round. Climates, therefore, would likely be more extreme but would have less seasonal variation. (13-3)
4. **EXPLAIN:** Heavy vehicle traffic in cities as well as the energy used to heat and light buildings raises average temperatures. Rainfall over large cities may differ from that of surrounding areas because of skyscrapers blocking the movement of moisture in the air. Vehicle exhaust and industrial smokestacks in cities also add particles to the air, which leads to increased rainfall. (13-4)
5. **INFER:** Some animals can eat only certain plants and thus live in the biomes where those plants are found. (13-5)
6. **INFER:** Cacti live in deserts. Without stems that store water, the cacti would die. (13-5)

INTERPRETING VISUALS

7. tropical rain forest (13-5)
8. desert (13-5)
9. tundra, taiga, temperate forest, desert, grassland, and savanna (13-5)
10. Antarctica. Its climate is cold year-round. (13-3)
11. Possible answers: North America: taiga, temperate forest, scrublands/chaparral, tropical rain forest, grassland and savanna, and desert; Asia: tundra, taiga, temperate forest, tropical rain forest, grassland and savanna, and desert (13-5)
12. Students should use the correct biome name and list some characteristic organisms for the biome. (13-5)

Chapter Project Wrap-Up

TRAVEL BROCHURE

Have students explain where they obtained the information for their brochures. You can use the Individual Activity Rubric from the Teacher's Resources to assess students' contributions. Fill in the rubric with the additional information below. For this project, students should have:

- named some major geologic sites, plants, and animals for the area.
- described the weather in the area and suggested types of weather gear.

Concept Challenges TEST PREP

WRITTEN RESPONSE **Answer the following questions in complete sentences.**

1. **PREDICT:** What kind of climate would a coastal town located near a cold-water current have?
2. **INFER:** Why are many deserts located on the leeward sides of mountains?
3. **INFER:** If Earth had no tilt, how would this affect the world's climates?
4. **EXPLAIN:** How do the types of objects found in a city help to create a microclimate?
5. **INFER:** Why would vegetation affect the types of animals that can live in a biome?
6. **INFER:** Why is having a stem that stores water an important adaptation for a cactus?

INTERPRETING VISUALS **Use Figure 13-19 below to answer the following questions.**

7. Which biome is found most often along the equator?
8. Which biome covers most of Australia?
9. What are the biomes found in North America?
10. Where is polar ice found? Why?
11. Which continent has the most biomes? What are they?
12. Which biome do you live in? What are some common plants and animals that are found in this biome?

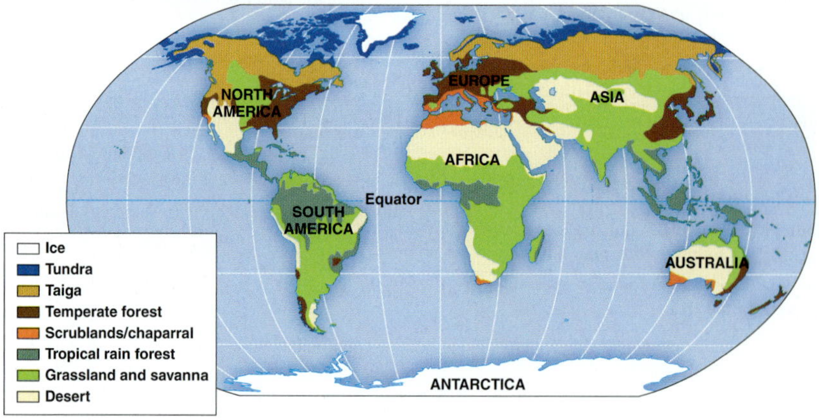

▲ Figure 13-19

324

Teaching Options

• TEACHING RESOURCES • Online or Print

Teacher's Resources

Key Term Review: Chapter 13, p. 10

Chapter 13 Test: pp. 11–13

Individual Activity Rubric: Teaching Tools, p. 5

Chapter Self-Check: Teaching Tools, p. 11

Big Idea Planner: Study Tools, pp. 11–12

Weekly Journal: Study Tools, p. 19

Portfolio Assessment Rubric: Teaching Tools, p. 10

PLANNING GUIDE

◆ **TEACHING THE CHAPTER** This chapter should take approximately 8–11 days to complete instruction and assessment.

	Skills and Features	Projects and Activities	Achieve Science Literacy	Meet Individual Needs
Chapter 14 Opener p. T325	MODEL	• Chapter Project		
Lesson 14-1 **What are the motions of Earth?** pp. T326–T327 Standards: D3a, D3b	CONTRAST, INFER, PREDICT • Investigate • Web InfoSearch • How Do They Know That?		• Study Hint • Reading Strategy • Ongoing Assessment	• Reteach Teacher's Resources* Concept Map Fishbone Map
Lesson 14-2 **What are time zones?** pp. T328–T329	HYPOTHESIZE, CALCULATE, ANALYZE • Web InfoSearch	• Cooperative Learning • Extending the Lesson	• Study Hint	• ESL/ELL Strategy Teacher's Resources* Enrichment Activity Time Zones
Lesson 14-3 **What causes the seasons?** pp. T330–T331 Standards: D3b, D3d	OBSERVE, HYPOTHESIZE, CONCLUDE • Building Math Skills • Hands-On Activity	• Extending the Lesson • More on the Web • Active Art	• Study Hint • Writing Hint • Ongoing Assessment	Teacher's Resources* Enrichment Activity
Lesson 14-4 **What are solstices and equinoxes?** pp. T332–T333 Standards: D3b, D3d	NAME, IDENTIFY, DESCRIBE • Building Reading Skills • How Do They Know That?	• More on the Web • Lab Challenge	• Study Hint	• Reteach
Big Idea **How does the changes of seasons affect animal behavior?** pp. T334–T335 Standard: D3d	RESEARCH, COMMUNICATE, ANALYZE • Science Log: Writing Activity	• Big Idea • Extending the Lesson • Close Activity	• Writing Hint • Tips for Using Technology	Teacher's Resources* Big Idea Planner
Lesson 14-5 **What are the motions of the Moon?** pp. T336–T337 Standards: D3a, D3b	CALCULATE, EXPLAIN, ANALYZE • Building Reading Skills • People in Science	• Cooperative Learning • Extending the Lesson • More on the Web	• Study Hint • Reading Strategy	• ESL/ELL Strategy

*www.conceptsandchallenges.com, password: ccearth09

Standards: For details on the correlation to National Science Standards see pages *xx–xxi*.

Planning Guide continues on next page.

	Skills and Features	Projects and Activities	Achieve Science Literacy	Meet Individual Needs
Lesson 14-6 **What are the phases of the Moon?** pp. T338–T339 Standard: D3b	CONTRAST, PREDICT, COMPARE • Building Science Skills • Integrating Physical Science	• Extending the Lesson • More on the Web	• Study Hint • Ongoing Assessment	• ESL/ELL Strategy
Lab Activity **Modeling the Phases of the Moon** pp. T340–T341 Standard: D3b	OBSERVE, ANALYZE, INFER • Set-Up Time: 5–10 min • Lab Time: 25–30 min	• Extend the Lab Activity	• Tips for Using Technology	• ESL/ELL Strategy Teacher's Resources* Lab Activity Report
Lesson 14-7 **What causes an eclipse of the Moon?** pp. T342–T343 Standard: D3b	DESCRIBE, HYPOTHESIZE, EXPLAIN • Hands-On Activity	• Integrating Physical Science • More on the Web • Active Art	• Study Hint • Reading Strategy	• Reteach Teacher's Resources* Enrichment Activity
Lesson 14-8 **What causes an eclipse of the Sun?** pp. T344–T345 Standard: D3b	ANALYZE, EXPLAIN, COMPARE • Health and Safety Tip • How Do They Know That?		• Study Hint • Writing Hint • Ongoing Assessment	Teacher's Resources* Enrichment Activity
Chapter 14 Challenges pp. T346–T348	• Chapter Summary • Key Term Challenges • Content Challenges • Concept Challenges	• Chapter Project Wrap-Up	• Study Hint • Preparing Students for Standardized Tests	• ESL/ELL Strategy Teacher's Resources* Chapter Self-Check Weekly Journal

*www.conceptsandchallenges.com, password: ccearth09

Teacher Notes

Chapter 14 The Motions of Earth, the Sun, and the Moon

▲ **Figure 14-1** Sunsets are enjoyed for their many colors.

The passing of a single day can be seen by watching the Sun in the sky. The Sun appears each morning in the east. This is sunrise. During the daytime, the Sun appears to move westward across the sky. Finally, at sunset, it disappears below the western horizon. This is what we call a day. A few hours later, the Sun seems to rise again in the east to begin a new day. However, the Sun never actually rises or sets. It is just an illusion.

► How do you think the motions of Earth makes it seem as if the Sun is moving across the sky?

Contents

Teaching Options

ASSESSMENT PLANNER

For assessment in the Student Edition, see the following pages:

Content Assessment
Checking Concepts: pp. 327, 329, 331, 333, 337, 339, 343, 345
Thinking Critically: pp. 327, 329, 331, 333, 337, 339, 343, 345
Interpreting Visuals: p. 343
Chapter Challenges: pp. 346–348

Alternative Assessment
Building Skills: pp. 331, 333, 337, 339
Web InfoSearch: pp. 327, 329
Science Log: p. 334

Performance-Based Assessment
Hands-On Activity: pp. 331, 343,
Lab Activity: pp. 340–341

• TEACHING RESOURCES • Online or Print

Teacher's Resources
Lesson Review: Ch. 14, pp. 2, 3, 4, 6, 8, 9, 10, 11
Enrichment: Ch. 14, pp. 5, 12
Key Term Review: Ch. 14, p. 13
Chapter 14 Test: pp. 14–16
Laboratory Manual: pp. 73–76

Chapter 14 The Motions of Earth, the Sun, and the Moon

Chapter Overview

In this chapter, students will learn about standard time zones and identify eight U.S. time zones. They will see what causes the seasons. They will note the Moon's motions and phases. Finally, they will learn about lunar and solar eclipses.

About Figure 14-1 Earth's rotation period in relation to the Sun is the standard method of measuring time. One day lasts on average 24 hours and is called the *mean solar day*, which is the time interval from one noon to the next. Noon is when the Sun has reached its highest point in the sky. However, Earth's rotation can also be based on the time it takes Earth to rotate in relation to a star other than our Sun. This method of keeping time is called *sidereal time*, and it is used mostly by astronomers.

Answer Possible answer: An observer on Earth does not feel the motion, so the Sun seems to be moving instead.

Linking Prior Knowledge

For this chapter, students should recall:

- the meaning of longitude and positions of the equator, poles, and hemispheres (Lesson 1-6).
- how climate affects living organisms (Lesson 13-5).

Chapter Project

CREATING A MOTION MODEL

Divide the class into five groups. Have each group create a small working model or simulation of one of the five motions discussed in this chapter: Earth's rotation, which causes day and night; Earth's revolution around the Sun; the phases of the Moon; or a lunar and solar eclipse. Suggest that the models have movable parts that can be used to demonstrate the motion. A poster illustrating the motion should accompany each model as well as a brief explanation of it. **learning styles: kinesthetic, auditory**

14-1 *(pp. 326–327)*

1 Introduce

INVESTIGATE

TIME: 10 minutes

PURPOSE: Students will model night and day.

MATERIALS: wooden stick, plastic-foam ball, marker, modeling clay, flashlight or desk lamp, protractor (optional)

PROCEDURE: Have students work in pairs or small groups. Explain that the ball represents Earth and the flashlight represents the Sun. The line of dots on the ball represents the equator. If necessary, provide students with protractors to help them estimate a 25° angle. **learning styles: visual, tactile**

THINK ABOUT IT: The light illuminates half the ball. When the ball rotates, a different part of it lights up.

STUDY HINT A process can be very difficult to understand from a diagram. Have students act out the actual motions to better grasp what is happening.

Linking Prior Knowledge Ask students what they think makes day differ from night. Also, ask them why the amount of daylight and darkness changes throughout the year.

2 Teach

Discussion Differentiate between Earth's revolution and its rotation. Remind students that Earth revolves (travels around the Sun) from west to east, or in a counterclockwise direction, while it rotates (spins on its axis) in the same direction. Ask students to picture a carnival ride in which the riders sit in seats that spin at the periphery of a large wheel. As the wheel turns, the seats revolve around the center of the wheel. At the same time, each seat rotates on its axis. **learning style: auditory**

READING STRATEGY Have students create concept webs in their notebooks. (They can use the Fishbone Map or Concept Map from the Teacher's Resources.) The topic should be "Earth's motions." As students read the lesson, they can add arms to their webs on which they record Key Terms or other details about Earth's motions.

14-1 What are the motions of Earth?

INVESTIGATE

Modeling Day and Night

HANDS-ON ACTIVITY

1. Push a wooden stick through a plastic foam ball. Mark a line of dots around the ball.
2. Place the stick upright in modeling clay. Tilt it at about a third of the distance to the desk. Turn on a flashlight or desk lamp and darken the room.
3. Point the angled stick toward the light source. Rotate it and observe.
4. Move the ball so the stick points away from the light source and observe.

THINK ABOUT IT: What happens to the light when it hits the ball? What do you observe when you rotate the ball on its stick?

STEP 3

Objective

Explain Earth's motions.

Key Terms

revolution (rehv-uh-LOO-shuhn)**:** movement of a planet or other body orbiting another body

axis (AK-sihs)**:** imaginary line through the center of a planet or other body around which that body spins

rotation (roh-TAY-shuhn)**:** spinning of a planet or other body on its axis

Revolution Although you cannot feel it, Earth is always moving. It moves in an orbit around the Sun. This movement is called **revolution.** One complete revolution around the Sun takes 365¼ days, or one year. Earth's path as it goes around the Sun is called its orbit. As Earth travels around the Sun, Earth's orbit forms a slightly flattened circle, or oval, shape.

1 **DESCRIBE:** What is the motion of Earth around the Sun called?

Rotation Besides revolving, Earth also spins on its axis. An **axis** is an imaginary line through the center of a planet or other body around which that body spins. Earth's axis is tilted at an angle of 23½ degrees relative to an imaginary line drawn perpendicular to its orbit. The North Pole of Earth always points toward Polaris, the North Star.

The spinning of a planet or other body on its axis is called **rotation.** Our planet rotates once every 24 hours. This gives us a daily cycle of daylight and darkness, or day and night. The side of Earth's surface that faces the Sun has daylight. The side that faces away from the Sun has darkness. At any moment, half of Earth is experiencing daylight. The other half is experiencing darkness.

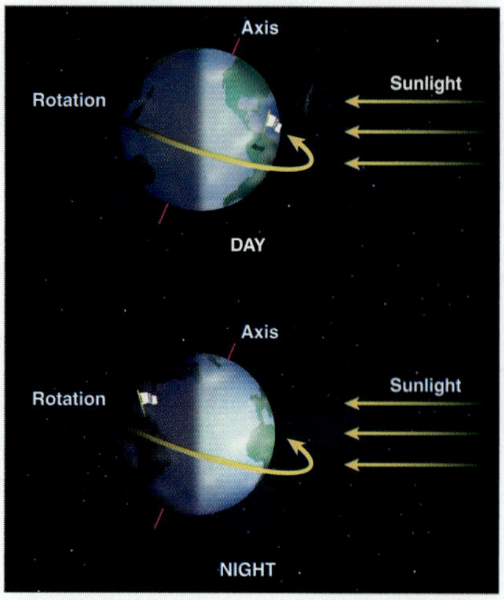

▲ **Figure 14-2** Earth rotating on its axis causes day and night.

2 **IDENTIFY:** What is the motion of Earth on its axis called?

Teaching Options

FAST FACTS *Ancient Sun Worship*

Many cultures have regarded the Sun either as a god or as the symbol of a god. In ancient Egypt, the sun god Ra (also called Re) was depicted with the body of a human and the head of a falcon. Ra was believed to be reborn each day at sunrise. The worship of Ra flourished during the Old Kingdom in Egypt. Later, the Egyptian pharaoh Akhenaton established the worship of another sun god, Aton (also called Aten).

The ancient Greeks worshiped the Sun god Helios, who was believed to ride a golden chariot across the sky each day. A major temple honoring Helios was built on the island of Rhodes, where the Colossus of Rhodes, an enormous statue of Helios, came to be regarded as one of the Seven Wonders of the World. Sun worship was also practiced by the Babylonians, the ancient Persians, and many indigenous peoples of North and South America.

Sunrise and Sunset When viewed from above the North Pole, Earth rotates on its axis from west to east. This makes the Sun appear to rise in the east and set in the west. As Earth rotates, the Sun seems to move across the sky from east to west. This is called the Sun's apparent motion.

3 **RELATE:** In which direction does Earth rotate?

CHECKING CONCEPTS

1. How long does Earth take to make one complete revolution around the Sun?
2. What causes the change from day to night?
3. Why does the Sun appear to rise in the east?
4. At what angle is Earth's axis tilted?

THINKING CRITICALLY

5. **CONTRAST:** How is rotation different from revolution?
6. **DESCRIBE:** What is the shape of Earth's orbit?
7. **PREDICT:** What would happen if Earth did not rotate?
8. **INFER:** Like the Sun, the Moon seems to rise in the east and set in the west. Why?

Web InfoSearch

The 24-Hour Day Since ancient times, people have had an interest in measuring and recording the passage of time. The ancient Egyptians were the first to use a 24-hour day. Their system was based on observing a series of 36 stars called decan stars. The decan stars rise and set in the sky every 40 to 60 minutes.

SEARCH: Use the Internet to find out more about decan stars. What are they? How did they lead the Egyptians to create the 24-hour day? Start your search at www.conceptsandchallenges.com. Some key search words are **decan stars, Egyptian astronomy,** and **ancient calendars.**

How Do They Know That?

EARTH ROTATES ON ITS AXIS

A pendulum is a weight that swings freely back and forth. It can be used to show that Earth rotates.

French physicist Jean Bernard Leon Foucault (1819–1868) provided proof of Earth's rotation with a pendulum. Using an early camera, he photographed the Sun. The camera took pictures on a light-sensitive plate. Leaving the camera focused on the Sun for some time, Foucault showed that the Sun's position relative to Earth changed. He invented a pendulum-driven device to keep the camera in line with the Sun. However, he noticed that his pendulum tended to swing in the direction from which it was first released. If he tried to turn it, it always returned to its original path. From this experiment, he concluded that Earth was rotating.

▲ **Figure 14-3** Foucault's pendulum

Foucault set up a demonstration to prove his conclusion. He released a giant ball over a pile of sand. The ball scratched a straight line in the sand. Over the course of the day, that line shifted again and again to the right. Eventually, it came full circle.

Thinking Critically What did it mean that the pendulum was not changing course?

Answers to questions

1 **DESCRIBE:** revolution

2 **IDENTIFY:** rotation

3 **RELATE:** west to east

3 *Assess*

CHECKING CONCEPTS ANSWERS

1. $365\frac{1}{4}$ days, or 1 year
2. Earth's rotation
3. because Earth rotates from west to east
4. $23\frac{1}{2}°$

THINKING CRITICALLY ANSWERS

5. **CONTRAST:** The spinning of Earth on its axis is rotation, whereas Earth's orbit around the Sun is revolution.
6. **DESCRIBE:** oval or elliptical
7. **PREDICT:** One side of Earth would always have daylight and the other would always have night.
8. **INFER:** Earth rotates on its axis from west to east.

4 *Close*

Web InfoSearch

The 24-Hour Day Have students use the Internet to find out about the 24-hour day. Ask: *What are the decan stars? How did they lead the Egyptians to create the 24-hour day?* Some key search words are **Egyptian astronomy, decan stars,** and **ancient calendars.** Students can begin their search at www.conceptsandchallenges.com.

How Do They Know That?

EARTH ROTATES ON ITS AXIS

Real-Life Connection At the Smithsonian Institute in Washington, D.C., there is a large Foucault pendulum. It appears to be constantly changing its direction of swing. Under the Foucault pendulum are many pegs in a circle. The pendulum knocks another peg down every few minutes as Earth rotates beneath it.

Discussion Ask students: *What is a pendulum?* (a weight attached to a wire or string that swings back and forth) *What hypothesis did Foucault use a pendulum to prove?* (Earth rotates.)

Thinking Critically Answer
Earth was rotating beneath it.

RETEACH

Demonstrating Earth's Tilt Using a Trippensee Planetary Model, show students that Earth's axis is tilted about $23\frac{1}{2}°$ from the perpendicular to the path of its orbit. Have students observe that Earth's axis always tilts in the same direction as Earth revolves around the Sun. Trippensee Planetary Models are available at most scientific supply houses.
learning style: visual

ONGOING ASSESSMENT

Have students describe the Sun's apparent motion around Earth and explain why this motion is only apparent.

• TEACHING RESOURCES •
Online or Print

Teacher's Resources

Lesson Review: Chapter 14, p. 2

Concept Map: Study Tools, p. 4

Fishbone Map: Study Tools, p. 5

14-2 *(pp. 328–329)*

1 Introduce

STUDY HINT Have students summarize the meanings of both time zone charts in the lesson, using their own words. Ask students to use the captions of both charts to create a summary that contains a topic sentence and supporting details.

Linking Prior Knowledge Review with students the meaning of *longitude* (Lesson 1-6).

2 Teach

Demonstration Set up a sundial in the school yard and demonstrate how it can be used to tell time. See the Web site of the North American Sundial Society (http://sundials.org) for suggestions on where to buy a sundial or how to make one yourself (see Lab Video, Segment 14). Caution students never to look directly at the Sun.
learning style: visual

Discussion Discuss solar time and the establishment of time zones. Point out that solar noon is determined by the location of the Sun and that different places have solar noon at different times. Discuss why time zones are important and the confusion that would result without the establishment of time zones.
learning style: auditory

Reinforcement Using a globe or map, review what is meant by longitude. Guide the class to see that Earth moves through 360° of longitude during one complete rotation on its axis. Point out that a day is divided into 24 hours. Develop the idea that there is one hour difference for each 15° of longitude ($360°/24\text{ h} = 15°/\text{h}$).
learning style: visual

Answers to questions

1. **HYPOTHESIZE:** Places a short distance apart would have different times.
2. **DESCRIBE:** a system by which all places within a time zone have the same time
3. **CALCULATE:** 3:00 P.M.
4. **INFER:** You move the calendar back one full day.

14-2 What are time zones?

Objectives

Use standard time zones to compare times around the world. Identify eight U.S. time zones.

Key Terms

solar noon: time of day when the Sun is highest in the sky

standard time: system whereby all places within a time zone all have the same time

International Date Line: boundary formed where the first and 24th time zones meet

Solar Time The Sun can be used to tell time. Using the Sun to measure time is called solar time. When the Sun is highest in the sky, the time is **solar noon.** Because Earth rotates, different places have solar noon at different times. For example, when it is solar noon in Philadelphia, it is 4 minutes after solar noon in New York and 17 minutes after solar noon in Boston.

1. **HYPOTHESIZE:** What problem is likely to arise using solar time?

Standard Time Even though clocks allowed people to keep track of time, there was still a problem. Each region set its own time, and there was no way to coordinate among regions. This problem grew when railroads were built and people traveled more. A system was needed on which to base a train schedule. This system would also help travelers schedule meetings in other towns.

In 1884, an international agreement was reached that divided Earth's surface into 24 time zones. Each of these time zones was to be 15 degrees of longitude wide. All places within a time zone would have the same time. This time was called **standard time.**

2. **DESCRIBE:** What is standard time?

U.S. Time Zones The United States has eight time zones. These are Eastern Standard, Central Standard, Mountain Standard, Pacific Standard, Alaska Standard, Hawaiian Standard, Samoa Standard, and Atlantic Standard.

As you move west, the time in each zone is one hour earlier than in the previous time zone. As a result, when it is noon in New York, it is 11:00 A.M. in Chicago.

Look at Figure 14-4. Notice that time zones do not have straight boundaries. The boundaries were drawn this way to keep whole states or large neighboring cities in the same time zone.

3. **CALCULATE:** If it is 12:00 noon in California, what time is it in Virginia?

World Time Zones There are 24 standard time zones and 24 hours in a day. The first time zone is on one side of the boundary called the **International Date Line.** The twenty-fourth time zone is on the other side. As you cross over the International Date Line, you either gain or lose 24 hours. If you were to travel west, you would have to move your calendar ahead one full day when you crossed the International Date Line.

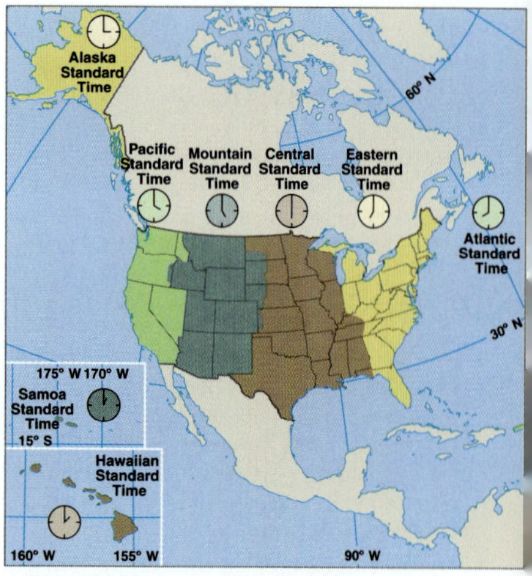

▲ **Figure 14-4** Time zones of the United States

4. **INFER:** What happens to the date if you cross the International Date Line going east?

Teaching Options

! FAST FACTS
Daylight Saving Time

Explain that daylight saving time extends daylight one hour later in the evening by artificially beginning the day one hour later. The idea of daylight saving was mentioned in an essay by Benjamin Franklin in 1784. It has been used in the United States and much of Europe since World War I, when the system was adopted to conserve fuel needed to produce electrical power. Federal legislation set the period of daylight saving time in the United States as the first Sunday in April to the last Sunday in October.

COOPERATIVE LEARNING

Telling Time Using Figure 14-4, locate and identify the time zones in the continental United States. Display a large U.S. map and have students take turns reading aloud the names of any two places in the United States in different time zones. Tell half the students to write a sample time for the first place on the board. Have the rest of the class determine the time for the second place.
learning styles: visual, auditory

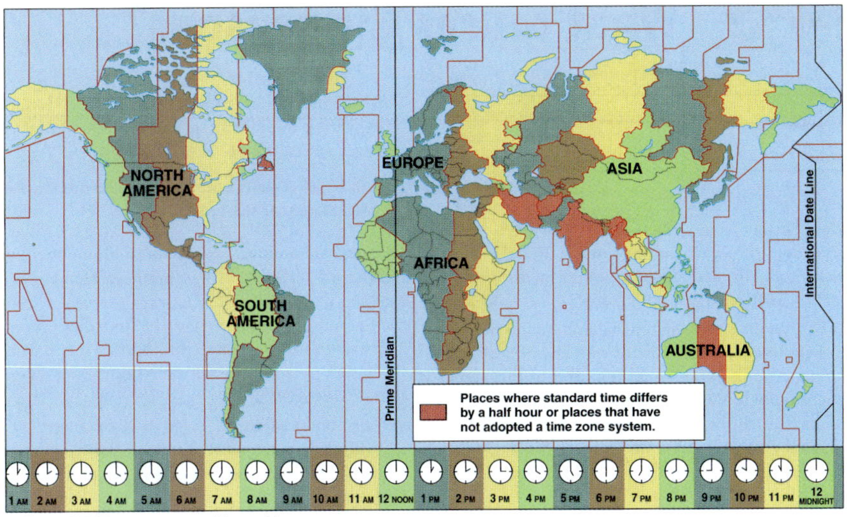

▲ **Figure 14-5** Time zones around the world

✓ CHECKING CONCEPTS

1. There are __________ time zones in the world.
2. Using the Sun to measure time is __________.
3. The 48 states in the continental United States have __________ time zones.
4. If you cross the International Date Line going east, you would __________ 24 hours.
5. As you travel west, you move the clock __________ one hour for each time zone.

THINKING CRITICALLY

Use Figures 14-4, 14-5, and Appendix B in this book to help you answer the following questions.

6. IDENTIFY: In which time zone is Texas located?
7. ANALYZE: If it is 3:00 P.M. in Florida, what time is it in Oregon?
8. CALCULATE: Suppose you leave California at 12:00 noon Pacific Standard Time and take a six-hour plane trip to Boston. When you land in Boston, what would the time be there?
9. IDENTIFY: In which time zone do you live?
10. ANALYZE: Earth rotates from west to east. Would a town to your west have solar noon before or after you? Explain.

Web InfoSearch

Jet Lag *Jet lag* is a term given to the tired feeling people get when they travel across three or more time zones. Jet lag occurs because your body is adjusted to the time zone where you live and not adjusted to the time zone to which you have just traveled. It is easier for your body to adjust to a longer day traveling west than to a shorter day traveling east. For this reason, jet lag usually affects people who are flying east more than it affects people who are flying west.

SEARCH: Use the Internet to find out more about jet lag. How does it affect choice of flying times? What methods are there for overcoming the effects of jet lag? Start your search at www.conceptsandchallenges.com. Some key search words are **time zone, jet lag,** and **avoiding jet lag.**

3 Assess

✓ CHECKING CONCEPTS ANSWERS

1. 24
2. solar time
3. 4
4. gain
5. back

THINKING CRITICALLY ANSWERS

6. IDENTIFY: Central Standard
7. ANALYZE: noon
8. CALCULATE: 9:00 P.M.
9. IDENTIFY: Accept any one of the following: Eastern Standard Time, Central Standard Time, Mountain Standard Time, Pacific Standard Time, Alaska Standard Time, Hawaii-Aleutian Standard Time, Samoa Standard Time, or Atlantic Standard Time.
10. ANALYZE: After; As you move west, clock time is earlier.

4 Close

Web InfoSearch

Jet Lag Jet lag is a term given to the overall tired feeling people get when they travel across three or more time zones. Jet lag occurs because your body is adjusted to the time where you live, not to the time zone to which you have traveled. Have students use the Internet to find out more. Ask students: *How does jet lag affect choice of flying times? What methods are there for overcoming the effects of jet lag?* Some key search words are **jet lag, time zone jet lag,** and **avoiding jet lag.** Students can begin their search at www.conceptsandchallenges.com.

EXTENDING THE LESSON

Daylight Saving Time Pro and Con Have students find out why some states, such as Arizona, Hawaii, and sections of Indiana, do not use daylight saving time, while most states do use it. Students can report their findings to the class in an oral report.
learning style: auditory

• TEACHING RESOURCES •
Online or Print

Teacher's Resources

Lesson Review: Chapter 14, p. 3

Enrichment: Chapter 14, p. 5

Time Zones: Visuals, p. 11

ESL/ELL STRATEGY

Identifying Time Zones Have English-language learners identify the time zones in their country of birth. Then, have them convert the local time to the current time in this time zone.

14-3 *(pp. 330–331)*

1 Introduce

STUDY HINT Have students preview the lesson by reading the title, objective, subheads, and figure captions. Then, have students write questions that they would like to have answered during their reading. Encourage students to write the answers to these questions as they read.

Linking Prior Knowledge Review with students the positions of the equator, poles, and hemispheres (Lesson 1-6). Have students locate them on a map or globe.

2 Teach

Demonstration Using a globe to represent Earth and a student to represent the Sun, demonstrate Earth's position relative to the Sun during the four seasons.
learning style: visual

Discussion Use Figure 14-6 to discuss Earth's tilt and seasons. Explain the relationship between direct rays, slanted rays, and the seasons. Point out that the angle of the Sun's rays and the length of day determine the seasons.
learning styles: visual, auditory

Answers to questions

1. **DESCRIBE:** the tilt of Earth's axis and Earth's movement around the Sun
2. **IDENTIFY:** winter

WRITING HINT Remind students that an apostrophe and an *s* are used to show ownership, or possession. Point out the phrase *Earth's axis* on page 330. Explain that the apostrophe and *s* goes with the owner (in this case, Earth) and what follows is the thing owned (axis). Have students write two sentences about Earth that show possession, using an apostrophe and an *s*.

14-3 What causes the seasons?

Objective
Explain what causes the change of seasons.

Why the Seasons Change As you learned earlier, Earth's axis is tilted at 23 1/2 degrees. The North Pole of Earth points toward Polaris, the North Star. The axis always points in the same direction as Earth moves in its orbit.

The seasons are caused by a combination of the tilt of Earth's axis and Earth's movement around the Sun. At one point in Earth's orbit of the Sun, the North Pole of Earth is tilted toward the Sun. Six months later, the North Pole is tilted away from the Sun. This change in position changes the angle at which the Sun's rays strike Earth's surface and is responsible for the change of seasons.

When the North Pole is tilted toward the Sun, the Northern Hemisphere receives the Sun's rays almost directly. The more direct the rays are, the better that part of Earth's surface is heated.

The tilt of Earth's axis also causes the total number of daylight hours to change. When the North Pole is tilted toward the Sun, the Northern Hemisphere has more hours of daylight than it does of darkness.

The combination of more direct rays and longer days warms the Northern Hemisphere. It causes summer in that region.

1 **DESCRIBE:** What combination of factors causes the change of seasons?

Opposite Seasons When it is summer in the Northern Hemisphere, the Southern Hemisphere tilts away from the Sun. The rays it receives from the Sun are less direct. They are spread out more. There are also fewer daylight hours in that part of the world. In fact, there are more hours of darkness than of daylight.

The combination of fewer daylight hours and less direct rays causes that part of Earth's surface to cool. The lower temperatures cause winter in that part of the world.

The Northern and Southern hemispheres have opposite seasons. Look at Figure 14-6. Notice that when the Northern Hemisphere is tilted away from the Sun, it has winter.

At the same time it is winter in the Northern Hemisphere, the Southern Hemisphere is tilted toward the Sun and has summer. When summer comes to the Northern Hemisphere, it is winter in the Southern Hemisphere.

Go Online **active art**
For: The Seasons activity
Visit: PHSchool.com
Web Code: cfp-5012

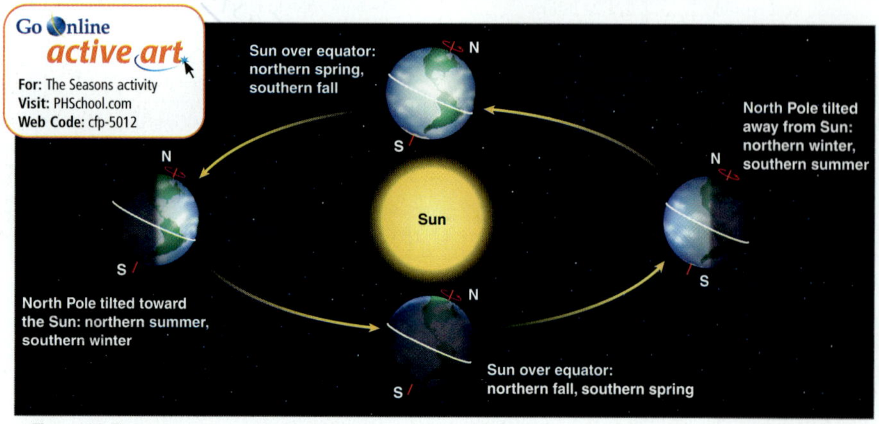

▲ **Figure 14-6** The seasons are caused by Earth's tilt and its revolution around the Sun.

330

Teaching Options

EXTENDING THE LESSON

Signs of the Changing Seasons Because Earth's orbit around the Sun progresses at a regular rate of one complete revolution every 365 1/4 days, the "official" beginning of each season occurs on or very close to the same date every year. However, because of yearly or monthly variations in weather, the appearance of the visible signs of the changing seasons often occur before or after the official beginning of a season. For example, leaves turning color is generally a sign of autumn, and the sprouting of the first wildflowers or the return of migrating birds are typical signs of spring.

Ask students to recall some recent seasonal events and decide whether their appearance occurred close to the actual beginning of the season, before it, or after it. Have students speculate on why the seasonal events occurred when they did. **learning styles: visual, auditory**

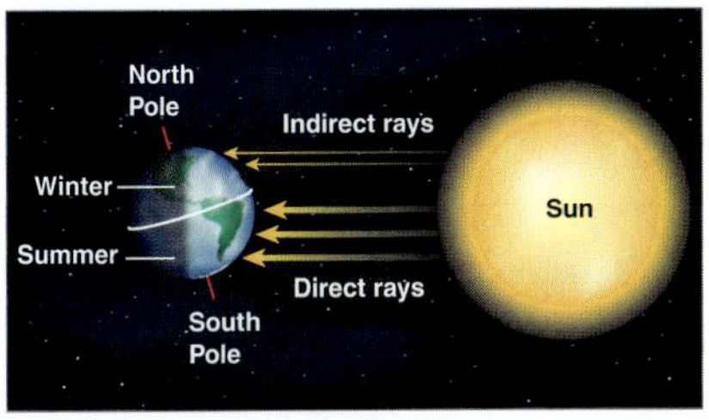

▲ **Figure 14-7** December in New York City

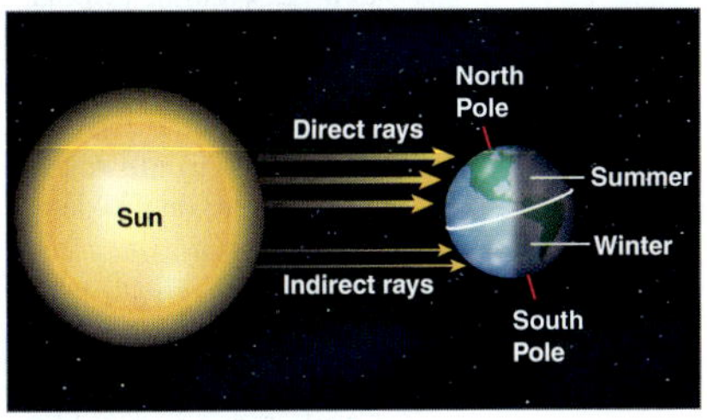

▲ **Figure 14-8** June in Australia

2 IDENTIFY: If it is summer in the Northern Hemisphere, what season is it in the Southern Hemisphere?

CHECKING CONCEPTS

1. The North Pole points __________ the Sun during the summer in the Northern Hemisphere.
2. Seasons are caused by the __________ of Earth's axis.

THINKING CRITICALLY

3. **ANALYZE:** If it is summer at the North Pole, what season is it at the South Pole?
4. **INFER:** Why do you think direct sunlight produces more heat than indirect sunlight?

BUILDING MATH SKILLS

Measuring Light from the Sun strikes Earth's surface at different angles. Angles are measured in degrees. A full circle has 360 degrees. When the Sun is directly overhead, near the equator, it is at an angle of 90 degrees to Earth's surface. A 90 degree angle is called a right angle. It is one-fourth of a circle. When the Sun is near the horizon, it is at an angle of 0 degrees to Earth's surface. Earth's axis is tilted at an angle of 23½ degrees. About what fraction of a right angle is this?

Hands-On Activity

MODELING DIRECT AND INDIRECT RAYS

You will need a flashlight and graph paper.

1. Hold a flashlight about 20 cm above a sheet of graph paper. Shine the flashlight straight down. On the paper, trace around the outer edge of light. Count how many boxes are inside the circle you made. Record the number.
2. Tilt the flashlight, making sure you keep it at the same height. Again, trace around the edge of the light. Count the number of boxes inside this. Record the number.

▲ **STEP 2** Again trace around the outline of light made by the flashlight.

Practicing Your Skills

3. **OBSERVE:** **a.** Which light rays were spread out over a larger area, the direct rays or the slanted rays? **b.** Which rays were brighter, the direct rays or the slanted rays?
4. **HYPOTHESIZE:** Which rays do you think heat better, direct rays or slanted rays?
5. **CONCLUDE:** When do you think Earth's surface in the Northern Hemisphere gets the more direct rays of the Sun, in the summer or in the winter? Explain.

3 *Assess*

CHECKING CONCEPTS ANSWERS

1. toward
2. tilt

THINKING CRITICALLY ANSWERS

3. **ANALYZE:** winter
4. **INFER:** The light and heat from the Sun are more concentrated when sunlight is more direct.

4 *Close*

BUILDING MATH SKILLS ANSWER

Measuring Since 90 ÷ 23.5 = 3.8, or about 4, the tilt angle is about one-quarter of a right angle.

Hands-On Activity

MODELING DIRECT AND INDIRECT RAYS

TIME: 15 minutes

PURPOSE: modeling, comparing, hypothesizing, concluding

DISCUSSION: Make sure that students understand that the flashlight represents the Sun and the graph paper represents Earth's surface.

ALTERNATIVE MATERIALS: A gooseneck lamp can be used instead of a flashlight.

SAFETY TIP: If you use a gooseneck lamp, caution students to avoid touching the lamp's shade or bulb.

Practicing Your Skills Answers

3. **OBSERVE:** **a.** the slanted rays **b.** the direct rays
4. **HYPOTHESIZE:** direct rays
5. **CONCLUDE:** in summer, because temperatures are higher in the summer and direct rays produce more heat

MORE ON THE WEB

The Land of the Midnight Sun
You can have students visit www.conceptsandchallenges.com to read about Alaska, which is sometimes called the Land of the Midnight Sun. Then, you can ask students to create a tourism flyer for Alaska.

ONGOING ASSESSMENT

Have students explain to a partner why the Northern and Southern hemispheres have opposite seasons.

• TEACHING RESOURCES • Online or Print

Teacher's Resources

Lesson Review: Chapter 14, p. 4

Enrichment: Chapter 14, p. 5

Go Online active art

For: The Seasons activity
Visit: PHSchool.com
Web Code: cfp-5012

Students can interact with the art of the seasons online.

14-4 (pp. 332–333)

1 Introduce

STUDY HINT Before beginning this lesson, read the list of Key Terms aloud so that students can hear their pronunciations.

Linking Prior Knowledge Remind students that they learned about how climate affects living organisms (Lesson 13-5). Discuss how climate relates to seasons and length of day.

2 Teach

Demonstration Using a globe to represent Earth and a student to represent the Sun, demonstrate Earth's position relative to the Sun during the solstices and equinoxes.
learning style: visual

Discussion To introduce this lesson, write the following dates on the board: March 21, June 21, September 21, December 21. Ask students what these dates signify. Point out that in the Northern Hemisphere, June 21 is called the summer solstice and December 21 is called the winter solstice. Describe the positions of Earth on the solstices. Refer students to Figure 14-9 to illustrate your descriptions. Identify March 21 and September 21 as the equinoxes and describe the position of Earth on these days.
learning styles: auditory, visual

Reinforcement Ask how long day and night are at the equator during an equinox. (12 hours of darkness and 12 hours of daylight.) Ask how long day and night are at the South Pole, and how long in your community. (12 hours of darkness and 12 hours of daylight) Ask whether there is anywhere on Earth that does not have 12 hours of darkness and 12 hours of light during an equinox. (There are 12 hours of darkness and 12 hours of daylight everywhere on Earth.)
learning style: auditory

Answers to questions

1 **NAME:** winter

2 **IDENTIFY:** June 21

3 **NAME:** September 21 and March 21

14-4 What are the solstices and equinoxes?

Objective
Define solstice and equinox.

Key Terms
perihelion (per-uh-HEE-lee-uhn)**:** point in a planet's orbit at which it is closest to the Sun
aphelion (uh-FEE-lee-uhn)**:** point in a planet's orbit at which it is farthest from the Sun
solstice (SAHL-stihs)**:** day of the year the Sun reaches its highest or lowest point in the sky
equinox (EE-kwih-nahks)**:** day the Sun shines directly on the equator

Near and Far During the year, the distance between Earth and the Sun changes. Earth is at **perihelion,** or closest to the Sun, in early January. Earth and the Sun are then about 147 million km apart. The Northern Hemisphere has winter. In early July, Earth is at **aphelion,** or its farthest point from the Sun. This distance is about 152 million km. Yet, this is summer in the Northern Hemisphere. The seasons do not depend on how far Earth is from the Sun.

1 NAME: During which season is the Northern Hemisphere closest to the Sun?

The Solstices The beginning of the summer and winter seasons is marked by a day called a solstice. The word **solstice** means "Sun stop."

In the Northern Hemisphere, the first day of summer is on or about June 21. The Sun seems to travel in its highest path across the sky. This day is called the summer solstice. On this day, Earth is very close to aphelion. The North Pole has 24 hours of daylight. At the same time, the South Pole points away from the Sun. The South Pole has 24 hours of darkness. Winter begins on this date in the Southern Hemisphere.

▲ **Figure 14-10** The period of daylight reaches a full 24 hours at the polar zones in summer. This phenomenon is known as the midnight Sun.

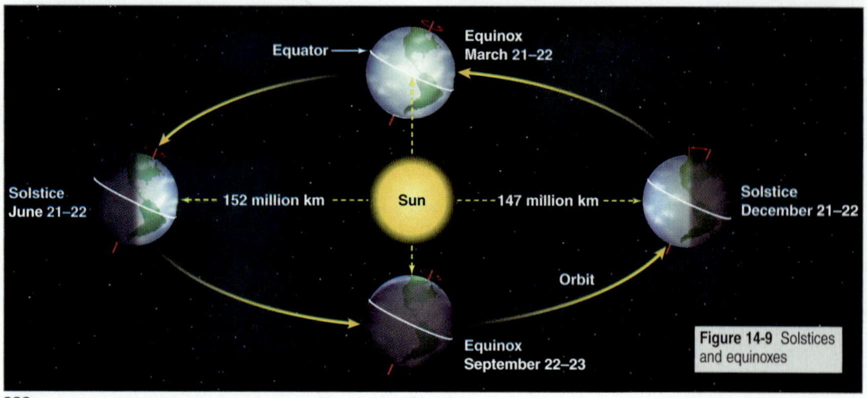

Figure 14-9 Solstices and equinoxes

332

Teaching Options

RETEACH

Understanding Seasons and Distance Use Figure 14-9 to point out to students that differences in the distance between Earth and the Sun do not cause the seasons. Have students use tracing paper and a pencil to trace Figure 14-9. Have them mark with an *X* Earth's approximate position in its orbit when it is closest to the Sun at perihelion (in January, or close to its December position). Point out that in this position, the Northern Hemisphere has winter. **learning styles: visual, tactile**

QUICK CONVERSIONS		
	Metric/SI	**English**
Earth-Sun distance		
Perihelion	147 million km	91.3 million mi
Aphelion	152 million km	94.5 million mi
Stonehenge		
Sandstone blocks	4 m	13 ft
Circle diameter	30 m	98 ft
Flat stone	5 m	16 ft
Distance to stone marker	73 m	239 ft

The first day of winter in the Northern Hemisphere is on or about December 21. This day is called the winter solstice. On the winter solstice, Earth is very close to perihelion. The North Pole points away from the Sun. The Sun seems to follow its lowest path across the sky. On this day, the South Pole has 24 hours of daylight.

2 IDENTIFY: What day marks the first day of winter in the Southern Hemisphere?

The Equinoxes Because of the tilt of Earth's axis, the Sun's position relative to Earth's equator is constantly changing. Most of the time, the Sun is north or south of the equator at solar noon. Two times during the year, the Sun is directly over the equator at solar noon. These are the **equinoxes.** The equinoxes mark the beginning of the spring and fall seasons. The spring, or vernal, equinox occurs around March 21. The fall, or autumnal, equinox occurs around September 21. During an equinox, there are 12 hours of darkness and 12 hours of daylight everywhere on Earth.

3 NAME: On which days of the year does the North Pole have 12 hours of daylight and 12 hours of night?

CHECKING CONCEPTS

1. What is a day with 12 hours of daylight and 12 hours of darkness called?
2. What date marks the beginning of winter in the Northern Hemisphere?
3. Are the days longer or shorter in the Northern Hemisphere right after the autumnal equinox?
4. *Equinox* means "equal night." How many hours of daylight and darkness are there during the fall and spring equinoxes?

THINKING CRITICALLY

5. DESCRIBE: What is winter like in your area? Explain why.
6. HYPOTHESIZE: On the winter solstice, the North Pole has 24 hours of darkness. How might this affect your life that day if you lived there?

BUILDING READING SKILLS

Using Prefixes The prefix *equi-* means "equal." Find five words in the dictionary that contain this prefix. Write down these words and their definitions on a sheet of paper. Circle the part of the definition that relates to the prefix.

How Do They Know That?

PREDICTING THE CHANGE OF SEASONS

Ancient peoples made calendars based on astronomical events that occurred the same time each year. Some of these calendars were monuments. One such monument is called Stonehenge.

Stonehenge stands on a plain in Salisbury, England. It was built around 1848 B.C. The monument is made of huge blocks of sandstone about 8 m high. They form a circle 30 m across. Inside the circle, there are two horseshoe-shaped sets of stones. A flat block of sandstone 5 m long rests inside the inner horseshoe. About 73 m away from the flat sandstone block is a stone marker.

Stonehenge is a giant calendar that can predict the change of seasons. It can also predict the solstices and equinoxes. A shadow is cast by the stone marker onto the flat sandstone block. This marks the direction in which the Sun rises and sets on the longest day of the year, the summer solstice.

▲ **Figure 14-11** Stonehenge is a giant calendar that predicts solstices and equinoxes.

Thinking Critically What might ancient peoples have needed calendars for?

MORE ON THE WEB

Creating Equinox Festivals
Have the students go online to learn about festivals held during the equinoxes. Students can then create a festival for either the fall or spring equinox and invent a history for their festival.

• TEACHING RESOURCES • Online or Print

Teacher's Resources
Lesson Review: Chapter 14, p. 6

Laboratory Manual
Lab. Challenge: How can you track the Sun's movement across the sky? pp. 73–76

3 Assess

CHECKING CONCEPTS ANSWERS

1. equinox
2. December 21
3. shorter
4. 12 hours each

THINKING CRITICALLY ANSWERS

5. DESCRIBE: Possible answer: The days are short but get longer as the winter passes because the Sun's position is constantly changing relative to the equator, and this changes length of day in other places.
6. HYPOTHESIZE: Possible answer: I would probably spend more time indoors in well-lighted areas or maybe sleep more.

BUILDING READING SKILLS ANSWER

Using Prefixes Possible answer: *Equidistant* means "equally distant from a point." *Equilibrium* means "a state of equality between two forces." An *equilateral* triangle has three equal sides. An *equivocal* statement can have several meanings that are equally likely to be true. *Equimolecular* means "having an equal number of molecules."

4 Close

How Do They Know That?

PREDICTING THE CHANGE OF SEASONS

Geology Connection Prior to the development of modern methods of dating, scientists could only speculate about how old Stonehenge was. After analyzing charcoal that was found near Stonehenge, using the carbon-14 dating method, scientists concluded that Stonehenge was built around 1848 B.C. Carvings found on the stones also indicate a similar age.

Discussion Ask students: *On what date is a shadow cast onto the flat sandstone block at Stonehenge?* (on or about June 21, the summer solstice) Discuss how ancient peoples may have transported the huge blocks of sandstone that make up Stonehenge.

Thinking Critically Answer
Possible answers include to plan special events and ceremonies, to decide when to plant crops, or to keep track of the movement of objects in the sky.

The Big Idea

(pp. 334–335)

1 *Introduce*

Objective Students will look at the seasonal behavior of several animals. Students will choose one migrating animal to research and write about in their Science Logs.

Linking Prior Knowledge Ask students to recall how Earth's tilt causes seasons (Lesson 14-3). Then, have students review the effect of seasonal changes on climate (Lesson 13-2). Finally, ask students to remember how climate affects living things (Lesson 13-5).

2 *Teach*

Discussion Point out that the migratory animals discussed in this feature are air creatures and sea creatures. Ask students: *What special features do these animals have that allow them to travel great distances?* (strong wings or fins) *What limitations prevent modern land animals from migrating in the same way?* (obstacles such as large bodies of water and buildings and highways that block migratory routes)

Use the background information in Fast Facts to help guide students in choosing a topic for their Science Logs.

THE BIG IDEA

How does the change of seasons affect animal behavior? Have students start their research for their Science Log Writing Activity at www.conceptsandchallenges.com. Be sure students use search engines and websites that are reliable and acceptable. You may also wish to provide copies of the *Big Idea Planner* and *Big Idea Science Log Entry* study tools found in the Teacher's Resources.

Integrating Life Science

THE Big IDEA

How does the change of seasons affect animal behavior?

Does the weather in your town change with the seasons? If so, you probably respond to this by wearing different clothing and participating in different activities. Animals respond to environmental changes, too. Some shed thick coats to prepare for warm summer weather. Others store food to prepare for the time cold weather makes food scarce. Animals that migrate take more drastic actions. They leave their homes and travel thousands of miles in search of better living conditions for the season. Many breed in their winter homes.

California gray whales are migratory animals. During the summer and early fall, the whales live in the cold waters of the Chukchi and Bering seas. In late fall, they migrate southward. They travel more than 8,000 km to the calm, warm waters of Baja California, in Mexico. In the shallow waters of their winter home, the whales breed and produce young. In the late spring, they return to their Arctic summer home.

Monarch butterflies migrate in the same general direction as California gray whales. The butterflies spend the summer in the northern United States. In the fall, they migrate to the California coast, Mexico, and Florida.

Of all migrating animals, Arctic terns travel the greatest distance. These birds spend the summer breeding in northern Canada. In the fall, they travel more than 17,600 km southward to their Antarctic winter home. The arrival of spring triggers a return trip northward.

Look at the illustrations and photos that appear on these two pages. Then, follow the directions in the Science Log to find out more about "the big idea." ✦

BERING SEA

PACIFIC OCEAN

California Gray Whales

During their southward journey in fall, California gray whales travel along the West Coast. For this reason, whale watching is a popular fall activity in California.

WRITING ACTIVITY

Science Log

Choose a migrating animal. You may choose one that is mentioned here or another one that you know about. Write a report about the different environments in which the animal you have chosen lives. Tell why it migrates, as well as where it migrates to. Start your search for information at www.conceptsandchallenges.com.

334

Teaching Options

! FAST FACTS *Meet the Migrators*

Arctic Terns
Arctic terns are related to gulls. They are about 45 cm (1.5 ft) long, with a bluish gray back, white breast, and black crown. Arctic terns spend almost their entire adult lives in the air. They can hover in midair, which helps them spot and swoop down to catch fish, shrimp, and krill, which make up most of their diet.

Gray Whales
Gray whales belong to the group of whales called baleen whales. Instead of teeth, baleen whales have long, thin, fiberlike plates, called baleen, that extend down from the sides of their mouths. To feed, the gray whale dives to the ocean floor, where it sucks water, mud, sand, and tiny organisms into its mouth. It uses its tongue to force the water through the baleen. The tiny creatures are trapped in the whale's mouth and swallowed.

Monarch Butterflies
When monarch butterflies arrive at their winter resting places, they gather in large groups on trees or bushes. Clasping a twig or leaf with their sharp claws, the butterflies hang upside down, with one of their wings overlapping the wing of the butterfly below it. This formation gives the cluster of monarchs protection from cold nighttime temperatures, as well as from wind and rain.

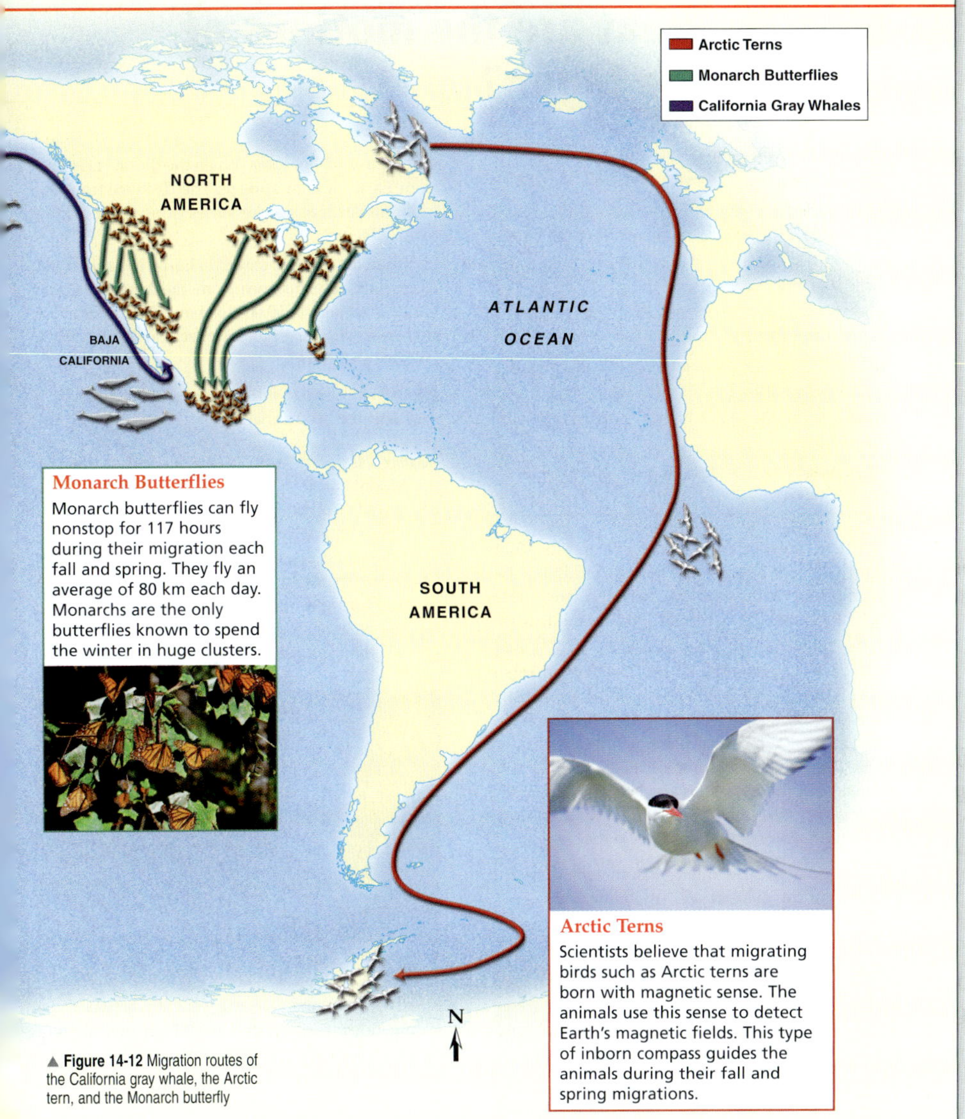

Monarch Butterflies
Monarch butterflies can fly nonstop for 117 hours during their migration each fall and spring. They fly an average of 80 km each day. Monarchs are the only butterflies known to spend the winter in huge clusters.

Arctic Terns
Scientists believe that migrating birds such as Arctic terns are born with magnetic sense. The animals use this sense to detect Earth's magnetic fields. This type of inborn compass guides the animals during their fall and spring migrations.

▲ **Figure 14-12** Migration routes of the California gray whale, the Arctic tern, and the Monarch butterfly

3 Assess

Use the Writing Activity Rubric from the Teacher's Resources to assess students' Science Logs. Fill in the rubric with the additional information below. For this assignment, students should have:

- chosen and researched a migrating animal.
- written a report about the two environments in which the animal lives.

4 Close

Writing Travel Stories Have students write a fictional story about a migrating animal. Their story should include possible experiences of an animal on its seasonal journey. Encourage students to be creative but insist on scientific accuracy.

Alternatively, students can research the survival behaviors of animals that neither migrate nor hibernate during cold winter months, such as deer, foxes, wolves, and other mammals. Students can prepare short oral or written reports on how these animals stay active all winter despite the cold, the shortage of food, and the effects of snow and ice.
learning style: auditory

WRITING HINT Encourage students to write several drafts of their report or story. For the first draft, students should be sure that they have included an introduction, supporting details, and a conclusion. By a second or third draft, they should pay special attention to grammar, punctuation, and spelling.

Tips for Using Technology

Using Presentation Software for Reports Have interested students use presentation software to complete their Science Log activity on a migrating animal. Presentation software allows them to use a computer to put text and art together in a slide show. Students can then print out their slide show and use it as a handout to follow during their presentations. Students may find the Thumbnail Sketch graphic organizer in the Teacher's Resources useful in planning their presentations.

• TEACHING RESOURCES • Online or Print

Teacher's Resources

Big Idea: Chapter 14, p. 7

Big Idea Planner: Study Tools, pp. 11–12

Big Idea Science Log Entry: Study Tools, p. 13

EXTENDING THE LESSON

Researching Seasonal Changes Have students find out how hibernation helps animals such as woodchucks survive cold winters.

14-5 *(pp. 336–337)*

1 Introduce

STUDY HINT Have students write all four Key Terms on separate index cards. As they read the lesson, have them write the definition of each term on the back of the card.

Linking Prior Knowledge Have students recall what they learned about tides (Lesson 10-6). Ask them to describe evidence of tides that they have observed at a beach or harbor. (Answers may include exposed rocks or pilings covered with seaweeds and marine creatures such as barnacles and mussels.)

2 Teach

Demonstration Use a large globe to guide students to understand why people see only one side of the Moon. With masking tape, divide the globe into four equal sections from pole to pole. Shine a light on the globe and demonstrate that each part is illuminated at one time or another. Walk around the room and slowly rotate the globe so that the same side faces the class throughout your "revolution."
learning style: visual

Discussion Describe the rotation and revolution of the Moon. Refer students to Figure 14-13 to observe the positions of the Moon at apogee and perigee.
learning styles: visual, auditory

Answers to questions

1. **CALCULATE:** 3,500 km/h × 24 h/day = 84,000 km/day
2. **INFER:** because the Moon's orbit is elliptical
3. **INFER:** twice a month
4. **EXPLAIN:** Earth must rotate an extra 50 minutes each day to "catch up" with the Moon.

READING STRATEGY Write the following Key Terms on the board: *apogee* and *aphelion*, *perigee* and *perihelion*. Point out the letters that each pair have in common, then compare their definitions.
learning style: visual

14-5 What are the motions of the Moon?

Objectives

Describe and identify two motions of the Moon.
Explain how the Moon's motions affect the tides.

Key Terms

apogee (AP-uh-jee): point at which the Moon is farthest from Earth

perigee (PER-uh-jee): point at which the Moon is closest to Earth

spring tide: tide that is higher or lower than a normal tide

neap tide: tide that is not as high or as low as a normal tide

Rotation and Revolution Earth and its Moon travel together around the Sun. In addition, the Moon orbits Earth. Its journey around Earth takes about a month. The Moon also rotates on its axis.

The Moon rotates more slowly on its axis than Earth does. It takes the Moon 27⅓ days to rotate once on its axis. The Moon revolves around Earth at a speed of about 3,500 km/h. It takes the Moon 27⅓ days to make one complete revolution around Earth. This is the same time it takes the Moon to rotate. All of these motions are constantly changing the relative positions of the Sun, Earth, and Moon and how the Moon appears to us in the sky.

1 **CALCULATE:** How many kilometers does the Moon travel around Earth in one day?

▼ **Figure 14-13** The revolution and rotation of the Moon

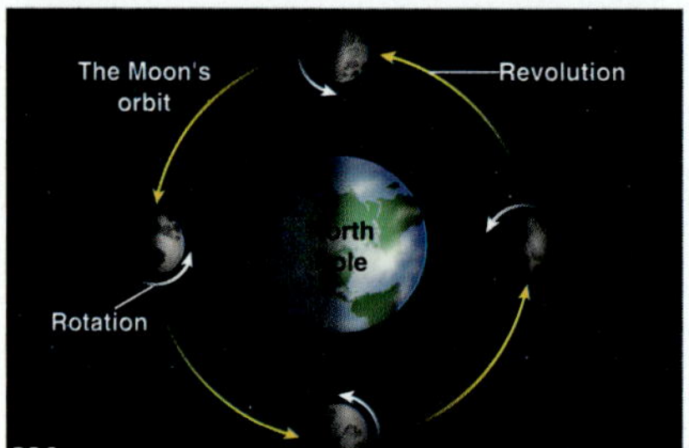

336

Apogee and Perigee When the Moon is farthest from Earth, it is at **apogee.** The distance from the Moon to Earth at apogee varies from about 404,000 to 407,000 km.

When the Moon is closest to Earth, it is at **perigee.** At perigee, the distance from the Moon to Earth varies from about 350,000 to about 370,000 km. This change in distance has an effect on Earth's tides, among other things.

2 **INFER:** Why do the distances between Earth and the Moon change?

Moon's Gravity and Tides At certain times in the Moon's orbit of Earth, the Sun and the Moon are lined up in their orbits. The combined gravitational pull of the Sun and the Moon results in higher high tides and lower low tides on Earth. Therefore, the daily tidal range is greatest at these times. During these periods, which occur in a regular cycle twice a month, the tides become what are called **spring tides.**

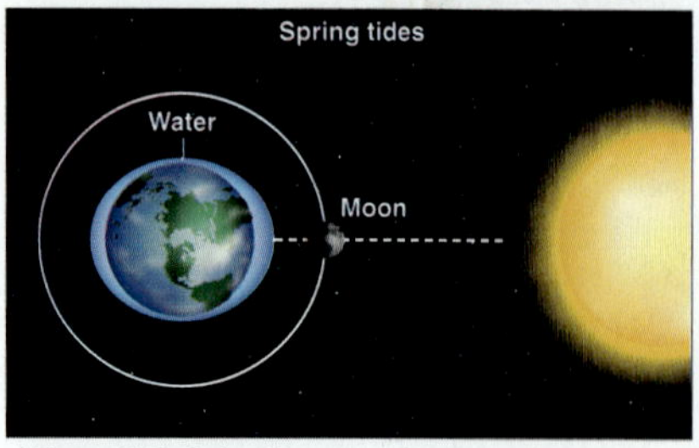

▲ **Figure 14-14** Spring tides

At certain times of the year, in between the spring tides, the Sun and the Moon pull at right angles to each other in relation to Earth. Because the Sun and the Moon pull at right angles to each other, their gravity forces do not combine or act together on Earth's oceans. Instead, they pull against each other. As a result, the daily tidal range is small. Tides that occur during this time are called **neap tides.**

Teaching Options

EXTENDING THE LESSON

Collecting and Organizing Data Have students collect newspaper articles giving the times the Moon rises for the next week. Tell students to organize the data in a table, graph the results, and compare the times. Have them make a general statement about how the times vary. (Moonrise occurs 50 minutes later each day and so the Moon can sometimes be seen during daylight hours.) **learning style: visual**

QUICK CONVERSIONS

	Metric/SI	English
Revolution	3,500 km/h	2,170 mi/h
Apogee	405,000–407,000 km	252,000–253,000 mi
Perigee	360,000–370,000 km	226,000–230,000 mi

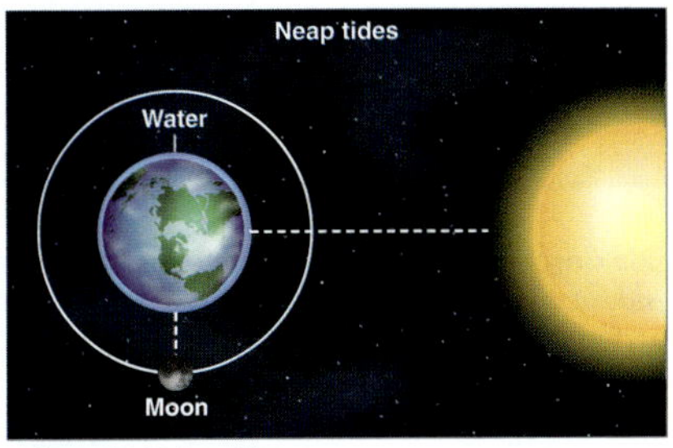

▲ **Figure 14-15** Neap tides

3 INFER: How often do spring and neap tides occur?

Moonrise Like the Sun, the Moon appears to rise in the east and set in the west. As the Moon revolves around Earth, Earth must go through more than one rotation to "catch up" with the Moon. Earth must rotate 24 hours and 50 minutes to bring the Moon back into view. As a result, the Moon comes into view at moonrise about 50 minutes later each day.

4 EXPLAIN: Why does the Moon rise about 50 minutes later each day?

CHECKING CONCEPTS

1. How long does it take the Moon to make one complete revolution around Earth?
2. What causes the tides?
3. Why does the Moon not always rise at the same time?

THINKING CRITICALLY

4. EXPLAIN: Why does the Moon appear to rise in the east and set in the west?
5. ANALYZE: How many times does the Moon rotate during one revolution around Earth?
6. CALCULATE: If the Moon rose at 6:45 P.M. on Tuesday, at about what time would it rise on Thursday?

BUILDING READING SKILLS

Using Prefixes The prefix *peri-* means "around," "surrounding," or "close." Perigee is when the Moon is closest to Earth. Using a dictionary, find five other words that contain this prefix. Explain how the prefix relates to the meaning of each word.

People in Science

ASTRONAUT

The first person to travel in space was a Soviet air force pilot named Yuri Gagarin. His one orbit around Earth in April 1961 took less than two hours. American astronauts first flew in space in the *Mercury* and *Gemini* missions. However, it was the dramatic landing of American *Apollo* astronauts on the Moon in July 1969 that really marked the beginning of a human presence in space.

▲ **Figure 14-16** An astronaut collects lunar samples from the Moon's surface.

An astronaut travels and works in space. The word *astronaut* usually refers to a person involved in the American space program. Astronauts help design and test spacecraft and other space equipment. While in space, they operate their spacecraft and analyze any problems that arise. They also carry out experiments in space. Astronauts work for the National Aeronautics and Space Administration, or NASA. Many are also military officers. To be an astronaut, you need a background in mathematics and science. Most astronauts have advanced degrees. Astronauts who are pilots also have many hours of flight experience.

Thinking Critically Which Earth science specialists might want to be astronauts?

3 Assess

CHECKING CONCEPTS ANSWERS

1. $27\frac{1}{3}$ days
2. the combined gravitational pull of the Sun and Moon
3. because Earth rotates slower than the Moon revolves

THINKING CRITICALLY ANSWERS

4. EXPLAIN: because Earth rotates from west to east
5. ANALYZE: once
6. CALCULATE: 6:45 P.M. + 100 minutes (50 minutes × two days) = 8:25 P.M.

4 Close

BUILDING READING SKILLS ANSWERS

Using Prefixes Possible answers: A *perianth* surrounds a flower. *Perinatal* refers to the time very close to birth. A *perimeter* is the boundary of an area. The *perihelion* is the point closest to the Sun. The *pericardium* is a membrane surrounding a vertebrate's heart.

People in Science

ASTRONAUT

Real-Life Connection On February 20, 1962, astronaut John Glenn became the first American to orbit Earth. In his spacecraft, *Friendship 7*, Glenn orbited Earth three times, and his trip lasted 4 hours and 55 minutes. Glenn came back to Earth a national hero. From 1974 until 1999, he served as a U.S. senator from Ohio. In 1998, Glenn returned to space aboard the space shuttle *Discovery*. At the age of 77, he was the oldest person ever to travel into space.

Discussion Explain to students that *astronaut* comes from two Greek words that mean "star" and "sailor." Discuss the appropriateness of the term *astronaut* to describe the profession.

Thinking Critically Answer

Possible answers include geologists, astronomers, and meteorologists.

MORE ON THE WEB

Viewing an Earthrise Have students use the Internet to view image galleries of an Earthrise, as seen from the Moon. Then have them write a report as if they were astronauts on the Moon seeing Earth for the first time.

COOPERATIVE LEARNING

Answering Questions Have pairs answer the Checking Concepts and Thinking Critically questions.

• TEACHING RESOURCES • Online or Print

Teacher's Resources

Lesson Review: Chapter 14, p. 8

ESL/ELL STRATEGY

Pairs of Opposites Point out to English-language learners that the Key Terms are two pairs of opposites and that if they can remember the meaning of one for each pair, they should be able to recall or figure out the other.

14-6 *(pp. 338–339)*

1 *Introduce*

STUDY HINT The diagram in Figure 14-17 is very important in understanding this lesson. Have students explain the sequence that occurs and how Earth's and the Moon's motions create the effect of phases on Earth.

Linking Prior Knowledge Ask students to remember that the side of Earth that faces the Sun has light and the side that faces away from the Sun has darkness (Lesson 14-1). Then, ask students if they have noticed whether the Moon looks the same or different from night to night over the course of a month.

2 *Teach*

Demonstration Demonstrate the phases of the Moon by shining a projector on a globe. Tell students that the projector represents the Sun. Attach a plastic-foam ball to a pencil to represent the Moon. Using Figure 14-17 as a guide, hold the ball at different positions around the globe. Tell students to note that only certain parts of the ball are lighted at the different positions. **learning style: visual**

Discussion Display a calendar and have students note the dates of the new, full, first-quarter, and third-quarter Moon. Point out that these are the phases of the Moon. Explain why the Moon appears to change shape. **learning style: visual**

Reinforcement Draw a diagram on the board similar to Figure 14-17 to represent various positions of the Moon in its orbit around Earth. Guide students to see why the Moon changes its appearance during the month. **learning style: visual**

Answers to questions

1. **IDENTIFY:** because of the way light from the Sun is reflected from the Moon as a result of the positions of the Sun, the Moon, and Earth
2. **DEFINE:** the phase during which more than half the Moon is visible
3. **COMPARE:** The waxing crescent phase is after the new Moon. The waning crescent phase is before the new Moon.

14-6 What are the phases of the Moon?

Objective

Identify the different phases of the Moon.

Key Terms

phases (FAYZ-uhz)**:** changing shapes of the Moon

crescent (KREHS-uhnt) **phase:** phase when less than half the Moon is visible

gibbous (GIHB-uhs) **phase:** phase when more than half the Moon is visible

Phases of the Moon From Earth, the Moon sometimes looks round. Other times, it looks like a thin sliver. The Moon appears to change shape because of the way it reflects light from the Sun. The changing shapes of the Moon are called **phases.** These phases depend on the positions of the Sun, the Moon, and Earth.

1. **IDENTIFY:** Why does the Moon appear to change shape?

Waxing Phases Look at Figure 14-17. When the side of the Moon facing Earth is dark, it appears as if there is no Moon at all. This is called a new Moon. As the Moon revolves around Earth, a small part of it becomes visible. As the visible part of it increases, the Moon is said to be waxing. The first phase is called the waxing **crescent phase.** During the crescent phase, less than half of the Moon is visible. When the Moon has moved one-quarter of the way around Earth, it enters the first quarter phase. This means that half of the side of the Moon facing Earth is visible.

As the Moon continues in its orbit, more and more of the side facing Earth becomes visible. This is called the waxing **gibbous phase.** During the gibbous phase, more than half of the Moon is visible. Finally, the Moon completes half of its trip around Earth. The whole side facing Earth is visible. This is the full Moon.

2. **DEFINE:** What is the Moon's gibbous phase?

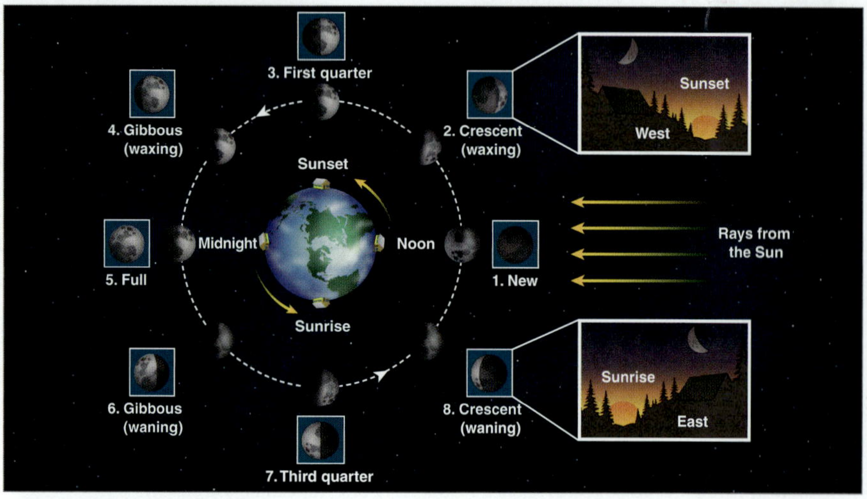

▲ **Figure 14-17** The phases of the Moon

Teaching Options

EXTENDING THE LESSON

Observing the Moon Assign a Moon observation for the next 4 weeks. Have students keep a record of the Moon's phases, and the time and date each phase occurs. Encourage them to make drawings of how the Moon looks each night. **learning style: visual**

ONGOING ASSESSMENT

Make flash cards that show a sketch or photograph of each phase of the Moon. Call on students to identify the phase shown as you hold up each card.

ESL/ELL STRATEGY

Going Through a Phase Ask English-language learners what they think the expression "going through a phase" means. Ask how this expression might relate to the Moon.

Waning Phases As the Moon continues to move around Earth, less and less of the surface becomes visible. The Moon is waning. After the full Moon, the Moon enters the waning gibbous phase. At the last quarter phase, only half of the Moon's surface facing Earth is visible. The last phase of the Moon is the waning crescent phase.

The Moon takes 29½ days to go through all of its phases. This is a little longer than the time for one revolution of the Moon around Earth. The reason for this is that as the Moon orbits Earth, the two bodies are also moving around the Sun. The Moon must travel a little farther to get directly between Earth and the Sun.

3 COMPARE: When do the waxing and waning crescent phases of the Moon take place?

CHECKING CONCEPTS

1. The Moon reflects light from the ___________.
2. The phases of the Moon depend on the positions of Earth, ___________, and the Sun.
3. During the ___________ phase of the Moon, half of the side of the Moon facing Earth is visible.
4. When the Moon has finished half of its revolution around Earth, the Moon is called a ___________ Moon.
5. The Moon goes through all of its phases in ___________ days.

THINKING CRITICALLY

6. CONTRAST: How do the crescent phase and the quarter phase of the Moon differ?
7. PREDICT: What would the Moon look like if it did not reflect light from the Sun?
8. SEQUENCE: List the eight phases of the Moon in order, beginning with the new Moon.

BUILDING SCIENCE SKILLS

Comparing The time period between one full Moon and the next is called a lunar month. On a calendar, identify each of the Moon's phases for the next two months. How does the length of the lunar month compare to the length of the calendar month?

Integrating Physical Science

TOPICS: radiation, light

HOW THE MOON IS LIT

Light travels in a straight line. Usually, it cannot go through an object. When light strikes an object, it may be totally absorbed, in which case it appears black. If the light is totally or partially reflected, then the object can appear white. Different colors of light might also be partially absorbed, with only some of them being reflected. The object will then appear in the reflected color.

We see the Moon because some of the light from the Sun is reflected off of the Moon rather than absorbed. The Moon appears white in the sky. Unlike stars, the Moon makes no light of its own. If it were not for the Sun, we would not be able to see the Moon. As the Moon revolves around Earth, the Sun's light strikes the Moon at different angles. This makes the Moon appear to change shape as viewed from Earth. We call this feature of the Moon "going through phases."

▲ **Figure 14-18** The crescent phase of the Moon

Thinking Critically Could we see the Moon if it absorbed all of the light instead of reflecting some of it?

3 *Assess*

CHECKING CONCEPTS ANSWERS

1. Sun
2. the Moon
3. quarter
4. full
5. 29 1/2

THINKING CRITICALLY ANSWERS

6. CONTRAST: During a crescent phase, less than half of one side of the Moon is visible. During a quarter phase, one-half of one side of the Moon is visible.
7. PREDICT: The Moon could not be seen if it did not reflect light from the Sun.
8. SEQUENCE: new Moon, waxing crescent, first quarter, waxing gibbous, full Moon, waning gibbous, third quarter, waning crescent

4 *Close*

BUILDING SCIENCE SKILLS ANSWER

Comparing A lunar month is 29 1/2 days. Calendar months range from 28 to 31 days.

Integrating Physical Science

TOPICS: radiation, light

HOW THE MOON IS LIT

Real-Life Connection *Albedo* is the term for the measure of how reflective the surface of a planet, a moon, an asteroid, or other celestial body is. A body that reflects no light at all has an albedo of zero. Because much of the Moon is covered by a thin layer of powdered rock that darkens its surface, it has a relatively low albedo. Only about 7 percent of the sunlight hitting it is reflected from its surface; the rest is absorbed.

Discussion Ask students why a flower appears yellow. (Only yellow light is reflected by the flower.) Ask why a blank sheet of paper or the Moon appears white. (Light is almost totally reflected by the paper and the Moon.)

Thinking Critically Answer
no, because when light is absorbed a surface appears dark

MORE ON THE WEB

Simulating Phases of the Moon
Have students use a Moon phase simulation to see what phase the Moon shows on any date between 1800 and 2199.

• TEACHING RESOURCES • Online or Print

Teacher's Resources
Lesson Review: Chapter 14, p. 9

Lab Activity

(pp. 340–341)

1 Prepare the Lab

Modeling the Phases of the Moon

SET-UP TIME:  LAB TIME: 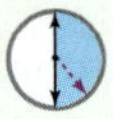

BACKGROUND: The phases of the Moon were of great importance to early sky watchers. The period between one full Moon and the next was designated a month. The appearance of certain full Moons marked the time of planting and harvest.

The Moon progresses through stages in which only parts of the side facing Earth are visible. It starts out as dark, goes to partly visible, full, partly visible again, and back to dark. This takes place over a period of 29 1/2 days. These phases are caused by the varying alignments of Earth, the Moon, and the Sun. At any particular time, we see all or parts of the Moon's near side, depending on the temporary alignment of these three bodies.

PURPOSE: Students will investigate the cause of the changing phases of the Moon.

ALTERNATIVE MATERIALS: Any white ball (such as a table-tennis ball or a baseball) can be used in place of the plastic-foam ball.

SCIENCE SKILLS: Students will **model** lunar phases and **record** their observations in a diagram.

ADVANCE PLANNING: Elevate a 150-watt lightbulb in a clip lamp or a table lamp without the shade in a corner of the room. Pull down the shades to make the room dark. If necessary, darken the windows further with black construction paper.

2 Guide the Lab

PROCEDURE: Discuss the procedure and explain how data will be recorded in a phase diagram. Divide students into pairs. Each pair will work together to model Moon phases and observe what is taking place.

Provide students with a copy of the Lab Activity Report found in the Teacher's Resources for students to record their observations and conclusions.

LAB ACTIVITY
Modeling the Phases of the Moon

Materials
- Metric ruler
- Plastic foam ball with toothpick
- Table lamp with 100-watt bulb
- Paper
- Marker

BACKGROUND

Moonshine is actually light from the Sun reflecting off the Moon's surface. When viewed from Earth on different nights throughout the month, the Moon appears to have different shapes. These shapes are called phases.

PURPOSE

In this activity, you will investigate why the Moon has phases.

PROCEDURE

1. Work with a partner in a darkened room. Place the lamp on a shelf or bookcase so that the bulb is about 120 cm above the floor. The bulb represents the Sun, the ball is the Moon, and your head is Earth.

2. Sit facing the lamp and hold the ball level with the bulb, between you and the lamp. Have your partner stand behind you and draw three circles to represent the Sun, Moon, and Earth in this setup.

3. Now have your partner shade in the part of the Moon that is not lighted by the Sun. Label this drawing Position 1.

4. Turn your body and the Moon about 45 degrees to the left. With the Moon in this position, have your partner draw the shape of the lighted part of the Moon that can be seen from Earth. Label this drawing Position 2.

▲ **STEP 2** The ball should be held level with the bulb, between you and the lamp.

▲ **STEP 4** Turn your body and the Moon about 45 degrees to the left.

Teaching Options

! FAST FACTS About Phase

Although *phase* is most often used to describe the Moon as seen from Earth, in astronomy, *phase* refers to how much of the lighted surface of any planet or satellite can be seen at a distance from that celestial body. Since their orbits pass between Earth and the Sun, Mercury and Venus are the only planets in the solar system that, when viewed from Earth, show all the phases that the Moon shows. The other planets only show either gibbous or full phases.

ESL/ELL STRATEGY

Summarizing Procedures English-language learners may benefit from stating the procedures in their own words. Have English-language learners explain the procedures to English-proficient students.

▲ **STEP 5** Turn your body and the Moon another 45 degrees.

5. Turn your body and the Moon to the left another 45 degrees and repeat Step 4. Label this drawing Position 3.

6. Continue turning the ball in the same direction, 45 degrees at a time, until you come back to your original position. Have your partner draw and label the Moon shape at each position.

7. Copy the chart in Figure 14-19. Using the information from your partner's diagrams, draw each phase of the Moon in its appropriate section in the chart.

Moon's Phases

New	First quarter	First half	Three-quarters
Full	Three-quarters	Last half	Last quarter

▲ **Figure 14-19** Use a copy of this chart to organize your diagrams.

CONCLUSIONS

1. OBSERVE: In what position is the Moon when it is in its first quarter phase?
2. OBSERVE: What position of the Moon is opposite its new Moon phase?
3. ANALYZE: Why does the Moon have phases?
4. INFER: What happens when the Moon is directly between Earth and the Sun?

TROUBLESHOOTING: Make sure that students hold the Moon slightly above their heads when it is on the opposite side of the Sun so that they can see the full Moon. If they are in direct line with the Sun, a lunar eclipse occurs.

SAFETY TIP: Remind students to move carefully in the dark corners of the room and to avoid touching the hot bulb.

EXPECTED OUTCOME: Students will learn that the Moon's phases are created by the varying alignments of the Moon, Earth, and the Sun.

LAB HINT: Make sure that students understand that their heads and eyes represent Earth in this model.

3 Conclude the Lab

1. OBSERVE: off to the side of Earth
2. OBSERVE: on the other side of Earth from the Sun
3. ANALYZE: When the Moon orbits Earth, the Sun lights up different portions of the Moon. About once a month, we see its lighted side, and about 2 weeks later, we see its shadowed side. In between, we see the Moon partially lighted and partially in shadow.
4. INFER: The Moon blocks the Sun's light from reaching Earth. (A solar eclipse occurs.)

Use the Laboratory Activity Rubric from the Teacher's Resources to assess students' lab activities. Fill in the rubric with the additional information below. For this activity, students should have:

- worked with a partner and modeled the phases of the Moon.
- drawn and labeled each phase of the Moon.

4 Extend the Lab

Have students research different kinds of full Moons to answer questions such as *Is the Moon actually blue when there is a blue Moon?* or *What are the harvest Moon and hunter's Moon?* As a class, discuss the relationship of the Moon's phases to the length of the months in a year. If all months were 29.5 days long, there would be 12.4 months a year. In just a few year's time, all the months would be shifted to different seasons.

Tips for Using Technology

Using Graphics Software for Diagramming the Lab Have interested students use graphics creation software to complete their diagrams for the Lab Activity. The pencil, or drawing tool, allows students to draw in freehand. This can be useful if they want to sketch something. The text tool allows a student to enter text into an illustration. The text tool can be used to label parts of the illustration, to give it a title, or for any other means.

• TEACHING RESOURCES • Online or Print

Teacher's Resources

Lab Activity Report: Study Tools, pp. 16–17

Laboratory Activity Rubric: Teaching Tools, p. 8

14-7 (pp. 342–343)

1 Introduce

STUDY HINT Have students look at the diagrams in this lesson carefully to be sure that they understand the differences, which are subtle but important.

Linking Prior Knowledge Ask students who have observed a lunar eclipse to share their experiences with the class.

2 Teach

Demonstration Place an object on a sheet of white paper and hold it in front of a lamp. Have students observe the shadow that forms. Point out the umbra and penumbra of the shadow.
learning style: visual

Discussion An eclipse occurs whenever one celestial body is partially or totally obscured by the shadow of another. Observations of such events have provided important contributions to the physical sciences. For example, in 1675, Danish astronomer Ole Rømer used eclipses of Jupiter's satellites to calculate the speed of light. Also, observations of starlight passing near the Sun during a solar eclipse helped validate Einstein's general theory of relativity. Using Figure 14-20, describe a lunar eclipse. Have students describe the positions of the Sun, the Moon, and Earth during a lunar eclipse.
learning styles: visual, auditory

Answers to questions

1 **IDENTIFY:** the umbra and penumbra

2 **DEFINE:** the passing of the Moon through Earth's shadow

3 **DESCRIBE:** The entire face of the Moon is darkened.

READING STRATEGY Suggest that students read the lesson straight through, without worrying about understanding everything they read. Instruct them to keep reading even if they encounter a concept they do not understand. Once they have read everything, they can go back and reread anything that they did not understand the first time.

14-7 What causes an eclipse of the Moon?

Objectives

Describe a lunar eclipse. Distinguish between a total and a partial lunar eclipse.

Key Terms

umbra (UHM-bruh): center or dark part of a shadow

penumbra (pih-NUHM-bruh): light part of a shadow

lunar eclipse (ih-KLIHPS): passing of the Moon through Earth's shadow

Casting Shadows When you walk outside on a sunny day, you can see your shadow. A shadow is formed when an object blocks a light source. A shadow has two parts. The center of a shadow is very dark. The dark part of a shadow is called the **umbra.** Around the outside of a shadow, you will see a lighter part. The light part of a shadow is called the **penumbra.**

People and objects are not the only things that can cast shadows. Earth, the Moon, and other bodies in space also cast shadows.

1 **IDENTIFY:** What are the two parts of a shadow?

Eclipse of the Moon As the Moon revolves around Earth, it usually passes above or below Earth's shadow. Sometimes the Moon passes directly through Earth's shadow. As a result, sunlight is blocked from reaching the Moon. When the sunlight is blocked from the Moon, a **lunar eclipse** occurs. A lunar eclipse can occur only during the full-Moon phase.

2 **DEFINE:** What is a lunar eclipse?

Total or Partial Eclipses Sometimes the Moon moves entirely into Earth's umbra. When this happens, all of the Sun's light is blocked. The entire face of the Moon darkens. This is called a total lunar eclipse.

Sometimes, only part of the Moon moves into Earth's umbra. Sunlight can still reach the Moon. As a result, only part of the Moon darkens. This is called a partial lunar eclipse.

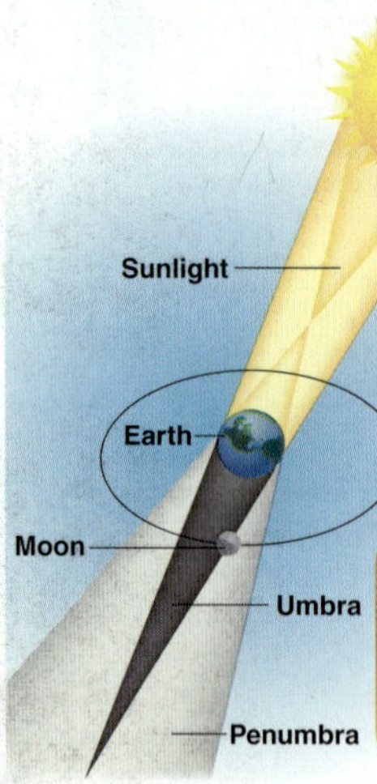

Go Online active art
For: Moon Phases and Eclipes activity
Visit: PHSchool.com
Web Code: cfp-5013

▲ **Figure 14-20** A partial lunar eclipse

Total lunar eclipses are rare. Also, they do not occur at regular intervals. They may occur as often as six months apart or as much as 2½ years apart.

▲ **Figure 14-21** A partial lunar eclipse

3 **DESCRIBE:** How does the Moon look during a total lunar eclipse?

Teaching Options

INTEGRATING PHYSICAL SCIENCE

Refracting Light Rays of light refract, or bend, as they pass from one substance to another. During a total lunar eclipse, Earth's atmosphere refracts some light into the umbra. As a result, the Moon, instead of appearing totally black, ranges in color from a dull gray to copper, depending on the amount of dust in Earth's atmosphere. This same phenomenon occurs all the time and is called earthshine. Have students find out in which phases earthshine is visible (crescent phases), and have them draw diagrams of the path of the light that causes earthshine. **learning style: visual**

Go Online active art
For: Moon Phases and Eclipes activity
Visit: PHSchool.com
Web Code: cfp-5013

Students can interact with the art of moon phases and eclipses online.

CHECKING CONCEPTS

1. The dark part of a shadow is the __________.
2. A lunar eclipse occurs when Earth is between the __________ and the Moon.
3. A __________ lunar eclipse occurs when part of the Moon is in Earth's umbra.
4. During a __________ lunar eclipse, all of the Moon is dark.

THINKING CRITICALLY

5. HYPOTHESIZE: When would a partial lunar eclipse be difficult to see?
6. INFER: Can a lunar eclipse occur during the quarter-Moon phase? Why or why not?
7. INFER: Can Earth's shadow have only an umbra and not a penumbra? Explain your answer.

INTERPRETING VISUALS

Use Figure 14-22 to answer the following questions.

8. EXPLAIN: The Moon in Figure 14-22 is in a total eclipse. What would happen if the Moon were to move into the penumbra?
9. HYPOTHESIZE: If Earth and the Moon were closer to the Sun, how would the size of the umbra change?
10. EXPLAIN: Why can a lunar eclipse occur only during a full Moon?

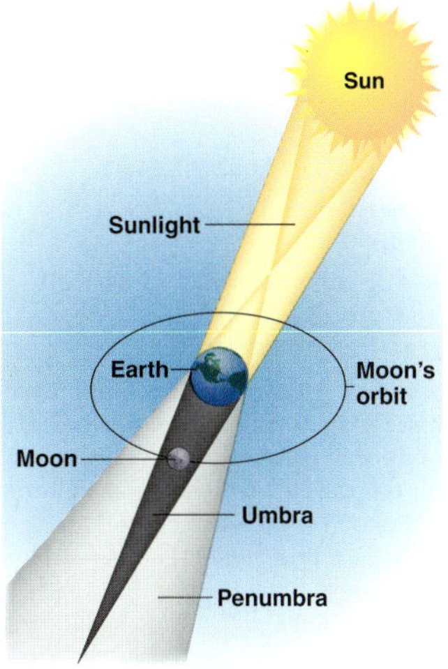

▲ **Figure 14-22** A total lunar eclipse

Hands-On Activity

MODELING LUNAR ECLIPSES

You will need tracing paper, a ruler, and a pencil.

1. Figure 14-23 shows the Moon in three different positions as it revolves around Earth. Copy the diagram onto a sheet of paper.
2. Indicate on your diagram which position represents a total eclipse, which represents a partial eclipse, and which represents no eclipse of the Moon.

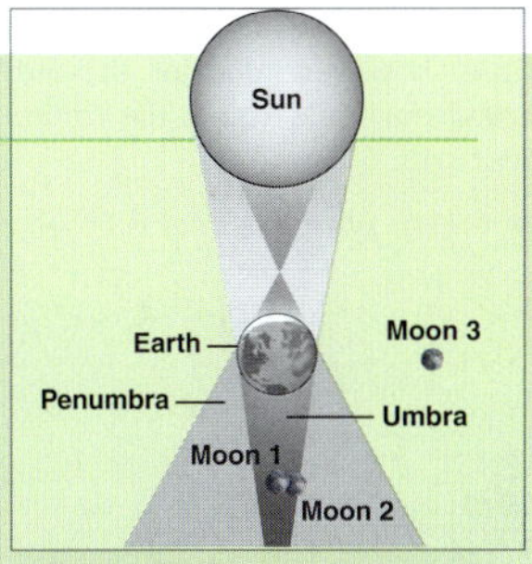

▲ **Figure 14-23** The Moon revolving around Earth

Practicing Your Skill

3. ANALYZE: Does Position 1 represent a total eclipse, a partial eclipse, or no eclipse of the Moon? Explain.
4. ANALYZE: Does Position 2 represent a total eclipse, a partial eclipse, or no eclipse of the Moon? Explain.
5. ANALYZE: Does Position 3 represent a total eclipse, a partial eclipse, or no eclipse of the Moon? Explain.
6. MODEL: Draw a model of a total eclipse of the Moon.

3 Assess

CHECKING CONCEPTS ANSWERS

1. umbra
2. Sun
3. partial
4. total

THINKING CRITICALLY ANSWERS

5. HYPOTHESIZE: when only a tiny part of the Moon is in Earth's umbra
6. INFER: No, a lunar eclipse can only occur during a full Moon, when Earth is between the Sun and the Moon.
7. INFER: No, a shadow always has an umbra and a penumbra.

INTERPRETING VISUALS ANSWERS

8. EXPLAIN: It would be light again.
9. HYPOTHESIZE: It would be wider.
10. EXPLAIN: The Moon has to be on the side of Earth away from the Sun.

4 Close

Hands-On Activity

MODELING LUNAR ECLIPSES

TIME: 15 minutes

PURPOSE: modeling, analyzing

DISCUSSION: Ask students: *During what phase does a lunar eclipse occur?* (full Moon)

Practicing Your Skills Answers

3. ANALYZE: total eclipse; The Moon is entirely in Earth's umbra.
4. ANALYZE: partial eclipse; The Moon is partly in Earth's umbra.
5. ANALYZE: no eclipse; The Moon is not in Earth's umbra.
6. MODEL: Drawings should have the Moon, Earth, and the Sun lined up in a straight line.

RETEACH

Modeling a Lunar Eclipse

Divide the class into small groups. Distribute a softball, a small plastic-foam ball, and a light source to each group. Direct students to shine the light on the softball (Earth) and then place the smaller ball (the Moon) in the umbra of the softball. Caution students not to look directly into the light. Darken the room as much as possible and check that the light from one group does not interfere with that of other groups.

learning styles: visual, kinesthetic

• TEACHING RESOURCES • Online or Print

Teacher's Resources

Lesson Review: Chapter 14, p. 10

Enrichment: Chapter 14, p. 12

MORE ON THE WEB

Predicted Lunar Eclipses Have the students use the Internet to find a schedule of predicted lunar eclipses. If any lunar eclipses are predicted to happen in the near future, suggest to students that they make an attempt to view them.

14-8 *(pp. 344–345)*

1 *Introduce*

STUDY HINT Have students read the lesson feature on page 345 before beginning the lesson. Discuss the concepts presented in the feature. Ask students how they think this information relates to topics they have already studied or to the lesson they are about to study.

Linking Prior Knowledge Ask students if they can remember seeing a movie in which a solar eclipse was shown or described. Ask them if any of the characters in the movie were aware ahead of time that an eclipse was going to happen. Explain that, like lunar eclipses, solar eclipses are predictable.

2 *Teach*

Demonstration Darken the room as much possible. Use a globe, a small plastic-foam (or table-tennis) ball, and a light source to model a solar eclipse. Caution students not to look directly into the light.
learning style: visual

Discussion Refer students to Figure 14-25. Explain how solar eclipses occur and why a solar eclipse varies at different places on Earth. **learning styles: visual, auditory**

Answers to questions

1. **DEFINE:** the passing of the Moon directly between Earth and the Sun
2. **DESCRIBE:** the Moon blocking the entire face of the Sun
3. **IDENTIFY:** a partial solar eclipse

WRITING HINT Remind students that if they are asked to compare two things, such as a lunar and a solar eclipse or a partial and a total solar eclipse, they should describe how the things are similar. Explain that they can use words or phrases such as *just as*, *in the same way*, *alike*, and *similarly* to do this. If they are asked to contrast, they should describe differences between two things. To help them contrast, they can use words or phrases such as *however*, *in contrast*, and *differ*.

14-8 What causes an eclipse of the Sun?

Objective

Explain how a solar eclipse occurs.

Key Terms

solar eclipse: passing of the Moon between Earth and the Sun

corona (kuh-ROH-nuh): outer layer of the Sun's atmosphere

Casting Shadows An eclipse of the Sun is called a **solar eclipse.** A solar eclipse occurs when the Moon passes directly between Earth and the Sun. During a solar eclipse, the Moon casts a shadow on Earth. Figure 14-24 shows a solar eclipse. During a solar eclipse the Sun looks like it is covered by a black circle. This circle is the Moon.

Like Earth, the Sun has an atmosphere. The outer layer of the Sun's atmosphere is called the **corona.** The corona is like a halo around the Sun. Usually, the corona cannot be seen from Earth because the Sun itself is so bright. During a solar eclipse, however, most of the Sun's surface is blacked out. As a result, the corona can be seen from Earth.

1 **DEFINE:** What is a solar eclipse?

Kinds of Solar Eclipses Like lunar eclipses, solar eclipses can either be total or partial. A total solar eclipse occurs when the entire face of the Sun is blocked by the Moon. Only the outer atmosphere of the Sun still shows. A partial solar eclipse happens when only part of the Sun's face is blocked.

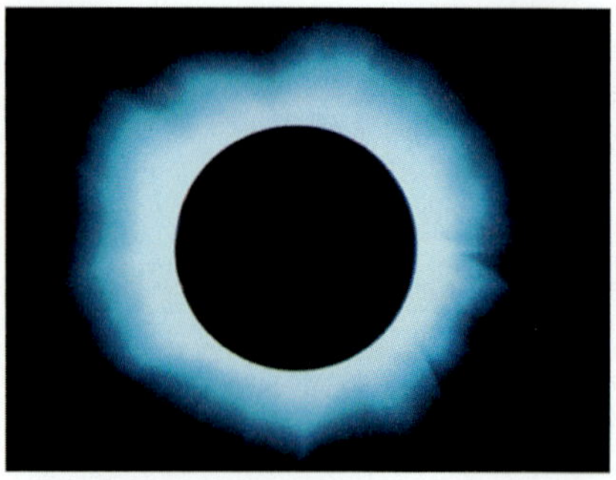

▲ **Figure 14-24** A total solar eclipse

2 **DESCRIBE:** What causes a total solar eclipse?

Viewing Solar Eclipses Look at Figure 14-25 below. When the Moon's umbra touches Earth, people within it see a total solar eclipse.

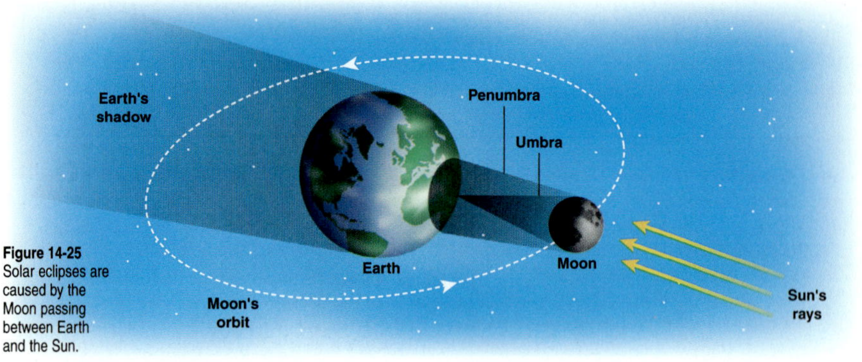

Figure 14-25 Solar eclipses are caused by the Moon passing between Earth and the Sun.

Teaching Options

FAST FACTS *Solar Eclipses in History*

Throughout history, solar eclipses have had significant effects on people and governments. In A.D. 628, the death of Empress Suiko of Japan was blamed on the total eclipse that occurred the day she died. In A.D. 840, Emperor Louis I of Bavaria died, reportedly as a result of witnessing 5 minutes of a total eclipse. His three sons fought over his empire, which was eventually divided into France, Germany, and Italy. A solar eclipse in Japan in A.D. 975 led authorities to declare a general amnesty.

During the American Revolutionary War, scientists at Harvard University organized an expedition to observe a total eclipse that was to occur in October 1780. The destination of the expedition, western Penobscot Bay in Maine, was then under the control of British forces. An agreement was negotiated between the Americans and the British that allowed the scientists to carry out their expedition without fear of attack.

The umbra of the Moon is very small. Therefore, a total solar eclipse is visible from only a small area of Earth. People who are in the Moon's penumbra see a partial solar eclipse.

The penumbra of the Moon's shadow is much larger than the umbra. As a result, a partial solar eclipse can be seen over a larger area of Earth than a total solar eclipse can. Partial solar eclipses are seen more often than total solar eclipses.

3 IDENTIFY: Which kind of solar eclipse is seen more often, a total or a partial solar eclipse?

CHECKING CONCEPTS

1. When does a solar eclipse occur?
2. What is a partial eclipse of the Sun?
3. What happens when the Moon's umbra touches Earth?
4. Which part of the Moon's shadow is the largest?

THINKING CRITICALLY

Use Figure 14-25 to answer the following questions.

5. ANALYZE: What part of the Moon's shadow are you in if you see a total solar eclipse?
6. EXPLAIN: What can people outside the area of a total solar eclipse see?
7. COMPARE: How are total eclipses of the Moon and the Sun similar?
8. INFER: Why is a partial solar eclipse seen more often than a total solar eclipse?

HEALTH AND SAFETY TIP

Never look directly at the Sun, especially during a total solar eclipse. Doing this can cause permanent eye damage. Some people use a special filter to view the Sun. However, poorly made filters can also cause eye damage. The best way to watch a solar eclipse is with a pinhole box. Go online to learn how to make one. Use the key search words **pinhole box.**

How Do They Know That?

PREDICTING SOLAR ECLIPSES

People have always been fascinated by solar eclipses. In some cultures, solar eclipses were connected with superstition, mystery, and fear. Some ancient cultures thought the darkened sky caused by a solar eclipse was a sign of the displeasure of certain spirits. The ancient Chinese thought that solar eclipses happened when a dragon in the sky tried to swallow the Sun.

Descriptions of solar eclipses have been found dating back many centuries. In Babylon, a record of solar eclipses was kept from 747 B.C. on. In China, 36 solar eclipses were recorded between 720 B.C. and 495 B.C. Scientists have calculated the exact dates of many past solar eclipses. The records of eclipses in ancient writings have been used to pinpoint the dates of historical events.

▲ **Figure 14-26** This moment during a solar eclipse is known as the diamond ring effect. Can you see why?

Today, scientists can accurately predict solar eclipses. This is important in the study of the Sun and the Moon.

Thinking Critically Why do you think solar eclipses were once feared?

3 Assess

CHECKING CONCEPTS ANSWERS

1. when the Moon passes between Earth and the Sun
2. when only part of the Sun's face is blocked by the Moon
3. People within the umbra see a total solar eclipse.
4. the penumbra

THINKING CRITICALLY ANSWERS

5. ANALYZE: the umbra
6. EXPLAIN: If they are within the penumbra, they can see a partial eclipse.
7. COMPARE: The entire Moon or Sun is within the umbra.
8. INFER: A total solar eclipse is visible over only a small area of Earth at one time.

HEALTH AND SAFETY TIP

Have interested students make a pinhole box. They can show their finished products to the class.

4 Close

How Do They Know That?

PREDICTING SOLAR ECLIPSES

Literature Connection Have students read Mark Twain's *A Connecticut Yankee in King Arthur's Court* and describe in a book report the incident involving a solar eclipse.

Discussion Ask students: *How are solar eclipses used to date historical events?* (Scientists can calculate the exact dates of solar eclipses and correlate the dates to records of eclipses in ancient writings.)

Thinking Critically Answer Possible answers include that people did not understand what caused solar eclipses and thought that the Sun might stop shining forever or that it was an omen of something bad about to happen.

ONGOING ASSESSMENT

Have students use three different-sized balls to model the positions of celestial bodies involved in a solar eclipse.

• TEACHING RESOURCES • Online or Print

Teacher's Resources

Lesson Review: Chapter 14, p. 11

Enrichment: Chapter 14, p. 12

Challenges (pp. 346–348)

Chapter Summary

Review Give students index cards and assign each student a chapter summary point. Have them turn their summary point into a question and write it on the card, then write the answer on the other side of the card. Collect the cards and have students take turns drawing a card and answering the question.

STUDY HINT Have pairs of students take turns quizzing each other on the definitions of the Key Terms. Remind them to look in the lessons for any terms they cannot define.

Key Term Challenges

MATCHING

1. solar noon (14-2)
2. phases (14-6)
3. spring tide (14-5)
4. axis (14-1)
5. lunar eclipse (14-7)
6. rotation (14-1)
7. revolution (14-1)
8. aphelion (14-3)

IDENTIFYING WORD RELATIONSHIPS

9. Rotation is the spinning of a body on its axis. Revolution is the movement of a body in an orbit around another body. (14-1)
10. A meridian is a line that runs from pole to pole on the surface of Earth. The International Date Line runs from pole to pole and is the boundary between the first and twenty-fourth time zones. (14-2)
11. An equinox is one of two days in the year when the Sun is directly over the equator at solar noon. A solstice is one of two days in the year when the Sun reaches its highest or lowest point in the sky. (14-4)
12. The apogee is the point at which the Moon is farthest from Earth. The perigee is the point at which the Moon is closest to Earth. (14-5)
13. A penumbra is the light part of a shadow. An umbra is the dark part of a shadow. (14-7)

Chapter 14 Challenges

Chapter Summary

Lessons 14-1 and 14-2

- **Revolution** and **rotation** are movements of Earth. Earth's rotation causes day and night. The number of daylight hours changes because Earth's **axis** is tilted.
- The Sun appears to rise in the east and set in the west because of Earth's rotation. When the Sun is highest in the sky, the time is **solar noon.**
- Earth is divided into 24 time zones. There are eight **standard time** zones in the United States. As you cross the **International Date Line,** you either gain or lose one full day.

Lesson 14-3

- The seasons are caused by Earth's tilt and orbit around the Sun. The Northern and Southern hemispheres have opposite seasons.

Lesson 14-4

- During the year, the distance between Earth and the Sun varies.
- **Solstices** and **equinoxes** mark the changes in seasons. The North Pole tilts toward the Sun on the summer solstice. The North Pole tilts away from the Sun on the winter solstice. The Sun shines directly on the equator during the vernal and autumnal equinoxes.

Lesson 14-5

- The Moon revolves around Earth in the same way Earth revolves around the Sun. The Moon rotates once every $27\frac{1}{3}$ days.
- **Apogee** and **perigee** describe the Moon's position in relation to Earth.
- The motions of the Moon cause changes in tides. Two kinds of tides are **spring tides** and **neap tides.**

Lesson 14-6

- The Moon has waxing and waning **phases.**

Lessons 14-7 and 14-8

- The **umbra** and the **penumbra** are the two parts of a shadow.
- A **lunar eclipse** occurs when the Moon passes through Earth's shadow. A **solar eclipse** occurs when the Moon passes between Earth and the Sun. Eclipses can be partial or total.

Key Term Challenges

aphelion (p. 332)	perigee (p. 336)
apogee (p. 336)	perihelion (p. 332)
axis (p. 326)	phases (p. 338)
corona (p. 344)	revolution (p. 326)
crescent phase (p. 338)	rotation (p. 326)
equinox (p. 332)	solar eclipse (p. 344)
gibbous phase (p. 338)	solar noon (p. 328)
International Date Line (p. 328)	solstice (p. 332)
lunar eclipse (p. 342)	spring tide (p.336)
neap tide (p. 336)	standard time (p. 328)
penumbra (p. 342)	umbra (p. 342)

MATCHING **Write the Key Term from above that best matches each description.**

1. time when the Sun is highest in the sky
2. changing shapes of the Moon
3. daily tidal range is the largest
4. imaginary line through the center of a planet or other body on which that body spins
5. passing of the Moon through Earth's shadow
6. movement of Earth on its axis
7. movement of Earth in its orbit
8. point in its orbit when Earth is farthest from the Sun

IDENTIFYING WORD RELATIONSHIPS **Explain how the words in each pair are related. Write your answers in complete sentences.**

9. rotation, revolution
10. International Date Line, meridian
11. equinox, solstice
12. apogee, perigee
13. penumbra, umbra
14. crescent phase, gibbous phase
15. solar eclipse, lunar eclipse

346

Teaching Options

PREPARING STUDENTS FOR STANDARDIZED TESTS

Reading Strategy: Remind students to scan a question or written passage to locate keywords that can help them answer the question about the passage.

Writing Strategy: Tell students to make sure that they use capital letters for the first word in a sentence, for place names, and for the names of days of the week, months, and holidays.

Interpreting Visuals Strategy: Remind students that numbering or an alphabetical sequence in illustrations often indicates the order of events.

ESL/ELL STRATEGY

Restating Events English-language learners will benefit from restating events in their own words. Have pairs of students work together to explain such events as phases, eclipses, and tides.

Content Challenges TEST PREP

FILL IN **Write the term or phrase that best completes each statement.**

1. When viewed from above the North Pole, Earth rotates on its axis from __________.
2. Earth's axis is tilted at an angle of __________.
3. Because of Earth's tilted axis, the North Pole of Earth always points toward __________.
4. In the Northern Hemisphere, the summer solstice occurs on __________.
5. During an equinox, the number of daylight hours and nighttime hours are __________.
6. The 48 states of the continental United States have __________ standard time zones.
7. The Moon takes __________ days to make one complete revolution around Earth.
8. The Moon takes __________ days to make one complete rotation on its axis.
9. As the visible part of the Moon decreases, the Moon is __________.
10. The waxing crescent phase of the Moon occurs __________ the full Moon.
11. When the daily tide range is largest, this is called a __________ tide.
12. When the daily tide range is smallest, this is called a __________ tide.
13. Because Earth rotates, the Sun appears to set in the __________.
14. Earth is closest to the Sun during the month of __________.
15. The darkest center part of a shadow is called the __________.

TRUE/FALSE **Write *true* if the statement is true. If the statement is false, change the underlined term or phrase to make the statement true.**

16. The change from day to night is caused by Earth's revolution.
17. The number of daylight hours is not equal all year because of Earth's tilted axis.
18. The change in seasons is caused by Earth's rotation.
19. In the Southern Hemisphere, the first day of summer is June 21.
20. The vernal equinox marks the first day of spring.
21. Earth is divided into 24 time zones, each 15 degrees of longitude wide.
22. As you move east, each time zone is one hour earlier than the previous time zone.
23. When the Moon is closest to Earth, it is at apogee.
24. The Moon rises 50 minutes earlier each day.
25. As the part of the Moon that is visible increases, the Moon is waxing.
26. The Moon appears to change shape because of the way it reflects light from the Sun.
27. During a total lunar eclipse, the entire face of the Moon is darkened.
28. When the Moon moves entirely into Earth's umbra, people within see a solar eclipse.
29. The difference between high tide and low tide is greatest during spring tides.
30. The Sun is farthest from Earth at perihelion.

14. The crescent phase is a phase of the Moon when less than half the Moon is visible. The gibbous phase is a phase of the Moon when more than half the Moon is visible. (14-6)
15. A solar eclipse occurs when the Moon passes between Earth and the Sun and casts a shadow on Earth. A lunar eclipse occurs when the Moon passes through Earth's shadow. (14-7, 14-8)

Content Challenges

FILL IN

1. west to east (14-1)
2. $23\frac{1}{2}^\circ$ (14-1)
3. Polaris or the North Star (14-1)
4. June 21 (14-4)
5. equal (14-4)
6. four (14-2)
7. $27\frac{1}{3}$ (14-5)
8. $27\frac{1}{3}$ (14-5)
9. waning (14-6)
10. before (14-6)
11. spring (14-5)
12. neap (14-5)
13. west (14-1)
14. January (14-4)
15. umbra (14-7)

TRUE/FALSE

16. false; rotation (14-1)
17. true (14-1)
18. false; tilted axis (14-3)
19. false; December 21 (14-3)
20. true (14-4)
21. true (14-2)
22. false; west (14-2)
23. false; farthest (14-5)
24. false; later (14-5)
25. true (14-6)
26. true (14-6)
27. true (14-7)
28. false; lunar (14-8)
29. true (14-5)
30. false; aphelion (14-4)

ALTERNATIVE ASSESSMENT

Explaining the Seasons Have students draw and label a diagram to explain why Earth experiences seasons.

PORTFOLIO ASSESSMENT

Making Student Portfolios
Students can demonstrate their comprehension of the concepts in this chapter by making a portfolio. The Chapter Self-Check, the Big Idea Planner, and the Weekly Journal are some of the reproducibles in the Teacher's Resources that they can include in their portfolios. Encourage students to select their own work, such as the diagram of the Moon's phases, for inclusion in their portfolios. You can use the Portfolio Assessment Rubric also found in the Teacher's Resources to assess students' portfolios.

Concept Challenges

WRITTEN RESPONSE

1. **CALCULATE:** The Moon rises 50 minutes later each day, so 6:30 P.M. + 50 minutes = 7:20 P.M. (14-5)
2. **COMPARE:** It takes the Moon slightly longer (29 1/2 days) to go through a complete series of phase changes than to make one complete revolution around Earth (27 1/3 days). (14-6)
3. **HYPOTHESIZE:** The gravitational attraction is slightly stronger at perigee because gravitational attraction is strongest when objects are closest, and perigee is when the Moon is closest to Earth. (14-5)
4. **COMPARE:** They are equal. (14-2)
5. **CONTRAST:** During a lunar eclipse, Earth is between the Moon and the Sun. During a solar eclipse, the Moon is between Earth and the Sun. (14-7, 14-8)
6. **INFER:** The Moon's gravitational pull is slightly stronger at perigee and slightly weaker at apogee. Thus, at perigee, high tides will be higher and low tides will be lower. At apogee, high tides will be lower and low tides will be higher. (14-5)

INTERPRETING VISUALS (14-6)

7. waxing crescent
8. waxing gibbous
9. third quarter
10. new Moon
11. because the part of the Moon visible from Earth is decreasing
12. waning gibbous
13. 29 1/2 days
14. waxing crescent, first quarter, waxing gibbous, full, waning gibbous, third quarter
15. the way the Moon reflects light from the Sun and the change in the relative positions of the Sun, the Moon, and Earth

Chapter Project Wrap-Up

CREATING A MOTION MODEL

Invite other classes or family members to see students' models. Students can answer questions about their simulations. You can use the Individual Activity Rubric found in the Teacher's Resources to assess students' work. Fill in the rubric with the additional information below. For this project, students should have:

- created a small working model or simulation of one of the five main motions.
- provided an appropriate poster and a one-page description of the model.

Concept Challenges TEST PREP

WRITTEN RESPONSE **Answer each of the following questions in complete sentences.**

1. **CALCULATE:** If the Moon rises at 6:30 P.M. on Thursday, what time will it rise on Friday? Explain your answer.
2. **COMPARE:** How does the number of days between one full Moon and the next compare to how long it takes the Moon to revolve once around Earth?
3. **HYPOTHESIZE:** Gravitational attraction decreases with distance. Is the Moon's gravitational attraction to Earth stronger at apogee or perigee? Explain.
4. **COMPARE:** How does the number of time zones on Earth compare with the number of hours in a day?
5. **CONTRAST:** How do the position of the Sun, the Moon, and Earth during a solar eclipse differ from their positions during a lunar eclipse?
6. **INFER:** How do the tides on Earth change during the Moon's apogee and perigee?

INTERPRETING VISUALS **Use Figure 14-27 below to answer the following questions.**

7. Which phase of the Moon does the picture labeled *A* represent?
8. Which phase of the Moon does the picture labeled *C* represent?
9. Which phase of the Moon does the picture labeled *F* represent?
10. Which phase of the Moon is not shown?
11. Why is the phase of the Moon labeled *E* called a waning phase?
12. Which phase of the Moon follows a full Moon?
13. How long does it take the Moon to go from one new Moon phase to the next new Moon phase?
14. Name the phases of the Moon in the correct order, beginning with the one labeled *A*.
15. What causes the Moon to change phases?

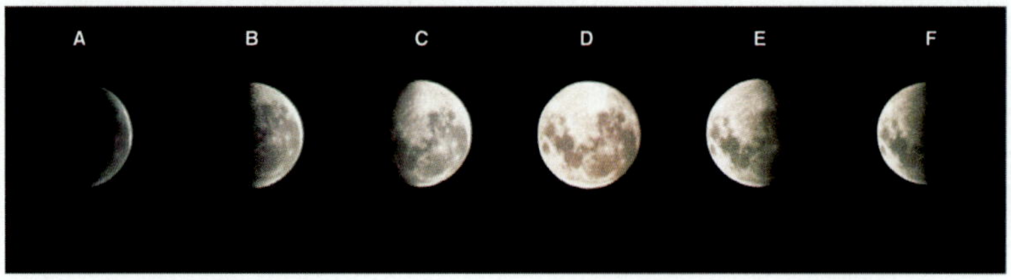

▲ **Figure 14-27** Phases of the Moon

Teaching Options

• TEACHING RESOURCES • Online or Print

Teacher's Resources

Key Term Review: Chapter 14, p. 13

Chapter 14 Test: pp. 14–16

Individual Activity Rubric: Teaching Tools, p. 5

Chapter Self-Check: Teaching Tools, p. 11

Big Idea Planner: Study Tools, pp. 11–12

Weekly Journal: Study Tools, p. 19

• TEACHING RESOURCES • Online or Print

Teacher's Resources

Portfolio Assessment Rubric: Teaching Tools, p. 10

Chapter 15 Human Impact on the Environment

PLANNING GUIDE

◆ TEACHING THE CHAPTER This chapter should take approximately 11–14 days to complete instruction and assessment.

	Skills and Features	Projects and Activities	Achieve Science Literacy	Meet Individual Needs
Chapter 15 Opener p. T349	RESEARCH	• Chapter Project		
Lesson 15-1 **What is the environment?** pp. T350–T351 Standard: D1k	DESCRIBE, EXPLAIN, OBSERVE • Web InfoSearch	• Extending the Lesson • Integrating Chemistry	• Study Hint	• Reteach
Lesson 15-2 **What are natural resources?** pp. T352–T353 Standards: D1f, D1k	DEFINE, CONCLUDE, CLASSIFY • Web InfoSearch • Integrating Life Science		• Study Hint • Reading Strategy	• ESL/ELL Strategy Teacher's Resources* Enrichment Activity
Lesson 15-3 **What are our main energy sources?** pp. T354–T355 Standard: D1k	NAME, DESCRIBE, CLASSIFY • Investigate • Building Math Skills • Science and Technology	• Extending the Lesson • More on the Web • Active Art	• Study Hint	• ESL/ELL Strategy Teacher's Resources* Venn Diagram
Lesson 15-4 **What are some alternate energy sources?** pp. T356–T357	DEFINE, ANALYZE, IDENTIFY • Web InfoSearch • Hands-On Activity	• Cooperative Learning	• Study Hint	• ESL/ELL Strategy
Lesson 15-5 **How can natural resources be conserved?** pp. T358–T359 Standard: D1e	IDENTIFY, INFER, DESCRIBE • Real-Life Science	• Extending the Lesson • More on the Web	• Study Hint • Ongoing Assessment	• Reteach Teacher's Resources* Enrichment Activity
Lesson 15-6 **What is wildlife conservation?** pp. T360–T361 Standard: D1k	RESEARCH, IDENTIFY, CONCLUDE • Investigate • Building Science Skills • Integrating Life Science	• Extending the Lesson • More on the Web	• Reading Strategy	Teacher's Resources* Enrichment Activity

*www.conceptsandchallenges.com, password: ccearth09

Planning Guide continues on next page.

Standards: For details on the correlation to National Science Standards see pages *xx–xxi*.

Chapter 15 Human Impact on the Environment *(Planning Guide continued)*

	Skills and Features	Projects and Activities	Achieve Science Literacy	Meet Individual Needs
Lesson 15-7 **What causes air pollution?** pp. T362–T363 Standard: D1h	IDENTIFY, INFER, HYPOTHESIZE • Health and Safety Tip • Hands-On Activity	• Cooperative Learning	• Study Hint	
Big Idea **How can chemical reactions make air deadly?** pp. T364–T365 Standard: D1h	RESEARCH, COMMUNICATE, ANALYZE • Science Log: Writing Activity	• Big Idea • Close Activity	• Writing Hint • Tips for Using Technology	• ESL/ELL Strategy Teacher's Resources* Big Idea Planner Thumbnail Sketch
Lesson 15-8 **What causes water pollution?** pp. T366–T367 Standard: D1f	EXPLAIN, DESCRIBE, NAME • Investigate • Designing an Experiment • Integrating Life Science	• Extending the Lesson • More on the Web	• Study Hint • Ongoing Assessment	• Reteach Teacher's Resources* Enrichment Activity Designing an Experiment
Lab Activity **Saving Water and Energy** pp. T368–T369 Standard: D1f	CALCULATE, ANALYZE • Set-Up Time: 5–10 min • Lab Time: 25–30 min	• Extending the Lesson • Extend the Lab Activity	• Tips for Using Technology	Teacher's Resources* Lab Activity Report
Lesson 15-9 **How is water purified?** pp. T370–T371 Standard: D1f	INFER, IDENTIFY, ANALYZE • Health and Safety Tip • People in Science	• Extending the Lesson • More on the Web • Lab Challenge	• Study Hint • Reading Strategy • Ongoing Assessment	• Reteach Teacher's Resources* Flowchart
Lesson 15-10 **What causes land pollution?** pp. T372–T373	DESCRIBE, ANALYZE, APPLY • Designing an Experiment • Real-Life Science	• Cooperative Learning • Extending the Lesson • More on the Web • Active Art	• Study Hint	• ESL/ELL Strategy Teacher's Resources* Designing an Experiment
Lesson 15-11 **What are some global effects of pollution?** pp. T374–T375	EXPLAIN, APPLY, INFER • Science and Technology	• Cooperative Learning • Integrating Chemistry	• Reading Strategy	
Chapter 15 Challenges pp. T376–T378	• Chapter Summary • Key Term Challenges • Content Challenges • Concept Challenges	• Chapter Project Wrap-Up	• Study Hint • Preparing Students for Standardized Tests	• ESL/ELL Strategy Teacher's Resources* Chapter Self-Check Weekly Journal

*www.conceptsandchallenges.com, password: ccearth09

Chapter 15 The Human Impact on the Environment

▲ **Figure 15-1** Cutting down trees threatened the northern spotted owl.

The northern spotted owl is a small bird found mostly in the old-growth forests of the Pacific Northwest. Because of its limited range, the population of spotted owls has always been a small one. However, the species is becoming even rarer and is now on the endangered species list. To help save the birds, the U.S. Forest Service banned the cutting down of trees on government land in the area. This angered the timber industry.

▶How have humans had an impact on the spotted owl's environment?

Contents

Teaching Options

ASSESSMENT PLANNER

For assessment in the Student Edition, see the following pages:

Content Assessment
Checking Concepts: pp. 351, 353, 355, 357, 359, 361, 363, 367, 371, 373, 375
Thinking Critically: pp. 351, 353, 355, 357, 359, 361, 363, 367, 371, 373, 375
Interpreting Visuals: pp. 359, 371, 375
Chapter Challenges: pp. 376–378

Alternative Assessment
Building Skills: pp. 355, 361
Web InfoSearch: pp. 351, 353, 357
Science Log: p. 365

Performance-Based Assessment
Hands-On Activity: pp. 357, 363
Designing an Experiment: pp. 367, 373
Lab Activity: pp. 368–369

• TEACHING RESOURCES • Online or Print

Teacher's Resources
Lesson Review: Ch. 15, pp. 2, 3, 4, 5, 6, 8, 10, 12, 14, 15, 16
Enrichment: Ch. 15, pp. 7, 9, 13
Key Term Review: Ch. 15, p. 17
Chapter 15 Test: pp. 18–20
Laboratory Manual: pp. 77–80

Chapter 15 The Human Impact on the Environment

Chapter Overview

In this chapter, students will learn about the nutrient cycles that support life on Earth. Students will identify renewable and nonrenewable natural resources and discuss resource conservation. Students will also learn about wildlife conservation and the activities that cause pollution.

About Figure 15-1 The northern spotted owl lives primarily in the Pacific Northwest. It nests in broken treetops and hollows of old or dead standing trees. Most nesting sites are found in large, dense, old-growth forests made up of trees that are more than 100 years old. The old-growth forests in which these birds live are an important source of valuable lumber.

Answer Logging has reduced the living space available to the owl, making it difficult for the species to survive and reproduce.

Linking Prior Knowledge

For this chapter, students should recall:

- the three parts of Earth (Lesson 1-2).
- how fossil fuels form (Lesson 4-6).
- soil chemistry (Lesson 7-5).
- how water is recycled (Lesson 9-1).
- gases in the atmosphere (Lesson 11-1).

Chapter Project

CONSERVATION OF RESOURCES DEBATE

Have students prepare for a debate. Divide the class into four groups: one on preserving nature versus land development; another on the value of saving endangered species; a third on energy development versus strip mining; and a fourth on landfills versus incineration. Each group should research the pros and cons of the issue and participate in a 10-minute debate, with the whole class voting on which side gave the better argument. Each group should also provide written documentation of their research findings.

15-1 *(pp. 350–351)*

1 *Introduce*

STUDY HINT Before beginning this lesson, write the title of each section on the board. Have students copy these titles in their notebooks, using an outline format. As students read each section, they should make up a topic sentence for each paragraph under each section head.

Linking Prior Knowledge Review the parts of Earth: the hydrosphere, lithosphere, and atmosphere. Elicit from students that the hydrosphere, lithosphere, and atmosphere also make up the environment in which all life on Earth depends for survival (Lesson 1-2).

2 *Teach*

Discussion Students probably have heard the term *environment* many times. Point out that the environment is everything that surrounds a living thing. Have students describe the environment in which they live and how the environment supplies them with the things they need for survival. Ask students what living things get from the environment. Students should recognize that living things get everything they need from the environment. Also, discuss how different types of pollution might affect living things.
learning style: auditory

Reinforcement Explain that the materials provided by the environment are recycled, or used over and over again. On the board, draw diagrams illustrating the carbon cycle and the nitrogen cycle. Use Figures 15-2 and 15-3 as guides. Point out that both of these nutrient cycles involve the lithosphere and the atmosphere.
learning styles: visual, auditory

15-1 What is the environment?

Objectives
Describe the cycles in nature. Define pollution.

Key Terms
environment (ehn-VY-ruhn-muhnt): everything that surrounds a living thing
pollution (puh-LOO-shuhn): anything that harms the environment
pollutant (puh-LOOT-uhnt): harmful substance in the environment

A Support System for Life Living things get everything they need from their environment. The **environment** is everything that surrounds a living thing. The atmosphere, the hydrosphere, and the lithosphere are parts of the environment. For example, the atmosphere supplies living things with important gases. The hydrosphere provides water. The lithosphere provides important materials such as minerals and soil.

Some of these resources pass through cycles. For example, the water cycle allows Earth's water to be used again and again. Carbon dioxide, oxygen, and nitrogen also cycle through the environment.

1 DESCRIBE: What is the environment?

Pollution Anything that harms the environment is called **pollution.** Pollution occurs when harmful substances, or **pollutants,** are released into the environment. Pollutants harm Earth's air, water, and land. They upset nature's cycles.

An example of how humans upset nature's cycles is thermal pollution. Thermal pollution raises water temperatures in aquatic systems. The problems result mostly from industry using water to cool equipment or materials during various processes. The electric power industry, steel mills, oil refineries, and paper mills all use huge amounts of water for cooling. After the equipment has cooled, the "waste" water is often returned to where it came from. However, its temperature is higher than it was before. This temperature change can upset nature's cycles and the balance of the environment.

Farming also causes pollution through pesticide use. Pesticides are meant to protect food supplies from pests such as insects, rodents, weeds, molds, and bacteria. However, pesticides can kill birds and fish, as well as pests. They can also poison the foods people eat.

2 EXPLAIN: How does pollution upset nature's cycles and the balance of the environment?

The Nitrogen Cycle One important cycle in the environment is the nitrogen cycle. In the nitrogen cycle, bacteria take nitrogen gas from the air and make it into compounds. Both plants and animals use these compounds. Bacteria that make these compounds are called nitrogen-fixing bacteria.

Nitrogen-fixing bacteria live in the soil and in the roots of plants such as legumes. Farmers who grow crops often plant legumes in their fields to add nitrogen compounds to the soil.

When plants and animals die, the nitrogen compounds are broken down by different bacteria.

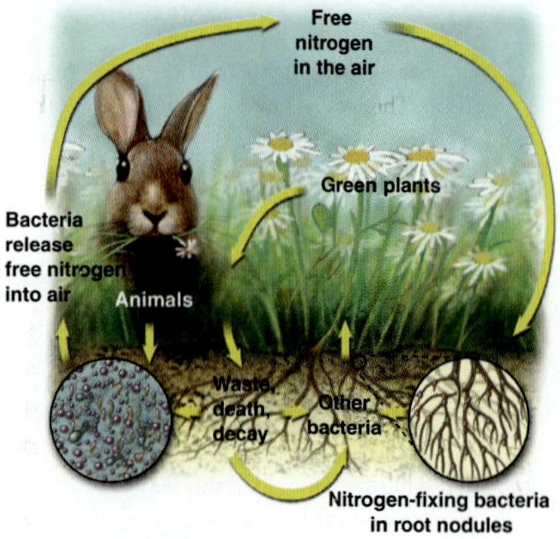

▲ **Figure 15-2** The nitrogen cycle

Nitrogen is then released back into the air or soil.

3 DESCRIBE: What happens in the nitrogen cycle?

Teaching Options

RETEACH

Re-creating the Nitrogen Cycle Divide the class into small groups. Ask students to draw pictures of plants, animals, soil, the Sun, clouds, and so on. Have students cut out the pictures and assemble them in a way that illustrates the nitrogen cycle. Have students glue the pictures on posterboard and add arrows to show the flow of nitrogen through the environment.
learning styles: tactile, visual

INTEGRATING CHEMISTRY

Researching Fertilizer Gardeners and farmers use fertilizers to encourage the growth of flowers, vegetables, and crops. Most fertilizers include the three most important elements for plant growth: nitrogen (N), potassium (K), and phosphorus (P). Have pairs of students use library resources and the Internet, or talk to local gardening or farming experts, to find out more about these elements and why they are important to plant growth.
learning style: auditory

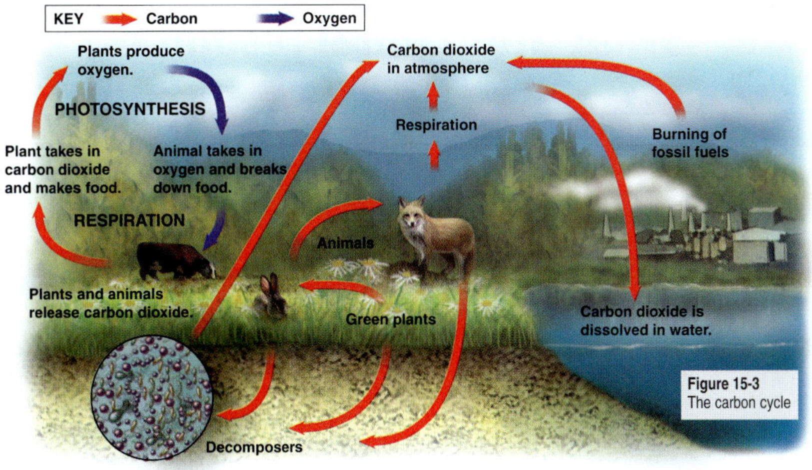

Figure 15-3 The carbon cycle

The Carbon Cycle Living things are made up of organic compounds. These contain carbon. Carbon is found in the atmosphere in the form of the gas carbon dioxide. The process by which carbon is recycled is called the carbon cycle. It is the repeated movement of carbon between Earth's atmosphere and living things. Green plants are producers. They use carbon dioxide to make food in photosynthesis. Some carbon dioxide is returned to the atmosphere when the food is used for energy during cellular respiration. The rest is stored as sugar. Animals are consumers. They eat producers or other consumers and use the food for energy. Some stored carbon dioxide is released when they breathe out.

All organisms eventually die. Their bodies are broken down by decomposers. Decomposers also release carbon dioxide into the atmosphere.

Fossil fuels are formed from the remains of ancient organisms. When fossil fuels are burned to release energy, more carbon dioxide is released into the atmosphere. This affects the carbon cycle.

4 **IDENTIFY:** What gas do animals release when they breathe out?

CHECKING CONCEPTS

1. What resources cycle through the environment?
2. What part of the environment supplies water?
3. What gas do animals that have lungs take in from the air when they breathe in?
4. What happens to the nitrogen compounds used by plants and animals when they die?
5. When does pollution occur?

THINKING CRITICALLY

Use Figure 15-2 to answer the following questions.

6. **ANALYZE:** How do animals get nitrogen compounds?
7. **OBSERVE:** What living things make nitrogen compounds?

Web InfoSearch

Rachel Carson Rachel Louise Carson (1907–1964) worked for the U.S. Fish and Wildlife Service. She was a marine biologist and science writer. Carson wrote about the ocean and pollution. Her most famous book, *Silent Spring*, was published in 1962. In this book, Carson warned that pesticides were harmful to the environment.

SEARCH: Use the Internet to find out more about Rachel Carson. What pesticide was banned because of her book? Are pesticides still used widely today? Start your search at www.conceptsandchallenges.com. Some key search words are **Rachel Carson** and **banned pesticides.**

Answers to questions

1 **DESCRIBE:** everything that surrounds a living thing

2 **EXPLAIN:** Pollution changes the environment, upsetting the balance between living and nonliving things.

3 **DESCRIBE:** Bacteria that live in the soil capture nitrogen from the atmosphere and change it into solid compounds that plants and animals use. When the plants and animals die, the nitrogen is released back into the air through decay.

4 **IDENTIFY:** carbon dioxide

3 Assess

CHECKING CONCEPTS ANSWERS

1. water, carbon dioxide, oxygen, and nitrogen
2. the hydrosphere
3. oxygen
4. The compounds break down, and nitrogen gas is released into the atmosphere.
5. when harmful substances are released into the environment

THINKING CRITICALLY ANSWERS

6. **ANALYZE:** by eating plants or other animals
7. **OBSERVE:** bacteria

4 Close

Web InfoSearch

Rachel Carson Rachel Louise Carson (1907–1964) worked for the U.S. Fish and Wildlife Service. She was a marine biologist and science writer. Carson wrote about the ocean and pollution. Her most famous book, *Silent Spring*, was published in 1962. In this book, Carson warned that pesticides were harmful to the environment. Have students use the Internet to find out more about Rachel Carson. Ask students: *What pesticide was banned because of her book? Are pesticides still used widely today?* Some key search words are **Rachel Carson, banned pesticides** and **Rachel Carson pesticides.** Students can begin their search at www.conceptsandchallenges.com.

EXTENDING THE LESSON

Looking for Environmentalists
Have interested students find out about scientists other than Rachel Carson who have contributed to our knowledge of the environment. Some examples are marine biologist Sylvia Earle, who has worked to preserve ocean environments, especially coral reefs; biologist Barry Commoner, who was one of the first U.S. environmentalists; and biologist Paul Ehrlich, who has written widely on the harmful effects of human overpopulation on the environment. Have students record their findings in a written report. **learning style: linguistic**

• TEACHING RESOURCES • Online or Print

Teacher's Resources
Lesson Review: Chapter 15, p. 2

15-2 *(pp. 352–353)*

1 Introduce

STUDY HINT Have students write three facts in the lesson related to the map of natural resources shown in Figure 15-4. Then, ask them to tie the facts together in a way that would satisfy the lesson objective.

Linking Prior Knowledge Ask students to name what they think are natural resources.

2 Teach

Demonstration Have students look around the classroom and identify the kinds of natural resources they see. Ask what natural resources are used in objects such as desks and other furniture (wood, plastic from petroleum products, metals), lights (glass, metals, electricity), heaters or air conditioners (air, energy), and books (wood or other plant fibers used to make paper). **learning styles: visual, tactile**

Discussion Ask students why they think it is important for people to conserve natural resources. Make sure that students understand the necessity of conserving both renewable and nonrenewable resources. **learning style: auditory**

Reinforcement Write *Natural Resources* on the board. From this head, draw arrows to each of the following heads: *Renewable Resources* and *Nonrenewable Resources*. Under *Renewable Resources*, list *air*, *water*, *soil*, and *living things*. Under *Nonrenewable Resources*, list *fossil fuels* and *minerals*. Have students copy the charts in their notebooks. **learning styles: auditory, visual**

READING STRATEGY Ask students how adding the prefix *non-* changes the word *renewable*. (reverses the meaning of the word) Ask students to think of other words in which *non-* reverses the meaning.

Answers to questions

1. **DEFINE:** material from Earth that is used by living things
2. **LIST:** air, water, soil, and living things
3. **DEFINE:** a natural resource that is not easily replaced
4. **ANALYZE:** It increases.

15-2 What are natural resources?

Objective

Distinguish between renewable and nonrenewable natural resources.

Key Terms

natural resource: material from the earth that is used by living things

conservation (kahn-suhr-VAY-shuhn)**:** wise use of natural resources

Natural Resources When a spaceship travels into space, it carries everything its crew needs to live. A spaceship must have a supply of food, air, water, and fuel. Earth can be compared to a spaceship. Earth has everything living things need to survive. Materials from the earth that are used by living things are called **natural resources.**

1 **DEFINE:** What is a natural resource?

Renewable Resources Natural resources that can be reused or replaced are called **renewable resources.** Air, water, soil, and living things are renewable resources.

The water cycle allows Earth's water to be used over and over. New soil is formed to replace soil that has been carried away by wind and water. New plants grow to replace those that die. Animals are born to replace animals that die.

▲ **Figure 15-5** Freshly cut timber, considered a renewable resource, is often floated down a river to reach factories in distant locations.

2 **LIST:** What are four renewable resources?

Nonrenewable Resources Oil, coal, and natural gas are fossil fuels. They were formed from plants and animals that lived long ago.

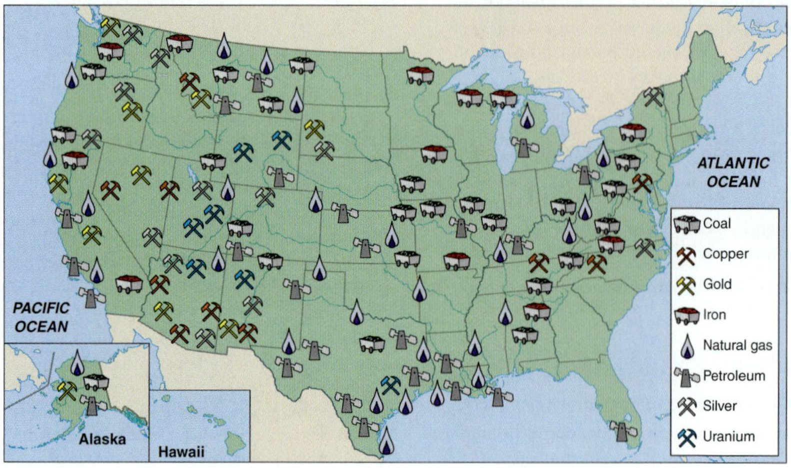

▲ **Figure 15-4** Natural resources in the United States

352

Teaching Options

! FAST FACTS *Conserving Soil*

Like other natural resources, soil needs to be conserved. Soil is necessary for growing crops and natural vegetation, supporting wildlife, filtering impurities from water, and preventing flooding. The National Cooperative Soil Survey estimates that almost 2 billion tons of valuable topsoil have been lost every year since 1995. This loss is largely a result of water and wind erosion. Buildup of salts from irrigation also ruins topsoil.

Between 1985 and 1995, soil conservation practices reduced soil erosion by about 30 percent. Such practices include planting grass and trees on bare soils, terrace farming, contour farming, and practicing conservation tillage. Conservation tillage is a method of growing crops in which the stubble from the previous crop is not plowed completely under. The crops are planted into the stubble, which protects the soil underneath. However, since 1995, use of such conservation methods has been reduced while loss of cropland has increased because of urban construction. Today, the soil on an estimated 30 percent of America's cropland is eroding faster than it can be replaced.

Fossil fuels and minerals are called **nonrenewable resources**. Nonrenewable resources are natural resources that are not easily replaced by nature or by human effort. Nonrenewable resources can take millions of years to form. Once nonrenewable resources are used up, they are basically gone forever.

3 **DEFINE:** What is a nonrenewable resource?

Conservation The wise use of a natural resource is called **conservation.** As the number of people on Earth gets larger, the need for natural resources increases. People must learn how to use natural resources wisely to help them last longer.

4 **ANALYZE:** What happens to the need for natural resources as Earth's population grows?

CHECKING CONCEPTS

1. Water, soil, and air are __________ natural resources.
2. Oil is a __________ natural resource.
3. Natural resources come from __________.
4. Fossil fuels __________ be reused or replaced.
5. The wise use of natural resources is called __________.

THINKING CRITICALLY

6. **CLASSIFY:** Classify each of the following as a renewable or a nonrenewable resource: trees, coal, nitrogen, people, diamonds, water, natural gas, soil, oil, and iron.
7. **CONCLUDE:** Why is it important to conserve even renewable resources?

Web InfoSearch

National Parks The U.S. National Park Service protects natural scenery, wildlife, and historic sites for the public to enjoy. Yellowstone National Park was the first national park in the world.

SEARCH: Use the Internet to find out more about the U.S. national park system. What parks are part of the system? Also, how are national parks a resource? Start your search at www.conceptsandchallenges.com. Some key search words are **Yellowstone Park, National Park System,** and **National Park Service.**

Integrating Life Science

TOPICS: endangered species, predation

THE WHALING AGREEMENT

In nature, whales have very few natural enemies. However, they have been preyed upon by people for centuries. People have invented ways, such as better harpoons and faster ships, to make the hunting and killing of many whales at once easier.

By 1946, many whale populations had been listed as endangered. In response, the International Whaling Commission (IWC) was formed. It set limits on the killing of whales. However, whale populations kept decreasing. So in 1986, the IWC asked countries to voluntarily stop killing whales altogether. Japan, Norway, and Iceland refused and still hunt whales. So do Alaska Natives.

▲ **Figure 15-6** Whales were almost driven to extinction because people viewed them as a renewable resource. However, the whales could not breed fast enough to replace lost individuals.

Thinking Critically What human actions have endangered the whales?

3 Assess

CHECKING CONCEPTS ANSWERS

1. renewable
2. nonrenewable
3. Earth
4. cannot
5. conservation

THINKING CRITICALLY ANSWERS

6. **CLASSIFY:** *Renewable*: trees, nitrogen, people, water, and soil; *Nonrenewable*: coal, diamonds, natural gas, oil, and iron
7. **CONCLUDE:** so people do not use them up before they can be replaced

4 Close

Web InfoSearch

National Parks Have students use the Internet to find out more about national parks. Ask: *What parks are part of the U.S. National Park System? How are national parks a resource?* Some key search words are **National Park System, Yellowstone Park,** and **National Park Service.** Students can begin their search at www.conceptsandchallenges.com.

Integrating Life Science

TOPICS: endangered species, predation

THE WHALING AGREEMENT

Background Baleen whales have filterlike structures, called baleen, instead of teeth. The baleen strain tiny organisms, called plankton, from the water. Baleen whales include blue whales and humpback whales. Toothed whales include sperm whales, orcas, dolphins, and porpoises. You may want to have students research numbers for total current populations of these mammals and identify which ones are increasing due to conservation efforts.

Ask students to find out the kinds of natural resources people obtained from hunting whales. (Whale oil was used in lamps until kerosene became widely available. Whale meat was used as food. Ambergris, a substance found in the stomachs of sperm whales, was used in perfumes.)

Thinking Critically Answer
hunting and pollution

ESL/ELL STRATEGY

Natural Resources Discuss with English-language learners the concept of natural resources. Explain that the word *natural* is related to the word *nature*, which means "the world and all things except those made by humans." The word includes anything that is made by nature, so natural resources are resources that are supplied by Earth.

• TEACHING RESOURCES • Online or Print

Teacher's Resources

Lesson Review: Chapter 15, p. 3

Enrichment: Chapter 15, p. 7

15-3 *(pp. 354–355)*

1 Introduce

INVESTIGATE

TIME: 10–15 minutes

PURPOSE: Students will observe the characteristics of a piece of coal.

MATERIALS: anthracite or lignite coal, hand lens

PROCEDURE: Have pairs of students write a description of their observations and draw one or more sketches of what they see.

THINK ABOUT IT: Students may be able to discern some of the layers of sediment that were compressed to form the coal. They may also observe fossil imprints that show coal's organic origins.

STUDY HINT Have students compare and contrast the fuels discussed in this lesson. Use a Venn Diagram from the Teacher's Resources or create a chart with the information.

Linking Prior Knowledge Review what students learned about fossil fuels (Lesson 4-6). Remind students that oil, coal, and natural gas are fossil fuels. Ask how fossil fuels formed, and why they are considered a nonrenewable resource. (Fossil fuels take millions of years to form and are being used up faster than they can be made.)

2 Teach

Discussion Emphasize that fossil fuels are the main sources of energy used in the United States. Refer students to the circle graph in Figure 15-7. Ask students which energy source is used to produce the most energy. (oil) Stress the importance of conserving fossil fuels.
learning style: auditory

Answers to questions

1. **NAME:** coal, oil, and natural gas
2. **DESCRIBE:** They do not pollute the air, and moving water is a renewable resource.
3. **IDENTIFY:** It produces deadly radiation that could be released in an accident. Nuclear wastes remain dangerous for thousands of years and are difficult to dispose of safely.

15-3 What are our main energy sources?

INVESTIGATE

Observing Coal
HANDS-ON ACTIVITY

1. Observe a lump of hard coal (anthracite) or soft coal (lignite). Look at its color, texture, and shape. Record your observations in detail.
2. Use a hand lens to examine the coal more closely. Look for fossils or imprints of plant or animal remains.

THINK ABOUT IT: What more did you notice about the coal when you used the hand lens? What does this tell you about what coal is made of?

Objective
Name and describe the main sources of energy.

Key Terms
hydroelectric (HY-droh-ee-LEHK-trihk) **power:** electrical energy produced from moving water
nuclear (NOO-klee-uhr) **energy:** energy produced by splitting or combining atoms

Fossil Fuels **Fossil fuels** are the world's main energy sources. Oil, coal, and natural gas are fossil fuels. Fossil fuels can generate electricity and heat. They can also run car engines and other types of machinery.

Go Online
active art
For: Hydroelectric Power activity
Visit: PHSchool.com
Web Code: cfp-3025

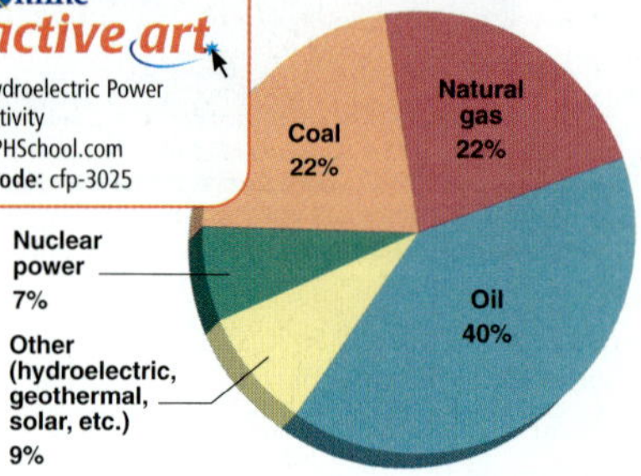

▲ **Figure 15-7** Current energy sources in the United States

However, most of our current energy sources have problems associated with them. Fossil fuels are rapidly being used up. Their use can also harm the environment.

People need to find ways to cut down on the use of fossil fuels. Some ways to do this include using energy-efficient cars, carpooling, lowering the heat, and turning off unneeded lights.

1 **NAME:** What are three fossil fuels?

Hydroelectric Power Water stored behind a dam has potential energy. When this water flows downhill, its stored potential energy changes to the energy of motion, which is called kinetic energy. The electrical energy power plants produce from running water is called **hydroelectric power.** At hydroelectric power plants, the moving water turns the blades of a machine called a turbine. The turbine drives an electric generator that produces electricity.

▲ **Figure 15-8** Hydroelectric plant on the Tocantins River in Brazil

Teaching Options

! FAST FACTS

Nuclear Energy

Nuclear energy is released during a nuclear reaction as a result of the conversion of mass into energy. The conversion involves the binding energy of the nucleus of an atom. Nuclear energy is released in two ways: by fission and fusion. Fission is how today's nuclear reactors produce commercial energy. It was first discovered by Otto Hahn, Lise Meitner, and Fritz Strassmann in the late 1930s. In 1942, the first sustained nuclear chain reaction was achieved by Enrico Fermi in the United States.

ESL/ELL STRATEGY

Power and Energy Although the words *power* and *energy* are often used as though they mean the same thing, they do not. Tell English-language learners that in physics, energy is the capacity for doing work or making something move. Types of energy include mechanical, electrical, and chemical. Power refers to the *rate* of doing work or of producing or consuming energy. It is a measure of the output of an engine or other power source.

Today, falling water is used to produce about 20 percent of the world's electricity. Hydroelectric power does not pollute the air. Also, its energy source, moving water, is renewable. However, dams can affect the wildlife living on or near a river by changing the environment. Dams can also alter the natural flow of rivers.

2 DESCRIBE: What are two advantages of hydroelectric power plants?

Nuclear Energy Energy produced by splitting atoms is **nuclear energy.** The heat given off when atoms are split can be used to produce steam. The steam powers generators to make electricity. Uranium is commonly used as fuel.

There are drawbacks to using nuclear energy. Deadly radiation is produced in nuclear reactors. An accident could release this radiation. Also, a way must be found to safely store the wastes, which remain deadly for thousands of years.

Energy can also be released by combining atoms. However, this kind of nuclear energy is very difficult and costly to produce.

3 IDENTIFY: What are some problems with using nuclear energy?

CHECKING CONCEPTS

1. What is the most commonly used fossil fuel?
2. How can people produce electricity from moving water?
3. What kind of energy does a dam store?
4. How is nuclear energy produced commercially?

THINKING CRITICALLY

5. HYPOTHESIZE: How might building a dam affect the living things in an area? Give an example.
6. CLASSIFY: Classify each of the following items as having potential or kinetic energy: **a.** a book on a shelf; **b.** milk in a bottle; **c.** a book falling to the floor; **d.** a parked car; **e.** juice being poured; **f.** a moving car.

BUILDING MATH SKILLS

Researching and Graphing Find out how energy sources are used in the United States. Then, create a pie graph comparing the different uses.

Science and Technology

CLEANING UP OIL SPILLS

Every year, there are thousands of crude oil spills. These spills harm aquatic life and seashores. Within hours, oil slicks—large bodies of oil floating on the surface of the water—are formed. The oil is spread quickly by waves and sea animals moving through the water. Swift action must be taken to prevent serious damage to the environment.

▲ **Figure 15-9** The damage from some oil spills can be limited by using special equipment.

Scientists have developed some technologies for cleaning up oil spills. None is perfect. Special nets called containment booms keep the oil in an enclosed area, where it can be pumped out later. Controlled burning of oil slicks is another method. Some chemicals can break up the grease. These allow the oil to rise above the water. The oil can then be moved away from the spill site. Certain bacteria can also be brought in to break down the complex chemicals in oil. Fertilizers and minerals can also encourage other bacteria already in the water to break down the oil.

Thinking Critically How do sea animals spread the oil?

MORE ON THE WEB

Hydroelectric Power Have students use the Internet to learn about hydroelectric power plants. Have them begin their search at www.conceptsandchallenges.com.

EXTENDING THE LESSON

Conserving Energy Invite a speaker from a local utility company to discuss energy conservation measures and how energy is supplied to your community.

• TEACHING RESOURCES • Online or Print

Teacher's Resources

Lesson Review: Chapter 15, p. 4

Venn Diagram: Study Tools, p.10

Go Online active art

For: Hydroelectric Power activity
Visit: PHSchool.com
Web Code: cfp-3025

Students can interact with the art of hydroelectric power online.

3 Assess

CHECKING CONCEPTS ANSWERS

1. oil
2. The potential energy of water held behind the high wall of a dam is converted to kinetic energy when the water is released. The kinetic energy of the falling water turns the turbines of a generator to produce electricity.
3. potential energy
4. Nuclear power plants split atoms, releasing heat energy that is used to produce steam. The steam turns the turbine blades of an electrical generator.

THINKING CRITICALLY ANSWERS

5. HYPOTHESIZE: Possible answer: Water builds up behind the dam, flooding areas that were previously dry land and destroying wildlife habitats.
6. CLASSIFY: potential: a, b, d; kinetic: c, e, f

4 Close

BUILDING MATH SKILLS ANSWER

Researching and Graphing U.S. Department of Energy figures for the year 2000 were residential 15%, commercial and industrial 37%, transportation 19%, and electric power production 29%.

Science and Technology

CLEANING UP OIL SPILLS

Life-Science Connection The fur and feathers of ocean mammals and birds help insulate these animals from cold ocean waters. Animals contaminated with oil remove it by eating it, which poisons them. Detergent used to clean up the oil also temporarily removes the water repellency of feathers and fur.

Discussion Discuss the kinds of ocean and weather conditions that might make cleanup difficult. (storms that hamper equipment and spread oil faster; seaweed or other floating debris that clogs machinery) Point out that because cleanup is so difficult, oil-spill prevention has been stressed, such as better tanker-hull construction.

Thinking Critically Answer Possible answer: Animals with oil-soaked fur or feathers move from place to place, taking the oil with them.

15-4 *(pp. 356–357)*

1 Introduce

STUDY HINT Have students summarize the process shown in Figure 15-10, from the collecting of solar energy using the solar panel to the production of heat to warm the house.

Linking Prior Knowledge Lead a discussion of energy resources and the current energy-supply situation. Ask students what they know about energy sources other than fossil fuels, hydroelectric power, and nuclear power. Many students will likely know something about solar energy.

2 Teach

Demonstration Use a simple solar cell-powered motor, calculator, radio, or flashlight to show students how the energy from sunlight can be converted into electrical energy. Point out that the Sun's energy can be used in two ways: to heat materials and to power solar cells, which transform sunlight directly into electricity.
learning style: visual

Discussion Show students pictures of volcanoes, geysers, and hot springs. Describe these physical features and remind students that there are very high temperatures within Earth. Relate this to geothermal energy.

Reinforcement Write *Alternative Energy Sources* on the board. From this head, draw arrows to each of the following heads: *Solar Energy*, *Geothermal Energy*, *Wind Energy*, and *Tidal Energy*. Ask volunteers to describe each of these energy resources. Have students copy the information in their notebooks under the correct heads.
learning styles: visual, auditory

Answers to questions

1. **DEFINE:** devices that change sunlight into electricity
2. **IDENTIFY:** deep inside Earth
3. **EXPLAIN:** to pump water and produce electricity
4. **APPLY:** places where high tides are very high and low tides are very low, which in the United States occur only in Maine and Alaska

15-4 What are some alternative energy sources?

Objective

Describe five alternative energy sources.

Key Terms

solar energy: energy from the Sun

solar cell: device that converts sunlight into electricity

geothermal (jee-oh-THUHR-muhl) **energy:** energy produced from heat inside Earth

Solar Energy Alternative energy refers to energy not produced by burning fossil fuels or wood. Like fossil fuels, all alternative energy sources have problems related to their use.

Energy from sunlight, or **solar energy,** is one alternative energy resource. Solar energy is already used to heat some homes. It can also be used to produce electricity. Solar collectors collect the Sun's energy, and devices called **solar cells** on the collector convert the sunlight into electricity.

Solar energy is a renewable resource. It does not pollute. However, solar energy is only a good source of energy in places that receive above-average amounts of sunlight all year.

1 **DEFINE:** What are solar cells?

Geothermal Energy Heat produced deep inside Earth is called **geothermal energy.** This heat melts rocks and warms groundwater. In some places, steam is produced and can generate electricity for heating homes. In other places, the heated water is piped directly into homes. In Iceland, about 80 percent of the homes get their hot water and heat from geothermal energy.

2 **IDENTIFY:** Where is geothermal energy produced?

Wind Energy The energy from moving air, or wind energy, is another renewable resource. Windmills have been used for many years to pump water and produce electricity. In California, big windmill farms are used to make electricity in some areas. Windmills do not pollute the air. However, there are few places on Earth where the wind blows steadily enough for windmills to be efficient sources of energy.

3 **EXPLAIN:** What are windmills mostly used for?

Energy from the Tides Tidal energy is the energy of rising and falling tides. Tidal energy can be used to produce electricity. However, it can be used only near a shoreline where the tides vary from very high to very low.

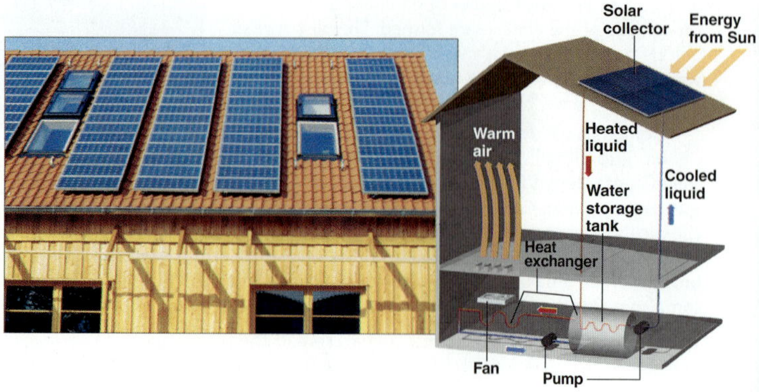

Figure 15-10 ▶ Shown are two different kinds of solar heating systems. On the left, panels of solar cells are used to generate electricity. On the right, a solar collector heats water on the roof. It then pumps the heated water down through a heat exchanger. A fan circulates the warm air around the house.

Teaching Options

! FAST FACTS More on Solar Collectors

If you leave a closed car in bright sunshine, the temperature of the interior rises above the outdoor air temperature. The car collects and stores the Sun's heat. Similarly, radiant heat from the Sun can be collected by solar panels. In a type of collector called a liquid collector, an array of closely spaced, thin black tubes encased in clear glass or plastic is attached to a heat-absorbing plate. These tubes are filled with water or another heat-absorbing fluid. As the fluid flows through the tubes, heat from the Sun raises the fluid temperature. Fluid heated in a solar collector can be piped under floors or into radiators to heat rooms in a building. If clean household water is pumped through the tubes, it can be used for drinking, bathing, and doing laundry.

ESL/ELL STRATEGY

***Explaining* Geothermal** To help English-language learners understand the meaning of *geothermal*, explain that *geo* means "earth" and *thermal* means "heat." Ask English-language learners to describe to you how these meanings relate to the term *geothermal energy*.

▲ **Figure 15-11** A tidal energy plant in Nova Scotia, Canada

APPLY: Where in the United States could tidal energy be produced?

CHECKING CONCEPTS

1. What devices change sunlight into electricity?
2. What is geothermal energy?
3. Why can we not rely on windmills for energy?

THINKING CRITICALLY

4. **ANALYZE:** Which alternative energy source do you think is best? Why?
5. **INFER:** If you were building a house, where would you put your solar collectors? Why?
6. **EXPLAIN:** How does a pinwheel or a kite show that wind has energy?

Web InfoSearch

Biomass Fuels Material from living things is called biomass. Biomass can be a fuel source. Wood and solid animal waste are biomass fuels. Methane gas from sewage is also a biomass fuel. Ethanol from sugarcane is already used heavily in Brazil.

SEARCH: Use the Internet to find out more about biomass fuels. How are they produced? What are they used for? Start your search at www.conceptsandchallenges.com. Some key search words are **biomass fuels, biomass fuel uses,** and **biomass alternative energy.**

Hands-On Activity

MEASURING SOLAR ENERGY ABSORPTION

You will need black construction paper, white construction paper, and two thermometers.

1. Place two thermometers in a sunny place. ⚠ CAUTION: Be careful if using glass thermometers.
2. Cover one thermometer with black construction paper and the other with white construction paper.
3. After 15 minutes, check the temperatures of both thermometers.

▲ **STEP 2** Cover the thermometers with paper.

Practicing Your Skills

4. **OBSERVE:** What were the temperature readings of the two thermometers?
5. **COMPARE:** Which thermometer had the higher reading?
6. **RELATE:** Which color takes in more heat energy from the Sun, black or white?
7. **CONCLUDE:** What color do you think is used in solar collectors? Why?

COOPERATIVE LEARNING

Comparing Energy Sources Ask small groups of students to work together to list the similarities and differences among geothermal energy, wind energy, and tidal energy. Have students share their findings with the class. **learning style: auditory**

• TEACHING RESOURCES • Online or Print

Teacher's Resources

Lesson Review: Chapter 15, p. 5

3 Assess

CHECKING CONCEPTS ANSWERS

1. solar cells
2. heat produced deep inside Earth
3. Few places on Earth have winds that blow steadily enough to provide a constant supply of energy.

THINKING CRITICALLY ANSWERS

4. **ANALYZE:** Whatever students choose, they should give logical reasons for their choice.
5. **INFER:** Possible answer: on the roof, facing the direction from which the most sunlight comes
6. **EXPLAIN:** Wind makes the pinwheel turn. Wind carries the kite into the air.

4 Close

Web InfoSearch

Biomass Fuels Material from living things is called biomass. Have students use the Internet to find out more about biomass fuels. Ask students: *How are they produced and stored? What are they used for?* Some key search words are **biomass fuels, biomass fuel uses,** and **biomass alternative energy.** Students can begin their search for information at www.conceptsandchallenges.com.

Hands-On Activity

MEASURING SOLAR ENERGY ABSORPTION

TIME: 20–30 minutes

PURPOSE: observing, comparing, relating, concluding

COOPERATIVE LEARNING: Have students work in pairs to complete this activity. Ask each pair to contribute their temperature readings to a class chart.

Practicing Your Skills Answers

4. **OBSERVE:** The thermometers should show two different readings.
5. **COMPARE:** the thermometer covered with black paper
6. **RELATE:** black
7. **CONCLUDE:** black, because it absorbs more of the Sun's heat energy than white or other colors

15-5 *(pp. 358–359)*

1 Introduce

STUDY HINT As students read through the lesson, have them make a table of conserved materials and conservation methods.

Linking Prior Knowledge Ask students to discuss what they currently do to conserve natural resources. Many students may be familiar with the recycling of glass, paper, or metal. Students may also be familiar with energy conservation measures such as turning off lights, TVs, and other devices when they are not being used.

2 Teach

Demonstration Show students examples of materials that households in their area are being encouraged to recycle. These probably include newspapers, aluminum beverage cans, and some types of plastic. Ask students to identify the renewable and nonrenewable resources used to make these materials. Discuss the importance of recycling each resource.
learning styles: auditory, visual

Discussion Ask students to think about what their lives would be like if there were no metals. Point out that there would be no computers, bicycles, cars, telephones, televisions, stereos, jewelry, appliances, and so on. Ask students to think about life without trees, paper products, or glass. Identify recycling and replacement as two ways to conserve natural resources.
learning style: auditory

Answers to questions

1. **DESCRIBE:** It allows resources to be used again.
2. **LIST:** the oceans, the Moon, and asteroids
3. **STATE:** Possible answers include turning off water when brushing teeth, taking showers instead of baths, and fixing leaky faucets promptly.
4. **NAME:** contour plowing, terracing, strip-cropping, and planting windbreaks

15-5 How can natural resources be conserved?

Objective
Describe ways to conserve natural resources.

Key Term
recycling: using a natural resource over again

Recycling Natural resources can be conserved by **recycling,** or by using them over again. Aluminum, glass, paper, and plastics can all be recycled. For example, when paper is recycled, fewer trees need to be cut down. Recycling also uses less energy than finding new resources. This conserves fossil fuels.

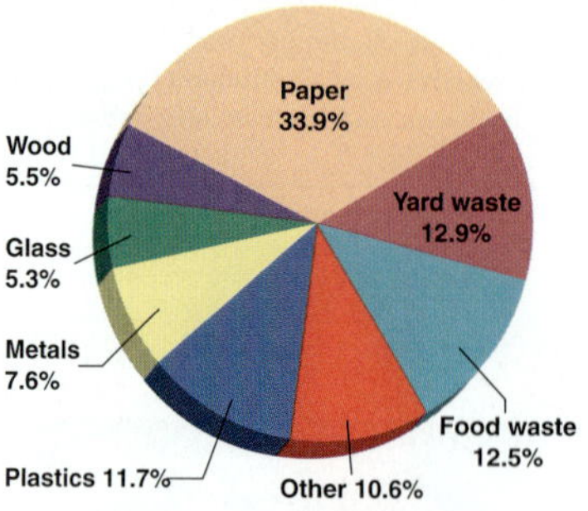

▲ **Figure 15-12** Types of household waste

Another way to conserve resources is to use other materials in their place. For example, some iron and aluminum parts in engines can be replaced with plastic. Some steel cables can be replaced by nylon ones.

1. **DESCRIBE:** How does recycling conserve natural resources?

New Mineral Sources Mineral resources on land are running out. However, the oceans may be a new source of minerals. Many minerals, such as sodium, sulfur, potassium, magnesium, and bromine, are dissolved in ocean water. Scientists are looking for less expensive ways to take minerals from the ocean.

Another new source of minerals might be space. Moon rocks contain aluminum, magnesium, titanium, and silicon. Asteroids may contain billions of tons of nickel and iron. Spacecraft have been sent to study asteroids with this in mind. Astronomers use special instruments to determine an asteroid's mineral makeup.

2. **LIST:** What are some possible new sources of minerals?

Water Conservation The supply of clean, fresh water on Earth is limited. One way to conserve water is not to waste it. Do not run the water when you brush your teeth. Take showers rather than baths because showers use less water. Fix leaky faucets quickly.

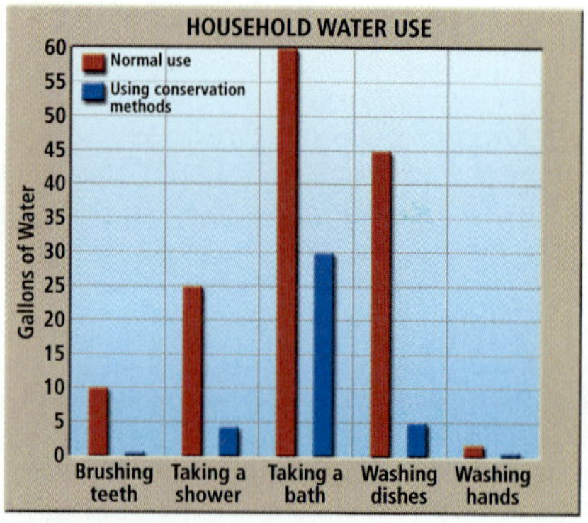

▲ **Figure 15-13**

3. **STATE:** What are two ways to conserve water?

Soil Conservation Some farming methods help conserve soil. For example, in contour farming, farmers plow across the slope of the land. In terracing, flat areas, or terraces, are created on the side of a hill. Both these methods help prevent water from washing away soil. In strip-cropping, grasses are planted between rows of crops to hold water and help stop erosion.

Teaching Options

QUICK CONVERSIONS

	English	Metric/SI
Taking a shower	10 gal 1 gal	38 L 3.8 L
Brushing teeth	24 gal 4 gal	91 L 15 L
Taking a bath	59 gal 30 gal	223 L 113 L
Washing dishes	43 gal 3 gal	163 L 11 L
Washing hands	2 gal 1 gal	7.6 L 3.8 L

EXTENDING THE LESSON

Composting Composting can be done indoors or outdoors in a large bin or garbage can. Besides the items listed in the text, other suitable composting items include grass clippings and other yard waste, ashes, paper, eggshells, nut shells, and coffee grounds. Dairy products, meat scraps, and oils should not be added to a compost bin. The compost material should be mixed with soil, kept moist, and turned over with a shovel about once a week.
learning style: auditory

▲ **Figure 15-14** Strip-cropping can help conserve soil.

Windbreaks such as rows of trees help stop wind from blowing soil away. Fertilizers can replace the nutrients used up by plants. So can planting legumes.

4 NAME: What farming methods help conserve soil?

CHECKING CONCEPTS

1. What process reuses natural resources?
2. How can iron and aluminum engines be updated to conserve resources?
3. What do fertilizers add to soil?
4. What farming methods help stop water from washing away soil?

THINKING CRITICALLY

5. EXPLAIN: How does recycling reduce waste?
6. CALCULATE: About 90 new aluminum cans can be made from 100 recycled cans. If each person in your class recycled one can each day, how many new cans could be made from what your class recycled in a month?

INTERPRETING VISUALS

Use Figure 15-13 to answer the following questions.

7. IDENTIFY: Which household activity uses the most water?
8. INFER: What could you do to save water while doing this activity?
9. IDENTIFY: Which household activity shows the biggest difference between normal use and conservation use?
10. INFER: What do people do that might cause the difference mentioned in Question 9?

Real-Life Science

COMPOSTING

One way to help conserve natural resources is by composting. Composting turns food waste into fertilizer. It also reduces the amount of solid waste that must be put in landfills.

Composting is usually done outdoors, in a shaded area. In a compost pile, scraps of plant materials, such as peels and rinds, are mixed with soil. Microbes in the soil break down the plant material. This releases nutrients. These microbes eat both "brown" and "green" plant materials. Brown materials, such as straw and mulch, are from dead or dried plants. The microbes use them for energy. Green materials, such as leaves and other plant parts, provide nitrogen and other nutrients. As the microbes break down the plant materials, heat is given off. This warms the surrounding soil.

▲ **Figure 15-15** Composting helps the environment by recycling food waste. It also reduces the need for commercial fertilizers.

Thinking Critically What plant wastes from your dinner last night could you have used for composting?

3 Assess

CHECKING CONCEPTS ANSWERS

1. recycling
2. use plastics in place of metals
3. nutrients
4. contour plowing, terracing, and strip-cropping

THINKING CRITICALLY ANSWERS

5. **EXPLAIN:** Recycled materials are not thrown away.
6. **CALCULATE:** Possible answer: 30 students × 30 recycled cans/student = 900 recycled cans; 90 new cans are made for each 100 recycled cans, so 900 recycled cans × (90 new cans/100 recycled cans) = 810 new cans

INTERPRETING VISUALS ANSWERS

7. **IDENTIFY:** taking a bath
8. **INFER:** Possible answer: Take a shower instead of a bath.
9. **IDENTIFY:** washing dishes
10. **INFER:** Possible answer: Run water while washing and rinsing instead of using a basin full of water.

4 Close

Real-Life Science

COMPOSTING

Life-Science Connection Demonstrate composting for students. Fill a 1-gal container (or a 2-L plastic bottle with the top cut off) one-third full of moistened soil. Do not use sterilized potting soil. Bury a few dried leaves and pieces of apple in the soil. Cover the opening with several layers of cheesecloth secured with a rubber band. Place the jar in a dimly lit area and have students monitor the changes for the next 2 weeks. (The leaves and apple pieces will decay.) Every few days, gently shake the jar to remix the contents. If the soil becomes dry, add a few drops of water through the cheesecloth.

Discussion Explain to students that composting mimics the natural decay processes that take place in the environment. When leaves fall or a plant or an animal dies, the remains are decomposed by bacteria, fungi, insects, and other organisms. This process makes nutrients that were in the remains available to other organisms.

Thinking Critically Answer Students should list a variety of vegetable materials.

MORE ON THE WEB

Backyard Conservation
Have students use the Internet to read about different methods of resource conservation and write a short report on the resource conservation method of their choice.

• *TEACHING RESOURCES* • *Online or Print*

Teacher's Resources

Lesson Review: Chapter 15, p. 6

Enrichment: Chapter 15, p. 7

ONGOING ASSESSMENT

Have students create posters, newspaper advertisements, radio public service announcements, or other materials for a campaign to encourage recycling in their school or community.

RETEACH

Conserving Soil Have any students who have worked on a farm or garden describe some of the methods used to conserve soil. **learning style: auditory**

15-6 *(pp. 360–361)*

1 Introduce

INVESTIGATE

TIME: 10–15 minutes

PURPOSE: Students will learn that detergents, including those used to clean oil-soaked birds that have been caught in oil spills, can have harmful effects.

MATERIALS: mixing bowl, measuring spoons, water, vegetable oil, detergent powder, paper towels

PROCEDURE: Have students work in pairs to complete the activity. Students will observe that the oil floats on water. The detergent breaks up the oil, allowing it to sink.

THINK ABOUT IT: The detergent breaks up the oil coating, destroying the bird's waterproofing. The bird may no longer be able to float.

Linking Prior Knowledge Ask if students know of any wildlife that is endangered or threatened in their area. Students may know that some American wildlife, such as bison, gray wolves, and grizzly bears, are threatened or endangered because of hunting or habitat loss. Students may be familiar with the terms *threatened* and *endangered*, but they may not understand the difference between them.

2 Teach

Discussion Tell students the definition of the terms *extinct*, *endangered*, and *threatened* as they pertain to wildlife. Describe the main threats to wildlife today. Point out that certain species of seemingly common wildlife, such as woodpeckers and small fish, are endangered or threatened. Discuss the different measures being taken to conserve wildlife. Make sure that students understand that wildlife includes plants.
learning style: auditory

READING STRATEGY Review with students the definition, pronunciation, and spelling of the word *habitat*. Point out that spelling is made easier by remembering that the word *habitat* contains the word *habit*.

Answers to questions

1 **DEFINE:** all the plants and animals in an area

2 **IDENTIFY:** loss of habitat

15-6 What is wildlife conservation?

INVESTIGATE

Observing the Effects of Detergent on Wildlife

HANDS-ON ACTIVITY

1. Pour one cup of tap water into a medium-size mixing bowl.
2. Add 1 teaspoon of vegetable oil. Observe the surface of the water.
3. Sprinkle 2 teaspoons of laundry detergent powder over the surface of the water. Gently stir with a spoon to mix, but try not to produce bubbles. Again, observe the surface of the water.

THINK ABOUT IT: The oil on birds' feathers makes them waterproof. This helps the birds stay afloat. What could detergent do to bird feathers? How does this endanger the birds?

Objectives

Identify the main threat to wildlife today.
Describe two ways to conserve wildlife.

Key Term

wildlife: all the plants and animals that live in an area

Endangered Wildlife All of the plants and animals that live in an area make up the area's **wildlife.** Over the past few centuries, hundreds of plant and animal species have died out, or become extinct. Many more are endangered, or moving toward extinction. Once a species becomes extinct, it will never return. The African elephant is an endangered animal. There are very few African elephants left. Other species have been identified as threatened. This means their numbers are going down, although they are not yet considered to be endangered.

1 **DEFINE:** What is wildlife?

Loss of Habitat Loss of habitat is the main threat to wildlife. As the number of people on Earth increases, there is less space for other species. Land that was once covered with forest is now used for cities, farms, roads, and industry. The tropical rain forests are rapidly disappearing. More than half of all plant and animal species that live on Earth today live in tropical rain forests.

▲ **Figure 15-16** African elephants, though endangered, are still hunted for their ivory tusks.

2 **IDENTIFY:** What is the biggest threat to wildlife today?

Laws and Wildlife One way to conserve wildlife is to pass laws that protect it. Some of these laws regulate hunting and fishing. In recent years, some countries have passed laws stopping the use of certain pollutants. There are also new laws to stop the destruction of wildlife habitats.

3 **DESCRIBE:** How do laws help protect wildlife?

Wildlife Refuges A wildlife refuge is an area, like a national park, where animals and plants are protected. Setting up wildlife refuges helps conserve wildlife. Some zoos also help conserve wildlife. Breeding programs at these zoos have helped save some endangered species from extinction. In some cases, members of some species are returned to their natural environment.

Teaching Options

! FAST FACTS *The Endangered Species Act*

The U.S. government passed the Endangered Species Act in 1973. This act outlaws the hunting and collecting of endangered or threatened species of plants and animals. It also protects these species by preventing construction projects and other economic activities, such as logging, that would damage or destroy their habitats. The act even prohibits importing endangered or threatened species or products made from them. The U.S. Fish and Wildlife Service works to enforce the Endangered Species Act.

Soon after the act was passed, people began to challenge it. They claimed that the economic benefits of development projects can outweigh the benefits of protecting plant and animal habitats. The act provides for the formation of a committee to rule on such challenges. In 1992, for example, the committee ruled to allow logging in certain areas of old-growth forest, even though such activity would threaten the habitat of the endangered northern spotted owl.

▲ **Figure 15-17** Whooping cranes are an endangered species. This pair is at a national wildlife preserve in Arkansas.

In the United States, the National Wildlife Refuge System operates more than 550 wildlife refuges across the country. One example is Klamath Basin in Oregon and California. This vast region of shallow lakes and marshes protects many species of water birds and plants. Alaska has sixteen refuges, making up 83 percent of the entire National Wildlife Refuge System. On its 77 million protected acres are caribou, bears, moose, and wolves, among others.

4 **STATE:** What are two ways to conserve wildlife?

CHECKING CONCEPTS

1. Wildlife in danger of extinction is called ___________.
2. To conserve wildlife, many different countries have passed laws that include rules for fishing and ___________.
3. Once a kind of plant or animal becomes ___________, it will never return.
4. A wildlife ___________ is an area where wildlife is protected.

THINKING CRITICALLY

5. **INTERPRET:** What does the statement "Extinct is forever" mean?
6. **CONCLUDE:** Is it important to save the rain forests? Why?

BUILDING SCIENCE SKILLS

Researching and Writing The dodo bird is an extinct animal. So is the passenger pigeon. Find out about five other animals or plants that once thrived but are now extinct. In which countries did they live? When? What were they like? How did their loss affect the environment in which they lived? Write a report with your findings.

Integrating Life Science

TOPICS: habitats, endangerment, biodiversity

ENDANGERED ANIMALS

You have read about the northern spotted owl. Unfortunately, this owl is not the only endangered animal. There are hundreds of others. The main causes of endangerment are loss of habitat, hunting, and animals captured for the pet trade.

In recent years, there has been a big conservation effort to save the giant panda, which is native to China. For example, trees in panda habitats are no longer being cut down. However, some scientists fear that only a thousand or so pandas are left. Some other well-known animal species facing possible extinction are elephants, tigers, manatees, gorillas, whales, tortoises and marine turtles, rhinoceroses, and condors.

▲ **Figure 15-18** The giant panda, which is endangered, is native to China.

Thinking Critically How is saving species an act of conservation?

3 **DESCRIBE:** Possible answer: Laws help prevent overcollecting, polluting, and loss of habitat.

4 **STATE:** Accept any two of the following: setting up wildlife refuges, establishing breeding programs, or passing laws to stop pollution and conserve habitat.

3 *Assess*

CHECKING CONCEPTS ANSWERS

1. endangered
2. hunting
3. extinct
4. refuge

THINKING CRITICALLY ANSWERS

5. **INTERPRET:** Extinct organisms will never return.
6. **CONCLUDE:** Possible answer: Yes. Rain forests provide habitat for many species of wildlife, so preserving the rain forest helps preserve wildlife.

4 *Close*

BUILDING SCIENCE SKILLS

Researching and Writing Reports should be logical and accurate and refer to at least two sources.

Integrating Life Science

TOPICS: habitats, endangerment, biodiversity

ENDANGERED ANIMALS

Background Wildlife species often depend on one another for food, shelter, and other such needs. If one species disappears, others may become endangered or threatened. For example, the disappearance of the bamboo that giant pandas depend on for food could mean that pandas would not have enough food to survive.

Discussion Ask students: *Why do wildlife conservationists need to know how species depend on one another for food and sometimes even shelter?* (to better understand how to preserve threatened or endangered species)

Thinking Critically Answer
Possible answer: Wildlife is a natural resource, and saving a species is a way to conserve wildlife.

MORE ON THE WEB

Pelican Island Have students use the Internet to learn about the Pelican Island National Wildlife Refuge and write a short report on it.

• TEACHING RESOURCES • Online or Print

Teacher's Resources
Lesson Review: Chapter 15, p. 8
Enrichment: Chapter 15, p. 9

EXTENDING THE LESSON

Wildlife Up Close and Personal If possible, take students to visit a local zoo, wildlife refuge, or wildlife rescue center. You can usually make advance arrangements for a tour. Some wildlife organizations may be able to come to your school to make a presentation, and they may bring animals or plants for students to observe. Have students write about their experience. **learning style: visual**

15-7 *(pp. 362–363)*

1 Introduce

STUDY HINT Before beginning this lesson, write the title of each section on the board. Have students copy these titles in their notebooks, using an outline format. As students read the lesson, have them write down the two most important points from each section.

Linking Prior Knowledge Ask students what they know about air pollution and if they have observed its effects. Review the gases in Earth's atmosphere (Lesson 11-1). Point out that air pollution involves gases that are naturally found in the atmosphere, as well as gases that are artificially introduced by human activities.

2 Teach

Demonstration Bring in a used air-conditioner filter or air filter from a car. (Keep it in a closed plastic bag.) Ask what the filter does. (removes solid particles from the air) Point out that air contains particulate pollutants, such as dust and soot, and chemical pollutants.
learning style: visual

Discussion When discussing acid rain, remind students that acidity is measured in pH units. Explain that most plants grow best in soils with a pH of between about 6.3 and 6.8, and that other living things are also sensitive to pH. Point out that normal rain is slightly acidic, which means its pH is a little below 7.0. Acid rain has a pH of about 5.6 or below. Describe how acid rain harms both living and nonliving things.
learning style: auditory

Answers to questions

1. **IDENTIFY:** the burning of fossil fuels
2. **EXPLAIN:** Gases combine with water in the air to form acids.
3. **DESCRIBE:** a layer of warm air trapped between two layers of cold air
4. **RELATE:** Carbon dioxide traps heat energy from the Sun.
5. **EXPLAIN:** Laws put limits on the amount of pollution that may be released by cars and factories.

15-7 What causes air pollution?

Objectives

Identify causes of air pollution. Explain how air pollution harms the environment.

Key Terms

smog: mixture of smoke, fog, and chemicals

acid rain: rain containing acids produced by water chemically combining with certain gases

Air Pollution The burning of fossil fuels is the major cause of air pollution. When these fuels are burned, harmful substances enter the atmosphere. Thick, black smoke from chimneys contains dust and soot. Dust and soot are pollutants that can remain in the air for a long time. They can irritate eyes, throat, and lungs. Many cities have a smog problem. **Smog** is a mixture of smoke, fog, and chemicals. Smog also irritates the lungs of people, especially those with breathing problems.

▲ **Figure 15-19** Smog over a city pollutes the air.

1 **IDENTIFY:** What is the major cause of air pollution?

Acid Rain Cars and factories burn fossil fuels. This releases large amounts of carbon dioxide and other gases into the air. Some of these gases mix with water in the air and form acids. The acids then fall to Earth as **acid rain.** Acid rain is harmful to both living and nonliving things.

Acid rain that falls into lakes and streams kills fish. Acid rain also harms trees and causes brick, stone, and metal structures to weather, or break apart.

▲ **Figure 15-20** Scientists study trees affected by acid rain.

2 **EXPLAIN:** How does acid rain form?

Temperature Inversions Some large cities are ringed by mountains. As cool air flows into a valley, it pushes the warm air upward. This warm air becomes caught between two layers of cold air. The warm air cannot rise farther because of the cooler air pushing down on it. The mountains block the warm air from leaving the area. This potentially dangerous condition is known as a temperature inversion.

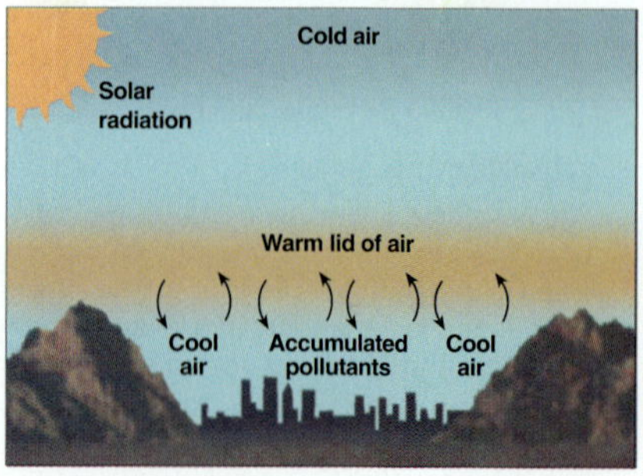

▲ **Figure 15-21** A temperature inversion traps polluted air.

During a temperature inversion, smog will continue to build up until a new weather system pushes the warm air up over the mountains. The cool air below can then circulate, reducing pollution.

3 **DESCRIBE:** What is a temperature inversion?

Teaching Options

! FAST FACTS *The pH of Common Liquids*

Solution	pH*
Human stomach acids	2
Lemon juice	2
Tomato juice	4–5
Milk	6
Pure water	7.0 (neutral)
Seawater	8–9
Milk of magnesia	10
Household ammonia	11
Drain cleaner	12

*The acidity of a solution is measured in pH units, which range from 0 to 14. A neutral solution (distilled water) has a pH of 7.0. A solution with a pH below 7.0 is acidic. A solution with a pH above 7.0 is alkaline. The table at the left shows the approximate pH values of some everyday liquids.

Carbon Dioxide Levels Fuels need oxygen to burn. When fossil fuels burn, they give off the gas carbon dioxide. The carbon dioxide traps heat energy produced by the Sun's rays. This is called the greenhouse effect. Carbon dioxide is a greenhouse gas.

Some scientists think that the increase of carbon dioxide in the air is causing the average air temperature to rise around the world. This rising temperature may cause dramatic climate changes.

4 RELATE: How can an increase of carbon dioxide in the air cause air temperature to rise?

Protecting the Air Many countries have laws to help control or reduce air pollution. In the United States, most vehicles must have antipollution devices. Factories, too, must have filters on their smokestacks. The filters reduce the number of solid particles, called particulates, released into the air. However, using less fuel is still the best way to reduce air pollution.

5 EXPLAIN: How do laws help reduce air pollution?

CHECKING CONCEPTS

1. Smoke contains tiny pieces of dust and __________.
2. Smog is a mixture of smoke, fog, and __________.
3. Carbon dioxide traps __________ energy.
4. Some pollutant gases combine with __________ in the air to form acid rain.

THINKING CRITICALLY

5. PREDICT: The rising heat in Earth's atmosphere could cause glaciers to melt. How could this affect sea level and coastal cities?
6. HYPOTHESIZE: Is acid rain more of a problem in wet or dry climates? Explain.
7. INFER: What two words form the word *smog*?

HEALTH AND SAFETY TIP

Cities with air pollution problems often warn residents when smog levels are very high. Find out why people in these cities should not exercise outdoors on these smog-alert days.

Hands-On Activity
OBSERVING AIR POLLUTANTS

You will need a hand lens, a microscope slide, a toothpick, and petroleum jelly.

1. Use the toothpick to coat one side of a glass slide with a thin layer of petroleum jelly.
2. Place the slide on a window ledge overnight.
3. Examine the slide with a hand lens the next day.

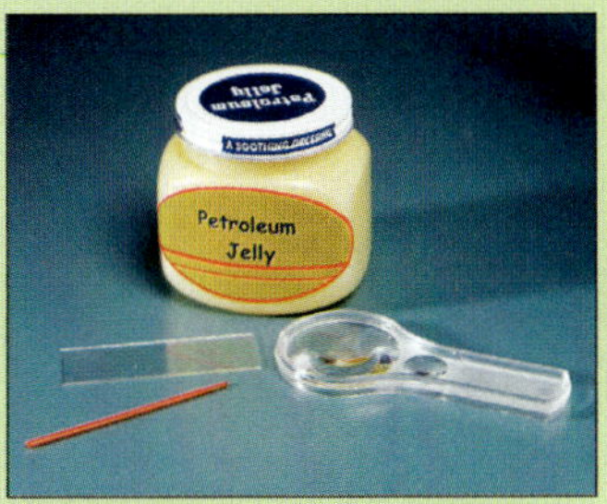

▲ **STEP 1** Coat one side of the slide with petroleum jelly.

Practicing Your Skills

4. OBSERVE: What are some of the things you observed on your slide?
5. INFER: What do you think these particles are?
6. IDENTIFY: Where do you think these particles came from?
7. HYPOTHESIZE: **a.** Where in your neighborhood do you think you could place your slide to trap fewer pollutants? **b.** Where in your neighborhood do you think you could place your slide to trap more pollutants?
8. ANALYZE: How could you prove both of your hypotheses?

COOPERATIVE LEARNING

Testing the Hypothesis Discuss the Practicing Your Skills questions from the Hands-On Activity as a class. Divide the class into three or four groups and have each group work out a way to actually test the hypothesis.
learning styles: auditory, tactile

• TEACHING RESOURCES • Online or Print

Teacher's Resources
Lesson Review: Chapter 15, p. 10

3 Assess

CHECKING CONCEPTS ANSWERS

1. soot
2. chemicals
3. heat
4. water

THINKING CRITICALLY ANSWERS

5. PREDICT: Sea level would rise and cause flooding in low-lying coastal cities.
6. HYPOTHESIZE: Possible answer: wet climates, because they have more rain
7. INFER: *smoke* and *fog*

HEALTH AND SAFETY TIP

People who are active outdoors when smog levels are very high can suffer shortness of breath, chest pain, coughing, and respiratory congestion. High smog levels can worsen diseases such as asthma and bronchitis.

4 Close

Hands-On Activity
OBSERVING AIR POLLUTANTS

TIME: 15 minutes each on two consecutive days

PURPOSE: observing, inferring, hypothesizing, analyzing

SAFETY TIP: Make sure that the slides cannot fall and create a hazard.

COOPERATIVE LEARNING: Have pairs of students work together to set up and observe their slides.

Practicing Your Skills Answers

4. OBSERVE: Students will probably observe dust particles, soot, and perhaps other airborne items such as seeds, fibers, or insects.
5. INFER: pollutants
6. IDENTIFY: things that float in the air
7. HYPOTHESIZE: Possible answers: **a.** areas away from auto exhaust and dust **b.** areas near building exhausts (vents and other outlets through which fumes from buildings are released) and construction sites
8. ANALYZE: Possible answer: by placing a slide in each area and comparing the amount of pollution collected on each one

The Big Idea

(pp. 364–365)

1 *Introduce*

Objective Students will describe how smog forms, identify the components of smog, and explain how smoke from burning coal can release sulfur dioxide into the air. Students will research air quality in their area and write a radio script announcing current conditions.

Linking Prior Knowledge Have students recall what *environment* means (Lesson 15-1). Ask them to list some life-forms that are impacted by a smoggy environment. (plants and animals)

2 *Teach*

Demonstration Smoke is a colloid. Its particles cannot be filtered from the air in which they are suspended. To demonstrate, cut a stick of incense so that it will fit inside a beaker. Place the wooden end of the incense into a small cork and place it upright inside the beaker. Light the incense and cover the beaker with a coffee filter, securing it firmly with a rubber band. As the incense burns, a small amount of the smoke will pass through the coffee filter and the aroma of the incense will be detectable. **learning style: visual**

Discussion After students have read the article and callouts, encourage them to use reference materials to find out more about health-related problems associated with air pollution. Students could find out about the types of air pollution that occur in their area as compared to other parts of the world before writing radio scripts for their Science Logs.

Integrating Chemistry

THE Big IDEA

How can chemical reactions make air deadly?

If you live in a large city, you have probably observed skies darkened by a blanket of smoke and fog. This is smog. Smog is a colloid. In a colloid, particles remain in suspension and do not settle out. Smog contains several chemical compounds, fog, and soot. If you have respiratory problems, smog might make you feel uncomfortable. Usually, it is not deadly.

The smog that blanketed London, England, in 1952 was a thick, gray haze that smelled like coal smoke. Then the haze turned yellow and quite deadly. The air smelled terrible. Taking even the smallest breath caused a sharp pain in the lungs. When the smog finally began to lift, nearly 4,000 Londoners were dead.

The cause of this deadly smog has been traced to the coal-burning plants around the city. The plants had released coal smoke, containing sulfur dioxide, and particles of soot into the air. Ordinarily, air contaminated by this mixture would soon be replaced by fresh air moving into the area. However, the air mass surrounding London remained in place for almost a week. Above the city, a layer of warm air was trapped between two layers of cool air. This stopped the normal air circulation.

Fuel oils burned for heat increased the amount of sulfur dioxide in the air. Eventually, the air became more acidic than lemon juice. With each breath, residents of the city drew sulfuric acid into their bodies.

Look at the illustrations that appear on these two pages. Then, follow the directions in the Science Log to find out more about "the big idea."✦

Soot

The solid particles of soot provided water vapor in the air with a surface upon which to condense.

Breathing Deadly Air

With each passing hour, the number of sulfur dioxide molecules in the air increased. Few people in London ventured outdoors. Those who did covered their mouths and noses with scarves. However, the cloths did not filter out the sulfur dioxide that was in the air.

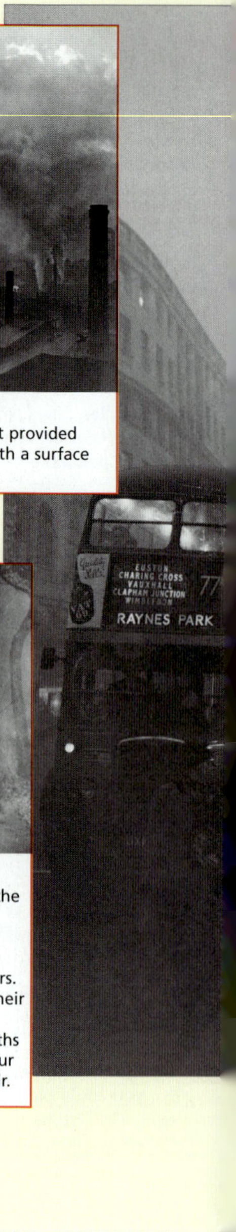

Teaching Options

! FAST FACTS *Pollution and the Industrial Revolution*

Although pollution has caused health problems ever since people first started to live in crowded cities during ancient times, it first became widespread and dangerous during the Industrial Revolution, which started in Great Britain during the late 1700s. The Industrial Revolution brought tremendous increases in the use of fuel-powered machinery to produce goods. The machinery was made of iron, which was manufactured by burning coal. Coal was also burned to run industrial machinery. Such dramatic growth in the burning of coal pumped enormous amounts of smoke and soot into the air of industrialized areas, such as large cities in England, Belgium, and the United States. The pollution problems that first emerged during the Industrial Revolution continue to impact the world today.

ESL/ELL STRATEGY

Mixing Up Mixtures Some students may mix up the meanings of the words *colloid*, *mixture*, and *smog*. Explain to them that the word *mixture* is the most general term, that a colloid is a type of mixture, and that smog is a type of colloid. Draw this in a diagram on the board.

Science Log

WRITING ACTIVITY

The Environmental Protection Agency (EPA) provides daily information about air quality. Do research to find out what the current conditions are where you live. Start your search for information at www.conceptsandchallenges.com. Then, write a script for a radio news broadcast explaining these conditions. Include facts such as smog levels and whether there is currently a temperature inversion.

▲ **Figure 15-22** Except for the soot it contains, smog, like most colloids, cannot be separated by filtering. In London, the smog contained molecules of sulfur dioxide from coal smoke. These molecules dissolved in the water droplets making up the fog and formed sulfuric acid.

Colloids Scatter Light

As the days passed, visibility dwindled to zero. The particles of the colloid scattered light from the headlights of cars. In the photo of beakers, you can see that light passes through a solution but is scattered by a colloid.

THE BIG IDEA

How can chemical reactions make air deadly? Have students start their research for their Science Log Writing Activity at www.conceptsandchallenges.com. Be sure students use search engines and websites that are reliable and acceptable. You may also wish to provide copies of the *Big Idea Planner* and *Big Idea Science Log Entry* study tools found in the Teacher's Resources.

3 Assess

Use the Writing Activity Rubric from the Teacher's Resources to assess students' Science Logs. Fill in the rubric with the additional information below. For this assignment, students should have:

- described air-quality conditions in their area and explained how air pollution is generated.
- communicated, in paragraph form, accurate information about air quality and how it could affect public health in their area.

4 Close

Writing Radio Scripts The scripts should include information about air pollution in the students' area, possible health effects, and what people can do to help reduce air pollution. Make sure that students produce a written script. Ask volunteers to read their scripts aloud as if they were talking on the radio.

WRITING HINT Remind students that a script is meant to be read aloud. They should write sentences that are short and easy for a listener to understand.

Tips for Using Technology

Using Presentation Software to Show Environmental Conditions

Have students use presentation software to add visuals to their scripts. Presentation software allows them to use a computer to put text and art together in a slide show. They can use a scanner to incorporate their own art in their slide show. Students can also use presentation software to share their data with the class. Have students use the Thumbnail Sketch graphic organizer from the Teacher's Resources to plan their presentations.

• *TEACHING RESOURCES* • *Online or Print*

Teacher's Resources

Big Idea: Chapter 15, p. 11

Big Idea Planner: Study Tools, pp. 11–12

Big Idea Science Log Entry: Study Tools, p. 13

Writing Activity Rubric: Teaching Tools, p. 9

Thumbnail Sketch: Study Tools, p. 24

15-8 *(pp. 366–367)*

1 Introduce

INVESTIGATE

TIME: 10–15 minutes

PURPOSE: Students will observe the dilution of a water pollutant.

MATERIALS: water, 1-gal clear container, red food coloring, measuring cup

PROCEDURE: Have students work in small groups to complete this activity. As students add more water, make sure that they take time to observe the results. Have students keep track of the amount of water they add. Tell students to stop when they can no longer see any red.

THINK ABOUT IT: Students' answers will vary. The pollution does not go away just because it cannot be seen. Even a small amount of water pollution is easily spread throughout a lake or stream.

STUDY HINT Have students summarize Figure 15-24 in a short paragraph and draw some conclusions from what the chart shows.

Linking Prior Knowledge Ask students to recall their study of the water cycle (Lesson 9-1). Point out that pollution entering any waterway, even a drainage ditch beside a road, will be carried away when it rains and can eventually end up in streams, lakes, groundwater, wetlands, or the ocean.

2 Teach

Discussion Remind students that all living things require water for survival. Explain that water pollution can come from many sources. The air, soil, sewage, and chemicals can all pollute water. Also, mention oil spills as an additional source of water pollution. Discuss incidents such as the *Exxon Valdez* oil spill in Alaska in 1989, which was the worst oil spill in U.S. history. **learning style: auditory**

Answers to questions

1. **EXPLAIN:** They are polluted.
2. **DESCRIBE:** bacteria, soaps, and chemicals
3. **NAME:** farming and industry
4. **EXPLAIN:** It becomes less harmful.

15-8 What causes water pollution?

INVESTIGATE

Observing Water Pollution

HANDS-ON ACTIVITY

1. Pour $\frac{1}{2}$ cup of tap water into a one-gallon glass container.
2. Add and stir in two drops of red food coloring.
3. Add 1 cup of tap water at a time to the jar until the red food coloring disappears from view.

THINK ABOUT IT: How many cups of water did it take to make the red food coloring disappear? Is the red food coloring gone entirely just because it can no longer be seen? What does this tell you about the effects of pollution on streams and wildlife?

Objective

Describe the major sources of water pollution.

Key Term

sewage: wastes usually flushed away in water

Water Pollution Water pollution occurs when harmful substances enter water sources. Some pollutants dissolve in the water. Others float. Many lakes and rivers were once used for drinking and swimming. Today, many of these are polluted. Not even fish can live in some of these waters.

▲ **Figure 15-23** The foam in this river is from detergents.

1. **EXPLAIN:** Why are many lakes and rivers not used for drinking water today?

Sewage Wastes that usually are flushed away in water, such as biological wastes, are called **sewage.** Sewage is a source of water pollution. Bacteria and other disease-causing organisms live in sewage. Sewage also contains soaps and chemicals.

2. **DESCRIBE:** What harmful things are in sewage?

Chemical Pollutants Many different chemicals pollute the water. Fertilizers are chemicals that help plants grow. Pesticides are used to kill animal pests and weeds. Farmers use fertilizers and pesticides on their food crops.

Many of the chemicals that make up fertilizers and pesticides are poisonous to living things. They can be washed off farm fields and seep into the groundwater. This water is then carried to lakes and rivers. The chemicals in the water may harm organisms that live in the water.

FRESHWATER POLLUTANTS		
Kinds of Pollutants	**Examples**	**Sources**
Disease-causing organisms	*Giardia*, *Cryptosporidium*, bacteria	Human wastes, runoff from livestock pens
Pesticides and fertilizers	DDT, nitrates, phosphates	Runoff from farm fields, golf courses
Industrial chemicals	PCBs, carbon tetrachloride, dioxin	Factories, industrial waste, disposal sites
Metals	Lead, mercury, copper	Factories, waste disposal sites
Radioactive wastes	Uranium, carbon-14	Medical and scientific disposal sites, nuclear power plants
Petroleum products	Oil, gasoline	Road runoff, leaking underground storage tanks

▲ **Figure 15-24**

Chemical pollutants also come from industry. Some industries illegally bury their wastes in barrels, or metal drums, in the ground. If the drums rust or break apart, the wastes can leak into the groundwater.

Teaching Options

EXTENDING THE LESSON

Cleaning Up Toxic Waste
Common ways of cleaning up toxic wastes include burial, burning, and bioremediation. Burial often just hides the wastes, and burning may transform toxic wastes from a solid form into harmful gases and particles that enter the air. Bioremediation, the use of microorganisms to break down toxic wastes, is probably the most effective cleanup method. The microorganisms eliminate toxins by turning them into harmless compounds. Have students research efforts being made to clean up toxic-waste sites. Ask them to present their findings in an oral report.

RETEACH

Water Pollution Sources Bring to class images of polluted waterways or newspaper and magazine articles related to water pollution. Show the images to the class and discuss the pollution issues they raise. Read excerpts from the articles as a starting point for discussion of different sources of water pollution. **learning styles: visual, auditory**

Industries also dump chemical wastes directly into water. These wastes are harmful to fish living in the water. They are also harmful to people who eat the fish.

Mercury is a poisonous element that is sometimes found in polluted water. Many fish cannot be eaten because they contain mercury. Figure 15-24 contains examples of different types of freshwater pollutants, including chemical pollutants.

3 NAME: What are two sources of chemical pollutants?

Protecting the Water Sewage treatment plants have been built in many cities and towns. These plants change sewage into less-harmful materials.

Laws also help fight water pollution. These laws require industries to clean their wastes before dumping them into lakes and rivers. Laws have also banned the use of certain pesticides and soaps.

4 EXPLAIN: What happens to sewage at sewage treatment plants?

CHECKING CONCEPTS

1. What are pesticides used for?
2. What causes water pollution?
3. What happens when metal drums storing industrial wastes rust?
4. Where do fertilizers and pesticides that seep into groundwater end up?
5. Why is sewage potentially harmful?

THINKING CRITICALLY

6. MODEL: Draw a flowchart that shows how a pesticide could end up in a person's body.
7. EXPLAIN: How could you help stop water pollution?
8. PREDICT: What might happen if drums containing harmful chemicals were buried on a beach? Explain your answer.

DESIGNING AN EXPERIMENT

Design an experiment to solve the following problem. Include a hypothesis, variables, a procedure with materials, and a type of data to study. Also, tell how you would record your data.

PROBLEM: Sewage increases the amount of algae in water. Algae are plantlike organisms. As algae die and decay, they use up the oxygen in the water. How can you show what will happen to fish living in the water as the oxygen in the water is used up?

Integrating Life Science

TOPICS: ecology, environment

EUTROPHICATION

Runoff carries nutrients into lakes. Farming, the cutting of timber, and sewage treatment plants can also change the chemical balance in lakes. The buildup of nutrients in lakes is called eutrophication (yoo-troh-fuh-kay-shuhn). It can turn lakes into swamps and then into dry land.

▲ **Figure 15-25** A buildup of nutrients caused the green algae to grow out of control in this body of water.

In nature, plants eventually die and add organic sediment to a lake's bottom. This sediment may combine with silt from erosion. Over time, the lake may gradually be filled in. However, this natural process can be speeded up by eutrophication from certain farming practices and the treatment of sewage in sewage plants. Healthy lakes can become choked with vegetation. Eutrophication is a widespread problem in U.S. lakes today.

Thinking Critically What human practices cause eutrophication?

ONGOING ASSESSMENT

Have students compare and contrast air and water pollution. Ask: *What are important sources of air and water pollutants? How can air and water pollution be reduced or avoided?* Have students present their answers and ideas to the class.

MORE ON THE WEB

Types of Water Pollution At www.conceptsandchallenges.com, students can read more about the different types of water pollution.

• TEACHING RESOURCES • Online or Print

Teacher's Resources

Lesson Review: Chapter 15, p. 12

Enrichment: Chapter 15, p. 13

Designing an Experiment: Study Tools, p. 14

Designing an Experiment Rubric: Teaching Tools, p. 7

3 Assess

CHECKING CONCEPTS ANSWERS

1. to kill insects that harm crops
2. harmful substances that enter the water
3. Wastes can leak into groundwater.
4. in lakes and rivers
5. because it contains bacteria and other disease-causing organisms as well as harmful chemicals

THINKING CRITICALLY ANSWERS

6. MODEL: Possible answer: pesticides being applied to a crop, rain washing the pesticide into a nearby stream, a fish absorbing the pollutant from the water, a person eating the fish
7. EXPLAIN: Possible answer: by not putting harmful substances into waterways
8. PREDICT: Waves could break up the drums, releasing pollutants into the ocean.

4 Close

DESIGNING AN EXPERIMENT

Use the Designing an Experiment Rubric from the Teacher's Resources to assess students' experiments. Fill in the rubric with the additional information below. For this assignment, students should have:

- mentioned that if the level of oxygen in the water is too low, the fish will die.
- measured oxygen levels in the water.

Integrating Life Science

TOPICS: ecology, environment

EUTROPHICATION

Background Runoff is rainwater or melted snow that ends up in lakes and streams. It often carries pollutants in the soil. Nutrients in runoff can come from fertilized farmlands, feedlots where cattle or other animals are kept, and sewage. They can also come from the natural decay of organisms. These nutrients encourage the growth of algae and plants.

Discussion Discuss some ways to cut down on eutrophication. (using less fertilizer and treating sewage properly before releasing it into the environment)

Thinking Critically Answer
using fertilizer, logging, land clearing, and improper sewage disposal

Lab Activity

(pp. 368–369)

1 Prepare the Lab

Saving Water and Energy

SET-UP TIME: **LAB TIME:**

BACKGROUND: Humans tend to take water for granted. People in some parts of the country experience periodic shortages of water, however, especially in the summer. These shortages tell us that we cannot count on unlimited supplies of this resource for the future. As worldwide demand for water and other resources continues to climb, the need for conservation becomes apparent.

PURPOSE: Students will estimate water loss from a leaky faucet and monitor energy loss from leakage of hot water.

SCIENCE SKILLS: Students will **estimate** the amount of water lost from a leaky faucet, then **calculate** the potential loss of water and energy in a day, a month, and a year.

2 Guide the Lab

PROCEDURE: Discuss the procedure with students and describe how fast the faucet should drip to get results. Possible results: If 50 mL of water are collected in 10 minutes, then 50 mL × 6 = 300 mL/hr, × 24 hr/day = 7,200 mL/day, × 365 days/yr = 2,628,000 mL/yr, or 2,628 L lost per year. If the temperature of the water is 23°C, then 2,628 L/yr × 23°C × 1 Cal/°C = 60,444 Cal/yr. Provide students with a copy of the Lab Activity Report found in the Teacher's Resources to record their observations and conclusions.

EXPECTED OUTCOMES: Students will see that a significant amount of water and energy for water heating can be lost from a leaky faucet. They can infer how much water they use when they turn on the faucet.

LAB ACTIVITY
Saving Water and Energy

Materials

Safety goggles, lab aprons, watch with second hand, graduated cylinder, thermometer, paper, cup, pencil, calculator, large glass

BACKGROUND

A leaky faucet can result in a big waste of water and energy. Each drop, though small, adds up. Over time, a lot of water can be wasted down the drain. If hot water is coming from the faucet, energy is wasted along with the water. Fixing a leaky faucet can help save valuable resources.

PURPOSE

You will calculate how much water and energy a leaky faucet wastes.

PROCEDURE

1. Turn on the cold water tap just enough to create a steady drip. Adjust the tap until the rate of drip is about 10 to 20 drips per minute. Place a large glass under the drip.
2. Copy the chart shown in Figure 15-26.
3. Predict how many liters of water you think will be wasted if the faucet drips for an hour. Write down your prediction in your data chart, as shown in Figure 15-26.
4. Collect the water dripping from the faucet for 10 minutes. Use the graduated cylinder to measure the amount. In your data chart, record in milliliters (mL) the amount of water collected.
5. To find out how many liters of water are wasted in one year, carry out the calculations in the chart as shown in Figure 15-27.

▲ **STEP 1** Adjust the tap until the rate of drip is correct.

▲ **STEP 4** Measure how much water was lost by pouring it into the graduated cylinder.

Teaching Options

QUICK CONVERSIONS

Water Temperature	Metric/SI	English
Hot water	40°C	104°F
Tap water	17°C	63°F
Temperature increase	23°C	41°F

EXTENDING THE LESSON

Conserving Water in Space

Water conservation on the International Space Station is especially important to the astronauts on board. Since it is very costly to transport water from Earth to the space station, water needs are supplied by an extensive water recycling system. Water used for washing and running fuel cells is purified and reused. Water is even collected from the water vapor breathed out by humans and animals on board and from their bodily wastes. Ask students: *What can be done to reduce water consumption in space?*

6. When a hot water faucet leaks, both water and energy are wasted. Heat energy is used to heat water. A calorie is the amount of heat energy needed to raise the temperature of 1 mL of water by 1°C. A kilocalorie is the amount of heat needed to raise the temperature of 1 L (1,000 mL) of water by 1°C. The water from the hot water tap is 40°C.

7. Take the temperature of the water coming from the cold water tap. Find out how much the temperature of the cold water has been raised.

8. To estimate the amount of heat energy wasted by a leaky hot water tap, multiply that number by the number you calculated would be lost in a year. Record this number in your data table.

▲ **STEP 7** Measure the temperature of the cold water.

My Data	
My prediction	
Amount of water collected	
Volume of water per year	
Temp. of cold water	
Energy to heat 1 L of water	
Amount of energy (in kilocalories) wasted per year	

▲ **Figure 15-26** Use a copy of this table to record your observations.

How to Find Volume of Water Wasted	
To get mL per hour	__ mL × 6 = __ mL/hr
To get mL per day	__ mL/hr × 24 = __ mL/day
To get mL per year	__ mL/day × 365 = __ mL/yr
To get liters per year	__ mL/yr ÷ 1,000 = __ L/yr

▲ **Figure 15-27** Use a copy of this table to do your calculations.

CONCLUSIONS

1. CALCULATE: Burning one barrel of fuel oil produces about 23,000,000 kilocalories of heat energy. How many barrels of oil would be wasted by a leaky faucet dripping hot water for a year?
2. CALCULATE: How many 2-liter bottles of water would be wasted in a year?
3. ANALYZE: What can you do to limit the amount of water that is wasted?
4. ANALYZE: Why is it important to conserve resources such as oil and water?

3 Conclude the Lab

1. CALCULATE: Possible answer: Using the previous figures, 60,444 Cal/yr ÷ 23,000,000 Cal/barrel of oil = 0.003 barrels of oil used per year.
2. CALCULATE: Possible answer: Using the previous figures, 2,628 L ÷ 2 L/bottle = 1,314 bottles.
3. ANALYZE: Possible answers include: Fix leaky faucets. Instead of running water, fill a basin full of water for shaving, washing hands, and washing or rinsing dishes. Run dishwashers only with a full load. Use hot water only when necessary. Take showers instead of baths.
4. ANALYZE: Possible answer: so that we will have them available when we need them

Use the Laboratory Activity Rubric from the Teacher's Resources to assess students' lab activities. Fill in the rubric with the additional information below. For this activity, students should have:

- calculated how much water a leaky faucet wastes.
- calculated how much energy a leaky faucet wastes.

4 Extend the Lab

Ask students to use similar methods to measure specific uses of water, such as brushing teeth, showering, washing dishes, washing a car, or watering a lawn. Have students add these up to estimate daily, weekly, or monthly water consumption. In households that meter water use, students can compare their measurements to the totals that appear on monthly water bills.

Tips for Using Technology

Developing Student Web Pages for Lab Findings Interested students can develop a Web page that displays their lab findings about saving water and energy. Developing a basic Web page can be easy, even if students are not familiar with HTML. Web sites such as www.Geocities.com, www.Angelfire.com, and www.Tripod.Lycos.com offer free Web site hosting. They also offer tools that make creating a Web page easier. These tools allow students to create basic Web pages with little or no use of HTML.

• TEACHING RESOURCES • Online or Print

Teacher's Resources

Lab Activity Report: Study Tools, pp. 16–17

Laboratory Activity Rubric: Teaching Tools, p. 8

15-9 *(pp. 370–371)*

1 Introduce

STUDY HINT Have students draw simple sketches or diagrams of the water-purification methods listed on this page. Doing this can help students both understand and recall the methods later.

Linking Prior Knowledge Ask how many students drink bottled water. Ask why they do and if they think it is really better or safer. Have students describe what they think happens to water before it is pumped to homes or schools.

2 Teach

Demonstration Demonstrate sedimentation by adding a teaspoon of soil to a clear container of water and allowing it to settle during class. Demonstrate filtration by pouring the water with soil into a coffee filter set inside a second clear container. Pour the water carefully so that the sediment that has settled to the bottom remains in the first container. **learning style: visual**

Discussion Display a clear glass of water. Ask students where their drinking water comes from. (Most municipal water supplies come from wells, springs, streams, and reservoirs.) Ask students what materials the water in wells, springs, streams, and reservoirs may contain. (minerals, sand, dirt, twigs, bacteria, and so on) Point out that water must be purified, or cleaned, of these substances. Describe how water is purified before it ends up in a person's drinking glass. **learning style: auditory**

READING STRATEGY Have students turn the information in Figure 15-28 into a flowchart to help them understand and remember it. Have them use the Flowchart graphic organizer from the Teacher's Resources.

Answers to questions

1 **IDENTIFY:** calcium and magnesium

2 **NAME:** sedimentation, coagulation, filtration, aeration, and chlorination

3 **DESCRIBE:** distillation, freezing, and filtering

15-9 How is water purified?

Objectives

Explain why water must be purified. Identify five ways to purify water.

Key Terms

hard water: water containing a lot of minerals, especially calcium and magnesium

soft water: water containing few or no minerals

desalination: process of getting freshwater from salt water

▲ **Figure 15-29** A water treatment plant

Drinking Water Drinking water comes mostly from reservoirs, springs, or wells. A reservoir is a lake made by people. Reservoirs store large amounts of freshwater. Pipelines carry the water to homes and businesses.

Some water contains dissolved minerals such as calcium and magnesium. Water that has high levels of calcium and magnesium in it is called **hard water.** It is difficult for soap to form suds in hard water. Water with few or no minerals in it is called **soft water.**

1 IDENTIFY: What are two minerals in hard water?

Purifying Water Water for drinking and bathing must be purified, or cleaned. Water is usually purified in water treatment plants. Water can be purified in the following ways.

- **Sedimentation** (sehd-uh-muhn-TAY-shuhn): The water is allowed to stand for long periods of time. Heavy particles, such as sand and dirt, settle to the bottom and are removed.
- **Coagulation** (koh-ag-yoo-LAY-shuhn): Alum and other chemicals that cause particles to clump together are added to the water. The clumps of particles settle to the bottom. The particles are then removed from the water.
- **Filtration** (fihl-TRAY-shuhn): Water is passed through a filter to remove small particles.
- **Aeration** (air-AY-shuhn): Water is sprayed into the air. Oxygen from the air dissolves in the water. The oxygen kills some of the harmful microorganisms in the water.
- **Chlorination** (klawr-uh-NAY-shuhn): Chlorine is added to the water. This chemical kills any harmful microorganisms that are in the water.

2 NAME: What are five ways to purify drinking water?

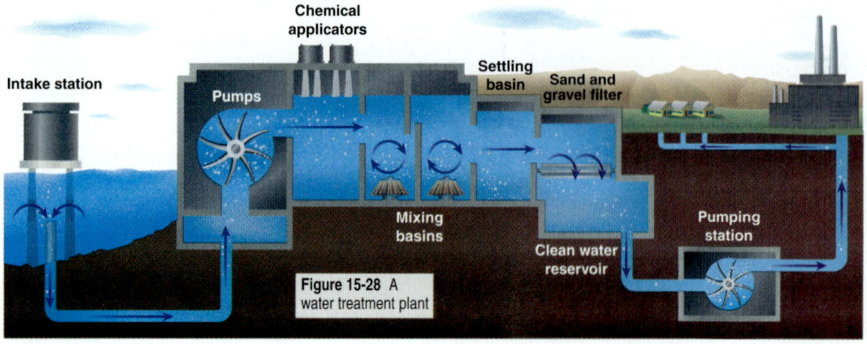

Figure 15-28 A water treatment plant

Teaching Options

EXTENDING THE LESSON

Observing Water Hardness

Demonstrate the difference between hard and soft water. Add a few drops of liquid soap to two test tubes, one containing distilled water and one containing hard water. (You can make hard water by dissolving a small amount of calcium bicarbonate or table salt in water.) Shake both test tubes well. Students will observe the lack of suds in the test tube containing hard water. Have students write their observations. Invite interested students to look up more information on the chemistry of hard water and why it does not produce many suds.

RETEACH

Observing Coagulation

Demonstrate coagulation by adding some soil to a 100-mL graduated cylinder containing 80 mL of water. Agitate the cylinder to mix the soil. Add 15 mL of saturated aluminum solution and 5 mL of concentrated ammonium hydroxide.
learning style: visual

ONGOING ASSESSMENT

Have students use their own words to define the terms *sedimentation*, *coagulation*, *filtration*, *aeration*, and *chlorination*.

Desalination The process of getting freshwater from salt water is called **desalination.** One method of desalination is to distill the salt water. The water is boiled until it evaporates. This leaves the salt behind. The water vapor is then condensed to produce liquid freshwater. Another method involves freezing the water. This also leaves the salt behind. A third method is to pump ocean water at high pressure through a fine filter. The filter separates out much of the salt in the water. Water still containing too much salt is returned to the ocean.

Desalination is a very costly operation. It uses large amounts of energy and a lot of equipment. However, some nations, such as those in the Middle East, do it anyway. These nations have very limited freshwater resources.

3 DESCRIBE: What are three ways to desalinate water?

CHECKING CONCEPTS

1. Hard water contains high levels of calcium and __________.
2. A human-made lake that stores freshwater is called a __________.
3. Desalination means to __________ salt from water.

THINKING CRITICALLY

4. ANALYZE: Explain the water purification method described in each statement. **a.** Chemicals that cause particles to clump together are added to water. **b.** Water is sprayed into the air. **c.** Water stands for long periods of time.

INTERPRETING VISUALS

Use Figure 15-28 to complete the following questions.

5. IDENTIFY: What materials are used for filtration?
6. ANALYZE: Where does the water in reservoirs come from?
7. INFER: What substance is added to water to kill microorganisms?

HEALTH AND SAFETY TIP

Many "safe-looking" products found around the home are really dangerous chemicals. They can contaminate water supplies. Examples of such products are bleaches, detergents, and other cleaning agents. Read the labels on products you find around your home and see how to safely dispose of them.

People in Science

WATER PURIFICATION TECHNICIAN

A water purification technician works at a water treatment plant. He or she controls and maintains the equipment at the plant and also takes water samples during the purification process. The water samples are tested for impurities. The technician also checks the amount of chemicals in the water. It is part of the technician's job to make sure that the proper amounts of chlorine and other chemicals are being used.

To become a water purification technician, you must have a high school diploma. Water purification technicians may learn their trade through on-the-job training. Some water purification technicians take two-year programs in water purification technology.

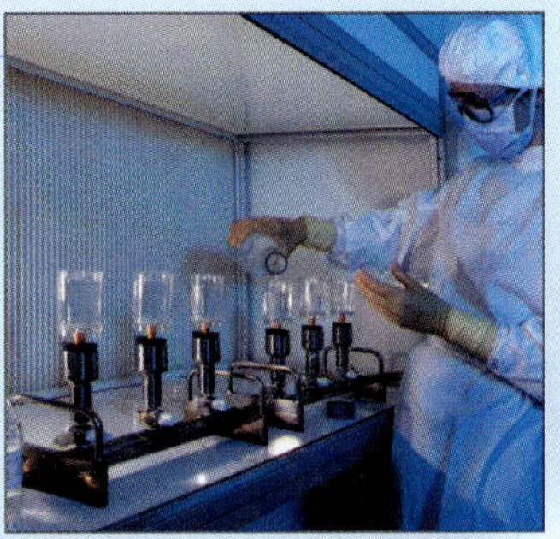

▲ **Figure 15-30** A water purification technician tests water samples.

Thinking Critically What is the purpose of a water treatment plant?

3 Assess

CHECKING CONCEPTS ANSWERS

1. magnesium
2. reservoir
3. remove

THINKING CRITICALLY ANSWER

4. ANALYZE: **a.** Coagulation causes particles to clump together and settle to the bottom. **b.** Aeration uses the oxygen in air to destroy some harmful microorganisms in the water. **c.** Sedimentation allows large particles to settle to the bottom.

INTERPRETING VISUALS ANSWERS

5. IDENTIFY: sand and gravel
6. ANALYZE: a large water source such as a lake, which is fed by rain (indicated by clouds in the illustration)
7. INFER: chlorine

4 Close

People in Science

WATER PURIFICATION TECHNICIAN

Real-Life Connection Bring a pool-tester kit to class. Show students how it is used to check the level of chemicals in a pool. Relate the demonstration to the work of a water purification technician.

Discussion Ask students whether they would consider becoming a water purification technician. Have them explain why or why not. Point out that water purification plants clean up the water that goes into the water supply for people to drink. These technicians do not treat wastewater at a sewage treatment plant. **learning style: auditory**

Thinking Critically Answer
to purify water from wells, springs, streams, and reservoirs that supply drinking water

MORE ON THE WEB

Hard Water Primer Have students use the Internet to learn more about hard water. Then, have students share their findings with the class.

• TEACHING RESOURCES • Online or Print

Teacher's Resources

Lesson Review: Chapter 15, p. 14

Flowchart: Study Tools, p. 6

Laboratory Manual

Lab. Challenge: How can hard water be softened? pp. 77–80

15-10 (pp. 372–373)

1 Introduce

STUDY HINT Have students write the subhead names on a sheet of paper and put a summarizing sentence under each. Then, have students underline the important terms in each sentence and write definitions for them.

Linking Prior Knowledge Ask students to recall what they learned about leaching in soils (Lesson 7-5). Ask what they think would happen to oil or other chemicals spilled on bare soil.

2 Teach

Demonstration Calculate how much garbage students generate in 1 week. On a Monday morning, weigh the empty garbage can in the classroom. Record the weight. For the next 5 days, weigh the garbage can and its contents at the end of each day. Subtract the weight of the garbage can from the total weight to get the weight of the garbage. At the end of the week, add up the total amount of garbage produced by students. **learning style: tactile**

Discussion Ask students to describe the different items they throw away every day. Ask what they think happens to these items. (They are probably trucked to landfills.)

Explain the difference between biodegradable and nonbiodegradable wastes. Write *Biodegradable* and *Nonbiodegradable* on the board as the heads for two lists. Ask students to name commonly discarded items and identify which list they belong to. Write these on the board. Discuss the differences in handling these two types of waste. Ask for suggestions about how to encourage the recycling or reuse of nonbiodegradable wastes. **learning style: auditory**

Answers to questions

1. **IDENTIFY:** litter and garbage
2. **DESCRIBE:** They are buried in drums in the ground.
3. **ANALYZE:** Possible answers include recycling more materials, burning some for energy, and using some as fertilizer.

15-10 What causes land pollution?

Objective

Describe the causes of land pollution.

Key Terms

litter: materials that are carelessly thrown away

biodegradable: material that breaks down easily

Litter and Garbage Have you ever seen cans, bottles, papers, and plastic materials that were thrown on the ground? These carelessly thrown-away materials are called **litter.** Litter is one of the causes of land pollution. Litter harms the land and destroys the beauty of many areas.

Garbage, mainly paper wastes, also causes land pollution. Each year people create billions of tons of garbage. The garbage builds up at dumps or is put in landfills. There, the garbage is buried between layers of soil. Many materials in garbage are **biodegradable.** This means they break down easily. Some, though, take many years to break down. Other materials contain harmful substances that may seep into the soil.

1 **IDENTIFY:** What are two causes of land pollution?

Industrial Wastes Chemical wastes from industry pollute the water. They also pollute the land.

Some industrial wastes are buried in drums in the ground. Many of these toxic wastes contain harmful metals such as mercury and lead. In many places, these wastes are leaking from the drums into the ground. They then pollute the land and can harm living things.

▲ **Figure 15-31** Toxic waste site in Houston, Texas

2 **DESCRIBE:** How are some industrial wastes disposed of?

Protecting the Land Getting rid of garbage and chemical wastes is a growing problem. Cities have taken some steps to reduce land pollution. Garbage is collected regularly. Streets are cleaned of litter. However, much more needs to be done.

Go Online **active art**
For: Sanitary Landfill activity
Visit: PHSchool.com
Web Code: cep-5042

◀ **Figure 15-32** Overflowing trash cans ruin the landscape of city streets.

372

Teaching Options

COOPERATIVE LEARNING

Observing Litter Divide the class into small teams. Assign each team an area of the school grounds or a local park. Have them observe and list the types of litter they find in their area on a given day. Ask them to create a graph or chart showing the relative amounts of cans, bottles, etc., and present them to the class. **learning style: visual**

Go Online **active art**
For: Sanitary Landfill activity
Visit: PHSchool.com
Web Code: cep-5042

Students can interact with the art of a sanitary landfill online.

EXTENDING THE LESSON

Modeling a Landfill More than half of all garbage produced in the United States ends up in large solid-waste disposal pits called landfills. Landfills are lined with thick plastic, clay, or other materials that prevent the garbage from contaminating the groundwater and soil. Each day's worth of garbage is covered with a layer of soil. Landfills are also equipped with drain pipes. These help prevent rainwater from washing harmful materials into the surrounding land. Invite students to find out more about their local landfill and make a model of it. Have them display their models in class.

One solution to the problem of land pollution is to make use of certain wastes. Some wastes can be used as fertilizers. Others can be burned as energy sources. Still other wastes, such as plastics, glass, paper, bottles, and cans, can be recycled.

The cleanup of chemical wastes that leak into the soil is a more difficult problem. Cleaning up these wastes is expensive and takes years.

3 ANALYZE: What are some solutions to the problem of land pollution?

CHECKING CONCEPTS

1. Materials that are carelessly thrown away are called ___________.
2. Some wastes can be burned to produce ___________.
3. Garbage put between layers of land creates a ___________.
4. Cleaning up chemical wastes from the ___________ is expensive and takes a lot of time.

THINKING CRITICALLY

5. ANALYZE: What are some steps you can take to reduce the amount of litter in your neighborhood?
6. APPLY: How can industrial wastes pollute both land and water?

DESIGNING AN EXPERIMENT

Design an experiment to solve the following problem. Include a hypothesis, variables, a procedure with materials, and a type of data to study. Also, include a way to record your data.

PROBLEM: Laws have been passed in many locations around the country that place a small deposit, such as a dime, on beverage containers. If the containers are taken back to a store, deposits are returned. Besides promoting recycling, deposits may reduce litter. Design an experiment that would prove whether deposits do reduce litter. Put your findings in a report. Include a graph or chart.

Real-Life Science

GETTING RID OF GARBAGE

The United States produces over 250 million tons of garbage each day. For years, solid waste was put into landfills. Most of that waste was made up of biodegradable compounds. As it decayed, more room became available in the landfill. However, in the last 50 years, new, longer-lasting materials have been put in landfills. Our landfills are becoming full.

What can we do with our garbage? Some common disposal methods are incineration, or burning, and ocean dumping. Incineration saves landfill space and can be used to produce heat and electricity. However, it also adds pollutants to the air. Special filters must be used to trap these pollutants. The dumping of sewage and other wastes in the ocean threatens ocean environments.

▲ **Figure 15-33** A landfill is one way to get rid of waste. However, our landfills are rapidly filling up.

Reusing and recycling can reduce our waste problem. So can composting. In composting, decayed organic waste is used as a natural fertilizer.

Thinking Critically Only a small percentage of our solid waste that can be recycled is recycled. Why do you think this is so?

3 Assess

CHECKING CONCEPTS ANSWERS

1. litter
2. energy
3. landfill
4. soil

THINKING CRITICALLY ANSWERS

5. ANALYZE: Possible answers: Recycle paper, glass, and metals. Use reusable items, such as cloth grocery bags or sturdy plastic picnic plates instead of disposable items.
6. APPLY: Buried wastes can leak into soil and leach into groundwater.

4 Close

DESIGNING AN EXPERIMENT

Use the Designing an Experiment Rubric from the Teacher's Resources to assess students' experiments. Fill in the rubric with the additional information below. For this assignment, students should have:

- described how they would compare the data in a chart or graph.
- included, in the procedure, a way to monitor the amount of litter discarded in a specific area over a given amount of time.

Real-Life Science

GETTING RID OF GARBAGE

Background Recycling has several stumbling blocks. The first is getting people to collect their reusable materials so that they can be processed. A second problem is finding creative ways to use the recycled material. A third vital component is developing a market for recycled goods and materials. The products have to represent a sensible choice for many consumers and manufacturers, not just those who are dedicated to environmental ideals.

Thinking Critically Answer Possible answers include: People are too busy to recycle their garbage. Some people have no place to put a compost pile for kitchen wastes. Community governments have not offered enough incentives or opportunities to encourage people to recycle.

ESL/ELL STRATEGY

Dividing Words Divide the word *biodegradable* into word parts for students who are learning English. (*Bio* means "living organism," and *degrade* means "to break down.")

• TEACHING RESOURCES • Online or Print

Teacher's Resources

Lesson Review: Chapter 15, p. 15

Designing an Experiment: Study Tools, p. 14

Designing an Experiment Rubric: Teaching Tools, p. 7

MORE ON THE WEB

Landfills Have students visit www.conceptsandchallenges.com to learn more about landfills. Have students explain why the United States does *not* have a growing reliance on landfills.

15-11 (pp. 374–375)

1 Introduce

Linking Prior Knowledge Ask students how they think pollutants might travel to another city, state, or country.

2 Teach

Demonstration Use pH indicator strips to test the pH of distilled water, tap water, and vinegar. Show students the color of each test strip and which pH value it matches on the comparison chart. Use a pH scale to chart the results on the board, showing that distilled water is neutral, tap water is very close to neutral, and vinegar is acidic. Place small pieces of chalk in the container with the vinegar and let students observe how the acidic solution breaks down the chalk. Point out that acid rain is sometimes as acidic as vinegar and can wear away stone and other substances. **learning styles: visual, auditory**

Discussion Show students a world map or a large map of North America. Point to the political borders and explain that borders do nothing to stop wind or flowing water or atmospheric air currents. Tell students that air pollution created in factories in the midwestern United States travels north and east into other regions and Canada. Discuss the causes and effects of acid rain, radiation, and ozone depletion. **learning styles: auditory, visual**

READING STRATEGY Have students read the summary questions before they read the main text and try to answer them. Then, have students reread these questions later and try to answer them again.

Answers to questions

1. **ANALYZE:** Pollution created in one country can reach other countries.
2. **HYPOTHESIZE:** It can damage forests and food crops, poison fish, and contaminate drinking water.
3. **IDENTIFY:** by winds and flowing water
4. **EXPLAIN:** It protects people from the damaging ultraviolet light that comes from the Sun.

15-11 What are some global effects of pollution?

Objective
Explain how pollution spreads across borders.

A World with No Borders The lines that divide countries on maps do not exist in reality. From space, you see just one world, with large areas of land surrounded by even larger areas of ocean.

Pollution does not stay in just one place. Fumes from automobiles, homes, and factories are carried by the wind around the globe. Pollutants dumped into rivers gradually make their way into the oceans. Ocean currents carry them around the world. Pollution is not just one country's problem. It is a global problem.

1 **ANALYZE:** Why is pollution a global problem?

Acid Rain All rainwater is slightly acidic. Certain gases, such as sulfuric oxide and nitric oxide, can mix with rainwater. This makes the drops more acidic. These gases are produced by power plants, industries, and motor vehicles. Global winds spread the gases across borders. There they mix with rainwater. Many countries other than the ones producing the gases are affected by acid rain.

Acid rain damages forests. It also collects in lakes and rivers, poisoning fish and water plants. It breaks down rock and mud and releases harmful chemicals into the water. About 10,000 lakes in Sweden contain so much mercury that the government has discouraged people from eating fish caught there. Some bodies of water contain so much dissolved copper that the hair of persons drinking the water turns green.

2 **HYPOTHESIZE:** How does acid rain affect the lives of people?

Radiation In 1986, a poorly designed nuclear reactor in Chernobyl, Ukraine, exploded. The explosion and fires that followed released huge amounts of radiation into the atmosphere. It was the world's worst nuclear energy accident.

Thirty people working near the reactor died shortly after the accident. Ten more people died later from thyroid cancer brought on by the accident. Winds and water carried radioactive dust around the world. The worst affected areas were near the plant. However, Poland, Hungary, and Sweden were also affected. The dust destroyed rich farmlands, contaminated cow milk, and exposed up to 600,000 people to radiation.

▲ **Figure 15-34** Chernobyl in the Ukraine was the site of the world's worst nuclear reactor accident to date.

3 **IDENTIFY:** How is radioactive dust spread around the world?

Ozone Holes Ozone is a special form of oxygen. Ozone is created from oxygen by lightning and ultraviolet (UV) light from the Sun. Most of this ozone ends up in a thin layer about 35 km above Earth's surface. Ozone is very important to life on Earth because it absorbs most of the UV light coming from the Sun. Large amounts of UV light cause eye damage and skin cancer in humans.

Some industrial chemicals, such as CFCs, release gases that rise into the upper atmosphere. These gases destroy the ozone layer, creating holes in it. In recent years, satellites have found lower than normal levels of ozone above Earth's poles, particularly in the South Pole region.

Teaching Options

INTEGRATING CHEMISTRY

How Ozone Forms Molecules of oxygen gas are made up of two oxygen atoms bonded together. Ozone gas molecules are made up of three oxygen atoms bonded together. When ultraviolet radiation from the Sun strikes an ozone molecule high in the atmosphere, one of the oxygen atoms splits away, turning the ozone molecule into an oxygen molecule. Ozone's ability to absorb ultraviolet radiation prevents its damaging effects from reaching Earth's surface. Have pairs or small groups of students work together to draw and cut out illustrations representing a number of oxygen molecules, the Sun, and ultraviolet radiation. Have students manipulate the illustrations to show ultraviolet radiation from the Sun splitting ozone molecules into oxygen molecules. **learning styles: visual, tactile**

QUICK CONVERSIONS

	Metric/SI	English
Height above Earth	20 km	12 mi
Smoke dispersal	60 km	37 mi

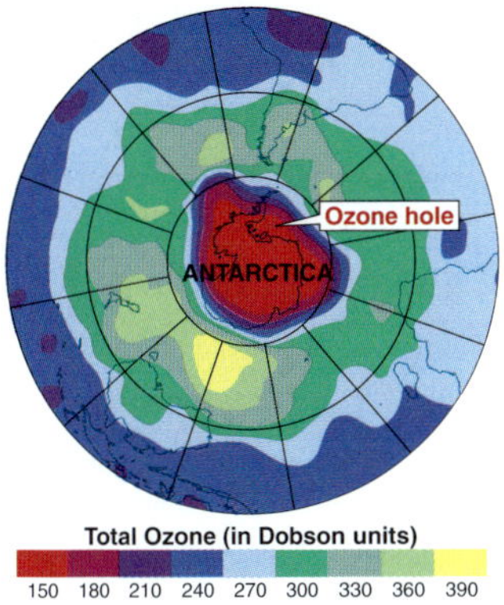

▲ **Figure 15-35** An enhanced satellite image of the ozone hole over the South Pole. The thinnest part is shown in red.

More ultraviolet light is reaching Earth's surface than in the past. Many industries have begun using ozone-safe chemicals. However, the damage already done may take decades to repair.

4 **EXPLAIN:** Why is the chemical known as ozone important to people?

CHECKING CONCEPTS

1. Why are various kinds of pollution considered to be a global problem?
2. How does pollution spread around the world?
3. How is acid rain spread?

THINKING CRITICALLY

4. **INFER:** How will population increases affect global pollution?
5. **APPLY:** What can individuals do to reduce global pollution?

INTERPRETING VISUALS

Use Figure 15-35 to answer the following questions.

6. What part of Earth are you looking at?
7. Where is the ozone hole the thinnest?
8. What does the yellow represent?
9. What does this map reveal to us about ozone in the atmosphere?

Science and Technology

USING SPACE TO TRACK POLLUTION

Looking at Earth from space, astronauts can see thousands of square kilometers at a time. This gives them a special view of Earth's surface.

From space, astronauts can see smoke from forest fires set in South America to clear land for farming. The smoke stretches out for thousands of kilometers across the Atlantic Ocean. During the Desert Storm conflict in the Middle East in 1991, astronauts saw giant black, smoky smudges from the burning of Kuwaiti oil fields. The smoke spread out for 60 km southward over the Persian Gulf.

▲ **Figure 15-36** Smoke from the burning Kuwaiti oil fields spread out for many kilometers.

Astronauts are not the only ones to use space to track pollution. Orbiting satellites use cameras and computers to map changes in world temperatures. They also track ozone levels over the poles. Data collected by satellites and astronauts are made available to scientists who study pollution. Noting changes in the world environment year after year helps scientists predict the future. Currently, they are predicting a worsening trend in global pollution.

Thinking Critically Why is space a good place to study global pollution?

COOPERATIVE LEARNING

Seeing the Whole Earth *Landsat* images use different colors to show features of Earth's surface, such as natural vegetation and cropland, water, snow or ice cover, bare ground, and urban areas. On the Internet, *Landsat* images of the same area from different years are available. Invite students to obtain such images of nearby cities or of other parts of the world. Have them compare past and present images and analyze the differences. Then, have them propose possible reasons for the changes and any impact on the environment.

• TEACHING RESOURCES • Online or Print

Teacher's Resources
Lesson Review: Chapter 15, p. 16

3 *Assess*

CHECKING CONCEPTS ANSWERS

1. Pollution is carried around the world by wind, water, and ocean currents.
2. by wind and water
3. by global winds

THINKING CRITICALLY ANSWERS

4. **INFER:** Possible answer: More people will burn more fuel, drive more cars, build more factories, and take part in other activities that can cause pollution.
5. **APPLY:** Possible answer: Find ways to use less energy and replace CFCs with chemicals that do not damage the ozone layer.

INTERPRETING VISUALS ANSWERS

6. Antarctica
7. the red area
8. the thickest part of the ozone layer
9. Possible answer: Certain parts of the ozone layer are more easily damaged.

4 *Close*

Science and Technology

USING SPACE TO TRACK POLLUTION

Real-Life Connection NASA's Earth Observing System uses satellites for long-term observation of the atmosphere. The *Terra* satellite circles Earth from pole to pole, observing carbon monoxide and other pollutants. Major sources of carbon monoxide and other greenhouse gases are fires, the burning of fossil fuels, and some industries. Carbon monoxide in the atmosphere is largely caused by human activities. The *Terra* satellite tracks the flow of pollution throughout Earth's atmosphere.

Discussion Display a map of the world. Point to the Amazon River basin in South America and tell students that rain forests in this region are being burned to create farmland. Show students how the smoke could spread. Next, point to the Middle Eastern countries of Iraq and Kuwait. Mention how fires from burning oil fields spread south along the Persian Gulf during the Desert Storm conflict in 1991. **learning styles: visual, auditory**

Thinking Critically Answer
Space gives a good view of many parts of Earth at the same time.

Challenges *(pp. 376–378)*

Chapter Summary

Review Before students begin the Challenges, go over the summary with them. Have them ask any questions they have about it or raise any issues of concern.

STUDY HINT Divide students into pairs. Have students quiz each other on the Key Terms for the chapter. Begin by having one student read the definition. The second student should identify the Key Term from memory. Then, have one student read the Key Term and the other provide the definition from memory.

Key Term Challenges

MATCHING

1. environment (15-1)
2. smog (15-7)
3. sewage (15-8)
4. hard water (15-9)
5. geothermal energy (15-4)
6. wildlife (15-6)
7. hydroelectric power (15-3)
8. nuclear energy (15-3)
9. pollution (15-1)
10. conservation (15-2)

FILL IN

11. acid rain (15-7)
12. litter (15-10)
13. natural resources (15-2)
14. recycling (15-5)
15. soft water (15-9)

Chapter 15 Challenges

Chapter Summary

Lessons 15-1 and 15-2

- The **environment** is everything that surrounds a living thing. Oxygen, nitrogen, and carbon dioxide cycle through the environment. **Pollution** harms the environment.

Lessons 15-3 and 15-4

- The main sources of energy are oil, coal, and natural gas. Fossil fuels are nonrenewable.
- **Hydroelectric power** is produced from quickly flowing water.
- The splitting and combining of atoms produces **nuclear energy.** Nuclear energy has serious drawbacks.
- **Solar energy** can heat homes and provide electricity.
- **Solar cells** convert solar energy to electricity.
- **Geothermal energy** is energy from heat within Earth. Tidal energy comes from tides. Wind, too, is used to produce electricity.

Lessons 15-5 and 15-6

- **Natural resources** are materials from Earth used by living things. Some can be reused or replaced. Others cannot. **Recycling** allows resources to be reused.
- Many kinds of **wildlife** are becoming extinct due to loss of living space.

Lesson 15-7

- Burning fossil fuels causes air pollution and increases the amount of carbon dioxide in the air. Dust and soot are air **pollutants. Smog** is a mixture of smoke, fog, and chemicals.
- **Acid rain** harms both living and nonliving things.

Lessons 15-8, 15-9, and 15-10

- **Sewage** and industrial waste produce water pollution.
- Drinking water comes from reservoirs, wells, or springs. Water contains dissolved minerals.
- **Litter,** garbage, and chemicals from farms and industry pollute land. Cleaning up chemical wastes from land is very difficult.

Lesson 15-11

- Pollution spreads around the world by wind and ocean currents. Acid rain, radiation, and ozone holes are some global pollution issues.

Key Term Challenges

acid rain (p. 362)
biodegradable (p. 372)
conservation (p. 352)
desalination (p. 370)
environment (p. 350)
geothermal energy (p. 356)
hard water (p. 370)
hydroelectric power (p. 354)
litter (p. 372)
natural resource (p. 352)
nuclear energy (p. 354)
pollutant (p. 350)
pollution (p. 350)
recycling (p. 358)
sewage (p. 366)
smog (p. 362)
soft water (p. 370)
solar cell (p. 356)
solar energy (p. 356)
wildlife (p. 360)

MATCHING Write the Key Term from above that best matches each description.

1. everything that surrounds a living thing
2. mixture of smoke, fog, and chemicals
3. waste that is usually flushed away in water
4. water that contains many minerals
5. heat energy from inside Earth
6. plants and animals living in nature
7. electricity produced from moving water
8. energy produced by splitting atoms
9. anything that harms the environment
10. wise use of natural resources

FILL IN Write the Key Term from above that best completes each statement.

11. Gases in the air mix with water and fall to Earth as __________.
12. Materials, such as paper and cups, that are left on the ground are called __________.
13. Materials that come from Earth and are used by living things are called __________.
14. The use of natural resources over and over again is called __________.
15. Water with few or no minerals is called __________.

376

Teaching Options

PREPARING STUDENTS FOR STANDARDIZED TESTS

Reading Strategy: Remind students to read true/false questions two or three times before answering. In some types of true/false questions, guessing will not be much help.

Writing Strategy: Explain to students that, when writing complete sentences, each sentence must express a complete thought. Give them examples of sentences that do and do not meet this guideline.

Interpreting Visuals Strategy: Instruct students to carefully compare information on a graph or chart with the information a question is asking for. They may need to look at the numbers and callouts on the graph more than once.

ESL/ELL STRATEGY

Classifying to Remember Have students group as many of the Key Terms as they can, to reinforce their meanings. For example, they can put *nuclear energy, geothermal energy, hydroelectric energy,* and *solar energy* into one group called *energy sources*. They can put *litter, acid rain, sewage,* and *smog* into another group called *pollutants* (another Key Term).

Content Challenges TEST PREP

MULTIPLE CHOICE Write the letter of the term or phrase that best completes each sentence.

1. Oil, coal, and natural gas are three different kinds of
   a. fossil fuels.
   b. water pollutants.
   c. acid rain.
   d. nuclear fuels.
2. Soap, chemicals, and bacteria are commonly part of
   a. acid rain.
   b. litter.
   c. garbage.
   d. sewage.
3. Sedimentation and coagulation are two kinds of
   a. water purification methods.
   b. farming methods.
   c. energy resources.
   d. air pollutants.
4. Pesticides and fertilizers are two kinds of
   a. sewage.
   b. natural resources.
   c. chemical pollutants.
   d. farming methods.
5. A species of plant that is no longer found alive is
   a. threatened.
   b. extinct.
   c. endangered.
   d. polluted.
6. Devices that convert sunlight into electricity are called
   a. nuclear reactors.
   b. solar cells.
   c. fossil fuels.
   d. geothermal panels.
7. The major threat to wildlife is
   a. air pollution.
   b. water pollution.
   c. loss of living space.
   d. overhunting.
8. Minerals and fossil fuels are examples of
   a. air pollutants.
   b. nonrenewable resources.
   c. renewable resources.
   d. land pollutants.
9. Electricity produced from heat inside Earth is
   a. geothermal energy.
   b. solar energy.
   c. nuclear energy.
   d. tidal energy.

TRUE/FALSE Write *true* if the statement is true. If the statement is false, change the underlined term to make the statement true.

10. The burning of wood is the major cause of air pollution.
11. Acid rain can cause brick to weather.
12. More carbon dioxide in the air may drive temperatures down.
13. Soap does not form suds easily in hard water.
14. Industrial wastes cause both water and land pollution.
15. Aluminum cans and newspapers are replacement materials.
16. Contour farming prevents water from eroding soil.
17. Farmers plant rows of trees on their land to prevent water from carrying away soil.
18. Uranium is a common fuel for nuclear energy.
19. Geothermal energy comes from the rising and falling of tides.
20. A species that is endangered is gone forever.

Content Challenges

MULTIPLE CHOICE

1. a (15-2)
2. d (15-8)
3. a (15-9)
4. c (15-8)
5. b (15-6)
6. b (15-4)
7. c (15-6)
8. b (15-2)
9. a (15-4)

TRUE/FALSE

10. false; fossil fuels (15-7)
11. true (15-7)
12. false; up (15-7)
13. true (15-9)
14. true (15-10)
15. false; recyclable (15-5)
16. true (15-5)
17. false; wind (15-5)
18. true (15-3)
19. false; Tidal (15-4)
20. false; extinct (15-6)

ALTERNATIVE ASSESSMENT

Connecting Ideas Have students write a few paragraphs explaining how the importance of natural resources to human life relates to the importance of finding ways to conserve and not pollute natural resources.

PORTFOLIO ASSESSMENT

Making Student Portfolios Students can demonstrate their comprehension of the concepts in this chapter by making a portfolio. The Chapter Self-Check, the Big Idea Planner, and the Weekly Journal are some of the reproducibles in the Teacher's Resources that they can include in their portfolios. Students may wish to include individual contributions to group work in their portfolios. You can use the Portfolio Assessment Rubric also found in the Teacher's Resources to assess students' portfolios.

Concept Challenges

WRITTEN RESPONSE

1. **EXPLAIN:** If wastes were burned instead of fossil fuels, less waste would be sent to landfills and fewer fossil fuels would be burned. (15-10)
2. **ANALYZE:** It reduces the amount of living space for wildlife and may pollute the land. (15-6)
3. **ANALYZE:** Polluted water may kill fish or make them unfit to eat. (15-8)
4. **INFER:** It is in less danger because gravity causes more water erosion on a slope. (15-5)
5. **ANALYZE:** Wind energy is effective only in areas that experience high winds on a regular basis. (15-4)
6. **INFER:** The water cycle provides us with clean drinking water. (15-9)
7. **LIST:** Accept any three of the following: oil, coal, natural gas, and minerals. (15-2)
8. **INFER:** Tidal energy would be good because coastlines are near oceans. (15-4)
9. **HYPOTHESIZE:** It reduces the amount of available food and living space. (15-6)
10. **ANALYZE:** Possible answers include air pollution from factories that can be spread by wind and water pollution from industries or farmlands that can be spread by flowing water or ocean currents. (15-11)

INTERPRETING VISUALS (15-3)

11. oil
12. 84%
13. 16%
14. natural gas
15. 26%

Chapter Project Wrap-Up

CONSERVATION OF RESOURCES DEBATE

Each group's written research material may be photocopied and distributed to the rest of the class. Alternatively, the documents could be displayed in the classroom.

You can use the Individual Activity Rubric from the Teacher's Resources to assess students' projects. Fill in the rubric with the additional information below. For this project, students should have:

- prepared a research report listing the facts and summarizing the arguments.
- participated in the debate.

Concept Challenges TEST PREP

WRITTEN RESPONSE Answer each of the following questions in complete sentences.

1. **EXPLAIN:** How can burning wastes both reduce land pollution and conserve fossil fuels?
2. **ANALYZE:** How does the growth of cities affect wildlife?
3. **ANALYZE:** How does water pollution affect the fishing industry?
4. **INFER:** Why is flat land in less danger of erosion by water than hilly land?
5. **ANALYZE:** Would wind energy be a good energy source in your area? Explain.
6. **INFER:** What cycle provides us with fresh supplies of drinking water?
7. **LIST:** What are three natural resources that are not renewable?
8. **INFER:** What alternative energy source might be good for coastline cities? Why?
9. **HYPOTHESIZE:** How does the destruction of a habitat affect a wildlife population living there?
10. **ANALYZE:** What forms of pollution other than acid rain, radiation, and ozone depletion could spread around the globe? How would they be spread?

INTERPRETING VISUALS Use Figure 15-37 to answer each of the following questions.

11. What energy resource accounts for the most energy use?
12. What percentage of the world's energy use do fossil fuels make up?
13. What percentage of the world's energy use do other fuels make up?
14. Which type of fuel accounts for more energy use, all nonfossil fuels combined or natural gas?
15. About what percentage of fossil fuel use does coal account for?

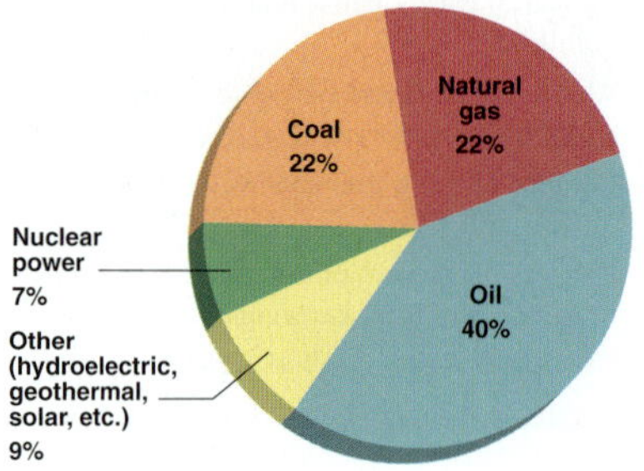

▲ **Figure 15-37** Energy resources

378

Teaching Options

• TEACHING RESOURCES •
Online or Print

Teacher's Resources

Key Term Review: Chapter 15, p. 17

Chapter 15 Test: pp. 18–20

Individual Activity Rubric: Teaching Tools, p. 5

Chapter Self-Check: Teaching Tools, p. 11

Big Idea Planner: Study Tools, pp. 11–12

Weekly Journal: Study Tools, p. 19

Portfolio Assessment Rubric: Teaching Tools, p. 10

Chapter 16 The Exploration of Space

PLANNING GUIDE

◆ **TEACHING THE CHAPTER** This chapter should take approximately 10–13 days to complete instruction and assessment.

	Skills and Features	Projects and Activities	Achieve Science Literacy	Meet Individual Needs
Chapter 16 Opener p. T379	RESEARCH	• Chapter Project		
Lesson 16-1 **What is astronomy?** pp. T380–T381 Standard: D3a	EXPLAIN, DESCRIBE, INFER • Investigate • Building Reading Skills • How Do They Know That?	• Extending the Lesson • More on the Web	• Study Hint	Teacher's Resources* Enrichment Activity
Lesson 16-2 **How is space explored?** pp. T382–T383 Standard: D3a	DEFINE, HYPOTHESIZE, CONCLUDE • Web InfoSearch • Real-Life Science	• Cooperative Learning • Extending the Lesson	• Writing Hint	• ESL/ELL Strategy • Reteach
Lesson 16-3 **How does a refracting telescope work?** pp. T384–T385	COMPARE, HYPOTHESIZE, OBSERVE • Building Science Skills • Hands-On Activity	• More on the Web • Integrating Physical Science	• Writing Hint • Ongoing Assessment	Teacher's Resources* Enrichment Activity Refracting Telescope
Lesson 16-4 **How does a reflecting telescope work?** pp. T386–T387	CONTRAST, ANALYZE, RESEARCH • Investigate • Building Science Skills • Integrating Environmental Science	• Cooperative Learning • More on the Web	• Study Hint • Ongoing Assessment	Teacher's Resources* Enrichment Activity Reflecting Telescope
Lesson 16-5 **What is a radio telescope?** pp. T388–T389	EXPLAIN, COMPARE, CONTRAST • Web InfoSearch • Science and Technology	• Extending the Lesson • Integrating Life Science	• Study Hint	• ESL/ELL Strategy
Big Idea **How do telescopes detect electromagnetic waves in space?** pp. T390–T391	RESEARCH, COMMUNICATE, ANALYZE • Science Log: Writing Activity	• Big Idea • Integrating Physical Science • Close Activity	• Tips for Using Technology	Teacher's Resources* Big Idea Planner

*www.conceptsandchallenges.com, password: ccearth09

Planning Guide continues on next page.

Chapter 16 The Exploration of Space *(Planning Guide continued)*

	Skills and Features	Projects and Activities	Achieve Science Literacy	Meet Individual Needs
Lesson 16-6 **How do astronomers measure distance?** pp. T392–T393 Standard: D3a	OBSERVE, CALCULATE, SEQUENCE • Building Math Skills • Hands-On Activity	• Extending the Lesson		• ESL/ELL Strategy Teacher's Resources* Enrichment Activity
Lesson 16-7 **How does a rocket work?** pp. T394–T395 Standard: D3c	EXPLAIN, ANALYZE, PREDICT • Building Math Skills • Hands-On Activity	• Extending the Lesson • More on the Web • Integrating Physical Science • Lab Challenge • Active Art	• Study Hint • Reading Strategy	• Reteach Teacher's Resources* Enrichment Activity
Lab Activity **Modeling Newton's Third Law of Motion** pp. T396–T397 Standard: D3c	OBSERVE, ANALYZE, COMPARE • Set-Up Time: 15–20 min • Lab Time: 40–45 min	• Extending the Lesson • Extend the Lab Activity	• Tips for Using Technology	• Reteach Teacher's Resources* Lab Activity Report Cause-and-Effect Diagram
Lesson 16-8 **What is the space shuttle?** pp. T398–T399	ANALYZE, INFER, IDENTIFY • Science and Technology	• Cooperative Learning • More on the Web	• Study Hint • Reading Strategy • Ongoing Assessment	Teacher's Resources* Enrichment Activity
Lesson 16-9 **What are satellites and space probes?** pp. T400–T401	NAME, DESCRIBE, CLASSIFY • Building Science Skills	• Extending the Lesson • More on the Web	• Study Hint	• ESL/ELL Strategy
Lesson 16-10 **How are space stations used?** pp. T402–T403	EXPLAIN, HYPOTHESIZE, INFER • Designing an Experiment • People in Science	• More on the Web	• Writing Hint • Ongoing Assessment	• Reteach Teacher's Resources* Designing an Experiment
Chapter 16 Challenges pp. T404–T406	• Chapter Summary • Key Term Challenges • Content Challenges • Concept Challenges	• Chapter Project Wrap-Up	• Study Hint • Preparing Students for Standardized Tests	• ESL/ELL Strategy Teacher's Resources* Chapter Self-Check Weekly Journal

*www.conceptsandchallenges.com, password: ccearth09

Chapter 16 The Exploration of Space

▲ **Figure 16-1** The space shuttle lifts off from its launch pad in Cape Canaveral, Florida.

The space shuttle is made up of an orbiter, a fuel tank, and booster rockets. The orbiter carries the crew and cargo into space and lands back on Earth like an airplane. The space shuttle is the world's first reusable spacecraft. It can fly missions and return home, then return to space later for more missions. The first shuttle mission, with the orbiter *Columbia*, was launched on April 12, 1981.
▶What do you think are the advantages in having a reusable spacecraft system?

Contents

Teaching Options

ASSESSMENT PLANNER

For assessment in the Student Edition, see the following pages:

Content Assessment
Checking Concepts: pp. 381, 383, 385, 387, 389, 393, 395, 399, 401, 403
Thinking Critically: pp. 381, 383, 385, 387, 389, 393, 395, 399, 401, 403
Interpreting Visuals: p. 399
Chapter Challenges: pp. 404–406

Alternative Assessment
Building Skills: pp. 381, 385, 387, 393, 395, 401
Web InfoSearch: pp. 383, 389
Science Log: p. 391

Performance-Based Assessment
Hands-On Activity: pp. 385, 393, 395
Designing an Experiment: p. 403
Lab Activity: pp. 396–397

• TEACHING RESOURCES • Online or Print

Teacher's Resources
Lesson Review: Ch. 16, pp. 2, 3, 4, 5, 7, 9, 11, 12, 14, 15
Enrichment: Ch. 16, pp. 6, 10, 13
Key Term Review: Ch. 16, p. 16
Chapter 16 Test: pp. 17–19
Laboratory Manual: pp. 81–82

Chapter 16 The Exploration of Space

Chapter Overview

In this chapter, students will learn about astronomy, the study of the universe. Students will also learn about the tools that astronomers use to study space, including different types of telescopes and spacecraft.

About Figure 16-1 NASA has constructed six space shuttles: *Columbia, Enterprise, Challenger, Discovery, Atlantis,* and *Endeavour*. America's fleet of orbiter vehicles are named after ships that made discoveries about Earth and its oceans. The first orbiter, *Enterprise*, was a prototype and was not designed to fly in space.

Answer A reusable spacecraft can save money, time, and resources.

Linking Prior Knowledge

For this chapter, students should recall:

- distances on a map (Lesson 1-7).
- the layers of Earth's atmosphere (Lesson 11-2).
- the motions of the Moon (Lesson 14-5).

Chapter Project

SPACE EXPLORATION TIMELINE

MATERIALS: paper rolls, markers, tape

Students can use the roll of paper to create a space exploration timeline. Divide the class into four or five groups, giving each a block of years. For example, Group 1 can do early developments (Galileo and Newton). Group 2 can do early modern space science (the V-2 rocket, *Sputnik, Explorer 1*). Group 3 can do more recent history (*Gemini, Apollo, Viking,* and *Voyager*), and Group 4 can show what is going on today. Have students gather magazine pictures of telescopes and spacecraft or download images from NASA and other space Web sites. Each group should also prepare a brief summary of their time period, making some general statements about it. The timelines can then all be taped together to form a large classroom display.
learning styles: visual, tactile

16-1 *(pp. 380–381)*

1 Introduce

INVESTIGATE

TIME: 10–15 minutes

PURPOSE: Students will calculate the length of a light-year.

MATERIALS: calculator, pencil, paper

PROCEDURE: Check calculations for each step.

1. 300,000 × 60 = 1,800,000 km/min
2. 1,800,000 × 60 = 1,080,000,000 km/hr
3. 1,080,000,000 × 24 = 25,920,000,000 km/day
4. 25,920,000,000 × 365 = 1,460,800,000,000 km/year

THINK ABOUT IT: Possible answer: It helps astronomers estimate distances to stars.

STUDY HINT Before beginning this lesson, have students scan the lesson subheads for an idea of what the lesson will be about. Have them write the kind of information that they think they will read. Tell students to later compare what they learn with what they thought they would learn.

Linking Prior Knowledge Have students review the motions of the Moon (Lesson 14-5). Ask students if they have ever been outside at night to look at the Moon, stars, and planets. Have them describe what they saw. Ask: *What could you have used to better see the Moon and planets?* (binoculars, telescope)

2 Teach

Discussion Ask students to define *astronomy*. (the study of stars, planets, and other objects in space) Have students suggest areas of astronomy that they would like to investigate. List all the topics on the board.

Answers to questions

1. LIST: Mercury, Venus, Mars, Jupiter, and Saturn
2. EXPLAIN: Farmers used astronomy to decide when to plant crops. Sailors used astronomy to find their way at sea.
3. DESCRIBE: planetologists and cosmologists

16-1 What is astronomy?

INVESTIGATE
Calculating the Length of a Light-Year
HANDS-ON ACTIVITY

1. Light travels about 300,000 km per second. How many kilometers does light travel per minute? Write this number down.
2. How many kilometers does light travel per hour? Multiply your number from Step 1 by 60. Write this number down.
3. How many kilometers does light travel in a day? Multiply the number you got in Step 2 by 24. Write this number down.
4. Multiply the number you got in Step 3 by 365, for the number of days in a year. This will give you the distance in kilometers that light travels in one year.

THINK ABOUT IT: Why would knowing the speed of light be important to astronomers?

Objective

Explain what is meant by astronomy.

Key Terms

astronomy (uh-STRAHN-uh-mee): study of stars, planets, and other objects in space

solar system: Sun and all the bodies that orbit it

Ancient Astronomy People have wondered about the skies for thousands of years. The study of space and all the objects in it is called **astronomy.** Astronomy is one of the oldest sciences. Ancient records predicting eclipses of the Sun date back to the first century B.C.

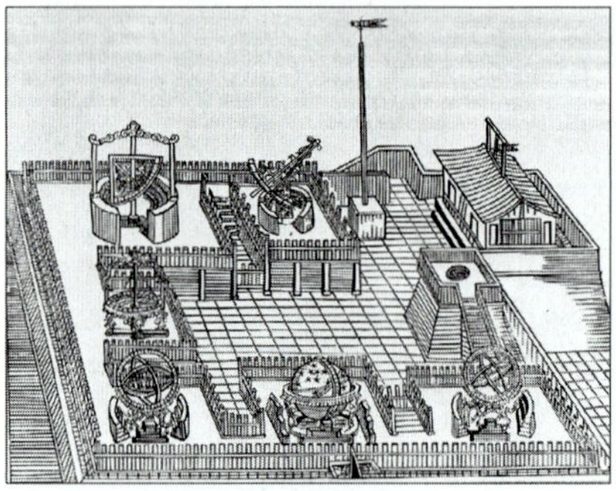

▲ **Figure 16-2** Ancient astronomers studied the skies with special tools.

The planets Mercury, Venus, Mars, Jupiter, and Saturn were all known more than 5,000 years ago. No new planets were found until 1781, when Uranus was discovered.

1. LIST: Which planets did the ancients know?

Uses of Astronomy Ancient peoples solved many problems using astronomy. Most ancient societies depended on the changing of the seasons. Farmers had to know when to plant their crops. Astronomers were able to predict the coming of spring for them. Sailors, too, used the positions of the stars to navigate at sea.

2. EXPLAIN: How did the ancients use astronomy?

Modern Astronomy Ancient astronomers made observations using only their eyes. Over time, new tools helped astronomers to see farther into space. In 1609, the Italian scientist Galileo (gal-uh-LAY-oh) Galilei used a simple telescope to look at the Moon and other objects. Today, astronomers use telescopes, satellites, and space probes to observe the planets and study stars many trillions of kilometers from Earth.

What do modern astronomers study? They study the universe. The universe is everything that exists. Astronomers called planetologists study the solar system. The **solar system** includes the Sun and all of the bodies in space that orbit it. Astronomers called cosmologists study how the universe began and how it will end.

Teaching Options

! FAST FACTS *Ancient Chinese Astronomy*

Early Chinese astronomers kept careful records of many astronomical events and are credited with creating some of the first known maps of the universe. These rough charts may date back to 1500 B.C. The Chinese also recorded a comet sighting in 613 B.C., and within centuries of that sighting, they were routinely recording eclipses, meteor showers, sunspot cycles, and supernovas.

EXTENDING THE LESSON

Inferring Earth's Shape To demonstrate how early astronomers used observations to infer Earth's spherical shape, have students gather at one end of a table. Hold a pencil upright in your hand so that the pointed tip rests on the table. Move the pencil from the near to the far end of the table. Students should note that the pencil stays visible. Then, do the same over a curved surface, such as a globe or large ball. Students should note that the pencil appears to sink below the horizon of the curved surface. **learning style: visual**

▲ **Figure 16-3** Telescopes help astronomers study objects in space.

3 DESCRIBE: What are two kinds of astronomers?

✓ CHECKING CONCEPTS

1. Five thousand years ago, there were __________ known planets, not including Earth.
2. Sailors found their way using the __________.
3. The universe is __________ that exists.
4. Ancient astronomers made observations using only their __________.
5. Astronomers called __________ study how the universe began.
6. The first astronomer to use a telescope to view the Moon was __________.

THINKING CRITICALLY

7. **INFER:** Early astronomers observed that the planets changed their positions in the night sky. However, the stars kept the same positions. Why did sailors use the stars and not the planets to find their way at sea?

BUILDING READING SKILLS

Vocabulary On a sheet of paper, list the names of the following stars: Vega, Sirius, Aldebaran, Polaris, Algol, Capella, Betelgeuse, and Proxima Centauri. Find out what each name means in its original language. Write the meaning and identify the original language next to each name. Also, find out how the names are pronounced.

How Do They Know That?

EARLY IDEAS ABOUT EARTH

▲ **Figure 16-4** Nicolaus Copernicus

Astronomy is an ancient science. Records dating back to before 2000 B.C. show that there was a calendar based on movements of the Sun and Moon. At that time, most people believed Earth was flat, with the stars inside a giant sphere that moved around Earth.

In the sixth century B.C., Pythagoras (pih-THAG-uh-ruhs), a Greek mathematician, challenged the view that Earth was flat. He noted that the masts of ships disappeared below the horizon. He also saw that Earth's shadow on the Moon during a lunar eclipse was curved. From these observations, Pythagoras reasoned that Earth must be round. In the fourth century B.C., the Greek philosopher Aristotle (AR-ihs-taht-uhl) claimed that Earth was at the center of the universe. This claim went mostly unchallenged for almost 2,000 years. Then, in the early sixteenth century, the Polish mathematician Nicolaus Copernicus (koh-PUHR-nih-kuhs) suggested that the Sun, not Earth, was the center of the solar system. Later observations by Tycho Brahe, Johannes Kepler, Galileo, and Sir Isaac Newton supported Copernicus. This was the beginning of modern astronomy.

Thinking Critically What change in thinking led to modern astronomy?

MORE ON THE WEB

Constellations Have students use the Internet to learn about constellations. Then, have them report to the class what they discover.

• *TEACHING RESOURCES* • *Online or Print*

Teacher's Resources

Lesson Review: Chapter 16, p. 2

Enrichment: Chapter 16, p. 10

Star Name	Meaning	Star Name	Meaning	Star Name	Meaning
Vega	the swooping eagle	Polaris	polestar	Betelgeuse	the hand of Gemini
Sirius	burning	Algol	the demon	Proxima Centauri	closest star in the constellation Centaurus
Aldebaran	the follower	Capella	nanny goat		

3 *Assess*

✓ CHECKING CONCEPTS ANSWERS

1. five
2. stars
3. everything
4. eyes
5. cosmologists
6. Galileo

THINKING CRITICALLY ANSWER

7. **INFER:** The positions of the stars provided an unchanging "sky map" that the sailors could use to determine their position at sea.

4 *Close*

BUILDING READING SKILLS

Vocabulary Refer to the chart in the Teaching Options on this page.

How Do They Know That?

EARLY IDEAS ABOUT EARTH

Real-Life Connection Copernicus's theory directly challenged the widely held view that Earth was at the center of the universe. It was considered so controversial that it was not published until 1543, the year of his death. The publisher of the book included a disclaimer stressing that the theory was not meant to challenge Earth's place in the universe. For this reason, Copernicus was not posthumously charged by religious authorities with blasphemy.

Discussion Tell students that astronomers often refer to the period following the publication of Copernicus's theory as the Copernican Revolution. Ask students why *revolution* is a good word to describe this time. (It forever changed the way people viewed the universe.)

Thinking Critically Answer
the theory that the Sun was at the center of the solar system

16-2 *(pp. 382–383)*

1 Introduce

Linking Prior Knowledge Have students recall the four main layers of Earth's atmosphere (Lesson 11-2) (troposphere, stratosphere, mesosphere, and thermosphere). Tell them that Earth's atmosphere does not have a clear upper boundary. Instead, it gradually blends into outer space. The very highest region of the atmosphere, called the exosphere, contains few air molecules. Many space telescopes orbit in or above the exosphere.

2 Teach

Demonstration To demonstrate how Earth's atmosphere affects space observations, place a piece of cheesecloth over an aquarium. Have students note how the cloth blocks their view of the inside of the aquarium. Ask: *What does the cloth represent?* (Earth's atmosphere)
learning styles: visual, auditory

Discussion Tell students that the government spends billions of dollars annually on the space program. Have students suggest reasons for this investment. List students' responses on the board. Possible reasons are to learn more about the universe, including the Moon and planets; to be able to conduct experiments in space; to search for life on other planets; to improve methods of forecasting weather; to improve national defense; to test various scientific theories; to produce better methods of communications; and to develop a better understanding of how other bodies in space affect our planet.
learning style: auditory

Answers to questions

1 **INFER:** to answer questions about the universe

2 **DEFINE:** huge collections of stars, gas, and dust that travel together through space

3 **INFER:** It is above Earth's atmosphere.

WRITING HINT Tell students that writing an outline can help them to better organize and recall information.

16-2 How is space explored?

Objective

Explain how space exploration helps scientists learn about Earth and the universe.

Key Term

galaxy (GAL-uhk-see): huge collection of stars, gas, and dust that travels together through space

Answers from Space The Space Age really began in 1957 with the launch of the Russian satellite *Sputnik*. A year later, the United States launched *Explorer 1*. Since then, humankind has sent spacecraft all around the solar system and people to the Moon. Scientists hope that these missions will answer questions about the universe. They want to know, for example, how the solar system formed and whether intelligent life exists elsewhere in the universe.

▲ **Figure 16-5** *Explorer 1* was the first U.S. spacecraft to be successfully launched.

1 **INFER:** Why do scientists explore space?

How the Universe Formed Space exploration may provide clues as to how the universe was formed. Scientists think that the universe began about 13 billion years ago in a huge explosion called the Big Bang.

Soon after the Big Bang, galaxies began to form. A **galaxy** is a huge collection of stars, gas, and dust that travels together through space as one body. The most distant galaxies were probably formed soon after the Big Bang.

2 **DEFINE:** What are galaxies?

Hubble Space Telescope Astronomers are learning more about the universe by using a telescope that is orbiting Earth. The Hubble Space Telescope was launched into space in 1990. It is named after the American astronomer Edwin Hubble. This telescope has observed and taken pictures of many of the planets and distant galaxies that are hard or even impossible to see clearly from Earth.

When the Hubble began operating in space, it did not work properly. Images from deep space were blurry. The problem was corrected several years later during a mission of the space shuttle *Endeavour*.

3 **INFER:** How can the Hubble Space Telescope see so far into space?

◀ **Figure 16-6** The Hubble Space Telescope orbits Earth. It has provided images of many galaxies, such as the Large Magellanic Cloud, seen above.

382

Teaching Options

COOPERATIVE LEARNING

Life on Other Planets Divide the class into two groups. Have the groups debate whether life exists elsewhere in the universe. Point out that space exploration may someday answer this question.
learning style: linguistic

RETEACH

Talking About Spinoffs Bring in a spinoff product, such as a freeze-dried food product, mentioned in this lesson. Ask students to describe the product and its use on Earth and probable use in space.
learning styles: visual, linguistic

EXTENDING THE LESSON

Researching Space Tell students that the satellite *Sputnik* was launched as part of the International Geophysical Year, which lasted from July 1, 1957, to December 31, 1958. Have interested students research the astronomical observations made by scientists from different countries during that time.
learning style: linguistic

CHECKING CONCEPTS

1. Space exploration really began in the year __________.
2. Scientists think that the universe began about __________ years ago.
3. The explosion that scientists think might have started the universe is called the __________.
4. The __________ Telescope, which orbits Earth, is able to see very distant galaxies.
5. The space shuttle __________ fixed the Hubble's imaging problem.

THINKING CRITICALLY

6. **HYPOTHESIZE:** How might traveling deeper into space help scientists answer some of their questions about space?
7. **INFER:** Why is it easier for telescopes in orbit to view stars?
8. **CONCLUDE:** How has the Hubble Space Telescope increased our knowledge of space?

Web InfoSearch

Astrolabes The most common astronomical instrument until about 1650 was the astrolabe. It was used to observe the altitude of stars and to guide sailors. The Greek astronomer Hipparchus probably refined the first one. Astrolabes were often made of brass. The Adler Planetarium and Astronomy Museum in Chicago has the largest collection of astrolabes in North America.

▲ **Figure 16-7** Astrolabe from the seventeenth century

SEARCH: Use the Internet to find out more about astrolabes. Are they still being made today? Start your search at www.conceptsandchallenges.com. Some key search words are **astrolabe instrument** and **Hipparchus astrolabe.**

Real-Life Science

SPINOFFS FROM SPACE

How can space exploration help people on Earth? It produces spinoffs. A spinoff is a benefit or product that results from an unrelated activity or process.

Sending spacecraft into space is very costly. One way to save costs is by developing smaller, lighter equipment. Smaller radios, computers, and televisions made for use in space travel are now used on Earth. Many foods sent into space with the astronauts are freeze-dried to make them lighter and last longer. Many of these foods are available in supermarkets today. Lightweight, fireproof, durable clothing, nonstick cookware, and sunglasses that adjust to changes in light are all spinoffs from space.

▲ **Figure 16-8** Improvements in fireproof clothing are a spinoff from space exploration.

Other spinoffs help us recover from illnesses. Doctors on Earth need to check on astronauts in space. To meet this need, scientists developed tiny monitoring, recording, and transmitting devices now used on Earth.

Thinking Critically Why is freeze-dried food good for any type of travel?

3 *Assess*

CHECKING CONCEPTS ANSWERS

1. 1957
2. 13 billion
3. Big Bang
4. Hubble Space
5. *Endeavour*

THINKING CRITICALLY ANSWERS

6. **HYPOTHESIZE:** Possible answer: Scientists can observe astronomical events that cannot be seen from Earth.
7. **INFER:** because Earth's atmosphere does not distort or cloud the image
8. **CONCLUDE:** It has observed and created images of the planets and distant galaxies that cannot be seen from Earth.

4 *Close*

Web InfoSearch

Astrolabes Have students use the Internet to find out if astrolabes are still being used today. Some key search words are **astrolabe instrument** and **Hipparchus astrolabe.** Students can begin their search for information at www.conceptsandchallenges.com.

Real-Life Science

SPINOFFS FROM SPACE

Background Space exploration can create problems as well as benefits. One increasing problem is space junk. Since 1957, some 4,000 satellites have been launched into space. When they cease functioning, satellites either fall out of orbit and burn up in Earth's atmosphere or continue to orbit while disintegrating. Debris from spacecraft and space missions may pose a threat to future space travel.

Discussion Have students brainstorm the advantages and disadvantages of space exploration. List their responses on the board in a two-column table.

Thinking Critically Answer
It is lightweight, does not need refrigeration, and lasts a long time.

ESL/ELL STRATEGY

Working With Peers Working with students who are proficient in English can help English-language learners to communicate. Have students work in pairs. One student should write three questions about the lesson and read these aloud to the other student, who tries to answer them. Then, partners should reverse roles. Both partners should check each other for accuracy.

• *TEACHING RESOURCES* • *Online or Print*

Teacher's Resources
Lesson Review: Chapter 16, p. 3

16-3 *(pp. 384–385)*

1 Introduce

Linking Prior Knowledge Bring a camera to class. Review the parts of the camera and how they work. Draw a simple diagram on the board to trace the path of an image through the camera's lens and body onto the film. Tell students that the most important part of a camera is the lens.

2 Teach

Demonstration Introduce this lesson by displaying a magnifying glass. Ask students why people use magnifying glasses. (to make objects appear bigger) Point out that the magnifying glass is thicker at the center than at the edges. Tell students that this lens is a convex lens.
learning styles: visual, tactile

Discussion Refer students to Figure 16-10. Explain how a refracting telescope works. Have students identify the objective lens and eyepiece lens. Have them trace the paths of the light rays. Ask volunteers to describe how the lenses work.
learning styles: visual, auditory

Reinforcement If possible, bring to class a refracting telescope and demonstrate how it works. Allow students to study the telescope closely. Then, have a volunteer explain the function of each lens.

WRITING HINT Tell students that for the Building Science Skills activity on page 385, they should take notes, organize their thoughts before preparing a final draft of their work, and choose their punctuation carefully. Prepared lists can benefit from good organization and proper punctuation just as paragraphs and essays can.

Answers to questions

1 **NAME:** craters, plains, and hills

2 **DESCRIBE:** convex

3 **NAME:** objective lens, eyepiece lens

16-3 How does a refracting telescope work?

Objective
Explain how a refracting telescope works.

Key Terms
refracting (rih-FRAKT-ing) **telescope:** telescope that uses convex lenses to produce an enlarged image
convex (kahn-VEHKS) **lens:** lens that is thicker in the middle than it is at the edges

Galileo's Telescope Galileo was the first person to look at objects in space through a telescope. Have you ever looked through a telescope? If you have, you know that a telescope makes faraway objects appear much nearer than they are. If you look at the Moon through a telescope, you can see many features on its surface that you cannot see with your eyes alone. Galileo looked at the Moon soon after the telescope was invented. He was the first person to see that the Moon's surface is not smooth. Galileo saw craters, plains, and hills on the Moon.

▲ **Figure 16-9** Galileo's telescope, with Italy in the background

1 **NAME:** What features did Galileo see on the surface of the Moon?

Refracting Telescopes Galileo used a **refracting telescope.** It consisted of a tube with two lenses inside. A lens is a piece of glass or plastic that refracts, or bends, light.

The telescope shown in Figure 16-10 is a simple refracting telescope. The lenses in a refracting telescope are convex lenses. **Convex lenses** are lenses that bulge outward. They are thicker in the middle than they are at the edges.

When light passes through a convex lens, the light appears to bend inward, as shown in Figure 16-10. The light produces an image that is larger than the image you would see with your eyes alone.

2 **DESCRIBE:** What kinds of lenses are used in a refracting telescope?

How the Lenses Work Each of the lenses in a refracting telescope has a special job to do. The lens at the far end of the tube is called the objective lens. The objective lens collects light and brings the image into focus. The lens at the other end of the tube is called the eyepiece. This lens enlarges, or magnifies, the image formed by the objective lens.

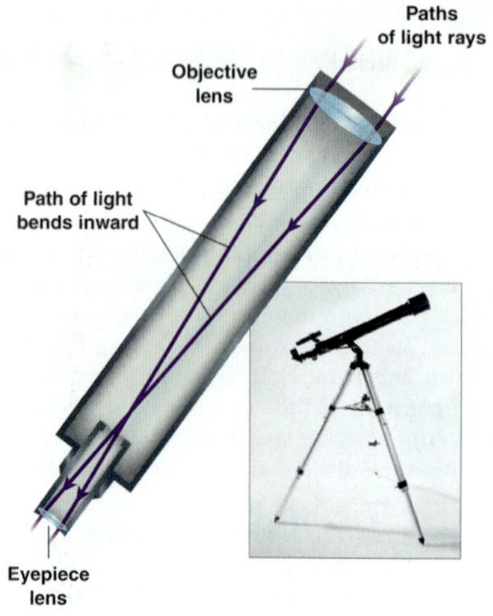

▲ **Figure 16-10** A simple refracting telescope

384

Teaching Options

! FAST FACTS *Eyeglass Lenses*

Farsighted people have a difficult time viewing close-up objects because of the shape of their eyeballs. Light rays cannot converge, or focus, directly on the light-sensing tissue at the back of the eyeball, called the retina. Light rays reach the retina before they converge. Convex lenses correct farsightedness by causing light rays to converge sooner.

Nearsighted people have a difficult time focusing on faraway objects. Light rays converge before they reach the retina. Concave lenses, which spread light rays apart, correct this problem by causing the light rays to converge later.

INTEGRATING PHYSICAL SCIENCE

Refracting Light To produce a clear image of a star, the light waves from it must be focused, or brought together in a single point. A refracting telescope uses a convex lens. Light waves from stars arrive on Earth at different times. As they pass through the lens, they slow down and bend. In the middle, where the lens is thickest, the waves slow down the most. Differences in speed cause the light waves to focus, or arrive at the same point at the same time.

The objective lens and the eyepiece lens work together. They can produce a sharp, clear image of a distant object.

3 NAME: What are the two lenses in a refracting telescope called?

CHECKING CONCEPTS

1. A telescope makes the Moon appear __________.
2. Galileo saw __________, hills, and plains on the Moon.
3. The __________ in a refracting telescope contains two convex lenses.
4. A convex lens appears to bend, or __________, light.
5. The lenses in a refracting telescope are __________ lenses.
6. The two lenses that are used in a refracting telescope are the objective lens and the __________ lens.

THINKING CRITICALLY

7. COMPARE: Which surface features of the Moon are also found on Earth?
8. INFER: Why does the image that is formed by a telescope appear to be nearer than the object really is?
9. HYPOTHESIZE: If light did not bend when it was passed through a lens, would the refracting telescope still work? Why or why not?

BUILDING SCIENCE SKILLS

Observing Many objects that you use every day contain convex lenses. Make a list of common objects that contain convex lenses. List as many of them as you can. Briefly describe how the lenses are used in these objects. Hint: What part of your body has a convex lens?

Hands-On Activity

MAKING A SIMPLE TELESCOPE

You will need 2 convex lenses or magnifying glasses.

1. Look at a distant object through a convex lens.
2. Move the lens back and forth slowly. Stop when you see the object clearly through the lens.
3. Without moving the first lens, hold a second convex lens close to your eye.
4. Move the second lens back and forth slowly. Stop when you can see the distant object clearly through both lenses.

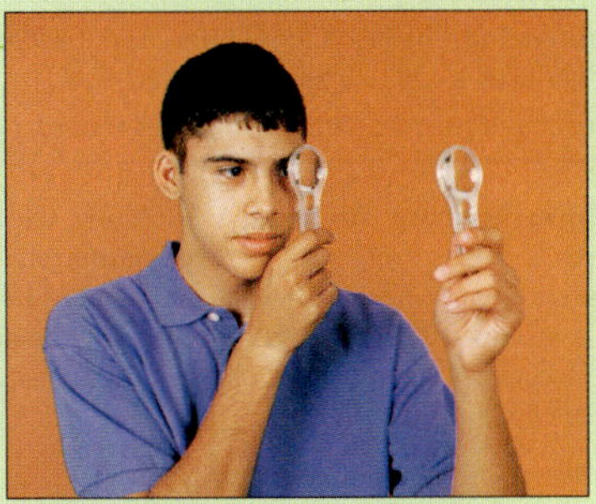

▲ **STEP 4** The distant object should show clearly through both lenses.

Practicing Your Skills

5. IDENTIFY: What part of a telescope does the first lens represent?
6. OBSERVE: How does the image appear through the first lens?
7. IDENTIFY: What part of a telescope does the second lens represent?
8. OBSERVE: How does the image appear when you look through both lenses?
9. ANALYZE: Would a telescope with a large objective lens be better than one with a small objective lens? Explain.

MORE ON THE WEB

Writing About Galileo
Have students learn more about Galileo by visiting www.conceptsandchallenges.com. Then, have students write a short report about Galileo and his accomplishments.

ONGOING ASSESSMENT

Have students draw a diagram of a simple refracting telescope in their notebooks. Tell students to label the objective lens and the eyepiece lens and to include brief descriptions of the function of each lens.
learning style: visual

• TEACHING RESOURCES • Online or Print

Teacher's Resources

Lesson Review: Chapter 16, p. 4

Enrichment: Chapter 16, p. 6

Refracting Telescope: Visuals, p. 12

3 Assess

CHECKING CONCEPTS ANSWERS

1. nearer
2. craters
3. tube
4. refract
5. convex
6. eyepiece

THINKING CRITICALLY ANSWERS

7. COMPARE: craters, plains, hills
8. INFER: The eyepiece lens magnifies the object.
9. HYPOTHESIZE: no, because the image would be out of focus

4 Close

BUILDING SCIENCE SKILLS

Observing Lists may include eyes, microscopes, telescopes, binoculars, cameras, and movie or slide projectors. In each of these objects, the lens collects the light and brings the image into focus.

Hands-On Activity

MAKING A SIMPLE TELESCOPE

TIME: 10–15 minutes

PURPOSE: identifying, observing, analyzing

Discussion Tell students that in this activity, they will make a model telescope. Tell students that models are copies or imitations of objects. They are used to understand how something works. You may want to have students list other models used in both everyday life and in science.

Practicing Your Skills Answers

5. IDENTIFY: the objective lens
6. OBSERVE: distorted and larger
7. IDENTIFY: the eyepiece lens
8. OBSERVE: larger and closer
9. ANALYZE: yes, because it would focus better

16-4 *(pp. 386–387)*

1 Introduce

INVESTIGATE

TIME: 10–15 minutes

PURPOSE: Students will model the way light travels.

MATERIALS: cardboard cut into 15 × 15-cm pieces, scissors, modeling clay, flashlight, large index card, metric ruler

PROCEDURE: Divide the modeling clay into four pieces. Have students work in pairs to measure and cut the cardboard.

THINK ABOUT IT: Light appears when the notches are lined up. Light travels in a straight line.

STUDY HINT Before beginning this lesson, have students read the lesson feature on page 387. Discuss the concepts presented in it. Ask students how they think this information relates to topics that they have already studied or to this lesson.

Linking Prior Knowledge Ask students to tell how lenses and mirrors are alike and how they are different. Write students' comments on the board. (Possible answer: Alike: Both help you see images. Different: You look through a lens to see the image, but you see the image reflected in a mirror.)

2 Teach

Discussion Use Figure 16-12 as a guide to draw a reflecting telescope on the board. Point out the locations of the mirrors and lenses in the telescope. Then, have students compare the diagram of this telescope with the refracting telescope in Figure 16-10.

Reinforcement Remind students that mnemonics can be used to remember the meanings of words. Ask: *What would help you to remember the meaning of the word* concave? Point out that *concave* includes the word *cave*, a structure that curves inward.

Answers to questions

1. **DEFINE:** a telescope that uses a mirror to collect light
2. **EXPLAIN:** so that a camera can be attached to it to take pictures
3. **EXPLAIN:** They have very large mirrors.

16-4 How does a reflecting telescope work?

INVESTIGATE

Modeling the Way Light Travels

HANDS-ON ACTIVITY

1. Cut three 15- by-15 cm cards from cardboard. On each card, in the middle of one side, cut a 2- by-2 cm notch. Use modeling clay to place the cards about 10 cm apart. Line up the notches so they are in a straight line with each other.
2. Shine a flashlight behind the cards. Use the modeling clay to hold up a large index card at the other end of the row of cards.
3. Darken the room. Observe any light on the index card.
4. Move the cards so that the notches are not in a straight line. Observe the light on the index card.

THINK ABOUT IT: When did light appear on the index card? What does this tell you about how light travels?

Objectives

Explain how a reflecting telescope works.
Compare it with a refracting telescope.

Key Terms

reflecting (rih-FLEHKT-ing) **telescope:** telescope that uses a concave mirror to collect light

concave mirror: mirror that curves inward

Using Mirrors to See A refracting telescope uses lenses to collect light. A **reflecting telescope** uses a mirror. This mirror is a **concave mirror.** It curves inward. In a reflecting telescope, the mirror collects light from distant objects and focuses it to form an image.

1 DEFINE: What is a reflecting telescope?

Newton's Telescope In 1668, Isaac Newton, an English scientist, made the first reflecting telescope. He used a concave mirror to collect the light and a convex lens as an eyepiece to magnify the image. He also used a flat mirror to reflect light from the concave mirror to one side. This prevented the eyepiece from blocking the incoming light.

Some reflecting telescopes have the eyepiece to one side of the telescope tube. A camera can then be attached to it to take pictures.

▲ **Figure 16-11** Newton uses his reflecting telescope.

2 EXPLAIN: Why is the eyepiece that is used in a reflecting telescope sometimes found on one side of the tube?

Modern Reflecting Telescopes Modern reflecting telescopes have very large mirrors. The Hale Telescope on Mount Palomar in California has a mirror 5 m across.

The twin Keck telescopes that sit on the summit of Mauna Kea in Hawaii have 36 mirrors that together measure about 10 m across. This gives a light-collecting area almost half the size of a tennis court. Composite telescopes are being developed today that will have mirrors of up to 16 m across.

Teaching Options

! FAST FACTS *Telescope Quality*

One of the most important features of a telescope is the diameter of its objective lens or mirror. The larger the lens or mirror, the more light the telescope will gather and the sharper the image will be. The steadiness of a telescope also determines its quality. Steadiness can be assessed by trying out a telescope before buying it.

COOPERATIVE LEARNING

Building Telescopes Have students work as a team to construct a simple reflecting telescope. Tell them to mount a screen at the center of a ruler and to point the ruler at some distant object. Using a concave mirror, demonstrate how to focus the image of the distant object on the screen. Have a volunteer describe the image formed on the screen. Then, extend the screen to one side of the ruler and place the mirror at a slight angle to form a better image. Have a volunteer describe the image formed by the mirror. Next, demonstrate how to view the image formed by the mirror through a convex lens. Have a student describe the image. **learning styles: visual, tactile**

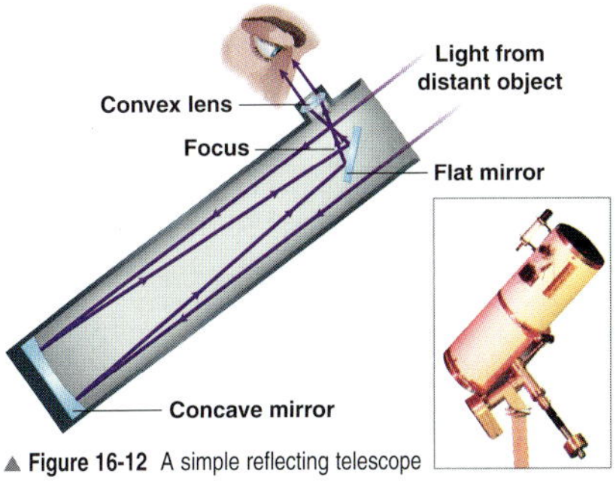

▲ **Figure 16-12** A simple reflecting telescope

3 **EXPLAIN:** What is the benefit of larger mirrors in telescopes?

✓ CHECKING CONCEPTS

1. What kind of telescope uses a mirror to collect and focus light?
2. Who made the first reflecting telescope?
3. What kind of lens is used as an eyepiece in a reflecting telescope?
4. Where is a reflecting telescope's eyepiece sometimes found?
5. Where are two modern reflecting telescopes with very large mirrors found?

THINKING CRITICALLY

6. **CONTRAST:** How do refracting and reflecting telescopes differ?
7. **ANALYZE:** Why is a reflecting telescope with a large concave mirror better than a reflecting telescope with a small mirror?
8. **INFER:** What might the pictures taken by a reflecting telescope be used for?

BUILDING SCIENCE SKILLS

Researching An observatory is a building or group of buildings with large telescopes. Find out what kinds of telescopes are used in the following observatories: Kitt Peak, Mauna Kea, Yerkes, and Mount Wilson. Write a report.

Integrating Environmental Science

TOPICS: pollution, overdevelopment

LIGHT POLLUTION

Because of streetlights, the night sky near cities and towns is never very dark. This is true even when there is no Moon showing and clouds hide the stars. The lights reflect off water droplets and other particles in the air. So much light makes it difficult for astronomers to observe stars and other objects using optical telescopes. Astronomers call this interference light pollution.

Street lamps cause a large percentage of light pollution. Businesses that leave on electric signs all night contribute, too. Many people keep lights on around their homes for safety and security. All of these lights brighten the nighttime sky and make viewing difficult. They can also upset the natural cycles of plants and animals in the area. Fewer lights, shields on streetlights that direct light down, and lights that are activated only when needed can all help reduce light pollution.

Thinking Critically What could you do to cut down on light pollution?

▲ **Figure 16-13** Satellite photos from 20 years ago (top) and today (bottom).

MORE ON THE WEB

Writing About Newton You can have students visit www.conceptsandchallenges.com to learn more about Newton. Then, have them write a short report on Newton.

ONGOING ASSESSMENT

Obtain catalogues or brochures from stores that sell telescopes. Point out the different kinds of telescopes. Have students note the features and costs. Then, have students write their own catalogue ads for refracting and reflecting telescopes.

• TEACHING RESOURCES • Online or Print

Teacher's Resources

Lesson Review: Chapter 16, p. 5

Enrichment: Chapter 16, p. 6

Reflecting Telescope: Visuals, p. 13

3 Assess

✓ CHECKING CONCEPTS ANSWERS

1. reflecting
2. Isaac Newton
3. convex
4. on the side of the tube
5. Mount Palomar in California, Mauna Kea in Hawaii

THINKING CRITICALLY ANSWERS

6. **CONTRAST:** A refracting telescope uses lenses to collect light. A reflecting telescope uses a mirror.
7. **ANALYZE:** A larger mirror allows the telescope to gather more light.
8. **INFER:** Possible answers: to study space and to see distant objects better

4 Close

BUILDING SCIENCE SKILLS

Researching Check students' reports. (Kitt Peak, 16 optical telescopes, the biggest of which is a 4-m reflecting telescope; Mauna Kea, the site of the Keck Observatory, which has the two largest reflecting telescopes in the world; Yerkes Observatory, the site of the world's largest refracting telescope; Mount Wilson Observatory, two solar telescopes and two reflecting telescopes)

Integrating Environmental Science

TOPICS: pollution, overdevelopment

LIGHT POLLUTION

Life-Science Connection Sea turtles lay their eggs on the beach and then bury them in the sand. The young turtles hatch at night and head instinctively toward the sea guided by moonlight reflecting off the water. Light pollution from streetlights and beach houses, however, can confuse the young turtles. They may turn from the sea and head for the street. In coastal areas such as the Atlantic and Gulf shores of Florida, volunteers stand guard over newly hatched turtles and guide them toward the sea.

Discussion Ask: *Where would you build a large telescope?* (away from city lights)

Thinking Critically Answer
Possible answers may include using fewer lights, installing shields on lights, and using lights that turn on only when needed.

16-5 *(pp. 388–389)*

1 *Introduce*

STUDY HINT Remind students to take neat, legible notes during class discussions so that they will find it easier to study for tests.

Linking Prior Knowledge To introduce this lesson, bring a portable radio to class. Turn on the radio. Ask students how a radio works. Point out that a radio collects radio waves. The receiver changes the radio waves to sound waves. Then, tell students that radio telescopes pick up radio waves from space.

2 *Teach*

Demonstration Stand outside the room and ring a bell. Then, go back in the room and ask students what object made the noise. (a bell) Point out that even though students could not see the bell, they could identify it based solely on its sound. Relate the demonstration to the use of radio telescopes to study stars.
learning style: auditory

Discussion Use Figure 16-14 to describe how radio telescopes are used to study stars. Discuss the advantages of radio telescopes. Ask students what kinds of things might disrupt radio signals. Discuss how these factors may affect data collected by radio telescopes.
learning style: auditory

Reinforcement Have students make a table in their notebooks that lists the characteristics of refracting, reflecting, and radio telescopes. Students should include the different parts of each telescope and explain the function of each.
learning style: linguistic

Answers to questions

1. **EXPLAIN:** The light from the Sun is brighter than the light from the stars and drowns them out.
2. **COMPARE:** The antenna of a radio telescope works like the mirror of a reflecting telescope.
3. **EXPLAIN:** Radio waves can travel through clouds.

16-5 What is a radio telescope?

Objective
Explain how a radio telescope works.

Key Term
radio telescope: telescope that can receive radio waves from sources in space

Star Search Stars are usually not visible during the day. They are there, but you cannot see them because of the Sun.

The Sun is also a star. It is the star closest to Earth. The bright light of the Sun outshines the dimmer light from the distant stars. On rainy nights, clouds hide the light from the stars. Not even a refracting or reflecting telescope can help you see the stars on a cloudy night.

1. EXPLAIN: Why are the stars not visible in daylight?

Radio Waves from Space Stars not only send out visible light. They also send out all forms of electromagnetic radiation.

A **radio telescope** can receive radio waves from space. It can find stars during the day or when there are clouds.

In 1932, an American engineer, Karl Jansky, heard the first radio signals from space. He found that the radio waves were coming from the center of our galaxy, the Milky Way Galaxy.

In 1937, Grote Reber built a radio telescope with a 9.5-m reflector dish, or antenna. With this, Reber was able to make the first radio map of the Milky Way Galaxy.

The antenna of a radio telescope works like a mirror does in a reflecting telescope. The antenna collects and focuses radio waves given off by stars and other objects in space. The radio waves are transmitted to a receiver. An astronomer can actually "listen" to the stars this way.

2. COMPARE: How is a radio telescope like a reflecting telescope?

Advantages of Radio Telescopes There are three main advantages to using radio telescopes. First, a radio telescope can detect some objects that refracting and reflecting telescopes cannot see. Many objects in the universe give off strong radio waves but very little visible light.

Second, a radio telescope can be used in any kind of weather. Radio waves can travel through clouds in Earth's atmosphere. Reflecting or refracting telescopes cannot be used on very cloudy nights.

Third, a radio telescope can be used during the day, when stars other than the Sun are not visible. Reflecting or refracting telescopes can only be used at night, when the Sun's light does not drown out the light from other, more distant stars.

3. EXPLAIN: How is a radio telescope able to function on a cloudy night?

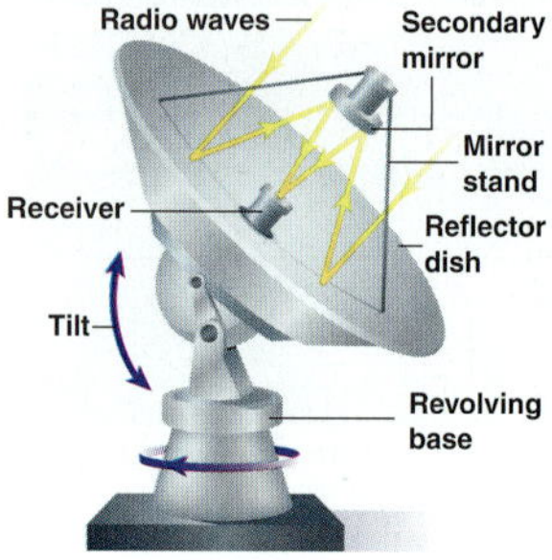

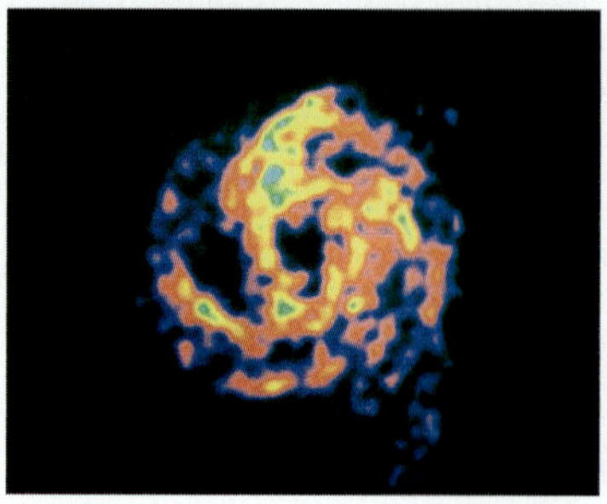

Figure 16-14 Left: The antenna system of a radio telescope. Right: A false-color image produced by a radio telescope.

Teaching Options

EXTENDING THE LESSON

Researching Radio Signals Have interested students find out more about Karl Jansky and his discovery of radio waves from outside the solar system. Tell students to write their findings in a report and share their reports with the class. **learning style: linguistic**

INTEGRATING LIFE SCIENCE

How the Ear Hears Just as a radio telescope's antenna collects radio waves and transmits them to a receiver, allowing astronomers to "listen" to stars, ears allow humans and other animals to hear by collecting sound waves and transmitting them to the brain. After sound waves enter the ear, they strike a thin, tightly stretched membrane called the eardrum, which begins to vibrate. The vibrations travel through tiny bones inside the ear, eventually reaching tiny nerve fibers in the cochlea. These nerve fibers turn the vibrations into nerve impulses, which travel to the cochlear nerve. The cochlear nerve transmits the nerve impulses to the brain. The brain, in turn, interprets the nerve impulses as sounds. Ask students to draw simple diagrams showing the path of a sound wave from a source to the brain. **learning style: visual**

CHECKING CONCEPTS

1. The star closest to Earth is __________.
2. You cannot see stars on a __________ night.
3. A __________ telescope can be used during the day.
4. The first radio map of the Milky Way Galaxy was made by __________.
5. The __________ of a radio telescope works like the mirror of a reflecting telescope.

THINKING CRITICALLY

6. CONTRAST: How do radio telescopes differ from reflecting and refracting telescopes?
7. LIST: What are three advantages of radio telescopes over optical telescopes?
8. INFER: How do you think astronomers can identify that a sound picked up by a radio telescope is from space?

Web InfoSearch

Interferometry A radio interferometer is two or more radio telescopes linked electronically. This greatly increases their power. Radio interferometers, such as the Very Large Array, can make much more detailed pictures than single radio telescopes can.

▲ **Figure 16-15** The Very Large Array

SEARCH: Use the Internet to find out more about this. Where is the Very Large Array found? What is it used for? Start your search at www.conceptsandchallenges.com. Some key search words are **radio telescope, interferometer,** and **Very Large Array radio telescope.**

Science and Technology

THE SEARCH FOR LIFE ELSEWHERE

Exobiology is a branch of biology that deals with the search for life, including intelligent life, outside of the solar system. In the 1950s, American scientists realized that radio signals could come from or be sent across the galaxy. The signals were, therefore, perfect for communicating with aliens, should any be talking or listening.

▲ **Figure 16-16** The radio telescope at Arecibo in Puerto Rico

Project SETI (Search for Extraterrestrial Intelligence) was formed in 1959 to listen for radio signals. In 1974, the radio telescope at Arecibo in Puerto Rico sent the first human message to the stars. It was made up of 1,679 on-off pulses aimed 25,000 light-years away, at a cluster of stars known as M13.

SETI was canceled in 1993 without having received any alien signals. However, in 1995, a new program was formed called Project Phoenix. Based in Greenbank, West Virginia, Phoenix used radio telescopes from all over the world to listen for artificial radio signals from space.

Thinking Critically If you could contact an alien, what would you tell it about Earth?

3 Assess

CHECKING CONCEPTS ANSWERS

1. the Sun
2. cloudy
3. radio
4. Grote Reber
5. antenna

THINKING CRITICALLY ANSWERS

6. **CONTRAST:** Radio telescopes have an antenna and pick up radio waves rather than light waves.
7. **LIST:** They can detect objects that give off very little light, can be used in any kind of weather, and can be used during the day.
8. **INFER:** Possible answers: Certain frequencies have been reserved for use by astronomers or the radio dish was aimed at the sky field.

4 Close

Web InfoSearch

Interferometry Have students use the Internet to find out about the Very Large Array. Some key search words are **interferometer, radio telescope,** and **Very Large Array radio telescope.** Students can begin their search at www.conceptsandchallenges.com.

Science and Technology

THE SEARCH FOR LIFE ELSEWHERE

Real-Life Connection In 1960, Dr. Frank Drake became the first astronomer to search for intelligent life elsewhere in the universe. He used a radio telescope in West Virginia to listen for signals. SETI (the Search for Extraterrestrial Intelligence) had begun, and soon both NASA and a private organization had established SETI programs.

Discussion Ask students: *What useful purpose does exobiology serve?* (Possible answers include to protect ourselves or to learn about nonhuman cultures and technologies.)

Thinking Critically Answer
Possible answers include telling the alien about the basis of life on Earth or sharing our knowledge of the universe.

ESL/ELL STRATEGY

Understanding Radio Maps Some English-language learners may have difficulty understanding the concept of a radio map. Find and distribute examples of radio maps in astronomy texts or on the Internet. You should also refer these students back to Chapter 1 to review the lessons on how maps can show many different types of information.

• *TEACHING RESOURCES* •
Online or Print

Teacher's Resources
Lesson Review: Chapter 16, p. 7

The Big Idea

(pp. 390–391)

1 *Introduce*

Objective Students will be able to identify the parts of the electromagnetic spectrum, compare types of modern telescopes, and understand how the invisible parts of the electromagnetic spectrum are used in many devices that study the universe.

Linking Prior Knowledge Have students recall what they have learned about space exploration (Lesson 16-2). Ask: *Which well-known telescope was launched in 1990 by the space shuttle* Discovery*?* (the Hubble Space Telescope) *What advantage does the Hubble Space Telescope have over Earth-based telescopes?* (It is above Earth's atmosphere and thus obtains clearer images of stars and other objects in space.)

Have students recall the types and origins of telescopes (Lessons 16-3, 16-4, 16-5).

2 *Teach*

Demonstration Because visible light can be seen, students may think of it as distinctly separate from other parts of the electromagnetic spectrum. Write the similarities among all electromagnetic waves on the board. For example, all parts of the electromagnetic spectrum

- transmit energy through space.
- travel at the speed of light in a vacuum.
- have wavelengths.

Encourage students to draw a picture to illustrate each similarity and write a caption for their drawings, using the above points. **learning styles: auditory, visual**

Discussion Have students discuss how they use the invisible parts of the electromagnetic spectrum in their lives. Challenge them to name examples not included in this feature. (cell phones, TV remote controls, motion detectors) **learning style: auditory**

Integrating Physical Science

THE Big IDEA

How do telescopes detect electromagnetic waves in space?

A rainbow shows the colors of the visible spectrum. Each color of the visible spectrum forms from light rays that have a particular wavelength. Violet has the shortest wavelength of the visible spectrum. Red has the longest wavelength.

Rays of light can have wavelengths shorter than violet and longer than red. We cannot see light at these wavelengths with our unaided eyes. However, special telescopes and other instruments are able to detect this light. Using these tools, scientists know that the visible spectrum is just one small part of the entire electromagnetic spectrum.

Astronomers use the invisible parts of the electromagnetic spectrum to study light rays with shorter or longer wavelengths than those of the visible spectrum. Using these tools, astronomers have gathered a great deal of information about distant stars and galaxies.

Note that infrared light, microwaves, and radio waves all have wavelengths longer than the colors of the visible spectrum. These invisible parts of the electromagnetic spectrum are used in everyday life. Microwaves cook foods. Infrared light keeps foods warm. Radio waves broadcast movies and music around the world. Ultraviolet light, X-rays, and gamma rays all have wavelengths shorter than the colors of the visible spectrum. Tanning booths use ultraviolet rays to darken the skin. X-rays take pictures of bones and teeth. Gamma rays can kill cancer cells.

Figure 16-17 ▶ Electromagnetic spectrum

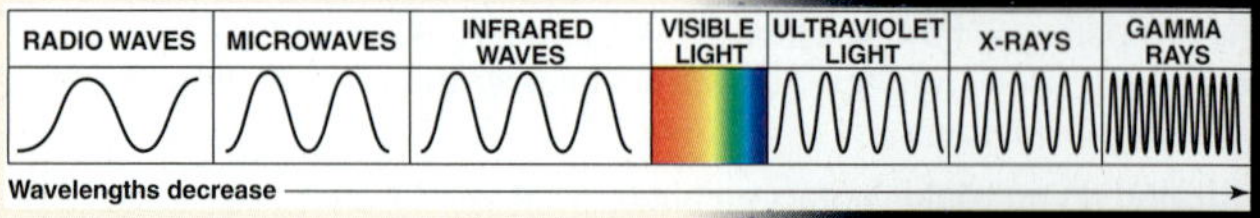

Look at the illustrations that appear on these two pages. Then, follow the directions in the Science Log to learn more about "the big idea." ✦

390

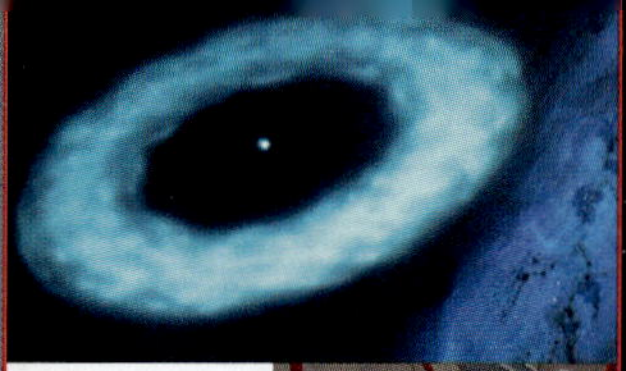

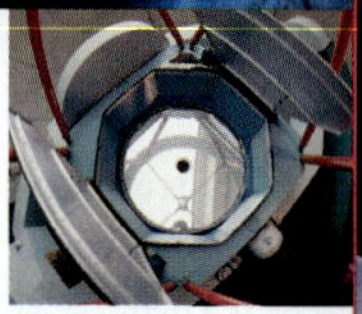

Infrared Telescopes
Infrared telescopes (right) are commonly placed high on satellites that travel through space. One infrared telescope detected rings of dust around the star Vega (shown above) and other nearby stars. Scientists believe these rings might be newly forming solar systems.

Gamma Ray Telescopes
The Compton Gamma Ray Observatory (above right) spent four years in space compiling an image of the Milky Way. The halo effect shown (above left) was made by gamma rays emitted from space.

Teaching Options

! Fast Facts — *Dark Matter and Black Holes*

The Chandra X-Ray Observatory, launched in 1999, was designed by NASA to study X-rays. Chandra has given scientists additional knowledge about dark matter, which, though currently undetected, makes up most of the universe. In addition, Chandra studies black holes, which are extremely high-gravity objects thought to exist at the center of galaxies. Matter, including light, that is sucked into a black hole is heated and emits X-rays.

INTEGRATING PHYSICAL SCIENCE

Placing Wavelengths on a Continuum Electromagnetic radiation can be classified according to wavelength or frequency. Wavelength refers to the distance between successive wave peaks. Frequency is the number of waves that occur each second. Have students study Figure 16-17. Ask: *Which type of electromagnetic radiation has the longest wavelength?* (radio waves) *Which type of electromagnetic radiation has the lowest frequency?* (radio waves) Have students write the following on a continuum from shortest to longest wavelength: red, radio waves, X-rays, violet. (X-rays–violet–red–radio waves)

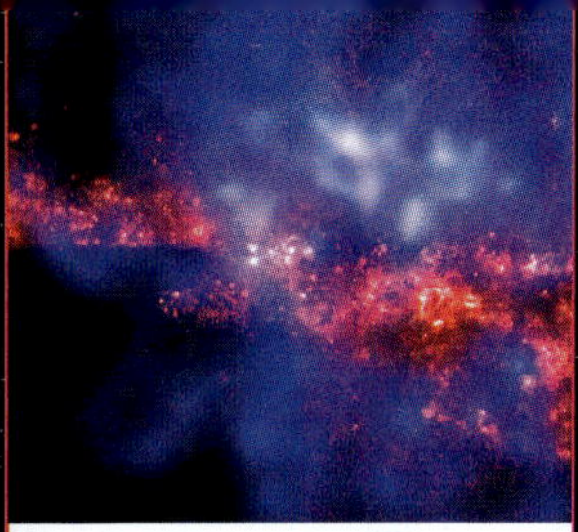

X-Ray Telescopes
Most X-ray telescopes, like the Chandra Observatory shown below, contain a reflector made up of a series of curved mirrors. Astronomers use X-ray telescopes to study hot gases (shown above), black holes, comets, and quasars.

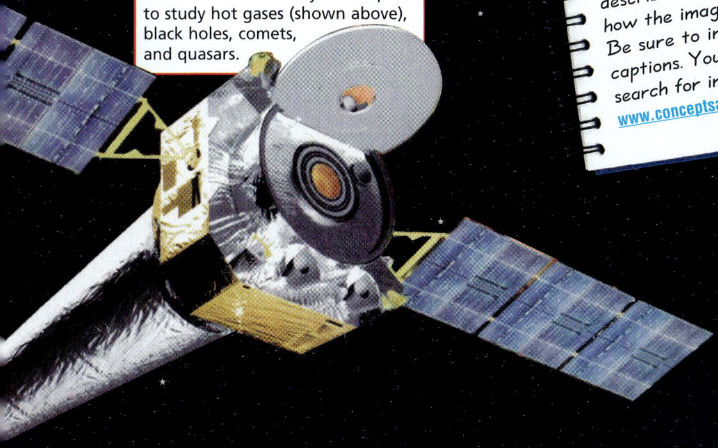

WRITING ACTIVITY

Science Log

At the peak of Mauna Kea, in Hawaii, is an infrared telescope. Do some research to find out where you can see the images this telescope produces. Choose three of the images. Write a paragraph about each image that describes what is shown and how the image was taken. Be sure to include photo captions. You can start your search for information at www.conceptsandchallenges.com.

Ultraviolet Light Telescopes
Ultraviolet telescopes are often located aboard satellites like the one shown here. The telescopes have gathered information about very hot space objects such as the Sun (right), quasars, and white dwarfs.

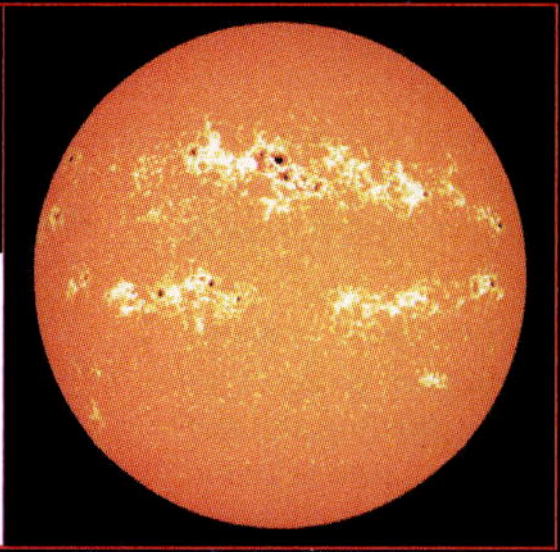

THE BIG IDEA

What do modern telescopes show us about the universe? Have students start their research for their Science Log Writing Activity at www.conceptsandchallenges.com. Be sure students use search engines and websites that are reliable and acceptable. You may also wish to provide copies of the *Big Idea Planner* and *Big Idea Science Log Entry* study tools found in the Teacher's Resources.

3 Assess

Use the Writing Activity Rubric from the Teacher's Resources to assess students' Science Logs. Fill in the rubric with the additional information below. For this assignment, students should have:

- identified the different parts of the electromagnetic spectrum and compared and contrasted different types of modern telescopes.
- communicated, in paragraph form, information about infrared telescope images.

4 Close

Space Booklet Have students search magazines or the Internet for examples of images taken in space by infrared, microwave, ultraviolet, X-ray, and gamma-ray telescopes. Tell students to cut out and glue the images or download and print the images to make a booklet. Students should label and describe each image and indicate what it taught astronomers about stars.

Tips for Using Technology

Saving Internet Images for Science Log Reports Demonstrate for students the ability to save graphic files while surfing the Internet for images taken by the telescope at Mauna Kea. Students may use the saved images in their Science Log report.

On a Macintosh®-based computer, click and hold the mouse button on the image you would like to save. A menu will appear. Move the mouse pointer to *Download Image to Disk*. Let go of the mouse button. A *Save* dialog box will appear. Choose the location into which you would like to save the file.

• TEACHING RESOURCES • Online or Print

Teacher's Resources

Big Idea: Chapter 16, p. 8

Big Idea Planner: Study Tools, pp. 11–12

Big Idea Science Log Entry: Study Tools, p. 13

Writing Activity Rubric: Teaching Tools, p. 9

16-6 *(pp. 392–393)*

1 Introduce

Linking Prior Knowledge Have students review distances on a map (Lesson 1-7). Remind students that distances on Earth can be measured in centimeters, meters, or kilometers. Tell them that distances in space are so great that much larger units are needed for measuring. Introduce the light year and AU (astronomical unit, which is equal to the mean distance from Earth to the Sun—about 150 million km) as units of measurements used by astronomers.

2 Teach

Demonstration The inner planets are much closer to one another than the outer planets are. To demonstrate this relationship, use plastic-foam balls or marbles to represent the planets. Place the planets in proper sequence on the floor, using a scale of 10 cm = 1 AU. Ask students: *Which two planets are farthest apart?* (Uranus and Neptune) *Which planets are closest together?* (Earth and Venus are 0.3 AU apart, as are Venus and Mercury.) **learning style: visual**

Discussion Define 1 AU as the mean average distance between Earth and the Sun, or about 150 million km. Have students study the chart in Figure 16-19 and discuss the distances of other planets from the Sun. Point out that Mercury and Venus are closer to the Sun than Earth is and that they are therefore less than 1 AU from the Sun. **learning styles: auditory, visual**

Reinforcement Randomly read aloud the names of the planets in Figure 16-19. Have students give each planet's distance from the Sun in AU.
learning styles: auditory, visual

Answers to questions

1 **EXPLAIN:** When measured in kilometers, the distances in space are so great that the numbers are too large to work with easily.

2 **INTERPRET:** about 700 light years

3 **OBSERVE:** 0.7 AU

16-6 How do astronomers measure distance?

Objective

Identify two methods astronomers use to measure distances in space.

Key Terms

light-year: unit of measurement equal to about 10 trillion km

parallax (PAR-uh-laks)**:** apparent change in the position of a distant object when seen from two different places

astronomical (as-truh-NAHM-ih-kuhl) **unit (AU):** unit of measurement based on the Sun's distance from Earth and equal to about 150 million km

Measuring Distance Most distances on Earth can be measured in meters or kilometers. How would you measure the distance from Earth to the stars? You might try to use kilometers. However, astronomers have found that the distances to stars are so great that the numbers are too large to work with easily.

For example, the star called Proxima Centauri (PRAHK-suh-muh sen-TAW-ree) is the closest star, other than the Sun, to Earth. Proxima Centauri is 40,000,000,000,000 km from Earth. As you can see, this is a very large number. Astronomers had to create special units to measure distances in space.

1 **EXPLAIN:** Why did scientists create special units to measure distances in space?

Light-Years Astronomers often measure the distance to an object in space using light-years. A **light-year** is equal to the distance light travels in one year.

Light travels through space at a speed of about 300,000 km per second. A light-year is equal to almost 10 trillion km. Light from the Sun reaches Earth in a little more than 8 minutes. Light from the North Star, Polaris, reaches Earth in about 431 years.

2 **INTERPRET:** How far, in light-years, is Polaris from Earth?

Parallax Astronomers can use parallax to find out distances to the closer stars. **Parallax** is the apparent change in the position of a distant object when seen from two different places.

Figure 16-18 shows how a nearby star seems to move against a background of more distant stars. By measuring how much the nearby star appears to move, astronomers can calculate how far away the star actually is. Nearby stars have a larger angle of parallax than distant stars do.

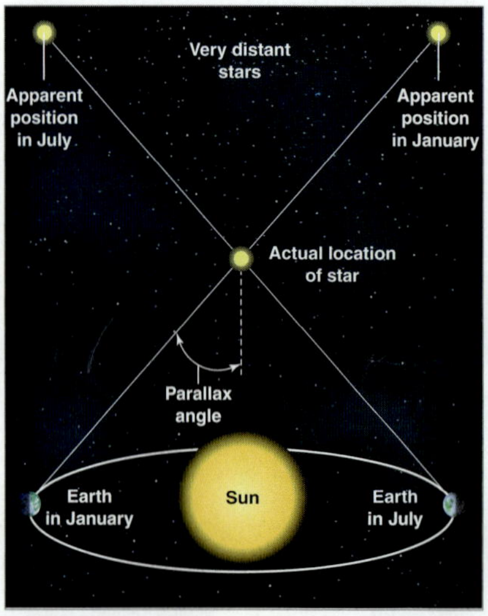

▲ **Figure 16-18** Parallax

Astronomical Units Although it varies, the average distance from Earth to the Sun is about 150 million km. Astronomers call this distance an astronomical unit. One **astronomical unit**, or 1 **AU**, is equal to 150 million km. An astronomical unit can also be used to measure distances in space, especially distances between the planets in the solar system. Figure 16-19 shows the distances of the planets from the Sun in astronomical units.

Teaching Options

ESL/ELL STRATEGY

AU to Light-Years Be sure that English-language learners understand why astronomers have developed two different means of measuring distance in space. One (AU) is better for nearby objects, such as the planets and moons of the solar system. The other (light years) is for objects that are very far away. Have students practice converting different distances from km to AU until they are relatively comfortable using this measuring system.

EXTENDING THE LESSON

Researching Other Measurement Units Have small groups of interested students use reference books or the Internet to find out how scientists measure distances between objects that are much too small to be seen, such as atoms. Students can also research other astronomical units of measurement, such as the parsec, which is equivalent to 3.26 light years. Have the groups share the results of their research with one another. **learning style: linguistic**

DISTANCES OF PLANETS FROM THE SUN			
Planet	Distance	Planet	Distance
Mercury	0.4 AU	Jupiter	5.2 AU
Venus	0.7 AU	Saturn	9.5 AU
Earth	1.0 AU	Uranus	19.2 AU
Mars	1.5 AU	Neptune	30.1 AU

▲ Figure 16-19

OBSERVE: In astronomical units, how far from the Sun is Venus?

CHECKING CONCEPTS

1. Why do astronomers not measure the distances to stars in kilometers?
2. What is the name of the star, other than the Sun, closest to Earth?
3. What is the average distance from Earth to the Sun in kilometers?
4. What is an astronomical unit?
5. What is the speed of light in space?
6. What is a light-year?

THINKING CRITICALLY

7. CALCULATE: Convert each of the distances shown in Figure 16-19 into kilometers.
8. SEQUENCE: Put the planets in order based on distance to each other, from greatest distance to least distance.

BUILDING MATH SKILLS

Calculating One light-year is equal to the distance light travels in one Earth year, or 10 trillion km. The diameter of Earth is only 13,000 km. The star known as Alpha Centauri is 4.4 light-years away. Figure out how many Earths would have to be placed side by side to reach Alpha Centauri. Procyon is 11.4 light-years away. How many Earths would be needed to reach the star known as Procyon?

Hands-On Activity

OBSERVING PARALLAX

You will need three sheets of continuous-feed computer paper, a metric ruler, a drinking straw, and tape.

1. Remove one of the side edges from the computer paper. Tape this strip horizontally to the wall.
2. Tape the straw upright on a metric ruler at the 15-cm mark.
3. Stand about 3 m from the wall. Hold the ruler level with the floor. Close your left eye and line up the straw with the left end of the paper strip.
4. Open your left eye and close your right eye. Count the number of holes the straw appears to move along the strip.
5. Move the straw to the ruler's 30-cm mark and repeat Steps 3 and 4.

▲ **STEP 3** Close your left eye and line up the straw with the left end of the paper strip.

Practicing Your Skills

6. IDENTIFY: What does the strip of paper with the holes represent?
7. OBSERVE: **a.** At the 15-cm mark, how many holes did the straw appear to move? **b.** How many did it move at the 30-cm mark?
8. ANALYZE: Why does the straw appear to change position?
9. COMPARE: Which has a greater parallax, a nearby star or a faraway star? Why?

QUICK CONVERSIONS		
	Metric/SI	English
1 AU	150 million km	93 million mi
1 light-year	10 trillion km	6.2 trillion mi

• TEACHING RESOURCES •
Online or Print

Teacher's Resources

Lesson Review: Chapter 16, p. 9

Enrichment: Chapter 16, p. 10

3 *Assess*

CHECKING CONCEPTS ANSWERS

1. If measured in kilometers, the numbers would be too large to work with.
2. Proxima Centauri
3. 150 million km
4. the average distance from Earth to the Sun
5. about 300,000 km/sec
6. the distance light travels in one year

THINKING CRITICALLY ANSWERS

7. CALCULATE: 0.4 AU = 60 million km; 0.7 AU = 105 million km; 1.0 AU = 150 million km; 1.5 AU = 225 million km; 5.2 AU = 780 million km; 9.5 AU = 1,425 million km; 19.2 AU = 2,880 million km; 30.1 AU = 4,515 million km; 39.4 AU = 5,910 million km
8. SEQUENCE: Uranus and Neptune (10.9 AU), Uranus and Saturn (9.7 AU), Jupiter and Saturn (4.3 AU), Mars and Jupiter (3.7 AU), Earth and Mars (0.5 AU), Venus and Earth (0.3 AU), Mercury and Venus (0.3 AU)

4 *Close*

BUILDING MATH SKILLS ANSWER

Calculating 4.4 light-years × 10,000,000,000,000 km/light-year = 44,000,000,000,000 km ÷ 13,000 km = 3,384,615 Earths; 11.4 light-years × 10,000,000,000,000 km/light-year = 114,000,000,000,000 km ÷ 13,000 km = 8,769,230,769 Earths

Hands-On Activity

OBSERVING PARALLAX

TIME: 15–30 minutes

PURPOSE: identifying, observing, analyzing

ALTERNATIVE MATERIALS: Punch holes in regular intervals in a strip of plain paper instead of using computer paper.

Practicing Your Skills Answers

6. IDENTIFY: stars
7. OBSERVE: Answers will vary. The number of holes moved should not exceed three or four.
8. ANALYZE: because it was viewed through one eye and then the other
9. COMPARE: a nearby star, because it has a larger angle of parallax

16-7 *(pp. 394–395)*

1 *Introduce*

STUDY HINT Have students link the main ideas in the three sections in this lesson into a paragraph that answers the lesson question.

Linking Prior Knowledge Students have likely seen a rocket launch on television or in a movie. Have students describe what they saw. Tell them that in this lesson, they will learn how a rocket overcomes the force of gravity. Remind them of what they learned about gravity (Lesson 8-2).

2 *Teach*

Demonstration Extend a ruler halfway over the edge of a table. Balance a plastic-foam ball on the end of the ruler resting on the table. Using a downward motion, strike the extended end of the ruler with your hand. The ball will fly into the air. Have students identify the action and reaction here. Point out that gravity causes the ball to fall back down. Flip the ball again, using a greater force. Guide students to see that the greater the downward force of your hand, the higher the ball rises. Explain that a great force is needed to move a rocket away from Earth's surface so that it is not pulled back by gravity.

Discussion Refer students to Figure 16-21. Point out the different parts of a rocket and describe a rocket's thrust.
learning styles: auditory, visual

Answers to questions

1. **STATE:** For every action, there is an equal and opposite reaction.
2. **DEFINE:** the reaction force that pushes a rocket forward
3. **EXPLAIN:** There is no air in space to supply the oxygen needed to burn fuel.

READING STRATEGY Make sure that students understand the scientific meaning of the word *law*. In science, a law describes a natural phenomenon. It is an unalterable fact. Have students make the connection to laws made by governments.

16-7 How does a rocket work?

Objective
Describe how a rocket works.

Key Term
thrust: forward force produced in a rocket engine

Newton's Third Law of Motion The Chinese invented rockets almost 800 years ago. They used them for fireworks and weapons.

To understand how a rocket works, you must know Newton's third law of motion. This law says that for every action, there is an equal and opposite reaction. For example, suppose you are sitting in a rowboat on a lake. You throw a rock into the lake. This is the action. At the same time, the rowboat moves backward slightly. This is the reaction.

1. **STATE:** What is Newton's third law of motion?

Rocket Engines The force that pushes a rocket forward is called **thrust.** The greater the thrust, the higher and faster the rocket will travel. What causes thrust? Fuel is burned inside a rocket engine. The fuel can be a solid or a liquid. As the fuel burns, inside the engine it produces hot gases that begin to expand. The expanding gases create pressure inside of the engine. This pressure forces the hot gases out of the back of the rocket. This is the action force. The rocket moves forward, in the opposite direction. This is called the thrust, or reaction force.

▲ **Figure 16-20** The *Saturn V* rocket was much more powerful than earlier rockets. It was used to launch the *Apollo* astronauts to the Moon.

2. **DEFINE:** What is thrust?

Escaping Earth's Gravity A lot of thrust is needed for a rocket to escape Earth's gravity. To get into Earth orbit, a rocket must reach a speed of more than 40,000 km/h. Large amounts of fuel are needed to produce enough thrust to reach this speed.

Rocket engines need oxygen to burn fuel. In space, there is no air to supply the oxygen. Rockets must carry their own.

As a rocket moves farther away from Earth, the pull of Earth's gravity on the rocket becomes weaker. Once the rocket is in space, there is little to slow it down. The rocket does not need to burn fuel to keep moving. It keeps moving in the same direction at a constant speed. Fuel is needed in space only to change the rocket's speed or direction.

Go Online **active art**
For: Multistage Rocket activity
Visit: PHSchool.com
Web Code: cfp-5021

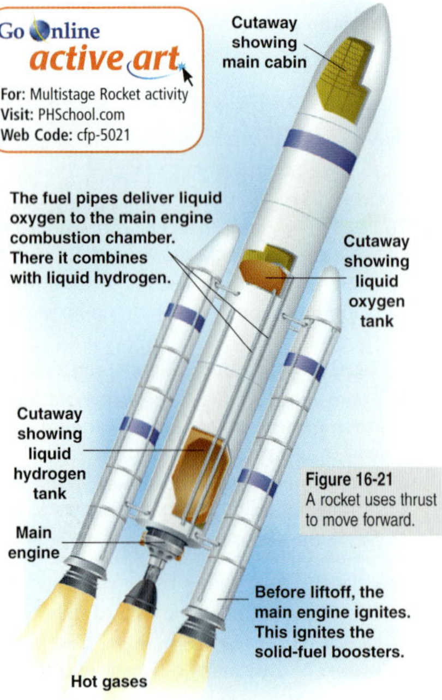

Figure 16-21 A rocket uses thrust to move forward.

3. **EXPLAIN:** Why must rockets carry oxygen?

Teaching Options

EXTENDING THE LESSON

Rocket Scientists at Work Have interested students research the work of Robert Goddard and Werner Von Braun, two scientists associated with the early development of rockets. Tell students to present their findings in a report. **learning style: linguistic**

INTEGRATING PHYSICAL SCIENCE

Other Newtonian Laws of Motion Sir Isaac Newton's first and second laws of motion also pertain to the movements of rockets.

- The first law of motion states that an object will remain at rest or continue to move in a straight line unless a net force acts on it. In space, a rocket will continue to move in the same direction at a constant speed unless thrust is used to change its speed or direction.
- The second law of motion states that the acceleration of an object is determined by the net force acting upon the object divided by the object's mass. The forces that act on a rocket are gravity and thrust. Have students list and categorize other examples of Newton's first and second laws of motion. **learning style: auditory**

CHECKING CONCEPTS

1. Rockets were invented by the __________ people.
2. Newton's third law of motion says that for every action, there is an equal and __________ reaction.
3. The force that pushes a rocket forward is called __________.
4. Rocket fuel needs __________ to burn.
5. As a rocket moves farther away from Earth, the pull of gravity becomes __________.

THINKING CRITICALLY

6. **ANALYZE:** What is fuel needed for in space?
7. **PREDICT:** Could a rocket traveling at a speed of 10,000 km/h get into space? Explain.
8. **INFER:** Why is thrust also known as the reaction force?
9. **EXPLAIN:** Why does a rocket not need to burn fuel to keep moving once it is in space?

BUILDING MATH SKILLS

Calculating Thrust is a force. The unit of force in the metric system is the newton (N). One newton (1 N) is equal to 4.5 lb. Below is a list of rockets and their thrusts at launch. The thrust is given in pounds. Convert each thrust into newtons. Then, list the rockets in order, from least thrust to greatest thrust.

Delta	205,000 lb
Saturn V	7,570,000 lb
Mercury-Atlas	367,000 lb
Space shuttle	6,925,000 lb
Vanguard	28,000 lb

For example:

1 N = 4.5 lb

105,000 lb ÷ 4.5 lb/N = 23,333 N

Hands-On Activity

DEMONSTRATING ACTION AND REACTION FORCES

You will need about 3 m of fishing line, a 3-cm piece of a drinking straw, tape, a balloon, and a twist tie.

1. Inflate a balloon. Close the end with a twist tie.
2. Thread the fishing line through the straw. Tape the straw across the top of the balloon.
3. Tie one end of the string to a chair. Hold the other end.
4. Move away from the chair. Hold the string tight. Remove the twist tie from the balloon and observe what happens.

▲ **STEP 2** Tape the straw to the balloon after you thread the fishing line through it.

Practicing Your Skills

5. **EXPLAIN:** What happened to the air in the balloon when you removed the twist tie?
6. **OBSERVE:** Where did the balloon go?
7. **INFER:** In which direction did the air from the balloon move?
8. **IDENTIFY:** **a.** What was the action force? **b.** What was the reaction force?

3 *Assess*

CHECKING CONCEPTS ANSWERS

1. Chinese
2. opposite
3. thrust
4. oxygen
5. weaker

THINKING CRITICALLY ANSWERS

6. **ANALYZE:** to change the rocket's speed or direction
7. **PREDICT:** No; A rocket must reach a speed of more than 40,000 km/h to escape Earth's gravity.
8. **INFER:** because thrust is a reaction to the pressure that forces the hot gases out of the back of the rocket
9. **EXPLAIN:** The pull of Earth's gravity becomes weaker, and the rocket continues in the same direction at a constant speed.

4 *Close*

BUILDING MATH SKILLS ANSWERS

Calculating *Delta*: 205,000 lb = 45,556 N; *Saturn V*: 7,570,000 lb = 1,682,222 N; *Mercury-Atlas*: 367,000 lb = 81,556 N; Space shuttle 6,925,000 lb = 1,538,889 N; *Vanguard*: 28,000 lb = 6,222 N; *Vanguard*, *Delta*, *Mercury-Atlas*, Space shuttle, *Saturn V*
learning style: logical/mathematical

Hands-On Activity

DEMONSTRATING ACTION AND REACTION FORCES

TIME: 15–30 minutes

PURPOSE: explaining, observing, inferring, identifying

Discussion Discuss reactions in general. Ask: *How would you react if you received a low grade in class?* Lead students to see that reactions occur in response to events or behaviors.

Practicing Your Skills Answers

5. **EXPLAIN:** It came out.
6. **OBSERVE:** toward the chair
7. **INFER:** in the opposite direction of the balloon's movement
8. **IDENTIFY:** **a.** the air escaping the balloon **b.** the forward movement of the balloon

MORE ON THE WEB

Designing a Rocket Car
Groups of students can use the Internet to design a rocket car of their own.

RETEACH

Newton's Third Law of Motion
Write on the board: *For every action, there is an equal and opposite reaction.* Ask students for examples that illustrate this law. (a recoiling rifle, jumping off a small boat, a rotating lawn sprinkler) Then, have students demonstrate this law of motion.
learning styles: auditory, kinesthetic

• *TEACHING RESOURCES* •
Online or Print

Teacher's Resources
Lesson Review: Chapter 16, p. 11
Enrichment: Chapter 16, p. 13

Laboratory Manual
Lab. Challenge: How does a heat shield work? pp. 81–82

Go Online
active art

For: Multistage Rocket activity
Visit: PHSchool.com
Web Code: cfp-5021

Students can interact with the art of multistage rockets online.

Lab Activity

(pp. 396–397)

1 Prepare the Lab

Modeling Newton's Third Law of Motion

SET-UP TIME: **LAB TIME:**

BACKGROUND: Outer space can be directly explored by using vehicles that can carry spacecraft and humans into orbit and beyond. The only vehicle capable of spaceflight is the rocket. The principles behind rocket flight are based on Newton's three laws of motion. The third law states that for every action, there is an equal and opposite reaction. The action in rocketry is the backward thrust of burning rocket fuel. The reaction is the movement of the rocket in the opposite direction. The other two laws of motion relate to inertia and factors that control the magnitude of the thrust.

PURPOSE: Students will investigate Newton's third law of motion by constructing and flying a model rocket.

ALTERNATIVE MATERIALS: A table tennis ball or plastic-foam ball can be used in place of the paper nose cone.

SCIENCE SKILLS: Students will **construct** a model rocket and **observe** its flight. They will **analyze** the factors that affected the rocket's flight and its performance.

ADVANCE PLANNING: Select a location for the rocket launch away from overhead light fixtures or wires. Obtain empty 35-mm film canisters from a local photo finisher. The caps on the canisters must snap into place. Caps that extend out over the edge of the canisters will not work properly.

2 Guide the Lab

PROCEDURE: Students may construct rockets individually or in pairs. Discuss the procedure with students. Demonstrate how to roll a nose cone from a semicircle of paper. Before adding the antacid tablet to the water, have students practice putting the caps on the canisters. When the cap is properly in place, they will hear a snap. Review the launch procedure with students. Provide students with a copy of the Lab Activity Report from the Teacher's Resources for students to record their observations and conclusions.

LAB ACTIVITY
Modeling Newton's Third Law of Motion

Materials

Safety goggles, 35-mm film canister with inner-locking cap, sheet of paper, cellophane tape, scissors, fizzing antacid tablet, water, tablespoon

BACKGROUND

The only way to reach space is to ride a rocket. The force driving most rockets is produced when rocket fuel burns. Hot gases shoot out from the engines of the rocket, and the rocket travels upward into space. However, not all rockets have to burn fuel to work. Thrust can be produced in many ways.

PURPOSE

In this activity, you will construct and fly a rocket.

PROCEDURE

1. Copy the chart in Figure 16-22.
2. Using half a sheet of paper, roll and tape a snug tube around a film canister. Make sure the cap end of the canister sticks out 12 mm. Tape the tube to the canister.
3. Cut out and tape rocket fins to the lower end of the tube. The fins can be of any shape. Make three or four fins. Cut out a half-circle of paper. Roll it into a nose cone. CAUTION ⚠ Be careful when using scissors.
4. Tape the nose cone to the upper end of the tube. Your rocket is now ready. Draw a diagram of the rocket in your chart.
5. Put on your safety goggles. Take your rocket to a launch site. Put a tablespoon of water inside the canister. Drop in one-half of an antacid tablet that fizzes when dropped into water.

▲ **STEP 2** Roll the paper around the film canister.

▲ **STEP 3** Tape rocket fins to the lower end of the tube. Prepare a nose cone.

Teaching Options

EXTENDING THE LESSON

Future Space Missions Have students use newspapers, magazines, or the Internet to research future space missions. Students can work in small groups to gather information about the type of spacecraft involved, its mission, and its planned launch date. Have students write newspaper articles describing the space mission of their choice.
learning style: linguistic

RETEACH

Reviewing Newton's Third Law of Motion Have students identify the action force in this activity. (the fizzing antacid tablet) Ask: *What is the reaction force?* (the flight of the rocket) *What is another name for reaction force?* (thrust) Have students write three cause-and-effect statements about the lab activity. Students can use the Cause-and-Effect graphic organizer from the Teacher's Resources.
learning styles: auditory, linguistic

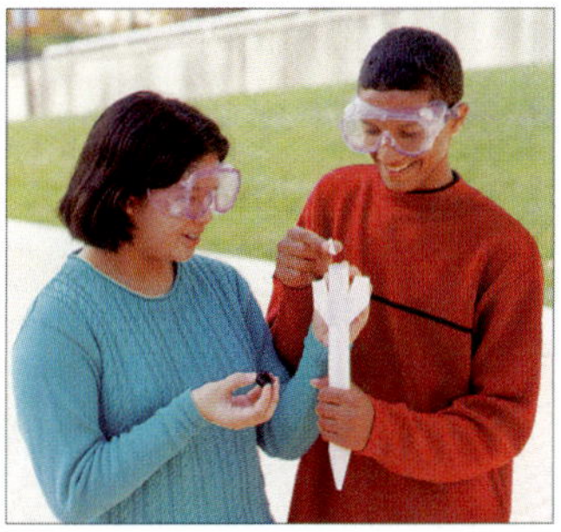

▲ **STEP 5** Drop half of your antacid tablet into the canister.

▲ **STEP 6** Set your rocket on the launch pad and step away.

6. Quickly snap the film canister cap in place. Set the rocket right side up on the launch pad and step away. Record your observations in your chart.

Diagram of your rocket	How well did your rocket perform?
Diagram of someone else's rocket	How well did that rocket perform?

▲ **Figure 16-22** Draw a diagram of your rocket and someone else's rocket. Record your observations.

CONCLUSIONS

1. **OBSERVE:** How high did your rocket go?
2. **OBSERVE:** Did any other rockets fly higher than yours?
3. **ANALYZE:** What do you think made some rockets fly higher than others?
4. **COMPARE:** Write a report comparing the two rockets you observed. Say why you think one worked better than the other.

TROUBLESHOOTING: Make sure that the body tube of the rocket is securely taped onto the canister. The exposed cap end should extend out of the bottom of the body.

SAFETY TIP: Make sure that students wear eye protection when launching the rockets.

EXPECTED OUTCOME: Students will observe that the film canister cap and the contents of the canister will pop out and fall down, and that the rocket will fly upward in the opposite direction.

LAB HINT: Place the rockets on plastic-foam plates to catch the liquid expelled by the launch. To estimate rocket heights, use an erasable marker to mark several measurements on a wall, then launch the rockets near this wall. To help identify the rockets, encourage students to decorate their rockets with markers.

3 Conclude the Lab

1. **OBSERVE:** Rockets should travel at least a meter into the air.
2. **OBSERVE:** It is possible that some of the rockets will achieve great heights.
3. **ANALYZE:** The reaction force was greater because of variations in amounts of water and fizzing action.
4. **COMPARE:** Check students' reports for an explanation that references Newton's third law of motion.

Use the Laboratory Activity Rubric from the Teacher's Resources to assess students' lab activities. Fill in the rubric with the additional information below. For this activity, students should have:

- constructed and flown a rocket.
- drawn conclusions from their recorded observations.

4 Extend the Lab

Have students repeat the activity to experiment with different amounts of antacid tablets, different amounts of water, and different water temperatures. Make sure that students test only one variable at a time and keep careful lab notes on each test.

After students have written their reports, have them discuss the factors that both improved and detracted from the rockets' performance. List their responses on the board. Based on this list, have students decide on the best rocket design.

Tips for Using Technology

Using Word Processing Software to Write Lab Reports

Have students use word processing software to write their lab reports for this activity. Students can print out their word processing document to share with the class. Tell students to check their spelling before printing, using the word processing program's spell checker. Caution them, however, that the spell checker will not recognize a misspelled word if it forms an acceptable word, even though the word is being used incorrectly in that context (for example, *to, too,* and *two*).

• TEACHING RESOURCES • Online or Print

Teacher's Resources

Lab Activity Report: Study Tools, pp. 16–17

Laboratory Activity Rubric: Teaching Tools, p. 8

Cause-and-Effect Diagram: Study Tools, p. 3

16-8 *(pp. 398–399)*

1 Introduce

STUDY HINT Have students list three design features of the space shuttle and three uses. Then, ask them to relate the shuttle's design to its uses if they can.

Linking Prior Knowledge Many students have either traveled on an airplane or seen one land. Have students describe the take-off and landing of an airplane. Tell them that the space shuttle lands much like an airplane but is launched into space like a rocket.

2 Teach

Demonstration If possible, obtain a model space shuttle for students to examine. Show them the main parts of the shuttle and demonstrate how the booster rockets and liquid-fuel tank separate from the orbiter. If you cannot obtain a model space shuttle, bring a model airplane to class and have students compare and contrast its design with the illustration of the space shuttle in Figure 16-23.
learning styles: visual, tactile

Discussion Refer students to Figure 16-23. Ask them to point out the three main parts of the shuttle and to describe the function of each part.
learning styles: auditory, visual

Answers to questions

1. **INFER:** It could be reused many times, which would be a cost savings.
2. **LIST:** solid-fuel booster rockets, liquid-fuel tank, and shuttle orbiter
3. **NAME:** Spacelab

READING STRATEGY After they have read the lesson, have students discuss in small groups whether or not they would want to ride on the space shuttle. Have students explain their reasoning, based at least partly on what they read.

16-8 What is the space shuttle?

Objective
Explain how the space shuttle works.

A Reusable Spaceship In the 1960s and 1970s, all spaceflights were made with spacecraft that could be used only once. This was very expensive and wasteful. The National Aeronautics and Space Administration (NASA) decided to build a spacecraft that could go into space and return to Earth many times. This new kind of spacecraft was known as the Space Transportation System, or the space shuttle.

1. **INFER:** How was the space shuttle an improvement over earlier spacecraft?

Design of the Shuttle The space shuttle has three main parts. Two of the parts are needed to get the shuttle into space. These are the solid-fuel booster rockets and the liquid-fuel tank. The third part is the shuttle orbiter. The orbiter is the only part that stays in space after launch.

How does the shuttle get into orbit? At launch, both the booster rockets and the orbiter's rocket engines are fired. After launch, the boosters separate from the orbiter. They fall back into the ocean and can be hauled out to be used again.

The large liquid-fuel tank provides fuel for the orbiter's engines. The tank drops away before the orbiter reaches orbit. It is not reusable.

Once the orbiter is in space, the engines are turned off. They are fired again to change the orbiter's speed and direction when it is ready to return to Earth. When its mission is over, the orbiter glides to a landing on a runway.

◀ **Figure 16-24** The orbiter uses a parachute to help slow its speed upon landing.

2. **LIST:** What are the three main parts of the space shuttle?

Uses of the Shuttle The shuttle's cargo bay is designed to carry between 10,000 and 50,000 kg of equipment into orbit. The weight depends partly on the shuttle's altitude in space. Satellites and space probes can be launched from the shuttle.

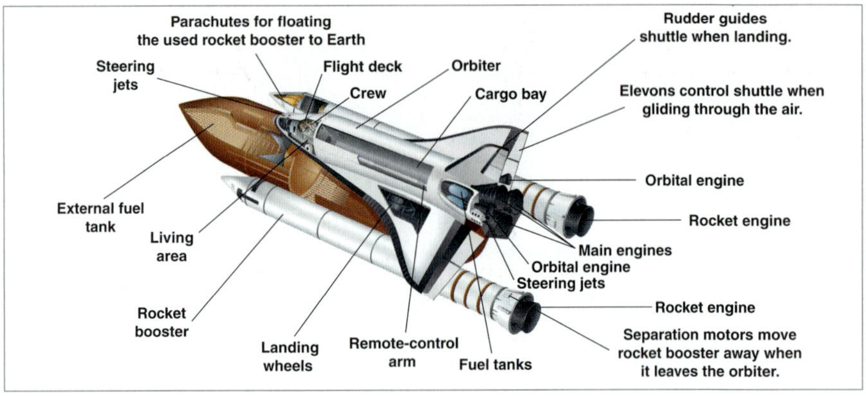

▲ **Figure 16-23** The space shuttle

Teaching Options

FAST FACTS *A Snowstorm in Space*

In 1988, the former Soviet Union tested its first space shuttle. The shuttle was named *Buran*, which is the Russian word for *snowstorm*. *Buran* was launched from a cosmodrome in the Soviet Union. The shuttle was boosted into orbit by a huge Energia rocket. After circling Earth twice, *Buran* made a remote-controlled landing. Many people who saw photographs of *Buran* said that it looked very much like American space shuttles. After this initial test flight, the *Buran* program was delayed by funding difficulties. The program was ended in 1993.

COOPERATIVE LEARNING

Jobs in Space Tell students that getting a spacecraft into orbit requires the effort of many dedicated workers. Assign students to small groups and have each group research a different career in the aerospace industry. Students should then give oral reports about the daily work tasks and the training or education needed to do the job. **learning styles: auditory, linguistic**

Shuttle astronauts can repair satellites in orbit. Satellites can also be returned to Earth for repair.

The shuttle will sometimes carry a flying laboratory called Spacelab in its cargo bay. In this laboratory, a variety of science experiments are performed. The shuttle is also used to ferry people and supplies into space.

▲ **Figure 16-25** Spacelab is used for experiments aboard the space shuttle.

NAME: What is the laboratory carried by the shuttle called?

CHECKING CONCEPTS

1. Another name for the space shuttle is __________.
2. The shuttle orbiter lands like an airplane on a __________.
3. Satellites are carried in the shuttle's __________.
4. Spacelab is a flying __________.

THINKING CRITICALLY

5. **ANALYZE:** How is the space shuttle useful to people on Earth?
6. **INFER:** Why is it important for the boosters to be separated from the orbiter before going into orbit?

INTERPRETING VISUALS

Use Figure 16-23 to answer the following questions.

7. **IDENTIFY:** How many main parts does the space shuttle have?
8. **INFER:** What part of the shuttle actually goes into space?
9. **ANALYZE:** How does the design of the shuttle make it a kind of space plane?

Science and Technology

THE NATIONAL AERONAUTICS AND SPACE ADMINISTRATION

The National Aeronautics and Space Administration, or NASA, started operations in 1958. NASA is a government agency that supports research. It also develops vehicles and programs for exploring Earth's atmosphere and space.

Over the years, NASA has achieved many technological breakthroughs. Project Mercury, NASA's first major project, proved that technology could help humans to survive in space. The *Apollo* missions put people on the Moon.

Space shuttles today are helping to build the International Space Station. Spacecraft are being sent to Mars and beyond. The Hubble Space Telescope takes pictures of the solar system and the stars beyond. Satellites are being launched into space for many different purposes. NASA technology has also been used to improve our health. *Apollo* technologies led to computerized aerial tomography, or CAT scans, and better kidney dialysis machines. Space shuttle technologies led to a new pump used in artificial hearts.

▲ **Figure 16-26** NASA technicians work on Earth and in space.

Thinking Critically What types of vehicles has NASA built?

3 *Assess*

CHECKING CONCEPTS ANSWERS

1. Space Transportation System
2. runway
3. cargo bay
4. laboratory

THINKING CRITICALLY ANSWERS

5. **ANALYZE:** Possible answer: It increases our knowledge of the universe and launches satellites that increase our knowledge about Earth.
6. **INFER:** They are empty of fuel and so would be useless extra weight in space. If they are ejected, they can be recovered and used again.

INTERPRETING VISUALS ANSWERS

7. **IDENTIFY:** three
8. **INFER:** the orbiter
9. **ANALYZE:** Possible answer: It can be flown like an airplane and land on a runway like an airplane, but it travels in space.

4 *Close*

Science and Technology

THE NATIONAL AERONAUTICS AND SPACE ADMINISTRATION

Real-Life Connection Sally Ride was the first U.S. woman to fly in space. Her first flight was aboard the seventh space shuttle flight in June 1983. On the second and third shuttle flights, she acted as capsule communicator, the person on the ground who talks to the astronauts in flight.

Discussion Have students brainstorm reasons why clear, steady communication between the ground and the shuttle crew is so crucial. Use the opportunity to discuss the importance of teamwork in general.

Thinking Critically Answer
Accept any of the following: space shuttles, land rovers, satellites, space probes, spacecraft, and space stations.

MORE ON THE WEB

Viewing Space Shuttles To view video clips of various space shuttles launching and landing, have students visit the Internet.

ONGOING ASSESSMENT

Have students take turns describing some characteristic of the space shuttle. Students may mention the design or function of a particular shuttle part, or a use for the shuttle itself. **learning styles: auditory, visual**

• TEACHING RESOURCES •
Online or Print

Teacher's Resources

Lesson Review: Chapter 16, p. 12

Enrichment: Chapter 16, p. 13

16-9 *(pp. 400–401)*

1 Introduce

STUDY HINT Have students write in their notebooks one fact about each of the visuals shown in the lesson, other than what is in the caption. This information can help them understand and remember the lesson.

Linking Prior Knowledge Remind students that artificial things are made by people. Have them give examples of both artificial and natural things. Write their responses on the board. Ask: *Is the Moon an artificial or a natural satellite?* (natural)

2 Teach

Demonstration Use a globe and a small ball to demonstrate the four different types of orbits shown in Figure 16-28. Hold the ball about 5 cm above the globe, then move the ball around the globe to represent a circular orbit. Orbit the ball about 15 cm above the poles on the globe to represent a polar orbit. Orbit the ball about 50 cm above the equator to represent a geostationary orbit. Vary the height of the orbit to represent an eccentric orbit. Have students identify each orbit.
learning style: visual

Discussion Ask students how they think artificial satellites and space probes are used to explore space. Discuss the advantages of space probes. (They can be sent to examine distant and dangerous places.) **learning style: auditory**

Reinforcement Remind students that the United States launched its first spacecraft, *Explorer 1*, a year after *Sputnik* was launched. Have students write a news article about one of the events mentioned in this lesson.

Answers to questions

1. **NAME:** an orbit
2. **DESCRIBE:** to collect information about Earth, the Sun, stars, other planets, comets, and other bodies in the solar system; for communications and navigation
3. **IDENTIFY:** Jupiter and Neptune

16-9 What are satellites and space probes?

Objective
Explain how artificial satellites and space probes are used to explore space.

Key Terms
satellite (SAT-uhl-eyet): natural or artificial object orbiting a body in space
orbit: curved path of one object around another object in space

Artificial Satellites

A **satellite** is any object, natural or artificial, that follows a curved path around another object in space. The curved path of the object is its **orbit.**

▲ **Figure 16-27** An *Intelsat VI* satellite

For thousands of years, astronomers were only able to study the skies from the surface of Earth. Then, on October 4, 1957, the Space Age began. On that day, the Soviet Union launched the first artificial satellite. It was called *Sputnik*. *Sputnik* circled Earth every 96 minutes.

Most satellites are launched into one of four main orbits. Nearly circular orbits are usually about 250 km above Earth's surface. Polar orbits are usually about 800 km high.

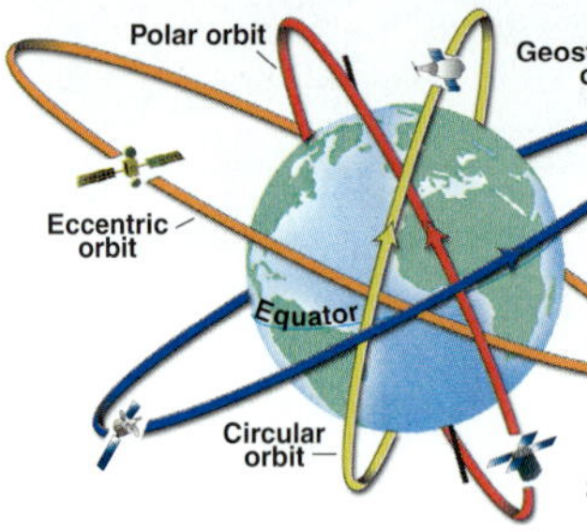

◀ **Figure 16-28** Satellite orbits

The height of a highly elliptical, or eccentric, orbit depends on where the satellite is in its orbit. A geostationary orbit is 36,000 km above Earth's surface. A satellite in geostationary orbit stays above the same spot on Earth. This is because it travels at the same speed as the planet rotates.

1 **NAME:** What is the curved path of a satellite called?

Uses of Satellites Since *Sputnik*, thousands of artificial satellites have been placed in orbit around Earth. These collect information about Earth, the Sun, the stars, other planets, comets, and other bodies in the solar system.

For example, the Global Positioning System (GPS) is a radio navigation system of about 24 satellites operating in circular orbits. GPS is a guidance service provided by the U.S. Air Force Space Command unit.

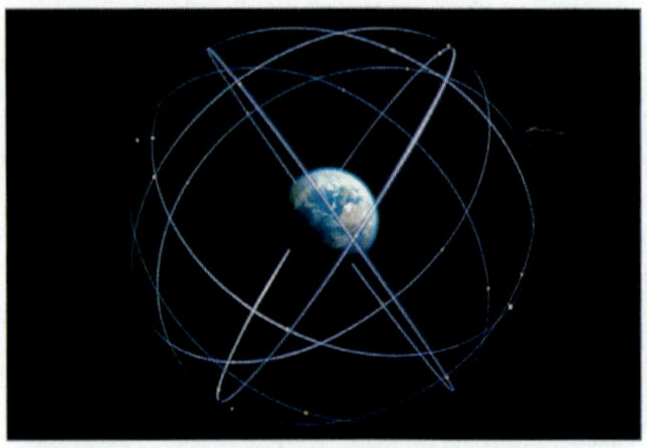
▲ **Figure 16-29** The 24 GPS satellites circle Earth using six different orbits.

The system relays positions on Earth's surface in degrees of latitude and longitude. GPS receivers with computers have improved surveying techniques. Surveyors are now able to monitor changes of Earth's crust caused by tectonic motion. Meteorologists are able to use GPS signals to measure the temperature and water content of the atmosphere. Geologists can accurately map changes in the Greenland ice sheet to help them understand climate changes.

Teaching Options

ESL/ELL STRATEGY

Telling Things Apart Distinguish for English-language learners the differences among *space probes*, *satellites*, and *spacecraft*. A spacecraft is a generic name for all kinds of hardware sent into space. A satellite is a spacecraft that is put into orbit usually for scientific study or communications. A space probe is a spacecraft usually sent on a one-way journey to visit another part of the solar system and send back data. Have students list some examples of each type to further clarify the differences.

EXTENDING THE LESSON

Viewing Spacecraft If possible, arrange a visit to a space museum where rockets, satellites, or space probes are on display. You might instead visit a virtual space museum on the Internet. Afterwards, have students report on what they have learned.
learning styles: visual, linguistic

QUICK CONVERSIONS

	Metric/SI	English
Circular orbit	250 km	155 mi
Polar orbit	800 km	500 mi
Geostationary orbit	36,000 km	22,000 mi

Computers are essential in turning satellite data or measurements into meaningful information. For example, computers help meteorologists translate weather satellite data into current temperatures, pressures, humidity readings, and wind speeds. This information is used to create a weather report.

2 DESCRIBE: What are satellites used for?

Space Probes Astronauts have been sent to the Moon to explore it and have returned safely to Earth. Space can also be explored by space probes, which do not carry people. In fact, many kinds of space exploration are best done with space probes. Space probes can go places that would be too far or too dangerous for astronauts to travel to.

For example, part of the *Galileo* space probe sent to Jupiter entered its atmosphere in 1995. It radioed important information to Earth before being destroyed by high temperatures and pressure. *Cassini*, launched in 1997, reached Saturn in 2004. It was scheduled to orbit the planet and its moons until June 2008.

▲ **Figure 16-30** The *Cassini* probe explored Saturn from orbit when it got there in 2004.

Space probes are usually sent on one-way missions. They do not return to Earth. Some space probes to the outer planets, such as *Voyagers 1* and *2*, have even been sent out of the solar system.

The *Voyager* space probes were launched in 1977. In 1989, *Voyager 2* became the first spacecraft from Earth to reach Neptune. Both space probes sent back stunning images of all the giant planets.

Scientists believe that *Voyager 2* will continue traveling for thousands of years. They hope to continue receiving signals from both space probes until at least the year 2020.

3 IDENTIFY: What are two planets that have been visited by space probes?

CHECKING CONCEPTS

1. What is a satellite?
2. What was *Sputnik*?
3. What was the first space probe to reach Neptune?
4. What did *Cassini* do when it got to Saturn?

THINKING CRITICALLY

5. NAME: What are two advantages that space probes have over spaceships and human crews?
6. INFER: *Telstar* was one of the first artificial satellites. It was launched in 1962. *Telstar* was a communications satellite. What do you think was the function of *Telstar?*
7. INFER: How does GPS help ships and aircraft?

BUILDING SCIENCE SKILLS

Classifying Below is a list of some important artificial satellites. Find out the function of each. Then, classify each according to its function. On a piece of paper, write the words *Communications Satellite*, *Weather Satellite*, *Navigation Satellite*, and *Scientific Satellite*. Place each satellite under its correct heading.

Early Bird	*Explorer I*	*Intelsat*	*Tiros*
Echo	*Landsat*	*Nimbus*	*Transit*

◀ **Figure 16-31** *Tiros*

3 Assess

CHECKING CONCEPTS ANSWERS

1. a natural or an artificial object orbiting a body in space
2. the first artificial satellite
3. *Voyager 2*
4. orbit the planet and its moons for four years

THINKING CRITICALLY ANSWERS

5. NAME: Space probes can go to places that would be too dangerous for humans. Space probes can also be sent on one-way missions.
6. INFER: It was to relay television signals. Accept any thoughtful responses related to communications.
7. INFER: It relays their position in degrees of latitude and longitude.

4 Close

BUILDING SCIENCE SKILLS ANSWER

Classifying

Communications Satellites: *Early Bird, Echo, Intelsat*
Weather Satellites: *Tiros, Nimbus, Landsat*
Navigation Satellite: *Transit*
Scientific Satellites: *Explorer 1, Landsat*

MORE ON THE WEB

The Golden Record Students can learn about the Golden Record on the Internet. They can view and hear some of the media included on the record.

The Parts of Voyager Students can also visit the Internet to see a diagram of the *Voyager* space probe. Then, have them write a short report on why they believe each component was necessary.

• TEACHING RESOURCES • Online or Print

Teacher's Resources
Lesson Review: Chapter 16, p. 14

16-10 (pp. 402–403)

1 Introduce

Linking Prior Knowledge Have a volunteer describe the layout of his or her home. Point out that most homes have a kitchen, one or more bathrooms, one or more bedrooms, and so on. Tell students that a space station is a temporary home for scientists. Thus, a space station also has some of the same features as students' homes.

2 Teach

Demonstration Obtain a sample of freeze-dried food from your local grocery store or camping supply store. Have students observe the food before you prepare it. Cook the food according to directions. Have students observe the cooked food. Discuss the characteristics of freeze-dried food that make it suitable for space travel. **learning style: visual**

Discussion The idea of living and working on a space station can be used to stimulate discussion. Ask students if they would like to work on the International Space Station and have them give reasons why or why not. **learning style: auditory**

Answers to questions

1. **EXPLAIN:** for living and working in space for long periods of time
2. **NAME:** *Skylab*
3. **DESCRIBE:** laboratories, living quarters, docking bays, solar panels, and a satellite repair shop
4. **HYPOTHESIZE:** Possible answers include dealing with medical emergencies or with psychological problems because of living in close quarters or being isolated from friends and family on Earth.

WRITING HINT Tell students that they can improve their spelling skills by keeping a dictionary on hand when writing. Encourage them to refer to the dictionary when doing any type of work that requires writing, even just answering questions.

16-10 How are space stations used?

Objective

Explain the role of space stations.

Space Stations Artificial satellites and space probes are very useful in space exploration. They can gather information and send it back to scientists on Earth. The scientists do not have to leave Earth. Sometimes, however, scientists prefer to make their observations directly. In a space station, scientists can live and work in space for long periods of time.

1 **EXPLAIN:** What is a space station useful for?

Early Space Stations In 1971, the former Soviet Union launched the first space station, *Salyut* ("salute"). The first U.S. space station, *Skylab*, was launched in 1973. It was about the size of a small house. Three teams of astronauts visited *Skylab*. Eventually, *Salyut* and *Skylab* wandered from their orbits, causing them to fall to Earth. They had shown, though, that people could safely live and work in space.

▲ **Figure 16-32** *Skylab* was the first U.S. space station.

The former Soviet Union later constructed in space its second space station, called *Mir* ("peace"). The first part of *Mir* was launched in 1986. Some Russian cosmonauts stayed on *Mir* for more than a year, setting new records. *Mir* fell out of orbit and back to Earth in 2001.

2 **NAME:** What was the first U.S. space station?

The International Space Station The United States and Russia launched the first parts of a new space station, called the International Space Station, or ISS, in 1998. Other countries are involved in this project, as well. The ISS may be complete by 2010. Many parts have already been carried into orbit by space shuttles or other rockets.

People have been working and living on the ISS since the year 2000. They use the space shuttle to travel to and from Earth. The completed ISS will have laboratories, living quarters, docking bays for shuttles, and solar panels for energy. It will also have a satellite repair shop.

▲ **Figure 16-33** The International Space Station (ISS) is scheduled to be finished by the year 2010.

3 **DESCRIBE:** What are some parts of the ISS?

Teaching Options

RETEACH

Reviewing Kinds of Spacecraft Write *Spacecraft* on the board. Write the following subheads under the main head: *Satellite*, *Space Probe*, *Space Shuttle*, and *Space Station*. Have students take turns identifying characteristics of each type of spacecraft. **learning style: linguistic**

ONGOING ASSESSMENT

Have students work in groups of four or five to design a blueprint for a space station. Tell them to reread the lesson for hints on what to include in their designs. Specifically, refer students to the subhead Living in Space on page 403. Afterwards, have the groups present their designs to the class. Students should identify the parts of their space station and explain the function of each part. **learning style: visual**

Living in Space Astronauts on the ISS must live and work in conditions that are very different from those on Earth. Life-support systems on board must provide oxygen and remove carbon dioxide. The air has to be pressurized. Food and water must be supplied from Earth. The water must be recycled.

Because of the very low gravity in space, the body does not have to work as hard in space as it does on Earth. Muscles tend to weaken. Therefore, exercise is essential to keep the astronauts' bodies in shape. Also, sleeping areas in space must have straps to prevent astronauts from floating off. Eyeshades must be worn because the Sun rises and sets every hour and a half on a spacecraft in near-Earth orbit.

4 **HYPOTHESIZE:** What are some other problems astronauts might have living in space?

CHECKING CONCEPTS

1. In a __________, scientists can live and work in space.
2. The first space station was called __________.
3. Parts for the ISS are carried into orbit mostly by the __________.
4. The ISS will be used to repair __________.

THINKING CRITICALLY

5. **INFER:** Astronauts living on a space station have to adjust to the weightlessness of space. Ordinary chores are often more complicated than they are on Earth. Identify three things that you do every day on Earth, such as take a shower, that might be very difficult to do on a space station.

DESIGNING AN EXPERIMENT

Design an experiment to solve the following problem. Include a hypothesis, variables, a procedure with materials, and a type of data to study. Also, tell how you would record your data.

PROBLEM: Earth's climate may be getting warmer partly because of the burning of fossil fuels. This could melt polar ice caps and create new deserts. How can scientists study this problem from the ISS?

People in Science

AEROSPACE WORKER

Space probes, satellites, and space stations are manufactured by the aerospace industry. This industry employs about one million workers. About half of the people in the aerospace industry actually work on putting spacecraft together. Many inspectors check the quality of each job as it is completed.

Because the aerospace industry uses the latest technologies, many aerospace workers have a background in science and engineering. The aerospace industry also employs lawyers, accountants, and clerical workers. If you are a high school graduate and a graduate of a technical school or college, you might want to find a job in the aerospace industry.

▲ **Figure 16-34** Many aerospace workers help assemble rockets.

Thinking Critically Why is it important to have many inspectors checking spacecraft?

MORE ON THE WEB

The International Space Station Have students visit www.conceptsandchallenges.com to learn more about the ISS. Students can search the Internet for flight schedules or images and video clips of space walks and construction on the station.

• TEACHING RESOURCES • Online or Print

Teacher's Resources

Lesson Review: Chapter 16, p. 15

Designing an Experiment: Study Tools, p. 14

Designing an Experiment Rubric: Teaching Tools, p. 7

3 Assess

CHECKING CONCEPTS ANSWERS

1. space station
2. *Salyut*
3. space shuttle
4. satellites

THINKING CRITICALLY ANSWER

5. **INFER:** Possible answers may include brushing teeth, taking out the trash, and setting the table for a meal.

4 Close

DESIGNING AN EXPERIMENT

Use the Designing an Experiment Rubric in the Teacher's Resources to assess students' experiments. Fill in the rubric with the additional information below. For this assignment, students should have:

- demonstrated an understanding of global warming.
- identified a way that ISS scientists could study the problem.

People in Science

AEROSPACE WORKER

Technology Connection NASA was created in 1958 by President Dwight D. Eisenhower. Its main launch center is located in Kennedy Space Center in Florida. This coastal location has generally fine weather and is near the ocean, making it easy to retrieve rocket fuel tanks and other equipment that fall into the sea after launch. Rockets and space shuttles are constructed and stored in Kennedy Space Center's Vehicle Assembly Building. Also on site is a 4.6-km (2.9-mi) runway used for space shuttle landings.

Discussion After reading this feature, have students identify the different kinds of jobs involved in the aerospace industry. On the board, make a list of the jobs. Have students discuss the advantages and disadvantages of each job.

Thinking Critically Answer
Possible answer: to make sure that no mechanical problems arise during the spacecraft's mission that might put astronauts or expensive materials at risk

Challenges (pp. 404–406)

Chapter Summary

Review Before students begin the Challenges, review the objectives of each lesson with them. If time permits, have students work in pairs to identify at least one relevant fact related to each objective.

STUDY HINT Have pairs of students take turns quizzing each other on key concepts in this chapter.

Key Term Challenges

MATCHING

1. orbit (16-9)
2. thrust (16-7)
3. astronomical unit (16-6)
4. concave mirror (16-4)
5. convex lens (16-3)
6. galaxy (16-2)
7. solar system (16-1)
8. satellite (16-9)

FILL IN

9. astronomy (16-1)
10. refracting telescope (16-3)
11. radio telescope (16-5)
12. light year (16-6)
13. reflecting telescope (16-4)
14. parallax (16-6)
15. astronomical units (16-6)

Chapter 16 Challenges

Chapter Summary

Lessons 16-1 and 16-2

- **Astronomy** helped ancient societies solve practical problems. Modern astronomers use telescopes, satellites, and space probes to study the **solar system** and the rest of the universe.

Lessons 16-3 and 16-4

- Galileo used an early **refracting telescope** to look at the Moon. A refracting telescope uses two **convex lenses.**
- A **reflecting telescope** usually uses a **concave mirror** to collect light. Newton made the first reflecting telescope.

Lesson 16-5

- A **radio telescope** picks up radio waves from objects in space.

Lessons 16-6 and 16-7

- **Light-years** are used to measure distances in space. One light-year is the distance light travels in one year, or about 10 trillion km.
- One **astronomical unit,** the distance from Earth to the Sun, is equal to 150 million km.
- Newton's third law of motion explains how rockets work. **Thrust** pushes a rocket forward.
- Rocket engines carry oxygen to burn fuel in space for changing speed or direction.

Lesson 16-8

- The space shuttle is reusable. Both the booster rockets and the liquid-fuel tank drop off before the orbiter reaches space. The orbiter returns to Earth and lands on a runway.
- The shuttle has many uses, including the launching and repairing of satellites.

Lessons 16-9 and 16-10

- Thousands of artificial **satellites** are in **orbit** today around Earth.
- Space probes can go to dangerous places. They can be sent on one-way missions.
- A space station allows scientists to live and work in space for long periods of time.
- The ISS is being built in orbit from parts carried by space shuttles. It will have a full-time research laboratory and satellite repair station.

Key Term Challenges

astronomical unit (p. 392)	parallax (p. 392)
astronomy (p. 380)	radio telescope (p. 388)
concave mirror (p. 386)	reflecting telescope (p. 386)
convex lens (p. 384)	refracting telescope (p. 384)
galaxy (p. 382)	satellite (p. 400)
light-year (p. 392)	solar system (p. 380)
orbit (p. 400)	thrust (p. 394)

MATCHING Write the Key Term from above that best matches each description.

1. curved path of an object around another object
2. force that pushes a rocket forward
3. distance of 150 million km
4. mirror that curves inward
5. lens thicker in the middle than at its edges
6. large system of stars
7. the Sun and all the bodies that circle the Sun
8. natural or artificial object that orbits a body in space

FILL IN Write the Key Term from above that best completes each statement.

9. The study of stars, planets, and other bodies in space is called __________.
10. A telescope that uses two convex lenses to produce an enlarged image of an object is a __________.
11. A telescope that is used to study radio waves coming from space is a __________.
12. A unit of distance equal to 10 trillion km is the __________.
13. A telescope that uses mirrors to form an image is a __________.
14. An apparent change in the position of a distant object when seen from two different places is called __________.
15. To measure distances to the planets, astronomers usually use __________.

Teaching Options

PREPARING STUDENTS FOR STANDARDIZED TESTS

Reading Strategy: Tell students that they should try their best to ignore distractions during a test and to remain focused on the questions and answers.

Writing Strategy: For questions that require an answer in paragraph form, tell students to first create a topic sentence. They can then build their answer around this sentence.

Interpreting Visuals Strategy: Tell students that all elements of a diagram, including colors, labels, arrows, and captions, should be closely examined when interpreting visuals.

ESL/ELL STRATEGY

Grouping Key Terms Have English-language learners group the Key Terms into meaningful categories, such as space hardware or tools (satellites, telescopes, and mirrors) and astronomy concepts (solar system, orbit, and parallax). Studying the terms as groupings can help students recall the meanings of unfamiliar terms.

Content Challenges TEST PREP

MULTIPLE CHOICE Write the letter of the term or phrase that best completes each sentence.

1. The first artificial satellite launched into space was
   - a. *Apollo*.
   - b. *Voyager*.
   - c. *Telstar*.
   - d. *Sputnik*.
2. A concave mirror
   - a. curves inward.
   - b. curves outward.
   - c. is flat.
   - d. curves inward and outward.
3. The telescope used by Newton was a
   - a. refracting telescope.
   - b. reflecting telescope.
   - c. radio telescope.
   - d. scanning telescope.
4. The closest star to Earth, other than the Sun, is
   - a. Io.
   - b. Proxima Centauri.
   - c. Polaris.
   - d. *Voyager*.
5. An astronomical unit is equal to the distance from
   - a. Earth to the Moon.
   - b. the Sun to the Moon.
   - c. Earth to the Sun.
   - d. Earth to Pluto.
6. To get into space, a rocket must reach a speed of more than
   - a. 3,000 km/h.
   - b. 4,000 km/h.
   - c. 30,000 km/h.
   - d. 40,000 km/h.
7. The space station that is currently in orbit is called
   - a. the space shuttle.
   - b. *Saturn V*.
   - c. *Apollo*.
   - d. the ISS.
8. Muscles can weaken in space because of the
   - a. high temperatures.
   - b. low air pressure.
   - c. low gravity.
   - d. change of diet.

TRUE/FALSE Write *true* if the statement is true. If the statement is false, change the underlined term to make the statement true.

9. The first artificial satellite was launched in 1957.
10. Scientists think that the universe began when a huge explosion called the Big Bang occurred.
11. The telescope used by Galileo was a radio telescope.
12. The first person to study the surface of the Moon with a telescope was Newton.
13. The lens in a refracting telescope that collects light and focuses the image is the objective lens.
14. A telescope that uses mirrors instead of lenses is a refracting telescope.
15. Stars can be studied on a cloudy night with a radio telescope.
16. The *Galileo* space probe is studying the atmosphere of Earth.
17. The only part of a space shuttle that goes into space is the liquid-fuel tank.
18. Newton's third law of motion states that for every action, there is an equal and opposite reaction.

Content Challenges

MULTIPLE CHOICE

1. d (16-2)
2. a (16-4)
3. b (16-4)
4. b (16-6)
5. c (16-6)
6. d (16-7)
7. d (16-8)
8. c (16-10)

TRUE/FALSE

9. true (16-9)
10. true (16-2)
11. false; refracting (16-3)
12. false; Galileo (16-3)
13. true (16-3)
14. false; reflecting (16-4)
15. true (16-5)
16. false; Jupiter (16-9)
17. false; orbiter (16-8)
18. true (16-7)

ALTERNATIVE ASSESSMENT

Space Probe Broadcast Have small groups of students each research a different space probe. Students can use magazines, newspapers, or the Internet as reference materials. Have students find out the launch date of their space probe. Have them also research five other significant events that occurred on that date. Groups can then use their findings to prepare a radio broadcast for that date in history. Have each group present its broadcast to the class.
learning styles: linguistic, auditory

PORTFOLIO ASSESSMENT

Making Student Portfolios Students can demonstrate their comprehension of the concepts in this chapter by making a portfolio. The Chapter Self-Check, the Big Idea Planner, and the Weekly Journal are some of the reproducibles in the Teacher's Resources that they can include in their portfolios. They may also wish to contribute their Chapter Project or a radio script they have written to include in their portfolios. You can use the Portfolio Assessment Rubric also found in the Teacher's Resources to assess students' portfolios.

Concept Challenges

WRITTEN RESPONSE

1. **EXPLAIN:** Possible answer: It helps scientists learn more about how the solar system formed and whether life exists elsewhere in the universe. (16-2)
2. **CONTRAST:** A reflecting telescope uses a concave mirror to collect light. A refracting telescope uses convex lenses. (16-3, 16-4)
3. **INFER:** They can go on one-way trips and to places that are too dangerous for humans. (16-9)
4. **EXPLAIN:** Scientists measure how much the star appears to move against a background of stars farther away. From this, they calculate the distance. (16-6)
5. **EXPLAIN:** The law says that for every action, there is an equal and opposite reaction. In a rocket, the pressure of hot gases is the action force. The thrust of the rocket is the reaction force. (16-7)
6. **ANALYZE:** It can be reused. (16-8)
7. **INFER:** There is little to slow down a space probe in space, so it does not need to burn fuel to keep moving. (16-7, 16-9)
8. **INFER:** Possible answer: Uranus could not be seen with the available technology. (16-1)
9. **ANALYZE:** Possible answers may include space probes, space telescopes, Earth-based telescopes, space stations, space shuttles, and space missions. (16-2)
10. **INFER:** One natural satellite, the Moon, orbits Earth. (16-9)
11. **CALCULATE:** There are about 66,667 astronomical units in one light year. (16-6)

INTERPRETING VISUALS (16-6)

12. Alpha Centauri A, B, C
13. Procyon
14. three
15. No. The information is not in the chart, and the stars are not lined up neatly in the universe, one behind the other. You would need to know angles and use trigonometry to figure out the answer.
16. They might be located on opposite sides of our galaxy.

Concept Challenges TEST PREP

WRITTEN RESPONSE **Answer each of the following questions in complete sentences.**

1. **EXPLAIN:** Why is space exploration important?
2. **CONTRAST:** What is the difference between a refracting telescope and a reflecting telescope?
3. **INFER:** What are the advantages of unmanned space probes?
4. **EXPLAIN:** How do scientists use parallax to calculate the distances to stars?
5. **EXPLAIN:** How does Newton's third law of motion describe how a rocket works?
6. **ANALYZE:** What is the advantage of the space shuttle over earlier spacecraft?
7. **INFER:** How are space probes able to travel to the outer reaches of our solar system without a huge fuel reserve?
8. **INFER:** Before the discovery of the planet Uranus, no planets had been discovered for 5,000 years. Why did it take such a long time for astronomers to discover Uranus?
9. **ANALYZE:** If you were an astronomer today, what types of tools could you use to explore space?
10. **INFER:** How many natural satellites orbit Earth?
11. **CALCULATE:** One astronomical unit is equal to 150 million km. One light-year is equal to 10 trillion km. How many astronomical units are in one light-year?

INTERPRETING VISUALS **Use Figure 16-35 to answer each of the following questions.**

12. Which of these star systems is closest to Earth?
13. Which of these star systems is farthest from Earth?
14. How many stars are in the Alpha Centauri star system?
15. Can you tell from this chart how close these stars are to one another? Explain your answer.
16. How could two stars that are about the same distance from Earth be far apart from each other?

NEAREST STARS AND STAR SYSTEMS

Name of Star or Star System	Distance (in light years)
Alpha Centauri A, B, C	4.4
Barnard's Star	5.9
Lalande 21185	8.3
Sirius A, B	8.6
Ross 154	9.7
Epsilon Eridani	10.5
HD 217987	10.7
Ross 128	10.9
61 Cygni A, B	11.4
Procyon	11.4

▲ Figure 16-35

406

Teaching Options

Chapter Project Wrap-Up

SPACE EXPLORATION TIMELINE

Students' projects can be exhibited around the classroom. Students can present their posters and explain what each picture represents.

You can use the Individual Activity Rubric from the Teacher's Resources to assess students' projects. Fill in the rubric with the additional information below. For this project, students should have:

- labeled pictures of each event or historical person that appears on the timeline.
- have set up reasonable intervals between events.

• TEACHING RESOURCES • Online or Print

Teacher's Resources

Key Term Review: Chapter 16, p. 16

Chapter 16 Test: pp. 17–19

Individual Activity Rubric: Teaching Tools, p. 5

Chapter Self-Check: Teaching Tools, p. 11

Big Idea Planner: Study Tools, pp. 11–12

Weekly Journal: Study Tools, p. 19

Portfolio Assessment Rubric: Teaching Tools, p. 10

Chapter 17 The Solar System

PLANNING GUIDE

◆ **TEACHING THE CHAPTER** This chapter should take approximately 10–13 days to complete instruction and assessment.

	Skills and Features	Projects and Activities	Achieve Science Literacy	Meet Individual Needs
Chapter 17 Opener p. T407	RESEARCH, INTEGRATE	• Chapter Project		
Lesson 17-1 **What is the solar system?** pp. T408–T409 Standards: D3a, D3c	CLASSIFY, SEQUENCE, RESEARCH • Building Science Skills • How Do They Know That?	• Extending the Lesson • Integrating Chemistry • Lab Challenge	• Study Hint • Reading Strategy	• Reteach Teacher's Resources* The Solar System
Lesson 17-2 **What do we know about orbits?** pp. T410–T411 Standard: D3c	DESCRIBE, HYPOTHESIZE, ANALYZE • Hands-On Activity		• Study Hint	• Reteach
Big Idea **What keeps the planets and moons in orbit?** pp. T412–T413 Standard: D3c	RESEARCH, COMMUNICATE, ANALYZE • Science Log: Writing Activity	• Big Idea • Close Activity	• Tips for Using Technology	• ESL/ELL Strategy • Reteach Teacher's Resources* Big Idea Planner
Lesson 17-3 **What do we know about Earth's moon?** pp. T414–T415 Standards: D3a, D3c	IDENTIFY, CALCULATE, OBSERVE • Building Science Skills • Hands-On Activity	• Extending the Lesson • More on the Web	• Reading Strategy	
Lesson 17-4 **What are the other moons in the solar system?** pp. T416–T417 Standard: D3a	SEQUENCE, INFER, RESEARCH • Building Science Skills	• Extending the Lesson • More on the Web	• Study Hint • Ongoing Assessment	• ESL/ELL Strategy
Lab Activity **Looking Back in Time** pp. T418–T419 Standard: D3a	OBSERVE, ANALYZE, INFER • Set-Up Time: 15–20 min • Lab Time: 40–45 min	• Extend the Lab Activity	• Tips for Using Technology	Teacher's Resources* Lab Activity Report

*www.conceptsandchallenges.com, password: ccearth09

Planning Guide continues on next page.

Standards: For details on the correlation to National Science Standards see pages *xx–xxi*.

	Skills and Features	Projects and Activities	Achieve Science Literacy	Meet Individual Needs
Lesson 17-5 **What do we know about Mercury, Venus, and Earth?** pp. T420–T421 Standard: D3a	DESCRIBE, INFER, COMPARE • Web InfoSearch • Integrating Life Science	• Cooperative Learning	• Study Hint	• ESL/ELL Strategy Teacher's Resources* Enrichment Activity Venn Diagram
Lesson 17-6 **What do we know about Mars?** pp. T422–T423 Standard: D3a	CONTRAST, IDENTIFY, APPLY • Investigate • Web InfoSearch • Integrating Life Science	• Extending the Lesson	• Reading Strategy	• ESL/ELL Strategy • Reteach Teacher's Resources* Enrichment Activity
Lesson 17-7 **What are asteroids and meteoroids?** pp. T424–T425 Standard: D3a	LOCATE, EXPLAIN, CONTRAST • Web InfoSearch • Real-Life Science	• Extending the Lesson	• Study Hint • Writing Hint	• ESL/ELL Strategy
Lesson 17-8 **What do we know about Jupiter and Saturn?** pp. T426–T427 Standard: D3a	COMPARE, RESEARCH, MODEL • Building Science Skills • Science and Technology		• Study Hint • Writing Hint • Ongoing Assessment	Teacher's Resources* Enrichment Activity
Lesson 17-9 **What do we know about Uranus, Neptune, and the dwarf planets?** pp. T428–T429 Standard: D3a	IDENTIFY, EXPLAIN, HYPOTHESIZE • Web InfoSearch • People in Science	• Cooperative Learning • Integrating Chemistry		• ESL/ELL Strategy • Reteach Teacher's Resources* Enrichment Activity
Lesson 17-10 **What is a comet?** pp. T430–T431 Standards: D3a, D3c	DESCRIBE, EXPLAIN, INFER • Investigate • Designing an Experiment • How Do They Know That?	• Extending the Lesson • More on the Web	• Study Hint • Ongoing Assessment	• Reteach Teacher's Resources* Designing an Experiment
Chapter 17 Challenges pp. T432–T434	• Chapter Summary • Key Term Challenges • Content Challenges • Concept Challenges	• Chapter Project Wrap-Up	• Study Hint • Preparing Students for Standardized Tests	• ESL/ELL Strategy Teacher's Resources* Chapter Self-Check Weekly Journal

*www.conceptsandchallenges.com, password: ccearth09

Chapter 17 The Solar System

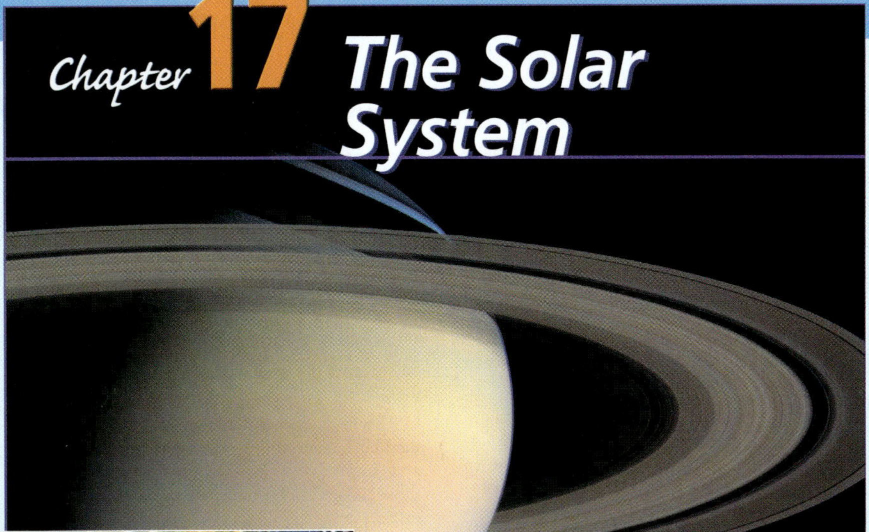

▲ **Figure 17-1** Saturn's majestic rings, shown here in false colors, were first seen by Galileo in 1610.

Saturn is the second largest planet in our solar system. Its beautiful rings make it the easiest planet to recognize. Like Jupiter, Saturn is a huge ball made mostly of gas. It is surrounded by clouds. Saturn is ten times farther from the Sun than Earth is. To the naked eye, it looks bright yellow. Computers often enhance the images taken of Saturn using false colors. This allows its atmosphere and its many individual rings to be seen.

▶ What do you think Saturn's rings might be made of?

Contents

Teaching Options

ASSESSMENT PLANNER

For assessment in the Student Edition, see the following pages:

Content Assessment
Checking Concepts: pp. 409, 411, 415, 417, 421, 423, 425, 427, 429, 431
Thinking Critically: pp. 409, 411, 415, 417, 421, 423, 425, 427, 429, 431
Interpreting Visuals: p. 411
Chapter Challenges: pp. 432–434

Alternative Assessment
Building Skills: pp. 409, 415, 417, 427
Web InfoSearch: pp. 421, 423, 425, 429
Science Log: p. 412

Performance-Based Assessment
Hands-On Activity: pp. 411, 415
Designing an Experiment: p. 431
Lab Activity: pp. 418–419

• TEACHING RESOURCES • Online or Print

Teacher's Resources
Lesson Review: Ch. 17, pp. 2, 3, 6, 7, 8, 10, 11, 12, 13, 16
Enrichment: Ch. 17, pp. 9, 14
Key Term Review: Ch. 17, p. 17
Chapter 17 Test: pp. 18–20
Laboratory Manual: pp. 83–86

Chapter 17 The Solar System

Chapter Overview

In this chapter, students will learn about the orbits and features of the eight planets that make up the solar system. Students will also explore natural satellites, including Earth's Moon, as well as asteroids, meteoroids, and comets.

About Figure 17-1 Saturn's seven rings are made of lumps of frozen water and bits of rock that range in size from tiny particles to large boulders. The frozen water also varies in size. It may be in the form of hail, snowflakes, or large icebergs. Only three rings are easily visible: the large outer A ring, the central B ring, and the inner C ring.

Answer Possible answer: dust and ice

Linking Prior Knowledge

For this chapter, students should recall:

- the factors that affect the rate of weathering (Lesson 7-3).
- the definition of *erosion* (Lesson 8-1).
- the greenhouse effect (Lesson 13-4).
- the cause of Earth's seasons (Lesson 14-3).
- the movement of Earth around the Sun (Lesson 14-4).
- the missions of satellites and space probes (Lesson 16-9).

Chapter Project

BIG IDEA POSTER

Have students create their own "Big Idea" two-page spread based on lessons in this chapter. They can choose from life or physical science. Some ideas for physical science are the elements that make up the planets, what makes each planet the color it is, or where planetary magnetic fields come from. Some ideas for life science are the possibility that life could develop elsewhere in the solar system or that humans could live on other planets. Fully designed pages can be put on posterboards for display.

17-1 (pp. 408–409)

1 Introduce

STUDY HINT Have students copy the section titles in their notebooks, using an outline format. As students read each section, they should write the topic sentence for each paragraph in the section under the appropriate title.

Linking Prior Knowledge Have students write their family's last name and the first names of family members in their notebooks. Point out that the Sun also has a family. Tell students that the eight planets have some features in common, yet each is unique, just as family members are.

2 Teach

Demonstration To illustrate the great size of the Sun in relation to the planets, draw a circle on the board about 1 m (39 in.) in diameter. Inside this circle, draw a smaller circle 1 cm (0.39 in.) in diameter. Tell students that the large circle represents the Sun and the small circle represents Earth. **learning style: visual**

Discussion Have students discuss ways of observing the planets. Ask if students have a telescope or have ever looked through a telescope and what they saw. Broaden the discussion to include the types of technology that scientists use to study the planets. **learning styles: auditory, linguistic**

Answers to questions

1. **DEFINE:** a spinning cloud of gases and dust
2. **IDENTIFY:** Jupiter, Saturn, Uranus, and Neptune
3. **CLASSIFY:** Mercury, Venus, Earth, and Mars

READING STRATEGY To help students make connections while reading, ask them to think about the word *system* and the types of things a system can describe. (school system, railroad system, digestive system) Lead students to understand that a system is made up of interacting parts. Then, make the connection to the solar system.

17-1 What is the solar system?

Objectives

Name the planets that make up the solar system. Describe how they are grouped.

Key Terms

solar system: the Sun and all the objects that orbit the Sun

nebula (NEHB-yuh-luh)**:** cloud of gas and dust in space

orbit: curved path of one object around another object in space

Formation of the Solar System The **solar system** is the Sun and all the objects that orbit it. Scientists are not sure how the solar system formed. However, several theories have been developed. One popular theory states that the solar system formed from a spinning cloud of gas and dust called a **nebula.**

Scientists think that gravity caused the nebula to shrink, or contract, to form the Sun. After the Sun formed, the leftover gas and dust in the nebula formed the other objects in the solar system. This took many millions of years.

The Sun contains more than 99.8 percent of the mass in our solar system. The planets make up most of the rest of the mass.

A planet is a body in orbit around the Sun. It must have enough mass to assume a round shape and have an orbit that is clear of other bodies.

1. **DEFINE:** What is a nebula?

The Sun's Family The Sun's family consists of eight planets with at least 162 moons. There are four inner planets: Mercury, Venus, Earth, and Mars. These planets are small and rocky. There are four outer planets: Jupiter, Saturn, Uranus, and Neptune. These planets are large and composed of gases. They are often called the gas giants. There are also three dwarf planets: Ceres, Pluto, and Eris. They have a total of four moons. Together these objects exist on a plane that is over 10 trillion km in diameter.

The inner and outer planets are separated by the Main Asteroid Belt. The belt lies between Mars and Jupiter and is composed of hundreds of thousands of small rocky bodies. These are often called planetoids.

Beyond the orbit of Neptune is another region called the Kuiper Belt. It is a disk-shaped region containing many small icy bodies. Many comets come from this belt. Far beyond the Kuiper Belt is another vast cloud of comets called the Oort Cloud.

2. **IDENTIFY:** Which planets are known as the gas giants?

Planet Pathways Ancient peoples observed that the planets changed their positions among the stars. The planets seemed to wander in the sky. The word *planet* comes from a Greek word meaning "wanderer."

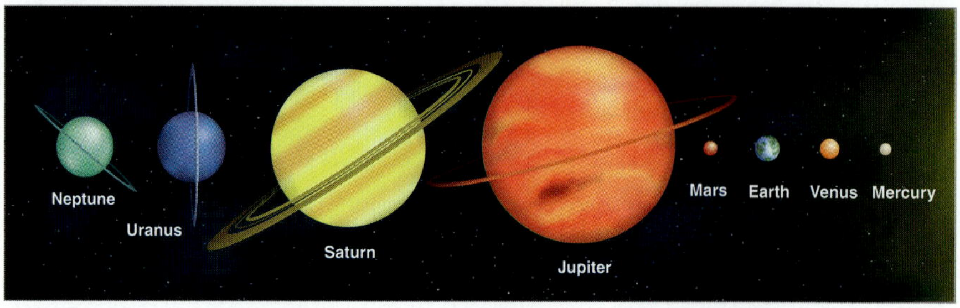

▲ **Figure 17-2** The solar system contains eight planets. They are divided into the inner planets and the outer planets.

Teaching Options

INTEGRATING CHEMISTRY

Elements in the Early Solar System As the early solar system slowly cooled, various substances began condensing into liquid and solid particles that eventually merged to form planets. Elements such as iron condensed at high temperatures. These condensed elements helped to form the cores of the inner planets that were close to the hot Sun. Gases such as hydrogen condensed at cooler temperatures. These condensed gases eventually formed the outer gaseous planets that were farther from the Sun. Have students list the planets that formed when elements condensed at high temperatures. (Mercury, Venus, Earth, and Mars)

RETEACH

Comparing Planets Refer students to Figure 17-2. Have students make a two-column chart in their notebooks. Tell them to head the columns *Inner* and *Outer*. Say the names of the planets in a random order. Have students locate the planet in Figure 17-2 and write its name in the correct column in their notebooks. **learning styles: visual, auditory**

Earth is one of the eight planets in the solar system. Like the other planets, Earth moves in a curved path around the Sun. This path is the planet's **orbit.** All of the planets orbit the Sun in the same direction.

3 CLASSIFY: Which are the inner planets?

CHECKING CONCEPTS

1. The solar system may have formed from a __________.
2. The solar system contains __________ planets.
3. A nebula is a __________.
4. The path of a planet around the Sun is the planet's __________.
5. The planets can be divided into two groups, the inner planets and the __________ planets.
6. The __________ separates the inner and outer planets.

THINKING CRITICALLY

7. SEQUENCE: List the planets in order from nearest to the Sun to farthest from the Sun.
8. HYPOTHESIZE: Why is the solar system often referred to as the Suns family?

BUILDING SCIENCE SKILLS

Researching As seen from Earth, some planets seem to have unusual motions in their orbits. For example, early astronomers were puzzled to discover that for part of the year, Mars and some other planets seemed to move backward in their orbits compared with the movements of Earth. Do research to find out why Mars and some other planets seem to sometimes exhibit these retrograde motions. Draw a diagram of these motions and write a report explaining them.

▲ **Figure 17-3** Mars, along with some other planets, exhibits retrograde motion.

How Do They Know That?

PLANETS BEYOND THE SOLAR SYSTEM

Are there planets going around other suns? Are there other solar systems? Planets located outside our solar system are known as extra solar planets.

It isn't easy to find planets around other suns. Planets are very small compared with stars and usually have less than one-billionth the brightness. Still, it is possible to find planets orbiting other stars. Today, powerful telescopes may be able to detect other extrasolar planets if they are large enough.

Planets have a gravitational effect on stars the same way that stars have a gravitational effect on planets. Planets cause a star to wobble. Telescopes can detect this wobble. Using this method, a planet was discovered in orbit around the star 51 Pegasi in 1995. More than 300 extrasolar planets have been discovered so far.

Thinking Critically Why are planets that orbit other stars called extrasolar planets?

▲ **Figure 17-4** An artist's drawing of 51 Pegasi and some of its planets.

EXTENDING THE LESSON

Discovering Planets Have students use astronomy texts to research when and how each planet was discovered. Students can create a poster or present the information as a chart.
learning style: linguistic

• TEACHING RESOURCES •
Online or Print

Teacher's Resources
Lesson Review: Chapter 17, p. 2
The Solar System: Visuals, p. 14

Laboratory Manual
Lab. Challenge: How can a scale model of the solar system be made? pp. 83–86

3 *Assess*

CHECKING CONCEPTS ANSWERS

1. nebula
2. eight
3. spinning cloud of gas and dust
4. orbit
5. outer
6. Asteroid Belt

THINKING CRITICALLY ANSWERS

7. SEQUENCE: Mercury, Venus, Earth, Mars, Jupiter, Saturn, Uranus, Neptune
8. HYPOTHESIZE: Possible answer: The eight planets and the Sun together make up the solar system, just as siblings and parents make up a family.

4 *Close*

BUILDING SCIENCE SKILLS

Researching All planets exhibit retrograde motion as observed from Earth. The planets that are farther from the Sun than Earth is orbit more slowly, while planets closer to the Sun orbit faster. When Earth catches up with and passes a slower-moving planet, or when a faster-moving planet passes Earth, the other planet appears to move backwards in the sky.

How Do They Know That?

PLANETS BEYOND THE SOLAR SYSTEM

Life-Science Connection The probability of life on other planets has long been a debate among astronomers. In the 1960s, scientist Frank Drake formulated an equation that provides an estimate of the number of advanced civilizations that may exist in the Milky Way Galaxy. Drake's equation is based on his estimates of variables that would need to exist in order for such civilizations to exist. He estimated that about 10,000 advanced civilizations may exist in the Milky Way alone.

Discussion Have students list the conditions necessary for life as we know it. Point out that the building blocks of life on Earth (carbon, oxygen, nitrogen, and hydrogen) are relatively common throughout the universe. Have students debate the probability of life on extrasolar planets.

Thinking Critically Answer
The prefix *extra-* means "outside or beyond." The planets that orbit other stars are called extrasolar because they are beyond or outside our solar system.

17-2 *(pp. 410–411)*

1 *Introduce*

STUDY HINT Have students link all of the Key Terms in the lesson in a paragraph that summarizes the lesson. Remind students to write complete sentences and include a topic sentence.

Linking Prior Knowledge Before beginning this lesson, have students review the movement of Earth around the Sun (Lesson 14-4). Have students describe the path of a runner or a race car around an oval track. Tell students that the planets follow similar paths around the Sun. Each planet, however, follows its own track or orbital path.

2 *Teach*

Demonstration Toss a small, unbreakable object into the air. Ask students why the object falls to the ground. (Gravity pulls the object toward Earth.) Point out that gravity acts on all objects on Earth, including people. **learning styles: visual, auditory**

Discussion Call attention to Figure 17-5. Ask students to describe the difference between a circle and an ellipse. Explain that Earth's orbit around the Sun is elliptical. Using Figure 17-6 as a guide, point out that the distance between Earth and the Sun is not always the same.
learning styles: visual, auditory

Reinforcement Tell students that Earth is sometimes farther and sometimes closer to the Sun. Have students find the average distance of Earth to the Sun, using the distances shown in Figure 17-6. (152,000,000 km + 147,000,000 km = 299,000,000 km; 299,000,000 km ÷ 2 = 149,500,000 km; 149,5000,000 km = average distance of Earth to the Sun) Have students find the difference between the average and perihelion distances. (149,500,000 km − 147,000,000 km = 2,500,000 km) Also, have students find the difference between the average and aphelion distances. (152,000,000 km − 149,500,000 km = 2,500,000 km)
learning style: logical/mathematical

17-2 What do we know about orbits?

Objective
Describe the shape of Earth's orbit.

Key Terms
ellipse (eh-LIPS): flattened circle, or oval
perihelion (per-uh-HEE-lee-uhn): point in a planet's orbit at which it is closest to the Sun
aphelion (uh-FEE-lee-uhn): point in a planet's orbit at which it is farthest from the Sun

Gravity Every object in the universe pulls on every other object. This pull is the force of gravity, or gravitational attraction. There is gravitational attraction among all objects in the universe. For example, there is gravitational attraction between the Sun and the planets. This gravitational attraction pulls the planets toward the Sun as they move through space. Instead of flying off into space, the planets move in orbits around the Sun. Gravity also pulls all objects on Earth toward the center of the planet.

1 STATE: What force pulls all nearby objects toward the center of Earth?

Earth's Orbit A circle is perfectly round. All lines drawn from the center of the circle to its rim are the same length. An **ellipse** looks like a slightly flattened circle or oval. Lines drawn from the center to different points on its rim are different lengths.

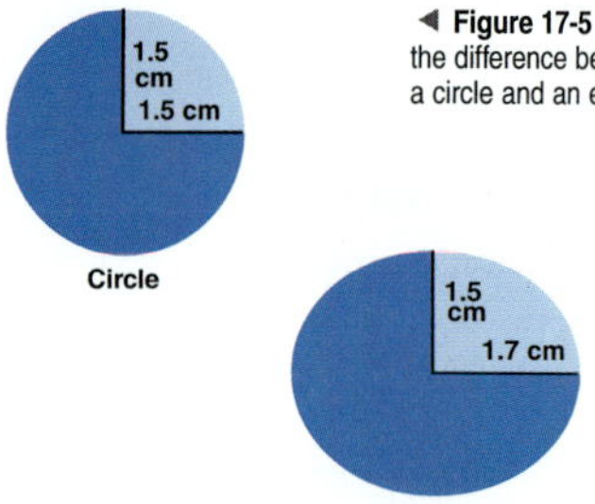

◄ **Figure 17-5** Note the difference between a circle and an ellipse.

Earth and all the other planets travel around the Sun in elliptical orbits. This means the planet is not always the same distance from the Sun. In January, for example, Earth reaches its perihelion. **Perihelion** is the point at which a planet is closest to the Sun. Earth is about 147 million km from the Sun at perihelion. In July, Earth reaches its aphelion. **Aphelion** is the point at which a planet is farthest from the Sun. Earth is about 152 million km from the Sun at aphelion.

2 DESCRIBE: What is the shape of Earth's orbit?

Orbital Velocity The speed at which a planet travels in its orbit is called its orbital velocity. The closer a planet is to the Sun, the greater is its orbital velocity. A planet moves fastest at perihelion. It moves slowest at aphelion.

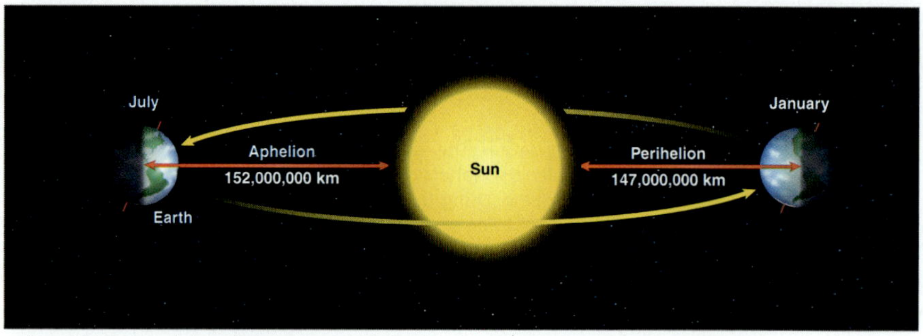

▲ **Figure 17-6** Earth travels in an elliptical orbit around the Sun. It is closer to the Sun at perihelion than at aphelion.

Teaching Options

QUICK CONVERSIONS		
	Metric/SI	**English**
Perihelion	147,000,000 km	91,300,000 mi
Astronomical unit	150,000,000 km	93,000,000 mi
Aphelion	152,000,000 km	94,500,000 mi

RETEACH

Illustrating Ellipses Have a student draw an ellipse on the board. Have another student draw the Sun within the ellipse and two circles to represent Earth at perihelion and aphelion. Have a third student write in the distance from Earth to the Sun at perihelion and aphelion.
learning style: visual

The closer two objects are to each other, the greater the gravitational attraction between them. As a planet gets closer to the Sun, the gravitational attraction between the planet and the Sun increases. As a result, the planet moves faster in its orbit.

As a planet moves farther from the Sun, the gravitational attraction between the planet and the Sun decreases. The planet slows down. This is why a planet moves fastest at perihelion and slowest at aphelion. The difference in gravitational attraction also explains why the planets closer to the Sun move faster than those farther from the Sun.

3 STATE: When does Earth move fastest in its orbit?

CHECKING CONCEPTS

1. An ellipse has an __________ shape.
2. Earth's path around the Sun is its __________.
3. Earth moves in an __________ orbit.
4. The point at which Earth is closest to the Sun is __________.
5. A planet's speed in its orbit is its __________.

THINKING CRITICALLY

6. HYPOTHESIZE: The planets are different distances from the Sun. Mercury is closest to the Sun. Neptune is farthest from the Sun. Which of the two planets has the greater orbital velocity? Explain.

INTERPRETING VISUALS

Use Figure 17-6 to answer the following questions.

7. ANALYZE: What month is it when Earth is closest to the Sun?
8. ANALYZE: What month is it when Earth is farthest from the Sun?
9. CALCULATE: How much farther from the Sun is Earth at aphelion than at perihelion?

Hands-On Activity

DRAWING THE SHAPE OF EARTH'S ORBIT

You will need paper, cardboard, two pins, string, a pencil, and a metric ruler.

1. Place a sheet of paper on the cardboard. Draw a horizontal line 1 cm long near the center of the paper.
2. Stick one pin through the paper and cardboard at the end of each line. Label one of the pins Sun.
3. Tie the ends of a piece of string together to make a loop about 10 cm long. Place the loop around the pins.
4. Hold a pencil point inside of and against the string. At the length of the string loop, move the pencil around the pins, marking the paper.

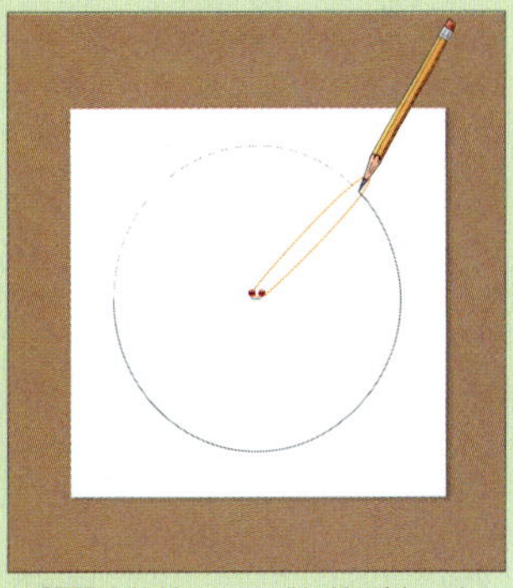

▲ **STEP 4** Hold your pencil inside of the string and move it around.

Practicing Your Skills

5. INFER: You have created a slight ellipse. Why do people assume Earth's orbit is a circle?
6. IDENTIFY: Label the points of aphelion and perihelion.
7. IDENTIFY: Measure and label the distances to aphelion and perihelion.
8. IDENTIFY: Label the months Earth reaches aphelion and perihelion.

• *TEACHING RESOURCES* •
Online or Print

Teacher's Resources
Lesson Review: Chapter 17, p. 3

Answers to questions

1 STATE: gravity

2 DESCRIBE: elliptical

3 STATE: at perihelion

3 *Assess*

CHECKING CONCEPTS ANSWERS

1. oval
2. orbit
3. elliptical
4. perihelion
5. orbital velocity

THINKING CRITICALLY ANSWER

6. HYPOTHESIZE: Mercury has the greater orbital velocity. The closer a planet is to the Sun, the greater is its orbital velocity.

INTERPRETING VISUALS ANSWERS

7. ANALYZE: January
8. ANALYZE: July
9. CALCULATE: 5,000,000 km

4 *Close*

Hands-On Activity

DRAWING THE SHAPE OF EARTH'S ORBIT

TIME: 15–30 minutes

PURPOSE: inferring, identifying

DISCUSSION: To help students distinguish between *aphelion* and *perihelion*, explain the roots of these words. The root *helio* means "pertaining to the Sun." The prefix *ap-* means "away from." The prefix *peri-* means "near."

Practicing Your Skills Answers

5. INFER: it is only slightly elliptical
6. IDENTIFY: Students' drawings should be clearly labeled, with *perihelion* at the point nearer the Sun and *aphelion* at the point farther from the Sun.
7. IDENTIFY: Students' drawings should be labeled *147,000,000 km* at perihelion and *152,000,000 km* at aphelion.
8. IDENTIFY: aphelion—July; perihelion—January

The Big Idea

(pp. 412–413)

1 *Introduce*

Objective Students will recognize the effects of gravity on planetary orbits. Students will also compare and contrast the different orbital speeds of the eight planets.

Linking Prior Knowledge Ask students to recall what they learned about orbits (Lesson 17-2). Have them describe the shape of the planets' orbits. (elliptical)

Have students review the definitions of *aphelion* and *perihelion* (Lesson 17-2). Ask: *When does a planet move fastest in its orbit?* (at perihelion) *Why?* (The gravitational attraction between the planet and the Sun increases.)

2 *Teach*

Demonstration Help students calculate how many times faster Earth's orbital speed is compared to Saturn's. (30 km/s ÷ 9.6 km/s = 3.1 times faster) Have students do the same calculation for the other planets. In each case, students should compare the orbital speed of the planet to Earth's orbital speed. Remind students that Venus and Mercury have faster orbital speeds than Earth.
learning style: logical/mathematical

Discussion Tell students that the famed scientist Isaac Newton (1642–1727) was the first to relate the force of gravity to orbital movements. He recognized that gravity was a universal force and developed physical laws that explained the gravitational attraction between objects.

Refer students to the diagram of the solar system in Figure 17-7. Ask: *Which planet has the fastest orbital speed?* (Mercury) *Which planet is closet to the Sun?* (Mercury) *Which planet has the second-fastest orbital speed?* (Venus) *What is the second planet from the Sun?* (Venus) Help students grasp the relationship between distance from the Sun and orbital speed. Planets that are closer to the Sun have faster orbital speeds because of the stronger gravitational attraction of the Sun. **learning styles: auditory, visual**

Integrating Physical Science

THE BIG IDEA

What keeps the planets and moons in orbit?

When you throw a ball, you give it a forward motion. At the same time, gravity pulls the ball toward the center of Earth. As a result, the ball has two motions. It has a forward and a downward motion. These two motions cause the ball to follow a curved path.

Suppose that you tied a string around the ball and swung the ball around your head. You would feel an outward pull on the string. If you were to let go of the string, the ball would fly away from you. However, as long as you hold onto the string, the ball will keep moving in a curved path around your head. This curved path is the ball's "orbit."

The motion of the planets around the Sun is similar to the motion of the ball. The planets, which are in motion, keep moving forward in a straight line. However, there is also a force pulling them inward, toward the Sun. This is the force of gravity. Together, gravity and forward motion help determine a planet's orbital motion.

When you hear the word *gravity*, you probably think of the force that pulls objects toward Earth. However, gravity is not limited to our planet. It happens in space, too. Every object in space pulls on every other object. That means that gravity occurs between the Sun and each planet. This attraction toward the Sun keeps the planets from spinning off into space. Gravity also keeps moons orbiting planets.

Look at the illustrations that appear on these two pages. Then, follow the directions in the Science Log to learn more about "the big idea." ✦

412

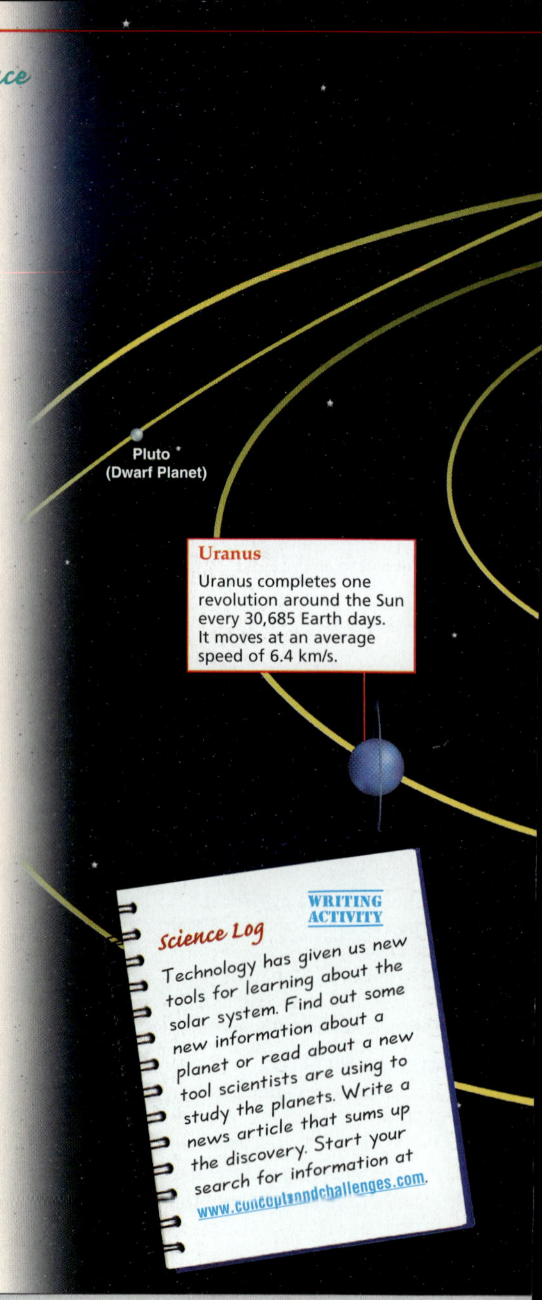

WRITING ACTIVITY

Science Log

Technology has given us new tools for learning about the solar system. Find out some new information about a planet or read about a new tool scientists are using to study the planets. Write a news article that sums up the discovery. Start your search for information at www.conceptsandchallenges.com.

Teaching Options

! FAST FACTS *Planet Data*

All the planets in the solar system are unique. Here are some interesting facts about some of them:

- A day on Mercury lasts two-thirds of its year.
- Jupiter rotates so fast that its day lasts only about 10 hours.
- Saturn's winds blow at speeds exceeding 1,600 km/h (1,000 mph).
- Uranus's ocean of water and ammonia is a steamy 2,200°C (4,000°F).

ESL/ELL STRATEGY

Gravity Tell students that the word *gravity* comes from a Latin term meaning "heavy." Ask them how the definition of this word is related to its Latin origin.

RETEACH

Classifying Planets Write the names of the eight planets on the board in random order. Have students rewrite the names in sequence from closest to farthest from the Sun.

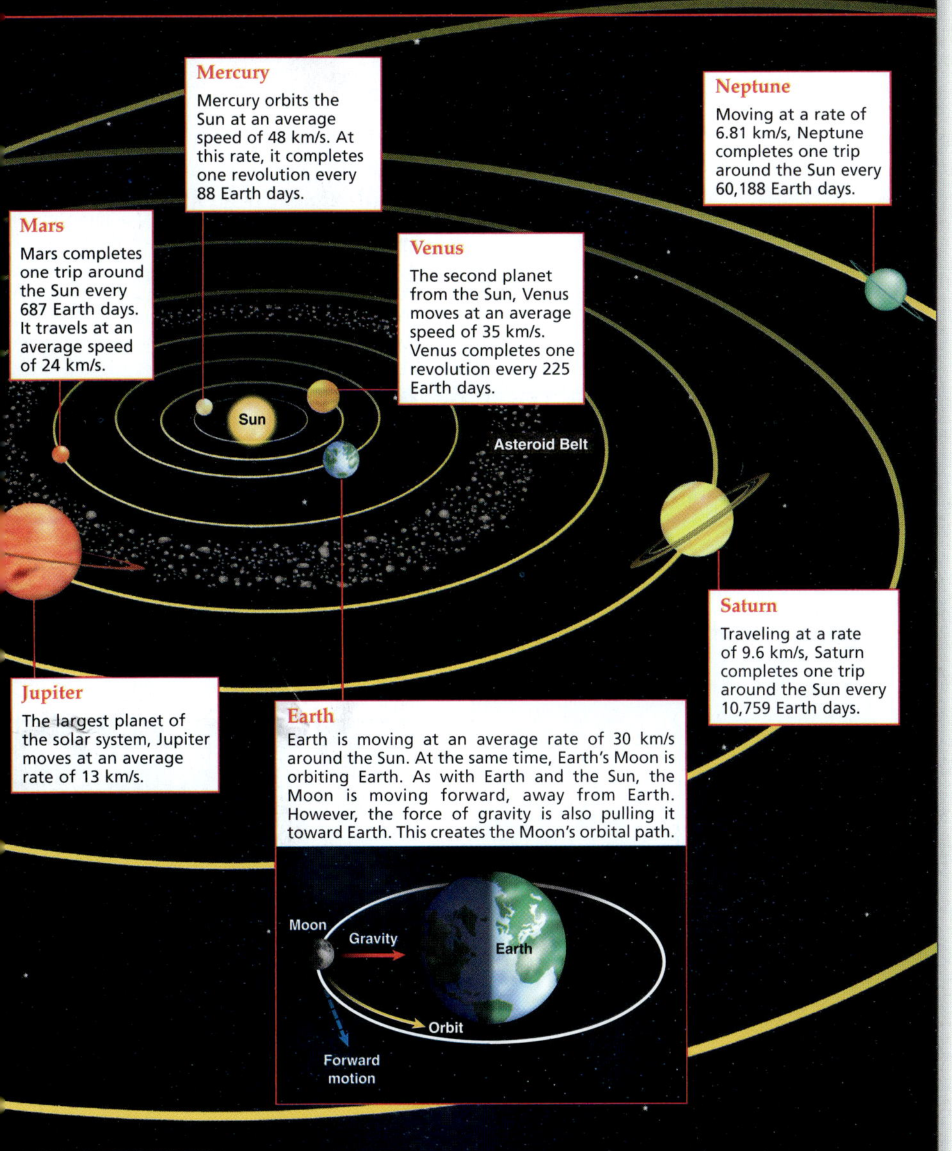

Figure 17-7 The solar system

THE BIG IDEA

What keeps the planets and moons in orbit? Have students start their research for their Science Log Writing Activity at www.conceptsandchallenges.com. Be sure students use search engines and websites that are reliable and acceptable. You may also wish to provide copies of the *Big Idea Planner* and *Big Idea Science Log Entry* study tools found in the Teacher's Resources.

3 Assess

Use the Writing Activity Rubric from the Teacher's Resources to assess students' Science Logs. Fill in the rubric with the additional information below. For this assignment, students should have:

- recognized the effects of gravity on planetary orbits.
- compared and contrasted the different orbital speeds of the planets.

4 Close

Creating a Model Have students use the information in this feature to create a model of the solar system. Students can use plastic-foam balls, modeling clay, or other materials to represent the planets. Encourage students to use materials or colors to highlight the unique characteristics of each planet. For example, Earth is largely covered by water and thus should be blue. Students' models should show the planets in their proper sequence from the Sun. Have students do research and explain to the class something interesting or unusual about the solar system, using the model as a visual aid in the presentation.
learning styles: tactile, visual

Tips for Using Technology

Using Desktop Publishing to Create Newspaper Stories Have students use desktop publishing software to publish their newspaper articles. Desktop publishing software allows students to combine text and art on a page. Most word processing software programs include tools for desktop publishing. Students can use these tools to give text special formats, to make a special border around the text, and to plan a page layout. In addition, students can add clip art into a document. Students can also use a scanner to incorporate their own art into their documents.

• *TEACHING RESOURCES* • *Online or Print*

Teacher's Resources

Big Idea: Chapter 17, p. 5

Big Idea Planner: Study Tools, pp. 11–12

Big Idea Science Log Entry: Study Tools, p. 13

Writing Activity Rubric: Teaching Tools, p. 9

17-3 *(pp. 414–415)*

1 *Introduce*

Linking Prior Knowledge Students are familiar with landscape features on Earth, such as mountains and plains. Tell them that the same types of features are found on the Moon. Then, have students recall what they learned about weathering (Lesson 7-3) and erosion (Lesson 8-1). Tell them that because the Moon has no atmosphere, weathering or erosion has not changed the Moon's surface.

2 *Teach*

Demonstration To demonstrate the lack of erosion and weathering on the Moon, place some soil on two plates. Put one plate outside to expose it to the elements. Put the other plate in a closet or sheltered interior location. After a week, have students observe the differences between the two plates of soil. Relate the demonstration to the absence of an atmosphere on the Moon. **learning style: visual**

Discussion Display photographs of the full Moon. Point out features of the Moon, such as mountains, craters, and plains. Have students identify the features in the photographs. **learning style: visual**

READING STRATEGY Point out to students that "seas" on the Moon are not water, even though that is the Latin meaning of the term *maria*.

Answers to questions

1 **STATE:** July 20, 1969

2 **NAME:** Possible answers include that the Moon is Earth's only natural satellite; the Moon is 3,400 km (2,100 mi) in diameter; and the Moon has less mass than Earth, so its gravitational attraction is not as strong.

3 **IDENTIFY:** Galileo

17-3 What do we know about Earth's Moon?

Objective

Describe some features of the Moon.

Key Terms

mare (MAH-ray), *pl.* **maria:** broad, flat plain on the Moon's surface

crater (KRAYT-uhr)**:** round hole on the Moon's surface

Moon Landing On July 20, 1969, an American astronaut, Neil Armstrong Jr., stepped onto the surface of the Moon. He was the first human to do so. He and his crew had traveled 384,000 km from Earth to the Moon. The trip took five days. Since 1969, 12 astronauts have been to the Moon and returned to Earth. *Apollo 17* in 1972 was the last mission to send astronauts to the Moon.

▲ **Figure 17-8** Astronauts walked on the Moon for the first time in 1969.

1 **STATE:** When did the first person walk on the Moon?

Moon Facts Earth has only one natural satellite, called the Moon. The Moon is much smaller than Earth. It has a diameter of about 3,400 km. Because the Moon has less mass than Earth does, its gravity is less than Earth's gravity.

The gravity on the Moon's surface is only one-sixth as strong as the gravity on Earth. The Moon's weaker gravity means that you can jump much higher on the Moon than you can on Earth.

The Moon has no liquid water, and it has an extremely thin atmosphere. Temperatures on the Moon can range from over 100°C to below –200°C. Astronauts need to wear space suits to survive on the Moon.

2 **NAME:** What are three important facts about the Moon?

The Moon's Surface Galileo saw and named the three main types of features on the Moon's surface. He named the smooth, dark areas that he saw **maria.** The word *maria* means "seas" in Latin. The first astronauts to land on the Moon landed in the Sea of Tranquility. Today, scientists know that the Moon's maria are not seas but broad, flat plains. Galileo also saw light areas on the Moon. These light areas are mountains, or highlands. Some mountains on the Moon are higher than the highest mountains on Earth.

▲ **Figure 17-9** The Moon has no water to drink and no air to breathe. Visitors must bring their own supplies.

The third feature Galileo saw on the Moon's surface were its many **craters.** Large objects striking the Moon's surface caused most of the craters. Erupting volcanoes may have caused the rest.

414

Teaching Options

! FAST FACTS *The Race for Space*

During the late 1950s and throughout the 1960s, the United States and the former Soviet Union were engaged in a race for space. As a matter of national pride, both countries wanted to be the first to land on the Moon. In 1957, it appeared that the Soviet Union was pulling ahead in the race with the successful launch of *Sputnik*, the first artificial satellite in space. Four years later, Soviet cosmonaut Yuri Gagarin became the first person to orbit Earth.

During those early years, both the U.S. and Soviet space programs suffered numerous setbacks, including a fire on the launch pad that killed three *Apollo* astronauts. However, by 1961, American scientists had designed a plan for a crewed mission to the Moon. In this respect, they were far ahead of the Soviets, who did not finalize plans for their crewed mission until 1967. Although both space programs accomplished spectacular feats, the United States was the first and only country to land people on the Moon.

▲ **Figure 17-10** Craters on the Moon

IDENTIFY: Who named the main features on the Moon's surface?

CHECKING CONCEPTS

1. The distance from Earth to the Moon is __________ km.
2. The Moon has a __________ of 3,400 km.
3. The word *maria* means __________ .
4. Three features on the Moon's surface are maria, __________, and highlands.

THINKING CRITICALLY

5. **CALCULATE:** How much would a 120-lb person weigh on the Moon?
6. **CALCULATE:** How much would you weigh on the Moon?
7. **INFER:** Astronauts left their footprints on the Moon's surface. These footprints may remain unchanged for millions of years. Why?
8. **ANALYZE:** Why did the astronauts who walked on the Moon have to wear boots with lead weights in them?

BUILDING SCIENCE SKILLS

Observing Many features on the Moon are visible to the unaided eye. Others can be seen clearly through binoculars. Observe a full Moon. First use only your eyes. Then use binoculars. Compare how well you are able to see certain features with your unaided eyes and with binoculars. Make a drawing of the features you were able to identify and label those features.

Hands-On Activity

MODELING CRATER FORMATIONS

You will need a shoebox, plaster of Paris, a metric ruler, and three rocks of different sizes.

1. Mix the plaster of Paris according to the directions. Make enough to fill the shoebox one-third of the way up.
2. Pour the plaster of Paris into the shoebox. Just before the plaster hardens, drop one of the rocks into the box from a height of 25 cm. Then, drop the other rocks from the same height.
3. Remove the rocks and drop them from a height of 10 cm.
4. Remove the rocks and let the plaster of Paris harden. Measure the rocks and craters in as many ways as you can.

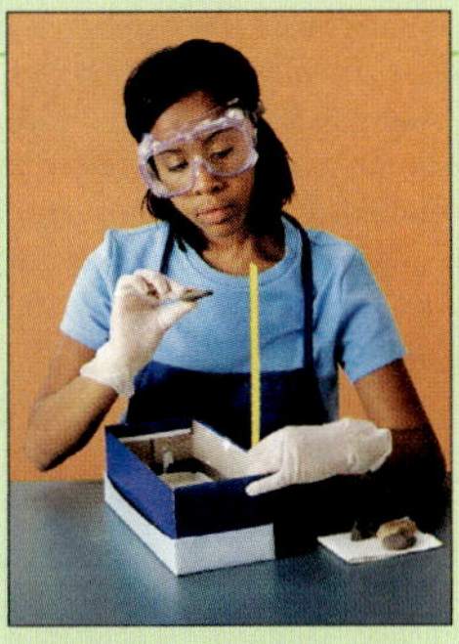

▲ **STEP 3** Drop the rocks a second time from a lower height.

Practicing Your Skills

5. **EXPLAIN:** What do the rocks represent?
6. **COMPARE:** Which rock made the deepest crater? Which rock made the widest crater?
7. **APPLY:** How do you think craters were formed on the Moon?

3 *Assess*

CHECKING CONCEPTS ANSWERS

1. 384,000
2. diameter
3. seas
4. craters

THINKING CRITICALLY ANSWERS

5. **CALCULATE:** 20 lb (1/6 of 120 lb)
6. **CALCULATE:** Students' answers should equal 1/6 of their weight on Earth.
7. **INFER:** The Moon has no wind or water to erode the footprints.
8. **ANALYZE:** The force of gravity on the Moon is much less than it is on Earth. The lead weights helped the astronauts stay on the ground.

4 *Close*

BUILDING SCIENCE SKILLS

Observing Students' drawings should be clear representations of the Moon's surface as seen from Earth and should include any of the three main features: maria or plains, mountains or highlands, and craters.

Hands-On Activity

MODELING CRATER FORMATIONS

TIME: 15–30 minutes

PURPOSE: explaining, comparing, applying

DISCUSSION: Tell students that astronomers can determine the relative ages of craters by observing how the craters overlap. In satellite images of moons and planets, an older crater will usually appear to be underneath a younger crater. Have students examine how their own craters overlap and determine which craters are oldest and which are youngest.

Practicing Your Skills Answers

5. **EXPLAIN:** large objects striking the Moon's surface
6. **COMPARE:** Possible answer: The largest rock dropped from the greatest height made the deepest crater. The largest or widest rock made the widest crater.
7. **APPLY:** by large objects striking the Moon or by erupting volcanoes

EXTENDING THE LESSON

Exploring the Moon If possible, show students a video of the first Moon landing. Then, have interested students find out more about Project *Apollo*. Have students share the results of their research with the class.

learning styles: visual, linguistic

• TEACHING RESOURCES •
Online or Print

Teacher's Resources

Lesson Review: Chapter 17, p. 6

MORE ON THE WEB

Studying Earth's Moon Have students use the Internet to learn more about Earth's Moon.

17-4 *(pp. 416–417)*

1 Introduce

STUDY HINT Have students create flash cards with moon names on one side and important features, such as planet orbited, on the other side. Have them decorate their cards with drawings or pictures cut out from magazines and then test one another using the cards.

Linking Prior Knowledge Have students recall what they learned about the movement of Earth around the Sun (Lesson 14-4). Remind them that moons orbit planets just as planets orbit the Sun.

2 Teach

Demonstration Use two students to demonstrate the orbits of moons around planets and the orbits of planets around the Sun. Stand in the middle of a cleared space. Have one student slowly walk around you in a large oval path. At the same time, have a second student circle the first student in a smaller orbital path. Ask students what you represent (the Sun), what the first student represents (a planet), and what the second student represents (a moon).
learning styles: visual, kinesthetic

Discussion Discuss the relationship between number of moons and planet size. Point out that the larger planets generally have numerous moons, while the smaller planets have few or none. Relate the discussion to the greater gravitational attraction of the larger planets.

Have students note that space probes have helped people learn about all the moons in the solar system. Ask: *How many moons have* Voyager 1 *and* Voyager 2 *discovered?* (35)
learning styles: auditory, linguistic

Answers to questions

1. **DESCRIBE:** Earth, Mars, Jupiter, Saturn, Uranus, and Neptune
2. **IDENTIFY:** Earth and Mars
3. **IDENTIFY:** Saturn

17-4 What are the other moons in the solar system?

Objective
Compare the moons of the different planets in the solar system.

Key Term
satellite (SAT-uhl-eyet)**:** natural or artificial object orbiting another body in space

Natural Satellites A **satellite** is any natural or artificial object that orbits another object in space. People have observed the only natural satellite of Earth, the Moon, since ancient times.

For thousands of years, Earth was believed to be the only planet in the solar system to have a moon. Then, in 1610, Galileo discovered four of the moons of Jupiter. Today, astronomers know that six of the planets (Earth, Mars, Jupiter, Saturn, Uranus and Neptune) have moons.

Until the late 1970s, almost all of the known moons had been discovered using Earth-based telescopes. Then, two space probes, called *Voyager 1* and *Voyager 2*, traveled beyond the Asteroid Belt to the outer planets. They sent back to Earth the first detailed, close-up photographs of the many moons that orbited Jupiter, Saturn, Uranus, and Neptune. Together, *Voyagers 1* and *2* discovered at least 35 new moons.

1 **DESCRIBE:** What planets have natural satellites?

Moons of the Inner Planets Mercury and Venus are the only planets in the solar system without at least one moon. Earth has only one moon.

Mars has two moons, named Deimos and Phobos. These were discovered in 1877. They are made of dark, carbon-rich rock. Because of their makeup, both reflect very little light from their surface. Both moons are small and lumpy. Deimos has a peanut shape. Both moons also have many craters. One crater on Phobos is 10 km across.

Both moons orbit close to Mars. Phobos takes less than a Martian day to go once around the planet. Deimos takes a little more than a day.

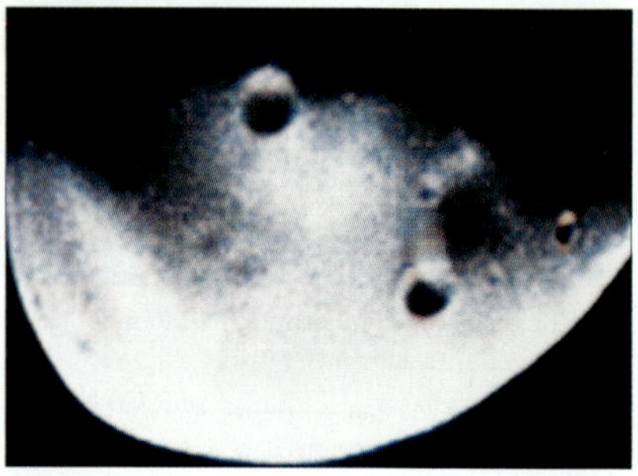

▲ **Figure 17-11** Mars's moon Deimos is shaped like a peanut.

2 **IDENTIFY:** Which of the inner planets have one or more moons?

Moons of the Outer Planets Jupiter has at least 63 known moons. Jupiter and its moons have been described as a minisolar system.

Jupiter's moons are very varied. Some are rocky. Others are icy. The four moons first viewed by Galileo are often referred to as the Galilean moons. They are the largest of Jupiter's moons.

The Galilean moons travel around Jupiter in a nearly circular orbit, almost exactly around the planet's equator. It is possible to track the movements and positions of these moons using a good pair of binoculars.

The Galilean moons played an important role in the history of astronomy. When their orbits were first observed, it was proven that not everything in space revolved around Earth.

In 1995, a space probe, appropriately named *Galileo*, arrived at Jupiter to study the Galilean moons. The largest Galilean moon is Ganymede, followed by Callisto, Io, and Europa. Ganymede is the largest moon in the solar system, with a diameter of 5,268 km. Callisto has a dark, icy

Teaching Options

ESL/ELL STRATEGY

Moon and Moons Be sure that English-language learners understand the proper use of the term *moon.* Point out that other planets have natural satellites, or moons, but that Earth is the only planet with a satellite actually named Moon. To help students remember, have them make a table that lists the planets and the number of moons each planet has.
learning style: logical/mathematical

EXTENDING THE LESSON

Modeling A Planet and Its Moon(s)
Have students research a planet or dwarf planet (other than Earth) and its satellite system. Tell them to determine relative sizes by finding diameters and creating ratios. Moon diameter divided by planet diameter times 100 gives a percent that can be used as a scale to make proportional models of the planet and moon(s). For example, for dwarf planet Pluto:

Diameter Charon	1205 km x 100 = 52.4%
Diameter Pluto	2300 km
Diameter Nix	130 km x 100 = 5.7%
Diameter Pluto	2300 km

learning style: visual, tactile

surface with many white craters. Impacts from asteroids and other bodies probably exposed clean ice beneath the dirtier top layer. Io has active volcanoes on it. Europa is covered with ice that may have liquid water below the surface.

◀ **Figure 17-12** Io is a large, volcanically active moon of Jupiter.

Saturn has 59 known moons. Five of these moons are very large. Saturn's largest moon is Titan. Titan is the second largest moon in the solar system.

Huygens is a probe sent to Titan by the European Space Agency. It arrived there in 2004, as part of the *Cassini* mission to Saturn. It parachuted toward the surface and reported on conditions beneath the moon's orange clouds.

▲ **Figure 17-13** This is an artist's concept of Saturn and six of its moons as seen from a seventh moon.

Uranus has 27 known moons. Titania is the largest, with Oberon second. The moons of Uranus are very varied. Some have deep canyons and long scars on their surface. Others have large, smooth areas between areas riddled with craters. Miranda may have more types of landforms than any other body in the solar system. Ten of Uranus's moons were discovered by *Voyager 2* in 1986.

Neptune has 13 known moons. Only two, Triton and Nereid, are visible from Earth. Triton orbits in the opposite direction from the rest of Neptune's moons. It has the coldest surface in the solar system, at –235°C. Scientists think Triton is covered with frozen nitrogen and methane.

The dwarf planet Pluto has three moons. The largest is Charon. It was discovered in 1978. Charon is about half the size of Pluto. Some scientists suspect its surface to be covered with water ice and impact craters. One theory suggests that an icy fragment of Pluto was knocked off when Pluto collided with another object. This fragment could be Charon.

The dwarf planet Eris has one natural satellite.

3 IDENTIFY: Which outer planet has the most moons?

CHECKING CONCEPTS

Match each planet or dwarf planet with its correct moon.

1. Jupiter	**a.** Charon
2. Mars	**b.** Triton
3. Uranus	**c.** Io
4. Pluto	**d.** Titania
5. Neptune	**e.** Titan
6. Saturn	**f.** Phobos

THINKING CRITICALLY

7. **SEQUENCE:** List the planets in order from the one with the most moons to the ones with the fewest. Include planets that have no moons.
8. **INFER:** Why are the four largest moons of Jupiter called the Galilean moons?

BUILDING SCIENCE SKILLS

Researching The names given to moons are chosen by the International Astronomical Union. Newly discovered moons are numbered first, with the year they are discovered, then named later. The names chosen come from many different sources. For example, recently discovered moons of Uranus were named after characters in the plays of Shakespeare. Choose the moons of one planet. Find the source of each moon's name. Write a report of your findings.

3 Assess

CHECKING CONCEPTS ANSWERS

1. c	**3.** d	**5.** b
2. f	**4.** a	**6.** e

THINKING CRITICALLY ANSWERS

7. **SEQUENCE:** Saturn (30), Jupiter (28), Uranus (21), Neptune (8), Mars (2), Earth (1), Venus (0), and Mercury (0)
8. **INFER:** because they were first viewed by Galileo, and they were given his name to honor him

4 Close

BUILDING SCIENCE SKILLS

Researching You may want to have students work in small groups and assign a different planet to each group. Also, consider dividing the satellites of some planets among different groups. Three groups, for example, can each research the ten moons of Saturn. Have the groups give brief oral presentations to the class. See the chart of moon name origins below. **learning style: interpersonal**

MOON NAMES

Planet	Origins of Names of Major Moons (Unless noted otherwise, all names are of Greek mythological characters.)
Mars	*Phobos*, *Deimos*: Both are sons of Ares (god of war) and Aphrodite (goddess of love). *Phobos* is the Greek word meaning "fear," and *deimos* is the Greek word for "panic."
Jupiter	*Io*, *Europa*, *Callisto*, *Ganymede*: The first three are females with whom Zeus (king of the gods) fell in love. Ganymede is a boy that Zeus chose to be cup-bearer to the gods.
Saturn	*Titan*: The Titans are a family of giants.
Uranus	*Titania*, *Oberon*, *Miranda*: Titania and Oberon, respectively, are the queen and king of the fairies in Shakespeare's *A Midsummer Night's Dream*. Miranda appears in Shakespeare's *The Tempest*.
Neptune	*Triton*: Triton, son of Poseidon (god of the seas), is a demigod having a human head and upper body and the tail of a fish.
Pluto (dwarf planet)	*Charon*: Charon ferries souls across the River Acheron (River of Woe) into Hades (the underworld).

MORE ON THE WEB

Deimos and Phobos Have students use the Internet to learn more about Deimos and Phobos, the two moons that orbit Mars.

• TEACHING RESOURCES • Online or Print

Teacher's Resources
Lesson Review: Chapter 17, p. 7

ONGOING ASSESSMENT

Write the names of the moons mentioned in this lesson on slips of paper. Put the slips into a box. Have students take turns withdrawing a slip and identifying which planet the moon orbits. **learning style: visual**

Lab Activity

(pp. 418–419)

1 Prepare the Lab

Looking Back in Time

SET-UP TIME: **LAB TIME:**

BACKGROUND: Scientists can estimate the age of the surface of a moon by examining the number of craters present. Jupiter's moon Io has almost no craters on its surface. This indicates that the surface of Io is relatively young. Io is a volcanic moon, and sulfurous lava flows continually rework its surface. In contrast, Jupiter's moon Callisto is heavily pocked with craters. This indicates that Callisto's surface has been stable for billions of years, with the exception of meteorite impacts. The surface of the far side of Earth's Moon is similar to Callisto's surface. It is marked by thousands of craters. The surface of the near side of Earth's Moon has some craters, too, but it is also covered by large, relatively flat areas called maria. Maria are broad basins filled in by basalt lava flows.

PURPOSE: Students will model the formation of craters on the surfaces of moons.

ALTERNATIVE MATERIALS: A shallow dish with a rim can be used in place of a petri dish. Droppers can be used to make the craters.

SCIENCE SKILLS: Students will **model** the formation of craters on the surfaces of moons. They will **observe** the models of other students and **estimate** the age of the model craters.

ADVANCE PLANNING: Obtain images of cratered moons, including Earth's Moon. Have students study the images before beginning the lab. Images of moons can be found at NASA Web sites.

2 Guide the Lab

PROCEDURE: Organize students into small groups. Discuss the procedure. After students have completed their models, have groups exchange models with one another for age analysis. Provide students with a copy of the Lab Activity Report from the Teacher's Resources to record their observations and conclusions.

LAB ACTIVITY
Looking Back in Time

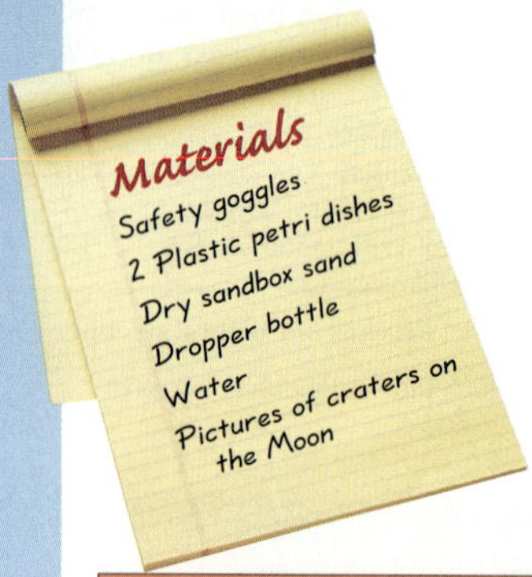

▲ **STEP 1** Fill the petri dish with sand and the dropper bottle with water.

▲ **STEP 3** Move your dish around as you squeeze out drops of water from the bottle.

BACKGROUND

The solar system has more than 90 moons. Meteorite impacts have altered the terrain of most of these moons by creating craters. Many moons, such as Earth's Moon and Jupiter's Callisto, are heavily pitted with meteor craters. If a plain on the surface of a moon has craters, it is likely that the plain formed first and was later altered by impacts. The number of impacts helps scientists estimate the age of the plain. If the plain has very few or no craters, it probably formed very recently. If the plain is heavily cratered, the plain is probably ancient. Scientists can often determine the sequence of when the impacts occurred that created the craters. They do this by looking for overlapping craters.

PURPOSE

In this activity, you will create a moon with many craters on its surface and infer its history.

PROCEDURE

1. Put on your safety goggles. Fill one petri dish almost to the top with sand. Fill the dropper bottle with water.

2. Copy the chart in Figure 17-14. Hold the dropper bottle about 25 cm above the dish and squeeze out a drop of water. Observe what happens when the water strikes the surface of the sand. Record your observations in your chart.

3. For the next minute, squeeze a drop from the bottle every second. Move the dish around as you release the drops every second. Have some drops fall by themselves. Have other drops fall on top of each other or close together. Observe the raindrop pattern you produced on the sand. Record your observations.

4. Trade your dish for a dish made by another group. Try to determine the sequence in which the craters were made. Record your observations.

Teaching Options

! Fast Facts — *Origin of the Moon*

Prior to the *Apollo* missions, there were three theories about the Moon's formation.

- **Co-accretion theory:** Earth and the Moon were once a double planet.
- **Escape theory:** The Moon formed from a large piece of Earth that broke away from the planet, which, at the time, spun much faster.
- **Capture theory:** The Moon formed in some distant part of the solar system and was eventually captured by Earth's gravity.

These three theories have been largely proved false by evidence gathered during the *Apollo* missions. Currently, the most widely accepted theory of the Moon's formation proposes that a Mars-sized object collided with Earth approximately 4.5 billion years ago. Dust and debris from this collision were ejected into space and eventually coalesced into the Moon.

5. In your chart draw a picture of the dish with circles to represent the craters. Look at the overlapping craters. Mark older craters with the number 1. Mark newer craters with the numbers 2, 3, and so on.

6. Fill a second petri dish with sand. Use your thumb to make a large crater in the center.

7. Squeeze drops from the bottle in and around the big crater. Observe and record what happens.

8. Look at pictures of the craters on the Moon. Can you tell which craters are younger and which are older? Note down your comments below the chart.

▲ **STEP 7**
Squeeze drops of water around the large crater you made with your thumb.

Step	What You Observe	Written Description
1		
2		
3		

▲ **Figure 17-14** Use a copy of this chart to write down your observations and draw your pictures.

CONCLUSIONS

1. **OBSERVE:** What happened when the first drop of water hit the sand?
2. **ANALYZE:** How can you tell which crater is newer when two craters overlap?
3. **OBSERVE:** What happened to the large crater you created with your thumb when it was hit with more drops of water?
4. **ANALYZE:** What is the difference between an old crater and a young one?
5. **INFER:** How do craters tell scientists if the surface of a moon is very young or very old?

TROUBLESHOOTING: Make sure that students do not create too many craters. Otherwise, the surface of the moon will become soupy.

SAFETY TIP: Remind students to wear safety goggles at all times during the experiment. If glass dishes are used in place of plastic petri dishes, then caution students to handle the dishes with care. If a dish should break, have students tell you immediately.

EXPECTED OUTCOME: Students will observe how successive craters overlap earlier craters and use this information to estimate the relative ages of the craters.

LAB HINT: Make sure that students understand that the drops they make represent meteorites striking a solid planetary or moon surface.

3 Conclude the Lab

1. **OBSERVE:** It formed a crater.
2. **ANALYZE:** The newer crater lies on top of or cuts into the older crater.
3. **OBSERVE:** The large crater flattened and filled in.
4. **ANALYZE:** Older craters are overlapped by younger craters. Older craters are also less distinct than younger craters.
5. **INFER:** Old surfaces have many craters. Young surfaces have fewer craters.

Use the Laboratory Activity Rubric from the Teacher's Resources to assess students' lab activities. Fill in the rubric with the additional information below. For this activity, students should have:

- created a moon with many craters.
- inferred the moon's history.

4 Extend the Lab

Discuss craters on Earth. Ask: *Why doesn't Earth have as many craters as the Moon?* (Erosion and weathering on Earth wear away craters.) Allow students to examine images of craters on various moons in the solar system. Based on the skills learned in this lab, have students try to estimate the relative ages of the craters.

Have interested students research the names, sizes, and locations of at least three impact basins on Earth's Moon. For example, the largest impact basin, the South Pole Aitken Basin, is about 2,500 km (1,600 mi) wide. Students can use astronomy texts, the Internet, or other references for their research.

Tips for Using Technology

Searching the Internet for Information on Impact Basins
Remind students that one important tool for finding information on the Internet is a search engine. A search engine allows the user to type in a keyword or phrase. The engine then searches thousands of sites on the Internet for those that contain the keyword or phrase.

Suggest to students that they surround phrases in quotation marks, such as "impact basins." This will instruct the search engine to search for sites that have that exact phrase.

• TEACHING RESOURCES •
Online or Print

Teacher's Resources

Lab Activity Report: Study Tools, pp. 16–17

Laboratory Activity Rubric: Teaching Tools, p. 8

17-5 *(pp. 420–421)*

1 *Introduce*

STUDY HINT Have students fill in a Venn Diagram from the Teacher's Resources, using the concepts and facts presented about the three innermost planets.

Linking Prior Knowledge Have students recall what they learned about the greenhouse effect (Lesson 13-4). Review the definition of global warming, which was also discussed in the same lesson.

2 *Teach*

Demonstration Make a model of the Earth's magnetic field by placing a bar magnet on a sheet of paper or plastic. Tell students that, like Earth, the bar magnet has a north pole and a south pole. Sprinkle iron filings around the magnet and explain that they represent the charged particles that are trapped in Earth's magnetic field. Gently tap the edge of the paper or plastic until the filings are aligned in a pattern around the magnet. Have students compare the pattern of the filings with the lines of Earth's magnetic field shown in Figure 17-17. **learning style: visual**

Discussion Before beginning this lesson, have students list the names of the inner planets. (Mercury, Venus, Earth, and Mars) Tell students that they will learn about the characteristics of the innermost planets in this lesson. Based on what they know, ask students to brainstorm similarities and differences among Mercury, Venus, and Earth. Write a list of students' responses on the board. After completing the lesson, have students revise the list to incorporate new knowledge.
learning styles: auditory, visual

Answers to questions

1 **DESCRIBE:** It is a dry, rocky planet with no atmosphere.

2 **INFER:** because Venus is similar in size, mass, and density to Earth

3 **LIST:** moderate temperatures, an atmosphere that contains oxygen, and liquid water

17-5 What do we know about Mercury, Venus, and Earth?

Objective
Identify the basic features of the three innermost planets.

Mercury Mercury is the planet closest to the Sun. Of all the planets, Mercury travels the fastest around the Sun. However, it rotates slowly on its axis. This dry, rocky planet has no atmosphere. Temperatures on Mercury range from 430°C during daylight hours to –170°C during Mercury's night. Because Mercury is so close to the Sun, astronomers cannot see it easily from Earth. However, astronomers learned a lot when, in 1974, the space probe *Mariner 10* visited Mercury and sent back photographs. In the 1990s, radar was used to study Mercury's surface. We now know that the surface of Mercury is covered with craters.

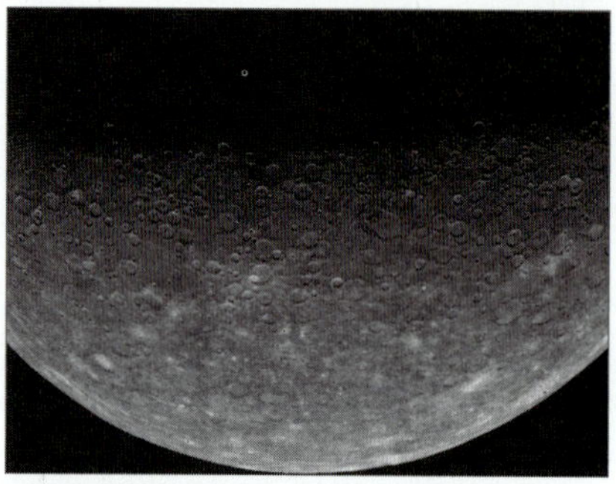

▲ **Figure 17-15** Mercury is very similar to Earth's Moon. It has no atmosphere, and its surface is covered with impact craters.

1 **DESCRIBE:** What is Mercury like?

Venus The planet Venus is similar to Earth in size, mass, and density. However, Venus is a very hostile world. Its average temperature is higher than the average temperature of any other known planet. The air pressure on Venus is 90 times more crushing than that of Earth's air at sea level.

Astronomers think these conditions are related to Venus's carbon dioxide atmosphere and thick clouds of sulfuric acid. From Earth, we can only see Venus's cloudtops.

▲ **Figure 17-16** The pressure on the surface of Venus, seen here, is equal to that of the pressure on Earth deep down in the ocean.

Venus has retrograde rotation. This means that it spins slowly in the opposite direction from most of the other planets.

Many space probes have orbited Venus or landed on it. These probes have revealed smooth plains, mountains, and valleys. Like the Moon, Venus goes through phases. We see only varying amounts of its sunlit side.

2 **INFER:** Why was Venus once called Earth's twin?

Earth The third planet from the Sun and the fifth largest in the solar system is Earth. Earth is the only planet known to have oceans of liquid water. It is also the only planet known to support life. Life occurs on Earth because of moderate temperatures, an atmosphere containing oxygen, and liquid water.

Earth has a magnetic field around it because of the iron in its core. Electrically charged particles from the Sun, mostly electrons and protons, bounce around in the atmosphere above Earth. They are trapped in Earth's magnetic field.

Occasionally, some of these particles escape from the magnetic field and rain down on Earth. They strike atoms and molecules in the upper atmosphere and cause them to glow. In the Northern Hemisphere, this glow is known as the aurora borealis, or the northern lights. In the southern hemisphere, it is the aurora australis, or southern lights.

Teaching Options

! FAST FACTS *Viewing Venus*

Venus's thick clouds are good reflectors of sunlight. This is why Venus usually appears far brighter than other planets in the night sky. Its dense cloud cover, however, makes it impossible to view the planet's surface using visible light. Venus's surface has been analyzed and mapped using Earth-based radar observations, probes that landed on the planet's surface, and probes that circled the planet and bounced microwaves off its surface.

COOPERATIVE LEARNING

Developing Questions After students have read through the lesson, have them write at least three questions based on the content. The questions should be different from those already in the text. Then, have students exchange questions with a partner and write answers for the questions they receive. **learning style: linguistic**

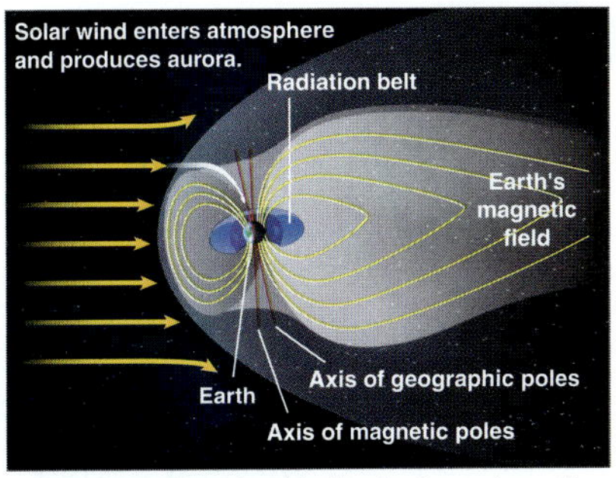

▲ **Figure 17-17** Charged particles from the Sun are trapped in Earth's magnetic field.

3 LIST: What conditions make life on Earth possible?

CHECKING CONCEPTS

1. What are the three innermost planets?
2. What does the surface of Mercury look like?
3. What gas in the atmosphere makes the surface of Venus so hot?
4. Which inner planet has liquid water?

THINKING CRITICALLY

5. COMPARE: How is the surface of Mercury similar to the surface of Earth's Moon?
6. ANALYZE: Venus has rocks similar to basalt. On Earth, these rocks are usually found near volcanoes. What does this suggest?

Web InfoSearch

More About Venus Sunlight striking the surface of a planet warms the ground. This releases heat radiation. Like glass in a greenhouse, the atmosphere traps some of the heat. In some places, such as on Venus, the heat keeps building up. This is called a runaway greenhouse.

SEARCH: Use the Internet to find out what the temperature is on Venus and why. Could Earth someday become like Venus? How could it? Start your search at www.conceptsandchallenges.com. Use the key search words **runaway greenhouse effect, Venus, planet profile,** and **carbon dioxide.**

Integrating Life Science

TOPICS: origin of life, molecular biology

HOW LIFE ON EARTH BEGAN

Nobody knows for sure how life on Earth began. Scientists think they know when it happened—about 3.5 billion years ago.

Many scientists think that the first living things were certain molecules, or strings of atoms. These molecules floated in water, alongside billions of unattached atoms. As these molecules bumped into other molecules, they traded atoms. Eventually, a chain of atoms came together that could actually make copies of itself using the free-floating atoms.

▲ **Figure 17-18** Stromatolites are among the oldest fossils. They were formed in sedimentary rock by blue-green algae.

How can a chain of atoms make copies of itself? It happens all the time, right inside your body. It is how your body grows. Molecules inside your body grab onto atoms to create more molecules. When molecules first did this billions of years ago, they began to show some of the characteristics of living things.

Over time, life forms that were especially good at making copies of themselves outnumbered other life forms. These life forms were our very distant ancestors.

Thinking Critically What parts of your body can make copies of themselves?

ESL/ELL STRATEGY

Earth's Magnetic Field Help students learning English to understand Earth's magnetic field by discussing and explaining magnetism, showing how a magnet works by using a bar magnet and relating this to a much larger magnet, Earth.

• TEACHING RESOURCES • Online or Print

Teacher's Resources

Lesson Review: Chapter 17, p. 8

Enrichment: Chapter 17, p. 9

Venn Diagram: Study Tools, p. 10

3 *Assess*

CHECKING CONCEPTS ANSWERS

1. Mercury, Venus, and Earth
2. It is covered with craters.
3. carbon dioxide
4. Earth

THINKING CRITICALLY ANSWERS

5. COMPARE: Both Mercury and the Moon are dry and have many craters.
6. ANALYZE: Venus has, or once had, volcanoes.

4 *Close*

Web InfoSearch

More About Venus Venus's surface gets very little sunlight. Light that does get through warms the ground. This releases heat radiation. Like glass in a greenhouse, Venus's atmosphere traps the radiation. The heat continues to build up. Have students use the Internet to find out more about Venus, such as what the temperature is on Venus and if Earth could become like Venus. Some key search words are **Venus planet profile, greenhouse Venus,** and **runaway greenhouse effect.** Students can begin their search at www.conceptsandchallenges.com.

Integrating Life Science

TOPICS: origin of life, molecular biology

HOW LIFE ON EARTH BEGAN

Background The earliest life-forms on Earth were simple bacteria. These single-celled organisms photosynthesize, and thus release oxygen into the air. Scientists hypothesize that these early bacteria may have played a large role in increasing the amount of oxygen in Earth's early atmosphere. This made it possible for more complex organisms to evolve.

Discussion Ask students what types of life-forms first inhabited Earth. Lead students to understand that most scientists believe that the earliest life-forms were simple one-celled organisms; more complex organisms evolved later.

Thinking Critically Answer
cells

17-6 *(pp. 422–423)*

1 Introduce

INVESTIGATE

TIME: 10–15 minutes

PURPOSE: Students will construct 3-D glasses to observe images of Mars.

MATERIALS: stiff paper, scissors, red filter, blue filter, *Pathfinder* pictures

PROCEDURE: Tell students to mark the positions of their eyes on the paper before cutting out the holes.

THINK ABOUT IT: Possible answers: Pictures would show mountain heights, depths of valleys, and so on.

Linking Prior Knowledge Have students recall what they learned about Earth's seasons (Lesson 14-3). Relate this to seasons on Mars.

2 Teach

Discussion Ask any students who have been to the Grand Canyon in Arizona to describe the canyon system or show photographs. Tell students that Mars also has complex canyon systems.
learning styles: visual, auditory

READING STRATEGY A good strategy to use with struggling readers is making connections. Most students are familiar with features on Earth such as valleys and volcanoes. Have students describe these features as they are on Earth. Ask students to keep in mind what they know as they read about Mars.

Answers to questions

1. **CONTRAST:** Mars is smaller and its surface is reddish. It also has a much thinner atmosphere and is extremely dry.
2. **IDENTIFY:** many space probes and two *Viking* landers
3. **EXPLAIN:** Mars's gravity is weaker.
4. **INFER:** the Martian south pole or beneath the surface

3 Assess

17-6 What do we know about Mars?

INVESTIGATE

Seeing Mars in 3-D

HANDS-ON ACTIVITY

1. Make a viewer by cutting out two eyeholes from a rectangle of stiff paper. The holes should match your eye positions.
2. Tape a piece of red filter over one hole. Tape a piece of blue filter over the other hole.
3. Hold the viewer in front of your eyes so that your right eye looks through the blue filter and your left eye looks through the red filter. Examine pictures of Mars's surface taken by the *Pathfinder* mission supplied by your teacher.

THINK ABOUT IT: Why might scientists find 3-D pictures useful for studying Mars?

STEP 2

Objective

Describe features on the planet Mars.

Key Term

rift: valley caused by a crack in the crust of a planet

The Red Planet Mars is the fourth planet from the Sun. Its orbit period is 687 days. Its rate of rotation is 24 hours and 37 minutes. Therefore, the lengths of a day on Mars and on Earth are almost the same. Mars has seasons similar to Earth's seasons because of the similar tilt of its axis. However, Mars is half the diameter of Earth. Its surface is reddish in color. Mars has volcanoes, valleys, polar ice caps, craters, and river channels. It has a thin carbon dioxide atmosphere.

The air pressure on Mars is about the same as the air pressure on Earth at an altitude of 35 km. Except for some icy spots at the poles, no water is visible on the Martian surface. Mars is a dry world. It can be colder than Antarctica in winter and as warm as a spring day in the American Midwest.

1 **CONTRAST:** How do Mars and Earth differ?

The Martian Surface Many space probes, including two *Viking* landers, have studied Mars. Photographs show that the surface has many craters and is covered with loose rocks. Winds of up to 100 km/h raise giant dust storms that cover the planet. Scientists think that Mars probably once had rivers or lakes on its surface.

2 **IDENTIFY:** What spacecraft visited Mars?

▲ **Figure 17-19** The surface of Mars was photographed by the *Viking* lander. It showed a landscape with rocks scattered all around.

Teaching Options

FAST FACTS *Mission Misses*

Early missions to Mars met with failure. In September 1999, the *Mars Climate Orbiter* was lost before it reached the planet. The probable cause was faulty calculations on the part of NASA engineers. Three months later, the *Mars Polar Lander* was lost just before touchdown on the planet's surface. The spacecraft is thought to have malfunctioned. More recent missions have had greater success, but not without problems of their own. In 2002, the rover *Opportunity* repeatedly experienced motor problems with its robotic arm.

EXTENDING THE LESSON

Exploring Mars Have interested students research and write about the missions of the rovers *Opportunity* and *Spirit*, which landed on Mars in 2004, or the *Phoenix Lander*, which arrived on the red planet in 2008. Each student can focus on one aspect of their missions, such as type of data gathered by the robot rover *Sojourner*. Have students share their reports with the class. **learning style: linguistic**

Giant Volcanoes and Canyons The largest known volcano in the solar system is called Olympus Mons. Olympus Mons is found on Mars. It is 21 km high. Had Olympus Mons formed on Earth, it would not have been nearly as high. This is because Earth's gravity is stronger. At the summit of the volcano is a large crater. Mars has many other large volcanoes. Mars also has a large **rift,** or crack, in its crust that forms a complex canyon system. The canyon, called Vallis Marineris, stretches 4,000 km. On Earth, this canyon would reach across the United States.

3 EXPLAIN: Why are volcanoes on Mars much larger than volcanoes on Earth?

Water on Mars Scientists have used robot spacecraft to search for water on Mars. Finding water could mean that Mars has life. There is some water ice at the Martian south pole. There are also many dry channels on Mars that look as though they were carved by running water. Where is the water today? It may be frozen beneath the surface or have boiled away in the thin atmosphere.

4 INFER: Where might water be on Mars?

CHECKING CONCEPTS

1. What causes the huge dust storms on Mars?
2. How are Mars and Earth similar?
3. How do the Vallis Marineris on Mars and the Grand Canyon on Earth compare in size?
4. Why do we think water might exist on Mars?

THINKING CRITICALLY

5. INFER: If Mars once had water, what might have happened to it?
6. APPLY: Could humans live on Mars? How?

Web InfoSearch

Martian Canals In 1877, an Italian astronomer said that he had seen "channels" on Mars. This word became "canals" when the report was translated into English. Some people, including the astronomer Percival Lowell, assumed that the canals must have been built by Martians.

SEARCH: Use the Internet to find out more about the "canals" on Mars. Do they exist? Start your search for information at www.conceptsandchallenges.com. Use the key search words **Martian canals, Mars canals Lowell,** and **Percival Lowell.**

Integrating Life Science

TOPICS: bacteria, fossils

FOSSILS FROM MARS?

In 1984, a meteorite was found in Antarctica that scientists think came from Mars. Like other planets, Mars has been struck by asteroids often in its past. Mars was struck by one such asteroid about 16 million years ago. Pieces of the surface broke off and went flying into space. Scientists believe a meteorite from this asteroid strike eventually fell to Earth.

Some scientists studying the rock found unusual, rodlike structures in it. They suggested that these tiny structures might be bacteria fossils. If the rodlike structures are from Mars, it could mean that life existed on Mars many years ago. However, these structures might have become part of the meteorite as it fell through the atmosphere to Earth.

▲ **Figure 17-20** Possible bacteria fossils (right) were found on an Antarctic ice sheet similar to the one above.

Thinking Critically What evidence of possible life was in the meteorite?

CHECKING CONCEPTS ANSWERS

1. strong winds
2. Mars has seasons similar to Earth's and a similar rate of rotation. Mars also has volcanoes, polar ice caps, and river channels.
3. Vallis Marineris is much larger.
4. There is some water ice at the Martian south pole and evidence of erosion from flowing water on the planet's surface.

THINKING CRITICALLY ANSWERS

5. INFER: It might be frozen beneath the surface or have boiled away in the thin atmosphere.
6. APPLY: Possible answer: Yes. Humans could live on Mars if they had adequate life support for air and water, and if they had protection from extreme cold.

Web InfoSearch

Martian Canals Have students use the Internet to find out more about the "canals" on Mars and if they exist. Have students start their search at www.conceptsandchallenges.com. Students can use the key search words **Martian canals, Mars canals Lowell,** and **Percival Lowell.**

4 *Close*

Integrating Life Science

TOPICS: bacteria, fossils

FOSSILS FROM MARS?

Background Some scientists argue that the markings in the meteorite, which is known as ALH 84001, are not actually fossils. They point out that the shapes could have just as likely been formed by nonorganic processes.

Discussion Students will learn about asteroids and meteorites in Lesson 17-7. However, you may want to preview the Key Terms on page 424. An *asteroid* is "a large chunk of rock or metal that orbits the Sun." A *meteorite* is "a piece of rock or metal that falls on a planet or moon's surface."

Thinking Critically Answer
fossils of bacteria

RETEACH

Identifying Characteristics of Mars Write the word *Mars* and the following subheads on the board: *Orbital Period*, *Rate of Rotation*, *Color of Surface*, and *Atmosphere*. Ask students to write on the board information about each subhead.
learning styles: visual, linguistic

• *TEACHING RESOURCES* • *Online or Print*

Teacher's Resources

Lesson Review: Chapter 17, p. 10

Enrichment: Chapter 17, p. 14

ESL/ELL STRATEGY

Martian Channels Place a layer of sand in a large pan. Fill a clean dishwashing soap bottle with water. Open the cap and turn the bottle upside down so that water flows over the sand, making channels like those on Mars. Be sure that students understand that water no longer flows on Mars's surface. learning style: visual

17-7 *(pp. 424–425)*

1 Introduce

STUDY HINT Have students practice stringing together the Key Terms in meaningful sentences that will help them remember their individual meanings. For example, "A meteoroid becomes a meteor if it flashes briefly in the sky or a meteorite if it crashes into Earth."

Linking Prior Knowledge Ask students if they have seen a shooting star. Tell them that shooting stars are not really stars at all. They are meteors that burn up as they enter Earth's atmosphere.

2 Teach

Demonstration Refer students back to page 387 in Chapter 16. Tell them that light pollution is a glow in the night sky caused by streetlights, brightly lit signs, and security lights on buildings. Light pollution can make it difficult to view the night sky. To demonstrate, have students observe as you shine a flashlight on the ceiling of a brightly lit classroom. Then, turn the overhead light off and shine the flashlight on the ceiling again. Tell students that the flashlight represents a meteor shower and the overhead light represents city lights. **learning style: visual**

Discussion Ask students where the best place for viewing a meteor shower would be. (far from city lights) Discuss why meteor showers can only be seen at night. (Sunlight keeps people from seeing the light of meteors.) Discuss whether meteor showers occur only at night or 24 hours a day. (24 hours a day) Remind students that meteor showers are particles from passing comets. **learning style: auditory**

Answers to questions

1. **LOCATE:** between Mars and Jupiter
2. **NAME:** a meteor
3. **EXPLAIN:** Earth passes through the dust left by a passing comet, and the dust burns up in Earth's atmosphere.
4. **IDENTIFY:** in Antarctica

17-7 What are asteroids and meteoroids?

Objective

Compare asteroids, meteoroids, meteors, and meteorites.

Key Terms

asteroid (AS-tuhr-oid)**:** large chunk of rock or metal that orbits the Sun

meteoroid (MEET-ee-uhr-oid)**:** small piece of rock or metal that travels through space

meteor (MEET-ee-uhr)**:** rock or metal that enters Earth's atmosphere

meteorite (MEET-ee-uhr-eyet)**:** piece of rock or metal that falls on a planet or moon's surface

Asteroids There is a large gap in the solar system between the orbits of Mars and Jupiter. For many years, astronomers thought the gap must contain a planet. They searched the gap but did not find the missing planet. Then, in 1801, an Italian astronomer spotted a tiny dot of light that orbited the Sun, like the planets. The astronomer had discovered the first asteroid. Ceres, as it was named, is 940 km across. It is now considered a dwarf planet.

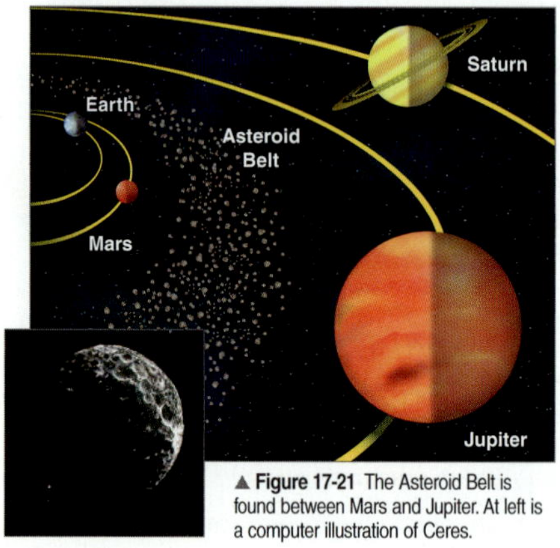

▲ **Figure 17-21** The Asteroid Belt is found between Mars and Jupiter. At left is a computer illustration of Ceres.

Ceres and other **asteroids** are leftover debris from the formation of the solar system. They are made up of pieces of rock or metal or a combination of rock and metal. Today, thousands of asteroids are known. Most orbit the Sun in the so-called Asteroid Belt between Mars and Jupiter.

1. **LOCATE:** Where is the Asteroid Belt?

Meteoroids Asteroids can collide, breaking off small pieces from both asteroids. These pieces, called **meteoroids,** scatter around the solar system. Most pieces are smaller than a sand grain. If a meteoroid enters a planet's atmosphere, friction will cause all or some of the meteoroid to burn. A bright streak will then be seen in the sky. Meteoroids that produce that bright streak are called **meteors.** Large pieces of meteoroids that reach a planet's surface are called **meteorites.** A crater can result from the impact.

◀ **Figure 17-22** A meteorite much bigger than the one below struck Arizona. It formed the crater known as Meteor Crater, seen at left.

2. **NAME:** What is a meteoroid that enters Earth's atmosphere called?

Meteor Showers Sometimes many meteor flashes will be seen in a night. These displays are called meteor showers. The meteoroids that fall in meteor showers are tiny, dust-sized particles from comets. Comets travel in long orbits that regularly take them close to the Sun. As a comet approaches the Sun, some of the ice in it melts. This releases trapped dust particles. When Earth passes through the dust left by a passing comet, the dust burns up in the atmosphere. A meteor shower occurs.

3. **EXPLAIN:** How are meteor showers produced?

424

Teaching Options

! FAST FACTS *How the Dinosaurs Died*

The hypothesis that linked the extinction of dinosaurs to an asteroid impact was first put forth by noted physicist Luis Alvarez. Alvarez was helping his son, a geologist, analyze layers of clay in Italy. The clay was deposited around the time of the dinosaurs' disappearance. When analyzing the clay, the Alvarezes unexpectedly found high levels of the element iridium, which is rare on Earth but relatively common in asteroids and comets. This led Luis Alvarez to propose that the iridium layer was deposited by a devastating asteroid strike that caused mass extinctions on Earth.

EXTENDING THE LESSON

Meteorology The word *meteorology* has the same Greek root as the words *meteoroid*, *meteor*, and *meteorite*. The meaning of the Greek root *meteor* is "high in the air." Have students explain why the word *meteorology* aptly describes the study of Earth's atmosphere. (The study of Earth's atmosphere is concerned with processes that occur high in the air.)

Meteorites on Earth More than 30,000 meteorites have been found on Earth's surface. However, most meteorites fall into the sea. Some meteorites have crashed through the roofs of houses and slammed into cars. Fortunately, these events are very rare.

Once in a while, a large meteorite strikes Earth's surface. One such impact took place in Arizona 50,000 years ago. A 1-km-wide crater, shown in Figure 17-22, was produced when the meteorite struck.

The best place to look for meteorites today is in Antarctica. When meteorites fall on ice, they are easy to see. Some four-billion-year-old meteorites have been found there.

4 IDENTIFY: Where are meteorites easy to spot?

CHECKING CONCEPTS

1. Where are meteoroids found?
2. What is a meteoroid?
3. How big are the particles in a meteor shower?
4. Where do most meteorites land on Earth?
5. What particles are burned up in meteor showers?

THINKING CRITICALLY

6. CONTRAST: How are meteoroids, meteorites, and meteors different?
7. INFER: How might the surface of a meteorite look after going through Earth's atmosphere?

Web InfoSearch

How the Moon Formed One thing scientists wanted to learn from the *Apollo* program was how the Moon was created. Moon rocks provided scientists with important clues. Today, some scientists believe that the early Earth was struck by a very large object, very likely an asteroid. Some of the debris from the collision flew off into space. Eventually, this debris came together to become the Moon.

SEARCH: Use the Internet to find out what evidence led scientists to this theory. Go to www.conceptsandchallenges.com to begin your search. Some key search words are **Moon rocks** and **asteroids.**

Real-Life Science

NEAR-EARTH OBJECTS

Not all asteroids are in the Asteroid Belt. Thousands cross or come close to Earth's orbit every year. These are named near-Earth objects (NEOs). About 918 NEOs are considered a possible danger to Earth. These objects are all being monitored by several groups.

About 65 million years ago, an asteroid about 10 km wide struck Earth. The huge crater it is believed to have formed, called Chicxulub, is found in the Gulf of Mexico and on Mexico's Yucatán peninsula. This is the impact that many scientists now think might have ended the rule of the dinosaurs. However, no known NEO is currently a serious threat to Earth.

▲ **Figure 17-23** Artist's concept of an asteroid hitting Earth

What if such an object should appear? Can anything be done to stop a catastrophe? Some people have suggested blowing the object up with a nuclear bomb. However, this might create many small pieces that could strike with the same deadly force. It would be better to nudge the asteroid in another direction. However, no one knows exactly how to do this yet.

Thinking Critically How might you force an asteroid to change direction?

ESL/ELL STRATEGY

Distinguishing Key Terms Have students learning English look up the suffixes *-oid*, *-or*, and *-ite* and use the information to write definitions for *meteoroid*, *meteor*, and *meteorite*.
learning styles: linguistic, auditory

• *TEACHING RESOURCES* •
Online or Print

Teacher's Resources
Lesson Review: Chapter 17, p. 11

WRITING HINT Suggest that students use complete sentences when answering the Thinking Critically questions. Explain that good reasoning requires clear, ordered, complete thoughts.

3 *Assess*

CHECKING CONCEPTS ANSWERS

1. in space
2. a small piece of rock or metal that travels through space
3. dust-sized
4. in the sea
5. dust particles from comets

THINKING CRITICALLY ANSWERS

6. CONTRAST: Meteoroids are small pieces of rock or metal that travel through space. Meteors are meteoroids that enter the atmosphere and burn up. Meteorites are meteors that strike the surfaces of planets or moons.
7. INFER: Possible answer: A meteorite might look melted or charred, since part of it may have burned as it passed through the atmosphere.

Web InfoSearch

How the Moon Formed Have students use the Internet to find out more about what led scientists to the current theories of how the Moon was formed. Ask students: *What clues did Moon rocks obtained from the* Apollo *mission provide?* Some key search words are **Moon rocks** and **asteroids.** Students should begin their search at www.conceptsandchallenges.com.

4 *Close*

Real-Life Science

NEAR-EARTH OBJECTS

Discussion Have students discuss how an asteroid impact could lead to mass extinctions. (The impact would send huge clouds of dust and debris into the atmosphere, blocking out sunlight. Many plants and animals would die.)

Thinking Critically Answer
Possible answer: A soft tap from an object launched from Earth might knock an asteroid off a collision course.

17-8 *(pp. 426–427)*

1 *Introduce*

STUDY HINT Before beginning this lesson, have students scan the subheads and look at the illustrations. Have them write questions that they would like to have answered during their reading. Encourage students to write the answers as they read the lesson.

Linking Prior Knowledge Review cloud formation with students (Lesson 12-5) and explain that Jupiter is covered with bands of clouds. Tell students that the Great Red Spot on Jupiter is similar to a cyclone on Earth. Hot spots on Earth produce eruptions of magma. The hot spot of Jupiter produces eruptions of hot gases.

2 *Teach*

Demonstration Partially freeze water in a cake pan so that a thin crust of ice forms on the surface of the liquid. Then, have students observe while you gently move the pan back and forth so that the water sloshes under the ice. Have students note how the ice breaks up. Relate this to the movement of water under the ice that covers Europa's surface.
learning style: visual

Discussion Describe the characteristics of Jupiter and Saturn. Have students observe the photographs of Jupiter and Saturn on pages 426 and 427. Tell them to note the Great Red Spot of Jupiter and the distinctive rings around Saturn.
learning styles: auditory, visual

Answers to questions

1. NAME: hydrogen and helium
2. DESCRIBE: a huge storm on Jupiter
3. IDENTIFY: hydrogen and helium

WRITING HINT Tell students that the best way to write a research report is to take good notes. Have them pick out important points and summarize them in their own words on separate note cards. Then, have students put the cards in a logical sequence and think of good transitions between ideas.

17-8 What do we know about Jupiter and Saturn?

Objective
Identify some features of Jupiter and Saturn.

A Gas Giant Jupiter is the largest planet in the solar system. It is the fifth planet out from the Sun. Because of its size, Jupiter can be seen without a telescope. Its mass is twice that of all the other planets combined. Jupiter has a diameter of 143,000 km. Earth's diameter is less than 13,000 km.

Jupiter is a gas giant with a rocky core. It is made up mostly of the light gases hydrogen and helium. Its density is only one-fourth that of Earth's. Colorful bands of clouds cover the entire planet.

All of the gas giants, including Jupiter, have rings around them. These rings are made up of small particles of dust and ice. Each particle is a tiny satellite. The rings circle the planet around the equator. A thin, faint ring around Jupiter was discovered by *Voyager 1* in 1979.

▲ **Figure 17-24** The planet Jupiter (upper right) is big enough to hold more than 1,300 Earths. The other objects in the picture are four of Jupiter's moons.

1. NAME: What two gases make up most of Jupiter?

The Great Red Spot The largest and best-known feature of Jupiter is its Great Red Spot. Astronomers believe this is a huge storm, similar to a cyclone. It was probably created by steam and ammonia rising from the atmosphere below Jupiter's cloudtops.

In 1994, Jupiter was hit by comet Shoemaker-Levy 9. The comet was pulled apart by Jupiter's gravity. Twenty-one pieces crashed into the thick atmosphere of Jupiter. Astronomers studied the impacts to learn more about Jupiter's atmosphere.

▲ **Figure 17-25** The Great Red Spot is a huge storm on Jupiter.

2. DESCRIBE: What is the Great Red Spot?

The Ringed Planet The second largest planet in the solar system and the sixth out from the Sun is Saturn. The diameter of Saturn is about 121,000 km.

Like Jupiter, Saturn is a gas giant made up mostly of hydrogen and helium. Saturn has colorful bands of clouds like Jupiter. However, Saturn is less dense than Jupiter is. It is even less dense than water.

Saturn's rings are its most distinctive feature. Galileo thought they were ears or handles when he saw them through his telescope. There are three main rings. Many thousands of smaller rings orbit inside the main rings. There is a large gap between two rings called the Cassini division. The *Cassini* probe began studing Saturn's rings in 2004.

Teaching Options

! FAST FACTS *Jupiter and Gravity*

A computer simulation model indicates that Jupiter's great gravitational force may serve as a sort of shield that protects the inner planets, including Earth, from frequent comet and asteroid strikes. When the computer program simulates a solar system without Jupiter, the small, rocky, inner planets are regularly bombarded by comet impacts. When the program simulation includes Jupiter, however, the strong gravitational pull of the gas giant draws away many comets and asteroids that would otherwise reach the inner planets.

QUICK CONVERSIONS		
Planet Diameters		
	Metric/SI	English
Earth	13,000 km	8,000 mi
Jupiter	143,000 km	89,000 mi
Saturn	121,000 km	75,000 mi

▲ **Figure 17-26** The orange bands in this picture are really clouds racing from east to west around Saturn.

IDENTIFY: What is Saturn mostly made up of?

CHECKING CONCEPTS

1. The largest planet in our solar system is __________.
2. The second largest planet in our solar system is __________.
3. Jupiter's mass is more than __________ the mass of the other planets put together.
4. Jupiter is made up mostly of __________ and helium.
5. The Great Red Spot is caused by steam and __________ rising into Jupiter's atmosphere.

THINKING CRITICALLY

6. **COMPARE:** What are two ways in which Jupiter and Saturn are alike?
7. **INFER:** What makes Jupiter and Saturn gas giants?

BUILDING SCIENCE SKILLS

Researching and Modeling Two small moons of Saturn, Pandora and Prometheus, orbit on either side of what is called the F ring. These moons are known as shepherds. *Pioneer 11* first saw them in 1979. Find out what these moons do and how they do it. Report your findings. Then, draw a sketch of Saturn's ring system and label the shepherd moons and the F ring.

Science and Technology

PROBING THE SECRETS OF JUPITER

The *Voyagers* gave us incredible images of Jupiter and its moons as the probes flew by. Then, in December 1995, the space probe *Galileo* went into orbit around Jupiter. This began a lengthy and still ongoing scientific exploration of Jupiter and its moons.

Galileo's first task involved separating a smaller probe and releasing it. This probe, slowed by a parachute, fell through Jupiter's atmosphere. Instruments on the probe measured the amounts of gases. Scientists found much less helium and water vapor than they had expected. They were also surprised to learn that the winds grew stronger instead of weaker as the probe descended.

▲ **Figure 17-27** The *Galileo* space probe

Galileo also provided new information about Jupiter's largest moons. For example, it appears that sections of ice covering Europa are moving away from each other. This movement might be caused by water beneath the ice. Another surprising discovery was a magnetic field around Ganymede.

Thinking Critically Why might finding water on Europa be exciting to scientists?

• TEACHING RESOURCES • Online or Print

Teacher's Resources

Lesson Review: Chapter 17, p. 12

Enrichment: Chapter 17, p. 14

ONGOING ASSESSMENT

Have students use astronomy texts or reference sources to research the diameters of the nine planets. Then, have students use wadded-up sheets of newspaper to make scale models of the planets. The models should accurately show the relative sizes of the planets in relation to one another.

learning styles: visual, tactile

3 Assess

CHECKING CONCEPTS ANSWERS

1. Jupiter
2. Saturn
3. twice
4. hydrogen
5. ammonia

THINKING CRITICALLY ANSWERS

6. **COMPARE:** Both are gas giants. Both also have rings.
7. **INFER:** They are huge planets made up mostly of hydrogen and helium gases.

4 Close

BUILDING SCIENCE SKILLS

Researching and Modeling The small moons of Saturn, Pandora and Prometheus are examples of shepherd satellites. The moons orbit on either side of the F ring, and their gravitational forces keep the particles that make up the ring from straying from their orbit. Thus, the F ring remains narrow and its edges sharp. Check students' sketches for inclusion of rings and the correct labeling of the two shepherd moons, Pandora and Prometheus, and the F ring.

Science and Technology

PROBING THE SECRETS OF JUPITER

Life-Science Connection The ice that covers the surface of Europa appears to be cracked. If water does exist under the ice, the cracks may allow light from the Sun to reach the water. This, in turn, may spark life processes such as photosynthesis.

Discussion Tell students that the space probe *Galileo* was launched in 1989. Ask: *How long did* Galileo *take to get to Jupiter?* (about 6 years) *Based on information gathered by* Galileo, *what inferences did scientists make about Europa?* (Water might be present on the planet.)

Thinking Critically Answer
Possible answer: It might indicate the possibility that life as we know it could exist on Europa.

17-9 *(pp. 428–429)*

1 Introduce

Linking Prior Knowledge Students first learned about satellites and space probes in Lesson 16-9. Tell them that when the *Voyager* space probes were launched in 1977, scientists only expected the mission to last 5 years. Now, scientists expect to receive signals from the space probes until about the year 2020. At that point, *Voyager 2* is expected to leave the solar system and pass into interstellar space.

2 Teach

Demonstration Use a tennis ball and marbles to demonstrate distances among the planets. Place the tennis ball in an open area. This represents the Sun. Then, using a scale of 1 AU (astronomical unit) = 10 cm, place the marbles in their appropriate position and the correct distance from the model of the Sun. Use these scale distances:
Mercury, 3.9 cm
Venus, 7.2 cm
Earth, 10 cm
Mars, 15.2 cm
Jupiter, 52 cm
Saturn, 95.4 cm
Uranus, 191.8 cm
Neptune, 300.6 cm
Have students identify each planet.
learning style: visual

Discussion Ask students: *Which gas giant planet might be most interesting to see up close?* Have them explain their choices, using the information they learned in this chapter about the characteristics of these planets. **learning styles: auditory, linguistic**

Answers to questions

1 **LIST:** hydrogen, helium, and methane

2 **IDENTIFY:** 1845

3 **IDENTIFY:** a small, planetlike object that orbits the Sun and shares its orbit with similar objects

4 **EXPLAIN:** rocks or ice fragments

17-9 What do we know about Uranus, Neptune and the dwarf planets?

Objective
Identify some features of Uranus, Neptune and the dwarf planets.

Key Terms
dwarf planets: small, planetlike objects that orbit the Sun and share their orbits with similar objects

Uranus The seventh planet from the Sun and the third largest planet in the solar system is Uranus. The diameter of Uranus is about 51,000 km.

In 1781, Uranus became the first planet to be discovered by using a telescope. Very little was known about Uranus until 1986, when the *Voyager 2* space probe flew past it. *Voyager 2* sent back many images and discovered ten new moons.

◀ **Figure 17-28** Uranus may have collided with another large body early in its history and been knocked over. This image was taken by an infrared camera. The planet's real color is blue-green, and its ring is very thin.

Uranus is made mostly of hydrogen and helium. The blue-green color of the clouds shows that the atmosphere also contains methane.

Oddly, the axis of Uranus is horizontal. The planet appears to be lying on its side. Its poles point toward the Sun. This unusual tilt may have been caused by a collision with another planet-size body early in its history.

Uranus spins fast and has a very strong magnetic field. It also has a faint ring.

1 **LIST:** What are three gases found in the atmosphere of Uranus?

Neptune The eighth planet from the Sun is Neptune. Neptune is similar to Uranus in size and mass. Its diameter is about 49,500 km, and it also has a faint ring around its equator. Neptune is so far from the Sun that it takes 165 Earth years to revolve once around it.

Neptune was discovered in 1846. It was the last planet to be visited by *Voyager 2*. Photographs taken by *Voyager 2* show that Neptune has a Great Dark Spot, similar to Jupiter's Great Red Spot. Neptune is also a gas giant. Its upper atmosphere is made mostly of clouds of frozen methane, which gives the planet its blue-green color. The lower atmosphere contains mostly hydrogen and helium. Neptune is the windiest planet. Winds can blow east to west at over 1,500 km/h.

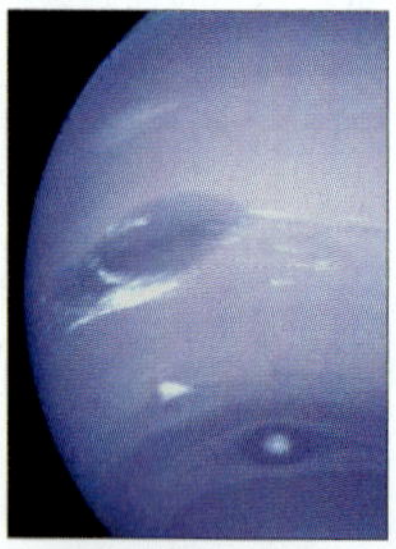

▲ **Figure 17-29** Through a telescope, Neptune is blue-green and shows few features.

2 **IDENTIFY:** When was Neptune discovered?

Dwarf Planets In 2006, the International Astronomical Union decided on a new definition of a planet. Pluto no longer fit that definition, so scientists reclassified Pluto as a dwarf planet. **Dwarf planets** are planetlike objects that orbit the Sun. Unlike planets, however, dwarf planets share their orbits with other, similar objects. For example, Pluto has an irregular orbit that overlaps the orbit of Neptune. Ceres and Eris are other dwarf planets. Dwarf planets are smaller than the other planets in the solar system.

▲ **Figure 17-30** The Sun looks very small from Pluto's orbit.

428

Teaching Options

ESL/ELL STRATEGY

Pluto's Orbit To help students understand Pluto's orbit, have them look at the solar system illustration on pages 412 and 413. Have them use their fingers to trace Pluto's orbit. Point out where Pluto's orbit crosses Neptune's orbit.

INTEGRATING CHEMISTRY

Methane Methane, which is found in the atmosphere of Uranus, is an organic compound. Each molecule of methane contains one atom of carbon and four atoms of hydrogen. Have students use plastic-foam balls to make models of methane compounds. **learning style: tactile**

COOPERATIVE LEARNING

Traveling Through Space Students have now learned about all eight planets. Have students develop a class skit in which students will act as tour guides, one for each planet, as they take passengers on a trip through the solar system. The passengers can ask the tour guide questions. Any questions that cannot be answered should be researched by volunteers and later presented to the class.

Pluto has a frozen surface of methane, nitrogen, and carbon dioxide. In 2006, NASA launched the *New Horizons* spacecraft. It will travel to Pluto and its moon, Charon. It should arrive in 2015, becoming the first spacecraft to visit there.

3 IDENTIFY: What is the definition of a dwarf planet?

The Kuiper Belt Between Neptune and the outer regions of the solar system is the Kuiper Belt. This belt was discovered in 1992. It was named after the Dutch astronomer who suggested that it might exist. The Kuiper Belt may contain as many as 100 million fragments of ice and rock. These fragments could be from a planet that never formed. Pluto, Eris, and their moons are classified as Kuiper Belt objects.

4 EXPLAIN: What kinds of objects are in the Kuiper Belt?

CHECKING CONCEPTS

1. What is the most unusual feature of Uranus?
2. How is Neptune similar to Uranus?
3. How long does Neptune take to complete one orbit around the Sun?
4. Which was the last planet that *Voyager 2* visited?

THINKING CRITICALLY

5. HYPOTHESIZE: Pluto may have once been a moon of Neptune. What might have caused Pluto to go into its own orbit?

Web InfoSearch

Neptune's Wobble A century ago, astronomers thought they observed a strange wobble in Neptune's orbit. Many thought the wobble was caused by a distant, unseen planet. Today, we know that the wobble does not exist at all.

SEARCH: Use the Internet to find out more about this. How did the search for another gas giant lead to the discovery of Pluto? Start your search at www.conceptsandchallenges.com. Some key search words are **Clyde Tombaugh, discovery of Pluto,** and **Pluto Charon.**

People in Science

JET PROPULSION LAB WORKER

The Jet Propulsion Laboratory (JPL) in Pasadena is run by the California Institute of Technology for NASA. Since 1960, it has been involved in exploring the solar system using space probes. Today, it serves as command center for missions such as *Voyager, Magellan,* and *Galileo*.

JPL employs many scientists, engineers, and technicians. They are involved in the launching and tracking of satellites, as well as space probes. The satellites study the oceans, the ozone, and other physical features of Earth. Scientists at JPL constantly analyze the data sent back to Earth. They must also check that the instruments are working and try to fix them remotely if they are not. JPL instruments also scan the skies for near-Earth objects that might collide with Earth. Most JPL scientists have a strong background in science and mathematics.

▲ **Figure 17-31** A Jet Propulsion Lab scientist is examining a volcanic Moon rock.

Thinking Critically What are some activities engineers at JPL might take part in?

RETEACH

Reviewing Planet Characteristics
Write *Outermost Planets* on the board. From this head, draw arrows to each of the following heads: *Uranus and Neptune*. Describe and list the important features of each planet under the appropriate head. Have students copy the information in their notebooks. **learning style: visual**

• TEACHING RESOURCES •
Online or Print

Teacher's Resources
Lesson Review: Chapter 17, p. 13
Enrichment: Chapter 17, p. 14

3 *Assess*

CHECKING CONCEPTS ANSWERS

1. its tilt, or horizontal axis
2. in size and mass
3. 165 Earth years
4. Neptune

THINKING CRITICALLY ANSWER

5. **HYPOTHESIZE:** It might have been knocked out of Neptune's reach by an asteroid or other large body.

4 *Close*

Web InfoSearch

Neptune's Wobble Have students use the Internet to find out more about this and how the search for another gas giant led to the discovery of Pluto. Some key search words are **Clyde Tombaugh, discovery of Pluto,** and **Pluto Charon.** Students should begin their search for information about this at www.conceptsandchallenges.com.

People in Science

JET PROPULSION LAB WORKER

Technology Connection When they were launched in 1977, the *Voyager* spacecraft were programmed to explore only Jupiter and Saturn. However, controllers at the Jet Propulsion Lab were able to reprogram the spacecraft so that they used the gravitational energy of the planets they passed to reach the other planets. At the time of their launching, the *Voyager* spacecraft were considered cutting-edge technology. Today, however, they are technological antiques. Many common laptop computers are more powerful than the *Voyager* computers.

Discussion Discuss the relationship between technology and space exploration. Help students understand that space exploration leads to new technologies. As technology improves, the ability to explore space improves.

Thinking Critically Answer
the building and launching of satellites and space probes, analyzing data, or the checking and repairing of instruments

17-10 *(pp. 430–431)*

1 Introduce

INVESTIGATE

TIME: 10–15 minutes

PURPOSE: Students will make scale models of a comet.

MATERIALS: pen, 5- and 10-in. paper plates, cotton balls, glue, streamers, tape, whole peppercorn or pea

PROCEDURE: Students can cut streamers from large sheets of paper or plastic.

THINK ABOUT IT: Possible answer: The comet is bigger than Earth but smaller than the Sun.
learning styles: visual, tactile

STUDY HINT Have students scan the lesson feature on predicting comets before they read the lesson. Understanding what is in the feature may help them make more sense of the rest of the lesson.

Linking Prior Knowledge Tell students that a comet is sometimes compared to a dirty snowball. Ask students what makes up a dirty snowball. (ice, dirt, water, dust, and grit) Ask students what happens when a dirty snowball is exposed to sunshine. (It melts, the water evaporates, and dirt is left behind.) Also, remind students what they learned about orbits (Lesson 17-2).

2 Teach

Discussion Refer students to Figure 17-33. Point out that a comet has a nucleus, a coma, and a tail. Describe the composition of each part. **learning style: auditory**

Reinforcement Make sure that students understand that comets orbit the Sun just as planets do. However, the comets' orbits are even more elliptical. Astronomers use a comet's orbital path to predict when it will next appear. Tell students that Halley's comet returns about every 76 years. It was last seen in 1986. Have students calculate when it will be seen next. (1986 + 76 = 2062)

Answers to questions

1 **DESCRIBE:** ice, frozen gas, and dust

17-10 What is a comet?

INVESTIGATE

Making a Scale Model of a Comet

HANDS-ON ACTIVITY

1. Using a pen, put a dot in the center of a 5-inch paper plate. This dot represents a comet's nucleus. Glue a cotton ball to the plate to cover the dot. The cotton ball represents the gas cloud that surrounds the nucleus of a comet. This is called the coma.
2. Attach five streamers to the back of the plate with tape. The streamers represent the comet's tail. Make sure all the streamers face in the same direction.
3. Compare your comet with a whole peppercorn or a small pea, which can represent Earth.
4. Compare your comet with a 10-inch paper plate, which can represent the Sun.

THINK ABOUT IT: How does your comet model compare to the size of Earth? How does it compare to the size of the Sun?

Objectives

Define comet and identify the features of a comet.

Key Terms

comet: lump of ice, frozen gas, and dust that orbits the Sun
nucleus: head or solid part of a comet
coma: gas cloud that surrounds the nucleus of a comet
tail: long, ribbonlike trail of comet dust and gas

Comet Parts Like planets and asteroids, comets are also members of the solar system. A **comet** is a lump of ice, frozen gas, and dust that orbits the Sun. Comets orbit the Sun in very long ellipses that often take them beyond the orbit of Pluto.

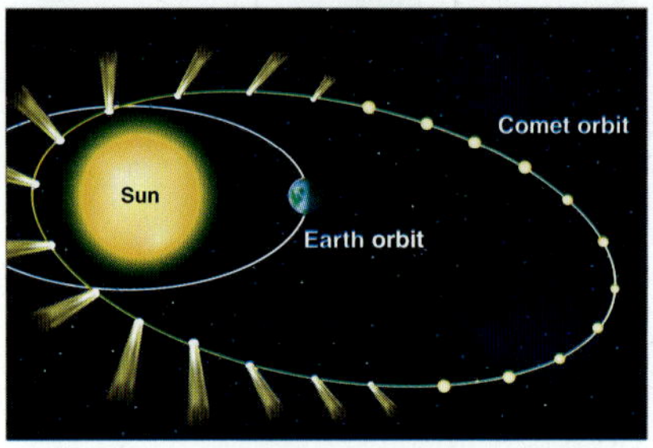

▲ **Figure 17-32** A comet's orbit typically takes it near the Sun and then back to the outer reaches of the solar system.

A comet has three parts. The core, or **nucleus**, is basically a dirty ice ball. It is made of water ice, frozen gas, and dust. The cloud of gas that surrounds the nucleus is called the **coma.** Stretching out from the coma is one or more **tails.** A stream of particles coming from the Sun pushes the tail or tails away from the center of the solar system.

▲ **Figure 17-33** The structure of a comet

1 **DESCRIBE:** What are comets made of?

Ghostly Travelers Comets travel in deep space, where it is very cold. For most of their trip around the Sun, they are invisible. As a comet approaches the Sun, it begins to warm up. Some of the ice begins to melt. Gas and dust are released from the nucleus and spread out to form the coma. Soon, the tail begins to form. Tails can stretch out for millions of kilometers.

From Earth, a comet looks like a ghostly patch in the sky. It suddenly appears one night as sunlight begins reflecting off it.

Teaching Options

! FAST FACTS *The Oort Cloud*

Astronomers theorize that the radius of the Oort Cloud ranges from 50,000 to 100,000 AU. Thus, the Oort Cloud may reach halfway to the Sun's nearest neighbor, Alpha Centauri. Comets that originate in the Oort Cloud are classified as long-period comets, or comets that take longer than 200 years to complete one orbit. Comets from the Kuiper Belt are short-period comets, which have orbital periods of 200 years or less.

RETEACH

Comet Tails Demonstrate how solar wind causes the tail of a comet to point away from the Sun. Obtain one of the models constructed in the Investigate activity. Hold it before a fan so that its "coma" faces the fan.
learning style: visual

EXTENDING THE LESSON

Researching Comets Have interested students research a comet other than Halley's, such as Encke, Tuttler, or Tempel 2. Have them find out when the comet last appeared and where it was most evident. Students can present their findings in a poster.

Over time, the coma expands, and the tail stretches out into space. After rounding the Sun, the comet moves back into deep space and disappears from view.

2 EXPLAIN: What causes a comet's tail to grow?

Where Comets Come From Astronomers think that many comets come from the Oort Cloud. This cloud, named after the Dutch astronomer Jan Oort, is far beyond Neptune's orbit. The Oort Cloud may contain trillions of inactive comets. A second cloud of comets or cometlike material called the Kuiper Belt exists beyond the orbit of Neptune.

Comets were formed billions of years ago when the solar system was young. After the Sun formed, gas and dust surrounding the new Sun formed into small pieces of ice, rock, and metal. These pieces eventually stuck together to form larger objects. The largest bodies became planets and moons. The smaller bodies became asteroids and comets.

3 INFER: What materials became comets?

CHECKING CONCEPTS

1. The parts of a comet are the nucleus, the coma, and the ________.
2. Comets orbit the Sun in long, ________ orbits.
3. The Oort Cloud is found outside the orbit of ________.
4. The heaviest materials, such as rock and ________, formed the planets, moons, comets, and asteroids.

THINKING CRITICALLY

5. **INFER:** Why are scientists interested in comets?
6. **INFER:** What happens to the size of the nucleus of a comet each time it orbits the Sun?

DESIGNING AN EXPERIMENT

Design an experiment to solve the following problem. Include a hypothesis, variables, a procedure with materials, and a type of data to study. Be sure to also include a way to record your data.

PROBLEM: Astronomers believe that the nucleus of a comet begins melting as the comet nears the Sun. Design an experiment that will prove this. Your experiment will be carried on a spacecraft that will travel with the comet as it orbits the Sun.

How Do They Know That?

PREDICTING COMETS

▲ **Figure 17-34** Halley's comet as seen in 1960

Scientists know that comets are coming days or months before they become visible to the naked eye. The comets appear in the sky right where scientists predict they will be. How do they do that?

Most comets travel around the Sun many times before they melt or leave the solar system. British astronomer Sir Edmund Halley was the first person to predict a comet's return, in 1758. The comet was later named after him.

Old comets, ones that have passed by the Sun before, are easy to predict. Today, comet orbits are calculated by computers. To find new comets, astronomers take pictures of the same regions of the night sky several days apart. If one of the stars in the first picture is in a different place in the second picture, it could be a comet. Powerful telescopes are then trained on the object. Eventually, astronomers identify it as a comet or some other object, like an asteroid. If it is a comet, it will begin releasing gas as it nears the Sun.

Thinking Critically Why do comets become visible as they get closer to the Sun?

2 EXPLAIN: The tail grows as the comet nears the Sun and warms up. The melting comet releases gas and dust that form the tail.

3 INFER: leftover materials from the formation of the solar system

3 *Assess*

CHECKING CONCEPTS ANSWERS

1. tail
2. elliptical
3. Neptune
4. metal

THINKING CRITICALLY ANSWERS

5. **INFER:** Possible answers may include that studying comets can help us learn more about how the solar system formed.
6. **INFER:** It gets smaller.

4 *Close*

DESIGNING AN EXPERIMENT

Use the Designing an Experiment Rubric from the Teacher's Resources to assess students' experiments. Fill in the rubric with the additional information below. For this assignment, students should have:

- identified a hypothesis that could, at least in theory, be tested.
- attempted to prove that a comet's nucleus begins to melt when it approaches the Sun.

MORE ON THE WEB

Comet Halley Have students use the Internet to learn more about Halley's comet. Then, have them share their findings.

ONGOING ASSESSMENT

Have students draw a comet and label its three parts. (nucleus, coma, tail) Have them describe what each part is made of.

learning styles: visual, linguistic

• TEACHING RESOURCES • Online or Print

Teacher's Resources

Lesson Review: Chapter 17, p. 16

Designing an Experiment: Study Tools, p. 14

Designing an Experiment Rubric: Teaching Tools, p. 7

How Do They Know That?

PREDICTING COMETS

Real-Life Connection Using relatively inexpensive telescopes, amateur astronomers can probe the depths of space. Amateur astronomers often form sky-watching groups. These dedicated amateurs are sometimes the first to notice changes in the night sky. In fact, amateur astronomers have discovered many new asteroids and comets.

Discussion Ask students: *How have amateur astronomers contributed to scientific research?* (They have discovered many new asteroids and comets.)

Thinking Critically Answer
Sunlight reflects off the gas and dust released from the nucleus.

Challenges (pp. 432–434)

Chapter Summary

Review Before students begin the Challenges, review the objectives of each lesson with them. If time permits, have students work in pairs. Each pair should come up with at least one relevant fact related to each of the objectives.

STUDY HINT Have pairs of students take turns quizzing each other on key concepts in this chapter.

Key Term Challenges

MATCHING

1. orbit (17-1)
2. ellipse (17-2)
3. comet (17-10)
4. satellite (17-4)
5. nebula (17-1)
6. rift (17-6)
7. meteor (17-7)
8. nucleus (17-10)
9. solar system (17-1)
10. aphelion (17-2)

APPLYING DEFINITIONS

11. An asteroid is a large chunk of rock that orbits the Sun. A meteoroid is a small piece of rock or metal that travels through space. (17-7)
12. A meteor is rock or metal that enters Earth's atmosphere. A meteorite is a meteor that strikes the surface of a planet or moon. (17-7)
13. Aphelion is the point in a planet's orbit when it is farthest from the Sun. Perihelion is the point in a planet's orbit when it is closest to the Sun. (17-2)
14. Craters are round holes on the surface of a planet or moon. Maria are broad, flat plains on the Moon. (17-3)
15. A nebula is a spinning cloud of gas and dust. The solar system probably formed from a nebula. (17-1)
16. A satellite is a natural or an artificial object that orbits another body in space. A moon is a natural satellite. (17-4)

Chapter 17 Challenges

Chapter Summary

Lessons 17-1 and 17-2

- The **solar system** may have formed from a spinning cloud of gases and dust called a **nebula.** There are eight planets in the solar system. The planets can be divided into inner and outer planets.
- An **ellipse** has an oval shape. Most planets travel around the Sun in elliptical **orbits.**
- Earth's orbital velocity is greatest at **perihelion** and least at **aphelion.**

Lesson 17-3

- The Moon is Earth's only natural **satellite.** Humans first landed on it in 1969. There is no air or water on the Moon. The Moon has **maria,** highlands, and **craters.**

Lessons 17-4 and 17-5

- Mercury and Venus have no moons. Mars has 2, Jupiter has at least 28, Saturn 30, Uranus 21, and Neptune 8.
- Mercury is the closest planet to the Sun. It has many craters. Venus has a blanket of sulfuric acid clouds. Earth supports life.

Lessons 17-6 and 17-7

- Mars is reddish. It might have had running water in the past.
- **Asteroids** orbit the Sun in a belt between Mars and Jupiter. Space objects called **meteoroids** may enter Earth's atmosphere. If they burn up, they are called **meteors.** If they strike Earth's surface, they are **meteorites.**

Lessons 17-8 and 17-9

- Jupiter is the largest, most massive planet. It is a gas giant. Its Great Red Spot is a huge storm. Saturn is the least dense planet and has distinctive rings. Rings are particles of ice and dust that orbit a planet around its equator.
- Uranus is a gas giant that appears to be lying on its side. Neptune, also a gas giant, has an upper atmosphere of frozen methane.

Lesson 17-10

- **Comets** are dirty ice balls. They have long, elliptical orbits.

Key Term Challenges

aphelion (p. 410)	meteoroid (p. 424)
asteroid (p. 424)	nebula (p. 408)
coma (p. 430)	nucleus (p. 430)
comet (p. 430)	orbit (p. 408)
crater (p. 414)	perihelion (p. 410)
ellipse (p. 410)	rift (p. 422)
mare (p. 414)	satellite (p. 416)
meteor (p. 424)	solar system (p. 408)
meteorite (p. 424)	tail (p. 430)

MATCHING **Write the Key Term from above that best matches each description.**

1. curved path of one object around another object in space
2. flattened circle, or oval
3. lump of dust and ice that orbits the Sun and has a gaseous tail
4. natural or artificial object orbiting another body in space
5. cloud of gas and dust in space
6. valley caused by a crack in the crust of a planet
7. rock or metal that enters Earth's atmosphere from space
8. head or solid part of a comet
9. Sun and all the objects that orbit the Sun
10. point in a planet's orbit at which the planet is farthest from the Sun

APPLYING DEFINITIONS **Explain the difference between the words in each pair. Write your answers in complete sentences.**

11. asteroid, meteoroid
12. meteor, meteorite
13. aphelion, perihelion
14. craters, maria
15. nebula, solar system
16. satellite, moon

432

Teaching Options

PREPARING STUDENTS FOR STANDARDIZED TESTS

Reading Strategy: Tell students to recheck their answers for any careless mistakes.

Writing Strategy: For answers that require several sentences, tell students to mentally organize their thoughts before writing them down.

Interpreting Visuals Strategy: If students are presented with data in a table, tell them that they can often recognize relationships among the variables by plotting the data in a graph.

ESL/ELL STRATEGY

Tell students to look for keywords such as *all*, *none*, or *not* in questions. For example, in the multiple-choice question 7, a keyword to note is *sometimes*. Have students make a poster of such clue words and continue to collect others such as *never* and *always*.

Content Challenges TEST PREP

MULTIPLE CHOICE **Write the letter of the term or phrase that best completes each statement.**

1. The planet closest in size to Earth is
   - **a.** Venus.
   - **b.** Mercury.
   - **c.** Uranus.
   - **d.** Jupiter.
2. The Great Red Spot is the best-known feature of
   - **a.** Mercury.
   - **b.** Neptune.
   - **c.** Uranus.
   - **d.** Jupiter.
3. The planet that is the farthest from the Sun is
   - **a.** Venus.
   - **b.** Neptune.
   - **c.** Uranus.
   - **d.** Jupiter.
4. The only planets in the solar system without at least one moon are
   - **a.** Jupiter and Mars.
   - **b.** Earth and Venus.
   - **c.** Jupiter and Mercury.
   - **d.** Mercury and Venus.
5. American astronauts first set foot on the Moon in
   - **a.** 1957.
   - **b.** 1969.
   - **c.** 1972.
   - **d.** 1975.
6. Earth is closest to the Sun in
   - **a.** July.
   - **b.** September.
   - **c.** March.
   - **d.** January.
7. The gas giants of our solar system are Jupiter, Saturn, Uranus, and
   - **a.** Earth.
   - **b.** Venus.
   - **c.** Neptune.
   - **d.** Mars.
8. The light-colored areas on the Moon are
   - **a.** mountains.
   - **b.** rivers.
   - **c.** plains.
   - **d.** oceans.
9. The first planet to be discovered by using a telescope was
   - **a.** Jupiter.
   - **b.** Uranus.
   - **c.** Mars.
   - **d.** Mercury.

TRUE/FALSE **Write *true* if the statement is true. If the statement is false, change the underlined term to make the statement true.**

10. Some scientists think a nebula expanded to form the solar system.
11. The core of a comet is made of dust and ice.
12. Earth moves fastest at perihelion.
13. The largest planet in the solar system is Neptune.
14. Earth has only one ring.
15. Uranus is tipped on its side.
16. The region between Mars and Jupiter is called the Meteoroid Belt.
17. Saturn's density is less than the density of Jupiter and even water.
18. The first asteroid found was named Ceres.

Content Challenges

MULTIPLE CHOICE

1. a (17-5)
2. d (17-8)
3. a (17-9)
4. d (17-4)
5. b (17-3)
6. d (17-2)
7. c (17-9)
8. a (17-3)
9. b (17-9)

TRUE/FALSE

10. false; contracted (17-1)
11. false; nucleus (17-10)
12. true (17-2)
13. false; Jupiter (17-8)
14. false; moon (3-4)
15. true (17-9)
16. false; Asteroid (17-7)
17. true (17-8)
18. true (17-7)

ALTERNATIVE ASSESSMENT

Summarizing Information Have students draw a diagram of the solar system that shows the eight planets and the Asteroid Belt arranged according to their distance from the Sun. Each item in the diagram should be labeled. Have students indicate on the diagram which planets are small and rocky, which are gas giants, which have moons, and which have rings.

PORTFOLIO ASSESSMENT

Making Student Portfolios Students can demonstrate their comprehension of the concepts in this chapter by making a portfolio. The Chapter Self-Check, the Big Idea Planner, and the Weekly Journal are some of the reproducibles in the Teacher's Resources that they can include in their portfolios. You may wish to make certain assignments planned portfolio items, such as the Chapter Project. Tell students ahead of time so that they put in their best effort. Use the Portfolio Assessment Rubric also found in the Teacher's Resources to assess students' portfolios.

Concept Challenges

WRITTEN RESPONSE

1. **INFER:** The planets may appear to wander across the sky, but in fact they do not. They move in predictable orbits. (17-1)
2. **INFER:** The Moon has no water or atmosphere to act as agents of erosion. (17-3)
3. **ANALYZE:** The inner planets are rocky. The outer planets are gas giants. (17-1)
4. **HYPOTHESIZE:** Possible answers may include that the Moon's lack of atmosphere and lack of light pollution make it a good place to study space or that the Moon can be mined for minerals. (17-3)
5. **INFER:** Water may indicate the presence of life. (17-6)
6. **ANALYZE:** It proved that not everything in space revolved around Earth. (17-4)
7. **EXPLAIN:** The dust trail left by the passing comet burns up in Earth's atmosphere. (17-7)
8. **INFER:** A stream of particles coming from the Sun pushes the tail away from the center of the solar system. (17-10)
9. **EXPLAIN:** The gravitational attraction between the Sun and Earth varies, depending on the distance from Earth to the Sun. (17-2)
10. **COMPARE/CONTRAST:** Possible answer: All eight planets orbit the Sun. The inner planets are small and rocky. The outer planets are gas giants. (17-1)

INTERPRETING VISUALS

11. 8 (17-9)
12. 2 (17-4)
13. Saturn (17-8)
14. no (17-5, 17-6)

Chapter Project Wrap-Up

BIG IDEA POSTER

Students' posters can be exhibited around the classroom. Students can present them and explain what each picture represents.

You can use the Individual Activity Rubric from the Teacher's Resources to assess students' projects. Fill in the rubric with the additional information below. For this project, students should have:

- created their own Big Idea spreads.
- included information that answers the question posed.

Concept Challenges TEST PREP

WRITTEN RESPONSE **Answer each of the following questions in complete sentences.**

1. **INFER:** Why is the term *wanderer* inaccurate for describing a planet?
2. **INFER:** Why do rocks on the Moon not weather or erode as rocks on Earth do?
3. **ANALYZE:** What is one trait that the inner planets have in common? The outer planets?
4. **HYPOTHESIZE:** What are some reasons astronauts may go back to the Moon?
5. **INFER:** Why are scientists looking for water on Mars?
6. **ANALYZE:** Why was Galileo's discovery of four moons orbiting Jupiter important to the history of astronomy?
7. **EXPLAIN:** How can a comet passing through Earth's orbit cause a meteor shower?
8. **INFER:** Why does a comet's tail always point away from the Sun?
9. **EXPLAIN:** Why does Earth's orbital velocity change as it orbits the Sun?
10. **COMPARE/CONTRAST:** How are the inner planets and outer planets alike and how are they different?

INTERPRETING VISUALS **Use Figure 17-35 to answer each of the following questions.**

11. How many moons does Neptune have?
12. How many planets have no moons?
13. What planet has the most known moons?
14. Do any of the inner planets have rings?

PLANETS, MOONS, AND RINGS		
Planets	**Number of Known Moons**	**Rings**
Mercury	0	No
Venus	0	No
Earth	1	No
Mars	2	No
Jupiter	62	Yes
Saturn	59	Yes
Uranus	27	Yes
Neptune	13	Yes

▲ Figure 17-35

Teaching Options

• TEACHING RESOURCES •
Online or Print

Teacher's Resources

Key Term Review: Chapter 17, p. 17

Chapter 17 Test: pp. 18-20

Individual Activity Rubric: Teaching Tools, p. 5

Chapter Self-Check: Teaching Tools, p. 11

Big Idea Planner: Study Tools, pp. 11–12

Weekly Journal: Study Tools, p. 19

• TEACHING RESOURCES •
Online or Print

Teacher's Resources

Portfolio Assessment Rubric: Teaching Tools, p. 10

PLANNING GUIDE

◆ **TEACHING THE CHAPTER** This chapter should take approximately 11–14 days to complete instruction and assessment.

	Skills and Features	Projects and Activities	Achieve Science Literacy	Meet Individual Needs
Chapter 18 Opener p. T435	MODEL	• Chapter Project		
Lesson 18-1 **How are stars formed?** pp. T436–T437 Standards: B3a, B3e	SEQUENCE, INFER, CLASSIFY • Building Science Skills • Science and Technology	• Cooperative Learning • Extending the Lesson • More on the Web	• Study Hint	
Lesson 18-2 **How is spectroscopy used to study stars?** pp. T438–T439 Standards: B3a, B3c	APPLY, ANALYZE, CALCULATE • Web InfoSearch • Hands-On Activity	• Extending the Lesson • Lab Challenge	• Study Hint • Reading Strategy	Teacher's Resources* Enrichment Activity
Lesson 18-3 **What is magnitude?** pp. T440–T441 Standard: D3d	DESCRIBE, HYPOTHESIZE, COMPARE • Building Math Skills • Hands-On Activity	• Cooperative Learning • Extending the Lesson	• Study Hint	• ESL/ELL Strategy
Lesson 18-4 **How are stars classified?** pp. T442–T443 Standard: D3a	CLASSIFY, IDENTIFY, ANALYZE • Building Math Skills • How Do They Know That?	• Extending the Lesson • More on the Web	• Study Hint • Ongoing Assessment	Teacher's Resources* Enrichment Activity Concept Map
Lesson 18-5 **What is a star's life cycle?** pp. T444–T445 Standards: B3a, B3e	DESCRIBE, EXPLAIN, SEQUENCE • Web InfoSearch • How Do They Know That?	• Cooperative Learning • Active Art	• Study Hint • Reading Strategy	• ESL/ELL Strategy • Reteach Teacher's Resources* Flowchart
Lesson 18-6 **What kind of star is the Sun?** pp. T446–T447 Standards: D3a, D3d	COMPARE, INFER, ANALYZE • Health and Safety Tip • Hands-On Activity	• Cooperative Learning • Integrating Physical Science		• ESL/ELL Strategy

*www.conceptsandchallenges.com, password: ccearth09

Planning Guide continues on next page.

Standards: For details on the correlation to National Science Standards see pages *xx–xxi*.

The Sun and Other Stars
(Planning Guide continued)

	Skills and Features	Projects and Activities	Achieve Science Literacy	Meet Individual Needs
Lesson 18-7 **What is the surface of the Sun like?** pp. T448–T449 Standard: D3d	EXPLAIN, DEFINE, MODEL • Web InfoSearch • Integrating Physical Science	• Extending the Lesson	• Study Hint • Ongoing Assessment	• Reteach Teacher's Resources* Enrichment Activity
Lesson 18-8 **How does the Sun produce energy?** pp. T450–T451 Standard: D3d	STATE, IDENTIFY, ANALYZE • Integrating Life Science	• Extending the Lesson • More on the Web • Integrating Health	• Study Hint • Reading Strategy • Ongoing Assessment	Teacher's Resources* Enrichment Activity
Lesson 18-9 **What are constellations?** pp. T452–T453	DEFINE, INFER, RESEARCH • Building Science Skills • Real-Life Science	• Extending the Lesson	• Study Hint	• Reteach
Lesson 18-10 **What are galaxies?** pp. T454–T455 Standard: D3c	COMPARE, DESCRIBE, CLASSIFY	• Cooperative Learning • More on the Web	• Study Hint • Writing Hint	• Reteach
Lab Activity **Counting Galaxies** pp. T456–T457	OBSERVE, ESTIMATE, INFER • Set-Up Time: 5–10 min • Lab Time: 25–30 min	• More on the Web • Extend the Lab Activity	• Tips for Using Technology	• ESL/ELL Strategy Teacher's Resources* Lab Activity Report Spreadsheet
Lesson 18-11 **What else do we know about the universe?** pp. T458–T459 Standards: B3a, D3b	EXPLAIN, ANALYZE, INFER • Investigate • Designing an Experiment • Integrating Physical Science	• Extending the Lesson	• Reading Strategy	• ESL/ELL Strategy Teacher's Resources* Designing an Experiment
Big Idea **How can we estimate the size of the universe?** pp. T460–T461	RESEARCH, COMMUNICATE, ANALYZE • Science Log: Writing Activity	• Big Idea • Cooperative Learning • Close Activity	• Tips for Using Technology	Teacher's Resources* Big Idea Planner Thumbnail Sketch
Chapter 18 Challenges pp. T462–T464	• Chapter Summary • Key Term Challenges • Content Challenges • Concept Challenges	• Chapter Project Wrap-Up	• Study Hint • Preparing Students for Standardized Tests	• ESL/ELL Strategy Teacher's Resources* Chapter Self-Check

*www.conceptsandchallenges.com, password: ccearth0

Chapter 18 The Sun and Other Stars

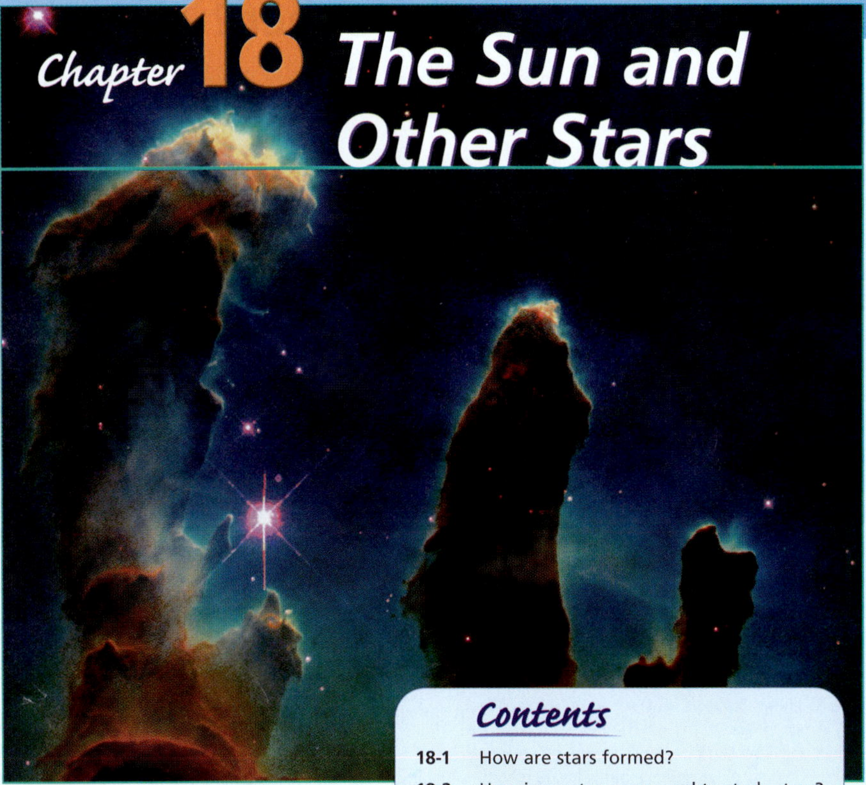

▲ **Figure 18-1** The Eagle Nebula is also known to astronomers as M16.

These dark structures may look like cave formations or undersea coral. However, they are actually columns of cool hydrogen gas and dust in a nebula. Nebulae are the birthplaces of new stars. The Eagle Nebula, above, is 6,500 light-years away from Earth. The tallest column is about one light-year tall. This picture was taken with a camera on the Hubble Space Telescope. The image reveals the various elements that make up the matter in the nebula.

▶ What do you think are some of the elements that make up stars?

Contents

Teaching Options

ASSESSMENT PLANNER

For assessment in the Student Edition, see the following pages:

Content Assessment

Checking Concepts: pp. 437, 439, 441, 443, 445, 447, 449, 451, 453, 455, 459
Thinking Critically: pp. 437, 439, 441, 443, 445, 447, 449, 451, 453, 455, 459
Interpreting Visuals: pp. 451, 455
Chapter Challenges: pp. 462–464

Alternative Assessment

Building Skills: pp. 437, 441, 443, 453
Web InfoSearch: pp. 439, 445, 449
Science Log: p. 461

Performance-Based Assessment

Hands-On Activities: pp. 439, 441, 447
Designing an Experiment: p. 459
Lab Activity: pp. 456–457

• TEACHING RESOURCES • Online or Print

Teacher's Resources

Lesson Review: Ch. 18, pp. 2, 3, 4, 5, 7, 8, 9, 10, 12, 13, 14
Enrichment: Ch. 18, pp. 6, 11
Key Term Review: Ch. 18, p. 16
Chapter 18 Test: pp. 17–19
Laboratory Manual: pp. 87–90

Chapter 18 The Sun and Other Stars

Chapter Overview

In this chapter, students will learn how stars, including the Sun, form. Students will find out how stars are classified and how they change over time. They will learn that groups of stars make up galaxies and that galaxies make up the universe.

About Figure 18-1 The Eagle Nebula in the constellation Serpens is in one of the spiral arms of the Milky Way. The gas and dust of the nebula are illuminated by a cluster of stars that the nebula is creating. The stars are about 5.5 million years old, and new ones are still forming.

Answer hydrogen and helium

Linking Prior Knowledge

For this chapter, students should recall:

- what elements are (Lesson 2-1).
- how telescopes are used to explore space (Lessons 16-3 to 16-5).
- a description of the solar system (Lesson 17-1).
- why objects in the solar system stay in orbit (Lesson 17-2).

Chapter Project

LIFE CYCLE OF A STAR MOBILE

MATERIALS: fake spider webbing, gray lint from a dryer, sparkles and spangles, plastic-foam balls of different sizes, markers, ice-cream sticks, hanger, yarn, cotton balls, silver stick-on stars

Have students create a mobile entitled *Evolution of a Star*, from the birth of a nebula (spider webbing mixed with lint and sparkles) to protostar, to main sequence star, to red giant or supergiant, to supernova, neutron star, white dwarf, or black hole. Divide the class into groups to prepare the parts of the mobile. The class can later put it all together and hang it from a hook in the ceiling or on a wall. Encourage students to be creative in designing the parts of the mobile, which can be joined by yarn, or clear fishing line. **learning style: tactile**

18-1 *(pp. 436–437)*

1 Introduce

STUDY HINT Have students read the Key Terms and their definitions. As they study this lesson, have students sequence the terms from smallest to largest: nebula, protostar, star, binary stars, cluster.

Linking Prior Knowledge Have students recall that elements are substances made up of only one kind of atom (Lesson 2-1). Ask students to use the information on this page to list the most common elements in stars. (hydrogen, helium, sodium, calcium, and iron)

2 Teach

Demonstration Have a volunteer recite the children's verse "Twinkle, Twinkle, Little Star." Ask students to hypothesize why stars appear to twinkle. Demonstrate this phenomenon by crumpling a piece of aluminum foil into a ball. Open the ball and spread it on a desktop. Place a transparent bowl filled with water over the wrinkled foil. Shine a flashlight into the water as a volunteer gently sets the water in motion using a pencil or a finger. Have the volunteer describe what happens to the light that hits the foil. (The light twinkles because it bends as it travels through the water.) Stars twinkle in the night sky because the light they emit bends as it travels through Earth's atmosphere. **learning style: visual**

Discussion After students have read this page, have them paraphrase the information describing the steps involved in the formation of a star. (Gas and dust combine to form a spinning cloud called a nebula. Gravity makes the cloud contract and flatten. Material condenses at the center of the disk to form a protostar. When the protostar gives off light and heat, it is a true star.) **learning style: auditory**

Answers to questions

1. **NAME:** hydrogen
2. **NAME:** a nebula
3. **DEFINE:** pairs of stars that travel through space together and revolve around each other

18-1 How are stars formed?

Objectives

Describe how stars form. Name some characteristics of stars.

Key Terms

star: ball of gases that gives off light and heat

nebula (NEHB-yuh-luh)**:** cloud of gas and dust in space

protostar: dense material in the center of a nebula that is about to become a star

binary stars: two stars that revolve around each other

star cluster: large group of stars that travel together through space

▲ **Figure 18-2** The Sun is only one of billions of stars. However, the unaided eye sees only a few thousand stars.

Tiny Points of Light The Sun is a star. A **star** is a big ball of gases that gives off heat and light. The Sun is only one of billions of stars that make up our galaxy, and there are billions of galaxies. Between the stars and galaxies are vast stretches of nearly empty space.

On a clear night, you can see thousands of stars in the sky. Most of these look to us like tiny points of light. This is because they are so far away. Most stars appear white to the unaided eye. However, a closer look reveals that stars come in different colors. For example, Rigel is a blue star. Betelgeuse (BEET-uhl-jooz) is red. The Sun, of course, is yellow. Stars also show a huge range in size, brightness, and surface temperature. How large and massive a star is determines what will eventually happen to it.

Different stars have different elements. Most stars are made up of the gases hydrogen and helium. Other elements often found in stars include sodium, calcium, and iron. The most common element in stars is hydrogen. All of these elements are found in varying amounts in stars.

1. **NAME:** What is the most common element in stars?

Formation of Stars A star forms from a cloud of gas and dust in space called a **nebula.** Gravity causes the nebula to contract and start spinning. This flattens the nebula into a disk. Material at the center of the disk forms a **protostar.**

As the nebula continues to contract, temperature and pressure build. Eventually, nuclear reactions begin, and the protostar starts to give off light and heat. A new star is born.

2. **NAME:** What is the first stage in star formation?

▲ **Figure 18-3** A star begins as a cloud of gas and dust that forms into a protostar. When it begins to give off light and heat, the protostar has become a true star.

436

Teaching Options

EXTENDING THE LESSON

Researching Star Clusters Have interested students find out about open star clusters such as the Pleiades or globular clusters such as Omega Centauri and M13. Have them find out what constellations the star formations are in, their ages, how big they are, and how they formed. Then, have students write a brief report and tell the class the most interesting facts about these star formations. **learning style: visual**

COOPERATIVE LEARNING

Looking at Star Births Have pairs of students work together to quiz each other on star formation. Have a student from each pair make a simple sketch of each stage on a separate card: nebula, spinning nebula, disklike nebula, protostar, star. The other student should put the sketches in the proper order or orally describe what is happening in each sketch.
learning styles: visual, linguistic

Double Stars and Star Clusters Unlike the Sun, most stars are double, or binary, stars. **Binary stars** are pairs of stars that travel through space together and revolve around each other.

Some stars are part of multiple-star systems. They contain three or more stars. Many stars move through space in large groups called **star clusters.** Some clusters are globular, or round, in shape. Others are open, or loosely arranged.

▲ **Figure 18-4** A globular cluster can contain millions of stars.

3 DEFINE: What are binary stars?

CHECKING CONCEPTS

1. What is the star closest to Earth?
2. What two gases make up most stars?
3. What is a nebula?
4. What causes a spinning nebula to contract?
5. What two shapes do star clusters form?

THINKING CRITICALLY

6. SEQUENCE: What are the stages in the formation of a star?
7. INFER: Why are rounded clusters called globular clusters?

BUILDING SCIENCE SKILLS

Classifying Stars are grouped according to color, or spectral classes. The letters *O, B, A, F, G, K,* and *M* represent spectral classes. Blue stars are in spectral class O or B. Red stars are M stars. Do some research on the following stars: Canopus, Arcturus, Sirius, Rigel, the Sun, Betelgeuse, Altair, and Capella. Then, put them in their correct spectral class.

Science and Technology

OBSERVATORIES

Most professional astronomers today work at observatories. Instead of using just their eyes to look through telescopes, they use sensitive electronic cameras to build up an exposure over many minutes or hours. Then, computers analyze the data. Astronomers study the pictures and the computer data.

Most observatories use reflecting telescopes with large mirrors. The Special Astrophysical Observatory in Russia has a mirror 6 m in diameter. However, large mirrors often sag under their own weight, blurring the image. This has led scientists to build reflecting telescopes that use many small mirrors acting as one. The Keck Telescope at Mauna Kea, in Hawaii, has a mirror made of 36 thin glass segments. Together, these segments make a mirror 10 m in diameter.

Many observatories in the United States, such as Palomar near San Diego, California, are run by cities or nonprofit groups. Kitt Peak National Observatory, near Tucson, Arizona, is the largest U.S. observatory.

Thinking Critically What activities are conducted inside observatories?

▲ **Figure 18-5** Kitt Peak National Observatory

QUICK CONVERSIONS		
Mirror Diameters	**Metric/SI**	**English**
Russian mirror	6 m	20 ft
Keck mirror	10 m	33 ft

MORE ON THE WEB

Viewing Star Births Have the students use the Internet to view some images of stars being created. Then, have students explain how they believe newly forming stars can produce such visual effects.

• TEACHING RESOURCES • Online or Print

Teacher's Resources

Lesson Review: Chapter 18, p. 2

3 *Assess*

CHECKING CONCEPTS ANSWERS

1. the Sun
2. hydrogen and helium
3. a spinning cloud of gas and dust
4. gravity
5. globular and open

THINKING CRITICALLY ANSWERS

6. SEQUENCE: gas and dust, nebula, protostar, and star
7. INFER: They are shaped like globes.

4 *Close*

BUILDING SCIENCE SKILLS ANSWERS

Classifying Canopus is an F star, Arcturus is a K, Sirius is an A, Rigel is a B, the Sun is a G, Betelgeuse is an M, Altair is an A, and Capella is a G.

Science and Technology

OBSERVATORIES

Physical-Science Connection A reflecting telescope is an optical instrument that uses concave and plane mirrors (which have a flat, or plane, surface) to collect and focus light. Light that enters such a telescope is reflected by the concave mirror onto the plane mirror. The plane mirror reflects the light to form an inverted image in the instrument, which is transmitted to an eyepiece, camera, or other device.

Discussion Ask students: *Why is using cameras and computers better than looking directly through telescopes to study the night sky?* (Cameras and computers can gather and process much more information than a person can in the same time period.)

Thinking Critically Answer Possible answer: Scientists study and analyze the data gathered by telescopes and computers to find out more about the universe.

18-2 *(pp. 438–439)*

1 Introduce

STUDY HINT Have students use their dictionaries to separate *spectroscopy* and *spectrograph* into word parts. Challenge students to use the meaning of each word part to write their own definition of each term. Have pairs of students check each other's definitions for accuracy.

Linking Prior Knowledge Have a volunteer describe what causes a rainbow to form. (Sunlight passes through raindrops and refracts, or bends. This bending separates light into the colors of the visible spectrum.)

2 Teach

Demonstration Explain that all stars emit a continuous spectrum of light—the visible light spectrum. Demonstrate this by shining sunlight through a prism. Have students observe that the white light from the Sun is made up of red, orange, yellow, green, blue, indigo, and violet light.
learning style: visual

Discussion Ask students: *How are fingerprints sometimes used to solve crimes?* (Because the fingerprints of every person are unique, they can be used to identify the person who committed a crime.) Ask how a star's spectrum is like a person's fingerprints. (Each element gives off a unique spectrum. The total spectrum of a star indicates the elements that are present in the star.) **learning style: auditory**

READING STRATEGY Tell students that the word root *scop* comes from Greek and means "to see." Ask them how they can use this knowledge to understand what a spectroscope is. Then, ask what other *scope* words they know. (Possible answers: telescope, microscope)

Answers to questions

1. **DEFINE:** a pattern of different colors of light coming from an object
2. **EXPLAIN:** the chemical makeup of a star
3. **APPLY:** yellow

18-2 How is spectroscopy used to study stars?

Objective

Describe how astronomers use spectroscopy to study stars.

Key Terms

spectrum, *pl.* **spectra:** pattern of different colors of light coming from an object

spectrograph (SPEHK-truh-graf)**:** device that measures the spectrum of an object

spectroscopy (spehk-TRUH-skohp-ee)**:** study of light coming from objects in space

Colors of Starlight White light is a mixture of different colors of light. When you aim a beam of white light at a prism, the prism separates or spreads out the light into its colors. This pattern of colors is called a **spectrum.**

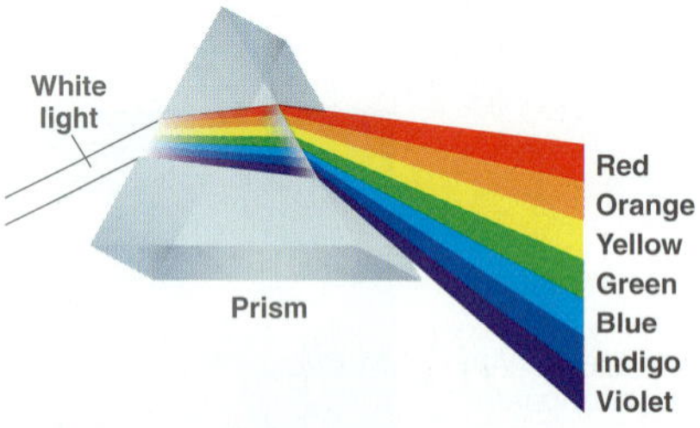

▲ **Figure 18-6** A prism breaks up white light into a spectrum.

The spectrum of white light contains red, orange, yellow, green, blue, indigo, and violet. These are the colors you see in a rainbow.

Astronomers learn about stars by studying their light. They use a **spectrograph,** which contains a prism that separates light from the star into bands of colors. The spectrograph also contains a camera to focus the spectrum and record it on photographic or electronic film. The study of light coming from objects in space is called **spectroscopy.**

1. **DEFINE:** What is a spectrum?

Color and Composition A spectrum reveals the chemical makeup of a star. Stars are very hot. When elements are heated enough, they give off light. Each element gives off its own spectrum of light. Astronomers can study this spectrum to find out what elements make up the star.

2. **EXPLAIN:** What does a spectrum reveal?

Color and Temperature The surface temperature of a star is related to its color. Stars come in different colors. Blue stars are the hottest, at around 29,000°C to 40,000°C. Red stars, at around 2,100°C to 3,200°C, are the coolest. A yellow star, like our Sun, is about 6,000°C. Scientists are able to determine the approximate temperature of a star from its spectrum.

3. **APPLY:** If a star is medium hot, what would be its color?

◄ **Figure 18-7** The spectra for hydrogen and helium

Teaching Options

! FAST FACTS *The Spectral Sequence*

Most stars belong to one of the seven main classes of the spectral sequence, introduced on page 437. Each class is defined according to its stars' colors and temperature range, though the divisions between classes are not precise.

- Blue stars include type O stars, which have surface temperatures of about 25,000K–40,000K (45,000°F–72,000°F); type B stars, which have surface temperatures of about 11,000K–25,000K (20,000°F–45,000°F); and type A stars, whose temperatures measure from 7,500K–11,000K (13,000°F–20,000°F). Type F stars, which measure about 6,000K–7,500K (10,000°F–13,000°F), are blue to white.
- White-to-yellow stars are classified as type G and have surface temperatures of 5,000K–6,000K (8,500°F–10,000°F).
- Orange-to-red type K stars are about 3,500K–5,000K (6,000°F–8,500°F). The coolest stars are type M stars. They shine red and have surface temperatures lower than 3,500K (6,000°F).

CHECKING CONCEPTS

1. Different colors of light make up a __________.
2. When an element is __________, it gives off a spectrum of light.
3. Each __________ in a star gives off its own spectrum.
4. The __________ of a star is related to its color.
5. The hottest stars are the color __________.

THINKING CRITICALLY

6. **ANALYZE:** A star has a very high surface temperature. What color does this star appear?
7. **CALCULATE:** What is the difference in degrees Celsius between the hottest stars and the coolest stars?

Web InfoSearch

The Doppler Effect Have you ever noticed that a siren moving toward you sounds different from one moving away from you? This is known as a Doppler effect. Both sound and light waves are affected by the Doppler effect.

SEARCH: Use the Internet to find out more about the Doppler effect. How is it important in astronomy? What are redshift and blueshift? How can scientists find out how fast an object is moving from its Doppler effect? Begin your search at www.conceptsandchallenges.com. Some key search words are **Doppler effect astronomy, Doppler redshift,** and **Doppler blueshift.**

Hands-On Activity

MAKING A SPECTROSCOPE

You can use a spectroscope to separate white light into a band of colors. You will need a cardboard tube, a diffraction grating, a sheet of black construction paper, scissors, tape, a light bulb, and colored crayons.

1. Tape a diffraction grating to one end of a cardboard tube.
2. Cut a thin slit in the center of a sheet of black construction paper. ⚠ CAUTION: Be careful when using scissors.
3. Cover the other end of the cardboard tube with the sheet of black paper. Use tape to hold the paper in place.
4. Hold the end of the tube with the diffraction grating up to your eye. Look through the tube at a light bulb. Slowly turn the tube until you see a spectrum. Draw the spectrum.
5. Use your spectroscope to look at other light sources, such as a fluorescent tube. ⚠ CAUTION: Do not look at the Sun with your spectroscope.

▲ **STEP 4** Look through your tube at a light bulb.

Practicing Your Skills

6. **DESCRIBE:** What does a spectroscope do?
7. **COMPARE:** a. Were the spectra from different light sources the same or different? b. How can you explain the similarities or differences of the spectra?

EXTENDING THE LESSON

Investigating Spectra and Energy Have interested students repeat the hands-on activity with a light that has an adjustable voltage. Have the students determine what happens when the voltage of the light is reduced (the blue end of the spectrum becomes fainter) and increased (the blue end of the spectrum becomes stronger). Help students conclude that the colors of the visible spectrum are related to the energy produced by the light source. **learning style: visual**

• *TEACHING RESOURCES* • *Online or Print*

Teacher's Resources

Lesson Review: Chapter 18, p. 3

Enrichment: Chapter 18, p. 6

Laboratory Manual

Lab. Challenge: How can light from stars be studied? pp. 87–90

3 *Assess*

CHECKING CONCEPTS ANSWERS

1. spectrum
2. heated
3. element
4. surface temperature
5. blue

THINKING CRITICALLY ANSWERS

6. **ANALYZE:** blue
7. **CALCULATE:** 26,900–36,800°C

4 *Close*

Web InfoSearch

The Doppler Effect Have students use the Internet to find out more about the Doppler effect and how it is important in astronomy. Have students answer the following questions: *What are redshifts and blueshifts? How can scientists find out how fast an object is moving from its Doppler effect?* Some key search words are **Doppler effect astronomy, Doppler redshift,** and **Doppler blueshift**. Students can begin their search at www.conceptsandchallenges.com.

Hands-On Activity

MAKING A SPECTROSCOPE

TIME: 20 minutes

PURPOSE: describing, comparing

SAFETY TIP: Remind students to be careful when using scissors. Stress the extreme danger of looking directly at the Sun.

COOPERATIVE LEARNING: You can have students work in groups of three. One student can tape the grating to the tube. One can cut the slit, and the third can secure the black paper to the tube. Encourage students to observe the spectrum given off by the light source.

Practicing Your Skills Answers

6. **DESCRIBE:** It breaks light into bands, or a spectrum, of color.
7. **COMPARE:** **a.** Different types of light will produce different spectra. **b.** No two light sources have identical spectra. They may be quite different or somewhat similar.

18-3 *(pp. 440–441)*

1 Introduce

STUDY HINT Have students write the lesson title in their notebooks. As they read the lesson, instruct students to find sentences that answer the question.

Linking Prior Knowledge Have students use a dictionary to find the meanings of the terms *apparent*, *absolute*, and *magnitude* as they are used in everyday language. Ask them to relate these meanings to how the terms are used in this lesson.

2 Teach

Demonstration Use a nickel and chromium wire, laboratory tongs, a heat-resistant glove, and a Bunsen burner to demonstrate how temperature can affect brightness. Hold the wire in a flame and ask students to observe its brightness. (The longer the wire is held in the flame, the hotter and brighter it becomes.)
learning style: visual

Discussion Have two volunteers describe and differentiate between the apparent and absolute magnitudes of a star. Prompt them, if necessary, to mention that distance from the observer is one distinguishing factor of these two measurements.
learning style: auditory

Reinforcement Have students refer to Figure 18-8 to answer the following questions: *Which object in the diagram is the brightest?* (the Sun) *What kind of body is this?* (a star) *Which planet is the brightest?* (Venus) *Which is brighter—the Sun or the faintest star observable with a telescope?* (the Sun) *Which planet has a greater apparent magnitude—Jupiter or Venus?* (Venus) *About how much brighter is Polaris than Pluto on this scale?* (about six times brighter)
learning styles: visual, logical/mathematical

Answers to questions

1. **NAME:** magnitude
2. **COMPARE:** a third-magnitude star
3. **DESCRIBE:** the actual brightness of a star

18-3 What is magnitude?

Objective

Compare apparent magnitude and absolute magnitude.

Key Term

magnitude (MAG-nuh-tood): way to measure a star's brightness

Brightness of Stars In some places you can see about 2,000 stars in the night sky without a telescope. Some stars appear brighter than others. One way to measure a star's brightness is by **magnitude.**

How bright a star appears to us depends on its temperature, size, and distance from Earth. A hot star is usually brighter than a cool star. A large star is usually brighter than a small star. The closer a star is to Earth, the brighter it usually appears to us.

1 **NAME:** What is the measure of a star's brightness called?

Apparent Magnitude The brightness of a star as seen from Earth is called the star's apparent magnitude. Planets can be classified this way, too, although they shine only with reflected light.

Astronomers have developed a scale for apparent magnitude. On this scale, a star with a low number appears brighter than a star with a high number. Bright stars usually have an apparent magnitude of 1. These are called first-magnitude stars. Very bright stars can have negative magnitudes. The dimmest stars you can see without a telescope are sixth-magnitude stars. You can see the magnitudes of some stars and other bodies in Figure 18-8 below.

2 **COMPARE:** Which appears brighter, a third-magnitude star or a sixth-magnitude star?

Absolute Magnitude Astronomers can also find the absolute magnitude of a star. The absolute magnitude of a star is its actual brightness. Absolute magnitude describes how bright a star would appear to us if all the stars were the same distance from Earth.

▼ **Figure 18-8** Comparison of apparent magnitudes

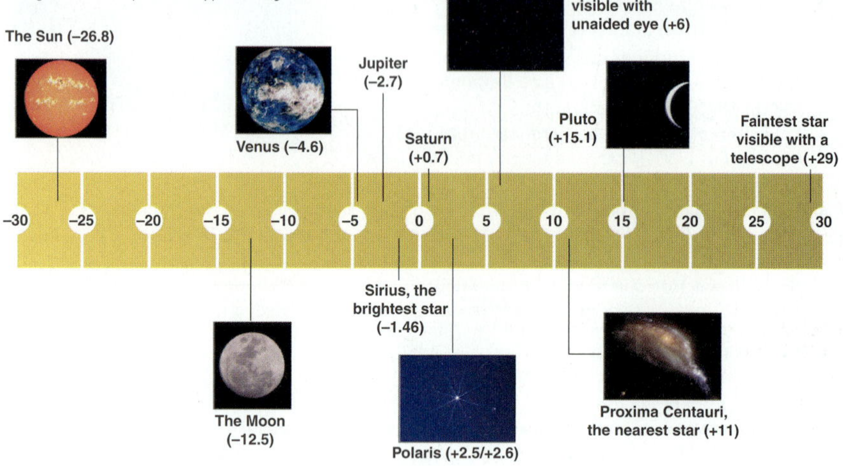

Teaching Options

COOPERATIVE LEARNING

Finding Apparent Magnitudes Have students find the apparent magnitudes of the following stars: Pollux (1.1), Sirius B (8.7), Canopus (−0.72), Ross 154 (10.6), Procyon (0.34), and Arcturus (–0.04). Have partners plot the findings on a number line similar to the one in Figure 18-8.
learning style: visual

EXTENDING THE LESSON

Connecting Magnitude and Distance Have interested students find out the absolute magnitude and distance from Earth of the stars in Figure 18-8. Provide graph paper for them to diagram the distance and brightness of each star.
learning style: visual

ESL/ELL STRATEGY

Apparent vs. Absolute Help English-language learners understand the difference between *apparent* and *absolute* by using the words in a different context. For example, apparent age is how old a person might look. Absolute age is based on the age written on the person's birth certificate. The word *apparent* comes from a French verb that means "to appear."

For example, the Sun appears very bright to us because it is so near Earth. If the Sun were farther away, it would not appear as bright to us. A dim star that is close to Earth may appear brighter than a bright star that is far away. All stars have both an apparent magnitude and an absolute magnitude.

DESCRIBE: What is absolute magnitude?

CHECKING CONCEPTS

1. Magnitude is a measure of a star's __________.
2. Magnitude depends on a particular star's size, __________, and distance from Earth.
3. A fourth-magnitude star appears __________ than a first-magnitude star.
4. The brightness of a star as seen from Earth is its __________ magnitude.
5. The actual brightness of a star is its __________ magnitude.

THINKING CRITICALLY

6. **HYPOTHESIZE:** Two stars have the same absolute magnitude but different apparent magnitudes. Explain how this can be true.
7. **COMPARE:** The Sun is the star closest to Earth. It has an absolute magnitude of 5.4. The star Altair has an absolute magnitude of 2.2. Which star is really brighter? Explain.

BUILDING MATH SKILLS

Calculating First-magnitude stars are about 2.512 times brighter than second-magnitude stars. These are about 2.512 times brighter than third-magnitude stars, and so on. Using this information, you can figure out how faint a star is compared to some other star. If two stars are five magnitudes apart, you would multiply 2.512 to the fifth power (2.512^5) to get 100. This means that a first-magnitude star is 100 times brighter than a sixth-magnitude star. Use this formula and a calculator to compare a star with a magnitude of 25 to a star with a magnitude of 5.

Hands-On Activity

OBSERVING MAGNITUDE

You will need one large flashlight and one small flashlight.

1. Work in groups of three.
2. Have one partner hold a large flashlight and another partner hold a small flashlight. Have both partners stand the same distance away from you and turn on their flashlights.
3. Compare the brightness of the two flashlights.
4. Now, have the person holding the small flashlight move closer to you. Compare the brightness of the two lights.

▲ **STEP 2** Have your partners stand the same distance away from you and turn on their flashlights.

Practicing Your Skills

5. **COMPARE:** **a.** When they were the same distance away, which appeared brighter, the large flashlight or the small flashlight? **b.** Which had the greater apparent magnitude? **c.** Which had the greater absolute magnitude?
6. **OBSERVE:** **a.** What happened when the small flashlight was moved closer? **b.** Which had the greater apparent magnitude? **c.** Which had the greater absolute magnitude?

• *TEACHING RESOURCES* • *Online or Print*

Teacher's Resources
Lesson Review: Chapter 18, p. 4

3 *Assess*

CHECKING CONCEPTS ANSWERS

1. brightness
2. temperature
3. dimmer
4. apparent
5. absolute

THINKING CRITICALLY ANSWERS

6. **HYPOTHESIZE:** Apparent magnitude depends on a star's temperature, size, and distance from Earth, so if the star is farther from Earth, it will look dimmer than a star of the same absolute magnitude (same temperature and size).
7. **COMPARE:** Altair is brighter than the Sun by more than 6 magnitudes. The lower the numbers, the higher is the absolute magnitude or brightness of a star.

4 *Close*

BUILDING SCIENCE SKILLS ANSWER

Calculating The stars are 20 magnitudes apart. Therefore, the brighter star (the one with the magnitude of 5) is 2.512^{20}, or about 100,090,000 times brighter than the dimmer star (the one with the magnitude of 25).

Hands-On Activity

OBSERVING MAGNITUDE

TIME: 15–20 minutes

PURPOSE: comparing, observing

COOPERATIVE LEARNING: Make sure that each student in a group tries out the experiment for himself or herself.

Practicing Your Skills Answers

5. **COMPARE:** **a.** the large flashlight **b.** the large flashlight **c.** the large flashlight
6. **OBSERVE:** **a.** It appeared brighter. **b.** the small flashlight **c.** the large flashlight

18-4 (pp. 442–443)

1 Introduce

STUDY HINT Have students make up questions on different elements of the H-R diagram that is shown in Figure 18-9. Then, have them quiz each other using these questions.

Linking Prior Knowledge Tell students that much of the information about stars comes from telescopic observations. Have students recall how telescopes are used to observe objects in space (Lessons 16-3 to 16-5).

2 Teach

Demonstration Use a long wooden match to demonstrate how color relates to temperature. Light the match and have students, at a safe distance, observe the different parts of the flame. Inform students that the blue part of the flame, which is closest to the head of the match, is the hottest part. The yellow part of the flame is cooler than the blue part. Extinguish the flame and have students note that the match head briefly glows orange or red. Have students compare this demonstration to the H-R diagram: blue stars are the hottest; yellow stars are cooler than blue stars but hotter than red stars; red stars have the lowest surface temperatures. **learning style: visual**

Discussion Have students refer to Figure 18-9 to answer the following questions: *Where are the brightest stars on the H-R diagram?* (the upper right) *What color are the hottest stars on an H-R diagram?* (blue) *Which stars are the coolest?* (red giants) *What is the absolute magnitude of a yellow star with a surface temperature of about 5,500K?* (about +2.5) *Are white dwarfs hotter or cooler than red giants?* (hotter) **learning style: visual**

Reinforcement Once you have finished teaching this lesson, have students use the Concept Map from the Teacher's Resources to explain star classification. In the box labeled *Concept*, have them write a statement summarizing how stars are classified. Then, in the other boxes, they can write short descriptions of each type of star, including its main characteristics. **learning style: linguistic**

18-4 How are stars classified?

Objective

Explain how the Hertzsprung-Russell diagram is used to classify stars.

Key Terms

main sequence star: star that falls within a long, narrow, diagonal band across the H-R diagram

red giant: large, bright star that is fairly cool

supergiant: very large and very bright star

white dwarf: very small, hot star

The H-R Diagram In the early 1900s, astronomers Ejnar Hertzsprung and Henry Russell each made a separate but similar discovery about stars. They found that there is a relationship between a star's absolute magnitude and its surface temperature and color. Together, Hertzsprung and Russell developed a chart called the Hertzsprung-Russell, or H-R, diagram. The H-R diagram shows that the brightness of most stars increases as the star's surface temperature increases.

1 **DESCRIBE:** What does the H-R diagram show?

Main Sequence Stars A star's position on the H-R diagram depends on its absolute magnitude and its surface temperature, or color. Temperature is in degrees Kelvin (K). Suppose that a star has a blue color and a low absolute magnitude. This star would be placed in the upper left corner of the diagram. A red star with a high absolute magnitude would appear in the lower right corner. Most stars fall in a narrow diagonal band that runs from the upper left to the lower right corner of the diagram. Stars that fall in this band are called **main sequence stars.** The Sun and most stars that you can see at night are main sequence stars.

2 **CLASSIFY:** What kind of star is the Sun on an H-R diagram?

Other Stars Some stars do not fall within the main sequence. They may be bright but not very hot. Many of these stars are red, orange, or yellow in color. Because they are not very hot, they should not be very bright. However, because these stars are very large, they give off a great deal of light. They have large absolute magnitudes. These stars are called **red giants.** Red giants appear in the upper right corner of an H-R diagram.

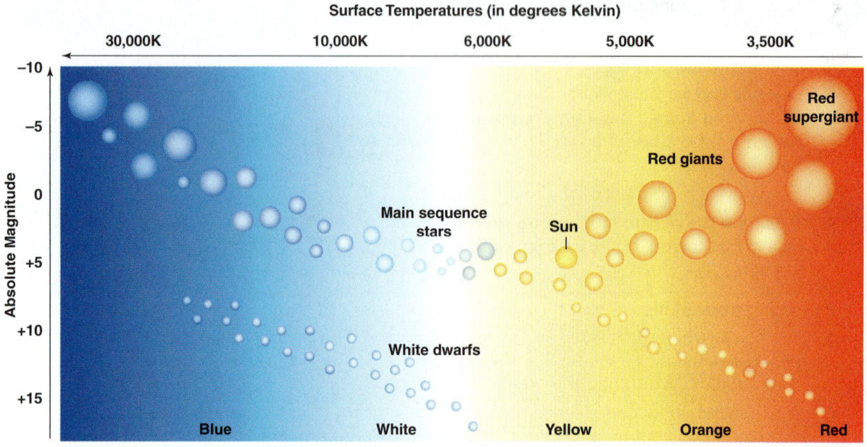

▲ **Figure 18-9** An H-R diagram shows the relationship between the brightness of stars and their surface temperatures.

Teaching Options

FAST FACTS

The Diameters of Stars

The diameters of stars can be estimated by using their luminosity. Stars in the main sequence are about 0.25 to 20 solar units in diameter. A solar unit is based on the diameter of the star in relation to the Sun. For example, a star with a diameter of 0.25 solar units would have a diameter 0.25 times that of the Sun. Giants are 10 to 80 solar units in diameter. Supergiants are up to 1,000 solar units in diameter. White dwarfs are less than 0.02 solar units in diameter. Note that the diameter of the Sun is approximately 1.39×10^6 km (865,000 mi).

ONGOING ASSESSMENT

Have students write a paragraph that summarizes the relationship between magnitude, color, and surface temperature.

MORE ON THE WEB

Simulating the H-R Diagram
You can have students use an online simulator to see how stars change location on an H-R diagram as they grow older.

Some stars are even larger and brighter than red giants. These stars are called **supergiants.**

Other stars that fall outside of the main sequence are hot but very small. These stars are blue or white. They are called **white dwarfs.** White dwarfs are found in the lower part of an H-R diagram, below the main sequence stars.

3 DEFINE: What are white dwarfs?

CHECKING CONCEPTS

1. The H-R diagram shows the relationship between __________ magnitude and temperature.
2. The Sun is a __________ sequence star.
3. A very large, bright star is a red __________.
4. Red giants are found in the upper __________ of an H-R diagram.
5. White dwarfs are small, __________ stars.

THINKING CRITICALLY

Use Figure 18-9 to answer the following questions.

6. COMPARE: Which star has a higher absolute magnitude, a red giant or a white dwarf?
7. COMPARE: Which star has a greater surface temperature, a red giant or a white dwarf?
8. ANALYZE: What is the average surface temperature of a white dwarf?
9. IDENTIFY: What stars are even larger than red giants?

BUILDING MATH SKILLS

Graphing Create a line graph that shows where each kind of star appears on the H-R diagram. Plot the absolute magnitude along the vertical axis. Plot the temperature along the horizontal axis. Label each point with the kind of star it represents. Do the points form a pattern? Write a few sentences describing how the pattern formed compares to the pattern in Figure 18-9.

How Do They Know That?

DISTANCES TO THE STARS

How do astronomers measure distances to stars? The distance of close stars can be found by observing the star's parallax. Parallax is based on how the star seems to shift position over time. However, the distances to stars that are very far away seemed almost impossible to measure. Then, an American astronomer discovered a way to find the distances to stars in other galaxies.

▲ **Figure 18-10** Henrietta Swan Leavitt used Cepheid stars to show that a star's brightness is related to its distance.

Henrietta Swan Leavitt (1868–1921) was born in Lancaster, Massachusetts. She graduated from what is now Radcliffe College and soon went to work at the Harvard Observatory. She was studying a kind of star called a Cepheid (SEHF-ee-id) variable. A Cepheid variable is a star that changes its brightness in a regular pattern over a set period of time.

In 1912, Leavitt identified 25 Cepheid variables in a nearby galaxy. She observed the period of each star, or how long the star took to go from bright to dim and back to bright again. She showed that the longer the period of a star is, the brighter the star. A star's brightness is related to its distance. Astronomers used this relationship to calculate the distance to faraway stars.

Thinking Critically Why do you think parallax is not useful for measuring all stars?

Answers to questions

1 DESCRIBE: the relationship between a star's absolute magnitude and its surface temperature and color

2 CLASSIFY: a main sequence star

3 DEFINE: blue or white stars that are very hot but small

3 *Assess*

CHECKING CONCEPTS ANSWERS

1. absolute
2. main
3. giant
4. right corner
5. hot

THINKING CRITICALLY ANSWERS

6. COMPARE: a red giant
7. COMPARE: a white dwarf
8. ANALYZE: about 9,500K
9. IDENTIFY: supergiants

4 *Close*

BUILDING MATH SKILLS ANSWER

Graphing Students' graphs should display a relatively smooth line that trends upper left to lower right. As surface temperature increases, absolute magnitude decreases.

How Do They Know That?

DISTANCES TO THE STARS

Technology Connection Until very recently, scientists were able to estimate distances to stars that were up to 300 light-years away. Satellites now enable astronomers to estimate distances up to about 1,500 light years.

Discussion Explain that parallax is an apparent or relative shift in position as observed from Earth. Have students hypothesize what causes this apparent shift in the position of a star. (Earth's revolution around the Sun) Inform students that the closer a star is to Earth, the larger its apparent shift in position. The distance to the star can be estimated from its parallax shift.

Thinking Critically Answer
Many stars are too far from Earth to be measured using parallax shifts.

EXTENDING THE LESSON

Variable Stars Variable stars are stars that vary in brightness. Explain that there are two types of variable stars. In one type, the changing brightness is caused by changes in the star's internal forces that cause the star to swell and shrink. This, in turn, causes the star's brightness to change. Another type of variable star is a member of a binary star system. As the fainter star of the pair passes in front of the brighter star, the brightness of the latter appears to decrease. Have students compare and contrast a solar eclipse and the eclipse of a binary star. **learning style: auditory**

• TEACHING RESOURCES • *Online or Print*

Teacher's Resources

Lesson Review: Chapter 18, p. 5

Enrichment: Chapter 18, p. 6

Concept Map: Study Tools, p. 4

18-5 (pp. 444–445)

1 Introduce

STUDY HINT Ask students why they think stars are described in terms of life cycles and death. After students read the lesson, have them write the reason and see if it is correct.

Linking Prior Knowledge Ask for examples of cycles with which students might be familiar, such as plant and animal life cycles, the water cycle, and the change of seasons. Stress that, unlike most cycles, the life cycle of a star can take billions of years.

2 Teach

Demonstration Pin a brightly colored circle (about 11 cm [5 in.] in diameter) to the back of a volunteer's shirt. Ask the volunteer to quickly rotate as the rest of the class observes the circle. Inform students that this is similar to the movement of a pulsar. Ask what the circle represents. (the light given off by the pulsar) Have them determine the speed of the rotating star by counting the number of times the circle is visible over a given period of time.
learning styles: visual, kinesthetic

Discussion Have students take turns describing the events that take place in the life cycle of a star. Make sure that they reference Figure 18-11 in their descriptions.
learning style: auditory

READING STRATEGY Have students find the compound words used in this lesson (*supergiant* and *supernova*). Have them divide each word into two parts. Challenge students to make a list of at least ten other compound words that contain *super-*.

Answers to questions

1. **EXPLAIN:** its mass
2. **DESCRIBE:** the period when hydrogen in a star's center continues changing into helium
3. **DESCRIBE:** It loses mass and becomes a white dwarf.
4. **INFER:** a large star

18-5 What is a star's life cycle?

Objective
Describe the stages in the life cycle of a star.

Key Terms
nova: explosion where the outer layers of a star are blown off
supernova: violent explosion where a star is blown apart

Life Cycle of a Star Stars change over time. This is called a life cycle. A star's complete life cycle takes many millions of years. As time passes, the star changes its mass into energy. The energy is given off as light and heat. Eventually, most of the mass is used up, and the star dies. By studying many different stars, astronomers have learned how stars change during their life cycles.

1 **EXPLAIN:** Where does a star's energy come from?

Protostar to Giant Stars are formed from nebulae. Gravity pulls the dust and the gas in the nebula together, forming a protostar. If the protostar becomes hot enough, nuclear reactions start to take place. In these reactions, hydrogen is changed into helium. Large amounts of energy are produced. These reactions turn the protostar into a star.

Now the star is on the main sequence. During this stage, hydrogen at the star's center continues to change into helium. When this process ends, small stars will stop shining and contract to become white dwarfs. Larger stars, however, will start new nuclear reactions that turn helium into carbon. The energy produced by these reactions expands the star and makes it cooler. It is now a red giant.

2 **DESCRIBE:** What is the main sequence stage?

Death of a Star The next stage in a star's life cycle depends on the star's mass. A medium-size star loses mass and begins to contract. The outer parts create a glowing cloud called a planetary nebula. The core of the star becomes a white dwarf. For a brief time, the white dwarf becomes very hot. The end stage of a white dwarf is thought to be a small, cold, dark object called a black dwarf. Scientists think it takes a trillion years for a black dwarf to form.

Sometimes a white dwarf will blow off its outer layers in a huge, bright explosion. This explosion is called a **nova.** A very massive star may blow itself apart in a **supernova.** During a supernova, or right after it, the star may collapse to become a neutron star or a black hole.

3 **DESCRIBE:** What happens when a medium-size star runs out of hydrogen?

Go Online **active art**
For: The Lives of Stars activity
Visit: PHSchool.com
Web Code: cfp-5043

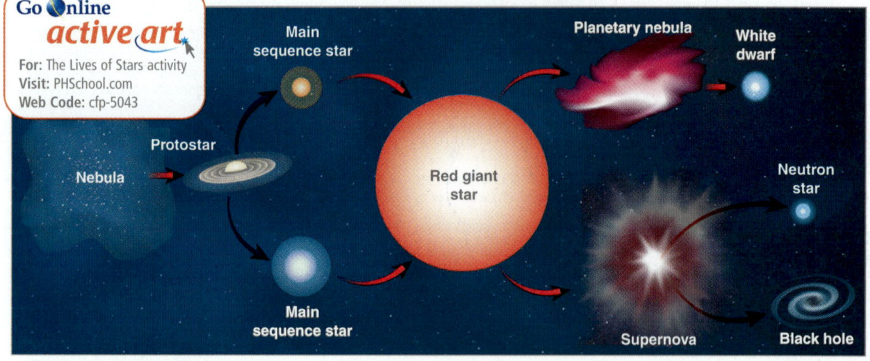

▲ Figure 18-11 The life cycle of a star

Teaching Options

RETEACH

Following a Star's Life Cycle Use a balloon to show how a typical star might change during its life cycle. Inflate the balloon and ask where the energy for this model star comes from. (from warm gases that you have exhaled) When the balloon is completely inflated, ask which part of the star's life cycle is represented. (the main sequence stage) Tie the balloon and prick a small hole in it so that the air slowly leaks out. Ask why the "star" is "dying." (Its source of energy is being depleted.) Ask students to describe how this model could have been used to show a nova. (The balloon could have been inflated until it burst.) **learning style: visual**

ESL/ELL STRATEGY

Cycles Point out that this lesson describes a cycle chronologically, step by step. Tell students that, when they write a description of a cycle, a good approach is to list the steps in the cycle in chronological order, and then write about the steps in that order. Have students use the Flowchart from the Teacher's Resources to help organize their notes.

Supernova 1987A On February 24, 1987, a supernova was seen in the Large Magellanic Cloud. This is more than 160,000 light years from Earth. The so-called Supernova 1987A was the first to be seen by the naked eye from Earth in almost 400 years. Astronomers studied the light coming from the supernova as it slowly dimmed. Viewing the supernova helped them to learn more about the evolution of stars.

4 **INFER:** What was the likely energy source of Supernova 1987A?

CHECKING CONCEPTS

1. A star changes __________ into energy.
2. The __________ stage is the briefest in a star's life cycle.
3. A main sequence star becomes a red giant or __________ when __________.
4. A white dwarf may blow off its outer shell in an explosion called a __________.
5. A supernova may occur in a very __________ star.

THINKING CRITICALLY

Use Figure 18-11 to complete the following exercise.

6. **SEQUENCE:** A flowchart shows a sequence of events. Create a flowchart showing the stages in the life cycle of a star. Begin with a protostar.

Web InfoSearch

Tycho Brahe Tycho Brahe was a 16th-century Danish astronomer. After studying law all day, he would go out at night to observe the stars. Brahe found that most of the astronomical information of the time was not accurate. He decided to devote his life to making accurate observations of the stars.

SEARCH: Use the Internet to find out more about Tycho Brahe's achievements. What did the king of Denmark give Brahe and why? Start your search at www.conceptsandchallenges.com. Key search words are **Tycho Brahe, Brahe King Denmark,** and **Tycho Brahe biography.**

How Do They Know That?

PULSARS

After a very massive star explodes, its core collapses. This collapse squeezes protons and electrons together until they become neutrons. This is how a neutron star is formed.

Neutron stars are incredibly dense. One teaspoonful of neutron star material would weigh over 2 billion tons. As they rotate, neutron stars send out radio waves and light that reach Earth in pulses. This is how pulsars got their name. Pulsars spin like the light atop a lighthouse. They emit radio or light waves or both that appear and then disappear in a regular pattern. The radio waves can be detected using a radio telescope.

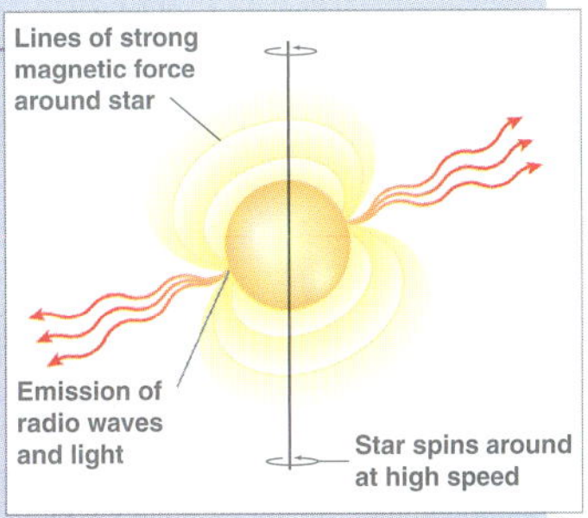

▲ **Figure 18-12** A pulsar spins like a light on a lighthouse.

The pulse of a pulsar tells us how fast the pulsar is spinning. The fastest pulsars discovered so far spin about 642 times per second. Some pulsars spin much more slowly, only once every few seconds.

Pulsars were first discovered in 1967 by Anthony Hewish and his graduate student, Jocelyn Bell. Hewish later received a Nobel Prize for identifying pulsars as a new class of stars.

Thinking Critically How did neutron stars get their name?

3 Assess

CHECKING CONCEPTS ANSWERS

1. mass
2. white dwarf
3. supergiant; helium changes to carbon
4. nova
5. large

THINKING CRITICALLY ANSWER

6. **SEQUENCE:** Rotating gas and dust contract to form a protostar. → Nuclear reactions in a protostar give off heat and light. → A star forms. → The star enters the main sequence stage. → All hydrogen is used up, and the star enters the red giant stage. → A red giant can contract to become a dwarf. → A dwarf can explode in a nova or supernova explosion. → A supernova can result in a neutron star.

4 Close

Web InfoSearch

Tycho Brahe Tycho Brahe devoted his life to making accurate observations of the stars. Have students use the Internet to find out more about Tycho Brahe's achievements. Some key search words are **Tycho Brahe, Brahe king Denmark,** and **Tycho Brahe biography.** Have students begin their search for information at www.conceptsandchallenges.com.

How Do They Know That?

PULSARS

Technology Connection In the late 1990s, a unique bursting pulsar was discovered by astronomers. This pulsar, which has been observed with the Proportional Counter Array on the *Rossi X-ray Timing Explorer (RXTE),* is the only pulsar known to generate both regular pulses and frequent enormous bursts of radiation.

Discussion Review that protons, neutrons, and electrons are parts of an atom. Protons and neutrons are found in the nucleus, or core, of an atom. Electrons spin around the core in definite orbits.

Thinking Critically Answer When very massive stars collapse, their protons and electrons are squeezed together until they become neutrons, forming a neutron star.

COOPERATIVE LEARNING

Demonstrating the Life Cycle of a Star Have groups of three students work together to create skits that show the stages in the life cycle of a star. Each group should accurately and creatively demonstrate at least the formation of a protostar, the main sequence stage, and the dwarf stage.

• *TEACHING RESOURCES* • *Online or Print*

Teacher's Resources

Lesson Review: Chapter 18, p. 7

Flowchart: Study Tools, p. 6

Go Online active art

For: The Lives of Stars activity
Visit: PHSchool.com
Web Code: cfp-5043

Students can interact with the art of the lives of stars online.

18-6 *(pp. 446–447)*

1 *Introduce*

Linking Prior Knowledge Ask students if they knew that the Sun is a star. Ask if they were surprised by this, and why or why not. Remind students that the Sun is at the center of the solar system (Lesson 17-1) and that objects stay in orbit around the Sun because of gravity.

2 *Teach*

Demonstration In order for students to get a sense of the size of the Sun when compared to that of Earth, draw both of these bodies on the board, using a scale of 1 mm to 1,000 km. At this scale, the Sun's diameter would be 1,392 mm (1.4 m, or about 4.5 ft). The diameter of Earth would be 12.75 mm (0.5 in.).
learning style: visual

Discussion To begin a discussion on the structure of the Sun, ask: *What is the best model of the structure of the Sun: a basketball, a peach, or a chocolate-covered candy-coated peanut? Why?* (A peach is a good model of the Sun. The skin represents the corona, the fleshy part of the fruit represents the chromosphere, the peach pit represents the photosphere, and the seed inside the pit represents the Sun's core.)
learning style: auditory

Reinforcement To summarize the information presented on the Sun's structure, ask: *Which part of the Sun is the source of this star's energy?* (the core) *How hot is the Sun's core?* (about 15,000,000°C) *What are the three parts of the Sun's atmosphere from the inside out?* (the photosphere, the chromosphere, and the corona) *How much cooler is the corona than the Sun's core?* (about 13,500,000°C)
learning style: auditory

Answers to questions

1. **NAME:** the core and the atmosphere
2. **IDENTIFY:** in the center
3. **LIST:** photosphere, chromosphere, and corona

18-6 What kind of star is the Sun?

Objective
Describe the parts of the Sun.

Key Terms
core: center of the Sun
photosphere (FOHT-oh-sfeer)**:** inner layer of the Sun's atmosphere
chromosphere (KROH-muh-sfeer)**:** layer of the Sun's atmosphere above the photosphere
corona (kuh-ROH-nuh)**:** outer layer of the Sun's atmosphere

Structure of the Sun The Sun is one of about 100 billion stars in our part of the universe. However, it is the only one close enough for us to study in detail. The Sun is an average star in size, mass, and temperature. Its diameter is 1,400,000 km. Because the Sun is made mostly of gases, it has no distinct boundaries. However, it has two main parts. These parts are the core and the atmosphere.

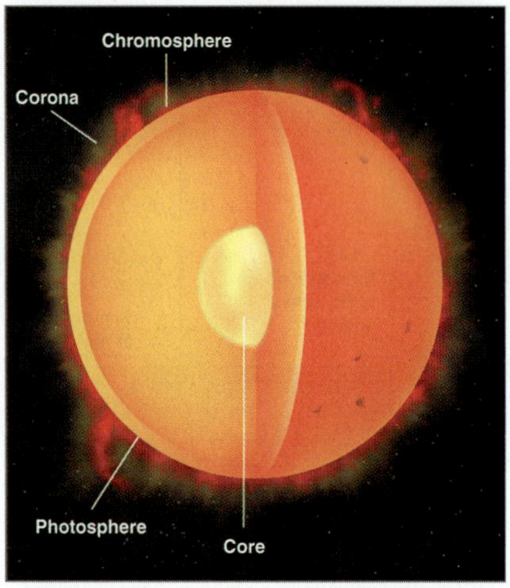

▲ **Figure 18-13** The atmosphere of the Sun is made up of the photosphere, the chromosphere, and the corona.

1 **NAME:** What are the two main parts of the Sun?

The Sun's Core The center and hottest part of the Sun is the **core.** The core makes up about 10 percent of the Sun's diameter. It is where the Sun's energy is produced. The temperature of the core is about 15,000,000°C.

2 **IDENTIFY:** Where is the Sun's core?

The Sun's Atmosphere Since the Sun is made up almost entirely of gas, its atmosphere takes up most of its volume. There are three thin layers at the outer edge of the atmosphere. One layer is the **photosphere,** or light sphere. The photosphere is the visible surface of the Sun. The gas in the photosphere glows and gives off light.

Above the photosphere lies another layer of the atmosphere, called the **chromosphere,** or "color sphere." This part of the Sun gives off a weak red glow that can be seen only under special conditions, such as during a solar eclipse. The chromosphere extends for thousands of kilometers.

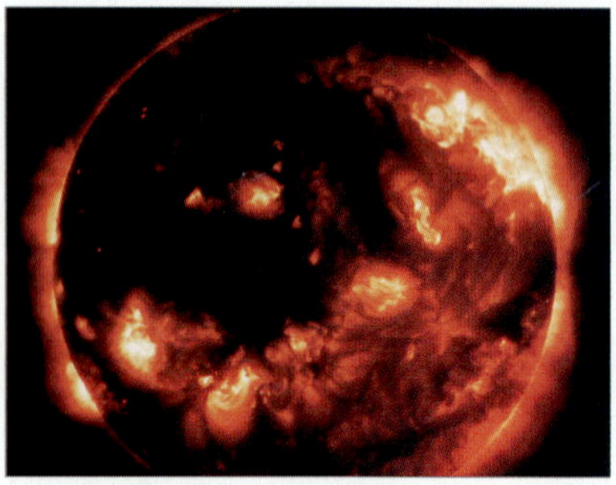

▲ **Figure 18-14** The chromosphere is seen more clearly in this photograph of the Sun taken by a special telescope.

The **corona** is the outermost layer of the Sun's atmosphere. This envelope of gases normally extends 1 million km from the Sun. It produces a glow about half as bright as a full Moon. The temperature of the corona is about 1,500,000°C. Like the chromosphere, the corona can usually be seen only during a solar eclipse.

3 **LIST:** What are three layers of the Sun's atmosphere?

446

Teaching Options

INTEGRATING PHYSICAL SCIENCE

Explaining Photons The Sun gives off energy in the form of electromagnetic radiation. Light and heat are two forms of electromagnetic radiation, which is made up of particles called photons. The rays of light emitted by the Sun consist of countless photons traveling in straight lines. Photons, like all forms of electromagnetic radiation, travel through space at the speed of light, which is 299,792 km/s. Ask students to find out how many kilometers the Sun is from Earth and calculate how long a photon takes to travel from the Sun to Earth. (Earth to Sun: 149,600,000 km; about 8 minutes)
learning style: logical/mathematical

ESL/ELL STRATEGY

Describing the Sun Have an English-language learner work together with a student who is proficient in English to make a chart that summarizes the characteristics of each layer of the Sun.

COOPERATIVE LEARNING

Modeling the Sun Have pairs of students find out how large each layer of the Sun is. Have them use these values to make a scale model of the internal structure of the Sun. **learning style: tactile**

CHECKING CONCEPTS

1. What is at the center of the Sun?
2. What is the approximate temperature of the Sun's core?
3. Which part of the Sun is the source of the Sun's energy?
4. How many layers make up the Sun's atmosphere?
5. Which layers of the Sun's atmosphere can be seen only during a solar eclipse?

THINKING CRITICALLY

6. **DESCRIBE:** What is the basic structure of the Sun?
7. **COMPARE:** Which is the hottest part of the Sun?
8. **INFER:** Why is the part of the Sun you named in Question 7 the hottest?
9. **ANALYZE:** The word *corona* comes from a Latin word meaning "crown." Why do you think the outer layer of the Sun is called the corona?

HEALTH AND SAFETY TIP

Earth's atmosphere filters out most of the harmful rays the Sun produces. However, some of the rays that get through can be harmful. For example, the Sun is far too bright to look at directly without damaging your eyes. Astronomers use special telescopes to study the surface of the Sun. Also, you know that sunlight can cause sunburn and can damage skin. Some kinds of skin cancer are caused by overexposure to the Sun. You should always take care to protect your skin and your eyes from overexposure to sunlight.

Hands-On Activity

OBSERVING THE SUN

You will need a ring stand, a clamp, a 20- by 28-cm sheet of cardboard, binoculars, and a sheet of white paper.

1. Clamp a pair of binoculars to a ring stand.
2. Cut a hole in the cardboard so that it fits over one eyepiece; the other eyepiece is covered. Tape the cardboard on securely.
3. Point the binoculars toward the Sun. ⚠ CAUTION: Never look directly at the Sun. Looking directly at the Sun could damage your eyes.
4. Hold the sheet of white paper a short distance from the eyepiece. An image of the Sun should appear on the sheet of white paper. Move the paper back and forth until your image of the Sun comes into sharp focus.
5. Quickly trace your image of the Sun onto the paper. Then, move the paper away to prevent scorching it. ⚠ CAUTION: If you hold the paper there for too long, it may ignite. Be sure to quickly remove it and the binoculars after you have drawn your image.

▲ **STEP 4** Hold the paper a short distance from the eyepiece and look for an image of the Sun on it.

Practicing Your Skills

6. **IDENTIFY:** When you look at your image of the Sun, what layer are you seeing?
7. **INFER:** Why do you think you cannot see the other layers of the Sun's atmosphere?
8. **INFER:** Why is this a good method to use to observe a solar eclipse?

QUICK CONVERSIONS		
Sun statistics	Metric/SI	English
Diameter	1,392,000 km	865,000 mi
Core temperature	15,000,000°C	27,000,000°F
Corona thickness	1,000,000 km	627,000 mi
Corona temperature	1,500,000°C	2,700,000°F

• TEACHING RESOURCES •
Online or Print

Teacher's Resources
Lesson Review: Chapter 18, p. 8

3 Assess

CHECKING CONCEPTS ANSWERS

1. the core and the atmosphere
2. about 15,000,000°C
3. the core
4. three
5. the chromosphere and the corona

THINKING CRITICALLY ANSWERS

6. **DESCRIBE:** The innermost core is surrounded by the atmosphere, which includes the photosphere, the chromosphere, and the corona.
7. **COMPARE:** the core
8. **INFER:** Nuclear reactions in the core make this part of the Sun the hottest.
9. **ANALYZE:** Possible answer: The corona forms a crown of light that surrounds the Sun.

4 Close

HEALTH AND SAFETY TIP

Ask students to list ways that they might protect their skin and eyes from damaging rays of the Sun. (by using sunscreen and wearing protective clothing and sunglasses)

Hands-On Activity

OBSERVING THE SUN

TIME: 15–20 minutes

PURPOSE: observing, inferring

SAFETY TIP: Remind students several times during this activity of the extreme danger of looking directly at the Sun.

Practicing Your Skills Answers

6. **IDENTIFY:** the photosphere
7. **INFER:** because the light from the photosphere drowns out the light from the other parts
8. **INFER:** Ultraviolet rays from the Sun are very harmful, so you should never look directly at the Sun.

18-7 *(pp. 448–449)*

1 Introduce

STUDY HINT As students read this lesson, have them write a sentence that paraphrases the information in each paragraph. Review students' sentences for accuracy.

Linking Prior Knowledge Have students who have observed the northern lights describe what they saw, or obtain a photo of the aurora and show it to the class. Have students review what they learned about the auroras (Lesson 11-2).

2 Teach

Demonstration Completely inflate a large, round, yellow balloon. Use a permanent marker to draw several sunspots on this model of the Sun. Slowly rotate the balloon (from east to west) to demonstrate that sunspots move in groups across the Sun as it rotates on its axis.
learning style: visual

Discussion Differentiate among sunspots, solar flares, and prominences. Have students refer to Figures 18-15 to 18-17 as you discuss these features of the Sun's surface. **learning styles: visual, auditory**

Reinforcement Check students' understanding of sunspots, solar flares, and prominences by asking the following questions: *Which of the Sun's surface features form as the result of cooling gases?* (sunspots) *Which of the Sun's surface features form when energy is given off suddenly from the Sun's atmosphere?* (solar flares) *What are solar prominences?* (streams of flaming gas that shoot out from the Sun's surface)
learning style: auditory

Answers to questions

1 **EXPLAIN:** because the Sun spins on its axis

2 **DEFINE:** streams of flaming gas that shoot out from the surface of the Sun

3 **DESCRIBE:** Solar flares can cause electrical outages and disrupt communications. They also cause the auroras.

18-7 What is the surface of the Sun like?

Objective
Describe sunspots, prominences, and solar flares.

Key Terms
sunspot: dark, cool area on the Sun's surface
prominence (PRAHM-uh-nuhns)**:** stream of gas that shoots high above the Sun's surface
solar flare: eruption of electrically charged particles from the surface of the Sun

Sunspots Some areas on the surface of the Sun are cooler than the areas around them. The gases in these cooler areas do not shine as brightly as the areas around them. As a result, these areas appear dark. The dark, cooler areas on the Sun's surface are called **sunspots.** Sunspots usually appear to move in groups across the Sun in the same direction. The appearance of movement is caused by the spinning of the Sun on its axis. Sunspots may last for days or even months.

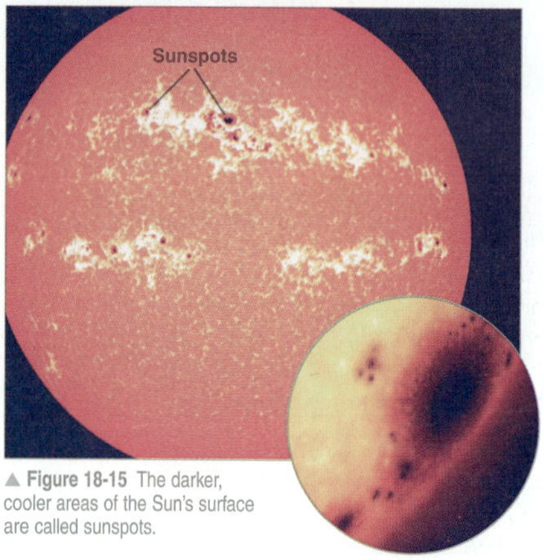

▲ **Figure 18-15** The darker, cooler areas of the Sun's surface are called sunspots.

1 **EXPLAIN:** Why do sunspots all appear to move in the same direction?

Prominences Streams of flaming gas shoot out from the surface of the Sun. These streams of gas are called **prominences.**

Prominences most often form in the chromosphere or photosphere. They can reach many thousands of kilometers above the Sun's surface. Then, they fall back into the Sun, forming huge arches. Prominences are best seen during a solar eclipse. They can last for weeks or months.

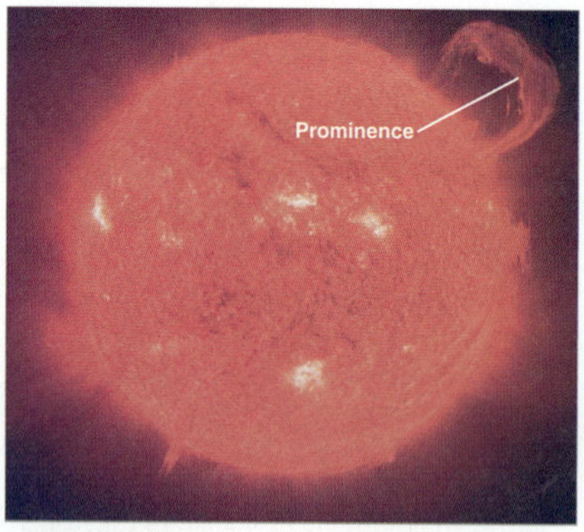

▲ **Figure 18-16** Prominences form in the chromosphere or photosphere.

2 **DEFINE:** What are prominences?

Solar Flares Energy sometimes builds up in the Sun's atmosphere. This buildup of energy usually happens near a group of sunspots. If the energy is given off suddenly, a **solar flare** is formed.

Solar flares usually do not last for more than an hour. Some may last for only a few minutes. In that time, though, they release enormous amounts of energy.

Solar flares send streams of electrically charged particles out into space. When these particles reach Earth's surface, they can cause electrical outages and disrupt communications. Solar flares are also the cause of the auroras, also called the northern and southern lights, on Earth.

448

Teaching Options

RETEACH

Depicting Auroras Have students use pastel-colored chalk or watercolors to make realistic depictions of auroras. Display the artwork in the classroom or in a school corridor.
learning styles: visual, tactile

EXTENDING THE LESSON

Measuring Sunspots Have students use rulers and the diameter of the Sun (1.39×10^6 km or 865,000 mi) to estimate the true sizes of the sunspots shown in Figure 18-15.
learning styles: tactile, visual

▲ **Figure 18-17** Solar flares send streams of charged particles out into space.

3 **DESCRIBE:** How do solar flares affect Earth?

CHECKING CONCEPTS

1. Sunspots appear dark because they are __________ than surrounding areas.
2. Streams of gas from the surface of the Sun are __________.
3. Solar flares release streams of __________ particles.

THINKING CRITICALLY

4. **MODEL:** Draw and label a picture of the Sun. Show all of the features of the Sun that were discussed.

Web InfoSearch

Sunspots and Ice Ages Astronomers have found that sunspot activity seems to build up and decrease in an 11-year cycle. This cycle has been linked to climate changes on Earth. For example, the Little Ice Age of the 1600s has been associated with a long period of low sunspot activity.

SEARCH: Use the Internet to find out more about sunspots. Why do they move across the Sun in an 11-year cycle? Could magnetic activity be a part of it? If so, how? Start your search for information at www.conceptsandchallenges.com. Some key search words are **sunspots, sunspots magnetic,** and **sunspot ice ages.**

Integrating Physical Science

TOPICS: magnetism, solar energy, ions

THE SOLAR WIND

Electrons and charged particles called ions move off the surface of the Sun. They slip through gaps in the magnetic fields of the Sun's corona and travel through space in all directions. This movement of charged particles through space is called the solar wind. It is not related to the air that blows across Earth's surface.

The ions heading toward Earth are drawn to Earth's magnetic poles. As the solar wind passes Earth, the charged particles that are moving at very high speeds may interact with the magnetic field of Earth. Once in Earth's atmosphere, they react with the oxygen and nitrogen atoms there. This reaction causes the light shows in the sky known as the auroras, or the northern and southern lights.

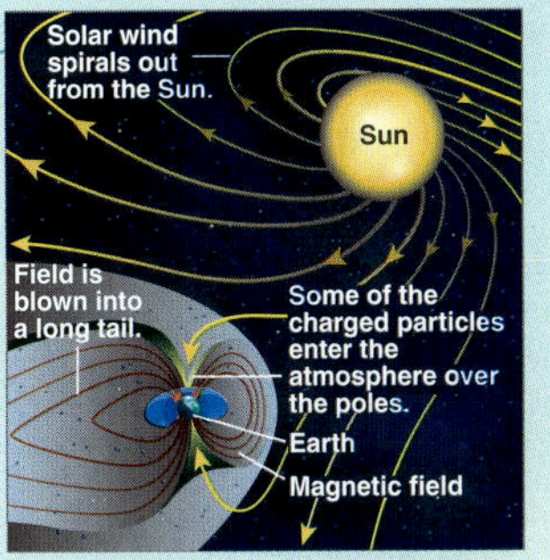

▲ **Figure 18-18** The solar wind spirals out from the Sun and enters the atmosphere over the poles.

The tail of a comet also shows evidence of the solar wind. The solar wind sweeps the gases evaporating from the comet nucleus into a tail. Because of the solar wind, a comet's tail always points away from the Sun.

Thinking Critically Why do you think ions are drawn toward the magnetic poles of Earth?

ONGOING ASSESSMENT

Have students write a short poem about the Sun, using the following terms: *the Sun, star, core, atmosphere, sunspot,* and *solar flare.* Students can also include some of the other Key Terms from this chapter. Ask several volunteers to read their poems aloud to the class.

• TEACHING RESOURCES • Online or Print

Teacher's Resources

Lesson Review: Chapter 18, p. 9

Enrichment: Chapter 18, p. 11

3 Assess

CHECKING CONCEPTS ANSWERS

1. cooler
2. prominences
3. electrically charged

THINKING CRITICALLY ANSWER

4. **MODEL:** Students' pictures should resemble Figure 18-13, with solar flare, sunspots, and preminences as well.

4 Close

Web InfoSearch

Sunspots and Ice Ages Astronomers have found that sunspot activity seems to build up and decrease in an 11-year cycle. This cycle has been linked to climate changes on Earth. Have students use the Internet to find out more about sunspots. Ask students: *Why do sunspots move across the Sun in an 11-year cycle? Could magnetic activity be a part of it? If so, how?* Some key search words are **sunspots, sunspots magnetic,** and **sunspots ice age.** Students can begin their search at www.conceptsandchallenges.com.

Integrating Physical Science

TOPICS: magnetism, solar energy, ions

THE SOLAR WIND

Background An ion is an atom that has gained or lost one or more electrons. Ions can be positively or negatively charged. Positive and negative ions attract each other to form ionic compounds. This attractive force is known as an ionic bond.

Discussion Ask students: *Why do you think there are so many colors displayed in an aurora?* (Possible answer: The gases that make up Earth's atmosphere vary, causing different colors to be given off in an aurora.)

Thinking Critically Answer
Possible answer: because the poles produce a magnetic field that attracts the particles

18-8 *(pp. 450–451)*

1 Introduce

STUDY HINT Have students read the objective and the boldface subheads to preview the information in the lesson.

Linking Prior Knowledge Have volunteers take turns explaining an atom (Lesson 2-1).

2 Teach

Demonstration Make a circle about 20 cm (8 in.) in diameter out of modeling clay. Press it firmly against a flat tabletop. Place wooden matches upright in the circle about 1 cm (less than half an inch) apart from one another. Keeping students a safe distance away, light one of the matches and have students observe the chain reaction that takes place. Relate this to the nuclear reactions that take place in stars and nuclear reactors.
learning style: visual

Discussion Write Einstein's formula, $E = mc^2$, on the board. Explain that the superscript 2 means that the speed of light is multiplied by itself. Carry out this calculation on the board so that students can see how large this value is.
learning style: auditory

```
      300,000
    × 300,000
      000000
     000000
    000000
   000000
  000000
 900000
90,000,000,000 = 9 × 10^10
```

READING STRATEGY Have students find the terms *nuclear*, *nucleus*, and *nuclei* in the first paragraph on this page. Have them use a dictionary to identify the part of speech of each term. Ask them to correctly use each word in a sentence.

Answers to questions

1 **DESCRIBE:** a nuclear reaction

2 **EXPLAIN:** Atoms combine, or fuse, to produce a new element and energy.

3 **STATE:** $E = mc^2$

4 **IDENTIFY:** hydrogen atoms

18-8 How does the Sun produce energy?

Objective

Describe the process by which energy is produced in the Sun.

Key Terms

nucleus (NOO-klee-uhs), *pl.* **nuclei** (NOO-klee-eye): center, or core, of an atom

fusion (FYOO-zhuhn): reaction in which atomic nuclei combine to form larger nuclei

Solar Energy The Sun gives off energy in many forms, including heat and light. Where does this energy come from? You know that burning produces heat and light. However, burning does not produce the heat and the light of the Sun. The Sun produces energy by nuclear reactions. The **nucleus** is the center of an atom. In a nuclear reaction, the nuclei of atoms are changed.

1 **DESCRIBE:** What kind of reaction produces the heat and the light of the Sun?

Nuclear Fusion The Sun is about 71 percent hydrogen and 27 percent helium. Other elements make up the remaining 2 percent. Deep inside the Sun, the temperature is more than 15,000,000°C. At these high temperatures, the nuclei of hydrogen atoms combine, or fuse. This kind of reaction is called hydrogen **fusion.** In this reaction, four hydrogen nuclei combine to form one helium nucleus. The mass of a helium nucleus is less than the mass of all four hydrogen nuclei put together. The missing mass has been changed into energy.

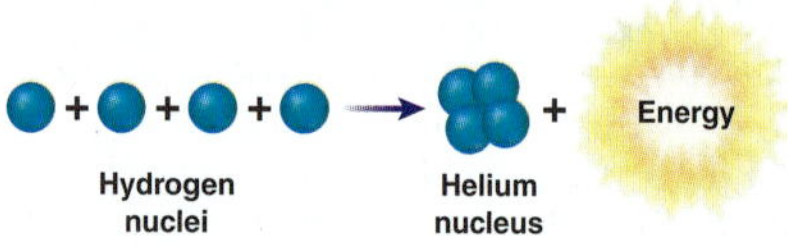

▲ **Figure 18-19** Hydrogen fusion

2 **EXPLAIN:** What happens in a fusion reaction?

Matter and Energy Albert Einstein helped explain how the Sun produces its energy. Einstein said that matter could be changed into energy. His equation $E = mc^2$ explains how a small amount of matter can be changed into a large amount of energy. In this equation, E is energy, m is mass, or the amount of matter, and c is the speed of light. The speed of light is 300,000 km/s. Astronomers have used Einstein's equation to calculate how much energy the Sun produces.

▲ **Figure 18-20** Albert Einstein

3 **STATE:** What is Einstein's equation?

Fusion Energy Fusion energy is one possible alternative source of nuclear energy. Fusion reactors may someday be a source of economical, clean, and safe energy. Hydrogen atoms can be combined in a fusion reaction to produce helium plus a great deal of energy. Because water contains hydrogen, the oceans could be an almost unlimited source of fusion energy.

▲ **Figure 18-21** The Tokamak fusion test reactor in Princeton, New Jersey

Extremely high temperatures and pressures are needed for fusion reactions to take place. Scientists are experimenting with different ways to bring about these reactions.

Teaching Options

EXTENDING THE LESSON

Comparing Fission and Fusion Have students find out about nuclear power plants that generate electricity in some parts of the world. Students will find that this type of reaction is fission, the splitting of atoms to generate energy. Make sure that students recognize that the reactions that take place in stars are fusion reactions—the joining of atoms to produce energy. Have students make illustrations that show how these two processes differ. **learning style: visual**

QUICK CONVERSIONS

	Metric/SI	English
Temperature of the Sun's core	15,000,000°C	27,000,000°F
Speed of light	300,000 km/s	186,000 mi/s

INTEGRATING HEALTH

Researching Vitamin D Have interested students find out why the nutrient vitamin D is also called the sunshine vitamin (because the body uses ultraviolet rays to manufacture it) and why it is important to the body. Students can present their findings in an oral report. **learning style: auditory**

One experimental fusion reactor uses a magnetic trap, or "magnetic bottle," to hold the fusion reaction. It also uses high-powered laser beams.

4 **IDENTIFY:** What would be the main fuel for fusion reactors?

CHECKING CONCEPTS

1. What are two forms of solar energy?
2. What is the core of an atom called?
3. What two elements make up 98 percent of the Sun?
4. What happens to the missing mass in a hydrogen fusion reaction?
5. What does *E* stand for in Einstein's equation?

THINKING CRITICALLY

6. **INFER:** Why are the nuclear reactions that take place in the Sun called fusion reactions?
7. **ANALYZE:** Explain how $E = mc^2$ shows that a small amount of matter can be changed into a very large amount of energy.

INTERPRETING VISUALS

Ordinary hydrogen nuclei contain one proton. Heavy hydrogen nuclei contain one proton and one neutron. Two pairs of heavy hydrogen nuclei fuse with two stray protons to form two light helium nuclei. These then fuse to form one ordinary helium nucleus. This reaction releases two protons, or ordinary hydrogen nuclei, and energy. Use Figure 18-22 to answer the following questions.

8. **IDENTIFY:** What kinds of atoms are produced during the Sun's fusion reactions?
9. **ANALYZE:** What symbol in the diagram represents the energy released?

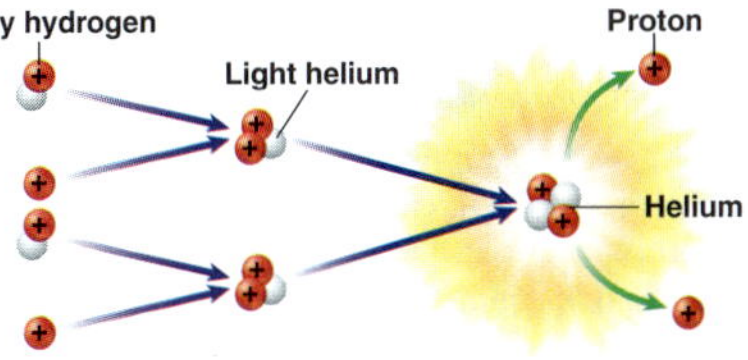

▲ **Figure 18-22** Fusion

Integrating Life Science

TOPICS: photosynthesis, food chain

FOOD FROM THE SUN

▲ **Figure 18-23** All animals depend on photosynthesis for food.

The Sun's energy reaches Earth as heat and light. This energy causes water on Earth to evaporate and rise into the air. The water then falls to Earth as rain. Rain helps plants to grow. Green plants use sunlight to make food. Without rain and sunlight, green plants would die. People and other animals could not survive.

The foodmaking process in green plants is called **photosynthesis**. The plants take in carbon dioxide and water. Sunlight supplies the energy needed by plants to make food. Some of the food is then used by the plants. The rest is stored. People and other animals use the stored food in plants when they eat the plants. Animals that eat other animals are taking in the energy from plants stored in the cells of the animals that are eaten.

During photosynthesis, plants give off oxygen. The oxygen goes into the air. People, other animals, and plants need oxygen to survive. Not all animals breathe air, but they do respire and get their oxygen this way. Without plants covering Earth and releasing oxygen, animals could not breathe.

Thinking Critically How is the oxygen produced by plants a kind of food animals must take in to survive?

ONGOING ASSESSMENT

Ask students to use what they learned in this chapter to explain the Sun's importance to Earth and its inhabitants. For example, the Sun provides light and heat for Earth and its inhabitants and is the energy source for photosynthesis.

MORE ON THE WEB

Learning About Solar Cells
Have students use the Internet to learn about solar cells and share with each other what they discover.

• TEACHING RESOURCES • Online or Print

Teacher's Resources

Lesson Review: Chapter 18, p. 10

Enrichment: Chapter 18, p. 11

3 Assess

CHECKING CONCEPTS ANSWERS

1. heat and light
2. the nucleus
3. hydrogen and helium
4. It is converted into energy.
5. energy

THINKING CRITICALLY ANSWERS

6. **INFER:** because atoms fuse, or join together
7. **ANALYZE:** The c^2 is a large number, so even if it is multiplied by a small number for mass, *E* will have a large value.

INTERPRETING VISUALS ANSWERS

8. **IDENTIFY:** helium
9. **ANALYZE:** the burst of white light

4 Close

Integrating Life Science

TOPICS: photosynthesis, food chain

FOOD FROM THE SUN

Background Chlorophyll is a pigment that traps light energy for green plants to use during photosynthesis. Photosynthesis produces a nutrient and oxygen from light energy, carbon dioxide, and water. The basic equation for this chemical reaction is $6CO_2 + 6H_2O + \text{light} \rightarrow C_6H_{12}O_6 + 6O_2$. Plants use some of the oxygen produced during photosynthesis for respiration. The rest of the gas is released via the plant's leaves. Some of the nutrient is used immediately for the plant's life processes. The rest is stored by the plant as sugar, starch, or other substances.

Discussion Ask students how photosynthesis benefits animals. Point out that virtually all food comes from photosynthesis, directly or indirectly. Discuss how the production of oxygen affected the evolution of life on Earth. (It generated an atmosphere that could support large, oxygen-using animals.)

Thinking Critically Answer
Possible answer: Oxygen is an essential substance, like food and water, for animals, and it must be taken into the bodies of animals.

18-9 (pp. 452–453)

1 Introduce

STUDY HINT To test themselves, have students turn all the questions in this lesson into fill-in-the-blank statements and then complete the statements.

Linking Prior Knowledge Ask students why stars rise and set throughout the night and why different constellations and star groups appear overhead at different times of the year. (because of the movements of Earth) Refer students back to Lesson 14-1 to review how Earth's rotation and revolution cause the changing views of the sky hour by hour and day by day.

2 Teach

Demonstration Use a large coffee can with both ends removed to trace three circles on white paper. Cut out the circles and sketch one of these constellations—Orion, the Big Dipper, and Leo—on each circle. Use a paper punch to indicate the stars that make up each constellation. Tape a circle over one end of the coffee can, darken the room, and use a flashlight to project the star pattern onto a screen. Repeat the procedure with the other two constellations.
learning style: visual

Discussion Explain and demonstrate to students how to use the star charts in Figure 18-24. Ask if a chart would be easier to use if you were looking down at it in front of you or overhead while you were looking at the stars. Discuss the differences in the chart that the two ways of looking would require.
learning style: auditory

Reinforcement Instruct students on how to find the Big Dipper and Little Dipper in the sky early in the evening and later at night. Ask students to do this over a weekend and to describe their observations. Ask: *Why do these constellations appear to change positions during a single night?* (Earth's rotation on its axis causes constellations to change positions.) Challenge students to explain why constellations change with the seasons. (Earth revolves around the Sun.)
learning styles: visual, auditory

18-9 What are constellations?

Objectives
Explain what constellations are. Name some familiar constellations.

Key Term
constellation (kahn-stuh-LAY-shuhn): group of stars that form a pattern in the sky

Star Patterns When you look at the night sky, you can see many stars. Some of these stars seem to form patterns. These patterns are called **constellations.**

Ancient peoples thought they could see animals and people in constellations. They named the constellations and made up stories about them. Today, astronomers recognize 88 constellations. Constellations are often used to locate individual stars.

Stars seem to move slowly across the sky. Different constellations and star groups appear overhead at different times of the year. The constellations visible in the Northern Hemisphere are different from those visible in the Southern Hemisphere. However, the stars are relatively stationary. It is the movements of Earth that are making it look like the stars are changing position in the sky.

1 DEFINE: What is a constellation?

Orion the Hunter In the Northern Hemisphere, Orion the Hunter is a constellation seen in the evening sky from November to March. It rises in the east.

Three bright stars make up Orion's belt. Betelgeuse, a red giant star, is under Orion's arm. Rigel, a giant blue-white star, is in Orion's knee.

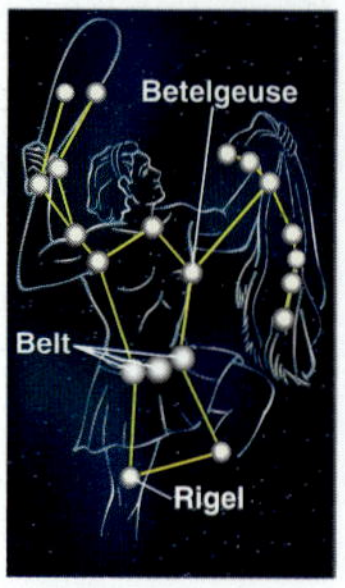

▲ **Figure 18-25** Orion

2 NAME: What are two stars in Orion?

The Bear Constellations One of the easiest constellations to find in the northern sky is Ursa Major. Its name means "big bear."

Ancient peoples thought they could see the shape of a large bear in the constellation. The Big Dipper is part of Ursa Major. It is made up of seven stars, three in the handle, four in the cup. Two bright stars in the cup of the Big Dipper are called the pointers. They point to Polaris, which is also known as the North Star.

Ursa Minor, meaning "little bear," is another constellation in the northern sky. Ursa Minor is also called the Little Dipper. Polaris is the first star in the handle of the Little Dipper.

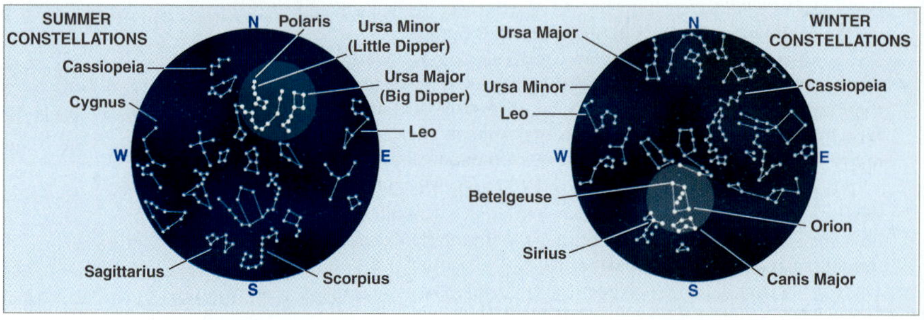

▲ **Figure 18-24** The summer and winter skies in the Northern Hemisphere

452

Teaching Options

RETEACH

Making Star Art Provide students with colored, gummed stars and black construction paper. Instruct students to drop, from a height of about 10 cm (4 in.), ten gummed stars on the black paper. Tell students to stick the stars on the paper exactly where they fell. Have students name this new constellation and write a short story about the object or character they see in the star pattern.
learning styles: visual, tactile

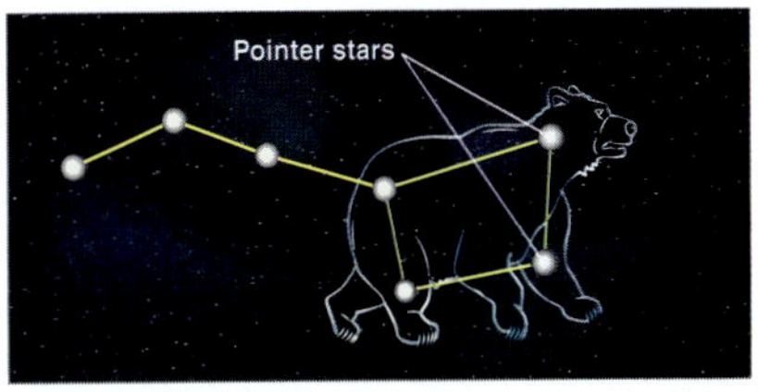

▲ **Figure 18-26** The Big Dipper is in the constellation Ursa Major.

3 NAME: What does the name Ursa Major mean?

CHECKING CONCEPTS

1. What are patterns of stars called?
2. How many constellations are there?
3. What is the Big Dipper?
4. What is the name of the North Star?
5. What is the name of the red star in Orion?

THINKING CRITICALLY

6. INFER: Why are the two bright stars in the cup of the Big Dipper called the pointers?
7. INFER: Although other stars appear to rotate in the night sky, Polaris does not seem to change position over the North Pole. Why do you think this is?
8. ANALYZE: Why do stars in constellations appear close even though they may be very far apart?

BUILDING SCIENCE SKILLS

Researching The zodiac is a band of constellations that lie along the ecliptic. Over a year's time, the Sun seems to trace a path in the sky along the ecliptic. What is the ecliptic? How many different constellations make up the zodiac? Do research on one of the constellations. What is the story behind it? Write your findings in a report.

Real-Life Science

IDENTIFYING THE CONSTELLATIONS

The oldest known astronomical texts were written by the ancient Sumerians. They recorded the names of constellations still known today as the lion, the bull, and the scorpion. The ancient Greek and Chinese peoples had different names for some of the same constellations. The ancient Egyptians named about 25 constellations, including the crocodile and the hippopotamus.

Astronomers continued naming the constellations right up until the twentieth century. In 1930, an official listing of 88 constellations was published by the International Astronomical Union. One of these is Leo the Lion.

▲ **Figure 18-27** The constellation called Leo is in the zodiac between Cancer and Virgo. Its brightest star is called Regulus.

To find the constellations, go out on a dark night and scan the sky. If you do not already know geographical directions in your area, learn them. This will help you use the sky charts published in books and magazines. To orient yourself, try to find the brightest stars and identify them. They can help you locate other star groups.

Orion is one of the easiest constellations to see in the winter sky. Look for Betelgeuse and Rigel. The three stars in Orion's belt point to Sirius in Canis Major (the Dog Star) and Aldebaran in Taurus, the bull constellation.

Thinking Critically How could you find out geographical directions in your area?

EXTENDING THE LESSON

Researching Constellations and Mythology Have interested students research one of the following constellations: Cygnus, Pegasus, Cassiopeia, Hercules, Cetus, Auriga, Monoceros, Centaurus, Aquila, or Lacerta. Have students draw the constellation and write a paragraph on the myth behind it. Display the drawings and paragraphs around the classroom.
learning style: visual

• TEACHING RESOURCES • Online or Print

Teacher's Resources

Lesson Review: Chapter 18, p. 12

Answers to questions

1 DEFINE: a group of stars that form a recognizable pattern in the sky

2 NAME: Possible answer: Betelgeuse and Rigel

3 NAME: big bear

3 Assess

CHECKING CONCEPTS ANSWERS

1. constellations
2. 88
3. a constellation made up of seven stars that resemble a large ladle
4. Polaris
5. Betelgeuse

THINKING CRITICALLY ANSWERS

6. INFER: They point to Polaris, the North Star.
7. INFER: It is located directly above the North Pole, which is one end of Earth's axis and so does not rotate.
8. ANALYZE: All stars are very far from Earth. At a great distance, two separate objects can appear to be close to each other.

4 Close

BUILDING SCIENCE SKILLS

Researching The path of the Sun among the stars as seen from Earth is the ecliptic. The Zodiac is a band of constellations that appear 9 degrees on either side of the ecliptic. The constellations of the Zodiac are Aquarius, Capricorn, Sagittarius, Scorpio, Libra, Virgo, Leo, Cancer, Gemini, Taurus, Aries, and Pisces.

Real-Life Science

IDENTIFYING THE CONSTELLATIONS

Geography Connection Help students identify some landmarks in their area that might help them determine geographical directions.

Discussion Tell students that stars can be used for navigation. Have students brainstorm to compile a list of people who might navigate by the stars. (Possible answers include hikers, sailors, and people crossing a desert.)

Thinking Critically Answer
by checking a map or looking in an atlas

18-10 *(pp. 454–455)*

1 Introduce

STUDY HINT Have students refer frequently to Figure 18-28 as they read this lesson.

Linking Prior Knowledge Have students review the nature of orbits (Lessons 17-2, 17-7, and 17-10) and the Chapter 17 Big Idea. Then, ask a student to define the term *ellipse*. Ask another student to tell the parts of speech for *ellipse* (noun) and *elliptical* (adjective).

2 Teach

Demonstration Make an amoeba-shaped figure from a piece of stiff cardboard. With a hole punch, make a hole near the edge of the shape in three places. You will need a small nail, a piece of corkboard, a weight, and a string that is as long as the longest dimension of the shape. Tie the weight to the piece of string.

Inform students that all galaxies rotate around a fixed point called the gravitational center. Ask where the gravitational center, or balancing point, of an object is. (In regularly shaped objects, such as circles and squares, the balancing point is the center.) Ask how one might find the center of an irregularly shaped object, such as an irregular galaxy. Use the shape you constructed to show students how. Put the small nail through one of the holes into the corkboard. Tie the unweighted end of the string to the nail and let it hang down. Carefully trace a straight line along the string. Repeat this procedure for the other two holes. The point of intersection of the three lines is the gravitational center, or balancing point, of the shape.
learning style: visual

Discussion Say aloud a type of galaxy: spiral, elliptical, or irregular. Have students find phrases from the text that describe that type of galaxy. You may wish to have students use the responses to this activity to make a table that contrasts the three main types of galaxies.
learning style: auditory

18-10 What are galaxies?

Objective
Describe the three main types of galaxies.

Key Terms
galaxy (GAL-uhk-see): huge collection of stars, gas, and dust that travel together through space
elliptical galaxy: galaxy shaped like a ball or slightly flattened ball
irregular galaxy: galaxy with no definite shape
spiral galaxy: galaxy shaped like a flattened disk with spiral arms

Galaxies Stars appear as small points of light. Among these points of light, you can also see some fuzzy patches. Some of these patches are nebulae, or clouds of gas and dust. Others are galaxies. A **galaxy** is a huge collection of stars, gas, and dust that travel together through space. Using the newest telescopes, astronomers can see billions of galaxies. Galaxies are the building blocks of the universe.

Galaxies have different shapes. Astronomers classify, or group, galaxies based on shape. There are three kinds of galaxies: elliptical, irregular, and spiral.

1 **DEFINE:** What are galaxies?

Elliptical and Irregular Galaxies Some galaxies have rounded shapes. These galaxies are known as **elliptical galaxies.** Elliptical galaxies can also look like slightly flattened balls. The stars in an elliptical galaxy are usually older than the stars in other kinds of galaxies. Elliptical galaxies are larger than irregular galaxies but smaller than spiral galaxies.

Galaxies with no regular shape are called **irregular galaxies.** Irregular galaxies are smaller and fainter than elliptical and spiral galaxies. They may be the most common kind of galaxy.

2 **COMPARE:** Which is more common, an elliptical galaxy or an irregular galaxy?

Spiral Galaxies **Spiral galaxies** are shaped like flattened disks. They usually have one or more spiral arms that branch out from their centers. One type of spiral galaxy is called the barred spiral. The arms of barred spirals branch out from the end of a short bar made up of stars and gas.

3 **DESCRIBE:** What special feature do spiral galaxies have?

The Milky Way Galaxy All the stars you can see in the night sky with your unaided eye are part of our home galaxy, called the Milky Way Galaxy. Our solar system is in the Milky Way Galaxy.

▲ **Figure 18-28** Galaxies come in three basic shapes: elliptical, irregular, and spiral.

454

Teaching Options

COOPERATIVE LEARNING

Classifying Galaxies Have students work in pairs to research the classifications for each of the following galaxies: Andromeda (spiral), the Whirlpool galaxy (spiral), NGC 205 (elliptical), NGC 221 (elliptical), NGC 3193 (elliptical), the Large Magellanic Cloud (irregular), the Small Magellanic Cloud (irregular), and NGC 5195 (irregular). Note that *NGC* stands for the New General Catalog, which lists galaxies by number.
learning style: interpersonal

RETEACH

Modeling the Milky Way Take some chalk, metersticks, and a calculator outside to construct with students a scale model of the Milky Way Galaxy. Remind students that the Milky Way is about 100,000 light-years in diameter. Using a scale of 1 mm to 1 light-year (or less, if you do not have a large school yard), have students compute how big to make the model galaxy (100,000 mm or 100 m [328 ft]). Challenge students to determine the approximate position of the Sun in this model of the galaxy. (67 m [109 ft] from the center of the galaxy) **learning style: tactile**

The Milky Way Galaxy is part of a group of more than 20 other galaxies. Together, these galaxies are known as the Local Group.

The Milky Way Galaxy is a spiral galaxy. It contains about 100,000 billion stars. At its center is a huge bulge of stars about 10,000 light-years across. The distance from edge to edge of the spiral arms is about 100,000 light-years. The Sun is in a spiral arm about two-thirds of the way from the center of the Milky Way Galaxy.

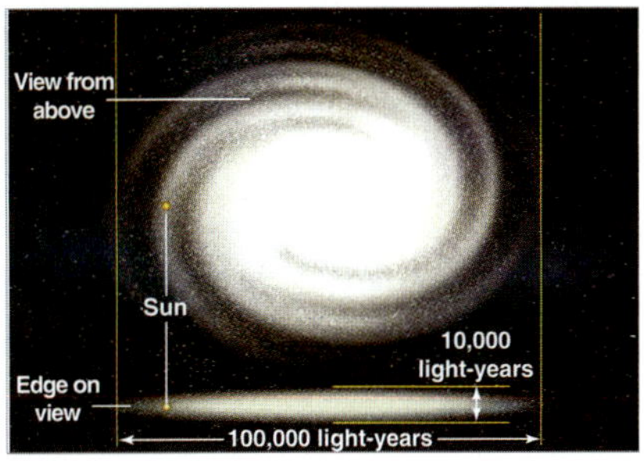

▲ **Figure 18-29** The Milky Way Galaxy

4 CLASSIFY: What kind of galaxy is the Milky Way Galaxy?

The Milky Way Band There is a band of stars stretching across the sky. Ancient peoples called this band the Milky Way. They thought it looked like a river of milk. When you look at the Milky Way band of stars, you are seeing the flattened disk of the Milky Way Galaxy, where the spiral arms lie.

▲ **Figure 18-30** Part of the Milky Way band of stars in the sky

5 IDENTIFY: What is the Milky Way band?

Galactic Neighbors Most galaxies are millions of light-years from Earth. For example, the great spiral galaxy in the constellation Andromeda, seen on the far right in Figure 18-28, is 2 million light years away. The Andromeda Galaxy is much larger than our own galaxy. You can see it without a telescope.

The closest galaxies to Earth are the Large Magellanic Cloud and the Small Magellanic Cloud. These irregular galaxies are 160,000 and 200,000 light years away, respectively.

6 IDENTIFY: Which galaxies are closest to the Milky Way Galaxy?

CHECKING CONCEPTS

1. Galaxies are made up of billions of __________.
2. Astronomers classify galaxies on the basis of their __________.
3. There are __________ main types of galaxies.
4. The oldest stars are found in __________ galaxies.
5. Stretching across the sky is the __________ band of stars.

THINKING CRITICALLY

6. LIST: What are the three main types of galaxies?
7. CALCULATE: If all the galaxies in the Local Group have about the same number of stars, about how many stars are in the Local Group?

INTERPRETING VISUALS

Use Figure 18-31 to complete the following exercise.

8. CLASSIFY: Classify each of the galaxies shown as spiral, elliptical, or irregular.

▲ **Figure 18-31**

WRITING HINT When students are calculating sums of numbers in the millions, billions, or trillions, tell them to write out all of the zeros so that they can translate accurately from the millions to the billions or trillions.

Answers to questions

1 DEFINE: huge collections of stars, gas, and dust that travel together through space

2 COMPARE: probably an irregular galaxy

3 DESCRIBE: two or more spiral arms that branch out from the center

4 CLASSIFY: a spiral galaxy

5 IDENTIFY: the flattened disk of the Milky Way Galaxy, where its spiral arms are found

6 IDENTIFY: the Large and Small Magellanic Clouds

3 Assess

CHECKING CONCEPTS ANSWERS

1. stars
2. shape
3. three
4. elliptical
5. Milky Way

THINKING CRITICALLY ANSWERS

6. LIST: elliptical, spiral, and irregular
7. CALCULATE: 100 billion × 20 = 2 trillion

INTERPRETING VISUALS ANSWER

8. CLASSIFY: A is irregular, B is spiral, and C is elliptical.

4 Close

Reinforcement Obtain a dozen or so photos of the three kinds of galaxies from *Sky and Telescope* or another astronomical magazine. Have students classify each galaxy either aloud or on the board.
learning style: visual

MORE ON THE WEB

Writing About Dark Matter
You can have students visit www.conceptsandchallenges.com to learn about dark matter. Then, have them write a report explaining the different theories that scientists have about dark matter.

• TEACHING RESOURCES • Online or Print

Teacher's Resources
Lesson Review: Chapter 18, p. 13

Lab Activity

(pp. 456–457)

1 Prepare the Lab

Counting Galaxies

SET-UP TIME: **LAB TIME:**

BACKGROUND: Above Earth's atmosphere, the Hubble Space Telescope can see deeper and more clearly into space than Earth-based telescopes. In 1995, the spacecraft was pointed at a tiny fragment of the sky. Over 10 days, hundreds of pictures were taken and combined to make a picture of galaxies in the deepest reaches of outer space. The area covered by the picture is equivalent to the size of a grain of rice held at arm's length. More than 30 million of these pictures would be needed to cover the entire sky.

PURPOSE: Students will estimate how many galaxies are in this Hubble Deep Field picture.

ALTERNATIVE MATERIALS: This activity can be conducted with printed copies of the picture or directly on a computer screen. Other pictures of dense star fields in our galaxy can also be used for estimating.

SCIENCE SKILLS: Students will **count** the number of galaxies that are present in a representative sample and **estimate** the total number of galaxies found in the entire picture.

ADVANCE PLANNING: Make copies of the picture in the text and enlarge it if you can or obtain the Hubble Space Telescope Hubble Deep Field picture from the Internet. (See the More on the Web in Teaching Options on this page.) Divide the pictures into quarters and identify them for reassembly later.

2 Guide the Lab

PROCEDURE: Organize students into four groups and give out the pictures. Discuss the procedure. Explain how representative samples can be used to estimate very large quantities. Provide students with a copy of the Lab Activity Report from the Teacher's Resources to record their observations and conclusions.

LAB ACTIVITY
Counting Galaxies

Materials
Hubble Deep Field picture
Calculator

BACKGROUND

The universe is very large. It has a vast number of galaxies that are each home to billions of stars like our Sun. The Hubble Deep Field picture represents only a tiny part of the whole sky. More than 30 million of these pictures would be needed to cover the entire sky.

PURPOSE

In this activity, you will estimate how many galaxies are found in a tiny fragment of the sky.

PROCEDURE

1. Your teacher will divide you into four groups. Examine the picture given to your group by your teacher. It was taken by the Hubble Space Telescope.

2. Copy the chart in Figure 18-33. Then, count the number of galaxies in your picture and record the number in your chart.

3. Your picture is a part of a larger picture. The larger picture is four times the size of your picture. The other groups are looking at the other three parts. How many galaxies do you think the entire picture contains? Estimate how many and record this number in your chart.

4. Trade the number of galaxies you counted in your picture with numbers arrived at by the groups that are counting the other three pictures. Record the numbers in your chart.

STEP 1 Examine the picture given to you by your teacher.

456

Teaching Options

FAST FACTS *The Hubble Space Telescope*

The Hubble Space Telescope is about the size of a tractor-trailer truck. It is 13.2 m (43.3 ft) long with a maximum diameter of about 4.2 m (13.8 ft). It weighs about 11,000 kg (24,000 lb) on Earth. The telescope is a low Earth-orbiting (LEO) satellite. At its current altitude of about 600 km (320 nautical mi), the telescope orbits Earth every 100 minutes at approximately 27,200 km/h (17,000 mph).

MORE ON THE WEB

Finding the Hubble Deep Field Image Have students go online to find the Hubble Deep Field picture.

ESL/ELL STRATEGY

Summarizing Procedures English-language learners may benefit from stating the procedures in their own words. Have English-language learners explain the procedures in their own words to English-proficient students.

▲ **Figure 18-32** The Hubble Deep Field image

My Data Chart	
Number of galaxies in your picture	
Estimate the total number of galaxies. (Your number times 4.)	
Number of galaxies in group 2's picture	
Number of galaxies in group 3's picture	
Number of galaxies in group 4's picture	
Total of all pictures	

▲ **Figure 18-33** Use a copy of this chart to record your calculations.

CONCLUSIONS

1. **OBSERVE:** How many galaxies did you count?
2. **ESTIMATE:** How many galaxies are there in the total picture?
3. **COMPARE:** How did your estimate compare with the actual count when you totaled the counts from the four groups?
4. **INFER:** For what other things might you use this estimating technique?

TROUBLESHOOTING: Several objects in the picture are stars in the Milky Way Galaxy. Tell students not to count objects that are round with small spikes of light coming out of them.

EXPECTED OUTCOME: Students will make reasonable estimates of the total number of galaxies shown in the picture.

LAB HINT: Point out that some galaxies may appear to be joined but are actually at different distances. They just lie in the same direction.

3 Conclude the Lab

1. **OBSERVE:** Answers should be similar for all groups.
2. **ESTIMATE:** Check students' math. Their answers should be in the 375 to 400 star range.
3. **COMPARE:** Discuss any gross differences among estimates.
4. **INFER:** Possible answer: estimating the number of marbles, pieces of candy, or other objects in a glass bowl

Use the Laboratory Activity Rubric in the Teacher's Resources to assess students' lab activities. Fill in the rubric with the additional information below. For this activity, students should have:

- estimated how many galaxies are found in a fragment of the sky.
- compared their estimates with the actual count.

4 Extend the Lab

Help students estimate how many galaxies are found in the entire universe by multiplying students' estimates for the Hubble Deep Field picture by 30 million. Write this value on the board. Have a volunteer correctly place the commas between digits.

Tips for Using Technology

Using Spreadsheets to Record Data Have students use spreadsheet software to record their estimations. Spreadsheets are tables that display data such as numbers, text, dates, or a combination of all three. Students can do math to analyze numerical data and make graphs and charts to present data in a colorful way. Students can even use spreadsheet software to present their data to the class. Have students use the Spreadsheet graphic organizer from the Teacher's Resources to plan their spreadsheets.

• TEACHING RESOURCES • Online or Print

Teacher's Resources

Lab Activity Report: Study Tools, pp. 16–17

Laboratory Activity Rubric: Teaching Tools, p. 8

Spreadsheet: Study Tools, p. 23

18-11 *(pp. 458–459)*

1 Introduce

INVESTIGATE

TIME: 10–15 minutes

PURPOSE: Students will measure distances between points on a model of the expanding universe.

MATERIALS: large round balloon; paper clip; small, thin rubber band; permanent marking pen; string (about 20 cm [8 in.] long); metric ruler

PROCEDURE: Draw a sample data table on the board so that students can see the number of columns and rows needed to record the data.

SAFETY TIP: Remind students to inflate the balloon completely, but to be careful that it does not pop.

THINK ABOUT IT: The distances increased as the balloon filled with air. The distance between some dots increased more than others because the balloon expanded in all directions as it was inflated. This model shows that, as the universe expands, galaxies move apart.

2 Teach

Discussion Ask students if they think the Big Bang is a good name for the theory that explains how the universe formed. Ask them to give reasons for their answers.
learning style: auditory

READING STRATEGY As students read this lesson, have them make sketches of the different concepts, even though some are rather abstract. Have them include whatever scientific details are given in the text and invent the rest.

Answers to questions

1 **EXPLAIN:** the Sun, the solar system, galaxies—everything in space

2 **EXPLAIN:** the explosion thought to have started the expansion of the universe

3 **ANALYZE:** Black holes can suck in nearby matter. When this happens, there are huge bursts of light and energy, which scientists call quasars.

4 **EXPLAIN:** When galaxies meet, gravity sometimes pulls them together to form a single, bigger galaxy.

18-11 What else do we know about the universe?

INVESTIGATE

Modeling the Expanding Universe

HANDS-ON ACTIVITY

1. Partially inflate a round balloon. Twist the opening and hold it shut with a paper clip or a rubber band.
2. With a marking pen, draw four dots on the surface of the balloon in a square pattern. Label them *A*, *B*, *C*, *D*.
3. Using a string, measure the distance between each of the four dots (*A* to *B*, *A* to *C*, *A* to *D*, *B* to *C*, *B* to *D*, and *C* to *D*). Hold the string next to a ruler and record each measurement in a table.
4. Inflate the balloon to its fullest and tie the end.
5. Repeat your measurements and record the distances.

THINK ABOUT IT: What happened to the distances between the dots as you inflated the balloon? Did some dots move farther apart than others did? Why? Suppose the balloon represents the universe. How does this activity show what is happening to the universe and to the galaxies in it?

Objective

Describe some unusual features of the universe.

Key Terms

Big Bang: explosion that may have begun the universe about 14 billion years ago

black hole: massive star that has collapsed and whose gravity is so powerful that it pulls in everything, even light

quasar: continuous burst of brilliant light and enormous energy from a very massive black hole

How Big Is the Universe The universe is everything, including the Sun, its solar system, and the many billions of stars in all the galaxies. It is everything that exists in space, even space itself. The distances between stars and galaxies are vast. If you could travel at the speed of light (300,000 km/s), it would take you more than 10 billion years to cross the part of space astronomers can see with their telescopes.

1 EXPLAIN: What does the universe include?

The Big Bang Scientists think that the entire universe was once contained in a single, very hot, very dense point. About 14 billion years ago, a huge explosion caused it to start expanding rapidly. This explosion is called the **Big Bang.**

As the universe expanded, it cooled. Atoms and molecules formed. Gradually, the atoms and molecules collected into the objects that make up today's universe, which is still expanding.

▲ **Figure 18-34** The radio waves caused by the Big Bang are represented in this picture by the pink ripples. These waves can still be detected today.

2 EXPLAIN: What is the Big Bang?

Black Holes and Quasars The universe is filled with many strange objects. For example, a black hole can form when a very massive star collapses in on itself. A **black hole** is an object so dense that nothing pulled into it by gravity can come out.

A black hole appears black because not even light can escape from its surface. All of the matter at the center of a black hole is squeezed into an infinitely tiny point.

Teaching Options

FAST FACTS

Glowing Black Holes

In October 2001, astronomers in the United States and Europe observed a glow about 10 billion times more intense than that of the Sun just outside a black hole in the constellation Centaurus. Astronomers hypothesize that this glow may be caused by friction between the black hole, which is spinning at nearly the speed of light, and the surrounding gases. The friction is probably the result of magnetic fields associated with the black hole.

EXTENDING THE LESSON

Explaining Gravity Wells Tell students that Albert Einstein first described gravitation as a distortion, or bending of space itself, caused by very massive objects. These objects make indentations in space, called gravity wells, much like a bowling ball would make an indentation on a mattress. Any smaller objects nearby are naturally drawn into the indentation. This is why planets orbit the Sun. Gravity wells formed by black holes are steeper, more like a funnel than a dent. Any object, even light, that passes near the gravity well of a black hole is sucked into it at the speed of light, never to escape.

Black holes are often found at the centers of galaxies. These black holes can be millions of times heavier than stars. When matter falls into these extremely massive black holes, tremendous bursts of light and energy are released. Astronomers call these bursts **quasars.** Quasars glow brighter than the light of a thousand galaxies.

3 ANALYZE: How are black holes and quasars related?

Colliding Galaxies The whole universe is in motion. Planets swirl around stars. Stars swirl around galaxies. Galaxies spread out into deep space. Sometimes two or more galaxies approach each other like colliding cars. The distance between stars is so great that the two galaxies pass right through each other like ghosts. However, gravity will sometimes pull galaxies like these together. They form into a single, bigger galaxy.

4 EXPLAIN: How can galaxies merge?

CHECKING CONCEPTS

1. About how many stars are in the universe?
2. What happens when stars fall into a black hole?
3. What are colliding galaxies?
4. How long ago did the universe form?

THINKING CRITICALLY

5. EXPLAIN: Why are the galaxies moving outward?
6. INFER: Do you think the galaxies will continue to move outward?
7. ANALYZE: If you had a spaceship that could travel at the speed of light, could you visit another galaxy?

DESIGNING AN EXPERIMENT

Design an experiment to solve the following problem. Include a hypothesis, variables, a procedure with materials, a type of data to study, and a way to record your data.

PROBLEM: The gravity field near a black hole is very strong. Any matter that gets near the black hole is sucked in by its powerful gravity. Design an experiment that models what might happen to matter as it passes near a black hole. Will it fall straight in or spiral around it first? Indicate why you chose one or the other.

Integrating Physical Science

TOPICS: forces, waves

GRAVITATIONAL WAVES

▲ **Figure 18-35** A gravitational wave spreads out a little like ripples in a pond.

Energy travels freely across the universe. We only see light waves, but scientific instruments can see radio waves, X-rays, ultraviolet light, and other forms of radiation. The energy in gravity also travels across the universe. Unless you are very close to an object, the gravity, or the attraction, you feel from it is very faint. Very massive objects have a stronger gravitational attraction than less massive objects do.

If all objects in the universe stayed still, the attraction of gravity would remain the same at all times. However, stars and galaxies do not stand still. When they move, they are believed to produce ripples in the gravitational field that travel across the universe as gravity waves. The waves would spread out in a manner similar to the water waves in a pond when a pebble is thrown in. However, instead of just moving across a flat surface like the pond waves, gravity waves would likely move in all directions.

Thinking Critically How could gravitational attraction cause waves?

3 Assess

CHECKING CONCEPTS ANSWERS

1. many billions
2. Tremendous bursts of light and energy, called quasars, are released.
3. galaxies that have merged
4. between 10 and 15 billion years ago

THINKING CRITICALLY ANSWERS

5. EXPLAIN: The universe is expanding because of an enormous explosion.
6. INFER: Possible answers: The expansion will never stop. The expansion will someday stop and contraction will take over. The expansion will eventually cease but the universe will not contract.
7. ANALYZE: No. There is no galaxy close enough to ours that could be reached in an ordinary human lifetime, even traveling at the speed of light.

4 Close

DESIGNING AN EXPERIMENT

Use the Designing an Experiment Rubric from the Teacher's Resources to assess students' experiments. Fill in the rubric with the information below. For this assignment, students should have:

- mentioned that black holes absorb matter.
- included a vacuum cleaner or some other object to simulate a black hole.

Integrating Physical Science

TOPICS: forces, waves

GRAVITATIONAL WAVES

Background Newton's law of universal gravitation states that every object in the universe is attracted to every other object with a force that is directly proportional to the products of its mass and inversely proportional to the square of the distance between the two objects:

$$F = G\left(\frac{m_1 m_2}{r^2}\right)$$

Discussion Ask which has a greater gravitational effect on Earth—the Sun or a larger star in another galaxy. Students should be able to deduce that the Sun, even though it is smaller, has a greater pull on Earth because it is closer.

Thinking Critically Answer
Matter moving through the universe could cause ripples called gravity waves in the surrounding gravitational field.

ESL/ELL STRATEGY

Quasar To help students remember the term *quasar*, inform them that it is short for "quasi-stellar radio source." *Quasi* means "to some degree." *Stellar* means "related to stars." A quasar is, to some degree, a star, but it is not an ordinary star.

• TEACHING RESOURCES •
Online or Print

Teacher's Resources

Lesson Review: Chapter 18, p. 14

Designing an Experiment: Study Tools, p. 14

Designing an Experiment Rubric: Teaching Tools, p. 7

The Big Idea

(pp. 460–461)

1 Introduce

Objective Students will explore how scientists think they can estimate how big the universe is. Students will also be introduced to some of the Milky Way Galaxy's nearest neighbors in space.

Linking Prior Knowledge Have students recall that galaxies are spread out in the universe and that there are three different types of galaxies (Lesson 18-10).

2 Teach

Discussion After students have read the information on these two pages, ask if they think that the universe will continue to expand forever. Tell students that this subject is still being debated and that data from the Hubble Space Telescope and other sources are contributing evidence to evaluate the theories.

In the **Do The Math** section, Galaxy 8 is larger. The numerator in the formula is four times larger for Galaxy 8.

THE BIG IDEA

How can we estimate the size of the universe? Have students start their research for their Science Log Writing Activity at www.conceptsandchallenges.com. Be sure students use search engines and websites that are reliable and acceptable. You may also wish to provide copies of the *Big Idea Planner* and *Big Idea Science Log Entry* study tools found in the Teacher's Resources.

Integrating Mathematics

THE Big IDEA

How can we estimate the size of the universe?

The universe has been expanding since the Big Bang. It is not expanding into anything, however. Instead, space itself is stretching out, and carrying the galaxies with it. The American astronomer Edwin Hubble was the first to present this idea. In 1929, he showed that the galaxies were rushing away from each other.

Hubble studied the brightness of a class of stars known as Cepheid stars. Using these stars, he was able to measure how far away certain galaxies were. He was also able to figure out how fast the galaxies were moving by examining their redshift. Hubble discovered that the galaxies closest to us are moving slower than the galaxies farther away. He also discovered that a galaxy's speed, or velocity (V), is directly related to its distance (D) from us. This relationship can be expressed using a number called the Hubble constant (H).

Knowing that the universe is expanding is very useful to astronomers. By measuring the velocity of a distant galaxy, they can determine its distance. Since the universe must be large enough to hold every galaxy, astronomers can use the most distant galaxies known to estimate the size of the universe.

However, astronomers have discovered that it is very difficult to give an exact value for Hubble's constant. This number could be as low as 15 and as high as 30 km/s/mly (kilometers per second per million light-years). Unless we can determine this value exactly, we can only give a range of possible sizes for the universe.

Look at the photos on these two pages. Then, follow the directions in the Science Log to learn more about "the big idea." ✦

The Milky Way
Earth and the other planets of our solar system are located about 50,000 light-years from the center of the Milky Way Galaxy, on one of its spiral arms.

The Small Magellanic Cloud
About 200,000 light years from Earth lies the Small Magellanic Cloud. By the 1920s, astronomers had realized that this fuzzy cloud of light was actually a distant galaxy.

The Large Magellanic Cloud
The Large Magellanic Cloud is an irregular galaxy visible only in the Southern Hemisphere. It is named after Portuguese explorer Ferdinand Magellan. Magellan's crew were the first Europeans to describe this cluster of stars that is more than 160,000 light-years from Earth.

460

Teaching Options

! FAST FACTS
Hubble's Law

Edwin Hubble was an American astronomer who discovered, among other things, that large-scale galaxies lie beyond the Milky Way and that these galaxies are distributed almost uniformly in all directions. Hubble was the first astronomer to offer evidence to support the theory of the expanding universe. The mathematical relationship known as Hubble's law states that the greater the distance between any two galaxies, the greater is their relative speed of separation.

COOPERATIVE LEARNING

Modeling Galaxies Have small groups of interested students use metric rulers, calculators, and adding machine tape to make scale models that show the distances between the galaxies mentioned in the callouts in the Big Idea. Display the models in the classroom.

learning styles: tactile, visual

Andromeda

The Andromeda Galaxy is the most distant galaxy that can be seen from Earth without a telescope. This spiral galaxy lies about 2 million light-years from Earth.

With the **Big Bang** explosion, the universe was created and began to expand outward.

DO THE MATH!

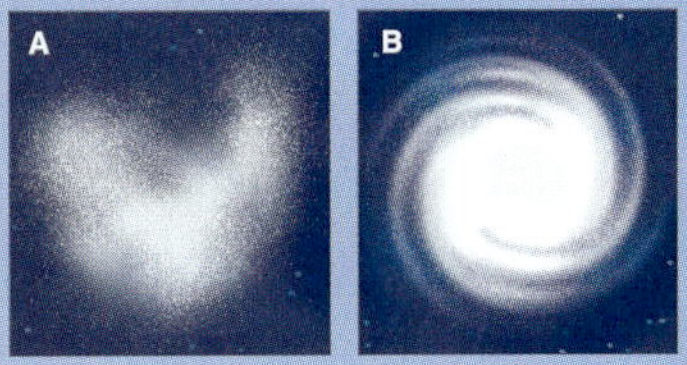

Figure 18-36

The two galaxies shown in Figure 18-36 appear to be the same size. They are not. One is simply farther away. You can tell which galaxy is bigger than the other by measuring how far each galaxy is from Earth.

STEP 1 Use the following formula to determine each galaxy's distance:

$$D = \frac{V}{H}$$

Distance (D) is equal to the velocity (V) of the galaxy divided by the Hubble constant. (H). Velocity, or speed, is measured in kilometers per second (km/s). Let Hubble's constant here be as 20 km/sec/mly.

STEP 2 Find out how far Galaxy A is from Earth. The velocity for Galaxy A is 1,000 km/s.

Galaxy A:

$$D = \frac{1{,}000 \text{ km/s}}{20 \text{ km/s/mly}} = \underline{\qquad\qquad} \text{ mly}$$

STEP 3 Find out how far Galaxy B is from Earth. The velocity for Galaxy B is 4,000 km/s.

Galaxy B:

$$D = \underline{\qquad\qquad} = \underline{\qquad\qquad} \text{ mly}$$

So, which galaxy is farther from Earth, A or B? Explain your choice.

WRITING ACTIVITY

Science Log

How would you like to take a trip through the universe and visit other parts of the Milky Way Galaxy, the neighboring Magellanic Clouds, and the Andromeda Galaxy? Research these terms online at www.conceptsandchallenges.com. Write a story for a travel magazine. Tell others how to prepare for the trip and what to look for on the way.

3 Assess

Use the Individual Activity Rubric from the Teacher's Resources to assess students' Science Logs. Fill in the rubric with the additional information below. For this assignment, students should have:

- accurately, but imaginatively, explained some of the differences among the galaxies discussed in the text.
- communicated, using a typical travel-magazine format, information about different galaxies.

Students may want to use desktop publishing software to create their Science Log travel article. See Tips for Using Technology in the Teaching Options on this page for more information.

4 Close

Have students work in pairs to find out why Hubble's constant is also called the law of red shifts.

Tips for Using Technology

Using Desktop Publishing Software to Create a Travel Article

Desktop publishing software allows users to combine text and art on a page. Most word processors include tools for desktop publishing. Students can use these tools to format their text, make special borders, and plan a page layout. In addition, students can bring art into a document. Clip art is art already drawn. Students can also use a scanner to incorporate their own art into their documents. Have students use the Thumbnail Sketch graphic organizer from the Teacher's Resources to plan their layouts.

• TEACHING RESOURCES •
Online or Print

Teacher's Resources

Big Idea: Chapter 18, p. 15

Big Idea Planner: Study Tools, pp. 11–12

Big Idea Science Log Entry: Study Tools, p. 13

Individual Activity Rubric: Teaching Tools, p. 5

Thumbnail Sketch: Study Tools, p. 24

Challenges (pp. 462–464)

Chapter Summary

Review Before students begin the Challenges, have pairs of students quiz each other on the boldface terms and any other unfamiliar words in the chapter summary.

STUDY HINT Have students refer to the chapter summary to help them answer the questions in the Challenges.

Key Term Challenges

MATCHING

1. photosphere (18-6)
2. constellation (18-9)
3. nebula (18-1)
4. magnitude (18-3)
5. sunspots (18-7)
6. star cluster (18-1)
7. white dwarf (18-4)
8. nucleus (18-8)
9. spectrograph (18-2)
10. irregular galaxy (18-10)

IDENTIFYING WORD RELATIONSHIPS

11. Both terms refer to the death of a star. In a nova, a star blows off its outer shell. In a supernova, the star blows itself completely apart. (18-5)
12. A protostar is an early stage in star formation. When nuclear reactions begin and the object begins to shine with its own light, it has become a true star. (18-1)
13. A binary star is a star system with two stars revolving around each other. A star cluster is a group of stars that travel together through space. (18-1)
14. A red giant is a star that has reached maturity and its full size. A white dwarf is a red giant star that has passed its maturity and is dying. (18-4)
15. An elliptical galaxy is oval-shaped. A spiral galaxy is pinwheel-shaped and has spiral arms. (18-10)

Chapter 18 Challenges

Chapter Summary

Lesson 18-1

- Most stars are hot, glowing balls of hydrogen and helium gases. The Sun is one of many billions of stars. A star forms from a **protostar** in a cloudy **nebula.**

Lessons 18-2 and 18-3

- **Spectroscopy** is the study of light from objects in space. A **spectrum** helps us identify a star's elements.
- The measure of a star's brightness is its **magnitude.** There is apparent magnitude and absolute magnitude.

Lessons 18-4 and 18-5

- An H-R diagram shows how a star's absolute magnitude is related to its surface temperature.
- Most stars, like the Sun, are **main sequence stars** on the H-R diagram.
- **Red giants** and **supergiants** are huge, bright stars. **White dwarfs** are small, hot stars.
- Stars begin life as protostars. Then, they enter the main sequence stage. Small and medium-size stars become white dwarfs, then black dwarfs. Very massive stars explode as **supernovas** and become neutron stars or black holes.

Lessons 18-6, 18-7, and 18-8

- The Sun has a **core** and an atmosphere. The Sun's atmosphere consists of the **photosphere,** the **chromosphere,** and the **corona.**
- **Sunspots** are dark, cooler areas on the Sun's surface. **Prominences** are streams of gas that arch above the Sun's surface. **Solar flares** send out energy and charged particles.
- The Sun has nuclear reactions called **fusion.** During fusion, hydrogen atoms combine to form helium.

Lessons 18-9, 18-10, and 18-11

- **Constellations** are groups of stars that form patterns in the sky.
- **Galaxies** are **elliptical, irregular,** or **spiral** in shape. The Milky Way Galaxy is a spiral galaxy.
- The Milky Way band of stars is in the Milky Way Galaxy.
- The universe is all that exists. **Black holes** pull in even light. **Quasars** emit tremendous energy.

Key Term Challenges

Big Bang (p. 458)	photosphere (p. 446)
binary stars (p. 436)	prominence (p. 448)
black hole (p. 458)	protostar (p. 436)
chromosphere (p. 446)	quasar (p. 458)
constellation (p. 452)	red giant (p. 442)
core (p. 446)	solar flare (p. 448)
corona (p. 446)	spectrograph (p. 438)
elliptical galaxy (p. 454)	spectroscopy (p. 438)
fusion (p. 450)	spectrum (p. 438)
galaxy (p. 454)	spiral galaxy (p. 454)
irregular galaxy (p. 454)	star (p. 436)
magnitude (p. 440)	star cluster (p. 436)
main sequence star (p. 442)	sunspot (p. 448)
nebula (p. 436)	supergiant (p. 442)
nova (p. 444)	supernova (p. 444)
nucleus (p. 450)	white dwarf (p. 442)

MATCHING **Write the Key Term from above that best matches each description.**

1. inner layer of the Sun's atmosphere
2. grouping of stars that form a pattern in the sky
3. cloud of dust and gas in space
4. measure of a star's brightness
5. dark, cooler areas on the Sun's surface
6. large group of stars in a galaxy
7. very small, hot star
8. center, or core, of an atom
9. device that measures the spectrum of an object
10. galaxy with no definite shape

IDENTIFYING WORD RELATIONSHIPS **Explain how the words in each pair are related. Write your answers in complete sentences.**

11. nova, supernova
12. star, protostar
13. binary star, star cluster
14. red giant, white dwarf
15. elliptical galaxy, spiral galaxy

Teaching Options

PREPARING STUDENTS FOR STANDARDIZED TESTS

Reading Strategy: Tell students to carefully read each question and each answer before deciding on a response in a multiple-choice test.

Writing Strategy: Tell students that, before they compose an answer to any written response question, they should organize their thoughts by writing key terms to use.

Interpreting Visuals Strategy: Remind students to read all of the labels associated with an illustration or photograph and to identify to which part of the diagram the labels are pointing.

ESL/ELL STRATEGY

Improving Communications Skills Working with peers encourages English-language learners to communicate. Have students work in pairs. One student should make up three questions about the chapter and read them aloud to the other student. The partner should answer the questions. Then, partners should reverse roles. Each pair should check each other to make sure that the correct answers were given.

Content Challenges TEST PREP

MULTIPLE CHOICE Write the letter of the term or phrase that best completes each statement.

1. The Sun is only one of billions of
   - **a.** nebulas.
   - **b.** stars.
   - **c.** galaxies.
   - **d.** constellations.
2. In Einstein's equation $E = mc^2$, c is
   - **a.** energy.
   - **b.** mass.
   - **c.** the speed of light.
   - **d.** the amount of matter.
3. The Milky Way Galaxy is
   - **a.** a spiral galaxy.
   - **b.** an elliptical galaxy.
   - **c.** an irregular galaxy.
   - **d.** a spherical galaxy.
4. The hottest stars are
   - **a.** red.
   - **b.** blue.
   - **c.** yellow.
   - **d.** orange.
5. The Sun's energy is produced in the
   - **a.** corona.
   - **b.** photosphere.
   - **c.** chromosphere.
   - **d.** core.
6. Supernova 1987A occurred in
   - **a.** the Milky Way Galaxy.
   - **b.** the Large Magellanic Cloud.
   - **c.** the Andromeda Galaxy.
   - **d.** the star cluster M13.
7. The actual brightness of a star is its
   - **a.** apparent magnitude.
   - **b.** first magnitude.
   - **c.** sixth magnitude.
   - **d.** absolute magnitude.
8. The spinning of the Sun on its axis causes
   - **a.** solar flares.
   - **b.** sunspot movement.
   - **c.** prominences.
   - **d.** fusion.
9. The outermost layer of the Sun's atmosphere is the
   - **a.** photosphere.
   - **b.** chromosphere.
   - **c.** corona.
   - **d.** core.
10. Betelgeuse is a star found in
    - **a.** the Big Dipper.
    - **b.** the Little Dipper.
    - **c.** Ursa Major.
    - **d.** Orion.

TRUE/FALSE Write *true* if the statement is true. If the statement is false, change the underlined term to make the statement true.

11. The gravity of a black <u>dwarf</u> is so great that nothing can escape.
12. The main features of the universe are <u>galaxies</u>.
13. The most common element in most stars is <u>helium</u>.
14. Sunspots are dark, <u>cooler</u> areas on the surface of the Sun.
15. The brightness of most stars increases as the star's surface temperature <u>decreases</u>.
16. The Sun and all the planets in the solar system are in the <u>Milky Way Galaxy</u>.
17. Each <u>element</u> gives off its own spectrum of light.
18. To separate white light into the colors of the rainbow, scientists can use a <u>prism</u>.
19. Galaxies with no regular shape are called <u>elliptical</u> galaxies.
20. Streams of gas shooting from the Sun's surface are called <u>prominences</u>.

Content Challenges

MULTIPLE CHOICE

1. b (18-1)
2. c (18-8)
3. a (18-10)
4. b (18-2)
5. d (18-6)
6. b (18-5)
7. d (18-3)
8. b (18-7)
9. c (18-6)
10. d (18-9)

TRUE/FALSE

11. false; hole (18-11)
12. true (18-10)
13. false; hydrogen (18-1)
14. true (18-8)
15. false; increases (18-4)
16. true (18-10)
17. true (18-2)
18. true (18-2)
19. false; irregular (18-10)
20. true (18-7)

ALTERNATIVE ASSESSMENT

Relating Terms Have students write a Science Log entry that explains the relationship among the following terms and phrases: *the Sun*, *stars*, *H-R diagram*, *fission*, and *a star's life cycle*.

CUMULATIVE ASSESSMENT

Assessing Students' Progress You may wish to administer the final examination from the Teacher's Resources at this point. Distribute copies of the Scantron Sheet, also from the Teacher's Resources, for students to record their answers.

PORTFOLIO ASSESSMENT

Making Student Portfolios Students can demonstrate their comprehension of the concepts in this chapter by making a portfolio. It is important that students include work that they are proud of or that they have put great effort into completing. The Chapter Self-Check, the Big Idea Planner, and the Weekly Journal are some of the reproducibles in the Teacher's Resources that they can include in their portfolios. They can also include the travel magazine article from the Big Idea Science Log. You can use the Portfolio Assessment Rubric also found in the Teacher's Resources to assess students' portfolios.

Concept Challenges (see p. T464)

WRITTEN RESPONSE

1. **COMPARE:** Novas and supernovas are both explosions that create a lot of light. (18-5)
2. **EXPLAIN:** Gravity contracts the nebula so that the material in the center becomes a protostar. (18-1)
3. **INFER:** Possible answer: Nuclear fusion in the core builds up heat in the atmosphere. When a certain amount of energy accumulates, the gases in the atmosphere expand explosively and shoot out from the Sun's surface. (18-6, 18-7)

Concept Challenges (cont.)

4. INFER: Gravity holds stars together in a galaxy. (18-10)
5. INFER: Nuclear fusion in the Sun's core produces energy in the form of heat and light. (18-8)
6. ANALYZE: The particles are from the photosphere and the chromosphere. (18-6, 18-7)
7. CONTRAST: A supergiant is larger than a red giant star. (18-4)
8. INFER: The core is the densest part of the Sun. (18-6)
9. ANALYZE: Astronomers use a spectrograph, which contains a prism that separates light into its characteristic colors, to find out the chemical makeup of a star. (18-2)
10. INFER: A star is born when gravity causes dust and gases to collapse and form an enormous, glowing sphere. A star dies when its source of fuel becomes exhausted. (18-5)
11. INFER: Fossil fuels are the remains of once-living things. While they were alive, the organisms (plants and other organisms containing chlorophyll) either produced their own food from sunlight or ate plants and animals that stored the sunlight as food. (18-8)
12. ANALYZE: During the life cycle of a star, hydrogen is converted to helium via nuclear processes. As a star ages, the amount of hydrogen decreases and the amount of helium increases. (18-5)

INTERPRETING VISUALS (18-6)

13. the core
14. the photosphere, chromosphere, and corona
15. 2,000,000°C
16. 0°C to 2,000°C

Chapter Project Wrap-Up

LIFE CYCLE OF A STAR MOBILE

Review the life cycle of a star (Lesson 18-5) and relate it to the mobile they have created. If students have questions about a specific stage in the formation of a star, ask the group that produced the model of that stage to explain. You can use the Individual Activity Rubric found in the Teacher's Resources to assess students' projects. Fill in the rubric with the additional information below. For this project, students should have:

- created a workable model of part of the mobile.
- described that part of the mobile for the report.

Concept Challenges TEST PREP

WRITTEN RESPONSE **Answer each of the following questions in complete sentences.**

1. COMPARE: How is a nova similar to a supernova?
2. EXPLAIN: What happens to material in the center of a nebula?
3. INFER: What forces cause streams of gases to shoot high above the Sun's surface?
4. INFER: What holds the stars together in a galaxy?
5. INFER: How does the Sun produce energy?
6. ANALYZE: Where are the electrically charged particles that stream out from the Sun created?
7. CONTRAST: What is the difference between a red giant star and a supergiant star?
8. INFER: Which is the densest part of the Sun?
9. ANALYZE: How do astronomers find out the chemical makeup of a star?
10. INFER: What is meant by the "birth" and "death" of a star?
11. INFER: Why can fossil fuels be called "stored sunlight"?
12. ANALYZE: Why is an old main sequence star made up of a larger percentage of helium than a young main sequence star?

INTERPRETING VISUALS **Use Figure 18-37 to answer the following questions.**

13. What is the center part of the Sun called?
14. What three layers make up the Sun's atmosphere?
15. What is the temperature of the corona?
16. About how many degrees Celsius hotter is the chromosphere than the photosphere?

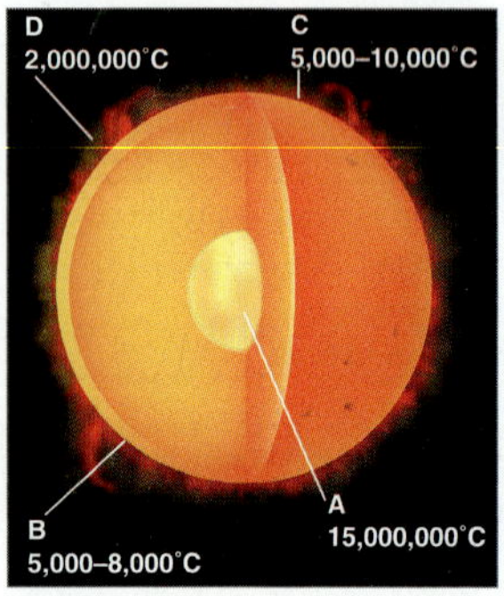

▲ Figure 18-37

464

Teaching Options

• TEACHING RESOURCES •
Online or Print

Teacher's Resources

Key Term Review: Chapter 18, p. 16

Chapter 18 Test: pp. 17–19

Individual Activity Rubric: Teaching Tools, p. 5

Chapter Self-Check: Teaching Tools, p. 11

Big Idea Planner: Study Tools, pp. 11–12

Portfolio Assessment Rubric: Teaching Tools, p. 10

• TEACHING RESOURCES •
Online or Print

Teacher's Resources

Scantron Sheet: Teaching Tools, p. 12

Appendix A Metric System

The Metric System and SI Units

The metric system is an international system of measurement based on units of ten. More than 90% of the nations of the world use the metric system. In the United States, both the English system and the metric system are used.

The *Système International*, or SI, has been used as the international measurement system since 1960. The SI is a modernized version of the metric system. Like the metric system, the SI is a decimal system based on units of ten. When you want to change from one unit in the metric system to another unit, you multiply or divide by a multiple of ten.

- When you change from a smaller unit to a larger unit, you divide.
- When you change from a larger unit to a smaller unit, you multiply.

METRIC UNITS		
LENGTH	SYMBOL	RELATIONSHIP
kilometer	km	1 km = 1,000 m
meter	m	1 m = 100 cm
centimeter	cm	1 cm = 10 mm
millimeter	mm	1 mm = 0.1 cm
AREA	SYMBOL	
square kilometer	km^2	1 km^2 = 1,000,000 m^2
square meter	m^2	1 m^2 = 1,000,000 mm^2
square centimeter	cm^2	1 cm^2 = 0.0001 m^2
square millimeter	mm^2	1 mm^2 = 0.000001 m^2
VOLUME	SYMBOL	
cubic meter	m^3	1 m^3 = 1,000,000 cm^3
cubic centimeter	cm^3	1 cm^3 = 0.000001 m^3
liter	L	1 L = 1,000 mL
milliliter	mL	1 mL = 0.001 L
MASS	SYMBOL	
metric ton	t	1 t = 1,000 kg
kilogram	kg	1 kg = 1,000 g
gram	g	1 g = 1,000 mg
centigram	cg	1 cg = 10 mg
milligram	mg	1 mg = 0.001 g
TEMPERATURE	SYMBOL	
Kelvin	K	
degree Celsius	°C	

▲ Figure 1

COMMON METRIC PREFIXES			
micro-	0.000001 or 1/1,000,000	deka-	10
milli-	0.001 or 1/1,000	hecto-	100
centi-	0.01 or 1/100	kilo-	1,000
deci-	0.1 or 1/10	mega-	1,000,000

▲ Figure 2

METRIC-STANDARD EQUIVALENTS	
SI to English	English to SI
LENGTH	
1 kilometer = 0.621 mile (mi)	1 mi = 1.61 km
1 meter = 1.094 yards (yd)	1 yd = 0.914 m
1 meter = 3.28 feet (ft)	1 ft = 0.305 m
1 centimeter = 0.394 inch (in.)	1 in. = 2.54 cm
1 millimeter = 0.039 inch	1 in. = 25.4 mm
AREA	
1 square kilometer = 0.3861 square mile	1 mi^2 = 2.590 km^2
1 square meter = 1.1960 square yards	1 yd^2 = 0.8361 m^2
1 square meter = 10.763 square feet	1 ft^2 = 0.0929 m^2
1 square centimeter = 0.155 square inch	1 $in.^2$ = 6.452 cm^2
VOLUME	
1 cubic meter = 1.3080 cubic yards	1 yd^3 = 0.7646 m^3
1 cubic meter = 35.315 cubic feet	1 ft^3 = 0.0283 m^3
1 cubic centimeter = 0.0610 cubic inch	1 $in.^3$ = 16.39 cm^3
1 liter = 0.2642 gallon (gal)	1 gal = 3.79 L
1 liter = 1.06 quarts (qt)	1 qt = 0.946 L
1 liter = 2.11 pints (pt)	1 pt = 0.47 L
1 milliliter = 0.034 fluid ounce (fl oz)	1 fl oz = 29.57 mL
MASS	
1 metric ton = 0.984 ton	1 ton = 1.016 t
1 kilogram = 2.205 pounds (lb)	1 lb = 0.4536 kg
1 gram = 0.0353 ounce (oz)	1 oz = 28.35 g
TEMPERATURE	
Celsius = 5/9(°F – 32)	Fahrenheit = 9/5°C + 32
0°C = 32°F (Freezing point of water)	72°F = 22°C (Room temperature)
100°C = 212°F (Boiling point of water)	98.6°F = 37°C (Human body temperature)
Kelvin = (°F + 459.67)/1.8	Fahrenheit = (K × 1.8) – 459.67

▲ Figure 3

Appendix B Map of the United States

▲ Figure 4

Appendix

A
Lake Superior
Superior Upland
WISCONSIN
St. Paul
Wisconsin River
Milwaukee
Madison
Lake Michigan
Green Bay
Straits of Mackinac
Lake Huron
MICHIGAN
Lansing
Lake St. Clair
Detroit
Chicago
Gary
Des Moines
Illinois River
Wabash River
Central Plains
ILLINOIS
Springfield
INDIANA
Indianapolis
Cincinnati
Missouri River
St. Louis
Jefferson City
MISSOURI
Ozark Plateau
ARKANSAS
Little Rock
Ohio River
KENTUCKY
Frankfort
Cleveland
OHIO
Columbus
Scioto R.
Lake Erie
Pittsburgh
Lake Ontario
Buffalo
Finger Lakes
Allegheny Plateau
Allegheny R.
PENNSYLVANIA
Harrisburg
Susquehanna R.
NEW YORK
Adirondack Mts.
Albany
Lake Champlain
Montpelier
VT.
N.H.
Concord
Connecticut R.
Hudson River
St. Lawrence River
MAINE
Penobscot R.
Augusta
Penobscot Bay
Boston
MASS.
Cape Cod
Providence
R.I.
Nantucket
Martha's Vineyard
Hartford
CONN.
Long Island Sound
Long Island
New York
Trenton
N.J.
Philadelphia
DELAWARE
Delaware Bay
Dover
Baltimore
Washington, D.C.
Annapolis
MARYLAND
WEST VIRGINIA
Charleston
Kanawha R.
VIRGINIA
James R.
Richmond
Norfolk
Roanoke R.
Appalachian Mountains
Blue Ridge
Cumberland Plateau
Cumberland R.
Nashville
TENNESSEE
Tennessee River
Memphis
Mississippi River
Pee Dee R.
Raleigh
NORTH CAROLINA
Hartwell Lake
SOUTH CAROLINA
Columbia
Savannah River
Altamaha R.
Charleston
Savannah
Atlantic Coastal Plains
Atlanta
GEORGIA
Chattahoochee R.
Birmingham
ALABAMA
Alabama R.
Tombigbee R.
Montgomery
MISSISSIPPI
Jackson
Pearl R.
LOUISIANA
Baton Rouge
Lake Pontchartrain
New Orleans
Mobile
Mobile Bay
Pensacola Bay
Tallahassee
Mississippi Delta
Galveston Bay
FLORIDA
Cape Canaveral
Tampa Bay
Lake Okeechobee
The Everglades
Miami
Cape Sable
Florida Keys
Straits of Florida
CUBA
BAHAMAS
ATLANTIC OCEAN
Gulf of Mexico
N
65°W
45°N
40°N
25°N
75°W
85°W
0
200
400 mi.
0
200
400 km
National capital
State capital
Major city
Mountain peak
Elevation
13,124 ft 4,000 m
6,562 ft 2,000 m
4,921 ft 1,500 m
3,281 ft 1,000 m
1,640 ft 500 m
820 ft 250 m
0 0
ATLANTIC OCEAN
San Juan
PUERTO RICO
CARIBBEAN SEA
18°N
67°W
66°W

Appendix C Science Terms

Analyzing Science Terms

You can often unlock the meaning of an unfamiliar science term by analyzing its word parts. Prefixes and suffixes, for example, each carry a meaning that comes from a word root. This word root usually comes from the Latin or Greek language. The following list of prefixes and suffixes provides clues to the meaning of many science terms.

WORD PART	MEANING	EXAMPLE
astr-, aster-	star	astronomy
bar-, baro-	weight, pressure	barometer
batho-, bathy-	depth	batholith, bathysphere
circum-	around	circum-Pacific, circumpolar
-cline	lean, slope	anticline, syncline
eco-	environment	ecology, ecosystem
epi-	on	epicenter
ex-, exo-	out, outside of	exosphere, exfoliation, extrusion
geo-	earth	geode, geology, geomagnetic
-graph	write, writing	seismograph
hydro-	water	hydrosphere
hypo-	under	hypothesis
iso-	equal	isoscope, isostasy, isotope
-lith, -lithic	stone	Neolithic, regolith
magn-	great, large	magnitude
mar-	sea	marine
meso-	middle	mesosphere, Mesozoic
meta-	among, change	metamorphic, metamorphism
micro-	small	microquake
-morph, -morphic	form	metamorphic
neo-	new	Neolithic
paleo-	old	paleontology, Paleozoic
ped-, pedo-	ground, soil	pediment
peri-	around	perigee, perihelion
-ose	carbohydrate	glucose, cellulose
seism-, seismo-	shake, earthquake	seismic, seismograph
sol-	sun	solar, solstice
spectro-	look at, examine	spectroscope, spectrum
-sphere	ball, globe	hemisphere, lithosphere
strati-, strato-	spread, layer	stratification, stratovolcano
terra-	earth, land	terracing, terrane
thermo-	heat	thermosphere
top-, topo-	place	topographic
tropo-	turn, respond to	tropopause, troposphere

▲ Figure 5

Appendix D Guide to Common Rocks and Minerals

COMMON ROCKS		
Rock Name	**Rock Type**	**Comments**
Anthracite coal	Sedimentary	Dark brown to black in color; shiny, hard, scaly, and dense; used as fuel; also used in producing coal gas and in the iron, steel, synthetic rubber, and dye industries
Basalt	Igneous	Fine-grained; dark in color; often rings like a bell when struck with a hammer; used for road paving
Compact limestone	Sedimentary	Fine-grained, soft, and porous; normally pale in color; used in building stones and in making lime
Conglomerate	Sedimentary	Very colorful; texture varies; consists of pebbles cemented together; often contains useful mineral grains; used as a building stone
Gneiss	Metamorphic	Coarse-grained; layers of different minerals often give a banded appearance; some varieties used as building material, both rough and polished
Granite	Igneous	Coarse-grained; mostly light in color, in shades of pink, gray, and white; important building stone, either polished or rough
Marble	Metamorphic	Coarse-grained; reacts with acid; color varies; often patterned; important building stone
Mica schist	Metamorphic	Medium- to coarse-grained; sparkly gray to black in color; contains important minerals
Obsidian	Igneous	Light to dark in color; glassy luster; sometimes translucent; brittle; used industrially as raw material; in ancient times was used for tools and sculptures
Pumice	Igneous	Light to dark in color; spongy appearance; many holes; lightweight, may float in water; used as building insulation
Sandstone	Sedimentary	Coarse-grained; usually sand color; often contains deposits of oil and gas; used for building
Shale	Sedimentary	Microscopic grains; clay composition; smooth surface, hardened mud appearance; some varieties, when processed, yield oil
Slate	Metamorphic	Find-grained; cleaves into thin flat plates; shiny dark gray color; used as shingles for roofing, for flooring, and blackboards

▲ **Figure 6**

Appendix D Guide to Common Rocks and Minerals (continued)

MINERALS			
Mineral	**Chemical Formula**	**Color**	**Streak**
Apatite	$Ca_5(F,Cl)(PO_4)_3$	Green, brown, red	White
Augite	$(Ca,Na)(Mg, Fe, Al)(Si,Al)_2O_6$	Dark green, black	Green to gray
Beryl	$Be_3Al_2Si_6O_{18}$	White, yellow, blue, green	White
Biotite (Mica)	Complex substance containing Fe, Mg, Si, O and other elements	Black, brown, dark green	White to light brown
Calcite	$CaCO_3$	Gray, white	White
Chalcopyrite	$Cu_2Fe_2S_4$	Brass, yellow	Greenish black
Copper	Cu	Copper red to black	Copper red
Corundum	Al_2O_3	Various colors	White
Diamond	C	Colorless to black	Colorless
Dolomite	$CaMg(CO_3)_2$	Pink, white, gray, green, brown, black	White
Feldspar	$(K, NA, CA)(AlSi_3O_8)$	Colorless, white, various colors	Colorless, white
Fluorite	CaF_2	Light green yellow, bluish green, and other colors	White
Galena	PbS	Lead gray	Lead gray
Gold	Au	Gold	Gold
Graphite	C	Black to gray	Black
Gypsum	$CaSO_4 \cdot 2H_2O$	Whitish gray	White
Halite	NaCl	Colorless and various colors	Colorless
Hematite	Fe_2O_3	Reddish brown to black	Light to dark red
Hornblende	Complex substance containing Ca, Na, Mg, Ti, and Al	Dark green, black, brown	Colorless
Magnetite	Fe_3O_4	Iron black	Black
Olivine	$(Mg,Fe)_2SiO_4$	Olive green	White
Pyrite	FeS_2	Brass, yellow	Greenish, brownish, black
Quartz	SiO_2	Colorless, white, any color when not pure	White
Serpentine	$Mg_3Si_2O_5(OH)_4$	Green, yellow, brown, or black	White
Silver	Ag	Silver, black	Silver
Sphalerite	ZnS	Brown, black, red, and other colors	White, yellow, brown
Sulfur	S	Yellow	White
Talc	$Mg_3Si_4O_{10}(OH)_2$	Gray, greenish white	White

▲ Figure 7

Appendix D Guide to Common Rocks and Minerals (continued)

MINERALS				
Mineral	**Luster***	**Hardness**	**Specific Gravity**	**Fracture/Cleavage**
Apatite	Vitreous	5	3.1–3.2	Fracture
Augite	Vitreous	5–6	3.2–3.6	Cleavage
Beryl	Vitreous	8	2.65–2.90	Fracture
Biotite (Mica)	Vitreous	2.5–6	2.8–3.2	Cleavage
Calcite	Vitreous	3	2.7	Cleavage
Chalcopyrite	Metallic	3.5–4	4.1–4.34	Fracture
Copper	Metallic	2.5–3	8.5–8.93	None
Corundum	Vitreous to adamantine	9	4.02	Fracture
Diamond	Adamantine	10	3.52	Cleavage
Dolomite	Vitreous to pearly	3.5–4	2.85	Cleavage
Feldspar	Vitreous	6	2.55–2.75	Cleavage
Fluorite	Vitreous	4	3.18	Cleavage
Galena	Metallic	2.5	7.4–7.6	Cleavage
Gold	Metallic	2.5–3	15.3–19.3	None
Graphite	Metallic	1–2	2.3	Cleavage
Gypsum	Silky	2	2.3–2.4	Cleavage
Halite	Vitreous	2.5	2.1–2.6	Cleavage
Hematite	Metallic	5–6	5.26	Fracture
Hornblende	Vitreous	5–6	3.2	Cleavage
Magnetite	Metallic	6	5.18	Fracture
Olivine	Vitreous	6.5–7	3.2–4.2	Fracture
Pyrite	Metallic	6–6.5	4.9–5.2	Fracture
Quartz	Vitreous	7	2.65	Fracture
Serpentine	Vitreous	2–5	2.5–3.2	Fracture
Silver	Metallic	2.5–3	10.0–12.0	None
Sphalerite	Metallic, submetallic, adamantine, resinous	3.5–4	3.9–4.1	Cleavage
Sulfur	Adamantine	1.5–2.5	2.0–2.1	Cleavage
Talc	Vitreous	1	2.7–2.8	Cleavage

**Luster is the appearance of the light reflected from the surface of a mineral. This is another aid to identifying minerals. Adamantine describes a mineral that has a brilliant diamondlike luster. Vitreous describes a mineral that has a glassy luster.*

▲ Figure 8

Appendix E Periodic Table of Elements

	Group 1	Group 2	Group 3	Group 4	Group 5	Group 6	Group 7	Group 8	Group 9
Period 1	1 H Hydrogen 1.008								
Period 2	3 Li Lithium 6.941	4 Be Beryllium 9.012							
Period 3	11 Na Sodium 22.990	12 Mg Magnesium 24.312							
Period 4	19 K Potassium 39.098	20 Ca Calcium 40.080	21 Sc Scandium 44.956	22 Ti Titanium 47.900	23 V Vanadium 50.942	24 Cr Chromium 51.996	25 Mn Manganese 54.938	26 Fe Iron 55.847	27 Co Cobalt 58.933
Period 5	37 Rb Rubidium 85.470	38 Sr Strontium 87.620	39 Y Yttrium 88.905	40 Zr Zirconium 91.220	41 Nb Niobium 92.906	42 Mo Molybdenum 95.940	43 Tc Technetium (97.906)	44 Ru Ruthenium 101.070	45 Rh Rhodium 102.905
Period 6	55 Cs Cesium 132.905	56 Ba Barium 137.340	57 La Lanthanum 138.910	72 Hf Hafnium 178.490	73 Ta Tantalum 180.948	74 W Tungsten 183.850	75 Re Rhenium 186.200	76 Os Osmium 190.200	77 Ir Iridium 192.200
Period 7	87 Fr Francium (223.000)	88 Ra Radium (226.025)	89 Ac Actinium (227.000)	104 Rf Rutherfordium (260.000)	105 Db Dubnium (260.000)	106 Sg Seaborgium (263.000)	107 Bh Bohrium (262.000)	108 Hs Hassium (265.000)	109 Mt Meitnerium (266.000)

58 Ce Cerium 140.120	59 Pr Praseodymium 140.907	60 Nd Neodymium 144.240	61 Pm Promethium (145.000)	62 Sm Samarium 150.350
90 Th Thorium 232.038	91 Pa Protactinium 231.036	92 U Uranium 238.029	93 Np Neptunium (237.048)	94 Pu Plutonium 244.000

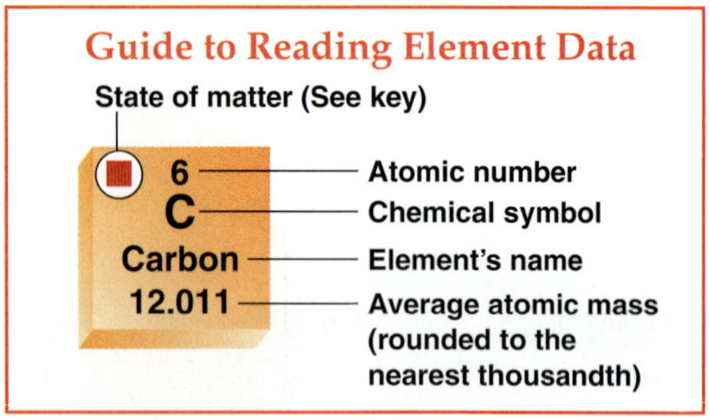

KEY

- Metal
- Metalloid
- Nonmetal
- To be discovered
- Gas
- Liquid
- Solid
- Synthetic elements

▲ Figure 9

Appendix

Group 10	Group 11	Group 12	Group 13	Group 14	Group 15	Group 16	Group 17	Group 18
								2 **He** Helium 4.003
			5 **B** Boron 10.811	6 **C** Carbon 12.011	7 **N** Nitrogen 14.007	8 **O** Oxygen 15.999	9 **F** Fluorine 18.998	10 **Ne** Neon 20.183
			13 **Al** Aluminum 26.982	14 **Si** Silicon 28.086	15 **P** Phosphorus 30.974	16 **S** Sulfur 32.064	17 **Cl** Chlorine 35.453	18 **Ar** Argon 39.948
28 **Ni** Nickel 58.710	29 **Cu** Copper 63.540	30 **Zn** Zinc 65.370	31 **Ga** Gallium 69.720	32 **Ge** Germanium 72.590	33 **As** Arsenic 74.922	34 **Se** Selenium 78.960	35 **Br** Bromine 79.909	36 **Kr** Krypton 83.800
46 **Pd** Palladium 106.400	47 **Ag** Silver 107.870	48 **Cd** Cadmium 112.400	49 **In** Indium 114.820	50 **Sn** Tin 118.690	51 **Sb** Antimony 121.750	52 **Te** Tellurium 127.600	53 **I** Iodine 126.904	54 **Xe** Xenon 131.300
78 **Pt** Platinum 195.090	79 **Au** Gold 196.967	80 **Hg** Mercury 200.590	81 **Tl** Thallium 204.370	82 **Pb** Lead 207.200	83 **Bi** Bismuth 208.980	84 **Po** Polonium (209.000)	85 **At** Astatine (210.000)	86 **Rn** Radon (222.000)
110 **Ds** Ununnilium (269)	111 **Rg** Unununium (272)	112 **Uub** Ununbium (277)	113 **Uut** (284)	114 **Uuq** Ununquadium (296)	115 **Uup** (288)	116 **Uuh** (298)	117 **Uus** (not yet recognized)	118 **Uuo** (294)

63 **Eu** Europium 151.960	64 **Gd** Gadolinium 157.250	65 **Tb** Terbium 158.924	66 **Dy** Dysprosium 162.500	67 **Ho** Holmium 164.930	68 **Er** Erbium 167.260	69 **Tm** Thulium 168.934	70 **Yb** Ytterbium 173.040	71 **Lu** Lutetium 174.970
95 **Am** Americium (243.000)	96 **Cm** Curium (247.000)	97 **Bk** Berkelium (247.000)	98 **Cf** Californium (251.000)	99 **Es** Einsteinium (254.000)	100 **Fm** Fermium (257.000)	101 **Md** Mendelevium (258.000)	102 **No** Nobelium (259.000)	103 **Lr** Lawrencium (262.000)

*Atomic masses in parentheses are of the most common form of the atom.

Appendix F Meteorological Scales

Saffir-Simpson Hurricane Scale

The Saffir-Simpson scale measures the wind speed of hurricanes. The category number tells the wind speed and amount of damage a hurricane can cause. Since the 1970s, the National Hurricane Center in the United States has used this scale to classify hurricanes. The scale was developed by Herbert Saffir, an engineer, and Robert Simpson, a former director of the National Hurricane Center. The National Hurricane Center is located on the campus of Florida International University, Miami, Florida. The Center is part of NOAA, the National Oceanic and Atmospheric Administration.

CATEGORY	WIND SPEED (mph)	DAMAGE
1	74–95	Minor
2	96–110	Moderate
3	111–130	Strong
4	131–155	Very Strong
5	More than 155	Devastating

◀ Figure 10

Beaufort Wind Speed Scale

This scale is named for the nineteenth-century British naval officer who devised it. The Beaufort Scale classifies wind speed according to its effects. It was originally used in 1806 as an aid for sailors. It has since been adapted for use on land. The scale has been modified through the years.

CODE	SPEED (MPH)	DESCRIPTION	EFFECTS ON LAND
0	Below 1	Calm	Smoke rises vertically
1	1–5	Light air	Smoke drifts slowly
2	6–11	Light breeze	Leaves rustle
3	12–19	Gentle breeze	Leaves and twigs move
4	20–28	Moderate breeze	Small branches move
5	29–38	Fresh breeze	Small trees sway
6	39–49	Strong breeze	Large branches sway
7	50–61	Near gale	Walking difficult
8	62–74	Gale	Twigs snap off trees
9	75–88	Strong gale	Minor damage
10	89–102	Whole gale	Significant damage
11	103–114	Storm	Widespread damage
12	Above 117	Hurricane	Widespread destruction

◀ Figure 11

The Fujita Intensity Scale

This scale provides a measure of the strength of a tornado. It was developed by tornado specialist Dr. Tetsuto Theodore Fujita, of the University of Chicago.

CATEGORY	WIND SPEED (mph)	DAMAGE
F0	40–71	Light
F1	72–112	Moderate
F2	113–157	Considerable
F3	158–206	Severe
F4	207–260	Devastating
F5	261–318	Incredible

◀ Figure 12

Appendix G Astronomy Tables

PLANETARY DATA				
Planet	Diameter (km)	Time for One Spin on Axis	Time for One Orbit of the Sun	Average Distance From the Sun (km)
Mercury	4,880	58.65 days	87.97 days	57.9 million
Venus	12,103.6	243.0 days	224.70 days	108.21 million
Earth	12,756	23.93 hours	365.26 days	149.6 million
Mars	6,794	24.66 hours	686.98 days	227.9 million
Jupiter	142,984	9.93 hours	11.86 years	778.57 million
Saturn	120,536	10.66 hours	29.4 years	1.43 billion
Uranus	51,118	17.24 hours	84.01 years	2.87 billion
Neptune	49,528	16.11 hours	164.78 years	4.5 billion

▲ Figure 13

KEY ROBOT PROBES TO PLANETS, COMETS, AND ASTEROIDS				
Probe	Type	Target	Encounter	Achievements
Pioneer 10	Flyby	Jupiter	1973	First to cross Asteroid Belt; took close-up photos of Jupiter
Mariner 10	Flyby	Mercury	1974–1975	Only probe to Mercury
Venera 9	Orbiter/lander	Venus	1975	First views of surface
Viking 1 and 2	Orbiters/landers	Mars	1976	Photographed surface; searched for life
Pioneer 11	Flyby	Saturn	1979	First detailed views
Voyager 1 and 2	Flybys	Jupiter Saturn Uranus Neptune	1979 1980–1981 1986 1989	Details of all four planetary systems; Voyager 2 was first probe to Uranus and Neptune
Giotto	Flyby	Comet	1986	First close-up view of nucleus
Galileo	Flyby/orbiter	Gaspra Jupiter	1991 and 1995	First flyby of an asteroid; first orbit of Jupiter
Mars Global Surveyor	Orbiter	Mars	1997	Mapped surface and searched for water
Mars Odyssey	Orbiter	Mars	2001	Mapped surface and searched for water
Cassini	Flyby/orbiter	Saturn	2004	Detailed studies of the planet and its rings

▲ Figure 14

SOLAR ECLIPSES THROUGH THE YEAR 2030		
Date	Duration of Totality (minutes)	Location
Nov. 23, 2003	4.4	Central America
April 8, 2005	0.7	South Pacific Ocean
March 29, 2006	4.1	Africa, Asia Minor, Russia
August 1, 2008	2.4	Arctic Ocean, Siberia, China
July 22, 2009	6.6	India, China, South Pacific
July 11, 2010	5.3	South America
Nov. 13, 2012	4.0	Australia, South America
Nov. 3, 2013	1.7	North and South America, Africa
March 20, 2015	2.78	Europe, Africa
March 9, 2016	4.15	Asia, Australia
August 11, 2018	2.7	Europe, Asia
July 2, 2019	4.5	Pacific Ocean, Chile, Argentina
Dec. 14, 2020	2.2	Pacific and Atlantic oceans, South America
Dec. 4, 2021	0.38	Antarctica, Southern Africa
April 20, 2023	1.3	Asia, Australia
April 8, 2024	4.5	North and Central America
August 12, 2026	2.3	North America, Europe
August 2, 2027	6.4	Africa, Asia
July 22, 2028	5.1	Asia, Australia
Nov. 25, 2030	3.7	Botswana, S. Africa, Australia

◀ Figure 15

Regularly Occurring Meteor Showers

Occasionally, 60 or more meteors flash across the sky in a single hour or two. These light displays are called meteor showers. They occur when Earth crosses a comet's orbit and groups of meteoroids from the comet's tail collide with Earth's atmosphere. The dates in the table below are peak dates. For most showers, the meteors can be seen for a few weeks, but the peak occurs somewhere around the middle of the time period.

METEOR SHOWERS		
Shower	Approximate Peak Dates	Associated Comet
Quadrantids	January 4–6	—
Lyrids	April 20–23	Thacher
Eta Aquarids	May 3–5	Halley's comet
Delta Aquarids	July 30	—
Perseids	August 12	Comet 1862 III
Draconids	October 7–10	Comet Giacobini-Zinner
Orionids	October 20	Halley's comet
Taurids	Nov. 3–13	Encke's Comet
Leonids	Nov. 18	Temple-Tuttle
Geminids	Dec. 4–16	3200 Phaeton

◀ Figure 16

Appendix H Star Charts

Spring Sky

To use this chart, hold it up in front of you and turn it so that the direction you are facing is at the bottom of the chart. This chart works best at 34° north latitude and at the following times: 10:00 P.M. on March 1; 9:00 P.M. on March 15; 8:00 P.M. on March 30. However, it can be referenced at other times and latitudes within the United States.

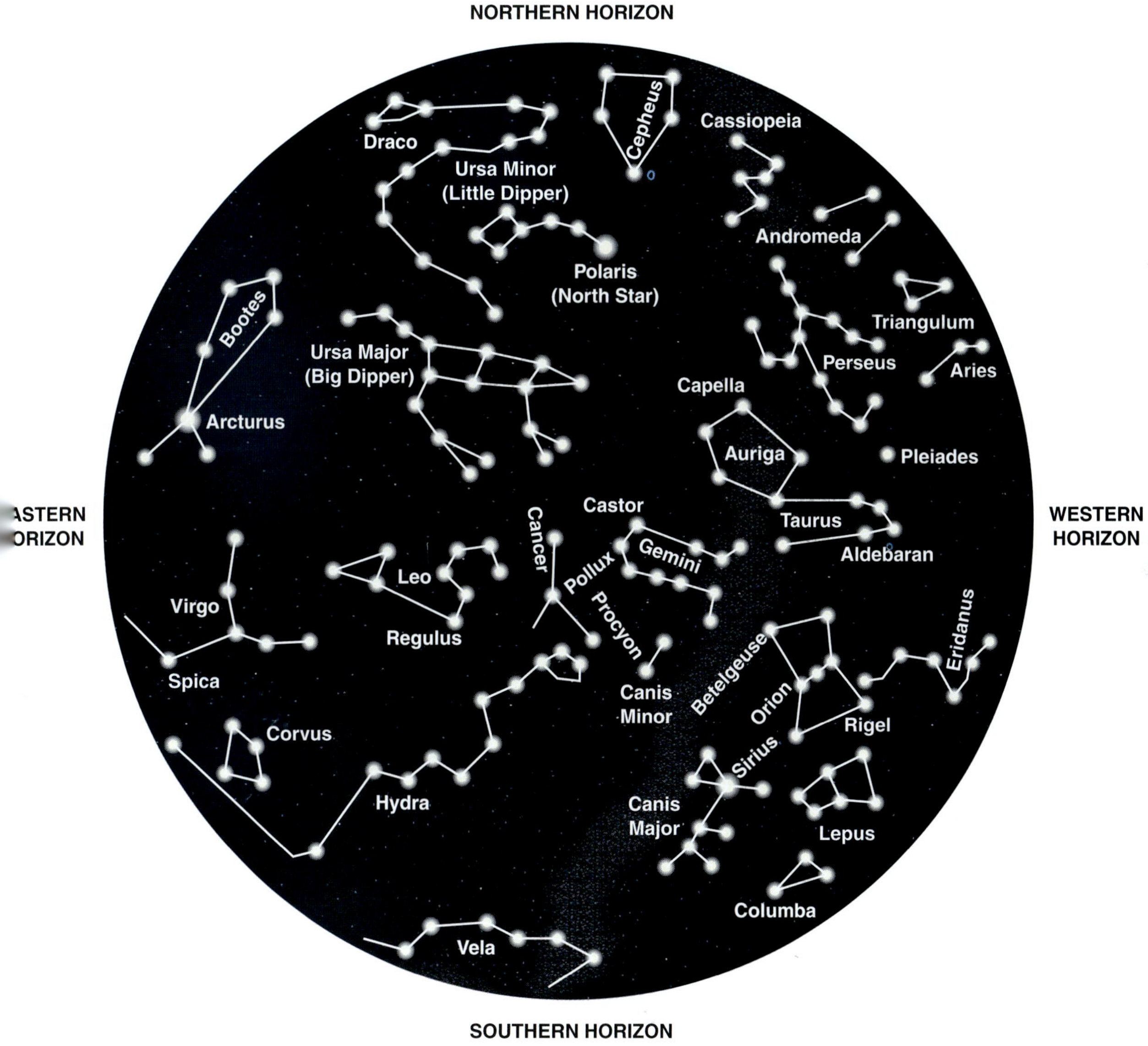

Summer Sky

To use this chart, hold it up in front of you and turn it so that the direction you are facing is at the bottom of the chart. This chart works best at 34° north latitude and at the following times: 10:00 P.M. on June 1; 9:00 P.M. on June 15; 8:00 P.M. on June 30. However, it can be used at other times and latitudes within the United States.

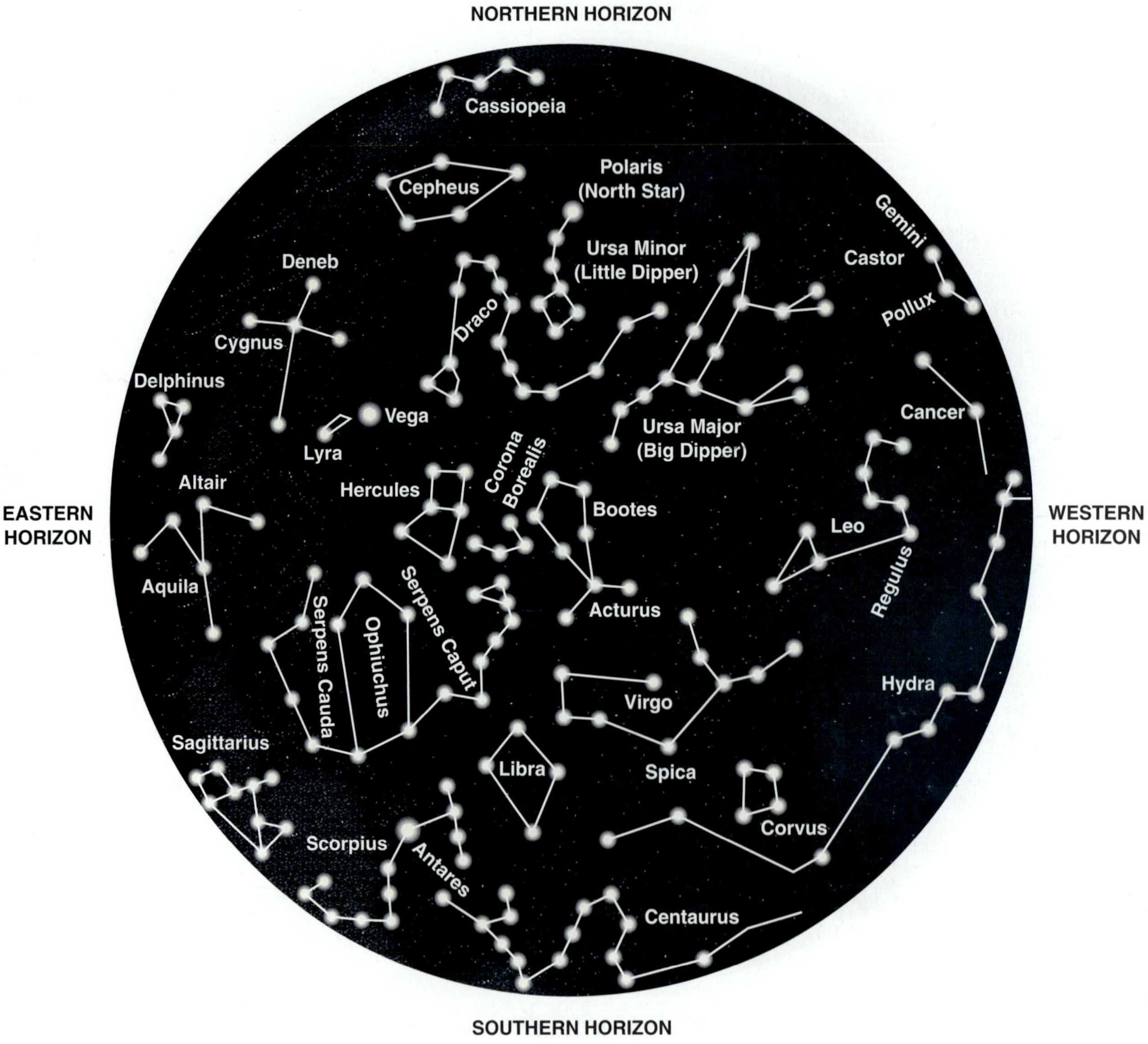

Appendix H Star Charts (continued)

Autumn Sky

To use this chart, hold it up in front of you and turn it so that the direction you are facing is at the bottom of the chart. This chart works best at 34° north latitude and at the following times: 10:00 P.M. on September 1; 9:00 P.M. on September 15; 8:00 P.M. on September 30. However, it can be used at other times and latitudes within the United States.

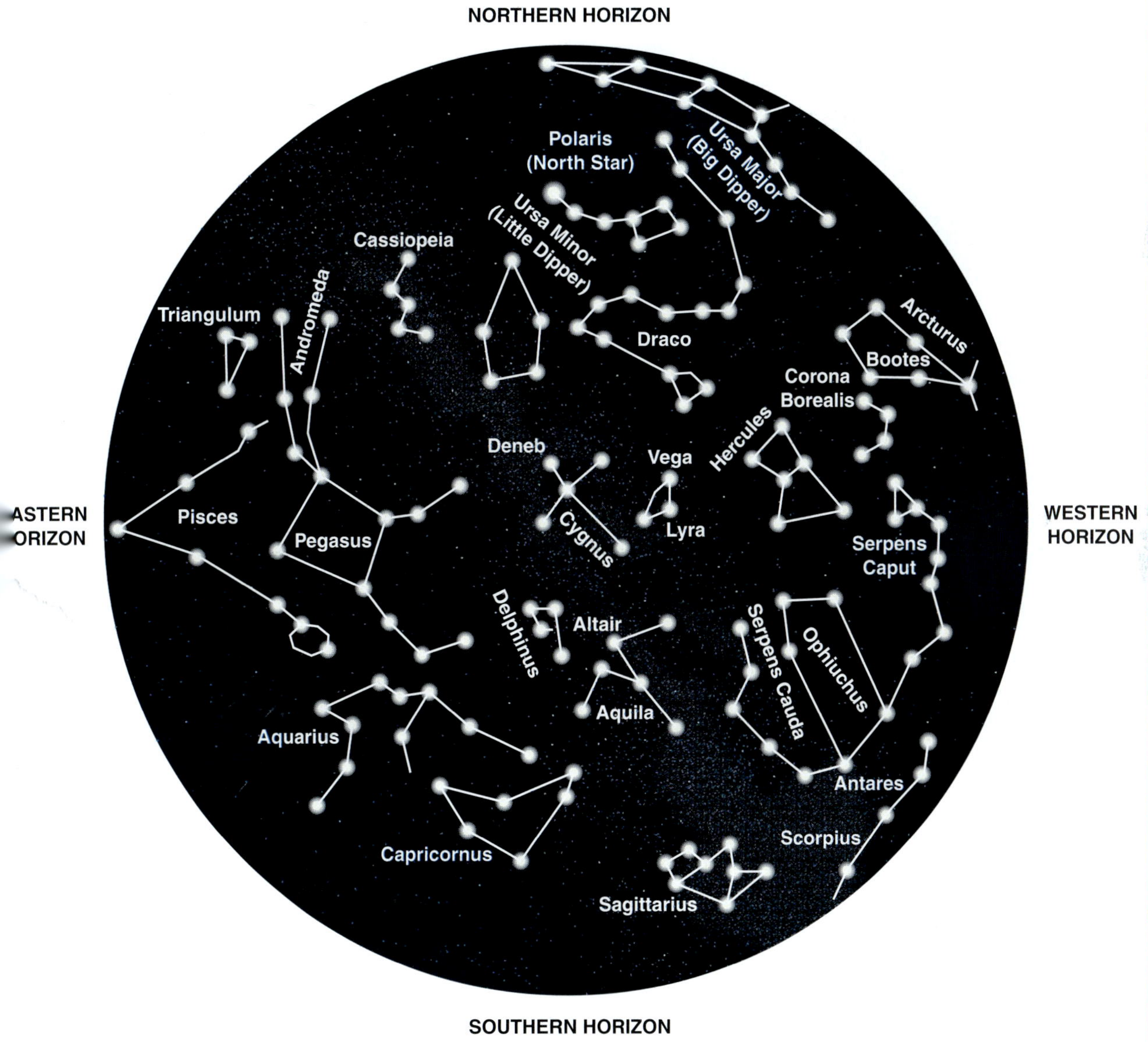

Appendix

Appendix H Star Charts (continued)

Winter Sky

To use this chart, hold it up in front of you and turn it so that the direction you are facing is at the bottom of the chart. This chart works best at 34° north latitude and at the following times: 10:00 P.M. on December 1; 9:00 P.M. on December 15; 8:00 P.M. on December 30. However, it can be used at other times and latitudes within the United States.

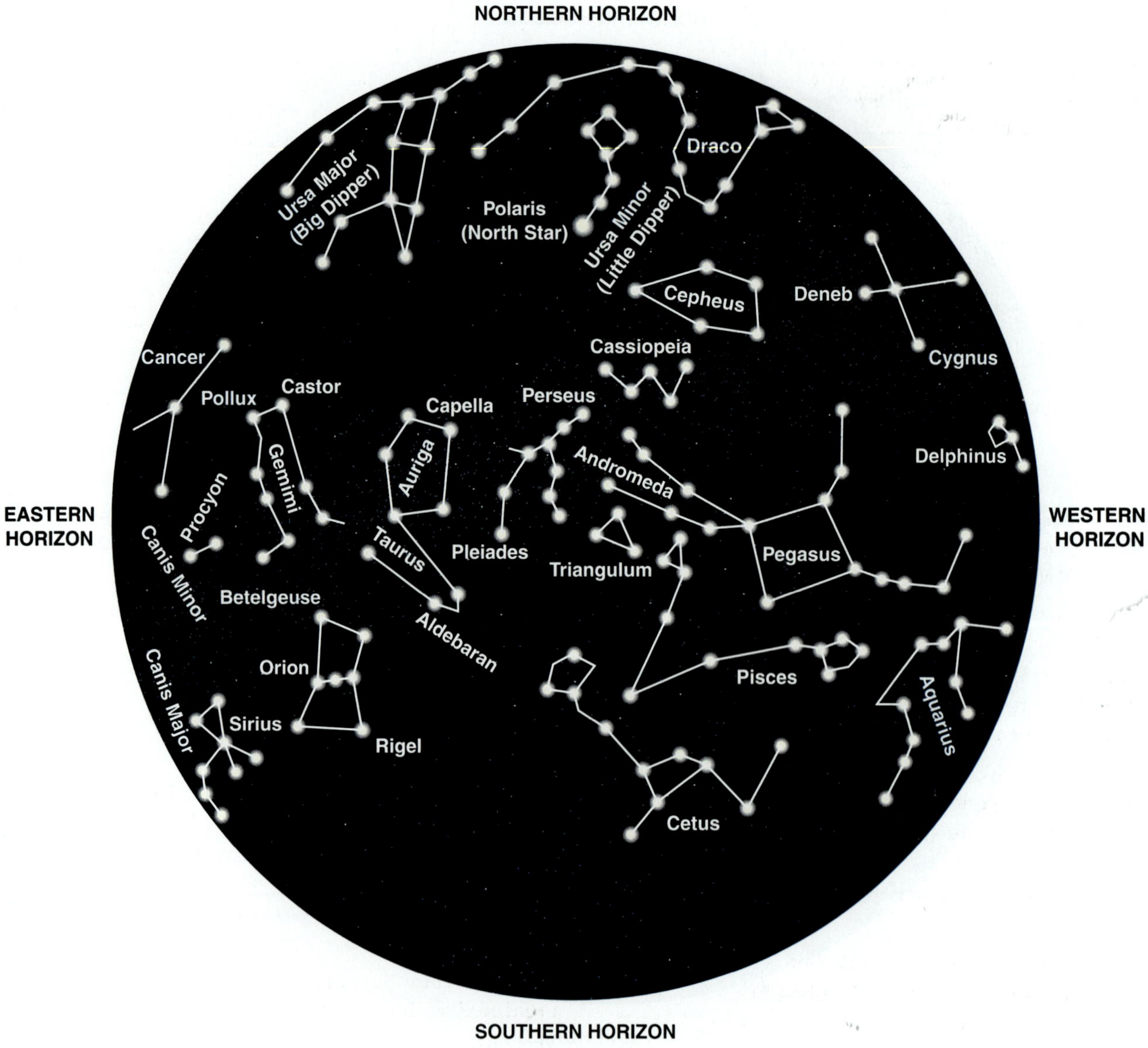

Glossary

Pronunciation and syllabication have been derived from *Webster's New World Dictionary*, Second College Edition, Revised School Printing (Prentice Hall, 1985). Syllables printed in capital letters are given primary stress. (Numbers in parentheses indicate the page number, or page numbers, on which the term is defined.)

PRONUNCIATION KEY					
Symbol	Example	Respelling	Symbol	Example	Respelling
a	transpiration	(tran-spuh-RAY-shuhn)	oh	biome	(BY-ohm)
ah	composite	(kuhm-PAHZ-iht)	oi	asteroid	(AS-tuhr-oid)
aw	atoll	(A-tawl)	oo	altitude	(AL-tuh-tood)
ay	abrasion	(uh-BRAY-zhuhn)	ow	compound	(KAHM-pownd)
ch	leaching	(LEECH-ing)	s	satellite	(SAT-uhl-yt)
eh	chemical	(KEHM-i-kuhl)	sh	specialization	(spehsh-uhl-ih-ZAY-shuhn)
ee	equinox	(EE-kwih-nahks)	th	thermocline	(THUR-muh-klyn)
f	hemisphere	(HEHM-ih-sfeer)	th	weathering	(WEHTH-uhr-ing)
g	galaxy	(GAL-uhk-see)	uh	volcanism	(VAHL-kuh-nihzm)
ih	anticline	(AN-tih-klyn)	y, eye	anticline, isobar	(AN-tih-klyn), (EYE-soh-bahr)
j	geologic	(jee-uh-LAHJ-ihk)	yoo	cumulus	(KYOOM-yuh-luhs)
k	current	(KUR-uhnt)	z	deposition	(dehp-uh-ZIHSH-uhn)
ks	axis	(AK-sihs)	zh	erosion	(e-ROH-zhuhn)

A

abrasion (uh-BRAY-zhuhn): wearing away of rock particles by wind and water (p. 184)

absolute age: true age of a rock or fossil (p. 98)

acid rain: rain containing acids produced by water chemically combined with certain gases (pp. 164, 362)

acid test: test that helps to identify minerals containing calcium carbonate (p. 50)

air current: up-and-down movement of air (p. 266)

air mass: large volume of air with about the same temperature and amount of moisture throughout (p. 292)

altitude (AL-tuh-tood): height above sea level (p. 310)

amber: hardened tree sap (p. 90)

anemometer (an-uh-MAHM-uht-uhr): instrument that measures wind speed (p. 272)

anticline (AN-tih-klyn): upward fold (p. 110)

aphelion (uh-FEE-lee-uhn): point in a planet's orbit at which it is farthest from the Sun (pp. 332, 410)

apogee (AP-uh-jee): point at which the Moon is farthest from Earth (p. 336)

asteroid (AS-tuhr-oid): large chunk of rock or metal that orbits the Sun (p. 424)

astronomical (as-truh-NAHM-ih-kuhl) **unit (AU):** unit of measurement based on the Sun's distance from Earth and equal to about 150 million km (p. 392)

astronomy (uh-STRAHN-uh-mee): study of stars, planets, and other objects in space (p. 380)

atmosphere (AT-muh-sfeer): envelope of gases surrounding Earth (pp. 18, 252)

atoll (A-tawl): ring-shaped coral reef around a lagoon (p. 244)

atom: smallest part of an element that can be identified as that element (p. 42)

axis (AK-sihs): imaginary line through the center of a planet or other body around which that body spins (p. 326)

B

barometer (buh-RAHM-uht-uhr): instrument used to measure air pressure (p. 262)

barrier reef: coral reef that forms around a sunken volcanic island (p. 244)

bedrock: solid rock that lies beneath the soil (p. 166)

Big Bang: explosion that may have begun the universe about 14 billion years ago (p. 458)

binary stars: two stars that revolve around each other (p. 436)

biodegradable: material that breaks down easily (p. 372)

biome (BY-ohm): large region with a characteristic climate and plant and animal communities (p. 318)

black hole: massive star that has collapsed and whose gravity is so powerful that it pulls in everything, even light (p. 458)

C

caldera (kal-DER-uh): large hole that forms when the roof of a magma dome collapses (p. 118)

capacity (kuh-PAS-ih-tee): amount of material something can hold (p. 280)

carbonation (kahr-buh-NAY-shuhn): chemical reaction that occurs when carbonic acid reacts with certain minerals (p. 160)

cave system: series of connected underground caves (p. 186)

cast: mold filled with hardened sediments (p. 90)

cellular respiration (rehs-puh-RAY-shuhn): process by which a cell releases energy from food molecules (p. 252)

chemical formula: formula that shows the elements that make up a compound (p. 44)

chemical (KEHM-ih-kuhl) **symbol:** shorthand way of writing the name of an element (p. 44)

chemical (KEHM-ih-kuhl) **weathering:** weathering that changes the chemical makeup of rocks (p. 158)

chromosphere (KROH-muh-sfeer): layer of the Sun's atmosphere above the photosphere (p. 446)

cinder cone: volcanic cone made up of rock particles, dust, and ash (p. 120)

cirrus (SIR-uhs) **cloud:** light, feathery cloud (p. 286)

clastic (KLAS-tihk) **rock:** sedimentary rock made up of fragments of rock (p. 74)

cleavage (KLEEV-ihj): tendency of some minerals to split along smooth, flat surfaces called planes (p. 52)

climate (KLY-muht): average weather conditions of an area over many years (p. 308)

cold front: forward edge of a cold air mass, formed when a cold air mass pushes under a warm air mass (p. 294)

coma: gas cloud that surrounds the nucleus of a comet (p. 430)

comet: lump of ice, frozen gas, and dust that orbits the Sun (p. 430)

communication: sharing information (p. 8)

composite (kuhm-PAHZ-iht) **cone:** volcanic cone made up of alternating layers of lava and rock particles (p. 120)

compound (KAHM-pownd): substance made up of two or more elements that are chemically combined (p. 42)

concave mirror: mirror that curves inward (p. 386)

condensation (kahn-duhn-SAY-shuhn): changing of a gas to a liquid (pp. 208, 284)

conduction (kuhn-DUHK-shuhn): transfer of heat through matter by direct contact (p. 258)

conservation (kahn-suhr-VAY-shuhn): wise use of natural resources (p. 352)

constant: something that does not change (p. 11)

constellation (kahn-stuh-LAY-shuhn): group of stars that form a pattern in the sky (p. 452)

continental (kahnt-ihn-EHNT-uhl) **drift:** theory that the continents were at one or more times a single landmass that broke apart and eventually moved into the positions they are in today (p. 138)

continental shelf: part of a continent that slopes gently away from the shoreline (p. 242)

continental slope: part of a continent between the continental shelf and the ocean floor (p. 242)

contour interval (KAHN-toor IHN-tuhr-vuhl): difference in elevation between one contour line and the next (p. 34)

contour (KAHN-toor) **line:** line drawn on a map that connects all points having the same elevation (p. 32)

controlled experiment: experiment in which all the conditions except one are kept constant (p. 11)

convection (kuhn-VEHK-shuhn): transfer of heat within a liquid or a gas (p. 258)

convection current: movement of a gas or a liquid caused by changes in temperature (p. 146)

convex (kahn-VEHKS) **lens:** lens that is thicker in the middle than it is at the edges (p. 384)

coprolite (KAHP-roh-lyt): fossilized dung or the stomach contents of ancient animals (p. 94)

coral: small animals found in warm, shallow ocean waters (p. 244)

core: innermost region of Earth (p. 20); center of the Sun (p. 446)

Coriolis effect: bending of Earth's winds and ocean currents by Earth's rotation (p. 232)

corona (kuh-ROH-nuh): outer layer of the Sun's atmosphere (pp. 344, 446)

crater (KRAYT-uhr): pit at the top of a volcanic cone (p. 118); round hole on the Moon's surface (p. 414)

crescent (KREHS-uhnt) **phase:** phase when less than half the Moon is visible (p. 338)

crest: highest point of a wave (p. 234)

crust: solid, thin outer layer of Earth (p. 20)

crystal (KRIHS-tuhl): solid substance with its atoms arranged in a regular, three-dimensional pattern (p. 52)

crystallization: formation of minerals caused by processes such as cooling and evaporation (p. 54)

cumulus (KYOO-myuh-luhs) **cloud:** big, puffy cloud (p. 286)

current: stream of water flowing in the oceans (p. 232)

cycle: series of events that happen over and over again (p. 80)

D

data (DAYT-uh): information you collect when you observe something (p. 3)

deflation (dee-FLAY-shuhn): removal of loose material from Earth's surface by wind (p. 198)

delta: triangular deposit of sediment located at the mouth of a river (p. 184)

density (DEHN-suh-tee): amount of matter in a given volume (p. 50)

density current: stream of water that moves up and down in ocean depths (p. 232)

deposition (dehp-uh-ZIHSH-uhn): process by which material carried by erosion is dropped in new places (p. 178)

desalination: process of getting fresh water from salt water (p. 370)

dew point: temperature to which air must be cooled to reach saturation (p. 284)

distortion (dih-STOR-shuhn): error in shape, size, or distance (p. 22)

dome mountain: mountain formed when upfolds in rocks created a rounded structure that looks like a bowl turned upside down (p. 114)

drumlin: oval-shaped mound of till (p. 192)

dwarf planet: small, planetlike objects that orbit the Sun and share their orbits with other bodies (p. 428)

E

Earth science: study of Earth and its history (p. 16)

earthquake: sudden, strong movement of Earth's crust (p. 124)

ebb tide: outgoing, or falling tide (p. 238)

element (EHL-uh-muhnt): simple substance that cannot be broken down into simpler substances by ordinary chemical means (p. 42)

elevation (ehl-uh-VAY-shuhn): height of a point on Earth above or below sea level (pp. 32, 112)

ellipse (eh-LIHPS): flattened circle, or oval (p. 410)

elliptical galaxy: galaxy shaped like a ball or slightly flattened ball (p. 454)

environment (ehn-VY-ruhn-muhnt): everything that surrounds a living thing (p. 350)

epicenter (EHP-ih-sehnt-uhr): place on Earth's surface directly above the focus (p. 124)

equator: imaginary line that runs around the middle of Earth's surface and divides it into the Northern and Southern hemispheres (p. 26)

equinox (EE-kwih-nahks): day the Sun shines directly on the equator (p. 332)

erosion (ee-ROH-zhuhn): process by which weathered material is removed and carried from a place (p. 178)

erratic (ih-RAT-ihk): boulder left behind by a retreating glacier (p. 190)

evaporation (ee-vap-uh-RAY-shuhn): process by which a liquid changes into a gas (p. 54); changing of a liquid to a gas (pp. 208, 278)

F

fault: break in Earth's crust along which movement occurs (p. 110)

fault-block mountain: mountain formed when normal faults uplift a block of rock (p. 114)

floodplain: flat area on the side of a river where sediments are deposited during floods (p. 184)

flood tide: incoming, or rising, tide (p. 238)

focus (FOH-kuhs)**:** point beneath Earth's surface where an earthquake starts (p. 124)

folded mountain: mountain formed by the folding of rock layers (p. 114)

foliated (FOH-lee-ayt-uhd)**:** texture of a metamorphic rock that has mineral crystals arranged in bands (p. 78)

fossil (FAHS-uhl)**:** remains or traces of an organism that lived long ago (p. 90)

fossil fuel (FYOO-uhl)**:** natural fuel that was formed from the remains of living things (p. 102); oil, coal, or natural gas (p. 354)

fracture (FRAK-chuhr)**:** tendency of some minerals to break into pieces with uneven surfaces (p. 52); break in a rock (p. 110)

fringing reef: coral reef that is directly attached to a shore (p. 244)

front: boundary between air masses of different densities (p. 294)

frost: ice formed from condensation below the freezing point of water (p. 284)

fusion (FYOO-zhuhn)**:** reaction in which atomic nuclei combine to form larger nuclei (p. 450)

G

galaxy (GAL-uhk-see)**:** huge collection of stars, gas, and dust that travel together through space (pp. 382, 454)

gastrolith (GAS-troh-lihth)**:** stone used to grind food (p. 94)

gem: stone that has been cut and polished (p. 58)

geologic (jee-uh-LAHJ-ihk) **time scale:** outline of the major divisions in Earth's history (p. 100)

geothermal (jee-oh-THUHR-muhl) **energy:** energy produced from heat inside Earth (p. 356)

geyser (GY-zuhr)**:** heated groundwater that erupts onto Earth's surface (p. 212)

gibbous (GIHB-uhs) **phase:** phase when more than half the Moon is visible (p. 338)

glacier (GLAY-shuhr)**:** moving river of ice and snow (p. 188)

global winds: large systems of winds around Earth (p. 268)

globe: three-dimensional model of Earth's surface (p. 22)

gram: basic unit of mass (p. 4)

gravity (GRAV-ih-tee)**:** force of attraction that exists among all objects in the universe (p. 410)

groundwater: water that collects in pores in soil and sinks into the ground (p. 210)

guyot (GEE-oh)**:** flat-topped, underwater seamount (p. 242)

H

half-life: length of time it takes for one-half the amount of a radioactive element to change into a stable element (p. 98)

hanging valley: small glacial valley above a main valley (p. 190)

hard water: water containing a lot of minerals, especially calcium and magnesium (p. 370)

hardness: property of a mineral that relates to how much the mineral resists being scratched (p. 48)

hemisphere (HEHM-ih-sfeer)**:** one-half of a sphere (p. 26)

horizon (huh-RY-zuhn)**:** soil layer (p. 170)

hot spot: place where magma reaches the surface of a tectonic plate (p. 150)

humidity: amount of water vapor in the air (p. 280)

humus (HYOO-muhs)**:** decaying remains of plants and animals (p. 166)

hurricane (HUR-ih-kayn)**:** tropical storm with very strong winds (p. 296)

hydroelectric (HY-droh-ee-LEHK-trihk) **power:** electrical energy produced from moving water (p. 354)

hydrolysis (hy-DRAHL-uh-sihs)**:** chemical reaction that occurs when minerals with little water content react with water (p. 160)

hydrosphere (HY-droh-sfeer)**:** part of Earth that is water (p. 18)

hypothesis: suggested answer to a question or problem (p. 10)

I

iceberg: large piece of a glacier that enters the ocean (p. 188)

ice cap: large sheet of ice found near Earth's poles (p. 188)

ice wedging: mechanical weathering caused by the freezing and melting of water (p. 158)

igneous (IHG-nee-uhs) **rock:** rock formed by the crystallization of hot melted rock or minerals (p. 66)

index fossil: fossil used to help determine the relative age of rock layers (p. 96)

indicator (IHN-dih-kayt-uhr) **kit:** used to test the acidity of soil; color will change (p. 172)

infer: to form a conclusion (p. 8)

International Date Line: boundary formed where the first and 24th time zones meet (p. 328)

irregular galaxy: galaxy with no definite shape (p. 454)

isobar (EYE-soh-bahr): line on a weather map that connects points of equal air pressure (p. 302)

isotherm: line on a weather map that joins places that have the same temperatures (p. 302)

K

karst topography: land that has sinkholes, caverns, and underground rivers (p. 186)

kettle lake: lake formed by a retreating glacier (pp. 192, 218)

L

L-wave: surface wave (p. 126)

lagoon: shallow body of water between a reef and the mainland (p. 244)

lake: low spot in Earth's surface filled with still water (p. 218)

landform: physical feature of Earth's solid surface (p. 112)

latitude (LAT-uh-tood): distance in degrees north or south of the equator in degrees (pp. 26, 310)

lava (LAH-vuh): magma that reaches Earth's surface (pp. 68, 118)

law of superposition: states that each undisturbed sedimentary rock layer is older than the layer above it (p. 96)

leaching (LEECH-ing): removing or washing away the minerals in soil (p. 168)

legend (LEHJ-uhnd): list of map symbols and their meanings (p. 28)

light-year: unit of measurement equal to about 10 trillion km (p. 392)

liter (LEET-uhr): basic unit of liquid volume (p. 4)

lithosphere (LIHTH-oh-sfeer): solid part of Earth (p. 18)

loess (LOH-ehs): thick deposits of windblown dust (p. 198)

longitude (LAHN-juh-tood): measurement in degrees east or west of the prime meridian (p. 26)

longshore current: movement of water parallel to a shoreline (p. 196)

lunar eclipse (ih-KLIHPS): passing of the Moon through Earth's shadow (p. 342)

luster: how a mineral's surface reflects light (p. 48)

M

magma (MAG-muh): molten rock inside Earth (pp. 68)

magma chamber: underground pocket of molten rock (p. 150)

magnetism (MAG-nuh-tihz-uhm): force of attraction or repulsion associated with magnets (p. 50)

magnitude (MAG-nuh-tood): way to measure a star's brightness (p. 440)

main sequence star: star that falls within a long, narrow, diagonal band across the H-R diagram (p. 442)

mantle (MAN-tuhl): thick layer of Earth below the crust (p. 20)

map: flat model of Earth's surface (p. 22)

map projection (proh-JEHK-shuhn): drawing of Earth's surface, or part of it, on a flat surface (p. 24)

mare (MAH-ray), *pl.* **maria:** broad, flat plain on the Moon's surface (p. 414)

mass: amount of matter in something (p. 4)

mass erosion: downhill movement of weathered materials caused by gravity (p. 180)

matter: anything that has mass and volume (p. 252)

meander (mee-AN-duhr): loop in a mature river (p. 214)

mechanical (muh-KAN-ih-kuhl) **weathering:** weathering in which the chemical makeup of rocks does not change (p. 158)

meniscus: curve at the surface of a liquid in a thin tube (p. 4)

meridian (muh-RIHD-ee-uhn): line on a map or a globe running from the North Pole to the South Pole along Earth's surface (p. 26)

mesosphere (MEHZ-uh-sfeer): third layer of the atmosphere (p. 254)

metamorphic (meht-uh-MOR-fihk) **rock:** rock formed when existing rocks are changed by heat and pressure (p. 66)

meteor (MEET-ee-uhr): rock or metal that enters Earth's atmosphere (p. 424)

meteorite (MEET-ee-uhr-eyet): piece of rock or metal that falls on a planet or moon's surface (p. 424)

meteoroid (MEET-ee-uhr-oid): small piece of rock or metal that travels through space (p. 424)

meter (MEET-uhr): basic unit of length or distance (p. 4)

microclimate (MY-kroh-kly-muht): very small climate zone (p. 314)

mid-ocean ridge: an ocean-floor feature resembling a mountain ridge on land (p. 140)

middle-latitude zone: region between 30° and 60° N and S latitude (p. 312)

millibar (MIHL-ih-bahr): unit of measurement for air pressure (p. 300)

mineral (MIHN-uhr-uhl): naturally occurring, inorganic solid formed from elements or compounds and having a definite chemical makeup and regular atomic structure (p. 46)

model: tool scientists use to represent an object or a process (p. 3)

mold: imprint or hollow in rock that is shaped like and made by an organism (p. 90)

molecule: (MAHL-ih-kyool): smallest part of a substance that has all the properties of that substance (p. 42)

molten (MOHL-tuhn) **rock:** melted minerals (p. 68)

monsoon: wind that changes direction with the seasons (p. 270)

moraine (muh-RAYN): ridge of till deposited by a melting glacier (p. 192)

N

natural resource: material from the earth that is used by living things (p. 352)

neap tide: tide that is not as high or as low as a normal tide (p. 336)

nebula (NEHB-yuh-luh): cloud of gas and dust in space (p. 436)

nekton (NEK-tahn): free-swimming ocean animals (p. 246)

newton: metric unit of force (p. 260)

nodule (NAHJ-ool): mineral lump found on the ocean floor (p. 240)

nonclastic rock: sedimentary rock made up of dissolved minerals or the remains of living things (p. 74)

nonfoliated: texture of a metamorphic rock that does not have mineral crystals arranged in bands (p. 78)

nonrenewable resource: natural resource that is not easily replaced by nature or human effort (p. 353)

nova: explosion where the outer layers of a star are blown off (p. 444)

nuclear (NOO-klee-uhr) **energy:** energy produced by splitting or combining atoms (p. 354)

nucleus (NOO-klee-uhs): head or solid part of a comet (p. 430); center, or core, of an atom (p. 450)

O

oceanography (oh-shuh-NAHG-ruh-fee): study of Earth's oceans (p. 226)

ooze: ocean sediment that contains the remains of many ocean organisms (p. 240)

orbit: curved path of one object around another object in space (pp. 400, 408)

ore: mineral that is mined because it contains useful metals or nonmetals (p. 58)

oxbow lake: curved lake formed when a bend in a river is cut off at both ends (p. 214)

oxidation (ahk-sih-DAY-shuhn): chemical change that occurs when oxygen reacts with another substance (p. 160)

P

P-wave: fastest earthquake wave (p. 126)

Pangaea (pan-JEE-uh): single, giant landmass, or continent, that later broke apart (p. 138)

parallax (PAR-uh-laks): apparent change in the position of a distant object when seen from two different places (p. 392)

parallel (PAR-uh-lehl): horizontal line on a map or globe that circles Earth from east to west at intervals starting at the equator (p. 26)

penumbra (pih-NUM-bruh): light part of a shadow (p. 342)

perihelion (per-uh-HEE-lee-uhn): point in a planet's orbit at which it is closest to the Sun (pp. 332, 410)

pH scale: number scale used to measure acidity (p. 172)

phases (FAYZ-uhz): changing shapes of the Moon (p. 338)

photosphere (FOHT-oh-sfeer): inner layer of the Sun's atmosphere (p. 446)

photosynthesis (foht-oh-SIHN-thuh-sihs): food-making process in green plants (p. 451)

physical property: observable characteristic that describes an object (p. 48)

phytoplankton (fyt-oh-PLANK-tuhn): source of food for floating animals in the ocean (p. 246)

plain: large, flat area just above sea level (p. 116)

plankton (PLANK-tuhn): floating organisms (p. 246)

plate boundary: place where two plates meet (p. 146)

plateau: large, flat area at a high elevation (p. 116)

pluton (PLOO-tahn): large body of igneous rock that can form into different shapes when magma cools inside Earth's crust (p. 68)

polar air mass: air mass that forms over cold regions (p. 292)

polar zone: cold region above 60° N and below 60° S latitude (p. 312)

pollutant (puh-LOOT-uhnt): harmful substance in the environment (p. 350)

pollution (puh-LOO-shuhn): anything that harms the environment (p. 350)

pond: body of water similar to a lake but usually smaller and shallower (p. 218)

pore: tiny hole or space (p. 210)

potable (POHT-uh-buhl): water that is fit to drink (p. 231)

precipitation (pree-sihp-uh-TAY-shuhn): process that occurs when elements and compounds leave a solution and crystallize out as solids (p. 54); water that falls to Earth's surface from the atmosphere (p. 208, 290)

predict: to state ahead of time what you think is going to happen (p. 8)

pressure: amount of force per unit of area (p. 260)

prominence (PRAHM-uh-nuhns): stream of gas that shoots high above the Sun's surface (p. 448)

protostar: dense material in the center of a nebula that is about to become a star (p. 436)

psychrometer (sy-KRAHM-uht-uhr): instrument used to find relative humidity (p. 282)

Q

quasar: continuous burst of brilliant light and enormous energy from a very massive black hole (p. 458)

R

radiant (RAY-dee-uhnt) **energy:** energy given off by the Sun that can travel through empty space (p. 256)

radiation (ray-dee-AY-shuhn): movement of the Sun's energy through empty space (p. 256)

radio telescope: telescope that can receive radio waves from sources in space (p. 388)

rain gauge (GAYJ): device used to measure rainfall (p. 290)

rapids: part of a river where the current is swift (p. 214)

recycling: using a natural resource over again (p. 358)

red giant: large, bright star that is fairly cool (p. 442)

reflecting (rih-FLEKT-ing) **telescope:** telescope that uses a concave mirror to collect light (p. 386)

refracting (rih-FRAKT-ing) **telescope:** telescope that uses convex lenses to produce an enlarged image (p. 384)

relative (REHL-uh-tihv) **age:** age of an object compared to the age of another object (p. 96)

relative humidity: amount of water vapor in the air compared with the amount of water vapor the air can hold at capacity (p. 282)

renewable resource: natural resource that can be reused or replaced (p. 352)

reservoir: artificial lake (p. 218)

residual (rih-ZIJ-oo-uhl) **soil:** soil remaining on top of the bedrock from which it formed (p. 168)

revolution (rehv-uh-LOO-shuhn): movement of a planet or other body orbiting another body (p. 326)

Richter (RIHK-tuhr) **scale:** scale that measures the energy released by an earthquake (p. 128)

rift: valley caused by a crack in the crust of a planet (p. 422)

rift valley: flat area between two ridges that is formed by spreading plates (p. 140)

Ring of Fire: major earthquake and volcano zone that almost forms a circle around the Pacific Ocean (p. 130)

rock: mixture of minerals, generally cemented together (p. 66)

rock cycle: series of natural processes by which rocks are slowly changed from one kind of rock to another kind of rock (p. 80)

rotation (roh-TAY-shuhn): spinning of a planet or another body on its axis (p. 326)

runoff: water from rain or snow that flows into streams and rivers from surface areas (p. 182)

S

S-wave: second earthquake wave to be recorded at a seismograph station (p. 126)

salinity (suh-LIHN-uh-tee): amount of dissolved salts in ocean water (p. 230)

sand bar: long, offshore underwater deposit of sand (p. 196)

satellite (SAT-uhl-eyet): natural or artificial object orbiting another body in space (pp. 400, 416)

saturated (sach-uh-RAYT-ihd): filled to capacity (p. 280)

scale: feature that relates distances on a map to actual distances on Earth's surface (p. 28)

sea arch: gap formed when waves cut completely through a section of rock (p. 194)

sea stack: column of rock remaining after the collapse of a sea arch (p. 194)

seafloor spreading: process that forms new seafloor (p. 140)

seamount: volcanic mountain on the ocean floor (p. 242)

sediment (SEHD-uh-muhnt): rock particles carried and deposited by water, wind, or ice (pp. 72, 184)

sedimentary (sehd-uh-MEHN-tuh-ree) **rock:** rock formed from pieces of other rocks that are pressed together (p. 66)

seismic (SYZ-mihk) **wave:** earthquake wave (p. 124)

seismograph (SYZ-muh-graf): instrument that detects and measures earthquakes (p. 124)

sewage: wastes usually flushed away in water (p. 366)

shield cone: volcanic cone made up of layers of hardened lava (p. 120)

simulation: computer model that usually shows a process (p. 3)

sinkhole: large hole in the ground formed when the roof of a cavern collapses (p. 186)

sling psychrometer (sy-KRAHM-uht-uhr): instrument used to find relative humidity (p. 282)

smog: mixture of smoke, fog, and chemicals (p. 362)

soft water: water containing few or no minerals (p. 370)

soil: mixture that includes silt, sand, and clay (p. 166)

soil profile (PROH-fyl): all the layers that make up the soil in an area (p. 170)

solar cell: device that converts sunlight into electricity (p. 356)

solar (SOH-luhr) **eclipse:** passing of the Moon between Earth and the Sun (p. 344)

solar energy: energy from the Sun (p. 356)

solar flare: eruption of electrically charged particles from the surface of the Sun (p. 448)

solar noon: time of the day when the Sun is highest in the sky (p. 328)

solar system: Sun and all the bodies that orbit it (p. 380)

solstice (SAHL-stihs): day of the year the Sun reaches its highest or lowest point in the sky (p. 332)

solution: mixture in which the particles of one substance are evenly mixed with the particles of another substance (p. 54)

sonar: echo-sounding system that bounces sound waves off the ocean floor (p. 228)

specialist (SPEHSH-uhl-ihst): person who studies or works on only one part of a subject (p. 16)

specialty: studying or working in only one part of a subject (p. 16)

specific humidity (hyoo-MIHD-uh-tee): actual amount of water in the air (p. 280)

spectrograph (SPEHK-truh-graf): device that measures the spectrum of an object (p. 438)

spectroscopy (spehk-TRUH-skoh-pee): study of light coming from objects in space (p. 438)

spectrum: pattern of different colors of light coming from an object (p. 438)

sphere (SFEER): round, three-dimensional object (p. 18)

spiral galaxy: galaxy shaped like a flattened disk with spiral arms (p. 454)

spit: long, narrow deposit of sand connected at one end to the shore (p. 196)

spring: natural flow of groundwater to Earth's surface (p. 212)

spring tide: tide that is higher or lower than a normal tide (p. 336)

standard time: system whereby all places within a time zone have the same time (p. 328)

star: ball of gases that gives off light and heat (p. 436)

star cluster: large group of stars that travel together through space (p. 436)

station model: record of weather information at a weather station (p. 300)

stratosphere (STRA-tuh-sfeer)**:** second layer of the atmosphere (p. 254)

stratus (STRAY-tuhs) **cloud:** sheetlike cloud that forms layers across the sky (p. 286)

streak: color of the powder left by a mineral (p. 48)

subduction (suhb-DUHK-shuhn) **zone:** place where old oceanic crust is forced back down into an ocean trench (p. 142)

submersible (suhb-MUHR-suh-buhl)**:** underwater research vessel (p. 228)

subscript (SUHB-skrihpt)**:** number in a chemical formula that shows how many of each atom are in one molecule of a compound (p. 44)

summit: highest point on a mountaintop (p. 112)

sunspot: dark, cool area on the Sun's surface (p. 448)

supergiant: very large and very bright star (p. 442)

supernova: violent explosion where a star is blown apart (p. 444)

symbol: drawing on a map that represents a real object (p. 28)

syncline (SIHN-klyn)**:** downward fold (p. 110)

T

tail: long, ribbonlike trail of comet dust and gas (p. 430)

talus: pile of rocks and rock particles that collects at the base of a slope (p. 180)

tectonic plate: large, solid piece of Earth's surface (p. 144)

temperature: measure of the amount of heat energy something contains (p. 4)

texture (TEHKS-chuhr)**:** size of crystals in an igneous rock (p. 70); size of soil particles (p. 168)

theory (THEE-uh-ree)**:** set of hypotheses that have been supported by testing over and over again (p. 10)

theory of plate tectonics (tehk-TAHN-ihks)**:** theory that Earth's crust is broken into plates that float on the upper part of the mantle (p. 144)

thermocline (THUR-muh-klyn)**:** layer of ocean water in which the temperature drops sharply with depth (p. 230)

thermosphere (THUR-muh-sfeer)**:** upper layer of the atmosphere (p. 254)

thrust: forward force produced in a rocket engine (p. 394)

thunderstorm: storm with thunder, lightning, and often heavy rain and strong winds (p. 296)

tide: regular change in the level of Earth's oceans (p. 238)

till: rock material deposited by a glacier (p. 190)

topography (tuh-PAHG-ruh-fee)**:** general form and shape of the land on Earth's surface (p. 32)

tornado (tawr-NAY-doh)**:** small, very violent, funnel-shaped cloud that spins (p. 296)

transpiration (trans-spuh-RAY-shuhn)**:** process by which plants give off water vapor into the air (p. 278)

transported (trans-POR-tihd) **soil:** soil moved away from the bedrock from which it was formed (p. 168)

trench: deep canyon (p. 142); deep canyon on the ocean floor (p. 242)

tributary (TRIHB-yoo-tehr-ee)**:** smaller stream that flows into the main stream of a river system (p. 182)

tropical (TRAHP-ih-kuhl) **air mass:** air mass that forms over warm regions (p. 292)

tropical zone: warm region near the equator (p. 312)

troposphere (TROH-puh-sfeer)**:** lowest layer of the atmosphere (p. 254)

trough (TRAWF)**:** lowest point of a wave (p. 234)

tsunami (tsoo-NAH-mee)**:** large ocean wave caused by an earthquake (p. 128)

U

umbra (UHM-bruh)**:** center or dark part of a shadow (p. 342)

nonfoliated: texture of a metamorphic rock that does not have mineral crystals arranged in bands (p. 78)

unit: amount used to measure something (p. 4)

V

variable: anything that can affect the outcome of an experiment (p. 11)

vegetation (vej-uh-TAY-shuhn)**:** plants (p. 318)

vent: volcano opening from which lava flows (p. 118)

volcanic mountains: mountains formed from lava or debris, such as ash or rocks, thrown out of a volcano (p. 114)

volcanism (VAHL-kuh-nihz-uhm)**:** movement of magma inside Earth (p. 118)

volcano (vahl-KAY-noh)**:** vent and the volcanic material around it (p. 118)

volume: amount of space an object takes up (p. 4)

W

warm front: forward edge of a warm air mass, formed when a warm air mass pushes over a cold air mass (p. 294)

water cycle: repeated pattern of water movement between Earth and the atmosphere (p. 208)

waterfall: steep fall of water, as of a stream, from a height (p. 214)

water table: upper layer of saturated rock and soil (p. 210)

wave: regular up-and-down movement of water (pp. 194, 234)

wave-cut terrace: flat section of rock formed by the erosion of a sea cliff (p. 194)

weather: day-to-day conditions of the atmosphere (p. 308)

weathering: breaking down of rocks and other materials on Earth's surface (p. 158)

well: hole dug below the water table that fills with groundwater (p. 212)

white dwarf: very small, hot star (p. 442)

wildlife: all the plants and animals that live in an area (p. 360)

wind: horizontal movement of air (p. 266)

wind vane: instrument that indicates wind direction (p. 272)

world ocean: body of salt water covering much of Earth's surface (p. 226)

Index

N

O

P

T

Photo Credits

Photography Credits: All photographs are by Pearson Education, Inc., John Serafin for Pearson Education, Inc., and David Mager for Pearson Education, Inc., except as noted below.

Cover: Earth Science
Chad Ehlers/Getty Images/Stone
Comstock Images/Jupiter Images

Back Cover: Comstock Images/Jupiter Images

Table of Contents: i t Comstock Inc.; v t Comstock Images/Jupiter Images; v b Charles D. Winters/Photo Researchers, Inc.; vi t Comstock Images/Jupiter Images; vi tr ChromoSohm/Photo Researchers, Inc.; vi b Jeff J. Daly/Fundamental Photographs; vii Comstock Images/Jupiter Images; viii t Comstock Images/Jupiter Images; viii tr Alistair Duncan/Dorling Kindersely Limited; viii b PhotoDisc/Getty Images; ix Comstock Images/Jupiter Images; x t Comstock Images/Jupiter Images; x b Philip Coblentz/Brand X/Alamy; xi Comstock Images/Jupiter Images; xii t Comstock Images/Jupiter Images; xii tr C.C. Lockwood/Animals Animals/Earth Scenes; xii b David Ducros/Science Photo Library/Photo Researchers, Inc.; xiii t Comstock Images/Jupiter Images; xiii tr NASA/Science Library/Photo Researchers, Inc.; xiv Comstock Images/Jupiter Images; xv Comstock Images/Jupiter Images; xvi Comstock Images/Jupiter Images; xvii Comstock Images/Jupiter Images; xviii Comstock Images/Jupiter Images; xviii Comstock Images/Jupiter Images

Front Matter P001 t Farrell Grehan/Photo Researchers, Inc.; P001 bl Brand X/Jupiter Images; P001 mr George Ranalli/Photo Researchers, Inc.; P001 br Comstock Images/Jupiter Images; P002 bl Jane Grushow/Grant Heilman Photography, Inc.; P002 br Eric Kamp/Phototake; P002 tr AFP/Stringer/Getty Images; P003 Comstock Images/Jupiter Images; P005 Comstock Images/Jupiter Images; P007 Comstock Images/Jupiter Images; P009 t SuperStock, Inc.; P009 bl Brand X/Jupiter Images/Alamy; P009 br Dwayne Newton/Photo Edit; P009 r Comstock Images/Jupiter Images; P010 Imageplus/Corbis Premium RF/Alamy; P011 Comstock Images/Jupiter Images; P013 Comstock Images/Jupiter Images

Chapter 1 P015 PhotoDisc/Getty Images; P016 tl Lloyd Sutton/Alamy; P016 bl Dwayne Newton/Photo Edit; P016 tr SuperStock, Inc.; P016 br Runk Schoenberger/Grant Heilman Photography, Inc.; P017 l NASA/Johnson Space Center; P017 r Tom Bean/DRK Photo; P018 Kim Heacox/DRK Photo; P021 The Granger Collection; P022 l Jeff Greenberg/Omni-Photo Communications; P022 tr Stephen VanHorn/Shutterstock; P023 CNES; Licensed by SPOT Image Corporation/Photo Researchers, Inc.; P025 Hulton Archive/Getty Images; P029 Chrislofoto/Shutterstock; P030 t PhotoDisc/Getty Images; P030 b The Granger Collection, New York; P031 tl United States Mint; P031 tr Tom Bean/Corbis; P031 bl Bettmann/Corbis; P031 br Bettmann/Corbis; P032 U.S. Geological Survey; P033 Ingram Publishing (Superstock Limited)/Alamy; P038 PhotoDisc/Getty Images; P039 PhotoDisc/Getty Images; P040 PhotoDisc/Getty Images

Chapter 2 P041 Jim Zuckerman/Corbis Premium RF/Alamy; P042 l Runk Schoenberger/Grant Heilman Photography, Inc.; P042 r The Natural History Museum/Alamy; P043 tl Martyn F. Chillmaid/Science Photo Library/Photo Researchers, Inc.; P043 tr Lawrence Migdale/Photo Researchers, Inc.; P043 br Bettmann/Corbis; P045 Charles D. Winters/Photo Researchers, Inc.; P046 Kent Foster/Photo Researchers, Inc.; P048 tr Fundamental Photographs; P048 bl Harry Taylor/Dorling Kindersley Limited; P048 bc Harry Taylor/Dorling Kindersley Limited; P048 br Paul Silverman/Fundamental Photographs; P050 l Breck P. Kent/Animals Animals/Earth Sciences; P050 r Gary Retherford/Photo Researchers, Inc.; P051 Gemaldegalerie, Dresden, Germany/Bridgeman Art Library; P051 inset Thessalonika Museum, Greece/Bridgeman Art Library; P052 t Colin Keates/Dorling Kindersley Limited; P052 b Charles D. Winters/Photo Researchers, Inc.; P053 Peter Horree/Alamy; P055 NASA; P055 inset NASA/Phototake; P058 t Biophoto Associates/Photo Researchers, Inc.; P058 bl Charles D. Winters/Photo Researchers, Inc.; P058 br M. Clayel Jacama/Photo Researchers, Inc.; P059 tl Hartmann/Sachs/Phototake; P059 tr Colin Keates/Dorling Kindersley Limited; P059 br Tobias Machhaus/Shutterstock; P060-61 bkgd Peter Essick/Aurora & Quanta Productions; P060 Dorling Kindersley Limited; P061 tl U.S. Geological Survey/U.S. Department of the Interior; P061 tm Stephen Oliver/Dorling Kindersley Limited; P061 tr Natural History Museum/Dorling Kindersley Limited; P061 m Ellen Beijers/Shutterstock; P061 b Dane Penland/Smithsonian Institution/AP Wide World Photos; P062 Jim Zuckerman/Corbis Premium RF/Alamy; P063 Jim Zuckerman/Corbis Premium RF/Alamy; P064 Jim Zuckerman/Corbis Premium RF/Alamy

Chapter 3 P065 Pixelite/Shutterstock; P066 Chromosohm/Photo Researchers, Inc.; P067 Jonathan Blair/Corbis; P068 Michael Nicholson/Corbis; P069 John Kay/Corbis; P070 l Phillip Hayson/Photo Researchers, Inc.; P070 r Breck P. Kent/Animals Animals/Earth Scenes; P073 tl Colin Keates/Dorling Kindersley Limited; P073 tr Runk/Schoenberger/Grant Heilman Photography, Inc.; P073 b Ric Ergenbright/Corbis; P073 b inset O.S.F./Animals Animals/Earth Scenes; P074 t Joyce Photographics/Photo Researchers, Inc.; P074 m Andrew J. Martinez/Photo Researchers, Inc.; P074 b Farrell Grehan/Photo Researchers, Inc.; P076 tl Phillip Hayson/Photo Researchers, Inc.; P076 tr Grace Davies/Omni-Photo Communications; P076 bl Phillip Hayson/Photo Researchers, Inc.; P076 br Grace Davies/Omni-Photo Communications; P077 Paul Jenkin/Animals Animals/Earth Scenes; P078 tl Andrew J. Martinez/Photo Researchers, Inc.; P078 tr Andrew J. Martinez/Photo Researchers, Inc.; P078 bl Tom Pantages/Phototake; P078 br Andrew J. Martinez/Photo Researchers, Inc.; P079 t PhotoDisc/Getty Images; P079 b Norbert Rehm/Shutterstock; P084 t Michael Nicholson/Corbis; P084 b PhotoDisc/Getty Images; P085 t Chris R. Sharp/Photo Researchers Inc.; P085 m Pixelite/Shutterstock; P085 b Emma Lee/PhotoDisc, Inc.; P086 Pixelite/Shutterstock; P087 Pixelite/Shutterstock; P088 Pixelite/Shutterstock

Chapter 4 P089 Ismael Montero Verdu/Shutterstock; P090 l Jan Kranendonk/Shutterstock; P090 r Carolina Biological Supply/Phototake; P091 tl Jonathan Blair/Corbis; P091 bl Tom McHugh/Photo Researchers, Inc.; P091 tr Jeff J. Daly/Fundamental Photographs; P094 Tom McHugh/Photo Researchers, Inc.; P095 Des Bartlett/Photo Researchers, Inc.; P099 John Reader/Science Photo Library/Photo Researchers, Inc.; P101 Steve Munsinger/Photo Researchers, Inc.; P102 Dorling Kindersley Limited; P103 PhotoDisc/Getty Images; P106 Ismael Montero Verdu/Shutterstock; P107 Ismael Montero Verdu/Shutterstock; P108 Ismael Montero Verdu/Shutterstock

Chapter 5 P109 Craig Hansen/Shutterstock; P110 Rolf Sziringer/Okapia Bild/Photo Researchers, Inc.; P111 Kevin Schafer/Peter Arnold, Inc.; P112 Jerry McCormick-Ray/Photo Researchers, Inc.; P113 The Granger Collection, New York; P114 t Craig Lovell/Corbis; P114 b Jim Brandenburg/Minden Pictures; P115 t Georg Gerster/Photo Researchers, Inc.; P115 b Galen Rowell/Corbis; P116 l Jim Steinberg/Photo Researchers, Inc.; P116 r George Ranalli/Photo Researchers, Inc.; P117 PhotoDisc/Getty Images; P119 t George Gerster/Photo Researchers, Inc.; P119 b David Weintraub/Photo Researchers, Inc.; P121 NASA/Science Source/Photo Researchers, Inc.; P122-123 bkgd David Weintraub/Photo Researchers, Inc.; P122 t David Muench/Corbis; P122 b Joel W. Rogers/Corbis; P123 t Layne Kennedy/Corbis; P123 b Maslov Dmitry/Shutterstock; P124 Ken M. Johns/Photo Researchers, Inc.; P125 b AFP/Stringer/Getty Images; P128 t David J. Cross/Peter Arnold, Inc.; P128 b Jasper Juinen/AP Wide World Photos; P131 Roger Rossmeyer/Corbis; P134 Craig Hansen/Shutterstock; P135 Craig Hansen/Shutterstock; P136 Craig Hansen/Shutterstock

Chapter 6 P137 Mikael Eriksson/Shutterstock; P138 The Granger Collection, New York; P141 David Rydevik/Wikipedia; P147 G. Brad Lewis/Getty Images; P151 Mikael Eriksson/Shutterstock; P152-153 Earth Imaging/Getty Images; P154 Mikael Eriksson/Shutterstock; P155 Mikael Eriksson/Shutterstock; P156 Mikael Eriksson/Shutterstock

Chapter 7 P157 Alistar Duncan/Dorling Kindersley Limited; P158 Mark Newman/Phototake; P159 t Barry L. Runk/Grant Heilman Photography, Inc.; P159 b PhotoDisc/Getty Images; P161 l PhotoDisc/Getty Images; P160 r eDesign/Shutterstock; P161 t Martin Bond/Science Photo Library/Photo Researchers, Inc.; P161 b Vera Bogaerts/Shutterstock; P164 Buddy Mays/Corbis; P165 Popshots/Omni-Photo Communications; P166 C. Allan Morgan/Peter Arnold, Inc.; P168 Sheila Terry/Photo Researchers Inc.; P170 Joy M. Prescott/Shutterstock; P172 t Dorline Kindersley Limited; P172 b Dorling Kindersley Limited; P173 l Dorling Kindersley Limited; P173 r Dorling Kindersley Limited; P174 Alistar Duncan/Dorling Kindersley Limited; P175 Alistar Duncan/Dorling Kindersley Limited; P176 Alistar Duncan/Dorling Kindersley Limited

Chapter 8 P177 NASA; P178 Larry Lefever/Grant Heilman Photography, Inc.; P179 t Grant Heilman/Grant Heilman Photography, Inc.; P179 b Doug Steley/Alamy; P180 l Chris R. Sharp/Photo Researchers, Inc.; P180 r Spencer Grant/Photo Researchers, Inc.; P181 G. R. Roberts/Omni-Photo Communication, Inc.; P182 Tom Bean/DRK Photo; P183 Donald Specker/Animals Animals/Earth Scenes; P184 l Jane Grushow/Grant Heilman Photography, Inc.; P184 r NASA/Grant Heilman Photography, Inc.; P186 t Runk/Schoenberger/Grant Heilman Photography, Inc.; P186 b Herb Segars/Animals Animals/Earth Scenes; P187 Sid & Shirley Rucker/DRK Photo; P188 l Grant Heilman/Grant Heilman Photography, Inc.; P188 r PhotoDisc/Getty Images; P189 t Science Photo Library/Photo Researchers, Inc.; P189 b Marine Biological Laboratory/Mark Marten/Photo Researchers, Inc.; P190 t G. R. Roberts/Omni-Photo Communications; P190 b Popshots/Omni-Photo Communications; P192 Eric Kamp/Phototake; P193 Simon Fraser/Science Photo Library/Photo Researchers, Inc.; P194 l Jim Corwin/Photo Researchers, Inc.; P194 r Francois Gohier/Photo Researchers, Inc.; P195 Alex Neauville/Shutterstock; P196 Larry Ulrich/DRK Photo; P197 Padre Island National Seashore, National Park Service; P198 PhotoDisc/Getty Images; P199 t Grant Heilman/Grant Heilman Photography, Inc.; P199 b Christopher Scott/Alamy; P202-203 bkgd AP Wide World Photo; P202 b Library of Congress; P203 t AP Wide World Photo; P203 b USDA/NRCS/NCGC/National Cartography and Geospatial Center; P204 NASA; P205 NASA; P206 NASA

Chapter 9 P207 PhotoDisc/Getty Images; P211 Stephen L. Alvarez/National Geographic Society; P212 PhotoDisc/Getty Images; P213 Sylvain Grandadam/Robert Harding Picture Library Ltd/Alamy; P214 Murray Wilson/Omni-Photo Communications; P215 James Sheffer/Photo Edit; P219 t Carr Clifton/Minden Pictures; P219 b Worldsat International, Inc./Photo Researchers, Inc.; P222 PhotoDisc/Getty Images; P223 PhotoDisc/Getty Images; P224 PhotoDisc/Getty Images

Chapter 10 P225 Morozova Tatyana/Shutterstock; P226 PhotoDisc/Getty Images; P227 Jeff Rotman/Alamy; P229 t Stuart Westmorland/Photo Researchers, Inc.; P229 b Patricia Jordan/Peter Arnold, Inc.; P231 Yann Arthus-Bertrand/Corbis; P238 t Jim Brandenburg/Minden Pictures; P238 b Jim Brandenburg/Minden Pictures; P239 Tom McHugh/Photo Researchers, Inc.; P240 tr Charles D. Winters, Photo Researchers, Inc.; P240 bl Jim Steinberg/Photo Researchers, Inc.; P240 br Jan Hinsch/Science Photo Library/Photo Researchers, Inc.; P243 Pavlovsky/Photo Researchers, Inc.; P244 l David Hall/Photo Researchers, Inc.; P244 r Jean-Marc Truchet/Getty Images; P245 t Douglas Faulkner/Photo Researchers, Inc.; P245 b Steve Wolper/DRK Photo; P246 t Jan Hinsch/Science Photo Library/Photo Researchers, Inc.; P246 b Doc White/Nature Picture Library; P247 Elisei Shafer/Shutterstock; P248 Morozova Tatyana/Shutterstock; P249 Morozova Tatyana/Shutterstock; P250 Morozova Tatyana/Shutterstock

Chapter 11 P251 Philip Coblentz/Brand X/Jupiter Images/Alamy; P253 Dennis Kunkel/Phototake; P255 Michael Giannechini/Photo Researchers, Inc.; P257 David DuCros/Science Photo Library/Photo Researchers, Inc.; P259 Darla Hallmark/Shutterstock; P262 l Bruce Heinemann/PhotoDisc/Getty Images; P262 l inset Yoav Levy/Phototake; P262 t Leonard Lessin/Peter Arnold, Inc.; P263 t Donna Ikenberry/Animals Animals/Earth Scenes; P263 b The Granger Collection, New York; P266 Bruno Herdt/Getty Image, Inc./Photo Disc/Getty Images; P271 PhotoDisc/Getty Images; P272 tr David Parker/Science Photo Library/Photo Researchers, Inc.; P272 bl Wil Blanche/Omni-Photo Communications; P272 br Mark C. Burnett/Photo Researchers, Inc.; P274 Philip Coblentz/Brand X/Jupiter Images/Alamy; P275 Philip Coblentz/Brand X/Jupiter Images/Alamy; P276 Philip Coblentz/Brand X/Jupiter Images/Alamy;

Chapter 12 P277 PhotoDisc/Getty Images; P278 l John Anderson/Alamy; P278 r Stuart Westmorland/Photo Researchers, Inc.; P279 Andrew Syred/SPL/Science Photo Library/Photo Researchers, Inc.; P280 Jim Steinberg/Photo Researchers, Inc.; P281 Dwayne Newton/Photo Edit; P284 t Rod Planck/Photo Researchers, Inc.; P284 b Richard Shiell/Animals Animals/Earth Scenes; P285 Hank Morgan/Photo Researchers, Inc.; P286 tr Bruce Watkins/Animals Animals/Earth Scenes; P286 bl John Spragens, Jr./Photo Researchers, Inc.; P286 mr Eastcott/Momatiuk/Animals Animals/Earth Scenes; P286 br Fred Whitehead/Animals Animals/Earth Scenes; P287 t Stephen Ingram/Animals Animals/Earth Scenes; P287 b Jim Brandenburg/Minden Pictures; P290 col. 1 Carson Baldwin, Jr./Animals Animals/Earth Scenes; P290 col. 2 C. C. Lockwood/Animals Animals/Earth Scenes; P290 col. 3 Johnny Johnson/DRK Photo; P290 col. 4 E. R. Degginger/Animals Animals/Earth Scenes; P291 Fogden, M. OSF/Animals Animals/Earth Scenes; P293 NASA/Science Photo Library/Photo Researchers, Inc.; P294 John Lemker/Animals Animals/Earth Scenes; p296 OAR/ERL/National Severe Storms Laboratory; P297 t PhotoDisc/Getty Images; P297 b Vstock/Alamy; P298 t Mary Evans Picture Library/Alamy; P298 b Keith Kent/Science Photo Library/Photo Researchers, Inc.; P299 Pasphotography/Shutterstock; P304 PhotoDisc/Getty Images; P305 PhotoDisc/Getty Images; P306 PhotoDisc/Getty Images

Chapter 13 P307 Kenneth Sponsler/Shutterstock; P307 inset Peter Chadwick/Dorling Kindersley Limited; P309 Wolfgang Kaehler/Corbis; P310 Michael Ellis/Shutterstock; P311 Jacques Jangoux/Alamy; P314 l Jim Schwabel/Index Stock Imagery/Jupiter Images; P314 r Reuters NewMedia Inc./Corbis; P318 Andy Lim/Shutterstock; P319 Robert Lubeck/Animals Animals/Earth Scenes; P319 inset Botanica/Jupiter Images; P320 t PhotoDisc/Getty Images; P320 m Francois Gohier/Photo Researchers, Inc.; P320 b L F Stewart/Shutterstock; P321 tl PhotoDisc/Getty Images; P321 tm Charlie Ott/Photo Researchers, Inc.; P321 tr Bios (Compost-Visage)/Peter Arnold, Inc.; P321 b Brad Thompson/Shutterstock; P322 Kenneth Sponsler/Shutterstock; P323 Kenneth Sponsler/Shutterstock; P324 Kenneth Sponsler/Shutterstock

Chapter 14 P325 Jhaz Photography/Shutterstock; P327 The Granger Collection, New York; P332 Galen Rowell/Corbis; P333 PhotoDisc/Getty Images; P334 Neil Rabinowitz/Corbis; P335 l Sol Sexto/Photo Disc/Getty Images; P335 r Smari/Getty Images; P337 NASA/Science Photo Library/Photo Researchers, Inc.; P339 Hashim Pudiyapura/Shutterstock; P342 John Chumack/Galactic Images/Photo Researchers, Inc.; P344 Brand X/Jupiter Images; P345 Roger Ressmeyer/Corbis; P346 Jhaz Photography/Shutterstock; P347 Jhaz Photography/Shutterstock; P348 t Jhaz Photography/Shutterstock; P348 b John Sanford/Science Photo Library/Photo Researchers, Inc.

Chapter 15 P349 Bob Hosea/Shutterstock; P349 inset U.S. Fish and Wildlife Service; P352 Jeff Greenberg/Omni-Photo Communications; P353 Brandon D. Cole/Corbis; P354 t Falk Kienas/Shutterstock; P354 b Jacques Jangoux/Peter Arnold, Inc.; P355 Joe Sohm/Pan America/Jupiter Images; P356 Chinch Grynieqicz/Ecoscene/Corbis; P357 Nova Scotia Power; P359 Alex S. MacLean/Peter Arnold, Inc.; P361 t C. C. Lockwood/Animals Animals/Earth Scenes; P361 b Mike Flippo/Shutterstock; P362 l Michael Ledray/Shutterstock; P362 r Will McIntyre/Photo Researchers, Inc.; P364-365 bkgd Bettmann/Corbis; P364 t Hulton-Deutsch Collection/Corbis; P364 b Bettmann/Corbis; P365 Bettmann/Corbis; P366 Karla Caspari/Shutterstock; P367 Josh Meyer/Shutterstock; P370 Werner H. Muller/Peter Arnold, Inc.; p371 Steve Allen/Alamy; P372 t Sam C. Pierson, Jr./Photo Researchers, Inc.; P372 b Jeff Greenberg/Alamy; P373 Paul Kern/Omni-Photo Communications; P374 Novosti/Science Photo Library/Photo Researchers, Inc.; P375 Yahya Ahmed/AP Wide World Photos; P376 Bob Hosea/Shutterstock; P377 Bob Hosea/Shutterstock; P378 Bob Hosea/Shutterstock

Chapter 16 P379 Brand X/Jupiter Images; P380 Private Collection/Bridgeman Art Library; P381 t Roger Ressmeyer/Corbis; P381 b LOC/Science Source/Photo Researchers, Inc.; P382 t NASA; P382 b NASA; P382 b inset NASA; P383 t Dr. Christian Brandstatter (d/b/a Austrian Archives)/Corbis; P383 b D. Falconer/PhotoDisc/Getty Images; P384 l James A. Sugar/Corbis; P384 r Jessica Wecker/Photo Researchers, Inc.; P386 Bettmann/Corbis; P387 t Jerry Lodriguss/Photo Researchers, Inc.; P387 m USGS/EROS Data Center; P387 b NASA/Corbis; P388 Dr. Steve Gull/Science Photo Library/Photo Researchers, Inc.; P389 t PhotoDisc/Getty Images; P389 b David Parker/Science Photo Library/Photo Researchers, Inc.; P390-391 bkgd Reuters NewMedia Inc/Corbis; P390 t Corbis; P390 m Roger Ressmeyer/Corbis; P390 bl Stocktrek Images/Getty Images; P390 br AFP/Corbis; P391 t AFP/Corbis; P391 bl NASA/Corbis; p391 br NASA/Goddard Space Flight Center; P394 Harvey Lloyd/Peter Arnold, Inc.; P398 PhotoDisc/Getty Images; P399 t NASA/Science Source/Photo Researchers, Inc.; P399 b PhotoDisc/Getty Images; P400 l NASA; P400 r ESA/CE/Eurocontrol/Science Photo Library/Photo Researchers, Inc.; P401 l David Ducros/Science Photo Library/Photo Researchers, Inc.; P401 r David A. Hardy/Science Photo Library/Photo Researchers, Inc.; P402 l NASA; P402 r Nasa/Science Photo Library/Photo Researchers, Inc.; P403 Bettmann/Corbis; P404 Brand X/Jupiter Images; P405 Brand X/Jupiter Images; P406 Brand X/Jupiter Images

Chapter 17 P407 NASA/JPL/Space Science Institute; P407 inset NASA/JPL/Space Science Institute; P409 t NASA; P409 b Lynette Cook/Photo Researchers Inc.; P414 l Nasa; P414 r NASA/Science Photo Library/Photo Researchers, Inc.; P415 NASA; P416 NASA; P417 t NASA/Jet Propulsion Laboratory; P417 b Brand X/Jupiter Images; P420 l NASA Headquarters - Greatest Images of NASA; P420 r NASA/Science Photo Library/Photo Researchers, Inc.; P421 Kathie Atkinson/Oxford Science/Jupiter Images; P422 NASA/Science Photo Library/Photo Researchers, Inc.; P422 inset NASA; P423 Gerald Cooyman/Animals Animals/Earth Scenes; P423 inset NASA/Science Photo Library/Photo Researchers, Inc.; P424 l MarkGlick/Science Photo Library/Photo Researchers, Inc.; P424 r David Parker/Science Photo Library/Photo Researchers, Inc.; P424 r inset Breck P. Kent/Animals Animals/Earth Scenes; P425 Chris Butler/Science Photo Library/Photo Researchers, Inc.; P426 l Brand X/Jupiter Images; P426 r NASA/Jet Propulsion Laboratory; P427 t Brand X/Jupiter Images; P427 b PhotoDisc/Getty Images; P428 l Space Telescope Science Institute/NASA/Science Photo Library/Photo Researchers, Inc.; P428 tr NASA/Jet Propulsion Laboratory; P428 br NASA; P429 Roger Ressmeyer/Corbis; P430 Derke/O'Hara/Getty Images; P431 NASA; P432 NASA/JPL/Space Science Institute; P433 NASA/JPL/Space Science Institute; P434 NASA/JPL/Space Science Institute

Chapter 18 P435 Jeff Hester and Paul Scowen (Arizona State University)/NASA HQ Audio Visual Branch; P436 Larry Gilpin/Getty Images; P437 t NASA/Omni-Photo Communications; P437 b Gary Ladd/Photo Researchers, Inc.; P440 t col 1 NASA Goddard Space Flight Center; P440 t col 2 NASA/Science Photo Library/Photo Researchers, Inc.; P440 t col 3 John Sanford/Science Photo Library/Photo Researchers, Inc.; P440 t col 4 Goddard Institute for Space Studies/NASA; P440 bl NASA Johnson Space Center; P440 bm John Sanford/Science Photo Library/Photo Researchers, Inc.; p440 br NASA/Goddard Space Flight Center; P443 The Granger Collection, New York; P446 Dr. Leon Golub/Science Photo Library/Photo Researchers, Inc.; p448 br NASA/Goddard Space Flight Center; P448 inset Mark Garlick/Science Photo Library/Photo Researchers, Inc.; P448 r NASA/Jet Propulsion Laboratory; P449 Detlev van Favenswaay/Photo Researchers Inc.; P450 t Library of Congress; P450 b Roger Ressmeyer/Corbis; P451 PhotoDisc/Getty Images; P454 l NASA/JPL-Caltech/CTIO; P454 m Anglo-Australian Observatory/Dorling Kindersley Limited; P454 r Anglo-Australian Observatory/Dorling Kindersley Limited; P455 Anglo-Australian Observatory/Dorling Kindersley Limited; P457 Robert Williams and the Hubble Deep Field Team (STScI)/NASA; P458 NASA/Jet Propulsion Laboratory; P459 Patrick Clark/PhotoDisc/Getty Images; P460 t J. Baum & N. Henbest/Science Photo Library/Photo Researchers, Inc.; P460 m Royal Observatory, Edinburgh/Science Photo Library/Photo Researchers, Inc.; P460 b Cerro Tololo Inter-American Observatory; P461 Audio Visual Branch/NASA Headquarters; P462 Jeff Hester and Paul Scowen (Arizona State University)/NASA HQ Audio Visual Branch; P463 Jeff Hester and Paul Scowen (Arizona State University)/NASA HQ Audio Visual Branch; P464 Jeff Hester and Paul Scowen (Arizona State University)/NASA HQ Audio Visual Branch

Appendix: P 465 Comstock Images/Jupiter Images; P467 Comstock Images/Jupiter Images; P469 Comstock Images/Jupiter Images; P471 Comstock Images/Jupiter Images; P473 Comstock Images/Jupiter Images; P475 Comstock Images/Jupiter Images; P477 Comstock Images/Jupiter Images; P479 Comstock, Inc.

Glossary/Idex: P481 Comstock Images/Jupiter Images; P491 Comstock Images/Jupiter Images

Back Cover: P491 Comstock Images/Jupiter Images

Teacher Notes

Teacher Notes

Teacher Notes

Teacher Notes

Teacher Notes

Teacher Notes

Teacher Notes

Teacher Notes

Teacher Notes